MW01118922

DRILLS

Science and Technology of Advanced Operations

For my friend Matt
Thank you for reading
this my book.
all the best

V. Astakhov

Manufacturing Design and Technology Series

Series Editor
J. Paulo Davim

PUBLISHED

Drills: Science and Technology of Advanced Operations
Viktor P. Astakhov

FORTHCOMING

Diamond Tools in Abrasive Machining
Mark Jackson

Manufacturing Design and Technology

Science and Technology of Advanced Operations

VIKTOR P. ASTAKHOV

CRC Press is an imprint of the
Taylor & Francis Group, an informa business

CRC Press
Taylor & Francis Group
6000 Broken Sound Parkway NW, Suite 300
Boca Raton, FL 33487-2742

Printed on acid-free paper
Version Date: 20130923

International Standard Book Number-13: 978-1-4665-8434-1 (Hardback)

Library of Congress Cataloging-in-Publication Data

Astakhov, Viktor P.
Drills : science and technology of advanced operations / author, Viktor P. Astakhov.
pages cm. -- (Manufacturing design and technology)
Includes bibliographical references and index.
ISBN 978-1-4665-8434-1 (alk. paper)
1. Drilling and boring machinery. I. Title.

TJ1260.A87 2014
621.9'52--dc23 2013036861

Visit the Taylor & Francis Web site at
http://www.taylorandfrancis.com

and the CRC Press Web site at
http://www.crcpress.com

Contents

Preface

Why don't you write books people can read?

Nora Joyce to her husband James (1882–1941)

Various studies and surveys indicate that hole making (drilling) is one of the most time-consuming metal cutting operations in the typical shop. It is estimated that 36% of all machine hours (40% of computer numerical controlled [CNC] machines) is spent performing hole-making operations, as opposed to 25% for turning and 26% for milling, producing 60% of chips. Therefore, the use of high-performance drilling tools could significantly reduce the time required for drilling operations and thus reduce hole-making costs.

Over the past decade, the tool materials and coatings used for drills have improved dramatically. New, powerful, high-speed spindles, rigid machines, proper tooling including precision workholding, and high-pressure, high-concentration metal working fluid (MWF) have enabled a significant improvement in the quality of drilled holes and an increase in the cutting speed and penetration rate in drilling operations. In modern machine shops, as, for example, in the automotive industry, the quality requirements for drilled holes today are the same as they used to be for reamed holes just a decade ago. The cutting speed over the same time period has tripled and the penetration rate has doubled.

Despite all these new developments, many drilling operations even in the most advanced manufacturing facilities remain the weakest link among other machining operations. Moreover, there is still a significant gap in the efficiency, quality, and reliability of drilling operations between advanced and common machine shops. This is due to a lack of understanding of not only the process and its challenges, but primarily of the design, manufacturing, and application methods of high-productivity (HP) drilling tools. It is totally forgotten that process capability, quality, and efficiency are primarily decided on the cutting edges of the drill as this tool does the actual machining while all other components of the drilling system play supporting roles presumably assuring drill best working conditions. Therefore properly designed and manufactured drilling tools for a given application is the key to achieving high efficiency in drilling operations. Such a tool is referred to as the high-productivity drill (HP drill) throughout the book.

WHY NOW?

Although the basic design and manufacturing principles of drilling tools have been studied and used since the time that the Great Pyramids were built, the implementation of the results of these studies has been rather modest. The most apparent cause for this is that these studies lacked a systemic approach; that is, one component, for example, tool life, was studied while other important parameters, for example, process efficiency, were not considered. Although this is true, it is not the real cause. The reality is that neither the machining system, as a whole, nor its components were ready for implementation of the advances made in the design and manufacturing of drilling tools.

In the not-too-distant past, the components of a machining system were far from perfect, and thus it was not possible to utilize the advantages of advanced drilling tools. Tool specialists were frustrated old machines with insufficient power and spindles that could be rocked by hands; part fixtures that clamped parts differently every time; part materials with inclusions and great scatter in essential properties; tool holders that could not hold tools without excessive runouts assuring their proper position; starting bushing and bushing plates that had been used for years without replacement; low-concentration often contaminated MWFs that were more damaging than beneficial to the cutting tool; manual sharpening and presetting of cutting tools; limited range of cutting speeds and feeds; low dynamic rigidity of machines, etc. The most advanced (and thus expensive) drilling tools, therefore, performed practically the same (or even worse) as basic tools made in a local tool shop. As a result, any further development in tool improvement was discouraged as leading tool manufacturers did not see any return on investment in such developments.

For many years, a stable though fragile balance was maintained between low-quality (and thus relatively inexpensive) drilling tools and poor machining system characteristics. Metal cutting research was attributed mainly to university labs, and their results were mostly of academic interest rather than addressing practical needs. It is clear that metal cutting theory and the cutting tool designs based on this theory were not requested as many practical specialists had not observed any application benefits of such tools, and thus the theory behind them.

This has, however, changed rapidly since the beginning of the twenty-first century as global competition forced many manufacturing companies, mainly automotive manufacturers, to increase the efficiency and quality of machining operations. To address these issues, leading tool and machine manufacturers have developed a number of new products—new powerful precision machines having a wide range of speeds and feeds, tool materials and coatings, new tool holders, automated workholding fixtures, advanced machine controllers, etc. These changes can be called the *silent* machining revolution as they are rather dramatic and happened in a short period of time. For example, the 2013 Hanover Fair, the world's largest trade show for industry, was intended to be the drive toward the fourth industrial revolution. Unfortunately, it was not noticed by many tool manufacturers or even researchers. Therefore, the following text lists a few significant changes that can barely be seen in university labs and machine shops.

MACHINE TOOLS

Dramatic changes in the design and manufacturing of machine tools can be summarized as follows:

1. *Machines with powerful digitally controlled truly high-speed motor spindles*: For example, machines with working rotational speeds of 25,000 rpm and 35 kW motor spindles are used in advanced manufacturing power train facilities in the automotive industry. New multiaxis CNC machines with an excess of power and spindles capable of 35,000 rpm rotational speed are also being rapidly introduced in the mold-making industry.
2. *New spindles that assure tool runout* <0.5 μm: These were implemented on many machines. High static and dynamic rigidity of such spindles and machines made with granite beds results in chatter (vibration)-free performance even at the most heavy cuts in truly high-speed machining conditions.
3. *High-pressure through-tool MWF supply*: New machines are equipped with a high-pressure (70 bar and more if needed) MWF (coolant) supply through the cutting tools to provide cooling and lubrication needed for high-speed operations. MWFs cleaned up to 5 μm are delivered at constant controlled temperatures suitable for a given machining operation.

Tool Holders and Tool Presetting Practice

Old-fashioned tool holders having 7/24 taper developed over half a century ago and sold today as CAT, BT, and ISO are being rapidly replaced with high-precision HSK, developed as a standard defined by DIN (German Institute for Standardization). Balanced hydraulic, shrink fit, and steerable tool holders have been developed and widely implemented for high-speed machining to minimize tool runout and to maximize tool holding rigidity. With shrink fit tool holders, vibration is reduced and cutting is noticeably faster and smoother due, in part, to the lack of set screws and component tolerance variances.

For years, tool presetting was one of the weakest links in assuring proper tool position and performance. No matter how good the tool and its tool holder are tool presetting accuracy determines the actual performance of the tool. Need for more accurate tool presetting, tool data transformability from the presetter to the machine tool controller, and tool performance traceability has led to the development of noncontact laser-automated tool presetting machines. Nowadays, tool presetting machines (e.g., Zollar and Kelch) are widely used in high-speed machining applications. Each tool contains an ID number to enable users to retrieve and use the data later on. The tool is mounted into the holder, and the CNC-driven tool length stop is set to ensure correct tool positioning—for example, tool tip to gage datum. Measurement results are transferred to the tool radio-frequency identification chip (e.g., the Balluff chip) embedded in the holders (see Chapter 2 for a detailed explanation). Such a presetting machine can provide accuracy within 3 μ on each tool, which, in turn, results in improved machining quality.

Advanced Cutting Process Monitoring

Many recent technologies offer tool and machine monitoring, from detecting tool presence to measuring the tool's profile. Some can even measure the power consumed by the spindle motor and use that information to control the feed rate and minimize machining time. The most common features of modern machine tool controllers developed for high-speed unattended manufacturing are as follows:

- *Detecting broken or absent tools*: Small, simple tool detectors that check for the presence of a drill or other cylindrical-shaft tool were developed and implemented on modern machine tools. Their small size and simple operation adapt them well for many environments, including machining centers, screw machines, and transfer machines.
- *Power monitoring*: Directly monitoring the power consumed by the spindle motor allows one to understand exactly what is happening with the tool in a drilling operation. Power monitoring systems take their data directly from the motor controller; others measure with transducers on the wiring to the motor. This is called the *sensory perception* of machines. The system obtains information that would not otherwise be detectable and notifies the operator as soon as possible when there is a problem.
- *Adaptation*: Not only can power-monitoring units detect broken or worn tools, but with an *adaptive control* option, they can also use them to control the feed rate, reducing machining time, yet extending tool life by keeping the tool load constant and well under maximum load.

Advances in Cutting Tool Materials

Improvements in the quality and consistency of the major groups of tool materials through implementing advanced tool materials technologies are of prime importance (see Chapter 3 for details). Improved quality of machining systems allows wide use of modern grades of polycrystalline diamond (PCD) tool material capable of milling, drilling, and reaming high-silicon aluminum alloys at speeds of 1,000–11,000 m/min. Modern grades of carbide tools combined with advanced coatings allow machining of alloyed steels at speeds of 300–600 m/min. Modern grades of polycrystalline cubic boron nitride (PCBN) allow hard machining operations, which substitute some grinding operations. New tool materials and advanced grades of existing tool materials, including nanocoatings, have also been introduced.

Advances in Cutting Tool Manufacturing

A number of significant advances in cutting tool manufacturing are taking place rapidly. The introduction of CNC tool grinders/sharpeners and CNC tool geometry inspection/measuring machines are probably the most significant.

For decades, manual tool grinding/sharpening machines were used in the cutting tool industry. It was not possible to maintain the geometry of ground/sharpened tools with reasonable accuracy, as this varied significantly from tool to tool. The exact tool geometry simulated by any advanced tool design program couldn't be reproduced with reasonable accuracy using such machines. It was also not possible to grind any complicated profile of the tool as might be required for optimal tool performance. Naturally, any advanced tool geometry suitable for optimal performance of a machining operation was bluntly rejected by machining practice as being *impractical* for a real world application.

This situation, however, has been changing rapidly since the beginning of the twenty-first century. Today's tool grinder is typically a CNC machine tool that usually has 4, 5, or 6 axes designed to produce drills from a carbide rod. Such machines are widely used in the cutting tool industry. High levels of automation, as well as automatic in-machine tool measurement and compensation, allow extended periods of unmanned production. Modern CNC tool grinding machines have the following distinctive features:

- Built-in 3D simulation. This wizard automatically simulates the tool in 3D, directly in the user interface, and displays the expected cycle time. This is accomplished directly on the machine so that an offline simulator is not needed.
- The traveling steady unit on the P-axis for accurate long-length tools. This provides three points of support to the tool. Each support applies continuous hydraulic pressure on the tool, even for reducing diameters, such as tools with a back taper.
- Vibration elimination techniques and process options that allow one to grind low surface roughness or mirror surface finishes.
- Consistent results of less than 3 μ runout of the tool holding.
- On-board measuring systems that are able to measure the ground tool in its original clamping inside the grinding machine.
- Robotic loading–unloading and automated grinding wheel changer.

No matter how good the fully optimized cutting tool geometry is (using, e.g., a FEM simulation software) and how well it is depicted in multiple section planes on the tool drawing made using a 3D CAD program, it is practically useless as such an optimized geometry cannot be reproduced and then inspected/measured with high accuracy. Until very recently, the most common practice of measuring tool geometry was manual inspection, which did not provide accurate results as it was dependent on how experienced the inspector was, the complexity of the tool, and many other factors that could not be controlled. Naturally, the accuracy of such an inspection was not sufficient to ensure the effective performance of HP drills.

To address this important issue, CNC tool inspection machines that are capable of inspecting cutting tools accurately have been developed (see discussion in Chapter 7). For example, the ZOLLER Genius 3 measuring and inspection machine is equipped with five CNC-controlled axes for measuring and inspecting virtually any tool parameter—it is fast, simple, precise to the micron, and fully automatic. Equipped with a 500-fold magnification incident light camera, Genius 3 can automatically inspect microtools. The machine includes measuring programs for virtually every parameter (radius contour tracing, effective cutting angle, clearance angle, helical pitch and angle, chamfer width, groove depth, tumble and concentricity compensation, step measurement, etc.) of cutting tools.

Drilling Tools as the Weakest Link

The preceding discussion suggests that many components of modern drilling systems are ready to fully support HP drills while the equipment available to tool manufacturers fully support their high-efficiency production with practically no restrictions. However, wide implementation of HP drilling tools is not yet the case. As a result low reliability of cutting tools and sporadic tool failures in advanced manufacturing facilities (i.e., in the automotive industry) are major hurdles in the way of wide use of efficient, unattended machining production lines and manufacturing cells to decrease direct labor costs and improve efficiency of machining operations. As such, significant downtime due to low reliability of drilling tools undermines the potential of highly efficient production lines and manufacturing cells, raising questions about the feasibility of unattended machining operations.

Readiness of Tool Users and Manufacturers to Address the Issue

The preceding discussion clearly indicates that all the common excuses of inferior tool performance, that is, subpar quality of machining operations, have been practically eliminated. Cutting tool users/machining operation designers/planners and cutting tool manufacturers have been pushed to the forefront to show their capability in producing and implementing advanced tools to address the new challenges in metal machining—high productivity rates, low-cost parts, great quality, and suitable tool reliability, particularly for unattended production lines and manufacturing cells.

Leading drilling tool manufacturers failed miserably to meet this new challenge. The root cause for this failure was insufficient understanding of drilling principles, including drilling tool geometry and drill manufacturing quality–related flaws. This is partially due to the lack of information in the literature where all aspects of drilling tools are explained. The few available papers on drill geometry present the results in differential geometry, matrix, and vectorial forms not comprehensible to many technical readers. It is not clear from the available literature sources why one should learn drilling tool geometry as no practical examples of geometry optimization are normally considered in the literature. In other words, the tooling industry is left with no proper scientific or even engineering support. The technical and reference literature present a mixture of old and new notions as technical data have been copied from one edition to the next for the last 50 years or more. It is therefore very difficult for a practical engineer to *separate the wheat from the chaff* as information is thoroughly mixed in the technical literature and on various technical websites.

AIM OF THE BOOK

The foregoing analysis suggests that the weakest link in the design of HP drilling operations is the readiness of cutting tool users/machining operation designers/planners and cutting tool manufacturers. The lack of relevant research work on modern hole-making operations just adds to the problem. This book aims to address the most important issues in drilling operations and thus to provide assistance with the design of such operations. It discards many old notions and beliefs and introduces scientifically and technically sound notions with detailed explanations.

UNIQUENESS OF THE BOOK

The major feature of this book is the introduction and development of the concept of high-productivity (HP) drill design and its manufacturing and application features. A frequently asked question concerns the difference between HP drills and normal drills. In other words, what makes a drill an HP drill? The answer to this is discussed in the system approach combined with the *VPA-Balanced*© design concept discussed in Chapter 1. This combination can be briefly described as follows.

The *VPA-Balanced* design concept is the methodology used to design and manufacture a drill following a set of rules:

1. The geometry of the drill point is designed for a particular work material so that it ensures
   a. Minimum axial force and drilling torque (Chapter 4)
   b. Balanced design of the major cutting and chisel edges to guarantee chip flows from the different cutting edges with no crossing/interference and to provide sufficient room to ensure there are no restrictions on these flows (Chapters 4 and 5)
   c. Perfect force balance of a VPA-Balanced drill due to proper designing and manufacturing tolerances (Chapter 7)
   d. Proper distribution of MWF flow in the machining zone (Chapters 5 and 6)

2. The drill material is selected to be
   a. Of high quality and consistency (Chapter 3)
   b. Application specific
3. The drill should be of high manufacturing quality (Chapter 2), including final inspection of the tool (Chapter 7).

Although an HP drill designed adhering to the *VPA-Balanced* design concept is fully capable of ensuring the highest possible productivity of a given drilling operation, this capability is realized if and only if it is fully supported by the drilling system and its components. Are HP drills universal? No, they are not. They can be compared to high-performance cars, which require good road conditions, highly qualified drivers, and careful maintenance. High-performance cars may not be suitable for country roads, subpar maintenance, or for carrying heavy loads while moving from one apartment to the next. Moreover, they are not meant for everyday commuting from home to the workplace and back at an average speed 40 km/h. However, when the conditions are ideal, they deliver great comfort, joy of driving, and incredible reliability. As more and more people appreciate these benefits, the market segment of these cars increases quickly. This is also true for HP drills. They are suitable for high-quality, well-maintained drilling systems. They require high-quality tool materials and specialized, application-specific coatings as well as intelligent tool manufacturing systems, including complicated inspection equipment and procedures and computerized presetting in high-quality drill holders. The benefits include a high-penetration rate (which in turn results in high productivity), prolonged tool life, tool reliability, and better quality of machined holes—often two- or even three-tool hole operations can be reduced to a single-tool hole operation. HP drills are the major trend for the future where unattended, fully computerized manufacturing lines with greatly reduced labor cost, high productivity, and high consistency of quality of the manufactured parts are the ultimate goals/requirements.

On reading the previous paragraph, the reader may think this to be a dream or a remote possibility. However, this is a reality today, especially in the automotive industry, where manufacturing cells and production lines deliver high productivity (true high-speed machining), with minimum attendance, and great quality. It is surprising why some people, even specialists, cannot see the obvious results. The reliability of cars has increased significantly, allowing bumper-to-bumper warranties to be extended to 100,000 miles (160,000 km) for many cars. There are also a number of other significant improvements such as power-saving lights, reduced weight, more efficient fuel consumption, and increased safety. However, in spite of these improvements, the cost of these cars has not increased when compared to the cost of other commodities in the market.

To conclude this discussion, the author would like to remind the following. Only 20 years ago, the practical utility of carbide drills was being debated as their cost was much more than high-speed steel (HSS) drills and their applications required more intelligent handling and resharpening, better tool holders and machine spindles, etc. Many specialists at that time thought that carbide drills would have very limited use in practical manufacturing. However, today these drills are common standard tools used universally, while HSS drill usage has reduced significantly. The author is convinced that this will also be the case with HP drills as the economy of modern manufacturing dictates so. This book has been written to accelerate wider acceptance of such drills.

The distinguishing features of this book are as follows:

1. *Drilling is considered in this book instead of drills as is customary in the existent literature on the subject.* Drilling in this book is considered at two levels: (1) as a metalworking operation and (2) as a process. The former covers the steps in the design of optimal drilling operations, and thus is meant for process/machine/manufacturing lines and cell designers/developers, while the latter considers the cutting process with the unique features particular to drilling, and thus is meant for tool/tool materials/CNC tool grinder researchers/developers/designers.
2. The concept of a drilling system is introduced, and the interinfluence of the components of this system such as a drill, drilling machine, tool holder, fixture, coolant, etc.,

are analyzed. The coherency law is formulated. The so-called component approach is a common manufacturing practice in today's environment, where different manufacturers produce various components of the drilling system but no one seems to be responsible for system coherency. Unpredictable/unexplainable tool failure is a direct result of such an approach because the cutting tool is normally the weakest link in the machining system.

3. A clear objective of a drilling operation is set out. The prime system objective is explained as an increase in the drilling tool's (drill, reamer, etc.) penetration rate, that is, in drilling productivity.
4. Unparalleled drilling tool failure analysis methodology and procedures are considered. Novel procedures such as *part autopsy* and *drill reconstruction surgery* are introduced and explained with detailed examples. The HP drill realistic wear mode and wear regions are also discussed.
5. Tool materials that are common for making drilling tools are considered from the point of view of cutting tool design and implementation. Old/irrelevant yet common notions about cemented carbide tool material are discarded. Instead, relevant properties of this tool material are introduced. For the first time, the technology, properties, and manufacturing technology of PCD as a tool material are considered.
6. Design, technology, and implementation practices of PCD drills and reamers are considered. Vein and sandwich (DPI) drills, PCD-tipped tools, and full cross PCD drills are also included.
7. The essentials of drill geometry parameters are considered in a simple yet practical manner for drill designers, users, and grinders. The essential design features of known and advanced drilling tools, including HP drills, are discussed in detail.
8. The roles, types, and application techniques of MWF (coolant) in drilling operations are considered. The physical foundations of high-pressure application techniques of MWFs are discussed. Classification of near-dry machining (NDM) or minimum quantity lubrication (MQL) is presented, discussing their advantages and drawbacks. The essential components of the NDM system using a 360° vision approach as the key to successful implementation of NDM are considered.
9. Drilling metrology is considered for the first time in the literature. The meaning of the designations diametric and form tolerances of drilled holes are explained with examples. Drill geometry measurements/inspection using modern machines are considered. The standard and suggested tolerances for HP drills on drilling tool geometry parameters as well as their inspection are discussed in detail.

HOW THIS BOOK IS ORGANIZED

The structure of this book is unusual for the literature in the field because its logic is governed by HP drill design, manufacturing, and implementation theory and practice. A summary of the contents of the chapters is listed in the following.

Chapter 1: Drilling System

This chapter consists of two logically connected parts. The first introductory part presents a short classification of drilling operations. It discusses the components of drilling regime: cutting speed, cutting feed, feed rate, and material removal rate with practical examples; that is, it sets the scene for the other chapters by introducing the correct terminology, and precise definitions of the parameters of the drilling regime. The second part introduces the basics of the system approach. The structure of the drilling system is defined and the coherency law is formulated. The prime objective of the drilling system is established. The case for HP drills is considered and discussed. The design procedure for drilling systems is considered with practical examples.

Discussion on the selection of the proper components of a drilling system for HP drilling is considered, with an emphasis on the best components used in the automotive industry. The chapter argues that the quality of the components and equipment available today in drilling systems and in drilling tool production systems has overgrown the quality of the design and manufacturing of many drilling tools.

Chapter 2: Tool Failure as a System Problem: Investigation, Assessment, and Recommendations

This chapter considers drilling tool failure to be a system issue. Analyzing the drawbacks of the traditional notions and approaches, a new system-related definition of tool failure is given. The chapter argues that any tool failure is a system-related event. The causes of drill failure and their proper identification are discussed. Detailed procedures and methodology of drill failure analysis are considered along with examples. For the first time in tool failure analyses, *part autopsy* and tool *reconstruction surgery* are introduced and discussed with examples.

Standard tool wear measures and tool life analysis using wear curves are covered with practical examples using the Bernstein distribution for the assessment of tool quality and reliability. Typical wear regions of drills are considered with an emphasis on the assessment of tool wear of HP drills. Real mechanisms of tool wear are explained, showing that abrasion-adhesive wear is common in HP drilling. Abnormal wear modes due to castings' defects and cobalt leaching are discussed. The influence of the cutting speed on drill life is discussed showing that how the optimal cutting speed law can be applied in the design of HP drills. Wear mechanisms of PCD tools are also considered.

Chapter 3: Tool Materials

The selection of a cutting tool material type and its particular grade is an important factor when planning for a successful drilling operation. This chapter differs considerably from other chapters/books written on the subject in that it provides the knowledge base and practical information on tool materials for drilling tool designers, manufacturers, and end users. The emphasis is on HSS, cemented (sintered) carbides, and PCD tool materials as more than 98% of modern drilling tools are made of these tool materials. The chapter also focuses on common errors and misconceptions in manufacturing and implementation of drilling tools made of these materials. It provides in-depth coverage of tool materials, with definitions of terms and notions, detailed explanations of properties related to drilling, and practical examples, which make the chapter self-contained.

For the first time in the literature, the proper utilization of PCD in drilling tools has been considered in detail. Relevant properties of PCD, its selection, and its technology are considered. Common manufacturing flaws in PCD drilling tools are discussed along with examples. Recommendations to avoid such flaws are also provided.

Each type of tool materials is discussed in detail, covering its most essential properties. These are never discussed in the literature on tool materials and are poorly discussed in materials-related sources. For example, for cemetery carbide, the carbon balance is considered while for PCD, the concept of thermal stability is explained.

Chapter 4: Twist and Straight-Flute Drills: Geometry and Design Components

This chapter begins with the classification of various drills and introduces and defines the basic terms involved. A detailed explanation of the major constraints on the penetration rate imposed by the drilling tool itself and the correlation of these constraints with drill design and geometry parameters are provided. The force balance is defined as the major prerequisite feature in HP drill design/manufacturing.

The importance and system consideration of the drill geometry are explained with examples. Drill geometry is considered in the tool-in-hand system (T-hand-S), in the tool-in-machine system (T-mach-S), and in the tool-in-use system (T-use-S). The relevance of these systems to drill geometry parameters indicated in the tool drawing and to tool performance is revealed. Straight-flute and helical-flute tools are considered in order to help practical tool/drilling process designers to make proper selection of tool geometry parameters for a given application. Drill design optimization based on the load distribution on the rake face over the length of the drill cutting edge is introduced and explained with a number of practical drill designs.

Chapter 5: PCD and Deep-Hole Drills

This chapter discusses the designs and technologies of PCD drilling tools developed to meet the challenges of both new work material and high-speed machining, primarily in the automotive and aerospace industries. It discusses the disadvantages of some of the PCD-tipped drills that have been in use for a long time. The chapter points out a number of attempts (with some great successes) that were made in the development of the so-called full-face (cross-PCD) drills, broadly divided into two major design (technological) approaches: (1) PCD sintered in a part of the drill body and (2) fully sintered PCD segment brazed into the drill body. The designs of the full face for HP drilling are also introduced.

The chapter further discusses classification, geometry, and design of deep-hole drills. It deals with force balance and defines the term *self-piloting tool* (SPT). The history, design and application particularities, and geometries of gundrills, single tube system (commonly referred to as BTA) drills, and ejector drills are considered, with an emphasis on HP drills. Particularities of the MWF supply for each of these drill types are considered. STS and ejector drilling principles are compared to discard old notions while presenting information that will be useful for drill users.

Chapter 6: Metalworking Fluid in Drilling

This chapter argues that there are three equally important pillars for the successful application of MWF: (1) selection of the proper MWF, (2) delivery of this MWF into point of application, and (3) MWF maintenance. The physical delivery of MWFs to the machining zone is considered in this chapter. The chapter deals with the physics and technicalities of high-pressure MWF applications in drilling. It proves that the MWF flow, more than its pressure, defines the efficiency of high-pressure MWF applications. Using gun drilling as a simple example, the chapter reveals major issues that should be considered in the design of hole-making tools with internal high-pressure MWF supply.

The chapter further discusses that the costs of maintaining and eventually disposing of MWF, combined with health and safety concerns, have led to heightened interest in either eliminating MWF altogether or limiting the amount of MWF applied. As the former is not feasible for many applications, the latter, referred to as near-dry machining (NDM) or minimum quantity lubrication (MQL), is a current trend in industry. This chapter presents a classification of NDM methods, discussing their advantages and drawbacks. It considers the essential components of the NDM system, arguing that a 360° vision approach is the key to successful implementation of NDM.

The last section of the chapter argues that to realize the full potential of HP drilling, application of MWF will have to be monitored carefully; if the MWF parameters deteriorate, it can easily turn successful HP drilling operations into disastrous ones in terms of efficiency and productivity. Other issues such as concentration, water quality, and filtration are also discussed.

Chapter 7: Metrology of Drilling Operations and Drills

This chapter consists of two closely related parts. The first part is related to the various tolerances on the hole being drilled. The second part deals with drill metrology. For centuries, the metrology of drilling operations was not considered seriously as the tolerances on drilled holes were wide open.

Primitive hand gages and eyeballing measurements relying more on common sense and experience than on results of accurate measurements were considered common practice in drill metrology. Nowadays, however, with extensive use of HP drills and modern drilling systems, the tolerances on drilled holes have become the same as they used to be recently for reamed or even for ground holes.

The chapter introduces a novel concept of the dimensional inspection (metrological) system for HP drills. The datum feature for drilling tools as the solid foundation of the drill design and metrology is defined clearly. The chapter identifies drill geometrical and dimensional parameters on tool drawings and defines the procedures of their measurement/inspection with various measuring equipment including CNC drill inspection machines. The standard tolerances on the drill parameters are discussed, showing that they are not suitable for HP drills. The tolerances for HP drills are also discussed.

Appendix A: Axial Force, Torque, and Power in Drilling Operations

This appendix discusses the common ways of determining the force factors of drilling and points out some obvious flaws in such a determination. The concept of the chip compression ratio is explained as this is used in the chapters.

Appendix B: Tool Material Fundamentals

This appendix provides the supporting material for better comprehension of Chapter 3. It includes basics of metallurgical notions/terminology needed to understand the components and heat treatment of HSS. The manufacturing processes involved and the formation of essential properties for cemented carbides are discussed. The history of the development, formation of essential properties, and manufacturing technology of diamond as a tool material are also considered.

Appendix C: Basics of the Tool Geometry

The major objective of this appendix is to familiarize potential readers with the basic notions and definitions used in the analysis of tool geometry and the correlation of tool geometry parameters with the cutting force. It provides the fundamentals and definitions of the involved terminology for better comprehension of Chapters 4 and 5. It explains with examples that among many angles of the cutting tools, the clearance angle is the major distinguishing feature. The influence of other angles on the cutting process and its outcome are also discussed. Comprehensive coverage of edge preparation and its metrology is included.

Acknowledgments

I express my gratitude to all the people who provided support, helped with the testing and implementation of advanced tools, shared their viewpoints and discussed technical issues, allowed me to use their lab, inspection and production equipment, and assisted in the editing this book.

I thank Professor J. Paulo Davim for encouraging and enabling me to publish this book and many other book chapters that I have written.

I thank all my former and present teachers, colleagues, and students who have contributed to my knowledge of the subject.

Above all, I thank my wife, Sharon, and the rest of my family, who supported and encouraged me in spite of all the time spent away from them. It was a long and difficult journey for them.

Author

Viktor P. Astakhov earned his PhD in mechanical engineering from Tula State Polytechnic University, Tula-Moscow, Russia, in 1983. He was awarded a DSc designation (Dr. habil., Docteur d'État) and the title "State Professor of Ukraine" in 1991 for the outstanding service rendered during his teaching career and for the profound impact his work had on science and technology. An internationally recognized educator, researcher, and mechanical engineer, he has won a number of national and international awards for his teaching and research. In 2011, he was elected to the SME College of Fellows (http://www.docstoc.com/docs/99252367/Ten-Join-2011-SME-College-of-Fellows).

Dr. Astakhov currently serves as the CEO of Astakhov Tool Service Co. He is also the tool research and application manager of the General Motors Business Unit of PSMI. As a professor, he has been involved in supervising graduate students at Michigan State University. He has published monographs and textbooks, book chapters, and many papers in professional journals as well as in trade periodicals. He has authored the following books: *Geometry of Single-Point Turning Tools and Drills: Fundamentals and Practical Applications, Tribology of Metal Cutting, Physics of Strength and Fracture Control, and Mechanics of Metal Cutting.* He also serves as the editor in chief, associate editor, board member, reviewer, and advisor for many international journals and professional societies.

1 Drilling System

Everything is designed. Few things are designed well.

Brian Reed, American graphic designer

1.1 FUNDAMENTALS

1.1.1 Basic Drilling Operations

Drilling is a hole-making machining operation accomplished using a drilling tool. Figure 1.1a shows a common drilling arrangement in a drilling machine. The workpiece is clamped on the machine table with a vice equipped with jaws that clamp against the workpiece, holding it secure. The drill is clamped in the machine spindle that provides the rotation and the feed motions. Figure 1.1b shows a common drilling arrangement on a lathe. The workpiece is clamped in a self-centering three-jaw lathe chuck installed on the machine spindle that provides rotation and the tool is installed on the tailstock engaged with the lathe carriage that provides the feed motion.

A drilling tool is defined as an end cutting tool indented for one of the hole-making operations. Such a tool has the terminal (working) end and the rear end for its location in a tool holder. In all drilling operations, the primary motion is rotation of the workpiece or the tool or both (counterrotation drilling) and translational feed motion (Figure 1.2), which can be applied either to the tool or the workpiece depending on the particular design of the machined tool used.

There are a great number of drilling operations used in modern industry. Figure 1.3 shows some of the most frequently used. Although all these operations use the same kinematic motions and generic drilling tool definition, the particular tool designs, machining regimes, and many other features of the drilling tools involved are operation specific. These basic operations are defined as follows:

1. Drilling is the making of a hole in a workpiece where none previously existed. In this case, the operation is referred to as solid drilling. If an existing hole (e.g., a cored hole in die casting) is drilled, then the operation is referred to as core drilling. A cutting tool called the drill enters the workpiece axially through the end and cuts a hole with a diameter equal to that of the tool. Drilling may be performed on a wide variety of machines such as a lathe/turning center and drilling/milling/boring machine.
2. Boring (Figure 1.4) is the enlarging of an existing hole. A boring tool enters the workpiece axially and cuts along an internal surface to form different features, such as steps, tapers, chamfers, and contours. Boring is commonly performed after drilling a hole in order to enlarge the diameter, making steps and special features or to improve hole geometrical quality (i.e., to obtain high-precision diameter and shapes in the transverse [e.g., roundness] and longitudinal [e.g., position deviation] directions). Nowadays, however, many modern boring tools are multi-edge tools allowing significant increase in boring productivity and accuracy.

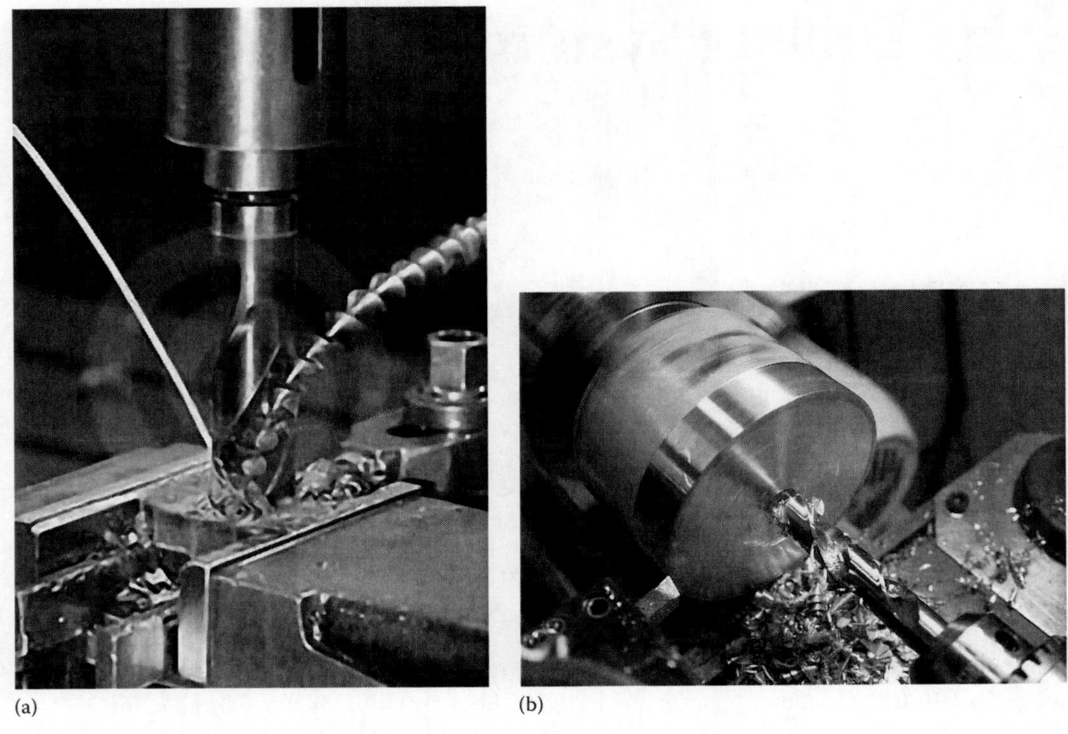

FIGURE 1.1 Generic drilling: (a) on a vertical drilling machine and (b) on a lathe.

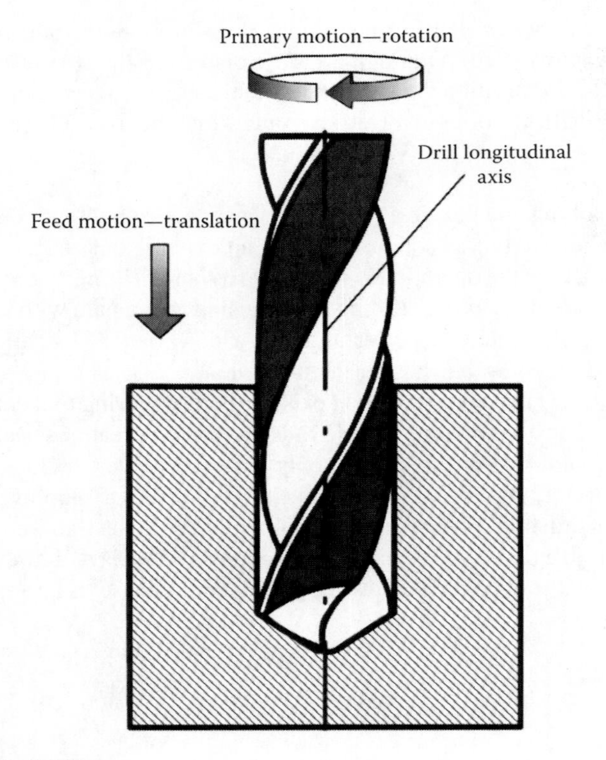

FIGURE 1.2 Motions in drilling.

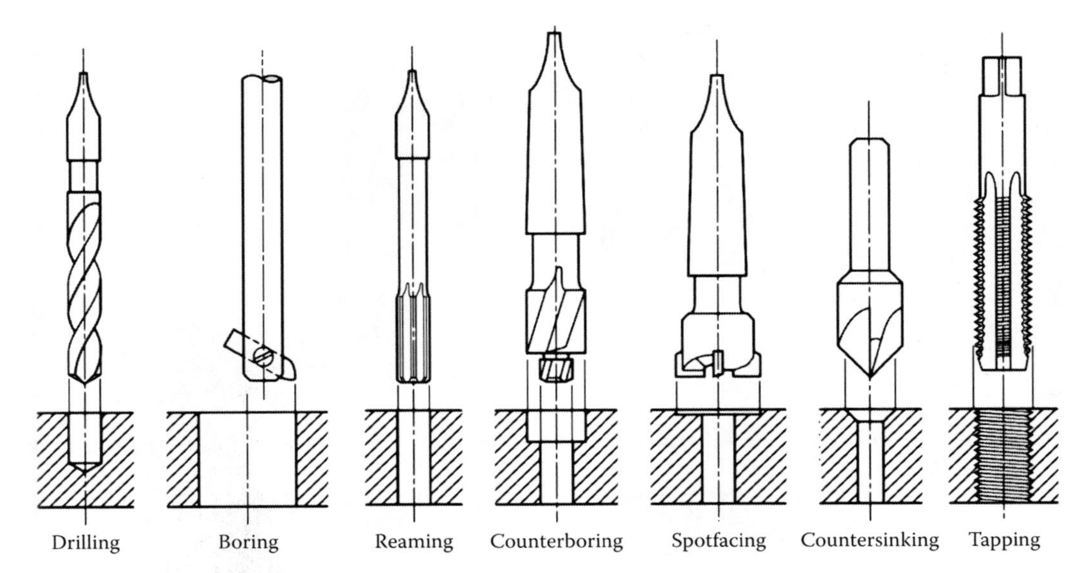

FIGURE 1.3 Basic drilling operations.

FIGURE 1.4 Boring.

3. Reaming (Figure 1.5) is the enlarging of an existing hole to accurate size and shape. An end cutting tool called the reamer enters the workpiece axially through the end and enlarges an existing hole to the diameter of the tool. Reaming is often performed after drilling or boring to obtain a more accurate diameter, better surface finish, and shape in the transverse direction.
4. Counterboring is flat-bottomed cylindrical enlargement of the mouth of a hole, usually of slight depth, as for receiving a cylindrical screw head. An end cutting tool referred to as the counterbore enters the workpiece axially and enlarges the top portion of an existing hole to the diameter of the tool. Counterboring is often performed after drilling to provide space for the head of a fastener, such as a bolt, to sit flush with the workpiece surface.
5. Countersinking is the process of making a cone-shaped enlargement at the entrance of a hole. An end cutting tool called the countersink enters the workpiece axially and enlarges the top portion of an existing hole to a cone-shaped opening. Countersinking is often

(a)

(b)

FIGURE 1.5 Reaming: (a) location of the part in the machine spindle, and (b) location of the reamer in the tailstock.

FIGURE 1.6 Tapping.

performed after drilling to provide space for the head of a fastener, such as a screw, to sit flush with the workpiece surface. Common included angles for a countersink include 60°, 82°, 90°, 100°, 118°, and 120°.

6. Spotfacing is a drilling operation performed where it is assumed that there will be a highly irregular face surface around a hole. This is common with castings. The spotface may be either below the surface of surrounding metal or placed on the top of a boss, as is typical with castings. The purpose of spotfacing can be either to provide flat surface to accommodate a screw head, nut, or washer or to make thru face to start other drilling operations. The spotface tool resembles an end mill cutter. A pilot in the center of the cutting surface is often added if the alignment of the existing hole and the spotface is important.
7. Tapping (Figure 1.6) is a drilling operation of cutting internal threads with an end form tool referred to as the threading tap. A tap enters the workpiece axially through the end and cuts internal threads into an existing hole. The existing hole is typically drilled by the required tap drill size that will accommodate the desired tap.

1.1.2 Machining Regime in Drilling Operations

The cutting speed and cutting feed are prime or basic parameters that constitute the machining regime in drilling operations.

1.1.2.1 Cutting Speed

In metric units of measure (the SI system), the cutting speed is calculated as

$$v = \frac{\pi d_{dr} n}{1000} \text{(m/min)} \tag{1.1}$$

where

$\pi = 3.141$

d_{dr} is the drill diameter in millimeters

n is the rotational speed in rpm or rev/min no matter which rotates, the drill or the workpiece

If both the drill and the workpiece rotate in opposite directions (the so-called counterrotation), then n is the sum of the rotational speeds of the drill, n_{dr}, and the workpiece, n_w, that is, $n = n_{dr} + n_w$.

For example, if $d_{dr} = 10$ mm and drill rotates with $n = 2170$ rpm while the workpiece is stationary, then $v = \pi d_{dr} n/1000 = 3.141 \times 10 \times 2170/1000 = 68.15$ m/min.

In the imperial units of measure, the cutting speed is calculated as

$$v = \frac{\pi d_{dr} n}{12} \text{(sfm or ft/min)} \tag{1.2}$$

where

$\pi = 3.141$

d_{dr} is the drill diameter in inches

n is the rotational speed in rpm or rev/min

For example, if $d_{dr} = ¾$ in. (19.05 mm) and the drill rotates with $n = 1220$ rpm while the workpiece is stationary, then $v = \pi d_{dr} n/12 = 3.141 \times 3/4 \times 1220/12 = 239.5$ sfm.

Although Equations 1.1 and 1.2 are exemplified for drills, they are perfectly valid for all drilling tools shown in Figure 1.3 having the basic motions shown in Figure 1.2. To calculate the cutting speed properly, the relevant diameter should be used in Equations 1.1 and 1.2 instead of d_{dr}. For example, for reaming, counterboring, spotfacing, and tapping, this diameter is equal to the outside tool diameter. For boring, diameter d_{br} equal to the finish diameter of the hole being bored should be used. When one deals with a multistage boring tool, then this diameter is equal to the largest finish diameter of the hole being bored. In countersinking, this diameter is equal to the largest diameter of the cone-shaped enlargement at the entrance of a hole.

Normally in the practice of machining, the cutting speed v is selected for a given tool design, tool material, work material, and particularities of a given drilling operation. Then the spindle rotational speed should be calculated using Equation 1.1 and the given diameter as

$$n = \frac{1000v}{\pi d_{dr}} \tag{1.3}$$

1.1.2.2 Feed, Feed per Tooth, and Feed Rate

The feed motion is provided to the tool or the workpiece, and when added to the primary motion leads to a repeated or continuous chip removal and the formation of the desired machined surface. In all drilling tools, the feed is provided along the rotational axis as shown in Figure 1.2.

Figure 1.7 provides visualization of the basic components of the drilling regime such as the cutting feed, depth of cut, and uncut chip thickness commonly referred as the chip load in professional literature. Designations of the components of the drilling regime are shown according to the International Organization for Standardization (ISO) Standard 3002/3.

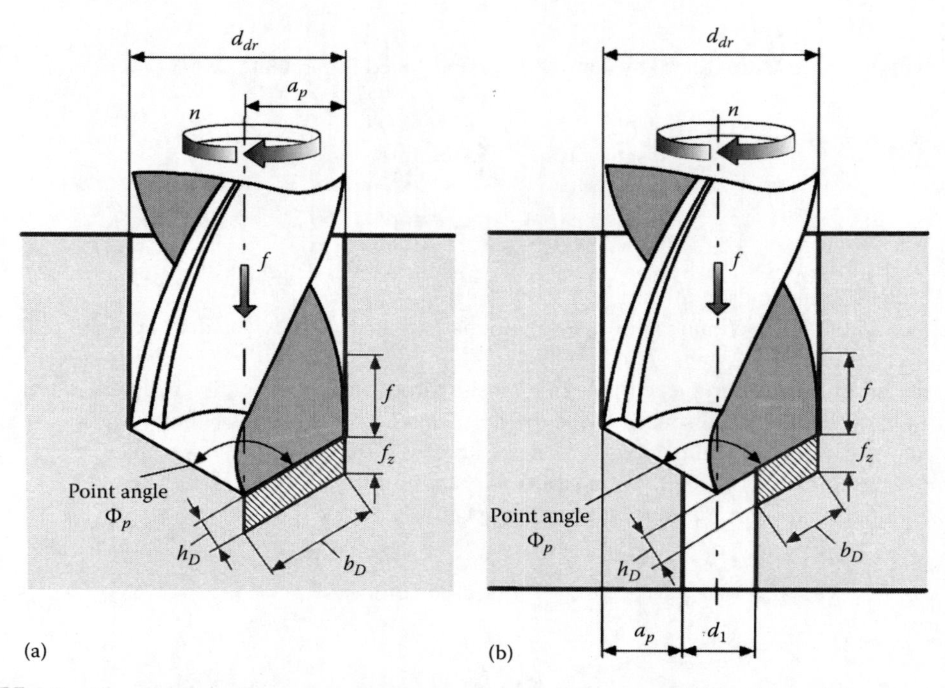

FIGURE 1.7 Visualization of the components of the drilling regime: (a) solid drilling and (b) core drilling.

The cutting feed, f, is the distance in the direction of feed motion at which the drilling tool advances into the workpiece per 1 rev. and thus, the feed is measured in millimeters per revolution (inches per revolution). The feed per tooth, f_z (the subscript z came from German *zahn*, i.e., a tooth), is determined as

$$f_z = \frac{f}{z} \tag{1.4}$$

where z is the number of cutting teeth.

The feed speed (ISO Standard 3002/3) commonly referred to in the literature as the feed rate, v_f, is the velocity of the tool in the feed direction. It measures in millimeters per minute (mm/min) or inches per minute (ipm) and calculates as

$$v_f = f \cdot n \tag{1.5}$$

where

f is the feed (mm/rev or ipm)
n is the rotational speed (rpm)

The feed speed (the feed rate) is often referred to as the penetration rate in the professional literature on drilling. It is used as a measure of drilling productivity. Substituting Equation 1.3 into Equation 1.5 and arranging the terms, one can obtain

$$v_f = k_{dt} f v \tag{1.6}$$

where $k_{dt} = 1000/(\pi d_{dr})$ is a constant for a given drill.

It directly follows from Equation 1.6 that the penetration rate depends equally on the cutting speed and feed. This fact should be kept in mind when designing a drilling operation/drill and selecting the tool material and components of a drilling system, for example, the metalworking fluid (MWF) (coolant) supply system.

Although Equations 1.4 and 1.5 are exemplified for drills, they are perfectly valid for all drilling tools shown in Figure 1.3 having the basic motions shown in Figure 1.2.

1.1.3 Depth of Cut and Material Removal Rate

The depth of cut in solid drilling is calculated as $a_p = d_{dr}/2$. In the case of core or *pilot* hole drilling shown in Figure 1.7b, the depth of cut is calculated as $a_p = (d_{dr} - d_1)/2$, where d_1 is the diameter of the pilot (core) hole.

The material removal rate is known as *MRR*, which is the volume of work material removed by the tool per unit time. Figure 1.8 presents visualization of the volume of the work materials removed in solid and core drilling. It directly follows from this figure that *MMR* (measured in mm^3/min) in solid drilling is calculated as

$$MMR = \frac{\pi d_{dr}^2}{4} v_f \tag{1.7}$$

Substituting Equation 1.6 into Equation 1.7, one can obtain

$$MMR = \frac{\pi d_{dr}^2}{4} v_f = \frac{\pi d_{dr}^2}{4}\frac{1000}{\pi d_{dr}} fv = 250 fvd_{dr} \tag{1.8}$$

Referring to Figure 1.8, one can calculate *MMR* (measured in mm^3/min) in core drilling as

$$MMR = \frac{\pi\left(d_{dr}^2 - d_1^2\right)}{4} v_f = \frac{\pi\left(d_{dr}^2 - d_1^2\right)}{4}\frac{1000}{\pi d_{dr}} fv = 250 fv\frac{d_{dr}^2 - d_1^2}{d_{dr}} \tag{1.9}$$

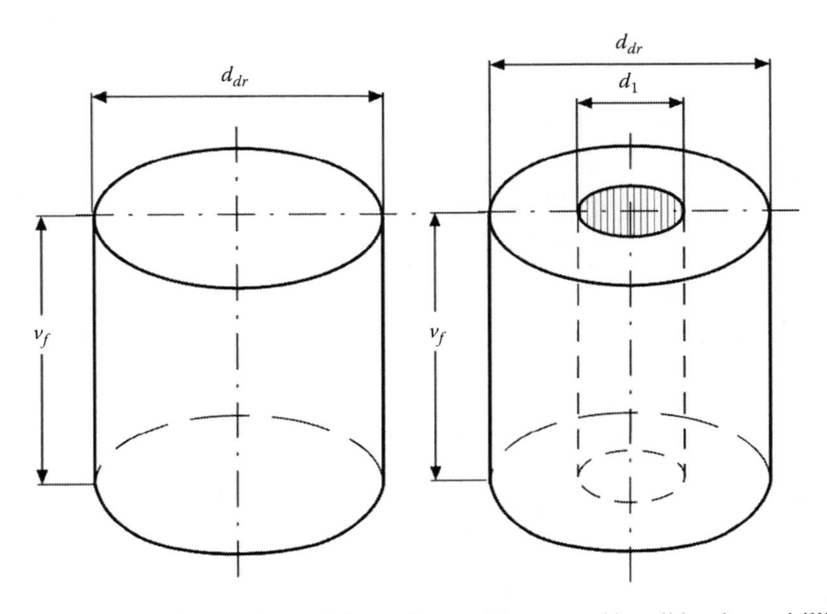

FIGURE 1.8 Visualization of the volume of the work materials removed in solid and core drilling.

1.1.4 Cut and Its Dimensions

Standard ISO 3002/3 defines the cut as a layer of the workpiece material to be removed by a single action of a cutting part. For a given cutting tooth, geometrical parameters of the cut are

- Nominal cross-sectional area, A_D
- Nominal thickness, h_D
- Nominal width, b_D

For solid drilling, these parameters are calculated referring to Figure 1.7a as follows:

The nominal thickness of cut known in the literature as the uncut (undeformed) chip thickness or chip load is calculated as

$$h_D = f_z \sin\left(\frac{\Phi_p}{2}\right) = \left(\frac{f}{z}\right)\sin\left(\frac{\Phi_p}{2}\right) \tag{1.10}$$

where Φ_p is the drilling tool point angle (discussed later in this book in Chapter 4).

The nominal width of the cut known in the literature as the uncut (undeformed) chip width

$$b_D = \frac{a_p}{\sin(\Phi_p/2)} = \frac{(d_{dr}/2)}{\sin(\Phi_p/2)} \tag{1.11}$$

The nominal cross-sectional area known in the literature as the uncut (undeformed) chip cross-sectional area is calculated as

$$A_D = h_D b_D \tag{1.12}$$

Substituting Equations 1.10 and 1.11 into Equation 1.12, one can obtain

$$A_D = \frac{f_z d_{dr}}{2} = \frac{f d_{dr}}{2z} \tag{1.13}$$

In the case of core or *pilot* hole drilling shown in Figure 1.7b, these parameters are calculated as

$$h_D = f_z \sin\left(\frac{\Phi_p}{2}\right) = \left(\frac{f}{z}\right)\sin\left(\frac{\Phi_p}{2}\right) \tag{1.14}$$

$$b_D = \frac{(d_{dr} - d_1)}{2\sin(\Phi_p/2)} \tag{1.15}$$

$$A_D = h_D b_D = \frac{f_z(d_{dr} - d_1)}{2} = \frac{f(d_{dr} - d_1)}{2z} \tag{1.16}$$

The foregoing considerations reveal that the material removal rate and undeformed chip cross-sectional area do not depend on the drill point angle while the uncut chip thickness and its width do.

Example 1.1

Problem

Determine the drill rotational speed, feed speed (the feed rate), depth of cut, material removal rate, nominal thickness of cut (uncut [undeformed] chip thickness or chip load) and width, and nominal cross-sectional area (the uncut [undeformed] chip cross-sectional area) for a drilling operation with a two-flute drill (z = 2) having Φ_p = 120° if the selected cutting speed v = 80 m/min, drill diameter is d_{dr} = 8 mm, and feed f = 0.15 mm/rev.

Solution

The spindle rotational speed is calculated using Equation 1.3 as

$$n = \frac{1000v}{\pi d_{dr}} = \frac{1000 \cdot 80}{3.141 \cdot 8} = 3184.7 \text{ rpm}$$

For practical purpose, n = 3185 rpm is adopted.

The feed speed (the feed rate) is calculated using Equation 1.5 as

$$v_f = fn = 0.15 \cdot 3185 = 477.75 \text{ mm/min}$$

The depth of cut is $d_w = d_{dr}/2$ = 6/2 = 4 mm.

The material removal rate is calculated using Equation 1.8 as

$$MRR = 250 f v d_{dr} = 250 \cdot 0.15 \cdot 80 \cdot 8 = 24{,}000 \text{ mm}^3\text{/min}$$

The nominal thickness of cut (uncut [undeformed] chip thickness or chip load) is calculated using Equation 1.10 as

$$h_D = \left(\frac{f}{z}\right) \sin\left(\frac{\Phi_p}{2}\right) = \frac{0.15}{2} \sin 60° = 0.065 \text{ mm}$$

The nominal width of the cut (the uncut [undeformed] chip width) is calculated using Equation 1.11 as

$$b_D = \frac{(d_{dr}/2)}{\sin(\Phi_p/2)} = \frac{8/2}{\sin 60°} = 4.619 \text{ mm}$$

The nominal cross-sectional area (the uncut [undeformed] chip cross-sectional area) is calculated using Equation 1.12 as

$$A_D = h_D b_D = 0.065 \cdot 4.619 = 0.300 \text{ mm}^2$$

Except for some special and form tools (e.g., the tap), the discussed formulae for calculating geometrical parameters of the cut are valid for all drilling tools. For counterboring and spotfacing, the point angle Φ_p = 180° should be used. For countersinking, these parameters vary over the cutting cycle so that the maximum nominal thickness and width of cut should be determined.

Example 1.2

Problem

Determine the drill rotational speed, feed speed (the feed rate), depth of cut, nominal thickness of cut (uncut [undeformed] chip thickness or chip load) and width, and nominal cross-sectional area (the uncut [undeformed] chip cross-sectional area) for a single-cutter boring tool shown in Figure 1.9. The tool has Φ_p = 190°, the selected cutting speed v = 240 m/min, bored hole diameter d_{br} = 40 mm, diameter of the hole to be bored d_1 = 35 mm, and cutting feed f = 0.10 mm/rev.

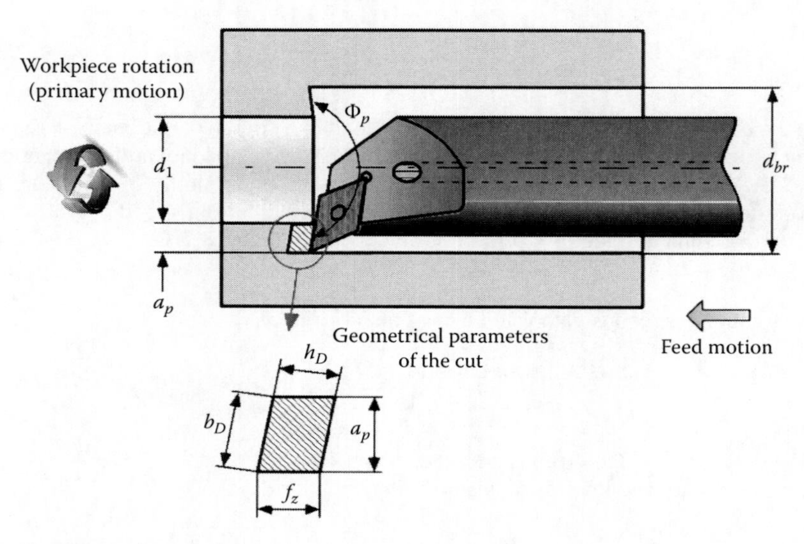

FIGURE 1.9 Boring operation.

Solution

The spindle rotational speed is calculated using Equation 1.3 as

$$n = \frac{1000v}{\pi d_{dr}} = \frac{1000 \cdot 240}{3.141 \cdot 40} = 1910.8 \text{ rpm}$$

For practical purpose, n = 1911 rpm is adopted.

The feed speed (the feed rate) is calculated using Equation 1.5 as

$$v_f = fn = 0.10 \cdot 1911 = 191.1 \text{ mm/min}$$

The depth of cut is $a_p = (d_{br} - d_1)/2 = (40 - 35)/2 = 2.5$ mm.

The nominal thickness of cut (uncut [undeformed] chip thickness or chip load) is calculated using Equation 1.10 as

$$h_D = \left(\frac{f}{z}\right)\sin\left(\frac{\Phi_p}{2}\right) = \frac{0.10}{1}\sin\left(\frac{190^\circ}{2}\right) = 0.10 \text{ mm}$$

The nominal width of the cut (the uncut [undeformed] chip width) is calculated using Equation 1.15 as

$$b_D = b_D = \frac{(d_{br} - d_1)}{2\sin(\Phi_p/2)} = \frac{40 - 35}{2\sin 190^\circ} = 2.49 \text{ mm}$$

The nominal cross-sectional area (the uncut [undeformed] chip cross-sectional area) is calculated using Equation 1.12 as

$$A_D = h_D b_D = 0.10 \cdot 2.49 = 0.249 \text{ mm}^2$$

1.1.5 Selecting Machining Regime: General Idea

Selecting the proper speed and feed rate for a particular drilling application is critical to reduce drill wear and breakage as well as to achieve high drilling efficiency in terms of cost per machined hole. In this author's opinion, the latter is the most proper measure of a drilling tool performance as well as the efficiency of the drilling operation. Therefore, the cutting speed and feed selection is not just technical as it used to be but rather is a process economy-driven issue to achieve the system objective. However, such a selection is not as straightforward as it used to be a few decades ago.

It used to be that speed and feed recommendations were selected as provided by the literature on the field. For example, one of the most popular resources is *Machinery's Handbook*, which celebrated with its 28th edition nearly 100 years as *The Bible of the Mechanical Industries*. The values selected in this way are always subject to specific job conditions so they were always considered as estimates to give the process designer/manufacturing specialist/operator an approximate starting point.

What makes selecting the right parameters so difficult is that there is little margin for error. Speeds and feeds that are too high, as well as speeds and feeds that are too low, can result in low efficiency of the whole operation and can cause drilling tool breakage. Moreover, the rapid change of tool material properties and tool coatings as well as drill design specifics, including the MWF application technique, *overrun* the recommendation provided in the reference literature so that, in many cases, the data provided can no longer be considered a good starting point.

Tables 1.1 and 1.2 give some recommendations for the selection of drilling speeds for the purpose of tool layout design (discussed later in this chapter). Once the design of the whole drilling operation

TABLE 1.1
Speed and Feed Recommendations: HSS Drills

Work Material	Hardness HB	HSS Grade	Cutting Speed (m/min)			Feed (mm/rev) for Drill Diameter (mm)			
			TiN	TiAlN	TiCN	9–12	13–17	18–24	25–35
Free machining steel	100–150	M4	60	85	72	0.17	0.22	0.30	0.35
(1112, 12L14, etc.)	151–200	M4	55	80	70	0.17	0.22	0.30	0.35
	201–250	M4	50	72	65	0.15	0.15	0.30	0.35
Low-carbon steel	85–125	M4	52	75	67	0.15	0.22	0.30	0.38
(1010, 1020, 1025, etc.)	126–175	M4	50	73	65	0.15	0.22	0.30	0.38
	176–225	M4	45	70	60	0.13	0.20	0.25	0.35
	226–275	M4	42	65	55	0.13	0.20	0.25	0.35
Medium-carbon steel	125–175	M4	50	73	65	0.15	0.15	0.30	0.38
(1045, 1140, 1151, etc.)	176–225	M4	45	70	60	0.13	0.13	0.25	0.35
	226–275	M4	42	65	55	0.13	0.13	0.25	0.35
	276–325	T15	40	60	52	0.10	0.10	0.22	0.33
Alloy steel (4130, 4140,	125–175	M4	45	65	60	0.15	0.20	0.25	0.35
4150, 5140, 8640, etc.)	176–225	M4	42	60	55	0.13	0.20	0.25	0.35
	226–275	M4	40	55	52	0.13	0.17	0.25	0.35
	276–325	T15	35	52	47	0.10	0.15	0.22	0.30
	326–375	T15	32	48	43	0.08	0.15	0.22	0.30
High-strength alloy	225–300	M4	42	58	55	0.15	0.25	0.30	0.35
(4340, 4330V, 300M, etc.)	301–350	M4	35	52	48	0.13	0.22	0.25	0.30
	351–400	T15	30	42	40	0.10	0.20	0.22	0.25
Structural steel A36, A285,	100–150	M4	42	58	55	0.15	0.25	0.30	0.35
A516, etc.	151–250	M4	35	52	48	0.13	0.22	0.25	0.30
	251–350	T15	30	42	40	0.10	0.20	0.22	0.25
High-temp. alloy	140–220	T15	9	12	10	0.07	0.15	0.20	0.25
Hastelloy, Inconel	221–310	M48	7	10	9	0.07	0.13	0.17	0.20
Stainless steel	135–185	M4	23	32	28	0.15	0.20	0.22	0.27
303, 416, 420, 17-4 PH, etc.	186–275	M4	18	28	25	0.13	0.17	0.20	0.25
Tool steel	150–200	T15	25	32	32	0.10	0.15	0.20	0.25
H-13, H-21, A-4, 0-2, 5-3, etc.	201–250	M48	18	28	26	0.10	0.15	0.20	0.25
Aluminum <8% Si		M4		180–250		0.20	0.33	0.40	0.50
>8% Si		M4		120–170		0.20	0.33	0.40	0.45

TABLE 1.2
Speed and Feed Recommendations: Carbide Drills

Work Material	Hardness HB	Grade	Cutting Speed (m/min)			Feed (mm/rev) for Drill Diameter (mm)			
			TiN	TiAlN	TiCN	9–12	13–17	18–24	25–35
Free machining steel (1112, 12L14, etc.)	100–150	P30	98	128	115	0.20	0.30	0.38	0.38
	151–200	P30	85	110	100	0.17	0.28	0.35	0.35
	201–250	P30	80	104	90	0.15	0.25	0.25	0.33
Low-carbon steel (1010, 1020, 1025, etc.)	85–125	P20	92	120	110	0.20	0.25	0.33	0.42
	126–175	P20	104	104	90	0.17	0.25	0.33	0.40
	176–225	P20	95	95	82	0.15	0.22	0.30	0.38
	226–275	P20	82	82	75	0.13	0.22	0.30	0.35
Medium-carbon steel (1045, 1140, 1151, etc.)	125–175	P20	80	104	90	0.17	0.25	0.30	0.40
	176–225	P20	73	95	84	0.15	0.22	0.28	0.38
	226–275	P20	64	82	72	0.15	0.22	0.28	0.38
	276–325	P20	55	70	63	0.13	0.22	0.25	0.35
Alloy steel (4130, 4140, 4150, 5140, 8640, etc.)	125–175	P20	76	100	87	0.17	0.25	0.33	0.40
	176–225	P20	70	92	80	0.15	0.22	0.30	0.38
	226–275	P20	64	82	72	0.15	0.22	0.30	0.38
	276–325	P20	60	76	69	0.13	0.20	0.27	0.35
	326–375	P20	52	67	60	0.10	0.17	0.25	0.33
High-strength alloy (4340, 4330V, 300M, etc.)	225–300	P20	48	60	55	0.15	0.22	0.25	0.30
	301–350	P20	43	55	48	0.13	0.20	0.22	0.27
	351–400	P20	37	48	43	0.10	0.17	0.20	0.25
						0.20			
Structural steel A36, A285, A516, etc.	100–150	P20	73	95	84	0.15	0.27	0.35	0.40
	151–250	P20	60	76	69	0.15	0.25	0.30	0.35
	251–350	P20	55	70	63	0.13	0.22	0.27	0.30
High-temp. alloy Hastelloy, Inconel	140–220	M10	80	105	90	0.10	0.17	0.22	0.27
	221–310	M10	60	85	70	0.10	0.15	0.20	0.25
Stainless steel 303, 416, 420, 17-4 PH, etc.	135–185	M30	48	64	56	0.17	0.22	0.30	0.35
	186–275	M30	37	48	43	0.15	0.20	0.27	0.20
Tool steel H-13, H-21, A-4, 0-2, 5-3, etc.	150–200	M10	48	67	58	0.10	0.17	0.22	0.27

is complete, and thus the parameters of the drill and drilling operation (tool holder, method of MWF supply, machine capabilities in terms of achievable speeds and feeds, etc.) have been selected, a standard or a special drill from at least two drill manufacturers should be quoted, asking them (besides the tool cost and lead time) to suggest speed and feed for the given designed operation as well as the estimation of tool life and tool reliability. If a special high-volume operation is to be designed, and thus a special drilling tool is to be used, a tool manufacturer should be involved in the design of the tool layout and then the drilling tool to achieve the maximum efficiency of this operation.

Although speeds and feed rates are determined by the type of material being drilled and the depth of the hole, there are two other important system considerations to keep in mind: tool holding and work holding. A frequent cause of drill breakage is a loose or poorly designed tool holder that imparts wobble to the drill. Even at slow speeds and feeds, a wobble will quickly break a drill. There are many types of tool-holding devices to choose from, but hydraulic and shrink-fit tool holders provide the most secure method of tool holding, because their use results generally in the least amount of runout. A precision collet tool holder is the next best option.

Workpiece clamping is also important. If the workpiece is not clamped properly, chatter or workpiece shifting due to cutting forces can be the case during drilling, which results in lower tool life, poor quality of the machined surface, and even breakage of the drill. If a drilling operation has been proceeding normally, and drills suddenly begin breaking, the first areas to check are the tool holder and the workpiece clamping.

1.1.6 Cutting Force and Power

1.1.6.1 Definition of Terms According to ISO Standard

While cutting, the tool applies a certain force to the layer being removed and thus to the workpiece. This force, known as the resultant cutting force **R**, is a 3D vector considered in the machine reference system (ISO 841) set in Figure 1.10a. The origin of this coordinate system is always placed at a point of the cutting edge. The y-axis is always in the direction of the primary motion, while the z-axis is in the direction of the feed motion. The x-axis is perpendicular to the y- and z-axes to form a right-hand Cartesian coordinate system.

For convenience, the cutting force is normally resolved into three components along the axis of the tool coordinate system. The main or power component of the resultant force $\mathbf{F}_c$ (known also as the tangential force) is along the y-axis. It is normally the greatest component. The force in the feed direction, which is the z-direction, is known as the feed or axial force $\mathbf{F}_f$. The component along the x-axis $\mathbf{F}_p$ is known as the radial component as it acts along the radial direction of the workpiece. The equal and opposite force **R** is applied to the cutting tool as a reaction force of the workpiece as shown in Figure 1.10b. This force is also resolved into three orthogonal components along the coordinate axis as shown in this figure. Additional component $\mathbf{F}_{xz}$ that acts in the xz-coordinate plane is also considered as it is essential for machining accuracy considerations.

1.1.6.2 Basis of the Cutting Force and Power Calculations

As known (Usachev 1915; Zorev 1966; Shaw 2004), power calculates as the product of the resultant force and the velocity in the direction of this force. In metal cutting, however, the magnitudes of the force components and the corresponding velocities can be considered. As the velocity in the

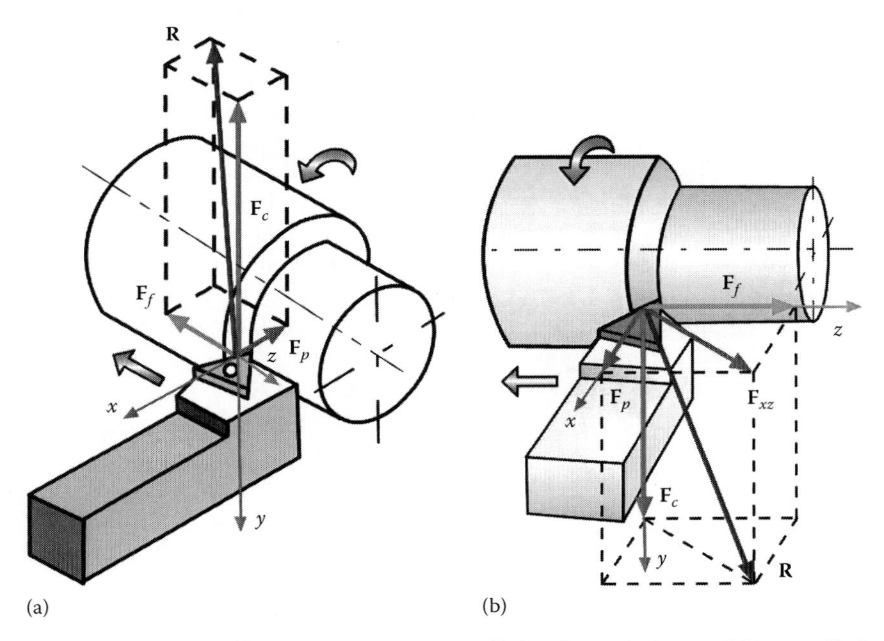

FIGURE 1.10 Cutting force and its components: (a) as applied to the workpiece and (b) as applied to the tool.

direction of the radial force F_p is zero, this component does not participate in power considerations. The axial force F_f is normally much smaller than the tangential force F_c. As discussed earlier, the velocity in the axial direction (the speed of feed) is negligibly smaller than the cutting speed. As a result, the contribution of the power due to the axial force F_f to the total cutting power is small. The greatest force component that acts in the direction of the cutting speed is F_c. Therefore, the cutting power is normally calculated as

$$P_c = F_c \cdot v \text{ (W)} \tag{1.17}$$

where
F_c is in newton (N)
v is in m/s

That is why F_c is often referred to as the power component of the cutting force.

$$P_{c-c} = \frac{F_z v}{MRR} = \frac{F_z v}{f v a_p} = \frac{F_z}{f a_p} \text{ (W/mm}^3\text{)} \tag{1.18}$$

where
the cutting feed f is in mm/rev
the depth of cut a_p is in mm

It is important to discuss here the common misconnects associated with Equation 1.18. As can be seen, the proper dimension of P_{c-c} is W/mm^3. Unfortunately, many specialists in the field do not realize the physical essence of this equation so they see only its second part where its formal dimension can be thought of as N/mm^2. As a result, P_{c-c} is often called specific cutting pressure (DeVries 1992; Anselmetti et al. 1995; Altintas 2000; Sreejith and Ngoi 2000; Boothroyd and Knight 2006) or even specific cutting force (symbol k_c) (Konig et al. 1972; Chang and Wysk 1984; Yoon and Kim 2004). In reality, it is not a true pressure or stress item.

Moreover, it is claimed that it is a kind of property of the work material that may characterize its machinability (Stenphenson and Agapiou 1996) and can be used to calculate the cutting force. The whole idea of the so-called mechanistic approach in metal cutting is based on this false perception. The role of tool geometry as the major contributor to the state of stress in the machining zone is totally ignored (Astakhov 2010).

It is discussed in Chapter 4 (Section 4.4.3) that the resultant force system in drilling can be represented by the axial force F_{ax} and drilling torque M_{dr} as shown in Figure 1.11. These parameters are used in the selection/design of the machine tool- and work-holding fixture design. The drilling torque and axial force vary in rather wide ranges depending upon the type of workpiece material, its metallurgical state and mechanical properties; tool design, geometry, and material; machining regime; MWF type and supply parameters; and many other particularities of the machining system. The power required in drilling operations P_{c-dr} is calculated using the drilling torque as follows:

In metric units of measure (the SI system),

$$P_{c-dr} = \frac{M_{dr} n}{9950} \tag{1.19}$$

where
P_{c-dr} is in kW
n is in rpm
M_{dr} is in N m

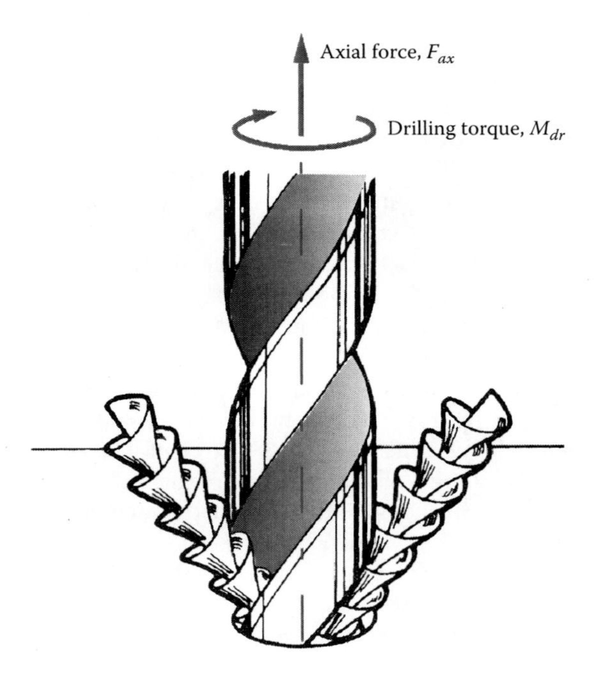

FIGURE 1.11 Total axial force and drilling torque.

In the imperial units of measure,

$$P_{c-dr} = \frac{M_{dr}n}{5252} \tag{1.20}$$

where
- P_{c-dr} is in hp
- n is in rpm
- M_{dr} is in ft-lb

Appendix A presents some common ways to determine the force factors in drilling.

1.2 DRILLING SYSTEM FOR HP DRILLS: STRUCTURE, PROPERTIES, COMPONENTS, AND FAILURE ANALYSIS

1.2.1 System Concept

Modern technological concepts make it possible to define the present stage of development as the system era. Management makes use of *system concept*, *system philosophy*, and *system approach*. Engineers and physical scientists speak of *system analysis*, *system engineering*, and *system theory*. Even in medicine or biology, the specialists speak of the *nervous system*, the *homeostatic system*, the *gene system*, etc. However, the picture is not as bright as it seemed to be in the 1960s when the system approach began to boom. Only in certain fields, for example, computer science, has the system concept been developing rapidly with great practical significance. As a result, only this field has system specialists (system analysts, system programmers, and system managers).

With the emergence of the concept of system engineering, the traditional role of specialization in engineering has been broadened or even completely changed. Traditionally, engineers specialized in a certain branch of engineering. At the system level, however, an engineer is not as much concerned with mechanics or even physics as he or she is with organization, information, and communication, with the mathematical, logical, or even phenomenological relationships among system components, whether they are physical or not. At this level, his or her principal enemy is always the complexity of a system under consideration so that wide knowledge not only in a certain engineering field but rather *broad-brush* education and experience is very useful in dealing with system problems.

System problems are often aptly described as a *can of worms*, because it is difficult to discriminate between the different elements of the problem such as the system's boundaries, the system's components and their levels, the system organization, and the interrelationships between the levels. The whole problem seems to be constantly in motion; the components are hopelessly intertwined, so much so that there may be only one indivisible component. It is difficult to grasp any one of the slippery components, and the problem is partly immersed in obscuring debris overshadowed by old beliefs, improper notions, and *experience-based* rules from the past often developed for considerably different systems or for older (sometimes really *ancient*) set of system components.

The process is somewhat as follows: the system engineer, faced with a problem derived from some system phenomenon, attempts to describe the structure of the system as a set of components. He or she assigns various relationships among the components and attempts to build the model of the system. Then he or she experiments with the model both mathematically and deductively, all the while checking the results of such comparisons and experiments with the requirements of the problem and experimental or heuristic evidence concerning the phenomenon itself. He or she modifies the model and experiments some more. Finally, he or she arrives at a satisfactory model and proceeds to analyze using various mathematical and computational techniques in order to arrive at an engineering decision. Though the previously discussed procedure looks relatively simple and logical, the chief problem here is to distinguish the system to be analyzed, its boundaries, and components. Intuition and experience at this stage are essential.

The systems approach integrates the analytic and the synthetic method, encompassing both holism and reductionism. It was first proposed under the name of *general system theory* by the biologist Ludwig von Bertalanffy (1969). Von Bertalanffy suggested that all systems studied by physicists are closed: they do not interact with the outside world. When a physicist makes a model of the solar system, of an atom, or of a pendulum, he or she assumes that all masses, particles, and forces that affect the system are included in the model. It is as if the rest of the universe does not exist. This makes it possible to calculate future states with perfect accuracy, since all necessary information is known.

However, as a biologist, von Bertalanffy knew that such an assumption is simply impossible for most practical phenomena. Separate a living organism from its surroundings and it will die shortly because of lack of oxygen, water, and food. Organisms are open systems: they cannot survive without continuously exchanging matter and energy with their environment. Manufacturing systems also fall into this category. The peculiarity of open systems is that they interact with other systems outside of themselves. This interaction has two components: input, that which enters the system from the outside, and output, that which leaves the system for the environment. In order to speak about the inside and the outside of a system, one needs to be able to distinguish between the system itself and its environment. The system and environment are, in general, separated by a boundary, which can be physical or conditional. For example, for living systems, the skin plays the role of the boundary. In many manufacturing systems, the boundary is conditional, that is, it can cover various components that may not be physically surrounded by a visible boundary. Moreover, the boundary may vary depending upon a particular system problem/issue to be considered/resolved.

The output of a system is in general a direct or indirect result from the input. What comes out needs to have gotten in first. However, the output is in general quite different from the input: the system is not just a passive tube but an active processor. For example, a steel bar stock that enters a manufacturing system transforms into a machined part and chips within this system, and electricity supplied to the system converts into mechanical work and then to heat. The transformation of input into output by the system is usually called throughput. This has given one all the basic components of a system as it is understood in systems theory (Figure 1.12).

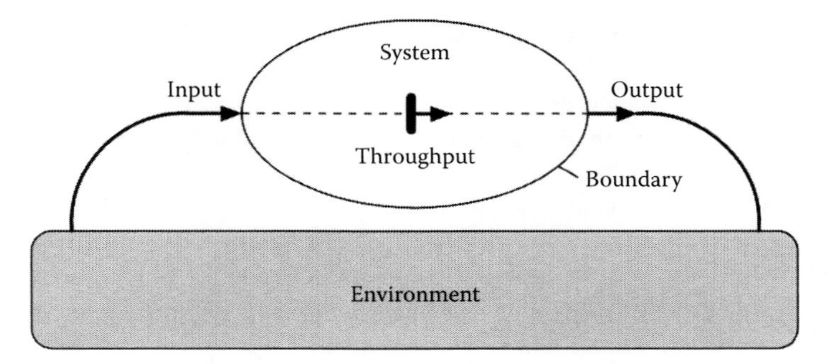

FIGURE 1.12 A system in interaction with its environment.

When one looks more closely at the environment of a system, he or she sees that it too consists of systems interacting with their environments. For example, the environment of a machine tool is full of other machines in the shop. If one now considers a collection of such systems that interact with each other, that collection could again be seen as a system. For example, a group of interacting machines form a production line, a machine shop, or a manufacturing company. The mutual interactions of the component systems in a way *glue* these components together into a whole. If these parts did not interact, the whole would not be more than the sum of its components. But because they interact, something more is added. With respect to the considered system, its components are seen as subsystems, that is, spindle system and MWF supply system. With respect to the components, the considered system is seen as a supersystem.

If the supersystem is considered as a whole, one doesn't need to be aware of all its components (subsystems). He or she can again just look at its total input and total output without worrying which part of the input goes to which subsystem. For example, if one considers a machine tool, he or she can measure the total amount of energy consumed in that machine (input), and the total number of parts manufactured (output), without knowing which components were responsible for which part of the input energy. This point of view considers the system as a *black box*, something that takes in input, and produces output, without one being able to see what happens in between (in contrast, if one can see the system's internal processes, it might be called as a *white box*). Figure 1.13 shows

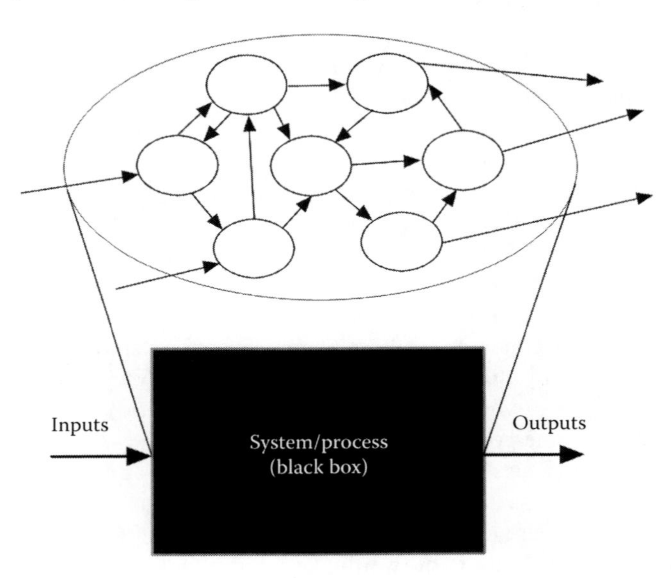

FIGURE 1.13 A system as a *white box*, containing a collection of interacting subsystems, and as a *black box*, without observable components.

a graphical representation of this concept. Although the black box view may not be completely satisfying, in many cases, this is the best one can get. For example, physics of many manufacturing processes is not exactly known or it is so complicated that it can be understood only by some university professors. Manufacturing specialists use the design of experiment techniques (Astakhov 2012) where some inputs are varied in the controlled manner and the system reacts in a certain way (output), for example, by producing more parts per shift, improved surface finish, and tool life. However, in most cases, they have little idea about the particular mechanisms that lead from the cause to the effect. Obviously, the variations of inputs, that is, the cutting speed, trigger a complex chain of interconnected reactions, involving different components and parts of the system, but the only thing that can be clearly established is the final result.

The black box view is not restricted to situations where specialists do not know what happens inside the system. In many cases, they can easily see what happens in the system, yet they prefer to ignore these internal details. For example, when quality of manufacturing is modeled using the most popular, for this purpose, Taguchi method, it does not matter which particular machine/operation/tool produces a particular parameter of the quality matrix. It is sufficient to know the whole quality matrix to estimate the influence of global inputs in the considered quality system. The *black box* view of the quality model is found to be much simpler and easier to use for quality improvements than the more detailed *white box* view, where individual machining operations and tools are considered. However, the black box view has its own severe limitations not normally revealed in the literature.

These two complementary views, *black* and *white*, of the same system illustrate a general principle: systems are structured hierarchically. They consist of different levels. At the higher level, one acquires a more abstract, encompassing view of the whole, without attention to the details of the components or parts. At the lower level, one can see a multitude of interacting parts but without understanding how they are organized to form a whole. The selection of the proper level of consideration and thus the proper approach to the study/design/problem-solving in the system is probably the most important step in system considerations, although little is known on such a selection when it comes to manufacturing system.

According to the analytical approach dominated thus far, the lowest level view is all one might need. If one knows the precise physical/chemical state of all the components and parts in the manufacturing system, he or she should be able to understand how that system functions. Traditionally, manufacturing is based on this reductionist view. Reductionism can mean either (1) an approach to understanding the nature of complex things by reducing them to the interactions of their parts, or to simpler or more fundamental things, or (2) a philosophical position that a complex system is nothing but the sum of its parts and that an account of it can be reduced to accounts of individual constituents. This can be said of objects, phenomena, explanations, theories, and meanings.

Reductionism strongly reflects a certain perspective on causality. In a reductionist framework, phenomena that can be explained completely in terms of relations between other more fundamental phenomena are called epiphenomena. Often there is an implication that the epiphenomenon exerts no causal agency on the fundamental phenomena that explain it.

Reductionism does not preclude the existence of what might be called emergent phenomena, but it does imply the ability to understand those phenomena completely in terms of the processes from which they are composed. This reductionist understanding is very different from that usually implied by the term *emergence*, which typically intends that what emerges is more than the sum of the processes from which it emerges.

Different alternative approaches to manufacturing systems have argued that such a view misses the most important thing: the system as a whole. For example, the spindle speed affects productivity and quality as well as heat generated, which in turn affects the state of the spindle. These interactions are not simple, linear, and straightforward cause and effect relations but complex networks of interdependencies, which can only be understood by their common purpose: maintaining the machining system in good conditions, that is, producing parts of required quality while assuring the needed productivity. This *common purpose* can only be considered at the level of the whole. It is meaningless at the level of an individual subsystem/system component or part.

One way to understand this is the idea of *downward causation*. According to reductionism, the laws governing the parts determine or cause the behavior of the whole. This is *upward causation*: from the lowest level to the higher ones. In emergent systems, however, the laws governing the whole also constrain or *cause* the behavior of the parts.

This reasoning can be applied to most of the surrounding things. Although the behavior of a transistor in a computer chip is governed by the laws of quantum mechanics, the particular arrangement of the transistors in the chip can only be understood through the principles of computer science. The structure of the DNA molecule, which codes our genetic information, is determined by the laws of chemistry. Yet, the coding rules themselves, specifying which DNA *triplet* stands for which amino acid, don't derive from chemistry. They constitute a law of biology. Each level in the hierarchy of systems and subsystems has its own laws, which cannot be derived from the laws of the lower level. Each law specifies a particular type of organization at its level, which *downwardly* determines the arrangement of the subsystems or components at the level below. When it is said that the whole is more than the sum of its parts, the *more* refers to the higher-level laws, which make the parts function in a way that does not follow from the lower-level laws.

Although each level in a hierarchy has its own laws, these laws are often similar. The same type of organization can be found in systems belonging to different levels. For example, all open systems necessarily have a boundary, an input, an output, and a throughput function. The components of a manufacturing system need energy in the same way that the system as a whole needs energy, even though the components might receive energy in different forms. The form is different, but the function is the same: to allow the components to function properly. Closed systems at different levels have many features in common as well. The binding forces that hold together the planets in the solar system, the atoms in a molecule, or the electrons in an atom, although physically different, have a very similar function. The embeddedness of a systems in supersystem holds for all types of systems: a manufacturing plant consists of shops, which consist of machines, which consist of components, which consist of materials, which consist of crystals/macromolecules, which consist of atoms/molecules, which consist of nucleons, which consist of quarks.

The modern general system theory is intended to reveal similarities in structures and functions for different systems, independent of the particular domain in which the system exists (Skyttner 2006). It is based on the assumption that there are universal principles of organization, which hold for all systems, be they physical, chemical, biological, mental, or social. The mechanistic worldview seeks universality by reducing everything to its material constituents. The systemic world view, on the contrary, seeks universality by ignoring the concrete material out of which systems are made, so that their abstract organization comes into focus. When it comes to manufacturing systems, a suitable fragile balance between these two approaches should be found depending on a particular objective of system analysis.

This chapter provides brief and simplified consideration of the drilling system and its components in terms of their suitability for high-efficiency drilling. It introduces basic system objectives and rules to help practical tool and manufacturing engineers in the intelligent selection of system components. The main attention is paid to the systemic selection of the cutting tool (drills) to achieve the system objective. This chapter introduces the novel system-based drill failure analysis procedure. Being practical and supported by many real-world examples, such a procedure should be considered as an inherent part of the drill system consideration as it aims to help practical manufacturing engineers, process planes, cutting tool designers, and practitioners improve performance/efficiency of drilling operations.

Although the following chapter mainly discusses the system approach to drilling, its main ideas are fully applicable to any cutting tool and tooling in modern high-efficiency machining environment. In this author's opinion, such an approach should be used in machining system design, retrofitting, and components and tool selection, that is, in any aspect of manufacturing.

1.2.2 Drilling System

1.2.2.1 Structure of the Drilling System

Significant progress in drilling has been achieved that resulted in the introduction of high penetration rate drilling. It has emerged during the past 10 years as the process that allows a penetration rate of more than 5 m/min for aluminum alloys, more than 2 m/min for cast irons, and more than 1 m/min for alloy steels. It became possible due to significant improvements in the manufacturing quality of drills, including the quality of their components, the implementation of better drilling machines equipped with advanced controllers as well as their proper maintenance, the application of better MWFs, better training of engineers and operators, and many other factors. However, the actual penetration rate and drilling process efficiency (the cost per unit length of drilled holes) vary significantly from one application to another, from one manufacturing plant to the next, depending on an overwhelming number of variables. Optimum drill performance in drilling is achieved when the combination of the cutting speed (rpm), feed, tool geometry, tool material grade, and MFW parameters is selected properly depending upon the work material (its hardness, composition, and structure), drilling machine conditions, and the quality requirements of the drilled holes (Astakhov 2001). To get the most out of a drilling job, one must consider the complete drilling system, which includes everything related to the operation (Figure 1.14). Such a consideration is known as the system engineering approach according to which the drilling system should be distinguished and analyzed for coherency of its components.

According to system engineering theory, it is improper to consider any component of a drilling system separately, thereby ignoring the system's properties. The so-called component approach is a common manufacturing practice in today's environment, where different manufacturers produce the various components of the machining system but no one seems to be responsible for system coherency. Low efficiency, subpar quality, and tool failure are direct results of such an approach. Any lack of coherence in the machining system normally leads to visible tool failure. Such failures can easily turn a drilling operation into the bottleneck operation in the automotive industry as a complete production line or a manufacturing cell can be down for a long time due

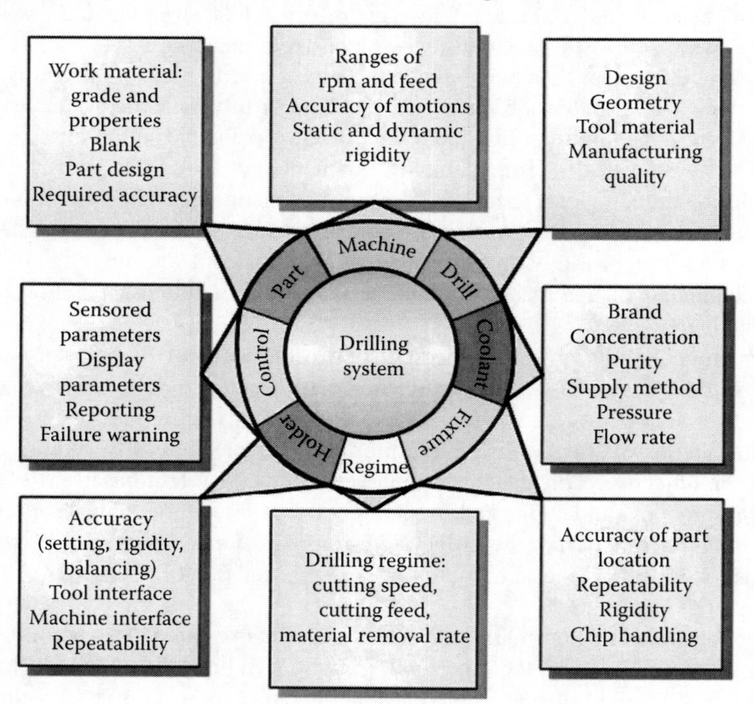

FIGURE 1.14 Drilling system structure.

to the failure of a single tool. The direct consequences are significant downtime, low efficiency of an operation, and poor quality of machined parts.

Reading these elaborations gained from everyday manufacturing practice, one can ask a logical question—what seems to be the problem? There are a number of drill manufacturers and even a greater number of drilling machine makers. Each manufacturing plant has trained personnel including engineers and operators, maintenance schedules, and re-sharpening services. However, our recent survey indicates that in the automotive and mold-making industries, the following are observed:

- The correct drill geometry is selected less than 30% of the time.
- The tool is used at the rated cutting regime only 48% of the time.
- Only 57% of the tools are used up to their full tool-life capability.
- The correct tool material is selected less than 30% of the time.
- The correct MWF (coolant) parameters (Figure 1.14) are used 42% of the time.
- The correct parameters of the drilling system are used less than 40% of the time.

To understand the performance of the drilling system and thus the root cause for many drilling-related problems, one should always consider the components of the drilling system shown in Figure 1.14 in a systemic way. One can appreciate the system properties of the drilling system if one realizes that the same drill used in different drilling machines shows a wide range of results from breakage to excellent performance; the same drill used on the same machine exhibits different results for different work materials; the same drill used on the same machine for drilling the same work material performs differently depending upon a particular brand of MWF used for the operation, the MWF flow rate, filtration, and temperature; and the performance of the same drill used on the same machine for drilling the same work material using the same MWF parameters would depend largely on the type and conditions of the tool holder. The same drill used on the same machine for drilling the same work material using the same MWF parameters and the same tool holder depends significantly on the machining regime. Moreover, the quality parameters of machined parts and drill performance are also affected by the part-holding fixture, namely, its accuracy, repeatability, and rigidity. The drill performance also depends on the extent of the operator's experience and training. The latter is particularly true if the control system provides relevant information to the operator and/or when this operator uses manual gages to inspect the quality of the machined hole. As seen, each individual system component can affect the system performance dramatically. The key here is to assure system coherency, that is, the condition when all the system components work as a *team* to achieve the ultimate system's objective.

Unfortunately, the drill manufacturer is often unfairly blamed for failures of the drilling system as the sole culprit because the drilling tool, as the weakest link in the drilling system, fails as a result of improper performance of various system components. For example, one manufacturer of gundrills for the automotive industry was blamed for gundrill breakage occurring at the tip–shank brazed joint. For over 5 years, this manufacturer tried to improve the strength of this joint. When this strength became sufficiently great, breakage of gundrill's carbide tips began to occur. An analysis of the root cause of this problem showed that the lack of the MWF flow rate supplied to the drilling zone caused the drill breakage. Because the root cause was not properly determined, the increased strength of the discussed brazed joint shifted the breakage to the carbide tip as a new weakest link.

1.2.2.2 Coherency Law

Although there are a number of important system laws, for example, the deformation and optimal temperature laws, the coherency law is always of prime importance in practice design and analysis of high performance (HP) drilling systems. The coherency law states that all components of the drilling (machining) system should be coherent, that is, logically connected and consistent in their quality to achieve the prime system objective. For example, if one buys and installs on an old car advanced sidewall design tires specially designed for use at the Indianapolis 500 car race, the performance of the car will be worse than with its old *native* tires although new tires come with the speedway's distinctive *Wing and*

Wheel official logo in full color. Although these tires are probably the best that the tire manufacturers can offer (not to mention their cost), they are not suitable for this old car. The same analogy can be made for cutting tools—the best and most expensive drill will not perform well if the machining system does not support its performance. In other words, a drill may have the best geometry and tool material and can be perfectly designed and manufactured, but it may not perform well for a given application. If, for example, the quality of the machined hole allows a drill with 20 μm runout and the drill is selected with 10 μm runout, then the total runout of the tool holder and spindle system should not exceed 10 μm, that is, a high-precision tool holder and spindle should be the case. If, for example, an HP drill is made of an advanced tool material, that is, submicrograin carbide with tantalum–hafnium additives (Ta_4HfC_5), and has internal channels for MWF supply, then the drilling system should be able (1) to run the drill at optimal cutting speed for this tool material, (2) to assure drill installation runout ≤10 μm, (3) to supply MWF with the pressure sufficient to achieve the flow rate needed for optimal drill performance, etc.

1.2.2.3 System Objective

In many books and research papers, drill tool life is of prime concern. In practical tool management, the cost of the drill, tool life, and other miscellaneous articles are of prime concern because these parameters are easy to measure and report. In manufacturing reality, this is not nearly the case.

The prime system objective is an increase in the drill penetration rate, that is, drilling productivity, which is entirely a combined system rather than a component characteristic. The major constraints are quality requirements (i.e., surface finish, straightness, shape) to drilled holes. Therefore, all system parameters should be selected so that they fully support this system objective, pushing the boundary of the major constraint as far as is physically possible. Unless such an understanding settles in the minds of researchers, tool manufacturers, and process designers/manufacturing professionals, no significant progress in drilling efficiency can be achieved.

As pointed out by Fiesselmann (1993), in all industries, on average, perishable cutting tools seldom represent more than 8% of the total direct/indirect product manufacturing costs. The author's experience shows that in the modern manufacturing plants in the automotive industry, the direct tooling cost is approximately 6% (if the most expensive broaching operations are excluded, then it is approx. 4%). For CNC machining centers and manufacturing cells where $1.00 is the benchmark, for 2,200 operating hours/year, $1.00 min means an operating cost of $132,000/year for just one machine (cell). Even factoring in 75% efficiency for loading/unloading, changing tools, and setup, an increase in the penetration rate by 50% amounts to a potential yearly savings of $24,750/CNC machining center/year. In the modern manufacturing plants in the automotive industry, where the time for robotic loading/unloading, inspection, and other auxiliary operations are minimized, a potential yearly savings can be as high as $35,000/machining cell/production line.

1.2.3 Case for HP Drills

Two things can be said about HP drills:

1. They drill better and faster, and last longer, than traditional high-speed, carbide, and polycrystalline diamond (PCD) drills.
2. They are much more expensive than traditional on-shelf drills due to special grades of tool materials, special grinds, closer tolerances on practically all design and geometry features, much higher requirements to surface roughness on the ground surface, etc. In other words, their design, manufacturing, and inspection require higher qualifications of the engineers, many extra operations, advanced equipment, well-qualified engineers, and well-trained operators.

Do the benefits justify the increased cost of HP drills? Not always. And not always in a way that can be seen just by analyzing short-term costs with no system considerations. Evaluating the economics of an HP drill involves a range of different factors. The bulk evaluation of drilling costs presented in the previous section does not account on many particularities of the drilling economy so that a more detailed analysis (at least, a methodology for such an analysis) is needed to understand the principal sources of HP drill efficiency and thus to carry out this analysis when a case for deployment of HP drills is developed/discussed.

To make the further analysis more detailed, a simple but realistic example of drilling work is considered as a model. The numbers used to analyze this job may not apply directly to a particular shop or application but they are realistic as based upon the current industrial practice. Moreover, the logic of the analysis can be applied to a wide variety of drilling operations. What this analysis shows is that the cost-effectiveness of a given drill may be determined by factors that are far removed from the tool's initial price. The prise often blinds the purchasing department/personnel as their objective is the purchasing cost reduction and not the total gains obtained by a company in the implementation of HP drills.

Table 1.3 presents the results of economy analysis. The analysis is based on a simple drilling job. Figure 1.15 shows a part where six holes are to be machined. Figure 1.16 shows particularities of the tool layout. The specifics of the job are listed as follows:

- The workpiece is made of high-silicon aluminum alloy, and it is run on a single manufacturing cell. Six cells are used in the shop.
- The production rate for all of the work on this part is 20 pieces/h/cell.
- The piece needs six drilled step holes having diameters as shown in Figure 1.16.
- The job runs three shifts for a total of 20 h/day, 5 days/week.
- The shop runs at a flat rate of $60/h including machine costs, labor, and overhead.

TABLE 1.3
Result of Economy Analysis/Comparison of the Standard and HP Drills

Item	Standard Drill	HP Drill
Feed rate (mm/min)	3,600	9,600
Total stroke length (mm)	26	26
Machining time (min)	0.072	0.027
Machining time cost per minute ($)	1	1
Machining time cost per hole ($)	0.072	0.027
Cost per part ($)	0.432	0.162
Drill life (holes)	5,000	7,000
Drill purchasing cost ($)	95	145
Drill cost per hole ($)	0.02	0.02
Combined cost per hole ($)	0.09	0.05
Combined cost per part ($)	0.55	0.29
Holes to drill per year	662,640.00	662,640.00
Number of re-sharpenings	8	8
Total drill life	40,000.00	56,000.00
Total drills needed	17	12
Average drill breakage per year due to casting and machining system defects	6	3
Actual total drills needed	23	15
Total drill purchase cost ($)	2,185.00	2,175.00
Re-sharpening cost per drill (S)	20.00	60.00
Total re-sharpening cost ($)	3,680.00	7,200.00
Total drills cost	5,865.00	9,375.00
Cutting cost per year ($)	**60,300.24**	**31,617.39**
Total cost ($)	**66,165.24**	**40,992.39**
Net cost per part ($)	0.60	0.37
Net cost per hole ($)	0.10	0.06

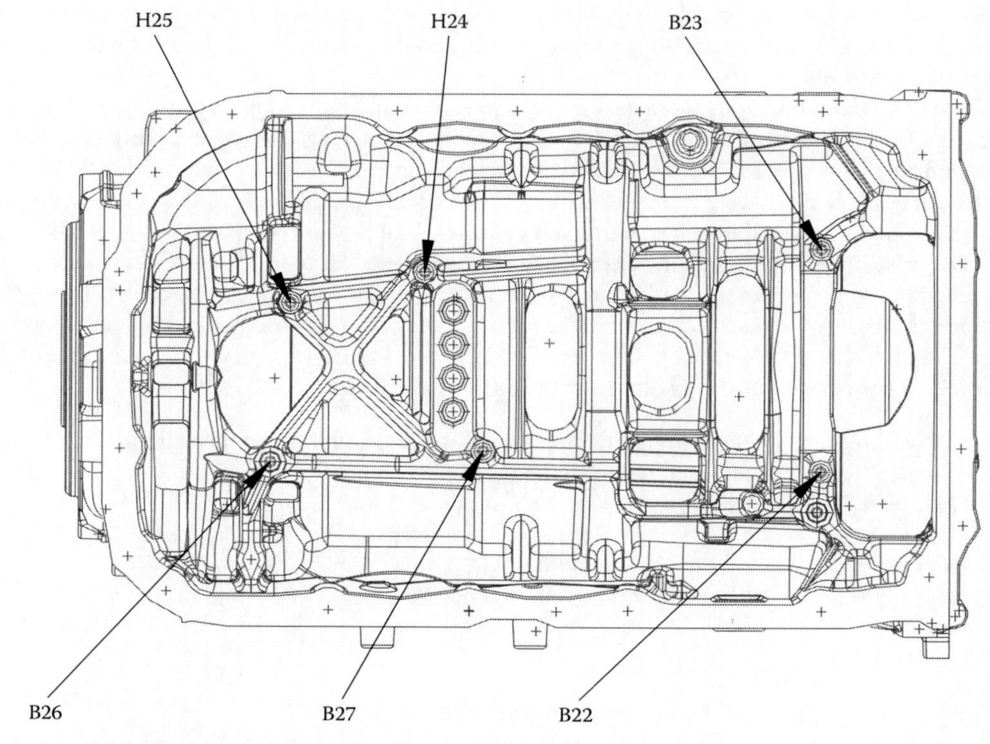

FIGURE 1.15 Six holes to be machined in each part.

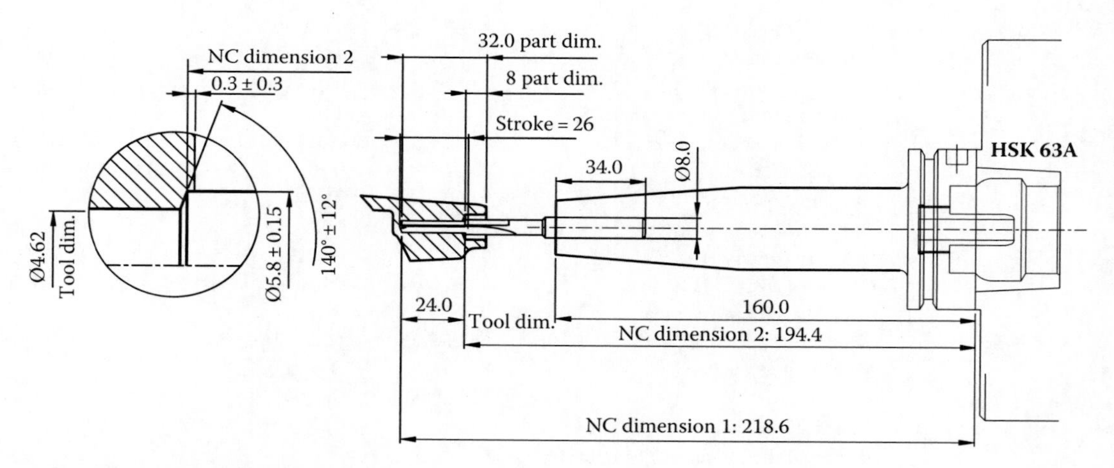

FIGURE 1.16 Particularities of the tool layout.

The analysis begins with looking at a basic measure of drill performance: the feed rate. The two tools compared are a typical, good-quality standard carbide drill and an HP carbide drill designed as the application-specific tool. The cutting regimes used for these drills were determined as follows:

1. The standard drill allows the spindle rotational speed n = 12,000 rpm and cutting feed f = 0.3 mm/rev. The speed is selected according to the drill manufacturer specification. It is limited by the grade of carbide and coating used. The cutting feed is limited by two major factors: (1) the maximum allowed axial force and (2) the ability of drill to reliably remove

the formed chips with no clogging of the chip flutes. Naturally, this drill was tried for high speed with no success as tool life reduced significantly due to accelerated drill corner wear (discussed in Chapter 2). Any attempt to increase the feed per revolution resulted in clogging of the chip flute and drill breakage.
2. The special drill is designed to the requirements of *VPA-Balanced*© design concept. It allows the spindle rotational speed n = 24,000 rpm and cutting feed f = 0.4 mm/rev.

The purchasing cost of the standard drill is $95 and its re-sharpening cost is $20 while those of the HP drill are $145 and $60, respectively. The difference in the initial cost absorbs the cost per development and testing, the higher cost of application-specific carbide grade (in the considered case, the generic HP carbide grade is changed to newly developed CERATIZIT grade with custom coolant hole diameters and location) (see Chapters 3 and 6 for details), and additional operations of finish grinding get a mirror surface finish. The higher re-sharpening cost is explained by additional grinding operations with specially dressed diamond wheels and detailed inspection after each re-sharpening.

The analysis presented in Table 1.3 does not account for the inventory cost and cost of the drill in the flow. These two will reduce saving margin. However, this analysis has not accounted for increased productivity, that is,

- The yearly program, while using HP drills, can be completed in 1/3 of the time needed if the standard drills are used. As a result, the CNC machine can be used for manufacturing other parts. For example, implementation of a few HP drills for the part shown in Figure 1.15 allowed sufficient cycle time so it was found that there is no need to purchase another manufacturing cell to handle the increased program. The resulted savings were much greater than that shown in Table 1.3.
- The reliability of the HP drill is at least twice greater than that of the standard drills. Moreover, these drills can handle casting porosity and small inclusions with no excessive wear and drill breakage. Saved downtime and cost of scrapped parts can be significant when the HP drills are used.

It is worth to mention that the automotive plant in the considered example runs six parallel manufacturing cells so the saving due to development and implementation of the single small HP drill shown in Table 1.3 is approx. $150,000/year. Note that the presented numbers are taken from a real-life case.

The real advantage of HP drills developed and manufactured using the *VPA-Balanced*© design concept is that it can be designed and made to address a particular need, for example, to increase productivity, to increase tool life, and to improve the quality of the drilled holes (surface roundness, diametric accuracy, true position, etc.). Moreover, three- or two-pass operations used to achieve high quality of drilled holes can be accomplished in one-pass saving machining time and tool costs.

1.2.4 Design of Drilling Systems

Although the design of each drilling system should follow its unique path depending on given practical conditions, the basic common features of the drilling system design are the same. As an example, Figure 1.17 shows a simplified flow chart for the HP drilling system design for the existing drilling machine, that is, for the most common practical case.

In this flow chart, the machine selection, fixture design, controller selection and programming, and verification stages are well covered in the literature (Campbell 1994; Parkesh 2003; Smid 2003; Nee 2004, 2010; Koening 2007). The selection of MWF parameters for drilling operations, design of the internal coolant channels, and chip removal parameters is discussed in Chapter 6. The tool layout, drill selection particularities, and particularities of some important components of the drilling system/drill manufacturing are discussed in this section.

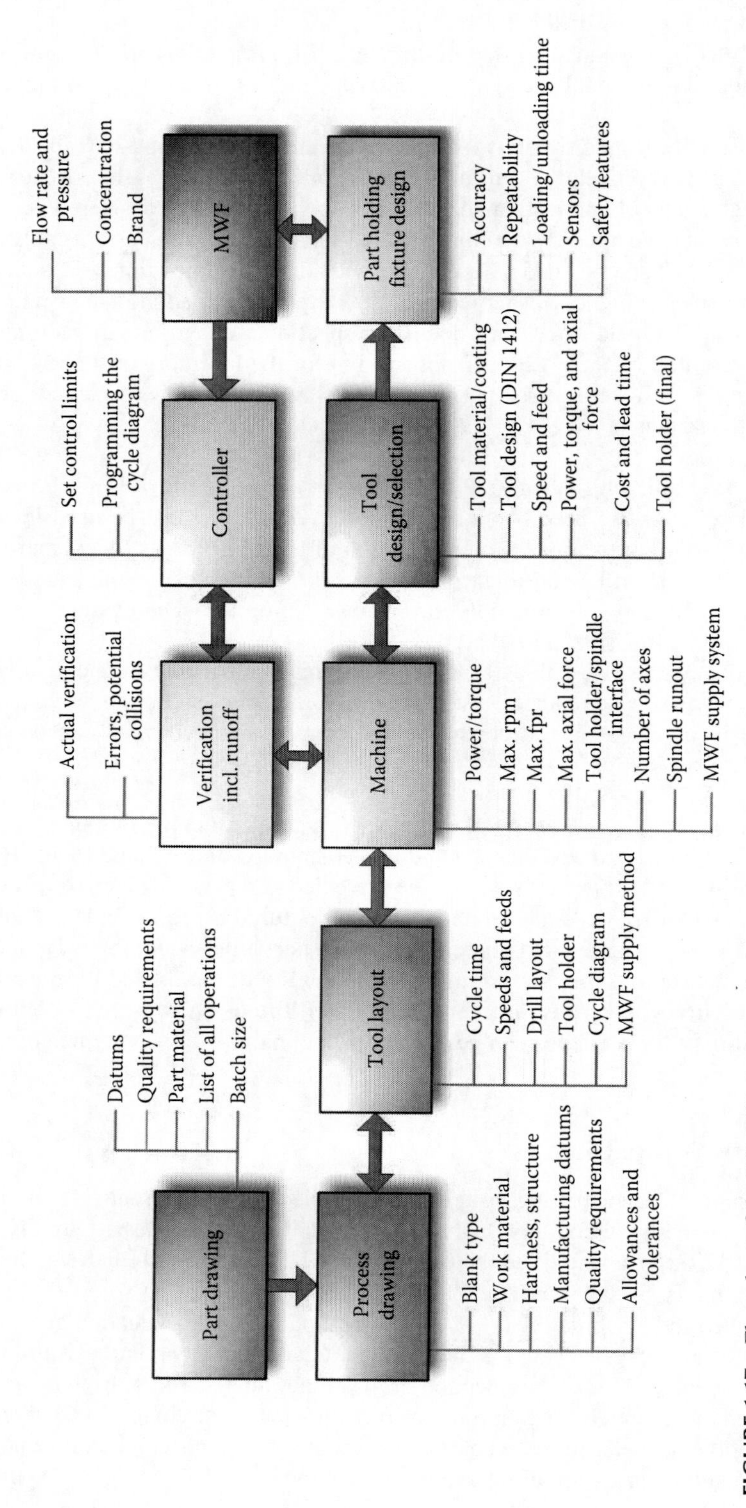

FIGURE 1.17 Flow chart of the HP drilling system design.

1.2.4.1 Part Drawing Analysis and Design of the Tool Layout

The design of a drilling operation should always begin with a part drawing analysis. In the considered example, it is assumed the batch size and the list of all manufacturing operations (preceding and subsequent to the considered drilling operation) are known. Moreover, it is assumed that a specific type of blank (casting, forging, bar stock, etc.) has been selected.

The analysis of the part drawing includes the following major steps:

1. Analysis of the part material: chemical composition, mechanical properties, and metallurgical state (including allowable inclusions, porosity, cavities). Special attention should be paid to allowable variation of these characteristics.
2. Datums and datum features as related to the hole to be machined.
3. Quality requirements to the considered hole that include size, shape, location tolerances, and surface roughness (or surface integrity if specified [Astakhov 2010]).

The results of this analysis are used in the constructing of the process drawing.

The next step is the design of the process drawing. This drawing is *derived* from the part drawing and contains essential information relevant to the hole-making operation. This information includes all the requirements to the drilling operation. Moreover, it contains suggestions on the proper part location to assure such requirements. Figure 1.18 shows an example of the process drawing. As can be seen, it includes the diameter of holes to be drilled and information on the location of this hole

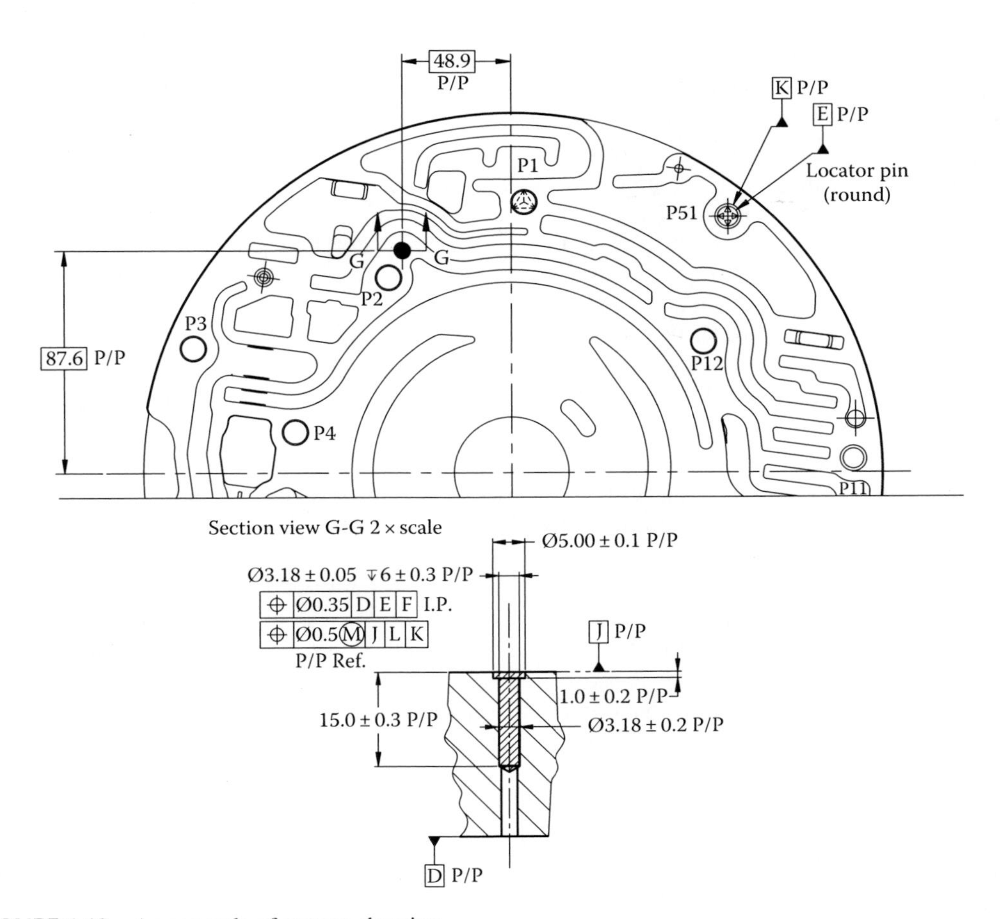

FIGURE 1.18 An example of process drawing.

with respect to the part design datums. Technological datums as, for example, location of the hole with respect to the locating pins, are also added to assure the required hole location with respect to the designed datum. The sense and meaning of GD&T designations are explained in Chapter 7 in details.

The next step is to design the tool layout. Usually, the tool layout is the handover document transferred from engineering to the shop floor. A tool layout captures the tool information in the language of engineering, consisting of drawings, bills of material, and parameter lists. A single-tool layout refers to a single-tool assembly for a certain operation performed with a specific spindle on a specific machine tool. The layout documents the components of the tool assembly, including spare parts.

An example of the tool layout in the automotive industry is shown in Figure 1.19 and the corresponding parameters are shown in Table 1.4. As can be seen, the following information is derived from the tool layout:

- Drill starting and end positions that define (1) the drilling length and (2) the length of the drill working part.
- The diameter of the drill (as derived from the diameter of the hole to be drilled) and drill point angle as equal to that requested by the part drawing.
- The drilling regime in terms of the cutting speed and feed. As the development of a tool layout is an iterative process, the initial assignment of these two regime parameters is normally based upon (1) cycle time available for the operation, (2) the data available in the company's tooling database (for similar applications), and (3) recommendation of the leading tool suppliers available in their online catalogs. The so selected cutting speed and feed then can be changed several times in the process of final revision of the drilling operation.
- The general tool holder that depends on the machine available for the operation. At this stage, the machine–holder interface is fixed (by HSK63-A in Figure 1.19) while the holder/tool interface is still open. In the considered case, it can be shrink-fit, hydraulic, collet, etc., interfaces depending upon the accuracy required, availability of a particular tool setting machine, and many other technical and logistic factors.
- The cycle diagram of the drilling operation as shown in Figure 1.20. This diagram allows one to calculate the machining cycle time (the time needed for the machining part of the drilling operations). When one adds this time to the time needed for loading–unloading of a part, he or she obtains the drilling cycle time.
- The drill sketch as shown in Figure 1.21. Note that this is not a drill drawing. Rather, it represents a general idea what kind of drill is needed for the drilling operation. Later on, the particular drill parameters (drill geometry, design, particular grade of tool material, coating, etc.) will be selected.

In the considered procedure, the machine is assumed to be known so that the cutting regime including the cutting power, cycle programming, tool holder, and stroke length is selected to be feasible for the chosen machine.

Tool layouts are a great way to communicate and brainstorm ideas back and forth between tool manufacturers (suppliers) and tool users. They are also used in programming the machine and tool presetting equipment. In continuous improvement efforts, on-site cost reduction teams improve cycle times—changing speed and feed rate or calling for alternate tooling—and thereby change the tool specification based upon information available in the tool layout.

1.2.4.2 Drill Selection/Design

According to the tool layout, the feed rate $v_f = 2757$ mm/min and the cutting speed $v = 250$ m/min are selected based upon the required (economy-justified) cycle time. At this stage, an important decision is to be made, namely, what tool to use: standard or custom. Although more than 70% of cutting tools in the automotive industry are custom, the use of relatively inexpensive standard drills seems to be an attractive option.

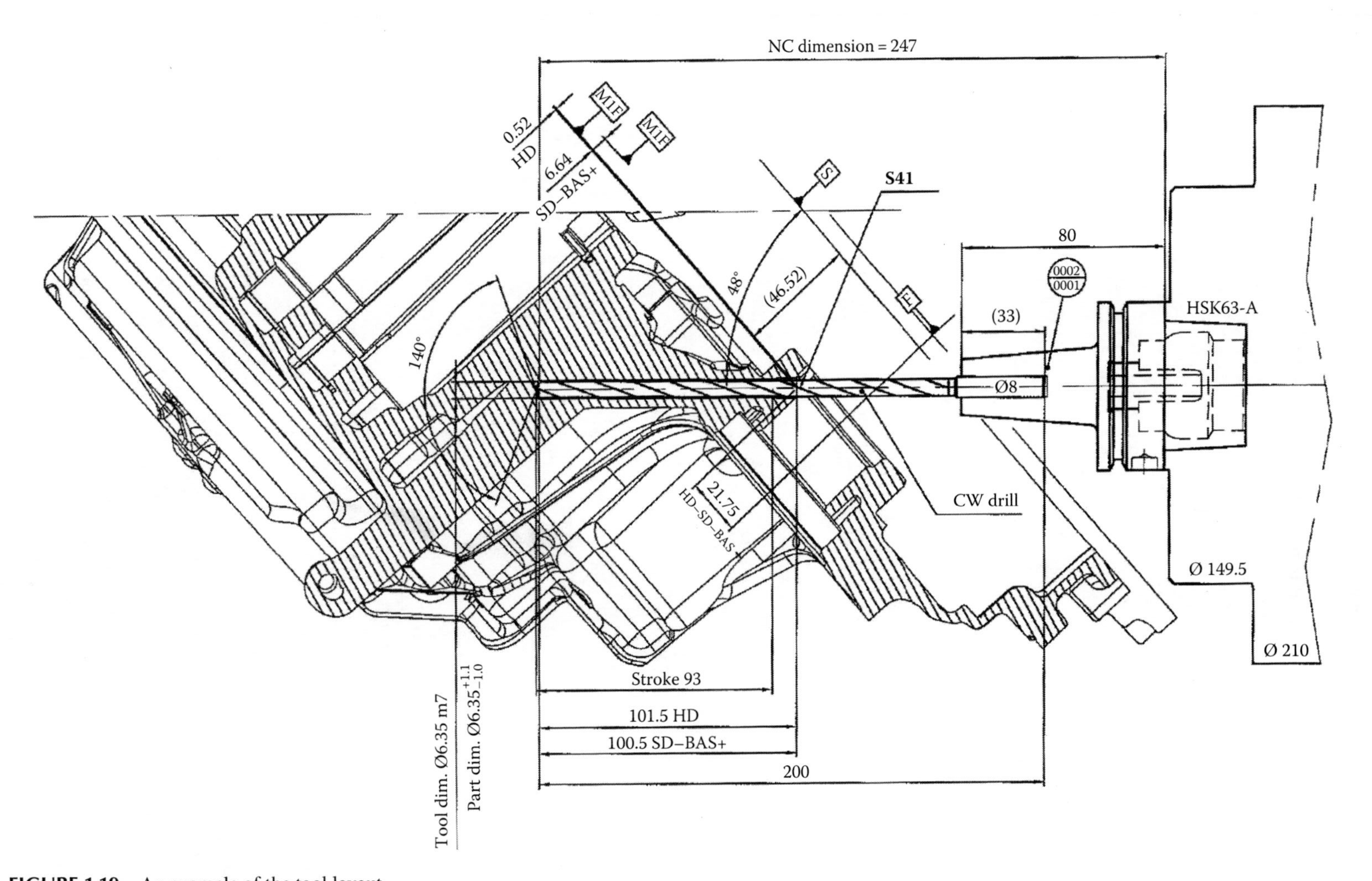

FIGURE 1.19 An example of the tool layout.

TABLE 1.4
Tool Layout Particularities

Work Material			Tool Material
Aluminum ANSI A380Mod			Sintered coated carbide
Si 7.5%–10%, Cu 2%–4%,			Tool type
Mg 0.3% max			Twist drill
Optional material			Tool holder
Aluminum ANSI A383Mod			Per layout
Si 12% max, Mg 0.3% max			
Blanks: die castings			
Machining regime as recommended by a leading cutting tool supplier			
Cutting diameter	d	mm	6.35
Cutting speed	v	m/min	250
Spindle rotational speed	n	rpm	12,532
Cutting feed	f	mm/rev	0.22
Feed per tooth	f_z	mm/rev/tooth	0.11
Uncut chip thickness (chip load)	h_D	mm	$h_D = t_z \sin(\Phi_p/2) =$ 0.11 sin(140/2) = 0.103
Feed rate	v_f	mm/min	2,757
Coolant			
Type			Water soluble
Concentration		%	8 min
Supply method			Internal through tool
Flow rate		l/min	18
Pressure		MPa	5.5

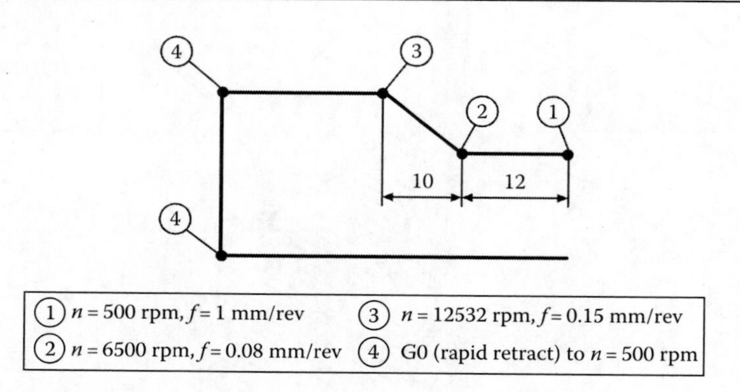

FIGURE 1.20 Drilling cycle diagram.

What exactly determines if a tool is special is a matter for discussion. The basic definition used in practice defines a custom drill as to be anything other than a jobber length drill. The custom tool is developed to address a particular need, for example, particular feature configuration in an engine or transmission part. However, when such a part goes into production, and the car manufacturer starts buying the tools in mass quantities, then this tool technically is no longer a custom tool. It can be referred to as a *standard custom tool* (Kennedy 2010).

Even when produced in large quantities, custom tools lack the economy of scale characteristic of standard tools. A part maker's willingness to pay a premium and take advantage of specials depends largely on the shop culture. Standard tooling is cheaper, but there is value in the more expensive custom tools; they offer more capability. If a shop uses the custom tools properly, their productivity can

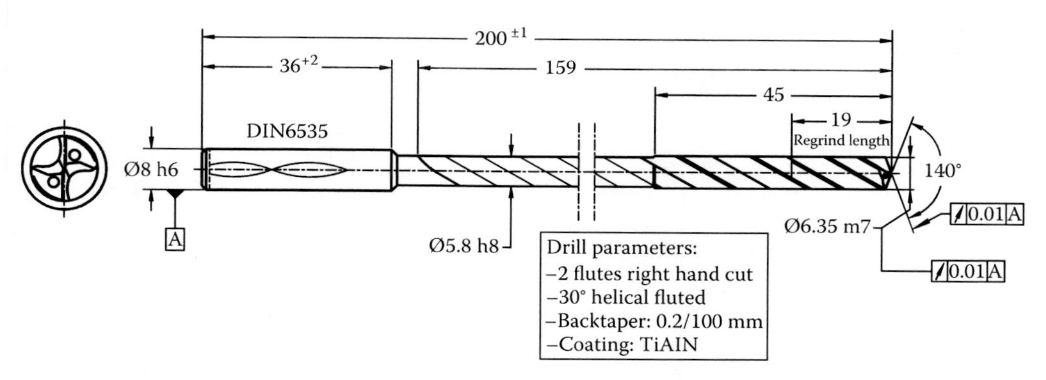

FIGURE 1.21 Drill sketch.

go way up and their profits can go way up too. If a shop is reluctant to make large capital purchases, a custom tool may enable this shop to increase productivity without purchasing new equipment.

Large aerospace and automotive manufacturers regularly use custom drills because they generally provide longer tool life and better hole quality, and the cost of reworking out-of-spec holes can easily exceed the custom tools' cost. Those manufacturers plan months ahead in most cases. Other shops may not be as proactive with inventory. A shop may deplete its supply of a certain diameter custom drill and then rush to find a standard drill in that diameter. This drill is simply issued to the shop floor, and the machine operator fights the consequences of such a decision because it is convenient.

Therefore, there are two possible outcomes of the considered step: (1) a standard drill is selected and (2) a custom-made drill is selected based upon the economy analysis similar to that presented in Section 1.2.3.

If a standard drill is selected, then the further consideration is the choice of a tool supplier as the selected standard drill can be manufactured by a number of companies. Selection of a particular supplier of the standard drill should be based on the comparison of the following items:

- Suitability of a standard drill for the considered operation in terms of work material and selected machining regime
- Track record of suppliers (if available) for supplying/supporting standard drills for other operations in the considered manufacturing facility or to the other manufacturing facilities of the company
- Quality rating of the suppliers
- Customer technical support provided by suppliers
- Lead time or on-shelf supply basis
- Cost of the drill and cost of re-sharpening
- Lead time for a re-sharpening including the location of the nearest re-sharpening facility

Often, however, the initial cost of the drill is the major driving rationale in the selection of the supplier as such a selection is carried out by the purchasing department. Fortunately, such a short-sighted strategy has been changing over the last decade so more and more direct manufacturing personnel have the decisive say in the selection of the tool supplier.

Once the supplier is chosen, the drill design, tool material, and coating are fixed by the chosen tool manufacturer. Technically, no further actions on the drill are needed. In the considered case, however, one problem, namely, the reliability of chip removal from the long hole, remains. This is because the drill is long so standard long drills have the thick web (discussed in Chapter 4) at the expense of reduced cross-sectional area of the chip flute. It may present a problem as the feed rate is high thus a great amount of chip is to be removed over the flutes.

Often, to deal with the problem of chip removal from long holes, the so-called peck drilling—a drilling operation that periodically retracts the tool to clear chips or flood the hole with MWF—is recommended by many literature sources on drilling so that the peck drilling cycle is a standard part of many drilling machine CNC controllers. Experienced practitioners, however, are of the opinion that peck drilling is the second most efficient way to destroy carbide drills. Putting them on a nice hard surface and then smashing them with a big hammer is the first way. There are two major reasons for that

1. Many work materials harden almost instantly if rubbed with tools, instead of cutting continuously. Even a couple of revolutions without continuous penetration can leave the bottom of the hole so hard, so shiny, and so perfectly armored and burnished that the drill has a hard time to start cutting again.
2. The loose chip left in the chip flute can drop back into the partially drilled hole as the drill retracts. Note that the chip is much harder than the work material due to severe plastic deformation during chip formation. When the drill returns, severe recutting takes place that is not beneficial to the tool wear.

When a custom-made drill is selected based upon the economy analysis similar to that presented in Section 1.2.3, a tool supplier capable to manufacture the special drill has to be chosen. Often, the special tool is considered as having some additional steps or other nonstandard features, not HP drills capable of high-efficiency drilling. When, however, the latter is the case, a small development project may be needed. The steps in such a project are considered in the next sections.

1.2.4.3 Drill Material Selection

To design a tool layout for an HP drilling operation, a drill material should be selected. It is an important decision to make at this stage because drilling accounts for a large percentage of the total cost of manufacturing, and one of the primary ways to cut drilling costs is to reduce drill breakage and premature wear.

Drill breakage is the most common problem associated with drilling operations, and it is a major cause of machine downtime. When a drill breaks, it must be removed from the workpiece using another drilling operation. This wastes valuable production time and can produce out-of-tolerance holes, scarred surface finishes, and other quality problems in the finished component.

Premature drill wear also reduces the efficiency of drilling operations. Typically, a drill wears too fast when for a given drilling operation (drilling system and properties of the work material), the tool material and cutting speed and feed rate are not selected properly for the application. The effect of an excessive cutting speed is normally easy to recognize as it causes the excessive wear of the drill periphery corners as discussed later in Chapter 2. In some cases, too slow speeds and feeds have the effect of burnishing the hole, creating unwanted bright spots on the hole walls, and cause tool wear. Premature wear also can lead to drill breakage as the drilling torque and axial force increase with increasing tool wear.

Chapter 3 of this book provides help in the proper selection of the tool material. In the considered case of drilling due to high cutting speed, sintered carbide or PCD is the material of choice. As the work material is high-silicon aluminum alloy then a special low cobalt carbide grade of high quality should be considered for the application. Alternatively, a PCD drill (see Chapter 5) can be used. A PCD drill is approximately five times more expensive and the lead time for such drills normally doubles that for special carbide drills. However, a PCD drill can offer at least twice greater cutting speed, and thus, the penetration rate combined with up to five times tool life increases. Although the listed advantages of a PCD drill are attractive, one needs to know the following before considering an implementation of such drills:

- Many leading drill manufacturers include PCD drills in their catalogs. The problem is that only a few of them are capable of manufacturing PCD drills of high quality and reliability.
- The drilling system for implementation of PCD drills should be of much greater quality than that for cemented carbide drills in terms of structural rigidity, available speeds and feeds, purity of MWF as well as its available flow rate at high-pressure, low casting inclusions and cavities count (discussed later in Chapter 2), etc.
- The tool re-sharpening should always be done by the tool supplier.

1.2.4.4 HP Drill Design/Geometry Selection

As pointed out in the Preface, the most frequently asked question is about the difference between HP drills and usual drills. The answer is in the combination of the system approach with *VPA-Balanced*© design concept to drill design, manufacturing, and application. This combination can be briefly described as follows.

VPA-Balanced© design concept is the methodology to design and manufacture a drill following a set of rules:

1. The geometry of the drill point is designed for a particular work material so that it assures
   a. Minimum axial force and drilling torque (Chapters 4 and 5)
   b. Balanced design of the major cutting and chisel edges to assure chip flows from the different cutting edges with no crossing/interference and providing sufficient room to assure that there are no restrictions on these flows (Chapters 4 and 5)
   c. Perfect force balance of *VPA-Balanced*© drill due to proper assigning of their design and manufacturing tolerances (Chapter 7)
   d. Efficient use of the MWF flow supplied to the drill through its proper distribution in the machining zone (Chapters 4 through 6)
2. The drill material is selected to be
   a. Of high quality and consistency (Chapter 3)
   b. Application specific (Chapter 3 provides the guidelines for making the proper choice of the tool material based upon its manufacturing practices and relevant properties)
3. High manufacturing quality (Chapter 2) including the tool final inspection (Chapter 7)

Although an HP drill designed following the *VPA-Balanced*© design concept is fully capable to assure the highest possible productivity of a given drilling operation, this capability is realized if and only if it is fully supported by the drilling system and its components.

Are HP drills universal? No, they are not. They can be compared to high-performance cars that require certain road conditions, high qualification of drivers, intelligent maintenance, etc. High-performance cars may not be suitable for country roads, low-cost maintenance, and for carrying heavy loads. Moreover, they are not meant for just everyday commuting *home–workplace and back* at average speed 40 km/h. However, when properly used, they deliver high comfort, joy of driving, and incredible reliability. As more and more people apposite these benefits, the market segment of these cars increases fast. Almost the same can be said about HP drills. They are suitable for high-quality, well-maintained drilling systems. They require high-quality tool materials and special, application-specific coatings, intelligent tool manufacturing including complicated inspection equipment and procedures, computerized presetting in high-quality drill holders, etc. The benefits of their use include high penetration rate (high productivity), long tool life and great tool reliability, and much better quality of the machined hole—often two- or even three-tool hole operations can be reduced to single-tool operation. HP drilling is the major trend for the future where unattended fully computerized manufacturing lines with greatly reduced labor cost, high productivity, and high consistency of quality of the manufactured parts is the ultimate goal.

Reading the previous paragraph, a casual reader may think that this is a dream or at least something from the distant future. In the author's opinion, it is today's reality, for example, in the automotive industry, where manufacturing cells and production lines deliver high productivity, with minimum attendance, and great quality.

To finish the introduction of HP drills, the author wants to remind potential readers that only 20 years ago, carbide drills were debatable tools as they cost much more than high-speed steel (HSS) drills; they require much more intelligent handling and re-sharpening, better tool holders and machine spindles, etc.; and many specialists at that time thought that carbide drills would have very limited use in practical manufacturing. Well, today these drills are common standard tools used universally while HSS drill usage has reduced. The author is deeply convinced that the same will happen with HP drills as the economy of modern manufacturing dictates so. That is why this book is written to accelerate wider acceptance of such drills.

1.2.4.5 Part-Holding Fixture Design/Selection

A part-holding fixture is a device for locating, holding, and supporting a workpiece during a manufacturing operation. Fixtures must correctly locate a workpiece in the intended orientation with respect to the machine axes. Such a location must be invariant in the sense that the device must clamp and secure the workpiece each and every time with the assigned accuracy. When the manufacturing drawing is properly designed as in the example shown in Figure 1.18, the design of the work-holding fixture does not present the problem as major manufacturing datums and their relations to the design datums are already defined. Figure 1.17 lists some important considerations in the work-holding fixture design.

When using standard drills, any type of work-holding fixture starting with the simplest shown in Figure 1.22 can be used. When one uses HP drills, work-holding fixtures should be efficient. In this context, the word *efficiency* implies the following:

- Effective loading/unloading of parts. When production lines or manufacturing cells are used, the robotic loading/unloading or rail transportation of pallets into machines is normally the case (Figure 1.23a). For stand-alone CNC machines, parts are preloaded in the fixture that can be robotically loaded into machine. Figure 1.23b shows an example where the fixture containing a number of parts is loaded into the CNC turntable installed in the CNC machine.

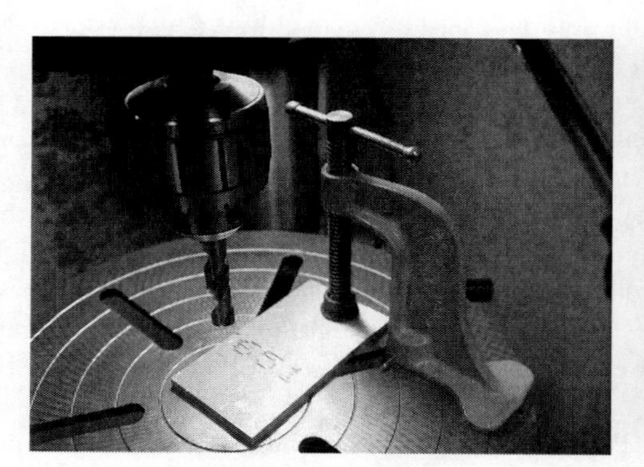

FIGURE 1.22 Simplest work-holding fixtures that can be used with standard drills.

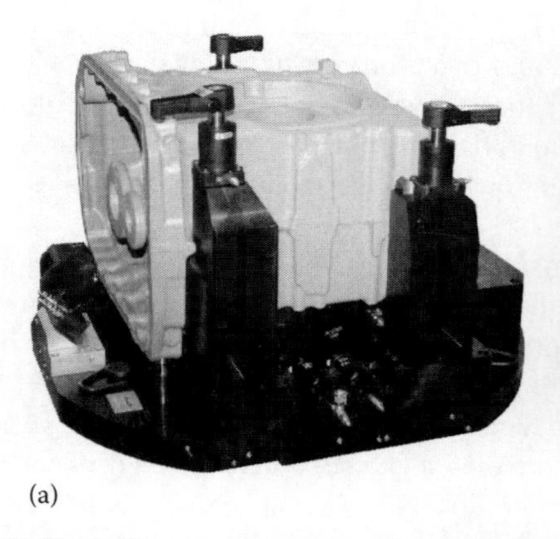

(a)

(b)

FIGURE 1.23 Examples of work-holding fixtures used in HP drilling operations; (a) automotive pallet-type fixture used in production lines and (b) fixture for a CNC stand-alone machine.

FIGURE 1.24 The dimensions inside the machine (*real-estate*) should be known in the fixture design/selection to accommodate the maximum number of parts machined in a single operation.

- The maximum parts to be machined simultaneously in a single operation for a given machine (Figure 1.24).
- Proper part position in the fixture according to the selected manufacturing datums (the home position). End switches or other signal transducers with communication with the machine controllers are used to indicate if the part is loaded/positioned properly.
- Structural rigidity of the fixture to assure vibration-free drilling operations.
- Additional MWF or compressed air flash to clean a fixture from the formed chips.

1.2.4.6 Metal Working Fluid

Chapters 5 and 6 of this book discuss particularities of MWF application in drilling tools. Figure 1.17 lists the major parameters of MWF to be considered at the stage of drilling operation design. The application of HP drills sets special requirements on MWF. Among them, the following are of prime importance:

1. The system should deliver the flow rate needed for efficient HP drill performance.
2. The purity of the MWF in terms of clearness (e.g., particle count), pH, special additives, etc. (Astakhov 2012), should be suitable for the application with HP drills.

Strict maintenance procedure including a detailed control plan should be developed to assure MWF quality.

1.2.4.7 Controller

Any modern drilling machine is actually a CNC machine. CNC stands for *computer numerically controlled*. It refers to any machine tool (i.e., mill, lathe, drill press) that uses a computer to electronically control the motion of one or more axes on the machine. CNC machine tools use software programs to provide the instructions necessary to control the axis motions, spindle speeds, tool changes, and so on. As such, multiple axes of motion simultaneously are controlled simultaneously resulting in 2D and 3D contouring ability. CNC technology also increases productivity and quality control by allowing multiple parts to be machined using the same program and tooling.

Once the part has been designed using conventional mechanical design methods (structural analysis, FEA, fatigue study, etc.), its 3D solid model of virtually any standard CAD format is imported into the *computer-aided manufacturing* (CAM) software (e.g., MasterCAM as the most popular). The raw material stock size, the part's coordinate origin, datums, and other necessary information for each tool used are imported from the process drawing. For standard tools, a tool library exists, which is simply a database of tools and their related parameters created earlier. As such, the appropriate tool is selected from the library and set the parameters necessary for machining that feature. Typical parameters include spindle speed, depth of cut, feed rate, number of passes, and tool path pattern. The use of a special tool may cause the need to create a library of special tools. Then the programmed tool path(s) is verified (using the cycle diagram on the tool layout) by running the CAM software's virtual machining cycle. The described routine is well known as presented in machine controller manuals with multiple examples (e.g., those by Siemens, Fanuc) and in many published books on the subject (e.g., Smid 2007; Kandray 2012). For HP drills, however, the capability of adequate process monitoring is of prime concern in the selection of a suitable controller as the part programming, tool change, and establishing speed and feed are nowadays standard functions of practically any controller.

The need to monitor drilling processes arises when there is a requirement for increasingly reliable, automated machinery and production cycles in order to satisfy the continual demands of the market to lower costs. Within this framework, it therefore becomes necessary to supervise HP drilling operations by systems that carry out process analysis of the machine tool.

Process monitoring is an absolute requirement, given the growing prevalence of unattended or semi-attended machining and the increasing cost of tooling, workpiece materials, and downtime. The price of not detecting something as simple as a broken tool is just simply too high to pay. One broken tool can trigger a chain reaction of failed tooling that may include breakage of other tools, machine damage, and many defective parts until the operator realizes the problem. If a tool breaks, the controller should be able to shut the machine off and stop the cycle.

Process monitoring encompasses three areas. Tool monitoring detects worn, broken, or missing tools. Machine condition monitoring uses sensors to detect excessive vibration and torque loads that can affect the machine itself. The third area is process optimization that includes adaptive, real-time control of cutting parameters. In addition to boosting productivity, adaptive control increases tool and machine life.

Modern process monitoring systems are based on a PC plug-in computer-integrated tool and machine monitoring card that is inserted into a slot in an open-architecture CNC. Via a sensor bus, the card can receive input from external power, force, torque, acoustic, and vibration sensors. It can also directly monitor torque signals generated by the machine's spindle and axis drives through a feature called digital torque adaptation. In adaptive control mode, the card employs a code that reads the information from the tool and machine and overrides the programmed machining parameters to increase or decrease the feed rate as needed. The feed rate increases when the tool is cutting air or drilling requirements are not high and decreases when the hole profile puts heavy loads on the drill and machine.

The following is the minimum list of the parameters that should be monitored in HP drilling:

1. Monitoring the power (as it is converted into the drilling torque) of the spindle has proven particularly effective in controlling the variations in drilling operations. Directly monitoring the power consumed by the spindle motor allows one to understand exactly what is happening with the tool. A new, sharp tool requires less power to cut than a worn, dull tool. Power monitoring systems are available that take their data directly from the motor controller; others measure with transducers on the wiring to the motor. It is called *sensory perception* for machines—the system obtains information that would not otherwise be detectable and notifies the operator as soon as possible when there is a problem. When first setting up a job, a *learn* cycle is run with new, sharp tools, during which the power monitor

takes data, which it *remembers* for future reference. The monitor sets a range of *normal* operating levels. While in production, the monitor notices when the power goes beyond one of the limits for the particular tool that is running. If the power goes up suddenly, that may mean a crash or a broken tool, and the monitor sends a stop signal to the machine controller. Also, as the tool wears, the power it requires goes up gradually. One can set the monitor to compare different tools/machining regimes and other particularities of machining operation, that is, such an operation can be optimized.

2. Detecting broken or missing tools. Post-process, positive-contact sensor (known as PCS) tool breakage detectors use a sensor arm to physically touch the tool tip at the end of each cycle. If the tool is broken or missed, the machine shuts off. More sophisticated post-process tool monitoring via lasers or coolant jets, ultrasound, and pressure sensors is also available.
3. Monitoring the axial force. Monitoring the force is important for assessment of tool performance. Force measurement is often made indirectly, by placing a sensor on the machine in an area that experiences strain proportional to the force to be measured. For example, Montronix, Inc. (Ann Arbor, MI) offered a three-axis load cell beneath the turret of a CNC lathe to measure all three standard components of the cutting force. The most advanced machines in the automotive industry have a load cell embedded in the spindle that gives the most precise measurement and monitoring of the cutting force. Moreover, the force history is stored in the memory of the machine tool controller and then can be analyzed to optimize cutting tool performance and/or compare cutting tools of different designs.
4. Monitoring a combination of MWF flow rate–pressure. The importance of this feature is discussed in Chapters 2 and 6.
5. Acoustic tool monitoring. Very small tools, 2 mm or less in diameter, can be difficult to monitor using force, torque, or power. For this situation, an acoustic monitoring technique was developed. A small acoustic sensor is installed in a location where it can *hear* differences due to breakage or wear. The monitoring system learns the sound/vibration characteristics of normal machining and sends an alert or stop signal when it detects a deviation outside the normal range.

For each listed parameter, the upper and lower limits are set in the controller so the controller issues a warning when a controlled parameter is approaching either limit and shuts off the machine when it passes the lower or upper limit. A clear message is issued by the controller explaining the reason for shutting off.

1.2.4.8 Machine Tools

A great variety of machine tools commonly referred to as machines are used to perform drilling operations. Besides the common use of vertical drilling machines and lathes to perform drilling operations as shown in Figure 1.1, drill presses and radial drilling machines (Figure 1.25) are also common machines for drilling used in job-shop production environment. Unfortunately, none of the listed machines can be used in HP drilling as their accuracy, rigidity, time needed for part loading–unloading, power and speed ranges, etc., are not in ranges required for such drilling. In other words, the introduced coherency law would be violated if an HP drill was used on the listed machines. As a result, the built-in efficiency of the HP drill cannot be realized due to machine limitations.

Suitable machines for HP drilling should be high-end modern CNC machines known as machining centers in industry. The term *machining center* describes almost any CNC milling and drilling machine that includes an automatic tool changer and a table that clamps the workpiece in place. On a machining center (as contrasted with a turning machine), the tool rotates, but the work does not. The most basic variety of this type of machine is also the most basic CNC machine tool—a vertical machining center. Examples of modern machining centers are shown in Figure 1.26. While vertical machining centers can be high-end machines because of

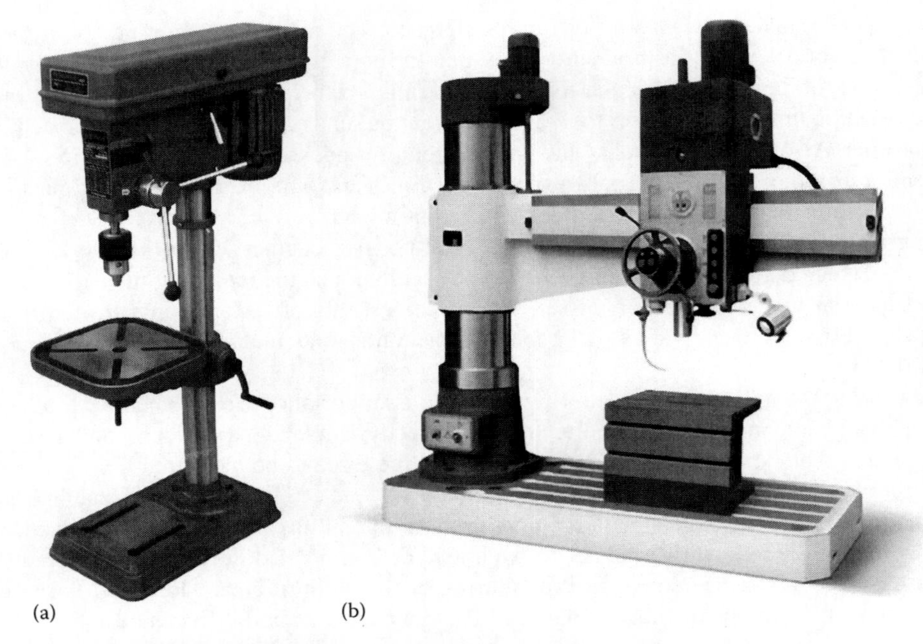

FIGURE 1.25 (a) Simple drill press and (b) radial drilling machine.

FIGURE 1.26 Examples of modern machining center.

their precision and/or their size, a small and simple vertical machining center is a relatively low-cost CNC machine tool that often represents a new machine shop's first machine tool purchase. The orientation of the spindle is the most fundamental defining characteristic of a machining center. Vertical machining centers and horizontal machining centers have (obviously) vertically and horizontally oriented spindles. Vertical machines generally favor precision while horizontal machines generally favor production—but these generalizations are loose, and plenty of machines break out of them. Other choices in machining center orientation include the universal machining center, which can change between vertical and horizontal spindle arrangement. More common than this is the five-axis machining center, which adds rotary motion to the machine's linear motion. The machine pivots the tool and/or the part not only to mill and drill at various angles but also to mill swept surfaces. Machining centers linked by an automated pallet system can form an automated machining cell. Such a cell can machine a queue of different parts without operator attention by shuttling the parts in and out of the various machines as appropriate. Related machines in this category include the boring mill, which generally describes a large machine for heavy and/or precise milling and hole making.

The selection of machine tools is largely a matter of judgment, based on the consideration of many variable factors. No fixed rules can be laid down, but the uses to which the machines are put are divided roughly into three classes that govern to a large extent the types of machines that should be purchased, whether they should be machines of a wide range of usefulness, standardized machines equipped with special tooling, or special-purpose machines of special design capable of very large and continuous production. To determine into which of these classes the requirement for new machines falls, an analysis should be made of the following factors: (1) quantity of production required and its duration, (2) method of machining and tolerances and finish required, (3) possibility of a change in design of the product, (4) cost of production, (5) when delivery of machine is required, and (6) money available for the purchase. The basic characteristics to be considered while selecting a machine for a particular drilling application are listed in Figure 1.17.

Many find it difficult to obtain unbiased information about machine tools. Obviously, people selling machine tools are skilled at pointing out the advantages of their products over those of the competition and just as skilled at downplaying the limitations inherent in their machines. Marketing brochures can effectively convey machine features and some level of technical performance data, but they do not provide true comparisons with competitive machines.

The first step is to develop a list of critical machine requirements for HP drilling applications. This list may include the following:

1. Size of working *envelope* (depending on the type of machine, this may include characteristics such as table size, tool clearance, tool holder size, and tool swing).
2. Ranges of available spindle speed and feed.
3. Machine horsepower/torque at the spindle and their *constancy* over the range of spindle speeds.
4. Spindle precision (runout) at high spindle speed. Spindle static and dynamic rigidity.
5. Machine structural rigidity. A matrix of machine tool specifications may leave out what is arguably the most critical factor in any high-speed operations—rigidity. Look for machine tools with rigid structures that dampen vibration, provide stiffness, and resist flexing even under high acceleration rates. Do not accept welded bases or frames. Machines should have heavily ribbed, cast iron construction that has been perfected through finite-element analysis.
6. Type of tool holders used.
7. Tool change time is an important consideration in HP drilling, where wringing seconds out of cycle times can lead to great cost savings.
8. Tool magazine capacity.
9. Type of machine control (see Section 1.2.4.7 for the requirements).
10. Compatibility with existing CAM software (or programs already written).
11. Number of available machining axes (generally between three and five).
12. High-pressure MWF (coolant) supply capability.

Once the list is developed, the next step is to have the group rank the importance of each of the critical machine requirements. The objective here is to achieve group consensus, not necessarily unanimous rankings, for each critical machine requirement. Any ranking scale can be used in this exercise, but generally a scale of 1–5 (with 5 being the highest ranking) works well. Ranking the critical requirements helps a company differentiate itself from the *typical* machine tool buyer by assigning more weight to specific machine tool features important in its operation.

Once the ranking of critical requirements has been established, this information should be loaded into a chart, or spreadsheet, along with the machine tool candidates to be evaluated. (A spreadsheet is recommended as it simplifies the calculations required and allows some *what-if* scenarios.) Each machine tool candidate is then rated against the ranked critical requirements. Once again, a scale of 1–5 is an effective rating method. Multiplying the critical requirement's ranking by the rating of each machine tool results in an overall numeric score for each machine tool alternative.

In the development of the chart, one should be aware that some manufacturers use Japanese Industrial Standard (JIS), others ISO, and some German VDI guidelines. Therefore, it is possible to have one machine brochure listing a machine's positioning accuracy at 2.5 μm, while another lists 5 μm. However, depending on what standard is used, the machine delivering 5 μm may be the more accurate. It is a good idea to ask the machine tool builder which standard they are using because the comparison of machines built on the same standards should be considered. Even when comparing machines built using the same standards, one should be aware that accuracy and repeatability may not always be truly accurate or reliable. A variety of factors can lead to variations between the specs in the sales brochure versus the specs of the machine installed on the shop floor.

In addition to accuracy specifications, shops should closely examine spindle specifications. Two spindles with seemingly identical motors can have significant differences in performance. Sales literature may not point out if specifications are for peak or continuous duty rating. Be aware that a 30,000 rpm spindle may only run a couple of minutes at a certain torque load. Check if you're comparing continuous power spindle ratings versus peak horsepower ratings.

Another important aspect to consider—and one that may not be apparent in the sales literature—is how machines deal with thermal issues. All high-speed machining centers have thermal issues. In the manufacturing process, heat is produced by the spindle axis drive systems and the cutting process. This heat can distort machine components and make it difficult to produce accurate parts. Spindle growth is of particular concern in HP drilling, where tolerances are especially tight.

Find suppliers that employ laser measuring equipment to monitor thermal growth of the spindle. Suppliers also may use temperature mapping techniques to measure thermal growth in the axes of their machining centers. Temperature sensors mounted on a machine casting and spindle head send dynamic data to the CNC, where it is processed on the fly. The control then calculates the thermal deviation using factory-developed algorithms and feeds new offsets to the relevant machine axes. Thermal compensation ensures real-time monitoring and feedback of temperature changes. Eliminating thermal variation ensures parts are machined to close tolerances with high repeatability.

Keep in mind that machine specifications can be misleading in terms of accuracy, spindle speeds, etc., and that a spreadsheet isn't always going to point you to the right machine for your shop. The best way to approach a machine tool purchase is to gather as much information as possible, ask suppliers educated questions about machine specifications, and talk to as many end users as possible. Finally, go beyond machine specifications and check into the service and support programs of your suppliers. Do they have parts and spindles on hand if you have a crash? Does the application department have the expertise to help you optimize your machining processes? Machine tool suppliers should be able to help shops engineer solutions that deliver outstanding and consistent results for their applications. Service and support may be hard to quantify on a spreadsheet, but they're vital and shouldn't be left out of the evaluation process.

1.2.4.9 Tool Holders

In the last 15 years, tool holders have changed significantly with more accurate methods of tool holding such as shrink-fit holders, hydraulic expansion chucks, and hydromechanical tool holders. Benefits of their use include more consistent and longer tool life, allowable speeds and feeds, shorter cycle times, and all-round improved productivity. Unfortunately, there are still a number of manufacturing companies that spend a lot of money on their machines but tend to have less consideration for the acquisition of adequate tool holders. Although in real life no one buys the cheapest tires for luxury high-performance cars, buying the chip tool holders for an expensive high-performance machining center is not considered as undermining the whole idea of the HP drilling project but rather a money-saving initiative.

It is explained in Chapter 7 (Section 7.7.5, Figure 7.59) that any tool holder has two interfaces that affect the system runout, namely, the spindle/tool holder interface and the tool holder/tool interface. Moreover, it is known from practice of machining that there are a number of possible combinations of design of these interfaces so the right combination for HP drilling should be considered. The further text is aimed to explain the meaning of the word *right* in this context.

A number of spindle/tool holder interfaces for drilling tools are available: tapered shank holders, ABS, FL holder, HSK, EPB, NTK, etc., systems. Factors to consider in making the decision for HP drilling are tool system rigidity, repeatability, durability, cost, ease of use, dependability, simplicity, and presettability (speed, accuracy, and repeatability). After careful consideration and based on many years' experience, HSK adapters are recommended for HP drilling because of inherent strength, repeatability, and durability, as well as their success in other applications. Such advantages come with a cost: the mating taper connection must be kept clean for accuracy and the spindle clamping mechanism must be cleaned and maintained regularly.

HSK tool holders are a German development. The abbreviation HSK stands for *hohl shaft kegel*, which, literally translated into English, means *hollow shank taper*. Among US end users, it is more commonly referred to as *hollow taper shank tooling*. The HSK design was developed as a non-proprietary standard suitable for both rotary and stationary applications. Its developers believed a single standard design was preferable to the growing number of tooling interfaces. The German DIN standards 69063 for the spindle receiver and 69893 for the shank are a set of standards that defined HSK tool holders for different applications. A total of six HSK shanks are defined. These shank styles are designated by the letters A through F. Each model is also identified by the diameter of the shank's flange in millimeters. Styles A, B, C, and D are for low-speed applications. E and F are for high speeds. The main differences between the styles are the positions of the drive slots, gripper location slots, coolant holes, and the area of the flange. Each style was defined for certain applications. It is critical that the proper type of HSK tooling be specified on high-speed machine tools, not just the correct size.

There are a number of tool holders with HSK spindle/tool holder interface. Not many of them can be used in HP drilling.

A self-centering chuck shown in Figure 1.27a includes jaws interconnected via a scroll gear (scroll plate) to hold onto a tool. Because such chucks most often have three jaws, the term three-jaw chuck without other qualification is understood by machinists to mean a self-centering three-jaw chuck. These chucks require a toothed key to provide the necessary torque to tighten and loosen the jaws.

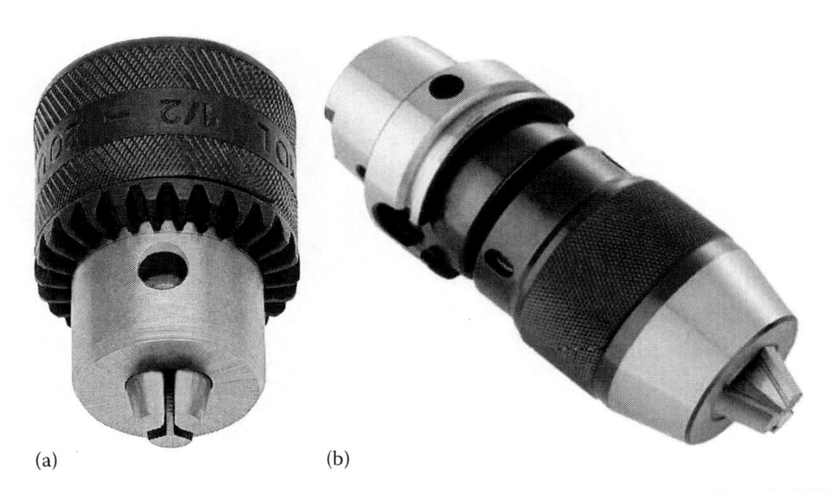

FIGURE 1.27 Self-centering universal drilling chuck: (a) general appearance and (b) with HSK.

This type of chuck is used on tools ranging from professional equipment to inexpensive hand and power drills for domestic use. Although it can be made with HSK tool holder/machine interface as shown in Figure 1.27b, and thus it seems that it can be used in HP drilling, it is not true in reality. The problem with such a chuck in HP drilling is excessive runout (±0.12 mm total indicated runout (TIR)—see Chapter 7 for definition) and lack of balance needed for high-speed applications.

At the low end of tool holder technology are side-lock holders widely used for CNC applications. Figure 1.28a shows VDI drill holder for index drills. These holders have one or two setscrews that press against the flat on the tool shank to secure it. Although side-lock holders are a pretty old technology in terms of precision, they are still common because it is an established technology, so the typical shop has a cabinet full of these holders to use up. As for quick tool change, side-lock holders require only loosening and tightening the screw(s) to change the cutting tool, and that takes less than a minute. Another quick-change benefit is that the holder can stay in the machine spindle while the tool is being changed. On the other hand, they are the least accurate of all types of holders as they create significant runout and are highly unbalanced. As a result, these tool holders should not be even considered for HP drilling although a side-lock holder can be made with HSK tool holder/machine interface as shown in Figure 1.28b, and thus it seems that it can be used in HP drilling.

In the middle of the technology spectrum are collet tool holders, or collet chucks (Figure 1.29). These consist of a collet holder body, a collet (usually made of spring steel) that is inserted into the body, and a nut that screws over the collet. The collet forms around the tool shank and exerts a

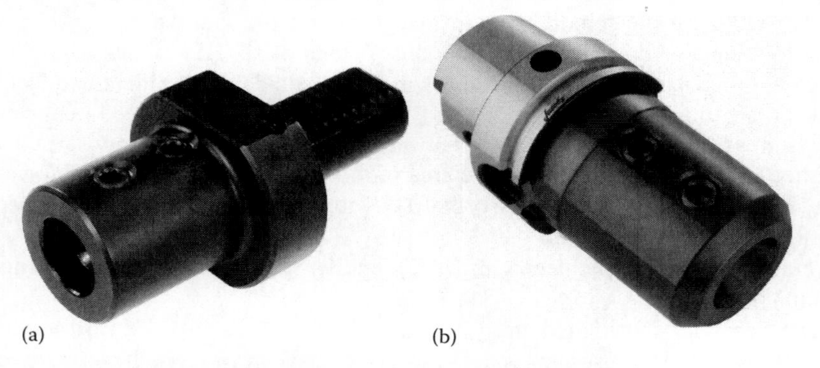

FIGURE 1.28 Side-lock holders: (a) VDI for index drills and (b) with HSK.

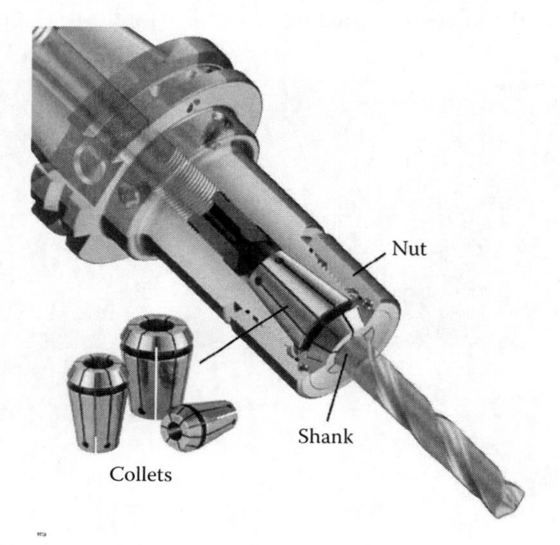

FIGURE 1.29 Collet tool holder.

strong clamping force when the nut is tightened. Pressing the tool from every direction toward the center line, the collet can effectively neutralize runout. And when this action alone doesn't provide enough centering, there is a simple method for improving concentricity. It involves an indicator and a mallet. The tool presetter measures the tool's runout in the tool holder—either in the spindle or out of it—then taps the high spot, and repeats both steps until the runout is appropriate to the cut. Using this approach, the tool presetter can achieve high level of tool concentricity that HP drilling may require. Tighter concentricity simply requires a more sensitive indicator, more patience and finesse, and unreasonably great presetting time. While it is important that all tool-holder shanks and bores be cleaned regularly, it usually just takes a quick wipe with a towel or shot of shop air. With collet holders, which should be cleaned thoroughly after every tool change, cleaning takes longer because there are more parts, and small chips get inside the slots in the collet and between the collet and the nut so it is required to take apart the whole assembly and clean each component carefully in between tool changes.

Although tool holders with precision collets can be used in HP drilling according to their advertised accuracy, the rigidity and repeatability of collet tool holders present some problems at high spindle rpms and great feed rates. Therefore, their use in HP drilling should be limited. An ideal collet would collapse by the same amount all the way around the tool. However, within a real-life collet that has, say, 0.4 mm of play in the ID before closing, tightening the nut may compress this play disproportionately to one side of the tool. If the resulting runout is too high for the cut, then it takes an indicator to diagnose this and a mallet to correct it.

Where purely mechanical tool holders may fail to achieve precise concentricity without manual intervention, at least two technologies—hydraulic and shrink-fit tool holders—can center the tool more repeatedly. Through the mechanisms of hydraulic pressure and thermal contraction, respectively, these tool holders close by essentially the same amount all the way around the shank of the tool. Thus, they take the principle of the collet tool holder one step further. Both hydraulic and shrink-fit tool holders essentially automate the tool centering process. By closing the tool-holder ID uniformly, both permit *hands-off* tool centering—no further adjustment is required.

The hydraulic tool holder works by using fluid to compress an internal expansion sleeve within the holder body. The hydraulic fluid delivers uniform pressure around this sleeve membrane allowing it to compress equally around the periphery of the cutting tool. Basically, that's how the hydraulic tool holder delivers its high concentricity specs. In operation, the concentricity of the hydraulic tool holder delivers runout accuracy of 0.003 μm or less at $2.5d_{sh}$ (d_{sh} is the shank diameter). It delivers high clamping torque of up to 900 N m (20 mm dia.) and 2000 N m (32 mm dia.) and vibration dumping as the shank hydraulic system absorbs vibrations. Another benefit of a hydraulic tool-holder system is rigidity. Rigidity plays an important role in predictable tool life as well as runout. The hydraulic tool holders are designed with through-the-tool MWF (coolant) capability. Air, MWF, and MWF/air mixture (for MQL applications—see Chapter 6) can be applied through the tool holder and cutting tool.

Figure 1.30 shows the working principle and particularities of hydraulic tool holders. As can be seen, first, the length of tool is set using the tool length-adjustable sleeve 7 actuated by setscrew 2. Then, the operator uses a key to rotate the actuation screw 5. The actuation screw 5 applies some force to actuation piston 6 that, in turn, creates pressure in the hydraulic medium 3. This pressure through the chamber system acts on the expansion sleeve 8 that firmly grips to shank 4. The enormous contact pressure over the expansion sleeve/tool shank contact squeezes the remaining oil, grease, or lubricant residues into the grooves 9 that assures dry contact.

As discussed in Chapter 7 (Section 7.7.5.5), when the high-precision shrink-fit tool holders are used, there are no means available for compensation of the total runout. Hydraulic tool holders allow some compensation mean when a hydraulic holder is used and the tool/holder assembly is preset in the spindle of a modern presetting machine. The total runout is measured and the tool is rotated to the position where the direction of the maximum runout of the spindle/holder assembly is opposite to that of the tool. Then the tool is locked in the holder in this position.

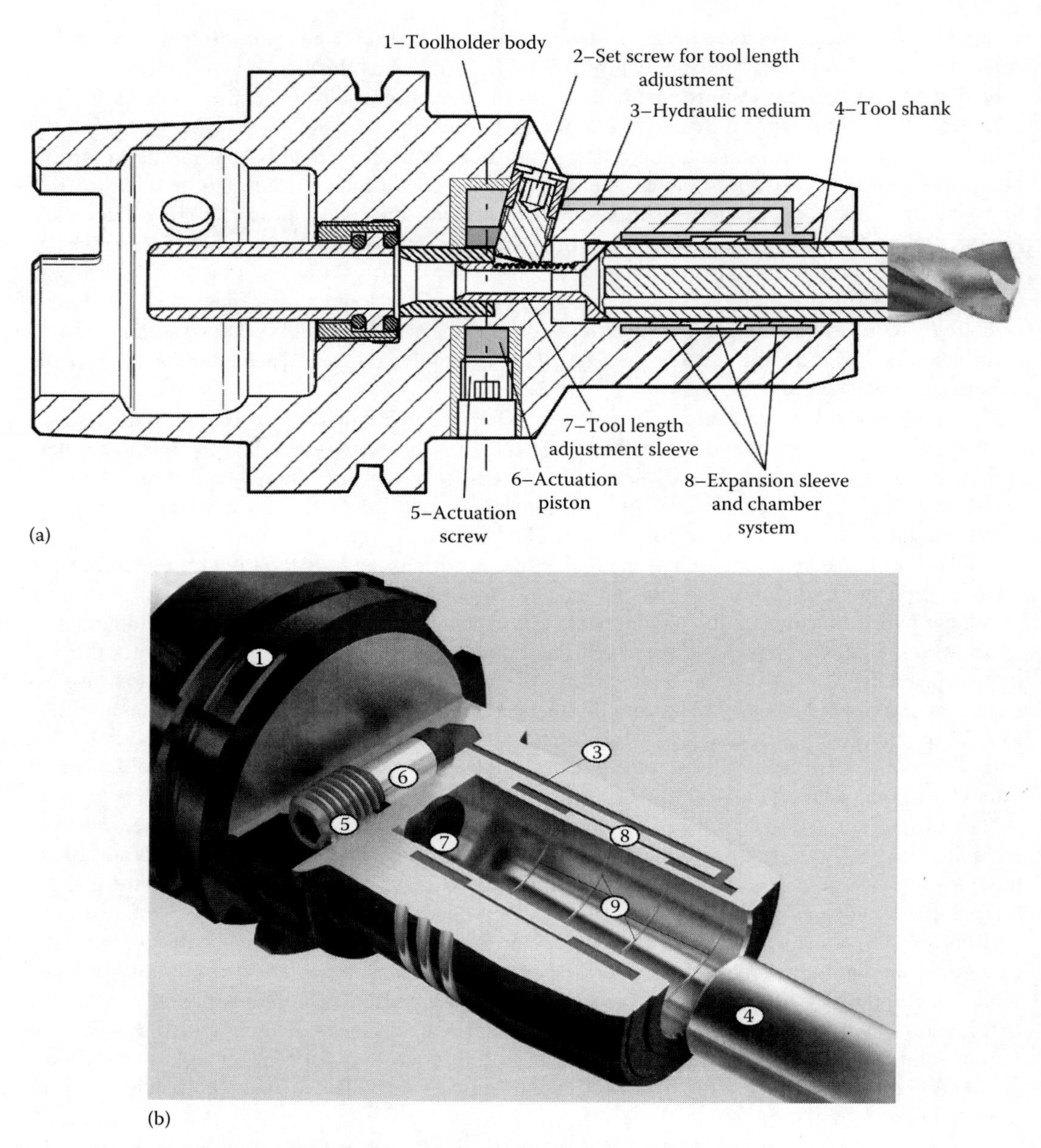

FIGURE 1.30 (a) Schematic and (b) particularities of hydraulic tool holder.

The shrink-fit tool-holder principle uses the expansion and contraction properties of metal to provide extremely powerful tool holding as shown in Figure 1.31. A shrink-fit tool holder's internal diameter that creates the tool/holder interface is deliberately undersized. When heated, it expands just enough to allow the tool to be inserted. When it cools, the contraction of the metal holds the tool shank firmly. Figure 1.32 shows a section view of the shrink-fit tool holder.

The key component of a shrink-fit tool-holding system is the induction-heating unit. Such a unit can be a part of the tool presetting machine or it can be a self-contained unit that is slightly smaller than a refrigerator and costs $15,000–$30,000. Being introduced first, the latter option is becoming less attractive as many modern tool presetting machines include built-in shrink-fit unit so that the tool can be set to the length and shrink fit on the same setting while proper information is written on its RFID read/write chip (see Section 2.4.3).

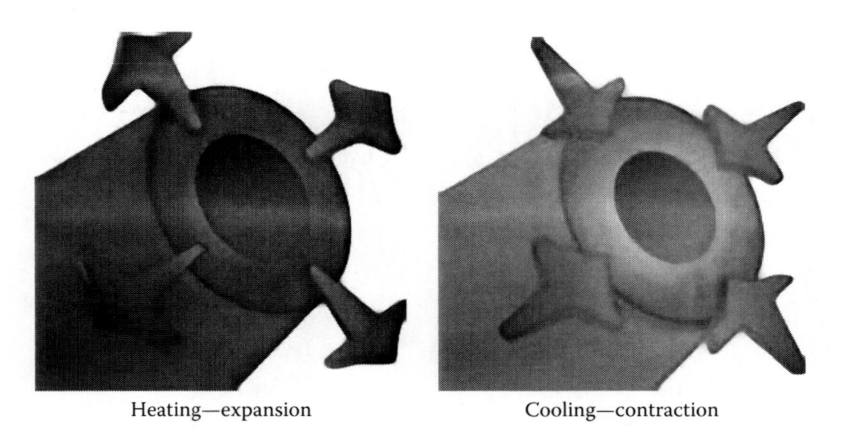

FIGURE 1.31 Principle of the shrink-fit tool holder.

FIGURE 1.32 A section view of the shrink-fit tool holder.

Figure 1.33 shows the tool removing sequence. To change tools, the operator places the tool holder in the tool presetting machine and recalls the tool number as the information about the heating time and position of the induction coil is stored there (Figure 1.33a). The operator initiates the tool removal sequence by pressing the corresponding key on the computer. The coil sets in the proper position and induction heat is applied (Figure 1.33b). In a few seconds, the heat expands the tool holder's bore enough for the operator to remove the dull tool manually. The heat is localized at the collar around the tool, so the operator wearing gloves can handle the tool and remove the tool mounted in the tool holder without fear of being burned. Once the tool is removed and placed on the cooling rake, the operator sets a radiator-type aluminum cooling block on the holder as shown in Figure 1.33c. Then the holder is put aside in the cooling fixture and a water-cooled unit is placed over the aluminum cooling block as shown in Figure 1.33d.

Although some promotion materials claim that the dull tool can be removed and a new one can be placed in the same cycle, it is not a good idea. This is because setting a new tool to the correct length requires some time so that the holder should be heated for extended time. As a result, the tool holder can be burned so it will not expand and contract to the desirable range (Figure 1.34). The induction-heating unit gives the operator total control over the heating process. Larger tool holders, because their ODs are relatively close to the coil, are heated more efficiently by the unit and, thus, require less power to heat them to the proper temperature than smaller tool holders do. The heating unit's controls allow the operator to vary power to regulate the amount of energy sent into the coil. The unit also allows the operator to control the heating of the tool holder by adjusting the cycle time. The automatic

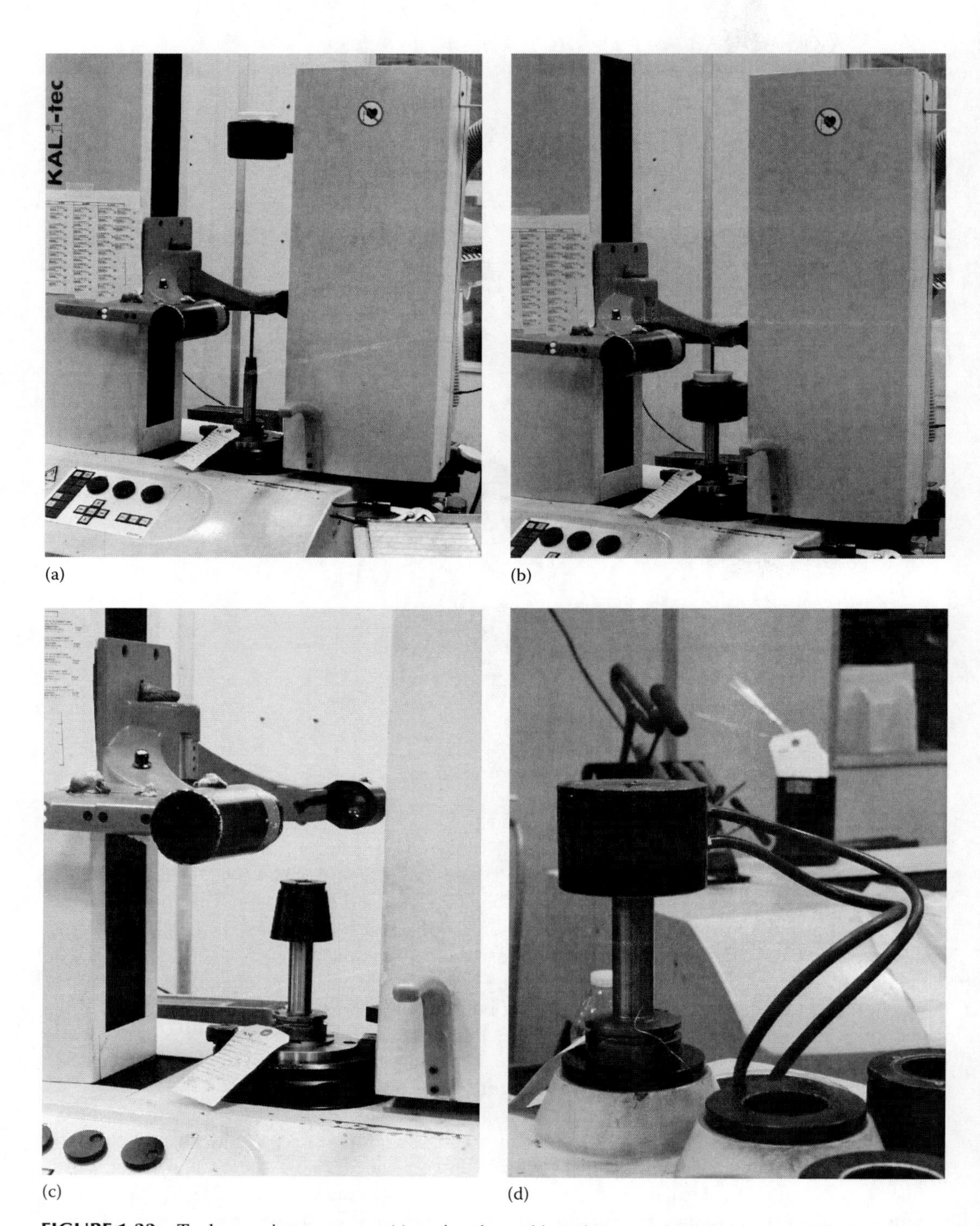

FIGURE 1.33 Tool removing sequence: (a) setting the tool into the presetting machine and selecting the corresponding program, (b) heating and removing the tool, (c) placing a radiator-type aluminum cooling block on the holder, and (d) holder in the cooling fixture with a water-cooled unit on it.

FIGURE 1.34 Burned tool holders.

control can be set to keep the heating unit on from 1 to 22 s. By setting the induction unit for a short cycle time, the operator can keep the heat concentrated in the clamping area. The tool holder will be out of the unit and in the cooling block before the heat can travel to other parts of the tool holder.

Once a shop has established the correct power and cycle settings for its tool holders, it can use the heating unit's internal control to regulate the power to the coil and the cycle time. When the internal control is engaged, the front-panel controls are disengaged. This feature prevents operators from sending too much power into the coil and overheating tools and tool holders, which may shorten the lives of these components. The operator can maintain even tighter control of the heating process by using an optional infrared sensor, which detects the amount of heat in the tool holder. When the tool holder reaches a preset temperature, the unit shuts off the power automatically.

Analysis of the suitability of hydraulic and shrink-fit tool holders for HP drilling reveals the following:

1. Both holders are fully suitable for HP drilling in terms of gripping power and accuracy.
2. The costs per unit of both holders are approximately the same. A good shrink-fit holder costs \$150–\$350 while a good hydraulic tool holder costs \$250–\$700 depending on holder size, supplier, batch, etc. However, shrink-fit tool holder requires more expensive presetting machines with the induction unit, trained operators, and specially written programs for heating cycles for each particular set of tools.
3. When using older machines and/or nonrigid part-holding fixtures, tool chatter may place a speed limit on the process, forcing the shop to accept a feed rate lower than it might otherwise achieve as the shrink-fit tool holder does not provide vibration damping.

Because of their standard configuration, and because of the slow rate of innovation of tool-holding systems relative to the machines that they serve, the lingering perception among some machining center users is that tool holders tend to be *commodity items*. Given the number of apparently successful tool-holder manufacturers competing in the marketplace today, the commodity conclusion is an easy one to draw. However, the commodity theory is wrong for HP drilling applications. While it is true that all tool holders of a specific type may look alike, definitely not all are created equal, not even to the first approximation. As a result, the cost of seemingly *similar* tool holders may differ two- to threefold. What makes any one thing better than another? In most metalworking applications, the difference between a good part and scrap is often a few microns on a critical dimension. Likewise, differentiating a high-precision tool holder is a matter of adherence to manufacturing tolerances.

Runout commonly wrongly attributed as concentricity in the tool-holding field is of prime concern. The cutting tool must rotate exactly on the rotational axis of the machine tool spindle. The means of accomplishing this near-perfect runout are also clear but complex as discussed in Chapter 7 (Section 7.7.5). To begin, the tool holder's tapered shank must sit within the corresponding spindle taper very precisely each and every time it is inserted. For this to happen, the mating surfaces must be matched to very tight cone angle tolerances. These tolerances are specified and published by national and international standards committees and are generally available for review by anyone. Well-made tool holders are measured for roundness and taper angle in gages that are calibrated by hard master gages. The methods involved in using production gaging vary from hard contact mechanical, hard contact/electronic analog, to noncontact analog techniques (such as air gages). All can be effective. The common denominator of these methods is the hard master gage used for calibration.

There is a definite variation between the master gages of the various tool-holder manufacturers. This strong assertion is based on the measurement of hundreds, if not thousands, of tool holders made by many different manufacturers over many years. Simply put, they vary. If it is assumed that all tool holders in the marketplace conform to their corresponding manufacturer's gages, then it follows that the master gages used by the various manufacturers are not the same. The problem is that this situation results in variations of spindle fit from manufacturer to manufacturer. The reason is easily understood. There is no *Grandmaster* gage for standard tapers. Although well-equipped and highly capable industrial metrology laboratories can measure tapers with adequate precision on rotary tables, there is no single reference hard gage that can readily authenticate other hard gages of the same size and taper rate.

Without a single source or master gage from which to trace all gages, it is understandable that there are variations in conformance to standard dimensions from part to part across the marketplace and that these variations can affect the quality of spindle fit as there is no national or international standard for reference. Although spindle and tool-holder taper fit and tool changing function have been specified by various standards such as ISO, American National Standards Institute (ANSI), JIS, or DIN, the particularities of the other end of the tool holder, where the drill shank fits the tool holder, are wide open. Tool-holder manufacturers alone set tolerances on the concentricity of these holding features relative to the tapered shank that enjoins the machine spindle. Only the tool-holder manufacturers can fix their machines so that the mating surfaces are machined with the tapered shank, with spindle interface as the surface of registry.

Because HP drilling requires spindle speeds of 10,000 rpm or higher, the tool holders used for such applications must be balanced as finely as the spindles in which they are installed. If they are not, the result will be unwanted vibrations that create chatter and will ultimately diminish surface finish and cutting tool life. In extreme cases, when imbalances are large, spindle damage can result. Much has been written about the physics of tool-holder balance in recent years. An important point, however, is that the force generated by an unbalanced condition within a tool/tool-holder assembly is proportional to the square of the spindle speed. Negligible forces generated at 1,000 rpm are 100 times greater at 10,000 rpm and 400 times greater at 20,000 rpm. The need for excellent concentricity is also more important at elevated spindle speeds because if the tool is not rotating on the spindle center line, it becomes a prime source of additional imbalance. For HP drilling operations, the hydraulic tool holders (HSK 63) are recommended to be balanced at 20,000 rpm to a 2.5 g rating.

Balanceable holders allow the end user to offset tool-holder imbalances by manipulation of radially located setscrews or balancing rings. In these cases, correcting imbalances in tool holders requires that the imbalance be located and quantified, which, in turn, requires the use of a sophisticated balancer. In some cases, a shop is better off using pre-balanced tool holders obtained from a reliable source and used in combination with high-quality cutting tools as a cost-effective alternative to the purchase and operation of a balancing machine.

1.2.5 Summary: Checklist of Requirements for the Drilling System for HP Drilling

The foregoing analysis allows to work out the following checklist of major requirements to the drilling system for HP drilling:

1. Application-specific design and geometry of HP drills that assures the minimum energy required for chip formation (see Section 1.2.4.4 and Chapters 4 and 5).
2. Application-specific tool material to support economically justified tool life and high quality of drilled surfaces (see Section 1.2.4.4 and Chapter 3).
3. Minimum runout. Low runout will provide dramatic payback by improving machining capability and reducing production costs (see Section 7.7.5).
4. Tool- and work-holding rigidity.
5. MWF (coolant) parameters control and maintenance (see Chapter 6).
6. Balancing.

Items 1–5 in this list are covered by the subsequent chapters, while item 6, balancing, has to be explained. HP drilling implies the use of high rpm spindles so that tool balancing plays the key role. Tool balance becomes an issue at 5000 rpm and up although it was not noticed as other issues, for example, with spindles static and dynamic rigidity, overshadow the impact of tool balancing. Typically associated with applications on machining centers, high-speed machining has been driven by improvements in machine technology alongside cutting tool advancements. As rpms have increased, so has the focus on the necessity of implementing balanced tooling assemblies.

Only with accurately balanced tools can the desired precision and surface quality be achieved repeatedly and reliably in high-speed machining and thus in HP drilling. There are two principal consequences of imbalance to consider:

1. Chatter, which causes deterioration of both machining quality and tool life
2. The short life of the bearings in the spindle, which results in major downtime and significant additional maintenance cost

It is easy to understand how balanced tools can reduce machine maintenance requirements and improve tool life and quality of machined parts when unbalance is quantified. Tool unbalance is the result of multiplying a tool's weight (mass) by its eccentricity (how much a tool's weight is off-center). Eccentricity can also be described as the distance between the tool's center of rotation and the actual center of the tool's mass. Commonly, the metric system is used when working with balance calculations. That means using micron for eccentricity and kilograms (kilos) for tool weight, which yield a measurement of unbalance in gram millimeters (g · mm).

The ISO has developed a quality standard for allowable unbalance. The standard is ISO 1940-1(2003) Mechanical vibration—Balance quality requirements for rotors in a constant (rigid) state—Part 1: Specification and verification of balance tolerances. The terms most relevant to tool balancing are defined as follows:

1. *Balancing* is a procedure by which the mass distribution of a rotor is checked and, if necessary, adjusted to ensure that the residual unbalance or the vibration of the journals and/or forces at the bearings at a frequency corresponding to service speed are within specified limits.
2. *Unbalance* is a condition that exists in a rotor when vibration force or motion is imparted to its bearings as a result of centrifugal forces.

3. *Initial unbalance* is unbalance of any kind that exists in the rotor (tool assembly) before balancing.
4. *Residual unbalance* is the final unbalance, that is, unbalance of any kind that remains after balancing.
5. *Amount of unbalance* is the product of the unbalance mass and the distance (radius) of its center of mass from the shaft axis. Units of amount of unbalance are gram millimeters (g · mm).
6. *State of a rotor* is determined by the unbalance behavior with speed, the types of unbalance to be corrected, and the ability of the rotor to maintain or to change the position of its mass elements and their centers of mass relative to each other within the speed range.

The standard points out that the response of the rotor to unbalance can change with the speed range and its bearing support conditions. The acceptability of the response is determined by the relevant balance tolerances. The speed range covers all speeds from standstill to the maximum service speed but can also include an overspeed as a margin for service loads (e.g., temperature, pressure, flow).

One and the same unbalance of a rotor in a constant (rigid) state can be represented by vectorial quantities in various ways in terms of resultant unbalance/resultant couple unbalance in one plane and in terms of a dynamic unbalance in two planes. The resultant unbalance vector may be located in any radial plane (without changing amount and angle); but the associated resultant couple unbalance is dependent on the location of the resultant unbalance vector. The center of unbalance (UC) is that location on the shaft axis for the resultant unbalance, where the resultant moment unbalance is a minimum.

Standard ISO 1940-1(2003) points out that rotors that need one correction plane only are typically disk-shaped rotors, provided that

- The bearing distance is sufficiently large
- The disk rotates with sufficiently small axial runout
- The correction plane for the resultant unbalance is properly chosen

Whether these conditions are fulfilled and single-plane balancing has been carried out on a sufficient number of rotors, the largest residual moment unbalance is determined and divided by the bearing distance, yielding a couple unbalance (pair of unbalances). If, even in the worst case, the unbalances found this way are acceptable, it can be expected that a single-plane balancing is sufficient. Figure 1.35a shows an example of representation of unbalanced vectors in the end planes. Reporting includes *a* (unbalance is 5 g · mm) and *b* (unbalance is 1.41 g · mm).

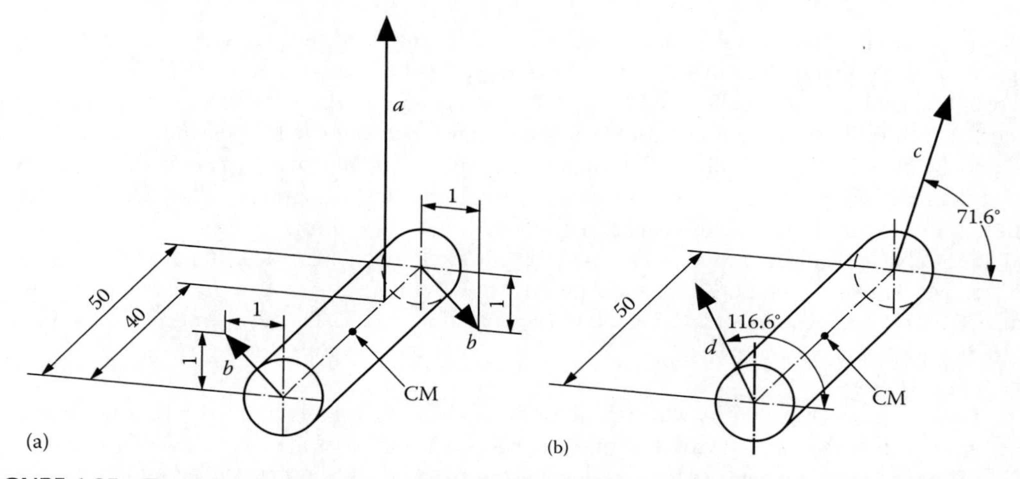

FIGURE 1.35 Example of representation of unbalance: (a) in terms of resultant unbalance vector together with an associated couple unbalance in the end planes, single-plane balancing, and (b) in terms of a dynamic unbalance in two planes, an unbalance vector in each of the end planes.

In most cases, rotational balancing machines are used (ISO 2953, Mechanical vibration—Balancing machines—Description and evaluation). The resultant unbalance can be determined and corrected to limits.

Naturally, most drilling tool assemblies are not close to disk-shaped rotors so this kind of balance is not sufficient for such tools. In practice, however, high-speed drilling tool assemblies are subjected to single-plane balancing. As such, the location of the balancing plane is chosen with no justification.

If a rotor in a constant (rigid) state does not comply with the earlier-listed conditions, the moment unbalance needs to be reduced as well. In most cases, resultant unbalance and resultant moment unbalance are assembled into a dynamic unbalance: two unbalance vectors in two planes (Figure 1.35b), called complementary unbalance vectors. For two-plane balancing, it is necessary for the rotor to rotate, since otherwise the moment unbalance would remain undetected. Reporting includes c (unbalance is 3.16 g·mm) and d (unbalance is 2.24 g·mm).

Rotors out of balance tolerance need correction. These unbalance corrections often cannot be performed in the planes where the balance tolerances were set but need to be performed where material can be added, removed, or relocated. Standard ISO 1940-1 (2003) defines the number of necessary correction planes and their location relative to the tolerance planes.

Standard ISO 1940-1 (2003) defines five methods to determine the balance tolerances. These methods are based on

1. Balance quality grades, derived from long-term practical experience with a large number of different rotors
2. Experimental evaluation of permissible unbalance limits
3. Limited bearing forces due to unbalance
4. Limited vibrations due to unbalance
5. Established experience with balance tolerances

The choice of method should be agreed upon between the manufacturer and user of the rotor.

On the basis of worldwide experience and similarity considerations, balance quality grades G have been established, which permit a classification of the balance quality requirements for typical machinery types. Balance quality grades G are designated according to the magnitude of the product $(e_{per} \cdot \Omega_t)$ expressed in millimeters per second (mm/s), where e_{per} permissible residual specific unbalance in kilogram meters per kilogram (kg·m/kg) or meters (m) and Ω_t is the numerical value (the magnitude) of the angular velocity of the service speed, expressed in radians per second (rad/s). If the magnitude is equal to 6.3 mm/s, the balance quality grade is designated G6.3. Balance quality grades are separated from each other by a factor of 2.5. A finer grading may be necessary in some cases, especially when high-precision balancing is required, but it should not be less than a factor of 1.6. Grade G16 is recommended for car and truck engine components, grade G6.3 is recommended for pumps and general machinery, grade 2.5 is recommended for machine tool drives, and grade G1 is recommended for most grinders.

According to ISO 1940-1 Standard, the permissible residual unbalance U_{per} can be derived on the basis of a selected balance quality grade G by the following equation:

$$U_{per} = 1000 \frac{(e_{per} \cdot \Omega_t) \cdot m_t}{\Omega_t} \tag{1.21}$$

where

U_{per} in gram millimeters
m_t is the tool mass in kilograms
e_{per} is the permissible residual specific unbalance in kilogram meters per kilogram (kg·m/kg) or meters (m)

A more practical unit is micrometers (μm) because many permissible residual specific unbalances are between 0.1 and 10 μm. The term e_{per} is useful especially if one has to relate geometric tolerances (runout, play) to balance tolerances. Obviously, $e_{per} = U_{per}/m_t$. Ω_t is the numerical value (the magnitude) of the angular velocity of the service speed, expressed in radians per second (rad/s), with $\Omega_t \approx n_t/10$ and the service speed n_t in revolutions per minute (rpm).

Using Equation 1.21, one can derive a practical formula for calculating the permissible residual unbalance U_{per} as

$$U_{per} = \frac{9549 m_t G}{n_t} \tag{1.22}$$

where

U_{per} in gram millimeters
m_t is the tool mass in kilograms
G is ISO 1940-1 grade in mm/s (i.e., G2.5)
n_t is maximum spindle speed the tool will be run in rpm

Example 1.3

Problem

Determine the amount of unbalance for a PCD monoblock (the drill is assembled in the tool holder) two-staged drill held in an HSK hydraulic tool holder. The weight of the monoblock is 1.2 kg. The monoblock is to be run at a maximum of 24,000 rpm and to be balanced to ISO G2.5 grade according to the recommendations of ISO 1940-1 Standard.

Solution

Using these data and Equation 1.22, one can calculate

$$U_{per} = \frac{9549 m_t G}{n_t} = \frac{9{,}549 \cdot 1.2 \cdot 2.5}{24{,}000} = 1.19\,\mathrm{g \cdot mm}$$

REFERENCES

Altintas, Y. 2000. *Manufacturing Automation. Metal Cutting Mechanics, Machine Tool Vibrations, and CNC Design*. Cambridge, U.K.: Cambridge University Press.

Anselmetti, B., Chep, A., and Mognol, P. 1995. Minimal database for the cutting parameters in con manufacturing systems. *International Journal of Computer Integrated Manufacturing* 8(4):277–285.

Astakhov, V.P. 2001. Gundrilling know how. *Cutting Tool Engineering* 52:34–38.

Astakhov, V.P. 2010. *Geometry of Single-Point Turning Tools and Drills. Fundamentals and Practical Applications*. London, U.K.: Springer.

Astakhov, V.P. 2012. Design of experiment methods in manufacturing: Basics and practical applications, Chapter 1. In *Statistical and Computational Techniques in Manufacturing*, Davim, J.P. (ed.). London, U.K.: Springer. pp. 1–54.

Bertalanffy, L. von. 1969. *General System Theory*. New York: George Braziller.

Boothroyd, G. and Knight, W.A. 2006. *Fundamentals of Machining and Machine Tools*, 3rd edn. Boca Raton, FL: CRC Press.

Campbell, P.G.Q. 1994. *Basic Fixture Design*. New York: Industrial Press.

Chang, T.C. and Wysk, R.A. 1984. *An Introduction to Automated Process Planning Systems*. Englewood Cliffs, NJ: Prentice Hall.

DeVries, W.R. 1992. *Analysis of Material Removal Processes*. New York: Springer-Verlag.

Fiesselmann, F. 1993. An option for doubling drilling productivity. *The Fabricator* 23(4):36–38.

Koening, D.T. 2007. Machine tool and equipment selection and implementation. In *Manufacturing Engineering: Principles for Optimization*, 3rd edn., Koening, D.T. (ed.). ASME, New York, Chapter 5.

Konig, W., Langhammer, K., and Schemmel, V. 1972. Correlations between cutting force components and tool wear. *Annals of the CIRP* 21(1):19–20.

Nee, J. (ed.). 2010. *Fundamentals of Tool Design*, 6th edn. Dearborn, MI: SME.

Nee, A.Y.C., Tao, Z.J., and Kumar, A.S. 2004. *An Advanced Treatise on Fixture Design and Planning*. Singapore: World Scientific Publications.

Parkesh, J. 2003. *Jigs and Fixtures Design Manual*. New York: McGraw-Hill.

Shaw, M.C. 2004. *Metal Cutting Principles*, 2nd edn. Oxford, U.K.: Oxford University Press.

Skyttner, L. 2006. *General Systems Theory: Problems, Perspectives, Practice*, 2nd edn. Singapore: World Scientific Publishing Company.

Smid, P. 2003. *CNC Programming Handbook*, 2nd edn. New York: Industrial Press.

Sreejith, P.S. and Ngoi, B.K.A. 2000. Dry machining: Machining of the future. *Journal of Materials Processing Technology* 101:287–291.

Stenphenson, D.A. and Agapiou, J.S. 1996. *Metal Cutting Theory and Practice*. New York: Marcel Dekker.

Usachev, Y.G. 1915. Phenomena occurring during the cutting of metals, (in Russian). *Izv. Petrogradskogo Politechnicheskogo Inst*. XXIII(1):321–338.

Yoon, M.C. and Kim, Y.G. 2004. Cutting dynamic force modelling of end milling operation. *Journal of Materials Processing Technology* 155–156:1383–1389.

Zorev, N.N. (ed.) 1966. *Metal Cutting Mechanics*. Oxford, U.K.: Pergamon Press.

2 Tool Failure as a System Problem

Investigation, Assessment, and Recommendations

Once you eliminate the impossible, whatever remains,
no matter how improbable, must be the truth.

Sherlock Holmes (by Sir Arthur Conan Doyle, 1859–1930)

As mentioned in the Introduction, drilling is the most common machining operation in the manufacturing industry. With the aid of advanced computer-aided design/computer-aided manufacturing (CAD/CAM) systems and open architecture controllers, production systems have reached the point where adaptive machining and agile manufacturing are on the verge of becoming practical for even relatively small machining enterprises and job shops; certainly with the trend toward customization and smaller production runs, this is essential (Rehorn et al. 2005).

Unfortunately, the issue of machine tool downtime continues to plague the industry. Downtime can be considered as any duration of time during which no machining operation is being performed on a given workpiece. This downtime can be thought of as consisting of two major parts. The first part, which can be referred as the scheduled downtime, is included in the cycle time and thus in efficiency analysis (termed as the unavoidable downtime). It includes the loading–unloading time, time needed for transportation between stations/machines, tool change and inspection time, spindle speed ramp up time, and maintenance time. The second part of the downtime, which can be referred as the uncontrolled downtime, is caused by tool failures.

The scheduled downtime reduction was in the center of attention of researchers and engineers for centuries, and thus, a number of advances to reduce this downtime have been made over a long time. Robotic loading–unloading and transportation, in-process gaging, state-of-the-art digital AC drives, and controllers are just few examples. All these advances dramatically reduced the scheduled downtime.

The uncontrolled downtime due to tool failures was not in the center of interest of researchers and engineers for years because (1) it was insignificant compared to the scheduled downtime; (2) speed and cutting feed were relatively low so that cutting tool failed by wear, which was periodically controlled; and (3) the costs of tool failure was low as cutting tools were relatively inexpensive and their failures did not bring much damage to the drilling system including the workpiece. However, the time has changed and this portion of the downtime has gradually become significant.

Tool failure is a major cause of unscheduled stoppage in a machining environment and is costly not only in terms of time lost but also in terms of capital destroyed (Rehorn et al. 2005). Some estimates state that the amount of downtime due to tool breakage on an average machine tool is on the order of 6.8% (Byrne et al. 1995), while when tool failures are considered, the figure is closer to 20% (Yeo et al. 2000). Even if the tool does not fail during machining, the use of excessively worn or damaged tools can put extra strain on the machine tool system and cause a loss of quality in the finished part.

Another important but routinely ignored aspect of the cutting tool failure is the failure effect. For example, a failure of a small, $100-worth carbide drill can result in a significant downtime in a high–production rate automotive plant. This downtime ranges from a half-hour downtime (for the removal of scrapped parts and resetting the manufacturing line/cell to its normal operating condition) to weeks when a special drill is the case (the common case in the automotive industry) and when all drills from the inventory were damaged (broken) due to drilling-system-related issue. This is because the lead time for such a drill can be significant (up to 8 weeks). One may wonder why not place an emergency order for such a drill in a nearby tool shop so the drill can be delivered in a day or so. Although apparently it can be done, in reality, it is difficult to achieve because

- A great portion of the lead time for a carbide drill is related with getting the proper tool material often with specific coolant hole location and of specific grade that the local tool shop may not have on shelf.
- The shop may not be an approved vendor, so the corresponding quality procedure requires a long time for tool testing and implementation before it can be used in actual part production.

Other issues directly correlated with a tool failure are the amount of scrap and cost of parts (or sometimes, units) containment including their part-by-part manual inspection and, sometimes, dissembling the units with potentially defected parts. This is explained as follows. When a tool breaks, the following losses are involved:

- The cost of the broken tool.
- The cost of the scrapped parts (material and previous handling/manufacturing).
- Downtime for removing scrapped parts, removing the broken tool and installing a new one, and resetting a manufacturing cell/production line.

When a tool fails to produce the parts with intended quality, the following is involved. In the job shop environment, each part is manufactured individually so there is a time assigned to control its parameters after each operation using in-process and post-process gages. If a drill fails to produce a proper hole, it should be detected and the part should be reworked. In the high–volume production environment, the losses are much greater. High-efficiency manufacturing facilities use unattended manufacturing lines/intelligent cells. For example, in the automotive industry, the part is measured when a tool is changed and then only each 40th–80th part is actually measured. If a drill fails to produce the proper hole (e.g., it drills holes with exit burr) as it was noticed during the periodic inspection, then (1) all parts manufactured since the previous inspection point should be contained and manually inspected for the similar defect(s) provided that they stay around in the movable rack or in the buffer, and (2) all parts that made their way to subassembly lines should be traced down, and when assembled, the subassemblies should be dissembled for possible defective parts.

As efficiency requires to keep the buffer size as small as possible, some parts with possible defects often find their way to final assembly and in the testing rigs. Such a containment becomes very costly and time-consuming. It undermines the whole idea and objective of unattended manufacturing as an army of people is needed to inspect the parts with possible defects manually.

In the author's opinion, the major problem is that the discussed downtime/reworking/containment/rejection of units in the final testing is not included in tooling cost saving even in the most advanced manufacturing plants. As a result, it is difficult to justify additional costs of tools (drills) having greater quality and thus reliability.

The foregoing analysis reveals the importance of tool failure analysis, finding the root cause of the failure and implementing the proper corrective actions. The sections to follow provide methodological help to carry out such an analysis and find the real root cause of the problem. Note that this is the first attempt to present such an analysis in the literature on metal cutting and cutting tool.

One may wonder, however, why the considerations of tool failure are placed before the chapters that introduce the major aspects of the tool materials, drilling tool geometry and design aspects, MWFs, and drilling metrology. The answer is rather simple—the author did it deliberately keeping in mind two major objectives:

1. To emphasize that a drilling tool failure is a system issue so it has to be treated as such if one really wants to find out the root cause of this failure and permanently fix it
2. To convince a potential reader that he or she needs to read other chapters of this book to design/select/use a drilling tool efficiently, that is, to reduce the machining downtime and cost per drilled hole associated with tool failure as each one of these chapters covers a subject associated with drill failure analysis

2.1 TRADITIONAL NOTIONS AND APPROACHES

In an ideal world, drills wear evenly at a predictable rate without possibility of catastrophic failure. As such, the tool management seems to be a continuous process, which can be run by business management-type personnel with some engineering training. Of course, that is not the reality of metal cutting where failures of drills do occur. As a result, any practical tool management process is a discrete process consisting of a great number of *problem-occurrence–problem-solving* dyads (couples). A problem occurrence can be thought of as failure, referred as tool failure. In this context, tool failure is understood in a broad sense. It starts from the selection of a tool that is not fully suitable for the application conditions, due to any one or the combination of the following factors: a wrong tool and/or a delivery at wrong time, incorrect tool inspection report, tool setting problems, tool delivery to the wrong tool rack (line), premature tool failure before achieving the intended tool life, quality problems with the machined parts (surface finish, dimensional and/or form and position tolerances, chatter marks, burr, spiraling, etc.), tool-handling failure (cracking, chipping, etc.), high CPU (cost per unit), and low reliability of the tools that causes unscheduled stoppage of the manufacturing lines for its replacement. In other words, the term *tool failure* includes all the facets of a tool management program.

The tool management process runs smoothly if and only if each and every *problem-occurrence–problem-solving* dyad is completed, that is, when for each problem that occurs on the daily basis, a proper, timely solution is provided. To accomplish that, a fast-response problem solution system should be an inherent part of any modern manufacturing business. Although there is no clear guidelines for establishing such a system as it is established mainly intuitively starting from a small manufacturing facility to the corporate level of global manufacturing companies, one underlying approach to establish a fast-response problem solution system is common.

According to this common approach, the supplier/manufacturer of the failed tool is asked to carry out the tool failure analysis. As such, the failed tool is sent back with minimum info on this failure. At best, a sales representative of the tool supplier picks up the failed tool asking some question surrounding the failure. The tool supplier should complete the 5Why form so that the root cause of the problem can presumably be found and thus corrected. In reality, however, it is not the case as most of the filled 5Whys forms do not even tackle the root cause of the problem unless there is an obvious to everyone manufacturing flaw(s) found in the failed tool or some obvious problem in the drilling system. This is because

- The information provided to the tool manufacturer to carry out a failure analysis is normally incomplete and often misleading.
- Many tool manufacturers do not have sufficiently educated and trained personnel and equipment to do such an analysis. Often, the top research and technical personnel of the overseas tool companies are located faraway so the lack of information and inability to see the *crime scene* and question the people involved result in a root cause analysis that is based on the second- or even thirdhand information about the failure. Such analyses are normally of little help to find the root cause of the problem. Moreover, they cannot be completed in a timely manner.

As a result, when a problem occurs, the common response is to fix what is perceived to be the problem. The major concern is to get back *on line* as soon as possible so that *quick-fix* solutions are common. Because the root cause of the problem is rarely found, the most proposed solutions are temporary pipe bandages. As a result, the same problem comes back again and again as any pipe bandage can only stop a leak for a while, but it cannot protect tube corrosion/damage.

2.2 FAILURE: A SYSTEM-RELATED DEFINITION

As discussed earlier, the drill is almost always the weakest link in the drilling system so it fails whenever something goes wrong in such a system. Besides few, all literature sources on metal cutting consider a tool failure as its wear (or sometimes its breakage). In the author's opinion, however, tool failure should be considered as a system-related issue, and thus, a new, system-related definition of the tool failure should be given and its potential effects should be considered.

In general failure analysis literature, a failure is an event when the equipment/machinery is not capable of producing parts at specific conditions when scheduled or is not capable of producing parts or performing scheduled operation to specification (Stamatis 2003). For drills, considered as a perishable (non-durable, consumable) commodity, this definition does not cover all aspects of drill failure so that a need is felt to provide a new definition of drill failure. It is given as follows:

Drill failure, in general, is an event when a given drill does not meet the predetermined requirement for the cost per drilled hole over its total life.

The introduced definition related the drill performance including its efficiency.

2.3 TOOL FAILURE PRIME SOURCES

Because any tool failure is a system-related event, the proper identification of the failure source known as the root cause is important. The system of the tool failures is rather complicated, but it is hierarchical. Figure 2.1 shows such a system for drill failure root cause analysis. As can be seen, this system consists of two subsystems, that is, the drilling-system-related and drill-related failures. The objective of the first stage of any root cause analysis is to identify to which of these two categories a given failure falls into. The second stage concentrates on finding the root cause within the identified category.

Figure 2.1 does not show drill handling failures, for example, shipping of a wrong drill, late orders, and incorrect labeling. Although this category of tool failures can be costly, failures associated with this category are not *technical* so they are investigated using management failure approaches.

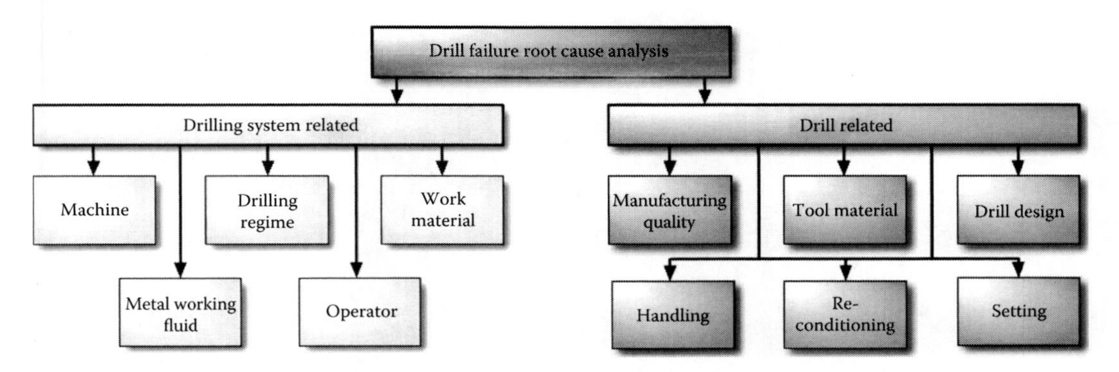

FIGURE 2.1 The system of causes of drill failure.

2.4 PREPARATION STAGE: COLLECTING INFORMATION

2.4.1 Knowing That a Failure Occurs

The most common ways for a tool engineer (hereafter referred to as the investigator) to learn about a tool failure are

1. *Actual failure*: From the tool (tag and/or tool-holder censor) brought from the shop floor preceded by e-mail and information given at the morning fast-response meeting
2. *Analytical failure*: From a cost-per-unit (CPU) statistical data analysis

The failure occurrence as found by either way is the formal reason/justification to start a tool failure investigation.

An analytical failure is normally obtained from the CPU analysis. For example, the tool may make the full tool life, but CPU is determined not only by this tool life but also by the number of the regrinds, that is, the total tool life. This type of failure can only be determined if the number of tool regrinds is tracked. This is particularly important for drills of high usage as some tool manufacturing companies remove too much tool material on each regrind (hoping to unfairly generate more business in the author's opinion). Because the tracked parameter as tool life does not change and if the number of regrinds is not tracked, it is impossible to determine high CPU associated with this drill.

2.4.2 Tool Tracking: The Tag System

The most common way to learn information about the failed tool (actual failure) is from its tool tag. An example of the properly filled tool tag is shown in Figure 2.2. This tool tag attached to the tool on its presetting contains valuable information on the following

1. The tool assembly number according to the tool layout. In the considered case, the part is TCH, operation 10, and tool 9.
2. When the tool was actually installed into the machine.
3. Machine/manufacturing cell number.
4. When the tool was prematurely pulled out from the machine.
5. Number of actual cycles it made.
6. The reason for removal—bad finish, which means that the roughness of the machined surface is out of specification.

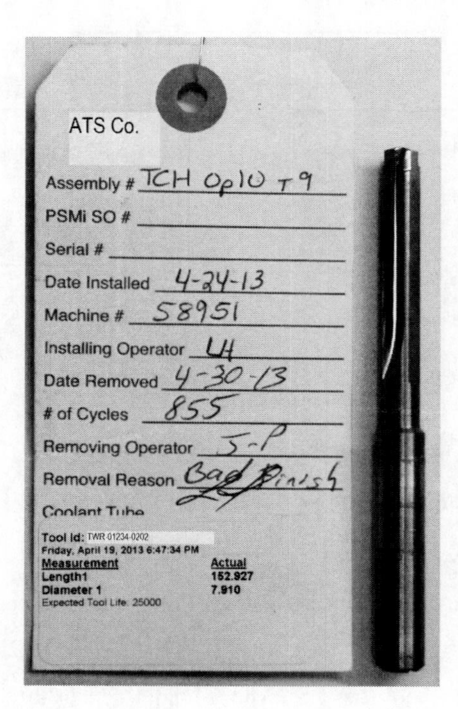

FIGURE 2.2 An example of the properly filled tool tag for the failed tool.

7. Printout from the tool presetting machine, which shows
   a. The intended tool life in the number of cycles and features per cycle. In the considered case, it is 25,000 cycles and each cycle includes 5 holes.
   b. Diameter of the drill as measured by the presetting machine. In the considered case, it is 7.910 mm.
   c. The gage length. In the considered case, it is 152.927 mm.
   d. The date when the tool was preset (important for tool tracking reasons, as a containment of a certain tool batch may be needed).

The tool tag system works well when the same parts in large quantities are routinely machined on the same machine as, for example, in the automotive industry when tool life (measured in the number of cycles) can be preset in the machine controller so the tool is automatically replaced when it reaches the end of its life, or, when pulled out prematurely, the controller informs the operator about the number of cycles made by this tool before failure.

In many other industries including rather advanced manufacturing facilities, machine tool operators have needed to manually input data regarding each tool as it is placed in a machine's *tool changer*. In addition, because a tool can be used only for a specific amount of time or a particular number of tasks, the operator has to keep track of how the tool has been used or examine it to determine if it is getting worn out. There is always the possibility of human error, causing a system to shut down because a tool either broke during the manufacturing process, was not operating properly due to being worn out, or was not installed in the correct location and was thus being used inappropriately. As such, no records of the machine malfunction and/or tool mishandling are available to the investigator.

2.4.3 Automated Tool Tracking with RFID

To automate a non-reliable procedure with tool tracking and thus to make tool tracking/performance evaluation easier, the identification technology that uses RFID is used. The tool holder is used as the transporter of data. The RFID chip (similar to the chips used to identify pet dogs) is attached to the tool holder, and the complete information on the tool presetting parameters are recorded on tool presetting.

Figure 2.3 shows how an RFID censor (Balluff chip) is installed in the tool holder. Normally, it is glued using epoxy in the corresponding pocket of a new tool holder. Figure 2.4 shows a typical location of the chip writer/reader on a tool presetting machine.

FIGURE 2.3 (a) RFID censor (Balluff chip) and (b) its installation in a tool holder.

FIGURE 2.4 Chip writer/reader on a tool presetting machine.

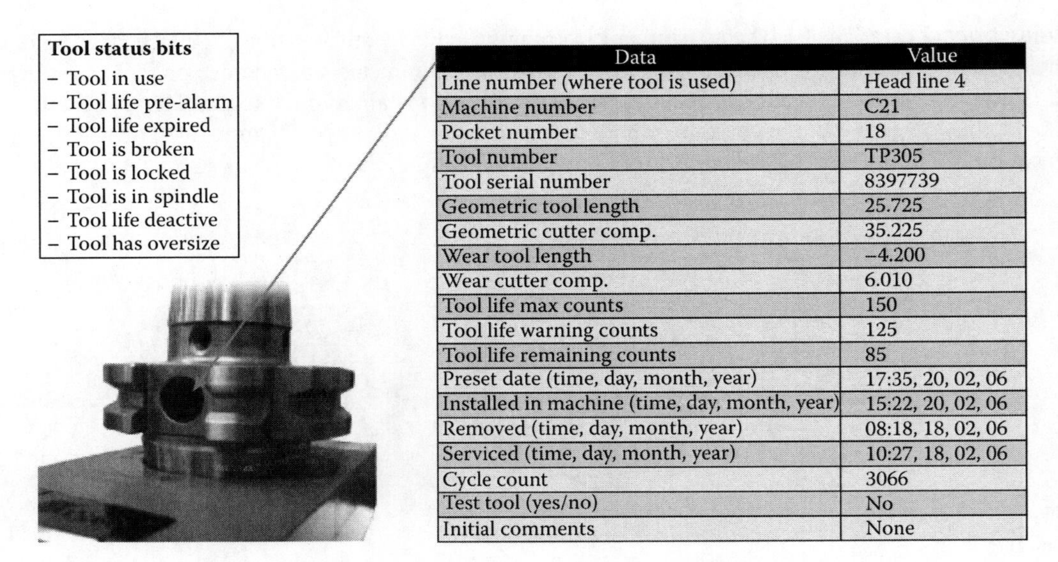

Data	Value
Line number (where tool is used)	Head line 4
Machine number	C21
Pocket number	18
Tool number	TP305
Tool serial number	8397739
Geometric tool length	25.725
Geometric cutter comp.	35.225
Wear tool length	–4.200
Wear cutter comp.	6.010
Tool life max counts	150
Tool life warning counts	125
Tool life remaining counts	85
Preset date (time, day, month, year)	17:35, 20, 02, 06
Installed in machine (time, day, month, year)	15:22, 20, 02, 06
Removed (time, day, month, year)	08:18, 18, 02, 06
Serviced (time, day, month, year)	10:27, 18, 02, 06
Cycle count	3066
Test tool (yes/no)	No
Initial comments	None

FIGURE 2.5 Parameters recorded in a Balluff chip.

Large amounts of information about the tool can be tracked using an RFID chip placed in the tool holder. This information can be dependably, repeatedly, cost-effectively, and accurately validated. In the example shown in Figure 2.5, 19 different parameters have been recorded on a simple, effective data carrier.

Figure 2.6 shows how an automated tool tracking system works. First, a tool is installed on a tool presetting machine (position 1), then the program for this tool is recalled and tool is set to

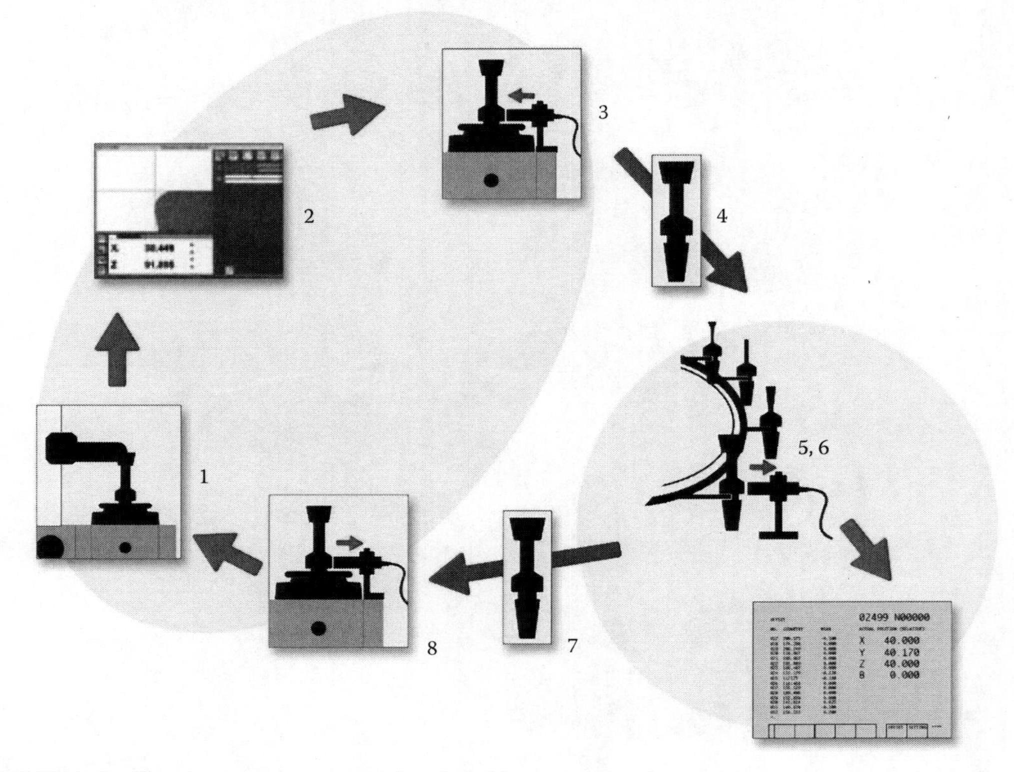

FIGURE 2.6 Showing how an automated tool tracking system works.

FIGURE 2.7 Chip reader/writer on a machine tool.

the intended dimensions and tolerances (position 2). The next step is recording the tool identification and presetting information onto a Balluff chip using the writer/reader of the tool presetting machine (position 3). Then, the preset tool is transported into the shop floor to the point of use (POU) (position 4). When required, the tool is installed into the tool magazine (position 5). An RFID reader/writer in the machine reads the information from the chip (Figure 2.7). Seamless communication between the tool holder and machine is assured. After completing the full tool life or pulling prematurely due to some problems, the tool is returned into the tool magazine where before the operator can remove it from the machine, RFID reader/writer in the machine writes the complete information associated with tool pull out including the data stamp (position 6). The tool is normally placed on the dull/failed tool cart/rack at POU. Then the tool is transported back to the presetting area (position 7). It is placed on the tool presetting machine and information is read from the Balluff chip by the writer/reader of this machine (position 8). This important information is stored in the corresponding database that can be accessed by the investigator whenever needed.

Many manufacturers are beginning to implant RFID read/write chips in every tool holder used in their shops. Data tracking and error proofing are two of the largest benefits to be realized when using RFID chips. RFID-based automatic tool identification and tracking can provide significant cost savings by minimizing expensive spindle and tool crashes. They generally provide more reliable results than either manual (tag) or bar code-based tool identification systems. In addition, the information gathered as a result of implementing RFID-based tool identification provides a foundation for automated tool room and reorder systems. With the capture of cutting tool wear and use information, replacement tooling can be ordered prior to breakages, reducing downtime. Accurate information about wear and use allows better management of replacement tooling reorders, supporting *just in time* operations and reducing unnecessary overstocking of expensive tooling.

2.4.4 Collecting Other Supporting Information and Evidences

The best starting point for evidence collecting is within the same shift when the failure occurred as the investigator can see the actual *crime scene* and talk to the *eye witnesses* of the tool failure.

The ability of the investigator to ask proper questions and the commitment to gather proper information and collect the supporting evidences are of enormous importance for the proper tool failure analysis and proper reporting of the root cause of this failure.

In talking to the *witness*, do not forget the following:

- The degree of one's emotions varies inversely with one's knowledge of the facts: the less one knows, the hotter he or she gets. In other words, the lack of facts inspires the fantasy.
- There are no facts, only interpretations/opinions. By providing adequate supporting evidences, the investigator should transform opinions into facts.
- The investigator is entitled to his/her own opinion, but not to his/her own facts.

The information provided by the tool tag of the failed tool and/or data stored on the RFID chip can be considered as the first hard evidence that allows the investigator to open the failure case. The next hard evidences are

- For broken and miserably failed tools, the investigator needs an actual part where the tool was broken/severely failed in. An old rule *no body, no case* should be clear to anyone involved as, with no actual part, the tool failure investigation becomes a guessing game.
- For quality (size, shape, surface finish) issue failures, the investigator needs an inspection report from CMM that shows the actual deviation from the allowable zone.

2.4.5 Assessment of the Collected Evidences: Obvious Root Causes

Obvious root causes of tool failure are those that

1. Are honestly reported in the tool tag by the shop floor, for example, *dropped* and *wreck*
2. Can be revealed by a simple visual or microscopic examination of the collected evidences with no special measurements

Most often, the obvious root causes are drilling-system-related although some tool-related causes are also found. The common machine-related obvious causes are

- *Tool dropping by the machine*: Machines can drop tools from time to time during a tool change. This usually results in painful consequences as, according to Murphy's law, the most expensive tool of greatest lead time with no backup is normally dropped. Moreover, when dropped, it hits as many obstacles as possible in the machine so it is almost completely destroyed.
- *Part mis-location in the work-holding fixture*: It happens surprisingly often even if the fixture is highly automated with end switches that presumably should assure the proper location of parts. Figure 2.8 shows some examples of drilling in mislocated parts. Often, the drilling tools break due to bending as a carbide drill can be bent only once.
- *MWF supply problems*: There is a wide range of such problems. Figure 2.9 shows the coolant hole(s) in the shank is(are) blocked by the sealing O-rings of the spindle and rotary union (a device that introduces the MWF flow to the rotating spindle). Figure 2.10 shows the coolant hole(s) in the shank is(are) blocked by chips. Figure 2.11 shows the coolant tube in the tool holder is blocked by chips. Figure 2.12 shows the coolant holes in the shank are blocked by the filter cartridge material. When two coolant holes or coolant tube in the holder is completely blocked, the MWF pressure increases beyond the upper limit preset in the machine so the controller should stop the drilling and the system message *high-pressure coolant fault* should be issued and written in the RFID chip. It may not happen, however, when the coolant holes (passages) are blocked only partially so the drilling tool continues to work with minimum MWF supply. Figure 2.13 shows a high-penetration rate PCD-tipped drill worked with no/not sufficient MWF supply. As can be seen, even aluminum rod is extruded through the coolant hole.

FIGURE 2.8 Examples of drilling in mislocated parts.

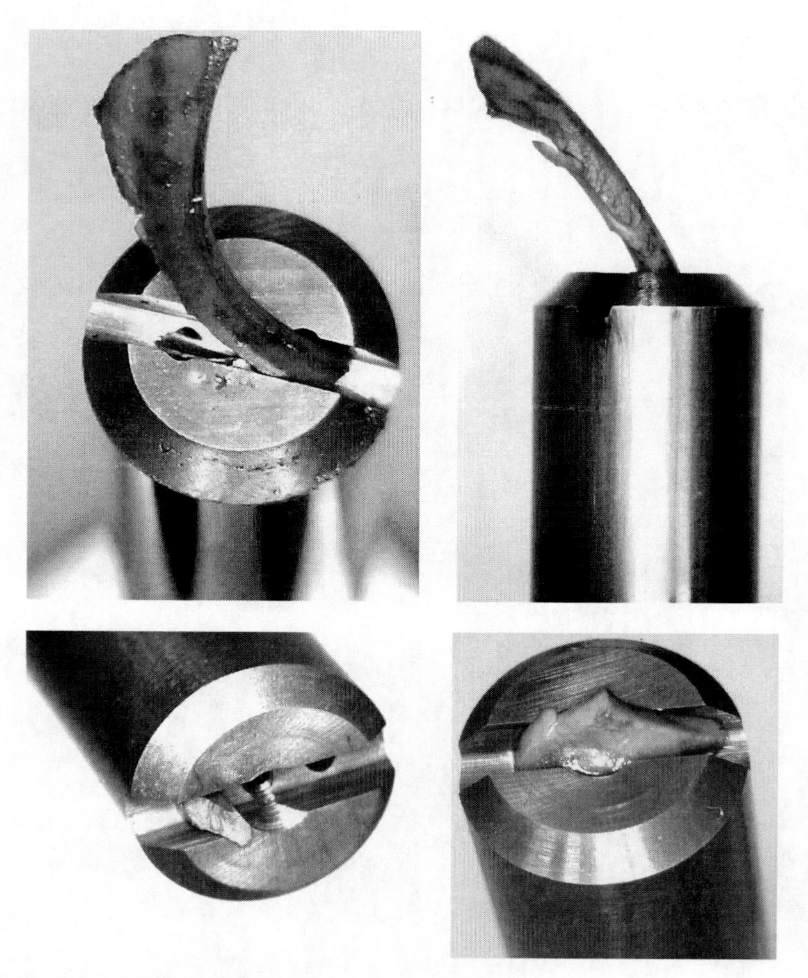

FIGURE 2.9 Coolant hole(s) in the shank is(are) blocked by the sealing O-ring.

FIGURE 2.10 Coolant hole(s) in the shank is(are) blocked by chips.

FIGURE 2.11 The coolant tube in the tool holder is blocked by chips.

FIGURE 2.12 The coolant holes in the shank are blocked by the filter cartridge material.

FIGURE 2.13 High–penetration rate PCD-tipped drill worked with no/not sufficient MWF supply.

FIGURE 2.14 Carting cavities at the drill entrance.

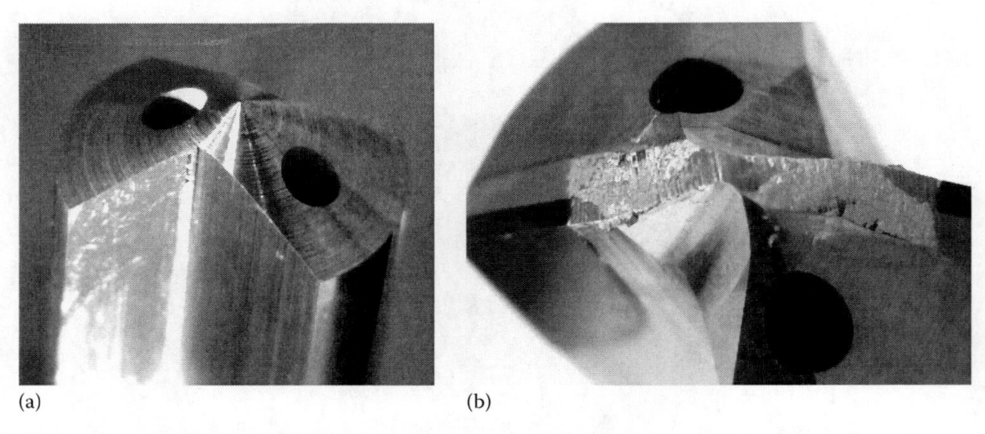

(a) (b)

FIGURE 2.15 Drill with missed flank faces (a) and a drill with *negative clearance* flank faces (b).

- *Workpiece problems*: Among multiple obvious workpiece problems, casting problems are common. Figure 2.14 shows carting cavities at the drill entrance. Such defects normally lead to drilling tool breakage.
- *Tool problems*: There are a number of possible obvious tool problems with which a drill may pass the presetting stage as no fault is detected. The most common is missed or improperly applied features. As an example, Figure 2.15 shows drills with missed flank faces and with *negative clearance* flank faces. Figure 2.16 shows an example of a rather common case where the brazed insert is missed.

2.4.6 Additional Information Needed for Normal Tool Failure Analysis

The obvious root causes for tool failures constitute no more than 20% of total failures so that the investigator should be prepared to carry out a *normal* tool failure analysis. Before he or she even starts such an analysis, additional information is always needed. This may include

1. Tool drawing.
2. Tool layout.
3. Tool inspection report.
4. Part inspection report.
5. Drilling system background information, for example, reported problems with the machine (cell, line, etc.), MWF inspection backlog information, and reported problems with MWF for the machine where the failure happened.

FIGURE 2.16 Missed insert due to improper brazing.

6. If the problem was fixed and machine (cell, line) is running, then the information on how the problem was fixed is essential, that is, by replacing the tool, by repairing/servicing spindle/fixture/MWF supply, and/or by changing to different batch of castings.
7. Tool history.
8. Last purchasing order particularities.
9. Inventory count and delivery schedule for the next supply.

In the ideal modern e-world, all this information should be readily available to the investigator in just a few clicks of the computer mouse. Moreover, this information should be complete, authentic, and up-to-date. Unfortunately in the real world, the opposite is true even at the best presumably high-tech manufacturing facilities. Trying to collect the listed information, the investigator gradually develops an incredibly high guessing ability combined with strong intuition.

2.4.6.1 Tool Drawing

The problems begin with an attempt to obtain the actual tool drawing because

- No drawings for standard tool are normally kept. Instead, only reference into a tool supplier catalog number is given, which may not be even current as this number changes with a new catalog issued. Naturally, no particularities of the tool design/geometry, tool material, and other essential information that can be relevant to the tool failure analysis are given in the catalog.
- Drawings for special tools are normally not of great help either. Figure 2.17 shows an example of a typical drawing by a leading drill manufacturer for a special high-penetration rate drill. As can be seen, no particularities of the tool geometry and tool material are given. The situation is better for the drills re-sharpened in the house as the tool geometry particularities must be shown on such drawings.

2.4.6.2 Tool Layout

As discussed in Chapter 1, the tool layout contains the valuable information on the tool position and requirement to the drilled hole(s), and cutting speed and feeds are often out of date as they may not reflect the changes made to the process even in the most advanced manufacturing facilities.

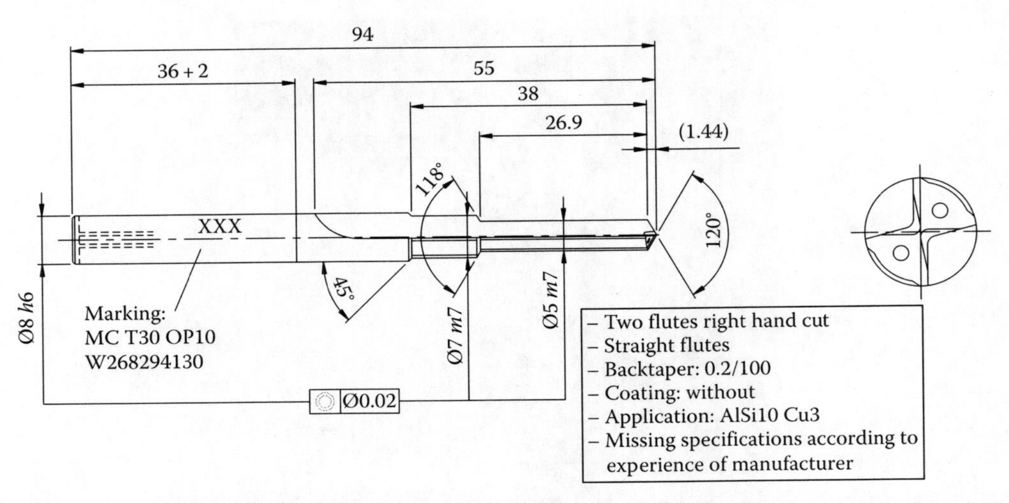

FIGURE 2.17 A typical drawing by a leading drill manufacturer for a special high-penetration rate drill.

The investigator, however, can contain the information about the speeds and feeds, cycle diagram, and tool offset used directly from the machine controller. In less advanced manufacturing facilities where the machine operator sets these parameters, obtaining the listed information presents a great challenge.

2.4.6.3 Tool Inspection Report

Tool inspection reports are supplied only for special drilling tools. In the author's experience, the most valuable parts of these reports are information on the measured drilling tool diameters and runouts (see Chapter 7) as shown in the examples presented in Figure 2.18. However, there are no particularities of the tool geometry (e.g., clearance angles) that are indicated although both reports are generated by one of the advanced tool inspection/measuring machines, Zoller/genius 3, which is fully capable of measuring the tool geometry of drilling tools. As discussed in Chapter 7, this is due to the lack of understanding of what and how to measure.

2.4.6.4 Part Inspection Report

It is essential when quality of the drilled hole is of concern, that is, the tool was prematurely pulled out due to a quality issue. Figure 2.19 shows an example of the part inspection report generated by a CMM machine. As can be seen, the drilled hole is oversized by 82 μm. In less advanced manufacturing facilities, a simple inspection report that lists the results of hand-gage measurements can be used. Regardless what type of the part inspection report is used, the quantified deviation of a quality feature from that assigned by the part/manufacturing drawing should be clear to the investigator.

2.4.6.5 Drilling System Background Information

Although there can be a great number of items in such information, the reported problems with the machine (cell, line, etc.) and MWF inspection backlog information are the two most common to know because, as the experience shows, they often directly correlate with tool failures. For example, knowing the information presented in Figure 2.20, one can construct a logical chain: A tool was pulled off prematurely on March 27 due to poor roughness of the machined surface—tool was excessively worn—the supplied MWF had unusually high ppm solids during this period of time.

Inspection Protocol 1 / 1

ZOLLER »genius Standard« 11/28/2012 **ABC Tool**

Inspected by: zoller 5:11:13PM

Master No. VNC675-gf23457

Description Multi-stage PCD tipped drill

Measurement Tool No. 1

Company Logo

Customer Best Tracks

ABC Tool **Job No.**

Reference No.

Tool No. TR012M2356

SteResult	Nom. value	U. tol.	L. tol.	Act. value	Diff. value	Tolerance
2 Run-out radial	0.000			0.004	0.004	
3 Diameter 1	7.500	0.000	-0.015	7.498	-0.002	
4 Step Length	-51.000	0.100	-0.100	-50.992	0.008	
5 D2 Run-out radial	0.000			0.002	0.002	
6 Diameter2	17.100	0.000	-0.018	17.082	-0.018	
8 Step Length	-27.800	0.100	-0.100	-27.809	-0.009	
9 Diameter 3	20.500	0.000	-0.020	20.491	-0.009	

Measure protocol "All edges" 1 / 1

»genius Standard« 9/28/12 **DEF Tools**

User zoller 10:11:45AM

ID no. MEASURE

Desc.

Remark

Stufe 1 / 3

Summary

Measure value	Nominal	Upper tolerance	Lower tolerance	Actual	
Lengthways dimension	264.405			264.405	✓
Crossways dimension	8.411	-0.029	-0.045	8.436	
Angle 1				0.12	✓
Run-out radial	0.000			0.002	✓
Run-out axial	0.000			0.000	✓

Details

Cutter	Z	X	A1
1	(x) 264.405	8.432	0.10
2	(x) 264.405	(x) 8.436	(x) 0.12

Statistic

Statistical value	Z	X	A1
Average	264.405	8.434	0.11
Difference	0.000	0.004	0.02
Maximum dimension	264.405	8.436	0.12
Minimum dimension	264.405	8.432	0.10
Standard deviation	0.000	0.002	0.01

FIGURE 2.18 Examples of tool inspection report by two leading drilling tool suppliers.

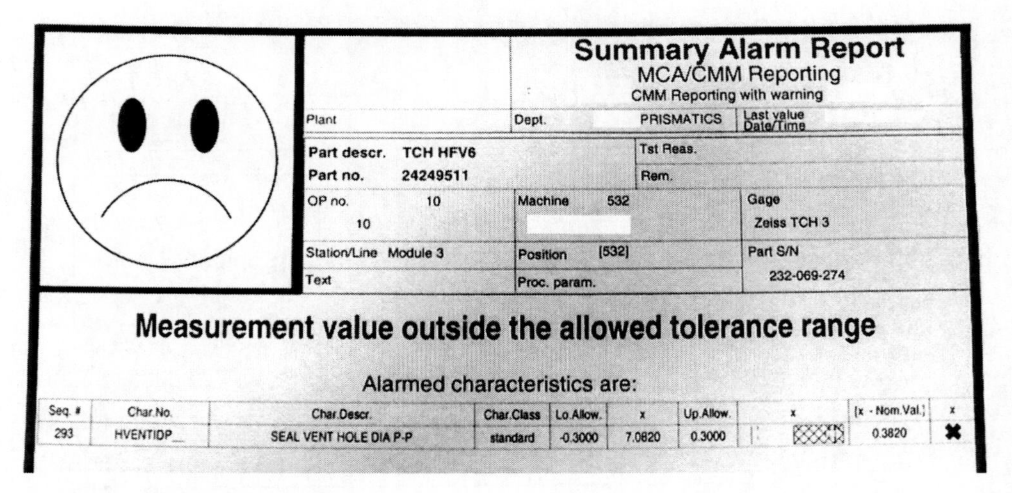

Summary Alarm Report

MCA/CMM Reporting

CMM Reporting with warning

Plant	Dept.	PRISMATICS	Last value Date/Time
Part descr. TCH HFV6		Tst Reas.	
Part no. 24249511		Rem.	
OP no. 10 10	Machine 532	Gage Zeiss TCH 3	
Station/Line Module 3	Position [532]	Part S/N 232-069-274	
Text	Proc. param.		

Measurement value outside the allowed tolerance range

Alarmed characteristics are:

Seq. #	Char.No.	Char.Descr.	Char.Class	Lo.Allow.	x	Up.Allow.		x	|x - Nom.Val.|	x
293	HVENTIDP__	SEAL VENT HOLE DIA P-P	standard	-0.3000	7.0820	0.3000			0.3820	✖

FIGURE 2.19 An example of the part inspection report.

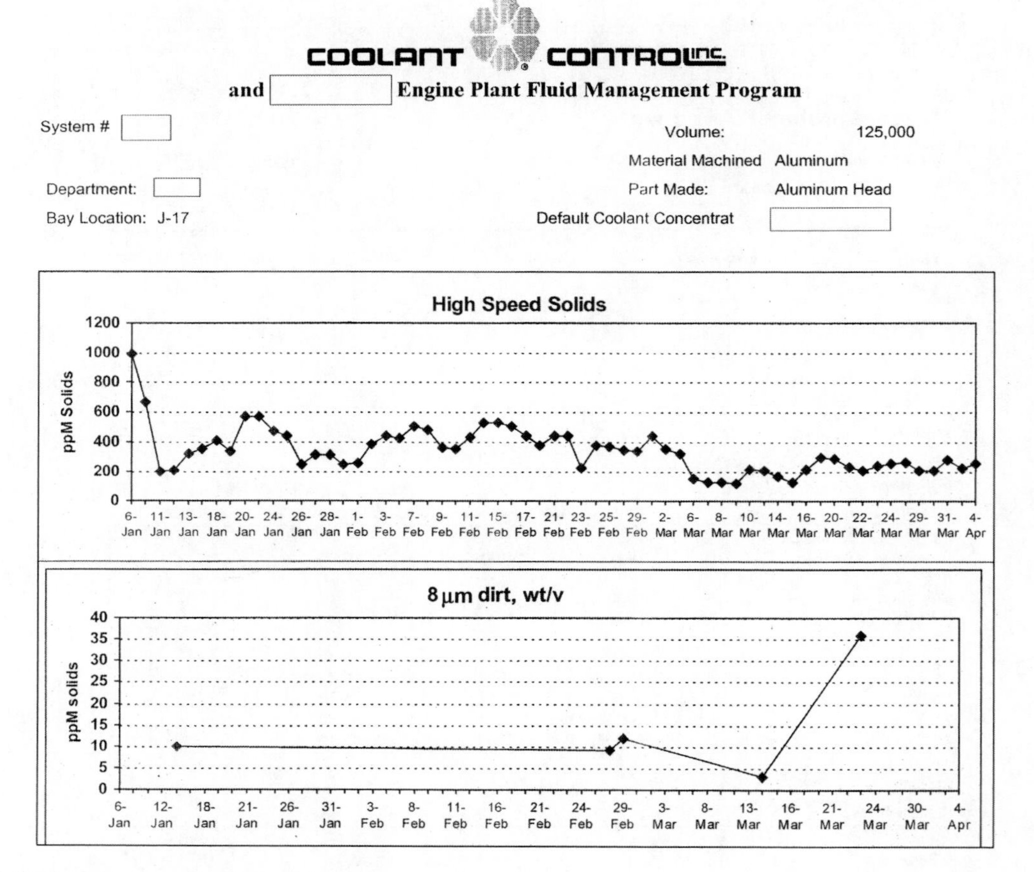

FIGURE 2.20 Coolant backlog information.

2.4.6.6 Additional Information if the Problem Was Solved

If the problem was solved in the shop flow, it is important for the investigator to know how it was solved. For example, if the problem was solved by replacing the failed tool by a new one from POU, then the following problems can be ruled out from further investigation:

- The problems related with the machine conditions
- The problems common for a batch of the blanks used as workpieces

The problems with MWF cannot be ruled out as these problems may relate to gradual tool wear, which may be revealed days later.

If, on the other hand, the problem was solved by replacing the failed tool preceded by fixing the problem with the machine (i.e., with the spindle runout, work-holding fixture), then the tool cause may be related to the drilling system rather than to the failed tool itself.

2.4.6.7 Tool History

The tool history helps the investigator to make a decision on the tool material. If, for example, a carbide drill was re-sharpened a number of times and each time in the past it made full tool life, then the tool material may safely be ruled out from further investigation. If the tool historically had a problem with making tool life, then knowing this problem can tremendously accelerate the investigation.

Last purchasing order particularities are also essential. It may be useful to know how many tools were supplied according to the last purchasing order and how many of them had problems.

2.4.6.8 Inventory Count and Delivery Schedule for the Next Supply

This information is useful if the problem with a particular tool batch is found, so containment of defective tools is needed. When it happens, all tools (including those in POU) from the same manufacturing batch are to be inspected and the inspection information is to be forwarded to the tool supplier. Often in the automotive industry, a *likelihood survival game* is played to select some *less-than-perfect tools* that may survive until the next tool supply has arrived. In some advanced manufacturing facilities where quality and productivity are of prime concern, a cutting tool containment procedure and a contingency plan are inherent parts of the tool management program.

2.5 PART AUTOPSY AND TOOL RECONSTRUCTION SURGERY

Two new terms particular to medicine are introduced in the tool failure analysis:

1. The first is the *part autopsy* defined as *postmortem* procedure to identify the root cause of the tool failure. It normally includes sectioning of a portion of the part containing fragments of the tool, the precut of the hole, and then manual opening of the hole in order to retrieve tool fragments. The part autopsy can be called also as tool autopsy as not only the part but also the tool is examined.
2. The second term is tool *reconstruction surgery* defined as the procedure of the reconstruction of the tool from the collected fragments using the edge cleaning putty as a binding Maximum Bond Krazy Glue® can also be of a great help.

The part autopsy and tool reconstruction surgery are special procedures performed only by highly trained engineering personnel and may require some special cutting and measuring equipment.

Moreover, they are expensive and time-consuming, so they are used only in special cases where the root cause of the drill failure cannot be determined using simpler procedures or when the tool is expensive so the root cause determination is a money-saving/persisting-problem-solving procedure.

To the best of the author's knowledge, nothing has been written on the discussed procedures simply because these procedures were not developed as the cost of drills was normally small. With gradual introduction of HP drills including, for example, PCD multistep drills, the cost of which can be as high as many thousands of dollars and lead time may be as long as 12–18 weeks, the development of the discussed procedures became of vital importance in the automotive industry. The following sections present a detailed example of the tool autopsy and reconstruction surgery (performed by Andrew Stanley, a qualified tool engineer).

2.5.1 Example: Step 1—Failure Information

A broken drill has been reported from the shop floor according to the tool tag shown in Figure 2.21 together with what is left from the drill. As indicated on the tool tag, the drill made 335 cycles out of 15,000 set as tool life. The part where the drill was broken was also supplied for the failure analysis (Figure 2.22). The part is made of high-silicon high-strength aluminum alloy A390. According to the information taken from the Balluff chip, the machining regime was set according to the tool layout, no abnormalities with MWF supply was reported, no excessive force/drilling torque prior to the fracture was recorded, and no part quality problems in the previously machined part were reported. The problem was solved, that is, normal part manufacturing was continued, by replacing the drill with one that was available at POU. The replacement drill was from the same manufacturing batch as the broken one.

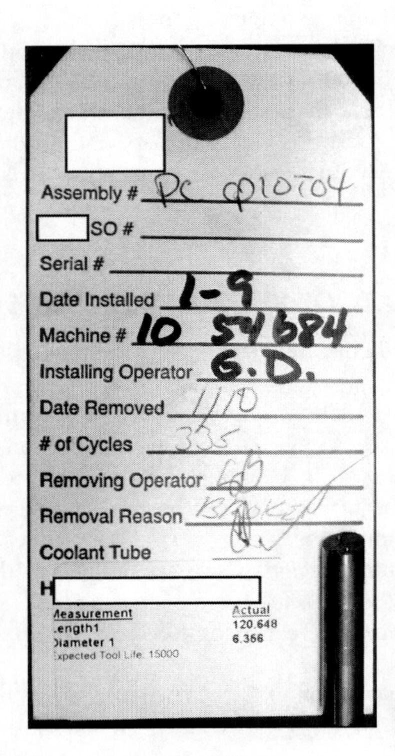

FIGURE 2.21 Tool tag and remaining of the broken drill.

FIGURE 2.22 Part with the remaining of the broken tool.

2.5.2 Example: Step 2—Analysis of the Collected Failure Information

The analysis of this preliminary information reveals

1. Tool breakage was not gradual tool deterioration (previous part quality, force components reading) but rather an abrupt event.
2. As the problem was solved by replacing the tool, the conditions of the machine and general conditions of die castings used as workpieces are not likely to have problems.
3. The tool inspection report received from the supplier with the tool was complete, that is, showed relevant information including important parameters of the tool geometry. The report indicated no deviation from the tool drawing tolerances.
4. Tool presetting information showed that the drill had no deviations from the parameters used in presetting, that is, its diameter, length, and runout were within the well-defined specifications.

These analyses narrow down the root cause of the problem to the particular machined part and the drill itself. Because of the high cost of the drill (more than $700) and high volume of production, the autopsy and reconstruction surgery procedures were found to be justifiable for the considered case.

A closer look at the tool drawing was taken to identify features that can potentially contribute to drill failure. Fortunately enough, the drawing was made according to the tool drawing requirements developed by the author (Astakhov 2010) and was established/reinforced by the engineering team. Figure 2.23 shows the relevant information taken from the tool drawing. The analysis of the geometry of the chisel edge reveals that the drill suffers from the known problem—a small gap between the chisel edge rake faces and the bottom of the hole being drilled discussed in Chapter 4 (Sections 4.5.8.2, Figure 4.108). The problem is enhanced by the fact that no web thinning (Section 4.5.7.1, Figure 4.74) is provided to ease the problem caused by the long chisel edge of the analyzed drill. As a result, high force acting on the brazed PCD insert should be expected. The drawing, however, does not provide any information on the required strength of the braze joint as the cost of a test to obtain and to verify this strength is enormously high.

2.5.3 Example: Step 3—Sectioning the Autopsy Specimen

Sectioning the autopsy specimen includes a number of stages. At the first stage, a decision is made about the sequence of precutting the hole with the broken tool and cutting off the piece with the precut

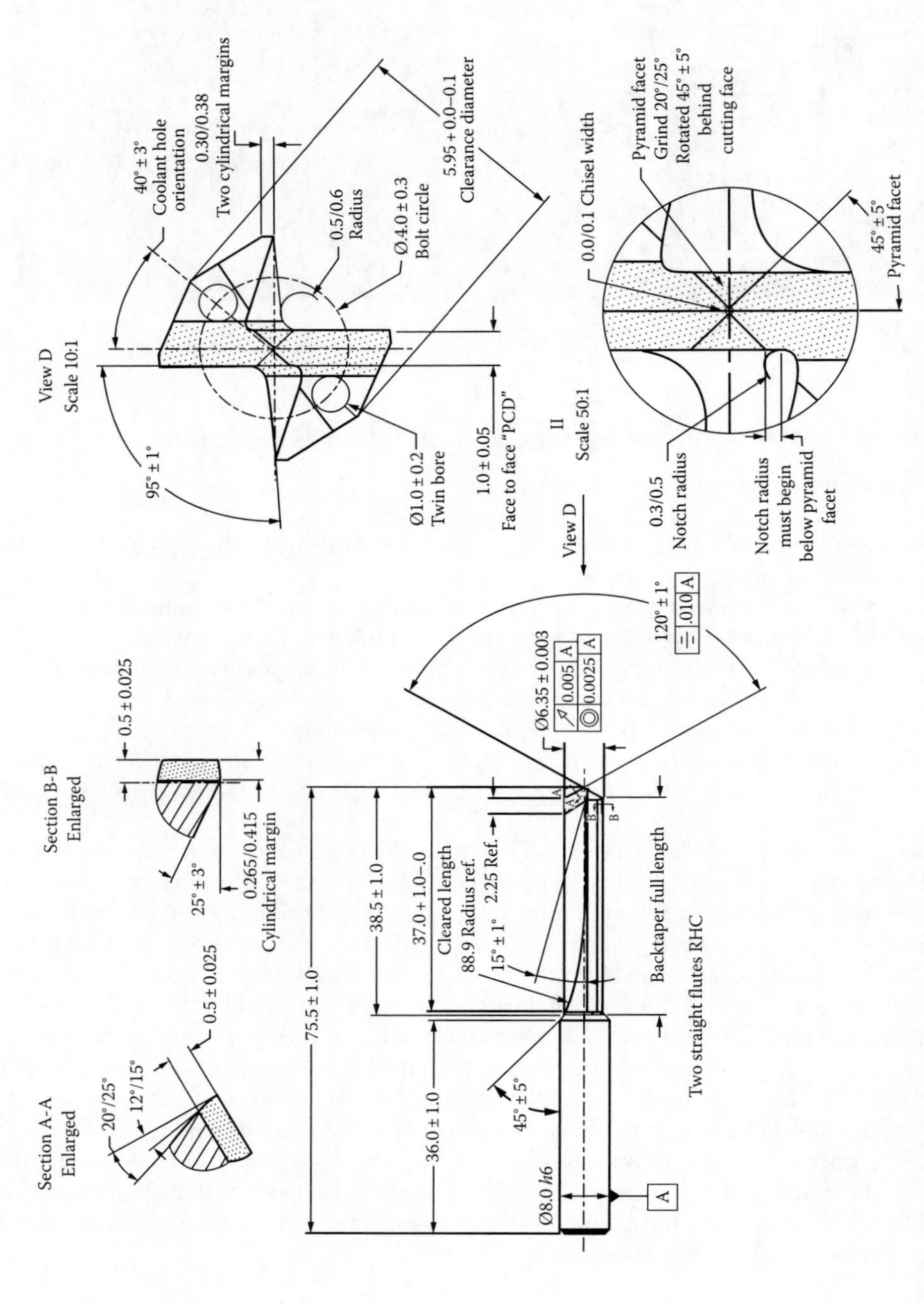

FIGURE 2.23 Relevant information taken from the tool drawing.

FIGURE 2.24 Showing how it was decided to cut the part.

hole from the rest of the part. It is made on the case-by-case basis depending upon the hole location and part configuration. Figure 2.24 shows how it was decided first to precut a quarter of the hole containing the broken tool and then the piece with this hole from the rest of the part. Corresponding lines are drawn using a sharpie marker pen.

At the second stage, the part is positioned on the table of a band saw with proper orientation along the corresponding marked lines. First, precutting of the hole with the broken tool is made leaving thin walls (~0.2–0.4 mm) to be broken manually in the course of the analysis (Figure 2.25a). Second, the piece with this hole from the part is cut off from the part (Figure 2.25b). Note that such sectioning of the part should be made only when the part is properly clamped to the table. Mechanical feed should be used, not held in hands and pushed against the blade. All safety measures/procedures should be followed.

(a)

(b)

FIGURE 2.25 Part cutting along the marked lines: (a) Precutting the hole with the broken tool and (b) cutting off the piece with this hole from the part.

2.5.4 Example: Step 4—First Microscopic Examination of the Sectioned Part

At this step, the sectioned part is cleaned using a light jet of compressed air and then is set under a microscope as shown in Figure 2.26 to reveal fractography of drill fracture and any other important feature particular to the breakage. Autopsy tools including a hammer, flat-nose screwdrivers, and various picks (similar to dental picks) are prepared.

Figure 2.27 shows the results of the first microscopic examination of the sectioned part. The following were revealed by varying the focus of the microscope:

- Fractography of the fractured surface reveals almost pure bending.
- Unusually big chips stuck in the chip flute indicate that this side of the drill actually failed.

2.5.5 Example: Step 5—Breaking the Precut Section and Separating Debris/Tool Fragments

At this step, the precut hole containing the broken tool is separated into two precut parts. A flat-nose screwdriver is used to break thin walls of the precut parts. Sometimes, light hammering may be needed for final separation (Figure 2.28). Then the precut quarter of the hole containing the broken

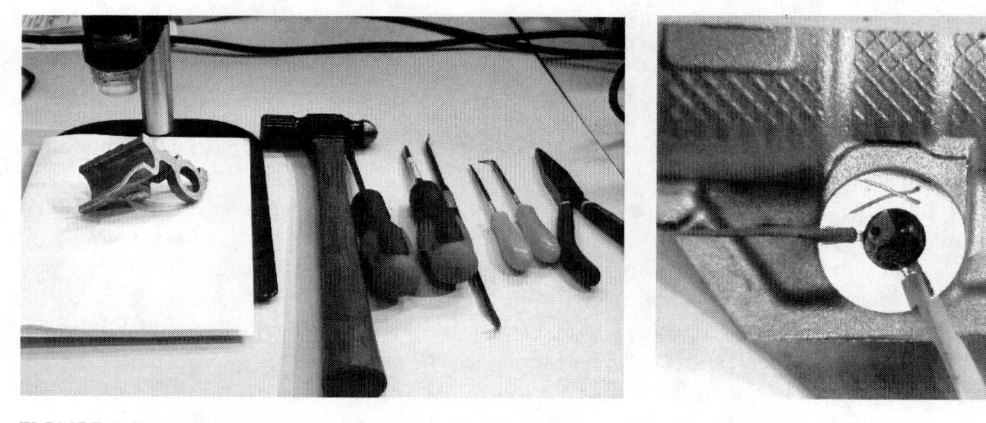

FIGURE 2.26 The sectioned part set under a microscope and autopsy tools are prepared.

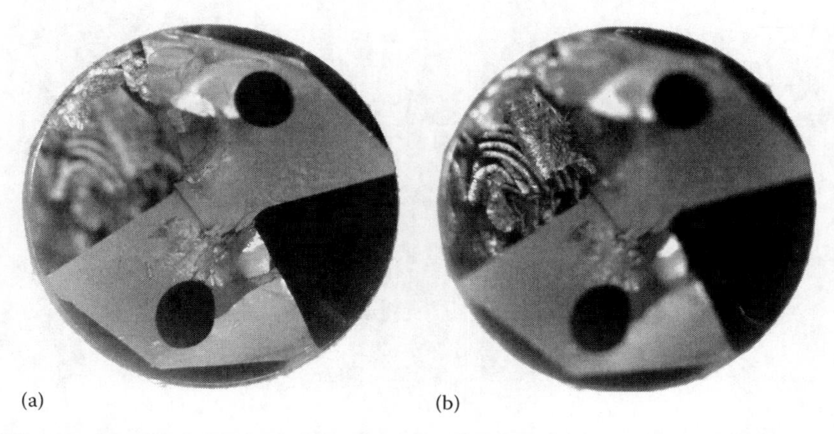

FIGURE 2.27 Results of the first microscopic examination of the sectioned part focusing on different features: (a) fractography of the drill fractured surface and (b) unusually large chips stuck in the chip flute.

FIGURE 2.28 Separating the precut parts.

FIGURE 2.29 Final separation of the precut parts.

tool is gently separated from the rest in the manner shown in Figure 2.29. The part containing the tool and debris is placed under the microscope and cleaning using a pick begins as shown in Figure 2.30.

The cleaning performed under the microscope actually consists of two parts:

1. Removing and separating debris as shown in Figure 2.31. All retrieved pieces of debris are analyzed to search for any abnormality in the chip shape/thickness, presence of inclusions, pieces of the tool material, etc.
2. Removing shattered pieces of the tool as shown in Figure 2.32. The rule of thumb at this stage is not to break the tool into more pieces than it was originally broken into. Removing the pieces of the broken tool, one should make the selection of the pieces that can be potentially used in the tool reconstruction surgery.

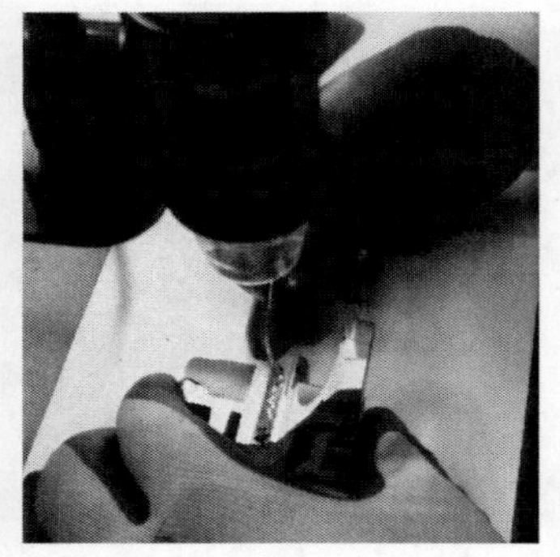

FIGURE 2.30 First look at the remaining tool/debris and the beginning of the cleaning using a pick.

FIGURE 2.31 Removing and separating debris.

FIGURE 2.32 Removing shattered pieces of the tool.

2.5.6 Example: Step 6—Examining the Surface of the Machined Hole

There are two prime objectives of such an examination:

1. To look for any abnormalities in the machined surface that can be potentially a cause of tool breakage. Casting defects (discussed later in this chapter) are common examples.
2. To look at the regularity of the feed marks left by the tool on the examined surface to assure that the feed (and thus feed rate) did not change in drilling.

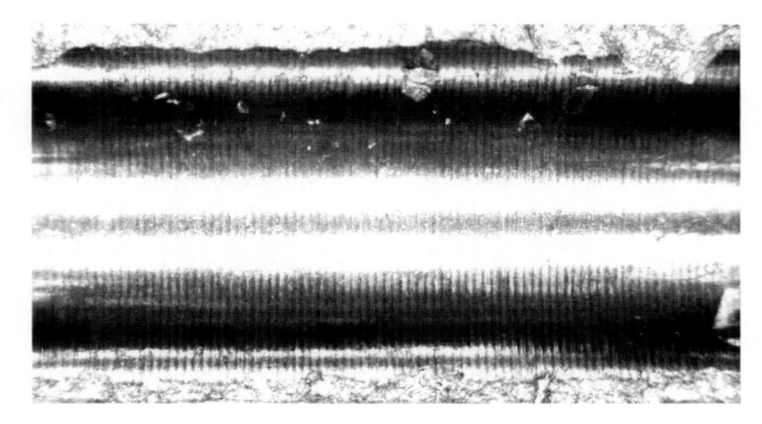

FIGURE 2.33 Examining the regularity of the feed marks left by the tool on the machined surface.

Figure 2.33 shows the surface of the machined hole in the considered case. A careful microscopic examination did not find any casting defects. The feed marks are regular so there was no change in the drilling regime until the drill was fractured.

2.5.7 Example: Step 7—Reconstruction of the Bottom of the Hole Being Drilled

The bottom of the hole being drilled is the most important surface of the partially drilled hole as actual cutting takes place there. Any irregularities in tool performance or shape/geometry of the cutting parts of the tool are inevitably reflected by the surface macro- and microtopology of this bottom. Therefore, its examination is an important stage of the failure analysis. Figure 2.34 shows that the bottom of the hole being drilled was reconstructed under the microscope by putting together the separated pieces of the hole and focusing the microscope on this bottom.

The observations show the following:

1. One cutting tooth was stalled while the other continued normal cutting for approximately 1/3 of revolution producing rather normal chip.
2. Drill just broke through the bottom of the hole being drilled so that the maximum load was shifted from the chisel edge to the major cutting edges.

FIGURE 2.34 Reconstruction of the bottom of the hole being drilled.

2.5.8 Example: Step 8—Reconstruction of the Drill and Root Cause Determination

The reconstruction (layer by layer) surgery of the broken drill is carried out under a microscope as shown in Figure 2.35. In the considered case, it begins with making a small *pedestal* of edge cleaning putty and placing it under the microscope.

In the considered case, a fragment of the tool body that supported the PCD insert is first examined as shown in Figure 2.36. The examination of this fragment reveals the possible root cause of the failure—the so-called dry brazing of PCD insert (discussed later in Section 2.6.10, Figure 2.71). As clearly seen, no wetting of the brazing filler material (BFM) is observed on the surface of brazing. It fully resembled the known appearance of the typical failure (fracture) of a PCD insert due to dry brazing as shown in Figure 2.37.

Figure 2.38 shows the full reconstruction of the drill. The following can be observed in this reconstruction:

- The PCD insert is separated into two fragments by a crack that runs from the cutting edge into the braze joint.
- Clear disengagement of the two PCD fragments from the braze joint.
- Traces of the built-up edge (BUE) on the rake faces of the major cutting edges and on the rake faces of the chisel edge.

The final analysis of the evidences presented in Figures 2.36 and 2.38 reveals that the dry brazing in the braze joint that should hold the PCD insert on the carbide body is the root cause of the problem.

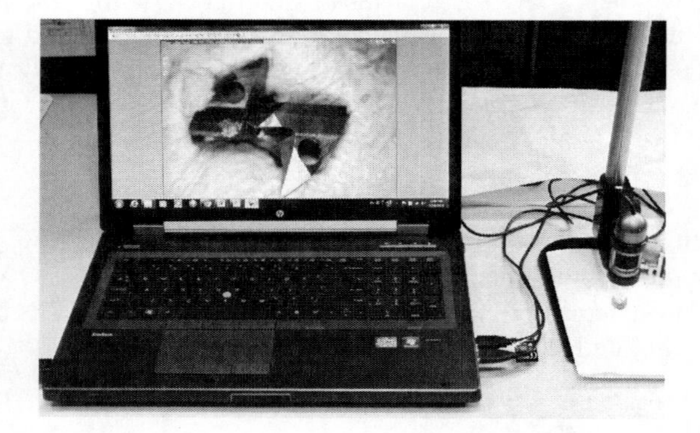

FIGURE 2.35 Setup for the drill reconstruction surgery.

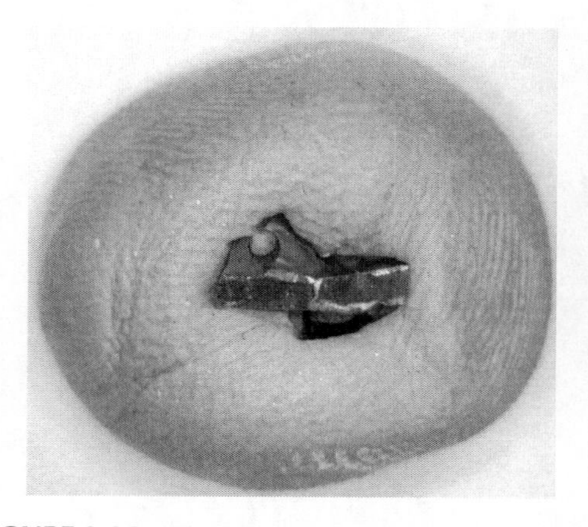

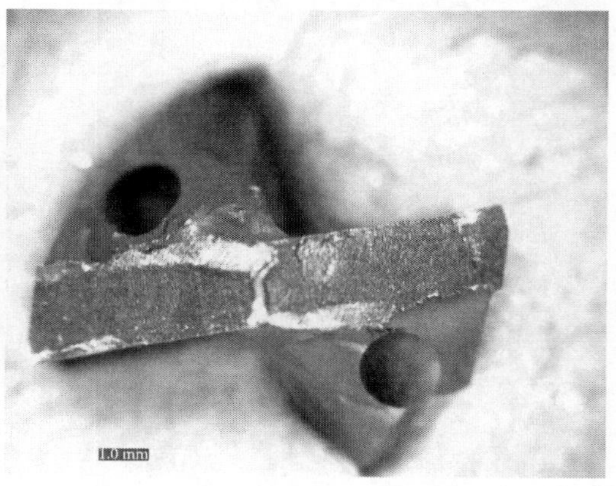

FIGURE 2.36 Placing and examining a fragment of the tool body that supported the PCD insert.

FIGURE 2.37 Example of *dry* brazing of a PCD insert.

FIGURE 2.38 Full reconstruction of the drill.

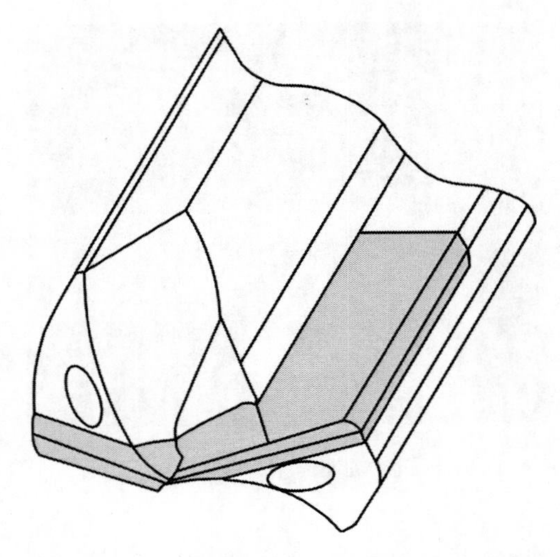

FIGURE 2.39 Proposed modifications of the drill geometry based on the *VPA-Balanced* drill concept.

However, a high cutting force due to inferior drill design/geometry is the major contributing factor to the failure of the braze joint. Therefore, the following recommendations were made:

1. The design of the rake face of the major cutting edges should be improved by web thinning (Chapter 4).
2. The geometry of the chisel edge has a major problem, namely, a small gap between the chisel edge rake faces and the bottom of the hole being drilled discussed in Chapter 4 (Figures 4.107 and 4.108). It may cause drill radial vibration and chattering of the drill margin.
3. The brazing technology should be improved to avoid dry brazing.

The proposed modifications of the drill geometry made according to *VPA-Balanced* drill concept are shown in Figure 2.39. As can be seen, web thinning is added. The double-gash design assures proper chip removal from the region of the split-point chisel edge. As such, the chevron shape of the braze joint changes to a simple straight joint as it should be sufficient to withstand a smaller and fully balanced cutting force.

2.5.9 Example: Step 9—Archiving the Evidence and Writing a Report

The last step is properly archiving the evidences and writing a report to be presented to the tool supplier. The evidences are placed in a plastic zip bag and stapled to the tool tag. Then it is properly stored with the reference to the written report. This archiving has two objectives:

1. Should similar failure occur again, the evidences can then be compared with those collected from a new failure to reduce the time needed for the analysis.
2. Tool suppliers do not always agree with the conclusions made in the report. If a supplier (normally the sales force at the meeting) does not agree with the root cause finding, the hardware evidences are made available to this supplier to carry out its own investigation on the matter. In the author's experience, however, such a validation investigation is never carried out because of the following:
   a. If these suppliers have the trained technical personnel and corresponding equipment, then the findings made in the original report are just confirmed through reading the report and briefly reexamining the evidences.
   b. If this supplier has no highly trained personnel, the problem *dies* itself as reading and understanding the report provided require some technical skills and experience.

2.6 TOOL WEAR

2.6.1 Background Information

In deforming processes used in manufacturing, concern over the tool wear is often overshadowed by considerations of forces or material flow. Except for hot extrusion, die life is measured in hours and days or in thousands of parts in metal-deforming operations (Schey 1983). In metal cutting, however, tool wear is a dominant concern because process conditions are chosen to give maximum productivity or economy, often resulting in tool life (the time to achieve the maximum allowable tool wear) in minutes. Central to the problem are high contact temperatures at the tool–chip and tool–workpiece interfaces combined with high relative speeds over these interfaces.

Generally, wear is thought of as erosion or sideways displacement of material from its *derivative* and original position on a solid surface performed by the action of another surface. In other words, wear is related to interactions between surfaces and more specifically the removal and deformation of material on a surface as a result of mechanical action of the opposite surface (often referred to as the counterbody). The need for relative motion between two surfaces and initial mechanical contact between asperities is an important distinction between mechanical wear compared to other processes with similar outcomes.

Wear can also be defined as a process where interaction between two surfaces or bounding faces of solids within the working environment results in dimensional loss of one solid. Aspects of the working environment that affect wear include loads and features such as unidirectional sliding, reciprocating, rolling and impact loads, speed, and temperature. Apart from general mechanical engineering where the wear of both counterbodies is considered as affecting performance of machines, wear on only one of two counter bodies, namely, the cutting tool, is relevant and thus considered in manufacturing, terming it as tool wear.

The nature of tool wear, unfortunately, is not yet clear enough in spite of numerous investigations carried out over the last 100 or even more years. Although various theories have been introduced hitherto to explain the wear mechanism, the complicity of the processes in the cutting zone hampers the formulation of a sound theory of cutting tool wear. Cutting tool wear is a result of complicated physical, chemical, and thermomechanical phenomena. Because different *simple* mechanisms of wear (adhesion, abrasion, diffusion, oxidation, etc.) act simultaneously with predominant influence of one or more of them in different situations, identification of the dominant mechanism is far from simple, and most interpretations are subject to controversy (Schey 1983). These interpretations are highly subjective and based on the evaluation of the cutting conditions, possible temperature and contact stress levels, relative velocities, and many other process-related paramagnets and factors. As a result, experimental or post-process methods are still dominant in the known studies of tool wear (Loladze 1958; Gordon 1967; Usui and Shirakashi 1982; Schey 1983; Olson et al. 1988; Shuster 1988; Jawahir and Van Luttervelt 1993; Marinov 1996; Stenphenson and Agapiou 1996; Luttervelt et al. 1998; Childs et al. 2000; Trent and Wright 2000), and only topological or simply geometrical parameters of tool wear are selected and thus reported in tool wear and tool-life studies.

It is well known in metal cutting that wear of the cutting tool may take place on the rake face or on the flank face, that is, over the so-called tool–chip and tool–workpiece interfaces. Although in reality wear takes place on both interfaces, wear on one of these two is normally predominant compared to the other, so this one is considered as its wear determines the tool life. Flank wear is commonly given priority in many studies on metal cutting, while face wear receives much less attention. No explanation for this "discrimination" is given. A need is felt to clarify the issue.

Figure 2.40 shows a simple model that helps to compare the wear-related conditions on the tool face and flank interfaces. As discussed in Section 3.1, a positive clearance angle is the major distinguishing feature of the cutting tool, that is, it is the only feature that distinguishes the cutting tool from those tools used in closely related manufacturing process as in splitting and shearing (Astakhov 2010). This model shows that when the tool is new (sharp), the flank–workpiece interface should not even exist, so only the tool–chip interface is considered. However practically,

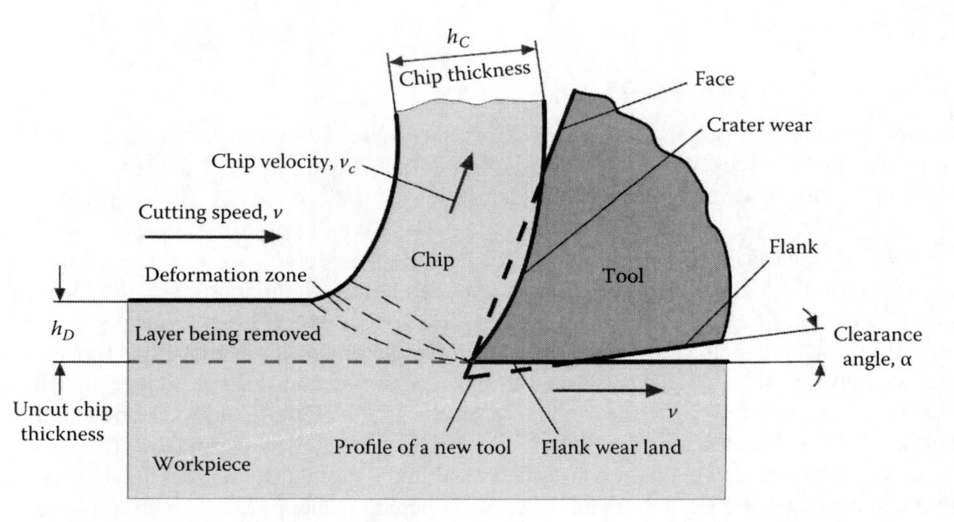

FIGURE 2.40 Notions of the rake and flank wear regions.

the tool–workpiece interface develops as soon as cutting begins due to two prime causes: work material springback and deformation of the cutting wedge (Astakhov 2012).

Comparison of the conditions over the tool–chip and the tool–workpiece interfaces reveals the following (Astakhov 2012):

1. The normal and shear stresses over the tool–chip interface are much higher than those at the tool–workpiece interface. Moreover, the normal stress maximum occurs not at the cutting edge but rather at a certain distance from the cutting edge.
2. The maximum temperature at the tool–chip interface is much higher than that at the tool–workpiece interface. Moreover, the location of this maximum is not at the cutting edge but rather at a certain distance from the cutting edge.
3. The location of the normal stress and temperature maxima and their distribution generally coincide with the shape of tool crater wear found behind the cutting edge as shown in Figure 2.40.
4. The sliding velocities at the interfaces are not the same. This velocity at the tool–workpiece interface is equal to the cutting speed v while that at the tool–chip interface is smaller as equal to the chip velocity v_c. The latter is calculated through the chip compression ratio ξ as $v_c = \xi v = v h_C / h_D$. The higher plastic deformation of the layer being removed takes place in metal cutting; the greater the chip thickness h_C compared to the uncut (undeformed) chip thickness h_D, the smaller is v_c compared to v.

As discussed in Chapter 1, one of the prime objectives in the cutting process optimization is the reduction of the amount of plastic deformation of the layer being removed to its physically possible minimum. When machining steel with an optimized using this criterion cutting tool, the chip compression ratio can be as low as 1.2 so that the chip velocity v_c is only 1.2 times smaller than the cutting velocity. Under this condition, crater wear takes place due to the previously listed reasons. In general turning of medium carbon steels, the chip compression ratio varies depending upon the optimality of the selected tool geometry and machining regime. Commonly, the chip compression ratio is found in the range of 1.6–2.2. Often, in such machining, crater wear takes place as the severity of the contact conditions at the tool–chip interface *overpowers* the relatively small difference in sliding velocities.

This is not the case, however, in the machining of a large group of difficult-to-machine materials and aluminum and titanium alloys in which wear at the tool–workpiece interface is predominant. Although each particular set of machining conditions should be treated on a case-by-case basis to understand why it happens, some underlying causes are general in such a consideration and thus can be of great help in any effort to reduce tool wear:

1. *Great chip compression ratios* in the machining of difficult-to-machine materials and aluminum alloys. Commonly, the chip compression ratio is found in the range of 3–8. As a result, the relative speed at the tool–workpiece interface is three to eight times greater than that at the tool–chip interface. A great difference is that the sliding velocities cannot be *overpowered* by the severity of the contact conditions at the tool–chip interface.
2. *Great springback* of the work material. The concept and causes for work material springback are considered in Chapter 4 (Section 4.6.9.5). As discussed, when the work material is being cut, the material is deliberately overstressed beyond the elastic limit in order to induce a permanent deformation and then separation of the stock to be removed. As the cutting edge passed a certain part of the machined hole, the load due to cutting is removed so that the applied stress returns to zero. As a result, the machined surface springs back because elastic deformation is recoverable deformation (elastic recovery). The higher the strength of the work material, the greater the springback; the lower the elasticity modulus, the higher the springback. Therefore, great springbacks and thus large contact area at the tool–workpiece interface are caused by the high strength of difficult-to-machine materials and low-elasticity modulus of light materials such as aluminum. Titanium alloys possess the worst combination of high-strength and low-elasticity modulus (see Section 4.6.9.5).
3. *Elastic deformation of the cutting wedge* due to high contact stress at the tool–chip interface. The surprisingly high contact stresses at the tool–workpiece interface cannot be explained even in principle by the common notion of springback of the work material (Astakhov 2006). The proper explanation includes the bending moment due to normal force acting at the tool–chip interface. A model of the cutting wedge (the part of the cutting insert between the tool–chip and tool–workpiece interfaces) deformation due to this moment is shown in Figure 2.41. As shown, the normal force F_N causes the deformation of the cutting wedge that results in the *diving* of the tool into the workpiece. The penetration δ_{pn} depends on the tool wedge angle, which in turn

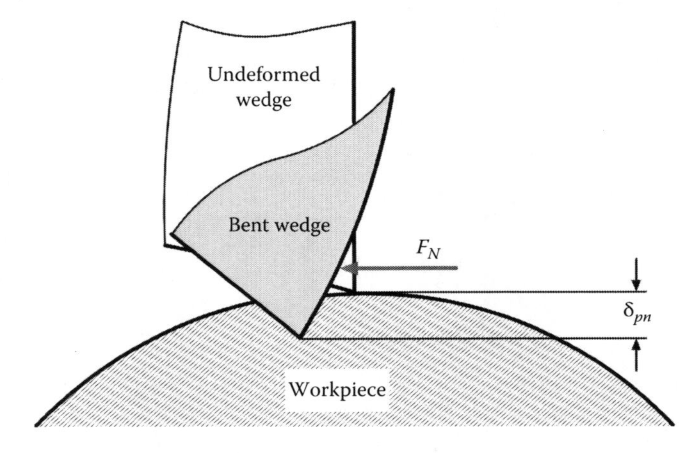

FIGURE 2.41 Model of the cutting wedge deformation.

is determined by the rake and clearance angles (see Appendix C). Depending on the cutting condition, δ_{pn} may reach as much as 5 μm in the stable cutting and may exceed 20 μm at the beginning of cutting when the normal force F_N is just applied (Astakhov 2012).

4. *Abrasion of the work material*: When work materials have highly abrasive solid phases in their structure, the flank wear is always predominant.

2.6.2 Standard Wear Assessment

In accordance with standard ANSI/American Society of Mechanical Engineers (ASME) Tool-Life Testing with Single-Point Turning Tools (B94.55M-1985), the principal types of tool wear, classified according to the regions of the tool they affect, are (Figure 2.42)

1. *Rake face or crater wear* produces a wear crater on the tool rake face. The depth of the crater *KT* is selected as the tool-life criterion for carbide tools. The other two parameters, namely, the crater width *KB* and the crater center distance *KM* are important if the tool undergoes re-sharpening.
2. *Relief face or flank wear* results in the formation of a flank wear land. For the purpose of wear measurement, the major cutting edge is considered to be divided into the following three zones: (a) Zone *C* is the curved part of the cutting edge at the tool corner; (b) Zone *N* is the quarter of the worn cutting edge of length *b* farthest away from the tool corner; and (c) Zone *B* is the remaining straight part of the cutting edge between Zones *C* and *N*. The maximum VB_Bmax and the average VB_B width of the flank wear are measured in Zone *B*, the notch wear VB_N is measured in Zone *N*, and the tool corner wear VB_C is measured in Zone *C*. As such, the following criteria for carbide tools are normally recommended: (a) the average width of the flank wear land $VB_B = 0.3$ mm, if the flank wear land is considered to be regularly worn in Zone *B*; (b) the maximum width of the flank wear land $VB_Bmax = 0.6$ mm, if the flank wear land is not considered to be regularly worn in Zone *B*. Besides, surface roughness for finish turning and the length of the wear notch $VB_N = 1$ mm can be used.

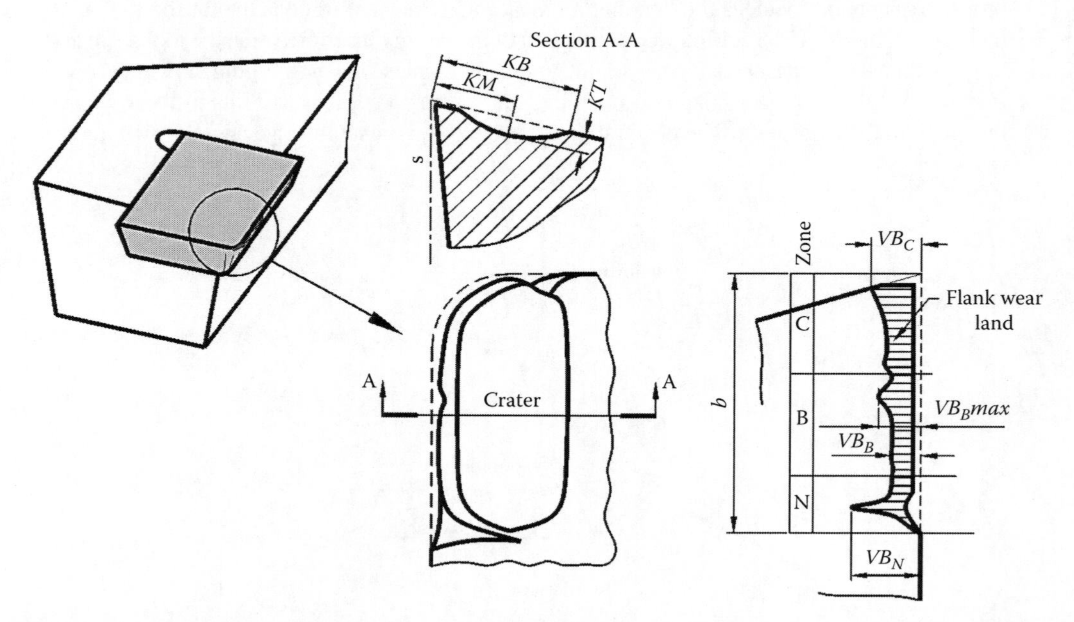

FIGURE 2.42 Types of wear on turning tools according to standard ANSI/ASME Tool-Life Testing with Single-Point Turning Tools (B94.55M-1985).

In the author's opinion, however, the listed geometrical characteristics of tool wear are subjective and insufficient because

- They do not account for the tool geometry (the clearance angle, the rake angle, the cutting edge angle, etc.) so they are not suitable to compare wear parameters of cutting tools having different geometries.
- They do not account for the cutting regime and thus do not reflect the real amount of the work material removed by the tool during the tool operating time, which is defined as the time needed to achieve the chosen tool-life criterion.

Tool wear curves illustrate the relationship between the amount of flank (rake) wear and the cutting time (τ_m) or the overall length of the cutting path (L_m). These curves are represented in linear coordinate systems using the results of cutting tests, where flank wear VB_Bmax is measured after certain time periods (Figure 2.43a) or after a certain length of the cutting path (Figure 2.43b). Normally, there are three distinctive regions that can be observed on such curves. The first region (I in Figure 2.43b) is the region of primary or initial wear. Relatively high wear rate (an increase of tool wear per unit time or length of the cutting path) in this region is explained by accelerated wear of the tool layers damaged during its manufacturing or re-sharpening. The second region (II in Figure 2.43b) is the region of steady-state wear. This is the normal operating region for the cutting tool. The third region (III in Figure 2.43b) is known as the tertiary or accelerated wear region. Accelerated tool wear in this region is usually accompanied by high cutting forces, temperatures, and severe tool vibrations. Normally, the tool should not be used within this region.

Tool wear types and patterns are well described in the literature on metal cutting (Shaw 1984; Astakhov 2004, 2006; Astakhov and Davim 2008). The assessment and proper reporting of tool wear are standardized by international (e.g., ISO 3685:1993, ISO 8688-1: 1989) and national (e.g., ANSI/*ASME* B94.55M-1985) standards. Tool wear in the standard assessments is always considered as a gradual process. The previously described two basic zones of wear in cutting tools, namely, flank wear and crater wear, are normally conserved, and thus, the corresponding parameters of tool wear are measured using a toolmaker's microscope (TMM) equipped with a video imaging system and having a resolution of approximately 0.001 mm. Stylus instruments similar to a profilometer and with laser interferometers are also used.

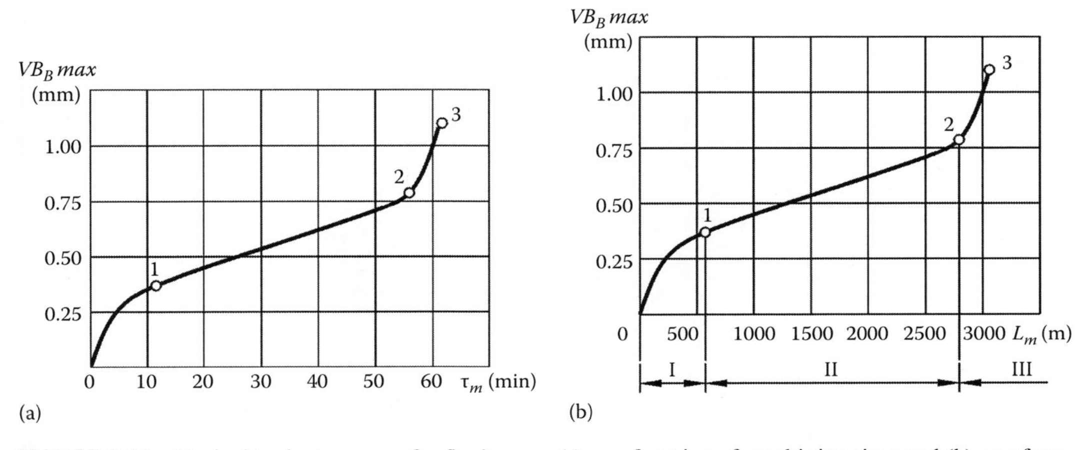

FIGURE 2.43 Typical tool rate curves for flank wear: (a) as a function of machining time and (b) as a function of the cutting length.

Standard tool-life testing and representation include Taylor's tool-life formula (Taylor 1907)

$$vT^n = C_T \tag{2.1}$$

where

v is the cutting speed
T is tool life in minutes
C_T and n are constants into which all cutting conditions affecting tool life must be absorbed

Although Taylor's tool-life formula is still in wide use today and is in the very core of many studies on metal cutting including the level of national and international standards, one should remember that it was introduced in 1907 as a generalization of many-year experimental studies conducted in the nineteenth century using work and tool materials and experimental technique available at that time. Since then, each of these three components underwent dramatic changes. Unfortunately, the validity of the formula has never been verified for these new conditions. Nobody proved thus far that it is still valid for any other cutting tool materials than carbon steels and HSSs, for cutting speeds higher than 25 m/min.

Figure 2.44 shows the procedure of experimental determining of Taylor's formula according to standard ANSI/*ASME* B94.55M-1985 for three cutting speeds $V_1 > V_2 > V_3$ (Astakhov and Davim 2008). Simple analysis of Taylor's tool-life formula shows that it actually correlates the cutting temperature with tool life as the cutting speed uniquely determines the cutting temperature.

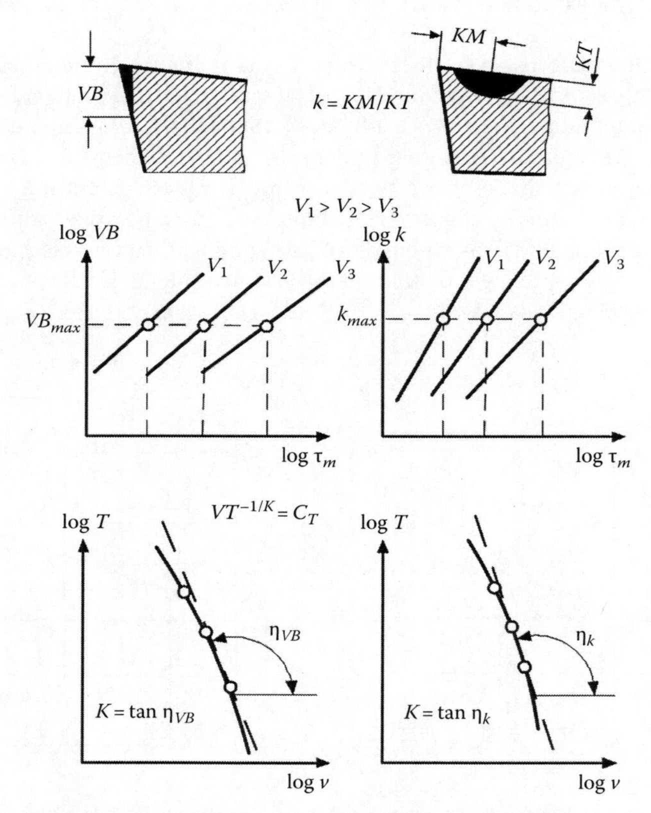

FIGURE 2.44 Procedure of experimental determining of Taylor's formula according to standard ANSI/*ASME* B94.55M-1985.

As can be seen, this formula states that the higher the cutting speed (temperature), the lower the tool life, which is in direct contradiction with well-known experimental studies and practice of metal cutting (Zorev 1966). Leading tool manufacturers clearly indicate the favorable range of the cutting speed (temperatures) for their tool materials. Deviation from the recommended speed (temperature) for a given tool material to either side lowers tool life. This, however, does not follow from Taylor's tool-life formula used in standards and practically all books (textbooks) on metal cutting.

The general mechanisms that cause tool wear normally described in the literature are (1) abrasion, (2) diffusion, (3) oxidation, (4) fatigue, and (5) adhesion. These mechanisms are explained by many authors, for example, by Loladze (1958), Shaw (1984), and Trent and Wright (2000).

In the author's opinion, however, out of the commonly referred mechanisms, only abrasion and adhesion are physically sound (Astakhov 2006), while others *borrowed* a long time ago from general machinery to describe tool wear do not provide any help in identifying the root cause of tool failures. This has been becoming obvious since earlier year when the conditions of the machining systems started to improve so the wear of the cutting tool can now be related to the tool design/material and its working conditions.

2.6.3 Statistical Analysis of Tool Wear Curves

A practical assessment of tool reliability requires a statistical analysis of tool wear curves obtained experimentally. As such, a proper statistical methodology is required to carry out such an analysis properly. This section aims to present an example of the tool-life statistical analysis methodology to help shop engineers in tool wear assessment.

As discussed in the previous section, wear curves are often used to explain phenomenology of tool wear. In full analogy with the classical wear curve, a cutting tool wear curve is thought of as consisting of three distinctive regions, namely, the region of initial wear, the region of steady-state wear, and the region of accelerated wear. The practice of tool wear testing does not always confirms this anticipation that is simply *borrowed* from general machinery.

Figure 2.45 shows a fragment of the results of a study dealing with the influence of the threading tap hardness on tool life. As seen, the hardness of the tool material affects not only tool life, which increases with this hardness, but also the appearance of the wear curves. For low hardness, the wear

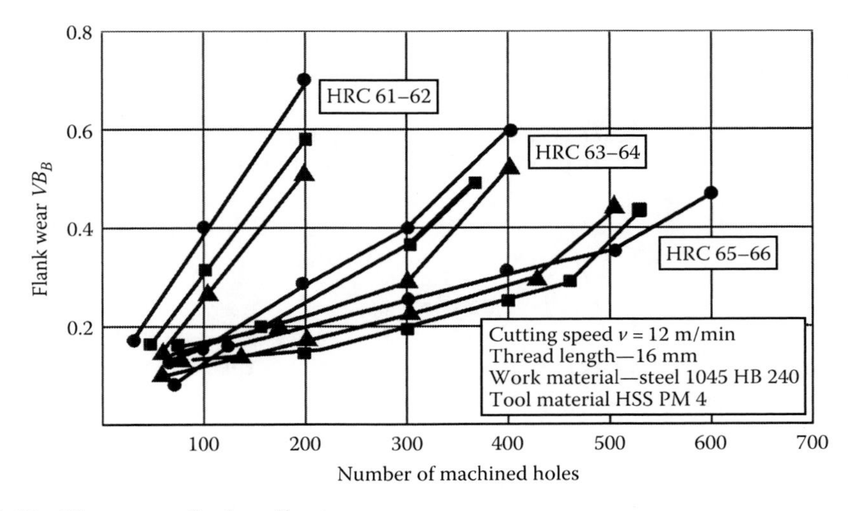

FIGURE 2.45 Wear curves for threading taps.

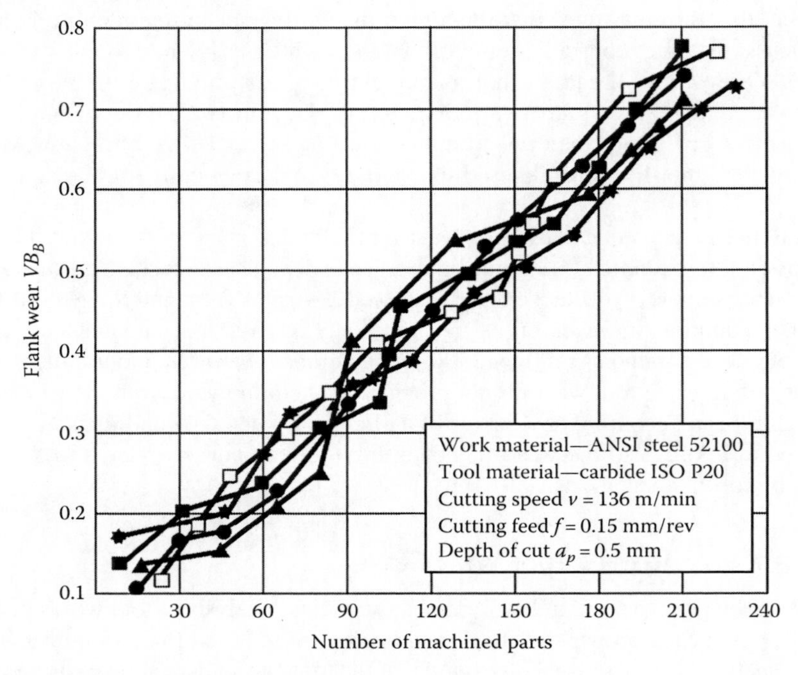

FIGURE 2.46 Wear curve in turning of bearing steel.

curves are practically straight lines so that there is only one region. For intermediate hardness, the region of accelerated wear becomes obvious so that the wear curves have two distinctive regions. For high harness, the wear curves resemble the classical wear curve.

Figure 2.46 shows an example of 5 wear curves (randomly selected out of 30 results) obtained in turning of ANSI 52100 bearing steel. As seen, there are no distinctive regions that can be distinguished on these curves. Moreover, the wear curves shown in Figures 2.45 and 2.46 intersect each other that indicates the rather stochastic nature of tool wear.

The foregoing analysis suggests that wear curve appearance is not a sufficient indicator for the analysis of cutting tool quality and reliability. Therefore, a more objective statistical methodology should be worked out to help researchers and practical engineers in the analysis not only tool life but also tool quality and reliability as such an analysis becomes important in modern manufacturing.

A suggested methodology of statistical analysis of the results of tool wear tests is carried out further using a practical example. Table 2.1 presents the results of a wear study of twist drills of 8 mm dia. made of M4 HSS. This table shows tool wear increments measured after each hundred drilled holes (14 min of machining time). The total machining time was 84 min, that is, the total number of measurements was $m = 6$. Table 2.2 summarizes the results of calculations.

The suggested methodology includes the following steps:

1. For each test drill, the average wear increment gained over time increment $\delta t = 14$ min is calculated as

$$\delta w_B^{(i)} = \frac{1}{m}\sum_{j=1}^{m} \delta w_j^{(i)} \tag{2.2}$$

TABLE 2.1
Tool Wear Increments

No. of Drill, i	Time Increment, j					
	t = 14 min	$2t$	$3t$	$4t$	$5t$	$6t$
	Tool Wear Increment $\delta w_j^{(i)}$ (mm)					
1	0.15	0.10	0.10	0.00	0.20	0.15
2	0.20	0.10	0.05	0.05	0.10	0.15
3	0.15	0.10	0.15	0.10	0.10	0.10
…	…	…	…	…	…	…
14	0.15	0.10	0.15	0.20	0.00	0.10
15	0.10	0.00	0.05	0.15	0.10	0.10
16	0.10	0.10	0.05	0.05	0.10	0.00
…	…	…	…	…	…	…
28	0.20	0.15	0.00	0.15	0.00	0.05
29	0.10	0.10	0.00	0.10	0.10	0.05
30	0.10	0.00	0.10	0.00	0.00	0.15

TABLE 2.2
Analysis of Wear Curves for Drills of 8 mm dia.

No. of Drill, i	$\delta w_B^{(i)}$	σ_i^2	$\log \sigma_i^2$
1	0.117	0.00466	$\bar{3}.6702$
2	0.108	0.00252	$\bar{3}.4014$
3	0.117	0.00068	$\bar{4}.8325$
…	…	…	…
14	0.117	0.00468	$\bar{3}.6702$
15	0.083	0.00268	$\bar{3}.4281$
16	0.067	0.00168	$\bar{3}.2253$
…	…	…	…
28	0.092	0.00742	$\bar{3}.8704$
29	0.075	0.00178	$\bar{3}.2504$
30	0.058	0.00442	$\bar{3}.6454$

$\delta w_B = 0.0903$ mm $\sum \sigma_i^2 = 0.1231$ $\log \sum \sigma_i^2 = -73.4581$

$\sum \left(\delta w_B^{(i)} - \delta w_B\right)^2 = 0.010035$

For example, for the first drill (i = 1),

$$\delta w_B^1 = \frac{1}{6}(0.15 + 0.10 + 0.10 + 0 + 0.20 + 0.15) = 0.117 \text{ mm} \tag{2.3}$$

2. The overall average wear increment gained over time increment δt = 14 min is calculated as

$$\delta w_B = \frac{1}{m}\sum_{i=1}^{n} \delta w_B^{(i)} = \frac{1}{30}\cdot 2.709 = 0.0903 \text{ mm} \tag{2.4}$$

This step deals with calculating the variances σ_i^2 of $\delta w_j^{(i)}$, their sum, and the average variance of wear increments $\sigma_{\delta w}^2$:

$$\sigma_i^2 = \frac{1}{m-1}\sum_{j=1}^{m}\left(\delta w_j^{(i)} - \delta w_B^{(i)}\right)^2 \quad \sum \sigma_i^2 = 0.1231 \tag{2.5}$$

$$\sigma_{\delta w}^2 = \frac{(m-1)\sum \sigma_i^2}{mn-1} + \frac{m}{mn-1}\sum_{i=1}^{n}\left(\delta w_B^{(i)} - \delta w_B\right)^2 \tag{2.6}$$

$$\sum_{i=1}^{n}\left(\delta w_B^{(i)} - \delta w_B\right)^2 = 0.010035 \tag{2.7}$$

$$\sigma_{\delta w}^2 = \frac{(6-1)0.1231}{6\cdot 30-1} + \frac{6}{6\cdot 30-1}0.010035 = 0.00377 \tag{2.8}$$

3. The next step is the verification of the initial assumption that the variances of the observations in the individual groups are equal. This situation is referred to as homogeneity of variance (and the absence of which is referred to as heteroscedasticity). The simplest yet statistically sound method for such verification is the Bartlett test (Bartlett 1937).

   The Bartlett test of the null hypothesis of equality of group variances is based on comparing the logarithm of pooled estimates of variance (across all the groups) with the sum of the logarithms of the variances of individual groups. The test statistics (Snedecor and Cochran 1989) as applicable for the considered case is based on the calculation of χ^2 number represented in the following form:

$$\chi^2 = \frac{2.3026}{1+((n+1)/3n(m-1))} n(m-1)\cdot\left[\log\frac{\sum_{i=1}^{n}\sigma_i^2}{n} - \frac{1}{n}\sum_{i=1}^{n}\log\sigma_i^2\right] \tag{2.9}$$

For the degrees of freedom $k = n - 1 = 30 - 1 = 29$,

$$\chi^2 = \frac{2.3026}{1+((30+1)/3\cdot 30(6-1))} 30(6-1)\cdot\left[\log 0.0041 - \frac{-73.4581}{30}\right] = 19.8 \tag{2.10}$$

This value is compared with the table critical value of chi-square at a 5% level of confidence (Snedecor and Cochran 1989). For $k = 29$ and $p = 0.05$, this value is $\chi_{cr}^2 = 42$. Because $\chi^2 < \chi_{cr}^2$, the variances of the observations in the individual groups are statistically equal.

4. The next step is to find if the difference between average increments of the tool wear is significant.

   The sum of squares of deviations between testing series is

$$Q_1 = m\sum_{i=1}^{n}\left(\delta w_B^{(i)} - \delta w_B\right)^2 = 6\cdot 0.01 = 0.06 \tag{2.11}$$

for degrees of freedom $k_1 = n - 1 = 30 - 1 = 29$.

The sum of square deviations within the series is

$$Q_2 = (m-1)\sum_{i=1}^{n}\sigma_i^2 = (6-1)\cdot 0.1231 = 0.616 \tag{2.12}$$

for degrees of freedom $k_2 = n(m-1) = 30(6-1) = 150$.

The F-criterion is then calculated as

$$F = \frac{Q_1/k_1}{Q_2/k_2} = \frac{0.06/29}{0.616/150} = 0.5 \tag{2.13}$$

This value is compared with the table critical value of the F-criterion at a 5% level of confidence $F_{cr} = 2.2$ (Snedecor and Cochran 1989). Because $F < F_{cr}$, the initial assumption (null hypothesis) is valid.

Because the experimental data passed statistical evaluation, they can be then used to determine parameters of the density function. The fitness of the obtained data for normal, lognormal, gamma, etc., distributions is consequently attempted (the test of goodness of fit), and the distribution function that fits the best to these data is then used in reliability analysis.

The difference between tool quality and the variance of tool life (Var[T]) can be distinguished. The simple yet accurate and practical assessment of tool quality is the value of constant C_v in the experimentally obtained correlation $v = f(T, f, a_p)$ (Astakhov et al. 1995; Astakhov 2006). The variance is the measure of the amount of variation of tool life with respect to its mean value. Table 2.3 shows the data for twist drills made of the same HSS manufactured by six different tool manufacturers.

As a rule, the higher the quality of tools, the lower the variance of their tool life. However, in the practice of manufacturing, there are a number of deviations from this general rule that allow distinguishing the quality of tool design including the suitability of the tool material and the quality of tool manufacturing. As can be seen in Table 2.3, the variance of the tool life for drills produced by tool manufacturer V clearly indicates good manufacturing quality of the drill. The detailed analysis of these tools showed that the clearance angle (see Chapter 4 for definition) of the lips assigned by the tool drawing was insufficient for the application, which caused relatively low tool life, while the manufacturing process stability and inspection practices were the best among other considered drill manufacturers.

TABLE 2.3
Quality and Variance of Tool Life for Drills from Different Tool Manufacturers

Supplier	Relative Constant C_{v}/C_{vi}	Variance $Var(T)$
I	1.00	0.22
II	0.69	0.42
III	0.48	0.47
IV	0.44	0.44
V	0.42	0.15
VI	0.40	0.56

The major issue with cutting tool testing and implementation is significant variation of their quality even within the same production batch that results in great scatter in tool life. Commonly, only a few cutting tools are used in laboratory testings carried out by universities or R&D departments. These tools are carefully selected from the same production batch, measured, and calibrated. Such a testing is suitable when one attempts to study the influence of a particular tool design, geometry features (e.g., the clearance angle), or machining regime (e.g., the cutting speed) on the outcomes of the machining process such as tool life, cutting force and power, machining quality, productivity, and efficiency. On the other hand, it is completely unacceptable in reliability studies results of which are to be used for assigning tool lives in unattended manufacturing operations (production lines and manufacturing cells), for example, in the automotive industry where a great number of tools are used and mass production of parts is the case.

To obtain adequate reliability results, a great number of tools should be tested. As such, all real-world imperfections such as the difference in machines and MWF supply and quality, variations in part-holding fixtures, tool holders, and controllers should be included in the tests. It makes it difficult to assure the statistical viability of the test results and even more difficult to assign proper tool reliability for unattended machining processes, particularly in the automotive industry, if the known *classical* distribution functions are used as suggested by the known studies on tool reliability. In the author's opinion, the Bernstein distribution should be used to evaluate experimental data for reliability testing in machining (Katsuki et al. 1992). A simple yet accurate and practical methodology for test data evaluation was developed using this distribution. This methodology is described as follows.

Consider a simple linear wear represented by wear curve 1 shown in Figure 2.47. Normally, it is approximated by a linear function $w(\tau_m) = a(\tau_m) + b$ where b is the initial wear, that is, $b = w(0) = w_0$, and a is the wear rate. It is understood that a is a random variable that depends upon the quality of a particular tool, as this quality directly affects the tool wear rate. If a and b both have normal distribution, then $a(\tau_m)$ is distributed normally with the following parameters of the normal distribution:

$$M\{w(\tau_m)\} = M\{a\}\tau_m + M\{b\} \quad (2.14)$$

$$\text{Var}\{w(\tau_m)\} = \tau_m^2 \text{Var}\{a\} + \text{Var}\{b\} \quad (2.15)$$

If w_M is the maximum allowed tool wear, then it follows from the normality of $w(t)$ and from

$$P\{\tau_m > T\} = P\{(aT + b) \leq w_M\} \quad (2.16)$$

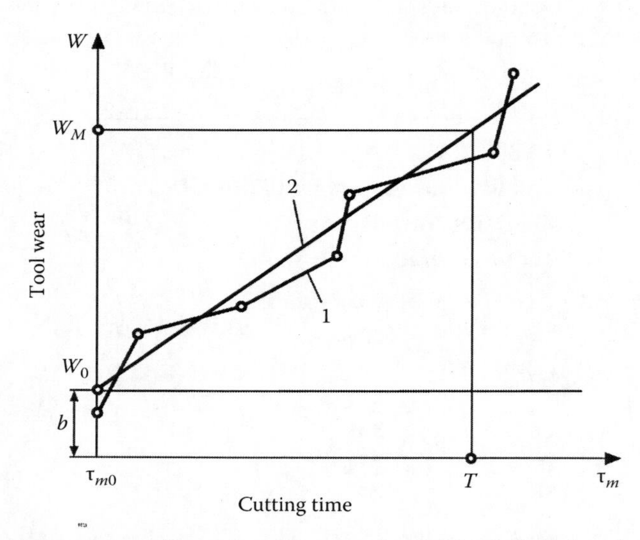

FIGURE 2.47 Wear curves: 1, experimental and 2, approximated for statistical analysis.

that

$$P\{\tau_m \leq T\} = 1 - P\{\tau_m > T\} = \Phi\left[\frac{T - ((w_M - M\{b\})/M\{a\})}{\sqrt{((\text{Var}\{a\}T^2 + \text{Var}\{b\})/M^2\{a\})}}\right] \quad (2.17)$$

This distribution is known as the Bernstein distribution (Babu et al. 2002). It differs from the normal distribution in that *the variance is time dependent.* The Bernstein distribution has three parameters:

$$\alpha = \frac{\text{Var}\{a\}}{M^2\{a\}}, \quad \beta = \frac{\text{Var}\{b\}}{M^2\{a\}}, \quad \delta = \frac{w_M - M\{b\}}{M\{a\}} \quad (2.18)$$

Substituting these parameters in Equation 2.16 and rearranging terms, one can obtain

$$F(T) = \Phi\left[\frac{T - \delta}{\sqrt{\alpha T^2 + \beta}}\right] \quad (2.19)$$

Tool reliability, that is, the probability of the tool working without failures over operational time T, is then calculated as

$$R(t) = P\{\tau_m > T\} = 1 - F(T) = 1 - \Phi\left[\frac{T - \delta}{\sqrt{\alpha T^2 + \beta}}\right] \quad (2.20)$$

The parameters of the Bernstein distribution defined by Equation 2.19 can be determined using tool wear curves similar to curve 1 shown in Figure 2.47. To determine these parameters, the following steps are recommended (Katsev 1974):

1. A tool wear test is carried out using N test tools. For each tool, the wear curve is plotted.
2. Using the least square method, each wear curve obtained experimentally is approximated by a straight line (curve 2 in Figure 2.47). For each line, coefficients a_i and b_i are determined.
3. Parameters of the Bernstein distribution are calculated using Equation 2.19.
4. Tool reliability is calculated using Equation 2.20.

As an example of use of the Bernstein distribution in tool testing, consider the results of the tests of twist drills of 10.5 mm dia. The split-point grind to assure drills' self-centering ability (see Chapter 4) and TiCN (see Chapter 3) to assure improved tool life were applied. Standard manufacturing, inspection, packing, and drill presetting procedures were used.

Initial wear w_0 was determined as that reached for $\tau_0 = 9$ min of drilling that corresponded to 50 drilled holes. This time period corresponds approximately to the region on initial tool wear on wear curves. The maximum allowed tool wear was $w_M = 0.5$ mm. The drilling time to achieve this wear is considered as tool life T. The tool wear rate is calculated as

$$r_w = \frac{w_M - w_0}{T - \tau_{m0}} \quad (2.21)$$

TABLE 2.4
Results of Testing of Drill of 10.5 mm

Drill Number	Initial Tool Wear, w_0 (mm)	Tool Life (min)	Wear Rate, r_w
1	0.10	117.0	0.0037
2	0.07	126.0	0.0037
3	0.07	93.6	0.0051
…	…	…	…
25	0.07	99.0	0.0048
26	0.05	93.6	0.0053
27	0.05	118.8	0.0041
…	…	…	…
43	0.10	77.4	0.0058
44	0.05	126.0	0.0038
45	0.08	77.4	0.0061
$M\{w_0\} = w_{0B} = 0.08$			$M\{r_w\} = r_{wB} = 0.0054$

The experimental data and calculations of the tool wear rate are shown in Table 2.4. Then the mean and variation of the wear rate and initial wear are calculated as

For the wear rate,

$$M\{r_w\} = \frac{\sum_{i=1}^{N} r_{wi}}{N} \tag{2.22}$$

$$\text{Var}\{r_w\} = \frac{1}{N-1}\sum_{i=1}^{N}\{r_{wi} - M\{r_{wi}\}\}^2 \tag{2.23}$$

For the initial wear,

$$M\{w_0\} = \frac{\sum_{i=1}^{N} w_{0i}}{N} \tag{2.24}$$

$$\text{Var}\{w_0\} = \frac{1}{N-1}\sum_{i=1}^{N}\{w_{0i} - M\{w_{0i}\}\}^2 \tag{2.25}$$

The results of the calculations presented in Tables 2.4 and 2.5 are then used to calculate the parameters of the Bernstein distribution as

$$\alpha = \frac{\sigma_r^2}{r_{wB}^2} = \frac{0.0000035}{0.0054^2} = 0.12 \tag{2.26}$$

$$\beta = \frac{\sigma_w^2}{r_{wB}^2} = \frac{0.00112}{0.0054^2} = 38.4 \tag{2.27}$$

$$\delta = \frac{w_M - w_{0B}}{r_{wB}} = \frac{0.5 - 0.008}{0.0054} = 77.7 \tag{2.28}$$

TABLE 2.5
Statistical Calculations

	Variance of the Wear Rate		Variance of the Initial Wear	
Drill Number	$(r_{wi} - r_{wB})$	$(r_{wi} - r_{wB})^2$	$(w_{01} - w_{0B})$	$(w_{01} - w_{0B})^2$
1	0.0017	0.00000289	0.02	0.0004
2	0.017	0.00000289	0.01	0.0001
3	0.003	0.00000009	0.01	0.0001
…	…	…	…	…
25	0.0006	0.00000036	0.01	0.0001
26	0.0001	0.00000001	0.03	0.0009
27	0.0013	0.00000169	0.03	0.0009
…	…	…	…	…
43	0.0004	0.00000016	0.02	0.0004
44	0.0016	0.00000256	0.03	0.0009
45	0.0007	0.00000049	0.00	0.0000
	$\sum_{i=1}^{N}(r_{wi} - r_{wB})^2 = 0.00015309$		$\sum_{i=1}^{N}(w_{01} - w_{0B})^2 = 0.0494$	
	$Var\{r_w\} = \sigma_r^2 = 0.0000035$		$Var\{w_0\} = \sigma_w^2 = 0.00112$	

The distribution function is calculated using Equation 2.19 as

$$F(T) = \Phi\left[\frac{T - 77.7}{\sqrt{0.12T^2 + 38.4}}\right] \tag{2.29}$$

For the degrees of freedom $k = N - 1 = 45 - 1 = 44$, $\chi^2 = 3.09$ and $P(\chi^2) = 0.22$ (Snedecor and Cochran 1989). Then the tool reliability at 90% level of confidence is calculated as

$$0.9 = 1 - \Phi\left(\frac{T_{0.9} - 77.7}{\sqrt{0.12T_{0.9}^2 + 38.4}}\right) \tag{2.30}$$

Using inverse Laplace table (Snedecor and Cochran 1989), one can find that $T_{0.9} = 53$ min.

2.6.4 Common Wear Regions of Drills

In full analogy with the wear region on a single-point tool shown in Figure 2.42, Figure 2.48 shows the most common types of the drill normal (gradual) wear classified according to the regions of the tool they affect. They are

1. *Rake face or crater wear* produces a wear crater on the tool rake face. The depth of the crater KT is selected as the tool life criterion.
2. *Relief face or flank wear* results in the formation of a flank wear land. For the purpose of wear measurement, the wear of the following regions is considered:
   a. The maximum width of the flank wear, VB_{max}, is measured on the primary drill flank.
   b. The corner wear width, VB_c, is measured on the primary drill flank at the drill corner.
   c. The chisel edge wear width, VB_{cl}, is measured on the flank face of the chisel edge.
   d. The margin wear, VB_{mg}, is measured on the flank face of the drill margin as shown in Figure 2.48.

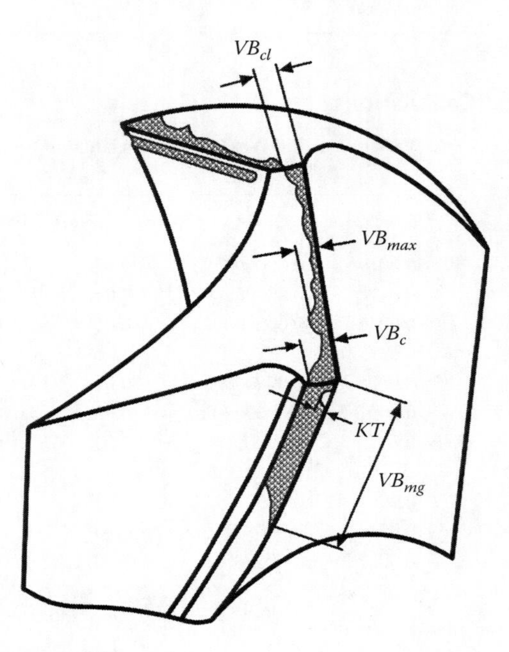

FIGURE 2.48 Types of wear on turning tools.

When the drill is designed and manufactured properly, when the drilling regime is optimized and the tool material of the drill is selected accordingly, and when MWF flow rate and its properties (particularly, concentration, pH, and purity) are suitable for the application, then the following wear regions should be the case in drilling of particular work materials:

- Crater wear in machining the majority of steels.
- Flank wear in machining of cast irons, difficult-to-machine materials and many other work materials. Maximum flank wear found at the distance over the cutting edge is approximately equal to (2/3…3/4) of drill radius from the drill center in machining of a wide group of work materials as, for example, difficult-to-machine materials and aluminum alloys. This is because the maximum drill temperature is found in this region under optimized drilling conditions (Granovsky and Granovsky 1985; Astakhov 1998/1999). Examples of such wear are shown in Figure 2.49.

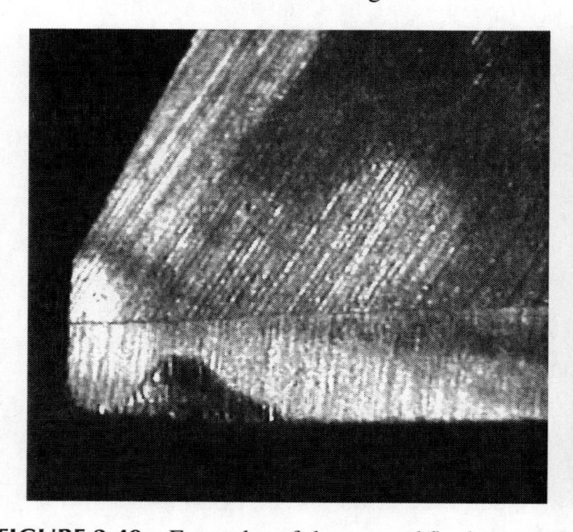

FIGURE 2.49 Examples of the *normal* flank wear of drills.

FIGURE 2.50 Examples of normal corner wear of a twist drill.

The listed flank wear is rarely found in the practice due to drill manufacturing tolerances discussed in Chapter 7. Due to drill runout and lip height wide-open tolerances, the drill corner normally has the greatest wear (Figure 2.50) and, therefore, is often selected as the criterion of tool life in practice. Normally, the corner wear $VB_c = 0.2–0.8$ mm (depending upon the drill diameter) is selected as the tool life criterion. Figure 2.51 shows typical tool life curves. Although for a given combination of the cutting feed, tool and work materials, and MWF type and parameters, the overall length of the cutting path (L_m) can be different, the general trend is the same under normal machining conditions. This trend is as follows:

- The tool life is the highest when the functional failure (e.g., tool breakage) is used as the tool-life criterion.
- The overall length of the cutting path increases as the chosen VB_c is greater.
- There is an optimal cutting speed where the tool life is the greatest for a given machining condition.

At low cutting speeds, an indication of excessive flank (corner) wear is drill *squeaking*, while at higher speeds, a high-pitched *squeal* coming from the drill itself may indicate a dull drill. A *squeaking* or a *squealing* drill should be unconditionally replaced.

The chisel edge wear has the same appearance for various drill materials. Typical chisel edge wear shown in Figure 2.52a includes the formation of radial grooves around this edge. As a result, the

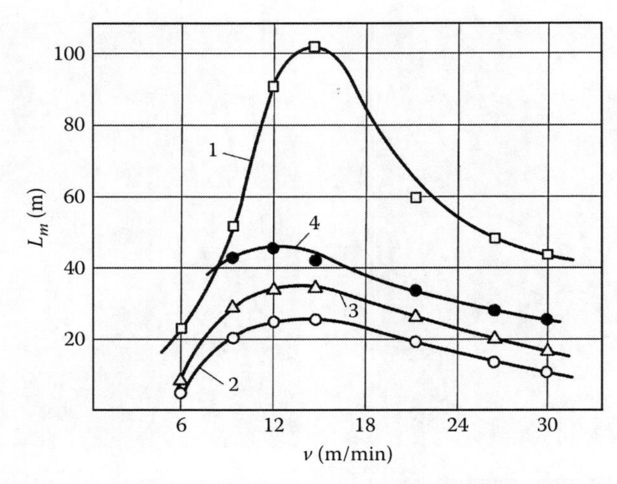

FIGURE 2.51 The overall length of the cutting path (L_m) depending upon the chosen tool life criterion: 1, functional failure; 2, corner wear VB_c = 0.3 mm; 3, corner wear VB_c = 0.4 mm; and 4, corner wear VB_c = 0.5 mm. HSS M4 drill of 10.2 mm dia., cutting feed f = 0.23 mm/rev. Work material AISI steel 1045, HB 185.

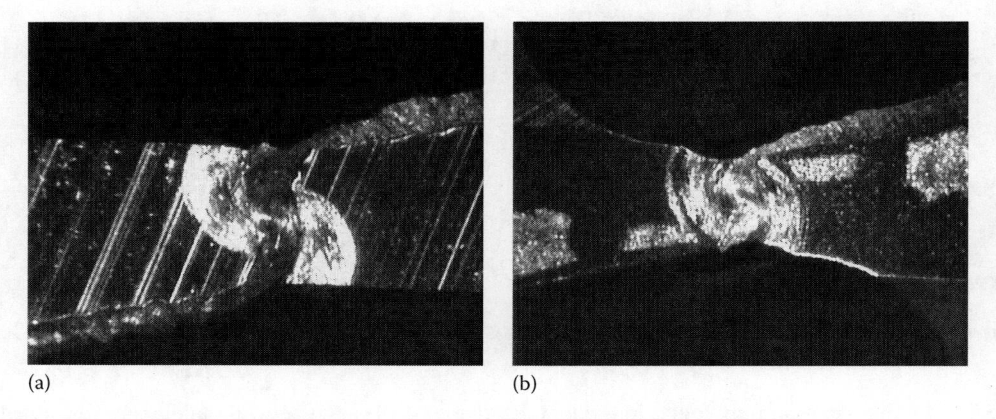

FIGURE 2.52 Typical chisel edge wear (a) and the BUE formed as a result of this wear (b).

BUE forms around the chisel edge as shown in Figures 2.52b and 4.109, 4.110, and 4.112. Although some researchers think that it actually helps drilling as drill point angle increases (e.g., Derval et al. 2011), in reality, the chisel edge wear has only harmful consequence as drill wandering at the hole entrance eventually may lead to breakage of the whole chisel edge region as shown in Figure 4.111.

The worst scenario, however, takes place when drill asymmetry due to both the manufacturing quality and accuracy of location is excessive (Chapter 7). This results in the wear of the drill margin assessed by length VB_{mg} (Figure 2.48). Figure 2.53 shows prevailing margin wear while the drill corner is not excessively worn. The appearance of margin wear is also shown in Figure 4.152.

Two most harmful consequences of drill margin wear are

1. Front taper forms on the drill instead of the intended back taper. This significantly increases the axial force and drilling torque. A drill with such wear may be jammed in hole and even can break.
2. A significant length of the drill has to be removed on its re-sharpening to restore the intended back taper and proper conditions of the drill corners. This reduces the total number of drill regrinds and thus the total drill life.

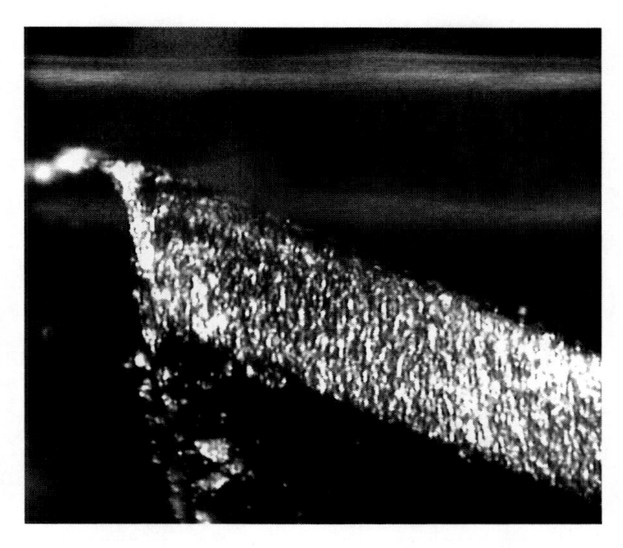

FIGURE 2.53 Prevailing margin wear while the drill corner is not excessively worn.

2.6.5 Assessment of Tool Wear of HP Drills

To analyze the performance of drilling tools on CNC machines, manufacturing cells, and production lines, the dimension tool life is understood as that time period within the cutting tool that assures the required dimensional accuracy of the machined parts. The dimension tool life can be represented by

- *The drill operating time* (T_{op}). The operating time is understood as that time within which there is no dimension compensation or tool change due to tool wear.
- *The number of machined parts/holes* (N_{mp}) during the tool operating time.
- *The overall length of drilled holes* (L_m) during the tool operating time.
- *The overall area of the machined surface* (A_{op}) during the tool operating time.
- *Linear relative tool wear* (h_{rs}) during the tool operating time.

All the listed characteristics of the dimension tool life are specific and thus, in general, are not fully suitable for the optimization of machining operations, comparison of various cutting regimes, assessment of various tool materials, and so on. For example, it is not possible to compare two different tool materials if two different cutting speeds (suitable respectively for each considered tool material) were used in the test. Among many specific wear characteristics suitable for automated machining introduced by the author earlier (Astakhov 2004, 2006), the surface wear rate is found to be most practical for the objective assessment of drill wear. For the drill corner, the surface wear rate is designated as I_{ec} and is measured as the radial wear per minute of the total machining time. For the drill margin, the surface wear rate is designated as I_{em} and is measured in an increase in the margin wear length per minute of the total machining time.

Figure 2.54 shows an example of the experimental assessment of surface wear rates of the drill corners I_{ec} and margin I_{em} (Derval et al. 2011). As can be seen, in the region of optimal cutting speed (see Figure 2.51), these curves are almost equidistant so the offered criteria can be used for reliable assessment of the tool wear without the full tool-life testing program, that is, at a fraction of cost and time of the standard procedure of tool-life assessment of drills. For laboratory analyses, *the specific dimension tool life is defined as* the volume of the workpiece machined by the drill per 1 μm of its radial corner wear. The surface wear rate and the specific dimension tool life are versatile tool wear characteristics because they allow to compare different tool materials for different combinations of the cutting speeds and feeds using different criteria selected for the assessment of tool life.

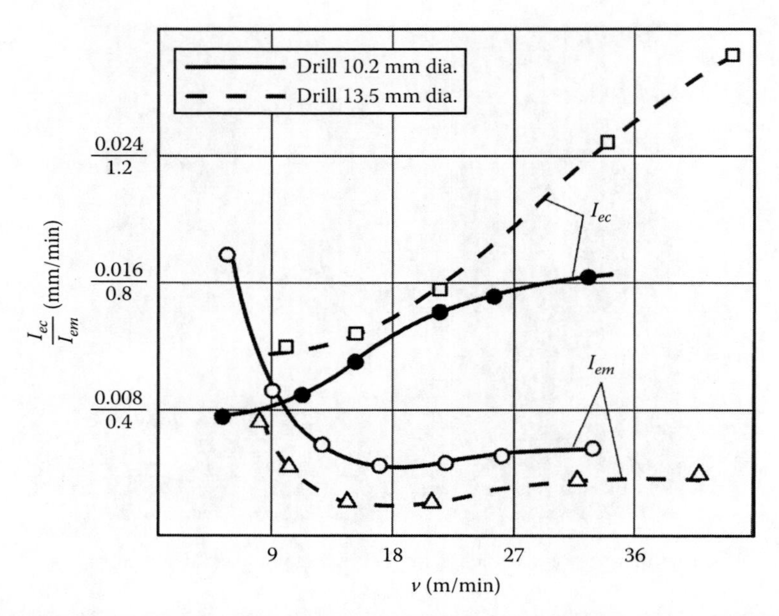

FIGURE 2.54 Surface wear rates of the drill corners I_{ec} and margin I_{em}. HSS M4 drill of 10.2 and 13.5 mm dia., cutting feed $f = 0.23$ mm/rev. Work material AISI steel 1045, HB 185.

2.6.6 Correlations of Drill Wear with Force Factors

Using an experimental setup similar to that shown in Figure 4.25, the increases in the axial force and drilling torque over the drill tool life starting from the sharp drill and finishing when the tool-life criterion is reached are shown in Table 2.6. As follows from these data, the drilling torque has the strongest correlation with the drill corner wear particularly at the optimal cutting speed. Moreover, the test results show that the drilling torque increases significantly at the end of tool life (Figure 2.55). This, combined with the simplicity of the torque measurement using modern drilling machine spindles with built-in torque sensors and advanced controllers, allows to recommend the drilling torque as the major parameter to control the end of drill life. In other words, the predetermined torque is set in the drill controller and this controller informs the machine operator about the end of drill life.

TABLE 2.6
Increases of the Axial Force and Drilling Torque When Tool Life Criterion Is Achieved Compared to a Sharp Drill

	Axial Force Increase (%)			Drilling Torque Increase (%)		
Cutting Speed (m/min)	Major Cutting Edges	Chisel Edge	Margins	Major Cutting Edges	Chisel Edge	Margins
9	23	47	30	40	10	50
15.6	28	50	22	58	10	32
27	71	25	4	76	13	11

Note: Cutting feed $f = 0.23$ mm/rev. HSS M4 drill of 10.2 mm dia. Work material AISI steel 1045, HB 185.

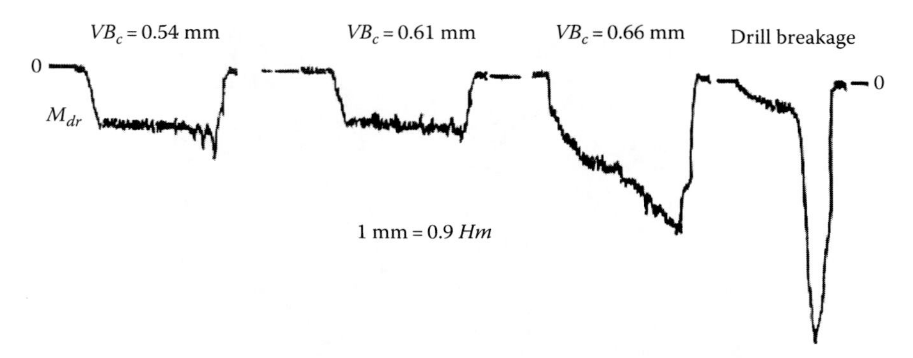

FIGURE 2.55 Change in the drilling torque (M_{dr}) with the corner wear up to drill breakage. Cutting speed $v = 31.8$ m/min and feed $f = 0.2$ mm/rev. HSS M4 drill of 10.2 mm dia. Work material AISI steel 1045, HB 185.

2.6.7 Assessment of Wear Resistance of Tool Materials

As discussed in Chapter 3 (Section 3.2.1), wear resistance is neither a defined characteristic nor a property of the tool material but is instead a system response arising from the conditions at the sliding interface. As mentioned earlier, the nature of tool wear is not yet sufficiently clear. On the other hand, wear resistance should be somehow assessed to compare different tool materials, thus helping a tool designer/end user in the selection of the proper tool material. The principal consideration in designing/using a wear test setup (rig, machine, methodology, and standard) is assuring that the test conditions, that is, the sliding configuration, conform to the practical situation.

ASTM (respective subcommittees such as Committee G2) attempted to standardize wear testing for specific applications issuing periodically updated standards. It developed a great number of standard wear tests starting with ASTM G40-10b *Standard Terminology Relating to Wear and Erosion*. The Society for Tribology and Lubrication Engineers (STLE) has also documented a large number of frictional wear and lubrication tests. In any testing, wear should be expressed as loss of material during wear in terms of volume. The volume loss gives a truer picture than weight loss, particularly when comparing the wear resistance properties of materials with large differences in density. For example, a weight loss of 14.8 g in a sample of tungsten carbide + cobalt (density = 14,800 kg/m^3) and a weight loss of 4.8 g in a similar sample made up of polycrystalline cubic boron nitride (PCBN) amborite *DBC50* (density = 4,800 kg/m^3) both result in the same level of wear (1 cm^3) when expressed as a volume loss. The inverse of volume loss can be used as a comparable index of wear resistance.

Their common test types are used to evaluate tool material wear resistance, namely, the block-on-ring and pin-on-ring Falex-type lubricity tests and the crossed-cylinder test. ASTM G99 *Standard Test Method for Wear Testing with a Pin-on-Disk Apparatus* defines a laboratory test procedure for determining the wear of materials during sliding using a pin-on-disk apparatus. Materials are tested in pairs under nominally nonabrasive conditions. The coefficient of friction can also be determined. Figure 2.56 shows Falex ISC tribometer specimen table. A test pin attached to a gimbaled arm is loaded against a test disk mounted on a motor-driven turntable. The friction and wear characteristics are quantified by measuring frictional forces, material removal, and displacement. Precise control and data acquisition provide accurate comparisons of material combinations. Interlaboratory test programs using the procedures of ASTM G77 have indicated that the coefficient of variation of the test results is typically 20%, while interlaboratory variations are larger, 30%.

The crossed-cylinder test basic configuration is shown in Figure 2.57. It has been used for a number of years in industry. Procedures and parameters used by different laboratories tend to vary although a standard practice has been developed and issued as ASTM G63 standard for this type of test.

FIGURE 2.56 Falex ISC tribometer specimen table.

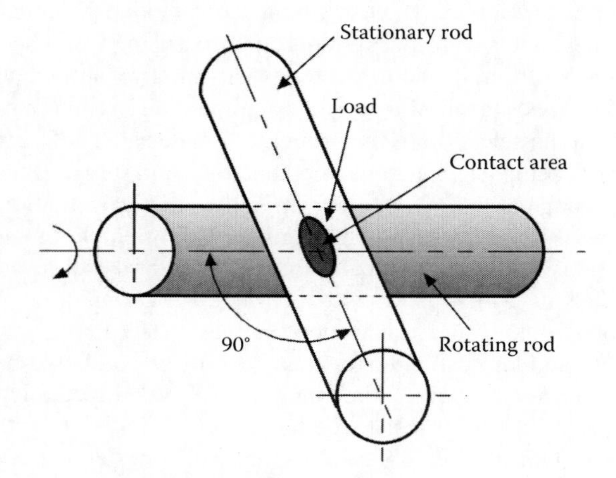

FIGURE 2.57 Basic configuration of the crossed-cylinder test.

One cylinder is held stationary and the other is pressed against it and rotated. The basic concept is to rank materials in terms of the wear produced after a fixed number of revolutions. Wear is directly measured by a mass loss technique but converted to volume for comparison.

The test is designed for unlubricated evaluation of test materials. The standard test has three procedures, which differ in terms of speed and duration to address different levels of wear behavior. Studies, which used the standard practice, have shown that the coefficient of variation for intralaboratory test is within 15% and for interlaboratory test, 30%.

Although these tests are widely used to evaluate wear resistance of different tool materials, one should clearly realize that it is difficult, if not impossible, to correlate bench test data with actual of the tested tool material in the cutting tool. This is because a number of crucial differences exist between the test conditions in rubbing tests and actual metal cutting:

- The major problem is that in rubbing tests, the continuous sliding contact occurs by cyclic reintroduction of the same surface element from the countermaterial. Although repeated, contact occurs between many machine elements, such as journal bearings, rotating seals, and engine pistons; in metal cutting, by contrast, tools generally slide against a freshly

formed, not previously encountered, surface. The physics and chemistry of this freshly formed surface (referred to as the juvenile surface) are considerably different from the contact surface in rubbing test.
- Tribological conditions in rubbing tests do not even remotely resemble those found in metal cutting. The level and distribution of the normal and shear stresses along the contact interfaces as well as the temperature and its distribution over these interfaces are principally different in these two processes. The average normal stress is 10–30 times and the contact temperature up to 20 times higher at the contact interfaces found in real metal cutting compared to those used in the ASTM tests. Due to these differences, the physical processes occurring at the rubbing test and metal cutting interfaces are highly dissimilar.

As a result of these differences, the available wear resistance data reported by many tool material suppliers cannot be considered as relevant in practical metal cutting applications.

2.6.8 Real Mechanisms of Tool Wear: Pure Abrasion, Adhesion, and Abrasion–Adhesive Wear

In the author's opinion, abrasion and adhesion are two basic mechanisms of cutting tool wear including drills although they are not properly addressed in metal cutting studies/tool applications.

Abrasive wear includes particle contamination and roughened surfaces that cause cutting and damage to a mating surface which is in relative motion to the first. There are two basic types of abrasive wear (Figure 2.58):

- Three-body abrasion occurs when a relatively hard contaminant (particle of dirt or wear debris) becomes imbedded in one metal surface and is squeezed between the two surfaces, which are in relative motion. When the particle size is greater than the MWF film thickness, scratching, plowing, or gouging can occur. This creates parallel furrows in the direction of motion, like rough sanding. Mild abrasion by fine particles may cause polishing with a satiny, matte, or lapped-in appearance.
- Two-body abrasion occurs when metal asperities (surface roughness, peaks) on one surface cut directly into a second metal surface. A contaminant particle is not directly involved. The contact occurs in the boundary lubrication regime due to inadequate lubrication or excessive surface roughness, which could have been caused by some other forms of wear.

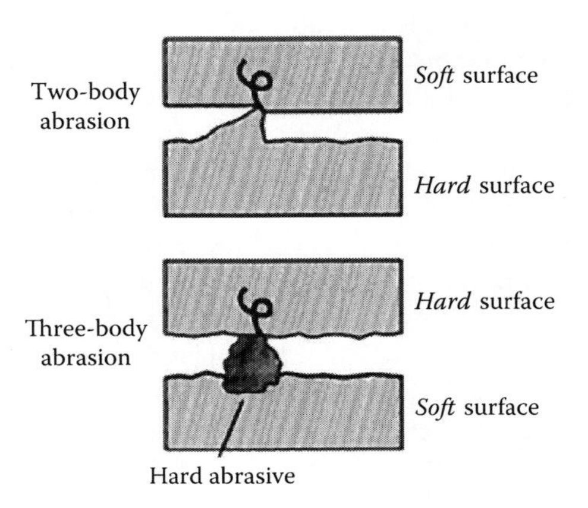

FIGURE 2.58 Two basic types of abrasive wear.

In the author's opinion, three-body abrasion takes place in drilling when MWF is dirty, that is, contains a great number of hard solid particles. To be specific, the words *dirty* and *hard* should be quantified. The following facts help such quantification:

- The grain size of modern sintered carbides used to make drilling tools is in the range of 0.6–1.5 μm (see Section 3.4) so that the same size of hard WC particles is found in MWF. In reality, clusters of carbide debris having size up to 25 μm are commonly found in MWF.
- The grain size of modern PCDs used to make drilling tools is in the range of 5–30 μm (see Section 3.5) so that the same sizes of hardest diamond particles are found in MWF.
- At best, 25 μm filters are used even in the advanced manufacturing facilities having the best MWF maintenance programs.
- The tolerance of many holes to be drilled in modern manufacturing facilities, that is, in the automotive industry, can be as tight as 10–15 μm.

These facts show that a drilling tool with internal high-pressure MWF supply is actually *bombarded* by a jet of MWF having hard *stones* of size that is greater or close to tolerance on the hole being drilled. As such, three-body abrasion takes place. The problem was noticed and solved in deep-hole drilling (see Chapter 5) where MWF is routed through multiple baffles in the tank and passed through multistage filtration to ensure fine filtration. The design of MWF tanks for deep-hole machining (DHM) allows fine particles to settle to the bottom of an MWF tank even after a series of magnetic and centrifugal filtrations. Two-tank systems including dirty and clean tanks are used. Unfortunately, this is not nearly the case for other types of drilling although the cost of modern drill and requirements to the drilled holes are high.

Two-body abrasion takes place when the work material has highly abrasive solid phases in its structure. Obvious examples are high-silicon–aluminum alloys widely used in the automotive industry. Less obvious are some solid phases in cast irons. Several phases can be present in iron, and their volume fraction and distribution have significant effects on tool life. Some of the phases that degrade machinability include (1) iron oxides and silicates formed during pouring; (2) carbides and ternary iron phosphides formed during eutectic solidification; (3) titanium, vanadium, and niobium carbides, nitrides, and carbonitrides formed by reactions in the iron; and (4) chromium and molybdenum carbides formed during cooling of the casting. Finely distributed carbides that form during solidification of cast iron have a detrimental effect on machinability. Carbon that remains in austenite grains after eutectic solidification (see Appendix B for terminology) must diffuse from the austenite and migrate to the graphite during cooling to the eutectoid temperature. High cooling rates and the presence of elements that either inhibit carbon diffusion or form stable carbides (e.g., molybdenum) reduce the rate of carbon transfer and can result in austenite that is supersaturated with carbon. (5) At or below the eutectoid temperature, the supersaturated austenite decomposes to produce abrasive micro-carbides distributed in the matrix.

Figure 2.59 shows the typical appearance of abrasive wear. As can be seen, the worn surface contains deep scratches in the direction of sliding left by abrasive particles/solid faces in the work material.

Generally, adhesive wear is thought of as the mechanical transfer of material from one contacting surface to another. It occurs when high loads, temperatures, or pressures cause the asperities on two contacting metal surfaces in relative motion to spot-weld together then immediately tear apart, shearing the metal in small, discrete areas. The surface may be left rough and jagged or relatively smooth due to smearing/deformation of the metal. In general machinery, adhesion occurs in equipment operating in the mixed and boundary lubrication regimes due to insufficient lubricant supply and/or its inadequate viscosity, insufficient internal clearances, incorrect installation, or misalignment. This can occur in rings and cylinders and bearings and gears. Normal break-in is a form of mild adhesive wear, as is frosting. Scuffing usually refers to moderate adhesive wear, while galling, smearing, and seizing result from severe adhesion. Adhesion can be prevented by lower loads, avoiding shock loading and ensuring that the correct oil viscosity grade is being used

FIGURE 2.59 Abrasive wear of the drill corners.

for lubrication. If necessary, extreme pressure and antiwear additives are added to the lubrication oil to reduce the damage due to adhesion.

In metal cutting, adhesion occurs due to mechanical bonding of juvenile freshly formed surface free of oxides. The harder the contact pressure and the rougher the tool contact surface, the stronger the bonding. Such an adhesion can be referred to as pure adhesion in metal cutting. It takes place at low cutting speeds when the built-up on the tool rake face is great. Due to high plastic deformation of the chip and high contact pressures at the tool–chip and tool–workpiece interface, extreme pressure and antiwear additives cannot penetrate into these interfaces (Astakhov 2006). As a result of adhesion of the work material converted into the chip and the tool rake face, the so-called built-up edge (commonly referred to as BUE in the literature on metal machining) is formed. Its appearance is shown in Figure 2.60. Such an appearance

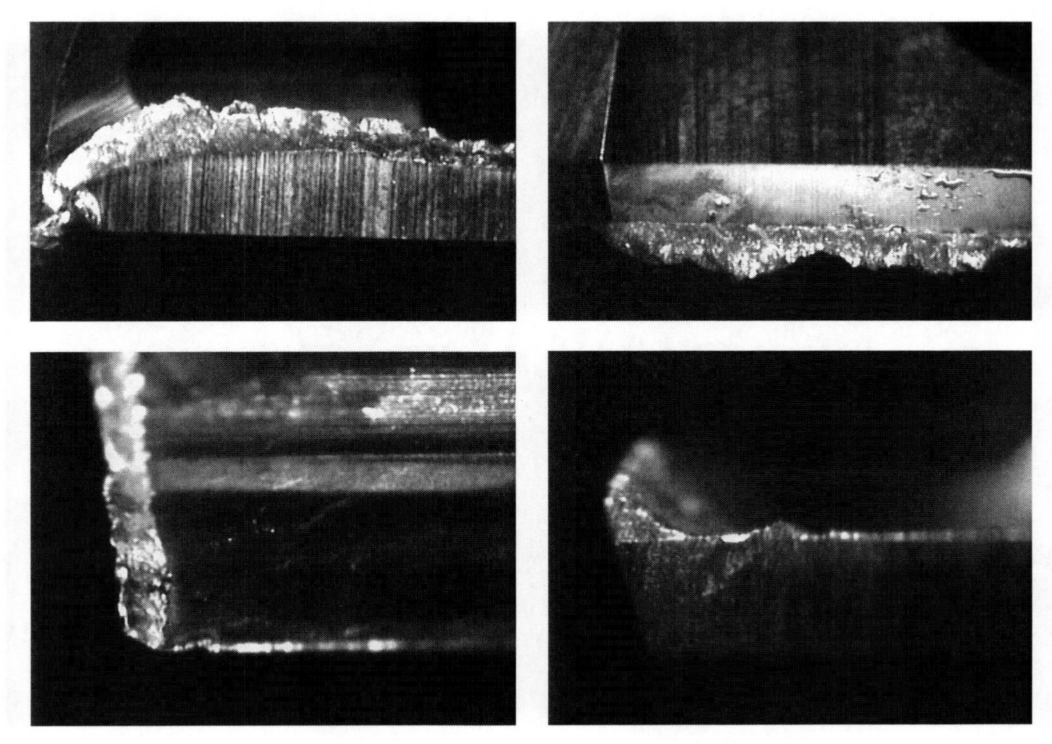

FIGURE 2.60 Adhesion of the work material to the drill rake face known as the BUE.

is deceptive so that many literature sources in the past and many specialists up to the present believe that BUE can protect the cutting edge from wear. Moreover, as it appears as a continuation of the cutting wedge, the *stable* BUE can practically eliminate the contact of the tool flank face and the workpiece. As discussed by the author earlier (Astakhov 1998/1999), this *looks-nice* story ends when one realizes that the tool life is lowest for the cutting conditions where BUE is greatest. In machining of steel, a reasonable tool life is achieved at high cutting speeds where BUE does not form at all (greater than 60 m/min for common steels).

Pretty pictures to support the *protective action of BUE* similar to those shown in Figure 2.60 appear in the literature on metal machining. However, the reality is much less pretty. Figure 2.61 shows the appearance of the drill rake face and the major cutting edges when BUE is partially removed (chemically). As can be seen, the drill's major cutting edges are completely destroyed by *the protector* BUE. This is because BUE is not stable in metal cutting so it changes within each cycle of chip formation (Astakhov 1998/1999). As BUE adheres to the rake face, the adhesion causes mechanical bonding (as glue with a piece of paper). When BUE is periodically removed by the moving chip (as its height becomes sufficient), it brings a small piece of the tool material with it (as the glue removed from the paper). The process repeats itself many many times so the tool becomes worn by this process. Note that no *welding* (the term often used in the literature) or other physical/chemical processes are involved in adhesion wear.

Pure adhesion in metal cutting significantly reduces with a cutting speed increase above approximately 60 m/min. Above this speed, no noticeable BUE exists on the tool rake face. While it is true for many other cutting tools, the cutting speed varies from its maximum to practically zero over the drill cutting edges so BUE is always a problem.

When the cutting speed becomes high, a considerable different mechanism of tool wear that can be referred to as abrasion–adhesive wear takes over. In the author's opinion, this is the prime mechanism of tool wear in modern manufacturing at high speeds. In short, its mechanism is as follows (Astakhov 2006). When the temperature on the tool rake face reaches 900°C–1200°C, a plastic

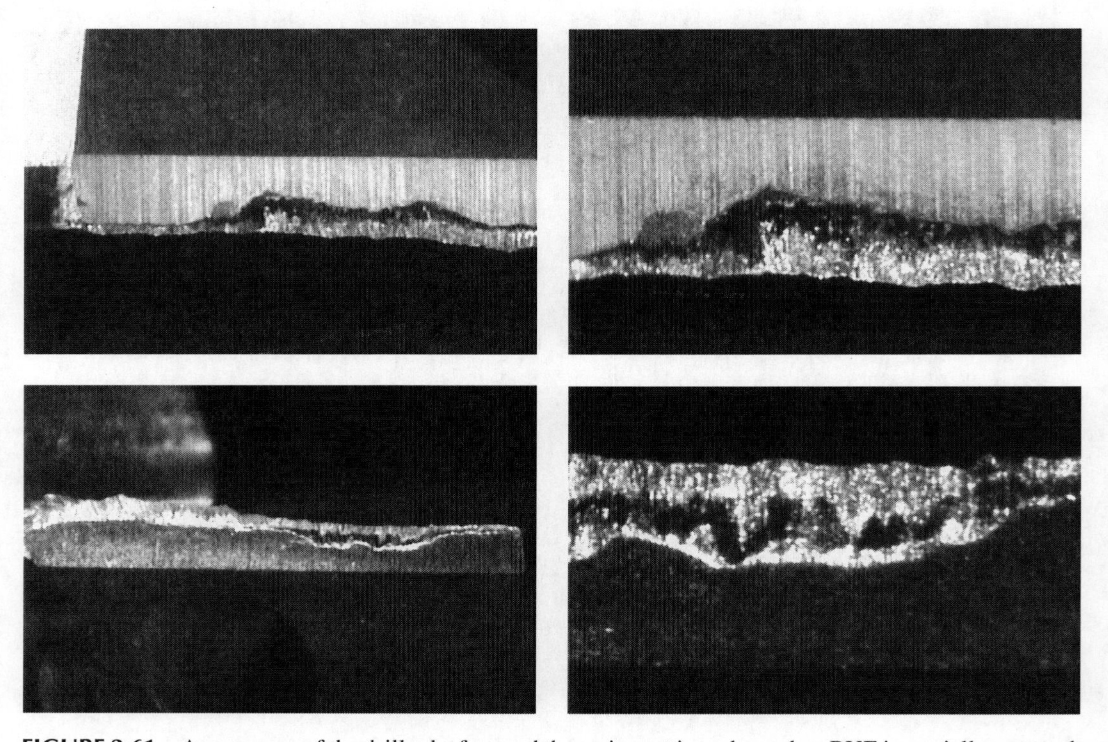

FIGURE 2.61 Appearance of the drill rake face and the major cutting edges when BUE is partially removed.

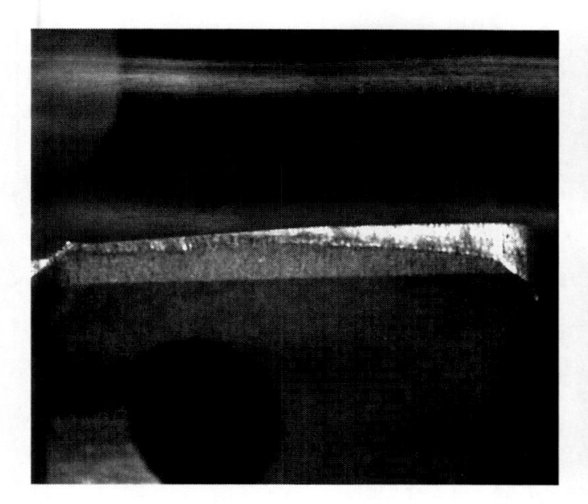

FIGURE 2.62 Appearance of abrasion–adhesive wear.

flow begins in volumes adjacent to this face. This flow takes place due to adhesion bonding between the rake face and the chip. In the case of carbide-tool materials, plastic deformation is greater in the cobalt matrix. This plastic deformation results in tearing off of carbide (WC and TiC) grains from *soft* cobalt layers (matrix) that undergo severe plastic deformation, *plowing* this *soft* layer by inclusions contacting in the work material, and *spreading* of the tool material on the chip and workpiece contact surfaces. Figure 2.62 shows the appearance of abrasion–adhesive wear on a drill. As can be seen, the wear marks are smooth compared to wear by pure abrasion shown in Figure 2.59 or by pure adhesion shown in Figure 2.61. This wear pattern should be regarded as normal as it indicates that the cutting conditions were close to optimum for a given drill material and geometry. As abrasion–adhesive wear is the basic mechanism of tool wear, modern tool coatings are designed to reduce its consequences although their *cover stories* do not mention this mechanism.

2.6.9 Special Wear Mechanisms: Reaction of the Cutting Tool on Increased Cutting Speed and the Optimal Cutting Temperature

2.6.9.1 Prevailing Concept

The concept of high-speed machining (HSM) was conceived by Dr. Carl J. Salomon during a series of experiments from 1924 to 1931. On April 27, 1931, Friedrich Krupp A.G. was granted the German Patent No. 523,594 referring to a *method of machining metal or of materials behaving similarly when being machined with cutting tools* based on metal cutting studies made by the inventor, Dr. Salomon, on steel, nonferrous, and light metals at cutting speeds of 440 m/min (1.444 ft/min) (steel), 1,600 m/min (5,250 ft/min) (bronze), 2,840 m/min (9,318 ft/min) (copper), and up to 16,500 m/min (54,133 ft/min) (aluminum). His contention was that the cutting temperature reaches a peak at a given cutting speed so that if the cutting speed is increased further, the temperature decreases.

Figure 2.63 shows a simplistic presentation of this concept (King 1985). As the cutting speed is increased from 0 to the normal mode v_1, the temperature will increase in a direct relationship until a peak value θ_{cr} is achieved. The cutting speed at θ_{cr} was termed as the critical cutting speed, v_{cr}. If the cutting speed is increased further, the cutting temperature declines according to this idea. The shape of the curve was thought to be dependent on the exact nature of the work material being cut. When the cutting speed was sufficiently increased, the resulting temperatures were reduced to those of the normal cutting temperature; thus, the normal cutting process can be carried out. The same cutting temperature θ_a corresponding to cutting speed v_1 found in the conventional speed range could possibly be reproduced at cutting speed v_2 found in the high-speed range.

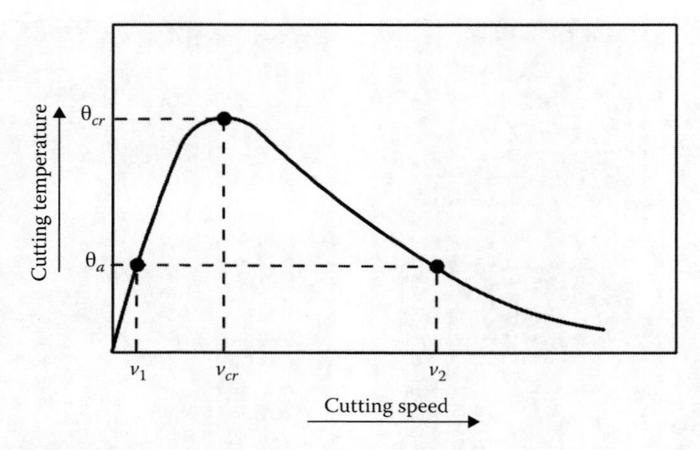

FIGURE 2.63 Idealized cutting speed versus cutting temperature plot.

There have been many versions of Salomon's curves used as reference by many researchers. Because much of the supporting data were lost during World War II and none of the participants in the research are alive to comment, the exact shape of the curves is left for speculations. However, the most commonly used version cited in most of the recent technical publications is shown in Figure 2.64. The solid lines represent data that Salomon was supposed to have been developing from experimental results. The broken lines indicate estimated results that were extrapolated by Salomon but not actually verified experimentally.

Although Salomon did not offer any rationale to support his so-called theory and because the details of his test are not known, Dr. Salomon is considered in the literature as the father of the so-called HSM. Many subsequent researchers tried to achieve the same results and a number of success stories have been reported in the literature (well summarized by King [1985]). The fundamental studies, machining tests, and practical applications of HSM attempted only in the early 1980s when high-speed spindles and thus machines became available. During the 1970s, a series of tests by the United States Navy with Lockheed Missiles Space Company showed that it was economically feasible to introduce HSM into the production environment in order to realize major improvement in productivity. In 1979, the US Air Force started a comprehensive research program in cooperation with

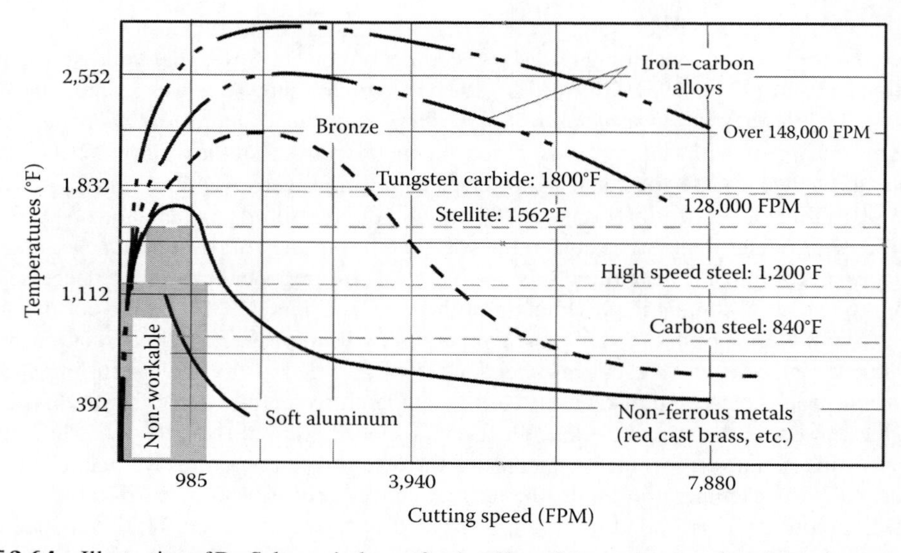

FIGURE 2.64 Illustration of Dr. Salomon's theory for the effect of the cutting speed on the cutting temperature.

General Electric (GE) to investigate the basic effective relationships and to exam the opportunities of integrating HSM into industrial application. The results of this program showed that the optimal cutting speed range in machining of aluminum alloys is 1500–4500 m/min. All tests were primarily concentrated on light metal alloys and in a few cases only on steel and cast iron. Other materials such as low-machinability steel, fiber-reinforced plastics, and alike materials were studied only to a small extent. Comprehensive and systematic scientific studies about fundamentals and investigations about the technical relationships between causes and effects as well as intensive consideration of the repercussions of this new metal cutting technology on the components involved in the metal cutting process as a whole were not available until the late 1970s. GE Company provided a database for machining aluminum alloys, titanium alloys, nickel-based superalloys, and steel (El-Hofy 2006).

The first industry-level HSM was attempted just 5 years ago in the automotive industry where the most advanced high rpm (up to 25,000) spindles with ceramic bearings and active control were used in the production lines and newly built power train plants. However, tool lives of many carbide and PCD tools were found to be one-third of those run at conventional cutting speeds for *the same* machining operations producing the same parts. Although the direct tooling cost at such plants does not normally exceed 5% so the reduced tool life at significantly decreased cycle time is fully justified by efficiency numbers, the idea depicted in Figures 2.63 and 2.64 did not find any experimental confirmation as the tool life of most of the cutting tools, mainly drilling tools, is limited by abrasion–adhesive wear greatly enhanced by high cutting temperatures.

The author's test results using HSM show that the cutting temperature does not reduce with the cutting speed as was expected. If the temperature increases even further, abrasion–adhesive wear becomes not the prime wear mode. A liquid layer forms between tool and workpiece due to diffusion leading to the formation of low-melting-point compound Fe_2W having a melting temperature T_m = 1130°C. This layer is quickly removed in cutting (Makarow 1976). At high cutting speeds, particularly in the machining of difficult-to-machine work materials, in parallel with plastic flow and spreading of the tool material over the chip and tool contact surfaces (that was conclusively proven by special tests with radioactive isotopes [Makarow 1976]), the plastic lowering of the cutting edge takes place. This plastic lowering is observed not only when carbide tools are used but also when PCD, metalloceramic, and ceramic tool materials are used. Often, in the machining of difficult-to-machine materials and in HSM, the plastic lowering of the cutting edge is the predominant cause of premature tool failure. The causes, mechanism, and appearance of the plastic lowering of the cutting edge were presented by the author earlier (Astakhov 2006).

The author is frequently asked if the data shown in Figures 2.63 and 2.64 are just simply fabricated or Dr. Salomon found (by chance, intuition, or *it suddenly dawned upon him*) *a sweet spot* so the presented data can be real. The short answer is it is physically feasible that Dr. Salomon found a certain combination of machining parameters for which the presented data are real. Immediately, two other questions arise: "Why was nobody else capable to obtain similar results?" and "If it is physically feasible, what has to be done to obtain similar results and thus to make HSM a common efficient operation?" A rather lengthy elaboration on the matter including a number of equation, graphs, and models can be reduced to the following conclusions:

1. According to the prevailed notion, new surfaces in metal cutting are formed simply by *plastic flow around the tool tip* (Shaw 1984). In other words, the metal cutting process is one of the deforming processes where the well-known single-shear plane model constitutes the very core of metal cutting theory, and thus, this process is thought of primarily as a cutting tool *deforming* a particular part of the workpiece by means of shearing. This notion combined with ignorance of the physics of material strength is the prime reason why Dr. Salomon's results have not been repeated despite all the efforts/time/resources spent. Neuroscientist Stuart Firestein in his recent book *Ignorance; How it Drives Science* (Firestein 2012) compared such attempts with an attempt to find a black cat in a dark room especially when there's no cat.

2. Unless an incorrect perception that metal cutting is one of the deforming process is totally abolished and the science of metal cutting comes back to the idea of one of the best minds of his time famed for his engineering studies, Franz Reuleaux of the Berlin Royal Technical Academy who as early as in 1890 provided that fracture occurs in metal cutting and thus cracks form ahead of the tool (Reuleaux 1900), no real progress in metal cutting is possible. Note also that the notion of metal cutting as the fracture of the work material was developed by founders of metal cutting studies as Time and Tresca (Time 1877; Time 1870; Tresca 1873) who pointed out a cyclic nature of this process. Frederic Taylor, a founder of experimental studies in metal cutting, I his most referred work (Taylor 1907) described crack creation and healing in each cycle of chip formation as the most distinguishing feature of metal cutting. Unfortunately, subsequent researchers, even those who tried to present the history of meta cutting, never mentioned these facts although thee routinely side works by the mentioned authors in introductions to their works.
3. The definition of the cutting process should be as follows. The process of metal cutting is defined as a forming process, which takes place in the components of the cutting system that are so arranged that the external energy applied to the cutting system causes the purposeful fracture of the layer being removed. This fracture occurs due to the combined stress including the continuously changing bending stress causing a cyclic nature of this process (Astakhov 1998/1999). The keywords are fracture and cyclic nature.
4. Since the beginning of time, human beings wanted to make their primitive tools/weapons and dwellings stronger. This approach was fully inherited by materials science and its multiple branches—to make parts, structures, machines, etc., stronger. In other words, materials, first natural and then man-made, were selected/designed/studied to withstand external loads as long as possible. Moreover, the beginning of plastic deformation known as yield where a part changes its shape irreversibly is considered as a failure because the machine or structure cannot continue its normal functioning after yielding. The lack of knowledge on the material behavior is covered by the so-called safety factor, that is, the theoretically obtained results are just simply multiplied by at least 2.6. In this context, metal cutting is a kind of black sheep because its main objective is to crash (fracture) the work material (physically separate the chip from the rest of the workpiece). Moreover, the goal is to do such a separation with minimum possible energy with no safety factors. Attempts to apply notions of materials with the traditional materials science having the opposite objective are fruitless.
5. Diamond is the hardest material, so it is used as the tool material to work under conditions where other tool materials normally fail. To cut a rough diamond down to a manageable size, the cutter must cleave it along the diamond's tetrahedral plane, where it is the weakest. Minerals cleave along particular crystallographic planes where the atomic bonding is weaker. It is similar to splitting a piece of wood—it splits fairly easily along the grain, but not across the grain. The number of cleavage planes and the angles between them are characteristic of specific minerals.
6. Most real-world work materials are polycrystalline, so they do not have the weakest crystallographic plane, which means that other physical phenomena should be found instead. This phenomenon is the state of stress in the deformation zone as their strength is sensitive to the state of stress and strain rate in this deformation zone (Astakhov 1998/1999). Adjusting the state of stress and interaction of the deformation and heat waves in the deformation zone (Astakhov 2006) plus assuring the strain rate needed, a particular material can be cut much easier than in traditional metal cutting. The high speed suggested by Dr. Salomon is needed for assuring the stain rate and energy wave propagation (Astakhov and Shvets 2010). In other words, the physics of metal cutting is much more complicated than it is presented in the known studies, most of which routinely violate basic laws of physics starting with the conservation law.

2.6.9.2 Optimal Cutting Temperature: Makarow's Law

To resolve the long-standing problem with the influence of the cutting temperature on tool life, the first metal cutting law (Makarow's law after A.D. Makarow who first pointed out the existence of the optimal cutting temperature) was formulated by Astakhov (2006) in the following form:

> For a given combination of the tool and work materials, there is the cutting temperature, referred to as the optimal cutting temperature θ_{opt}, at which the combination of minimum tool wear rate, minimum stabilized cutting force, and highest quality of the machined surface is achieved. Being a sole physical property of a given combination of the tool and work materials, this temperature is invariant to the way it has been achieved (whether the workpiece was cooled, preheated, etc.).

The Makarow's law, established initially for longitudinal turning of various work materials, was then experimentally proven for various machining operations. Therefore, the optimum cutting temperature for a given combination, *work material–tool material*, should be established and used as the only criterion for the suitability of this particular tool material for this particular work material. This temperature is a physical property and thus does not depend on the intrinsic details of tool design and geometry as well as on the parameters of a particular test setup. Any departure from this temperature in either side (lower or higher) unavoidably leads to reduced tool life. This cannot be reflected by Taylor's formula even in principle. Figure 2.65 shows a modern correlation curve *tool life versus cutting speed* where the optimal cutting speed is shown (Astakhov 2006).

In these considerations, the cutting temperature is the temperature measured by the tool–work thermocouple technique. It has the meaning of the average or integral temperature of the tool–chip interface. The huge advantage of such a representation of the cutting temperature is that the tool–work thermocouple can be used in any metal cutting studies and even on the shop floor for practical machines with intelligent controllers that can easily measure and then maintain this temperature. The idea of the tool–work thermocouple and its calibration procedure is well known (Shaw 1984; Astakhov 1998/1999, 2006; Trent and Wright 2000). Besides simplicity and feasibility of use on modern machine tools as the optimization criterion, a huge advantage of the optimal cutting temperature introduced this way is that it depends only on the work and tool material and thus should be established only once to their given combination. Then, it is valid for any tool and tool design made of this tool material and for this work material regardless of its heat treatment and particularities of the cutting operation.

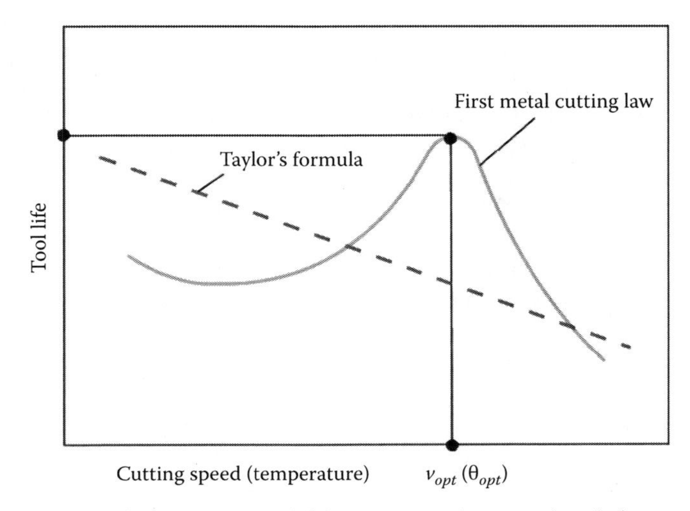

FIGURE 2.65 Modern correlation curve *tool life versus cutting speed* and that suggested by Taylor's formula.

Machining at optimal cutting temperature not only results in the minimum tool wear rate but also leads to obtaining the minimum cutting force and smallest roughness of the machined surface. Therefore, the proper *temperature gage* is developed for the first time to be used in simulations of the metal cutting process. Achieving the optimal cutting temperature is the second criterion (the first—the minimum plastic deformation of the layer being removed) for optimization of metal cutting (Astakhov 2006).

Reading this material about the optimal cutting temperature, one may wonder how it can be applied to drills as the cutting speed varies along the major cutting edges. Section 4.6.3 explains how to design drills having uniform temperatures equal to the optimal cutting temperature along their edges so that the maximum tool life and uniform tool wear along the cutting edge can be achieved.

2.6.10 Special Wear Mechanisms: PCD

Unfortunately, only phenomenological studies (the wear curves discussed earlier) of tool wear of PCD tool materials were attempted so far. It is pointed out (Stenphenson and Agapiou 1996) that microchipping of the cutting edge due to low toughness of PCD is the prime wear mode. This is particularly true in interrupted cutting of high-silicon–aluminum alloys where cutting edges chip readily, and tool consumption is unacceptably high. Chipping of the cutting edge commonly occurs during the initial stage of an interrupted machining process. As these chips accumulate, surface smoothness of the workpiece degrades and burring increases. The author's experience in high-volume production, however, shows that this is not the case when the drilling (machining) system is designed properly. When this is the case, HSM of heavily interrupted surfaces with PCD tools having even 10°–12° primary clearance angle takes place in face milling and drilling takes place with no microchipping of PCD inserts.

As discussed in Chapter 3, the composition of conventional PCD tools derives from the sintering process by which they are manufactured. Because this process yields unbalanced distributions of diamond particles and a metal catalyst (primarily cobalt), an inverse relationship exists between wear resistance and strength of the tool. Therefore, only very vague recommendations for the selection of PCD grades are available from leading PCD manufacturers. No recommendations on tool manufacturing practices using PCD are developed so that each PCD tool manufacturer developed such practices and procedures using mainly the experience gained in cemented carbide-tool manufacturing.

In the practice of the automotive industry, a 300%–400% scatter in tool life and unexplained breakage of PCD tools used for similar application is the case, and it seems that nobody can explain this scatter. Even if the failed PCD tool is sent to R&D departments of leading PCD manufacturers, their reports are of little help to find the root cause of the problem. A great disconnect exists between the PCD material and PCD cutting tool manufacturers as the former have insufficient knowledge in PCD tool design, manufacturing and implementation, while the latter have no trained staff and equipment to study the nature of PCD structure and thus to make proper assessment of PCD tool failures. Therefore, a need is felt to clarify a few long-standing issues providing some help to PCD tool manufacturers and users.

Contrary to common perceptions, the author's multiple observations of PCD tool wear patterns have shown that when the components of the machining system are coherent and the cutting regime is optimal, chipping does not occur at all in machining using PCD tools. Figure 2.66 shows the proper wear pattern of a PCD insert obtained when the tool was run at the optimal cutting conditions and tool life about 200,000 holes in high-silicon–aluminum alloy A380. As can be seen, there is no microchipping. Rather, a porosity-like structure of tightly bonded diamond crystals is developed as the catalysis (cobalt) is taken away by the sliding chip. In the author's opinion, achieving similar wear patterns is the goal in optimization of a machining operation with a PCD tool.

However, the discussed *ideal* wear of PCD tools occurs rather rarely. The problem is in the thermal stability of PCD tool material as discussed in Section 3.5.6. This problem is often found not only in PCD tool implementation but also in their manufacturing.

FIGURE 2.66 Proper (target) wear pattern of PCD insert.

The damage to the PCD structure can be classified as follows:

1. Pure mechanical, that is, due to high contact pressure on the tool–chip and/or tool–workpiece interfaces. Figures 2.67 and 2.68 show the appearance of these fractures respectively. As pointed out by Astakhov (2006), the fracture of the rake face happens when the feed per tooth exceeds the so-called breaking feed, while that on the tool–workpiece interfaces happens due to a variety of reasons, for example, an insufficient clearance angle (Astakhov 2010).
2. Pure thermal due to high temperatures. This can be further classified into two categories:
   a. Thermal damage imposed in tool manufacturing, that is, overaggressive electro-discharge machining (EDMing) and grinding as well as during brazing.
   b. Thermal damage due to high temperatures in cutting. High temperatures occur in machining particularly in HSM when the work material is highly abrasive, that is, creates a lot of heat due to friction at the tribological interfaces or when it is of low thermal conductivity as, for example, fiber-reinforced plastic materials.
3. Mechanical damage enhanced by high working temperature.

FIGURE 2.67 A crack developed on the tool rake face behind the tool–chip contact area due to high contact stresses on the rake face.

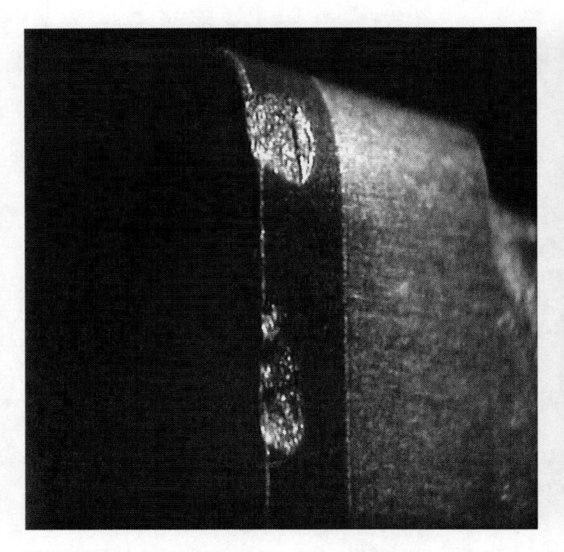
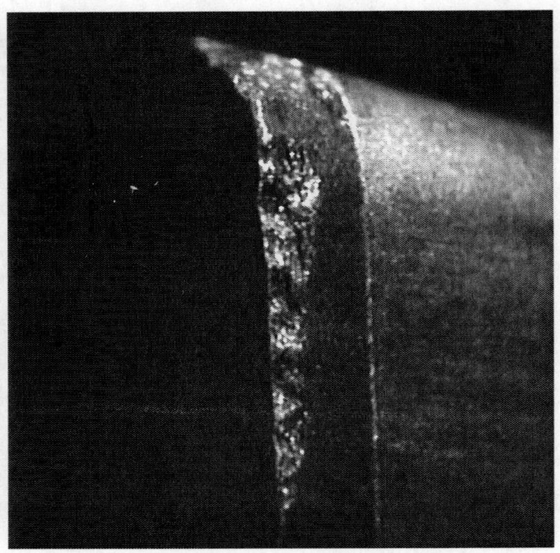

FIGURE 2.68 Chipping of the flank face of the tool due to high contact stress at the tool–workpiece interface (the flank contact area).

While high temperatures may occur in cutting at high cutting speeds and feeds, it is not likely that it occurs in many cases of machining of high thermal conductivity work material (e.g., aluminum alloys) with high flow rate water-based MWF supplied through tools. If this was the case, all PCD tools would fail. This is not the case as most of the tools complete tool life.

Therefore, thermal damage to PCD inserts during manufacturing is likely the case. Elevated temperatures affect the common PCD material in two principal ways:

1. Thermal degradation due to differential thermal expansion characteristics between the interstitial catalyst material and the intercrystalline-bonded diamond. As discussed in Chapter 3, such differential thermal expansion is known to occur at temperatures of about 400°C, causing ruptures to occur in the diamond-to-diamond bonding and resulting in the formation of cracks and chips in the PCD structure. Chipping and microcracking of cutting edges occur as a result (Miess and Rai 1996; Vandenbulcke and De Barros 2001; Chen et al. 2007).
2. Thermal degradation due to the presence of cobalt in the interstitial regions and the adherence of cobalt to the diamond crystals. Specifically, cobalt is known to cause an undesired catalyzed phase transformation in diamond (converting it to carbon monoxide, carbon dioxide, or graphite) with increasing temperature (Fedoseev et al. 1986; Coelho et al. 1995; Cook and Bossom 2000; Shimada et al. 2004).

The critical manufacturing operations in terms of thermal damage to PCD are brazing, EDMing, and grinding.

Figure 2.69 shows examples of poor brazing quality. Spilled brazing filler material (BFM) shows that excessive amount of the BFM was used, and thus, excessive heat was applied on brazing. As a result, heat-induced damage was done to the PCD tool material so that the tools failed prematurely. Figure 2.70 shows a direct correlation between overheating on brazing and PCD failure.

Even when the heat introduced on brazing is not excessive but the amount of the brazing filler is, as shown in Figure 2.69, the PCD tool quality deteriorates as the PCD insert rests on the relatively soft BFM rather than on the rigid tool body often made of sintered carbide for structural rigidity. In other words, such brazing undermines the whole purpose of the expensive carbide body. The rule of thumb for brazing is as follows—the proper brazing should not be distinguished with the naked eye

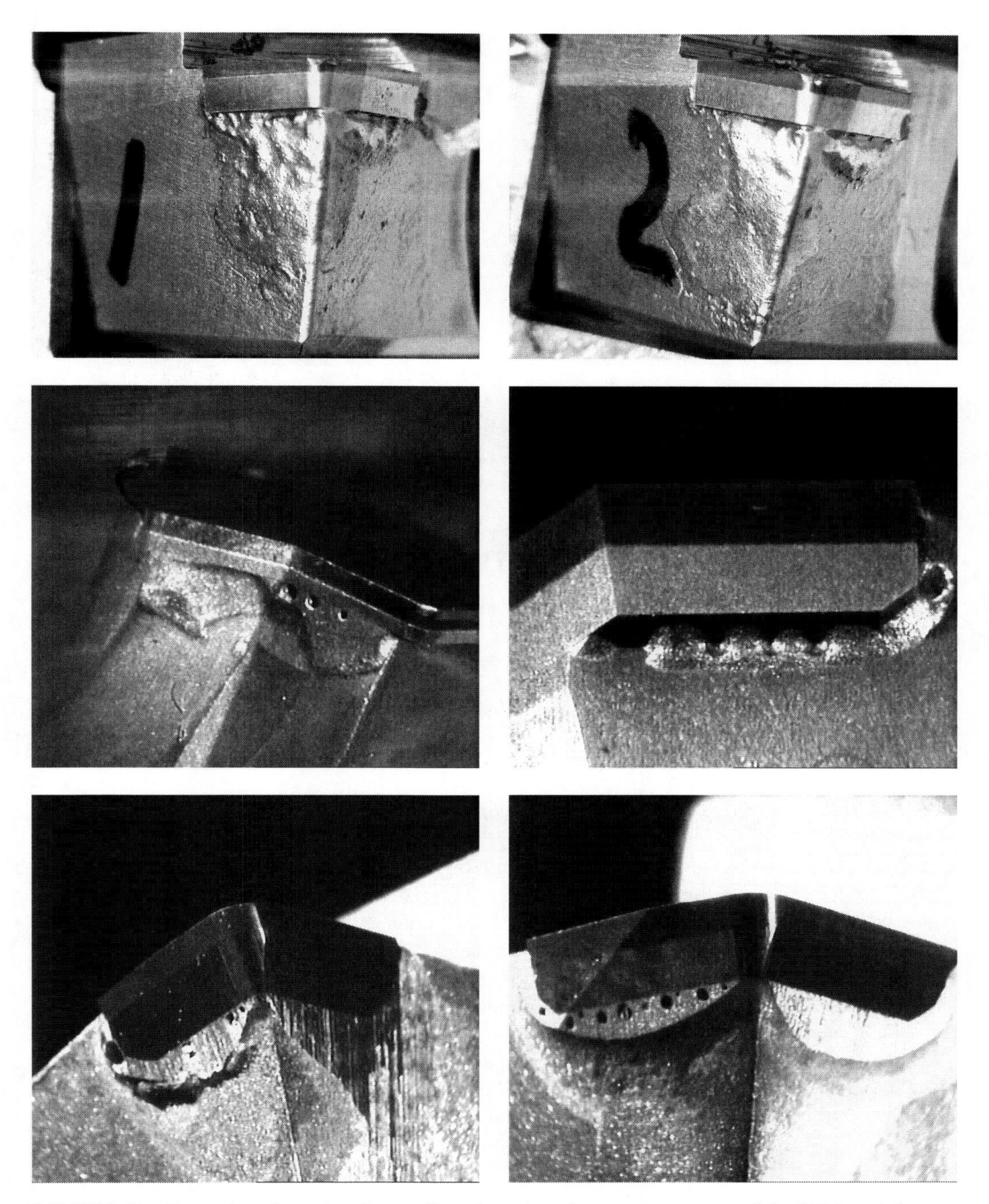

FIGURE 2.69 Examples of poor brazing quality where there is excessive amount of the BFM, and thus, too much heat was applied.

as no separate layer of BFM should be formed. Rather, BFM should only fill out the valleys on the surfaces of the PCD insert (the carbide substrate side) and those in the PCD pocket in the tool body.

On the other hand, when the heat applied on brazing is insufficiently combined with a low-quality (e.g., oxidized as left open for days) brazing flux and when the PCD insert and the pocket in the tool body are not freshly cleaned, plus the improper grade of BFM (a cheap, high melting

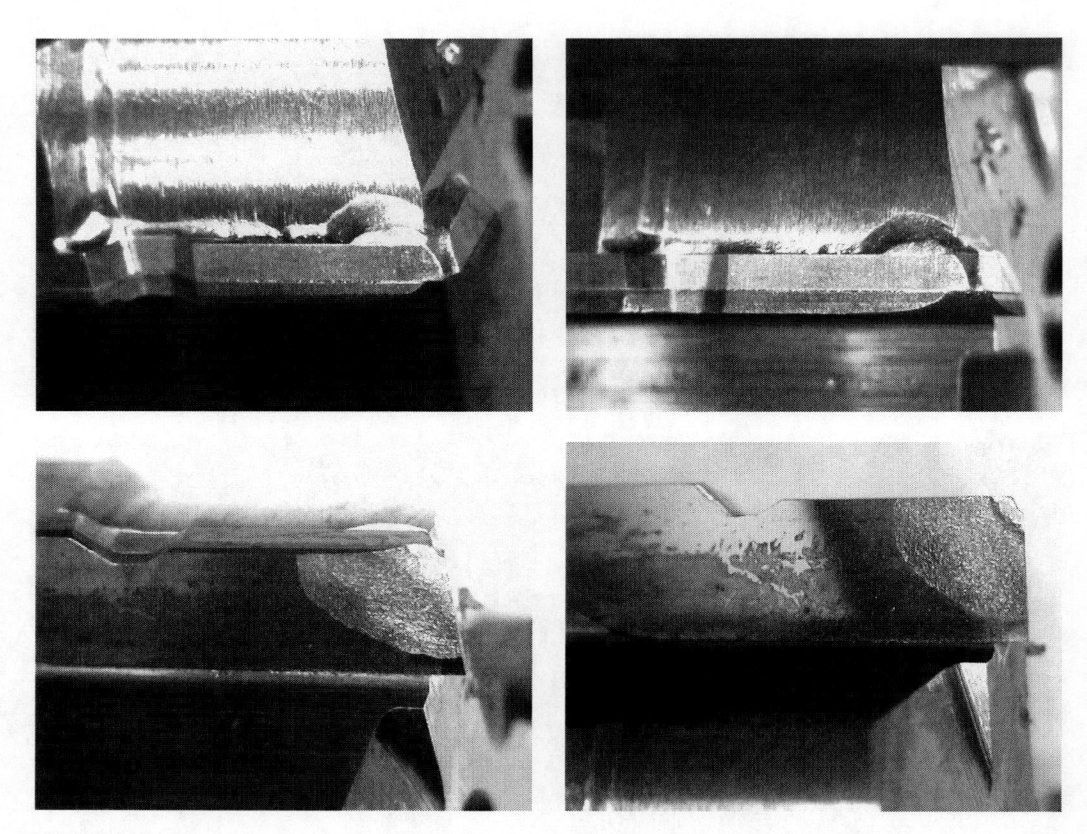

FIGURE 2.70 Showing a direct correlation between overheating on brazing and PCD failure.

temperature alloy) is used, then the so-called *dry brazing* is a result of the listed brazing conditions. It is already discussed in Section 2.5.8. Figure 2.71 shows a typical appearance of the pocket in the tool body and PCD insert that failed due to dry brazing.

In the author's opinion, the prime cause of the poor brazing is the use of induction or even torch brazing of PCD inserts with no temperature control. Experience shows that PCD inserts of any type cannot be brazed successfully without special processes, for example, vacuum (inert gas) brazing furnaces. Proper cleaning and fluxing procedures should be established and religiously maintained. Unfortunately, only a handful of PCD tool manufacturers conditionally justify these requirements. Proper PCD insert brazing for HP drilling tools is yet to be seen.

EDMing is another PCD tool manufacturing operation that requires close attention. As discussed in Chapter 3 (Section 3.5.2 and 3.53, Figures 3.79 and 3.83), a PCD disk (blank) consists of a layer of fine diamond powder sintered together into a dense uniform mass, approximately 0.5 or 2.0 mm thick, supported on a substrate of cemented carbide. The standard PCD blanks are made as disks of up to 74 mm in diameter. They are first typically cut with a wire EDM during the tool manufacturing process to obtain cutoff of shape and size close to the cutting insert in the tool (see Figure 3.76). Then these cutoffs are brazed to the tool body, often made of sintered carbide for drilling tools, and then trimmed with wire EDMing to the final dimensions or some stock is left for the finish grinding. If the latter is the case, then finish grinding operation is used to bring the tool to its final dimensions—normally thin margins and primary flank faces are ground on drilling tools.

EDMing PCD is difficult because diamond is not typically conductive and EDM is a process for vaporizing conductive materials. Cobalt, on the other hand, is conductive, and wire EDM machinability of PCD is a function of how uniformly the conductive metallic phase is dispersed throughout the diamond matrix. If cobalt is not evenly dispersed (as it is the case in low-quality PCD products),

FIGURE 2.71 Typical appearance of the pocket in the tool body and PCD insert that failed due to dry brazing.

an agglomeration of diamond grains can cause the wire to break or deviate as the wire tries to erode a material that it cannot. Additionally, coarse-grained PCD grades are inherently more difficult to cut because of their relatively large grain size. Cutting with EDM may also produce an undercut at the interface between the diamond and carbide substrate, which requires more grinding to finish the tool's periphery. Typically, an extra 0.25–0.40 mm stock of PCD is left after EDMing, and a tool is ground to final size. The quality of the ground edge that can be produced is crucial in achieving the optimal tool life and required surface finish. Grinding typical medium-grade PCD can cause larger grains to pull out from an edge, creating a rough cutting edge. Using high-quality PCD with tailored diamond particle distribution and improved packing density of the diamond grains, PCD tool manufacturers can achieve a higher-quality edge.

Both operations, EDMing and grinding, present a great challenge for PCD tools. Diamond is the hardest known substance, so when grinding PCD tools with a diamond grinding wheel, the amounts of material removed from the workpiece and wheel are about equal. This makes the grinding operation very slow and expensive. However, grinding makes a huge difference in the performance of

HP PCD tools as it produces much better quality of the cutting edges. Typical ground edges have a surface finish of *Ra* 0.04, whereas EDMed surfaces at best have a finish of *Ra* 0.20. Rougher surface after EDMing promotes the development of microcracks starting from the cutting edge as microvolumes of the work material adhere to cobalt in the serrations (valleys) left by EDMing.

Unfortunately, many PCD tool manufacturers and even some leading wire EDM machine suppliers try to deny this obvious fact observed by the author on a daily basis. They maintain, with no supporting evidence, that the difference in performance is small (5–15%) so that the higher cost and longer lead time required to grind PCD cannot be justified. They maintain that wire EDMs cut PCD efficiently and to an edge quality that is close to what grinding can achieve because of advances in the EDM generator, the circuitry that controls the impulses of electrical energy to create an effective spark. Developments in solid state technology provide a more uniform and reliable *square wave form* to the individual impulses. According to the author's inexpedience, it is not true. The latest EDM machines include a polishing capability that provides polishing after EDMing to improve the quality of the cutting edge, that is, reduce its micro-serrations.

The problem with wire EDMing, however, is that a great amount of heat is applied directly to microvolumes of the PCD insert being cut as the process relies on the cobalt evaporation. Before cobalt evaporates, it expands on heating at a much greater rate than that of PCD crystal, which creates interfacial cracks in the EDMed surface. On the QWD horizontal machines designed and made especially for PCD tools by Vollmer Co., a leading EDM machine manufacturer, repositionable jointed nozzles direct a flood of dielectric fluid to the cutting zone to reduce the damage to PCD. These machines also use a common grade of EDM oil as the dielectric fluid rather than deionized water. It is recommended to use the ten thou (0.25 mm) diameter zinc-coated brass electrode wire for almost all PCD and carbide cutting.

The major problem, however, arises when overaggressive EDMing is used. Figure 2.72 shows typical PCD failures due to overaggressive EDMing. The following can be observed in this figure:

1. Traces of wire *jumping* over PCD crystals when high feed rate of the wire is applied.
2. Deep groove between the PCD layer and the carbide substrate. As discussed in Chapter 3 and shown in Figure 3.93, the interface between the PCD layer and the carbide substrate is rich in cobalt. As cobalt is the only conductive face, overaggressive EDMing evaporates cobalt from this interface leaving a deep groove.
3. The collapsed cutting edge due to overheating on EDMing. The cutting edge becomes round instead of sharp that results in a number of quality problems with the machined parts.

Analyzing a great number of PCD tool premature failures, the author developed the following classification of these failures:

1. *Light* overheating of PCD. This kind of overheating causes the development of a network of microcracks in the PCD structure that made it structure weaker. When the stresses of cutting are applied, the chipping of the PCD structure takes place. It could range from chipping of relatively small portions of the PCD structure as shown in Figure 2.73 to severe chipping of whole PCD layer up to the carbide substrate as shown in Figure 2.74.
2. *Medium* overheating of PCD. Although there are a number of possible outcomes of this kind of overheating, the common denominator is PCD layer flaking of different sizes that range from micro-flaking (Figure 2.75), medium flaking (Figure 2.76), and severe layered flaking (Figure 2.77).
3. *Severe* overheating of PCD. Severe overheating of PCD leads to the plastification of the PCD layer, which can be local as shown in Figure 2.78 or even global as shown in Figure 2.79. As such, the PCD layer behaves as a highly plastic material with the corresponding wear pattern developed after just a few working cycles.

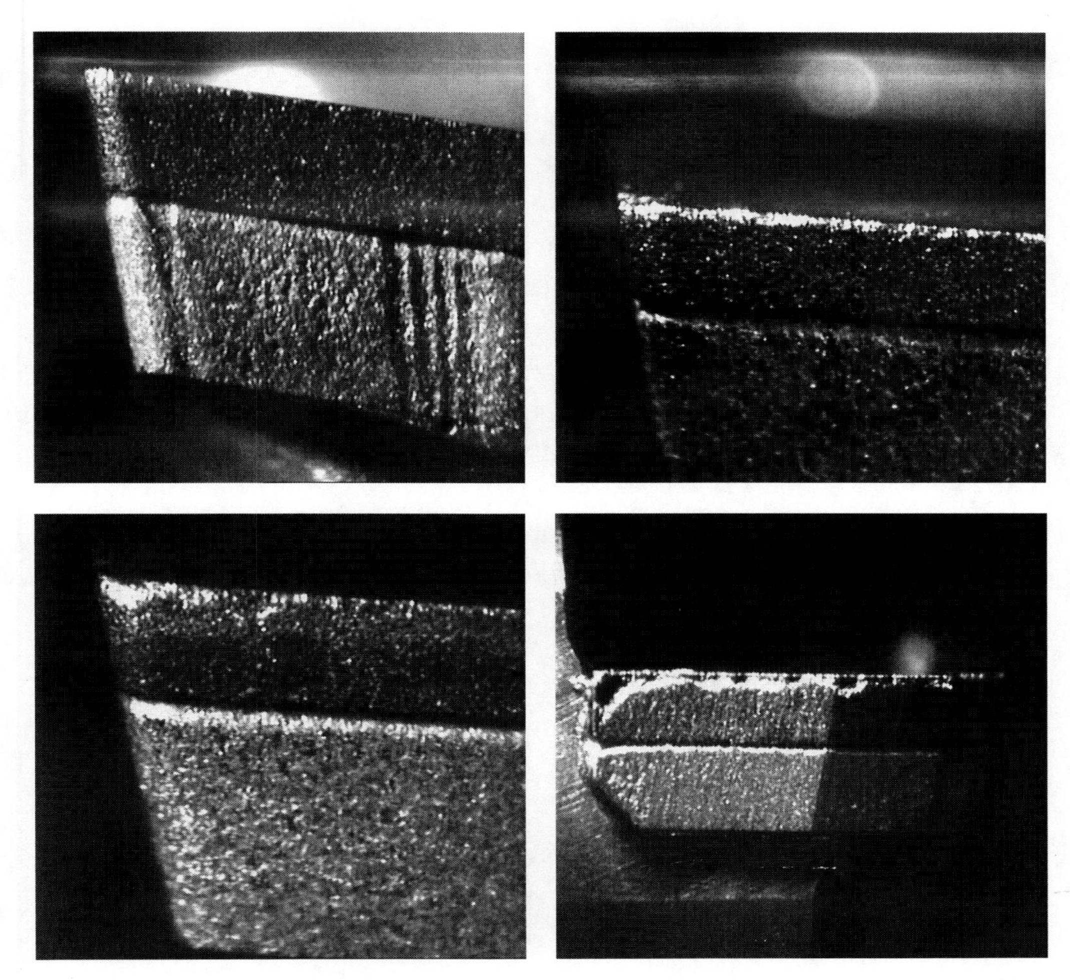

FIGURE 2.72 Typical PCD failures due to overaggressive EDMing.

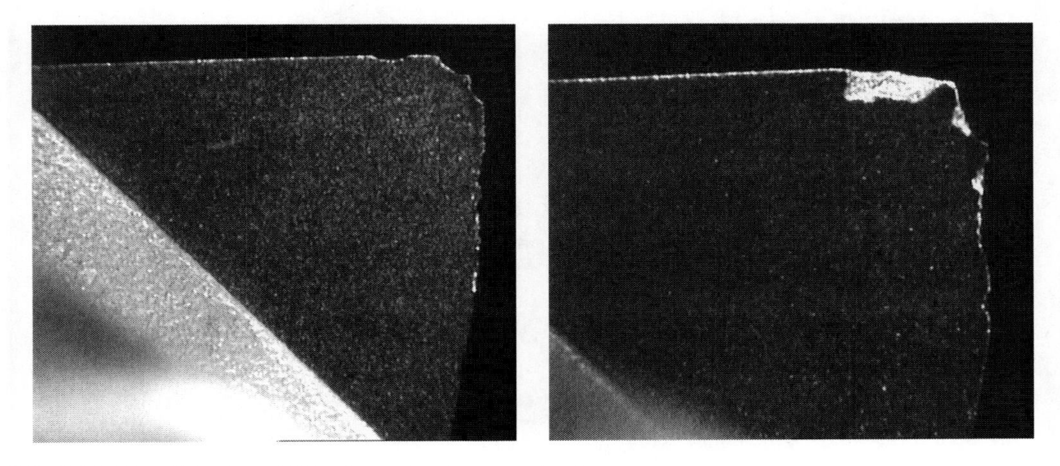

FIGURE 2.73 Chipping of PCD due to *light* overheating I.

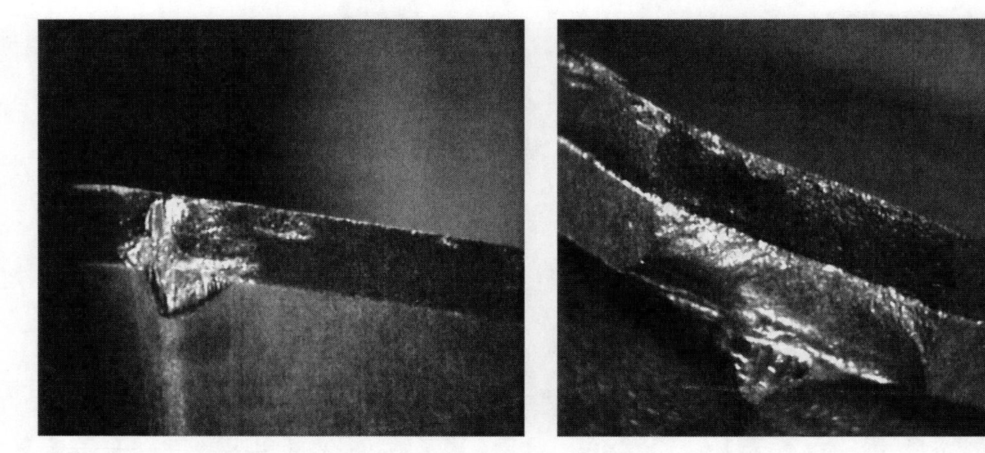

FIGURE 2.74 Chipping of PCD due to *light* overheating II.

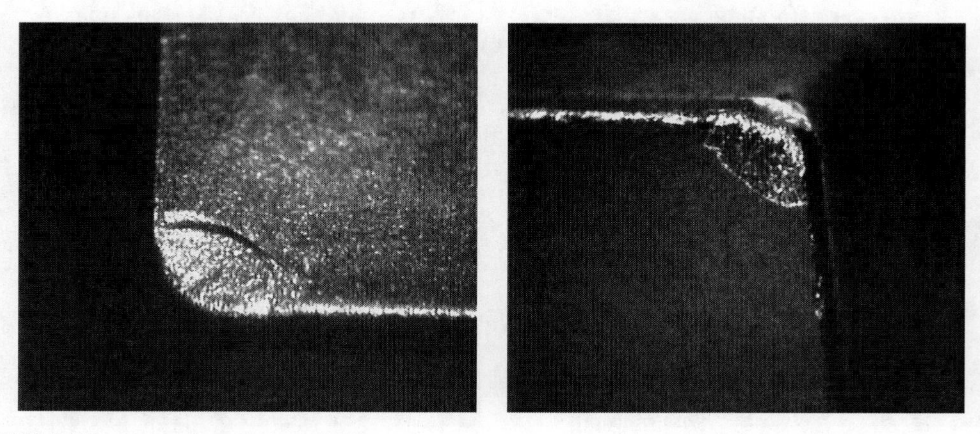

FIGURE 2.75 Examples of micro-flaking.

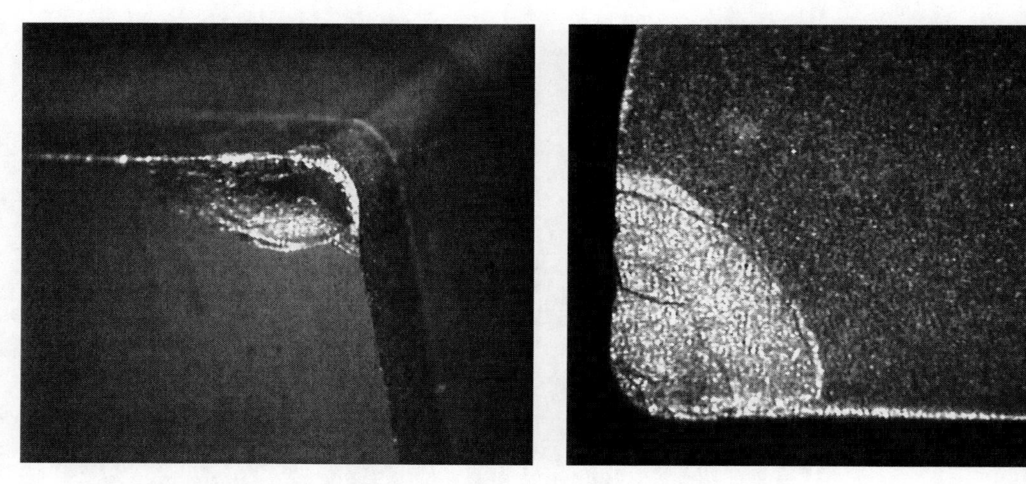

FIGURE 2.76 Examples of medium flaking.

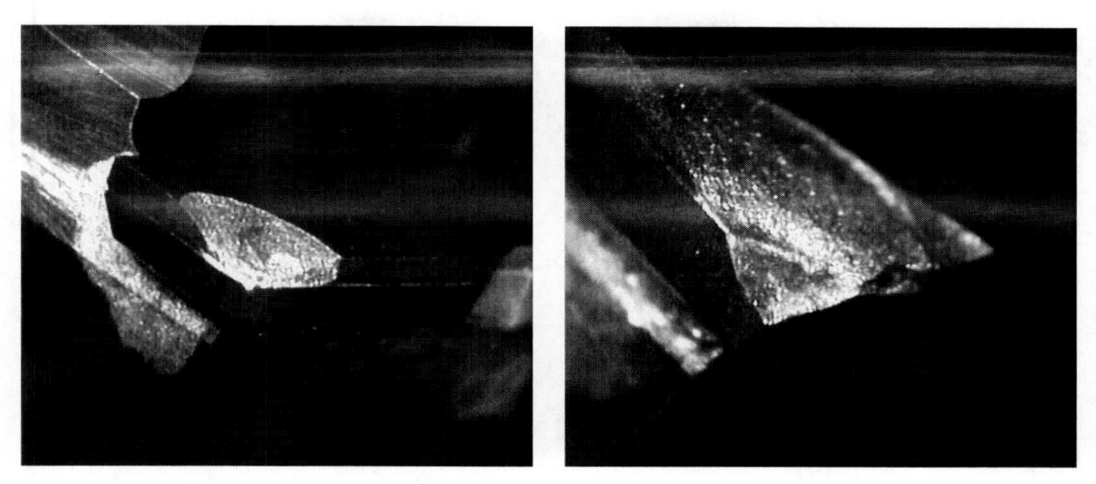

FIGURE 2.76 (continued) Examples of medium flaking.

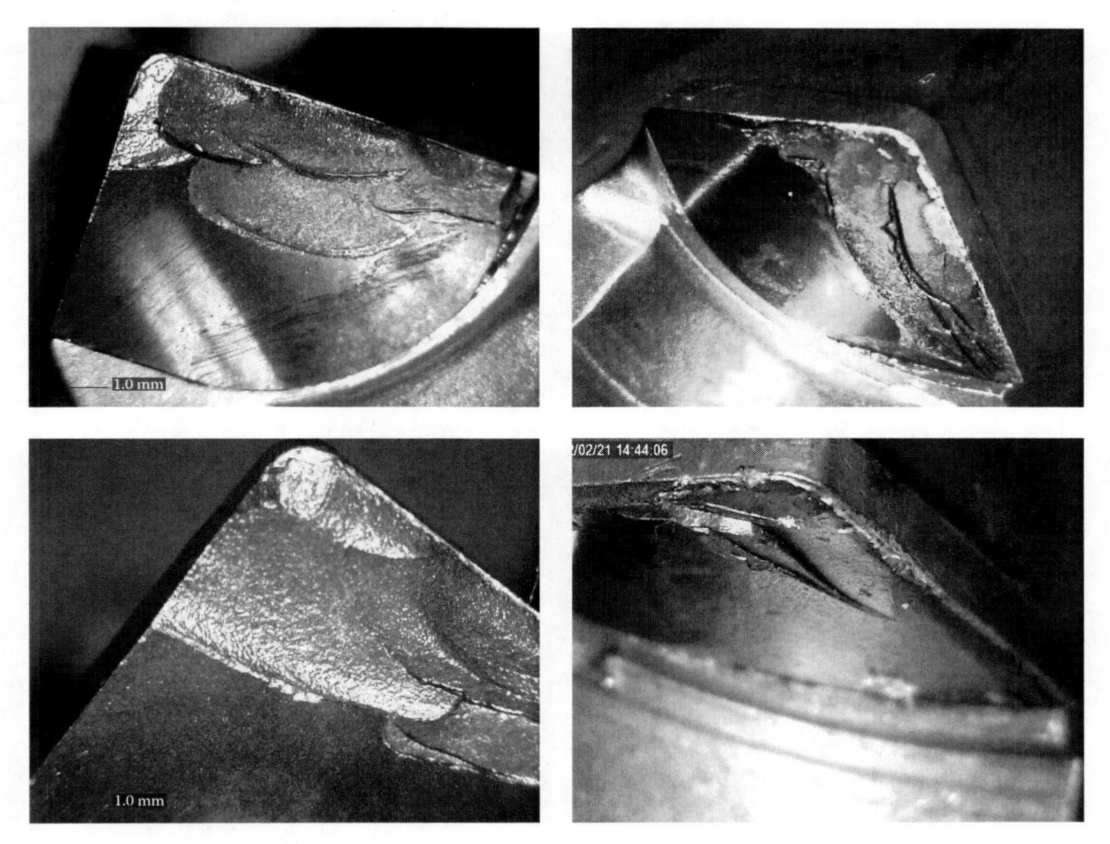

FIGURE 2.77 Examples of severe layered flaking.

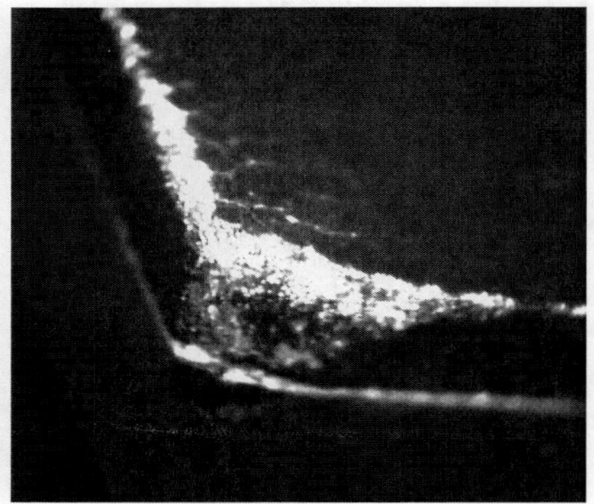

FIGURE 2.78 *Plastification* of PCD layer I.

FIGURE 2.79 *Plastification* of PCD layer II.

2.6.11 Casting Defects and Tool Wear/Failure

Casting defects are principal *enemies* of HP drills as such drills are seemingly much more sensitive to these defects than *normal* drills. The practice of HP drill application shows that the opposite is actually true as the survival rate of HP drills in the presence of casting defects is actually greater than that of *normal* drills due to variety of reasons starting with the application-specific tool material. Although it is true, casting defects may dramatically affect tool failure rate. This section attempts to classify the known casting defects and their effects on drills.

The casting defects affecting drill performance can be classified as follows:

1. Porosity and microinclusions
2. Casting cavities
3. Macroinclusions

Because the nature of porosity and content of inclusions depend on a particular material, method of casting, and many other metallurgical particularities, further consideration is restricted to aluminum–silicon alloys for clarity and particularity. These alloys are the materials of choice for many automotive and aerospace parts as well as domestic food and pump part castings. Other applications include components for food handling and marine castings.

Hydrogen is the only gas that is appreciably soluble in aluminum and its alloys. Porosity in aluminum is caused by the precipitation of hydrogen from the melt or by shrinkage during solidification, and most often by a combination of these defects (Zolotorevsky et al. 2007). Dissolution of hydrogen in large amounts in aluminum castings results in hydrogen gas porosity. For castings that are a few kilograms in weight, the pores are usually 0.01–0.5 mm (0.00039–0.020 in.) in size (Chen 2009). In larger castings, they can be up to a millimeter (0.040 in.) in diameter. Standard ISO 10049:1992 *Aluminium alloy castings—Visual method for assessing the porosity* specifies a method of inspection and describes the acceptance conditions, tabulates the severity levels, that is, number and size of pores, and the interpretation of the results.

Pores that form in the matrix of aluminum alloy castings lead to significant deterioration in casting quality. In brief, porosity is formed and distributed in the matrix of the casting. The pore count of aluminum–silicon alloys can be affected by the solidification mode of the alloy, the amount of oxide film and/or particle inclusions, the cooling rate, the atmospheric pressure, and the hydrogen level in the melt.

Porosity and porosity-size inclusions present a serious challenge in HSM of aluminum–silicon alloys as a myriad of pores should be thought of as a countless number of small sharp razor-type edges enhanced by hard SiC microinclusions. These literally shave the tool when their amount is excessive. Figure 2.80a shows an example of an excessively worn carbide drill and Figure 2.80b shows the result of an autopsy of the machined hole where porosity and microinclusions can clearly be seen. Unfortunately, there is no standard/procedure/methodology for microporosity and microinclusion assessment in terms of their influence on tool wear. As a result, the problem should be dealt with on a case-by-case basis with detailed autopsy of parts and collecting the relevant information.

Casting cavities are large shrinkage cavities (Figure 2.81). Normally, a drill breaks when it attempts to cut through such a cavity because its force balance is violated as one side of the drill cuts the air. As discussed in Chapter 4, proper drill performance completely relies on the drill force balance so the resultant axial force and drilling torque do not cause drill bending. If this balance is disturbed, however, the drill would bend. The greater violation of the force balance, the greater the bending, the higher the chance of dill breakage. However, this is only a part of

(a)

(b)

FIGURE 2.80 Excessive abrasive wear: (a) appearance of a worn drill and (b) result of the autopsy of the machined hole.

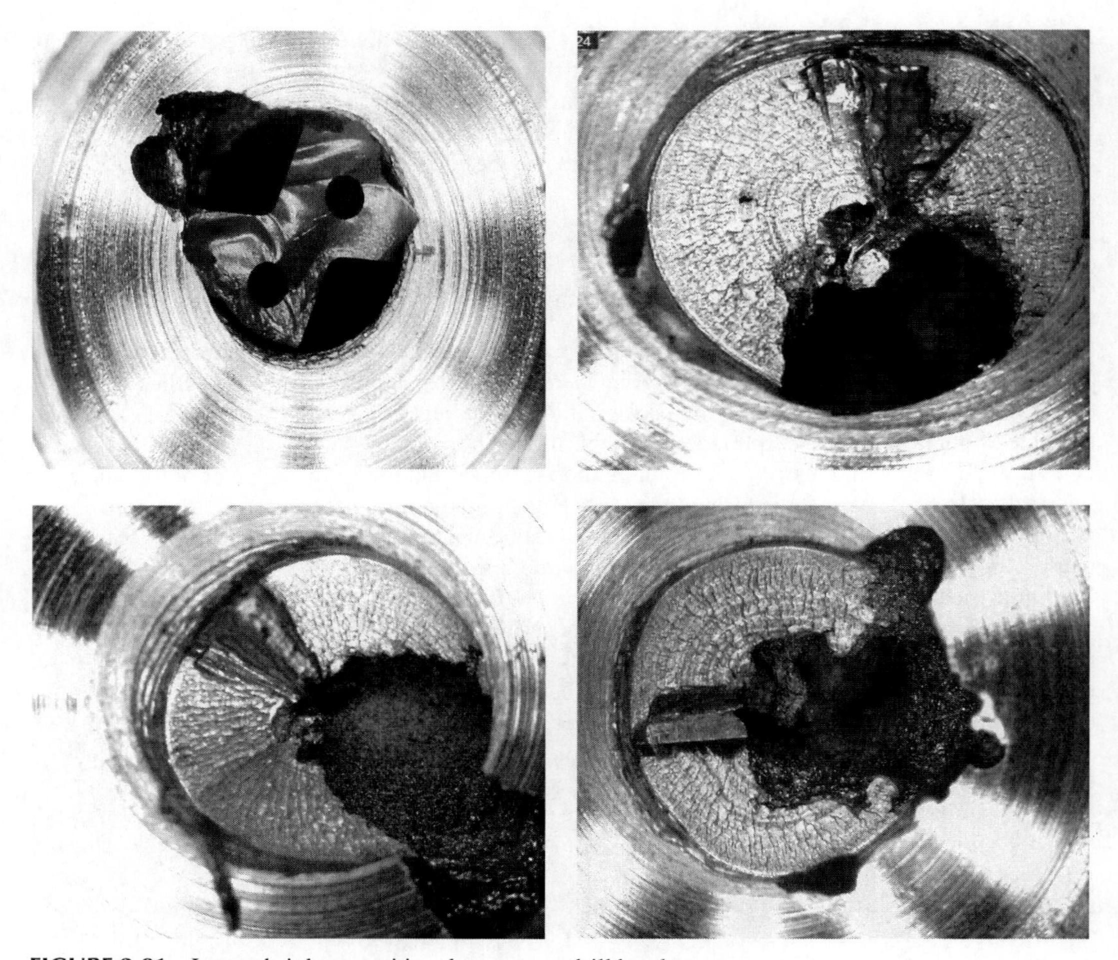

FIGURE 2.81 Large shrinkage cavities always cause drill breakage.

the problem. When a twist drill is compressed by the axial force, it winds becoming a bit shorter (see Section 4.4.3.5). If the axial force is suddenly removed from one of the cutting edges as this edge collapses into a cavity, the drill partially unwinds so one of its edges jumps ahead causing drill bending. This destroys the tool 100% of the time.

When a multi-edge, particularly helical-flute tool is used, then the cavity may not affect the force balance significantly. However, tool brakeage may occur for a number of other reasons. Figure 2.82 shows an example of such a case where a tap with an internal MWF supply having three helical flutes made the complete hole in the presence of a casting cavity. However, the tap was broken on its retraction as the chips formed on tapping accumulated in the cavity and then jammed the tool on its retraction. Part autopsy is the only means to reveal the root cause of the problem.

Nonmetallic inclusions are a particular concern in cast aluminum. Because of its reactivity, aluminum oxidizes readily form in liquid and solid states. Oxidation rate is greater at molten metal temperatures and increases with temperature and time of exposure. Magnesium in aluminum alloys oxidizes and, with time and temperature, reacts with oxygen and aluminum oxide to form spinel. Many oxide forms display densities similar to that of molten aluminum and sizes that reduce the effectiveness of gravimetric separation. Also, most oxides are wet by molten aluminum, reducing the effectiveness of mechanical separation methods. Inclusions occur as varying types with

FIGURE 2.82 Chip trapped by the tool on retraction.

differing sizes and shapes. Aluminum oxides are of different crystallographic or amorphous forms such as films, flakes, and agglomerated particles. Magnesium oxide is typically present as fine particulate. Spinels can be small, hard nodules or large, complex shapes. Aluminum carbide and aluminum nitride can be found in smelted aluminum but are of no significance due to their size and concentration in aluminum castings. Refractory and other exogenous inclusions may be identified by their appearance and composition.

Figure 2.83 shows appearance of small inclusions. Conditionally, inclusions are small when their size does not exceed 1/10 of the drill diameter. As these inclusions are very hard, they cause drill chipping in the manner shown in Figure 2.84. If not detected timely by the machine controller, this chipping then results in drill breakage.

Figure 2.85 shows an example of macroinclusion. When a drill comes into contact with such an inclusion, it unavoidably breaks. Part autopsy is required if one needs to establish the root cause of the problem. A report with such a root cause is then sent to the casting supplier, and countermeasures to avoid such microinclusions are requested as this type of inclusion affects not only machinability of castings but also the service properties of the part.

(a)

(b)

FIGURE 2.83 Small inclusions: (a) in the bottom of the holes being drilled and (b) in the bore wall.

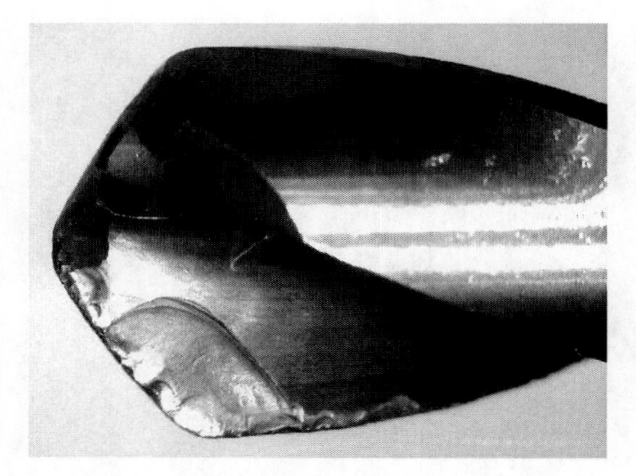

FIGURE 2.84 Drill chipping as a result of cutting through small inclusions.

FIGURE 2.85 Example of a macroinclusion.

2.6.12 Special Wear Mechanisms: Cobalt Leaching

As discussed in Chapter 3, cobalt is by far the most widely used binder metal or *cement* in cemented tungsten carbides because it most effectively wets tungsten carbide grains during carbide sintering. For this reason, cobalt is believed to be superior to other binder metals in terms of eliminating residual porosity and achieving high-strength and toughness values in sintered products. However, a great disadvantage of cobalt as a binder metal is its leaching caused by various reasons. Except for some simple cases, such reasons are not yet well determined. The simplest yet rather common is corrosion due to chemical reaction of cobalt with corrosive agents. The corrosion process involves the dissolution of the cobalt binder at exposed surfaces leaving a loosely knit skeleton of tungsten carbide grains having little structural integrity. This mechanism is often referred as cobalt *leaching* and is typically accompanied by flaking off of unsupported carbide grains in the affected surface areas. Figure 2.86 shows the essence of cobalt leaching.

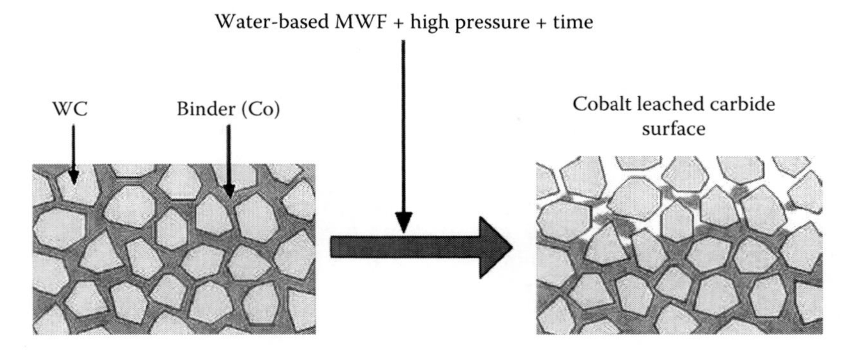

FIGURE 2.86 The essence of cobalt leaching.

Although WC + Co grades have fairly good resistance to attack by acetone, ethanol, gasoline, and other organic solvents as well as by ammonia, most bases, weak acids, and tap water, exposure to formic, hydrochloric, hydrofluoric, nitric, phosphoric, sulfuric, and other strong acids can result in a relatively rapid deterioration of the binder phase. Corrosion rates are affected also by temperature, by the concentration and electrical conductivity of the corrosive agent, and by other environmental factors such as high contact pressure and relative speed of the solvents. Figure 2.87 shows a typical appearance of cobalt leaching on a micrograph.

The carbide corrosion–enhanced leaching can be explained as follows. There is a chemical process called *chelation*. In chelation is (1) a chemical compound in which the central atom (usually a metal ion) is attached to neighboring atoms by at least two coordinate bonds in such a way as to form a closed chain or (2) to cause (a metal ion) to react with another molecule to form a chelate. Dissolving would mean that the cobalt would break up into individual cobalt-containing molecules in the water. Chelation means that it forms unique chemical compounds. This chelation causes the reddish or purplish coloration of a water-soluble MWF left around a failed tool. The observed colors on the cutting tools in many tool autopsies fully resemble colors (reddish and purple) of the cobalt salts. These reddish and purple colors have never been explained in tool analyses. The suggestions

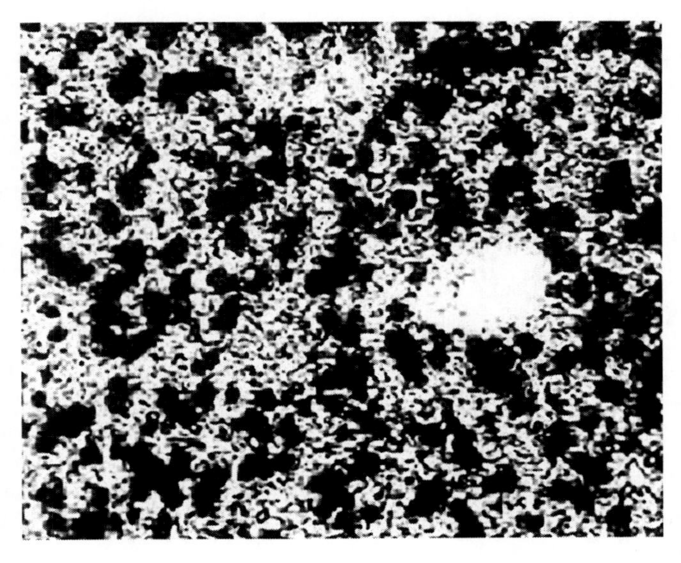

FIGURE 2.87 Structure of a cemented carbide with leached cobalt binder. In this condition, the skeleton of tungsten grains can break like glass.

TABLE 2.7
Resistance to Corrosion of Cobalt and Nickel Binder Cemented Carbides

<table>
<tr><th>pH</th><th>WC + Co Binder</th><th>WC + Ni Binder</th></tr>
<tr><th></th><th colspan="2">Resistance to Corrosion</th></tr>
<tr><td>12
11
10</td><td>Very good</td><td rowspan="4">Very good</td></tr>
<tr><td>9
8</td><td>Good</td></tr>
<tr><td>7</td><td>Fair</td></tr>
<tr><td>6</td><td>Poor</td></tr>
<tr><td>5
4</td><td rowspan="3">Poor to no resistance</td><td>Good</td></tr>
<tr><td>3
2</td><td>Fair</td></tr>
<tr><td>1
0</td><td>Poor</td></tr>
</table>

ranged from *tool bleeding* to *transmission fluid found its way to the coolant system.* The foregoing analysis provides a simple yet chemically sound explanation.

Table 2.7 shows the corrosion resistance details of tungsten cemented carbide having cobalt nickel binder as a function of the pH value of a water-soluble coolant, which roughly explains the carbide corrosion–enhanced leaching. Although practically all books on MWFs (coolants) point out that pH 7–9 should be the target range in MWF maintenance, no proper explanation of this experience-gained knowledge is provided.

Studies show (e.g., Jiang and Shang 2010; Xiaoming et al. 2011) that the oleic acid triethanolamine (TEA) ester solution could inhibit cobalt leaching to a certain degree, boric acid ester solution could suppress it even better, and composite boric acid ester containing benzotriazole has the best inhibition effect. Its inhibition mechanism is that composite boric acid ester containing benzotriazole inhibits cobalt leaching of the cemented carbide tool by a layer of complete and compact protective membrane that is generated on the surface of a cemented carbide tool. Cobalt leaching of the cemented carbide tool is effectively inhibited by adding the composite boric acid ester containing benzotriazole in water-based MWFs (coolants).

In the author's opinion and experience, the issue with cobalt leaching is much wider than that described earlier. The appearance of wear due to cobalt leaching resembles that due to abrasion so that it is physically impossible to attribute it to cobalt leaching. The analysis of the contact surface of worn tools showed severe cobalt leaching over the tool contact surface. It is caused presumably by strong bonding between the matrix material of the work composite material and cobalt in the cutting tool material. When sufficient microvolume of cobalt is leached, then collapse of the skeleton of tungsten grains takes place so that wear pattern appearance resembles abrasion.

Cobalt leaching can be observed even with no microscopes for some special tools. An example of such tools is PCD-tipped drills and reamers having a carbide body. Normally, tool life of such tools can be as high as 60,000 cycles so that the tool stays in the machine for as long as 4–5 months. Moreover, after tool life run, such tools are normally re-tipped, that is, the carbide body is still the same. Long exposure to a subpar quality MWF results in cobalt leaching that can be seen with the naked eye. Figure 2.88 shows two examples.

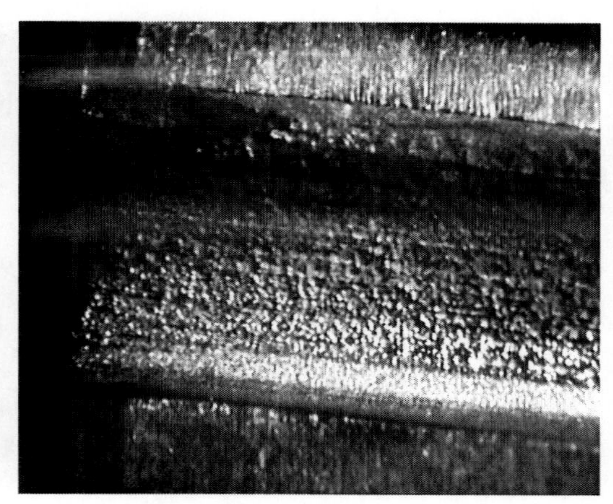

FIGURE 2.88 Appearance of severe cobalt leaching.

Experience shows that:

1. Cobalt leaching starts in carbide manufacturing:
   a. The prime cause is in carbide tool grinding/sharpening with water-soluble MWFs (coolants).
   b. The secondary important source of cobalt leaching in tool manufacturing/reconditioning is its cleaning/coating/recoating, which includes cleaning/stripping of old coating. It explains why recoated carbide tools generally show lower (up to 50%) tool life than new tools.
2. In actual part cutting, water-soluble MWFs are used. It is believed that pH of the water and TEA added to many MWFs are primarily responsible for cobalt leaching. When it happens, the wear pattern appears as a result of abrasion wear. The matter gets worse when: (a) high-pressure MWF is used so that extensive carbide leaching is enhanced by the mechanical action of MWF, and (b) a combination of cutting speeds and high contact temperatures enhance the kinetics of chemical reactions promoting cobalt leaching. Moreover, if the water used for MWF preparation contains a high amount of chlorine, cobalt leaching is enhanced even further. And the last, but not the least, factor to be considered is the time the tool stays in the machine.

2.6.13 Facts and Physics of the Wear of Tool Materials

2.6.13.1 Need for a New Theory of Tool Wear

The scatter in tool life observed in the practice of testing of tool materials of the same composition and bulk properties from different tool material suppliers may reach 300%–400%, which is at least an order greater than that for common engineering materials. As discussed in Chapter 3, the existing tool material classifications (within the same type, e.g., cemented carbide) are too vague and thus cannot explain this scatter even in principle. For example, the tools made of the same carbide grade according to ISO specification, for example, M10 manufactured by different carbide suppliers, may have considerably different tool life. For many years, this fact was not noticed in industry as the machining, in general, and drilling, in particular, systems were of subpar quality so the tool performance was determined by other components of the drilling system rather than the properties of tool materials. As discussed at the beginning of this book, the time has changed and so has the quality of the drilling system. Many common excuses for poor tool performance were eliminated. As a result, the difference in the tool materials of the same grade but from different tool material manufacturers has become more and more noticeable particularly under challenging applications where tool performance is closely monitored.

If one would attempt to find the difference in the mechanical, physical, chemical, etc., properties of a particular tool material to understand the discussed scatter in tool performance, then he or she finds the following:

- Only very few, mostly irrelevant to the cutting performance, properties of a particular grade of the tool material is actually available in colorful catalogs of various tool companies (see Chapter 3 for details)
- Contacting the tool material supplier does not help to answer the questions and/or to resolve the issue with the performance of a particular tool material because (a) the contactor is not sure what to ask for and (b) the supplier may not have, or may not want to reveal, the requested property or properties. As such the magic words *proprietary information* work each and every time.

Chapter 3 of this book provides help for specialists to understand the essence and manufacturing particularities of the basic tool materials. It is discussed that a great number of manufacturing processes and alternatives exist in tool materials manufacturing that may result in the wide range of their performance results although the ingredients of a particular tool material can be absolutely the same. The difference in the cost of apparently the same tool material from different tool materials suppliers can be significant. Moreover, often the bulk mechanical and physical properties of this tool material are the same as from various suppliers, so the difference in performance cannot be explained based on these properties and characteristics. As a result, it is rather difficult to justify higher cost of a tool material of presumably better quality if the reported composition and bulk properties of this material are the same as those of a cheaper one. Unfortunately, many companies do not have time and sufficiently trained personnel to carry out the full tool-life investigation to distinguish the difference in the performance between these two materials, so it often settles with the cheaper tool material.

In the author's opinion, this problem should be solved particularly because the tool materials and cutting tool become more and more expensive and because unattended manufacturing is used wider and wider. To do this, a fresh look at the physics of the wear of tool materials should be taken with no blinds imposed by the known physics of wear in general machinery. This section describes the foundation of such an approach presenting/explaining some known facts and setting the stage for further research and development on the matter.

Modern tool materials are composite materials, so their structure includes at least two distinctive solid phases: Extremely hard nonmetallic particles (WC, diamond, boron nitride, Al_2O_3, etc.) are held together by the matrix material (e.g., cobalt). As these materials are sintered/shaped, great deformation and temperatures are applied (see Appendix B for details); interfacial defects as microcracks, residual stress, and even defects of the atomic structure occur. Microcracking can be attributed to a mismatch of the thermal expansion coefficients between the phases, as well as from thermal expansion anisotropy of the phases.

These defects evolve upon cooling, starting from a stress-free high-temperature state and specifically in areas with large grain-to-grain misorientation (Komarovsky and Astakhov 2002). These defects (examples are shown in Figures 2.89 and 2.90) lower the strength, wear resistance, and other useful mechanical and physical properties of tool materials. Moreover, as the population of these defects is of random nature, the great scatter in the performance data is observed in the practice. This scatter presents a serious problem both in structural and in tool materials.

Further analysis is concentrated on PCD tool material (see Chapter 3 for the description of its composition and sintering). Although intergranular wear, grain cleavage, peeling, and spalling of grains have been suggested as possible wear mechanisms of the PCD (Miklaszewski et al. 2000;

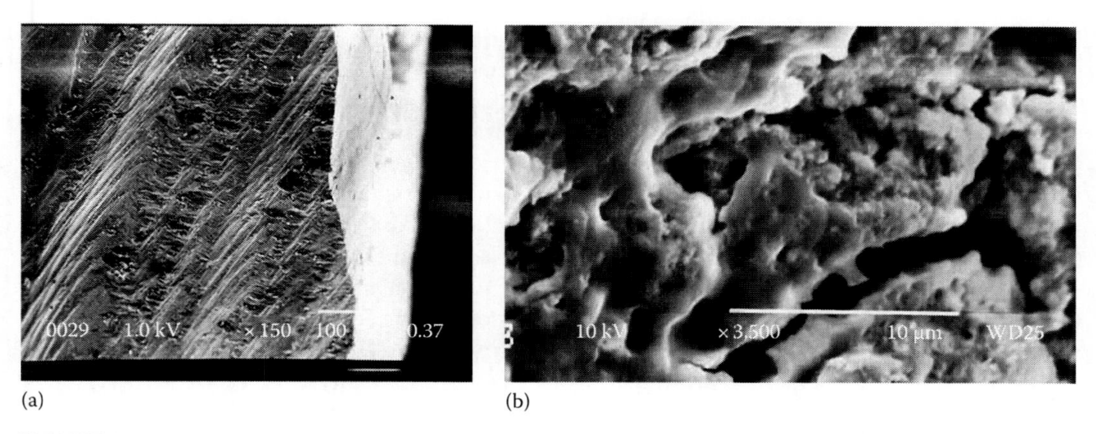

FIGURE 2.89 SEM images of surface and internal cracks in the PCD tool material: (a) surface cracks on a freshly ground surface (magnification ×150, scale bar is 100 μm) and (b) internal interfacial cracks after tool-life run (magnification ×3500, scale bar is 10 μm).

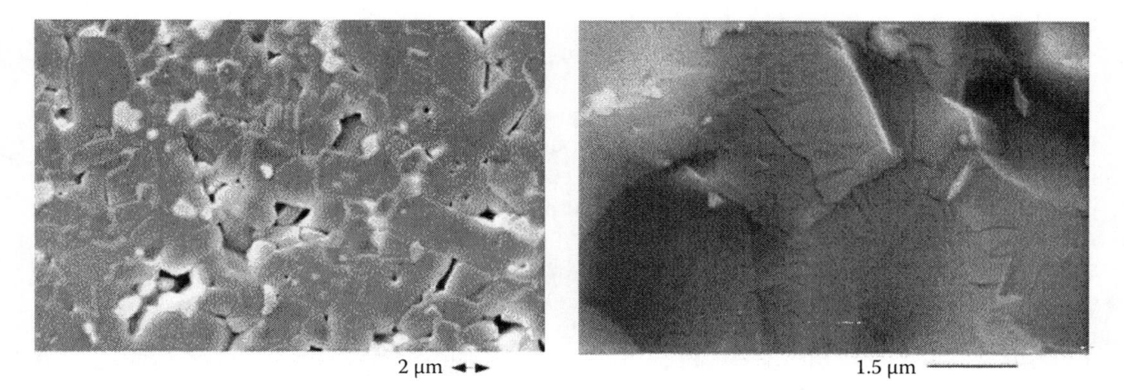

FIGURE 2.90 SEM images of surface and internal cracks in a ceramic tool material.

Bai et al. 2004; Philbin and Gordon 2005), only qualitative descriptions of suggestive nature are normally provided to explain these mechanisms. Moreover, these defects are considered to be formed due to high temperature and stresses in machining, while the microdefects formed on PCD sintering were not considered. No quantitative characterization of microcracks and their effect on the wear patterns were attempted.

The model of state of solids can be used as the basis to reveal the wear mechanism of PCDs and other SHM that deals with diamond/diamond interfacial defects (stresses and microcracks) is closely coupled in the processes of the generation and propagation of many small cracks occurred at the diamond/cobalt interfaces. However, the role of microcracking is complex and not well understood on a quantitative basis. Better understanding requires an ability to quantify the crack morphology and to reveal how that structure evolves due to stress and temperature fields found in metal cutting of high-silicon–aluminum alloys.

The complexity and strong coupling of microcracking to thermal and mechanical properties have been widely recognized (Evans 1978; Evans and Fu 1985; Laws and Lee 1989; Ortiz and Suresh 1993) but are still not well understood and thus correlated with tool wear. Improved understanding is hampered by (1) an inability to quantify microcracking, (2) limited observations of microcrack array responses to loads, and (3) limited evaluations of the mechanics models proposed to explain the role of microcracks and their role in tool wear mechanism (Kim et al. 1996; Kim and Wakayama 1997; Yousef et al. 2005). In the author's opinion, the first and foremost stage in further

PCD development should be concerned with the quantification of microcrack structure (Yurgartis et al. 1992). Special techniques to classify, count, and measure the length, spacing, and orientation of microcracks should be developed for PCDs and other tool materials using the developed methodology and test equipment available (e.g., the electron channeling contrast methodology using CamScan 44 FE SEM equipped with a Schottky thermal field emission gun [Crimp 2006]). Digital image processing should be applied to make the measurements, thereby providing time effective, reproducible, statistically significant estimates of crack morphology. A key feature of the image processing strategy is to abstract the crack structure to a set of medial axis lines.

Among the aforementioned general phenomenological mechanisms used to characterize tool materials (Astakhov and Davim 2008), namely, abrasion, diffusion, oxidation, fatigue, and adhesion, only abrasion–adhesive wear should be directly used in PCD tool wear analyses because no diffusion, oxidation, and fatigue are found in the known wear analysis of PCD tools. As such, for newly developed thermal stable PCD grades (see Chapter 3), abrasive wear is predominant. Abrasive wear is a phenomenon familiar to all; it is controlled by a subtle interplay of many effects. This makes the task of constructing a generalized model of the process from scratch extremely ambitious. In the author's opinion, this can now be attempted thanks to decades of painstaking research by many workers into different aspects of abrasion. The achieved results constitute the foundation on which a generalized model may be built.

The focus of a new model should be on the interatomic bonds of a solid and their consequences for wear behavior. The theory of the nonequilibrium thermodynamics of solids (Komarovsky and Astakhov 2002) can be used to understand the response of PCD to forces, heat, and other energy fields in metal cutting. The internal pressure or stress in PCD is defined as the vector representing the resistance to volume change, $\mathbf{P} = d\mathbf{F}/ds$, where $\mathbf{F}$ is the force of atomic interaction and s is the surface areas enclosing a volume. The thermodynamic equation of state (Komarovsky and Astakhov 2002) $PV = \mathbf{s}(N, V, T, \mathbf{P})T$, where $\mathbf{s}$ is the entropy vector and P is the magnitude of $\mathbf{P}$, should be used at the most general level of wear studies. The state of PCD is defined to be the shape of the rotors resulting from the diamond (not PCD) sintering process—that is why Chapter 3 and Appendix B explain this process in great detail. A rotor is a closed dynamic cell of solids. The equation of state relates the temperature of the solid structure to its ability to generate resistance forces. Compressions are those atoms located on the decreasing portion of a bond force minimum and provide resistance under heat adsorption, and dilatons are those located on the increasing portion to a force maximum and offer resistance in heat radiation. The compression–dilaton pattern of the bonds in a structure determines its response to loads, temperature, and other environmental conditions. Influence on the size effect, stresses, and aging in the wear-resistance response of PCD should be studied accounting for the number of interfacial defects (microcracks and voids) in PCD and the possibility of crack growth and crack coalescence that lead to micro- and then to macrochipping of PCD.

The description of dynamic loading and wear due to this loading can be developed from the equation of state to explain the physical nature of the time-dependent response. The increased resistance influences the initiation and propagation of microcracks that, according to this equation, is related to entropy. The thermodynamic potentials of PCD with varying numbers of interatomic bonds can then be derived to develop parameters of state using the periodic law of variations in state including mechanical hysteresis and its effects on the dilaton/compression transformations. The theory of strength in such a development is a generalization of the kinetic theory of strength that postulates that thermal fluctuations are key in breaking atomic bonds that can be directly correlated with wear. Wear, chipping, and fracture are attributed to what is called the Maxwell–Boltzmann factor (from the distribution of energy states), which describes the concentration density and energy of particles in a given region of the solid and which introduces stress concentrations. Breaking of bonds releases internal energy. Fracture and wear, however, are considered to be thermally dependent processes governed by the intensity of the potential field caused by stress and temperature anisotropy in the presence of dilaton traps.

2.6.13.2 Diffusion Self-Healing of Microcracks

Centuries-old buildings have been said to have survived these centuries because of the inherent self-healing capacity of the binders used for cementing building blocks together. Unfortunately, this self-healing capacity was never considered as the main course for such a survival. This is a well-known fact that was seen as the forgiveness of nature rather than self-healing of inherently smart building materials.

Today, the field of self-healing is considered a new area of materials research. It was in 2001 that White et al. (2001) published their results on self-healing in polymer-based systems by microencapsulated healing agents. This and related research in other fields of materials science were the result of an initiative by NASA launched among selected top institutes in the United States in 1996. Since then, the field is developing rapidly. Because healing presupposes the presence of a defect and a defect generally emerges at a very small scale, probably at the nanoscale, it is not surprising that self-healing is one of the promising application fields of nanotechnology.

Komarovsky and Astakhov (2002) presented the complete theory of diffusion healing of composite materials showing that practically all of these materials can be considered as self-healing materials. A number of advanced methods of self-healing are suggested in this publication. Today, our understanding of self-healing materials is as follows. Self-healing materials are man-made materials, which have the built-in capability to repair structural damage autonomously or with the minimal help of an external stimulus (energy fields [Komarovsky and Astakhov 2002]). The initial stage of failure in materials is often caused by the occurrence of small microcracks/microdefects (even at atomic–molecular scale) throughout the material. The proper characterization of such microdefects can help the ranging of the tool material's quality formed in their manufacturing so the cost of better tool materials can be easily quantitatively justified. The proper application of the external energy fields to enhance diffusion healing of such microdefects in composite tool materials to improve their quality should be the goal of the first stage of the development of better tool materials. The next stage can be the development of truly self-healing tool materials. In such materials, the occurrence of these microcracks is *recognized* in some way. Subsequently, mobile species, for example, atoms, have to be triggered to move to these places and perform their self-healing duty. These processes are ideally triggered by the occurrence of damage itself, in which case it is called an autonomous self-healing event. In practice, one could well imagine that self-healing that is triggered by an external stimulus greatly enhances the reliability and durability of tool materials.

A few examples of attempts to heal microcracks due to processing in solids are the following:

- One of the major problems in semiconductor manufacturing relates to the introduction of microcracks introduced during the form-shaping operations while producing the final thickness. The enhancement of the critical current density by using uniaxial pressing instead of cold rolling to produce the final thickness of the wire or tape, prior to annealing, is discussed in Haldar et al. (1992). US Patent No. 5,550,103, 1996, discusses the in-line pressing method that serves to improve the properties of the superconductor by healing the microcracks and lining up the flux pinning centers.
- Age-hardenable alloy parts having melting points in excess of 1000°C, in particular high-temperature superalloys, characterized by the presence of such structural defects as cast micropores, and/or grain boundary voids or internal microcracks resulting from high-temperature service, are improved in mechanical properties by subjecting these parts to hot isostatic pressure (HIP) (US Patent No. 4,302,256). Implementation of the HIP process in manufacturing of cutting tool carbides and ceramics (pioneered by ISCAR Co) allowed significant improvement in tool life and, what is more important, stabilization of the performance of such tool materials (scatter reduced by 75%). US Patent No. 6,544,458 discusses a similar process of preparation of a ceramic material with microcracks healing.

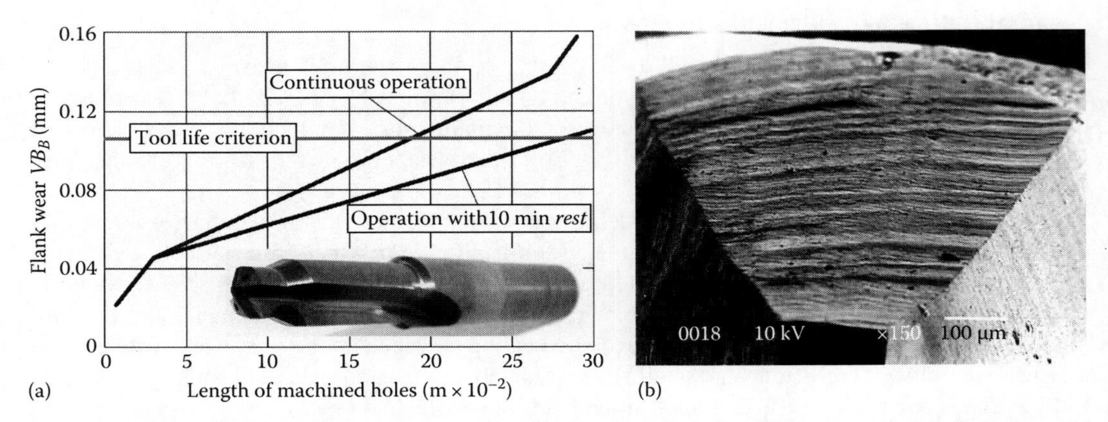

FIGURE 2.91 Comparison of tool life (normalized) for a PCD drill used in a continuous operation and a tool used in an operation with10 min *rest* (a) and SEM image (×150) of the PCD cutting insert at the end of tool life used in the operation with a 10 min *rest* (b).

- Brantley et al. (1990) observed that microcracks in quartz ~100 μm in length and <~10 μm in width heal in 4 h at 600°C and pressure of 200 MPa.
- Hung et al. (2008) found that grinding of lithium disilicate ceramics with diamond burs introduces flaws and microcracks. Subsequent heat treatments, veneer firing, or glazing heal these defects. As a result, manyfold increase in the performance was observed.
- Self-healing takes place in machining. For example, tool life of a PCD tool increases when the tool is allowed a 10 min *rest* between drilling successive holes compared to a PCD tool used in an uninterrupted (continuous) operation (5 s for loading/unloading) as it follows from Figure 2.91. Not one of the known wear mechanisms is able to explain this phenomenon.

Diffusion healing of tool materials has been performed since the 1950s although not one study or other literature source refer these known treatments as such. This is because such treatments are covered by the generic name *cryogenic treatment* (*CT*). Such a name is misleading as much more complicated thermomechanical processes are used including cooling and heating cycles. Section 3.3.7.3 discusses such a treatment for HSSs. Because this treatment is not recognized as diffusion healing, its regime is not selected properly based on the healing results, that is, the microdefect populations in the tool before and after treatment are not assessed thus the treatment results vary in a wide range as presented in Table 3.5. A very similar treatment is applied to cemented carbides as discussed in Section 3.4.8. The cycle diagram shown in Figure 3.72 shows that the tool material is cooled and then heated in a multistage operation. Because this treatment is not recognized as diffusion healing, the treatment results vary in a wide range as no clear technical, material properties-related objective of such a treatment is set.

The Diffusion© process was developed by Diffusion Ltd. (Windsor, Ontario, Canada) primarily for the use in the automotive industry for enhancing and stabilizing properties of the tool materials. This is because severe working conditions and the absence of the so-called safety factor used in the part design can easily contrast the difference in the performance of the original and treated by diffusion tools. The preliminary testing showed 50%–200% improvement in tool life of various tool materials. The Diffusion© process has also been used for improving the performance of structural components of the engines, transmissions, body parts, etc., including Canadian NASCAR engines. Preliminary test results showed that the tougher the working conditions of a structural component, the greater the performance results achieved due to the treatment.

REFERENCES

Astakhov, V.P. 1998/1999. *Metal Cutting Mechanics*. Boca Raton, FL: CRC Press.

Astakhov, V.P. 2004. The assessment of cutting tool wear. *International Journal of Machine Tools and Manufacture* 44:637–647.

Astakhov, V.P. 2006. *Tribology of Metal Cutting*. London, U.K.: Elsevier.

Astakhov, V.P. 2010. *Geometry of Single-Point Turning Tools and Drills: Fundamentals and Practical Applications*. London, U.K.: Springer.

Astakhov, V.P. 2012. Tribology of cutting tools. In *Tribology in Manufacturing Technology*, Davim, P.J. (ed.). New York: Springer, pp. 1–66.

Astakhov, V.P., Galitsky, V.V., and Osman, M.O.M. 1995. A novel approach to the design of self piloting drills. Part 1. Geometry of the cutting tip and grinding process. *ASME Journal of Engineering for Industry* 117:453–463.

Astakhov, V.P. and Shvets, S.V. 2010. Interaction between deformation and thermal waves in metal cutting. *International Journal of Advances in Machining and Forming Operations* 2:1–18.

Babu, G.J., Angelo, J., Canty, A.J., and Chaubey, Y.P. 2002. Application of Bernstein polynomials for smooth estimation of a distribution and density function. *Journal of Statistical Planning and Inference* 105(2):377–392.

Bai, Q.S., Yao, Y.X., Bex, P., and Zhang, D. 2004. Study on wear mechanisms and grain effects of PCD tool in machining laminated flooring. *International Journal of Refractory Metals and Hard Materials* 22(2–3):111–115.

Bartlett, M.S. 1937. Properties of sufficiency and statistical tests. *Proceedings of the Royal Society A* 160:268–282.

Brantley, S.L., Evans, B., Hickman, S.H., and Crerar, D.A. 1990. Healing of microcracks in quartz: Implications for fluid flow. *Geology* 18:136–139.

Byrne, G., Dornfeld, D., Inasaki, I., Ketteler, G., Konig, W., and Teti, R. 1995. Tool condition monitoring (TCM)—The status of research and industrial application. *Annals of the CIRP* 44(2):541–567.

Chen, Y., Zhang, L.C., and Arsecularatne, J.A. 2007. Polishing of polycrystalline diamond by the technique of dynamic friction. Part 2: Material removal mechanism. *International Journal of Machine Tool and Manufacturing* 47:1615–1624.

Chen, Y.-J. 2009. Relationship between ultrasonic characteristics and relative porosity in Al and Al–XSi alloys. *Materials Transactions* 50(9):2308–2313.

Childs, T.H.C., Maekawa, K., Obikawa, T., and Yamane, Y. 2000. *Metal Machining. Theory and Application*. London, U.K.: Arnold.

Coelho, R.T., Yamada, S., Aspinwall, D.K., and Wise, M.L.H. 1995. The application of polycrystalline diamond tool materials when drilling and reaming aluminium based alloys including MMC. *International Journal of Machine Tool and Manufacturing* 35:761–774.

Cook, M.W. and Bossom, P.K. 2000. Trends and recent developments in the material manufacture and cutting tool application of polycrystalline diamond and polycrystalline cubic boron nitride. *International Journal of Refractory Metals & Hard Materials* 18:147–152.

Crimp, M.A. 2006. Scanning electron microscope imaging of dislocations in bulk materials, using electron channeling contrast. *Microscopy Research and Technique* 69:374–381.

Derval, A.I., Ragring, N.A., and Samsonove, V.A. 2011. Failure of twist drills in automated manufacturing (in Russian). *Nauka i Obrazovanie (Science and Education*. pp. 4–15. http://engbul.bmstu.ru//search.html?word=%D0%BE%D1%82%D0%BA%D0%B0%D0%B7%D1%8B.

El-Hofy, H. 2006. *Fundamentals of Machining Processes: Conventional and Nonconventional Processes*. Boca Raton, FL: CRC Press.

Evans, A.G. 1978. Microfracture from thermal expansion anisotropy–I. Single phase systems. *Acta Metallurgica* 26:1845–1853.

Evans, A.G. and Fu, Y. 1985. Some effects of microcracks on the mechanical properties of brittle solids—II. Microcrack toughening. *Acta Metallurgica* 33(8):1525–1531.

Fedoseev, D.V., Vnukov, S.P., Bukhovets, V.L., and Anikin, B.A. 1986. Surface graphitization of diamond at high temperatures. *Surface and Coatings Technology* 28:207–214.

Firestein, S. 2012. *Ignorance: How It Drives Science*. New York: Oxford University Press.

Gordon, M.B. 1967. The applicability of binomial law to the process of friction in the cutting of metals. *Wear* 10:274–290.

Granovsky, G.E. and Granovsky, V.G. 1985. *Metal Cutting* (in Russian). Moscow, Russia: Vishaya Shkola.

Haldar, P., Hoehn, J.G., Rice, J.A., and Motowidlo, L.R. 1992. Enhancement in critical current density of Bi–Pb–Sr–Ca–Cu–O tapes by thermomechanical processing: Cold rolling versus uniaxial pressing. *Applied Physics Letters* 60:495–498.

Hung, C.Y., Lai, Y.L., Hsieh, Y.L., Chi, L.Y., and Lee, S.Y. 2008. Effects of simulated clinical grinding and subsequent heat treatment on microcrack healing of a lithium disilicate ceramic. *International Journal of Prosthodontics* 21(6):496–498.

Jawahir, I.S. and Van Luttervelt, C.A. 1993. Recent developments in chip control research and applications. *Annals of the CIRP* 42(2):659–693.

Jiang, Z. and Shang, C. 2010. Study on the mechanism of the cobalt leaching of cemented carbide in triethanolamine solution. *Advanced Materials Research* 97–101:1203–1206.

Katsev, P.G. 1974. *Statistical Methods for Cutting Tools* (in Russian). Moscow, Russia: Mashinostroenie.

Katsuki, A., Onikura, H., Sajima, T., and Akashi, T. 1992. Development of deep-hole boring tool guided by laser. *Annals of the CIRP* 41(1):83–87.

Kim, B.N., Naitoh, H., Wakayama, S., and Kawahara, M. 1996. Simulation of microfracture process and fracture strength in 2-dimensional polycrystalline materials. *JSME International Journal* 39(4):548–554.

Kim, B.N. and Wakayama, S. 1997. Simulation of microfracture process of brittle polycrystals: Microcracking and crack propagation. *Computational Materials Science* 8:327–334.

Komarovsky, A.A. and Astakhov, V.P. 2002. *Physics of Strength and Fracture Control: Fundamentals of the Adaptation of Engineering Materials and Structures*. Boca Raton, FL: CRC Press.

Laws, N. and Lee, J.C. 1989. Microcracking in polycrystalline ceramics: Elastic isotropy and thermal anisotropy. *Journal of the Mechanics and Physics of Solids* 37(5):603–618.

Loladze, T.N. 1958. *Strength and Wear of Cutting Tools* (in Russian). Moscow, Russia: Mashgiz.

Luttervelt, C.A., Childs, T.H.C., Jawahir, I.S., Klocke, F., and Venuvinod, P.K. 1998. Present situation and future trends in modelling of machining operations. Progress report of the CIRP Working Group 'Modelling of Machining Operations'. *Annals of the CIRP* 74(2):587–626.

Makarow, A.D. 1976. *Optimization of Cutting Processes* (in Russian). Moscow, Russia: Mashinostroenie.

Marinov, V. 1996. Experimental study on the abrasive wear in metal cutting. *Wear* 197:242–247.

Miess, D. and Rai, G. 1996. Fracture toughness and thermal resistance of polycrystalline diamond compacts. *Materials Science and Engineering A* 209:270–276.

Miklaszewski, S., Zurek, M., Beer, P., and Sokolowska, A. 2000. Micromechanism of polycrystalline cemented diamond tool wear during milling of wood-based materials. *Diamond and Related Materials* 9(3–6):1125–1128.

Olson, M., Stridh, B., and Söderberg, S. 1988. Sliding wear of hard materials—The importance of a fresh countermaterial surface. *Wear* 124:195–216.

Ortiz, M. and Suresh, S. 1993. Statistical properties of residual stresses and inter-granular fracture in ceramic materials. *Journal of Applied Mechanics* 60:77–84.

Philbin, P. and Gordon, S. 2005. Characterisation of the wear behaviour of polycrystalline diamond (PCD) tools when machining wood-based composites *Journal of Materials Processing Technology* 162–163:665–672.

Rehorn, A.G., Jin, J., and Orban, P.E. 2005. State-of-the-art methods and results in tool condition monitoring: A review. *International Journal of Advanced Manufacturing Technology* 26:693–710.

Reuleaux, F. 1900. Über den taylor whiteschen werkzeugstahl verein sur berforderung des gewerbefleissen in preussen. *Sitzungsberichete* 79(1):179–220.

Schey, J.A. 1983. *Tribology in Metalworking*. Metals Park, OH: American Society for Metals.

Shaw, M.C. 1984. *Metal Cutting Principles*. Oxford, U.K.: Oxford Science Publications.

Shimada, S., Tanaka, H., Higuchi, M., Yamaguchi, T., Honda, S., and Obata, K. 2004. Thermo-chemical wear mechanism of diamond tool in machining of ferrous metals. *CIRP Annals—Manufacturing Technology* 53:57–60.

Shuster, L.S.H. 1988. *Adhesion Processes at the Tool–Work Material Interface* (in Russian). Moscow, Russia: Mashinostroenie.

Snedecor, G.W. and Cochran, W. 1989. *Statistical Methods*, 8th edn. Ames, IA: State University Press.

Stamatis, D.H. 2003. *Failure Mode Effect Analysis: FMEA from Theory to Execution*. Milwaukee, WI: Quality Press (ASQ).

Stenphenson, D.A. and Agapiou, J.S. 1996. *Metal Cutting Theory and Practice*. New York: Marcel Dekker.

Taylor, F.W. 1907. On the art of cutting metals. *Transactions of ASME* 28:70–350.

Time, I. 1870. *Resistance of Metals and Wood to Cutting* (in Russian). St. Petersbourg, Russia: Dermacow Press House.

Time, I. 1877. *Memore sur le Rabotage de Métaux*. St. Petersbourg, Russia.

Trent, E.M. and Wright, P.K. 2000. *Metal Cutting*, 4th edn. Boston, MA: Butterworth-Heinemann.

Tresca, H. 1873. Mémores sur le Rabotage des Metaux. *Bulletin de la Société d'Encouragement pour l'Industrie Nationale* 15:585–685.

Vandenbulcke, L. and De Barros, M.I. 2001. Deposition, structure, mechanical properties and tribological behavior of polycrystalline to smooth fine-grained diamond coatings. *Surface and Coatings Technology* 146–147:417–424.

White, S.R., Sottos, N.R., Moore, J., Geubelle, P., Kessler, M., Brown, E., Suresh, S., and Viswanathan, S. 2001. Autonomic healing of polymer composites. *Nature* 409:794–797.

Xiaoming, J., Xiuling, Z., and Suoxia, H. 2011. Study of composite inhibitor on the cobalt leaching of the cemented carbide tool. *Advanced Science Letters* 4(4–5):1352–1256.

Yeo, S.H., Khoo, L.P., and Neo, S.S. 2000. Tool condition monitoring using reflectance of chip surface and neural network. *Journal of Intelligent Manufacturing* 11:507–514.

Yousef, S.G., Rodel, J., Fuller Jr., E.R., Zimmermannz, A., and El-Dasher, B.S. 2005. Microcrack evolution in alumina ceramics: Experiment and simulation. *American Ceramic Society* 88(10):2809–2816.

Yurgartis, S.W., MacGibbon, B.S., and Mulvaney, P. 1992. Quantification of microcracking in brittle-matrix composites. *Journal of Materials Science* 27:6679–6686.

Zolotorevsky, V.S., Belov, N.A., and Glazoff, M.V. 2007. *Casting Aluminum Alloys*. Oxford, U.K.: Elsevier.

Zorev, N.N. 1966. *Metal Cutting Mechanics*. Oxford, U.K.: Pergamon Press.

3 Tool Materials

When it comes to atoms, language can be used only as in poetry.
The poet, too, is not nearly so concerned with describing facts
as with creating images.

Niels Bohr (1885–1962), a Danish physicist, Nobel Prize laureate (1922)

The selection of a cutting tool material type and its particular grade is an important factor to consider when planning a successful drilling operation. A basic knowledge of each cutting tool material and its performance is therefore important so that the correct selection for each application can be made. Considerations include the type and properties of the material to be machined, the part/blank type and shape, machining conditions, and the required hole (bore) quality for the considered operation. Note that the cost per machined hole (the size of production lot, yearly program, existing machine/machining practice, etc.) should also be considered in the selection of the proper (technically and economically) tool material for a given drilling application.

The aim of this chapter is to provide important basic as well as advanced information on each type of cutting tool material used in drilling operations, its advantages/limitations, and the recommendations for its best use.

3.1 WORDS OF WISDOM

Starting considerations in the tool material selection:

1. The kind/grade of tool material accounts for one-third of the tool successful performance.
2. The tool design, its manufacturing quality, and proper implementation are two-thirds of the success.
3. Even the best salesman is not really an expert in your drilling operation and system. The selection of the proper tool material including coating is your job for standard and especially for special cutting tools.
4. Test various tool materials/coating until you find what works for your application then keep testing for what works better.
5. There is always a hardness–toughness trade-off. Settle for as much toughness of the tool material as you absolutely have to have.
6. Select the allowable amount of tool wear (the criterion of tool life, e.g., the maximum width of the wear land on the flank face as discussed in Chapter 2) as much as the quality of part (the component to be machined) and/or tool strength allows.
7. In automated production, fix-pocket brazed tools are more effective/reliable in performance and presetting.
8. High-hardness, high-performance tool materials cannot be brazed successfully without special processes, for example, vacuum (inert gas) brazing furnaces. The harder the tool material, the more sensitive it is to the brazing particularities (including brazing alloy and temperature control). Do not buy a tool if you see the spill of the brazing alloy (filler) and/or the brazing layer can be clearly seen. The brazed insert should rest on the rigid tool body and not on the soft brazing alloy. See Chapter 2 for detailed explanations.

9. Although listed in multiple catalogs/trade literature/books, often claimed by salesmen, maintained by widespread notions as equivalent, no two HSSs, cemented carbide grades, superhard tool materials, or anything else similar are exactly the same (besides the color and shape). The higher production/technological culture, the better benchmarking of tool performance is established, the greater the difference.

3.2 BASIC PROPERTIES

Many types of tool materials, ranging from high-carbon steels to ceramics and diamonds, are used as cutting tool materials in today's metalworking industry. In modern drilling operations, three types of tool material are primarily used: (1) HSSs, (2) cemented carbides, (3) diamond tool material including PCDs, and CVD diamond grades. In some special cases, for example, in hard boring, polycrystalline boron nitride (PCBN) is also used. It is important to be aware that differences exist among tool materials, what these differences are, and the correct application for each type of material (Davis 1995).

The three general properties of a tool material are as follows:

- *Hardness*: Defined as the resistance to indenter penetration. It directly correlates with the strength of the cutting tool material (Isakov 2000). The ability to maintain high hardness at elevated temperatures is called hot hardness. Figure 3.1 shows the hardness of typical tool materials as a function of temperature.
- *Toughness*: Defined as the ability of a material to absorb energy before fracture. The greater the fracture toughness of a tool material, the better it resists shock load, chipping and fracturing, vibration, misalignments, runouts, and other imperfections in the machining system. Figure 3.2 shows that, for tool materials, hardness and toughness change in opposite directions. A major trend in the development of tool materials is to increase their toughness while maintaining high hardness.
- *Wear resistance*: In general, wear resistance is defined as the attainment of acceptable tool life before tools need to be replaced. Although seemingly very simple, this characteristic is the least understood and, moreover, is a subject to misinterpretation/misunderstanding.

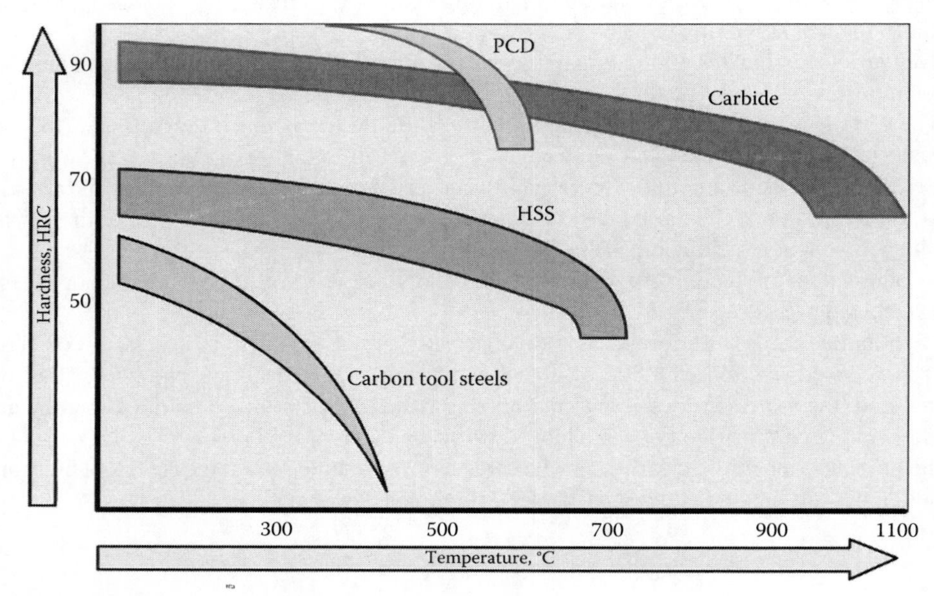

FIGURE 3.1 Hardness of tool materials versus temperature.

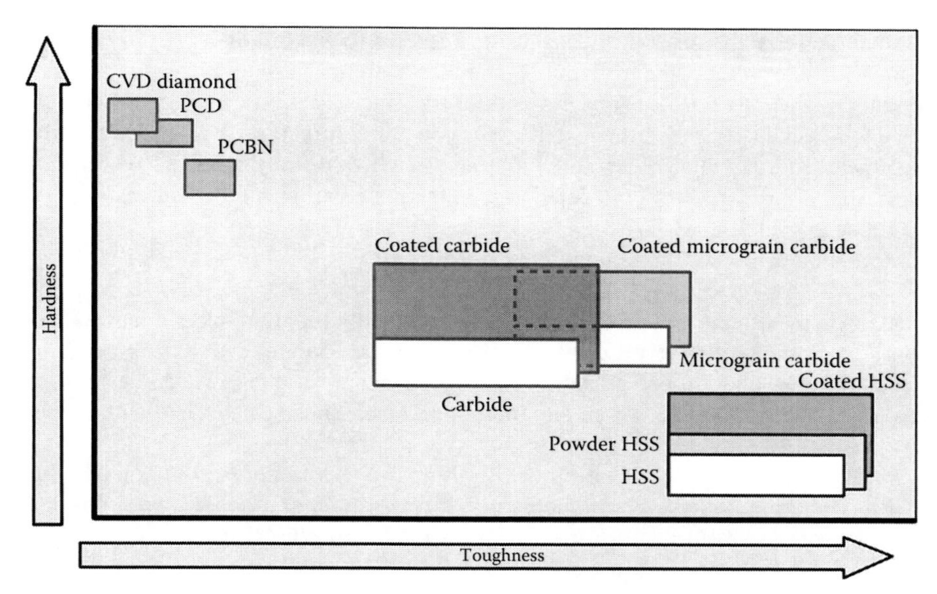

FIGURE 3.2 Hardness and toughness of tool materials used in drilling operations.

Among these three characteristics, hardness is the simplest as many people have natural perception of hardness even not knowing its proper definition. The other two basic properties require some explanation as they are explained considerably different in various sources of the professional literature. This should equip a specialist who tries to select a tool material with an ability to ask simple questions about the test methods used to obtain these characteristics and thus the relevance of the result in metal cutting.

3.2.1 Wear Resistance

Wear resistance is neither a defined characteristic nor a property of a tool material but is instead a system response arising from the conditions at the sliding interface. The nature of tool wear, unfortunately, is not yet sufficiently clear despite numerous theoretical and experimental studies. Cutting tool wear is a result of complicated physical, chemical, and thermomechanical phenomena. Because various simple mechanisms of wear (adhesion, abrasion, diffusion, oxidation, etc.) act simultaneously with a predominant influence of one or more of them in different situations, identification of the dominant mechanism is far from simple, and most interpretations are subject to controversy. Moreover, wear depends on the type of relative motion, normal stress, and sliding speed that brings a number of new variables in wear assessment. Because of these variations, different wear rates are commonly reported for the same combination of the tool and work material in the literature.

On the other hand, wear resistance should be somehow assessed to compare different tool materials and thus help a tool designer/end user in the selection of the proper tool material. The principal consideration in designing/using a wear test setup (rig, machine, methodology, and standard) is assuring that the test conditions, that is, the sliding configuration conform to the practical situation.

3.2.2 Toughness

The toughness of a hard tool material is an even less relevant characteristic bearing in mind the methods used for its determination. For carbides, the short rod fracture toughness measurement is common, as described in the ASTM standard ASTM B771-11 *Standard Test Method for Short Rod Fracture Toughness of Cemented Carbides*. The property K_{IcSR} determined by this test method is believed to characterize the resistance of cemented carbide to fracture in a neutral environment in the presence of a sharp crack under severe tensile constraint, such that the state of stress near the crack front approaches tri-tensile plane strain, and the crack-tip plastic region is small compared with the crack size and specimen dimensions in the constraint direction. A K_{IcSR} value is believed to represent a lower limiting value of fracture toughness. This value may be used to estimate the relation between the failure stress and the defect size when the conditions of high constraint described earlier would be expected.

This test method can serve the following purposes:

1. To establish the effects of fabrication variables on the fracture toughness of new or existing materials in quantitative terms significant to service performance
2. To establish the suitability of a material for a specific application for which the stress conditions are prescribed and for which maximum flaw sizes can be established with confidence

Figure 3.3 shows the basic configuration of the short rod specimen used in fracture toughness standard tests. The lengths a_1, a_2; diameter a_3; slot width a_4; and angle a_5 are defined by the standard. The short rod test to determine the short rod fracture toughness K_{IcSR} proceeds as follows. Increasing prying forces F are applied to the notched end. Due to sharp point at the tip of the chevron, a crack initiates (pops in) at a certain F. As this force increases, the crack advances stably until the critical length, a_c. The short rod fracture toughness K_{IcSR} is then calculated by the formula

$$K_{IcSR} = \frac{T_m^* F_{max}}{a_3^{3/2}} \tag{3.1}$$

where

F_{max} is the maximum prying force

T_m^* is the minimum value of the normalized stress intensity factor (a geometry specific parameter)

It has to be pointed out that the test method deals with linear elastic fracture mechanics used as the theoretical foundation for this test and nonlinear test conditions. Being understood by specialists

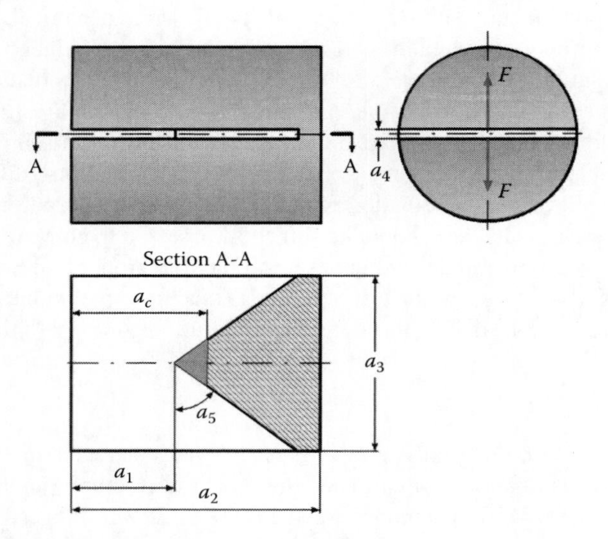

FIGURE 3.3 The short rod specimen basic configuration.

in linear elastic fracture mechanics, the significance of the short rod fracture toughness is not readily clear for specialists in metal cutting, although it routinely appears in practically all books and catalogs dealing with the tool materials in the list of important tool materials' properties. One may wonder if the same chevron configuration and/or the preexisting crack(s) exists in tool materials, if the loading condition in the test and in metal cutting are even similar, to what extent a great compressive force found in metal cutting may affect crack propagation, to what extent very high temperatures and energy density flux needed for cutting and thus transmitted through the tool material may affect crack initiation and propagation, etc.

3.3 HIGH SPEED STEELS

3.3.1 Why HSSs?

HSS today relates to a group of high-alloy, W-Mo-V-Co steels, designed to cut other materials efficiently at high speeds (compare to carbon tool steels). Once considered as an obsolete tool material, which application range should reduce to hand and woodcutting tools as new grades of other superior tool materials and new rigid machines are introduced, HSS survived as a high-performance tool material. Modern grades of HSSs combined with advanced coatings allow high cutting speeds that were considered for 10 years suitable only for carbide tools.

HSS has the following advantages:

1. Great bending strength that is significantly higher than that of any other cutting material as can be seen in Figure 3.4. It provides better resistance to cutting-edge chipping, increased feed per tooth, and greater depth of cut.
2. Compared to other tool materials, a sharp cutting edge can be achieved even in conventional grinding. As a result, the following service properties can be achieved:
   a. Less work hardening of the work material that is extremely important in machining of titanium alloys, austenitic stainless steels, and nickel alloys
   b. Better surface quality
   c. Closer tolerances
   d. Lower cutting forces that are particularly important in machining of thin-walled and nonrigid parts
   e. Lower cutting temperatures and thus smaller heat-affected layer in the machined surface

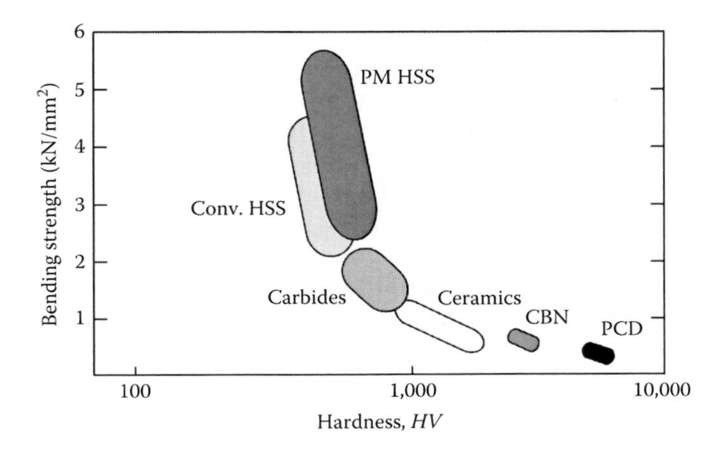

FIGURE 3.4 Bending strength of tool materials.

3. Much greater tolerance for nonrigid machining system including old machines and nonrigid part fixtures.
4. Much greater tolerance for specific work material/part conditions: nonhomogeneity of work material, cross-holes, welding joints, inclined bore entrance, etc.
5. Great resistance to thermal shocks and thus adaptation to practically all lubrication conditions.
6. Low cost per machined part.

3.3.2 Brief History

In 1868, the English metallurgist Robert Forester Mushet developed a tool steel adding tungsten (W) to a high-carbon steel. The composition of Mushet's steel, marketed as *R. Mushet's Special Tool Steel* through 1990, consisted of nominally 2% carbon (C), 2.5% manganese (Mn), 7% tungsten (W), often 0.5% of chromium (Cr), and 1.1% of silicon (Si) (Krauss et al. 1998). This steel had the remarkable capacity to harden during air cooling after forging or heating and is regarded as the first HSS.

A dramatic period in the development of HSS was precipitated by Frederick Winslow Taylor and Maunsel White, working with a team of assistants at the Bethlehem Steel Company at Bethlehem, PA, in the years around 1900. They discovered that very-high-temperature heating, much higher than typically applied at that time, and subsequent air cooling of chromium–tungsten steels, developed in that period by others in competition to the Mushet's tungsten steels, could produce and maintain exceptional hardness during machining at high speeds, even at red heat. The high-temperature heating, its remarkable effects, and the benefits of high-temperature tempering were truly innovative contributions by Taylor and White. Patents filed by Taylor and White were subjected to considerable legal argument because it was not clear whether they were claiming: a new steel or a new heat-treating process.

By 1910, production and use of HSS, containing 18% W, 4% Cr, and 1% vanadium (V) commonly referred to as 18-4-1, was firmly established. It is essentially the full analog of T1 HSS used today. However, shortage of tungsten during World War II spurred the development and commercial acceptance of the molybdenum (Mo) and molybdenum–tungsten grades. The HSSs are designated as group M or group T steels in the AISI classification system, depending on whether the major alloying approach is based on molybdenum or tungsten.

Although HSS and HSS tool manufacturers claim in their literature even at the ASTM level (Krauss et al. 1998) that the group T and M HSSs have effectively the same performance, it is not true in reality. The HSSs of group M, that is, the molybdenum HSSs, are much more widely used than the tungsten HSSs only because of cost advantages and tool manufacturability. The difference in cost is based in part on the fact that the atomic weight of molybdenum is about one-half that of tungsten; therefore, on a weight basis, less molybdenum than tungsten is required to provide *equivalent* performance. Molybdenum and tungsten atoms have similar atomic radii and form similar carbides. A replacement of 1.6–2.0 wt% W with only 1 wt% Mo has been found to produce markedly similar microstructure and properties in HSS tools.

In the author's experience, however, the HSSs of the group T outperform those in the group M in high-demanding applications provided that the tool design and manufacturing quality are proper and that the conditions of the machining system fully support these tools. The limiting factors of HSSs of the T group is high cost of tungsten and relatively poor grindability of HSSs of this group, which require much more intelligent selection of grinding wheels and grinding machines for their manufacturing. Naturally, the T group is overkill for hand drills, while in machining of difficult-to-machine materials where quality and cost-efficiency is of concern, this group shows its full advantages.

TABLE 3.1
Important Dates in the Development of HSSs

Date	Development
1868	Air-hardening tungsten alloy steel: Mushet
1898	HSS high-heat hardening: Taylor/White
1903	0.70 C–14 W–4 C prototype of modern HSS
1904	Alloying with 0.3% V
1906	Electric furnace melting introduced
1910	18 W–4 Cr–l V (18-4-1) steel (Tl) introduced
1939	High C high V super HSSs (M4 and T15)
1940	Start of substitution of molybdenum for tungsten
1953	Sulfurized free-machining HSS
1961	Rockwell C 70 HSS (M40 series)
1970	Introduction of powdered metal HSSs (PM HSS)
1973	Higher silicon and nickel contents of M7 to increase hardness
1980	Development to cobalt-free super HSSs
1980	Titanium nitride ceramic coating of tool steels
1982	Aluminum-modified high-speed tool steels

Table 3.1 lists important dates in the development of HSSs. Today, powder metallurgy (referred to as PM) HSSs are used exclusively for manufacturing of various cutting tools. The fine structures that result from rapid solidification in the PM process offer premium characteristics for both the manufacturers of cutting tools and their users. The more uniform distribution and the finer size of carbides in PM HSSs are especially evident in comparisons with larger diameter bars of conventionally produced HSSs, where carbide segregation is more of a problem. Thus, while the benefits pertain to cutters of all dimensions, they are more pronounced in larger tools.

There are four principal benefits of PM HSSs for tool users. The primary benefit is the availability of higher alloy grades, which cannot be manufactured by conventional steelmaking. These grades provide enhanced wear resistance and heat resistance for cutting tool applications. Second, the increased toughness of PM HSSs not only provides greater resistance to breakage (particularly valuable in intermittent cutting operations) but it also allows a tool to be hardened by 0.5–1.0 points higher on the Rockwell C scale without sacrificing toughness. Both longer tool life and higher cutting speeds can be realized. Third, PM HSSs have improved grindability with no reduction in wear resistance of the tool. This means reduced grinding wheel wear. Grinding can be done more quickly with less danger of damage to the cutter, and it leaves an edge that produces a smoother finish on the workpiece. Fourth, the greater consistency in heat treatment and uniformity of properties of PM HSSs increase the degree of predictability for scheduling tool changes. This factor is particularly advantageous in multi-spindle machines, where a single cutter failure affects several spindles and usually requires changing all cutters (including some that may have a lot of life left) for the sake of prudence.

3.3.3 Common Grades of HSS

Figure 3.5 shows the basic HSS grades and their chemical composition, properties, and availability. The large number of HSSs available today (partially listed in Figure 3.5) represents a number of choices based on performance and alloy modifications within the two major groups. The alloying

Designation AISI (ISO)	Normal chemical composition						Abrasion resistance	Red hardness	Toughness	Availability
	C	W	Mo	Cr	V	Co				
Group I—General purpose										
M1 (HS 1-8-1)	0.80	1.50	8.00	4.00	1.00	—				Very good
M2 (HS 6-5-2)	0.85	6.00	5.00	4.00	2.00	—				Excellent
M33	0.90	1.50	9.50	4.00	1.15	8.00				Fair
M34	0.90	2.00	8.00	4.00	2.00	8.00				Fair
M36	0.80	6.00	5.00	4.00	2.00	8.00				Good
T1 (HS 19-0-1)	0.75	18.00	—	4.00	1.00	—				Good
T2	0.80	18.00	—	4.00	2.00	—				Poor
T5	0.80	18.00	—	4.00	2.00	8.00				Good
Group II—Superior wear resistance										
M3 Type 1	1.05	6.00	5.00	4.00	2.40	—				Good
M3 Type 2	1.20	6.00	5.00	4.00	3.00	—				Very good
M7 (HS 2-8-2)	1.00	1.75	8.75	4.00	2.00	—				Good
Group III—High red hardness HSS										
M41	1.10	6.76	3.75	4.25	2.00	6.00				Fair
M42 (HS 2-9-1-8)	1.10	1.50	9.50	3.75	1.15	8.00				Very good
M43	1.25	1.75	8.75	3.75	2.00	8.25				Very good
Group IV—Super HSS										
M4	1.30	5.50	4.50	4.00	4.00	—				Poor
T15	1.50	12.00	—	4.00	5.00	5.00				Excellent

FIGURE 3.5 Basic HSS grades and their chemical composition, properties, and availability.

modifications are based not only on tungsten and molybdenum contents but also on carbon content, the total content of carbide-forming elements (including tungsten, molybdenum, chromium, and vanadium), and cobalt content.

Figure 3.5 lists not only the chemical composition of the commonly used HSS but also specifies the relative wear resistance, red hardness, toughness, and availability of 16 HSSs. These steels are classified in four groups:

3.3.3.1 Group I: General-Purpose HSSs

HSSs included in this group provide properties that permit efficient metal removal in 70% of drilling and milling applications. The *M* steels contain molybdenum as their chief alloying element. The *T* steels contain tungsten. M2 high-speed molybdenum steel is used on most applications. Its chemical composition provides balanced wear, red hardness, and strength qualities. It is readily available as a standard HSS and is stocked in blanks, forgings, and bar stock for special tools. It is economical as it possesses good grind and machinability. M33, M34, M36, and T5 have high cobalt content providing higher red hardness at the cost of toughness of the tool. They are not as readily available and are selected for special applications where the combination of these properties is advantageous.

3.3.3.2 Group II: Abrasion-Resistant HSSs

The Group II HSSs contain higher vanadium and carbon content. Higher content of vanadium carbide in M3 provides superior wear resistance than that available in the Group I general-purpose steels. M7 also has higher carbon content and is often selected for cutting tools where greater wear resistance is required.

3.3.3.3 Group III: High Red Hardness HSSs

HSSs of this group can be heat treated to 68–70 HRC but are generally heat treated to 66–68 HRC. The high-cobalt, high-carbon combination provides higher red hardness than that available in the other groups. The improved red hardness and wear resistance are at the expense of toughness. The M40 series steels are selected for machining of hardened materials as an alternate for T15 particularly for machining difficult-to-machine superalloys.

3.3.3.4 Group IV: Super HSSs

HSSs of this group are hardened between 66 and 68 HRC. They are high-tungsten, high-carbon, high-vanadium steels; T15 also contains cobalt. HSS M4 is slightly tougher than T15 but does not have the red hardness or wear resistance qualities of T15. T15 is used for machining of hard metals and alloys, particularly stainless steels and superalloys. Each cutting tool material possesses ingredients that impart cutting qualities that lend themselves to certain conditions.

Under common operating conditions, it is usually best to utilize standard HSS grades. If they do not perform, the cutter material selection chart should be used to determine the properties needed (abrasion resistance, red hardness, strength) for the applications.

3.3.4 Factors Affecting Intelligent Grade Selection of HSS

Although the information presented in the previous section is readily available and sufficient for many users, it is not nearly enough to make an intelligent selection of HSS. The common trial-and-error method suggested in the previous section is time-consuming and may not result in the selection of the optimal HSS grade.

There are two levels of selection of HSS grades: (1) selection by tool manufacturers for the tools they are producing and (2) selection by the end users for particular applications. These two selection procedures may not share the same objective because the objective of any tool manufacturer is to maximize the profit while the objectives of an end user can be to maximize tool performance, minimize cost per part, maximize productivity, improve quality, and many others. End users should clearly realize that selecting a particular grade of HSS at their level is like selection of a recipe of a cake from a cookbook (e.g., a charlotte cake). Depending on the quality of ingredients (the quality of the HSS which tool manufacturer buys from an HSS supplier) and the backing skill/experience (tool manufacturing including heat treatment of the purchased HSS), considerably different results are normally achieved. The problem with the assessment of the quality of the selected grade of HSS is not as simple as with the cake. The latter can be tested directly by the end user with the following results: *I like it—it is what I wanted or even better* or *I don't like it—it is not what I wanted.* Unfortunately, this is not nearly the case with an HSS cutting tool, for which performance is system-dependent, that is, as explained in Chapter 1, it depends on the tool design and geometry, machining regime, MWF parameters, machine conditions, and so on. If the System Coherency Law discussed in Chapter 1 is violated, that is, a selected HSS grade is better than the quality of other components of the drilling system, then no advantages of this grade can be gained in drilling operations. In this case, the selected expensive grade is a pure waste of money. If, on the other hand, a poor grade of HSS is selected for a high-quality machining system, no advantages of this quality can be gained either, the inadequate quality of the selected HSS will determine productivity and efficiency of this expensive machining system.

Not realizing that it is the case, HSS tool manufacturers have much greater challenge in the selection of HSSs for their tools besides very few who manufacture their own HSS grades. They have to take a much deeper dive into the properties of HSSs than that by cutting tool end users. This is because there are a number of important decisions to make.

First one is that tool manufacturers have to select the proper grade(s) of HSS for their product(s) as at the end of the day they are responsible for the performance of their HSS tools particularly today when a tool end user can select and test tools from various manufacturers globally. To help in such a selection, the properties of various grades of HSS should be quantified in the manner

TABLE 3.2
Performance Factors and Processing Information for Some T-Type HSS

Factor	T1	T2	T5	T6	T15
Major factors					
Wear resistance	7	8	7	8	9
Toughness	3	3	1	1	1
Hot hardness	8	8	9	9	9
Minor factors					
Usual working hardness, HRC	63–65	63–66	63–65	63–65	64–68
Depth of hardening	*D*	*D*	*D*	*D*	*D*
Finest grain size at full hardness, Shepherd standard	$9^1/_2$	$9^1/_2$	$9^1/_2$	$9^1/_2$	$9^1/_2$
Surface hardness as-quenched, HRC	64–66	65–67	64–66	64–66	65–68
Core hardness (25 mm diam. round), HRC	64–66	65–67	64–66	64–66	65–68
Manufacturing factors					
Availability	4	4	4	2	3
Cost	4	4	5	5	5
Machinability	5	5	2	1	1
Quenching medium	S, O, A	S, O, A	S, O, A	S, O, A	S, O, A
Dimensional change on hardening	M	M	M	M	M
Susceptibility to decarburization	M	M	H	H	H
Approximate hardness as-rolled or forged, HB	525	525	575	575	575
Annealed hardness, HB	217–255	217–255	217–255	217–255	217–255

Note: Quenching medium: A, air cool; O, oil quench; S, salt bath quench; M, medium; H, high.

as shown in Tables 3.2 and 3.3. Besides the major and minor factors affecting tool performance, manufacturing factors should also be considered. Availability, cost, and machinability directly affect the tool manufacturing process economy and thus the manufacturing cost of the tool. Annealed hardness defines machinability. The chief challenge, however, is to select an HSS grade(s) for standard tools. Being on-shelf catalog products, such tools should show satisfactory performance for a great application range including various work materials, drilling system properties, quality requirement, and so on. In most cases, HSS grades M1, M2, and M7 are selected for general-purpose HSS tools, M50 is used for hand tools, M35 (known in industry as *cobalt*) for higher performance, and M42 (known in industry as *super cobalt*) for most challenging applications.

The second tough challenge is where to buy the selected grade(s). There are a number of HSS suppliers offering the same HSS grades. However, for the same grade, there are a number of variables: quality of HSS, cost, lead time, and many other considerations. Quite often, unfortunately, the choice of the supplier of HSS is based on the cost and lead time rather than quality of HSS because the former two are readily understood by HSS tool manufacturers while quality of HSS is rather obscure for many HSS cutting tool design/manufacturing specialists.

The third challenge is about heat treatment of HSSs. Quenching medium, dimensional change on hardening, susceptibility to decarburization, and approximate hardness as-rolled or forged should meet the existing heat treatment facilities and practices. Common issues found in heat treatment of HSSs will be considered further in this chapter.

TABLE 3.3
Performance Factors and Processing Information for Some M-Type HSS

Factor	M1	M2	M3:1	M3:2	M4	M7	M10	M30	M33	M35	M36	M41	M42	M50
Major factors														
Wear resistance	7	7	8	8	9	8	7	7	8	7	7	8	8	6
Toughness	3	3	3	3	3	3	3	2	1	2	1	1	1	3
Hot hardness	8	8	8	8	8	8	8	8	9	8	9	9	9	6
Minor factors														
Usual working hardness, HRC	63–65	63–65	63–66	63–66	63–66	63–66	63–65	63–65	63–65	63–65	63–65	66–70	66–70	61–63
Depth of hardening	*D*	*D*	*D*	*D*	*D*	*D*	*D*	*D*	*D*	*D*	*D*	*D*	*D*	*D*
Finest grain size at full hardness, Shepherd standard	$9\frac{1}{2}$	$9\frac{1}{2}$	$9\frac{1}{2}$	$9\frac{1}{2}$	$9\frac{1}{2}$	$9\frac{1}{2}$	$9\frac{1}{2}$	$9\frac{1}{2}$	$9\frac{1}{2}$	$9\frac{1}{2}$	$9\frac{1}{2}$	$9\frac{1}{2}$	$9\frac{1}{2}$	$8\frac{1}{2}$
Surface hardness as-quenched, HRC	64–66	65–66	64–66	64–66	65–67	64–66	64–66	64–66	64–66	64–66	64–66	63–65	63–65	63–65
Core hardness (25 mm diam. round), HRC	64–66	65–66	64–66	64–66	65–67	64–66	64–66	64–66	64–66	64–66	64–66	63–65	63–65	63–65
Manufacturing factors														
Availability	4	4	3	3	3	4	4	2	1	1	2	2	3	2
Cost	3	4	4	4	4	3	3	4	5	4	5	5	5	4
Machinability	6	5	4	4	3	5	6	3	2	3	2	2	2	3
Quenching medium	S, O, A	S, O, A	S, O, A	S, O, A	S, O, A	S, O, A	S, O, A	S, O, A	S, O, A	S, O, A	S, O, A	S, O, A	S, O, A	S, O, A
Dimensional change on hardening	M	M	M	M	M	M	M	M	M	M	M	M	M	L
Susceptibility to decarburization	H	H	H	H	H	H	H	H	H	H	H	H	H	M
Approximate hardness as-rolled or forged, HB	525	525	550	550	575	525	525	575	575	575	575	575	575	575
Annealed hardness, HB	207–235	212–241	223–255	223–255	223–255	217–255	207–255	235–269	235–269	235–269	235–269	241–269	235–269	207–235

The forth issue is about coating. Coatings can be of general purpose or application specific and will be considered further in this chapter. The challenge is to select one of many possibilities: purchase own coating chamber and coat HSS tools in house, use local (if any) specialized coating facility, or use coating facility of another tool company. Although the first option looks attractive, it is not always feasible as it requires a considerable investment, justification of additional manufacturing personnel, preprocessing and post-processing equipment, for example, for tool deep cleaning before coating and coating thickness control. The second option is common as coatings are applied by professionals having vast experience. The drawbacks of this option, however, is the increased lead time as a certain number of HSS tool subjected to the same coating should be collected to justify a coating chamber-efficient run. It is not always an easy task for relatively small tool companies.

3.3.5 Formation of Properties

Formation of basic properties of HSSs takes place starting with casting and post-casting treatments. To understand the difference in quality (and thus in the cost) of HSS from various steel suppliers, one needs to know the basics of HSS casting and post-casting processes to be able to ask intelligent questions/proper information when selecting an HSS supplier and buying HSS. In doing that, one needs to realize clearly that the chemical composition of HSS (also known as the particular grade of HSS according to ISO and AISI standards) affects the quality of the tool made of this steel to a much smaller degree than the casting, post-casting, and heat-treating processes used. It is the same as with the charlotte cake mentioned previously where the final results (the appearance and taste) depend more on the experience of the baker (selecting and maintaining the proper *manufacturing parameters*) than on the quality of the ingredients or even a recipe.

Unfortunately, this important fact is commonly ignored in chapters on tool materials included in metal cutting and cutting tool books. Therefore, a need is felt to clarify the issue. This sections aims to present the information needed for a tool designer/user to understand HSSs in the depth required for proper tool design/application. Basic terminology and notions needed to understand the influence of the components in HSS casting are presented in Section B.1.1.

HSSs are complex, multicomponent alloy systems in which microstructures are very dependent on the kinetics of solidification and solid-state reactions and the nonuniform distributions of phases that accompany these reactions (Roberts et al. 1998). Following Hoyle (1988), the best way to rationalize the microstructural changes that occur during the processing of HSSs is to consider vertical sections through the multicomponent systems that make up these steels. The vertical sections plot regions of phase stability as a function of the temperature and carbon content, but do not give the compositions of the coexisting phases or the configuration of the phases. Nevertheless, the vertical sections indicate when various phases must form during processing, and this information considered together with other observations provides a good understanding of the evolution of microstructure in HSSs.

Figure 3.6 shows a vertical section through the Fe–W–Cr–C system at 18% W and 4% Cr as developed by several investigators and serves as the base to characterize phase relationships in Tl-type HSSs. Figure 3.7 shows a vertical section through the Fe–W–Mo–Cr–V–C system at 6% W, 5% Mo, 4% Cr, and 2% V and serves as the base for M2-type HSSs.

Superimposed on the vertical sections, the horizontal cross-hatched region in Figure 3.6 shows the normal range of carbon variations in tungsten HSSs, and the vertical line *AB* in Figure 3.7 shows a typical carbon concentration for M2 steel. In these compositions, starting from the liquid phase, dendritic or branched crystals of 8-ferrite nucleate and grow in the liquid upon cooling.

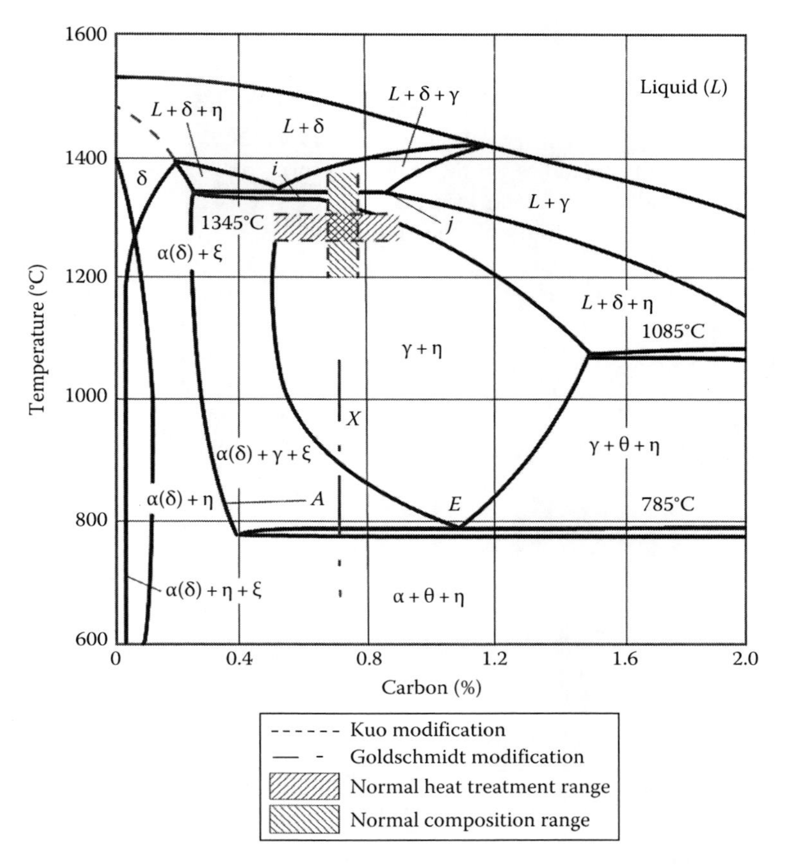

FIGURE 3.6 Phase diagram for T1-type HSS based on 18 W-4 Cr section through the quaternary Fe–W–Cr–C system. (From Roberts, G.A. et al., *Tool Steels*, 5th edn., ASM International, Metals Park, OH, 1998.)

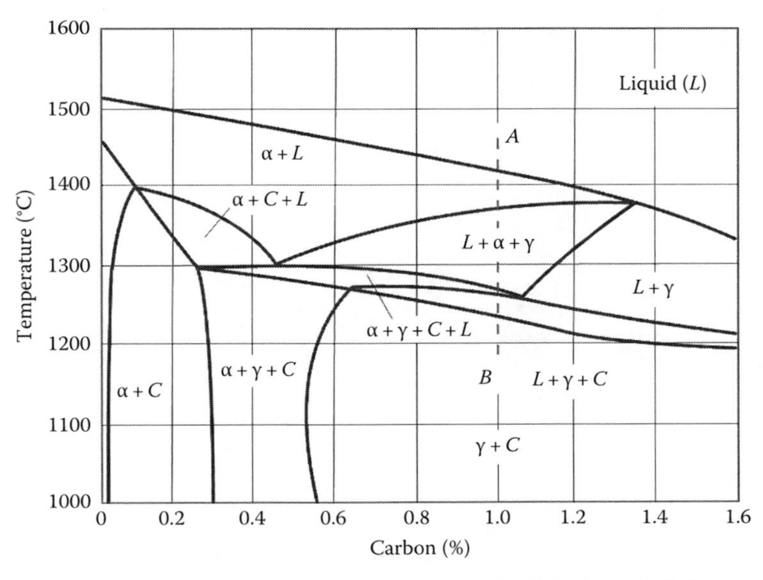

FIGURE 3.7 Phase diagram for M2-type HSS based on the 6 W-5 Mo-4 Cr-2 V section through the Fe–W–Mo–Cr–V–C system. (From Roberts, G.A. et al., *Tool Steels*, 5th edn., ASM International, Metals Park, OH, 1998.)

3.3.5.1 Casting of HSS

Traditionally, molten HSS is poured into the ingot mold as shown in Figure 3.8a. As such, progressive solidification starts from the walls and the base of the mold, moving inwards toward the thermal center or axis. The liquid phase of HSS decreases in volume during and after solidification, and there is insufficient solid metal to fill the shell first formed. The result is a body of steel with cavitation in the region of the last metal to solidify. The severity and distribution of this depend on several factors including the quality of steel, superheat at the time of pouring, method of pouring (whether direct or indirect), and the dimensions and taper of the mold. Such cavitation is also associated with impurities, which concentrate by segregation forming an undesirable distribution of undesirable elements in the final product. It is thus important to influence the amount and position of cavitation by reducing it to a minimum and locating it where it will be least harmful in the solidified ingot.

The grain macrostructure in ingots in most castings have three distinct regions or zones: the chill zone, columnar zone, and equiaxed zone. These zones are schematically shown in Figure 3.9.

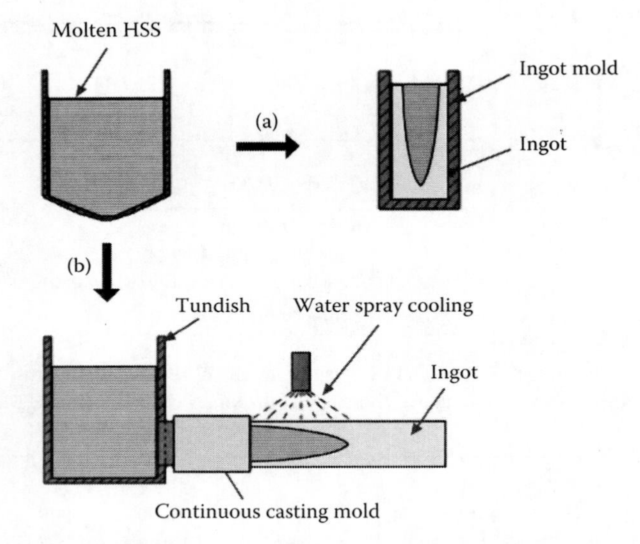

FIGURE 3.8 Schematic diagrams of the processes of HSS casting: (a) ingot mold casting and (b) continuous casting.

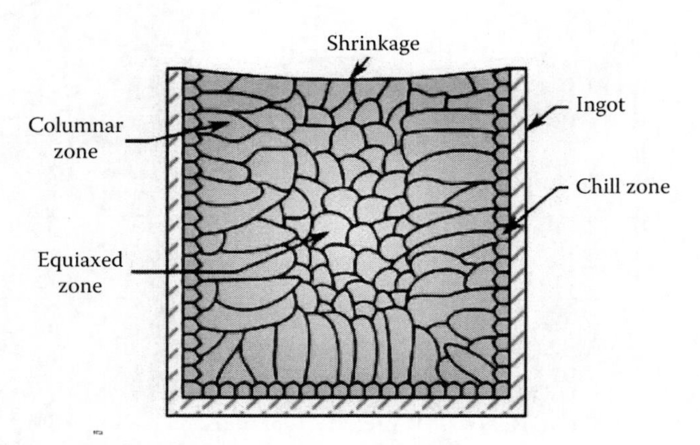

FIGURE 3.9 Distinctive zones in grain microstructure in ingots.

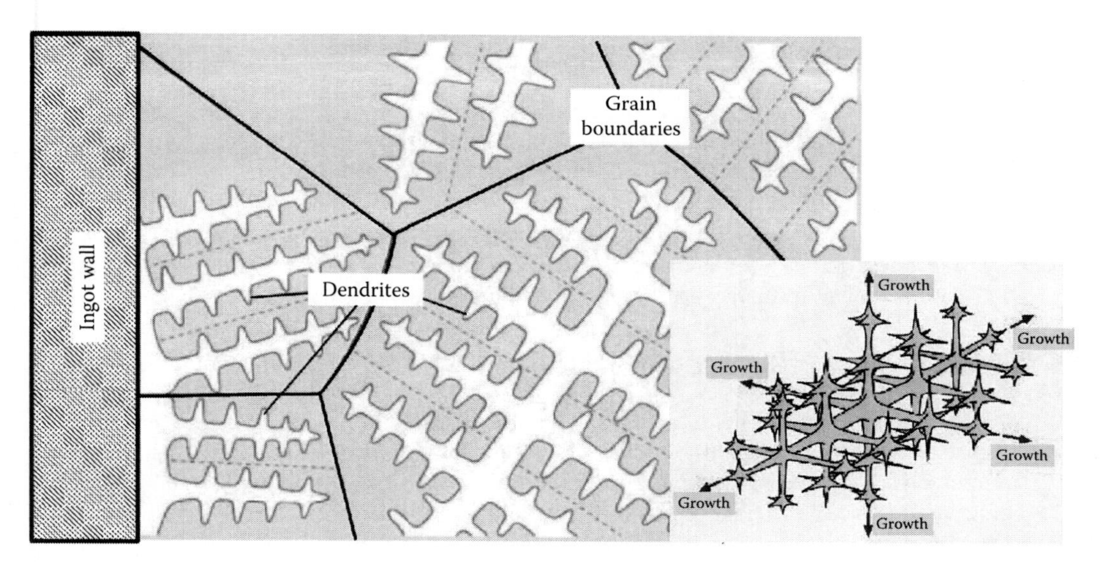

FIGURE 3.10 Appearance of dendrites.

The chill zone is named so because it occurs at the walls of the mold where the wall chills the material. Here is where the nucleation phase of the solidification process takes place. As more heat is removed, the grains grow toward the center of the casting. These are thin, long columns that are perpendicular to the casting surface, which are undesirable because they have anisotropic properties. Finally, in the center the equiaxed zone contains spherical, randomly oriented crystals. These are desirable because they have isotropic properties. The creation of this zone can be promoted by using a low-pouring temperature, alloy inclusions, or inoculants.

Crystallization occurs around small nuclei, which may be impurity particles. The first crystals have the crystal shape into which the metal would naturally solidify. However, as the crystal grows, it tends to develop spikes, and its shape changes into a tree-like form called a dendrite (Figure 3.10). As the dendrite grows, the spaces between its arms fill up. Spikes develop as the crystal grows in the directions in which the liquid is coolest. As these warm up in turn, so secondary, and then tertiary, spikes develop. Outward growth stops when growing arms meet others. Eventually, the entire liquid solidifies, and there is little trace of the original dendritic structure, only the grain into which the dendrites have grown.

The conventional method for manufacturing HSSs mainly comprises two steps, namely, the conventional casting and hot working. The as-cast microstructure of HSSs consists of the dendritic austenite decomposition products (generally martensite with some retained austenite) and the eutectic carbide networks, which are heterogeneously distributed in the interdendritic regions. However, in ingots, the only way to improve uniformity is to apply substantial amounts of hot work at high temperatures.

The dendritic crystal in HSS is cross-shaped and the liquid surrounding the dendrite has solidified as eutectic microstructures. The ferrite crystals are primarily iron-rich, and the other alloying elements are rejected into the surrounding liquid. With further cooling, the steel enters the three-phase ferrite–liquid–austenite phase field (Figures 3.6 and 3.7), and austenite forms around the ferrite dendrites. Finally, the remaining interdendritic liquid, highly enriched in the strong alloy carbide-forming elements, solidifies as eutectic structures consisting of colonies of closely spaced austenite and alloy carbide crystals. During these final stages of solidification, four phases coexist, namely, δ-ferrite, austenite, carbide, and liquid. Eventually, solidification is complete, and the ferrite is replaced by austenite, leaving a microstructure that consists of austenite and alloy carbides. Subsequent hot-work reductions and austenitizing heat treatments for hardening are performed in this phase field.

The interdendritic eutectic microstructures in as-cast HSS microstructures may take several morphologies and consist of various alloy carbides, depending on steel composition and solidification conditions. The primary alloy carbide in HSSs is M_6C or η-carbide, where M, the metal component, may consist of tungsten, molybdenum, and iron. Vanadium is typically present in MC carbides, and tungsten and molybdenum may be present in M_2C carbides.

Despite the diversity of compositions resulting from the alloy design development, the solidified microstructure of HSSs has maintained its characteristic basic feature, that is, dendrites surrounded by a more or less continuous interdendritic network of eutectic carbides. These microstructures have evolved over time, however, mainly as a result of changes in two main parameters:

1. *Alloy composition*: The most important changes caused by the progress in alloy design concern the type, morphology, and volume fraction of the eutectic carbides.
2. *Solidification process*: When considering the role of conventional solidification processing (cooling rate ranging from 10^{-3} to 10^2 K/s) on the solidified microstructure, the changes in microstructure caused by changing cooling rate are mainly with regard to dendrite size and eutectic colony distribution, both being more homogeneous as a result of the application solidification of electroslag remelting or refining treatments through minor additions. With rapid solidification processing (cooling rate ranging from 10^2 to 10^7 K/s), the resultant microstructure is mainly characterized by its very reduced scale, the dendrite arm spacing or cell size ranging from 0.01 to 1 μm, that is, two to four orders of magnitude smaller than that of normal solidified microstructures. Despite this remarkable difference in size, the more or less continuous interdendritic (or intercellular) network of eutectic carbides remains as a distinctive characteristic of the microstructure, even for cooling rates as high as 10^6 K/s, since short-range segregation still takes place. For extremely high cooling rates of more than 10^8 K/s, however, segregation is suppressed and a featureless microstructure is observed, probably as a result of diffusionless solidification.

3.3.5.2 Dealing with Cast Structure

The concentration of the alloying elements in the interdendritic eutectic colonies constitutes a very nonuniform structure unsuitable for hardening of HSSs. The dendritic and interdendritic regions increase in size as solidification rates decrease. To minimize the associated heterogeneity in structure and composition, the size of HSS ingots is limited to square sections of 300–400 mm on a side. Figure 3.11 is a panoramic sketch showing the evolution of the as-cast microstructure throughout the century of development of HSSs (Boccalini and Goldenstein 2001).

To assess the quality of HSSs properly, the solidified microstructures of ingot specimens should be dealt with professionally. One of the methods is to observe the microstructure under an optical

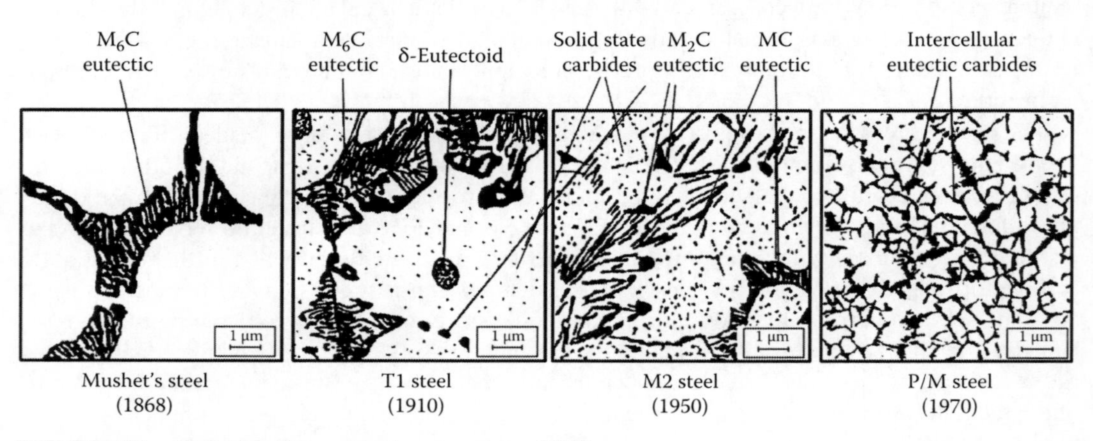

FIGURE 3.11 Evolution of as-cast microstructure of HSSs.

microscope, using Murakami etchant (3 g $K_3Fe(CN)_6$10 g NaOH + 100 mL H_2O) (Hetzner and Geertruyden 2008). To observe the 3D morphology of eutectic carbides, the specimens should be deeply etched in an etchant of 5 mL HF + 100 mL H_2O and then observed by a field emission SEM. The chemical compositions of carbides can be measured using an energy dispersive spectroscopy (EDS). The microstructures of carbides can be determined by electron backscatter diffraction (EBSD), transmission electron microscope (TEM), and x-ray diffraction (XRD). As even large tool manufacturing companies/HSS tool users may not have the listed equipment and correspondingly trained personnel, a professional service should be requested to control quality of HSSs.

The method used to the improve structure of HSSs can be broadly divided by those used in casting and those used after casting (solidification) is completed.

The traditional method involves improvement by hot reduction (severe plastic deformation at elevated temperatures) of ingots after casting. Figure 3.12 shows how hot reduction alters the solidification structure of HSS (Roberts et al. 1998). At moderate reductions, the as-cast network of interdendritic eutectic colonies are elongated in the direction of rolling in a still-interconnected pattern, sometimes referred to as a *hooky* pattern. After substantial amounts of hot work, the combined action of high temperature and deformation spheroidizes the lamellar or fibrous eutectic carbides and aligns the carbides in bands. Further increases in reduction bring the bands closer together. The net effect of the hot work is to approach a uniform dispersion of alloy carbides in the austenitic matrix. However, in order to completely eliminate banding of the dispersed carbides, hot reductions in excess of 97% have been shown to be necessary.

The type, amount, and dispersion of alloy carbides continue to be altered by thermal processing after solidification and hot work. Cooling from hot-work temperatures, annealing, and austenitizing for hardening all affect the primary carbide distributions in HSSs as microstructures attempt to approach equilibrium under changing thermal conditions. However, there is a limit on homogeneity of the primary carbide distributions achieved with this traditional method.

Other methods deal with improvement of the primary carbide distributions in HSS on its casting/solidification. Advanced primary processing approaches, such as continuous casting,

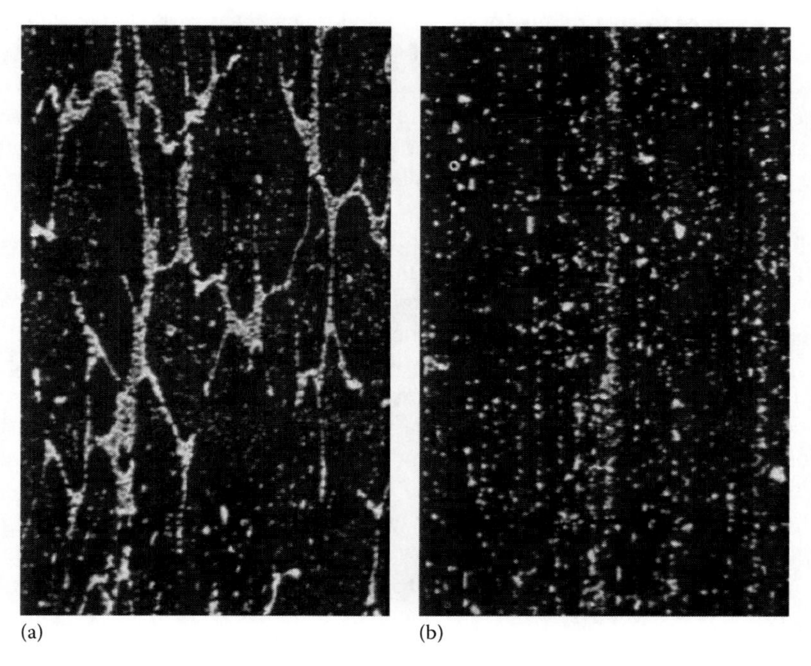

FIGURE 3.12 Microstructure of HSS showing alteration in carbide aggregates caused by forging, (a) longitudinal microstructure after moderate reduction, (b) longitudinal microstructure after more severe reduction.

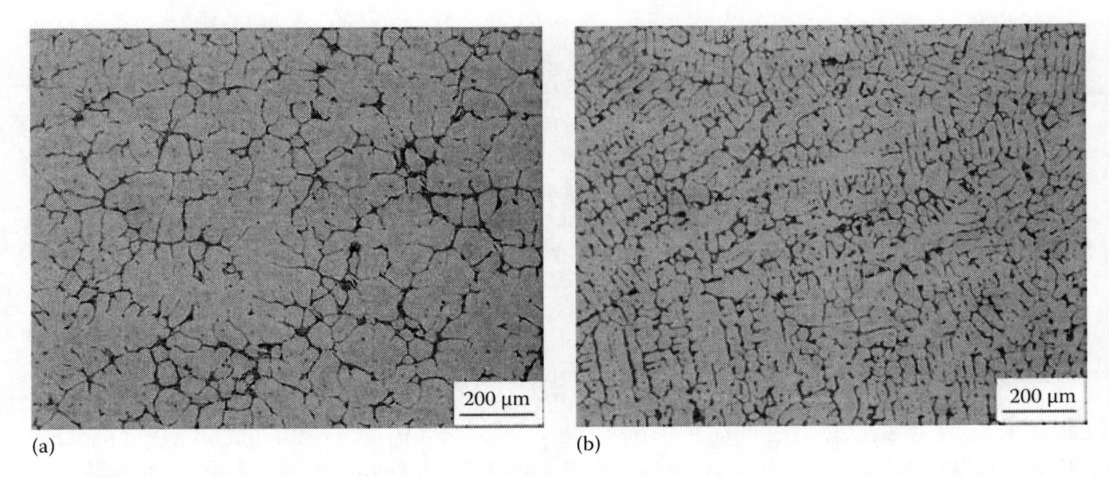

FIGURE 3.13 As-cast structure of M2 HSS ingots produced by (a) iron mold casting and (b) continuous casting.

powder production, and consolidation, spray deposition and others, which involve solidification at very high rates, have the capability to significantly reduce segregation and improve uniformity of HSSs.

The continuous casting process (Figure 3.8b) is an advanced fabrication technology widely applied in the production of steels and other metals. However, it is rarely used in manufacturing HSSs. This is because these steels contain a great amount of carbon and alloying elements that may cause some macrostructural defects in continuous casting ingots, such as severe composition segregation and shrinkage cavity. Besides, cracks are easy to form on the surface of ingots, due to the lower ductility of HSSs. Thus, little is known about the solidification microstructure of HSSs produced by continuous casting (Xuefeng et al. 2011).

Figure 3.13 shows the microstructure of M2 HSS ingot, which consists of dendrites of primarily austenite and networks of eutectic carbides distributed in the interdendritic regions (Xuefeng et al. 2011). It is noticed that the dendrites are refined and the networks of eutectic carbides are distributed more homogeneously in the continuous casting ingot (Figure 3.13b), compared to those in the iron mold casting ingot (Figure 3.13a). The average cooling rates in both ingots is about 0.6 and 11.5 K/s. However, the secondary water spray cooling results in a higher cooling rate in the ingot by continuous casting.

PM and spray forming (SF) have been reported as important alternative routes for HSS production. The ability to promote refined and more uniform microstructures is their main advantage, leading to improved properties and higher isotropy. While PM application is a completely established technology, the Osprey process may be considered as a not totally explored field.

Research and development on PM processing of HSSs started about 1965. The objective was to improve the functional properties and performance of HSSs in demanding applications and thus increase the competitiveness of HSSs versus cemented carbides primarily in machining operations. The R&D works were focused on the production of fully dense HSS billets and hot-worked semifinished products using the manufacturing route: *melting + inert gas (nitrogen) atomization + powder encapsulation + hot isostatic pressing (HIP)* (Grinder 2010).

The first stage in PM HSS manufacturing is powder manufacturing. The most suitable metal powders for HIP are produced by gas atomization. The principle of the process is shown in Figure 3.14. The gas atomization process starts with molten metal pouring from a tundish through a nozzle. The stream of molten metal is then hit by jets of neutral gas such as nitrogen or argon and atomized into very small droplets, which cool down and solidify when falling inside the atomization tower. Powders are then collected in a can.

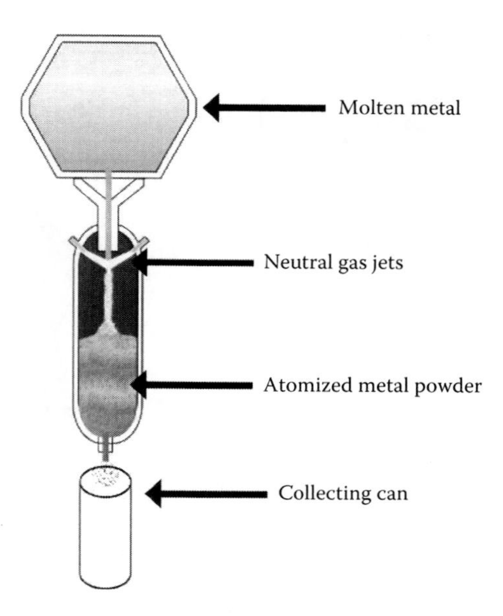

FIGURE 3.14 Principle of the gas atomization process.

This process has the following advantages:

- The perfectly spherical powder shape
- The high fill density due to the spherical shape and particle size distribution
- The excellent reproducibility of particle size distribution (Figure 3.15), ensuring consistent and predictable deformation behavior

HIP is a process to densify powders in a furnace at high pressure (100–200 MPa) and at temperatures from 900°C to 1250°C. The gas pressure acts uniformly in all directions to provide isostatic

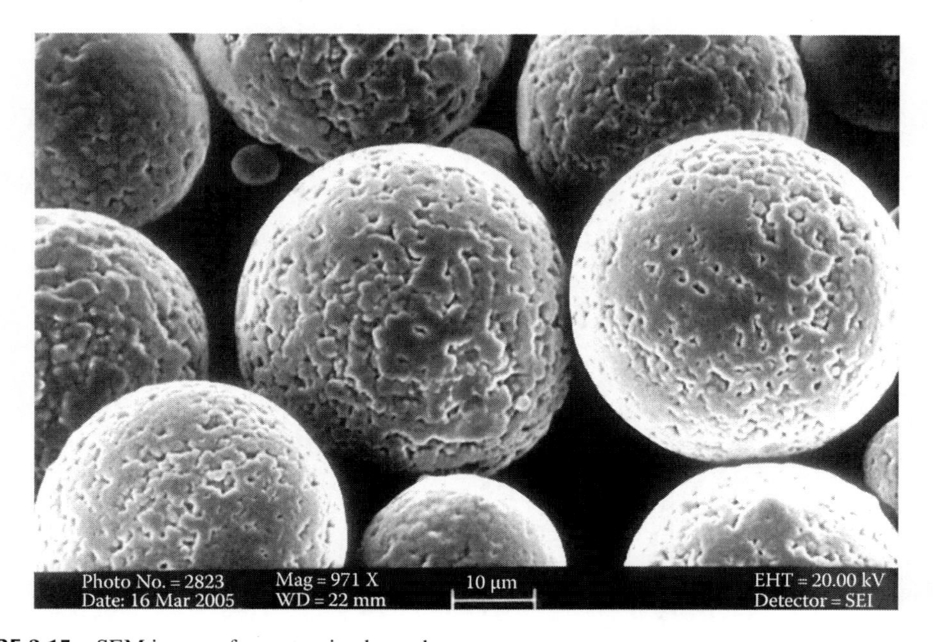

FIGURE 3.15 SEM image of gas atomized powders.

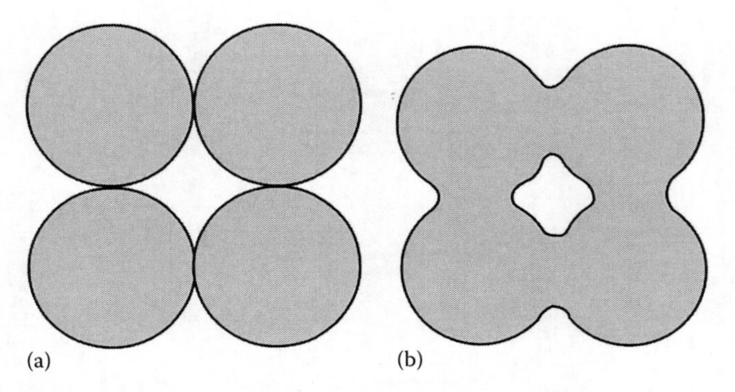

FIGURE 3.16 Representation of (a) initially unsintered powder particles and (b) sintered powder particles.

properties and 100% densification. It provides many benefits and has become a viable and high performance alternative to conventional processes such as casting and subsequent forging.

One should clearly realize that the temperature at which sintering is performed is lower than the melting point of the powdered material. Sintering consists of diffusion in solid state by which particles of compacted powder are bonded together (Figure 3.16). This is the basic working principle of powder technology.

A HIP unit consists mainly of a pressure vessel, a heating system, and an argon gas system as shown in Figure 3.17. Various HIP constructions are available, with or without a frame. For pressures above 100 MPa and HIP diameters above 900 mm, frame construction is chosen for safety reasons with or without top screw thread locking systems with different heating systems. Molybdenum furnaces are used for temperatures up to 1350°C and carbon graphite/tungsten furnaces up to 2200°C.

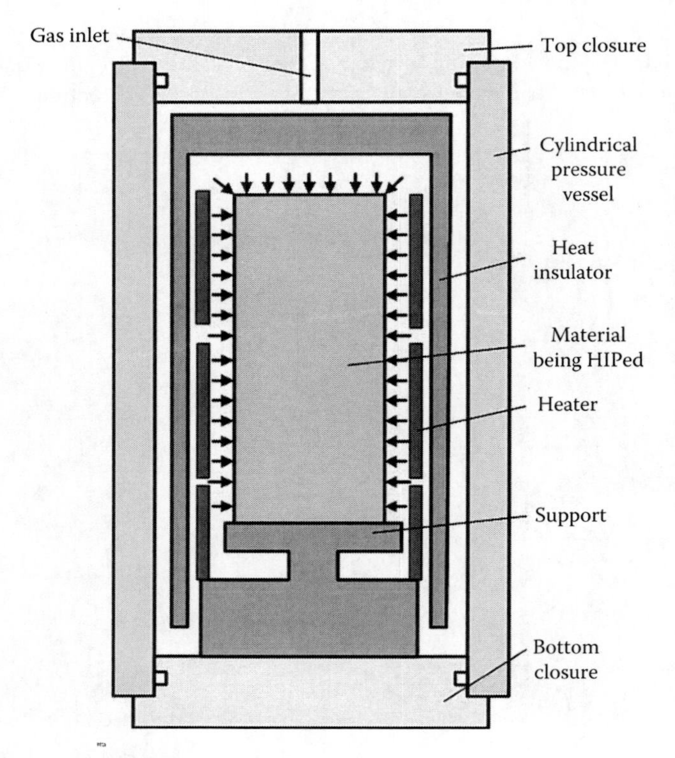

FIGURE 3.17 Schematic of HIP furnace.

Inside the pressure vessel, insulation (ceramic fibers and molybdenum sheets) is used to protect the steel pressure vessel against the temperature and to hold the high temperature inside the insulation. The bottom, cover, and pressure vessel are water cooled to protect the sealing ring and the vessel against temperature. In large HIP units, diameters can reach 2200 mm and height of more than 4000 mm, with a capacity of up to 30 tons.

The previously discussed problems with coarse and severe carbide segregation in conventional cast and forged HSSs are thus avoided in PM HSSs (Figure 3.18). It was found that PM HSSs exhibit several advantages as a result of the refined, homogeneous microstructure compared with their cast and wrought counterparts such as

- No macrosegregation
- Improved strength, toughness, and ductility at the same hardness
- Isotropic properties give less distortion at hardening of tools
- Improved grindability (extremely important in tool manufacturing)
- Improved hot workability permits higher carbide contents in PM HSSs and thus higher hardness at room and elevated temperature, which gives highly improved wear resistance
- In many cases large increase in tool life

Figure 3.19 shows the relationship between the bend fracture strength and the apparent size of the crack initiation site. In conventional HSSs, the initiation site is normally found at areas of

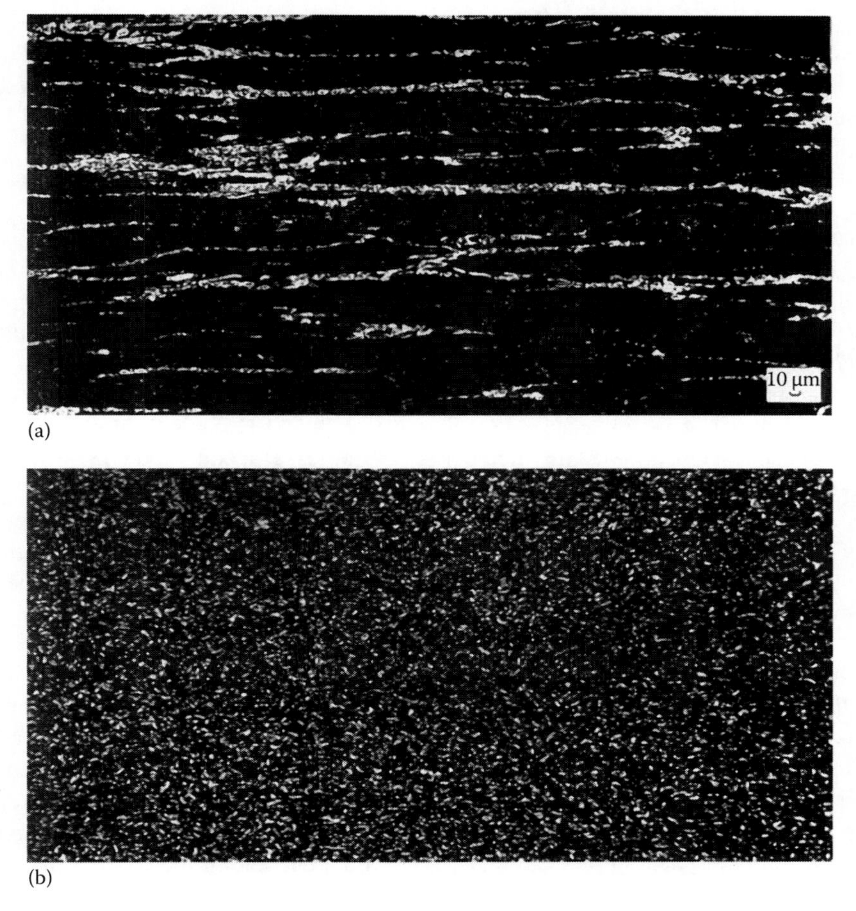

FIGURE 3.18 Microstructure of conventional (cast and forged) HSS (a) compared to that PM HSS (b) (black—iron matrix, white—carbides).

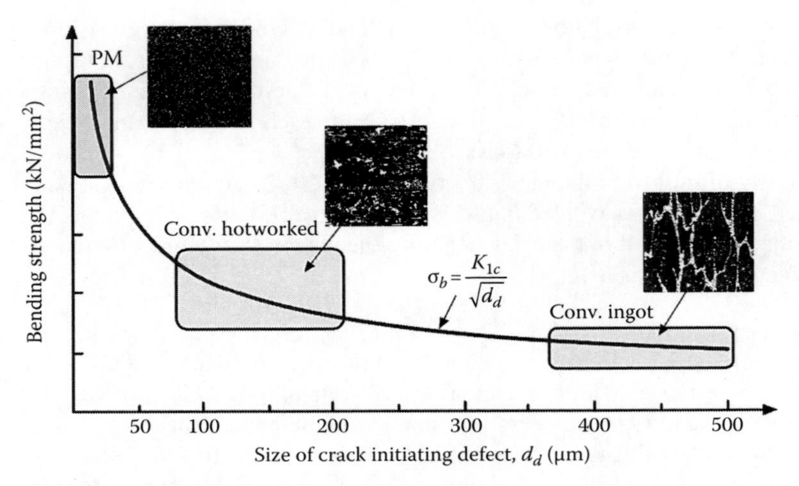

FIGURE 3.19 Strength versus defect size for HSS, HRC 65.

carbide segregation. Accordingly, the exogenous slag inclusions are seldom a problem for these steels. However, contrary situations exist for PM HSSs. As they have a uniform, fine carbide structure, it is often found that the cracks are initiated at exogenous inclusions. Major efforts have thus been made in order to decrease the amount and the maximum size of nonmetallic inclusions in PM HSSs.

PM HSSs may be produced by various processes, the most usual being the ASP, CPM, and APM processes. The differences in PM processes mainly due to HIP techniques (Mesquita and Barbosa 2003). The APM process has some advantages, since it is able to produce as-HIPed PM HSSs free from porosity and with no segregation of S, O, or C. This is possible due to a cold loaded mega-HIP system, where pressure and temperature are raised simultaneously. As APM steel is not subjected to any later forming process, it is considered the only truly isotropic PM HSS.

There is thus a direct relation between performance and functional properties of tools made of PM HSS and the microstructure, cleanliness, and inclusion contents of HSSs. This has motivated the major HSSs producers to develop and invest in highly modern and advanced metallurgical processes. PM production of HSSs is therefore nowadays high-tech comparable to, for example, the PM production of super alloys for aircraft engines.

PM has been applied in several situations due to success in refining HSS microstructure. In spite of its better performance in many cases, wide application of PM HSSs is limited by the relatively elevated cost of such products. The large number of operations, especially the HIP step, has considerably high cost, which impairs the total PM HSSs cost.

The advantage of SF process (also known as the Osprey process) in relation to PM is based on high cost of PM HSS (Figure 3.20). It occupies a position that is intermediate between technologies, which involve the refining of metals and the methods employed in PM. In PM or other rapid solidification techniques, the refined microstructure always relates to reduced sizes. On the other hand, SF is unique in combining a rapid solidification process (gas atomization) with a direct method for making bulk components. Since its development (Patent GB1972000026307 *Method for Making Shaped Articles from Sprayed Molten Metal*, priority—June 6, 1972), SF has been widely studied in several types of alloys. The effectiveness of the new process stems primarily from the fact that the high rates of crystallization of the melt in SF make it possible to form high-quality semifinished products that have a fine-grained structure but are free of macroscopic segregation-related defects. The SF process makes it possible to obtain commercial quantities of a wide range of high-quality products that are very efficient in service. Thus, in certain applications, the process can replace the production of high-quality alloy semifinished products by refining.

Figure 3.20, a stream of liquid metal is sprayed through a nozzle by an inert gas (nitrogen, argon). The droplets of metal formed as a result are accelerated to high velocities by rotating inlet

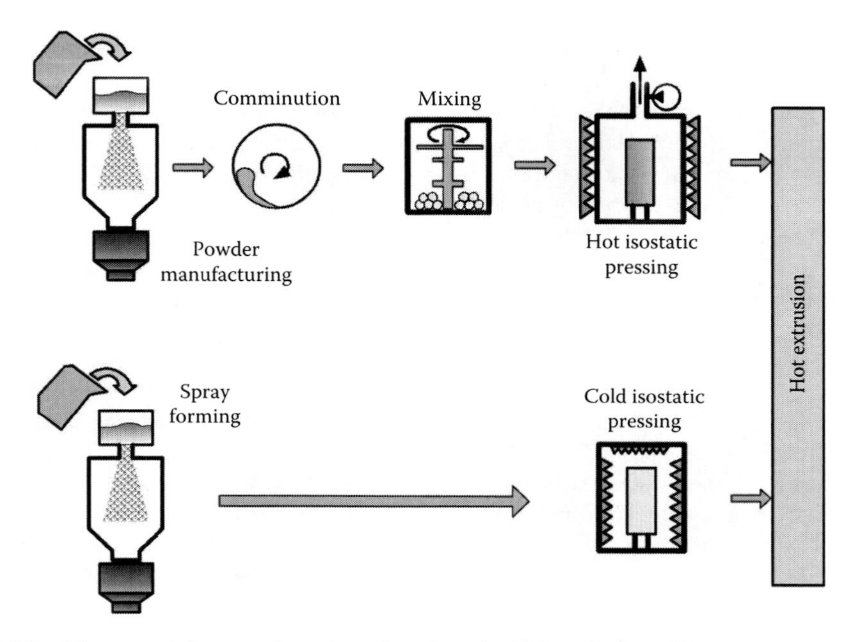

FIGURE 3.20 Diagram of the manufacturing of products by PM methods and SF.

disks and are then formed into a circular semifinished product. The droplets that do not end up in the semifinished product are removed from the chamber of the SF unit together with the outgoing gases and collected in the chamber of a gas-regeneration system. The regenerated product can be used again in conventional metal-refining. The end product of SF is a circular semifinished product that is subsequently deformed by forging or extrusion (Figure 3.21).

In SF, liquid metal from the crucible of the furnace used to make the steel enters an intermediate vessel with cylindrical nozzles in its bottom. If the height of the column of metal in the vessel is kept constant along with the diameter of the nozzles (5–7 mm), the unit can reach a productivity of 1000–2500 kg/h. It is also possible to use two systems of nozzles located on opposing sides of the vessel. The steel leaving the vessel is atomized in a jet of nitrogen or argon fed through high-speed

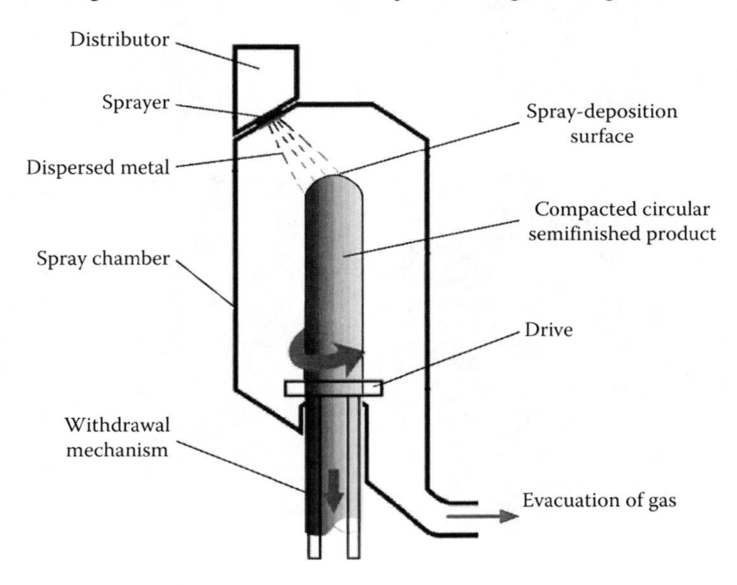

FIGURE 3.21 Schematic of the SF process.

nozzles and is directed onto the semifinished product at high velocities, thus forming the spray-compacted semifinished product (Lachenicht et al. 2011).

The semifinished product obtained by SF is distinguished by its low porosity and fine-grained structure. It is free of macroscopic and microscopic segregation-related defects and signs of a cast structure. The properties of the metal of the semifinished product are dramatically improved due to the high rate of crystallization. Here, the rate of crystallization in the process is higher than in continuous or ingot casting but lower than in PM methods. Among the advantages of SF of steel are the possibility of obtaining a homogeneous structure in the steel semifinished product and the fact that the manganese sulfides and carbides that are formed are smaller than in steels cast by conventional methods.

The as-cast microstructures of a conventional HSS, PM, and SF M3:2 can be compared in Figure 3.22 (Mesquita and Barbosa 2003). The finer microstructure of PM material is homogeneous (Figure 3.22b). This is a result of the higher cooling rates. The SF material microstructure shows considerable differences between the dense and porous regions (Figure 3.22c and d). In SF, it is well established that porous regions result from particles that solidified a large part of their volume during the flight, that is, in contact with the gas, and reach the substrate with just a small amount of liquid left. As a result, microstructures become finer, approaching that of PM material, which is fully solidified in gas atomization. Higher density is provided when particles reach the substrate with more liquid. However, the cooling rate is decreased and the carbide arrangements are coarser. Dense regions form a ringlike distribution near the surface.

In spite of the differences in some regions, the as-cast Osprey M3:2 presented a microstructure considerably finer than that of conventional material (Figure 3.22a). This is due to the better capacity of heat extraction during solidification in the SF process. The ability to produce such fine microstructures in a single process, without HIPing processes, is the main goal of the SF process. The SF material microstructure can be targeted between conventional and PM material ones. In the

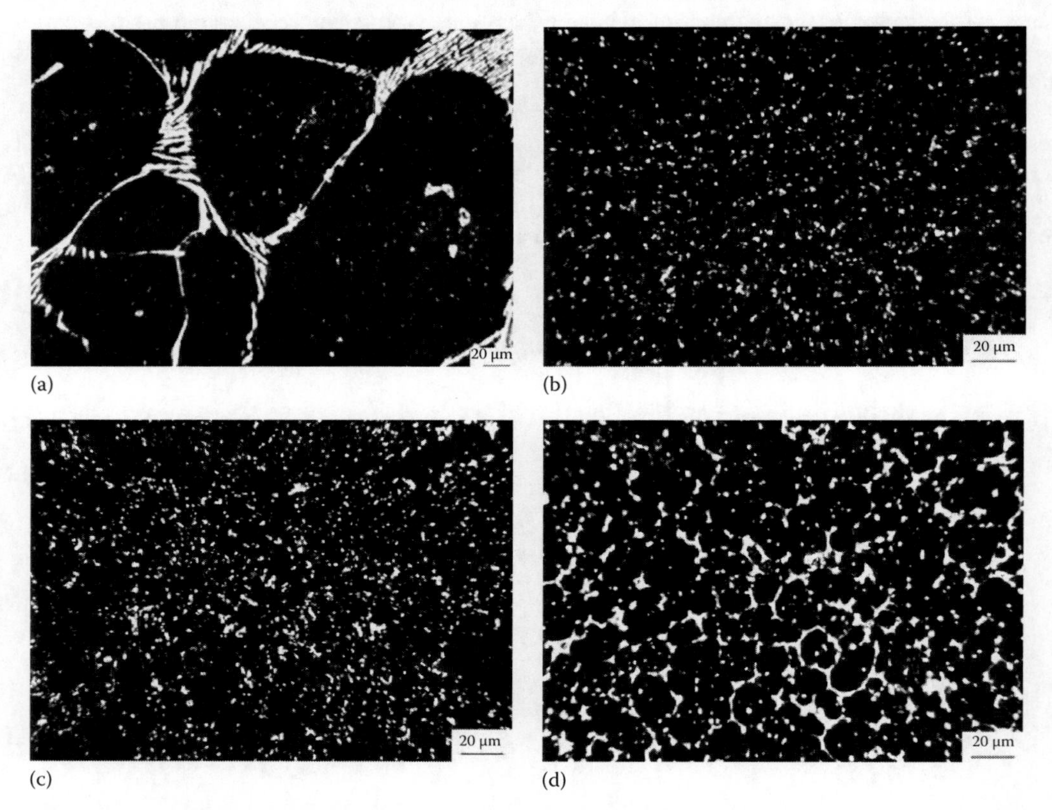

FIGURE 3.22 As-cast structures of (a) conventional HSS, (b) PM HSS, (c) porous region of SF HSS, and (d) dense region of SF HSS. HSS—M3:2.

opposite of conventional HSS, carbides in SF M3:2 do not form coarse morphologies. The majority of primary carbides is individualized and is finer. Besides, SF M3:2 HSS shows less variation in microstructure between core and surface. This fact is strictly related to the SF process. As considerable amounts of particles solidify during the *flight* period, final microstructure is less dependent of the section position than that of conventional HSS. Because of the relatively fine microstructure and small variation throughout each section, Osprey material is considered to be close to PM HSS. However, some differences still remain (Mesquita and Barbosa 2003).

3.3.6 Components in HSS

Understanding the influence of various alloying components in HSS and thus understanding why so many grades of HSSs are available to choose from is important. Basics of metallurgical notions/terminology needed to understand components of HSS are discussed in Section B.1.1.

HSSs are also steels but with large additions of refractory metals—tungsten, chromium, molybdenum, vanadium, and, in specialized cases, cobalt. The other element in steel, namely, *carbon*, forms *carbides* in carbon steels with just iron and in HSSs, with all the alloying additions except cobalt, which has other functions. So, in essence, HSS is a steel containing large amounts of refractory carbides, which provide hardness, high-temperature strength, wear resistance to tempering, with cobalt-enhancing high-temperature strength. Figure 3.23 lists the basic components of HSS and their influence on tool performance.

3.3.6.1 Tungsten and Molybdenum

Tungsten and molybdenum behave in the same manner by promoting red hardness and wear resistance. It is noted that the cutting performance of the steels increases in a linear manner as the percentage of either element increases. Molybdenum may be used to replace tungsten at the rate of around 1 wt% for every 1.6–2.0 wt% of tungsten.

3.3.6.2 Chromium

Addition of 4% chromium is made to all HSSs with the prime purpose of promoting depth hardening. The chromium in an annealed steel presents in the form of carbide, which dissolves into the austenite during the hardening cycle and hence becomes one of the primary sources of martensite in the

Cr	W	Mo	V	Co
Chromium	**Tungsten**	**Molybdenum**	**Vanadium**	**Cobalt**
Quantity - Approx. 4%	Quantity - Up to 20%	Quantity - Up to 10%	Quantity - 1%–50%, max. 10%	Quantity - 0%–10%
Role - Improves hardenability - Prevents scaling	Role - Cutting efficiency - Resistance to tempering	Role - Cutting efficiency - Resistance to tempering - Improves hardenability	Role - Forms very hard carbides for good abrasion wear resistance	Role - Improves heat resistance - Improves hot hardness - Slightly improves thermoconductivity
Origin - Various countries	Origin - Mainly China	Origin - By-product of copper and tungsten production	Origin - Present in many minerals	Origin - Mainly Canada, Morocco, and Zaire

FIGURE 3.23 Components of HSS and their influence.

quenched and tempered tool. Chromium in the absence of large quantities of retained austenite sharply retards the rate of softening in these steels, but itself does not produce a true secondary hardening peak.

3.3.6.3 Vanadium

This element is always present to a minimum of 1 wt% and generally up to 2 or 3 wt%. It can be higher in very highly alloyed grades. Vanadium forms extremely stable carbides such as VC or V_4C_3, which are virtually insoluble at normal hardening temperatures, and thus create a very effective means of limiting grain growth. The fact that the austenitizing temperatures encountered in the treatment of all HSSs approaches the solidus (see Figures B.1, 3.6, and 3.7) means that the retention of these complex vanadium carbides dispersed in the austenite matrix as it approaches the melting point restricts grain growth, which would otherwise be quite disastrous.

3.3.6.4 Cobalt

Not all HSSs contain cobalt, but possibly the newest and the best ones do. A few of the *super grades* have up to about 10 wt% maximum, although few special steels have higher additions. Why is cobalt in HSSs? A good question as it doesn't form carbides. The reasons that have brought cobalt to prominence in these latest alloys are the same as they always were. Cobalt dissolves in iron (ferrite and austenite) and strengthens it while at the same time imparting high-temperature strength (temperature on cutting surfaces can be 850°C). During solution heat treatment (to dissolve the carbides), cobalt helps to resist grain growth so that higher solution temperatures can be used, which ensures a higher percentage of carbides being dissolved. Steels are quenched after solution annealing, and the structure is then very hard martensite, including the retained high-temperature phase austenite plus carbides peppered throughout the structure.

Tempering precipitates the ultrafine carbides still in solution, and maximum hardness can be attained. Here, cobalt plays another important role, in that it delays their coalescence. This is important as it means that during cutting, the structure is stable up to higher temperatures. The addition of cobalt can raise the hardness by as much as 60 *HV*, depending on the specific grade of steel. Its prime purpose is to promote red hardness, which it does, however, at the expense of impact strength.

3.3.6.5 Carbon

As in all tool steels, carbon is essential to the hardenability of steel. Also, it is evident that, as the wearing properties and high hot hardness depend on the presence of massive amounts of complex alloy carbides, carbon is of prime importance. The carbon content of the eutectoid is reduced to a very marked extent by additions of W, Mo, and Cr, and therefore it is expected in most HSSs that the actual eutectoid composition is around 0.4 wt% of carbon. The usual carbon range for HSSs is 0.65–1.5 wt%, of which about 30 wt% is dissolved in the matrix. The hardness on the finished product increases rapidly up to about 1.0 wt% carbon. The higher carbon grades show a fairly marked fall off in ductility.

3.3.6.6 Sulfur

In its normal concentration of 0.03% or even less, sulfur has no effect on properties of HSSs. However, sulfur is added to certain HSSs to improve their machinability in the tool manufacturing process. Such HSSs are called resulfurized HSSs. The wide use of such HSSs is primary based on the finding by Berry (1970) who summarized comparative performance of M2 and resulfurized M2 HSSs as follows:

- Confirmation of beneficial effects of sulfur on machinability of tool steels of the M2 and also T1 grades.
- Examination of a variety of HSSs in turning operations concludes that sulfur additions are essential when good finish of machining surfaces is desired.
- Considerable success has been reported with several of the M series HSSs in terms of grindability. Sulfur levels have been reported as high as 1–3 wt%.
- Evidence has been presented that the effect of sulfur on grindability varies strongly with the amount of carbide present. Great improvement was reported with below about 8–11 wt% MC.

It should be pointed out, however, that no test condition are reported, and the whole work looks like promotional material by Climax Molybdenum Co., which had nothing to do with cutting tool design and application but rather manufactured *new* resulfurized HSSs that had great machinability.

Sulfur in any steel, other than when intentionally added as part of a two component system for free machining, is considered deleterious. Sulfur forms sulfide inclusions, which have a negative influence on the mechanical and functional properties of HSSs. It causes microchipping of the ground cutting edge that is particularly noticeable in high-performance HSS cutting tools designed, made, and applied properly.

The wrong perception of some tool manufacturers is that the situation is somewhat different in PM HSSs, as the sulfides precipitated are substantially smaller due to the high solidification rates during the atomization step. These small sulfides are evenly distributed throughout the steel and have less detrimental effect on, for example, hot workability. It has further been found that sulfur additions in the order of 0.10%–0.15% to PM HSS and also to tool steels, so-called resulfurized grades, improve the machinability resulting in higher productivity and surface finish in the manufacture of gear cutting tools and other products (Grinder 2010). In reality, the presence of sulfur in PM HSSs lowers tool life even greater.

Table 3.4 summarizes effects of other elements, generally found in trace quantities in HSSs. As can be seen, each element has its own purpose so that a meaningfully *designed* HSS has better tool manufacturing and service properties.

TABLE 3.4
Influence of Other Components in HSS

Element	Influence
Silicon Si	Up to about 1.0%, the influence of silicon HSS is slight. Increasing silicon content from 0.15% to 0.45% gives a slight increase in maximum attainable tempered hardness and has some influence on carbide morphology, although there seems to be a concurrent slight decrease in toughness. Some manufacturers produce at least one grade with silicon up to 0.65%, but this level requires a lower austenitizing temperature than that of the same grade of HSS with a lower silicon level to prevent overheating. In general, the silicon content is kept below 0.45% on most grades.
Manganese Mn	Present up to 0.35%. Above 0.4% induce cracking during heat treatment and grain growth at high temperatures.
Nickel Ni	Limited to 1%–2% max. Above 2% tendency to austenite stabilization. Ni tends to promote decarburization.
Aluminum Al	Used up to 1% to replace W and Mo but usuries can be caused by formation of alumina.
Tantalum Ta	Improves hot hardness, secondary hardening, and secondary hardening response, increases peak hardness stability to 650°C. Above 6% steel becomes unhardenable due to formation of stable carbides.
Niobium Nb	Improves secondary hardening response.
Titanium Ti	Added to refine grains and improve toughness. Also improves as-cast structure. Added by careful control ensuring that Ti + V does not exceed 6%–7%.
Nitrogen N	Added at about 0.035% to improve hot hardness and inhibition of grains growth. Nitrogen alloyed PM HSSs with 0.4%–0.6% N were developed in Japan by Kobe Steel Ltd in the early 1980. It has been reported that these nitrogen containing PM grades showed improved toughness and resistance to adhesive wear. New nitrogen alloyed PM tool steels with about 2% N are now introduced on market with potential large applications in tools for powder pressing, blanking, forming, and cold extrusion.
Boron B	Enhances the cutting performance but reduces forging qualities.
Selenium Se	Reportedly added to improve machinability.

3.3.7 Heat Treatment of HSS

Heat treatment of HSSs includes soft annealing, stress relieving, hardening, and subsequent tempering (Prabhudev 1988). Section B.1.1 includes explanations of basic terminology and definitions used in this section.

3.3.7.1 Soft Annealing and Stress Relieving

After forging and rolling or hot extrusion, HSS is subjected to *annealing treatment* in order to soften it and thus to increase its machinability. Annealing also aims to prepare the HSS structure for hardening. Annealing can be done either by an HSS manufacturer or by a tool manufacturer. Although annealing temperature is HSS grade specific, it is normally in the range of 880°C–890°C for the so-called soft annealing.

During annealing, HSS is slowly heated to the annealing temperature. As annealing is a diffusion process that aims to rearrange the HSS structure, the annealing time ranges from 4 to 6 h depending on the size of the HSS blank and its properties. Note that despite common practice, annealing time should be strictly controlled because if this time is too short, the hardness of the HSS blank would be too high so that its machinability is too low. On the other hand, if the annealing time is too long, the coagulation of carbides takes place with harmful influence on HSS hardenability. The cooling rate from the annealing temperature should be rather low, namely, 15°C–20°C/h to reach about 650°C in the annealing furnace. After that, HSS blank can be cooled to room temperature in the air.

As HSSs tend to decarburize during heat treatment, special attention is normally given to their heat treatment starting with soft annealing because the latter requires keeping HSS at high temperature for extended period of time. Normally, there are a number of possibilities.

The simplest is normally to heat treat HSS in house or in very local heat treatment facilities using usual furnaces. As such, decarburization can be avoided by box annealing, for example, by packing an HSS blank in burnt coke grids. The best results, however, are achieved when using furnaces with controlled atmosphere or vacuum. In a vacuum furnace, the product is surrounded by a vacuum. The absence of air or other gases prevents heat transfer with the product through convection and thus decarburization. Some of the benefits of a vacuum furnace are

- Uniform temperatures in a wide temperature range.
- Temperature can be tightly controlled within a small area.
- Low contamination of the product by carbon, oxygen, and other gases.
- Quick cooling (quenching) of product, which is important on HSS hardening.
- The process can be computer controlled to ensure metallurgical repeatability.

An inert gas, such as argon, is typically used to quickly cool the treated HSS back to nonmetallurgical levels (below 200°C) after the desired process in the furnace. This inert gas can be pressurized to two times atmosphere or more and then circulated through the hot-zone area to pick up heat before passing through a heat exchanger to remove heat. This process is repeated until the desired temperature is reached.

The hardness of properly annealed HSS for better machinability in tool making is normally in the range 240–300 HB depending upon a particular HSS grade and the parameters of the chosen annealing process.

If complicated high-precision multistaged HSS tools where a lot of machining (and thus machining residual stresses) is involved, it is desirable to apply *stress relieving heat treatment* before hardening. This is because if machining residual stresses are not properly removed before hardening, they may add to those induced on this hardening with a result of tool excessive distortion or even cracking. Note that stress relieving is not intended to make any modification to HSS metallurgical structure.

This treatment is done after rough machining. It includes heating to 550°C–650°C (1020°F–1200°F) either in a forced air circulation furnace or in a usual furnace. The material should be heated until its uniform temperature is achieved and held at this temperature for 1 h per 200 mm of tool thickness (diameter). After this, the tool is cooled down with the furnace. The correct work sequence is rough machining, stress relieving, and semifinish machining.

The excuse to avoid stress relieving because it takes too much time and associated with some additional cost is hardly valid. Correcting the dimensions and shape of a partially finished tool during semifinish machining of an annealed material is (with few rare exceptions) cheaper than making dimensional adjustments during finish machining of a hardened tool.

3.3.7.2 Hardening and Tempering

As the name implies, the major objective of hardening is to increase hardness of HSSs. To do that, one needs to understand what gives the desired harness to HSSs.

Typically, HSSs contain 0.25–0.30 volume fraction of carbides. The following carbides are found in HSSs: (1) MC, which is vanadium rich; (2) M_2C, which is a tungsten- or molybdenum-rich carbide; (3) M_6C such as Fe_3W_3C and Fe_4W_2C, where chromium, vanadium, and cobalt dissolve in Fe_3W_3C; and (4) $M_{23}C_6$, which is chromium rich and dissolves iron, vanadium, molybdenum, and tungsten. The hardness of these carbides is much greater than cementite or martensite as shown in Figure 3.24 (Campbel 2008).

The opinion of many specialists involved in HSS heat treatment is that an HSS tool is only as good as the heat treatment that it receives, and there is no such thing as an acceptable shortcut in the heat treating of HSSs. Heat treating is an inherently dangerous process and should be performed by trained professionals using proper procedures and adequate equipment whenever possible.

There are four basic steps that should be followed in any heat-treating process. They include in order preheating, austenitizing, quenching, and tempering. They are described as follows:

1. Preheating provides two important benefits. Since most HSSs are sensitive to thermal shock, a sudden increase from room temperature to the austenitizing temperature may cause tools to crack. Secondly, there is a phase transformation that the steel undergoes as it is heated to the austenitizing temperature that produces a change in density or volume. If this volume change occurs in a nonuniform manner, it can cause distortion of the tool. This problem is especially evident where differences in geometry or section size can cause some parts of the tool to transform before other parts have reached the target temperature.

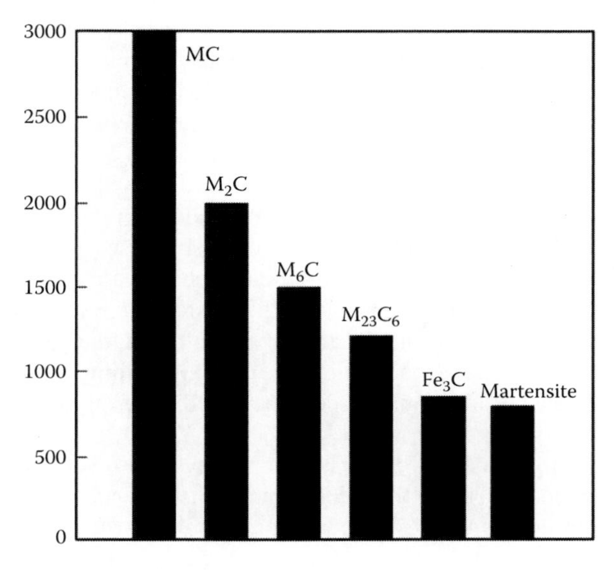

FIGURE 3.24 Relative hardness of alloy carbides in HSSs.

The material should be preheated to just below this critical transformation temperature and then held long enough for the entire cross section of the part to equalize. Once the part is equalized, then further heating to the austenitizing temperature will allow the material to transform while undergoing a minimum amount of distortion.

2. It is important that the tools are protected against oxidation and decarburization. The best protection is provided by a vacuum furnace, where the surface of the steel remains unaffected. Furnaces with a controlled protective gas atmosphere or salt baths also provide good protection. If an electric muffle furnace is used, the tool can be protected by packing it in spent charcoal or cast iron chips.
3. The austenitizing temperature that is selected depends strongly upon the alloy content of HSS. The target properties including hardness, tensile strength, and grain size also factor into the temperature that is chosen. In the annealed microstructure, the alloy content of the steel is primarily contained in the carbide particles that are uniformly distributed as tiny spheres. This condition is typically referred to as a spheroidized annealed microstructure. The idea behind austenitizing is to redistribute this alloy content throughout the matrix by heating the steel to a suitably high temperature so that diffusion can take place. Higher temperatures allow more alloy to diffuse, which usually permits a higher hardness. This is true as long as the temperature does not exceed the incipient melting temperature of the steel. If lower austenitizing temperatures are used, then less diffusion of alloy into the matrix occurs. The matrix is therefore tougher but may not develop high hardness. The hold times that are used depend upon the size of the tool and the temperature that is used.
4. Once the alloy content has been redistributed throughout the matrix, the steel must be cooled fast enough to fully harden it. This process is called quenching. By quenching the HSS properly, a new phase transformation occurs, and the microstructure changes from austenite to martensite. How rapidly this process must take place depends upon the chemical composition of the alloy. The best way to achieve the controlled cooling rate is in a vacuum furnace by inert gas pressure quenching in the range of 2–20 bar. There is an interest in the use of hydrogen for cooling in the 25–40 bar range due to its extremely high heat transfer rates. In gas quenching, part dimensional changes, although repeatable, are different than when quenching in oil. The trend today is to *dial in* the quench pressure. That is, they use only the highest pressure required to properly transform the material. The traditional way to quench HSS is in oil. Other mediums that are frequently used for quenching include water, brine, and salt bath. Whatever quenching process (medium) is used, the resulting microstructure is extremely brittle and under great stress. If the tool is put into service in this condition, it would likely shatter like glass. Some tools will even spontaneously crack if they are left in this condition. For this reason, tools that are quenched and cooled to hand warm should be tempered immediately.
5. Tempering is performed to soften the martensite that was produced during quenching. HSSs require several tempers before the tool can be put into service. This is because these alloys will retain a certain percentage of austenite when they are quenched, and during the first temper, some of this retained austenite will transform to untempered martensite. By performing a second temper, this new martensite is softened, thus reducing the chance of cracking. But by tempering a second time, some of the remaining austenite is transformed to untempered martensite, and so the process may need to be repeated several times.

Heat treatments used to harden HSSs first produce a mixture of austenite and alloy carbides. Initially in soft annealed HSSs, the microstructure is an array of ferrite grains and alloy carbides. When austenite begins to form on heating, the microstructure contains austenite, ferrite, and carbides. Both the ferrite grains and the carbide particles restrict grain growth of the austenite. Then, $M_{23}C_6$ begins to dissolve, and dissolution is complete when temperature 1095°C is reached on heating. Then, partial solution of the other carbides occurs until the austenite typically contains approximately 7–12 vol% alloy carbides of the types MC and M_6C. On cooling, there is immediate tendency for carbides to precipitate, especially on austenite grain boundaries. However, the diffusion rates of tungsten and molybdenum are very slow, and very slow cooling is required for significant carbide precipitation to occur on the grain boundaries and within the grains. The as-quenched steel contains 60–80 vol% martensite, 15–30 vol% retained austenite, and 5–12 vol% MC and M_6C. As mentioned previously, the as-quenched martensite is very hard and very brittle, has high residual stresses, and thus should be tempered immediately. The usual tempering range is 510°C–595°C. During tempering, precipitation of alloy carbides also occurs in the retained austenite, which is also supersaturated. This reduces the concentration of dissolved alloying elements in austenite, which rises the M_s temperature (i.e., the temperature of the alloy system at which martensite starts to form on cooling—see Appendix B for explanations). This causes more martensite to form when the steel is cooled after being tempered to reduce or to eliminate the retained austenite.

3.3.7.3 Cryogenic Treatment of HSSs

Over the years, the possibility of enhancing the properties of hardened steels by treating them at temperatures below room temperature has been of interest. Two types of low-temperature thermal treatments are subzero, or cold treatment, and cryogenic treatment (CT). Subzero treatment is carried out at −145°C (−230°F), while CT is carried out at −195°C (−320°F), the boiling point of liquid nitrogen (LN2). Interest continues to grow based on anecdotal evidence of the benefits of CT not only for steels but also in nonferrous metals like copper, aluminum, and titanium. The literature contains many claims about the advantages of CT (Kelkar et al. 2007; Baldissera and Delprete 2008; Sendooran and Raja 2011).

Subzero treatment has been applied in HSS processing for many years. This treatment consists of cooling the hardened component to the subzero temperature to transform the retained austenite into martensite. Reportedly, such a treatment increases high-hardness and wear resistance of HSS tools. Figure 3.25 illustrates the typical heat treatment curve normally adopted for subzero treatment (Prabhudev 1988). As can be seen, the subzero heat treatment is carried out after the quenching operation, before the phenomenon known as stabilization occurs that makes the retained austenite resistant to further transformation.

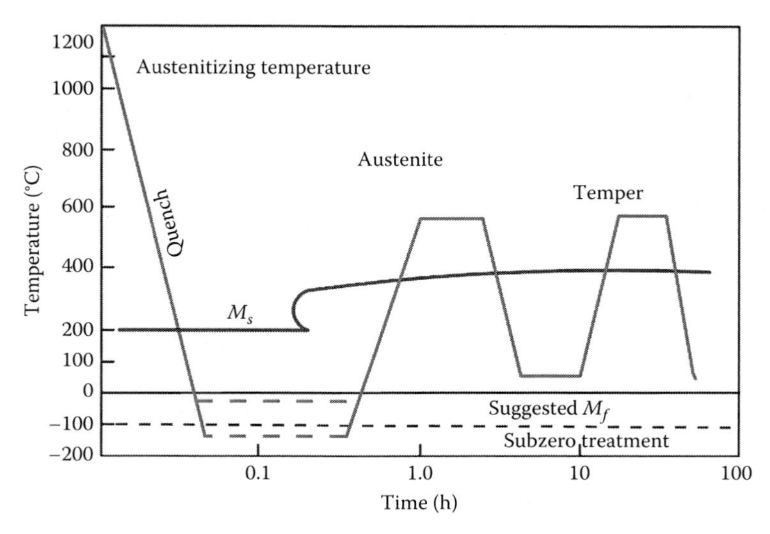

FIGURE 3.25 Schematic representation of a conventional hardening and tempering treatment and a subzero treatment superimposed on the isothermal transformation diagram for a M2-type HSS.

However, there is little information available in the scientific literature regarding understanding the mechanism behind property improvements beyond the obvious conversion of austenite to martensite, which only applies to usual tool steels where martensite is the hardest phase responsible for the cutting properties of a tool made of this steel. In HSSs, however, it is not so as directly follows from the data shown in Figure 3.24. As can be seen, the cutting properties of HSSs do not rely on martensite. Rather, they are determined by the carbides having much greater hardness than martensite so adding more martensite to the HSS structure seems to have little benefits.

Even though the mechanism behind improvement of the cutting properties of HSSs after CT has not been totally clarified even by most recent researches, different hypotheses coherently with microstructural observations have been suggested in literature (Baldissera and Delprete 2008).

A fundamental distinction among different CT is given by the parameters of the cooling–warming cycle. Two methods depending on the minimum temperature reached during the cycle are categorized (Baldissera and Delprete 2008):

- *Shallow cryogenic treatment (SCT) or subzero treatment*: The samples are placed in a freezer at 193 K, and then they are exposed to room temperature.
- *Deep cryogenic treatment (DCT)*: The samples are slowly cooled to 77 K, held down for many hours and gradually warmed to room temperature.

A great number of various SCT and DCT *recipes* have been developed. All of these recipes involve significant time of treatment starting from 1 and finishing with 168 h which adds to the cost and lead time in HSS tool manufacturing.

The resulting effect is somewhat contradictive with illusive explanations. Kelkar et al. (2007) carried out an extensive industrial-oriented thorough study on the properties of CT HSS M2. They concluded that cryogenically treated material should have an improved combination of hardness and toughness, but the effect is small and difficult to quantify. Table 3.5 shows DCT improvement of wear resistance reported in literature (Mohan Lal et al. 2001; Molinari et al. 2001; da Silva et al. 2006). As can be seen, great scatter in the results is the case.

The author's experience with CT of HSSs somehow confirms the reported results, that is, sometimes its application results in great improvement in tool life while the next batch of HSS tools made from the same grade of HSS from the same cutting tool manufacturer shows a little improvement or no improvement at all. This is completely in line with experience of many cutting tool professionals frustrated with a great scatter in the application results. Researches in the

TABLE 3.5
DCT Improvement of Wear Resistance Reported in Literature

Authors	HSS	Test Conditions	Reported Improvement
da Silva et al. (2006)	M2	Pin-on-disk	No significant changes
		Brandsma rapid facing test	+44% tool life
		Twist drills	+343% tool life (catastrophic failure end-life criterion)
		Shop-floor test of milling cutters	−22.8% machined part (appearance of burr end-life criterion)
Mohan Lal et al. (2001)	M2, T1	Pin-on-disk (M2)	+135% wear resistance
		Flank wear (T1, M2)	+110.2% wear resistance for T1 +86.6% wear resistance for M2
Molinari et al. (2001)	M2	Pin-on-disk	−51% wear rate

field cannot provide any help to understand the process and thus to reduce this scatter optimizing CT of HSS. As CT comes with additional cost and lead time, this scatter forced many practitioners to stop using this additional heat treatment as they feel that illusive gains cannot justify the additional cost.

In the author's opinion, a completely new look at HSS should be taken to understand the results of CT and other processes as HSSs so far are not treated by researchers as tool materials but rather as structural materials so that the corresponding unsuitable tests are used. In other words, researches do not pay much attention to what has made HSS a good tool material compared to high-carbon tool steels, that is, the data presented in Figure 3.24 are not analyzed properly. Therefore, a need is felt to clarify the issue.

Many steels can be heat treated to produce a great variety of microstructures and properties. Generally, heat treatment uses phase transformation during heating and cooling to change a microstructure in a solid state. In heat treatment, the processing is most often entirely thermal and modifies only structure. Thermomechanical treatments, which modify component shape and structure, and thermochemical treatments, which modify surface chemistry and structure, are also important processing approaches, which fall into the domain of heat treatment.

Although heat treatment is most widely used process today to enhance properties of materials, it has the known disadvantages and limitations. The major disadvantage is that nonequilibrium structures are created that causes significant internal stresses in materials and relaxation of the achieved enhancements over time.

Many engineering materials include distinctive solid phases in their structure. The higher material performance, the greater the difference in physical and mechanical properties of these phases. As these materials are shaped, great deformation and temperatures are applied. For example, for HSSs, hot rolling, extrusion, and heat treatment, are used. As there is a great difference in the behavior of the metallic and nonmetallic (carbide) solid phases under these severe conditions, interfacial defects as microcracks, residual stress, and even defects of the atomic structure occur (Komarovsky and Astakhov 2002). These defects lower the strength, wear resistance, and other useful mechanical and physical properties of HSS. Moreover, as the population of these defects is of random nature, the great scatter in the performance data is observed in the practice. For example, even when the cutting tool is of high quality and the machining system fully supports its performance, the scatter of the tool life can reach 40% due to microdefects in the tool material (Astakhov 2006). This scatter presents a serious problem in tool materials applications.

In the author's opinion, the only feasible and physically grounded way to improve the properties of the previously listed materials is to heal microcracks and microdefects as discussed in Section 2.6.13.2. This can be feasibly accomplished by their diffusion healing of a great energy flux and is introduced through the entire cross section of HSS tool. One of the possible ways to accomplish that is the application of the Diffusion© process developed by Diffusion Ltd. (Windsor, Ontario, Canada) primarily for the use in the automotive industry for enhancing and stabilizing properties of the tool materials.

3.3.8 Coating of HSS

HSS and HSS-PM are excellent substrates for all coatings such as TiN, TiAlN, TiCN, solid lubricant coatings, and multilayer coatings. Coatings considerably improve tool life and boost the performance of HSS tools in high-productivity, high-speed, and high-feed cutting or in dry machining and machining of difficult-to-machine materials (Soplop et al. 2009). Coatings provide (1) increased surface hardness, for higher wear resistance (abrasive and adhesive wear, flank or crater wear); (2) reduced friction coefficients that easies chip sliding, reduces cutting forces, prevents adhesion on the contact surfaces, reduces heat generated due to chip sliding, etc.; (3) reduced portion of the thermal energy that flows into the tool; (4) corrosion and oxidation of tool surfaces; and (5) improved surface quality of finished parts.

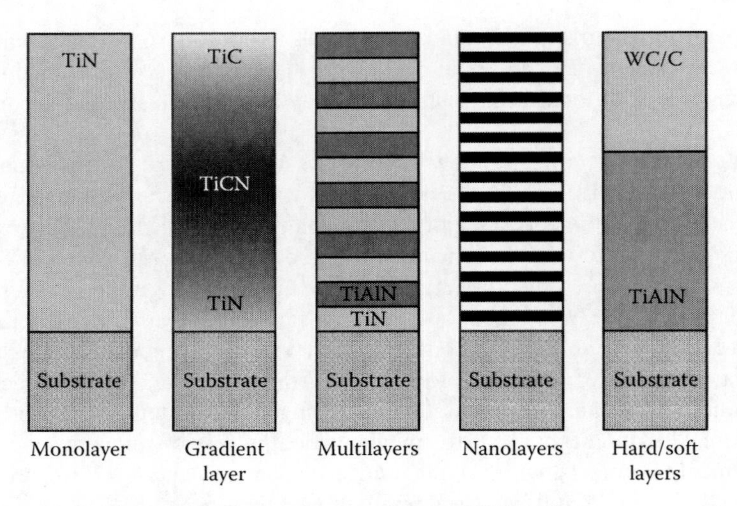

FIGURE 3.26 Representation of layers of modern coatings on HSS tools.

Common coatings for HSS applied in single or multilayers are shown in Figure 3.26. They are

- *Titanium nitride*: TiN—General-purpose coating for improved abrasion resistance. Color: gold; hardness *HV* (0.05): 2300; friction coefficient: 0.3; thermal stability: 600°C.
- *Titanium carbonitride*: TiCiN—Multipurpose coating intended for steel machining. Higher wear resistance than TiN. Available in mono and multilayer. Color: gray–violet, hardness *HV* (0.05): 3000; friction coefficient: 0.4; thermal stability: 750°C.
- *Titanium aluminum (carbo) nitride*: TiAlN and TiAlCN—High performance coating for increased cutting parameters and higher tool life. Also suitable for dry machining. Reduces heating of the tool. Multilayered, nanostructured, or alloyed versions offer even better performance. Color: black–violet, hardness *HV* (0.05): 3000–3500; friction coefficient: 0.45; thermal stability: 800°C–900°C. Some companies, however, discontinued to use of TiAlN as a standard coating. Instead, a much harder coating, AlTiN is recommended for applications (AlTiN: 4500 *HV* vs. TiAlN: 2600 *HV*).
- *WC–C and* MoS_2: Provides solid lubrication at the tool–chip interface that significantly reduces heat generation due to friction. Has limited temperature resistance. Recommended for high-adhesive work materials as aluminum and copper alloys and also for nonmetallic materials. Color: gray–black; hardness *HV* (0.05): 1000–3000; friction coefficient: 0.1; thermal stability: 300°C.
- *Chromium nitride*: CrN—The antiseizure properties of this coating make it preferred in situations where BUE is common. It is practically invisible coating when applied on HSS or sintered carbide. Color—metallic.
- *Aluminum titanium nitride + silicon nitride*: *n*ACo—Hard coating for difficult-to-machine materials.
- *Diamond-like carbon* (*DLC*): DLC has some of the valuable properties of diamond. When applied in pure form, it is as hard as natural diamond. In pure form, these diamond coatings offer extraordinary protection against abrasive wear and attack from atmospheric moisture and chemical vapors. Although smooth when seen with visible light, DLC actually has the form of a cobblestone street. In DLC, the cobbles are not crystalline; they are amorphous because they are made from random alternations between cubic and hexagonal lattices. The cobbles have no long-range order and so they have no fracture planes along which to break. The result is a very strong material.

Table 3.6 shows the coatings recommended by Melin Tool Co. for various work materials.

TABLE 3.6
Application of Coatings as Recommended by Melin Tool Co.

Work Material	Hardness	First Choice	Second Choice
Aluminum	HB 160–240	TiCN	TiN
Alloy steel	HRC 23–38	TiN	TiCN
Alloy steel	HRC > 38	AlTiN	AlTiN
Carbon steel	HB 160–240	TiN	TiCN
Carbon steel	HRC 23–38	TiN	TiCN
Carbon steel	HRC > 38	AlTiN	AlTiN
Hardened steel		*n*CAo	AlTiN
Low-carbon steel	HB 130–240	TiCN	TiN
Low-carbon steel	HRC 23–38	TiN	TiCN
Low-carbon steel	HRC > 38	AlTiN	AlTiN
Gray cast iron	HB 180–220	AlTiN	AlTiN
Nodular cast iron	HB 220–320	*n*ACo	AlTiN
Austenitic stainless steel	HB 180–220	TiCN	TiN
Martensitic stainless steel	HRC < 35	AlTiN	TiCN
Martensitic stainless steel	HRC ≥ 35	*n*ACo	AlTiN
Ni alloys		*n*ACo	AlTiN
PH stainless steel	HRC < 35	AlTiN	TiCN
PH stainless steel	HRC ≥ 35	*n*ACo	AlTiN
Ni-, Co-, and Fe-based superalloys		AlTiN	AlTiN
High Si aluminum		*n*ACo	AlTiN
High Si aluminum		*n*ACo	AlTiN

The first choice in thin-film coatings decades ago was TiN, the familiar yellow-gold coating still often used today. Some cutting tool end users still prefer old-fashioned TiN coating, unaware of the many choices now available. Dozens more have been developed in the meantime. Combined in multiple layers, coatings achieve a balance of properties not possible with a single-layer thick coating. As new thin-film ceramic coatings proliferate, the problem facing many engineers now may be one of too many choices, rather than too few.

Conventional coatings having rather rough surface are not inherently lubricious. As a result, applications of such coating solved the problem with tool protection but gave rise to another problem of higher heat generation on the tool–chip and tool–workpiece interfaces. New coatings have been developed to address this problem. Multilayered coatings combine lubricity and resistance to wear. This combination reduces friction at the tool–chip and tool–workpiece interfaces.

A noticeable trend in tool coatings is that they are becoming application specific. For example, Oerlikon Balzers (Amherst, New York), a supplier of coatings and surface technologies, now offers 24 different coatings. The Oerlikon Balzers developed application-specific coatings such as the Balinit Aldura for milling hardened tool steels with hardness exceeding HRC 50 and Balinit Helica, an AlCr-based coating designed specifically for twist drills. Balinit Aldura uses an aluminum–chromium–nitride (AlCrN)-based functional layer deposited onto a TiAlN support layer, all on a carbide substrate. The TiAlN ensures good adhesion and mechanical strength, while the AlCrN layer features excellent hot hardness and oxidation resistance (up to 1100°C) and insulates the tool from the heat of cutting.

Oerlikon Balzers has developed a new coating technology that should garner interest from tooling engineers: P3e (pulse-enhanced electron emission). Aluminum oxide-based coatings with high hardness, as well as thermal and chemical stability have been developed using this technology. With a deposition temperature below 600°C, substrates are coated with thermodynamically stable, alpha-aluminum oxide-based coatings without distorting the tool.

Nanocomposites are composed of typical coatings, such as TiAlN, embedded in an amorphous matrix of Si_3N_4. The resulting nanocomposite improves hardness, lubricity, and maximum operating temperatures. Platit (Grenchen, Switzerland), a coatings company, also patented a triple coating composed of an adhesion layer, an intermediate layer (mono or gradient), and a nanocomposite top layer for maximum wear resistance and toughness, for example, their *n*ACo3 triple coating, which has a bottom layer of TiN, a middle layer of AlTiN, and a nanocomposite top layer *n*ACo. *n*ACo is itself a nanocomposite of AlTiN embedded in an amorphous matrix of Si_3N_{4M}. This coating boasts a hardness of 38–45 GPa and a max operating temperature up to 1200°C, according to data provided by Platit.

Some important guidelines for coated tool users:

1. Rule No.1: Do not even try to solve your tooling problems with application of coating. Any tool coating is to enhance tool performance not to solve tooling problems.
2. The four equally important aspects of tool performance: tool design including its geometry, tool material (substrate), tool manufacturing quality, and coating. It is easy to have a new coating with a new color and higher hardness and maximum allowable temperature that reportedly enable performance improvements. However, you will not necessarily get such improvements without optimizing the other three listed aspects.
3. Should you decide to apply a coating on your cutting tool, a clear objective of coating application has to be first formulated. There are two feasible objectives of coating application: (a) increase tool life and (b) increase the allowable cutting speed (and thus productivity).
4. A clear benchmark should be established for the chosen objective, that is, the number of parts machined per tool life or the maximum speed allowed by the cutting tool to achieve the minimum required parts per tool life.
5. If you use a tool from a reputable tool company, then follow the recommendations on coating provided by this company for a given tool and application. Cutting tool/coating manufacturers are constantly working on new surface treatments that offer added protection against heat, friction, and abrasion. It is always best to check with your coating/tool manufacturer for the newest and greatest coatings available for your application.
6. A cost-effective coating can be dependent on many things, but there are usually one or more viable choices for every application. Choosing the correct coating and its attributes can mean the difference between a successful operation and one that shows little to no difference. Depths-of-cut, speeds, and MWFs (coolants) can all have an effect on the way a surface treatment may react.
7. Although it may bear the same name/chemical composition, for example, titanium carbonitride TiCN, the selected coating is not the same from different coating providers or even from different coating facilities of the same provider.
8. An important logistic/efficiency decision is to be made with regrindable drilling tools. First, it has to be determined the place of cutting tool prime wear, which can be the rake face crater, the wear land on the flank surface (often referred to as flank face), wear of margins, and so on (see Section 2.6.4). If the rake face wear is the case, then no recoating after grinding is needed. If, however, the flank or drill margin wear is the prime wear mode, then tool should be recoated after each regrinding. The problem arises with logistic as the proper procedure for recoating is coating stripping–regrinding–recoating. If the coating and grinding are carried out in house, then there are no problems. If the tool has to be sent for recoating, then a problem to maintain proper recoating sequence arises as many manufacturing facilities are not involved with coating and thus do not have the ability to apply a procedure of coating stripping as it may involve high hazard chemicals. Recoating by applying a second layer of coating on the existing one or grinding then stripping and recoating is not a good idea.
9. Although a number of coating properties are available, for example, hardness, the only feasible way to determine the effectiveness of the chosen coating is through trial and error, as this is due to the amount of variables while machining of a material.

Some important notes for tool manufacturers:

1. The dimensions shown on the tool drawing should be after coating. The tool shank should not be coated.
2. The surface roughness on the tool for coating should be the same as with no coating. Do not try to cover roughly ground tool surfaces by coating.
3. It is desirable to have a coating chamber in house although a learning curve and initial investments (buying a chamber, tool cleaning and coating stripping equipment, coating control equipment, training the personnel, etc.) are needed. However, the benefits are in logistic, consistency of coating (you know what you are applying), flexibility in terms of using application-specific coatings, etc.
4. Should you use an external service for coating, work closely with coating specialists making sure that you use the proper coating and coating regimes. For HSSs, it is important to make sure that the temperature on coating application does not overtemper your tool material with a result of decreasing its hardness. Analyze together the tool wear pattern to optimize the type of coating and coating regime.

3.4 CEMENTED CARBIDES

Carbide selection and its proper applications range seem to be straightforward when it comes to indexable carbide inserts as, knowing given machining conditions, a particular insert grade, coating, shape, holder, machining regime, and so on can be selected using colorful catalogs of leading tool manufacturers. The detailed, step-by-step selection procedure was presented by the author earlier (Astakhov 2011). In such a standard procedure, the tool end user intellectual input is in the selection of a particular tool manufacturer so then the rest is almost automatic. This is nearly the case with drilling tools for end users when they select a standard tool from a catalog of a tool company although small intellectual effort may be required to select a tool out of a few options.

The situation is quite different for drilling tool manufacturers and for end users of special drilling tools. Apart from HSSs, tool manufacturers do not affect the metallurgy of cemented carbides as they are not involved in heat treatment (at least intendedly) or other metallurgical procedures with this tool material. If a drilling tool is designed properly and made with decent manufacturing quality (in terms of its grinding, brazing, etc.), it should perform as intended provided that the carbide grade is selected properly for the application. A problem is how to select this proper grade out of thousand (combined with coating) available options accounting for the cost considerations (minimum cost per machined hole or maximum productivity whatever is more important for a particular drilling operation), lead time, grindability, and many other practical concerns.

Properties and application particularities of cemented carbides as tool material are well described in many literature sources and summarized by Upadhyaya (1998). Excellent methodological materials on the subject are those by Sandvik Coromant Co. (Iakovou and Koulamas 1996) and General Carbide Co. (Vagnorius and Sorby 2009). A great number of professional literature and textbooks on manufacturing contain at least a chapter on tool materials with extensive coverage of carbides (Davis 1995; DeGarmo et al. 2007; Nee 2010). It seems that all end users of carbides (both drilling tool manufacturers and tool users) need to know about this most widely used tool material is readily available. Moreover, if one needs training on carbides, a number of training courses, for example, offered by SME or the Tooling University, are widely offered. Therefore, a logical question why this book needs a section on cemented carbides (instead of referring multiple known sources) should be answered.

The major distinguished feature of this section is that it is written from the carbide user perspective not from the carbide manufacturers as many other widely available literature sources. It aims to shed some light on the darkest (yet still the most important for carbide users) corners of the carbide application business in drilling tools that have been somehow avoided in the known sources by variety of reasons.

3.4.1 What Is Cemented Carbide?

The cemented carbide is a range of composite materials. Cemented carbides consist of hard grains of the carbides of transition metals (Ti, V, Cr, Zr, Mo, Nb, Hf, Ta, and/or W) cemented or bound together by a softer metallic binder consisting of Co, Ni, and/or Fe (or alloys of these metals) known as the matrix. The grades of cemented carbide that include only tungsten carbides (WC) in their composition are known as straight grades. Because most of the commercially important cemented carbides contain mostly WC as the hard phase, the terms *cemented carbide* and *tungsten carbide* are often used interchangeably. Although the term *cemented carbide* is widely used in the United States, these materials are known internationally as hardmetals.

The process of combining tungsten carbide with cobalt is referred to as sintering. On sintering, the process temperature is higher than the melting point of cobalt while it is much lower than that of WC. As a result, cobalt is embedding/cementing the WC grains and thereby creates the metal-matrix composite (MMC) with distinct material properties. The naturally ductile cobalt metal serves to offset the characteristic brittle behavior of the tungsten carbide ceramic, thus raising its toughness and durability while lowering its hardness and thus abrasion wear resistance. Such parameters of tungsten carbide can be changed significantly by altering the cobalt content, grain size, addition of another (than WC) carbides, pre-sintering, sintering, and post-sintering processes.

The proportion of carbide phase is normally 70%–97% of the total weight of the composite and its grain size averages between 0.2 and 20 μm. Tungsten carbide (WC) and cobalt (Co) form the basic cemented carbide structure, and, as mentioned previously, the grades based on this concept are often referred to in simplified terms as straight grades. From this basic concept, many other types of cemented carbide have been developed. Thus, in addition to these simple WC–Co compositions, cemented carbide may contain varying proportions of titanium carbide (TiC), tantalum carbide (TaC) or niobium carbide (NbC), and others. Although they are called in the professional literature as, for example, titanium carbide (TiC), tungsten carbide (WC) is always the predominate portion of such cutting tool materials.

Although cobalt as the matrix material can be alloyed with or even completely replaced by other metals such as nickel (Ni), chromium (Cr), iron (Fe), molybdenum (Mo), or alloys of these elements, cobalt is the primary matrix material for cutting tool cemented carbide grades.

3.4.2 Brief History

Since the late 1800s when a French chemist, Henri Moissan, first synthesized it, tungsten carbide has been known as one of the hardest substances in existence, approaching diamond in this respect. In fact, he was seeking to produce man-made diamonds, but WC was the result. Since large solid pieces could not be produced, cast compositions containing tungsten carbide were tried but were too brittle and porous for use as an engineered material.

Carbide as a tool material was discovered in the search for a replacement of expensive diamond dies used in the wire drawing of tungsten filaments. Initiated by a shortage of industrial diamonds at the beginning of World War I in Germany, researchers had to look for alternatives. Regardless of a number of various testing and patents, it was not until the invention of sinterian when the first useful carbides were produced. At first, further melting experiments were performed, which failed as did sintering experiments with fine-grained WC powder. Also carburization of sintered tungsten compacts did not end up in useful materials. Finally, the first breakthrough was achieved by Heinrich Baumhauer through infiltrating a porous WC body with molten iron. Interestingly, this development was not performed at the laboratories of the study group but at the production works of the former Siemens lamp works in Berlin-Charlottenburg. Baumhauer applied for a patent on March 18, 1922 (German Patent No. 443,911). The research of Baumhauer stimulated further efforts in the study group. Karl Schröter, chief engineer at the *Osram study society for electrical lighting* improved the tungsten carburization. In a further step, WC powder was mixed with iron, nickel, or cobalt powder, then compacted and sintered. Based on these experiments, a patent was submitted by the Patent-Treuhand-Gesellschaft für elektrische Glühlampen m.b.H in Berlin, Germany, on March 30, 1923, which was granted on October 30, 1925 (German Patent No. 420,689) (Mills and Redford 1983). At that time, no one, even the most optimistic, could imagine the enormous breakthrough for this material in the tooling industry.

Not having resources to exploit this material on an industrial scale, Osram sold the license to Krupp at the end of 1925. In 1926, Krupp brought sintered carbide into the market. On June 10, 1926, the name of this product was entered into the register of trademarks under the name WIDIA (acronym for WIe DIAmant = like diamond). Its exceptional hardness and wear resistance represent a major breakthrough in tool engineering. The German word *Hartmetall* was used for the new product. In direct translation to English, this expression denotes *hardmetal*, a term which is also used internationally. The term *cemented carbide* was used first in the United States by researchers at GE/Carboloy, and it much better describes the nature of the metallic composite. The new product group performed very well, and WIDIA soon became a synonym for sintered carbide. The first product (WIDIA N—WC–6Co) was presented at the Leipzig Spring Fair in 1927. At first, only small amounts of WIDIA were produced at the Krupp works and even less in the rest of the world. Still the price of the material was extremely high—about $450 per pound, but even at that price its use could be justified economically. The practice of making only the tool tip out of cemented carbide was dictated as much by the cost of the material as it was by any other single consideration.

Following the introduction of WIDIA and its counterparts, Dr. Balke developed a tantalum carbide bonded together with metallic nickel. This material, called Ramet, resisted *cratering* and proved more successful in machining of steels than did tungsten carbide, which exhibited a tendency toward early cavitation near the cutting edge, where steel chips came into intimate contact with the tool face. It was this characteristic of the first carbides that impeded an advance in their application to machining steel comparable with the strides made in machining cast iron, nonferrous metals, and abrasive materials. Thus tantalum-bearing carbide opened new fields in steel cutting.

Single and multiple carbides of tantalum, titanium, and columbium were also to find use as crater preventives or resisters. In 1935, Philip McKenna, working at the time as a metallurgist for Vanadium Alloy Steel Company in Latrobe, PA, used a novel technique to manufacture crater-resisting carbides with improved strength and toughness and in 1937 introduced a tungsten–titanium carbide, which, when used as a tool material, proved to be effective in steel cutting. He received a patent for this formulation and went on to form the McKenna Metals Company, now known as Kennametal Inc.

The increasing demand for steel and tools in Germany at the beginning of the 1930's rapidly increased the sales from about 12 tons per year to 500 tons at the end of World War II. After World War II, sintered carbides became an international commodity and rapid advancements occurred in both science and technology. Constant progress by the carbide manufacturers in improvement of their products has lowered the cost of carbides from the original $450 per pound to such an extent that they are no longer regarded as precious metals. Depending upon grade, shape, and size, the cost, with greatly improved quality and capacity dramatically increased, is but a fraction of what it was in 1929.

According to the International Tungsten Industry Association, in 2008, roughly 50,000 tons of tungsten (W) were consumed worldwide in cemented carbides, which account for about 60% of the world's tungsten consumption (including recycled material). In terms of tonnages, stoneworking and machining of wood and plastics are the largest fields of application (26%), followed by metal cutting (22%), wear applications (17%), and chipless forming (9%). In contrast, the metal cutting group accounts for 65% of the turnover (due to its high degree of innovation), compared to stoneworking (10%), machining of wood and plastics (10%), wear applications (10%), and chipless forming (5%).

Today various grades of sintered carbides are the most common cutting tool materials. Although various carbide manufacturers may have different processing procedures, the final product is obtained by compacting the powder formulation by some technique and sintering the constituents into a solid mass in which cobalt, or a similar metal, bonds or cements the particles of carbide together.

3.4.3 Grade Classification

Despite official and unofficial *standards*, the true nature of carbide grades—how they are made and what they are made of—remains shrouded in mystery. The ISO system and the *C* system both offer individual manufacturers' recommendations for applications, but neither of them sets minimum requirements for the carbide's composition, properties, or performance. Is this good, bad, or irrelevant? It depends on whom you talk to (Craig 1997).

Tool manufacturers will tell you that today's carbide tools—with their vast array of compositions and coatings—are too complex to be covered by an all-encompassing standard. Certainly, the explosion of coatings in recent years (it is estimated that 80% of carbide tools sold today are coated) has complicated tool classifications. Naturally, coatings are no more standardized than substrates.

Tool manufacturers will further tell you that the current grading systems are adequate for most users' purposes. Some carbide-tool users, however, disagree, arguing that the lack of a carbide standard plays directly into the hands of the carbide-tool manufacturers by making it extremely difficult to switch from one brand's carbide grade to another brand's equivalent. A C-5 carbide tool from Producer *A* may vary drastically from Producer *B*'s C-5 tool. The existing classification systems tell users what inserts of a given classification should be used for, but they fail to describe the criteria for classifying carbide grades. The ISO and C systems leave users dependent upon the manufacturers' judgments of what grades are suited for a particular application, judgments that may or may not be based on sound analysis of a carbide grade's properties and cutting characteristics. The end result, these users say, is a costly, time-consuming reliance on trial and error to determine the best tool for a given job.

Some users have called for a descriptive standard that should delineate minimum requirements in terms of compositions and properties of a carbide material. The standard could still be used to prescribe certain uses for certain classifications, but users would be guaranteed that the tools meet certain criteria other than the manufacturer's opinion of how they should be used. Meeting certain physical criteria would also ensure a level of consistency from brand to brand that is currently lacking.

3.4.3.1 Earlier Standards

In its early years, carbide was a relatively simple material and, hence, easy to standardize. After World War II, allied investigators published detailed accounts of Krupp–Widia carbide compositions, production techniques, quality control methods, and the results of related research projects. This information formed the basis of an ersatz European standard. Continuing efforts to replace tungsten with less expensive materials led to a flood of carbide formulations based on materials such as titanium and tantalum. As a result, the neat, orderly German standard of tungsten-base grades was rendered obsolete. Nevertheless, the Krupp–Widia standard formed the basis for a system adopted by DIN, the German standards organization, based solely on application recommendations. The DIN system eventually was adopted as the ISO coding system we know today.

In the United States, Oscar Strand of the Buick Motor Division of General Motors developed an application-based classification code for nearly 100 grades in 1942. The grades were arranged into a simple system of only 14 symbols. This code evolved into the familiar *C* code, which became the unofficial American standard. While Buick went on much further to standardize carbide tools according to testable properties—such as chemical composition, tensile strength, compressive strength, thermal conductivity, density, average grain size (AGS), and grain size distribution—the C code as embraced nationwide was based solely on manufacturers' application recommendations and had no meaning in terms of testable properties. This classification is presented in Table 3.7 (Davis 1995).

TABLE 3.7
C-Grade Classification of Cemented Carbide

C-Grade	Application Category
Machining of cast iron, nonferrous, and nonmetallic materials	
C-1	Roughing
C-2	General-purpose machining
C-3	Finishing
C-4	Precision finishing
Machining of carbon and alloy steel	
C-5	Roughing
C-6	General-purpose machining
C-7	Finishing
C-8	Precision finishing
Nonmachining applications	
C-9	Wear surface, no shock
C-10	Wear surface, light shock
C-11	Wear surface, no shock
C-12	Impact, light
C-13	Impact. medium
C-14	Impact, heavy

Later, the US Department of Defense (DOD) attempted to rectify this situation by pressuring carbide-tool manufacturers to adopt a more descriptive system for government procurement. While the DOD's earlier attempts to standardize tool steels had at least partially succeeded, its efforts to standardize carbide failed.

The C code held sway in the United States until the late 1980s, when the ANSI adopted the ISO coding system—though references to C code classifications are still used informally and are, for most end users, still more easily recognizable. In practical terms, this move changed only the symbols used to codify carbide grades and the work material categories, not the basic concept of classifying carbide prescriptively rather than descriptively.

3.4.3.2 Current Standard

The ISO code has been given a higher profile by several major carbide-tool producers' adoption of the ISO color-coding system. The ISO grade for a given material is defined in the International Standard ISO 513-2004—Classification and application of hard cutting materials for metal removal with defined cutting edge—designation of main groups and group of application (Based on standard DIN 4990). According to this standard, the designation of groups of application for hard cutting materials includes the letter symbol (for carbides in accordance with Table 3.8), followed by a dash and the designation of the main group of chip removal and the group of application as shown in Table 3.9. Examples: HW—P10, HC—K20.

As shown in Table 3.9, there are six main groups of application. They are divided according to the different work materials that are to be machined. They are identified by a capital letter and an identifying color. Each main group of application is divided into application groups. The application groups are designated by the letter for the main group and a classification number. For example P01 stands for hardest grade and thus most wear-resistant grade in P application group whereas P40 is for toughest grade in this group.

TABLE 3.8
Letter Symbols for Carbides

Identification Letters	Materials Group
HW	Uncoated carbide, main content tungsten carbide (WC) with grain size ≥ 1 μm
HF	Uncoated carbide, main content tungsten carbide (WC) with grain size ≤ 1 μm
HT[a]	Uncoated carbide, main content TiC or TiN or both
HC	Carbide as mentioned earlier, but coated

[a] These grades are also called *cermets*.

TABLE 3.9
Application and Classification of Hard Cutting Materials

Main Groups of Application			Groups of Application		
Identification Letter	Identification Color	Material to be Machined	Hard Cutting Materials		
P	Blue	*Steel*:	P01	P05	↑ a ↓ b
		All kinds of steel and cast	P10	P15	
		steel except stainless steel	P20	P25	
		with an austenitic structure	P30	P35	
			P40	P45	
			P50		
M	Yellow	*Stainless steel*:	M01	M05	↑ a ↓ b
		Stainless austenitic and	M10	M15	
		austenitic/ferritic steel and	M20	M25	
		cast steel	M30	M35	
			M40		
K	Red	*Cast iron*:	K01	K05	↑ a ↓ b
		Gray cast iron, cast iron with	K10	K15	
		spheroidal graphite,	K20	K25	
		malleable cast iron	K30	K35	
			K40		
N	Green	*Nonferrous metals*:	N01	N05	↑ a ↓ b
		Aluminum and other	N10	N15	
		nonferrous metals,	N29	N25	
		nonmetallic materials	N30		
S	Brown	*Superalloys and titanium*:	S01	S05	↑ a ↓ b
		Heat-resistant special alloys	S10	S15	
		based on iron, nickel, cobalt,	S20	S25	
		titanium, and titanium alloys	S30		
H	Gray	*Hard materials*:	H01	H05	↑ a ↓ b
		Hardened steel, hardened	H10	H15	
		cast iron materials, chilled	H20	H25	
		cast iron	H30		

Note: a, Increasing speed, increasing wear resistance of cutting material; b, Increasing feed, increasing toughness of cutting material.

The standard points out that a group of application is not identical to a cutting tool material grade. Grades from different manufacturers that are in the same application group could be different as far as application range and performance level are concerned. In other words, the standard does not provide data for grade comparison chart.

The standard suggests that the manufacturers of hard materials should arrange, in proper order, their grades into an application group system according to the relative wear resistance and toughness of the grades. However, the major tool manufacturing companies were reluctant to follow this suggestion. Bernard North, currently the director of Kennametal India, expressed the position of many carbide manufacturers as (Craig 1997):

> We are concerned that, by trying to make the coding simple, you can end up misleading the customer. For example, stainless steels are treated as one material under these color codes. But there's a tremendous variation between different kinds of stainless steels and their machining characteristics. It may well be that some stainless steels are best machined with a tool that would ordinarily be used on cast iron. The actual application ranges of our tools really don't fit into a nice, neat split between workpiece materials. We have grades that are used successfully on gray cast irons, ductile cast irons, low-carbon steels, and stainless steels. How do you classify those grades according to the ISO system?

Therefore, the major tool manufacturers developed their own systems *generally* based upon requirements of ISO 513. One may only take a guess about the degree of such *generality*.

For example, Sandvik Coromant Co. has introduced the CoroKey system (Figure 3.27), which employs the ISO color designations as shown in Figure 3.28, but uses a different set of letter

FIGURE 3.27 CoroKey™ tool material identification system.

ISO	Material	ISO	Material
ISO P	**Steel** Reference material: Low alloy steel, CMC02.1/HB 180	ISO N	**Aluminum alloys** Reference material: Cast, non-aging, CMC 30.21/HB 75
ISO M	**Stainless steel** Reference material: Austenitic stainless steel, CMC 05.21/HB 180	ISO S	**Heat resistant alloys** Reference material: Ni-based, CMC 20.22/HB 350
ISO K	**Cast iron** Reference material: Gray cast iron, CMC 08.2/HB 220 Nodular cast iron, CMC 09.2/HB 250	ISO H	**Hardened steel** Reference material: Hardened and tempered, CMC 04.1/HRC 60

FIGURE 3.28 CoroKey™ tool material identification system: workpiece material.

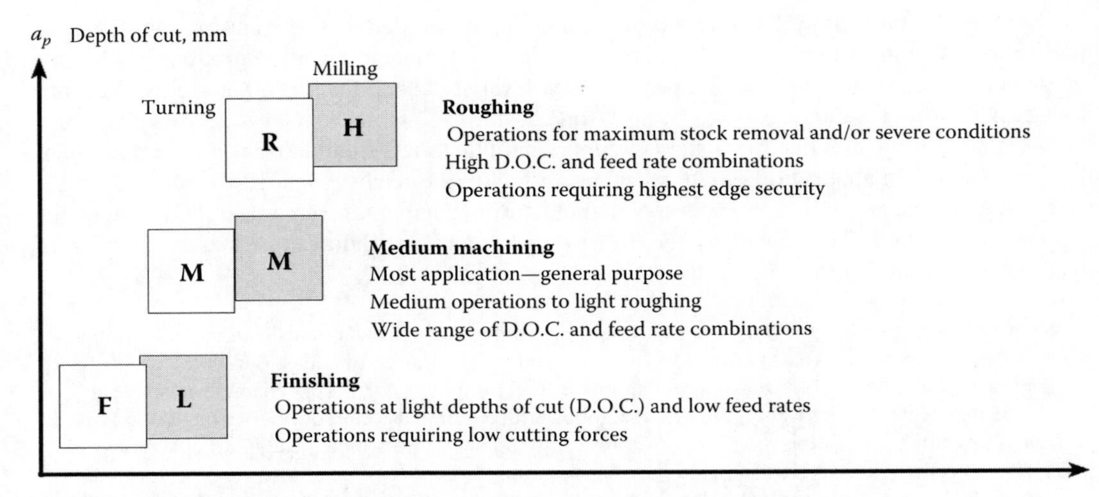

FIGURE 3.29 CoroKey™ tool material identification system: type of application.

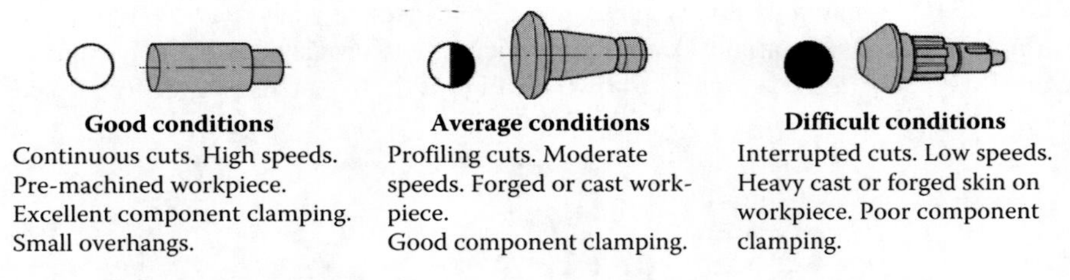

FIGURE 3.30 CoroKey™ tool material identification system: machining conditions.

prefixes—R for roughing, M for medium machining, and F for finishing (Figure 3.29). It also includes a graphical symbol for machining conditions (Figure 3.30).

Most carbide-tool manufacturers classify their products by subjecting the carbide material in question to performance tests run under a variety of conditions and parameters. Understanding the meaning behind a carbide-tool manufacturer's grade determination is not simple. One must read the manufacturer's literature carefully to see what it recommends in terms of machine type, speeds and feeds, workpiece materials, and maximum depth of cut (stock as referred to in catalogs) the tool can withstand. However, there is no guarantee that suggestions made by the carbide-tool manufacturers catalog and sales/technical representatives will work for a given application. This is because the in-house testing is very expensive for tool/carbide manufactures and physically cannot cover the extremely wide variety of applications so that their recommendations are mainly based on the field-test data. The latter is very subjective as the feedback information is normally far from complete as mainly provided by the sales force; thus, it is highly inconsistent.

With this lack of a one-to-one correlation between carbide materials and their classifications, combined with the subjectivity of the classification procedure, it's no wonder that carbide tools of the same grade classification from different manufacturers can vary widely in characteristics and performance. The fact that two brands have the same code doesn't mean anything. They could have completely different compositions and properties.

3.4.4 Problem

The main problem many end users have with the current C and ISO codes is that they seem to promote inconsistencies from one manufacturer to another and, hence, make switching from one brand of a given carbide grade to another brand of the same grade of similar performance a nontrivial business. That makes it hard to judge whether a switch of brands will actually lead to better, or poorer productivity. For example, if one buys a new brand of carbide tool, and it works better in its first application, it may happen because it is a better carbide grade, or it may happen because it is more suited for that particular operation. If the latter is true, the new tool may perform worse than the old tool once parameters of the operation are changed. Alternatively, it may happen because this new set of parameters is too close to carbide grade limits, where a small change can cause breakage, whereas the other tool may have a large safety factor where a small change won't cause a problem at all. Unless one knows exactly what he or she is working with, which implies either a lot of tests have been carried out (a time-consuming and costly option) or some kind of reliable standard should be made readily available. As a result, many users often reluctantly change a carbide supplier as it is very difficult to carry out a test program needed to change safely to another supplier. Some automotive companies have systems in place that discourage their production engineers from changing tool/carbide brands in the middle of a model, so that variants into the manufacture of a product that could affect the quality are not introduced.

An important issue is also with inconsistency of carbide quality from one manufacturing batch to the next. When indexable inserts are used in stable production environment (the same machine, MWF, and other variables), the inconsistency issue can easily be identified. For example, it may happen that a carbide supplier ships the wrong grade of inserts having the same geometry as the previous batch. As tool life suffers just at the beginning of using this new batch, the investigation of the root cause of the problem does not take much time and efforts although a lot of production time can be wasted particularly when the inserts are not on-shelf products. This, however, is not nearly the case when carbide blanks are used because it is rather difficult to identify the root cause of poor tool performance/breakage to inferior carbide grade as many other variable in tool manufacturing/application are involved.

3.4.5 Properties of Cemented Carbides

3.4.5.1 Introduction Notes

In mechanical engineering, the properties of a material play the key role in the selection of the proper material for a given application, calculations of the shape and dimensions of various parts, cost-related consideration, and many other design facets. As such, a practical engineer is equipped with standard and accepted methodologies to carry out this selection and calculations. The information on how to use the standard properties, for example, the tensile strength (both yield and ultimate) or thermoconductivity of a given material, is readily available. Moreover, many CAD commercial packages are available to assists engineers in such calculations. Therefore, it is expected that the properties of cemented carbide as a tool material should be at the same level of usefulness as those of structural materials. In other words, using these properties, a tool/process designer should be able to design/optimize the working part of a cutting tool for a given machining application and select the optimal machining regime bearing in mind minimization of the process cost. Unfortunately, this is not nearly the case in reality. There are two facets of the problem: (1) the listed properties' definitions (determination methods) and (2) their relevance (usefulness) in practical tool/process design.

As for the listed properties' definitions (determination methods), practically all properties are well defined by multiple national and international standards. For example, Davis (1995) presented a thorough description of physical and mechanical properties of cemented carbides with references to corresponding standards. He pointed out that evaluation of physical and mechanical properties of carbides is an important prerequisite to the selection of application-specific grades. How to do such a selection using the known properties in practice, however, is not mentioned by Davis (1995). Multiple company websites and catalogs present their versions of the properties definition, which are often rather far from those defined by relevant standards. In these sources, the methods of their determination are mentioned rarely (references to the corresponding standards) while proper standard designation, required to understand how a particular property was measured (e.g., see below transverse rupture strength [TRS] description), is not mentioned at all.

The relevance of the *standard* properties of cemented carbides, for example, TRS, hardness, and compressive strength, to the process of metal cutting is described only in very general qualitative manners. For example, the higher the hardness, the greater the wear resistance. As such, a simple question:

> Why 'hard' grade (for example, CeraTizit ultrafine grade CTU08 having 4% of Co binder and hardness HRA 95.2) has much greater wear resistance than 'soft' grade (for example, CeraTizit ultrafine grade TSF44 having 12% of Co binder hardness HRA 92.7) as their hardnesses differ only by 5% which very-well may be within the scatter in measurements and/or manufacturing quality variations?

makes sales and technical representatives speechless for a long time. Any further questions, for example, on TRS usefulness in cutting tool calculations (please show me a formula where I can plug in this TRS and calculate some practical tool parameters) should be avoided in order to keep them *in sane*.

In this section, an attempt is made to discuss the essence of the major properties of cemented carbides from the application point of view.

3.4.5.2 Groups of Properties

Major carbide cutting tool suppliers report some properties of cemented carbide blanks for drilling tools. For example,

- Sandvik Coromant Co. reports percentages of carbides and binder, qualitative description of grain size, hardness (*HV*), TRS, compressive strength, and density.
- CERATIZIT Co. reports grain size range, secondary carbide percentage, binder percentage, density, hardness (*HV*), TRS, and fracture toughness K_{1c}.
- Kennametal Co. reports percentage of binder, binder type, specific weight (in the wrong unit though), density, thermal and electrical conductivity, specific heat, mean coefficient of thermal expansion, hardness (HRA), TRS, modulus of elasticity including the method of its determination, compressive strength, Poisson's ratio and method of its determination, relative impact resistance, fracture toughness (ASTM B771-87), endurance limit (for rotating tool it relates to fatigue due to rotation under load), abrasion resistance factor, hot compressive yield stress at 1600°F, and modulus of rigidity.
- Mitsubishi Materials Co. reports the grain size, cobalt content, hardness (in HRA and *HV*), fracture toughness, TRS, density, and Young's modulus. Porosity rating is mentioned, but not reported.

Although the list of the properties is rather great, the question of proper carbide grade selection for a given application using these properties remains wide open. For even the most experienced tool designers, it is not clear how to correlate Poisson's ratio and specific heat of various carbide grades with their performance, that is, how to use these properties in the practice of tool design or tool failure analysis.

In general, a number of groups of properties are distinguished: chemical properties including material composition, mechanical properties, physical properties, metallurgical properties, manufacturing properties (which include, e.g., grindability, brazeability, coatability). In these groups, some properties are related to macro level while others are related to micro or even nanolevels. For example, the modulus of elasticity and hardness are example of properties related to the macro level, grain size relates to the micro level while the population of interfacial defects in the carbide structure is related to the nanolevel.

3.4.5.3 Formation of Properties: Basics

The formation of basic properties is described in this section in the manner suggested by Yao et al. (1995).

The properties of sintered WC–Co composites are critically dependent on their final composition and structure (Soleimanpour et al. 2012). Slight deviations from the ideal carbon content bring about the occurrence of either graphite or a ternary compound. Both of these phases are usually undesirable and result in degradation of mechanical properties and cutting performance. Therefore, the carbon content must be maintained within narrow limits to obtain the desired composite with optimum properties. It is now well established that two types of η phase can be obtained—$M_{12}C$ (Co_6W_6C) of substantially constant composition and an M_6C in which the composition can vary within the range of $Co_{3.2}W_{2.8}C$ and Co_2W_4C. The M_6C type of η phase is in equilibrium with the liquid phase and can nucleate and grow during the sintering process. This not only embrittles the structure by replacing the binder with a brittle phase but also reduces the effective contribution of WC to the strength of the composite. The $M_{12}C$ type is formed in the solid state (during cooling) with small grains distributed throughout the matrix and is therefore effectively less embrittling. Figure 3.31 shows an isothermal section of the WC–Co phase diagram at 1400°C (McHale 1994), which provides information on phase equilibria in the sintering range of commercial WC–Co composites.

Compared to other refractory carbides, the thermodynamic stability of tungsten carbide is relatively low, as is its room temperature hardness. At high temperatures, however, most cubic carbides rapidly lose their hardness, whereas the hardness of WC is quite stable. Coupled with this fact, the unique deformation characteristics of WC are the basis for its predominance as the hard refractory phase in cemented carbides. Other noteworthy properties of WC are its extremely high modulus of elasticity, second only to that of diamond, and its high thermal conductivity. The other refractory carbides have been used effectively as grain growth inhibitors in liquid-phase sintering of WC–Co. It has been found that the effectiveness of a transition metal carbide as a grain growth inhibitor is related to its thermodynamic stability, and they may be ranked as follows: VC > Mo_2C > Cr_3C_2 > NbC > TaC > TiC > Zr/HfC.

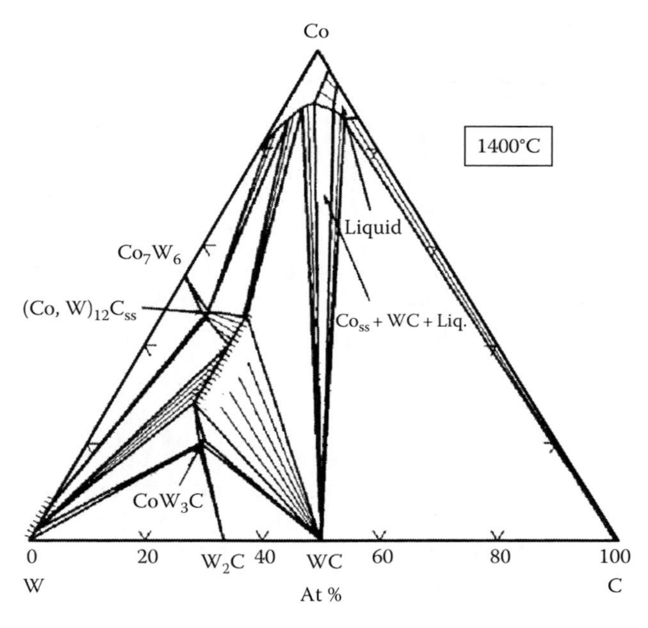

FIGURE 3.31 The isothermal section of WC–Co phase diagram at 1400°C.

It was also found that there is a maximum level above which no further grain growth inhibition occurs. This level is believed to correspond to the maximum solubility of the carbide phase in liquid cobalt. A liquid phase that is saturated with inhibitor carbide would reduce the solubility of WC and thereby reduce its coarsening rate.

The role of cobalt in cemented carbides is to provide a ductile bonding matrix for tungsten carbide particles. Cobalt is used as a bonding matrix because its wetting or capillary action during liquid-phase sintering allows the achievement of high densities. Because of the relatively high cost of cobalt, attempts have been made to design alternate materials, with iron and nickel as the predominant cobalt substitutes. However, sliding wear tests showed that cobalt content is very important for cemented carbides to have good wear resistance (Scussel 1992). Unfortunately, the tests were carried on under conditions that are very far from those found in metal cutting. Particularly, adhesion was not accounted for. This inferior testing result became a common notion that practically extinguished other matrix materials from the market.

Fracture in WC–Co systems with high Co contents has been found to occur mainly by the ductile rupture of Co through void nucleation and coalescence (Ravichandran 1994). Other fracture modes such as fracture along WC–Co interface and WC/WC grain boundary decohesion as well as cleavage across WC grains were also noted. These mechanisms occur especially at low volume fractions of Co binder in the composite at which the contiguity of WC grains begins to increase. The effect of the contiguity of WC skeleton on fracture toughness has also been demonstrated. In a fracture toughness experiment, in a given crack plane, the crack propagation is easy along the relatively weak WC–Co and WC/WC boundaries, and final fracture is primarily controlled by the area fraction of Co regions intact across the crack plane ahead of the tip. Since the WC/WC decohesion and WC–Co interface fracture energies are likely to be lower than the fracture energy absorbed in ductile fracture of the binder, ductile failure of Co can be considered as a primary mechanism manifesting the fracture resistance.

Several investigations attempted to correlate the microstructural parameters and mechanical properties of constituent phases to the experimentally-measured fracture toughness values. In particular, it has been universally found that the fracture toughness increases with volume fraction, the mean free path (MFP) length of the binder, and the size of WC grains. In addition, higher toughness has been suggested to result from the increased contiguity of Co binder, which minimizes the fracture along weak WC/WC boundaries. For a given volume fraction, geometrical arrangement of the ductile binder as a continuous thin matrix phase is beneficial for high toughness while retaining high strength. This arrangement could be the most desirable of several possible arrangements in which the deformation of ductile Co phase is highly constrained in the microstructure (Ravichandran 1994).

3.4.5.4 Some Important Properties

Hardness and fracture toughness are two the most important mechanical properties of cemented tungsten carbide and other cermets. Other mechanical properties, such as flexural strength, wear resistance, and impact resistance, are fundamentally dependent on the hardness and fracture toughness (Fang et al. 2009).

3.4.5.4.1 Hardness

Hardness is the resistance of cemented carbide to penetration by a diamond indenter under a specific load. Hardness is primarily a function of composition and grain size with higher binder metal contents and coarser tungsten carbide grain sizes producing lower hardness values. Conversely, low binder contents and fine grain sizes produce high-hardness values. Hardness is directly related to abrasive wear resistance. It is measured according to ASTM B-294 and ISO 3738 standards.

Hardness is normally determined using the Vickers indentation method according to ISO 3878 standard. The Vickers test uses a square-base pyramid shape diamond penetrator as shown in

Figure 3.32. The load may be varied: 5, 10, 20, 30, 50, or 120 kgf. *HV* 30 is preferred for testing of cemented carbides hardness. The force of a 30 kg, 294 N, is used to create a measurable indentation with minimal cracking at the corners. The load *F* is applied via the penetrator against the polished surface of the test specimen for 30 s. The resulting hardness reading depends on the load and the area of the pyramid penetrator's impression. It is calculated as

$$HV = \frac{1.8544F}{d^2} \tag{3.2}$$

where

load *F* is in kgf (30)

$d = (d_1 + d_2)/2$ is in millimeters

It is understood that any modern hardness tester similar to that shown in Figure 3.32 does this calculation automatically and, moreover, converts the obtained results in other hardness scales such as, for example, in the Rockwell A scale, which is more easily recognized (and thus commonly reported) in the United States.

As mentioned previously, hardness of cemented carbide directly correlates with the cobalt content and grain size. For any given constant cobalt content, when grain size is reduced, the hardness increases. It is shown in Figure 3.33 (Fang 2005).

3.4.5.4.2 Fracture Toughness

As discussed in Section 3.2.2, the toughness of a hard tool material is an even less relevant characteristic bearing in mind the methods used in its determination. For carbides, the short rod fracture toughness measurement is common, as described in the ASTM standard ASTM B771-11 *Standard Test Method for Short Rod Fracture Toughness of Cemented Carbides*. The property K_{IcSR} determined by this test method is believed to characterize the resistance of cemented carbide to fracture in a neutral environment in the presence of a sharp crack under severe tensile

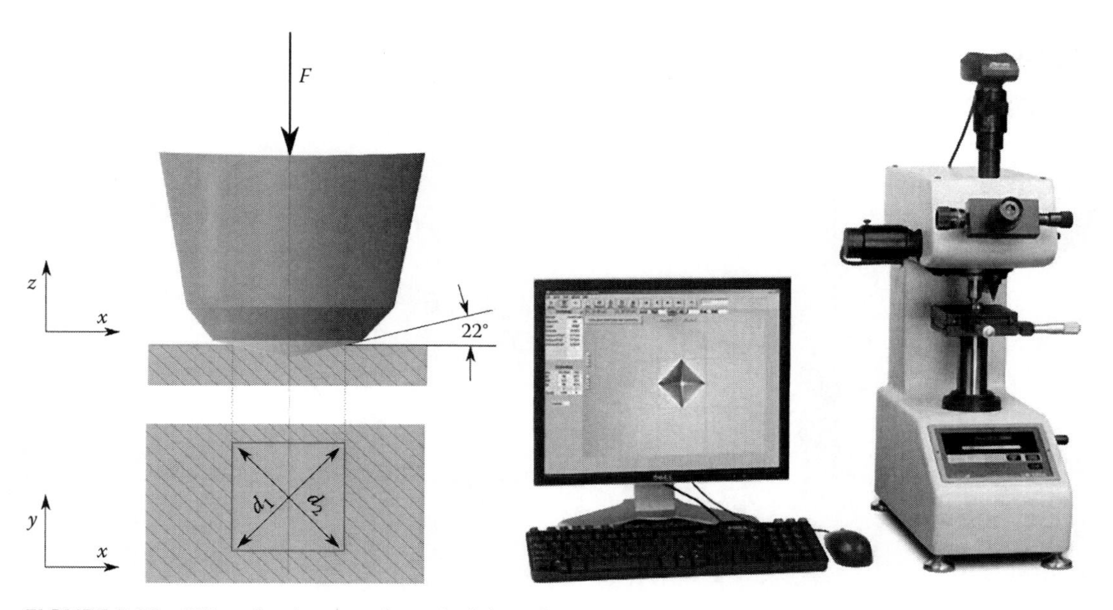

FIGURE 3.32 Vikers hardness testing principle and tester.

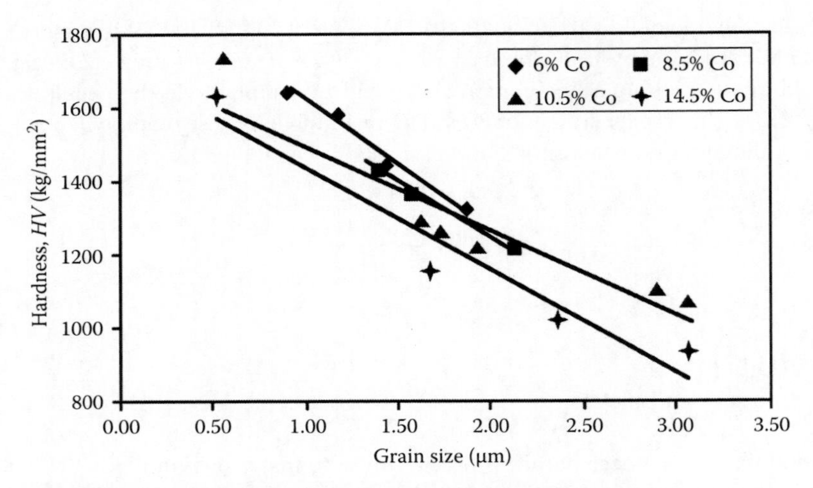

FIGURE 3.33 Hardness correlation with the cobalt content and grain size.

constraint, such that the state of stress near the crack front approaches tri-tensile plane strain, and the crack-tip plastic region is small compared with the crack size and specimen dimensions in the constraint direction.

Fracture toughness of cemented carbide directly correlates with the cobalt content and grain size. For any given constant cobalt content, when grain size is reduced, the fracture toughness decreases. It is shown in Figure 3.34 (Fang 2005).

Normally, fracture toughness and hardness are of prime concern in any cemented carbide grade/make selection for drilling tools as many tool/machining process designers have natural perception of these important characteristics in terms of their significance in maintaining a balance between wear resistance and tool breakage (chipping). Fracture toughness and hardness are well correlated for a given grade of cemented carbide. Figure 3.35 shows an example of correlation curve.

Unfortunately, the common notion in practice is opposite to the research facts shown in Figures 3.33 through 3.35. This common notion stems from multiple sales materials of various tool companies where micrograin carbide grades are promoted. According to this notion well

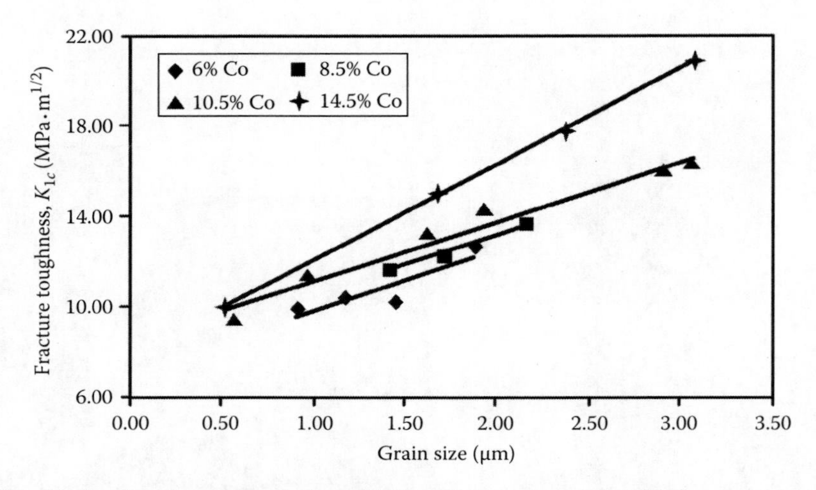

FIGURE 3.34 Fracture toughness correlation with the cobalt content and grain size.

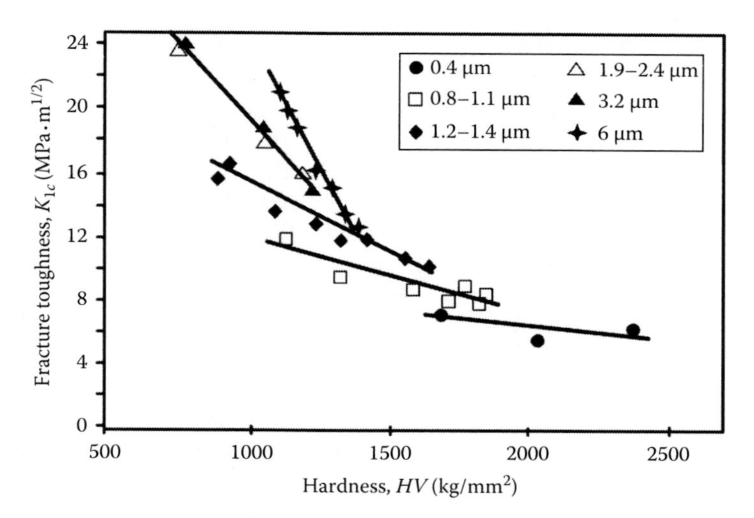

FIGURE 3.35 Relationship between the hardness and fracture toughness.

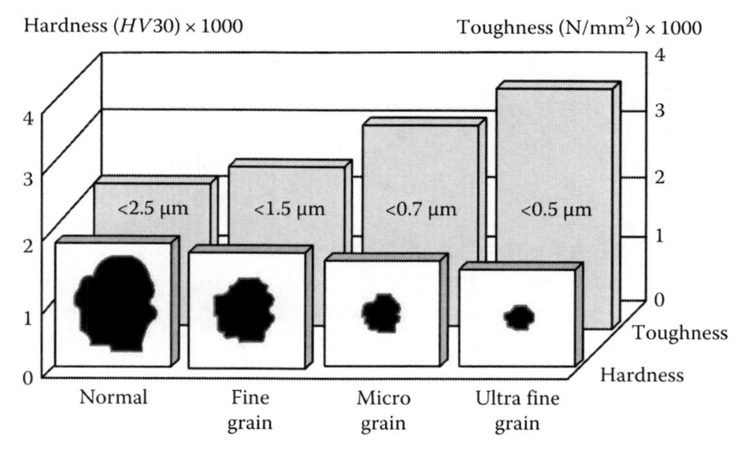

FIGURE 3.36 Hardness and toughness of cemented carbide of different grain size according to incorrect common notion common in practice.

explained in Zelinski (1998), "smaller-grain carbides deliver improved resistance to chipping (toughness) with little change in wear resistance (hardness)." It is shown in Figure 3.36 (Zelinski 1998). This is an example of how a sales pitch gradually becomes a *scientific fact* widely used in carbide selections in practice.

3.4.5.4.3 Transverse Rupture Strength

TRS is one of the most frequently used mechanical properties of cemented carbides (Gurland and Bardzil 1955; Gurland 1963; Exner and Gurland 1970). There are several reasons for its popularity in practice. First of all, TRS is very sensitive to porosity levels (Chatfield 1985). When porosity level is high, TRS values will not only be poor but also very inconsistent. Therefore, it has historically been used as an indicator of the quality of sintered materials in manufacturing. Today, however, due to advances in manufacturing technologies in the industry in the past two decades, the majority of commercial carbide materials are essentially porosity free (besides few exceptions when a sintered carbide material is made in a local *bakery* shop). While TRS continues to be an effective metric of the quality of cemented tungsten carbide products, it is more a true reflection of intrinsic strength of a sintered carbide material. This strength is a complex function of the cobalt content, grain size, carbon balance, and other chemical, compositional, and microstructural factors.

Secondly, because of its sensitivity to pores and other defects, TRS is often also viewed as a measure of *toughness* by tool application engineers and carbide sales force. When a carbide material contains significant porosity, the strong correlation between its TRS and the fracture toughness is easily understood as the pores are viewed as existing defects, of which the critical size is related to critical stress and the fracture toughness defined by Equation 3.1. But when the porosity level is very low or negligible, the relationship between TRS and the fracture toughness is not so straightforward. The flexural strength and fracture toughness are completely different concepts in the context of solid mechanics. TRS is a static tensile property while the fracture toughness is a measure of the resistance of a material to crack propagation. Unfortunately this fact is not a part of cemented carbide characterization found in many books on tool materials, catalogs of carbide-tool manufacturers, and corresponding websites where the properties of carbides are listed (Fang 2005).

TRS is determined in testing carried out according to ASTM B406-96 (2010) *Standard Test Method for Transverse Rupture Strength of Cemented Carbides* and ISO 3327:2009—*Hardmetals—Determination of TRS*. Note that both standards are very particular to the test conditions as a slightest deviation from the standard conditions may affect the test results significantly.

Test arrangement according to ISO 3327:2009 is shown in Figure 3.37. A rectangular test piece of Type A (35 ± 1 mm length, 5 ± 0.25 mm width, and 5 ± 0.25 mm height) or Type B (20 ± 1 mm length, 6.5 ± 0.25 mm width, and 5.25 ± 0.25 mm height) is used. A cylindrical test piece of 25 ± 1 mm length and 3.3 ± 0.25 mm diameter (Type C) can also be used. The standard requires that the test pieces must be free from visible surface cracks and structural defects.

The standard specifically points out that surface preparation of test pieces is an important variable and should be standardized to assure that consistent results are obtained. The test pieces must be ground on four faces, which are parallel to the length with a free-cutting diamond wheel, preferably resin bonded, using copious quantities of coolant. According to the requirement set by the standard:

> No pass shall exit 0.01 mm and all grinding marks shall be parallel to the length. The amount taken off each face shall not be less than 0.1 mm and the surface roughness shall be $R_a \leq 0.4$ μm. The four long edges shall be chamfered to 0.15 mm to 0.2 mm at an angle of 45° and all grinding marks shall be parallel to the length. Type C test pieces shall be centerless ground to a surface roughness $R_a \leq 0.4$ μm.

The standard also allows using test pieces in the as-sintered condition. Such test pieces shall have a chamfer of 0.4–0.5 mm at an angle of 45°, made before sintering to avoid flash. Bend strength results from as-sintered test pieces are generally significantly lower than those for ground test pieces.

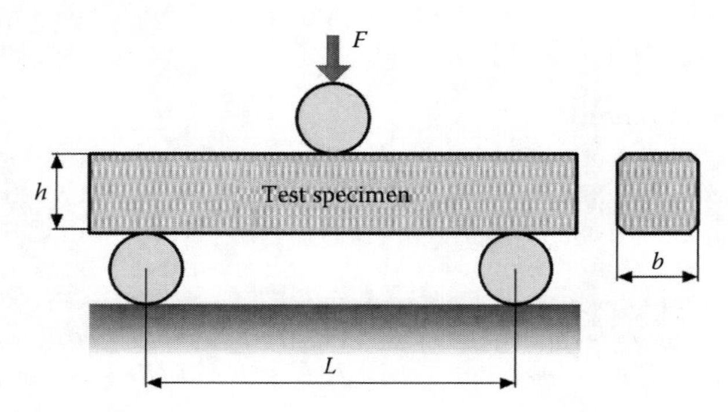

FIGURE 3.37 TRS test schematic.

The standard imposes the folloiwng requirements:

The deviation from parallelism of opposite longitudinal sides, in both the longitudinal and transverse directions, shall not exceed 0.05 mm for each 10 mm length for as-sintered test pieces and 0.01 mm for each 10 mm length for ground test pieces. For round test pieces, opposite sides shall be parallel within 0.015 mm.

Three cylinders (rollers), of which two are freely laying support cylinders with a fixed distance between them and one is freely laying force cylinder, shall be of equal diameter between 3.2 and 6 mm. Alternatively, the force may be applied by a ball having a diameter of 10 mm. Leading of the test piece can only be made via three cylinders if cylindrical test piece is used. Consequently, a ball is only applicable to test piece having plane surface.

The support cylinders and the force cylinder or ball shall be made of tungsten carbide hardmetal, which will not be visibly deformed by the applied force. The surface R_a of the cylinders and the ball shall not be greater than 0.63 μm.

The support cylinders shall be mounted parallel, with a span between them of 30 mm ± 0.5 mm for Type A test pieces and 14.5 mm ± 0.5 mm for Type B or Type C test pieces and to an accuracy of 0.2 mm for Type A test pieces. The mounting of the cylinders shall be such as to minimize deviations from parallelism of the support cylinders.

The test is carried out by applying a force to the force cylinder or ball. As such, the deviation of the line or point of the force application from the middle of the span shall not exceed 0.5 mm for Type A test pieces and 0.2 mm for Type B test pieces.

The TRS R_{bm}, expressed in newtons per square millimeter, for rectangular test pieces is calculated as

$$R_{bm} = \frac{3kFL}{2bh^2} \tag{3.3}$$

where k is the chamfer correction factor. For Type A test piece, $k = 1.03$ when the chamfer is in the range of 0.4–0.5 mm and 1.00 when the chamfer is in the range of 0.15–0.20 mm. For Type B test piece, $k = 1.02$ when the chamfer is in the range of 0.4–0.5 mm and 1.00 when the chamfer is in the range of 0.15–0.20 mm.

For round test piece, TRS R_{bm}, expressed in newtons per square millimeter, is calculated as

$$R_{bm} = \frac{8FL}{\pi d^3} \tag{3.4}$$

where d is the diameter of the test piece in millimeters.

TRS is reported as the arithmetical mean of at least five TRS determinations, round to the nearest 10 N/mm^2.

The standard points out that since the test piece geometry and surface preparation can significantly affect the values of TRS, it is important that the test report shall include the following information:

- A reference to this international standard.
- All details necessary for identification of the test sample.
- The type of test piece and the method of preparation of its surface.
- The method of applying force.
- The test results obtained. The following additional subscripts shall be added to the symbol indicating TRS to indicate the surface condition, that is, sintered (S) or ground (G):
  - For Type A test pieces: A30S or A30G
  - For Type B test pieces: B15S or B15G
  - For Type C test pieces: C15S or C15G

Examples: R_{bm30}(A30S), R_{bm30}(A30G).

The standard points out that, in general, Type B test pieces give strength values which are about 10%–20% higher than those obtained using Type A test pieces, depending on material tested and provided that they have the same surface conditions. Type C test pieces give strength values, which are about 5%–10% higher than Type B specimens whereas the increase of the strength values are material related.

The author's analysis of available information in TRS determination and the relevancy of this characteristic resulted in the following conclusions:

1. The requirements to the test particularities set by standard ISO 3327:2009 show that a small deviation from the test conditions affects the value of TRS. Particular attention is paid to surface roughness, which should not be worse than 0.4 μm. Note that many drilling tool used today do not have such smooth surface.
2. Practically all known to the author, tool/carbide manufacturers' catalogs, websites, and other technical material do not report the value of TRS in the manner required by standard ISO 3327:2009.
3. TRS should not be used as a fracture toughness measure.
4. The available data demonstrate that TRS is directly related to intrinsic mechanical properties of cemented carbides, namely, hardness and fracture toughness, when the effects of porosity are negligible.
5. The relationship between TRS and hardness is also not unidirectional but rather complex. For example, as shown in Figure 3.38, for cemented carbide having 10% Co, within a range of $800 < HV < 1500$, TRS appears to first increase and then decrease as the hardness increases. It reaches a peak value at approximately $HV = 1300$ kg/mm^2. When hardness is greater than 1500 kg/mm^2, TRS values could reach very high values, but the variations of TRS are also high in the high-hardness range. The dependence of TRS on cobalt content, grain sizes, and other microstructure parameters is embedded in its relation to the hardness and fracture toughness (Fang 2005).
6. Although TRS and fracture toughness K_{1c} are related, the relationship is not a linear one-to-one correlation as promoted by sales materials in the field. An example of such a correlation is shown in Figure 3.39. Although it generally follows a common notion that if TRS is high, the material must be *tough*; it could be misleading to assume a higher TRS would mean high fracture toughness.

The author's further analysis of the available data suggests that the use of TRS as a valuable property of cemented carbide should be discontinued until results of deeper studies dealing with the correlation of

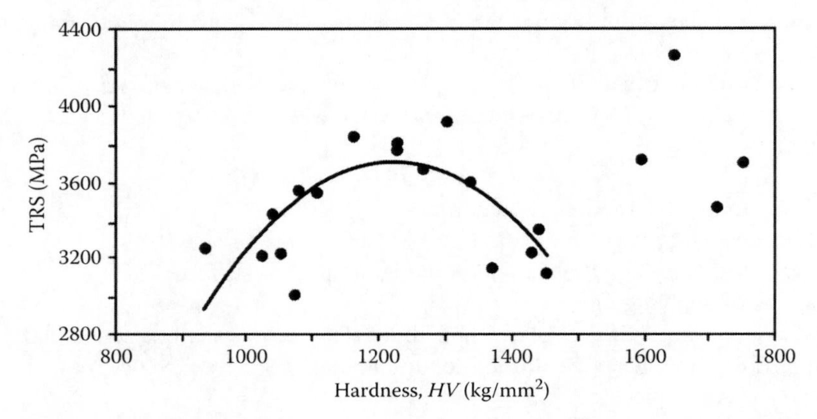

FIGURE 3.38 Correlation between TRS and hardness for 10% Co carbide obtained by Fang. (From Fang, Z.Z., *Int. J. Refract. Met. Hard Mater.*, 23, 119, 2005.)

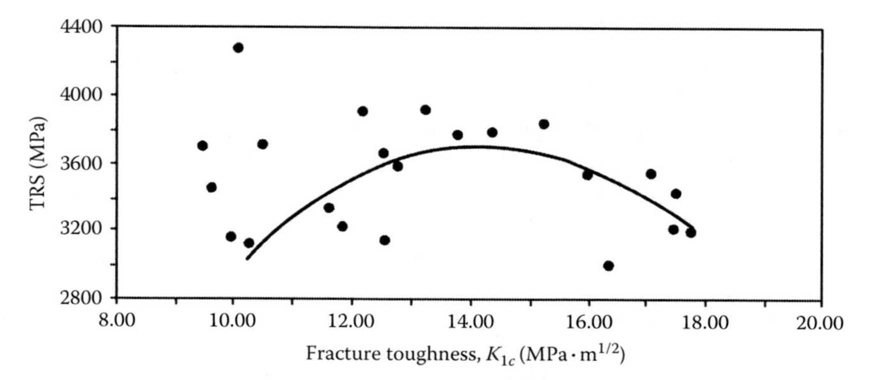

FIGURE 3.39 Correlation between TRS and fracture toughness for 10% Co carbide obtained by Fang. (From Fang, Z.Z., *Int. J. Refract. Met. Hard Mater.*, 23, 119, 2005.)

the TRS and microstructural features of cemented carbide will be available. Such result should explain a great scatter in the results when the standard method for TRS determination is performed even if the test is carried out in laboratory conditions with meticulous preparation of specimens, setting the setup, and carrying the tests.

3.4.5.4.4 Density

Density is the weight per unit volume of cemented carbide measured in grams per cubic centimeter (g/cm^3). It is essentially the weighted average of the densities of all of the components contained in the product and is therefore a check on its composition. For grades containing only tungsten carbide and a binder metal, the density of the composite decreases as the lighter binder metal content increases. Standards ASTM B-311 and ISO 3369 define procedures to measure carbide density. In the literature on carbide properties, density is often also called specific gravity although these two terms are not synonyms. Specific gravity is the density of a substance divided by the density of water. Since (at standard temperature and pressure) water has a density of 1 g/cm^3, and since all of the units cancel, specific gravity is usually very close to the same value as density (but without any units). Although it is probably interesting to know that the specific gravity of tungsten carbide is from 1½ to 2 times greater than that of carbon steel, it is very challenging to use this knowledge in making the proper choice of carbide grade.

3.4.5.4.5 Residual Porosity

Residual porosity is determined by visually examining the polished surface of a sintered sample at 100× or 200× magnification. Ratings for *A*-type porosity (pores less than 10 μm in diameter), *B*-type porosity (pores larger than 10 μm in diameter), and *C*-type porosity (carbon inclusions) are determined by comparing the size and frequencies of each pore type in the sample with those in standard photographs. Each standard photograph is associated with a numerical rating that is used to represent the porosity levels in the sample. In general, edge strength and toughness decrease as the level of residual porosity increases. At high levels of porosity, the wear resistance of the product may also be adversely affected. Standards ASTM B-276 and ISO 4505 define procedures to measure residual porosity by a visual examination of a polished surface. For example, a sample rated at A-02 B-00 C-00 porosity has pores less than 10 μm in diameter with a total pore volume of 0.02% of the sample. A rating of A-04 B-02 C-00 indicates a total pore volume of 0.06% for pores less than 10 μm across, 0.02% for pores between 10 and 40 μm, and an overall pore volume of 0.08%. Modern carbide grades from reputable carbide suppliers, however, have very little or practically no porosity so that this property is not even reported nowadays in the list of grade properties.

3.4.5.5 Nondestructive Testing of Carbide Properties Using Magnetic Measurements

Magnetic measurements are vital for controlling the consistency of carbide products. They comprise (1) moment (saturation) used for quality control and to estimate Co–W–C binder-phase composition and (2) coercivity used for quality control and to estimate WC grain size (Roebuck et al. 1999). The use of nondestructive methods to assess consistency of structure is very important because cemented carbides have microstructures on a very fine scale, which are difficult to inspect. Also, they are strong and not easy to test mechanically. However, cemented carbides contain cobalt which is ferromagnetic. Consequently, measurements of magnetic properties are widely used in industry to assess consistency. Both coercive force and magnetic saturation (moments) are measured for this purpose. If interpreted correctly, they can be linked to the cobalt composition and hence overall properties and performance. Although standards ASTM B-886, ASTM B-887, and ISO 3326 provide methods of magnetic saturation and coercive force measurement and reporting, the interpretation of the results obtained and how to establish their linkage with cemented carbides performance are not clear so that many companies and users developed their own sometime opposite interpretations.

The essence of magnetic saturation (moment) measurements can be explained as follows. Tungsten carbide–cobalt composites generally contain 5–20 wt% cobalt as a binder phase. The cobalt binder phase contains both tungsten and carbon in solution. The amount of tungsten dissolved in the cobalt binder phase can be assessed by measurement of the magnetic saturation or magnetic moment of the carbide because the saturation value of cobalt decreases linearly (Freytag et al. 1979; Roebuck and Almond 1988) with the addition of tungsten and is not affected by the amount of carbon in solution. The term *magnetic saturation* is an abbreviation for the saturation (or maximum) value of magnetic induction that can be obtained using a procedure given by ASTM B886-03 (2008) *Standard Test Method for Determination of Magnetic Saturation (Ms) of Cemented Carbides.* The values obtained give information that is related to the composition of the binder phase. For proper interpretation, it is necessary also to have additional data on the chemical composition, for example, wt% of Co and wt% of additional contaminant elements such as Ni and Fe in WC-based carbides.

Designations: I_s is magnetic moment/unit volume (intensity of magnetization/unit volume), tesla; $J_s = 4\pi I_s$ is saturation induction (polarization), tesla; σ is magnetic moment/unit weight (intensity of magnetization/unit wt), $T \cdot m^3/kg$; $\sigma = I_s/\rho$ where ρ is the density.

Table 3.10 provides examples of the different methods for expressing the magnetic moment data in SI units. In practical materials, there is a range of magnetic saturation and moment values associated with a variation in the carbon content of the carbide as shown in Figure 3.40.

TABLE 3.10
Typical Ranges of Magnetic Saturation and Moment Values

	5 wt% Co	10 wt% Co	15 wt% Co
Carbide			
Magnetic saturation, $4\pi\sigma$	8.0–10.1	16.1–20.2	24.1–30.3
Magnetic moment, σ	0.64–0.80	1.28–1.61	1.92–2.41
Percentage saturation	80–100	80–100	80–100
Binder phase			
Magnetic saturation, $4\pi\sigma$	160–202	160–202	160–202
Magnetic moment, σ	12.8–16.1	12.8–16.1	12.8–16.1
Percentage saturation	80–100	80–100	80–100

Note: SI units—$\mu T \cdot m^3/kg$.

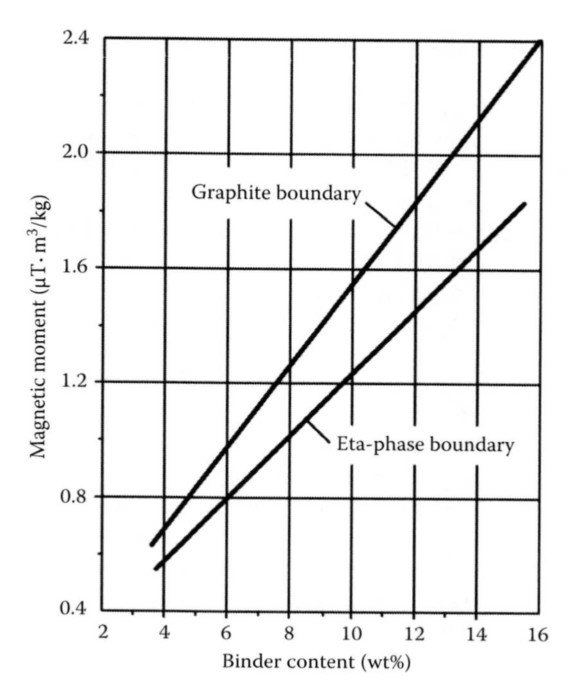

FIGURE 3.40 Magnetic moment of WC–Co carbide plotted against Co content.

In cemented carbides, the binder contains W and C in solution. Presence of C does not affect the cobalt magnetic saturation and moment values, but W decreases them. The decrease can be represented as

$$4\pi\sigma_B = 4\pi(\sigma_{Co} - 0.275m_w) \tag{3.5}$$

where

$4\pi\sigma_{Co}$ is 202 μT · m³/kg

m_w is the wt% W in the Co binder phase

When $4\pi\sigma_B$ attains a value of about 160 μT · m³/kg, eta phase begins to form and so the range 202–160 μT · m³/kg for the magnetic saturation of the cobalt phase defines the two-phase field in *WC–Co carbides*. This expression can be used to either calculate the W content of the binder phase from the magnetic moment or vice versa.

If a 6% Co cemented carbide has a magnetic saturation value of 11 μT · m³/kg, then $4\pi\sigma_B = 11/0.06 =$ 183.3 μT · m³/kg and therefore $m_w = (202 - 183.3)/(4\pi \cdot 0.275) = 5.4$ wt%.

The previous discussion has assumed that the binder phase in WC–Co carbides is a Co–W–C alloy. In practice, other elements can be present, particularly Fe, which can have an effect on the magnetic moment. This makes precise calculations difficult (Roebuck et al. 1999).

Magnetic coercivity measurements usually correlate with the grain size. There is a standard for coercivity measurements for hardmetals (cemented carbides), ISO 3326. Measurement of the WC grain size (d_{WC}) is not standardized, although ISO 4499 provides a method based on the intercept technique. The methods require care, and this is particularly the case for very fine-grained materials. The results of grain size measurement methods frequently depend on the method of preparation and examination of samples.

The use of linear intercept distributions (Heyn method) is a common method by which the grain size and distribution of the phases in composite hard materials is measured (Roebuck et al. 1999).

The distribution is generally plotted on cumulative probability paper (to linearize the plot) with the abscissa plotted as log (intercept distance). The distribution is obtained by counting individual intercept lengths (at least 200) on a randomly placed line laid across a polished and etched microstructural image. Area counting, by measurement of the number of grains per unit area, (Jeffries method) can also be used. This can be performed using optical microscopes for some coarser-grained WC carbides, but for many materials, the grain size is too fine to be resolved optically and so SEMs must be used. In very fine-grained carbides, the average grain size can be very small, less than 0.5 μm, for the arithmetic mean linear intercept, and good quality images are difficult to obtain in conventional SEMs. However, it has been shown that a high-resolution instrument that uses a field emission electron beam can provide acceptable images (Roebuck and Bennett 1986). Some research indicates that orientation imaging techniques based on EBSD provide more accurate measurements.

The most widely used nondestructive method for the assessment of grain size measures coercive force. The coercive force required to demagnetize a WC–Co carbide is primarily related to the WC–Co interphase area, because magnetic domain walls are pinned by these phase boundaries. The WC–Co interphase area is inversely related to the WC grain size. For smaller values of the latter, the interphase area increases and consequently so does the coercivity. However, coercivity also changes with variations in Co composition, and this should be allowed for in empirical equations relating WC grain size to coercivity (Roebuck 1995). Furthermore, difficulties can arise when trying to correlate coercivity with grain size due to effects of differences in cooling rate from the sintering temperature. This results in changes to the composition of the binder phase, which can have a significant effect on coercivity. For example, in low-carbon content carbides, it is possible to precipitate Co_3W in the binder phase on ageing of the carbide after accelerated cooling. This procedure can result in very significant changes in coercivity. Additionally, there may affect coercivity due to the phase boundaries between the hexagonal and face-centered cubic (FCC) phases in individual binder regions. Also, deformed carbides show a significant increase in coercivity due to deformation of the binder phase, including the creation of phase boundaries, twins, and stacking faults, which can be at least partially removed on subsequent annealing. Finally, the role of WC–WC contiguity must also be acknowledged since it can be seen that two materials of different contiguity, but similar grain size will have different interphase areas and hence different coercivities. Contiguity is a measure of the proportion of each WC grain that is in contact with adjacent contiguous WC grains (Roebuck et al. 1999).

Figure 3.41 shows a typical correlation between WC inverse grain size and coercivity for a range of WC–6 wt% Co carbides. It shows the relationship between magnetic coercivity and arithmetic mean linear intercept. These results can be characterized by a straight line given by

$$K = a + bd_{WC}^{-1} \tag{3.6}$$

where

K is the coercivity in kA/m
a and b are constants
d_{WC} is the WC arithmetic mean linear intercept in μm

The results in Figure 3.41 show more scatter than might be expected from the equipment used to make the coercivity measurements. It is possible that this scatter is due to a combination of the previously discussed microstructural effects on coercivity. Moreover, the linear regression fit for all the data did not give particularly good predictions for materials with grain sizes greater than about 2 μm.

Experimental results obtained for different Co content in the format of Equation 3.6 can be plotted as a property map. An example is shown in Figure 3.42, bearing in mind the constraint that it has not been extensively validated for coercivity values greater than about 15 kA/m, that is, equivalent to WC grain sizes (arithmetic linear intercept) less than about 0.8 μm. Additional work is needed to evaluate the finer-grained materials.

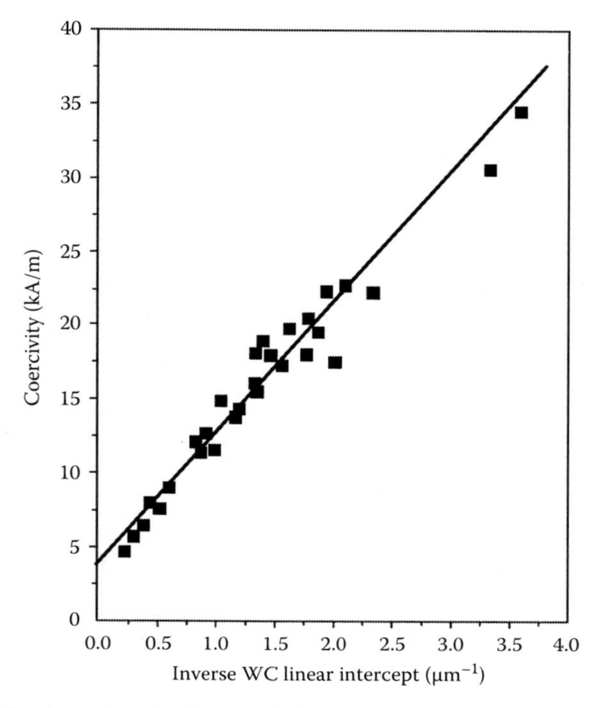

FIGURE 3.41 Coercivity–inverse grain size correlation.

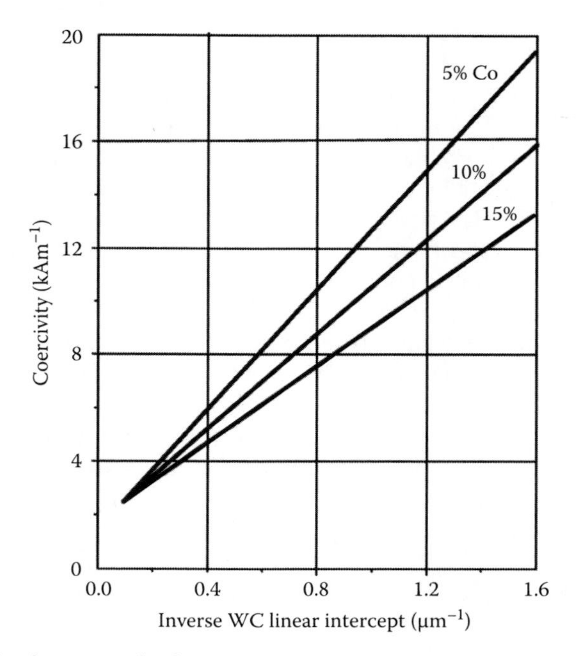

FIGURE 3.42 Coercivity–inverse grain size property map.

One needs to be aware, however, that the coercivity also varies with carbon content and W content of the binder phase. The materials in the evaluation exercise contained a range of carbon contents, and this has not been allowed for in the construction of the property maps. It has been assumed that the compositions are approximately in the center of the two-phase field. An estimate of uncertainty due to this source showed (Roebuck et al. 1999) that the calculated size varied by about ±20% across the two-phase field, whereas the measured variation is probably ±10%. The difference is thus probably due to variation in composition.

It is also possible that microstructural instabilities in low C carbides or residual stresses could affect measurements of coercivity through changes in internal strain and composition. These effects contribute added uncertainty to the correlation between grain size and coercivity and need to be systematically examined.

The dependence of mean grain size of cemented carbide on coercivity K within the grain size range of 1–5 μm allows the coercive force to be used as a parameter for nondestructive inspection of the mean size of carbide grains after sintering. As such, the most advanced correlation model developed by Gorkunov, Ulyanov, and Chulkina (Institute of Engineering Science, Russian Academy of Sciences) is

$$K = 10.2 + 21.7\left[\frac{\left(\langle D\rangle \rho_{WC} C_{Co}\right)}{\left(\rho_{Co}\left((1 - C_{Co}\right)\right)}\right]^{-4/3} \text{ A/cm} \tag{3.7}$$

where

$\langle D\rangle$ is the mean size of tungsten carbide grains

ρ_{WC} and ρ_{Co} are the densities of tungsten carbide and cobalt, respectively

C_{Co} is cobalt mass fraction

Note that the numerical coefficients in Equation 3.7 are determined empirically on experimental samples of hard tungsten–cobalt alloys. Although this model works well for laboratory-prepared specimens of carbide, for commercial hard alloys, the mean size of carbide grains can change within a wider range than in the case of experimental specimens; therefore, the coefficients 10.2 and 21.7 of Equation 3.7 may assume somewhat different values.

The strength and hardness of cemented carbides depend significantly on cobalt content. It can be quantitatively determined by measuring saturation magnetization. Magnetization curves for hard alloys have a peculiarity—linear magnetization increase in fields exceeding 1000 A/cm. The cobalt interlayer comprises thin ferromagnetic plates forming a continuous framework; in the general case, the plane of the plates can be both parallel and perpendicular to the direction of the field applied. Under demagnetization, shape anisotropy orients the magnetization vectors along the plane of the plates. The field applied first of all causes domain wall displacement in the plates whose planes are close to the direction of magnetization. In fields of ≈1000 A/cm, the domain wall displacement processes terminate for the most part. As the field increases, the further magnetization growth is due to the turn of the magnetization vectors to the direction of the field in the plates whose plane is perpendicular to the direction of the field. Consequently, the magnetic saturation of the composite material will occur in fields comparable with cobalt plate shape anisotropy fields. The evaluation of the shape anisotropy field for an infinitely thin cobalt plate in the direction perpendicular to its plane gives the order of 10 kA/cm. One should take this into consideration when developing magnetizers for nondestructive test instruments to be used for items made of cemented carbides.

The cobalt interlayer in hard alloys is not pure cobalt but a solid solution of tungsten and carbon atoms in cobalt (γ-phase) resulting from partial dissolution of tungsten carbide in cobalt. Thus, 8–10 wt% of tungsten carbide can dissolve in cobalt at 1340°C. As the temperature goes down to 1000°C, the solubility of tungsten carbide becomes 2–4 wt%. The physical properties of the γ-phase differ from those of pure cobalt, and, moreover, they depend on the amount of carbide dissolved.

Cobalt content in actual alloys is generally below 20–25 mass percent, the alloy structure being a mixture of two phases—tungsten carbide and the γ-phase (WC + γ). If carbon content deviates from the stoichiometric composition of the WC-phase beyond the homogeneity region of the γ-phase in sintering, yet another phase may appear.

With excess carbon, the isolation of free carbon (WC + γ + C) takes place, whereas with scarce carbon—the isolation of a η_1-phase (WC + γ + η_1) rich in tungsten and cobalt (Co_3W_3C). The isolations

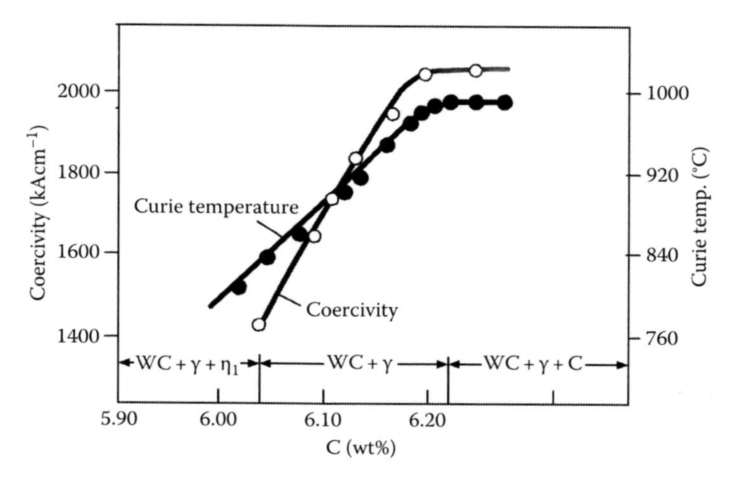

FIGURE 3.43 The effect of carbon content (C) in the cobalt binder on the saturation magnetization and the Curie temperature of the WC alloy—10 wt% Co.

of the η_1-phase make the cobalt binder brittle, this being explainable, firstly, by the brittleness of the η_1-phase and, secondly, by the fact that some cobalt from the composition of the γ-phase transforms to the η_1-phase. Since the strength characteristics of the alloys decrease drastically in this case, the presence of the η_1-phase in the alloy structure is impermissible. It follows from Figure 3.43 that the magnitude of saturation magnetization decreases significantly as carbon content decreases with respect to the stoichiometric composition of the WC-phase within the homogeneity region of the γ-phase.

The reason is that a lack of carbon causes excess atoms of free tungsten to dissolve in the cobalt binder. This results in greater parameters of the cobalt lattice and poorer exchange interaction of cobalt atoms as the interatomic distances grow, and this causes lower M_s of the alloy. When carbon content is excessive with respect to the stoichiometric composition of the WC-phase, some carbon atoms dissolve in the cobalt binder without changing the lattice parameters and the magnetic properties of cobalt, the others being isolated in the form of graphite.

The variation in the exchange interaction has a natural effect on the Curie temperature T_C of the alloys as follows from Figure 3.43. Curie temperature is a phase-sensitive characteristic, it is independent of the carbide phase grain size and of cobalt content, and it is governed only by the number of tungsten and carbon atoms dissolved in the cobalt binder within the two-phase region. The lack of carbon beyond the two-phase region accompanied by the formation of the η_1-phase will cause a further decrease in saturation magnetization, as part of the cobalt is bound in the non-ferromagnetic η_1-phase.

For practical measurement of magnetic properties of cemented carbide, a number of available equipment can be used or modernized for such measurements. For example, new AMH-5800 magnetometer is the latest in magnetic measurement technology for the measurement of magnetic properties of cemented carbides (WC in Co or Ni matrix) and semihard magnetic materials. It provides fast, repeatable, and accurate measurements.

The AMH-5800 measures the following parameters:

- Coercivity
- Magnetic moment
- Weight-specific saturation magnetization
- Magnetic polarization
- % Co or any other magnetic material in the alloy

As an example of the proper use of this sophisticated equipment, consider the results of study of tool life of carbide rock drills having multiple carbide cutting edges for drilling holes of large diameters (Salnikov 2009). This study found that tool life best correlates with the number of cycles achieved

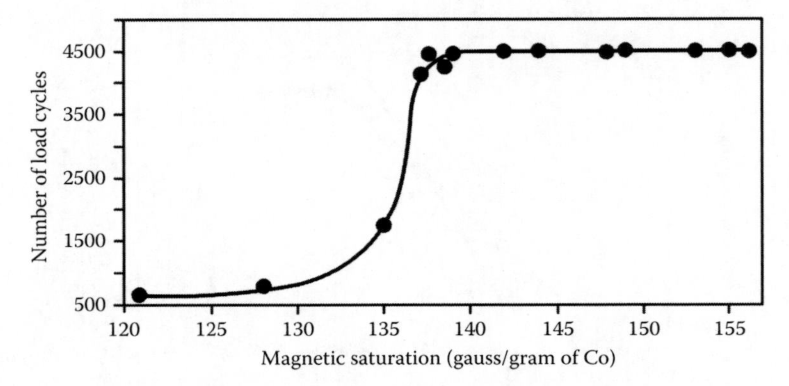

FIGURE 3.44 Number of load cycles achieved in shock test as a function of magnetic saturation for WC–10 wt% Co.

in shock load test. Figure 3.44 shows the number of load cycles achieved in shock test as a function of magnetic saturation for WC–10 wt% Co.

Magnetic saturation is a measure of the lump sum magnetism of pure cobalt. Pure cobalt in cemented carbides has normal magnetic properties. As discussed above, during sintering, some of the cobalt in the cemented carbides will alloy with either carbon or tungsten carbide. The maximum saturation of pure cobalt is 160 Gs/g Co. In cemented carbides, this saturation is always smaller. As can be seen in Figure 3.44, magnetic saturation of 145 Gs/g Co is considered as the lower acceptable limit, which results in at least 4000 cycles life. Using results of this study, magnetic saturation range 145–160 Gs/g Co was set in the acceptance test for cemented carbide for rock drills.

The next step in the study was to find out the correlation of magnetic saturation with an important component(s) (parameter(s)), which causes significant change in carbide performance. Detailed metallurgical analysis and tests based on the diagram shown in Figure 3.31 reveal that the carbon content is fully responsible in the difference in performance. Although some metallurgical sources (e.g., Upadhyaya 1998) point out that the maximum strength of cemented carbides is within the very narrow boundaries of the carbon (depending upon the cobalt content) of two-phase domain found in the phase diagram shown in Figure 3.31, the scatter of strength even within this domain is great. Table 3.11 shows the test results. As can be seen, insignificant change in the carbon content causes significant change in the cemented carbide performance.

TABLE 3.11
Life of Carbide Inserts (WC–10 wt% Co) in Cyclical Load Test as a Function of Carbon

Insert Code	Number of Shocks to Failure	Carbon Content (wt%)
R8812	256	5.37
R9629	2078	5.42
R3712	≥4000	5.49
R4418	1560	5.40
R5677	≥4000	5.51
R8652	345	5.39
R9314	157	5.36
R4875	≥4000	5.52
R4986	≥4000	5.50
R9963	1276	5.39
R6788	≥4000	5.48

The test results suggest that the optimal carbon content within the two-phase domain of the phase diagram shown in Figure 3.31 is from 6.08% to 6.13%, which assures the best combination of cemented carbide toughness and resistance to chipping for the discussed application. The use of a magnetometer provides fast and reliable quality control method for carbides.

For a given drilling application, the correlation similar to that shown in Figure 3.44 should be established for carbide chosen for the application and then used as a quality procedure. Moreover, many other useful correlations can be established using magnetic characteristics in magnetometer measurements. In other words, some tests, analysis, and knowledge of basic carbide metallurgy are needed to utilize the full capability of a modern magnetometer.

The discussed experimental findings suggest that a closer look on the carbon role and content in cemented carbide should be taken. Remarkably, there is a massive material available on the subject in the carbide-related metallurgical literature. For example, Upadhyaya (1998) pointed out that carbon content has major influence on grain growth in sintering. A slight increase in carbon content results in great grain growth rate so that the control of this content is crucial in carbide manufacturing.

Exner in his classical work on cemented carbides (Exner 1979) pointed out that the properties of WC–Co hardmetals (cemented carbides) are determined by three parameters, namely, the properties of the carbide phase, the cobalt binder, and the interaction between carbide and binder. Even a slight variation in the carbon content has a marked effect on the structure and mechanical properties of hardmetals. Gurland pointed out (Gurland 1954) that stoichiometrically, pure WC contains 6.17 wt% carbon. There is a very narrow range of ±0.1 wt% carbon where the hardmetal microstructure still remains in the two-phase region. Too high carbon content leads to the formation of uncombined (free) carbon in the microstructure, while too low content of carbon results in formation of the intermetallic eta phase. Both these features restrict quantity and uniformity of distribution of the binder phase, consequently, lowering the fracture toughness.

Upadhyaya et al. (2001) found that the major mechanical properties of cemented carbides depend on carbon content. For example, Figure 3.45 shows the variation of TRS as a function of carbon content of WC–6% Co alloy. Note that TRS is maximum in the two-phase $\alpha + \beta$ region (shown as γ phase in Figure 3.43). In the three-phase regions (WC + Co + η and WC + Co + C), the strength decreases rapidly. This result is in full agreement with that shown in Table 3.11 obtained under considerably different test conditions and for different grade of cemented carbide.

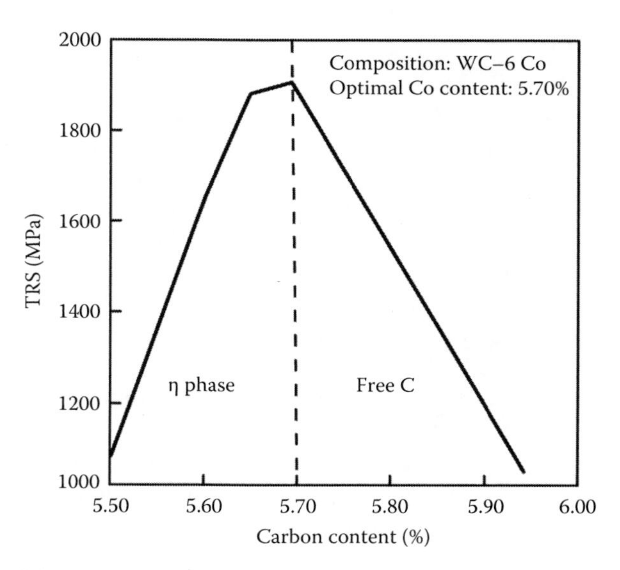

FIGURE 3.45 Effect of the total carbon content on the TRS of WC–6 Co alloy.

One explanation for the carbon phenomena is that the carbon content influences the solubility of tungsten in the cobalt binder of conventional WC–Co-based cemented carbides. In this way, the solid-solution hardening of the binder phase can be controlled (Zackrisson and Andren 1999).

The following conclusions can be drawn from the foregoing analysis:

1. Carbon content is one of the most important characteristics of cemented carbides. It affects a great variety of mechanical and microstructional properties of these materials.
2. The highest mechanical properties are achieved in the narrow zone of the two-phase region of the phase diagram shown in Figure 3.31. Note that such a diagram is grade and alloy content specific and thus should be known to make an intelligent decision on carbon content of a given grade of cemented carbides.
3. In metal cutting, in general, and in drilling, in particular, carbon content of various carbide grades are not reported or even mentioned in the list of properties of recommended carbide grades. In the author's opinion, this is the major hurdle in understanding the cutting performance of various cemented carbide grades and thus it should be corrected.

3.4.5.6 Nanoparticle Carbides: Research and Expectations

Nanoscale particle research has become a very important field in materials science. Nanoscale particles (1–100 nm) usually have physical properties different from those of large particles (10–100 μm) or the molecular/atomic species. It has been found that nanoparticles exhibit a variety of previously unavailable properties, depending on particle size, including magnetic, optical, and other physical properties as well as surface reactivity (Yao et al. 1995). Recent experiments have shown that consolidated nano-materials have improved mechanical properties, such as increased hardness of metals and increased ductility and plasticity of ceramics. The unique properties of nanoscale particles and nano-grain bulk materials can be attributed to two basic phenomena.

The first is that the number of atoms at the surface and/or grain boundaries in these materials is comparable to that of the atoms located in the crystal lattice; thus, the chemical and physical properties are increasingly dominated by the atoms at these locations. The second phenomenon is the *quantum-size effect* or quantum confinement effect. When particles approach the nanometer size range, their electronic and photonic properties can be significantly modified as a result of the absence of a few atoms in the lattice and the resulting relaxation of the lattice structure.

In conventional composites, the toughness decreases with increasing hardness, but the increase of hardness in nanostructured composites does not decrease their bulk fracture toughness. The difference between the fracture toughness of WC–Co using nanostructured powder and that using conventional ultrafine powder has also been shown to depend on the cobalt contents. For example, when a sample was sintered using nanostructured WC–10% Co powder, the sintered hardness was HRA 92.0 and the fracture toughness K_{1C} was 12.0 $\text{MPa}\cdot\text{m}^{1/2}$, while the fracture toughness of the conventional WC–10%Co carbide is approximately 9.0 $\text{MPa}\cdot\text{m}^{1/2}$ when HRA 92.0. This indicates over 30% increase in the fracture toughness.

The advances of technologies for the synthesis of nanosized particles, particularly nanosized WC and Co powders (Fu et al. 2001; Ban and Shaw 2002), raise prospects for superior mechanical properties of nanocrystalline WC–Co. The production of bulk nanocrystalline (grain sizes <100 nm) cemented tungsten carbide, however, remains a technological challenge because of the rapid grain growth during sintering. The rapid grain growth, or coarsening of nanosized particles, is an issue that affects not only the cemented tungsten carbide but also the manufacture of bulk nanocrystalline materials of a broad range of ceramic and metallic materials. Compared to the sintering of conventional micron-sized powders, the sintering of nanosized powders has an additional challenge of retaining nanoscaled grain sizes upon achieving full densification (Wang et al. 2008).

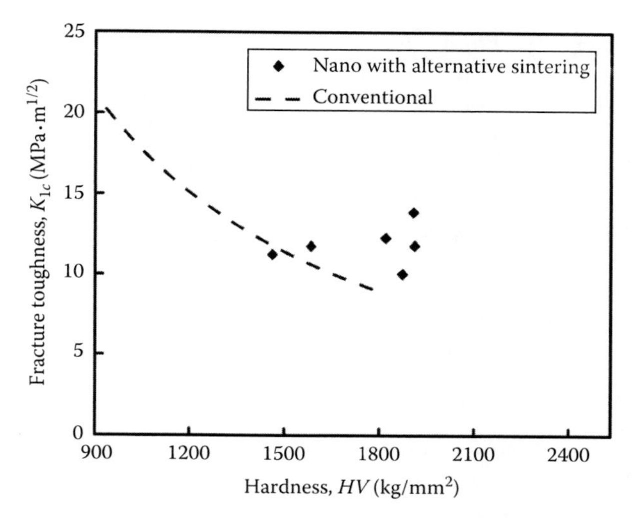

FIGURE 3.46 Fracture toughness versus hardness relationship of conventional WC–Co materials (dashed line) and the materials made from nanosized WC–Co powders by sintering via alternative techniques.

In the author's opinion, so far no one study points out the significant advantages of nanocarbides compared to conventional submicrograin-cemented carbides in quantitative manner related to metal cutting not to metallurgy (mechanical, structural, descriptive, and so on). For example, a report of many-year comprehensive study on the matter *Development of Bulk Nanocrystalline Cemented Tungsten Carbide for Industrial Applications* sponsored by DOE (Fang and Sohn 2009) suggests that the fracture toughness of a nanocarbide may not decrease in proportion to its hardness as in conventional carbides as shown in Figure 3.46. Being inconclusive as no grade of conventional and nanocarbides are indicated, these data were obtained using a different sintering technique so the claimed advantage can be very-well due to altered sintered technique. Although the results of wear test presented in the report showed that the wear resistance of nanostructure WC–Co is somehow higher than that of conventional cermets in proportion to their hardness, the test conditions are highly questionable. In this test, a series of grades of WC–Co with varying grain sizes and cobalt contents were made into pins with polished hemispherical shape. The radius of the tip was about 3.5 mm. During the tests, the pins slid under a modest load (9.8 N) on commercially ground silicon nitride plates. In reality, the stress found at the tribological interfaces in metal cutting are thousands-fold higher than that used in the test so it is highly unclear how to correlate the results obtained with the carbides performance in real metal cutting.

The available results on implementation of nanocarbides suggest that this material has great future in metal cutting. However, the tool design, its manufacturing quality, and implementation practice should correspond to the new properties of this tool material. In other words, more systemic studies are needed to find the optimal range of conditions for nanocarbide application in metal cutting.

3.4.6 Carbide Blanks

Carbide blanks are essentially unfinished carbide products that can be further modified or finished according to one's requirements. Practically all carbide drilling tools are made of carbide blanks. Carbide blanks are often finished into a final product by means of EDMing, grinding, polishing, or other industrial processes. Such blanks are available in various types of sizes and shapes. Most blanks are made according to various standards, including carbide companies standards, while others are custom-made products designed and manufactured according to particular custom specifications.

3.4.6.1 Blanks for Carbide-Tipped Drilling Tools

The simplest standard and custom-made blanks are those used for brazed carbide tools as, for example, for drilling tools shown in Figures 3.47 and 3.48. Such blanks can be made (EDMed, cut off, ground, etc.) from standard carbide strips similar to those shown in Figure 3.49 or can be purchased as made according to standards. There are some advantages and disadvantages of these alternatives, which should be taken into consideration in the tool design and manufacturing.

Making carbide blanks in-house from carbide strips (Figure 3.49) has the following advantages:

- Blanks can be made of the shape close to the net shape of the carbide inserts in the tool. As such, minimum finish grinding is needed. It is important when special tools similar to that shown in Figure 3.48 are designed and manufactured
- A special tool can be designed with intricate configuration of the cutting edges and optimal size of chip pockets
- Carbide strips are available in greater (than standard inserts) range of carbide grades including special grades
- Lead time of tools of special design can be reduced as carbide strips can be kept in inventory (the so-called on-shelf products).

These advantages, however, come with certain constraints. The major is that CNC EDM machines along with programming and experienced operators should be available in the shop. As modern EDM machines are highly productive (can be run unattended 24/7), it is not easy for a small shop to make efficient use of such a machine justifying its purchasing as such a machine could finish a month-worth lot of carbide blanks in 2 days.

The second alternative is to purchase standard blanks made to various available international (e.g., Standard ISO 242:1975) and national standards (e.g., German Standards DIN 4950, DIN 4966, DIN 8010, DIN 8011, and American National Standard for Cutting Tools ANSI B212.8-2002) as well as multiple company catalogs. For example, German Standards DIN 4950 defined the dimensions and tolerances for the shapes shown in Figure 3.50.

FIGURE 3.47 Carbide-tipped drill.

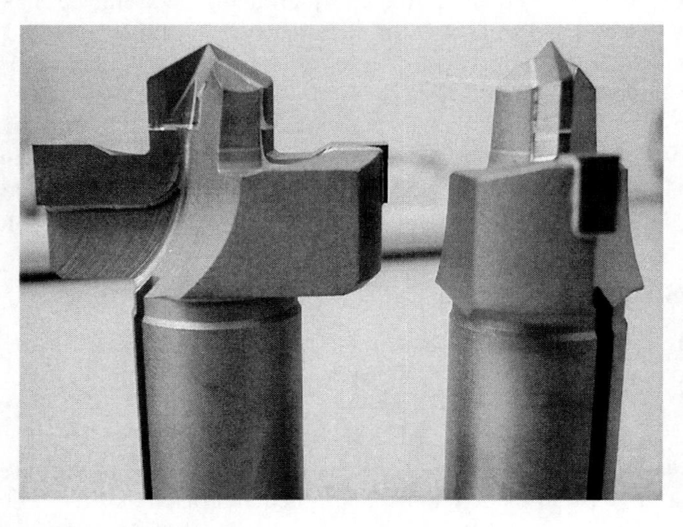

FIGURE 3.48 Carbide-tipped combined tool.

FIGURE 3.49 Carbide strips (rectangular bars).

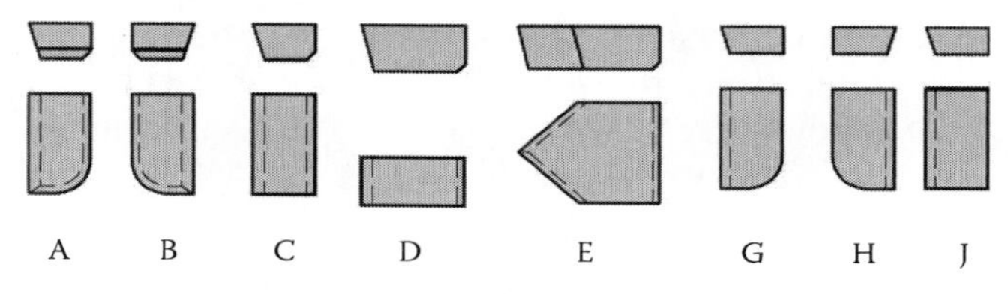

FIGURE 3.50 Carbide brazing blanks according to DIN 4950.

For example, American National Standard for Cutting Tools ANSI B212.8-2002 *Carbide Blanks for Twist Drills, Reamers, End Mills, and Random Rod* covers dimensional specifications and designations for carbide blanks for twist drills, reamers, end mills, and random rod including flat-type blanks for carbide-tipped twist drills, fluted-type blanks for carbide-tipped twist drills, radius-type blanks for carbide-tipped reamers, and rectangular-type blanks for carbide-tipped reamers. Figure 3.51 shows an example of carbide blank for drills.

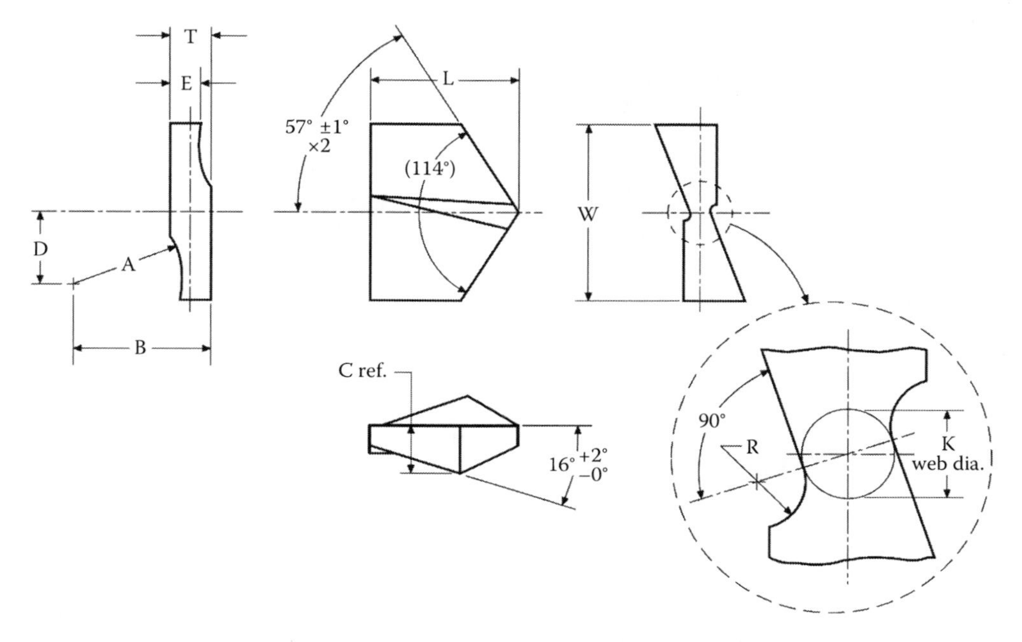

FIGURE 3.51 Some carbide blanks covered by standard ANSI B212.8-2002.

The use of such blanks makes sense for relatively simple designs of drilling tools as those shown in Figures 3.47 and 3.48 particularly when a great number of tools are to be produced so the lead time for blanks purchasing and the selection of their proper grade do not present problems. Often, some EDMing, cutoff, and/or rough grinding are required to bring a blank to the net shape (before and/or after brazing on the tool body) suitable for finish grinding.

3.4.6.2 Round Carbide Blanks

Round blanks of various design and dimensions are used in manufacturing of solid-carbide drilling tools. They are available from various carbide manufacturers. These blanks can be classified as generic round blanks of standards diameters and lengths, specialized round blanks made out of the generic blanks and special carbide preforms.

The simple blanks are carbide round bars (rods) widely available (Figure 3.52). The length and diameter of these bars can be made according to some national standards (e.g., ANSI B212.8-2002). They can be ordered as unground or centerless ground to tolerance h6, a common shank diameter tolerance for hydraulic and shrink-fit toolholders. All bars are normally available in standard lengths of 310 and 330 mm. They are also available as standard precut length and forms. For example, some round blanks according to standard ANSI B212.8-2002 are shown in Figure 3.53. Other than standard shape, round blanks can also be ordered from

FIGURE 3.52 Carbide round bars.

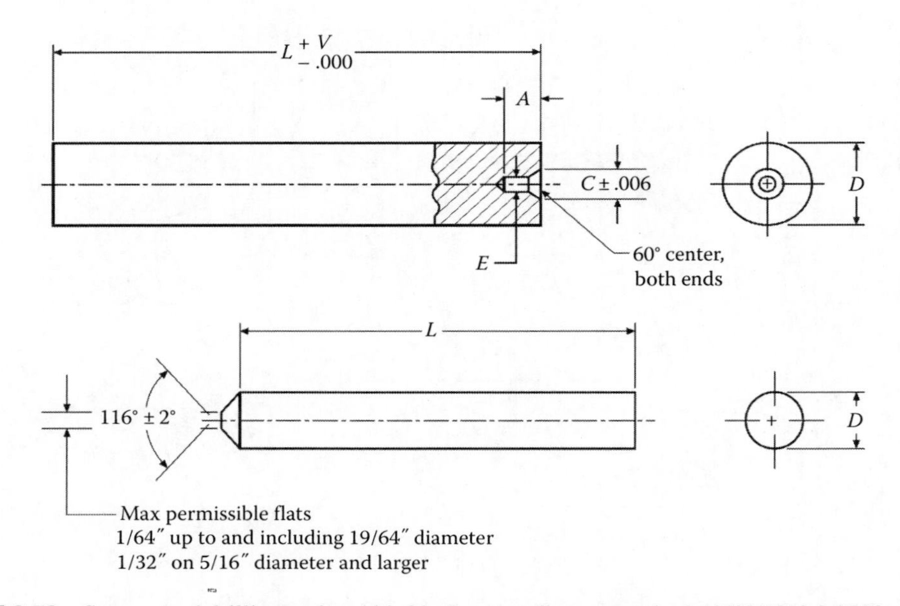

FIGURE 3.53 Some round drilling tool carbide blanks according to standard ANSI B212.8-2002.

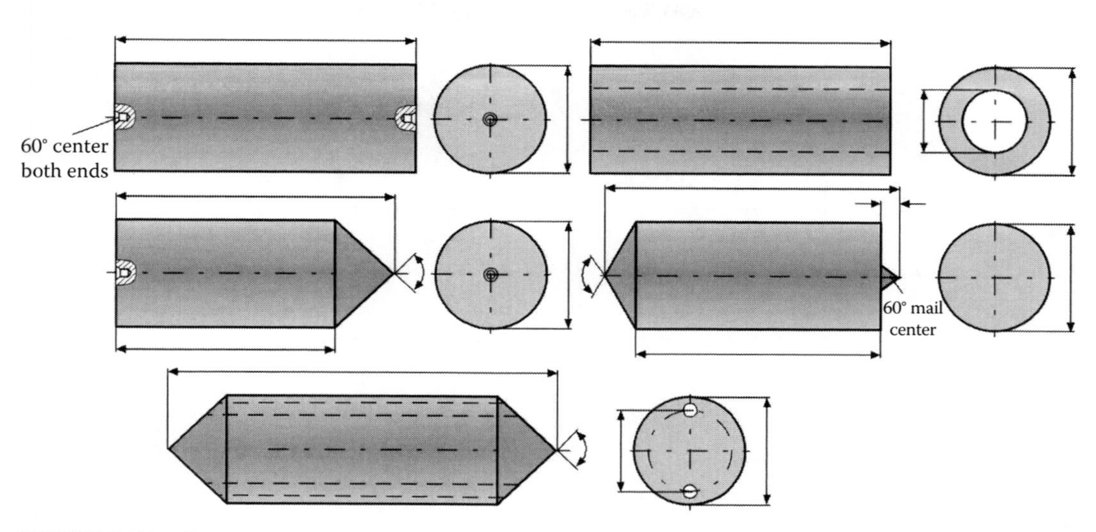

FIGURE 3.54 Some common nonstandard carbide round blanks.

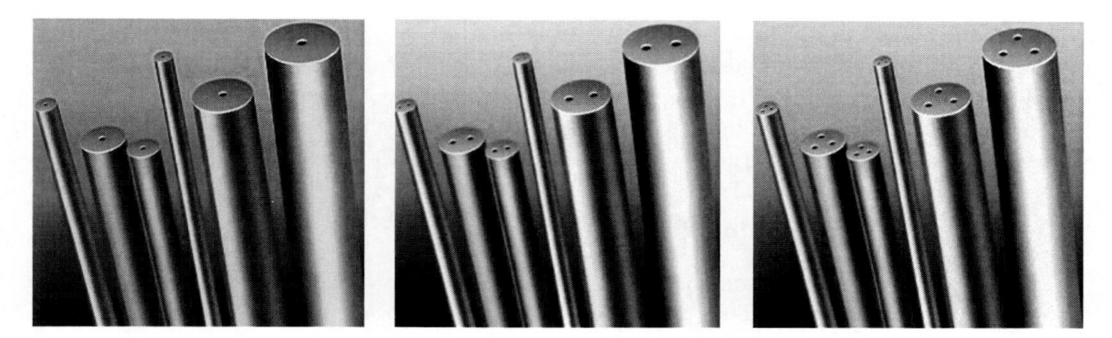

FIGURE 3.55 Carbide round bars with one, two, and three coolant holes.

various companies' catalogs or to be made to custom specifications. Some of common blanks in this category are shown in Figure 3.54.

Another common set of carbide round bars are those with coolant hole(s). Although they are considered in detail in Chapter 6, some important information and issues are briefly discussed here. Figure 3.55 shows that standard carbide bars may have one, two, or three coolant holes although special bars can have four and five holes.

The simplest method is to make a steel tool shank with a central drill hole. A carbide head with the same or similar diameter as the steel-shank is then brazed on its end as shown in Figure 3.56a. This carbide head may two or three holes leading in a starlike arrangement at an angle from the outside to the inside and which converge at the central hole of the shank. This combined tool can now feature spiral fluted or parallel chip flutes. Although similar carbide blanks have been used for years, this is not the best arrangement of the coolant holes as the location of their outlets changes with each tool regrind. As discussed in Chapter 6, such a location plays a key role in the proper introduction of MWF into the tool critical areas in which wear defines tool life.

Raw carbide bars produced in set lengths with a central hole from one side of the blank are a further development of this idea (Figure 3.56b). This central hole, which is drilled in a pre-sintered state, ends in the material just before the other end and depending on the number of chip flutes branches into two, three, or four holes drilled at the same angular pitch. In the case of set lengths produced

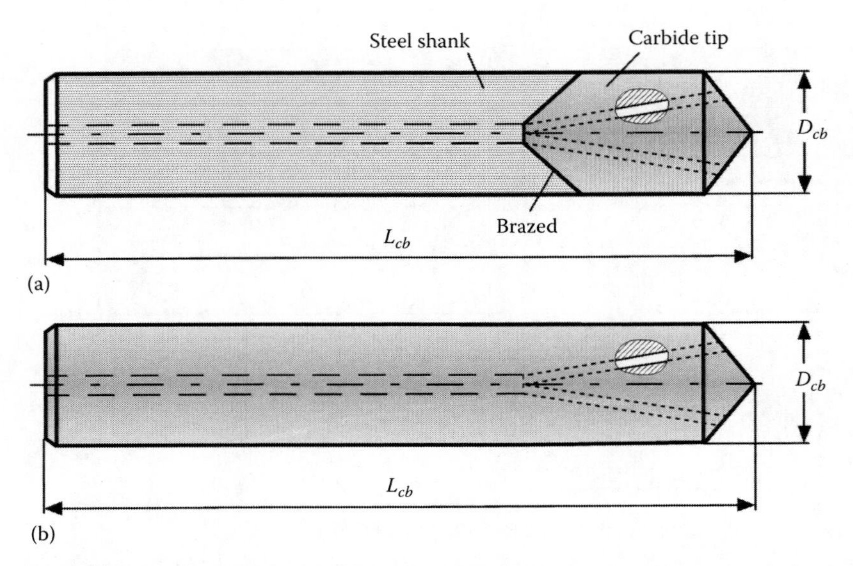

FIGURE 3.56 Blanks with a central coolant hole: (a) carbide tip brazed to the steel tool shank and (b) carbide bar with internal coolant holes.

in this manner, the spiral chip flutes can be ground with the desired helical angle. Naturally, such a blank inherited the discussed disadvantage of the brazed blank. In other words, changes in tool life can occur from one regrind to the next. Unfortunately, such variation is normally attributed to variation in carbide properties, quality of regrinding, and other factors, but never to the root cause—changing location of the outlet holes of the coolant channels. These locations are not normally indicated on tool drawings although, as discussed in Chapter 6, they belong to the category of the most important tool design parameters.

The configurations of common modern standard carbide bars with coolant holes are shown in Figure 3.57. The common method of their production is the extrusion press process using soft paraffin as a binder. The twisting process with an annular toothed nozzle is widespread in Europe. Twisted teeth, which twist the plasticized mass prior to its exit, are inserted in the end nozzle of the extrusion press.

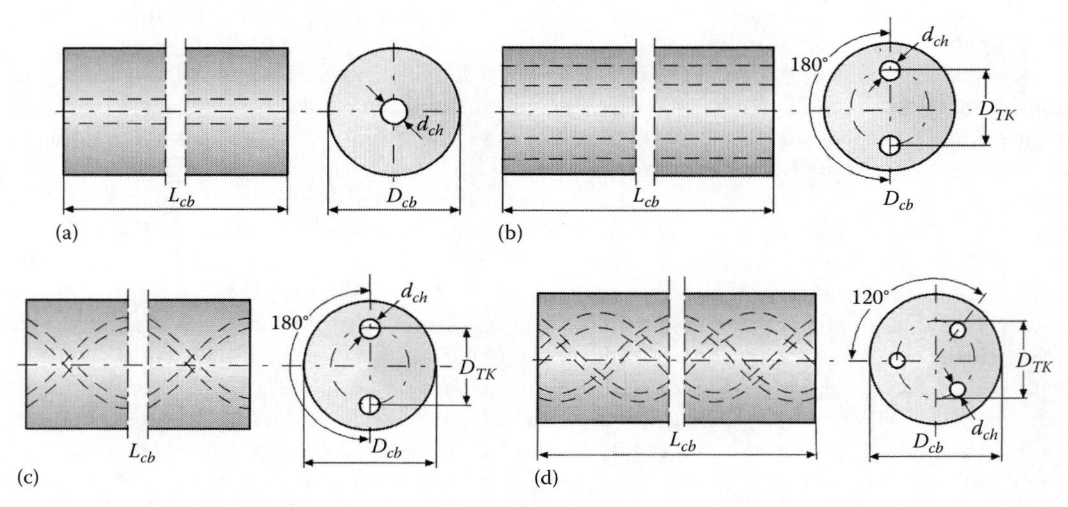

FIGURE 3.57 Geometrical parameters of carbide bars with coolant holes: (a) single central coolant hole, (b) two parallel coolant holes, (c) two helical coolant holes, and (d) three helical coolant holes.

Internal soft, elastic, or rigid twisted threads in the nozzle area produce the required coolant holes in the plastic rotating rod since these threads only reach up to the discharge of the mass from the nozzle. When the rotating rod-shaped mass exits the press, twisted helical coolant holes remain in the cylindrical plastic carbide rod instead of the threads. The rods are then placed on a special surface for removing the binder and for sinter-HIPing. The raw rods then have the typical twisted teeth from the nozzle on their outer shell, which produces the helix.

The production of annular teeth nozzles, whose teeth have been designed increasingly finer in the course of development, is expensive. New nozzles are required for each rod diameter, since there is a different pitch length for each rod diameter. Different toothed nozzles are also required for varying pitches with the same diameter.

It is particularly complicated when the temperature of the plastic mass to be pressed fluctuates or if the operating temperature of the press is fluctuating. This is often the case for piston-type presses working in batch operation. In this case, the plasticity of the material to be pressed can be detected easily using penetration and ductility measurements conducted dependent upon the temperature. As a result of these fluctuations, variations in pitch on the helical twisted sintered rods occur. A pitch correction on the pressed blanks is hardly possible because there is a rigid, specified twist in the ejection nozzle.

There are also processes in which a plastic rod is pressed out of the extrusion press with several parallel holes. After leaving the press, the plastic rod is taken up by a rolling unit, which forces it into a specific rotation, in order to produce the necessary twist. The handling of the relatively soft rod by a rotation unit after pressing is not simple. Marks and crushed spots left on blanks are highly disturbing factors, which must undergo a centerless grinding process following sintering during which the diameter of the blank must remain within a certain tolerance range (approximately 0.2–0.4 mm). Moreover, a rod that is twisted after leaving the high-pressure range of the press demonstrates significant changes during the subsequent thermal treatment process prior to sintering and during sintering.

Similar is a process of twisting sintered carbide rods with parallel coolant holes using local heating (usually inductive) and subsequent twisting of the traveling heated zone. Apart from the problems involved in repeated heat treatment following sinter-HIPing and all of the associated disadvantages such as grain size growth and bending, the handling of rods heated in this way is difficult. The leading manufacturers of extruded twisted carbide bars have mainly used piston-type extrusion presses in the past. The common piston type extrusion presses used to entail much less retooling effort than worm-type extrusion presses.

One disadvantage piston-type extrusion presses had compared with worm-type extrusion presses, which was not important when pressing solid rods, was the slow, creeping heating of both types of presses at the beginning of operation (approximately 1 h). This could not be completely eliminated by cooling measures because a certain processing temperature had to be set in any case. The pressing cycle is relatively short especially in the case of rods having diameters in excess of 10 mm because of the limited filling capacity of a piston extrusion press.

Carbide rods with helical coolant holes extruded using the worm-type press have much smaller pitch variations, which is important for long bars. Automatic cutoff and handling systems can be employed easily for continuous operation, and defective blanks can be immediately returned to the ongoing press operation. The long retooling times for worm-type presses feared in the past, especially in the case of small production quantities, no longer exist today. The worm-type extrusion press is particularly superior to the piston type when large diameter rods (>10 mm) are produced. This became possible with the introduction of new binders instead of soft paraffin. Initially, water-soluble plastics were used. As such, the water and the consequences of water use in the carbide powder had to be eliminated by a complicated drying and reduction process. Then the usual sintering process of the sinter-HIP process could be employed. As such, the remaining binder (approximately 1.5% of the overall pressed blank weight) is expelled without risk of the rods cracking. Modern binders consist of three organic components, which lend the pressed blanks from the extrusion press sufficient plasticity. Some 80% of the binder is easily dried out following extrusion in

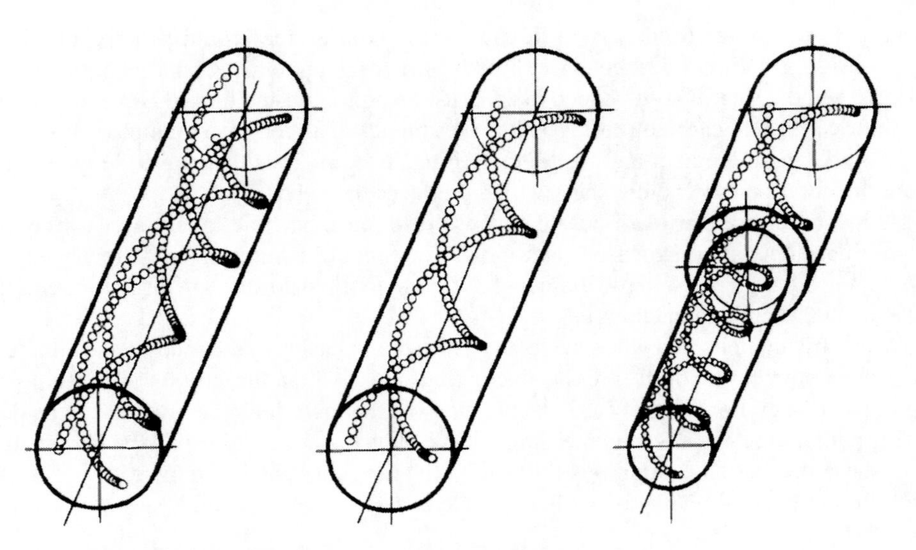

FIGURE 3.58 Common carbide blanks extruded using the worm-type press.

lacquer-drying ovens. Currently, press blanks up to a sinter diameter of 33 mm are dried in this manner. As opposed to the paraffin process, this is not an evaporation process, rather it is a drying process. The remaining binder (approximately 1.2% of the press blank weight) is removed in a simple, fast sintering cycle and condensed outside the oven.

A process has been developed and patented, which allows one to set any desired pitch for rods with helical twisted internal coolant holes in a very simple manner. Whereas in older processes the pitch had to be generated using components that demonstrate a set, rigid pitch and, therefore, can only produce this permanently set pitch, this new process allows any desired twist angle to be produced using a rotating nozzle. The ratio of the extrusion speed to rotational speed determined the pitch at which the plastic mass is twisted. Some common carbide bars produced using this process are shown in Figure 3.58.

A promising process of carbide blank manufacturing for drilling tools is composite grades carbide bar manufacturing. Traditionally, solid-carbide drilling tools have had a *monolithic* construction, meaning the substrate is made from a single grade of carbide. Such tools have limitations with regard to their performance for a couple of reasons (Mirchandani 2005).

3.4.6.3 Round Carbide Blanks Made of Advanced Carbides

In general, drilling operations are characterized by large differences in the cutting speed from the central region to the periphery. As discussed in Chapters 1 and 4, being practically zero at the axis of rotation, the cutting speed increases toward tool periphery corners proportionally to the tool radius. As the cutting feed is the same for all cutting parts of the tool, the loads on the central part of the tool are much greater than those in the parts close to the tool periphery. Therefore, considerable different properties of the tool material are needed in these regions. At the periphery, where the cutting speed is high, the tool material should be of maximum hardness to withstand abrasive wear while for the regions of the tool center, the tool material should be tough. As hardness and toughness for various grades of cemented carbides changes in opposite direction (Table 3.9), the selection of a particular carbide grade for a drilling tool always presents a problem. If a grade is selected *too hard* for maximum tool life (normally determined by the wear of tool corners), the regions of the cutting edge adjacent to the tool's center may fail because of microchipping, a result of the carbide grade being too hard. Alternatively, if this grade is selected to be tough, the cutting edge at the periphery fails prematurely because the grade is not hard enough. Thus, the grade of carbide selected for monolithic drills invariably represents a compromise between these competing failure modes.

Clearly, the performance of carbide drills could be improved if they were constructed from carbide that incorporated variations in physical properties from one location to another along the tool radius. Wear life would be substantially longer, for example, if the carbide grade were somewhat softer and tougher in the core region to prevent premature microchipping of the central part and harder in the peripheral region to prevent premature wear due to abrasion/adhesion (Mirchandani 2005).

Composite carbide grades have recently become available that have properties that vary from one location to another. Such grades consist of at least two distinct conventional carbide grades (e.g., Grade A and Grade B as shown in Figure 3.59a) arranged in a coaxial fashion. In this manner, it is possible to fabricate carbide substrates with distinctly different properties in the core region compared to the shell region. The performance ranges (cutting speed vs. material-removal rate) of drilling tools can be greatly expanded if they are made from a composite material. As shown in Figure 3.60, monolithic grades with a hardness greater than 94 HRA are typically applied at relatively high cutting speeds and low material-removal rates, such as when finishing, while softer grades (less than 90.0 HRA) are typically applied at lower cutting speeds and at higher feeds, such as when roughing. It is generally difficult to extend the operating range of either type of monolithic

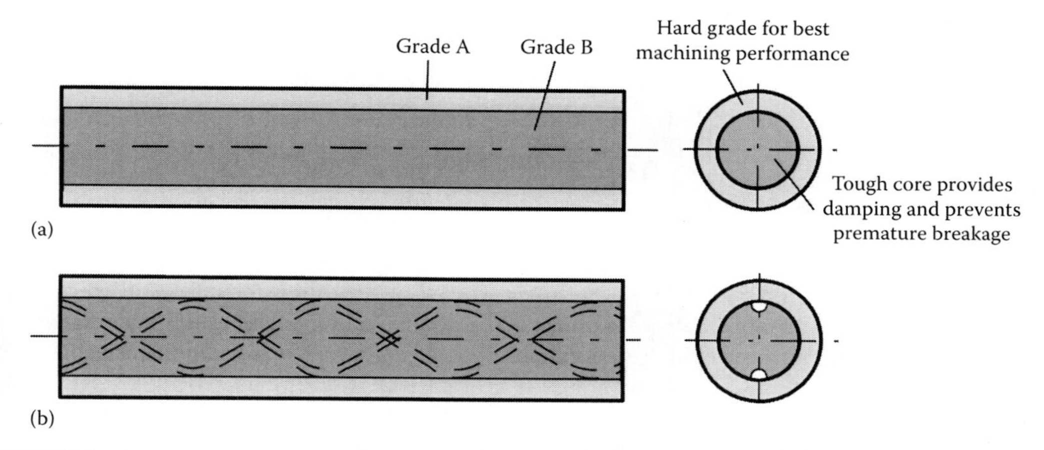

FIGURE 3.59 Structures of composite carbide bars: (a) with no coolant hole and (b) with coolant holes.

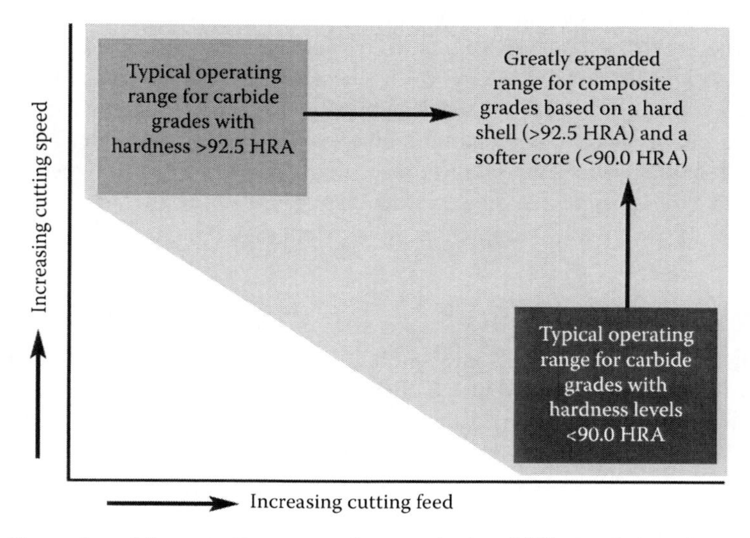

FIGURE 3.60 Expansion of the operating range when employing drilling tools based on composite carbide grades.

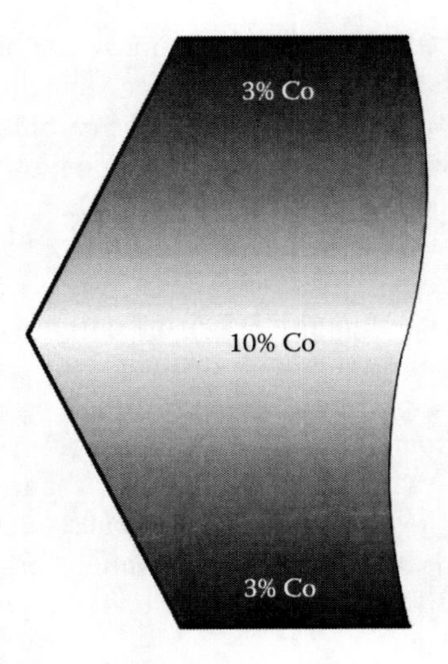

FIGURE 3.61 Diagram of the cemented gradient carbide.

grade without the risk of premature tool failure. For example, roughing with a tool made from a high-hardness grade is likely to fail prematurely by breaking. On the other hand, a softer grade run at a high cutting speed will experience deformation and excessive wear.

Composite construction helps to overcome the inherent limitations of monolithic grades. For example, a tool made of a composite grade with a high-hardness shell and a softer core can cut at speeds characteristic of high-hardness grades while removing material at rates usually seen with softer grades. This expands the operating range as shown in Figure 3.60. Obviously, the size of the operating range depends on the choice of the individual grades that make up the composite grade. Moreover, the coolant channels of any cross section and of any pitch can be made on the core in soft state (before sintering) as shown in Figure 3.59b.

The newest trend in the development of cemented carbide rods for drilling tool is the development of tool gradient materials (TGMs) fabricated with the use of PM methods. In these materials, a gradient change of binding cobalt phase and the reinforcing phase of WC aims to solve the problem involving the combination of high hardness and resistance to abrasive wear with high resistance to brittle cracking and, consequently, to ensure their optimal synergy with operating conditions. The diagram representing a structure of such materials is shown in Figure 3.61 (Dobrzański and Dołżańska 2010).

Note that composite carbide grades and TGMs can show their advantages only when the parameters of the drilling system fully support their application. Moreover, the drilling tool design and its manufacturing quality should correspond to the level of these tool materials. Otherwise, the implementation of these advanced materials may result in poorer tool performance compared to that made of *forgiving carbide grades*. In other words, the coherence system law should be followed in order to obtain the maximum advantages of these new tool materials.

3.4.7 Coating

Practically all that was set in Section 3.3.8 including rationale for selection, coating materials, and performance is fully applicable for carbide coating. This section briefly describes the basic methods used for coating application of carbide coating, particularities of coating management (logistics), as well as a new trend in carbide coating known as the tough coating.

3.4.7.1 Methods of Application

Chemical vapor deposition (CVD) coatings revolutionized cutting tools when introduced in the late 1960s. By the early 1980s, CVD had evolved into multilayer coating compositions of TiC,

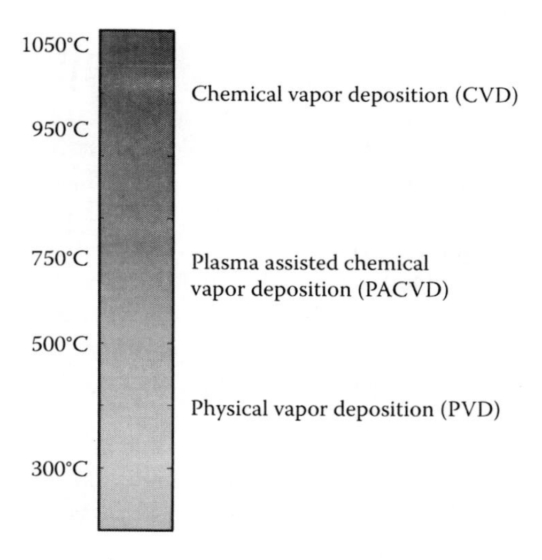

FIGURE 3.62 Process temperatures for common coating methods.

TiCN, and Al_2O_3. The CVD method involves heating up the substrate within a chemical reactor and exposing the substrate to a gas stream. The gases break down on the hot substrate surface, forming a coating layer. In general, the CVD method requires temperatures around 1000°C (Figure 3.62). A common coating uses the three gases $TiCl_4$ (titanium tetrachloride), H_2 (hydrogen), and N_2 (nitrogen) to produce TiN (titanium nitride) + HCl (hydrogen chloride). The HCl is a by-product of the process and must be disposed of according to strict environmental regulations. The advantages of the CVD method include optimal layer adhesion, as well as consistent layer distribution. The disadvantages of the CVD method are high process temperatures affecting the substrate, few suitable materials for coating as the coating material is fed in a gaseous form, and long cycle times.

Once considered as obsolete, CVD method has been significantly improved over the last decade becoming much appealing. These improvements include

1. Achievement of thicker multilayer coatings on carbide inserts, for example, up to 20 μm thick, medium-temperature CVD TiCN and Al_2O_3 multilayer designs for HSM of abrasive work materials. These coatings are nearly twice as thick as traditional multilayer CVD coatings and provide greater wear resistance and longer tool life in some specific applications.
2. Introduction of more sophisticated process control that includes CVD Al_2O_3 nucleation to obtain the desired α or κ crystalline phases of the coating. It is argued that α-phase Al_2O_3 is the most stable and also the most high-temperature-resistant and wear-resistant phase among several phases that can be CVD deposited. Precise control of coating temperatures and gas flow distribution resulted in the generation of exceptional coating thickness and adhesion consistency.
3. Achievement of about 50% less coating thickness variation from lot to lot than a decade ago.
4. Significant improvements in diamond film coatings produced by CVD technology. While this coating is still 100% crystalline diamond, the product has been improved from a single-layer polycrystalline coating as in 2001 to a multilayer nanocrystalline diamond coating. The multilayers of nanocrystalline diamond increase the fracture toughness of the coating and its fine grain structure of the submicron crystals and leave a smoother surface on the cutting edge for imparting finer part finishes. The multilayer structure is more resistant to cracking because each horizontal layer of diamond coating acts as a barrier that stops cracks from propagating further through the coating. This makes the coating stronger and helps it hold up better on cutting edges that experience mechanical shock from difficult-to-machine part materials or interrupted cuts.

After its introduction 30 years ago, PVD supplanted some CVD coatings. Since it was a lower-temperature process (Figure 3.62), this coating method can be used on heat-sensitive tool materials such as HSS and on variety of brazed tools with no damage to braze joints. Today, PVD and CVD are often considered as complementary processes. Each of these two has some special applications advantages. Combination CVD/PVD coatings are often utilized, with CVD comprising the first coating layer(s) and PVD comprising the smoother, finer top layer(s).

Recently, PVD coating development has focused on new compositions, nanocomposite coatings and Al_2O_3. AlTiN coatings applied via PVD have been called *the next best thing* to Al_2O_3 and until recently could only have been deposited on a commercial scale via CVD. PVD coatings that incorporate materials such as silicon and use new, nanotechnology-based materials, perform better, enabling the newest machine tools to machine faster and, in many instances, machine under dry or near dry (MQL) cutting conditions.

With PVD, there are two primary technologies used to coat the different substrates: the arc method (arc discharge) and the sputtering method (cathodic sputtering). Both methods share one additional advantage; the coating chambers are relatively easy to construct. The arc method involves an electrical power source (much like a lightning bolt) hitting the laminate material and transforming this material from a solid to a liquid and to a gaseous phase. The advantage of this process is high layer rates (in relation to sputtering). However, since the laminate material is in all three phases (solid, liquid, and gas), the potential for droplets (minute liquid particles) occurs. These droplets do not achieve the gaseous state. The sputtering method involves a thermal energy source, which transforms the solid laminate immediately to a gaseous state. No droplets occur as the material skips the liquid phase. However, the lower layer rates (in relation to arc) result in longer cycle times.

Plasma-assisted CVD coating method has been developed to combine the good adhesion of CVD and the low temperatures of PVD, while avoiding their typical drawbacks (high temperature with deformations and poor adhesion, respectively). Using microwaves or an electric field, a plasma is created. The energy for starting the chemical reactions generating the coating layer comes from the collision of the plasma ions and electrons. The process temperatures are lowered to 200°C–300°C (Figure 3.62), limiting distortions. Sometimes it is even necessary to raise a little the temperature to achieve better adhesion. PACVD is presently used mostly for deposition of diamond films.

There is much legitimate discussion—as well as hype—concerning nanotechnology use in coating. Such coatings became known as nano-coatings based on materials up to 100 nm in size. The key factor in nanotechnology is what the product does, rather than its size. Instead of looking at nanotechnology as a size range, it is more important to explore the point at which emergent properties can offer significant performance benefits, such as eliminating the defects that become failure mechanisms in tool coatings. These failure mechanisms can lead to cracks and tool stress. So far, however, not much conclusive evidence of their superior performance can be found. In the author's opinion, however, it does not mean that superior performance of nano-coating is not the case. The same as with nanocarbides, the tool design, its manufacturing quality, and implementation practice should correspond to the new properties of this class of coatings. In other words, more systemic studies are needed to find the optimal range of conditions for nano-coating application in metal cutting.

Although the listed coatings and the known coating application techniques seemingly work for indexable cutting inserts, which are not subjected to regrinds, a combined cost, logistic, and tool life problem arises for drilling tools that have to be reground. The only proper way to carry out regrinding includes stripping the old coating, regrinding the tool, and recoating the tool. It adds lead time and cost associated with regrinding. Moreover, reground tools often have lower tool life compared to new tools due to cobalt leaching on coating stripping.

To solve this problem, a considerably different way of coating, namely, the coating of carbide powder rather than coating the finish carbide product, can be used. The support for this technology

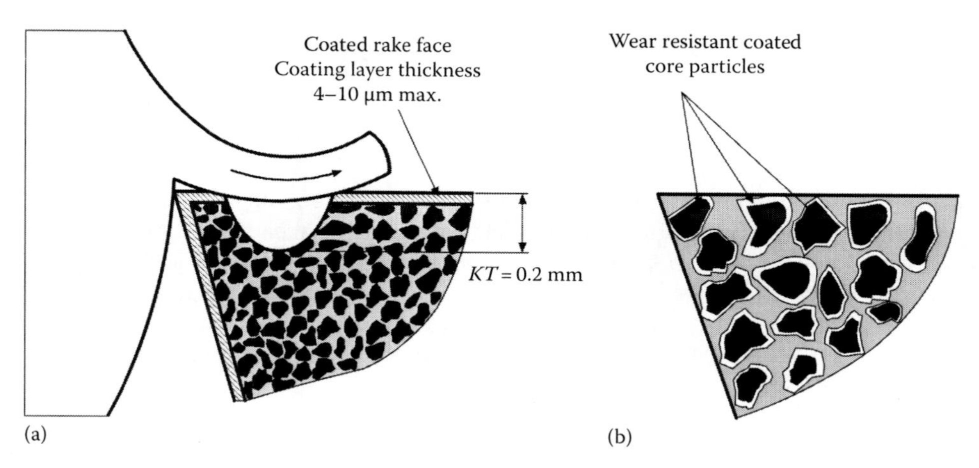

FIGURE 3.63 Representation of (a) conventional CVD coating and (b) TCHP.

stems for an undisputable fact shown in Figure 3.63a. As shown, when the wear of the rake face (the depth of the crater *KT*—see Section 2.6.2) reaches 0.2 mm, which is the most common criterion of tool life in finishing operations, no coating even in traces can be found on the rake face to protect the tool against wear as the thickness of the coating layer is 0.4–10 μm. The thickness of conventional coating is limited by delamination and cracking from different thermal expansion rates over large areas, and bending and surface loads severely limit coating thickness. The same can be said about the wear on the flank face where the average width of the flank wear land $VB_B = 0.3$ mm is commonly adopted as the tool life criterion (see Section 2.6.2) so that the depth of wear in the direction perpendicular to the flank face is approx. 0.1 mm for modern clearance angles. Therefore, the depth of wear is 20–40 times greater than that of the coating so one may wonder why any coating works at all.

The coating of carbide powder at the conceptual level has been known for a long time, but only recently its practical applications have become feasible. The complete literature review on the development history of this method can be found in German et al. (2005). One realization of this concept is tough coated hard powders (TCHPs), which are a new family of patented, high-performance metallurgical powders that incorporate unprecedented combinations of property extremes. They represent a class of engineered microstructure P/M-based hardmetals having combinations of critical properties that provide improvements in performance and productivity. These engineered property combinations include toughness, abrasive and chemical wear resistance, low coefficient of friction, light weight, and so on at levels not previously seen. TCHP powders can be fabricated into a multitude of industrial metal cutting inserts to leverage their key attributes to achieve manufacturing productivity improvements. These TCHP powders are created by incorporating hard particles in a tough matrix using proprietary manufacturing technologies. Engineered nanostructures are designed by encapsulating extremely hard *core* particles with a tough outer layer(s), for example, tungsten carbide and cobalt, which in the consolidation process becomes a contiguous matrix.

Figure 3.63b shows the structure of the hardmetal with the coated particles. As can be seen, TCHP-sintered microstructure is a cellular pseudoalloy of a contiguous tough tungsten carbide and cobalt mechanical support and binder phase containing chemically unadulterated wear-resistant core particles. These particles (such as TiN, TiC, TiB_2, ZrN, Al_2O_3, diamond, cBN, $AlMgB_{14}$, or B_4C) are dispersed evenly throughout the tough tool. When using multiple core particle materials to enhance different properties, each material and its unique properties are available simultaneously at the working surfaces and cutting edges of the tool throughout the entire substrate. This design multiplies the volume of wear-resistant material useable many times that in any possible coating, it provides a continuously renewed (self-healing) wear surface. The authors of the process/material claim that TCHP works all the time so that tool life can be extended many times.

TCHP powders and consolidated carbide blanks are manufactured and sold by Allomet Corporation (North Huntingdon, PA) as EternAloy®. Representative *core* particles include those traditionally used for extreme wear resistance (e.g., diamond, cBN, Ti(C,N), TiN, Al_2O_3). TCHP *composaloys* allow new combined levels of fracture toughness and hardness; resistance to abrasion, friction, wear, and corrosion; thermal conductivity; and impact resistance throughout the entire tool material. This innovation may extend uncoated tool life 10–30 times and coated tool life approximately 4–7 times. In addition, the tool material can be reground from one to five more time compare to conventional carbide tools before disposal. Naturally, tools made with TCHP *composaloys* do not require an external coating because hard and tough phases are already dispersed throughout the tool, resulting in a continuously renewed wear surface within a tough substrate.

The analysis of the available information on TCHP powders and consolidated carbide blanks, however, arises some issues:

1. The grain size of the modern sintered carbide grades is 0.6–0.8 μm. What should be the thickness of the coating on this grain that can provide wear resistance?
2. The micrographs of the carbide structure with TCHP powder shows that the cobalt content of more than 20 wt% versus 6 wt% is commonly used for high-performance carbide tools. The experimental results (Figure 3.33), however show that hardness being probably most important characteristic of the tool material decreases significantly with the cobalt content.
3. Hot hardness of sintered carbide is mostly affected by the cobalt content (Astakhov 2006). No data on hot hardness of the carbides with TCHP powder are available.
4. No comparison of tool lives of the best grades of sintered carbides and the carbides with TCHP powder is available.

3.4.7.2 Coating Strategies

Besides some drilling tools with indexable inserts, the vast majority of drilling tools are reground (re-sharpened) to restore their working conditions. The carbide drill is a good example: the cost of each regrinding/recoationg is 30%–50% of the price of a new tool and is reground up to 10 times. This adds up to more than three times the sales price of a new when the cost of all regrinds are added. Because most of carbide drills used today are coated, recoating strategy should be developed.

Two common strategies, 1 and 2, are illustrated schematically in Figure 3.64. In this figure, the common case for drilling tools is considered where tool predominant wear is flank wear *VB* and

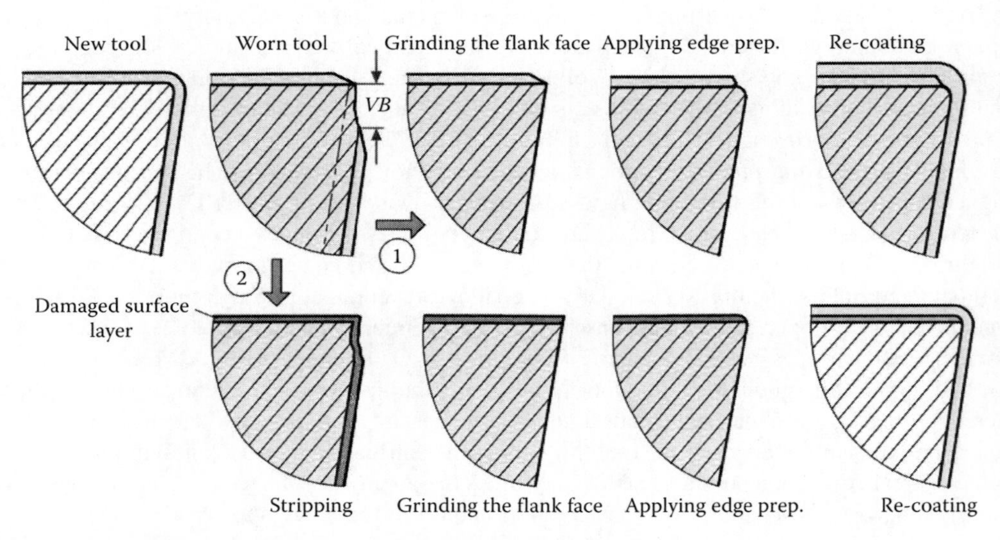

FIGURE 3.64 Two possible strategies in recoating of drilling tools.

the drilling tools are re-sharpened on the flank face to restore its original geometry. The coating is applied on the both the rake and flank faces, which could be planes or more complicated surfaces as helical, conical, etc.

According to the fist strategy, the tool is re-sharpened (the wear marks are removed as shown) and then the edge preparation is applied. After cleaning the tool, it is recoated. As can be seen, the old coating layer is completely removed from the flank face while that on the rake face is not removed. The adhesion of a second coating layer on a previous one decreases as the thickness of the preceding coating(s) increases. Although some coating companies recommend that the reground tools may be recoated two or three times without decoating (stripping), and still achieve a good tool life, it is not always the case particularly in HP drilling operation as the radius of the cutting edge increases with each successive coating application. It is discussed by the author earlier (Astakhov 2010), a difference in 3–4 μm in the cutting edge radius may result in a significant decrease in tool life. Therefore, this strategy can be selected for less demanding applications but it is totally unacceptable for HP drills.

The logistic of the first strategy is shown in Figure 3.65. As can be seen, three major transportations are involved providing that re-sharpening and recoating are done in different facilities than that of part drilling manufacturing facility. Although some larger manufacturing companies may have re-sharpening and coating facilities in house, the trend in industry, for example, the automotive industry is to send tools to outside re-sharpening and coating manufacturers. The best choice is to send the tool for reconditioning to its original tool manufacturer (OTM) capable to restore tool original geometry properly according to the original drilling tool manufacturing drawing not available to the costumer. Although a local tool reconditioning shop can be cheaper, the quality is normally higher with OTM. When OTM having no in-house coating facility handles tools reconditioning, the logistics shown in Figure 3.65 become a bit more complicated as the coating facility sends tools back to OTM and then OTM sends tools back to the drilling operation manufacturing facility.

The worst scenario, however, takes place when a tool having predominant flank wear is not recoated after sharpening, which is unfortunately common in manufacturing facilities having in-house re-sharpening service. In this case, tool life reduces as no coating protects the tool after re-sharpening. Some practitioners claim that they do not see any reduction of tool life. If this is the case, then the application of the coating on a new tool is waste of money as it does not affect tool life. The proper coating, however, affects tool life markedly and its grinding from the flank face reduces tool life noticeably.

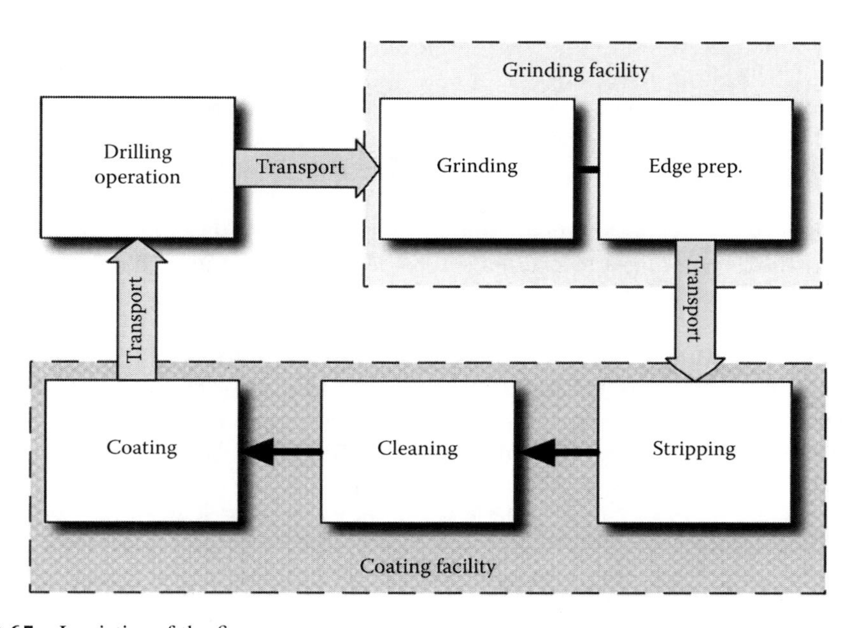

FIGURE 3.65 Logistics of the first strategy.

The second strategy is the proper way to be used for HP drills according to the author's experience. According to this strategy, the worn tool is fist decoated (coating is completely stripped), then the tool is properly reground, edge preparation is applied, and the tool is cleaned and then recoated. Although this sequence is logical and sounds simple, there are two important yet not well-known issues involved in this strategy that drilling tool users should be aware of. The first issue is coating stripping and the second one is logistic associated with the second strategy.

Most of modern decoating processes known as coating stripping rely on chemicals, which decompose the coating but also attack the underlying tool material. This may happen especially with cemented carbide. According to Platit AG coating company, the problem can be briefly described as follows. The stripping liquid may be either water-based or water-free (solvent-based). Although solvent-based liquids yield better results in laboratory tests, they are hardly used in industrial environment due to safety concerns (expensive reagents are required along with flammable and poisonous solvents, which have to be recovered at high cost). Water-based methods are thus the only ones, which come into question for large-scale decoating. Some of them require electrical current as well (electrolysis), in order to dissolve the coating. The chemical reactivity of the cobalt in the tool material is quite similar to that of titanium and/or chromium in the coating. This fact makes it a big challenge to dissolve the coating and at the same time preserve the cobalt-containing base material. This task is even more complicated when titanium carbide or chromium carbide are present in the cemented carbide. After a loss of 0.1–0.5 μm of base material, the surface may be, for example, polish-blasted so that the coating applied afterwards adheres well. Note that some stripping solutions attack not only cobalt but also the tungsten carbide of the cemented carbide, with very similar consequences to the adhesion of the coating applied afterwards.

As a result, a damaged layer forms on the surface of carbide as shown in Chapter 2, Figure 2.86. This layer lacks cobalt (cobalt leaching). Lack of binder results in decreased cohesion of the carbide grains, which remain loose on the base material. The surface becomes dull or even black. The depth of cobalt leaching is decisive. A depth from 0.1 to 0.5 μm may be *repaired* by posttreatment. Deeper cobalt leaching results in a mechanically unstable surface, and if a coating is applied, it shows weak adhesion or none at all. Therefore, the perfect stripping solution dissolves the coating but does not damage the cemented carbide. Real stripping solutions, however, do damage the cemented carbide. As a result, coating stripping should be done only by professionals to minimize the damage of carbide surface.

The time to accomplish coating stripping depends on the type of coating. It is a 6–28 h electrochemical process for Ti, Al–based (most popular) coatings. This adds the lead time and cost of tool reconditioning.

The logistic of the second strategy is shown in Figure 3.66. As can be seen, five major transportations are involved providing that re-sharpening and recoating are done in different facilities than that of part drilling manufacturing facility. Coating stripping cannot normally be a part of grinding facility as hazard chemicals are involved in stripping process. The coating facility normally has certifications to handle such chemicals, equipped rooms, and trained personnel. When OTM having no in-house coating facility that handles tools reconditioning, the logistics shown in Figure 3.66 becomes a lot more complicated. That is why the first strategy is often used although the second one is the proper.

3.4.7.3 Quality Control

Quality control of coatings assures their consistency in performance. Therefore, the tool manufacturer/coating provider needs to perform a quality inspection on the coating. The quality inspection procedure normally involves four areas of inspection layer thickness, layer adhesion, layer construction and structure, and layer composition and distribution.

3.4.7.3.1 Inspecting Layer Thickness

There are two primary methods for doing this: checking by calotte grinding or using x-ray florescent radiation. Calotte grinding involves using a small diamond grinding wheel (ball) that grinds away the coating to reveal the substrate below as shown in Figure 3.67. According to Platit AG

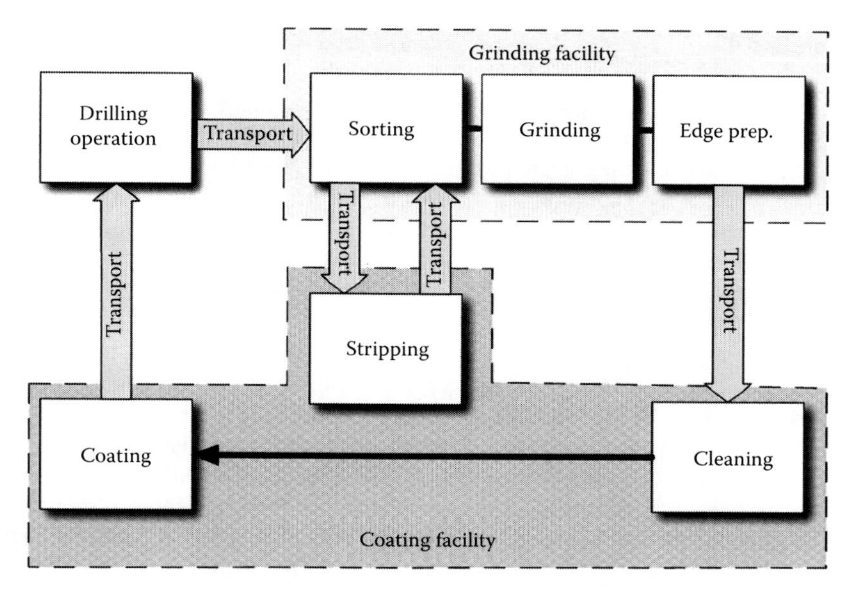

FIGURE 3.66 Logistics of the second strategy.

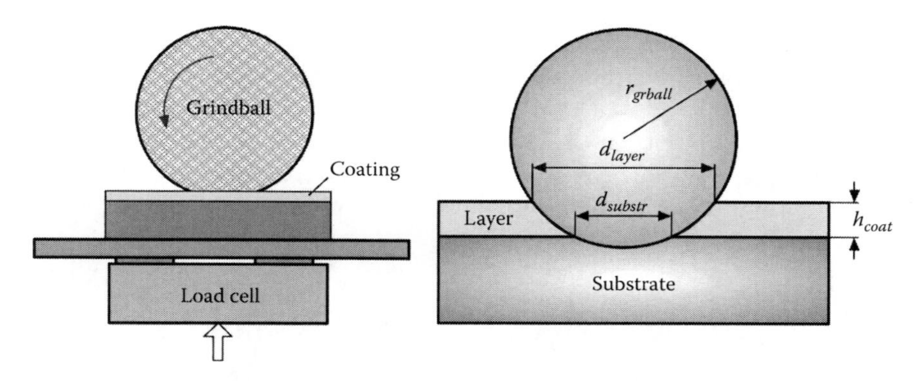

FIGURE 3.67 Measuring the coating thickness by spherical cap grinder (calotte grinding).

coating company, to prepare the calotte section samples, a grinding ball generates a calotte-shaped hollow in the coating of the sample. This results in circular (planar surface) or elliptical indentations (curved surface). Using a digital microscope for examining the calotte section sample, diameters d_{layer} and d_{substr} shown in Figure 3.67 the circular or elliptical indentations are determined by drawing in measurement circles. The coating thickness is calculated automatically as

$$h_{coat} = \frac{d_{substr}^2 - d_{layer}^2}{8r_{grball}} \tag{3.8}$$

Standard ASTM B568-98 (2009) *Standard Test Method for Measurement of Coating Thickness by X-Ray Spectrometry* covers test methodology for x-ray florescent radiation coating thickness measurements. This test method measures the mass of coating per unit area, which can also be expressed in units of linear thickness provided that the density of the coating is known. The method utilizes the x-ray fluorescence (XRF) effect. The proper selection of the operating parameters of the x-ray source and the filters of the detectors enables one to measure coating thickness for a various tool geometries. The energy of the fluorescence radiation is characteristic for the element that produced the radiation. The radiation produced in the substrate has a different energy as the fluorescence radiation produced in the coating. Using the method of differential filters, one or the other component of the radiation can be selected for measurement. The detector modules installed in the gage-head detect the selected radiation component in the backward directions, for example, the radiation from the coating. The measurement of the intensity of the fluorescence radiation allows a precise determination of the coating thickness.

3.4.7.3.2 Layer Adhesion

This is probably the most subjective of the quality control tests. Rockwell indentation and the scratch test are the two methods for checking layer adhesion.

Rockwell indentation involves exactly what it sounds. A diamond cone is indented in the sample. The coated substrate cracks, and the inspectors review the cracks at the indentation point via a microscope and evaluate the number and intensity of the cracks. Dependent upon the individual criteria set forth by the manufacturer, the inspector determines if it is acceptable or if it fails inspection. While this seems straightforward, the coating thickness also affects the number and intensity of the cracks, so a thicker coating would allow more cracks than a thinner coating. Platit AG coating company uses the following evaluation of the test results shown in Figure 3.68. The acceptable adhesion depends on customer application:

Normal HSS tools: HF1–HF4 acceptable
Normal HM tools: HF1–HF3 acceptable
High performance tools: HF1–HF2 acceptable.

The scratch testing method is a comparative test in which critical loads at which failures appear in the samples are used to evaluate the relative cohesive or adhesive properties of a coating or bulk material. During the test, scratches are made on the sample with a spheroconical stylus (generally

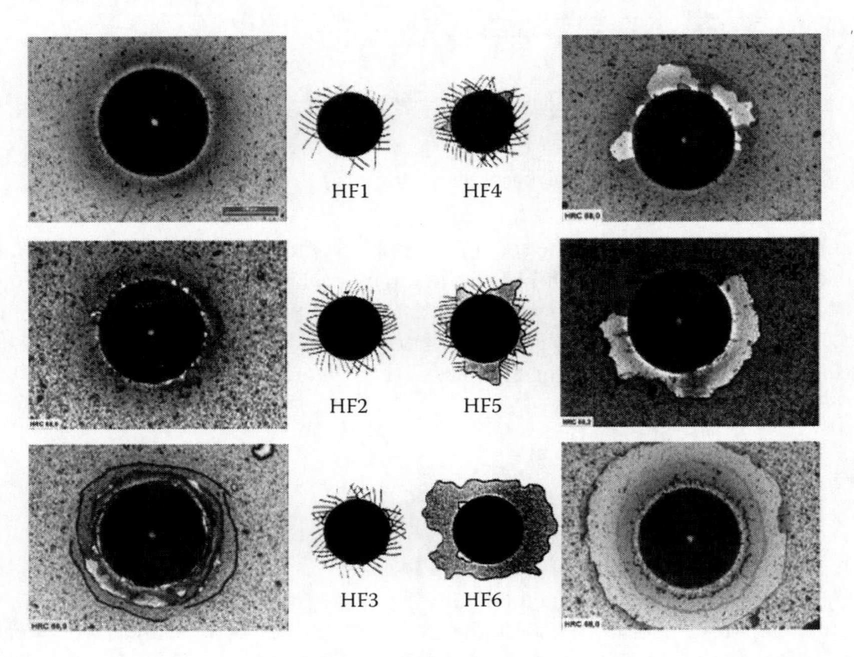

FIGURE 3.68 Possible results of the Rockwell indentation test according to classification suggested and used by Platit AG.

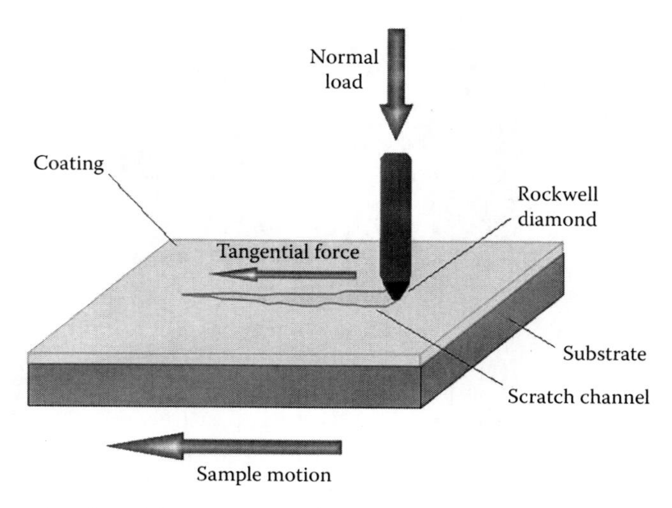

FIGURE 3.69 Schematic of scratch testing method.

Rockwell C diamond, tip radius ranging from 20 to 200 μm), which is drawn at a constant speed across the sample, under a constant load or, more commonly, a progressive load with a fixed loading rate (Figure 3.69).

When performing a progressive load test, the critical load (L_c) is defined as the smallest load at which a recognizable failure occurs. The driving forces for coating damage in the scratch test are a combination of elastic–plastic indentation stresses, frictional stresses, and the residual internal stresses. In the lower load regime, these stresses generally result in conformal or tensile cracking of the coating, which still remains fully adherent. The onset of these phenomena defines a first critical load. In the higher load regime, one defines another critical load, which corresponds to the onset of coating detachment from the substrate by spalling, buckling, or chipping.

The typical scratch tester uses three methods to detect thin-film coating failure: firstly, a load cell is used to measure the change in friction (Figure 3.70); secondly, acoustic emission detects cracking; and thirdly, after the test is complete, the scratch channel can be viewed using an optical microscope. The intensity of the acoustic emission is dependent on the type of thin-film coating failure

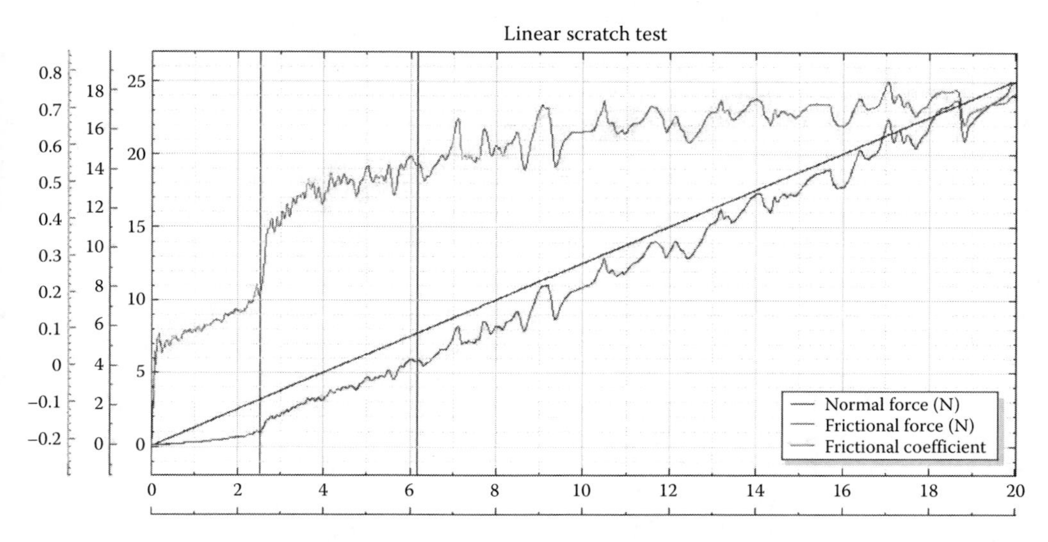

FIGURE 3.70 Record of a linear scratch test.

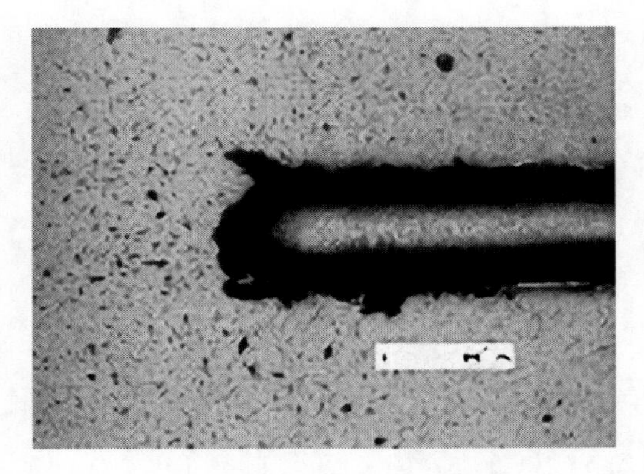

FIGURE 3.71 Example of the end of a scratch channel after a scratch test on a TiN thin film coating.

during the adhesion test, for example, cracking, chipping (cohesive failure), and delamination (adhesive failure). It is therefore good practice to always view the coating failure after the adhesion test using an optical microscope to confirm the critical load (Figure 3.71).

3.4.7.3.3 Layer Construction and Structure

After the thickness and adhesion of the coating were inspected, the next step is to review the layer construction and structure. This is reviewed in the same manner as the layer thickness by calotte grinding. Using the same ball, the coating is ground away and is viewed under a microscope. Here is where the different layer structures mentioned previously are visible and can be confirmed.

3.4.7.3.4 Layer Composition and Distribution

The last quality control check is the layer composition and distribution. This requires a SEM and energy-dispersive x-ray technology (EDX). The SEM produces extremely high magnification of images (as much as 200,000×) at high resolution combined with the EDX analysis to determine the materials and amounts of material in a small section of material (as small as 2 nm).

3.4.8 Cryogenic Treatment of Cemented Carbides

It was already pointed out in Section 3.3.7.3 that the possibility of enhancing the properties of hardened steels by treating them at temperatures below room temperature has been of interest for a long time. Two types of low-temperature thermal treatments are subzero, or cold treatment, and CT. Subzero treatment is carried out at −145°C (−230°F), while CT is carried out at −195°C (−320°F), the boiling point of LN2. Interest continues based on anecdotal evidence of the benefits of CT not only for steels but also in nonferrous metals like copper, aluminum, and titanium. The literature contains many claims about the advantages of cryogenics (Kelkar et al. 2007; Baldissera and Delprete 2008; Sendooran and Raja 2011).

Cemented carbides were also subjected to cryotreatments, and a wide diversity of experimental results was achieved in testing of cutting tools with cryogenically treated carbides as the tool material. Several different cryogenic processes have been tested by researchers. These involve a combination of deep freezing and tempering cycles. Generally, they can be described as a controlled lowering of temperature from room temperature to the boiling point of LN2(−196°C), maintenance of the temperature for about 24 h, followed by a controlled raising of the temperature back to room temperature. Subsequent tempering processes may follow (Yong et al. 2006). Note that the treatment cycle (soaking time, subsequent heating times, and so on) differ from one research to the next so no fair comparison of the available results can be made. For example, Gill et al. (2009) used a

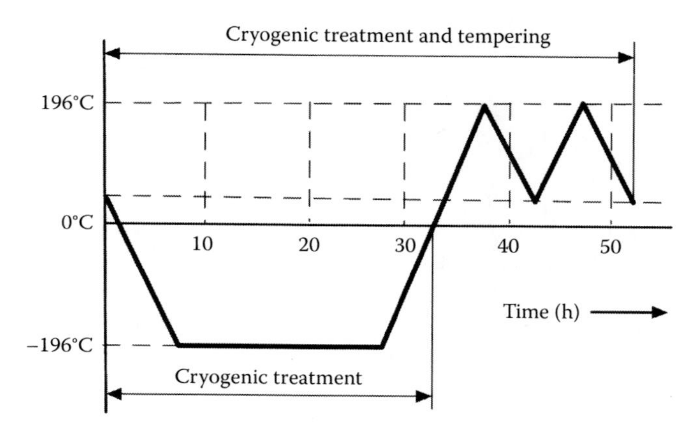

FIGURE 3.72 CT cycle used by Gill et al. (From Gill, S.S. et al., *Int. J. Mach. Tools Manuf.*, 49, 256, 2009.)

treatment cycle shown in Figure 3.72. As can be seen, the carbide inserts were kept under −196°C for almost 30 h while in the study by Yong et al. (2006), the carbide inserts were kept for 18 h at this temperature.

The known studies attempted to reveal two following outcomes of CT effect on cemented carbides:

1. How the properties of cemented carbides change (if any) after CT, and, if any change is found, and why this change occurred.
2. To what extent (if any) tool life (tool wear, cutting force, surface roughness of machined parts, etc.) is affected by such a treatment.

The analysis of the available literature shows that although some mechanisms of improvements of properties of cemented carbide are proposed, no solid evidences to support the proposed mechanisms have been provided so that it is still unclear how CTs alter the properties of cemented carbides and thus how to select the optimal regime for this treatment. For example, Bryson (1999) attributed the wear resistance, and hence the increase in tool life, of carbide tools to the improvement in the holding strength of the binder after CT. He believes that CT also acts to relieve the stresses introduced during the sintering process used in carbide manufacturing. However, Bryson also warned that under certain conditions, CT would have little or no effect on carbide tools. No evidence for the suggested mechanics is provided. Seah et al. (2003) found that the treated tools were superior to those of the untreated as-received inserts at high cutting speeds. From this study, they concluded that CT of tungsten carbide inserts increased the number of η-phase particles, a theory which they supported with photographs taken using a SEM. They assigned this as a reason for reducing TRS hence greater resistance to chipping, improved resistance to plastic deformation during cutting, and lower toughness. None of these properties as well as the actual amount of η-phase was actually investigated. Gallagher et al. (2005) analyzed the microstructural alterations of α (tungsten carbide), β (cobalt binder), γ (carbide of cubic lattice), and η (multiple carbides of tungsten and at least one metal of the binder) phases within the tungsten carbide tools caused by CT and linked these changes to the corresponding enhanced tool life. Thakur et al. (2008) used the help of XRD to demonstrate the formation of complex phases like W_3Co_3C and W_6Co_6C after CT. These complex phases result in the increase in hardness due to exposure of skeleton carbide matrix to later posttreatments.

The author's analysis of the available information on the performance change of cemented carbide after CT shows that there is a wide diversity of opinions and contradictive experimental results. For example, Yong et al. (2007) cryogenically treated tungsten carbide milling inserts and found 28.9%–38.6% increase in tool life. Reddy et al. (2009) reported that tool life of deep cryogenic-treated coated cemented tungsten carbide cutting tools increase by 27% when compared to untreated inserts for cutting speeds in the range between 200 and 350 m/min. They also observed that the cutting force decreases by 11% compared to untreated inserts for this range of cutting speeds. Gill et al. (2009) concluded that the deep CT has some effect on the performance of TiAlN-coated tungsten carbide inserts especially at lower cutting speeds. However, at higher cutting speeds, marginal gain in tool life can be obtained. It was also found that, deep CT weakens coating–substrate interfacial adhesion bonding. Overall, they do not recommend the deep CT for TiAlN-coated tungsten carbide inserts as the benefit gained is not significant. Yong et al. (2006) in their comprehensive study found that tools under mild cutting conditions stand to gain from cryogenic treatment, but heavy duty cutting operations with long periods of heating of the cutting tool will not benefit from it.

In the author's opinion, the following conclusions on CT of cemented carbides for drilling tool applications can be drawn:

1. The major gain that one can obtain by cryogenic treatment is an increased resistant to chipping.
2. Tool life may be increased for tools working at relatively low cutting speed. For HSM, no gains can be obtained.
3. CT may also interfere with coating adhesion and other essential properties so if it is used, it should be on a case-by-case basis.
4. Although the deep CT adds small cost to indexable carbide inserts as many of them can be treated simultaneously, this is not the case with drilling tools. In other words, the deep CT of special drilling tools could add significant cost.

3.4.9 Considerations in Proper Grade Selection

The discussion on the factors affecting intelligent selection of HSS presented in Section 3.3.4 is fully applicable to the considerations in the proper cemented carbide grade selection. In other words, although the commonly specified properties of commercial carbide grades are readily available and sufficient for many users, it is not nearly enough to make an intelligent selection of a particular carbide grade for drilling tools. The common trial-and-error method suggested earlier for the selection of the proper grade is time-consuming and may not result in the selection of the optimal grade.

As with HSS, there are two levels of grade selection: (1) selection by tool manufacturers for the tools they are producing and (2) selection by the end users for particular applications. These two selection procedures may not share the same objective because the objective of any tool manufacture is to maximize the profit while the objectives of an end user can be to maximize tool performance, minimize cost per part, maximize productivity, improve quality, and many others. End users should clearly realize that selecting a particular carbide grade at their level is only feasible when all other tooling and machining problems are solved so it is determined clearly that the most feasible way to improve the performance of a particular drilling tool is to use a better (for the application) carbide grade. The problem with the assessment of the quality of the selected grade is not as simple because the performance of any particular carbide grade is system-dependent, that is, it depends on the tool design and geometry, machining regime, MWF parameters, machine conditions, and so on. If the System Coherency Law discussed in Chapter 1 is violated, that is, a selected carbide grade is better than the quality of other components of the drilling system, then no advantages of this grade can be gained in drilling operations. In this case, such a selection is a pure waste of money. If, on the other hand, a poor carbide grade is selected for a high-quality machining system, no advantages of this quality can be gained, since an unsuitable carbide grade will determine productivity and efficiency of this expensive machining system. Saving pennies on carbides of subpar quality results in losing thousands on underutilization of capabilities of the expensive drilling system.

Not realizing that it is the case, carbide drilling tool manufacturers have much greater challenge in the selection of proper carbide grade for their tools besides very few who manufacture their own carbide grades. They have to take a much deeper dive into the properties of carbides than cutting tool end users. This is because there are a number of important decisions to make.

First one is that tool manufacturers have to select the proper carbide grade(s) for their product(s) as at the end of the day they are responsible for the performance of their tools particularly today when a tool end user can select and test tools from various manufacturers globally. To help in such a selection, the properties of various carbide grades are readily available. Although a number of mechanical, physical, chemical, and so on properties of a given grade, for example, its compressive strength, hardness, TRS, are available on the websites and colorful catalogs of carbide manufacturers, a selection of the proper carbide grade based only on these properties is virtually impossible as these properties are weakly correlated with the performance of a particular grade in cutting.

Selecting cemented carbide materials for cutting tools is a complex non–well-formalized process that consists of a series of compromises based mainly on the experience of the decision-making team (person). These compromises include balancing toughness and resistance to corrosion and shock, as well as to wear and abrasion resistance. Cemented carbide grain size, binder alloy percentage, additions of titanium and tantalum carbide, and other alloying elements along with considerations about design, applied loads, and stresses are combined to achieve the best compromise and result. Therefore, the good old *trial-and-error* method is widely used to determine suitability of a particular grade. As such, even a small R&D facility in the drilling tool company can provide significant help on the matter, particularly if research records are kept in the proper systematic format.

The second tough decision is where to buy the selected grade(s). There are a number of variables cemented carbide suppliers offering *similar* (e.g., C2) grades. However, for the same grade according to the previously discussed very general ISO or C-grade classification, there are a number of completely hidden from a casual eye variables. These variables can only be understood if the whole process (at least in general) of carbide manufacturing is understood. To help such understanding, Appendix B describes the carbide manufacturing process at the depth and level needed for carbide users. It is shown that a number of special processes and quality procedures can be involved in the manufacturing of cemented carbide. As a result, the quality and cost of *similar* carbides from different carbide manufacturers are not the same or even similar. The material presented in Appendix B should help end users and carbide drilling tool manufacturers ask intelligent questions to understand if the cost of the selected carbide grade is inflated or if it is justified by additional manufacturing operation in carbide manufacturing added to improve the quality and consistency of the end product.

Often, unfortunately, the choice of the carbide supplier is based on the cost and lead time rather than quality because the former two are readily understood by tool manufacturers while carbide quality is rather obscure for many cutting tool specialists.

The third choice to make is about coating. Coatings can be of general purpose or application specific. The common trend is to use a coating recommended in the literature, by local coating provider, or the same as *everybody uses*.

An intelligent carbide grade selection even for a given, well-defined application is not an exact science or even engineering. As discussed earlier, the situation with carbide grade performance and its inconsistency can be assessed using the trial-and-error method for indexable inserts used in stable production environment (the same machine, MWF, and other variables). Even with carbide indexable inserts, this is not the case in many practical applications when one insert can be used in a tool that performs various machining operations in, for example, a job shop environment. A completely different situation is with performance, and particularly with inconsistency, of carbide quality from one manufacturing batch to the next for drilling tools as there are a way greater number of aspects involved in the manufacturing and use of these tools. Often, it is next to impossible to determine if carbide quality and/or inconsistency is the root cause of a drilling tool failure problem.

The first step in selecting the proper grade of cemented carbide is to identify the failure mode for the considered operation, that is, the drilling system and to determine if a better carbide grade can be of any help. As with coatings, the selection of a better carbide grade should not be aimed on solving tooling or drilling system problems. In other words, no one grade can handle dirty MWF, inclusions/cavity of casings in the work material, improper tool holder, and so on. Therefore, if a common failure mode is, for example, drilling tool breakage, then other components of the drilling systems should be considered much before one is concerned with a better carbide grade. The common failure mode where a better carbide grade/make can help is tool wear and thus quality problems with the machined holes caused by this wear.

Selecting the proper carbide supplier is also a very important task. Some larger users have tried various strategies to address the perceived problem of carbide inconsistency from batch to batch. In the automotive and airspace industries, the cost of the carbide tools accounts for 5%–10% of the direct manufacturing costs while machine time makes up a much larger percentage. Therefore, these industries stress consistency and reliability of carbide tools over tool life to lower machining time.

One of the more intriguing strategies is the effort by some end users to push the quality issue back on the carbide-tool manufacturers where it belongs. For example, in an effort to improve the consistency of incoming carbide tools, Boeing Co. has a program that puts the burden of quality testing back on the tool suppliers. This company buys carbide tools solely from manufacturers that actually have labs where they perform acceptance testing on incoming carbide substrate raw materials. A certificate of carbide quality tested in the tool manufacturer's facility is required with each batch of carbide tools.

Some of the large companies developed personal standards for carbides quality and carbide-tool requirements. Boeing Co., for example, requires certification that the substrate material of the tools it buys meets Boeing's criteria and that the geometry generated matches the specifications Boeing has established in conjunction with the carbide-tool manufacturer. The rationale for such a move was that the company uses more and more carbide in HSM, and because there is no standard for carbide, Boeing worked with manufacturers to establish tool geometry guidelines and carbide specifications. According to these specifications, no recycled carbides are allowed in the tools they use. Particular attention is paid to the certification of raw carbides. Other industries have worked to create in-house standards to ensure the consistent performance of tools in given operations.

3.5 DIAMOND

3.5.1 Introduction

Progress in design and manufacturing led to the development of new engineering materials including a wide group of composites. One class of composite being looked at more and more by the aerospace and automotive industries is metal matrix composites (MMCs). MMCs include lightweight, and relatively low-strength alloys of aluminum, magnesium, or titanium reinforced by adding second-phase particles, whiskers, fibers, wires, or filaments. For the last 20 years, the focus of interest for the more common automotive and aerospace applications seems to be on aluminum reinforced with particles of silicon carbide or alumina fibers. Applications for these materials are being developed, in which heavier steel or iron components are replaced by lighter MMC substitutes. Examples in the automotive industry include engine blocks, transmission cases, brake disks, axles, or, in the leisure industry, items such as tennis racquets. Whether the application is a connecting rod or a tennis racquet, efficiency in use is obtained by reduction in weight, and acceleration is improved by reduction in the inertia of the moving mass.

Progress has been made both in the formulation of alloy compositions and in manufacturing routes. PM, co-spraying, low-pressure liquid metal infiltration, and die-casting techniques have been developed, and, in some of these, near net shape components can be made.

The wide implementation of MMCs particularly in HSM environment is being somewhat hampered by the fact that they are extremely abrasive and perceived as difficult-to-machine to close tolerances using conventional cutting tool materials. As components made from MMCs moved into mass production, machining costs/efficiency became of great importance, and thus a need to develop more efficient tool materials was fully realized. A new closer look at diamond as the tool material was taken as diamond is the hardest, most wear-resistant material in the world.

First discovered among alluvial (soil carried by water) deposits in the riverbeds of India, diamond has long been prized for its spectacular quality as a gem. Diamond's industrial use as tool material dates back to ca. 300 BC. Early uses of a diamond rock drill are found in Diderot's Encyclopedia in 1751.

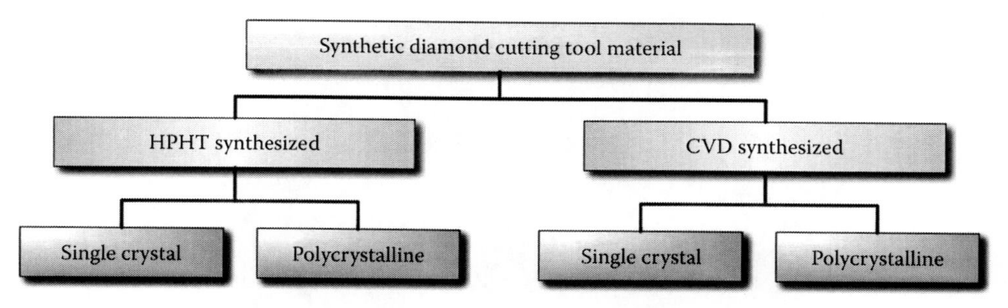

FIGURE 3.73 Simple classification of synthetic diamond-cutting tool materials.

Diamond-tipped precision single-point tools were used by J. Ramsden in 1771. Around 1900, large circular saw blades were set with diamonds to cut architectural stone. About the same time, grinding wheels were being developed by impregnating metal bodies with diamond particles.

The large-scale industrial use of diamonds in industry, however, begins with the development of synthetic diamonds. From a historical perspective, the real advancement of CVD diamond has evolved from work carried out from the 1950s onwards by researchers in the United States, Russia, and Japan. The first successful documented attempt to grow diamond at low pressures was by William G. Eversole of the Union Carbide Corporation (the United States) in 1952, making Eversole the first person to make diamond of any kind. In 1954, H. Tracy Hall (GE) successfully synthesized the first diamond using high-pressure high-temperature (HPHT) process. The commercial promise of diamond synthesized by HPHT processes was fulfilled when GE opened their first production units in 1956. Since this date, production of HPHT synthetic diamond has increased every year and is currently in excess of 300 tons per year. The properties of diamond and history of the development of synthetic diamond including description of the processes are presented in Appendix B.

A simple classification of synthetic diamond cutting tool materials is shown in Figure 3.73.

In machining where the wear mechanism is mainly abrasion, polycrystalline diamond (PCD) made using HPHT synthesis has already proved itself to be a superior tool material. Over the last 20 years, PCD has been accepted for volume production of hypereutectic aluminum–silicon alloy components in the automotive industry, the machining of nonferrous alloys of copper, the machining of abrasive plastics and plastic composites such as glass fiber and printed circuit boards, as well as volume machining of wood composites, such as chipboard.

3.5.2 Blanks for Drilling Tools: PDC and PCD Disks

Vast majority of PCD drilling tools are actually PCD-tipped tools made from PCD blanks manufactured using HPHT process over 30 years ago. There are two types of such blanks: polycrystalline diamond compacts (PDC) and PCD disks. PDC shown in Figure 3.74 are used to manufacture the cutting elements of drilling tools (called drill bit shown in Figure 3.75) for oil and gas drilling. PCD disks are used to make PCD cutting tips (Figure 3.76) for drilling tools similar to that shown in Figure 3.77 in metalworking industry. Although these two types of blanks were developed practically simultaneously, PDC and thus PDC drill bits experience much greater advancement rate due to high demands for advancement in oil and gas drilling compared to the metalworking industry.

PDC cutters, referred to at the time by the trade name Stratapax, were first developed by GE in 1973. They utilized GE's earlier invention, monocrystalline man-made (synthesized) diamond (see Appendix B), which was loaded into a pressure cell with cemented carbide substrate and repressed to produce a compact of 13 mm diameter and 3.3 mm length that incorporated a 0.5 mm thick diamond table. Over the next 30 years, a number of significant developments were made to improve both PDC design and materials properties as well as the design of drill bits.

The first difficulty was reliable mounting of PDC cutters in bit bodies. Brazing techniques and practice of the day frequently led to debilitating cutter loss and failed runs. Post-mount press-fit cutters that

FIGURE 3.74 PDCs.

FIGURE 3.75 Modern drill bit with PDC.

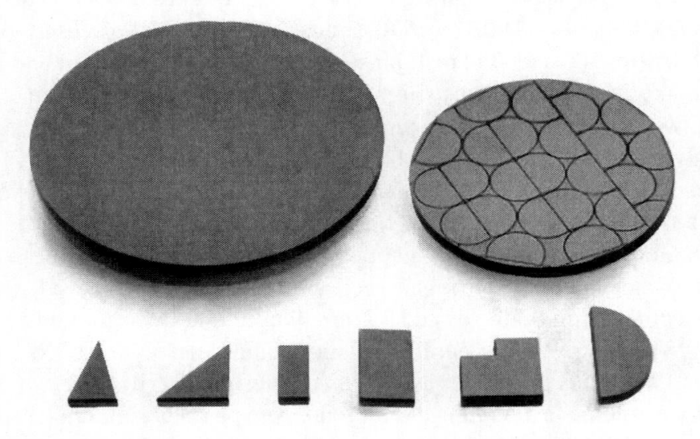

FIGURE 3.76 PCD disk blank and variety of PCD tips made out of this blank.

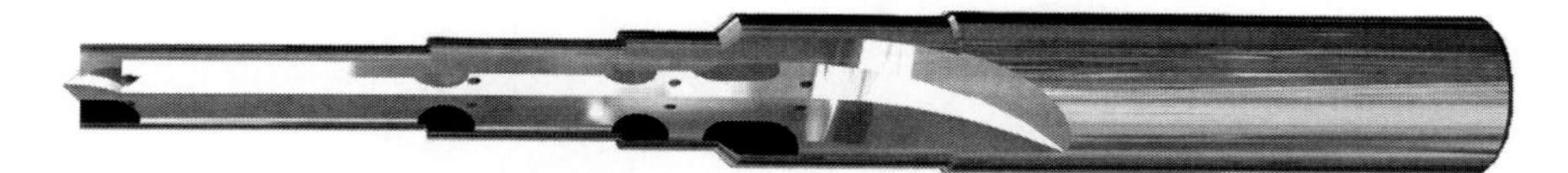

FIGURE 3.77 Typical PCD-tipped multistage drill.

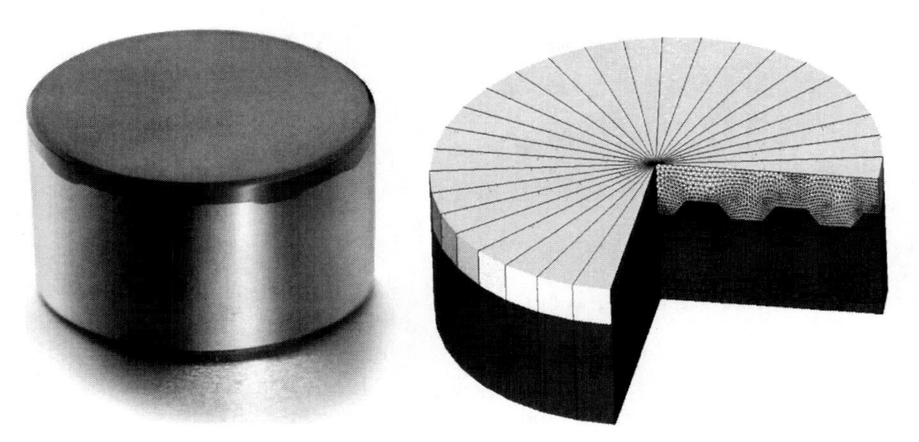

FIGURE 3.78 PDC with nonplanar diamond–substrate interface.

were deployed in steel-body bits were prone to fracture breakage of the post at the mounting point and to loss through erosion of the steel bit body. Improved brazing and mount pockets provided solutions.

The next major hurdle was reducing diamond table (layer) delamination from the substrate under impact loading. The initial solution to this problem was the development of nonplanar diamond–substrate interface by providing grooves and other patterns on the face of the cemented carbide substrate (e.g., as shown in Figure 3.78). At this stage of development, PDC bits had increased their market presence to about 15% of all footage drilled and were considered by many to have neared the peak of their potential development.

The major advancement, however, was made by reducing the so-called bit whirl, that is, self-regeneration off-center rotation condition as the prime source of the delamination. It was reduced by the design of force-balanced, symmetrical-bladed, spiral-edged drill bits. These improvements made by the late 1990s led to a significant increase in the use of such tools—by this time, PCD drill bits accounted for about 46% of all footage drilled in the oil field.

When design and manufacturing problems of PDC drill bits were solved, the coolant parameters and its delivery were optimized, and drilling cycle programming improved, common failures due to bit balling, cutter loss, and impact damage were mitigated. The next logical step in the further improvement of PDC drill bit performance was closer considerations of PCD layer properties as related to its performance in the normal wear mode, which is abrasive wear. It was realized that PDC cutters were subject to thermal damage and accelerated wear at the cutting tip, due to large amount of the residual cobalt catalysis remaining in the interstitial matrix of the cutters' PCD face. By reducing the cobalt content in the outermost layer of the diamond table, the cutters' abrasion resistance and thermal stability were significantly improved, allowing PDC bits to compete economically with roller-cone bits in even more applications.

In 2010, PDC bits accounted for an astonishing 65% of footage drilled in the oil and gas applications and still not do appear to have peaked in their development. More research than ever is going into PDC bits and especially into PDC cutters.

As mentioned previously, the development of metal cutting tools with PCD tips and thus PCD disks as blanks for such tips was much slower due to a number of reasons. The prime reason for that was

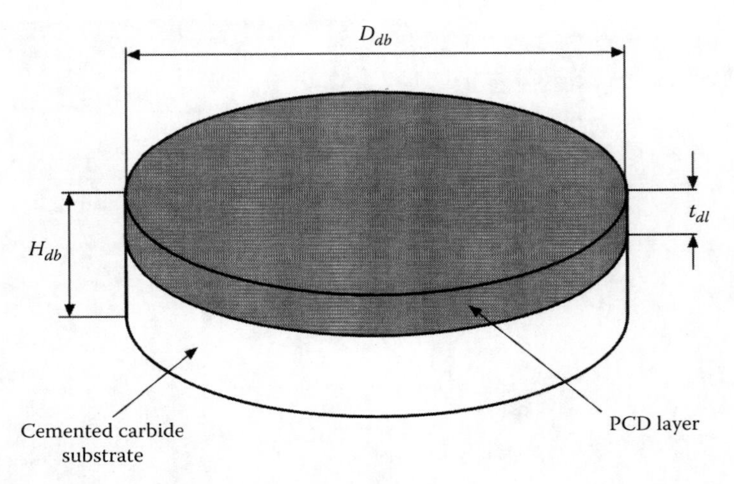

FIGURE 3.79 PCD disk and it geometrical parameters.

unavailability of drilling systems capable to run PCD drilling tools efficiently. Low available spindle speeds, low system rigidity, and excessively poor properties and maintenance of MWF were the prime reasons. Moreover, a widespread almost-religious belief is that PCD is only suitable for machining aluminum and other nonferrous materials because of the so-called affinity of diamond, which is essentially carbon in PCD and carbon in steels so that PCD *dissolves* in steels. As such, carbon in cemented carbides (WC) and that in steels mysteriously does not cause any *affinity* problems. As discussed in Appendix B, carbon as diamond or graphite is soluble in several metals (nothing to do with affinity), particularly Fe, Ni, Co when the temperature is high enough. However, this temperature is much higher than that leading to structure failures of PCDs as disucssed in Chapter 2. The development picked up its pace only when high volume of MMS (silicon–aluminums) and fiber-reinforced composite materials became the materials of choice in the automotive and in the aerospace industries, respectively.

Figure 3.79 shows a PCD disk (blank) and its geometrical parameters. Such a blank consists of a layer of fine diamond powder sintered together into a dense uniform mass, approximately 0.5 or 2.0 mm thick, supported on a substrate of cemented carbide. The expensive and technically difficult HPHT process used to produce the basic disk has been refined over the years, and it is now PCD blanks that are made as disks of up to 74 mm in diameter. By using different grain sizes of diamond and by changing the composition of the diamond and carbide layers, it has been possible to optimize the properties of PCD for specific applications. As a result, various grades are now synthesized by a number of companies around the world. The differences are in stability of sizes (H_{db} and h_{dl}), quality including consistency and residual stresses, and many others, which are not specified by PCD blank manufacturers. The blanks can be ground as that shown in Figure 3.76 or, which is common for high-quality blanks, mirror polished as that shown in Figure 3.80. The cost of discs of similar diameter can differ more than ten-fold.

3.5.3 Manufacturing of PCD Disks

3.5.3.1 Process

PCD disks are produced by high-pressure sintering techniques using a solvent–catalyst metal such as cobalt (Co) at pressure 6 GPa and temperature from 1673 to 1873 K. Mixed powder methods and infiltration methods are common sintering techniques. General Electric was the first company that developed one of the previous infiltration methods in 1972 and then industrialized PCD blanks. In GE's method, diamond powder is sintered and bonded to a WC–Co substrate by infiltration of Co from the substrate using presses similar to that described in Appendix B.

Figure 3.81 shows the assembly of the press for sintering, and Figure 3.82 shows details of a diamond powder disk and WC–Co substrate wrapped up in tantalum (Ta) shell. Such a wrapping is then placed in the assembly composed of a solid pressure salt (NaCl) surrounded by the crucible and graphite heater. Formation stages of a common PCD disk during HPHT sintering are shown in Figure 3.83. First, pressure is raised to its nominal level with little or no heating. During this stage, all the crystals are being pushed against each other with increasing force. Many diamond particles are sliding relative to each other, and many are cracking into two or more fragments with

FIGURE 3.80 Mirror-polished PCD disk.

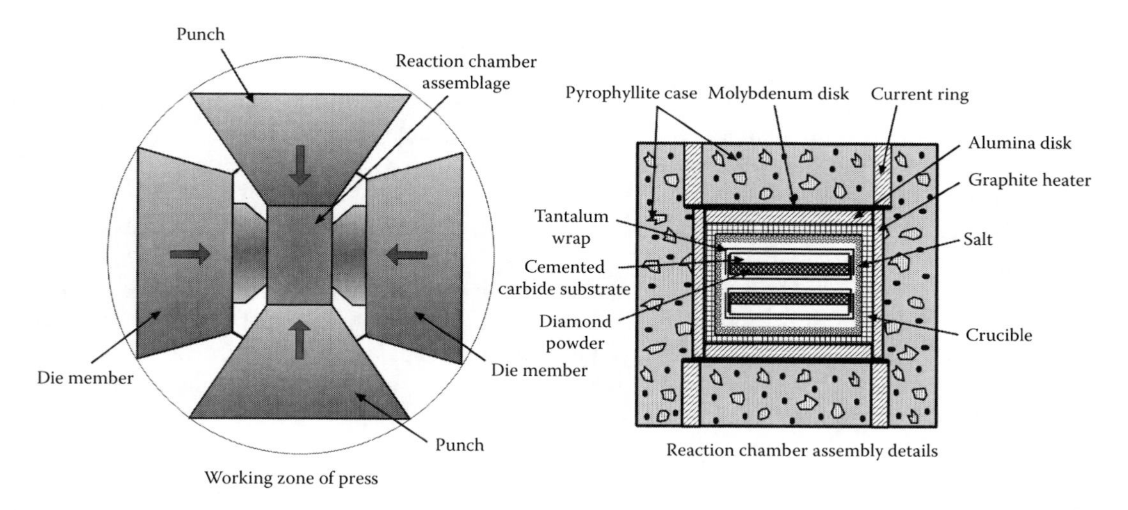

FIGURE 3.81 Apparatus used in PCD disks sintering.

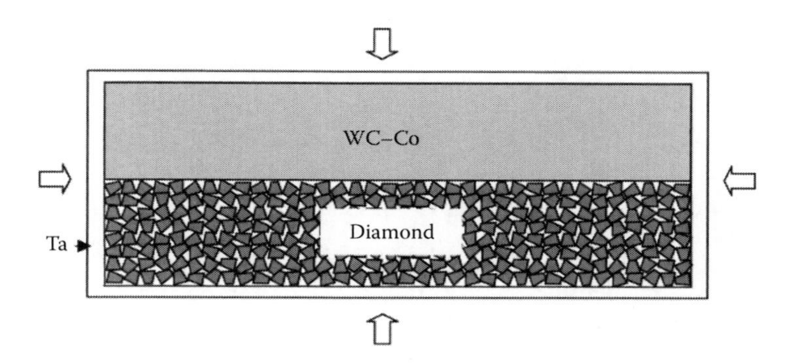

FIGURE 3.82 Diamond powder disk and WC–Co substrate wrapped up in tantalum (Ta) shell (the capsule).

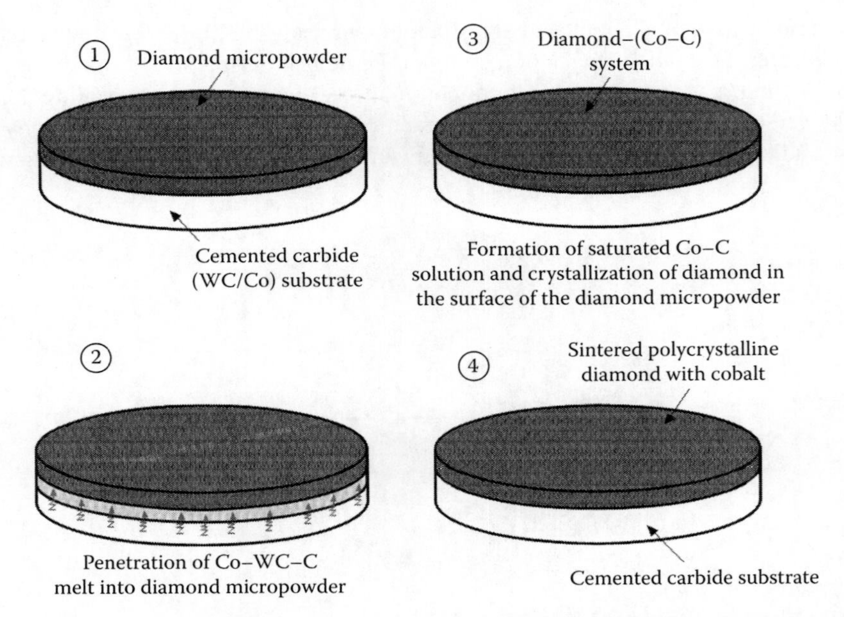

FIGURE 3.83 Formation stages of a common PCD disk during HPHT sintering: (1) start of the process with applying high pressure, (2) melting and infiltration stage, (3) phase formations, (4) finish of the process.

overall effect of increasing powder density (Uehara and Yamaya 1990). A coarser powder presents a higher degree of crushing than a fine one as the former includes much smaller average number of contact points and thus much higher contact stresses (compared to fine powders) that cause the described crushing.

After the crashed and compacted powder is under full pressure, the temperature is raised at a heating rate of 60°C/min to its nominal value (approx. 1600°C). The pressure–temperature condition is always kept in the thermodynamically stable region of diamond (see Figure B.24) above the eutectic temperature of carbon and Co. As the diamond powder is packed against a WC–Co substrate, cobalt is the source for the catalyst metal that promotes the sintering process. When the cobalt reaches its melting temperature of 1435°C at 5.8 GPa, it is instantaneously squeezed into the open porosity left in the layer of compacted diamond powder. At this point, the sintering process takes place through a mechanism of carbon dissolution and precipitation. Technically, this process is defined as a pressure-assisted liquid-phase sintering. The driving force for the densification under an extreme pressure is determined by the pressure itself and also by the contact area relative to the cross-sectional area of the particles. The reaction speed is proportional to the temperature and to the average effective pressure P_{ae}, which is the actual contact pressure between particles, as expressed in the equation (Belling et al. 2010a):

$$\log\left(\frac{d\rho_{ap}}{dt}\right) \approx -\frac{D_l P_{ae}}{RT} \tag{3.9}$$

where

ρ_{ap} is the powder apparent density
D_l is the carbon diffusivity in the molten metal catalyst
R is the ideal gas constant
T is the temperature

It follows from Equation 3.9 that the sintering process is faster if both the contact pressure and temperature are increased. On the other hand, when temperature and/or process time are insufficient, cobalt infiltration would not be completed resulting in weaker bonds of diamond grains.

The average effective pressure P_{ae} can also be expressed as (Bellin et al. 2010a)

$$P_{ae} \approx \frac{4a_{ps}^2}{Z_{sp}r_{sp}} P_{ex\text{-}s} \tag{3.10}$$

where

a_{ps} is the average particle size
r_{sp} is the radius of the contact area between two spherical particles
Z_{sp} is the number of surrounding particles
$P_{ex\text{-}s}$ is the external pressure applied to the system

It follows from Equation 3.10 that smaller grain size and better packing result in lower contact pressure. Therefore, in sintering PCD grades with small AGS, higher pressure and temperature are required to accomplish the sintering process.

The result of the process is a disk consisting of the WC–Co substrate and the diamond layer of certain thickness strongly bonded with this substrate. In this disc, PCD composite is a fully dense mass of randomly oriented, intergrown micron-size diamond particles that are sintered together in the presence of a metallic catalyst phase, usually cobalt. Small pockets of the catalyst phase, which promotes the necessary intergrowth between the diamond particles, are left behind within the composite material. Figure 3.84 shows SEM micrographs of the diamond grains before and after sintering.

It is believed that plastic deformation in the presence of catalyst is induced in the particles and diamond–diamond bonding occurs during sintering. It is explained by the Ostwald ripening mechanism. That is, some of the diamond particles dissolve into molten Co–C(–Zr) in the sintering stage (Co penetration stage) at high temperature and high pressure. The dissolved diamond precipitates again on the surface of the remaining diamond particles by the driving force caused by the difference in the solubility of diamond due to different curvatures of the particles and results in direct diamond–diamond bonding. It is considered that the previous solution and reprecipitation of diamond and Co penetration into the diamond powder are accelerated by the fine diamond particles generated in the pressure-rise stage.

Another explanation of the diamond sintering process is based on the formation of graphite during the heating stage. According to this theory, locally low-pressure regions are inevitably generated

(a)

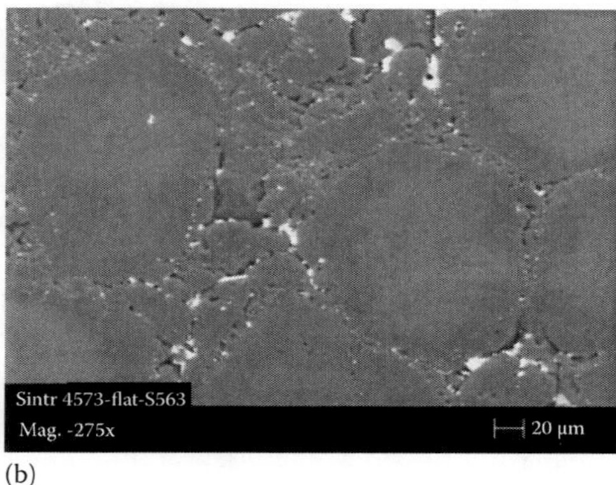

(b)

FIGURE 3.84 SEM micrographs of the diamond grains before (a) and after (b) sintering.

in the powder compact due to defects such as voids. Transformation of diamond into graphite occurs during the heating stage in the low-pressure regions. The graphite dissolves into molten Co more easily than diamond, and the direct bonding of diamond is formed by the precipitation of graphite as diamond. To confirm this process, some experiments were carried out at 6 GPa pressure and up to 1600°C using 10–15 and 2–4 μm grade powders. Neither Co nor graphite is detected by XRD or energy dispersive x-ray spectroscopy (EDS) in the diamond layers of either 10–15 or 2–4 μm grades sintered at 1500°C for 10 min. Therefore, it was concluded that transformation of diamond into graphite does not occur in the HPHT sintering process (Uehara and Yamaya 1990). However, this result is not conclusive as a very short sintering time (10 min instead of usual 1 h) and an excess of Co (additional Co plate placed on the top of the compact) were used in the test.

Therefore, it can be concluded that the phenomena of formation diamond–diamond bonds are not clearly demonstrated yet. More detailed and sophisticated studies are needed to understand and then to optimize the process, which in its present stage is fully empirical.

The Ta foil used as the wrapper in the process is changed into a TaC layer containing Co after sintering.

3.5.3.2 Powder Mix

The powder used in the sintering consists of the diamond powder and some other materials added for specific purposes. The presence of the latter is rarely mentioned in the literature as these additions in both the composition and the amount are closely guarded process secrets specific to a particular PCD blank manufacturer.

The quality of the feedstock diamond powder (powder mix) in this compact is of utmost importance in determining the final properties and consistency of the sintered product. Usually, the micronized diamond powder used in PCD manufacturing is a by-product of the industrial diamond powder synthesis process, where crystals of at least 100 μm size are produced mainly for stone cutting market (e.g., diamond saw blades, diamond wheels, loose diamond powder). There are a great variety of industrial diamond powders available on the market with an extremely wide quality range. Depending on the press cycle parameters (mainly pressure and temperature), crystals can be grown with different shapes (e.g., cubic rather than octahedral). Furthermore, if the crystals are growing too fast within the molten catalyst bath, it is possible to find metal inclusions buried deep inside the crystal at the end of the cycle.

Figure 3.85 shows SEM micrographs of starting diamond powders (Uehara and Yamaya 1990). As can be seen, a range of sizes rather than a particular size (grade) is normally assigned to the powder. Figure 3.86 shows particle size distribution in 2 and 4 μm graded commercial diamond powder of good quality (Shin et al. 2004). The better the quality of the diamond powder, the narrower the range of particle distributing in terms of their shapes and sizes.

Earlier researchers (e.g., Uehara and Yamaha 1990) reported that the grain size of diamond becomes smaller after sintering. They attribute this phenomenon to crushing of the starting powders by friction among the particles during the pressure-rise stage. To confirm this, diamond particles were observed after pressing at 5.5 GPa. As shown in Figure 3.87, all powers sizes used in the test were crushed down. The crushing was particularly remarkable in larger grades and a larger number of crushed fine particles smaller than 1 μm were observed in the crashed powders. This idea is fully accepted in the consideration of PDC for gas and oil drilling bits (Bellin et al. 2010a). Note that Uehara and Yamaha (1990) used the sample capsule with additional cobalt disk placed on the top of diamond powder and zirconium capsule while the tantalum foil was placed between the powder and substrate as shown in Figure 3.88.

Latter researchers, however, observed the opposite effect, that is, diamond grain growth, referred as the abnormal grain growth (AGG) in PCD during HPHT sintering (Shin et al. 2004). Some grains preferentially grow to a size of several hundreds of micrometers that later on makes it impossible for the PCD disk to be cut by the wire EDM. It has been generally accepted that AGG can occur in the presence of liquid phases (Kang 2005). The coarsening, or growth, of solid grains during

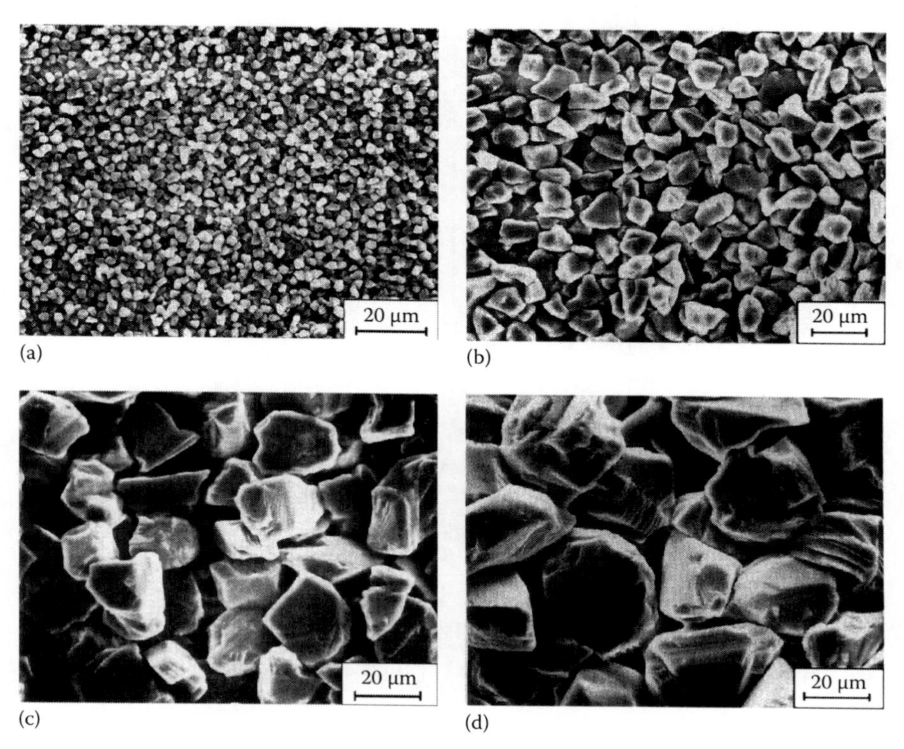

FIGURE 3.85 SEM micrographs of starting diamond powders. Grades, microns: (a) 3–6, (b) 10–15, (c) 22–36, (d) 40–60.

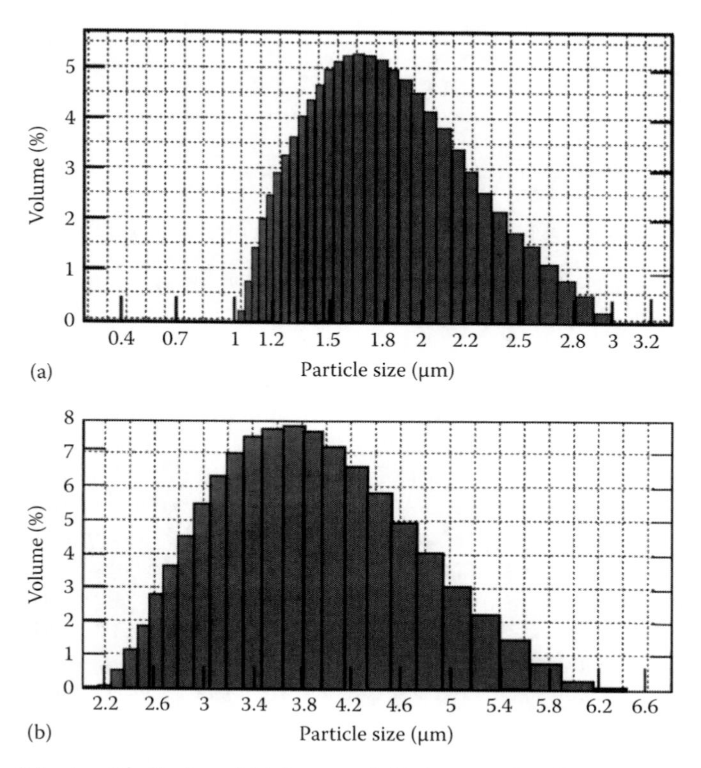

FIGURE 3.86 Particle size distribution of (a) 2 μm- and (b) 4 μm-graded diamond powder.

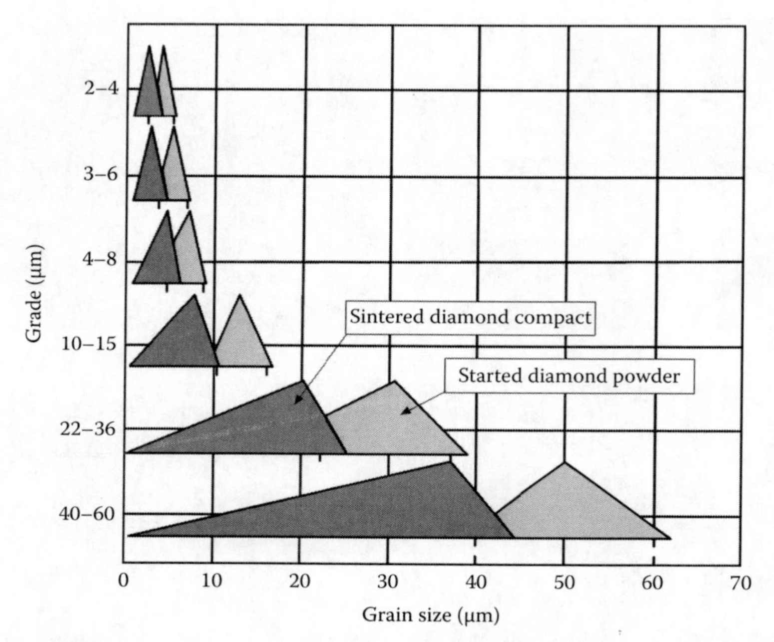

FIGURE 3.87 Schematic diagram showing the relationship between the grain size of diamond before and after sintering.

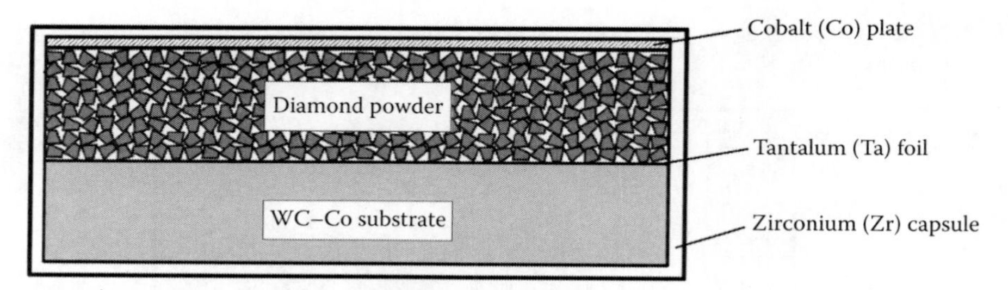

FIGURE 3.88 Sample capsule used by Uehara and Yamaha. (From Uehara, K. and Yamaya, S., High pressure sintering of diamond by cobalt infiltration, in: *Science and Technology of New Diamond*, Saito, O.F.S. and Yoshikawa, M. (eds.), KTK Scientific Publishers, Tokyo, Japan, 1990.)

liquid-phase sintering is usually explained in terms of the Ostwald ripening (Lifshitz, Slyozov and Wagner theory). Faceted particles are known to undergo AGG. Park et al. (1996) suggested a model in which, for grains dispersed in a liquid matrix being faceted with flat surface planes, the 2D nucleation would take place, resulting in AGG of certain crystals that are bigger than a specific minimum radius.

AGG in the fine-grained PCD can be controlled by addition of fine WC powder. Hong et al. (1988) reported behavior of cobalt infiltration and AGG during sintering of diamond on a cobalt substrate. Akaishi et al. (1991) studied AGG in the synthesis of fine-grained PCD compact and its microstructure. They found that addition of a small amount of cubic boron nitride powder suppressed AGG. Yu et al. discussed methods to suppress AGG of diamond in their sintering of 0.5–1.5 μm diamond powder using Ni–Zr alloy as a sintering aid under HPHT (Shin et al. 2004). The best reported results are achieved when the Co powder (approx. wt. 5%) of 1.5 μm in size and 0.8 μm WC powder as much as 0.35 wt% as a grain growth controlling agent are added while sintering took place using WC–10 wt% Co substrates (Shin et al. 2004).

Most available PCD blanks for drilling tools produced today have diamond grain size after HPHT sintering (*as-sintered*) of 1–30 μm. Finer, uniform, as-sintered diamond grain sizes, for example, of about 0.1–1 μm (referred to as submicron grades) have proven challenging to produce commercially using a common HPHT process where the compact of the diamond powder is placed on the top of a carbide substrate. This is not only because of AGG as described previously but primarily due to low packing density of submicron diamond particles that causes problems during loading of shielding enclosures and HPHT processing. Very fine pores between the submicron diamond grains in the initial diamond particle mass are difficult to uniformly penetrate with catalyst metal, leading to incomplete formation of strong bonds between diamond particles. As a result, the high surface areas of submicron diamond powders are not properly bonded. This is the actual cause for AGG.

One of many possible solutions to the problems is to blend diamond particles and particles of catalyst metal together. As such, the size of catalyst particles should be less than that of diamond particles (US Patent Application No; US 2007/0056778, Publ. date. Mar. 15, 2007). Then the resultant blend is sintered using a pressure and a temperature for a time sufficient to affect intercrystalline bonding between adjacent diamond particles. The catalyst metal, normally cobalt, may be about 0.5–15 wt% of diamond powder blend.

The foregoing consideration reveals that the diamond powder is a mix that may contain a lot of different technological additives to promote sintering (e.g., Co powder) and to prevent AGG (e.g., WC and/or Ni–Zr powder). The amount of these additives as well as their geometrical and physical characteristics is not disclosed in the PCD blank characterization. As a result, PCD cutting elements made of the same PCD grain size by different PCD blank manufacturers may exhibit considerably different machinability (e.g., grindability, sensitivity to heat, EDMability, etc.) in the cutting tool manufacturing and cutting performance in machining (e.g., tool wear, sharpness of the cutting edge, adhesion to the work material).

3.5.4 Grain Size

A single-crystal diamond is extremely hard and has very high abrasion resistance and thermal stability. At the same time, it is a highly anisotropic material, that is, the properties of a single-crystal diamond are not the same depending on the crystallographic plane in which they are measured. This allows the natural gemstones to be cut along specific *cleavage* planes where the energy required to split the crystal is at its minimum. In a diamond sintered compound, all the weak crystallographic planes are randomly oriented so, at a microscopic scale, this compound behaves isotropically (the properties do not depend on the load direction) with improved impact strength.

In selecting a particular grain size for a given application, it can be assumed that the smaller the size of the diamond crystals sintered together, the higher the abrasion resistance of a PCD cutting element. Moreover, such a cutting element can be easily EDMed and then finish ground (polished) to obtain a cutting edge of high quality with minimum serrations. When high requirement to surface roughness is the case, fine-grain PCDs (less that 1 μm) should be used. For normal PCD tipped reamers, 5–10 μm grain size is a common choice. Fine-grain PCDs are normally more expletive because:

- Initial diamond powder should be of good quality with a narrow distribution of particle sizes (Figure 3.86)
- Special recipes for powder compact and tight control of the sintering process parameters are required to prevent AGG
- Considerable higher pressure and temperature (compared to coarse-grained PCDs) are needed for their sintering.

However, these good properties come at the expense of lower impact strength compared to coarse-grained PCD grades.

Coarse grades are normally used for interrupted and/or high-impact cutting conditions, for example, in milling where PCDs with 30 μm or even higher grain sizes are recommended by PCD manufacturers. Tool manufacturing, particularly EDMing and grinding coarse-grained

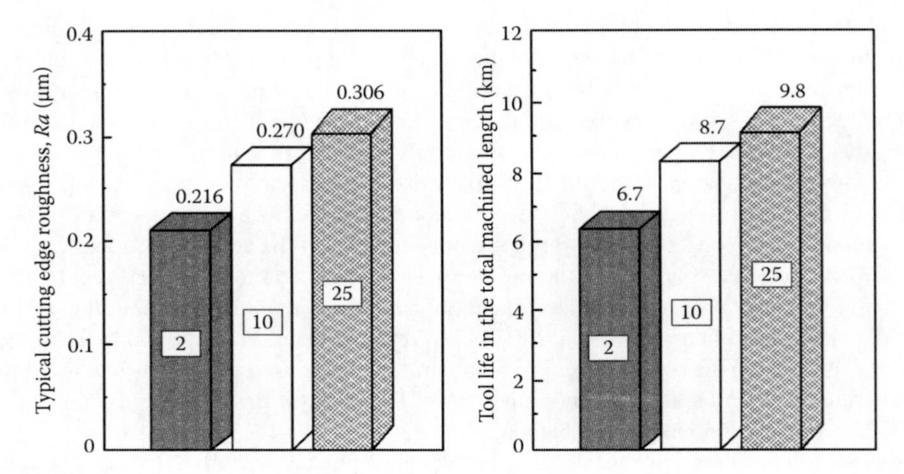

FIGURE 3.89 Relationship between PCD grain size, edge roughness, and abrasive wear rates.

PCDs present problems and generally require much more time to accomplish which also adds significantly to the tool manufacturing cost.

Figure 3.89 shows a typical abrasive wear rate and tool edge roughness values for fine (2 μm)-, medium (10 μm)-, and coarse (25 μm)-grained PCD tools with edges prepared by spark erosion and used to edge mill a ceramic-impregnated layer (Cook and Bossom 2000). When machining under harsh conditions, the differences in PCD wear resistance become more evident, therefore coarse-grained PCD tends to be used in such conditions. When machining under moderately harsh conditions, the PCD wear resistance is of less importance, and factors such as edge quality/surface finish must be considered; therefore, medium- and fine-grained PCD grades tend to be used.

The discussed relationship is not completely linear (Cook and Bossom 2000) even under harsh machining conditions; tool life of PCD can deteriorate if grain size is increased significantly beyond approx. 25 μm. Figure 3.90 shows the performance of various PCD grain size products in milling a ceramic-impregnated surface of a flooring board (Cook and Bossom 2000). As can be seen, the tool life achieved with an ultra-coarse-grained PCD (75 μm) is lower than that achieved with coarse and medium-grained products. While an ultra-coarse PCD has the theoretical abrasion resistance required for increased performance, the coarseness of the particles results in a substantially rougher cutting

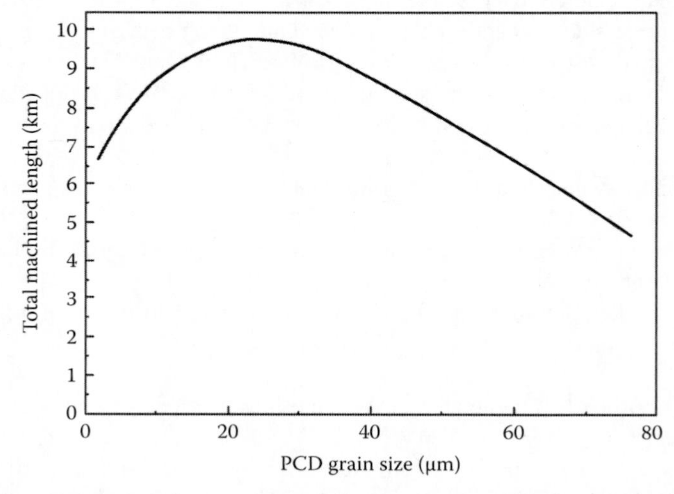

FIGURE 3.90 The performance of various PCD grades milling ceramic-impregnated high pressure laminates flooring board.

FIGURE 3.91 Secondary electron micrograph of PCD containing the grains of two distinctive sizes.

edge, which has a significant negative influence in contributing to overall tool performance. Therefore, by maintaining grain size at 25 μm and modifying the conditions of diamond synthesis, it has proven possible to produce a product with increased abrasion resistance but without reducing the edge quality characteristics.

When designing a PCD grade for specific application, it is possible to mix together diamond powders with considerably different average particle size and dimensional statistical distributions. It is believed that such mixing results in the achieving of a good degree of powder packing as it minimizes the empty spaces between crystals and favors a good sintering process during HPHT cycle, delivering a PCD with superior toughness and abrasion resistance. Figure 3.91 shows grain size distribution of the diamond phase consisting of a coarse fraction with an AGS of 35 μm with the intergranular space filled with a fine grain fraction with average size of 6 μm (Boland et al. 2010).

Element 6 Co., one of the major suppliers of PCDs for metal cutting industry, markets CTM 302 PCD grade. CTM contains a proprietary mix of micron diamond particle sizes between 30 and 2 μm. Figure 3.92 shows schematics of the packing density of CTM 302 and CTB025 PCD grades. A combination of this mix and carefully controlled sintering conditions increases the diamond packing density, which increases wear resistance. The improved packing density in turn results in a higher degree of contiguity between diamond grains, thereby enhancing the chipping resistance of

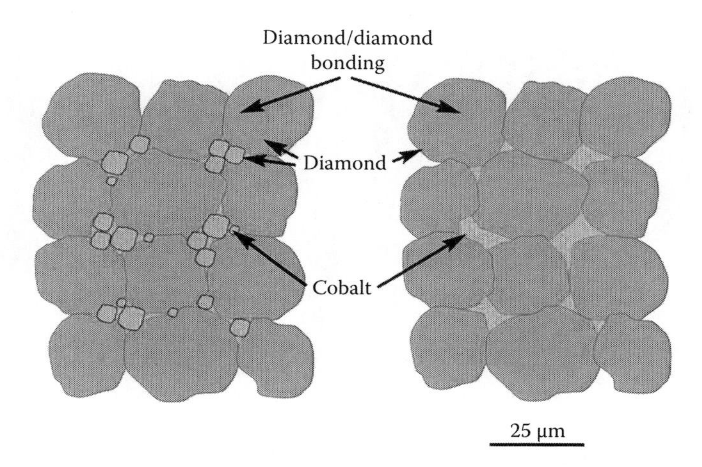

FIGURE 3.92 Schematics of the packing density of CTM 302 and CTB025 PCD grades (Element 6 Co.).

the PCD cutting elements. An added advantage of the increased packing density is the quality of the ground edge, which is superior to a normal coarse-grained PCD. Filling the areas between coarse diamond grains with finer diamond yields a continuous as opposed to a rougher more irregular cutting edge. Element 6 claims improvement the wear resistance of CTM 302 grade compared to usual CTB025 grade although the author's experience shows that this is not always the case.

The author's experience shows that the selection of the proper grain size is not a simple or straightforward task because the grain size affects the performance of a given cutting tool. This is because of the following:

- PCD blanks even of the same grain by different manufacturers are not the same in terms of composition (and thus properties/performance) of PCD layer. Consistency of PCD blank quality is of prime concern in many PCD applications for drilling tools.
- The quality of the cutting edge depends not only on the grain size but also on the manufacturing operations (and their regimes) used to make this edge, namely, EDMing, grinding, polishing, and lapping.
- The performance of PCD of a particular grain size depends on the work material, that is, its properties, chemical composition, inclusions, and porosity.
- The performance of PCD depends on the machining regime chosen for a drilling/reaming application and on the properties of the drilling system. Particularly, the cutting speed and the system rigidity are of prime concern.
- In HSM, the coolant properties and its application technique make significant contributions to PCD performance.

3.5.5 Interfaces

There are three interfaces in a typical PCD cutting element: (1) the top of the PCD layer that eventually becomes the rake faces of the tool, (2) the bottom of the carbide substrate that is used to attach this cutting element to the tool body, and (3) the interface between the sintered diamond layer and the cemented carbide substrate. In the literature, the prime attention is paid to the third interface while the other two are left aside although they play important role in PCD material performance in cutting tools.

The prime attention to the interface between the sintered diamond layer and the cemented carbide substrate is paid because this interface not only provides the necessary strength so that the cutting element can manage the static and dynamic shear loads that otherwise would cause the diamond layer to delaminate, but it also has to handle the residual stresses that arise within both the substrate and the PCD layer as a consequence of the HPHT sintering process. Moreover, as clearly seen in Figure 3.93, this interface is rich in cobalt, which creates some problem in EDMing PCD blanks as discussed in Chapter 2.

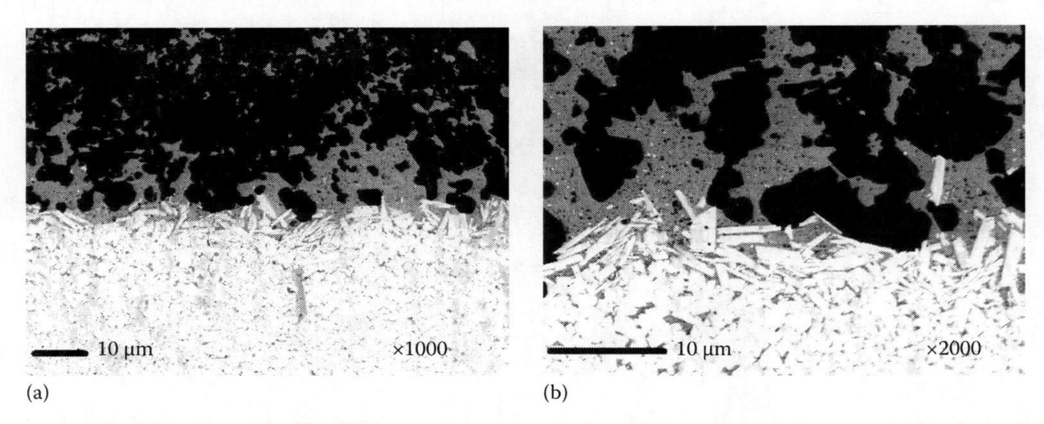

FIGURE 3.93 Micrographs of the PCD–carbide substrate interface at different magnifications: (a) ×1000, and (b) ×2000.

As discussed previously, PCD blanks are produced by the HPHT process. Temperatures from 1350°C to 1800°C and pressures from 5 to 9 GPa with an isostatic holding are used in this process that ensures the formation of a strong PCD layer and a reliable bonding of it with a carbide substrate. At the same time, the high thermobaric parameters cause considerable residual stresses in a finished product because of differences in bulk modulus and thermal expansion coefficients between PCD and carbide. Residual stresses are generated during the cooling stage when the PCD layer is already fully sintered. The thermal expansion coefficients mismatch between PCD layer and substrate causes the carbide to shrink more than the top PCD layer, forcing it to bend outward. The tensile state of stress within the diamond layer becomes worse as its thickness increases, leading in some instances to a spontaneous delamination of the top layer (Bellin et al. 2010a).

It is of importance for the sintering process optimization and PCD blanks processing to determine the level of residual stresses in the PCD layer and in the carbide substrate. Using the micro-Raman spectroscopy, Vohra et al. (1996) measured the compression residual stresses up to 1.3 GPa in the diamond layer and of tensile stresses from 0.87 to 0.95 GPa in the carbide substrate. Krawitz et al. (1999) measured the compression residual stress of 0.47 GPa in a diamond layer by the neutron diffraction. According to Paggett et al. (2002), the mean stress in a diamond layer plane ranges from 0.25 to 0.58 GPa.

It is evident that the previous data should be considered as the indicative ones in view of the difference in PCD blanks geometry and production technology as well as in the place, method, and accuracy of the stress measurements. However, the authors of the referred papers agree that residual thermal stresses bring about an essential decrease in the PCD blank mechanical strength. According to Chen and Xu (2010), up to 70% of PCD blank premature failures, including brittle fracture of the diamond layer and substrate, stratification, spallings, and cracking are caused just by residual stresses. Therefore, the analysis of residual stresses is a necessary stage in designing PCD blanks, and the development of methods to decrease these stresses is a line of investigation, which is promising for the improvement of the cutting insert quality. These methods include techniques (temperature gradient, annealing, stabilization, etc.) used both in sintering process and after its completion.

3.5.6 Thermal Stability

The previously described PCD tool material is known to be thermally unstable. In other words, once a PCD layer is formed, it starts its reverse transformation into carbon. However, the reverse reaction is extremely slow at room temperature. This is not the case, however, at high temperature involved in machining, particularly in HSM and when the work material is highly abrasive, that is, creates a lot of heat due to friction at the tribological interfaces or when it is of low thermal conductivity as, for example, fiber-reinforced plastic materials.

Elevated temperatures that occur in machining affect the common PCD material in two principal ways:

1. Thermal degradation due to differential thermal expansion characteristics between the interstitial catalyst material and the intercrystalline bonded diamond. Such differential thermal expansion is known to occur at temperatures of about 400°C, causing ruptures to occur in the diamond–diamond bonding and resulting in the formation of cracks and chips in the PCD structure. Chipping and microcracking of cutting edges occur as a result (Miess and Rai 1996; Vandenbulcke and De Barros 2001; Chen et al. 2007).
2. Thermal degradation due to the presence of cobalt in the interstitial regions and the adherence of cobalt to the diamond crystals. Specifically, cobalt is known to cause an undesired catalyzed phase transformation in diamond (converting it to carbon monoxide, carbon dioxide, or graphite) with increasing temperature (Fedoseev et al. 1986; Coelho et al. 1995; Cook and Bossom 2000; Shimada et al. 2004).

TABLE 3.12
Basic Properties of Standard PCD Inserts, Thermally Stable PCD Elements, Synthetic Monocrystalline Diamond, and Cemented Tungsten Carbide

Property	Standard PCD Insert	Thermal Stable PCD Element	Synthetic Monocrystalline Diamond	Cemented Tungsten Carbide (WC–15% Co)
Density, g/cm^3	3.90–4.10	3.42–3.46	3.51–3.52	13.80–14.10
Knoop microhardness, GPa	50–60	50–55	70–105	15–18
Compressive strength, GPa	7.5–8.5	4.0–5.5	7.0–9.0	12.0–14.0
Fracture toughness MPa m$^{1/2}$	9.0–10.0	7.0–8.0	3.0–4.0	11.0–12.0
Young's modulus, GPa	810–850	920–960	1000–1150	70–100
Thermal conductivity, W/m·K	550–750	120–250	700–1500	70–100
Wear resistance coefficient,[a] mm	0.15–0.20	0.20–0.25	—	>2.0
Thermal stability, °C	700	1200	1200	1400

[a] Value of linear wear of inserts after sandstone block testing.

Section 2.6.10 provides more information on the results of PCD overheating. These issues limit practical use of conventional PCD tool materials to about 700°C.

Although applications for PCD tools are rapidly developing, standard PCD grades have a substantial disadvantage with their low thermal stability. In oil and gas drilling, drill bits equipped with thermally stable polycrystalline diamond (TSPD) elements demonstrate high performance. On the basis of analysis of different publications (Akaishi et al. 1990, 1991; Wilks and Wilks 1991; Field 1992; Tomlinson and Clark 1992; German et al. 2005; Kang 2005; Boland et al. 2010; Chen and Xu 2010) in which there are descriptions of properties of the PCD inserts and TSPD elements, a comparison of the basic physical and mechanical properties of these superhard materials can be made. The data show (Table 3.12) that the most important operating mechanical properties of the diamond-containing layer of PCD inserts and TSPD elements are related to each other, such as hardness and wear resistance. Characteristic values of strength of PCD inserts are better than TSPD elements. However, the latter keep their own mechanical properties up to 1200°C while mechanical properties of PCD blanks quickly decrease after heating at temperatures more 700°C.

The technology to manufacture a two-layer composite of diamond-containing/cemented tungsten carbide substrate, which has a combination of high wear resistance and high thermostability, should be developed. Up to date, attempts to improve thermal stability of PCD can be generally classified as follows:

1. Removing the solvent–catalyst phase from the PCD material, either in the bulk of the PCD layer or in a volume adjacent to the working surface of the PCD tool (where the working surface typically sees the highest temperatures in the application because of friction events). This is commonly known as cobalt leaching.
2. Two-step sintering.
3. Use of a ceramic matrix material such as SiC instead of cobalt.

The essence of cobalt leaching is as follows (US Pat. Nos. 4,224,380 and 4,288,248). After finishing the PCD conventionally, the metallic phase is removed from the compact by acid treatment, liquid zinc extraction, electrolytic depleting, or similar processes, leaving a compact of substantially 100% abrasive particles that remove the cobalt phase though an acid etching process called leaching. For drilling applications, cobalt is removed up to 200 μm deep into the PCD layer.

As the cobalt phase remains inside the PCD layer, the loss of overall strength is not that significant as with other methods of TSP manufactured using other known methods.

Leaching a thin layer at the working surface dramatically reduces diamond degradation and improves the tool's thermal resistance. First, with no cobalt, the diamond–diamond bonds remain strong as little graphitization and cracking due to mismatch in thermal expansion properties occurs at high cutting temperatures. Second, the heat conduction of the contact diamond surface increases; thus, the heat is transmitted faster from the tool–chip and tool–workpiece that lowers the maximum cutting temperature.

Figure 3.94 shows wear curves for leached and non-leached PCD cutters subjected to a severe wear test (Bellin et al. 2010b). As can be seen, after sliding distance of about 5000 m, the volumetric wear rate of the unleached cutter increased dramatically. The leached PCD cutter, in contrast, maintained a relatively constant wear rate for about 15,000 m.

It has been found (US Patents Nos. 4,224,380 and 4,288,248) that the leached PCD can withstand exposure to temperatures up to 1200°C–1300°C without substantial thermal degradation. SEM analysis revealed that the non-leached samples exhibited many different characteristics when compared to the leached samples. The metallic phase began to extrude from the surface between 700°C and 800°C as viewed under 2000× magnification. As the temperature was increased to 900°C, the samples cracked radially from the rounded cutting edge to the center of the sample. The leached samples did not exhibit this behavior but were relatively unchanged until 1300°C. The diamond layers were clean at 1200°C, but at 1300°C, the edge looked rounded and fuzzy at 20× magnification. The images taken at 1000× magnification showed an etched surface with many exposed crystals due to thermal degradation of the surface.

There are several important issues associated with this approach (cobalt leaching) to achieving improved thermal stability. The prime concern is that a continuous network of empty pores result from the leaching, possessing a substantially increased surface area, which can result in increased vulnerability to oxidation (particularly at higher temperatures). This can then result in reduced strength of the PCD cutter at high temperatures. The second concern is the time needed for leaching, for example, according to US Patent No. 4,288,248 between 8 and 12 days, and strong chemicals used, such as hot concentrated acid solutions.

Two-step sintering results in the formation of a bi-layered sintered PCD disc, which includes a thermally stable top layer. According to US Patent No. 4,944,772, a leached PCD compact and cemented carbide support are separately formed. An interlayer of unsintered diamond crystals

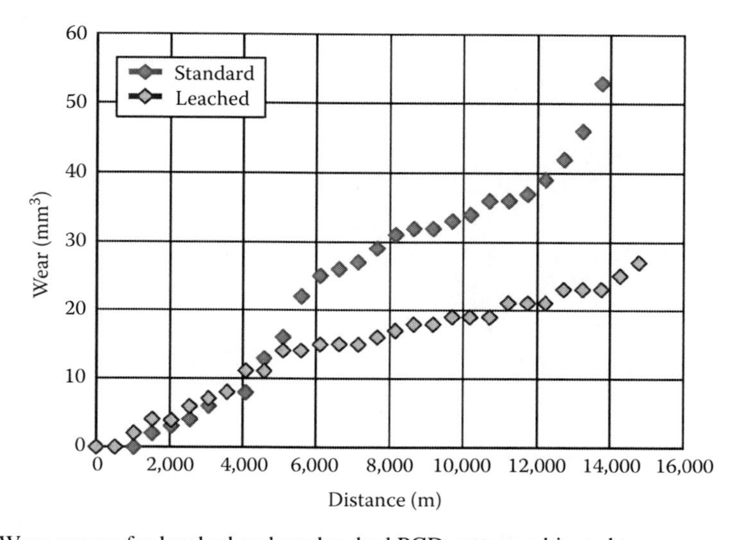

FIGURE 3.94 Wear curves for leached and nonleached PCD cutters subjected to a severe wear test.

(having a largest dimension of 30–500 μm) is placed between the carbide and thermally stable PCD (TSPCD) layer. A source of catalyst/sintering aid material is also provided in association with this layer of interposed crystals. This assembly is then subjected to the HPHT process, sintering the interlayer and bonding the whole into a bi-layered supported compact. In this application, appreciable re-infiltration of the TSPCD layer is not seen as advantageous, but the requirement for some small degree of re-infiltration is recognized in order to achieve good bonding.

The method according to US Patent No. 5,127,923 is an improvement to this approach, where a porous TSPCD layer is reattached to a carbide substrate during a second HPHT cycle, with the provision of a second *inert* infiltrant source adjacent a surface of the TSPCD compact removed from the substrate. Infiltration of the TSPCD body with this second infiltrant prevents significant re-infiltration by the catalyst metal of the carbide substrate. Where carefully chosen, it does not compromise the thermal stability of the previously leached body. A suitable infiltrant, such as silicon, for example, must have a melting point lower than that of the substrate binder.

The disadvantage of the bi-layer sintered PCD is the formation of high internal stresses because of the significant differences in properties between the leached/porous layer and the underlying sintered PCD and carbide substrate. This is exacerbated by the monolithic nature of the leached compact and often causes cracking at the PCD–substrate interface or through the PCD layer itself during the second HPHT cycle. Furthermore, the reattachment process itself can be difficult to control such that appreciable re-infiltration of the TSPCD layer does not occur during the second HPHT cycle. Moreover, multi-step sintering processes are both time-consuming and labor intensive. It is, therefore, desired that a thermally stable PCD material be developed in a relatively simple single-step manufacturing process.

This ambitious goal can be achieved if two catalyst metals are used simultaneously in the HPHT process. Figure 3.95 shows a model of the HPHT sintering method using double-sided infiltration of diamond micropowder by molten silicon and Co–WC–C melt from cemented tungsten carbide

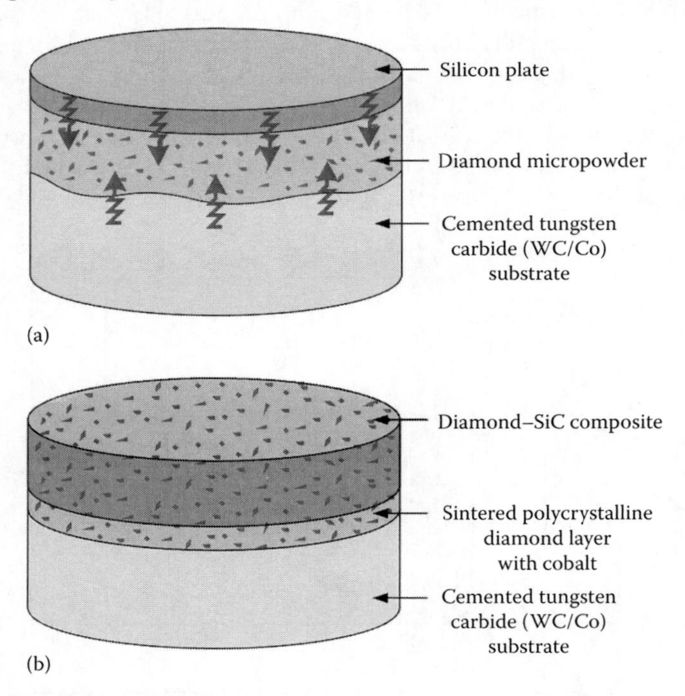

FIGURE 3.95 Starting and finishing stages of structure formation of thermally stable PCD blanks in HPHT sintering: (a) start of infiltration of the diamond layer from one side by liquid Si and from other side by Co–WC–C melt, (b) finished three-layered composite—top cutting layer is a thermally stable diamond–SiC composite.

substrate. By changing the pressure and heating rate, this method allows a change in the size of intermediate PCD–Co layer. This experimental method was developed at V.N. Bakul Institute for Superhard Materials of the National Academy of Sciences of Ukraine (Osipov et al. 2010).

All samples of thermally stable PCD blanks were produced using high-pressure apparatus (HPA) with profound anvil type. Within the high-pressure cell, pressures of 8.0–9.0 GPa and high temperatures of 1500°C–1800°C were created. The three-layer composite was sintered by the method of oncoming infiltration of diamond micropowder with AGS of 30–40 μm on the cemented tungsten carbide substrate (WC/15 mass. % Co). The experimental high pressure cell is shown in Figure 3.96.

The SEM and XRD microanalysis of the structure of the produced samples have shown that a thin transition zone containing 8%–12% cobalt forms between the carbide substrate and the diamond layer during the sintering as shown in Figure 3.97. The thickness of the transition zone

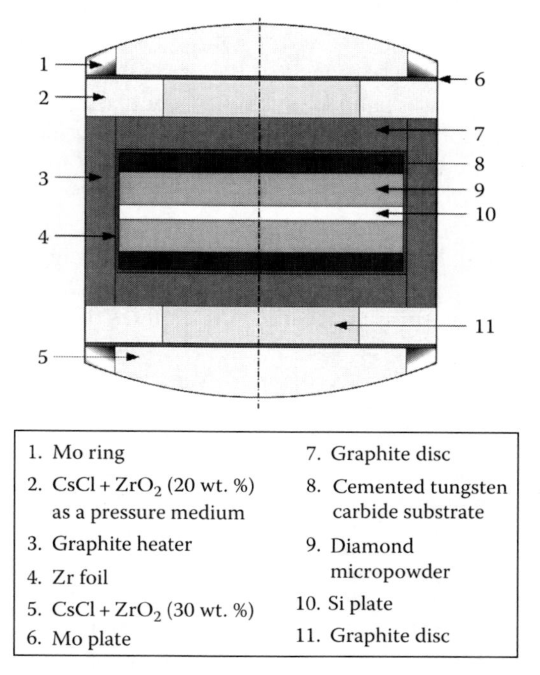

FIGURE 3.96 Cross section of the high-pressure cell assembly for HPHT sintering of thermally stable PCD blanks.

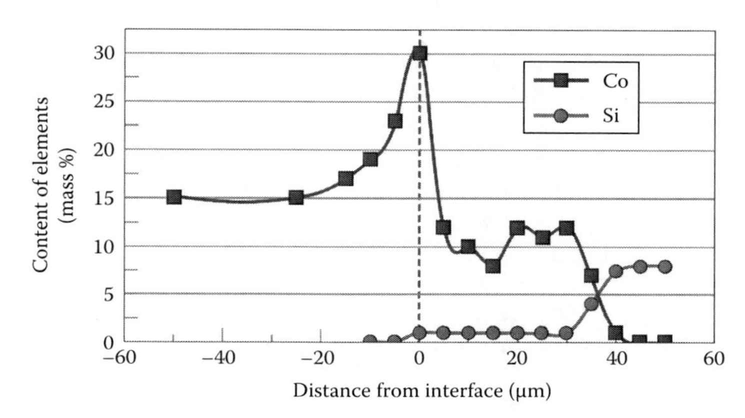

FIGURE 3.97 Concentration of chemical elements in the sample axial section at the interface between the WC–15% Co substrate and diamond layer.

approximately corresponds to the particle size of the initial diamond powder, that is, the cobalt melt penetrates into the diamond layer to a depth of no more than 40 μm. The amount of cobalt at the carbide-transition zone interface is essentially higher than in the substrate. At a distance of above 40 μm from the substrate, the diamond layer exhibits no cobalt. Free silicon is also not observed at the carbide-transition zone interface, and this is supported by XRD analysis.

The test results showed considerable increase in tool life (up to two times), allowable penetration rate (by 30%), and thermal stability (up to 1100°C) of the cutters made of the thermal stable inserts manufactured using this method.

3.5.7 PCD Grade Selection and Quality Inspection

3.5.7.1 Grade Selection Considerations

Since the first PCD blanks used for manufacturing cutting tools were introduced in the 1970s, tremendous technological advances have been made. Changes in HPHT technology have improved the sintering capability, yielding better diamond–diamond bonding and diamond–substrate bonding. Better bonding results in less tool breakage and delamination. Variations in diamond grain size and reduction of diamond–substrate interfacial residual stresses provided improved quality of cutting tools while adding the extra performance.

Almost as soon as PCD tool material was introduced, the dangers of thermal damage were recognized. In tool manufacturing, thermal damage is caused by overaggressive EDMing and grinding, while in machining it is caused by high cutting temperatures. As discussed in the previous section, the thermal degradation begins to occur at temperatures of about 400°C due to differential thermal expansion characteristics between the interstitial catalyst material and the intercrystalline bonded diamond that causes ruptures in the diamond–diamond bonding. This limits practical use of conventional PCD tool materials to about 650°C. The particular temperatures of the beginning of thermal damage and the maximum allowable temperature are diamond grade dependent, that is, they vary from one PCD manufacturer to the next depending on the exact PCD composition (the chemical composition of the powder mix) and parameters of HPHT sintering.

It is now established that the thermal damage of PCD tool material is actually the graphitization of boundaries of diamond grains that lead to weakening of diamond–diamond bonds. This issue is considered in Section 2.6.10 in details with the classification of the thermal damage and its appearance on the failed cutting tools. In the author's opinion, this issue is the most neglected in the whole PCD cutting tool business due to ignorance. There should be three simultaneously attempted directions in further improving this business:

1. Proper design of drilling systems accounting for the specific properties of PCD tool material. These systems should be capable of spindle speeds of 25,000–50,000 rpm, with minimum runout, rigid and dynamically stable structures and interfaces including tool holders and part-holding fixtures. High-pressure supply of extra-clean MWF of high concentration (9%–12%) should be an integral part of such systems. Tailored drilling tool designs accounting on thermal-sensitive operation in tool manufacturing and having higher back tapers, smaller margins, optimal clearance angles, edge preparations, etc., should be provided. Better quality of work materials, for example, die castings with minimum number of hard inclusion, casting cavities, and residual porosity should be considered.
2. Introduction of more intelligent operations capable of reduction of the thermal damage in the manufacturing of PCD tools. It includes but not limited to modern controlled EDMing, finish grinding, and vacuum brazing (the strictly controlled brazing temperature).
3. Much greater application of TSPCD in cutting tools.

There are too many choices and variables in the selection of proper PCD grade. The properties of PCD tool material can be varied by altering grain size, interface, and post-sintering treatment.

In addition, each PCD manufacturer has the variances of its specific grade powder composition, manufacturing processes, and experience (both PCD manufacturing and tool application). The result is a wide variety of PCD grades on the market. The tool geometry, that is, its rake, clearance, and inclination angles, can also affect a PCD tool performance in a somehow greater extent than that in HSS and cemented carbide drilling tools.

Unfortunately, many-year experience gained with tools made of other tool materials may not be applicable at all in the design and application of drilling systems and tools for PCD applications. Faced with the wide variety of choices, how does a tool specialist (manufacturing/process engineer) optimize PCD tool performance? How does a tool manufacturer (tool designer and tool manufacturer) manufacture an optimal tool and remain competitive? A logical move is to obtain the relevant information from the source, that is, PCD suppliers, as PCD tool manufacturers know much less about the properties of PCD and often offer to manufacture their tools with any PCD grade by the end user choice.

Unfortunately, the information on PCD grades for cutting tool provided by leading PCD suppliers allows making only a very general choice of a PCD grade for a given application. Figure 3.98 shows an example. As can be seen, the catalyst material (e.g., cobalt) is termed as the *binder*, the chemical composition is not provided, and the mechanical, physical, and chemical properties are not mentioned. It is not clear what is the difference between good and superior wear resistances—how this characteristic was measured and what units it has. The selection of the optimal machining regime and tool geometry is virtually impossible to find in the information provided by leading PCD manufacturers/suppliers. Direct calls and discussions on the matter have proven to be even less useful.

Application properties of the PCD tool material can be broadly categorized by their abrasion resistance, impact resistance (often referred to as chipping resistance), and thermal abrasion resistance. Manufacturing properties of the PCD tool material include EDMability (machinability by EDM), grindability, and brazeability (including the maximum allowable brazing temperature). Logistic properties may include availability and PCD supplier engineering support including possible failure analysis. The specialist must select the best grade for a given application, without risking other failure modes and without wasting money, that is, accounting for these three sets of properties. Unfortunately, not many supporting literature sources and reliable data are available as PCD is a relatively new tool material, and because the drilling system has undergone dramatic changes in the last 10–15 years so that the application data have been changing continuously, the selection process is not simple.

	Standard Grades				Special Grades		
	CC	CM	CF	CXL	W Grade	S Grade	L Grade
Type	Carbide backed	Carbide backed	Carbide backed	Carbide backed	Carbide backed	Carbide backed	Carbide backed
Grain size (mm)	25	10	4	25	(CC, CM, CF)	(CM)	(CM)
Diamond (vol. %)	94	92	90		90	80	50
Major binder	Co	Co	Co		WC	WC	WC
Characteristics	✓ Good wear resistance with strong diamond bonding	✓ For general purposes	✓ Good surface finish	✓ Specially designed to have superior wear resistance	✓ Easy WEDM cutting and good impact resistance	✓ Excellent impact resistance and easier tool fabrication	✓ Easier to grind and sharper cutting edge possible

FIGURE 3.98 PCD grade selection recommendation by a leading PCD tool material supplier.

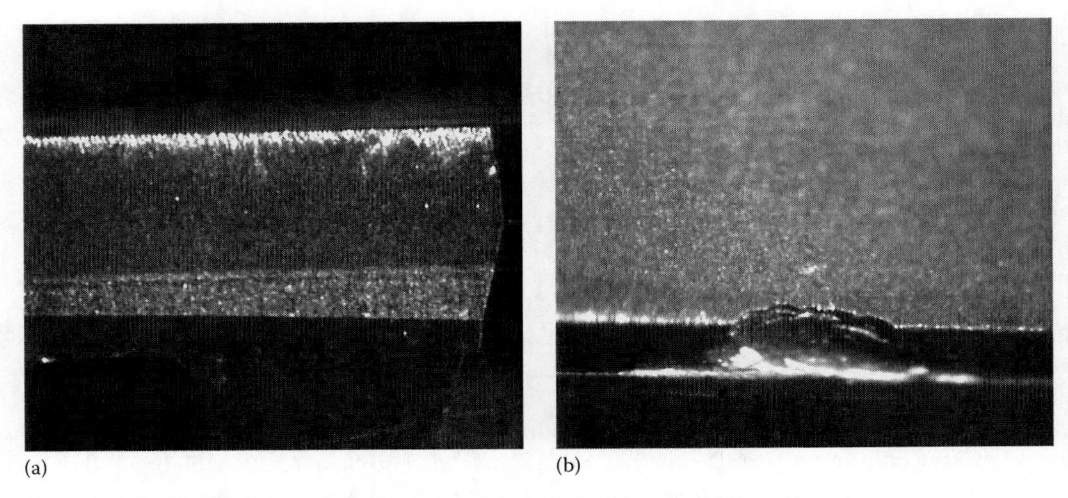

(a) (b)

FIGURE 3.99 Microchipping (a) and macro-chipping (b) of PCD cutting wedge.

PCD grades with high abrasion resistance normally have fine diamond grain size in a well-sintered blank. Such grades have great EDMability and grindability. The latter means that it is relatively easy to achieve the sharp cutting edge and good surface finish in grinding. Such grades are used for reamers and other finishing drilling tools where impact load, and thus damage is minimized.

PCD grades of coarse grain size normally have greater chipping resistance. In this context, chipping is understood as a range of PCD failures starting with small notches running perpendicular to the cutting edge (Figure 3.99a) and finishing with bulk fracture of the PCD layer (Figure 3.99b). Although chipping is normally attributed to the dynamic load of the cutting wedge (routinely wrongly referred as the cutting edge), for example, in milling, the chipping also often occurs in relatively *quiet* drilling operations due to variety of reasons. In the author's opinion based on observation of many chipped PCD wedges in various cutting tools, chipping is caused by high (for a given grade of PCD) contact stresses on the rake and flank faces (see Chapter 2, Figure 2.67) that, in turn, occur due to unsuitable PCD grade, excessively rough grind of PCD flank faces that leaves deep grinding marks on the flank faces acting as stress concentrators in drilling, improper choice of tool geometry (primarily the clearance angle), overaggressive EDM trimming and finish grinding that may damage the flank faces adjacent to the cutting edge, too great a feed per tooth, and many others.

HSM, where the cutting speed may easily reach 10,000 m/min, brought a new requirement to PCD tool materials—resistance to thermomechanical abrasion. PCD grades of this type should be highly abrasion resistant with an additional level of thermal stability. They are characterized by a fine diamond grain size in a well-sintered part, with the residual cobalt in the diamond table either partially or totally removed. Unfortunately such grades are not yet widely available for cutting tool manufacturing. Regarding the latter, the author cannot understand why this promising direction is not yet pursued for PCD blanks made for metal cutting tools although the same PCD manufacturers supply such materials for rock drilling (Bellin et al. 2010b). As described previously, thermal stable PCD grades should have very low or no cobalt in their final structure.

Historically, the first attempt to manufacture TSPCD was replacing the cobalt with silicon carbide. Because of the limitations associated with attaching this kind of PCD to the tool body, such as weakening of PCD, these attempts were only partially successful at the time. Nowadays with the further development in the technology, this direction seems to have gained some momentum at least at the research level.

As early as in 1993, two researchers from Sumitomo Co. leached a thin layer of the PDC diamond table of a regular PCD insert. In their subsequent test of this insert, they achieved dramatic increase in performance. However, their invention languished until about 2000. With more understanding on modern diamond layer composition and properties, diamond tool manufacturers applied this technique to PDC used in tools for rock drilling. PDC cutters treated in this manner, suitable for highly abrasive formations machining conditions where thermomechanical wear is expected. Nowadays such treated cutter are commonly referred to as leached cutters.

In another effort, researchers at US Synthetic applied a thermally stable layer to a carbide substrate. To make this cutter type, a thin PCD layer is leached of its cobalt and then placed on a carbide substrate and repressed in the HPHT press. This process yields a thermally stable layer attached to a carbide substrate. This cutter type also languished until recently when researchers came out with a more robust and reliable version of this technology. This process, known as the *two-step* process, now forms another category of premium cutters.

3.5.7.2 Quality Assessment of PCD Products

There are two levels of quality assessment of PCD product, namely, the laboratory assessment and manufacturing testing at the level of PCD and PCD tool production. While the former requires a well-equipped laboratory and highly qualified/trained personnel, the latter is performed at manufacturing level with much less sophisticated equipment and manufacturing personnel with no advanced degree in composite tool materials.

The laboratory level of quality assessment primarily includes microstructural assessment of PCD using SEM. SEM is a technique used to image the surface topography (secondary electron imaging—SEI) and composition of a sample (Backscatter Electron Imaging—BEI). In order to view and assess the microstructure of polycrystalline materials, backscatter imaging mode is utilized. In this imaging mode, diamond appears black, tungsten white with the binder phases being a mid-gray in color.

When investigating a failed PCD, the failed samples (returned by the customer) are cut and the cross section is polished in order to assess the microstructure of the PCD using SEM. Then the microstructure of this sample is compared to the *standard* structure in order to find out if the tool failed due to PCD quality, and if it did, then the root cause of failure should be determined.

The following are some basic characteristics determined using the image analysis technique:

- Diamond grain size (μm) (mean, standard deviation, distribution, and maximum size using the standard Saltikov method [Gublin 2008; Jeppsson et al. 2011])
- Cobalt pool size (μm) (Saltikov method [Gublin 2008; Jeppsson et al. 2011])
- Diamond content area (in % of area)
- Diamond MFP (μm)
- Cobalt MFP (μm)
- Diamond contiguity (in % of area)
- WC-phase area (in % of area)
- WC pool size (μm)
- Cobalt next neighbor distance (μm)

Being important to characterize PCD structure, these characteristics are not yet correlated with the performance of PCDs as the tool material. Nevertheless, they are supplied by PCD manufacturers to PCD tool manufacturers/users in the course of a failure analysis, which does not help to find the root cause of the failure as the latter cannot make any use of these characteristics.

Unfortunately, there are no standard manufacturing testing methods available for PCD tool materials used in the cutting tools. Perhaps the most convenient, cost-effective, and reliable test for

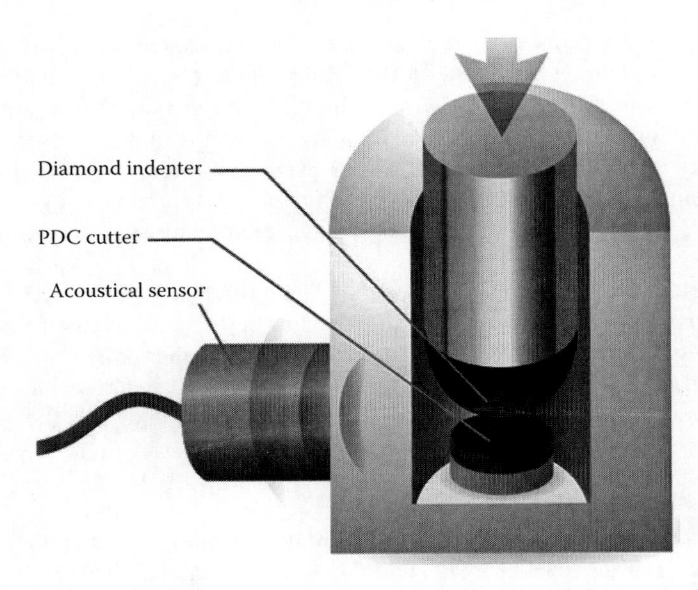

FIGURE 3.100 Varel's acoustic emission toughness test measures microcracking to qualify the strength of diamond–to-diamond bonds in the PCD cutter's diamond table.

the quality of PCD/CVD components is a monitored wear test. The pin-on-disk or pin-on-drum abrasive wear tests are unsuitable for such ultrahard materials because the tool wear is negligible. A more aggressive test has been devised in which the cutting element is monitored during a turning/cutting operation on an industry-standard corundum grinding wheel.

In the author's opinion, the acoustic emission toughness testing techniques (US Patent Applications, Pub. No. US 2011/0246096 (2011) and Pub. No. 2011/0239767 (2011)) developed by Varel International Co. for PDC used in rock drilling can be eventually adapted for PCD properties testing for cutting tool applications. This testing technology uses a domed PDC indenter to load a PDC as shown in Figure 3.100. The test cell is instrumented with a highly sensitive acoustic sensor. As the force applied to the PDC diamond layer is increased, it begins to experience microcracking. The acoustic sensor detects the microcracking events, which are then recorded on a hard drive and displayed in a monitor. The test accurately quantifies the strength of diamond–diamond bonds within the diamond table. Different types and grades of PDC cutter can be compared according to their resistance to load-induced microcracking to yield a highly predictive valuation of impact strength. In addition, this test can be applied as a process/quality control instrument for PDC cutter manufacturers (Bellin et al. 2010c).

Quality control procedures evaluate not only overall dimensional specifications but also grain structure, surface condition, and internal structural integrity. Advanced ultrasonic C-scan procedure is used to look deep into the structure of PCD materials to ensure that each PCD disk is free of internal defects and that the integrity of the interface between the diamond layer and the carbide substrate is optimal.

C-scan is an acoustic microscopy analytical testing. It is a noncontact screening method offering unique insight on the PDC cutter or blank. Its advantages include the detection of void formations, delaminations, cracks, fractures, as well as other hidden internal defects within inherently susceptible materials. The C-scan mode provides a planar view image at any specific depth. Much like the MRI machine used to examine patients at hospitals, the separation of the PDC layer can also be tested; illustrations of delaminations and void formations can be generated. Figure 3.101 shows a C-scan image to demonstrate an interfacial defect found between diamond layer and carbide (Bellin et al. 2010c). Smallest defects, down to even 5–10 μm, can be found using this technique.

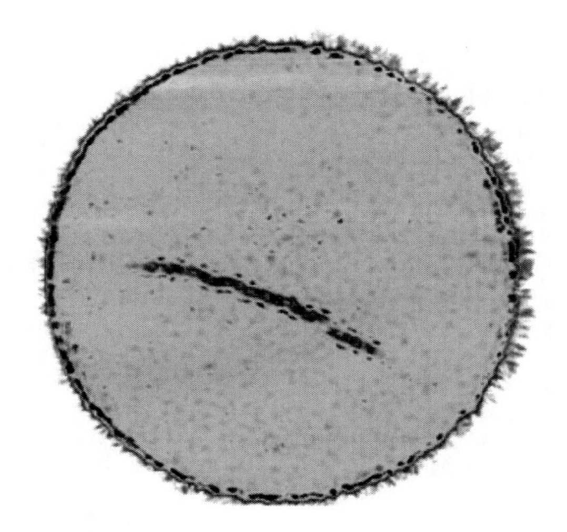

FIGURE 3.101 C-scan image of flawed diamond-to-carbide interface.

PCD blanks should also be subjected to ultraviolet dye penetrant examination, microscopic visual examination, and dimensional inspection using PCD-tipped micrometers and other specialized measuring techniques. Brazed products are inspected using ultrasonic inspection techniques to detect possible internal flaws.

REFERENCES

Akaishi, M., Kanda, H., and Yamaoka, S. 1990. Synthesis of diamond from graphite-carbonate systems under very high temperature and pressure. *Journal of Crystal Growth* 104:578–581.

Akaishi, M., Ohsawa, T., and Yamaoka, S. 1991. Synthesis of fine-grained polycrystalline diamond compact and its microstructure. *Journal of the American Ceramic Society* 74:5–10.

Astakhov, V.P. 2006. *Tribology of Metal Cutting*. London, U.K.: Elsevier.

Astakhov, V.P. 2010. *Geometry of Single-Point Turning Tools and Drills. Fundamentals and Practical Applications*. London, U.K.: Springer.

Astakhov, V.P. 2011. Turning. In *Modern Machining Technology*, Davim, J.P. (ed.). Cambridge, U.K.: Woodhead Publishing.

Baldissera, P. and Delprete, C. 2008. Deep cryogenic treatment: A bibliographic review. *The Open Mechanical Engineering Journal* 2:1–11.

Ban, Z.G. and Shaw, L.L. 2002. Synthesis and processing of nanostructured WC–Co materials. *Journal of Materials Science* 37:3397–3403.

Bellin, F., Dourfaye, A., King, W., and Thigpen, M. 2010a. The current state of PCD bit technology. Part 1: Development and application of polycrystalline diamond compact bits have overcome complex challenges from the difficulty of reliability mounting PDC cutters in bit bodies to accelerate thermal wear. *World Oil* September: 231:41–46.

Bellin, F., Dourfaye, A., King, W., and Thigpen, M. 2010b. The current state of PCD bit technology. Part 2: Leaching a thin layer at the working surface of a PCD cutter to remove the cobalt dramatically reduces diamond degradation due to friction heat. *World Oil* (October):53–58.

Bellin, F., Dourfaye, A., King, W., and Thigpen, M. 2010c. The current state of PCD bit technology. Part 3: Improvements in material properties and testing methods are being pursued to make PDC the cutter of choice for an increasing variety of applications. *World Oil* (November):67–71.

Berry, J.T. 1970. *High Performance High Hardness High Speed Tool Steels*. New York: Climax Molybdenum Co.

Boccalini, M. and Goldenstein, H. 2001. Solidification of high speed steels. *International Materials Reviews* 46(2):92–115.

Boland, J.N., Li, X.S., Rassool, R.P., and Hay, D. 2010. Characterisation of diamond composites for tooling. *Journal of the Australian Ceramic Society* 46(1):1–10.

Bryson, W.E. 1999. *Cryogenics*. Cincinnati, OH: Hanser Gardner Publications.

Campbel, F.C. 2008. *Elements of Metallurgy and Engineering Alloys*. Materials Park, OH: ASM International.

Chatfield, C.I. 1985. Comments on microstructure and the transverse rupture strength of cemented carbides. *International Journal of Refractory Metals & Hard Materials* 4(1):60.

Chen, F. and Xu, G. 2010. Thermal residual stress of polycrystalline diamond compacts. *Transactions of Nonferrous Metals Society of China* 20:227–232.

Chen, Y., Zhang, L.C., and Arsecularatne, J.A. 2007. Polishing of polycrystalline diamond by the technique of dynamic friction. Part 2: Material removal mechanism. *International Journal of Machine Tool and Manufacturing* 47:1615–1624.

Coelho, R.T., Yamada, S., Aspinwall, D.K., and Wise, M.L.H. 1995. The application of polycrystalline diamond tool materials when drilling and reaming aluminium based alloys including MMC. *International Journal of Machine Tool and Manufacturing* 35:761–774.

Cook, M.W. and Bossom, P.K. 2000. Trends and recent developments in the material manufacture and cutting tool application of polycrystalline diamond and polycrystalline cubic boron nitride. *International Journal of Refractory Metals & Hard Materials* 18:147–152.

Craig, P. 1997. Behind the carbide curtain. *Cutting Tool Engineering* 49(8):34–38.

da Silva, F.J., Franco, S.D., Ezugwu, E.O., and Souza, A.M., Jr. 2006. Performance of cryogenically treated HSS tools. *Wear* 261:674–685.

Davis, J.R. (ed.) 1995. *Tool Materials (ASM Specialty Handbook)*. Metals Park, OH: ASM.

DeGarmo, E.P., Black, J.T., and Kohser, R.A. 2007. *Materials and Processes in Manufacturing*, 10th edn. New York: John Wiley & Sons.

Dobrzański, L.A. and Dołżańska, B. 2010. Structure and properties of sintered tool gradient materials. *Journal of Achievements in Materials and Manufacturing Engineering* 43(2):711–733.

Exner, H.E. 1979. Physical and chemical nature of cemented carbides. *International Metals Reviews* 4:149–173.

Exner, H.E. and Gurland, J. 1970. A review of parameters influencing some mechanical properties of tungsten carbide–cobalt alloys. *Powder Metallurgy* 13:13–31.

Fang, Z.Z. 2005. Correlation of transverse rupture strength of WC–Co with hardness. *International Journal of Refractory Metals & Hard Materials* 23:119–127.

Fang, Z.Z. and Sohn, H.Y. 2009. Development of bulk nanocrystalline cemented tungsten carbide for industrial applications. Salt Lake City, UT: University of Utah.

Fang, Z.Z., Wang, X., Ryu, T., Hwang, K.S., and Sohn, H.Y. 2009. Synthesis, sintering, and mechanical properties of nanocrystalline cemented tungsten carbide—A review. *International Journal of Refractory Metals & Hard Materials* 27:288–299.

Fedoseev, D.V., Vnukov, S.P., Bukhovets, V.L., and Anikin, B.A. 1986. Surface graphitization of diamond at high temperatures. *Surface and Coatings Technology* 28:207–214.

Field, J.E. (ed.) 1992. *The Properties of Natural and Synthetic Diamond*. Cambridge, U.K.: Academic Press.

Freytag, J., Walter, P., and Exner, H.E. 1979. Charakterisierung bindephase von WC–Co hertmetallen uber magnetische eigenschaften. *Z. Metallkunde* 69(8):546–549.

Fu, L., Cao, L.H., and Fan, Y.S. 2001. Two-step synthesis of nanostructured tungsten carbide-cobalt powders. *Scripta Materialia* 44:1061–1068.

Gallagher, A.H., Agosti, C.D., and Roth, J.T. 2005. Effect of cryogenic treatments on tungsten carbide tool life: Microstructural analysis. *Transactions of NAMRI/SME* 33:153–160.

German, R.M., Smid, I., Campbell, L.G., Keane, J., and Toth, R. 2005. Liquid phase sintering of tough coated hard particles. *International Journal of Refractory Metals & Hard Materials* 23(4–6):267–272.

Gill, S.S., Singh, R., Singh, H., and Singh, J. 2009. Wear behavior of cryogenically treated tungsten carbide inserts under dry and wet turning conditions. *International Journal of Machine Tools and Manufacture* 49:256–260.

Grinder, O. 2010. PM production and applications of HSS. In *International Powder Metallurgy Directory 2010–2011*. Shrewsbury, England: Inovar Communications. International Powder Metallurgy Directory 14th Edition, pp. 125–130. http://www.pim-international.com/ipmd

Gublin, Y. 2008. On estimation and hypothesis testing of the grain size distribution by Saltykov method. *Image Analysis and Stereology* 27:163–174.

Gurland, J. 1954. A study of the effect of carbon content on the structure and properties of sintered WC–Co alloys. *Transactions of the AIME* 200:285–290.

Gurland, J. 1963. The fracture strength of sintered tungsten carbide–cobalt alloys in relation to composition and particle spacing. *Transactions of the Metallurgical Society of AIME* 227:1146–1150.

Gurland, J. and Bardzil, P. 1955. Relations of strength, composition, and grain size of sintered WC–Co alloys. *Transactions of the AIME, Journal of Metals* 203:11–315.

Hetzner, D.W. and Geertruyden, W.V. 2008. Crystallography and metallography of carbides in high alloy steels. *Materials Characterization* 59(7):825–841.

Hong, S.M., Akaishi, M., Kanda, H., Osawa, T., and Yamaoka, O. 1988. Behaviour of cobalt infiltration and abnormal grain growth during sintering of diamond on cobalt substrate. *Journal of Materials Science Letters* 23:3821–3826.

Hoyle, G. 1988. *High Speed Steels*. London, U.K.: Butterworth-Heinemann.

Iakovou, E.E. and Koulamas, C. 1996. Adaptive tool replacement policies in machining economics. *Journal of Manufacturing Science and Engineering* 118(4):658–663.

Isakov, E. 2000. *Mechanical Properties of Work Materials*. Cincinnati, OH: Hanser Gardener Publications.

Jeppsson, J., Mannesson, K., Borgenstam, A., and Ågren, J. 2011. Inverse Saltykov analysis for particle-size distributions and their time evolution. *Acta Materialia* 59(3):874–882.

Kang, S.J.L. 2005. *Sintering: Densification, Grain Growth and Microstructure*. Amsterdam, the Netherlands: Elsevier.

Kelkar, R., Nash, P., and Zhu, Y. 2007. Understanding the effects of cryogenic treatment on M2 tool steel properties. *Heat Treating Process* 7(8):57–60.

Komarovsky, A.A. and Astakhov, V.P. 2002. *Physics of Strength and Fracture Control: Fundamentals of the Adaptation of Engineering Materials and Structures*. Boca Raton, FL: CRC Press.

Krauss, G., Roberts, G., and Kennedy, R. 1998. *Tool Steels*, 5th edn. Metals Park, OH: ASTM International.

Krawitz, A.D., Winholtz, R.A., Drake, E.F., and Griffin, N.D. 1999. Residual stresses in polycrystalline diamond compacts. *International Journal of Refractory Metals & Hard Materials* 17:117–122.

Lachenicht, V., Scharf, G., Zebrowski, D., and Shalimov, A. 2011. Spray forming—A promising process for making high-quality steels and alloys. *Metallurgist* 54(9–10):656–668.

McHale, A.E. 1994. *Phase Equilibria Diagrams—Phase Diagrams for Ceramists*, vol. 10. Westerville, OH: The American Ceramic Society.

Mesquita, R.A. and Barbosa, C.A. 2003. Evaluation of as-hipped PM high-speed steel for production of large-diameter cutting tools. *Materials Science Forum* 416–418:235–240.

Miess, D. and Rai, G. 1996. Fracture toughness and thermal resistance of polycrystalline diamond compacts. *Materials Science and Engineering A* 209:270–276.

Mills, B. and Redford, A.H. 1983. *Machinability of Engineering Materials*. London, U.K.: Applied Science Publishers.

Mirchandani, P.K. 2005. Making a better grade. *Cutting Tool Engineering* 57(6):34–37.

Mohan Lal, D., Renganarayanan, S., and Kalanidhi, A. 2001. Cryogenic treatment to augment wear resistance of tool and die steels. *Cryogenics* 41:149–155.

Molinari, A., Pellizzari, M., Gialanella, S., Straffelini, G., and Stiasny, K.H. 2001. Effect of deep cryogenic treatment on the mechanical properties of tool steels. *Journal of Materials Processing Technology* 118:350–355.

Nee, J.G. (ed.) 2010. *Fundamentals of Tool Design*. Dearborn, MI: SME.

Osipov, A.S., Bondarenko, N.A., Petrusha, I.A., and Mechnik, V.A. 2010. Drill bits with thermostable PCD inserts. *Diamond Tooling Journal* 2(3):31–34.

Paggett, J.W., Drake, E.F., Krawitz, A.D., Winholtz, R.A., and Griffin, N.D. 2002. Residual stress and stress gradients in polycrystalline diamond compacts. *International Journal of Refractory Metals & Hard Materials* 20:187–194.

Park, Y.J., Hwang, N.M., and Yoon, D.Y. 1996. Abnormal growth of faceted (WC) grains in a (Co) liquid matrix. *Metallurgical and Materials Transactions A* 27(9):2809–2819.

Prabhudev, K.H. 1988. *Handbook of Heat Treatment*. New Delhi, India: McGraw-Hill.

Ravichandran, K.S. 1994. Fracture toughness of two phase WC–Co cermets. *Acta Metallurgica et Materialia* 42:142–150.

Reddy, T.V.S., Sornakumar, T., Reddy, M.V., Venkataram, R., and Senthilkumar, A. 2009. Machinability of C45 steel with deep cryogenic treated tungsten carbide cutting tool inserts. *International Journal of Refractory Metals & Hard Materials* 27:181–185.

Roberts, G.A., Crauss, G., and Kennedy, R. 1998. *Tool Steels*, 5th edn. Metals Park, OH: ASM International.

Roebuck, B. 1995. Terminology, testing, properties, imaging and models for fine grained hardmetals. *International Journal of Refractory Metals & Hard Materials* 13:265–279.

Roebuck, B. and Almond, E.A. 1988. Deformation and fracture processes and the physical metallurgy of WC/Co hardmetals. *International Materials Reviews* 32(2):90–110.

Roebuck, B. and Bennett, E.G. 1986. Phase size distribution in WC–Co hardmetal. *Metallography* 19:27–47.

Roebuck, B., Gee, M., Bennett, E.G., and Morrell, R. 1999. *Measurement Good Practice Guide No. 20: Mechanical Tests for Hardmetals*. Teddington, U.K.: National Physical Laboratory.

Salnikov, M.A. 2009. Development of WC cemented carbides with improved toughness and chipping resistance for rock drilling machines, Mechanical Engineering, Samara State University, Samara, Russia.

Scussel, H.J. 1992. *Friction and Wear of Cemented Carbides*, vol. 18. Metals Park, OH: ASM International.

Seah, K.H.W., Rahman, M., and Yong, K.H. 2003. Performance evaluation of cryogenically treated tungsten carbide cutting tool inserts. *Proceedings of the Institution of Mechanical Engineers Part B: Journal of Engineering Manufacture* 217:29–43.

Sendooran, S. and Raja, P. 2011. Metallurgical investigation on cryogenic treated HSS tool. *International Journal of Engineering Science and Technology* 3(5):3992–3996.

Shimada, S., Tanaka, H., Higuchi, M., Yamaguchi, T., Honda, S., and Obata, K. 2004. Thermo-chemical wear mechanism of diamond tool in machining of ferrous metals. *CIRP Annals—Manufacturing Technology* 53:57–60.

Shin, T., Oh, J., Oh, K., and Lee, D. 2004. The mechanism of abnormal grain growth in polycrystalline diamond during high pressure–high temperature sintering. *Diamond and Related Materials* 13:488–494.

Soleimanpour, A.M., Abachi, P., and Simchi, A. 2012. Microstructure and mechanical properties of WC–10Co cemented carbide containing VC or (Ta, Nb)C and fracture toughness evaluation using different models. *International Journal of Refractory Metals & Hard Materials* 31:141–146.

Soplop, J., Wright, J., Kammer, K., and Rivera, R. 2009. Manufacturing execution systems for sustainability: Extending the scope of MES to achieve energy efficiency and sustainability goals. Paper presented at *Industrial Electronics and Applications, ICIEA*, X'ian, China. pp. 3555–3559.

Thakur, D., Ramamoorthy, B., and Vijayaraghavan, L. 2008. Influence of different post treatments on tungsten carbide–cobalt inserts. *Materials Letters* 62:4403–4406.

Tomlinson, O.N. and Clark, I.E. 1992. Syndax3 pins—New concepts in PCD drilling. *Industrial Diamond Review* 52(3):109–111.

Uehara, K. and Yamaya, S. 1990. High pressure sintering of diamond by cobalt infiltration. In *Science and Technology of New Diamond*, Saito, O.F.S. and Yoshikawa, M. (eds.). Tokyo, Japan: KTK Scientific Publishers. pp. 203–209.

Upadhyaya, A., Sarathy, D., and Wagner, G. 2001. Advances in alloy design aspects of cemented carbides. *Materials and Design* 22:511–517.

Upadhyaya, G.S. 1998. *Cemented Tungsten Carbides: Production, Properties, and Testing*. Westwood, NJ: Noyes Publications.

Vagnorius, Z. and Sorby, K. 2009. Estimation of cutting tool failure costs. Paper presented at *IEEE International Conference on Industrial Engineering and Engineering Management*, Hong Kong, China. pp. 262–266.

Vandenbulcke, L. and De Barros, M.I. 2001. Deposition, structure, mechanical properties and tribological behavior of polycrystalline to smooth fine-grained diamond coatings. *Surface and Coatings Technology* 146–147:417–424.

Vohra, Y.K., Catledge, S.A., Ladi, R., and Rai, G. 1996. Micro-Raman stress investigations and x-ray diffraction analysis of polycrystalline diamond (PCD) tools. *Diamond and Related Materials* 5:1159–1165.

Wang, X., Fang, Z.Z., and Sohn, H.Y. 2008. Grain growth during the early stage of sintering of nanosized WC–Co powder. *International Journal of Refractory Metals & Hard Materials* 26:232–241.

Wilks, J. and Wilks, E. 1991. *Properties and Applications of Diamond*. Oxford, U.K.: Butterworth-Heinemann.

Xuefeng, Z., Feng, F., and Jianjing, J. 2011. Solidification microstructure of M2 high speed steel by different casting technologies. *China Foundry* 8(3):290–294.

Yao, Z., Stiglich, J.J., and Sudarshan, T.S. 1995. *Nano-Grained Tungsten Carbide–Cobalt (WC/Co)*. Fairfax, VA: Materials Modification, Inc.

Yong, A.Y.L., Seah, K.H.W., and Rahman, M. 2006. Performance evaluation of cryogenically treated tungsten carbide tools in turning. *International Journal of Machine Tools and Manufacture* 46:2051–2056.

Yong, A.Y.L., Seah, K.H.W., and Rahman, M. 2007. Performance of cryogenically treated tungsten carbide tools in milling operations. *International Journal of Advanced Manufacturing Technology* 32:638–643.

Zackrisson, J. and Andren, H.-O. 1999. Effect of carbon content on the microstructure and mechanical properties of (Ti, W, Ta, Mo)(C, N) ± (Co, Ni) cermets. *International Journal of Refractory Metals & Hard Materials* 17:265–273.

Zelinski, P. 1998. The fast track to high speed drilling. *Modern Machine Shop* (6):47–51.

4 Twist and Straight-Flute Drills
Geometry and Design Components

Good design goes to heaven; bad design goes everywhere.

Mieke Gerritzen (NL), a designer and co-founder of All Media Foundation

The major groups of drilling tools are *drills*, *reamers*, *counterbores*, and (threading) *taps* (*Metal Cutting Tool Handbook* 1989). The distinguished features of the drilling tools are the following:

1. All drilling tools are end cutting tools intended to make various types of holes (bores).
2. The prime cutting motion is rotation applied to the tool or to the workpiece or to both, and the feed motion is applied to the tool or to the workpiece along the longitudinal axis of the tool.

In high-volume operations, the drilling tool is application-specific. Different tool manufacturers offer a dozen or more recommended *best* tools for a particular job. Literature resources and patents also offer a great number of tool designs. As a result, a tool designer/process engineer is overwhelmed with the variety of available design/design features so that some methodological help is needed to steer a path clearly through this ocean of information toward major design concepts of HP drills. The first and foremost step in such a methodology is to understand the most important correlations between drilling tool design and geometry with its performance and drilling system requirements to assure this performance. Each drilling tool has features designed to solve specific problems—to extend tool life; to improve chip evacuation; to reduce drilling force, thereby increasing the allowable penetration rate; to improve the surface roughness or straightness of the machined hole; etc. Therefore, a clear system objective should be established before considering the design/purchase of a new drill for a given application. As mentioned earlier, achieving the maximum drill penetration rate while maintaining the required quality of machined holes is the most common system objective in high-volume production because it results in much greater manufacturing cost saving compared to other objectives such as improving tool life. Therefore, a clear understanding of the correlation between the drill/drilling system features and the allowable drill penetration rate is important.

The following sections discuss the most essential (to achieve the stated system objective) features of the drill that are not covered and/or not properly explained in the literature. The explanations are given in the simplest manner presenting the information needed to comprehend drill performance. For those involved in the development of a new drill and drilling process, the corresponding references are given for wider and deeper understanding of specific features.

4.1 CLASSIFICATION

A drill is an end cutting tool for machining holes having one or more cutting lips (major cutting edges) and having one or more helical or straight chip removal flutes. A great variety of drills are used in industry. They can be classified as follows:

Classification based on construction:

1. *Homogeneous drills*—those made of one piece of tool material such as carbide or HSS. Most of the HSS and carbide drills used in practice today are homogeneous drills.
2. *Tipped drills*—those having a body of one material with cutting lips (or their parts as the periphery corners) made of other materials brazed or otherwise bonded in place (see Figure 3.48). Cemented-carbide-tipped drill was used as an alternative to HSS for drilling difficult-to-machine materials, for woodworking drilling operations, and for drilling specific work materials such as concrete. Nowadays, however, the use of carbide-tipped drills is significantly reduced with wide implementation of homogeneous (often referred to as solid-carbide) drills. On the contrary, the use of PCD-tipped drills has been growing. The PCD tips in such tools are placed at the periphery regions of the major cutting edges. Alternatively, the full-face (cross) PCD drill can be made (see Chapter 5 for details).
3. *Indexable-insert drills*—those having cutting portions or indexable cutting inserts (cartridges) held in place. The indexable inserts can be installed in the drill body or placed in the cartridges that are installed into the drill body. The former designs allow smaller drill diameters but when severe breakage occurs, the whole drill body should be replaced. The latter design is more expensive and suitable for drills of larger diameters. However, when severe breakage occurs, only the cartridge is normally damaged. Moreover, much greater adjustment range in terms of setting the exact drill diameter is possible with this design. This type of drill can drill rapidly and produce holes up to five times the diameter of the drill length.
4. *Interchangeable (replaceable)-tip drills*—those made with replaceable drill head. An example of indexable (replaceable) drill head is shown in Figure 4.1. Clamping methods for interchangeable-tip drilling systems range from internal or external screws to turn-and-click systems. Compared to the typical ±0.12 mm accuracy of an indexable-insert drill, an interchangeable-tip drill can hold ±0.05 mm in most cases, compared to the ±0.025 mm typical with solid-carbide tools and ±0.015 mm typical with HP drills. The use of interchangeable-tip drills eliminates more than re-sharpening costs. During re-sharpening, the cost of having drills *in float* can be considerable. For example, one may have two or three

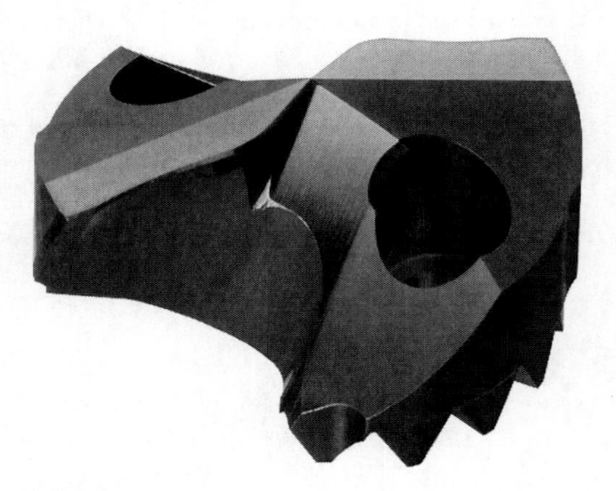

FIGURE 4.1 A replaceable drill head.

drills in operation and two or three drills as backups, but then he or she might have to have four more drills being sharpened, and four drills on the way back from the re-sharpener. Moreover, interchangeable-tip drills provide a further degree of flexibility in that the bodies can be fitted with tip geometries that maximize productivity for a specific workpiece material. For example, among the specialized geometries offered by Sumitomo for its SMD tools are MTL-style tips for steels and general-purpose applications and MEL-geometry tools designated for work materials like stainless steel and high-temperature alloys. It is also possible to combine the advantages of indexable-insert and interchangeable-head drill designs.

Classification based on shank configuration:

1. *Straight shank drills*—those having cylindrical shanks, which may have the same (Figure 4.2) or different diameter than the body of the drill. The shank can be made with or without driving flats, tang, neck, grooves, or threads.
2. *Taper shank drills*—those having conical shanks suitable for direct fitting into tapered holes in machine spindles, driving sleeves, or sockets (Figure 4.3). Tapered drills with Morse taper shanks are most common for *normal*-precision applications and generally have a tang meant exclusively to facilitate drill removal from the machine with a drift.

Classification based on the length-to-diameter ratio:

1. *Stub drills*—those having very short body length.
2. *Regular length drills*—drills having length-to-diameter ratio not exceeding 10. Jobber-length drills are the most common type of such drills. The length of the flutes is 10 times the diameter of the drill.
3. *Long drills*—drills having length-to-diameter ratio >10.

Classification based on number of flutes:

1. *Single-flute drills*—those having only one chip removal flute, for example, gundrills (see Chapter 5).
2. *Two-flute drills*—those having two chip removal flutes, for example, the conventional type of straight-flute and twist drills (Figure 4.4a).
3. *Multiple-flute drills*—those having more than two flutes. This drill type is commonly used for enlarging and finishing, drilled, cast, or punched holes (Figure 4.4b).

Classification based on MWF (coolant) supply:

1. *Drills with external coolant supply*—those having no special means for coolant supply.
2. *Drills with internal coolant supply*—those having internal coolant supply holes (Figure 4.5) or passages and those having coolant supply passages separated from the chip removal passages.

FIGURE 4.2 Drill with a cylindrical shank.

FIGURE 4.3 Drill with a conical (taper) shank.

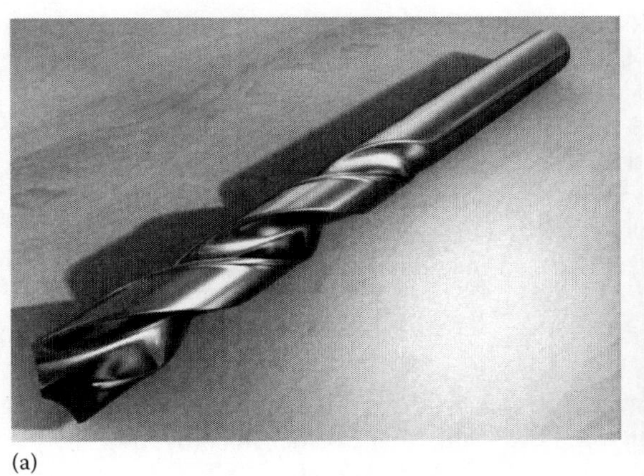
(a)

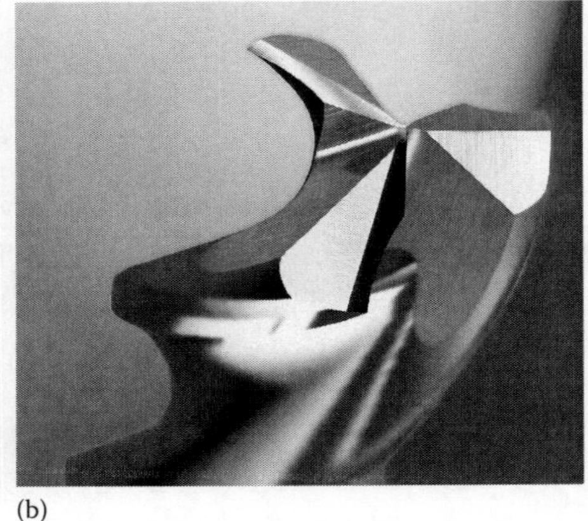
(b)

FIGURE 4.4 Drills: (a) two-flute and (b) three-flute.

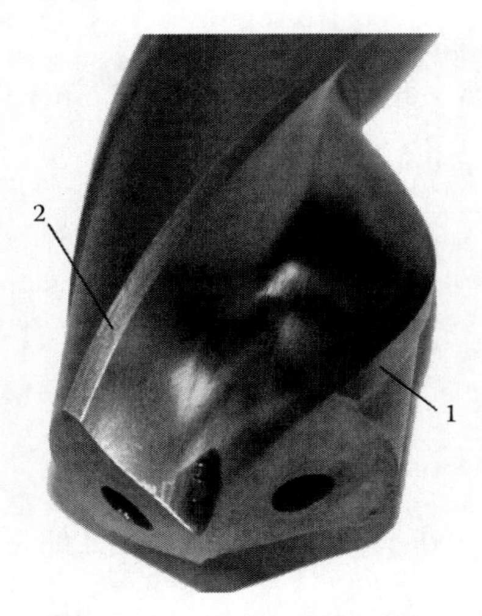

FIGURE 4.5 Twist drill with internal coolant supply holes and with (auxiliary) margins located on the heels: (1) side margin adjacent to the corner of the cutting edge and (2) additional margin made on the heel.

Classification based on assumed force balance:

1. *Transiently balanced drills*—those having only side margins adjacent to the corners of the major cutting edges as supporting means in the radial direction. The radial stability of such drills completely relies on the force balance in drilling.
2. *Transiently balanced drills with additional supports*—those relying on the complete force balance in drilling while having additional supporting margins normally located on the heels to improve drilling stability (Figure 4.5). Although the advantages of such a design in terms of improving drilling stability were known since the end of the nineteenth century, its wide use became feasible with improving drill manufacturing technology and drill setting accuracy. In other words, when CNC drill OD grinding became common and both runouts of spindles and drill holders were decreased dramatically.
3. *Self-piloting drills*—those drills designed so that the intentionally unbalanced radial force generated by the cutting edge in drilling acts on the supporting elements (often referred as guide pads). As a result, a self-piloting drill guides or steers itself during a drilling operation using the walls of the hole being drilled as the pilot surface. Chapter 5 discusses the self-piloting principle and various designs of self-piloting tools.

Classification based on functions and applications:

1. *Solid drills*—those making holes in a solid workpiece with no previously made holes (Figures 4.4 and 4.5).
2. *Spot drills*—short and rigid solid drills used to drill a starting hole (an indent) for the secondary (larger size) drill to enter, acting as a guide. An example of spot drills is shown in Figure 4.6.
3. *Center drills*—those for making a conical indentation in the end of a workpiece to mount it between centers for subsequent machining operations. An example is shown in Figure 4.7. Often, they are used as spot drills.
4. *Trepanning drills*—drill for cutting circular holes around a center. Treppaning, also known as trephination, is for making a burr hole like a compass that *cuts* only a small groove instead of solid core. Trepanning is a great alternative to solid drilling as it requires less cutting power, and in the majority of cases, the cores that remain can be used to produce other parts. Figure 4.8 shows examples of trepanning drills.
5. *Micro drills*—drills used for small holes mainly to drill circuit boards for electronic equipment. Often, micro drills are *pivot drills* as shown in Figure 4.9. *Pivot drills* were designed for watch and clock repair long time ago. The pivot is the small diameter cylindrical end on the arbor that carries a train wheel (gear) in a watch or clock. The pivot was formed by turning down the two ends of the hardened steel arbor or staff. Sometimes this small end breaks off or becomes badly worn. Complete replacement arbors have never been easily available or easy to make. So one way that repair people have coped is to drill a shallow hole in the end of the arbor and insert a piece of steel pivot wire. That process is called re-pivoting. The hole is drilled with a pivot drill, which is usually a short bit with a larger,

FIGURE 4.6 Spot drill.

FIGURE 4.7 Center drill.

FIGURE 4.8 Trepanning drills.

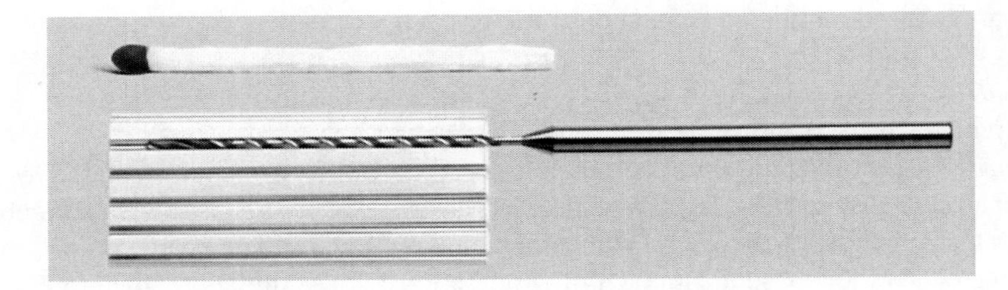

FIGURE 4.9 Microdrill.

long shank. The end of the bit is sharpened for optimum use in relatively hard steel. The bit may be a spade type or, if modern, it may be a twist type. It may be carbon steel or it may be HSS or carbide. Watchmakers could make their own pivot drills and replacement pivot inserts out of sewing needles. Many diameters of pivot drills are made, and most are probably not used in watch or clockwork nowadays.

6. *Combined drills*—although probably the simplest example of combined drills is the center drill shown in Figure 4.7, which combines the drill portion and an adjacent countersink portion, modern combined drills include the drill portion that can be combined with the reamer, cold rolling, thread, and other portions. Figure 4.10 shows two common examples. A drill reamer (Figure 4.10a) and, popular in the automotive industry, a thriller (Figure 4.10b) are used to reduce the cycle time. Thrillers enable drilling, chamfering, and thread milling to be performed with one tool, reducing the machining time.

Drills for Specific Work Materials

Although many of the drills meant for special work materials might have very distinctive appearance, they use the same principles as those for metals. Figure 4.11 shows example of woodworking drills. Figure 4.12 shows a modern hammer drill for machining holes in concrete. Figure 4.13 shows a drill for glass and ceramics.

(a)

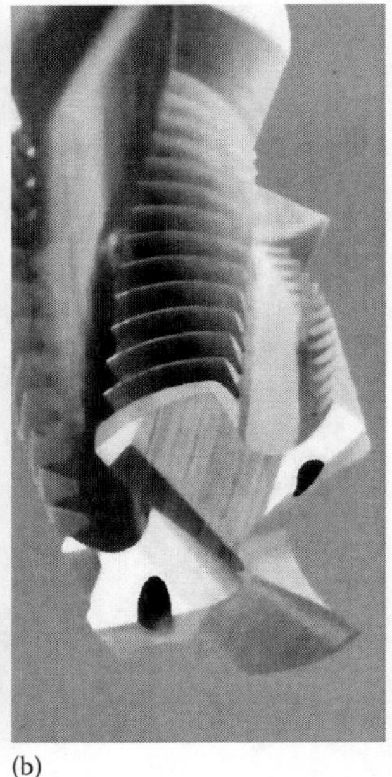

(b)

FIGURE 4.10 Combined drills: (a) drill reamer and (b) thriller (drill-thread mill).

FIGURE 4.11 Example of wood working drills.

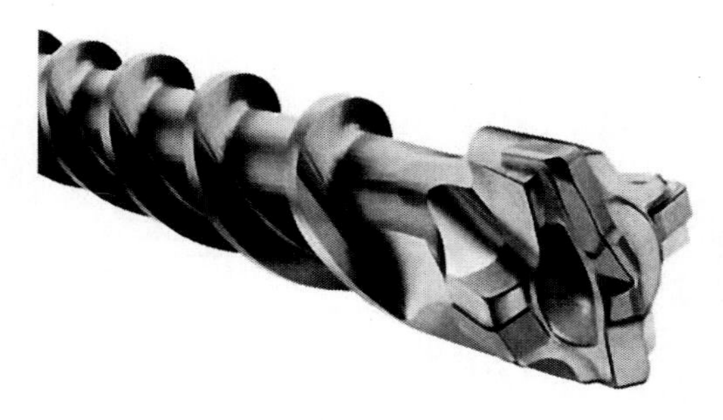

FIGURE 4.12 Hammer carbide drill for concrete.

FIGURE 4.13 Glass and tile drill.

4.2 BASIC TERMS

Understanding the proper terminology as related to drills is a key to understanding the drill design, manufacturing, and proper applications as all specialists involved in this chain should speak the same *language*. This is particularly true in drilling as many lay-language and shop-convenience terms developed over many years so that the same drill components may have multiple names and, what is more undesirable, understood differently by different specialists. Therefore, this section provides visualization and definitions of the basic terms.

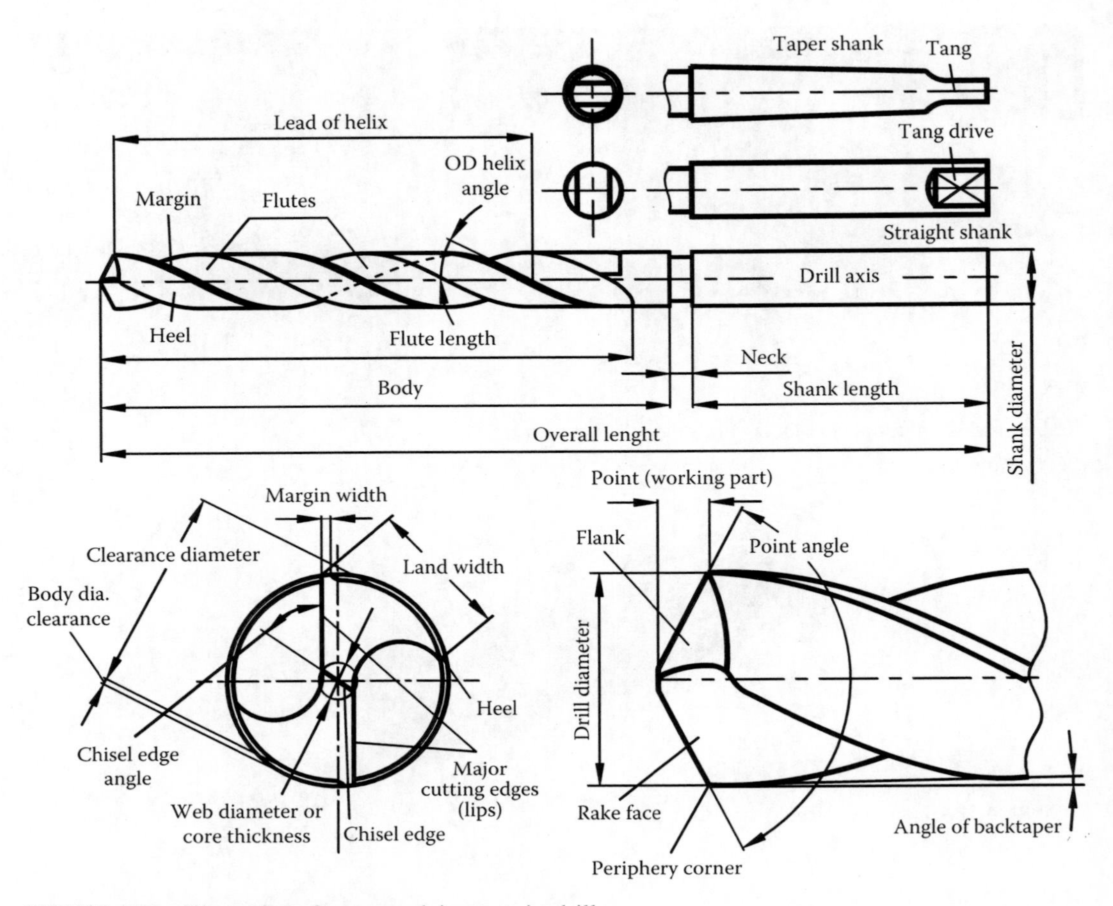

FIGURE 4.14 Illustration of terms applying to twist drills.

The basic terms used for straight-flute and for twist drills are the same so the basic terms related to the twist drill are considered in this section.

A twist drill is defined as an end cutting tool having one or more cutting teeth (often referred to as cutting lips in the trade literature) formed by the corresponding number of helical chip removal (transportation) flutes. A common twist drill is shown in Figure 4.14. It consists of the body, neck (optional), and shank.

The working part has at least two helical flutes called the chip removal flutes. The lead of the helix of the flute depends on many factors including the properties of the work material so it varies from 10° up to 45° for high-helix twist drills (the reason for this will be explained later). The flute profile and its location with respect to the drill longitudinal axis determine many facets of twist drill performance because (1) it determines the geometry of the drill rake face (the shape of the cutting edge (lip), the rake angle and its variation along this edge, the cutting edge inclination angle and its variation along this edge); (2) it determines the reliability of chip removal, that is, chip breakage into pieces (sections) suitable for transportation and the ease of such transportation; and (3) it determines the diameter of the web (the core thickness), that is, directly affecting the buckling stability of the drill. Moreover, together with the flute helix angle, it determines the torsional stability of the drill. As a result, a great number of various flute profiles have been developed, and many of them are available as applied to twist drills produced by various drill manufacturers (Astakhov 2010). Among them, the three shown in Figure 4.15 are basic ones.

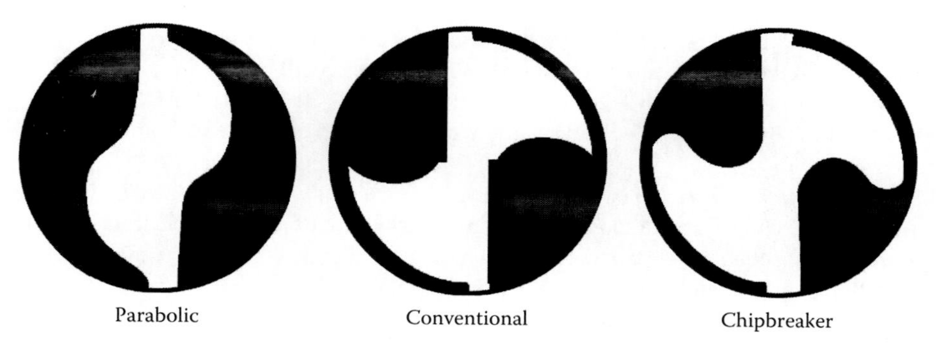

FIGURE 4.15 Common flute profiles.

Some important terms related to the twist drill design and geometry are defined as follows:

Back taper—a slight decrease in diameter from front to back in the body of the drill.

Body—the portion of the drill extending from the shank or neck to the periphery corners of the cutting lips.

Body diameter clearance—that portion of the land that has been cut away to prevent its rubbing against the walls of the hole being drilled.

Chip packing—the failure of chips to pass through the flute during the cutting action.

Chisel edge—the edge at the end of the web that connects the major cutting edges (a.k.a. the lips). Note that this definition is purely geometrical as in reality, there is more than one chisel edge. It will be explained later in this chapter.

Chisel edge angle—the angle included between the chisel edge and the cutting lip, as viewed from the end of the drill.

Clearance—the space provided to eliminate undesirable contact (interference) between the drill and the workpiece.

Cutter sweep—the section of the flute formed when the tool that is used to generate the flute exists the cut. Often, it is called the flute washout.

Cutting tooth—a part of the body bounded by the rake and flank surfaces and by the land.

Double margin drill—a drill whose body diameter clearance is produced to leave two margins on each land and is normally made with margins on the leading edge and on the heel of the land (Figure 4.5).

Drill axis—the imaginary straight line that forms the longitudinal center line of the drill.

Drill diameter—the diameter over the margins of the drill measured at the periphery corners.

Flute length—the length from the drill point to the extreme back end of the flutes. It includes the sweep of the tool that is used to generate the flutes and, therefore, does not indicate the usable length of flutes.

Flutes—helical or straight grooves cut or otherwise formed in the body of the drill to provide cutting lips, to permit removal of chips, and to allow cutting fluid (MWF) to reach the cutting lips if no other means of MWF delivery is provided.

Galling—an adhering deposit of nascent work material on the margin adjacent to the periphery corner of the cutting edge.

Helix angle—the angle made by the leading edge of the land with the plane containing the axis of the drill.

Heel—the trailing edge of the land.

Land—the peripheral portion of the cutting tooth and drill body between adjacent flutes.

Land clearance—see preferred term, body diameter clearance.

Land width—the distance between the leading edge and the heel of the land measured at right angles to the leading edge.

Lead—the axial advance of a helix for one complete turn or the distance between two consecutive points at which the helix is tangent to a line parallel to the drill axis.

Lip (major cutting edge)—a cutting edge that extends from the drill periphery corner to the vicinity of the drill center. The cutting edges of a two-flute drill extend from the chisel edge to the periphery.

Lip clearance (often referred to as relief)—the clearance made from the cutting edge to form the flank surface. There can be several consecutive flank surfaces with different clearance angles as the primary clearance, secondary clearance, etc., made to clear the lip as well as to prevent interference between the flank surface and the bottom of the hole being drilled.

Lip relief angle—obsolete term for the lip clearance angle. This angle is properly defined as the normal clearance angle at the periphery corner of the lip. This angle is often shown in twist drill drawings.

Margin—the cylindrical portion of the land that is not cut away to provide clearance.

Neck—the section of reduced diameter between the body and the shank of a drill.

Overall length—the length from the extreme end of the shank to the outer corners of the cutting lips.

Side rake angle—the axial rake angle as measured between the tangent to the rake face at the drill periphery point and an axial plane through this point.

Periphery—the outside circumference of a drill.

Periphery corner—the point of intersection of the major cutting edge (the lip) and the margin. In a two-flute drill, the drill diameter is measured as the radial distance between two periphery corners. Note that this diameter should be indicated in drill drawings, which is not always the case in reality.

Relative lip height—the difference in indicator reading between the cutting lips. Lips axial runout is another commonly used term (see Chapter 7).

Relief—the result of the removal of tool material behind or adjacent to the cutting lip and leading edge of the land to provide clearance and prevent interference (commonly called rubbing or heel drag) between the cutting tooth and the bottom of the hole being drilled.

Shank—the part of the drill by which it is held and driven.

Web—the central portion of the body that joins the lands. The extreme end of the web forms the chisel edge on a two-flute drill.

Web modification—modification of the web from its ordinary thickness, shape, and/or location to reduce drilling thrust, enhance chip splitting, and change chip flow direction. The simplest modification is web thinning.

Web thickness—the diameter of the web at the point, unless another specific location is indicated.

4.3 CONSTRAINTS ON THE DRILL PENETRATION RATE: DRILL

It was pointed out in Chapter 1 that the prime system objective is an increase in the drill penetration rate, that is, increase drilling productivity or reduce cost per unit. It was explained that this objective is a combined system rather than a component characteristic. The constraints imposed on the penetration rate by the major components of the drilling system are discussed in Chapter 1. This chapter introduces the constraints on the drill penetration rate imposed by the drilling tool itself and explains the correlations of the major tool geometry and design parameters with these constraints.

As discussed in Chapter 1, the feed rate (which is called the penetration rate in drilling) is calculated as the product of the cutting feed (mm/rev or ipr) and the spindle rotational speed (rpm). Therefore, this rate can be increased either by increasing the rotational speed or by increasing the cutting feed. There are some constraints and limits on each of these ways that should be understood.

The major constraint on the rotational speed is the cutting temperature primarily at the drill periphery corners as these have the highest linear (cutting) speed. The maximum allowable temperature is solely the property of the tool material (including its coating), while the maximum allowable rotational speed that causes this temperature is a function of many drill design and geometry variables (Astakhov 2006). This is because they define the state of stress in the deformation zone (the work of plastic deformation) and, thus, drilling force and torque, chip formation, and its sliding direction, as well as the sliding conditions on the tool margins and working conditions of the side cutting edges. Moreover, the tool design and geometry define to a large extent the self-centering of the drill and thus affect the drill transverse vibration, which is the prime cause of drill failure in many applications. Unfortunately, the listed factors and their intercorrelations are not well accounted for in the practice of drill design and implementation where the rotational speed for a given tool material is selected based only upon the work material (type and hardness).

Compared to the drill rotational speed, there are many more constraints (attributed to the drill itself) on the allowable cutting feed (feed per revolution). They include the drill buckling stability, excessive deformation, wear, and breakage. The force factors (drilling torque, axial force, and imbalanced forces) constitute this basis. Therefore, it is of importance in the design of HP drills and drilling systems to understand

1. The concept of the force factors in drilling
2. The drill-related constraints on the penetration rate imposed by the force factors
3. The basics of practical assessment of the force factors in drilling and their correlation not only with parameters of the drilling regime and properties of the work material as discussed in Appendix A but also with parameters of drill geometry and design.

4.4 FORCE BALANCE AS THE MAJOR PREREQUISITE FEATURE IN HP DRILL DESIGN/MANUFACTURING

4.4.1 Theoretical (Intended) Force Balance

Although in this section the concept of the force balance is discussed for a twist drill as the most common drill type used in industry, its essence and the way of assessment are fully applicable for any drill type. The essential features of the drill geometry and common designs are discussed further in the way they affect the force balance because this balance determines practically all the facets of tool performance, starting with the accuracy of the drilled hole and finishing with tool life.

For convenience and simplification of further considerations, the right-hand $x_0y_0z_0$ coordinate system, illustrated in Figure 4.16, is set as follows:

1. The z_0-axis along the longitudinal axis of the drill, with sense as shown in Figure 4.16, toward the drill holder.
2. The y_0-axis is perpendicular to the z_0-axis in the sense shown in Figure 4.16. The intersection of these axes constitutes the coordinate origin O as shown in Figure 4.16.
3. The x_0-axis is perpendicular to the y_0 and z_0 axes as shown in Figure 4.16.

Throughout all further considerations, this coordinate system is referred to as the original coordinate system in the tool-in-machine system (T-mach-S as defined in Appendix C). This system should also be considered as the datum system in drill and drilling machine accessories (drill holder, starting bushing, etc.) design. Any departure from this recommendation would result in the reduction of drill reliability considered as a complex parameter of its performance, including quality of machined holes, tool life, drill breakage, and chip removal problems.

As discussed in Chapter 1, the cutting force is a 3D vector that is commonly resolved in three components along the coordinate axis. In drilling, these components are termed the power, radial, and axial components. A simplified free-body diagram for a twist drill is shown in Figure 4.16.

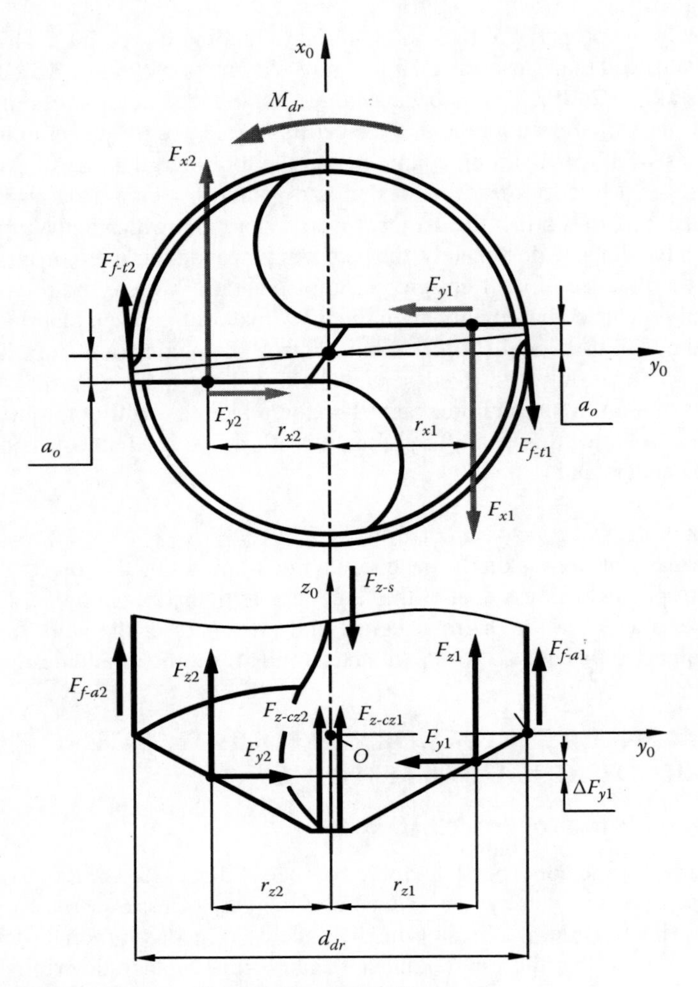

FIGURE 4.16 Simplified free-body diagram of a twist drill.

In this diagram, F_{x1} and F_{x2} are resultant power components, F_{y1} and F_{y2} are the radial components, and F_{z1} and F_{z2} are the axial components of the cutting forces acting on the first and the second major cutting edges (lips), respectively. The power and radial components of the cutting forces that act on the two parts of the chisel edge are not shown as these are small, while the axial components ($F_{z\text{-}cz1}$ and $F_{z\text{-}cz2}$) shown in Figure 4.16 are significant. The tangential $F_{f\text{-}t1}$ and $F_{f\text{-}t2}$ and axial $F_{f\text{-}a1}$ and $F_{f\text{-}a2}$ are components of the friction forces on the margins. The normal components of these forces are not shown.

The drilling torque applied through the spindle of the machine is calculated as

$$M_{dr} = F_{x1}r_{x1} + F_{x2}r_{x2} - F_{y1}a_o - F_{y2}a_o + F_{f\text{-}t1}\left(\frac{d_{dr}}{2}\right) + F_{f\text{-}t2}\left(\frac{d_{dr}}{2}\right) \tag{4.1}$$

and the axial force applied by the spindle is

$$F_{z\text{-}s} = F_{z1} + F_{z2} + F_{z\text{-}cz1} + F_{z\text{-}cz2} + F_{f\text{-}a1} + F_{f\text{-}a2} \tag{4.2}$$

The shown drill is in the static equilibrium in the x_0y_0 and z_0y_0 planes if and only if the following two equilibrium conditions are justified:

In the x_0y_0 plane

$$F_{x1}r_{x1} - F_{y1}a_o = F_{x2}r_{x2} - F_{y2}a_o \tag{4.3}$$

In the z_0y_0 plane

$$F_{z1}r_{z1} + F_{f\text{-}a1}\left(\frac{d_{dr}}{2}\right) = F_{z2}r_{z2} + F_{f\text{-}a2}\left(\frac{d_{dr}}{2}\right) \tag{4.4}$$

Equations 4.1 through 4.4 establish the full force balance, known as the theoretical or intended force balance. Unfortunately, this balance is not the case in practical drilling operations as many addition factors tend to disturb this balance.

4.4.2 Additional Force Factors in Real Tools

In practice, the conditions established by Equations 4.1 through 4.4 are rarely justified even to the first approximation. There are at least three principal domains of causes that may cause violation of this balance:

1. *Drill design/manufacturing accuracy*: For example, the major cutting edges (lips) may have so-called runout that stands for their inequality in terms of length, angular asymmetry (see Chapter 7 for details), etc. Moreover, these edges may have different elevation over the y_0-axis (the location distance of the major cutting edges a_o shown in Figure 4.16). The radial forces, for example, may not share the same line of action. Rather, these forces can be shifted by certain distance $\Delta_{F_{y1}}$ as shown in Figure 4.16. The same can be said about all the previously listed conditions of equilibrium.
2. *Drill mounting accuracy*: This includes both static and dynamic properties of the whole drilling system. Its static part includes the accuracy of drill location in the drill holder (chuck), the runout of the spindle of the machine, and static stiffness of the drilling system including workpiece fixture. The dynamic part includes many static characteristics considered under the action of the drilling force and torque plus dynamic stability of the drilling system.
3. *Workpiece issues*: These embrace geometrical and materials issues. The geometrical issues include drill entrance conditions (e.g., location and configuration of the previously made hole including cored holes in castings), cross holes, steps in the drilled holes and relation of their length to cross holes (in the automotive industry, the veins), wall-thickness difference in the drilled hole, and drill exit conditions (straight, inclined, thin-walled, etc.). The materials conditions include defects in the work material (considered in Chapter 2).

As a result, there is always unbalanced radial force $F_{xy\text{-}ud}$ acting in the x_0y_0 plane and the axial force $F_{z\text{-}ub}$ that acts not along the axis of rotation but its line of action is shifted by certain distance $e_{z\text{-}ub}$ from this axis. A model shown in Figure 4.17 illustrates these forces. The components $F_{x\text{-}ud}$ and $F_{y\text{-}ud}$ of the unbalanced radial force $F_{xy\text{-}ud}$ create the bending moments $M_{zx\text{-}ib}$ and $M_{zy\text{-}ib}$, respectively, in the corresponding z_0y_0 and z_0x_0 planes. The eccentric axial force $F_{z\text{-}ub}$ lowers buckling stability of the drill (discussed in the next section) restricting allowable penetration rate and causing drilled holes deviation from the true position.

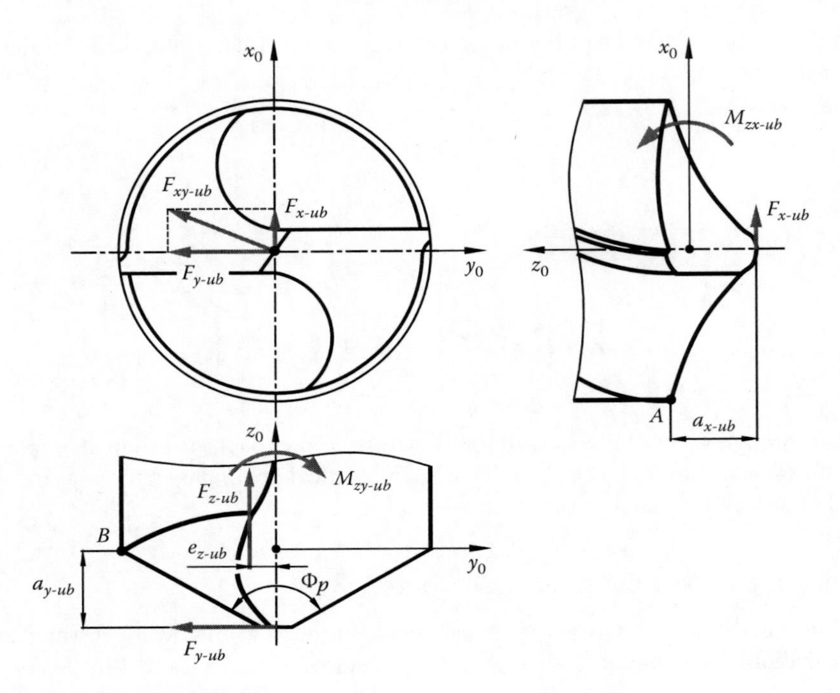

FIGURE 4.17 System of unbalanced loads.

It directly follows from the model shown in Figure 4.17 that arms of the components F_{x-ud} and F_{y-ud} of the unbalanced radial force F_{xy-ud}, namely, a_{x-ub} and a_{y-ub}, depend on drill design and geometry, primarily on the point angle Φ_p. The higher the point angle, the greater the arms, the greater the bending moments caused by the components of the unbalanced radial force. These bending moments cause a number of problems in drilling such as the shape distortions and diametric deviations of the hole being drilled.

It can be seen in Figure 4.17 that bending moments M_{zz-ib} and M_{zy-ib} have different effects on drill bending as point A belongs to the drill heel that has a certain body clearance while point B belongs to the drill margin formed with a cylindrical land that prevents drill shift in the z_0y_0 plane. This is the obvious case with straight-flute drills. One may argue, however, that the margins of a twist drill are helical so they should provide support for the whole cylindrical body of the drill. It would be true, however, if the drill had no back taper. The presence of back taper provides clearance of the drill body and thus some room for bending. To prevent drill bending in the z_0x_0 plane, the so double-margin drills were introduced (Figure 4.5). An example of such a design is shown in Figure 4.18. Sometimes, bore scrapers are used on the side of the additional margins to improve the surface finish of the machined holes. When the parameters and geometry of these additional design features are properly selected, they help to improve drill stability and the quality of the machined holes. However, the improvement is not always dramatic due to additional friction on the additional margins. In other words, their width, diameter, and back taper should be customized based on particular drilling conditions.

4.4.3 Resistance of a Drill to the Force Factors

Based upon the considerations made in the previous section, a model of a drill loaded by the force factors, namely, by the axial force F_{ax} (which is F_z force in the model of the force balance shown in Figure 4.16) and drilling torque M_{dr}, can be represented as shown in Figure 4.19. Two important length dimensions are considered in drill strength calculation: the maximum length L_{dr-1} (section A-A that corresponds to the face of the drill holder) and the maximum flute length L_{dr-s} where the ratio length/cross section is minimal (section B-B location that depends on the flute length and its profile).

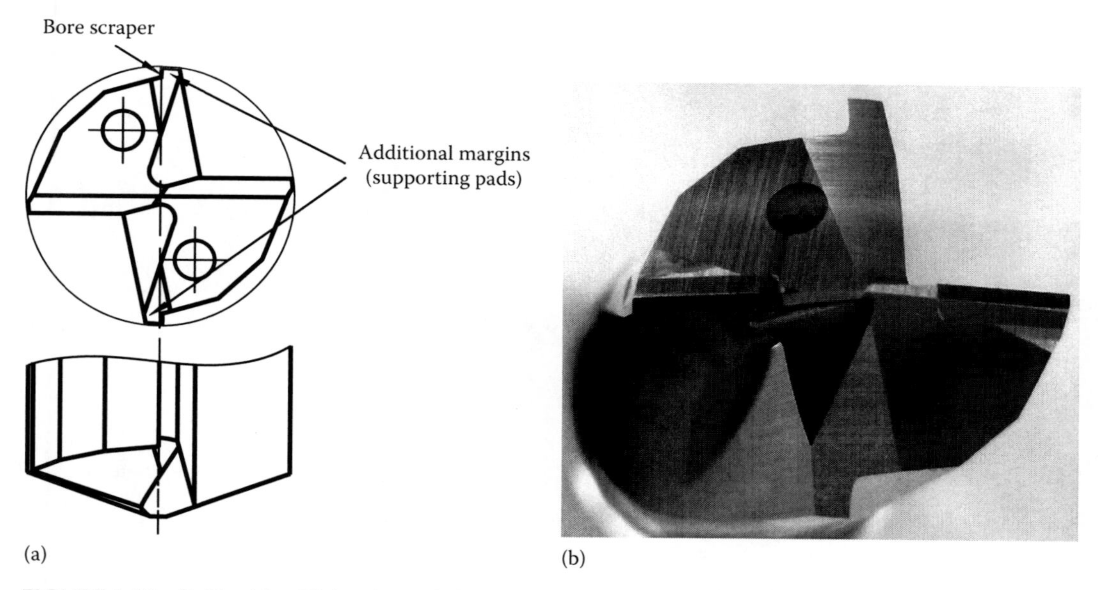

FIGURE 4.18 Drill with additional margins placed on the top of the heels: (a) the model and (b) a real drill.

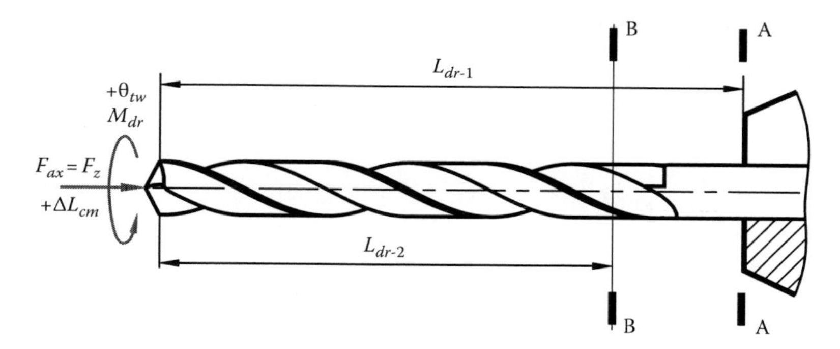

FIGURE 4.19 Drill loaded by the axial force and drilling torque.

4.4.3.1 Resistance to the Drilling Torque

In solid mechanics, *torsion* is the twisting of an object due to an applied torque. In drilling, torsion is a result of the drilling torque acting about the longitudinal axis as shown in Figure 4.19. Torsion, like a linear force, will produce both stress and strain. However, unlike linear stress and strain, torsion causes a twisting stress, called shear stress (τ_{st}), and a rotation, called shear strain (γ_{st}). It is important to be able to predict stresses and deformations that occur in the drill in this type of loading. Figure 4.20 shows a simple mode of torsion in drilling. In this model, a drill represented by a cylindrical body, which essentially is cantilever beam one end of which is rigidly fixed and the drilling torque is applied to the other (free) end.

Let's consider line AB drawn on the surface parallel to the beam longitudinal axis as shown in Figure 4.20a. When the drilling torque (or moment) is applied to the free end of drill represented by a circular beam as shown in Figure 4.20a, line AB becomes helical assuming position AB′, for which the angle of helix is γ_{tw}. The beam twists by an angle θ_{tw}. This angle is a function of the beam length, L, and stiffness represented by the shear modules G of the material the beam is made of modulus G. The twist angle starts at 0 at the fixed end of the beam and increases linearly as a function of the z-distance from this end. The change of angle, γ_{tw}, is constant along the length.

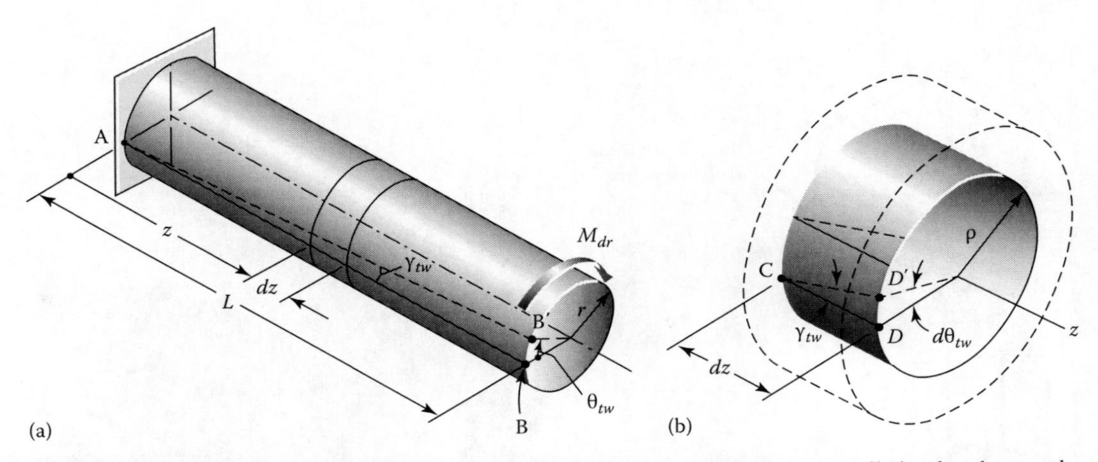

FIGURE 4.20 Model of torsion in drilling: (a) the whole circular beam and (b) a small circular element dz.

A small differential element, dz, is sliced from the beam as shown in Figure 4.20b. Because the cross sections bounded in this element are separated by an infinitesimal distance, the difference in their rotations, denoted by the angle $d\theta_{tw}$, is also infinitesimal. As the cross section undergoes the relative rotation $d\theta_{tw}$, straight line CD deforms into the helix CD′. By observing the distortion of the sliced element, one should recognize that the helix angle γ_{tw} is the *shear strain* of the element.

Two angles γ_{tw} and $d\theta_{tw}$ must be compatible at the outside edge (arc length $D - D'$). This gives the relationship,

$$\text{Arc length } D - D' = \rho d\theta_{tw} = \gamma_{tw} dz \tag{4.5}$$

from which the shear strain γ_{tw} is

$$\gamma_{tw} = \rho \frac{d\theta_{tw}}{dz} \tag{4.6}$$

The quantity $d\theta_{tw}/dx$ is the *angle of twist per unit length,* where θ_{tw} is expressed in radians. The corresponding shear stress (Figure 4.21a) is determined from Hooke's law as

$$\tau = G\gamma_{tw} = G\rho \frac{d\theta_{tw}}{dz} \tag{4.7}$$

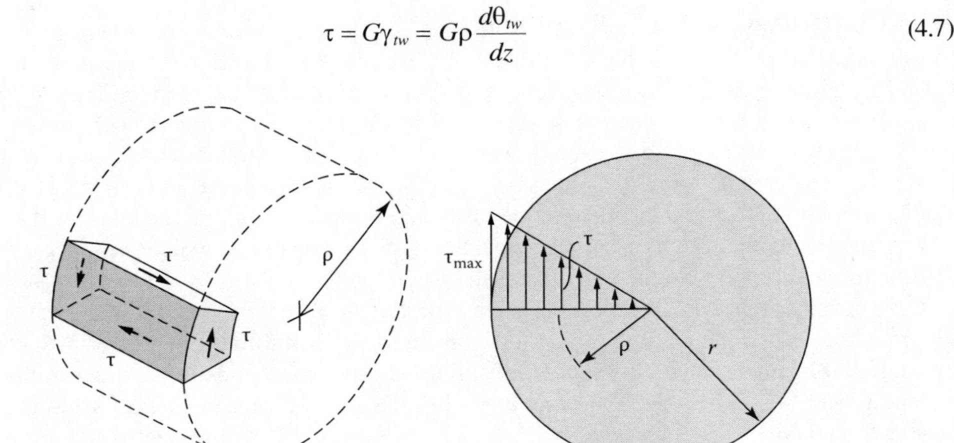

FIGURE 4.21 (a) Shear stress due to torsion and (b) shear stress distribution.

A simple analysis of Equation 4.7 reveals that the shear stress varies linearly with the radial distance ρ from the axis of the beam. This variation is shown in Figure 4.21b. As can be seen, the maximum shear stress, denoted by τ_{max}, occurs at the surface of the beam. Note that the previous derivations assume neither a constant internal torque nor a constant cross section along the length of the beam, that is, valid for a general case, for example, for a twist or straight-flute drill.

For practical calculation of τ_{max}, the following formula is normally used (Hibbeler 2010):

$$\tau_{max} = \frac{M_{dr} r}{J} \tag{4.8}$$

where J is the polar moment of inertia of the beam cross section.

Figure 4.22 shows formulae for calculating simple cross sections. For complicated cross sections as that of drills (e.g., those shown in Figure 4.15), any modern CAD program, that is, AutoCAD, does this calculation in a single click of a computer mouse. Once one has created a shape from lines and arcs (e.g., the cross section of the twist drill), he or she then creates a region from the shape by using the *region* command and selecting all the lines and/or arcs. Then the position of the coordinate system by location of its origin is defined. Once one has done this, he or she uses the *massprop* command and clicks the region being created. A text box pops up showing the moment of inertia results.

As follows from Figure 4.21, elements with faces parallel and perpendicular to the beam axis are subjected to shear stresses only. The normal stresses, shear stresses, or a combination of both may be found for other orientations. In a model shown in Figure 4.23, element a is in pure shear and its maximum shear stress is calculated using Equation 4.8. Element b shown in Figure 4.23 is located

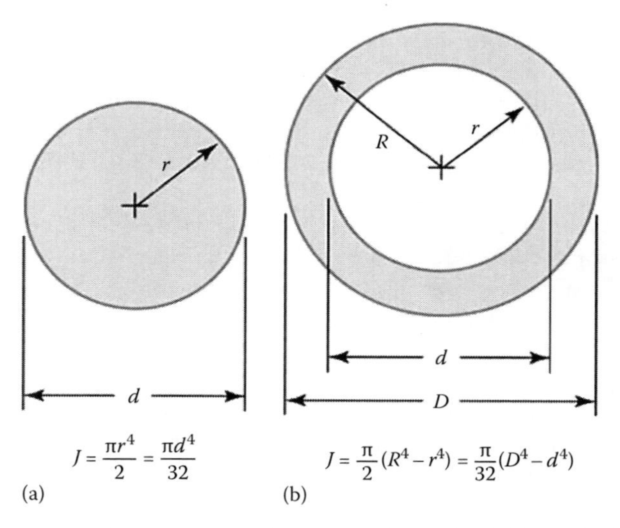

FIGURE 4.22 Polar moments of inertia for simple cross sections: (a) solid beam and (b) tubular beam.

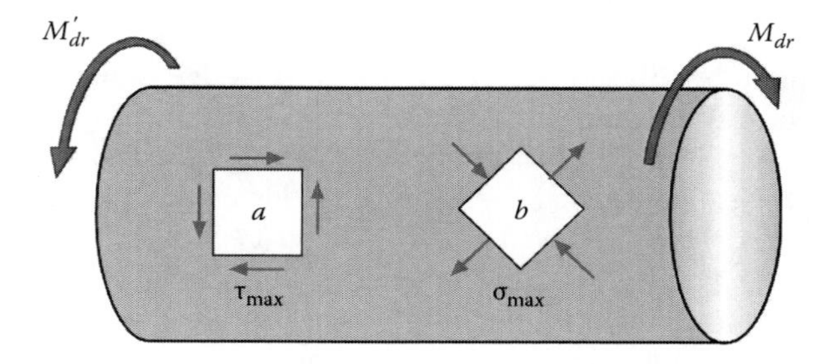

FIGURE 4.23 Maximum shear and normal stresses due to drilling torque.

FIGURE 4.24 Typical fracture of a drilling tool.

at 45° to the axis. It is subjected to a tensile stress on two faces and compressive stress on the other two. As well known (Hibbeler 2010),

$$\sigma_{max} = \sigma_{45^\circ} = \tau_{max} \tag{4.9}$$

Ductile materials generally fail in shear as many practical shafts. When subjected to torsion, a ductile specimen breaks along a plane of maximum shear, that is, a plane perpendicular to the shaft longitudinal axis. Brittle materials (tool materials) are weaker in tension than shear. When subjected to torsion, a brittle tool breaks along planes perpendicular to the direction in which tension is a maximum, that is, along surfaces at 45° to the drilling tool axis. Figure 4.24 shows an example.

Using the known tensile properties of the tool material, drilling torque (Equation 4.8), and tool cross-section geometrical properties (the polar moment of inertia), one can make a reasonably accurate assessment if the designed tool can withstand the drilling torque with no fracture.

4.4.3.2 Resistance to the Axial Force

According to Equation 4.2, the resultant axial force in drilling shown in Figure 4.19 is the sum of the axial forces on the major cutting edges (lips) and chisel edge and due to friction on the margins. The latter is small compared to the first two terms so that the contribution of the major cutting edges (lips) and the chisel edge are considered. It is very important to realize that the axial force produced by the unit length of the cutting edge is not a linear function of the location radius of this unit length. Rather, the contributions of the portions of the cutting edge located closer to the drill center are much greater than the periphery regions. That is why the so-called mechanist approach, which is when the unit cutting force is calculated and then its value is simply multiplied by the length of the drill cutting edge to obtain the cutting force, is incorrect in principle (see Appendix A).

To illustrate this statement, Figure 4.25 shows the principle (Figure 4.25a) and results (Figure 4.25b) of a simple axial force test. A pre-drilled test specimen made of gray cast iron (HB 200) is placed on a table dynamometer. An HSS twist drill of 29.5 mm dia. (d_{dr}) was used. As the drill progressed in the pre-drilled hole (of d_1 diameter), the contributions of different portions of the cutting edge into the resultant axial force can be assessed. Subtracting the axial force measured when drilled hole of and then 12, 18, and 24 mm dia.) from the resultant axial force measured when a solid specimen was drilled, one can obtain the contributions of the different parts of the major cutting edge

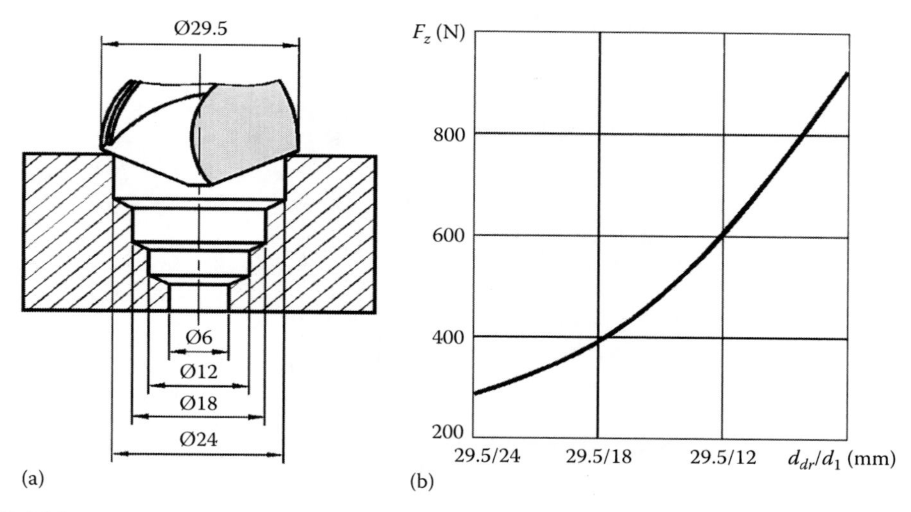

FIGURE 4.25 (a) Principle and (b) results of a simple axial force test. Cutting speed v = 59 m/min, feed f = 0.32 mm/rev.

and chisel edge into the resultant axial force. As can be seen in Figure 4.25b, different portions of the cutting edge contribute differently to the total axial force. For a *conventional* drill, the major cutting edges (lips) contribute approximately 30%–40%, minor cutting edges (margins) 10%, and chisel edge 50%–60% to the total axial force. Optimizing essential design parameters of the drill, one can not only achieve a reduction of the resultant axial force but also change substantially the relative contribution of the drill components to this force. This is one of the important stages in the development of the VPA-balanced drill concept or any other HP drill.

A significant axial force in drilling restricts the penetration rate because of the following:

- It affects the buckling stability of the drill. The compromising of this stability causes a number of hole quality problems. It also significantly reduces tool life causing excessive drill corner or even margin wear.
- Many machines used for drilling have insufficient thrust capacity that limits any increase in the penetration rate with standard drills.

Therefore, the reduction of the resultant axial force is vitally important when one tries to increase the allowable penetration rate of the drill. As the chisel edge is the major contributor to this axial force, one should (1) reduce the length of this edge and (2) improve the geometry of this edge. These two objectives can be achieved simultaneously as discussed later in this chapter.

The allowable axial force on a drill of a given design is restricted either by its buckling or by its compressive strength. Whenever a structural member is designed, it is necessary that it satisfies specific strength, deflection, and stability requirements. Typically strength (or in some cases fracture toughness) is used to determine failure, while assuming that the member will always be in static equilibrium. However, when certain structural members, for example, drills, are subjected to compressive loads, they may either fail due to the compressive stress exceeding the yield strength (which in case of brittle tool material is almost the same as the compressive strength), or they may fail due to lateral deflection (i.e., buckling). The maximum axial load that a drill can withstand when it is on the verge of buckling is called the critical load, $F_{z\text{-}cr}$. Any additional load greater than $F_{z\text{-}cr}$ will cause the drill to buckle and therefore to deflect laterally. Buckling is a geometric instability and is related to material stiffness, column length, and column cross-sectional dimensions. Strength does not play a role in buckling but does play a role in compression.

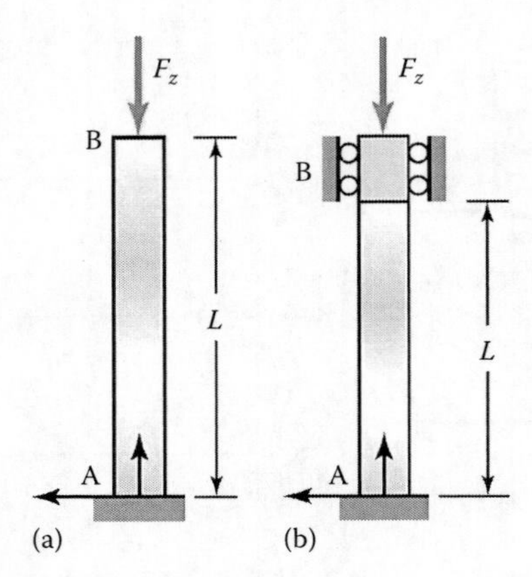

FIGURE 4.26 Two basic models for assessment the buckling stability of drills: (a) at the hole entrance and (b) in drilling.

Figure 4.26 shows two basic models for assessments of buckling stability of drills. Figure 4.26a shows a model for a drill as it starts drilling, that is, at hole entrance, while Figure 4.26b shows a model for a fully engaged drill, that is, when its margins are in full contact with the wall of the hole being drilled. The well-known equation (Vable 2012) allows to calculate $F_{z\text{-}cr}$ as

$$F_{z\text{-}cr} = \frac{\pi^2 EI}{L_{eff}^2} \tag{4.10}$$

where

L_{eff} is the effective length of the drill
E is the modulus of elasticity (sometimes called Young's modulus) of the drill material
I is the moment of inertia of the drill's cross section.

The effective length of the drill in Equation 4.10 is calculated as follows:

- For the case shown in Figure 4.26a: $L_{eff} = 2L$
- For the case shown in Figure 4.26b: $L_{eff} = 0.5L$.

For purposes of design, it is more useful to express the critical loading condition in terms of a stress such as

$$\sigma_{cr} = \frac{F_{z\text{-}cr}}{A_{d\text{-}cs}} = \frac{\pi^2 EI}{A_{d\text{-}cs} L_{eff}^2} \triangleright \sigma_{cr} = \frac{\pi^2 E}{(L_{eff}/k_{gr})^2} \tag{4.11}$$

where

$A_{d\text{-}cs}$ is the cross-sectional area of the drill
$k_{gr} = \sqrt{I/A_{d\text{-}cs}}$ is the smallest radius of gyration determined from the least moment of inertia I, i.e. for the least cross section.

The term, L_{eff}/k_{gr}, is known as the slenderness ratio and contains information about the length and the cross section of the drill.

It is important to understand the transition from compression failure of a drill under the applied axial force to that due to buckling the case where the applied stress, σ, is equal to the critical buckling stress, σ_{cr}, and the generalized compressive strength, σ_{cm}. At the point where such a transition takes place, $\sigma_{cr} = \sigma_{cm}$, that is,

$$\sigma = \sigma_{cr} = \sigma_{cm} = \frac{\pi^2 E}{(L_{eff}/k_{gr})^2} \tag{4.12}$$

If Equation 4.12 is solved for L_{eff}/k_{gr}, the resulting relation marks the combination of length and cross section at which the compressive behavior transitions from compression to buckling. This relation is known as the minimum slenderness ratio

$$\left(\frac{L_{eff}}{k_{gr}}\right)_{\min} = \sqrt{\frac{\pi^2 E}{\sigma_{cm}}} \tag{4.13}$$

This transition can be illustrated by plotting the relation between stress, σ, and slenderness ratio, L_{eff}/k_{gr}, as shown in Figure 4.27. Note that in reality, there is no sharply divided transition between yielding and buckling. Instead the σ versus L_{eff}/k_{gr} curve can be divided into three regions as shown in Figure 4.28. Region 1 is the short-length drill region in which failure occurs due to compression stress when $\sigma = \sigma_{cm}$. Region 2 is the intermediate-length drill region in which the drill either may fail due to compression or may buckle. In this region, empirical relations are used to approximate the resulting curve. Region 3 is for long drills. Buckling is the failure mode in this region.

Calculation of the drill strength in regions 1 and 3 is straightforward (the compression strength and critical stress, respectively), while to do so in region 2, material/geometry/loading-dependent

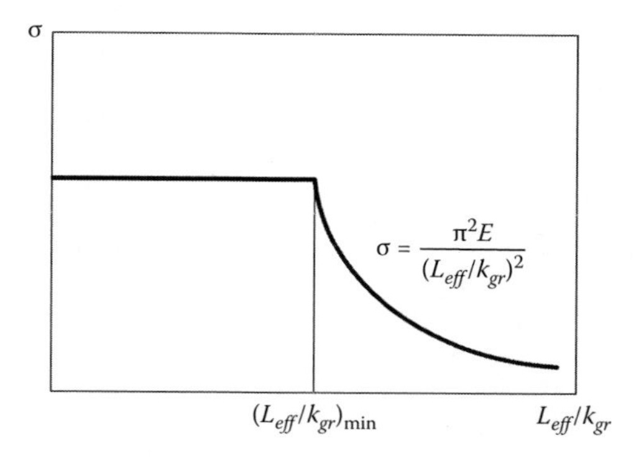

FIGURE 4.27 Stress versus slenderness ratio relation.

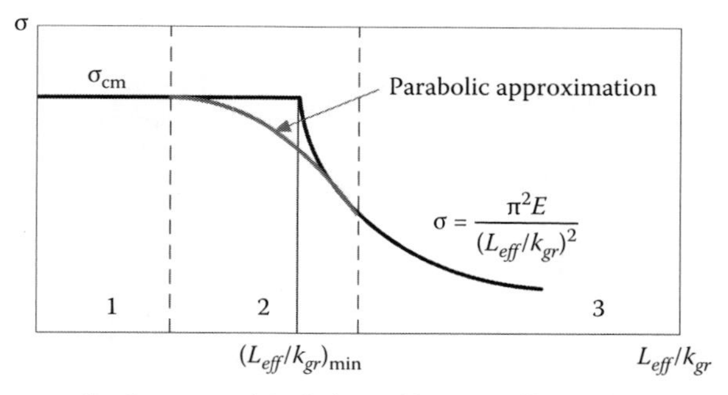

FIGURE 4.28 Stress versus slenderness ratio relation with parabolic approximation.

and empirical relations can be used. One common approach is to fit a parabola to the σ versus L_{eff}/k_{gr} curve from $\sigma = \sigma_{cm}$ to $\sigma = \sigma_{cm}/2$ as follows:

$$\sigma = \begin{cases} \sigma_{cm}\left[1 - \dfrac{(L_{eff}/k_{gr})^2}{((2L_{eff}/k_{gr})_{\min})^2}\right] & \text{for } 0 \le L_{eff}/k_{gr} < (L_{eff}/k_{gr})_{\min} \\ \dfrac{\pi^2 E}{(L_{eff}/k_{gr})^2} & \text{for } L_{eff}/k_{gr} > (L_{eff}/k_{gr})_{\min} \end{cases} \tag{4.14}$$

Example 4.1

An important example concerning buckling stability is considered next as a part of a drill premature failure analysis. The conditions of the drilling operations considered in the analysis were as follows:

1. Workpiece is a turbine shaft of an automatic six-speed rear-wheel drive transmission. A fragment of the shaft operation drawing is shown in Figure 4.29.
2. Work material is normalized steel SEA 1040 (equivalent AISI 1040) with the following mechanical properties: hardness 170 HB and tensile strength (ultimate 590 MPa, yield 374 MPa).
3. New and CNC-sharpened carbide twist drills having proper geometry, tool material grade, and suitable coating made by one of the most reputable tool companies were used. The drills were preset in drill holders using a Zoller machine so that their total runout did not exceed 25 μm. Figure 4.30 shows a drill preset in a holder.
4. CNC machines (Fanuc Robodrill machines) widely used in the automotive industry were used.
5. The drilling regimes used for the test drills are shown in the Table 4.1.
6. MULTAN B-400 MWF supplied at pressure of 3.1 MPa (450 psi) was used. The coolant concentration, pH, and clearness were kept within the limits assigned by the control plan.

Problem

Tool life of drills of 5.40 mm dia. (for drilling hole 5.500/5.300 mm dia. shown in Figure 4.29) is almost half (380 shafts) compared to that of the drills used for other holes (670 shafts) with no apparent reason.

Analysis

An analysis of the operational conditions that included all components of the drilling system showed that these components are selected properly including machining regimes shown in Table 4.1. Moreover, they are the same for all drills involved in the consideration. Therefore, these components were excluded from the further analysis. The only difference between the drills is their length. The drill of 5.40 mm dia. (for drilling hole 5.500/5.300 mm dia. shown in Figure 4.29) is a long drill, while other drills are regular-length drills (Jobber-length drills). As can be seen in Figure 4.31, the length-to-diameter ratio for this drill is 25.85 (for a new drill), which well exceeds 10 as with regular-length drills. This is because the hole 5.500/5.300 mm dia. is located near a flange made on the turbine shaft as clearly seen in Figure 4.29.

Analysis of drills of 5.40 mm dia. revealed the following:

- Their geometry (split-point self-started) and flute profile are suitable for the application.
- The overall number of regrind for these drills is almost half that for other drills. This is because severe wear of drill margins that requires much longer portion of the drill to be removed on the re-sharpening to restore proper geometry of these margins. Moreover, traces of cutting by the margins were observed as shown in Figure 4.32.

Part autopsy discussed in Chapter 2 was used to reveal the root cause of the problem. Following the autopsy methodology, a shaft that had completed the drilling operation was sectioned as shown in Figure 4.33 and the holes were examined. The examination revealed buckling instability of drills of 5.40 mm dia. Figure 4.34 shows the appearance of drilled hole when such instability takes place in drilling. As can be seen, the worst instability is observed at hole entrance as was

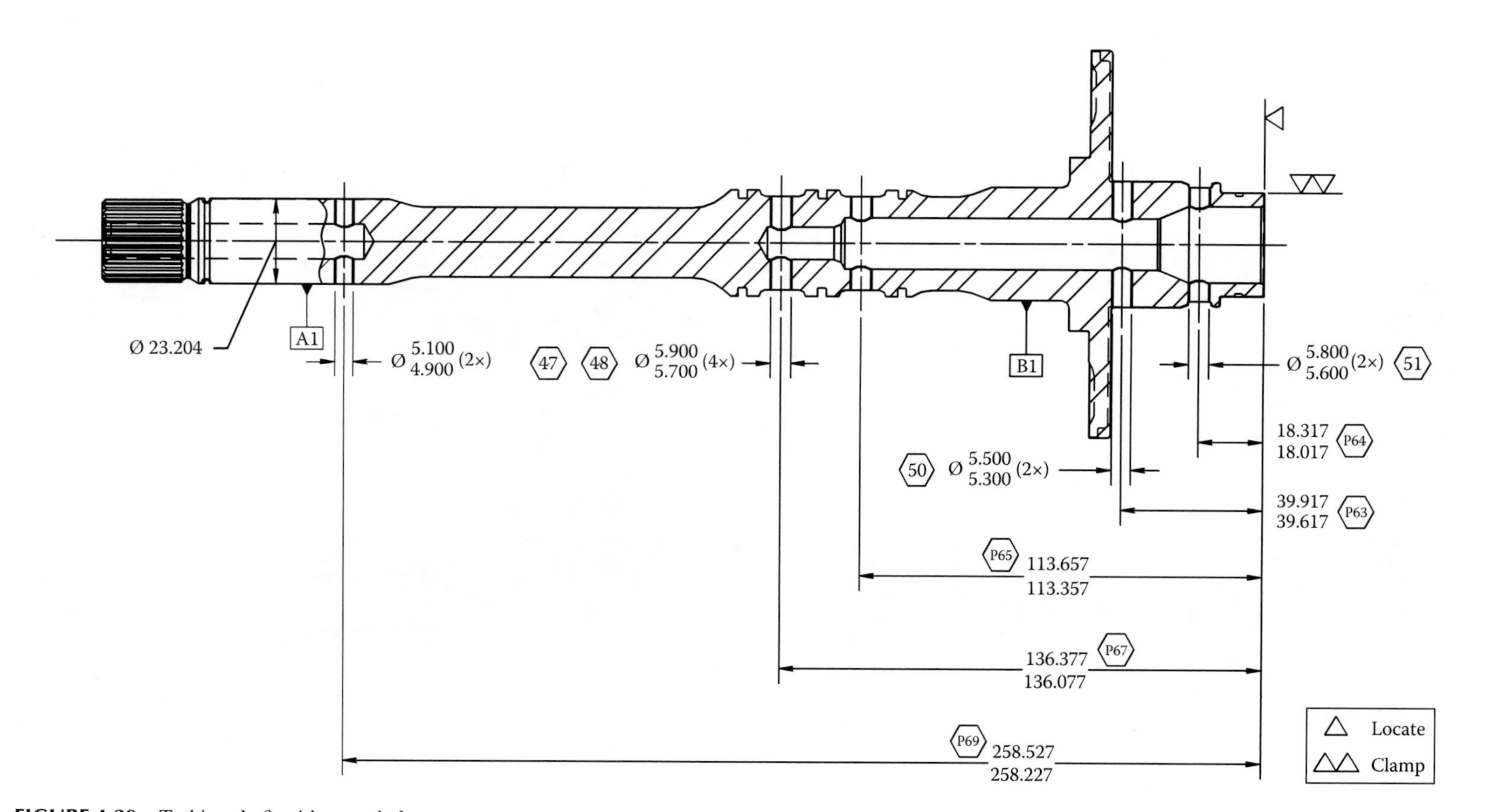

FIGURE 4.29 Turbine shaft with cross holes.

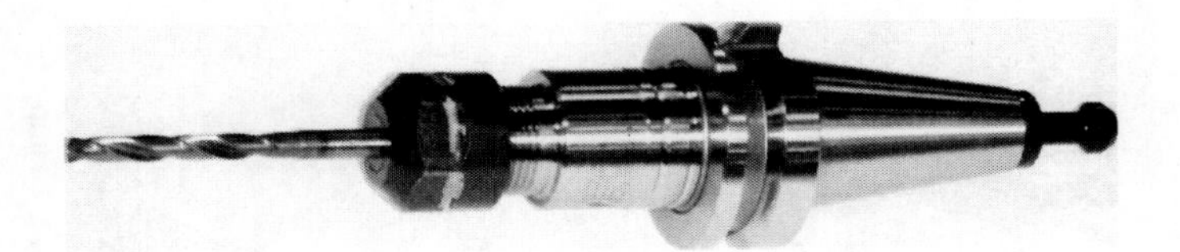

FIGURE 4.30 Drill preset in a holder.

TABLE 4.1
Drilling Regimes Used

Drill diameter, mm	5.00	5.40	5.70	5.80
Rotational speed, rpm	9315	8625	8130	8013
Cutting speed, m/min	146.3	146.3	145.6	146
Feed, mm/rev	0.13	0.10	0.13	0.13
Feed rate, mm/min	1211	862.5	1057	1042

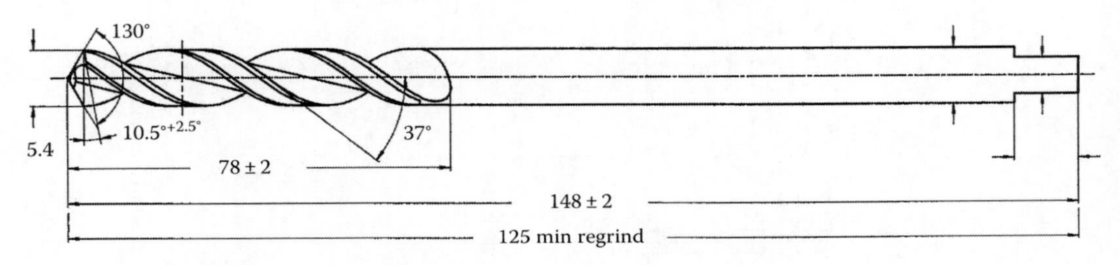

FIGURE 4.31 Length of the drill of 5.4 mm dia. as assigned by its drawing.

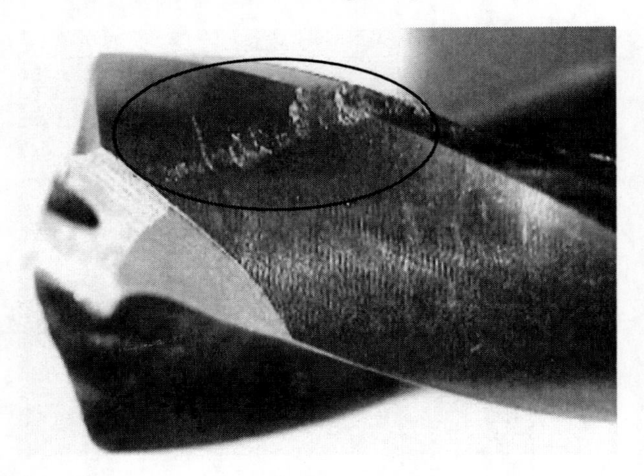

FIGURE 4.32 Showing an evidence of the cutting by drill margin.

discussed in analysis of the model shown in Figure 4.26a. When the drill is fully engaged in the hole being drilled, its deflection reduces as can be seen in Figure 4.34. The reason for this is discussed in the analysis of the model shown in Figure 4.26b.

Solution

A special thick-web drill (40%) with Din 1412 Form B point geometry and curved cutting edges (Astakhov 2010) was designed, manufactured, and implemented to solve the problem. Additionally, the cutting feed was reduced by 30% at the whole entrance while a higher feed that listed in Table 4.1 is then applied for the rest of the hole being drilled to meet the requirement of cycle time entrance.

FIGURE 4.33 Showing a sectioned part of the shaft.

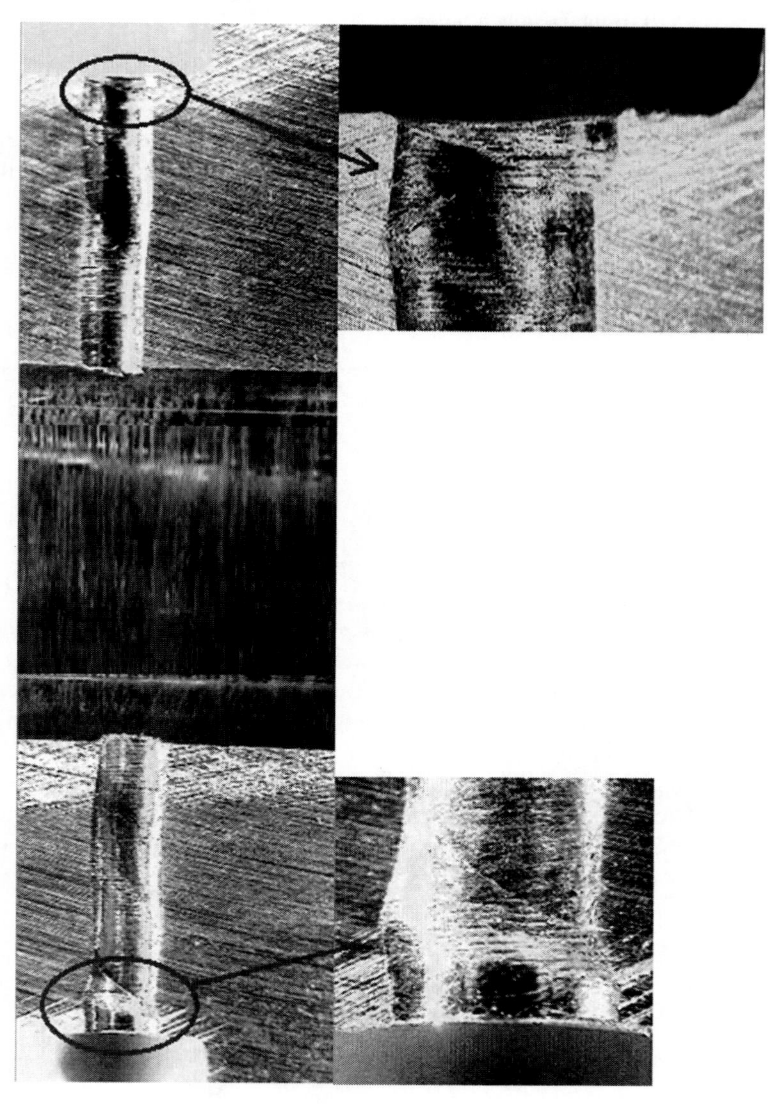

FIGURE 4.34 Showing the results of buckling instability.

4.4.3.3 Drill-Design/Process-Related Generalizations/Considerations Related to Resistance to the Force Factors

The foregoing analysis of the drill resistance to the axial force and drilling torque results in the following important drill-design/process-related generalizations to be used in the HP drill design:

1. The drilling torque is a function of the work material properties, drill diameter and geometry, and the drilling regime. Of these factors, the drill geometry and drilling regime can be varied to achieve optimal drill performance. As the cutting speed has a weak influence on the cutting force, it also has little influence on the drilling torque so that the cutting feed is the only factor to be considered. While for modern production CNC machines the drilling torque is not a limiting factor as these machines are equipped with powerful motors to deliver high torques, for relatively small machines, the drilling torque can be a constraint limited by the power of the drive motor. When the latter is the case, the feed per revolution is lowered, or the hole is drilled in two consecutive drilling operations using first a smaller drill and then a drill to the required hole size.
2. The ultimate tensile strength (UTS) is the proper mechanical characteristic of a tool material in its torsional resistance. High-quality HSSs have UTS up to 1000 MPa while 6%Co cemented carbide has 2200 MPa UTS. As a result, the allowable torque for the same tool made of cemented carbide is up to two times greater than that made of HSS.
3. The drilling torque is commonly the limiting force factor for regular-length drills (Jobber-length drills). A better accuracy of drill strength calculations is achieved if the Von Mises stress is used as this parameter accounts for the real state of stress in the body of a drill. Because in drilling only the axial force and torque are the force factors and stresses σ_z and τ_{zy} are present (as in combined torsion and bending/axial state of stress), the Von Mises stress is calculated as

$$\sigma_v = \sqrt{\sigma_z^2 + 3\tau_{xy}^2} \tag{4.15}$$

As such, the drill strength equation becomes

$$\frac{\sigma_{UTS}}{FS} \leq \sqrt{\left(\frac{F_z}{A_{d\text{-}cs}}\right)^2 + 3\left(\frac{M_{dr}d_{dr}}{2J_{d\text{-}cs}}\right)^2} \tag{4.16}$$

where FS is the safety factor.
4. Equation 4.13 and Figure 4.27 provide the idea of transition between compression and buckling, that is, about the region where the axial force is the limiting force factor. Figure 4.28 shows that there is a certain transition range rather than abrupt transition. Equation 4.14 shows how to calculate the critical stress in this range.
5. The easiest concept to grasp in analyzing buckling stability of a drill is that the maximum allowable axial force $F_{z\text{-max}}$ must be less than the critical buckling load $F_{z\text{-}cr}$ given by Equation 4.10. It follows from this equation that the maximum allowable axial force is calculated as

$$F_{z\text{-max}} \leq \frac{F_{z\text{-}cr}}{FS} = \frac{\pi^2 EI}{FSL_{eff}^2} \tag{4.17}$$

Therefore, the design equation can be written as

$$EI \geq \frac{\left(FSF_{z\text{-max}}L_{eff}^2\right)}{\pi^2} \tag{4.18}$$

In the design equation, everything on its right-hand side is known or set for the time being. Therefore, the left-hand side can be solved for either E or I, that is, either material or cross-section geometry. Usually, the tool material is already known for other reasons so that Equation 4.18 can be solved for I.

6. Three conclusions important for the drill design/application follow from the design equation (Equation 4.18):
   a. The maximum allowable axial force $F_{z\text{-max}}$ is directly proportional to the modulus of elasticity E. Therefore, for the same drill design (the moment of inertia I), drills made of cemented carbide allow higher penetration rate than those made of HSS as $E = 650$ GPa for cemented carbide while $E = 221$ GPa for HSS.
   b. The effective length of a drill has the strongest influence. As discussed in the analyses of two models shown in Figure 4.26, this length reduces four times when drill working end is restricted (Figure 4.26b). Therefore, the use of starting bushes or pilot holes for long drills is the most powerful mean to increase the allowable penetration rate in drilling. Another less powerful way to increase buckling rigidity is to reduce the feed rate at the hole entrance that, in turn, reduces $F_{z\text{-max}}$.
   c. The least effective but commonly used way to increase the drill buckling rigidity is a gradual increase of the moment of inertia from the drill's working end toward its shank by allying web thickness taper as shown in Figure 4.35. The American Society of Mechanical Engineers (ASME B94.11M-1993) and the Aerospace Industries Association of America, Inc. (NAS 907) standards define the conventional web thickness taper rate between 0.60 and 0.76 mm.

The section of the drill between the shank and the drill point is referred to as the body profile. Its design is wholly determined by the drill manufacturers and cannot be altered by the user. A lot has been written on this profile and its optimization (Spur and Masuha 1981; Thornley et al. 1987b; Agapiou 1993a,b; Chen 1997; Chen and Ni 1999). National and International standard organizations have come up with various recommendations on the forms suitable for given cutting conditions and workpiece materials. One of the common basic forms is shown in Figure 4.36. In this form, the central angle β_{fl} is selected to be equal or 2°–3° greater than that of the land. The web diameter is d_{wb} = (0.125–0.145)d_{dr} and increases toward the shank by up to 1.7 mm per 100 mm drill length; the body clearance diameter is d_{cl} = (0.125–0.145)d_{dr}; profile radii are r_{f1} = (0.75–0.90)d_{dr} and r_{f2} = (0.22–0.28)d_{dr}; and margin width is $b_m = (0.2\text{–}0.5)\sqrt[3]{d_{dr}}$. Back taper of 0.03–0.12 mm per 100 mm length is applied to the margins of HSS drill while that for cemented carbide drill is 0.1–0.2 mm per 100 mm length of the drill.

These earlier studies and recommendations were restricted by the ability of grinding machines to reproduce the optimized profiles with reasonable accuracy, and the computation techniques were not entirely suitable to calculate the geometrical parameters of the designed profile *in situ*, that is, within the design environment. As mentioned earlier, modern CAD programs, even the simplest as AutoCAD (Inventor), allow such calculation in one command so that any complicated profile can be assessed and its particularities can be accounted for in the exact determination of the geometrical characteristic needed for the previously discussed calculations. Moreover, modern CAD programs have a built-in FEM module that makes possible not only geometrical calculations

FIGURE 4.35 Web diameter (core thickness) increases from the tip toward the shank.

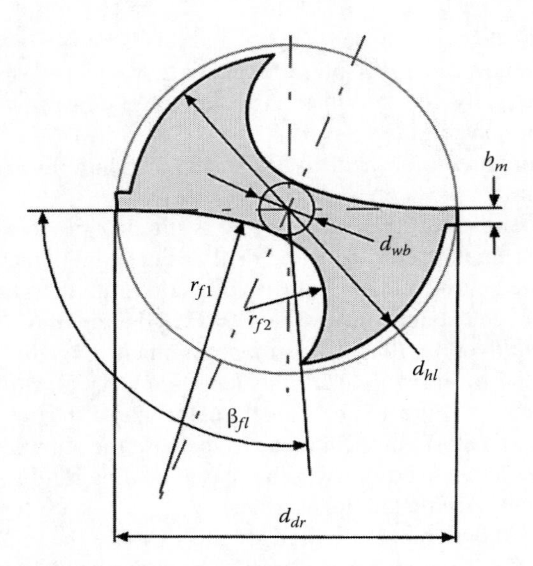

FIGURE 4.36 One of the basic forms of the body profile.

but also the strength calculations. Such an option has been already used in drill profile calculation and drill design (Abele and Fujara 2010). However, the clear objective of such a design is somehow lost in the development history.

Two main parameters influencing the design of the body profile are as follows:

1. The area of the flute. It is obvious that this area should be large enough to allow reliable chip transportation over this flute.
2. The drill cross section should be rigid enough to withstand the force parameters, that is, the drilling torque and the axial force.

As these are two competing factors, the optimization of the body profile is required to find a suitable compromise.

The body profile of the twist drill is shown in Figure 4.37. The angular land length (often considered in the direction perpendicular to the helix of the flute) L_{ld}, angular flute length L_{fl} or its central angle β_{fl}, and web diameter d_{wb} are normally considered in the most known optimizations although

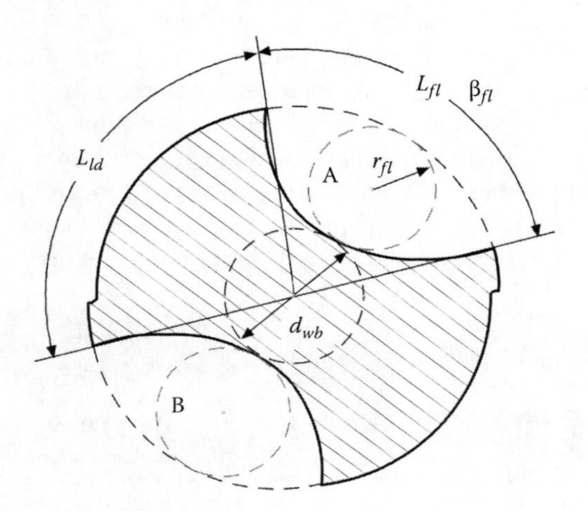

FIGURE 4.37 Body profile geometry parameters.

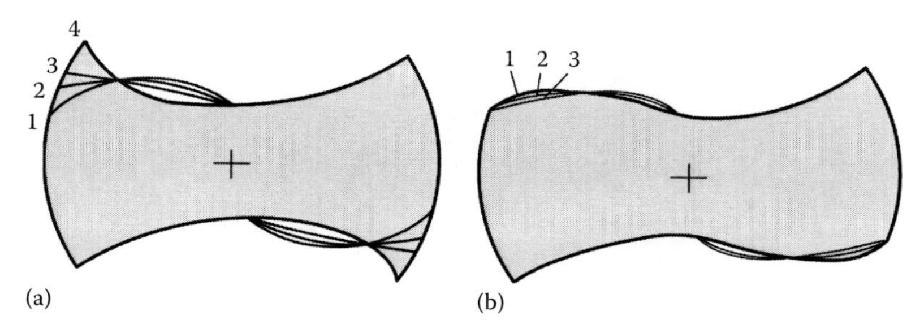

FIGURE 4.38 Body profiles with different (a) flute heel shapes and (b) protuberance locations.

other particularities of the body profile as, for example, different flute heel shapes and protuberance locations (Figure 4.38), can also be used (Chen and Ni 1999).

Thornley et al. (1987b) discussed that the total area of the chip flute is not the significant factor in chip removal because a typical drill chip is conical in shape and hence only the inscribed circle of the flute cross section it actually occupied is important. Consequently, the inscribed circle (In other words, circles A and B in Figure 4.37) of r_{fl} radius was proposed to be considered as the critical parameter. In other words, circles A and B in Figure 4.37 that could be inscribed in the flute space normal to the axis of the drill are considered to be a good index of flute disposal capacity since the curled-conical shape chip occupies only a circular part of the flute space. When the radius of the chip cone tends to be larger than r_{fl}, the chip flute is considered to be clogged. Although it was a great attempt of the body profile optimization, the proposed critical parameter of chip removal is not practical as the chip attains various shapes and sizes in drilling depending on the work material, machining regime, tool geometry, and body profile. There were many other attempts to optimize the body profile (its area, polar moment of inertia, and other cross-sectional characteristics).

In the author's opinion, besides increasing drill rigidity, the optimization of the flute profile should also aim to improve chip transportation in the flutes. The profile of a drill should be designed in such a way that the flutes provide the maximum space for the chip and facilitate chip removal while ensuring that the drill is capable to adequately withstand the drilling torque and axial force.

4.4.3.4 Improving Drill Rigidity

The so-called wide-web drills are more rigid and stronger than conventional drills; thus, higher penetration rates (up to 20%) and length-to-diameter ratio and better accuracy of machined holes are achieved with these drills. Wide-web drills are normally used to produce holes in difficult-to-machine and heat-treated work materials. Figure 4.39 shows the comparison of a conventional and a wide-web drill. Naturally, all the methods for the chisel edge modification as, for example, web thinning, split point, or even special shape of the cutting edges (Fuh and Chen 1995) are used for these drills as the *natural* length of their chisel edge is too great.

An apparent drawback to wide-web drills is the reduced cross-sectional area of the chip flutes. However, the penetration rate in drilling of difficult-to-machine and heat-treated work materials is rather low, which results in much smaller amount of chip produced and thus needed to be transported from the machining zone through these flutes.

Figure 4.40a shows a profile of a standard twist drill. As seen, the flute has two distinctive parts, namely, the so-called straight part 1, which in its intersection with the flank surface forms the major cutting edge (lip) and concavely curved surface 2 having a relatively large radius of curvature. Figure 4.40b shows the flute profile of a wide-web drill. As the web diameter 3 is great, the profile 4 is normally made using a single curve to enhance drill rigidity.

According to US Patents No. 4,744,705 (1988) and 5,230,593 (1993), the profiles of drills made of HSS and those made of cemented carbide are not the same particularly for heavy-duty drilling operations. The profile is usually characterized by the so-called flute-width ratio (FWR), which is the ratio

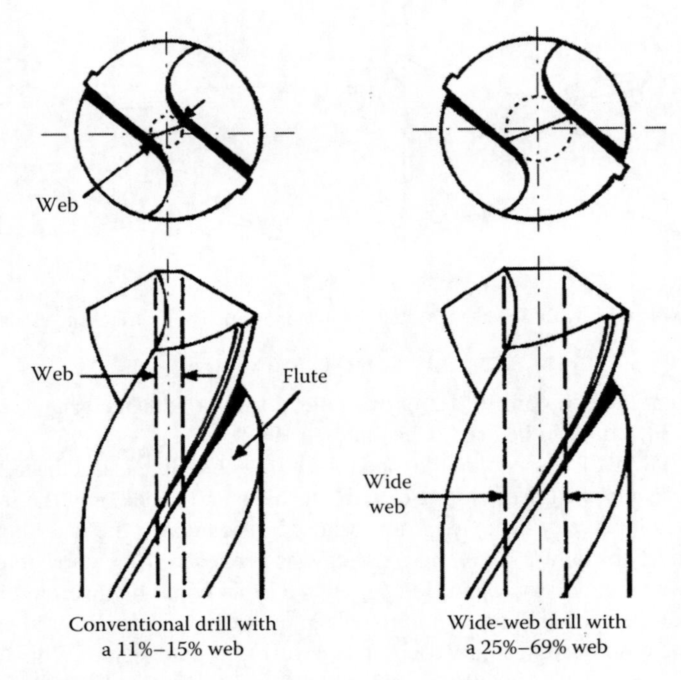

FIGURE 4.39 Conventional and wide-web drills.

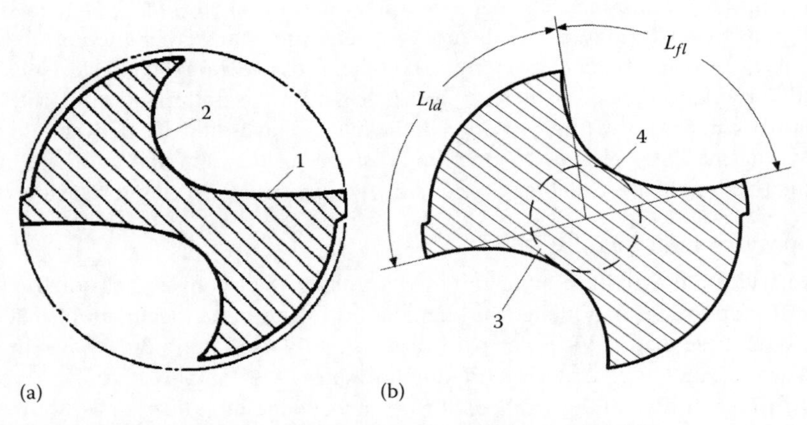

FIGURE 4.40 Profile of the (a) standard drill and that of a (b) wide-web drill.

of arc length L_{fl} of the flute to arc length L_{ld} of the lands (Figure 4.40b). For HSS drills having profile shown in Figure 4.40a, FWR is 0.7 at the forward end of the drill. To increase drill rigidity, FWR can be equal to 1.16 away from the forward end. For carbide drills, FWR is in the range of 0.4–0.8.

Figure 4.41 shows the relationships between the drill torsional rigidity and web thickness ratio for two different FWR according to the data presented in US Patent No. 4,583,888 (1988). As can be seen, an increase in the web thickness ratio and while a decreers in FWR lead to increase torsional drill rigidity. The chip evacuation, however, becoms more difficult. Thus, the web thickness ratio and FWR have their limits; generally, the web thickness ratio is set in the range 15%–23% and FWR is in the range of 1.0–0.76.

The quest for better shape of the chip flute and thus for more rigid profile of the drill body started a long time ago. As early as 1882, Hartshorn had proposed the profile (US Patent No. 262,588 [1882]) shown in Figure 4.42 that was appreciated and thus became popular over a century later.

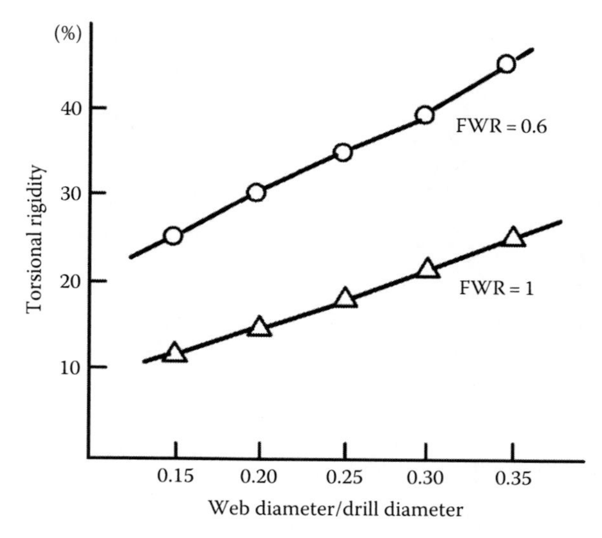

FIGURE 4.41 Drill torsional rigidity as a function of the web thickness ratio.

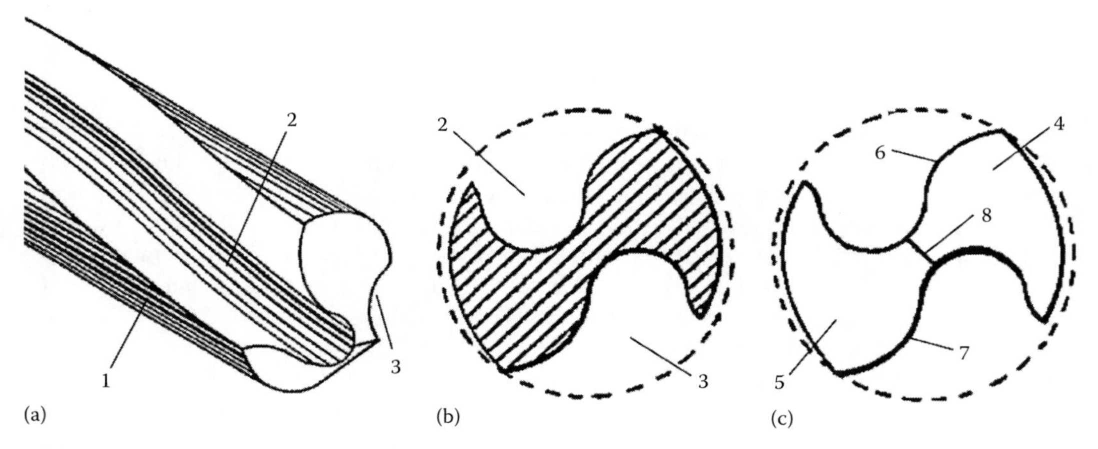

FIGURE 4.42 Drill design according to US Patent No. 262,588 (1882): (a) drill, (b) normal cross section, and (c) face view.

This drill has a drill body 1 and two straight or helical flutes (or grooves as in the text of the patent) 2 and 3 situated opposite to each other (Figure 4.42a and b). As can be seen in Figure 4.42b, the side of each flute that makes the cutting edge is of convex curvature along its entire length, so that when the forward end of the drill is ground away to form the point by providing the flank surfaces 4 and 5, the corresponding lips or the major cutting edges 6 and 7 (which form as the intersection lines between the flutes and the flank surfaces as shown in Figure 4.42c) of convex shape are produced. According to the patent, such lips' shape results in significant reduction of the drilling torque.

As modification of this flute profile, US Patent No. 4,065,224 (1976) claims that the shape of the convex cutting edges 6 and 7 shown in Figure 4.42c can be made so the constant rake angle along these edges is achieved for any desired combination of helix angle, point angle, and lip rake face shape required by the optimal drill performance.

One needs to realize that any increase in the web diameter requires the corresponding change in the tool geometry to keep the axial force at certain low level. Otherwise, any attempt to solve a problem with increasing drill stiffness will lead to another problem with an increase in the axial force that undermines the net result. Figure 4.43 illustrates how differences in drill web thickness

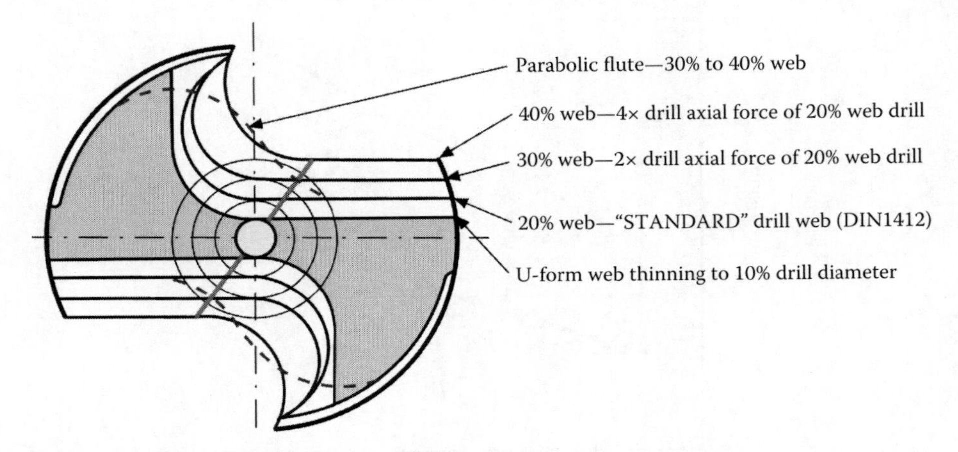

FIGURE 4.43 Effect of the web diameter on the axial force.

affect the axial (thrust) force requirements of drills. According to Fiesselmann (1993), it is seldom understood that a drill with 30% web requires almost twice the axial force of the 20% web drill. Further, the 40% web drill found in drills recommended for harder, tougher alloys, or which result from using parabolic flute drills for holes deeper than 10 times the drill diameter in depth, has an axial force requirement almost four times that of the 20% web drill.

4.4.3.5 Axial Force (Thrust)–Torque Coupling

Observations and tests comparative of straight-flute and twist drills showed that the latter allow greater critical axial force and torques. This result can easily be explained by the so-called torque–thrust coupling effect in twist drills. The coupling can be explained as follows. The body of a twist drill contains two helical flutes. If a torque is applied as shown in Figure 4.19, this torque tends to *unwind* the helix and thus increase the drill length. Conversely, if an axial force is applied to the end of the drill as shown in Figure 4.19, this force shortens the drill thus causes its *winding*. Therefore, the effects of the drilling torque and the axial force (thrust) on drill static stability partially compensate each other. This explains the results of observations. It is important to a drill designer/user to understand to which extent this compensation can go and how the design parameters of a twist drill to miximize this compensation.

Narasimha et al. (1987) proposed to assess torque/thrust coupling using the following coupling matrix:

$$\begin{Bmatrix} F_z \\ M_{dr} \end{Bmatrix} = \begin{bmatrix} K_{FF} & K_{MF} \\ K_{FM} & K_{MM} \end{bmatrix} \begin{Bmatrix} \Delta L_{cm} \\ \theta_{tw} \end{Bmatrix} \qquad (4.19)$$

where

K_{MM} is the axial stiffness under torsional restrain
K_{MF} is the torque-on-thrust coupling stiffness coefficient
K_{FM} is the thrust-on-torque coupling stiffness coefficient
K_{FF} is the torsional stiffness under axial restrain
ΔL_{cm} and θ_{tw} are deformations due to the force factors (Figure 4.19).

Table 4.2 shows the values of the stiffness coefficients obtained experimentally (Narasimha et al. 1987). Although unconventional, the units of the coefficient are meant for an easy quantitative comprehension of the results, that is, the coupling effect.

TABLE 4.2
Stiffness Coefficients

Drill Dia. (mm)	Helix Angle (deg)	Web Thickness (mm)	K_{FF} (N)	K_{MF} (N-m/rad)	K_{FM} (N-m)	K_{MM} (Nm²/rad)
12.7	32.76	1.905	2.47×10^6	1.91×10^3	2.00×10^3	0.599
	51.42	1.905	2.04×10^6	1.68×10^3	1.56×10^3	0.433
10.3	14.00	1.549	2.22×10^6	514.2	658.4	0.131
	31.94	1.549	2.41×10^6	1.18×10^3	1.42×10^3	0.229
	37.65	1.549	1.88×10^6	978.8	1359.1	0.165
9.5	12.96	1.473	2.09×10^6	437.9	535.0	0.100
	32.49	1.473	2.47×10^6	959.9	1.29×10^3	0.180
	36.80	1.473	2.47×10^6	765.1	800.8	0.146
6.4	13.67	1.219	2.47×10^6	185.9	151.7	0.019
	31.94	1.219	2.47×10^6	399.5	321.6	0.038
	35.48	1.219	2.47×10^6	339.7	187.8	0.032

The experimental results of several studies (De Beer 1970; Lorenz 1979; Narasimha et al. 1987) reveal the following:

- The flute helix angle, flute profile, and web thickness significantly affect the axial and torsional stiffness of drills.
- The pure torsional stiffness of drills is maximized for a helix angle of around 28°. Departure from this angle to either side lowers this stiffness significantly (see Table 4.2).
- The torque–thrust interaction (measured by K_{FM} and K_{MF}) has a distinct maximum at a helix angle of about 28°. As the same value of helix angle results in the largest increase in torsional stiffness, this explains a much higher allowable torque and axial force (thrust) for twist drills compared to straight-flute drills. This also provides an explanation to the fact that general-purpose drills are made with helix angle of 28°–30°.
- An increase in web thickness decreases the torque–thrust interaction, which means the benefit of the stiffening action of the axial force reduces.

The discussed results indicate that the penetration rate of a properly designed twist drill can be up to 20% higher than that of a straight-flute drill due to the coupling. This advantage is particularly important for long drills as the applied torque may significantly improve drill buckling stiffness. In the author's opinion, this is one of the most important advantages of twist drills rarely mentioned in the literature.

4.5 DRILL GEOMETRY

4.5.1 Importance of the Drill Geometry

Understanding the drilling tool geometry is a key to improving efficiency of drilling operations. This general statement is extensively elaborated in the author's earlier book on the drill geometry (Astakhov 2010). The drill geometry allows determining the following important parameters of any drilling operation:

1. *Uncut (undeformed) chip thickness (a.k.a. the chip load)*: Only when one knows and understands the tool geometry, he or she can properly determine the uncut chip thickness for each and every cutting element (wedge) involved.
2. *Chip compression ratio*: Measuring the chip thickness and dividing it by the uncut chip thickness, one can determine the chip compression ratio. Knowing this fundamental to

metal cutting theory and practice parameter, one can calculate practically all other process parameters and characteristics (Astakhov 2006; Astakhov and Xiao 2008). Methods of its practical determination in drilling are discussed in Appendix A.

3. *Direction of the chip flow*: The simplest yet very practical aspect of the drilling tool geometry is that this geometry defines the direction of chip flow. This direction is important to control chip breakage and evacuation.
4. *Cutting force on each cutting element as well as the total cutting force*: The cutting force is primarily determined by the mechanical properties of the work material, machining regime, and uncut chip thickness. Together with four other components of the cutting tool geometry, namely, the rake angle tool, cutting edge angle, tool minor cutting edge angle, and inclination angle, the uncut chip thickness defines the magnitudes of the orthogonal components of the cutting force. Knowing the correlation among the mentioned angles and force components, one can design HP drilling tools.
5. *Quality (surface integrity and machining residual stress) of machined surfaces*: The drilling geometry directly affects diametric shape and location accuracy of machined holes as well as their surface characteristics and parameters including surface roughness, machining residual stress, and topography.
6. *Tool life*: The geometry of the cutting tool affects tool life directly as this geometry defines the magnitude and direction of the cutting force and its components, sliding velocity at the tool–chip interface, the partition of the thermal energy released in machining, the temperature distribution in the cutting wedge, etc.

4.5.2 Tool Geometry Measures to Increase the Allowable Penetration Rate

There are four principle directions in increasing the allowable penetration rate, namely, (1) decrease the axial force and drilling torque by optimization of the drill geometry, (2) increase the torsional and buckling rigidities of the drill, (3) improve drill self-centering ability (its engagement or starting conditions), and (4) improve the shape of formed chips and their evacuation from the machining zone. Therefore, in the author's opinion, any HP drill design should be analyzed by its contribution in one (or more) to the listed directions. Such an approach significantly simplifies any analysis of a particular drill design and geometry as it provides clear understanding of an improvement made by any new design compared to the known. In this section, some important known designs are analyzed using the proposed *VPA-balanced* approach.

Decreasing the axial force can be achieved by

- Improving the tool rake geometry, which normally decreases the total drilling force and thus the axial force as its part
- Decreasing the length of the chisel edge
- Improving the geometry of the chisel edge
- Eliminating the chisel edge

Therefore, one needs to understand the geometry of the drill to apply these methods.

4.5.3 Straight-Flute and Twist Drills Particularities

Straight-flute drills find wide application in industry. The effectiveness of today's straight-flute drills changed traditional assumptions about the indispensability of spiral-flute drills known as twist drills. In some applications, a straight-flute drill is a better choice than a twist drill. Modern straight-flute drills were preceded by die drills: short, stiff tools with beefy tips for drilling hard steels. Run at a slow spindle speed and light feed rate, the die drill's strength and rigidity enabled it to make straight, round holes (Kennedy 2006).

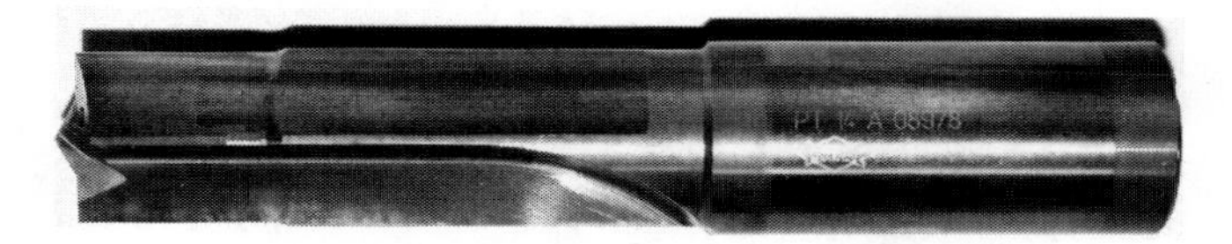

FIGURE 4.44 Modern straight-flute drill used in the automotive industry.

FIGURE 4.45 Twist drills.

Die drills worked well because the work materials they were engineered to machine typically produced short chips. Chip control is still the key issue in the application of straight-flute drills, which work best in materials that don't generate long, stringy chips. These include various grades of cast iron, powder metals, and medium- to high-silicon aluminum alloys widely used in the automotive industry. However, new machine tools with high rpm spindles and high-pressure coolant-delivery technologies, as well as enhanced tool geometries, have expanded the application range of straight-flute drills shown in Figure 4.44. These drills are cheaper than twist drills, their manufacturing and metrology are simpler, and multi-step constructions in one-pass tools for precession hole drilling are easy to implement. The use of straight-flute drills increased dramatically over recent years in the automotive and aerospace industries.

Twist drills (Figure 4.45) are the most widely used drills for general applications. When properly designed and made, the helical flute facilitates chip removal from the machining zone. The coupling of the axial force (thrust) and drilling torque enhances drilling stability and allowable penetration rate compared to straight-flute drills that are particularly important in drilling difficult-to-machine work materials. That is why twist drills dominate in such applications.

The twist drill bit was invented by Steven A. Morse who received US Patent No. 38,119 for his invention *Improvements of Drill-Bits* in 1863. The proposed original method of manufacture was to cut two grooves in opposite sides of a round bar, then to twist the bar to produce the helical flutes. This gave the tool its name, which stays although the proper name for this tool should be a drill with helical chip removal flutes or simply helical-flute drill. Nowadays, a flute is usually made by rotating the bar while moving it past a grinding wheel with its axis inclined at the helix angle to the axis of the bar and the profile of which corresponds to the flute profile in the normal cross section. For larger-diameter drills, other manufacturing methods of forming helical flutes in the drill body are also used.

Morse's aim in transforming a straight-flute drill into a twist drill was to enhance the chip transportation from the machining zone. The idea came from the Archimedes screw pump shown in Figure 4.46 developed around 250 BC (named after the Greek mathematician and inventor, Archimedes [c. 287–212 BC]) and used today not only for pumping water but also in screw conveyors for transporting solids, including the chip in manufacturing plants. The use of the Archimedes screw principle in drilling brought more advantages than Morse's initial thought, that is, the net result was much greater that he was bargained for.

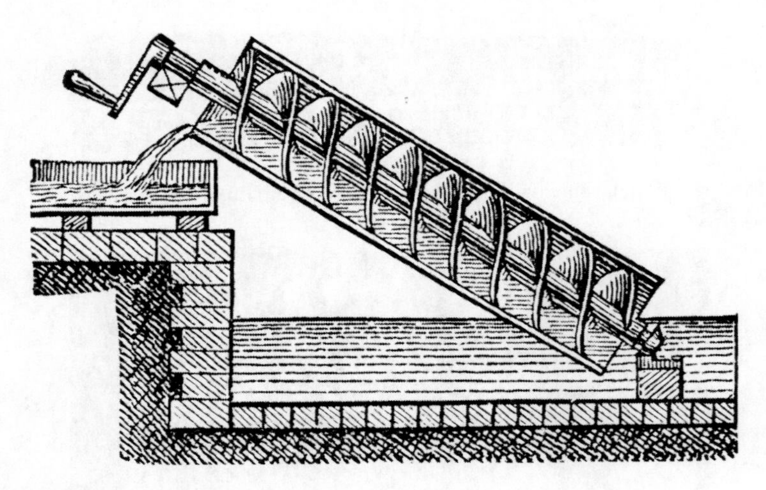

FIGURE 4.46 An Archimedes screw pump.

It is discussed later in this chapter that high periphery rake angles and chip shaping ability distinguish twist drills from straight-flute drills, allowing the machining of difficult-to-machine materials that produce the so-called long chips. Moreover, due to preferable chip shape, high feeds can be used. Reduction of the amount of plastic deformation due to high rake angle reduces the temperature at the drill periphery so that higher cutting speeds can be used for the same tool life. A higher penetration rate achieved with twist drills is due to the previously discussed torque–thrust coupling effect because its existence allows greater critical axial force and torques. One should understand, however, that all these advantages can be achieved if the following parameters are selected properly: (1) the web diameter, (2) flute shape, (3) flank angle distribution over the major cutting edge, (4) chisel edge geometry, and (5) surface roughness of the chip flutes and flank faces.

The major parameter affecting the previously discussed advantages of twist drills is the helix angle shown in Figure 4.14. In terms specific to twist drills, the helix angle can be found by unraveling the helix from the drill diameter, representing the section as a right triangle, and calculating the angle that is formed as shown in Figure 4.47. As can be seen

$$\omega_d = \arctan\frac{\pi d_{dr}}{p_{hl}} \tag{4.20}$$

where p_{hl} is the lead of the helix.

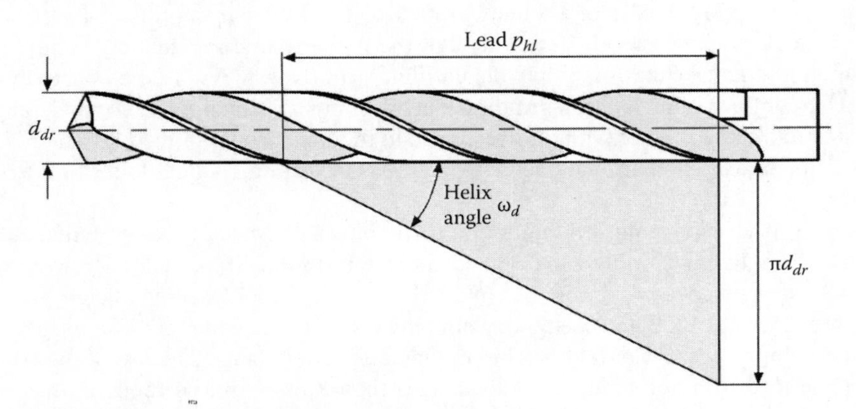

FIGURE 4.47 Unraveled helix corresponding drill diameter.

TABLE 4.3
Variation of ω_i with Radius r_i

	ω_i			
r_i/r_{dr}	$\omega_d = 15°$	$\omega_d = 30°$	$\omega_d = 45°$	$\omega_d = 60°$
1	15°	30°	45°	60°
0.8	12.1°	24.8°	38.7°	54.2°
0.6	9.1°	19.1°	31.0°	46.1°
0.4	6.1°	13.0°	21.9°	34.7°
0.2	3.1°	6.6°	11.3°	19.1°

Normally in tool drawings, the helix angle, ω_d corresponding to the drill outside diameter ($d_{dr} = 2r_{dr}$), is indicated. Knowing this angle, one can calculate the helix angle corresponding to any point i of the cutting edge located at radius r_i as

$$\omega_i = \arctan\left(\frac{r_i}{r_{dr}}\tan\omega_d\right) \tag{4.21}$$

It follows from this equation that the angle of helix reduces as a point of consideration is moved closer to the drill center as shown in Table 4.3.

The flutes of most twist drills have a standard helix angle (28°–32°), and this drill can be used to drill almost all work materials. For general-purpose work, twist drills having a nonstandard helix angle are not required. However, in cases where certain specific work materials must frequently be drilled, nonstandard helix angle drills may have significant advantages. Also called *fast-spiral* drills, high helix angle (34°–38°) drills have an excellent chip evacuation ability that is essential in drilling difficult-to-machine work materials and in deep-hole drilling. There are also slow helix angle (12°–22°) drills that are used to drill brass, soft bronze, and sheet materials. These materials are also drilled using straight-flute drills, having a zero-degree helix angle, because they will not tend to pull or run ahead of the feed and have less tendency to grab when opening up a hole. Reamers are normally provided with a slow helix angle.

Traditionally, besides helical flutes, two major differences in tool geometry (often referred as the point grind) between twist and straight-flute drills were the shape of the rake and flank surfaces. The rake face of a straight-flute drill is planar while that of a twist drill is helical. The flank face of a straight-flute drills is planar while for twist drills, hyperbolic, cylindrical, conical, and helical basic surfaces are used (Fujii 1970). However, these differences have been disappearing for the last 10 years with the rapid development of modern grinding fixtures and CNC grinders as well as introduction of better tool materials. Nowadays, the rake surface of twist drills is often modified (discussed later) so that the geometry particularities caused by the flute helical surface do not affect the rake geometry. The prime flank surface of even HSS twist drills is often made planar on the top of, for example, conical grind (Astakhov 2010), while many carbide twist drills are made with multifacet planar flank surfaces (Figure 4.48). Therefore, the further geometry analysis is equally applicable for straight-flute and twist drills as their geometrical features and particularities nowadays are easily interchangeable.

4.5.4 Systems of Consideration

Before we proceed any further, a potential reader is advised to comprehend the major concept of the systems of tool geometry considerations presented in Appendix C.

FIGURE 4.48 Planar flank face ground on multifacet carbide twist drills.

Three basic systems of consideration are used in the analysis and understanding of geometry of any drilling tool:

1. Tool-in-hand system (T-hand-S)
2. Tool-in-machine system (T-mach-S)
3. Tool-in-use system (T-use-S)

One can argue, however, that the proposed consideration may lead to overcomplicating of this seemingly simple subject. In the author's opinion, the opposite is actually the case as the proper drilling tool geometry cannot be considered without the mentioned systems of consideration. The necessity, significance, and physical relevance of using these three systems are explained in the following sections.

4.5.5 Drilling Tool Geometry in T-hand-S: Rake and Clearance Angles

Consideration of the basic system begins with T-hand-S. Figure 4.49 shows a cutting tooth of a drill having cutting edge *ab*. A point of consideration *i* is selected on this edge for the further analysis. The T-hand-S right-hand *xyz* current coordinate system is set as follows: its origin O is in point *i* selected on cutting edge *ab*, the *z*-axis is along the same direction as that of the assumed feed motion (vector $\mathbf{v}_f$ in), the *x*-axis is along the direction of the assumed direction of the cutting speed (the primary motion) (vector **v** in Figure 4.49), and *y*-axis is perpendicular to the *x* and *z* axes as shown in Figure 4.49.

In full analogy with standard definitions discussed in Appendix C, the basic planes in T-hand-S are shown in Figure 4.49. The tool reference plane P_r is defined as a plane through the selected point on the cutting edge and perpendicular to the assumed direction of primary motion; the assumed working plane P_f is defined as a plane through the selected point on the cutting edge, perpendicular to the reference plane P_r and containing the assumed direction of feed motion; the tool cutting edge plane P_s is perpendicular to P_r and contains the major cutting edge. If the major cutting edge is not straight, then the tool cutting edge plane should be determined for each point on the curved cutting edge thus being the plane that is tangent to the cutting edge at the point of consideration and that is perpendicular to the reference plane; the tool normal plane P_n is defined as a plane through the selected point on the cutting edge and normal to the cutting edge at the selected point on the cutting edge. The tool back plane, P_p, is defined as a plane through the selected point on the cutting edge and perpendicular both to the tool reference plane, P_r, and to the tool working plane, P_f.

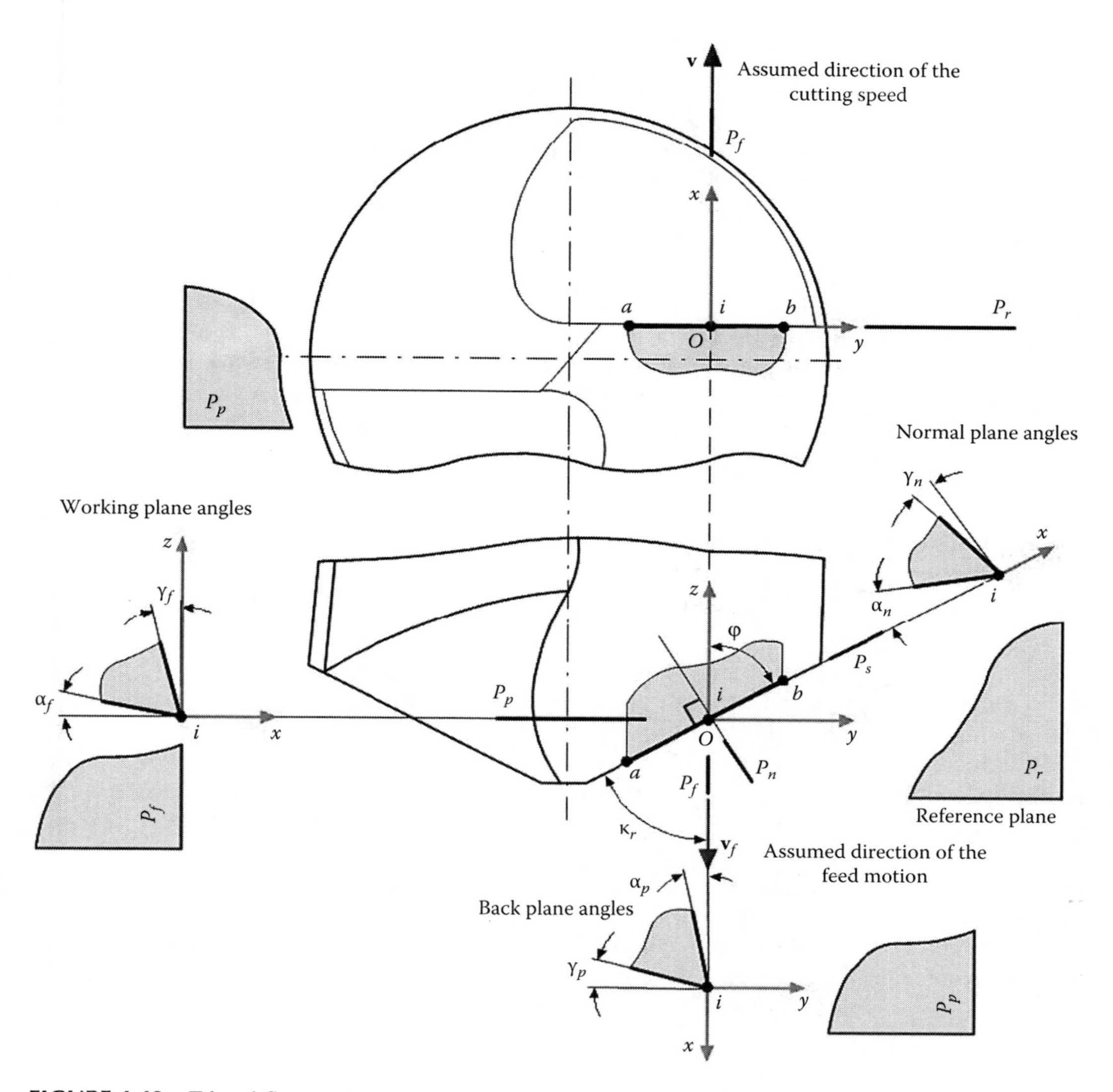

FIGURE 4.49 T-hand-S geometry.

The following three angles are of prime concern in T-hand-S:

1. The normal clearance angle is the angle between the tool cutting edge plane P_s and the tool flank plane as shown in Figure 4.49 measured in the normal plane P_n. It is clear that if the flank surface is not a plane, then the plane tangent to the curved flank surface at the considered point on the cutting edge is used instead of the flank plane.
2. The normal rake angle is the angle between rake face and the reference plane P_r measured in the normal plane P_n as shown in Figure 4.49. It is clear that if the rake face surface is not a plane, then the plane tangent to the curved rake face surface at the considered point on the cutting edge is used instead of the rake plane.
3. The half-point angle $\varphi = \Phi_p/2$ is the angle of the projection of the considered cutting edge into the reference plane P_r and the drill longitudinal axis. Although in many other tools, the tool cutting edge angle κ_r is considered (shown in Figure 4.49 and discussed in Appendix C) to determine edge orientation in the zx plane, the half-point φ and point Φ_p angles are found to be more convenient for drilling tool analyses. Obviously, $\varphi = \kappa_r$.

The great importance of the introduced angles is rather simple—these angles are indicated on the tool drawing and thus used in the manufacturing of the tool points, tool re-sharpening, and tool inspection. They, however, should not be used even to the first approximation to judge tool performance through application of the mechanics/physics of metal cutting.

The inspection of the normal rake and flank angles using modern tool geometry inspection machines is not always easy as many of them have some difficulties in measuring the normal rake and clearance angles directly. Instead, they are capable of measuring the so-called side (often referred as axial) and back (sometimes referred to as radial) angles. These angles shown in Figure 4.49 are in the working and back planes, respectively. To carry out a proper inspection, that is, to compare the results of inspection with those shown on the tool drawing, one needs to know the following simple relationships between the measured and the tool angles shown in the tool drawing. For example, if the rake γ_f and clearance α_f in the T-hand-S working plane are measured (the common case for many tool inspection machines), then the corresponding normal angles indicated on the drawing can be calculated as (Astakhov 2010)

$$\gamma_n = \arctan\left(\frac{\tan\gamma_f}{\sin\varphi}\right) \tag{4.22}$$

$$\alpha_n = \arctan(\tan\alpha_f \sin\varphi) \tag{4.23}$$

In reality, the normal rake and clearance angle are indicated in a drill drawing in the manner shown in Figure 4.49 only in some special cases.

Straight-flute drilling tools including drills are commonly ground with planar (flat) flank faces. As the flute is straight, the T-hand-S rake angle is assumed to be zero, so it is not indicated in the drawing. In this case, only normal primary (and secondary and tertiary, if applicable) clearance angle is indicated on the tool drawing. However, many straight-flute drills have the so-called gashes used for web thinning (discussed later). As such, a part of the rake face may have nonzero T-hand-S rake angle. Figure 4.50 shows a fragment of a tool manufacturing drawing by one of the largest drill manufacturers where the discussed parameters of the tool geometry in T-hand-S are indicated. It has to be pointed out, however, that

1. Nowadays the parameters of the drill geometry as those shown in Figure 4.50 rarely appear even on the tool manufacturing drawing as the tool designers/manufacturers rely on those set by the built-in programs of CNC drill grinders
2. These parameters rarely appear on customer drawings simply because many customers do not request this information completely relying on the tool manufacturer's expertise.

This makes an analysis and thus optimization of the tool geometry using the tool drawing almost impossible. Moreover, significant variations of the tool geometry parameters from one manufacturing lot/supply to the next are often observed when one tries to understand the root cause of the scatter in drilling tool performance.

Generally, a much more complicated indication system of the tool geometry parameters is used for twist drills. Moreover, this system varies significantly from one tool manufacturer to the next. To understand such a system, let's consider basic common cases possible in the defining of the rake face geometry: (1) rake face is not modified, that is, *natural* as determined by the helical-flute shape, (2) rake face is partially modified by gashes (the most common case), and (3) the rake face is fully modified.

When the rake face is not modified, this face is a part of the helical chip flute made on the tool. In this case, the rake angle is not indicated in the tool drawing. However, its determination is important for the analysis of the tool geometry. The senses of the T-hand-S rake and flank angles

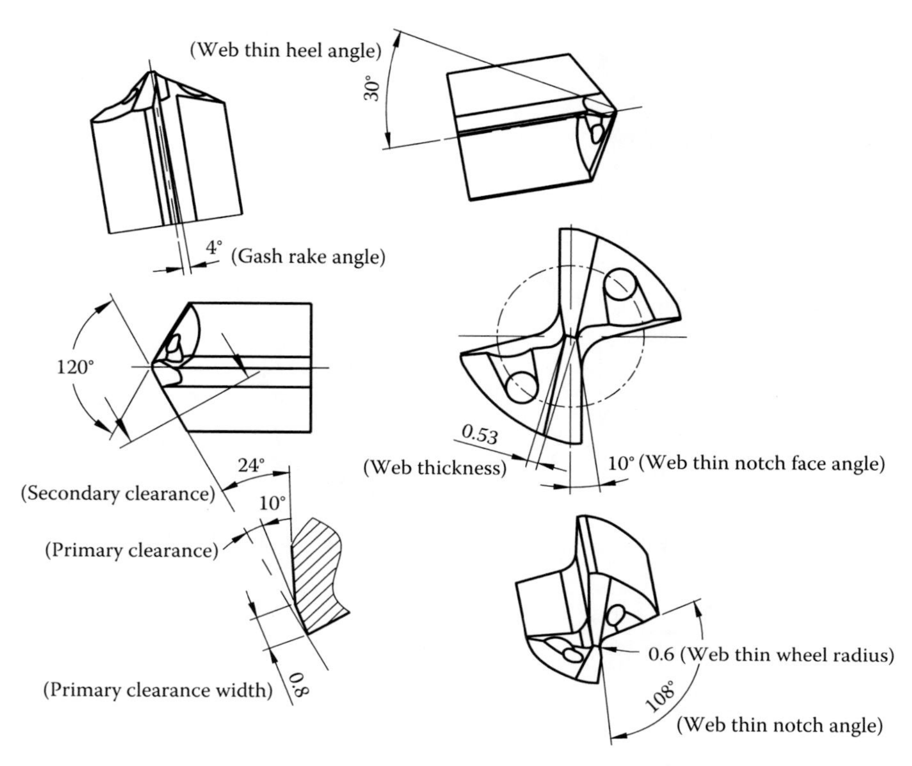

FIGURE 4.50 Representation of the tool geometry of a straight-flute drill with web thinning.

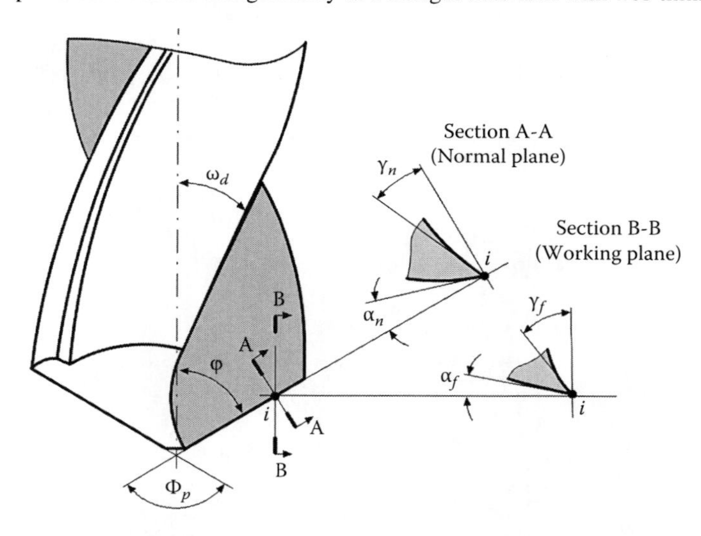

FIGURE 4.51 Rake and flank angles in the working and normal plane in T-hand-S for a twist drill.

in the working and normal reference planes are shown in Figure 4.51. As was shown by the author earlier (Astakhov 2010), the T-hand-S rake angle for a considered point i of the major cutting edge is calculated as follows:

In the working plane

$$\gamma_{f\text{-}i} = \arctan\left(\frac{r_i}{r_{dr}}\tan\omega_d\right) \quad (4.24)$$

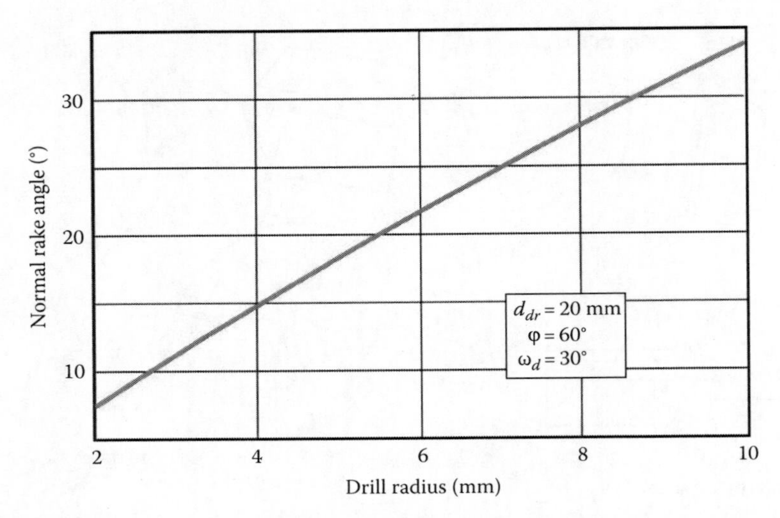

FIGURE 4.52 Variation of the T-hand-S normal rake angle over the cutting edge of a twist drill.

In the normal plane

$$\gamma_{n\text{-}i} = \arctan\left(\frac{r_i}{r_{dr}}\frac{\tan\omega_d}{\sin\varphi}\right) \tag{4.25}$$

Equation 4.25 shows that the T-hand-S normal rake angle varies over the cutting edge. Figure 4.52 shows an example. As can be seen, this angle decreases almost linearly from the drill corner to the inner end of the major cutting edge where this edge connects to the chisel edge.

The second basic case takes place when the rake face is modified by applying web thinning (gash). Nowadays, many drills have the so-called gashes used for web thinning. Figure 4.53 shows an example. As can be seen, a part of the rake face is helical having the *natural* rake angle calculated

FIGURE 4.53 A twist drill with web thinning by gashes.

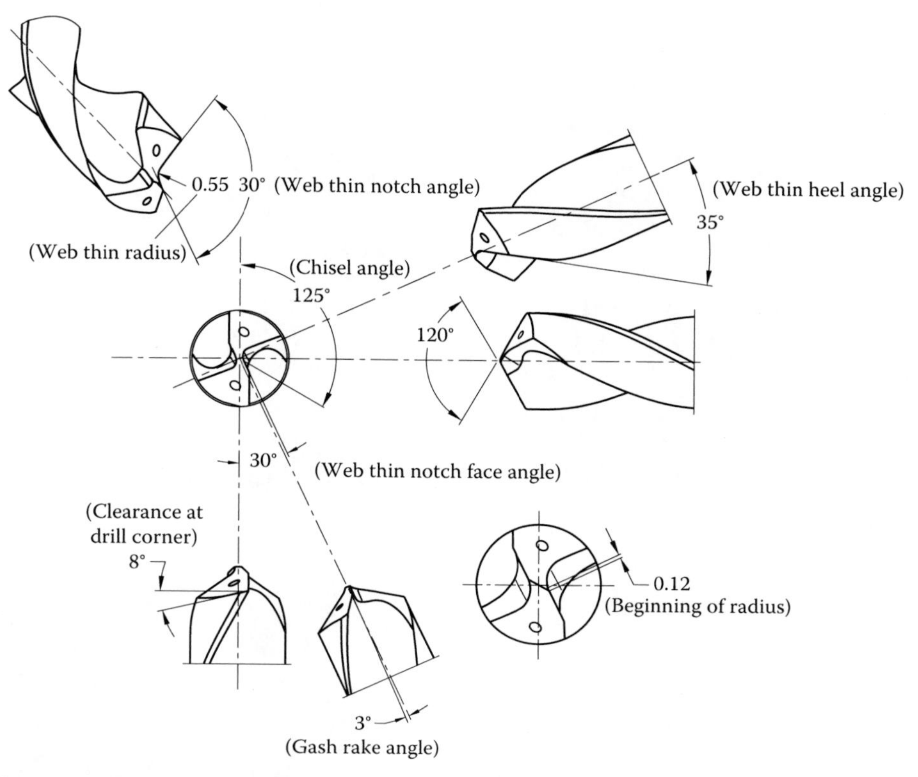

FIGURE 4.54 Representation of the tool geometry of the drill with web thinning shown in Figure 4.53.

using Equation 4.24, while the inner part is ground flat with a certain rake angle. Figure 4.54 shows a fragment of a tool manufacturing drawing by one of the largest drill manufacturers where the discussed parameters of the tool geometry in T-hand-S are indicated.

The same as was pointed out earlier for straight-flute drills, the parameters of the drill geometry as those shown in Figure 4.54 rarely appear even on the tool manufacturing drawing as the tool designers/manufacturers rely on those set by the built-in programs of CNC drill grinders. This makes an analysis and thus optimization of the tool geometry using the tool drawing almost impossible. Moreover, significant variations of the tool geometry parameters from one manufacturing lot/supply to the next are often observed when one tries to understand the root cause of the scatter in drilling tool performance.

The rake face of a twist drill can also be modified from the side of the drill corner to prevent drill chattering and/or grabbing at the machined hole exit. As discussed by the author earlier, one disadvantage of twist drills is that such drills have a tendency to *dig* or *grab* or run ahead of the feed particularly when breaking through the hole (Astakhov 2011b). It happens when nonrigid drilling system is used, for example, with hand drilling machines. It also happens when drilling very thin stock, owing to the tendency of the drill to *hook into* the work when breaking through a hole. In drilling soft work material, a chatter of the drill can also occur. The distinctive chatter marks on the hole bottom similar to that shown in Figure 4.55 often occur. The reason for that is explained in Section C.3.2.

To prevent this from happening, the high rake angle in the vicinity of the drill corners is reduced by grinding a flat rake face. An example of such a modified twist drill is shown in Figure 4.56. The T-hand-S rake angle in this case is application-specific and thus should be determined experimentally for a given application by maintaining the balance between the occurrence of chatter and increased cutting force factors. Commonly, however, the rake face in such a case is ground with a zero T-hand-S rake angle. The drill ground in this manner is also very effective for drilling unannealed steel or *hard spots* in cast iron.

FIGURE 4.55 Example of chatter marks on the bottom of the hole being drilled.

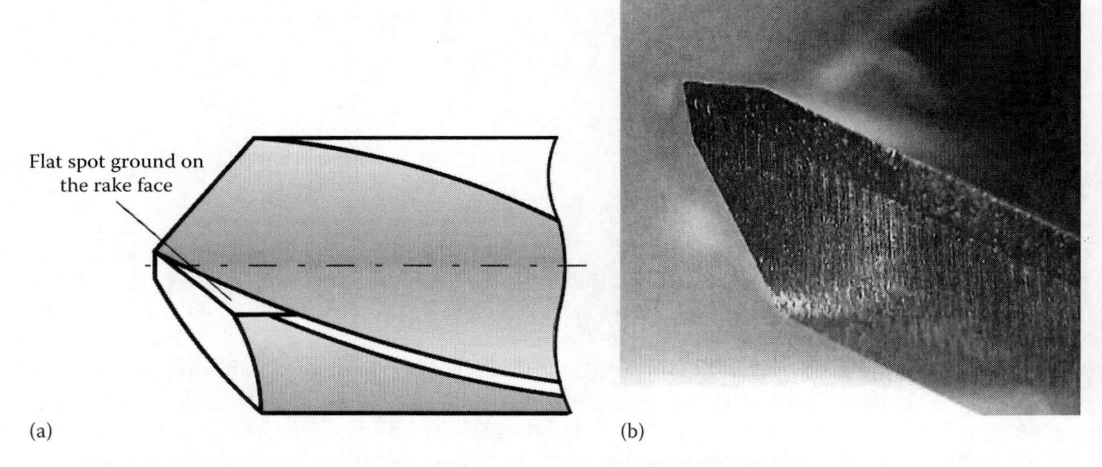

FIGURE 4.56 Modification to the rake face of a twist drill: (a) the model and (b) a real drill.

The third basic case when the rake face is fully ground to modify its *natural* geometry is not common, that is, used only for special applications. Although a number of various designs are known (Astakhov 2010), the plane rake faced (PRF) twist drill point geometry design developed by Armerego and Cheng (Armarego 1972a and b) seems to have a number of proven advantages. This design is shown in Figure 4.57. It has been shown that a positive normal rake angle over entire length of the major cutting edge is achieved. Moreover, the rake face modifications partially splits the chisel edge. The cutting mechanics analysis and experimental studies have confirmed the superiority of this drill point design over the conventional twist drills with significant reduction in the axial force (thrust) and drilling torque and an increase in tool life for both aluminum alloys and difficult-to-machine materials (Armarego 1972b; Wang and Zhang 2008).

A number of modifications to this design have been developed over the last 20 years. An analysis of such modification has been presented by Wang and Zhang (2008). One such modification to the general-purpose drills was attempted by grinding a PRF on the rake face of each lip using a narrow disk-shaped grinding wheel (Armarego 1972b). The essence of the geometry of the PRF drill point design is shown in Figure 4.58. As can be seen, to achieve such geometry, a conventional twist drill is modified by grinding PRF on the rake face of each lip using a disk-shaped grinding wheel.

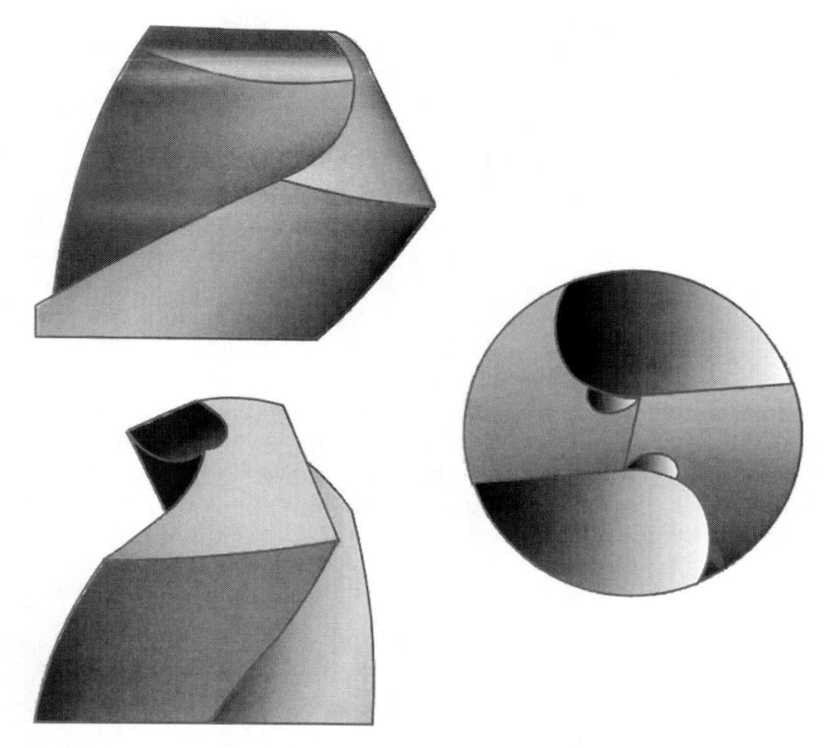

FIGURE 4.57 PRF twist drill point design.

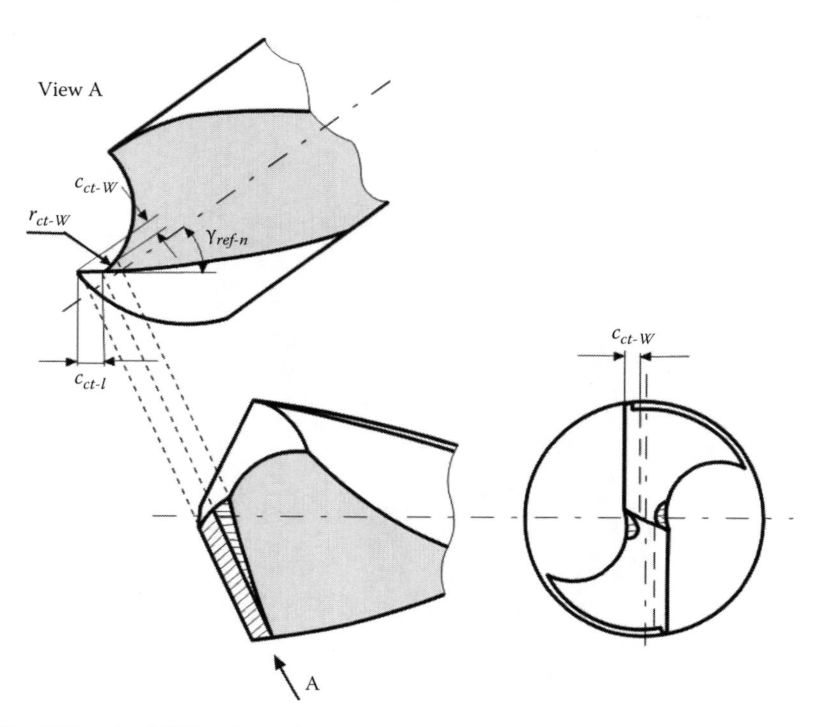

FIGURE 4.58 PRF twist drill developed by Wang and Zhang (2008).

According to the proposed geometry, a constant reference rake angle $\gamma_{ref\text{-}n}$ (the T-hand-S normal rake angle) is applied along each lip without changing the orientation of the major cutting edges. According to Wang and Zhang (2008), to achieve best performance, three geometry parameters of PRF drills should be selected as follows:

- Depth $c_{ct\text{-}w}$ should be equal to 20% of the web thickness.
- Size $c_{ct\text{-}l}$ for drill diameters of 5–30 mm is determined as $c_{ct\text{-}l} = 0.04267d_{dr} + 0.00364f + 0.00970$ (mm) where d_{dr} is the drill diameter (mm) and f is the cutting feed (mm/rev).
- Corner radius $r_{ct\text{-}w}$ is to be determined experimentally for the best drill performance. Too large a corner radius may reduce the drill strength, while too small $r_{ct\text{-}w}$ may cause sudden change in chip flow direction and thus may trap the formed chip from flowing smoothly into the flutes. A corner radius around 1 mm is considered reasonable for 7–13 mm drills.
- The reference rake angle at each lip has been selected to be equal to the reference rake angle at the outer corner of each lip and remains constant along each lip, such that the normal rake angle will remain positive along the whole lip although its value decreases as the radius decreases. Furthermore, in grinding PRF, the chisel edge is *point relieved* where each flank in the vicinity of the chisel edge corner is affected by the grinding wheel.

A few basic cases are possible in the defining of the flank face geometry in T-hand-S:

1. The primary flank is planar (flat). When the primary flank face is flat as shown in Figure 4.59, then regardless of a particular drill type, the primary clearance angle in T-hand-S is the same for all points of the major cutting edge. It is defined as in the model shown in Figure 4.49. Figure 4.50 shows an example of proper definition of the primary clearance angle in the tool drawing.
2. The primary flank is not planar. Commonly, quadratic and helical surfaces are used as the tool flank surface. In this case, the T-hand-S clearance angle is indicated in the tool drawing at the drill corner as shown in Figure 4.54. According to the author's many-year experience in the field, the type of flank surface and its parameters are never indicated in the tool drawing. This presents a great challenge in analyzing and optimizing drill geometry.

FIGURE 4.59 Twist drill with flat flank faces.

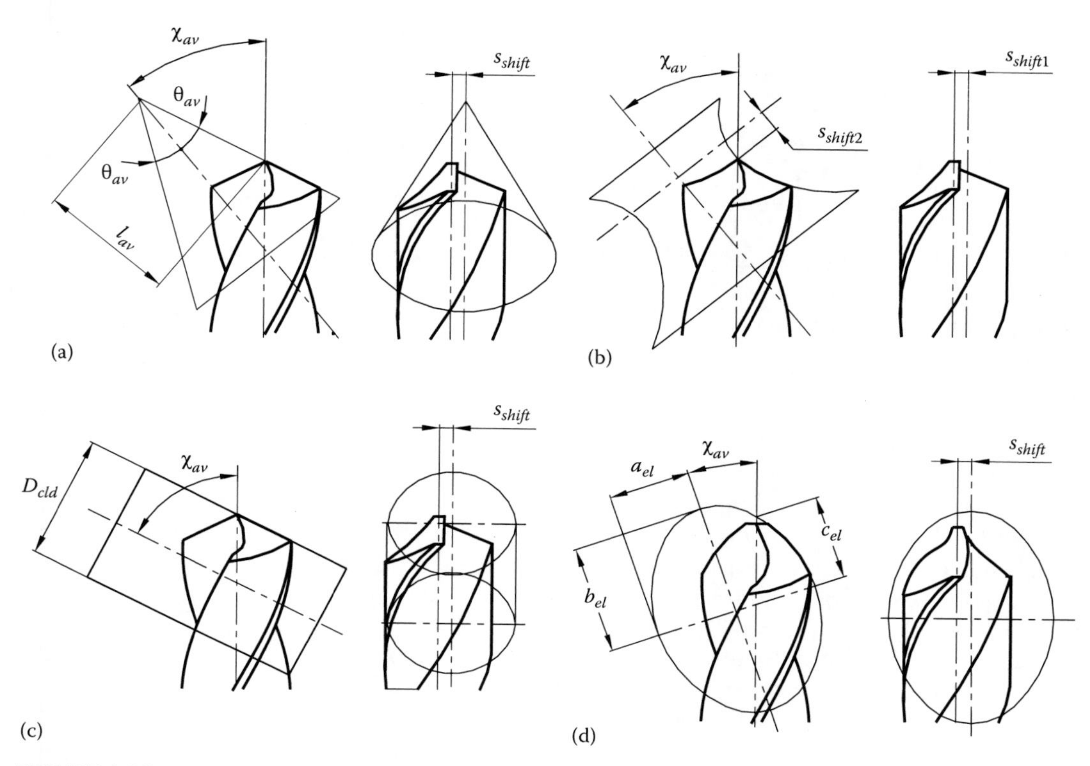

FIGURE 4.60 Simplified models for quadratic drill points: (a) conical, (b) hyperboloidal, (c) cylindrical, and (d) ellipsoidal.

Examples of quadratic surfaces include the cone, cylinder, ellipsoid, elliptic cone, elliptic cylinder, elliptic hyperboloid, elliptic paraboloid, hyperbolic cylinder, hyperbolic paraboloid, paraboloid, sphere, and spheroid. The geometry of these surfaces is well known (Beyer 1987; Mollin 1995) and can be utilized to achieve preferable cutting geometry of drills. Among these surfaces, the following have been used in the practice of drill design and manufacturing: (1) conical, (2) hyperboloidal, (3) cylindrical, and (4) ellipsoidal. Figure 4.60 also shows the relevant geometrical parameters that can be used to determine the grinding geometry knowing the desired geometry of the drill to be ground.

Although the geometric parameters of drills with quadratic-surface flank are well known, the most relevant studies of such drills were carried out a long time ego (Fujii 1970; Tsai and Wu 1979; Wright and Armarego 1983; Fugelso 1990). Two major drawbacks of these researches have to be pointed out: (1) no clear quantitative objective of the development/optimization of the tool geometry was specified and (2) insufficient experimental support of the theoretical, purely geometrical findings. Moreover, unavailability of grinding machines/fixtures to reproduce many of these complicated geometries with required accuracy at reasonable grinding time made these findings purely academical. As a result, only a few of them have passed reality check and are still sometimes used in industry.

The first one is known as the Racon® (US Trademark No. 73179521 Giddings & Lewis, Inc.; AMCA International Corporation; Registered on April 3, 1979, renewed on April 3, 1999) point grind. Its generic appearance is shown in Figure 4.61, and its basic idea is based on the point grind developed according to US Patent No. 1,309,706 (1917) (Astakhov 2010). It provides a continuously varying point angle, with the major cutting edges (lips) and margins blending together to form a smooth curve. Because the cutting edges are curved, there is less energy spent per unit length of the cutting edge and therefore less heat generated during drilling. Similar to the double-angle point grind (discussed later in this chapter), the outer periphery of the cutting lip is protected to reduce margin wear. Breakthrough burrs can be eliminated and tool life can be increased when drilling

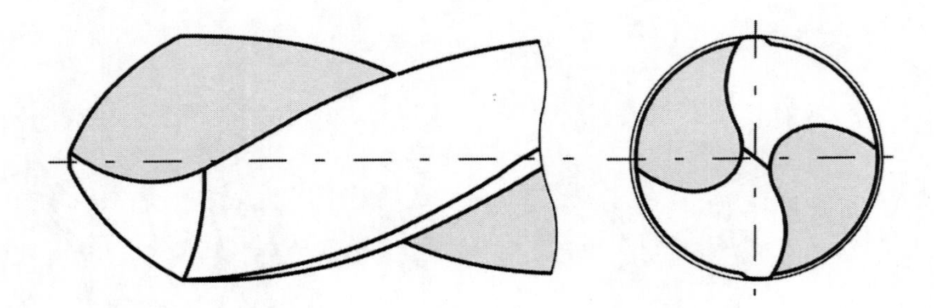

FIGURE 4.61 Racon® point grind.

abrasive work materials. The major limitation of this design is that it must be used through a guide bushing because it is not self-centering. To address this issue, the Bickford™ point was developed as a combination of the helical and Racon® points. It combines the self-centering feature of the helical point grind with the long life and burr-free breakthrough and higher feed capacity of the Racon point grind. The theory behind this point grind was presented by the author earlier (Astakhov 2010). In the author's opinion, these two point grinds *survived* only because Winslow Engineering Inc. (Fond du Lac, WI) produces machines capable of automatic grinding of such points.

The most common quadratic-surface flank point grind is the conical grind. Its appearance on a ground drill is shown in Figure 4.62. The results of an analysis of the conical grind presented by the author earlier (Astakhov 2010) reveal that the distribution of the flank angle along the cutting edge in cone grinding depends on the grinding parameters (four parameters shown in Figure 4.60a) so that it is insufficient to indicate the T-hand-S flank angle at the periphery point $\alpha_{n\text{-}1}$ and simple *conical grind* to know this distribution. This explains great scatter in the performance of the so-called standard twist drill with conical point. This also explains the popularity of the planar drill point where the tool geometry normally indicated in the drill drawing assures the same distribution of the flank angles along the major cutting edge (lip) and the chisel edge from one drill manufacture to another. Unfortunately, this issue is not well understood in the industry.

In helical point grind, the flank surface is formed as a part of helicoid coaxial to the drill longitudinal axis as shown in Figure 4.63. The resulting drill point is commonly referred to as a helical drill point. Figure 4.53 shows an example of a drill with the helical flank surface (grinders). When the diameter of the base cylinder is chosen properly, such a grind changes the flat blunt chisel to an *S* contour with a radiused crown effect that has its highest point at the center of the drill axis.

FIGURE 4.62 A drill having the conical point grind.

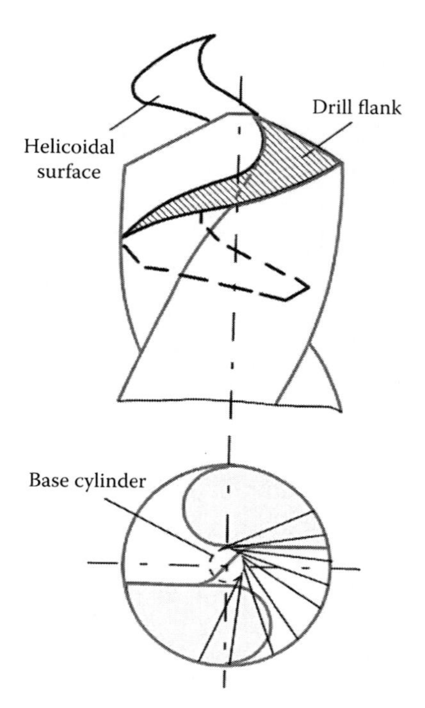

FIGURE 4.63 Helical flank surface.

This crown contour creates a continuous cutting edge from margin to margin across the web. The advantage is its self-centering ability that improves drill entrance stability.

The helical point grind is applied by many automatic and semiautomatic including desktop drill grinders. In T-hand-S, the flank angle is indicated in the tool drawing at the drill corner as shown in Figure 4.54. However, the actual distribution of the flank angles along the cutting edge depends on many setting parameters of a grinding machine and on the type of helical point grind (Hsieh 2005; Astakhov 2010) with no indication of these particularities on the tool drawing. This makes any analysis and optimization of the tool geometry extremely difficult.

4.5.6 Drilling Tool Geometry in T-mach-S and T-use-S: Clearance Angle

As mentioned earlier, the angles in T-hand-S are not *real*, that is, they cannot be used even to the first approximation for any tool performance analysis. Rather, any tool performance analysis should result in the determination of the optimal tool geometry in T-mach-S/T-use-S, and then the optimized angles should be *converted* into the corresponding angles in T-hand-S for tool manufacturing and inspection.

For simplification of understanding, T-mach-S and T-use-S are considered together as the contribution of the latter is just a *constant* to the tool angles in T-mach-S, while the angles in T-mach-S depend on the location of the considered point of the cutting edge with respect to the drill rotational axis. Figure 4.64 presents the basic idea of the kinematic clearance angle. In this figure, a point of consideration i is selected on the major cutting edge 1-2. The working plane (the same for T-mach-S and T-use-S systems) is drawn through the considered point and contains the vectors of the cutting speed, $\mathbf{v}$, and that of feed velocity, $\mathbf{v}_f$. The radius of the considered point is r_i. The drill rotates with n (rpm).

The right-hand T-mach-S $x_m y_m z_m$ is set according to the rules discussed in Section C.2.2, that is, its origin is in point i, the x_m-axis in the direction of the cutting speed $\mathbf{v}_i$, the z_m-axis in the direction of the feed velocity (along the longitudinal axis of the drill), and the y_m-axis is perpendicular to the

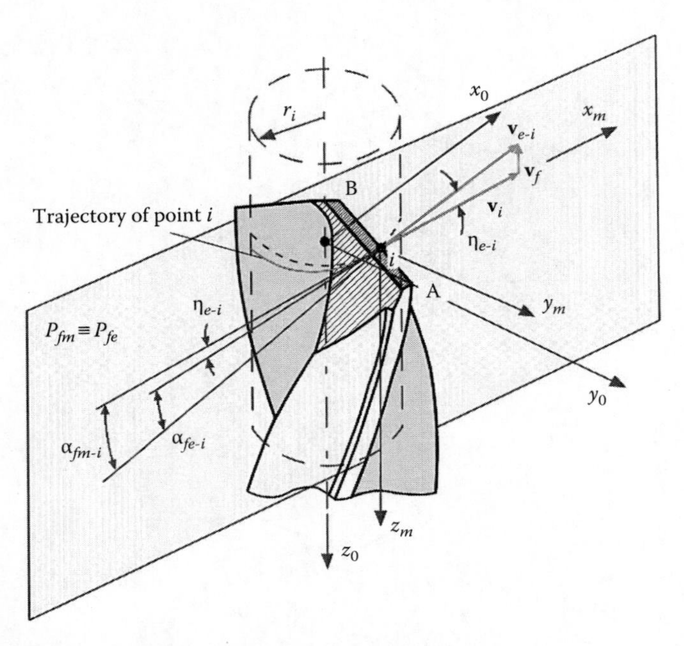

FIGURE 4.64 The basic idea of the kinematic clearance angle.

x_m and z_m axes as shown in Figure 4.64. As discussed in Section C.2.2, the T-mach-S coordinate system in general is not the same for different points of the cutting edge so it should be considered for each point of this edge individually.

As discussed in Chapter 1, as drill rotates with rotational speed n (rpm), the cutting speed for point i is calculated as $v_i = (\pi 2 r_i n)/1000$ and the feed velocity is calculated as $v_f = nf$. Note that the cutting speed depends on the radius of the point of consideration reducing with this radius, while the feed velocity is the same for any drill point. As discussed in Section C.2.3, the angle between the resultant cutting velocity $\mathbf{v}_e$ and that of cutting speed $\mathbf{v}$ for considered point i is the resultant cutting speed angle η_e. For a given point i of the cutting edge located at radius r_i, it is calculated as

$$\tan \eta_{e\text{-}i} = \frac{v_f}{v_i} = \frac{f}{\pi(2r_i)} \tag{4.26}$$

When drill is rotated and fed, point i generates a cylinder of radius r_i in its helical motion. Therefore, the most proper consideration of the clearance angle should be in the plane tangent to this cylinder in this point. The working reference plane $P_{fm} \equiv P_{fe}$ is the plane shown in Figure 4.64. Therefore, the working clearance angle in T-use-S is of prime importance in the tool design (assuring no interference of the flank face and the bottom of the hole being drilled) and cutting mechanics (as the sliding velocity is the highest in this direction) considerations and thus should be optimized. As follows from Figure 4.64 and the consideration presented in Appendix C, Section C.2.3, this angle in a given point i of the cutting edge is calculated as

$$\alpha_{fe\text{-}i} = \alpha_{fm\text{-}i} - \eta_{e\text{-}i} \tag{4.27}$$

where $\alpha_{fm\text{-}i}$ is the clearance angle in T-mach-S.

As discussed in Section C.2.3.2, in practical turning operations, angle η_e is very small and thus is taken into consideration only for some special cases where excessively high feeds are used, that is, thread cutting. Is it the same in considerations of the drilling tool geometry? To answer this question, let's consider a practical example using a carbide drill of 10 mm dia. The major cutting edge dimensions are as follows: radius of point A (Figure 4.64) is 5 mm and that of point B is 1.5 mm. The cutting speed is chosen to be 100 m/min and the feed per revolution is 0.4 mm/rev. Under these conditions, the drill rotational speed is $n = (1000v)/\pi d_{dr} = (1000 \cdot 100)/(3.141 \cdot 10) = 3183.6 \approx 3184$ rpm. The feed velocity is calculated (see Chapter 1) as $v_f = nf = 3184 \cdot 0.4 = 1273.6$ mm/min $\approx$ 1.27 m/min. The cutting speed for point A is equal to the drill cutting speed, that is, 100 m/min while that for point B is calculated as $v_B = (\pi d_B n)/1000 = (3.141 \cdot (2 \cdot 1.5) \cdot 3184)/1000 = 30$ m/min. Using Equation 4.26, one can calculate angle η_e for points A and B as $\eta_{e\text{-}A} = \arctan(1.27/100) = 0.73°$ and $\eta_{e\text{-}B} = \arctan(1.27/30) = 2.42°$.

The obtained results show that even in the most inner points of the major cutting edge and rather aggressive cutting feed, angle η_e is not that significant for the major cutting edge.

The first term in Equation 4.27 is the clearance angle in T-mach-S α_{fm} so its correlation with the corresponding clearance angle in T-hand-S should be established. To do this, a model of the T-mach-S geometry shown in Figure 4.65 is considered. As can be seen, the vector of the cutting speed at point i, $\mathbf{v}_i$ is perpendicular to the radius r_i of this point. Therefore, this actual direction of the cutting speed locates at angle μ_i to the direction of the assumed cutting speed in T-hand-S shown in Figure 4.49. The reference planes in T-mach-S are drawn through point of consideration i and defined as follows:

- *Tool reference plane* P_{rm} is perpendicular to the assumed direction of primary motion
- *Working plane* P_{fm} is perpendicular to the reference plane P_r and containing the vectors of the cutting speed and feed velocity
- *Tool back plane* P_{pm} is defined to be perpendicular to the drill rotational axis. This plane is perpendicular both to the tool reference plane, P_r and to the tool working plane, P_f.

Figure 4.65 shows the defined reference planes.

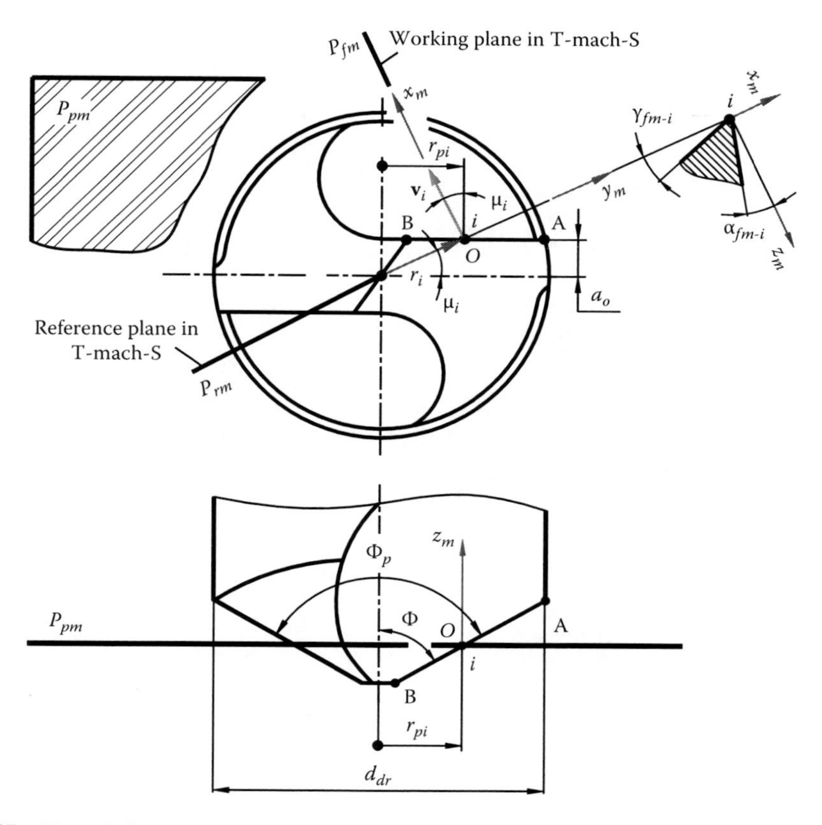

FIGURE 4.65 T-mach-S geometry.

The clearance angle α_{fm} is correlated with the normal clearance angle α_n in T-hand-S as follows:

$$\tan\alpha_{fm\text{-}i} = \frac{\cos\alpha_n \cos\mu_i + \cos\varphi \sin\mu_i}{\sin\varphi} \tag{4.28}$$

where the sense of angle μ_i is clearly seen in Figure 4.65 as the angle of the radius of point i.

Equation 4.28 can be rearranged for practical calculations as follows:

$$\alpha_{fm\text{-}i} = \frac{\arctan\left(\cos\alpha_n\sqrt{r_i^2 - a_o^2} + a_o\cos\varphi\right)}{r_i \sin\varphi} \tag{4.29}$$

Note also that $r_{pi} = \sqrt{r_i^2 - a_o^2}$.

As angle μ_i varies with location of the point of consideration i over the major cutting edge AB, clearance angle α_{fm} also changes correspondingly even if the normal clearance angle α_n in T-hand-S is kept constant as in the case with a planar flank face. As such, the calculations and thus optimization of α_{fm} do not present a problem. When the primary flank face of a drill is not planar, then the T-hand-S clearance angle α_n also changes with r_i so that Equation 4.29 should be modified correspondingly:

$$\alpha_{fm\text{-}i} = \frac{\arctan\left(\cos\alpha_{n\text{-}i}\sqrt{r_i^2 - a_o^2} + a_o\cos\varphi\right)}{r_i \sin\varphi} \tag{4.30}$$

where $\alpha_{n\text{-}i}$ is the T-hand-S clearance angle at point i of the cutting edge.

An analysis of Equation 4.29 and Figure 4.66 shows that the clearance angle α_{fm} increases with a_o (Figure 4.65) and decreases with the point angle $\Phi_p = 2\varphi$ particularly in the vicinity of the chisel edge. As can be seen, when a_o = 1.2 mm, when the point angle Φ_p increases from 120° (old drill standard)

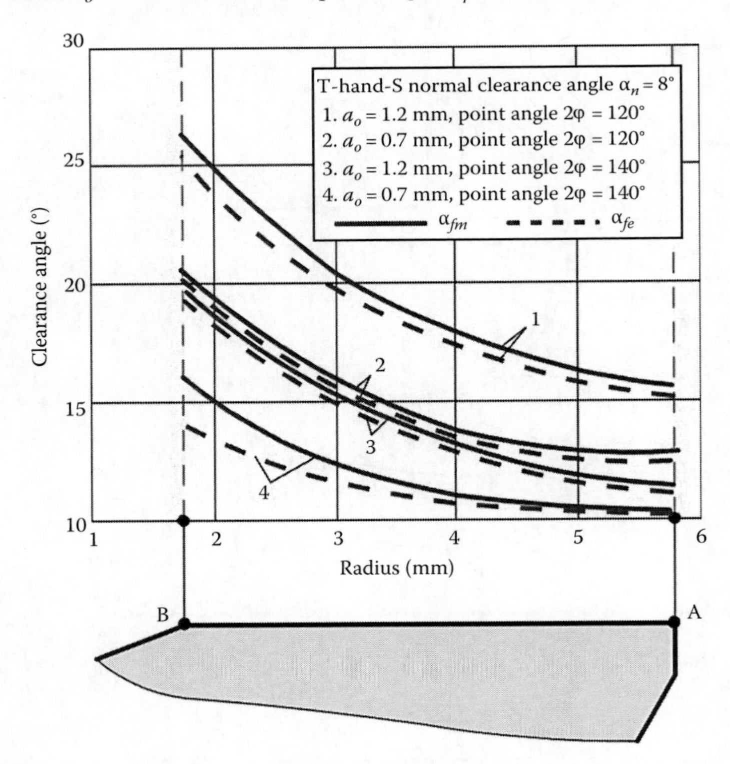

FIGURE 4.66 Variation of the clearance angles in T-mach-S α_{fm} and in T-use-S α_{fe} over the major cutting edge AB. A drill of 11.5 dia.

to 140° (recent standard for carbide drills), the clearance angle α_{fm} at point B decreases from 26.8° to 19.5°, that is, approximately by 1.4 times. The same variation has the clearance angle α_{fe}. This increases the strength of the cutting wedge in the vicinity of point B, which is essential in machining of high-strength steels. This also explains the change of the *old* standard point angle 120° into 140° for carbide drills by practically all major drill manufacturers.

Extremely important in the selection of the optimal clearance angle but never considered in the literature related to the cutting tool geometry is the conformity of the flank and machined surfaces. It is well known in tribology that this conformity is the prime factor affecting the contact process at the sliding interfaces and thus wear (Bhushan 2001).

To understand the idea of the conformity of two surfaces, let's consider two surfaces that make contact at point A. They are identified as follows: part surface P and generating (tool) surface T. Figure 4.67 shows sections of surfaces T and P. At the point of contact, these two surfaces have a common normal $\mathbf{n}_P$. The rate of conformity of surfaces P and T is a function of the radii of their normal curvature, that is, R_P and R_T (Radzevich 2010).

The case shown in Figure 4.67a corresponds to orthogonal cutting with the tool having non-planar flank face or to milling of a surface with a milling tool. In this case, $R_P \rightarrow \infty$. Figure 4.67b shows the case that corresponds to turning where both R_P and R_T are positive and the rate of conformity of surfaces P and T is much smaller than that in Figure 4.67a. The case shown in Figure 4.67c corresponds to hole-making operation, that is, drilling and boring. In this case, R_P is negative so surfaces P and T have the highest rate of conformity. Intuitively, one can realize that a higher rate of conformity leads to more close contact between the surface being machined P and the tool flank surface T and thus to a higher wear rate of the tool flank. Besides the contact condition, a higher rate of conformity results in a very narrow gap between the tool flank face and the surface being machined that make it difficult for MWF (coolant) to penetrate to regions located close to the tool–workpiece contact surface.

Figure 4.68 shows machined surface conformity in turning and boring. As can be seen, the machined surface in turning is convex with respect to point A, that is, it *escapes* from the tool

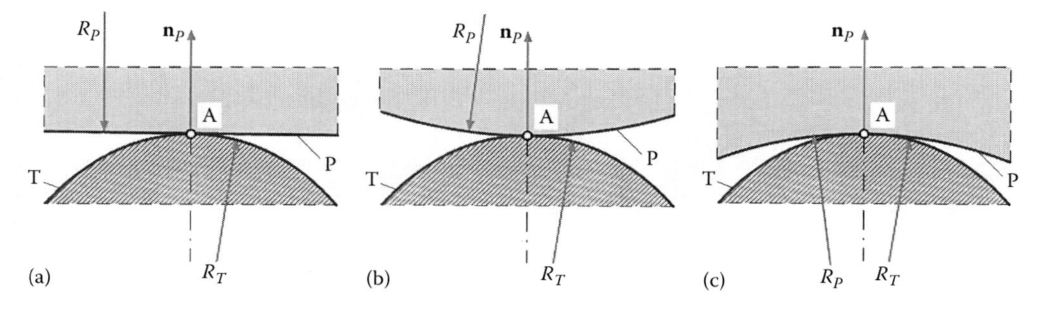

FIGURE 4.67 Normal sections of surfaces T and P: (a) "normal" rate of conformity as in orthogonal cutting, (b) smallest rate of conformity as in turning, and (c) high rate of conformity as in drilling.

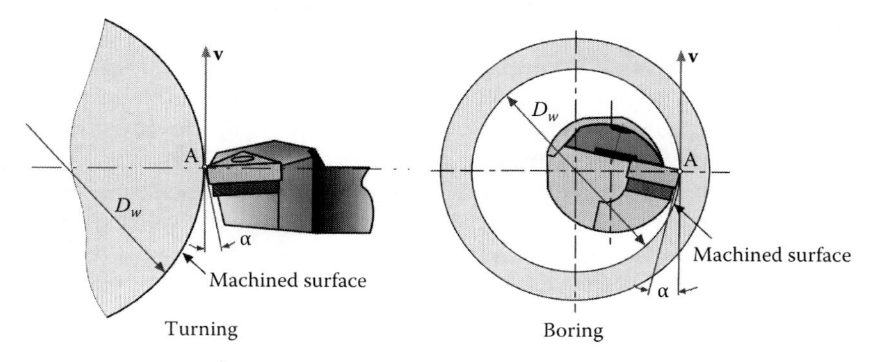

FIGURE 4.68 Machined surface conformity in turning and boring.

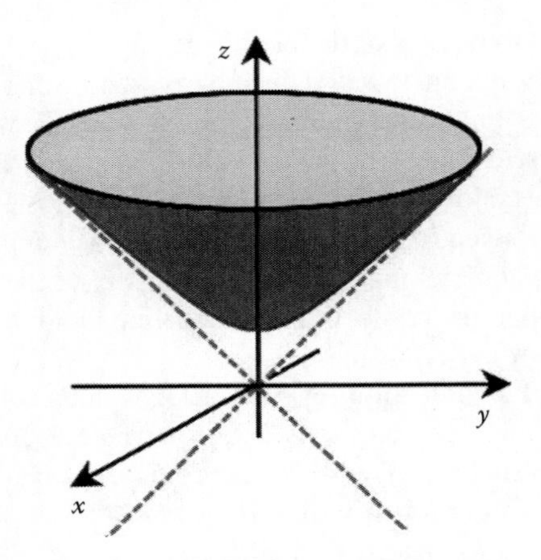

FIGURE 4.69 Surface being cut by a common drill is a hyperboloid.

flank face after the point of contact A. In hypothetical case where the work material springback (considered later in Section 4.5.9.5) is zero, machining is possible no matter how small the clearance angle is. The opposite picture is in boring where the machined surface is convex, that is, the machined surface moves toward the flank face. In a hypothetical case where the work material springback is zero, machining is only possible when the clearance angle is of some value to prevent interference of the flank face with the machined surface.

In drilling, however, the surface being machined is primarily not a cone surface as commonly depicted in the literature. As the major cutting edges (lips) locate ahead the centerline by distance a_o (see, e.g., Figure 4.65), the surface being cut is a hyperboloid as shown in Figure 4.69. Its strict equation is as follows:

$$-\frac{x^2}{a^2}-\frac{y^2}{b^2}+\frac{z^2}{c^2}=1 \tag{4.31}$$

As a hyperboloid of revolution is the case in drilling as drill rotates (also called a circular hyperboloid), then $a = b = a_o$ and $c = a_o/\tan\varphi$.

In practical calculations of the geometry of the major cutting edges and on the part drawing, such a hyperboloid is approximated by a cone as shown in Figure 4.70a. Let's consider some basic terminology. A *section* is the surface or outline of that surface formed by cutting a solid figure with a plane. If the solid figure is a right circular cone, the resulting curve is called a *conic section*. The diagram shown in Figure 4.70b shows such a cone. Four planes are shown, cutting through the cone at various angles, producing the curves shown in the following diagram. The intersection of each plane with the cone forms a conic section. The kind and shape of the conical section is determined by the angle of intersection of the plane with the axis and surface of the cone.

Angled views of a cone are produced with conic sections at different angles as shown in Figure 4.50b. Cutting at right angles to the axis produces a circle. Cutting at less than a right angle to the axis but more than the angle made by the side of the cone produces an ellipse. Cutting parallel to the side of the cone produces a parabola. Cutting more nearly parallel to the axis than to the side produces a hyperbola (the hyperbola in the diagram represents a cut parallel to the axis of the cone). Therefore, when considering the parameters of the tool geometry in the standard section planes in T-mach-S as shown in Figure 4.70b, the shape of the surface being machined differs considerably depending which one of these section planes is considered as shown in Figure 4.70c.

Considering Figure 4.70c, one can conclude that the best results in terms of balancing condition on the rake and flank surfaces are in the radial section where the surface being cut is a circle in this section as commonly depicted in the literature. In the normal section, the surface being cut is parabolic. As the aperture (the angle between two *generatrix* lines) for an *old standard* drill is 120° and

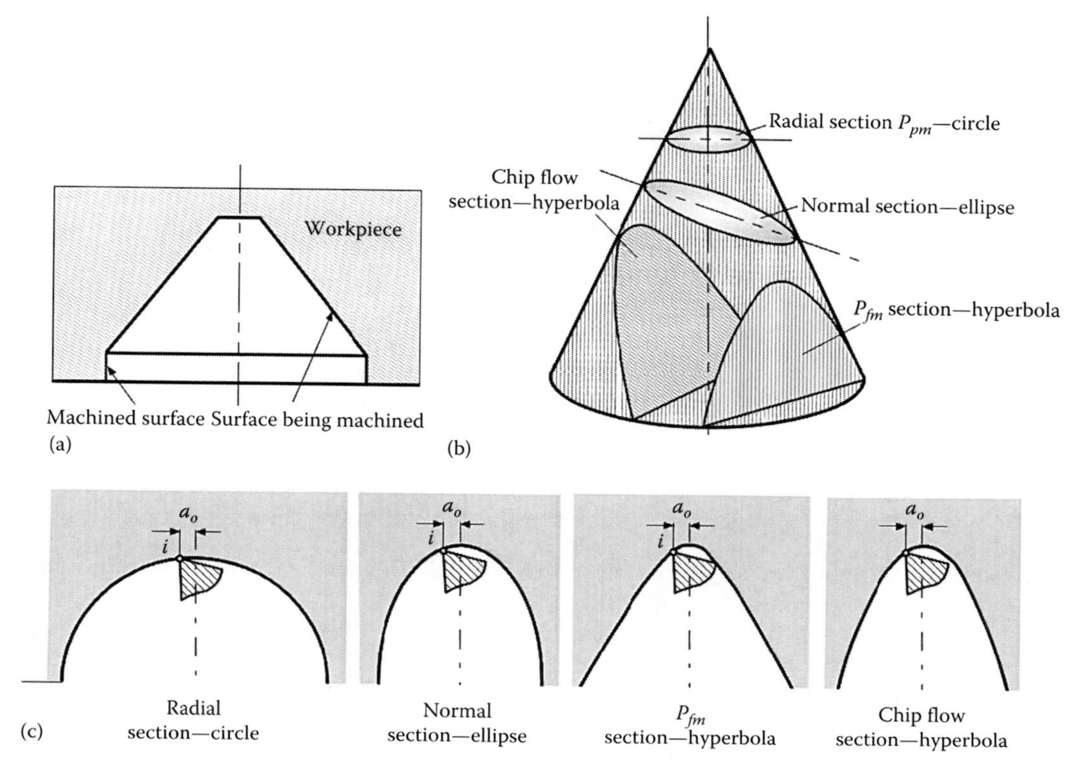

FIGURE 4.70 Conformity of the tool flank face and the surface being machined (the bottom of the hole being drilled): (a) approximation by a cone, (b) cone sections, and (c) appearance of the surface being cut in the standard cross-sectional planes.

new standard drill is 140°, then the angle between the radial and the normal section planes is 60° and 70°, respectively, that is, the ellipse in the normal section plane differs significantly from a circle.

In the P_{fm} section where the side (working) clearance angle is considered, the surface being cut is hyperbolic. Although the conditions on the flank face are improved in the immediate vicinity of point i as the surface of the cut curves away from this point, it does not last for long as this surface curves back with a small radius of curvature after a short distance. As can be seen, the greater the distance a_o (by which the major cutting edge is located ahead of the centerline), the better the conditions on the flank face and the lesser the chance of interference of the flank surface and the surface being cut. A wide class of drill design with an increased a_o became known as the crankshaft drill (point grind). An example of such a drill is shown in Figure 4.53. Much better conditions for MWF penetration on the vicinity of the flank face adjacent to the drill periphery corners, greater clearance angles over the cutting edge, and greater web diameters (increased allowable penetration rake) are major distinguishing features of such drills. Naturally, the resulting great length of the chisel edge requires its proper splitting, and the formed chip should be *smoothly* directed into the chip flute by properly grinding the transitional surfaces (known as gashes), which should be polished.

The foregoing analysis results in two important conclusions:

1. The smaller the drill diameter, the higher the clearance angle should be if the other conditions of a drilling operation are kept the same. This is because the conformity of the flank face and the surface being cut (the bottom of the hole being drilled) increase with decreasing drill diameter.
2. The greater the springback of the work material (the smaller its modulus of elasticity and the ultimate yield strength as considered later in Section 4.6.9.5), the greater should be the clearance angle.

TABLE 4.4
Recommended Flank Angles for Drills (for Twist Drill: The Periphery Clearance Angle)

	Suggested Flank Angle (°)		
Drill Diameter Range (mm)	**General-Purpose Drills**	**Drills for Tough and Hard Materials**	**Drills for Soft and Free Machining Materials**
0.35–1.00	24	20	26
1.05–2.50	21	18	24
2.55–3.00	18	16	22
3.05–6.50	16	14	20
6.55–8.95	14	12	18
9.00–13.00	12	10	16
13.10–20	10	8	14
>20	8	7	12

The recommended clearance angles are given in Table 4.4. These should be considered as good starting values in the optimization of the drill geometry parameters.

Unfortunately, many drill manufacturers and users do not follow these conclusions/recommendations, so that results in lower tool life, inferior quality of drilled holes, and drill breakage. According to the author's experience, optimization (making them close to the values shown in Table 4.4) of the clearance angles of HP drills, that is, meant for high-speed, high-penetration drilling of high-silicon automotive aluminum in the setting of the largest transmission plant, resulted in a two-time increase in tool life, three-time reduction of the drill breakage and their premature failures, and a 30% increase in the allowable penetration rate. Such an optimization first met great resistance from the largest drill manufacturing companies. When they were forced to manufacture such drills, the Oskar worth performance (the best dramatic acting)—"No it is impossible because it is impossible ever"—of their technical/sales personnel became a daily routine. The test and then implementation results, however, were rather stunning so that the optimized drills became standard for the application.

4.5.7 Drilling Tool Geometry in T-mach-S and T-use-S: Rake Angle

As discussed in Appendix C, the rake angle in T-use-S is calculated as (Astakhov 2010)

$$\gamma_{fe} = \gamma_{fm} + \eta_e \tag{4.32}$$

where angle η_e is calculated using Equation 4.26.

The first term in Equation 4.32 is the rake angle in T-mach-S γ_m so its correlation with the corresponding rake angle in T-hand-S should be established. To do this, the model of the T-mach-S geometry shown in Figure 4.65 is considered in the same way as in the previous section. The planes of the measurement, however, are not as well defined as those for the clearance angle. This creates a lot of controversial rules in the drill design. Unfortunately, these were never discussed in the literature, which adds even more confusion to this already complicated issue.

There are two principal approaches to the determination of the rake angle in T-mach-S γ_m. According to the first approach (Oxford 1955; Galloway 1957; Vinogradov 1985), this angle should be determined in the plane of chip flow as considered in Section C.3.5. According to this approach, it is termed the rake angle in the chip flow direction γ_{cf}. According to the second approach, γ_m should be determined in the normal plane P_{nm} (P_{ne}) (Fujii 1970; Astakhov 2010; Takashi and Jürgen 2010), that is, the rake angle γ_{nm} is considered.

4.5.7.1 Rake Angle in T-mach-S/T-use-S γ_{cf} Determination according to the First Approach

The first approach determines γ_{cf} as actually the rake angle not only in T-mach-S but rather in T-use-S as the chip appears only when the drill actually works. Using the approach discussed in Section C.3.5, one can determine the chip flow angle $\eta_{cf\text{-}i}$ (the angle between the *theoretical* chip flow direction and the normal to the major cutting edge) for a point of consideration i on the major cutting edge as

$$\sin\eta_{cf\text{-}i} = \sin\eta_{e\text{-}i}\cos\varphi_i + \cos\eta_{e\text{-}i}\sin\varphi_i\sin\mu_i \tag{4.33}$$

Then γ_{cf} is determined as (Vinogradov 1985)

$$\gamma_{cf\text{-}i} = \arcsin\left(\cos\lambda_{se\text{-}i}\sqrt{1-\cos^2\gamma_n\cos^2\eta_{cf\text{-}i}} - \sin\eta_{cf\text{-}i} + \sin\eta_{cf\text{-}i}\sin\lambda_{se\text{-}i}\right) \tag{4.34}$$

where $\lambda_{se\text{-}i}$ is the kinematic inclination angle at a point of consideration i on the major cutting edge. This angle is calculated as

$$\lambda_{se\text{-}i} = \arctan\left(\frac{\tan\lambda_{sm\text{-}i}}{\cos\eta_e}\right) \tag{4.35}$$

where $\lambda_{sm\text{-}i}$ is the T-mach-S inclination angle for a point of consideration i on the major cutting edge. The sense of this angle is shown in Figure 4.71. It follows from this figure that $\lambda_{sm\text{-}i}$ can be calculated through angle μ_i (Figure 4.65) as

$$\tan\lambda_{sm\text{-}i} = \tan\mu_i\sin\varphi \tag{4.36}$$

Figure 4.71 also shows an example of distribution of $\lambda_{sm\text{-}i}$ (and thus $\lambda_{se\text{-}i}$ as angle η_e [Equation 4.26] is small) for a drill with the web diameter equal to $0.2d_{dr} = a_o/2$ and half-point angle $\varphi = 58°$. As can be seen, the smallest $\lambda_{sm\text{-}A} = 9.97°$ is at the outer end of the major cutting edge (point A), while for the inner end B, this angle is $\lambda_{sm\text{-}B} = 50.50°$, that is, the greatest (Granovsky and Granovsky 1985).

According to an extensive experimental study carried out by Galloway (1957), the Stabler rule (Stabler 1964) is applicable for the drill geometry, that is, the chip flow angle $\eta_{cf\text{-}i} \approx \lambda_{sm\text{-}i}$. As such, Equation 4.34 can be significantly simplified to

$$\gamma_{cf\text{-}i} = \arcsin\left(\cos^2\lambda_{se\text{-}i}\sin\gamma_n + \sin^2\lambda_{se\text{-}i}\right) \tag{4.37}$$

Equation 4.37 establishes the correlation between the T-use-S γ_{cf} and T-hand-S γ_n rake angles through the drill design and geometry parameters. Figure 4.72 shows variation of γ_{cf} over the major cutting edge for a drill of 18 mm dia. with $a_o = 1.5$ mm and point angle $2\varphi = 120°$ for $\gamma_n = 0°$ and $\gamma_n = 10°$ (curves 1 and 2, respectively). As can be seen, the T-use-S γ_{cf} is greater than γ_n over the entire length of the major cutting edge. Moreover, γ_{cf} increases significantly for the region of this edge in the vicinity of its inner end adjacent to the chisel edge.

Equation 4.37 establishes direct correlation of γ_{cf} with the T-hand-S rake angle γ_n, point angle 2φ, and drill design parameter a_o. It can be assumed for practical calculation that $\lambda_{se\text{-}i} \approx \lambda_{sm\text{-}i}$ so Equation 4.37 can be rearranged for convenient use in the drill design as

$$\gamma_{cf\text{-}i} = \arcsin\frac{\sin\gamma_n\left(r_i^2 - a_o^2\right) + a_o^2\sin^2\varphi}{r_i^2 - a_o^2\cos^2\varphi} \tag{4.38}$$

The analysis of Equation 4.38 shows that γ_{cf} primarily depends on γ_n, a_o, and r_i while it weakly depends on the point angle $\Phi_p = 2\varphi$. For example, for a drill having $\gamma_n = 6°$, $a_o = 1.5$ mm: when $r_i =$ 4 mm, the rake angle $\gamma_{cf} = 12.33°$ when $\Phi_p = 118°$ and $\gamma_{cf} = 13°$ when $\Phi_p = 140°$. When $r_i = 10$ mm, the rake angle $\gamma_{cf} = 7°$ for $\Phi_p = 118°$ and $\Phi_p = 140°$.

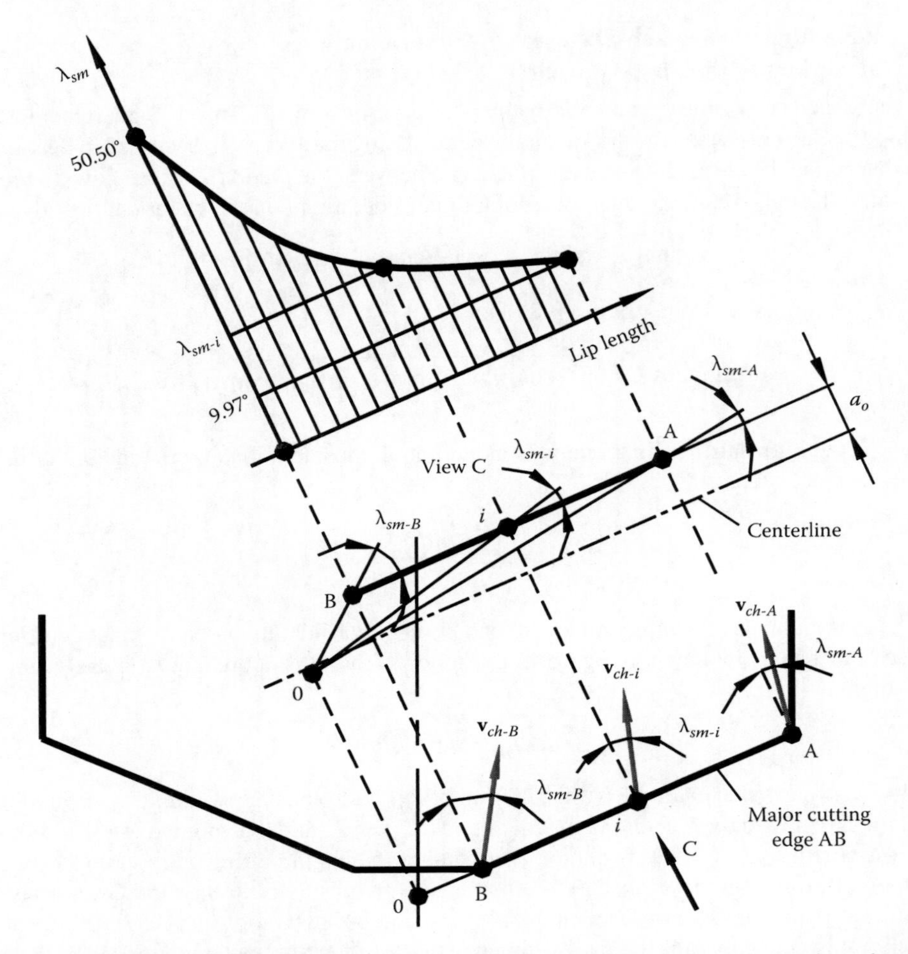

FIGURE 4.71 Inclination angle and the theoretical chip flow directions for the major cutting edge AB.

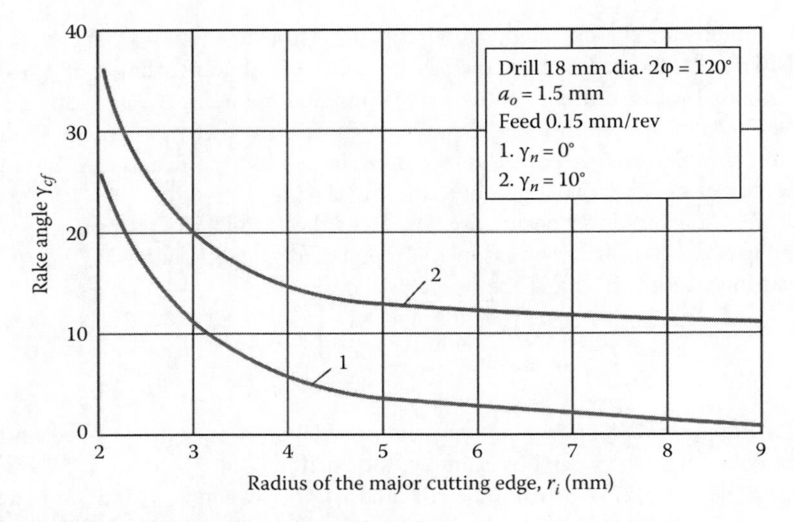

FIGURE 4.72 Variation of γ_{cf} over the major cutting edge.

To assess individual influence of the T-hand-S rake angle γ_n, point angle 2φ, and drill design parameter a_o, the following approximation of Equation 4.38 can be used (Vinogradov 1985):

$$\gamma_{cf\text{-}i} = \frac{0.08a^{2.13}\varphi^{1.2}}{r_i^{2.08}} + 0.81\gamma_n \tag{4.39}$$

where $a = 2a_o$.

The second term of Equation 4.39 includes only parameter γ_n, so by calculating the first term for various r_i, it is possible to determine the values of γ_n for the point of the major cutting edge to keep the same γ_{cf} over the whole major cutting edge, that is, to balance a drill in terms of direction of chip flow. For example, for a drill having the following parameters, $2\varphi = 120°$, $a = 3$ mm, and $\gamma_n = 0°$, the rake angle γ_{cf} is as follows:

r_i (mm)	2	3	5	7	9
γ_{cf} (°)	26.7	11.5	3.9	2.0	1.2

Therefore, to assure the condition $\gamma_{cf} = 0°$ over the entire cutting edge, the T-hand-S rake angle γ_n should vary according to $\gamma_n = -\gamma_{cf}/0.81$ as

r_i (mm)	2	3	5	7	9
γ_n (°)	−33.0	−14.2	−4.9	−2.4	−1.4

and thus the T-hand-S working rake angle as it is actually measured by many drill geometry inspection CNC machines (i.e., Helicheck) should vary according to Equation 4.22 as

$$\gamma_f = \arctan\left[\tan\left(\frac{\gamma_{cf}}{0.81}\right)\sin\varphi\right] \tag{4.40}$$

that is,

R_i (mm)	2	3	5	7	9
γ_f (°)	−29.3	−12.4	−4.3	−2.1	−1.3

The results of the presented calculations show that even the application of a gash having rake angle (the working [side] rake angle in T-hand-S) γ_f up to −29° on the drill rake face in the vicinity to its inner end adjacent to the chisel edge assures a neutral rake angle in T-mach-S. The detailed analysis of the gash geometry is presented by the author earlier (Astakhov 2010). Figure 4.73 shows

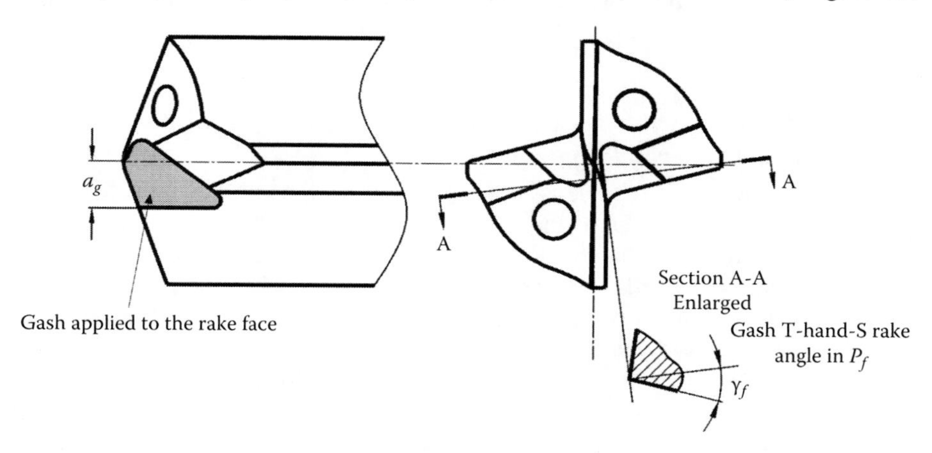

FIGURE 4.73 A gash applied to the rake face to modify its geometry (a straight-flute drill).

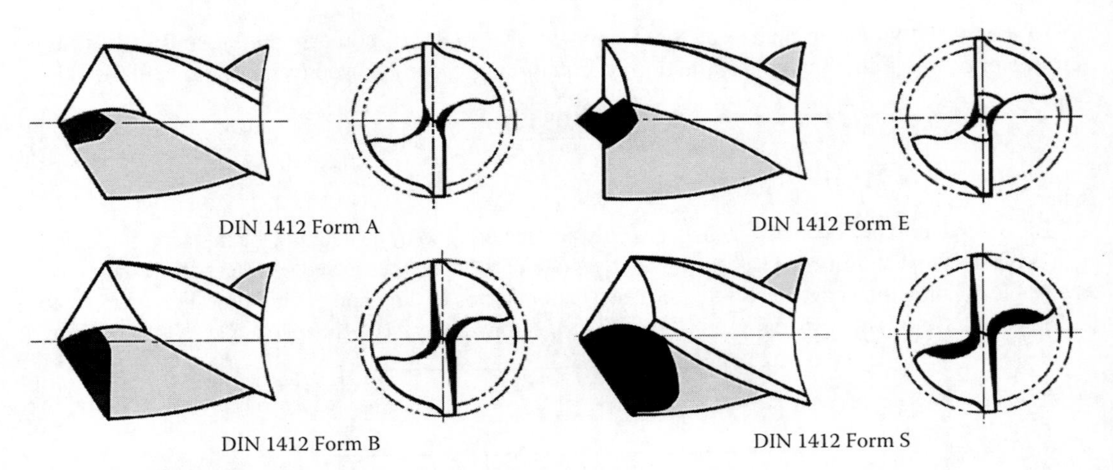

FIGURE 4.74 Gash extends according to DIN 1412 standard.

a fragment of the drill drawing where the gash is properly dimensioned. Its γ_f for straight-flute and twist drills should be as follows:

- For difficult-to-machine materials–26°
- For general-purpose drills–15°
- For high-penetration drills used for aluminum alloys–5°

Distance a_g should be approximately equal to $1/4d_{dr}$.

The application of a gash is known as web thinning as it reduces the length of the chisel edge (Astakhov 2010). Such an application is absolutely needed in machining of difficult-to-machine materials and for HP drills. Standard DIN 1412 defines the extent of gashes as shown in Figure 4.74. Type A point was initially intended for use on drills of over 20 mm to reduce the pressure on the web. Normally, the chisel edge is thinned up to 8% of diameter. Nowadays, with CNC grinding machines, this type becomes the most popular point for general applications. Type B point allows cutting edge runout correction and improving rake angle in the regions adjacent to the chisel edge. It was initially developed for brittle and difficult-to-machine work materials. Type D is known as cast iron point as its outer corners prevent frittering of the iron on breakthrough. Soon, it was found that this point grind is very useful for wide variety of work materials particularly when the exit burr is of concern. Type E was developed for use on sheet metal. It was soon found that various modifications of this grind are also useful for many applications. Point S is normally used on parabolic flute drills. Figure 4.75 shows practical realization of application of various gashes on common drills.

4.5.7.2 Rake Angle in T-mach-S/T-use-S γ_{ne} Determination according to the Second Approach

Although the determination of the T-use-S rake angle as γ_{cf} is predominant in the literature, it was noticed long time ago that it may not be the case (Oxford 1955) as the chip flow direction problem is complicated by the simple fact that the flowing chip is continuous across the major cutting edge. One should realize that in any drill, the length of the chip produced by the periphery point A (its nearest vicinity to assume some finite chip width) over one drill revolution is much greater than that by point B, which is at the inner end of the major cutting edge where this edge meets with the chisel edge. This is because the path traveled by point A in one drill revolution is r_A/r_B times greater than that by point B (where r_A and r_B are radii of point A and B, respectively). For example, for a standard 20 mm dia. drill, $r_A = 10$ mm and $r_B = 2$ mm, that is, the path passed by point A is five-fold greater than that by point B. The chip length is determined as the length of the path divided by

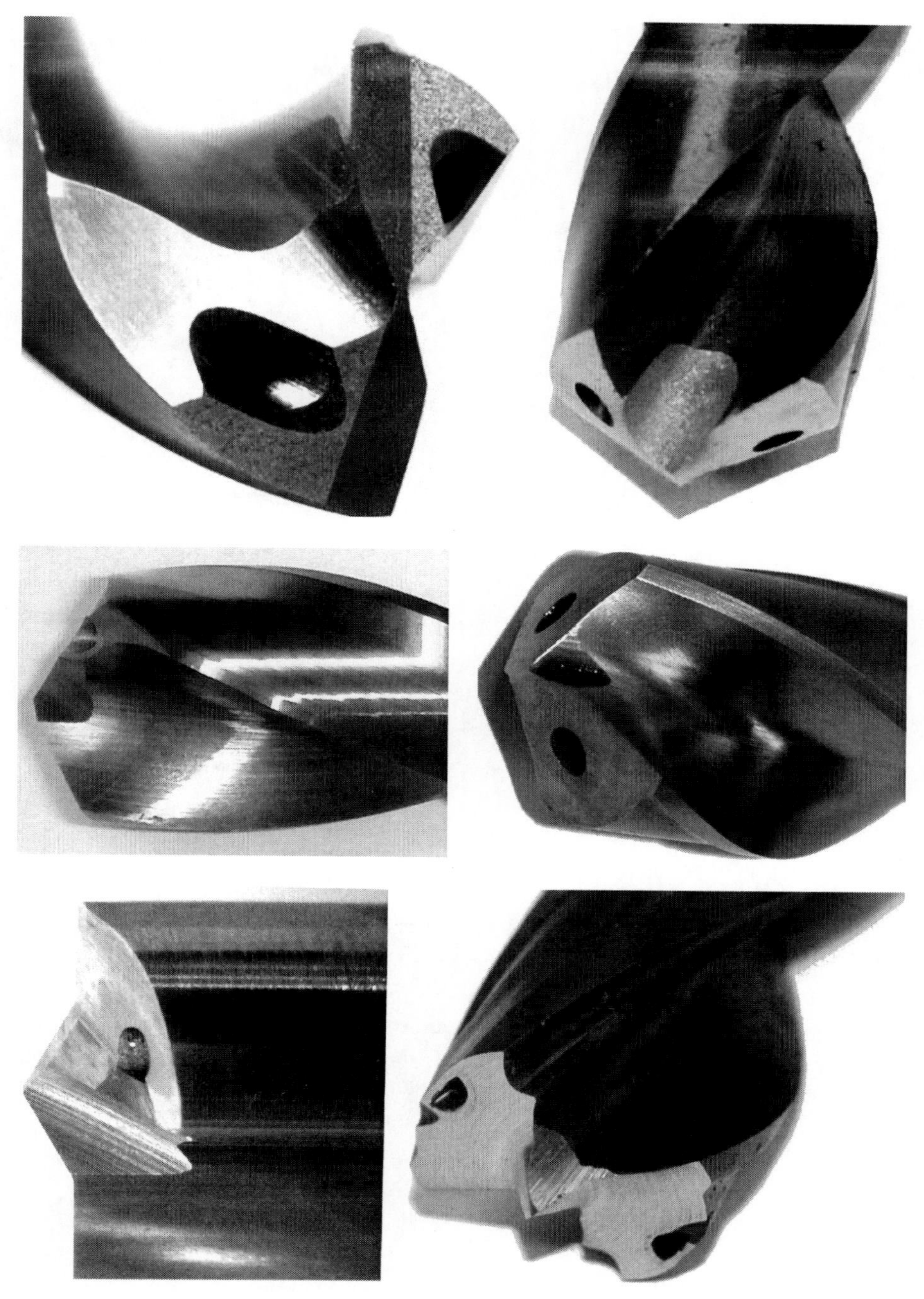

FIGURE 4.75 Common gashes applied to drills.

chip compression ratio (Astakhov 2004, 2006). Because this ratio is normally 50%–70% higher for point B because of its much smaller cutting speed and not favorable tool geometry, the total difference in the chip length produced by point A is normally seven to eight times greater than that by point B. Because the chip is continuous along the major cutting edge AB, the discussed difference causes chip curving into a cone shape as shown by a model in Figure 4.76a that closely resembles the reality (Figure 4.76b). In other words, the direction of actual chip flow deviates from that set by

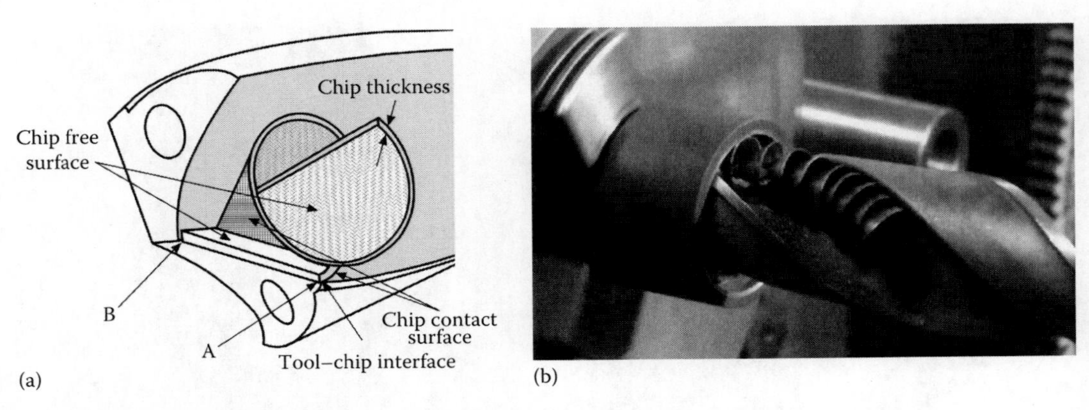

FIGURE 4.76 Chip flow: (a) model and (b) reality.

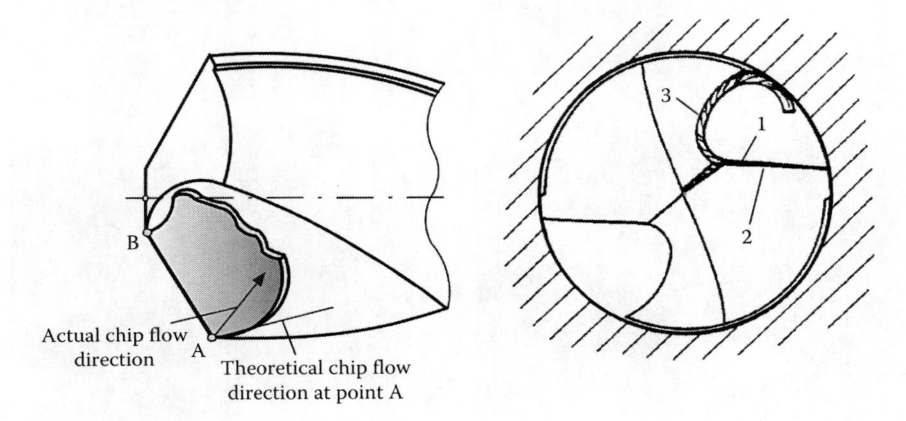

FIGURE 4.77 Simple model of chip flow.

the major cutting edge inclination angle as shown in Figure 4.77 (Astakhov 2010). As a result, the forming chip 1 flows along straight part 2 of the flute and then over its curved part 3 as shown in Figure 4.77 reaching the walls of the machined hole.

As the exact angle of chip flow is application-specific, the direction of chip flow according to this second approach is assumed to be perpendicular to the cutting edge (Fujii 1970; Astakhov 2010; Takashi and Jürgen 2010). As such, the T-use-S rake angle considered point i of the major cutting edge is calculated as

$$\gamma_{ne\text{-}i} = \gamma_{nm\text{-}i} - \upsilon_{e\text{-}i} \tag{4.41}$$

where

$\gamma_{nm\text{-}i}$ is the normal rake angle in T-mach-S
$\upsilon_{e\text{-}i} = \arctan(\tan \eta_{e\text{-}i} \sin \varphi)$ where η_e is determined according to Equation 4.26

As discussed by the author earlier (Astakhov 2010), when the major cutting edge of a drill is made with $\gamma_n = 0$ (the common case for straight-flute drills), then for a given point i on this edge, this angle is calculated as

$$\gamma_{nm\text{-}i} = -\arctan(\tan \mu_r \cos\varphi) = -\arctan\left[\tan\left(\arcsin\frac{a_o}{r_i}\right)\cos\varphi\right] = -\arctan\left(\frac{a_o}{r_{pi}}\cos\varphi\right) \tag{4.42}$$

Equation 4.42 defines the distributing of rake angles along cutting edge AB, that is, $r_B \le r_i \le r_A$. As can be seen, when $a_o > 0$, that is, when the major cutting edge is located ahead of the centerline, this rake angle is negative for any point of this cutting edge. Because $r_A > r_B$ and while angle φ and distance a_o are the same for all points of cutting edge AB, the absolute value of the rake angle at point B is greater than that at point A.

When the major cutting edge of a drill is made with modified rake surface having the normal T-hand-S rake angle $\gamma_n \neq 0$, for example, modified by applying additional design features as planar rake face (e.g., shown in Figures 4.58 and 4.74) then the distributing of the rake angle over cutting edge AB then is calculated as

$$\gamma_{nm}(r_i) = \gamma_n(r_i) - \arctan\left[\tan\left(\arcsin\frac{a_o}{r_i}\right)\cos\varphi\right] \tag{4.43}$$

or

$$\gamma_{nm}(r_{pi}) = \gamma_n(r_{pi}) - \arctan\left(\frac{a_o}{r_{pi}}\cos\varphi\right) \tag{4.44}$$

where $\gamma_n(r_i)$ determines the distribution of the rake face on the modified rake surface. Often, this surface is ground with a constant rake angle in T-hand-S so that this angle does not change over cutting edge AB, that is, $\gamma_n(r_i) = \text{Const.}$

When the rake face is helical, then calculations of the rake angle distribution over the major cutting edge should follow the following known procedure (Rodin 1971; Astakhov 2010):

$$\tan\gamma_{nm\text{-}i}(r_i) = \frac{1-\sin^2\varphi\sin\mu_i}{\sin\varphi\cos\mu_i}\tan\omega_i - \cos\varphi\tan\mu_i \tag{4.45}$$

where the sense of angle μ_i is shown in Figure 4.65 and ω_i is defined by Equation 4.21. As expected, when $\omega_i = 0$, Equation 4.45 coincides with Equation 4.42 obtained for a straight flute. Making simple geometrical rearrangement in Equation 4.45, one can obtain

$$\tan\gamma_{nm\text{-}i}(r_i) = \left(\frac{r_i}{r_{dr}}\tan\omega_d\right)\frac{1-(a_o/r_i)\sin^2\varphi}{\sin\varphi\cos(\arcsin(a_o/r_i))} - \cos\varphi\tan(\arcsin(a_o/r_i)) \tag{4.46}$$

An analysis of Equation 4.46 shows that when the rake surface is helical, the T-mach-S normal rake angle of a major cutting edge depends on the point angle Φ_p (as φ is a half of the point angle), on the distance a_o ($2a_o$ is often referred to in the literature as the web diameter although in general, the cutting edge may consist of a number of parts with individual a_o's or it can be inclined as per DIN 1214 Type B [shown in Figure 4.74]), and on the helix angle ω_d. Figure 4.78 shows the influence of the point angle for twist drill having the following parameters: $\omega_d = 30°$, $2a_o = d_{wb} = 0.2d_{dr}$ (Rodin 1971). As can be seen, in contrary to γ_{cf}, the point angle has significant influence on γ_{nm}. Small point angles cause a significant increase of the T-mach-S normal rake angle in the vicinity of the periphery point A with a sharp decrease of this angle along the cutting edge toward the drill axis. For a drill with $\Phi_p = 180°$, the normal rake angle varies along the cutting edge from 30° to 3°. For a drill with the standard point angle $\Phi_p = 120°$, the normal rake angle varies from +30° to –30°. Therefore, an increase in the point angle reduces the spread in the normal rake angle along the cutting edge.

The latter occurs because the point angle affects the shape and thus the curvature of the surface of the cut (the bottom of the hole being drilled). In general in drilling, this surface is hyperboloid as

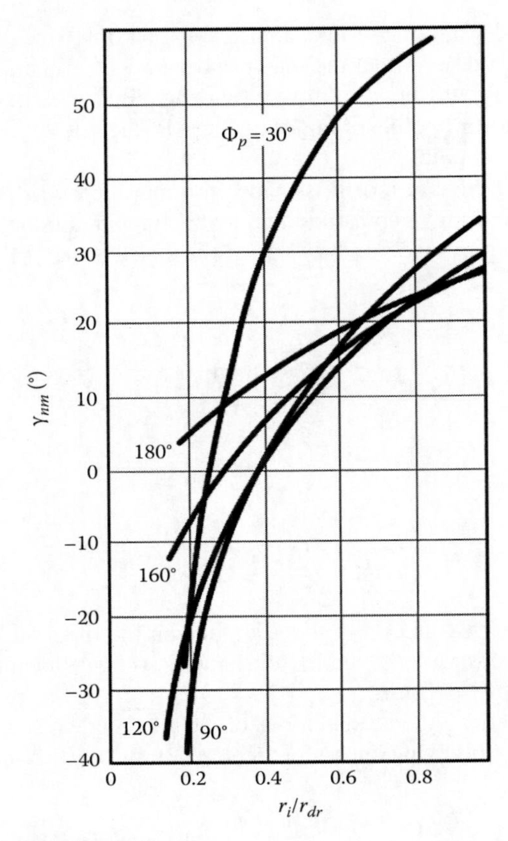

FIGURE 4.78 Influence of the point angle on γ_{nm}.

discussed earlier, which becomes a plane when $\Phi_p = 180°$. When it happens, the normal to the surface of cut does not change its direction along the cutting edge remaining to be parallel to the z_0-axis. As such, the location of the major cutting edge with respect to the centerline determined distance a_o ($2a_o = d_{wb}$) has only weak influence of the normal rake angle. Equation 4.46 can be modified for this case as

$$\tan \gamma_{nm}(r_i) = \left(\frac{r_i}{r_{dr}} \tan \omega_d \right) \cos \left(\arcsin \frac{a_o}{r_i} \right) \tag{4.47}$$

The influence of the distance a_o on the normal rake angle is shown in Figure 4.79 for standard drills with $\Phi_p = 120°$. As can be seen, an increase in a_o leads to decreasing of γ_{nm}. If the major cutting edge (lip) is located along drill radius, that is, $a_o = 0$, then $\mu_i = 0$ for any point of such a cutting edge. For this case, Equation 4.47 can be modified for this case as

$$\tan \gamma_{nm}(r_i) = \frac{r_i}{r_{dr}} \frac{\tan \omega_d}{\sin \kappa_r} \tag{4.48}$$

An analysis of Equation 4.48 shows that much smaller spread of the normal rake angle is achieved compared to standard twist drills. This is because if the cutting edge extends along drill radius, the surface of the cut (the bottom of the hole being drilled) is a conical surface. The normal to this surface at any point of the cutting edge has the same direction that improves the distribution of the normal rake angle over the cutting edge.

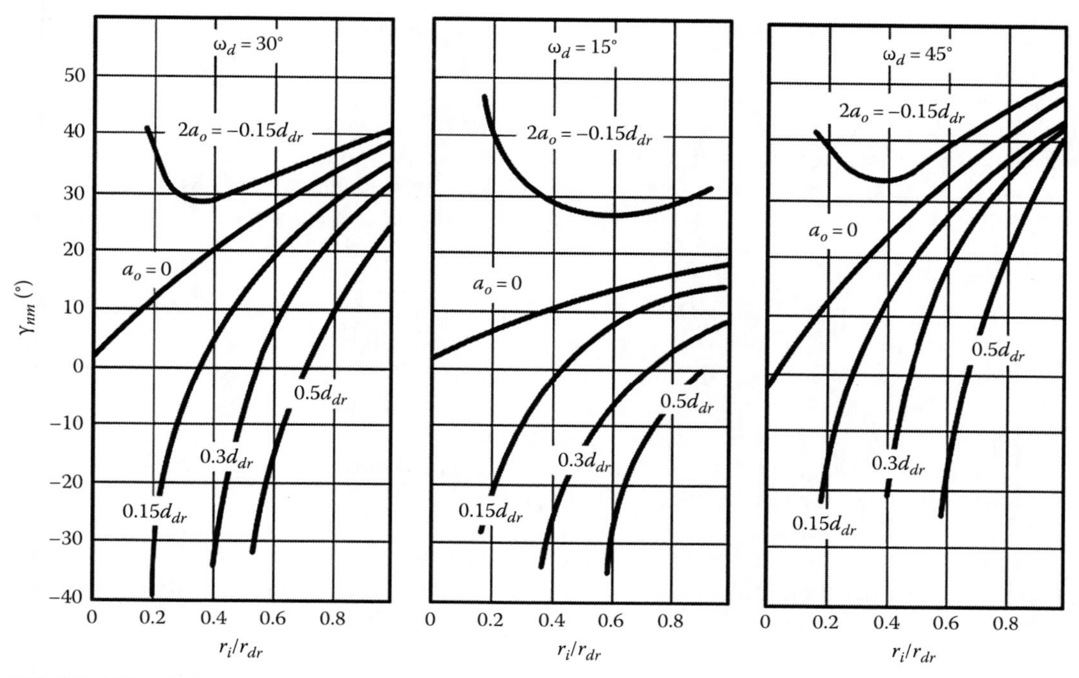

FIGURE 4.79 Influence of the distance a_o on the distribution of the normal rake angle.

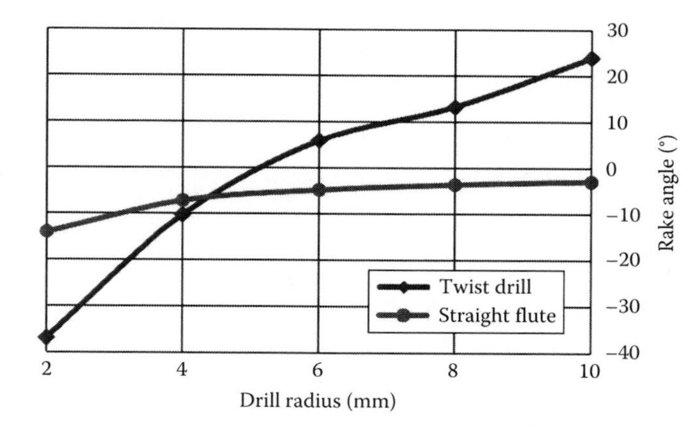

FIGURE 4.80 Rake angle distribution along the major cutting edge (lip) for a twist and a straight-flute drills.

Figure 4.80 shows a comparison of the distributions of the T-mach-S normal rake angle along the major cutting edge (lip) of straight-flute and twist drills (Equations 4.42 and 4.46, respectively). Drill diameter $d_{dr} = 20$ mm, web diameter (thickness) $d_{wb} = 2a_o = 4$ mm, and helix angle of the drill flute $\omega_d = 30°$. As can be seen, the rake angle for the twist drill varies from +24° at the drill periphery (point 1) to −37° at the inner end of the major cutting edge, while that for the straight-flute drill varies from −3° to −14°. Therefore, in terms of cutting conditions, the straight-flute drill has much more pleasurable rake angle distribution.

4.5.7.3 Comparison of the First and Second Approaches

Comparing Figures 4.72 and 4.80, one can conclude that the discussed approaches yield almost directly opposite results. In other words, according to the first approach (Figure 4.72), the T-use-S (T-mach-S) rake angle increases as the radius of the considered point of the major cutting edge decreases (Figure 4.72), while the opposite is true according to the second approach (Figure 4.72). This is an awkward situation for the one who tries to optimize the rake face drill geometry.

One of the ways to assess the chip flow direction is by analyzing the shape of the chip produced in drilling. Olson et al. (1998) pointed out that currently chip shapes are determined experimentally rather than predicted analytically or using FEM. The analytical or numerical determination of the chip shape and chip flow direction cannot be completed for drilling until all process variables, chip size, and its shape have been identified. The experimental program carried out by these authors resulted in the following conclusions:

- Six basic chip types were identified as produced in drilling. They are (1) conical chips formed when the chip follows flute shape, (2) fan-shaped chips formed when the chip cannot follow the flute shape, (3) chisel-edged chips formed by the chisel edge, (4) amorphous chips having wrinkled uncurled appearance, (5) needle chip causing severe upcurling, and (6) impacted chip that is not a primary chip form
- In each experiment, more than one chip type existed per sample. However, fan-shaped and corkscrew-shaped chips were the prime shapes.

In the author's opinion, however, the chip flow direction and thus its final shape depend on the properties of the work material, drill geometry and its material, drill manufacturing quality, machining regime, MWF supply conditions, etc. Therefore, there are a number of possibilities in existence.

Figure 4.81 shows the basic chip type in twist drilling at relatively low cutting speeds and feeds. It is the corkscrew-shaped chip. When this chip is the case, the chip flow direction is close to the theoretical chip flow direction provided that the shape of the chip flute is designed properly. To prove that this is the case, a special test was carried out using a pre-drilled workpiece shown in Figure 4.82a.

The author carried out a detailed study of the chip flow angle included in the analysis of the chip flow for different work materials, drill geometries, and machining regimes. Examples of the chip shapes obtained in this study are shown in Figure 4.83. The chip structure, its free and contact surfaces, chip deformation, and many other parameters of chip formation were analyzed. The results of

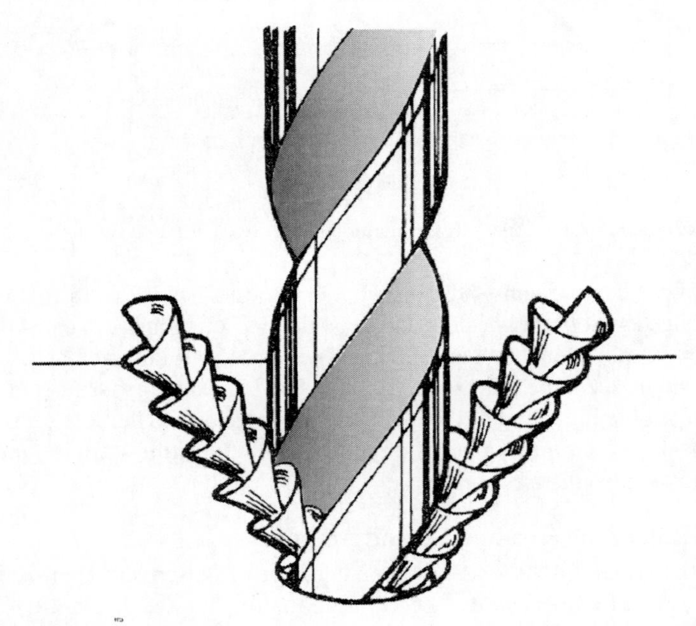

FIGURE 4.81 The basic chip type in twist drilling at relatively low cutting speeds and feeds.

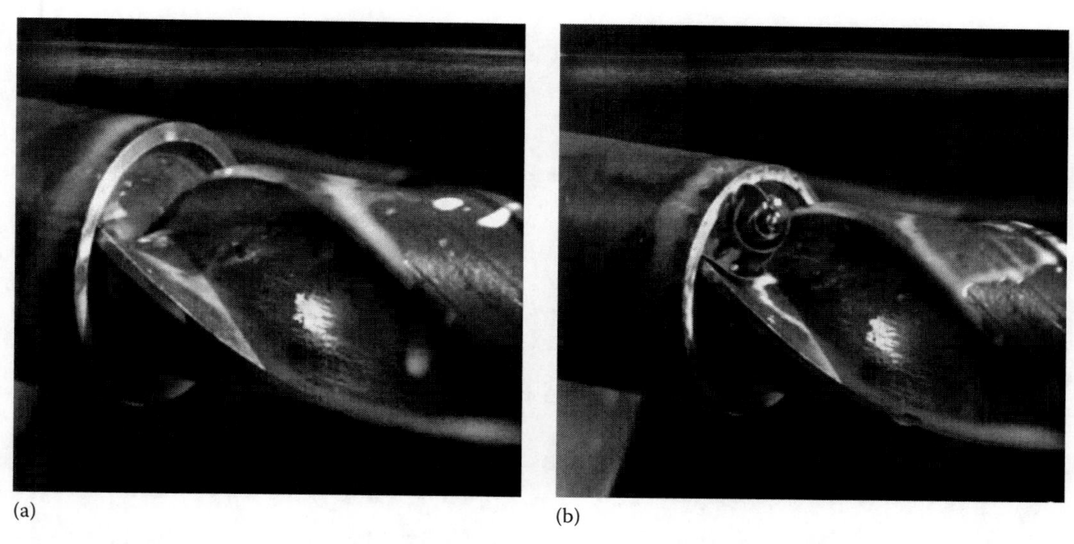

(a) (b)

FIGURE 4.82 Chip flow direction test: (a) predrilled workpiece and (b) observation of the chip flow direction.

the study reveal that, being application-specific, the chip flow direction in HP drill is always close to the direction of the normal to the cutting edge to the cutting edge regardless of the particular value of the drill point angle.

The observation of the chip flow direction for various drilling regimes and parts of the major cutting edge as well as for different work materials and drill designs results in the following conclusions:

1. The range of chip flow direction in general is often found between the theoretical chip flow direction and the direction of the normal to the cutting edge that differ by the inclination angle λ_{sm-i} (Figure 4.84). The chip flows in the direction of the normal to the cutting edge in HP drilling.
2. In drilling majority of steels with a twist drill using small cutting feeds, the chip flow on each individual section of the major cutting edge follows the theoretical direction as shown in Figure 4.85.
3. Although the chip flow on each individual section of the major cutting edge follows the theoretical direction, this is not the case when the whole cutting edge is engaged in cutting as the theoretical chip flow angle for each section is different as defined by Equation 4.36 and shown in Figure 4.71. As the theoretical chip flow angle varies along the major cutting edge, the unit chip flow from neighboring sections of this edge should cross each other. Moreover, as discussed in the previous section, the amount of chip produced by each section of the major cutting edge varies along its length as well as its deformation. As a result, for wide range of steel, the chip flow direction is between the normal and theoretical directions, while in HP drilling, it is in the normal direction.
4. The geometrical parameters of a common chip fragment shown Figure 4.86 vary along the length AB (the length of the major cutting edge). For given feed per tooth f_z and the cutting edge geometry that both define the uncut (undeformed) chip thickness (nominal thickness according to ISO Standard 3002/3) along this edge, h_D, the chip thickness h_C increased from A to B, that is, the chip deformation defined by the chip compression ratio ($\zeta = h_C/h_D$) normally increases from the drill periphery corner A toward the inner end B of the major cutting edge. As such, the radius of curvature of this chip fragment, R_{ch}, decreases from A to B due to the difference in the volume of the work material cut by section adjacent to points A and B.
5. When the feed and/or the cutting speed increase, the corkscrew-shaped chip falls apart into individual cones and small fans.

FIGURE 4.83 Studying the chip flow direction.

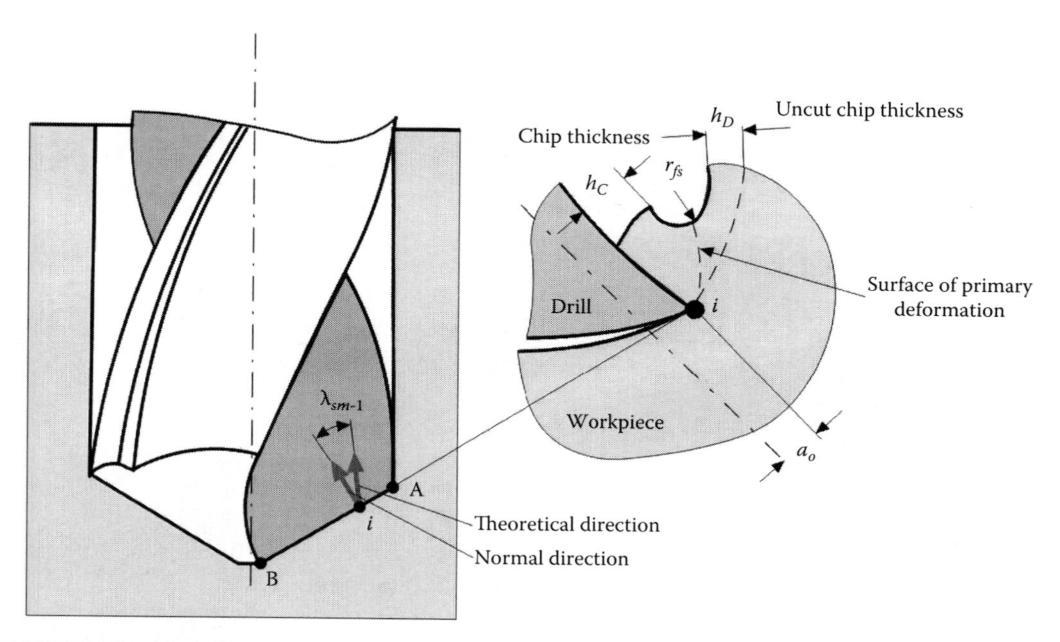

FIGURE 4.84 Possible range of chip flow directions between the theoretical and normal directions.

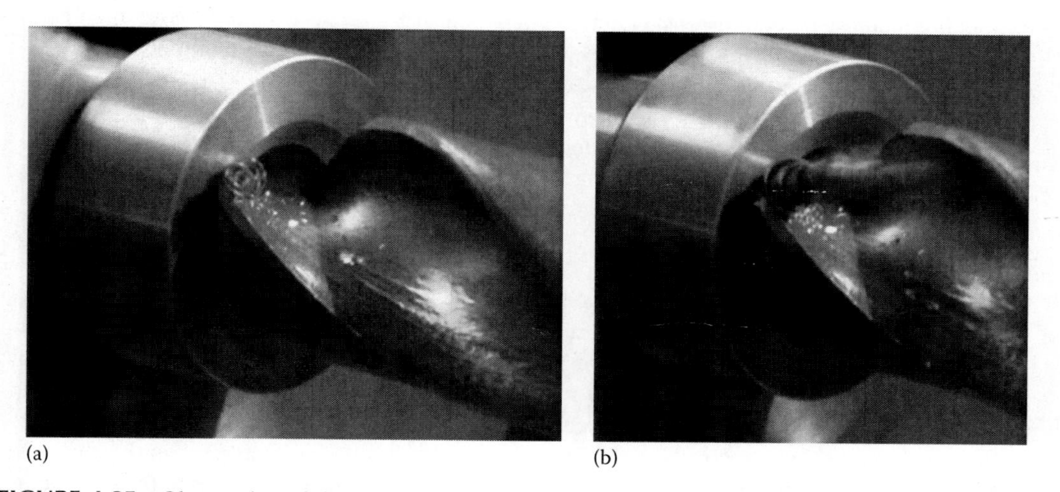

FIGURE 4.85 Observation of the chip flow direction for various drilling regimes and parts of the major cutting edge: (a) chip formation by a segment of the major cutting edge located close to the drill corner and (b) chip formation by a segment of the major cutting edge located in the mille of this edge.

6. The cutting speed has remarkable effect on the chip shape in machining of highly ductile materials. When the cutting speed is low (60 m/min), the string chips are produced. In this case, the direction of the chip flow is close to the theoretical direction. As the cutting speed increases to 200 m/min, the cone and fan chips are produced, as such, the chip direction is close to the normal direction.
7. In drilling with straight-flute drills at high speed, the shape of the chip and thus the direction of its flow are similar for wide range of work materials. Fanlike chips are produced in drilling different work materials with considerably different properties, namely, titanium alloy Ti6Al4V and aluminum alloy A380. The chip has distinctive deformation marks on its free side as it turns sharply.

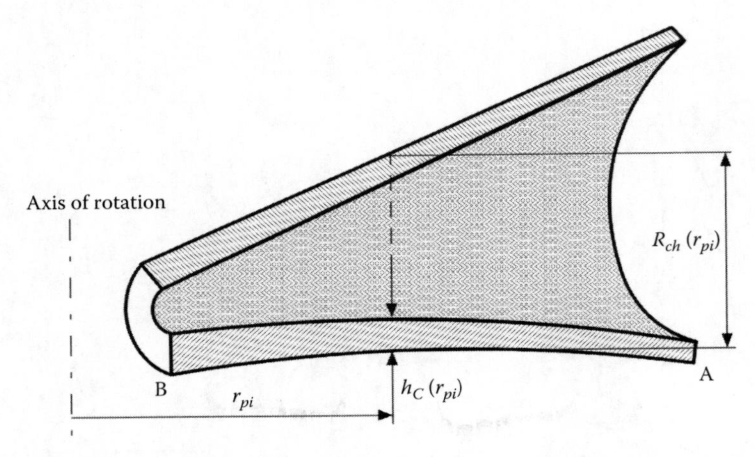

FIGURE 4.86 Parameters of a common chip fragment.

8. When a gash having a steep angle and great distance a_o is used in a drill meant for drilling a ductile work material, the chip continuity is violated at relatively low cutting feeds. As a result, the formed chip is heavily deformed in the vicinity of the gash. If the cutting feed is increased, however, the chip breaks into two separate flows and its pieces have a fanlike appearance. As such, the deformation of the chip, and thus the cutting torque, reduces and tool life increases. Therefore, as mentioned in Section 4.5.7.1, the gash distance a_g should be approximately equal to $1/4d_{dr}$ (Figure 4.73) to avoid unnecessary deformation of the chip.
9. Attempts to determine the chip shape and chip flow direction using commercial FEM metal cutting packages are too optimistic as any numerical simulation of the complicated chip deformation, flow, and curling processes are too complicated for such simulations as yet (Astakhov 2011a). To exemplify this statement, Figure 4.87 shows a comparison between the modeled and experimentally obtained chip shape (steel ANSI 1045, $v = 250$ m/min, $f = 0.25$ mm/rev).

4.5.7.4 Chip Breakage

In the author's opinion, the simplest and reliable method of chip breakage in drilling is modification of the rake face in the manner shown in Figures 4.57 and 4.58. In these designs, a chip-breaking step is ground on the drill rake face. However, as discussed in Section 4.5.5, the parameters of this step are functions of a particular drill design and the cutting feed. The latter make drills with a chip-breaking step on the rake face application-specific. The test of such drills has proven that not only particular feed makes them application-specific but also the cutting speed as well as the particular work and tool materials (including coating). As a result, a practical determination of the parameters of the chip-breaking step turns to be an expensive and time-consuming task.

To simplify finding the parameters of chip-breaking step, namely, its depth h_{st} and width b_{st} (Figure 4.88), the following practical procedure can be used. The procedure includes a simple turning test where a single-point cutter made of the same work material as the drill is set to cut the same work material as to be drilled. The cutter has a zero rake angle and is set with the tool cutting edge angle κ_r (see Appendix C) equal to the half-point angle of the drill φ. The cutting speed is chosen as equal to that in drilling and the cutting feed is equal to the feed per tool in drilling f_z (Chapter 1). The procedure includes the following steps:

1. Determination of the uncut (undeformed) chip thickness h_D as

$$h_D = f \sin \kappa_r \tag{4.49}$$

2. Measuring of the chip compression ratio (Appendix A) ζ

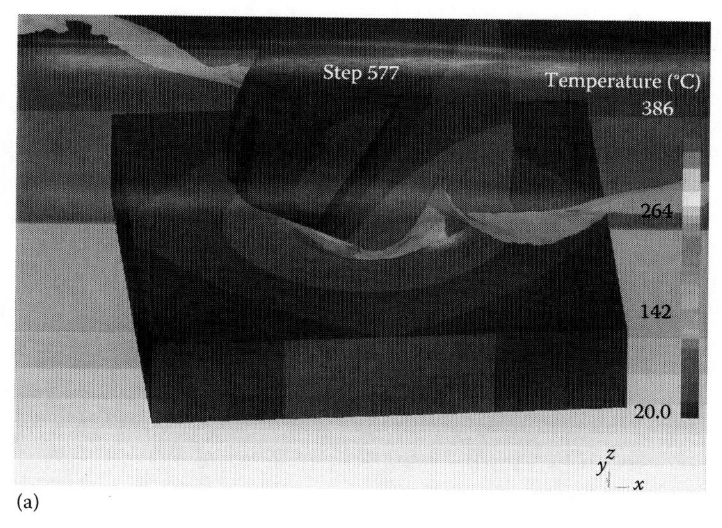

(a)

(b)

FIGURE 4.87 Showing a comparison between the (a) modeled and (b) experimentally obtained chip shape (steel ANSI 1045, $v = 250$ m/min, $f = 0.25$ mm/rev).

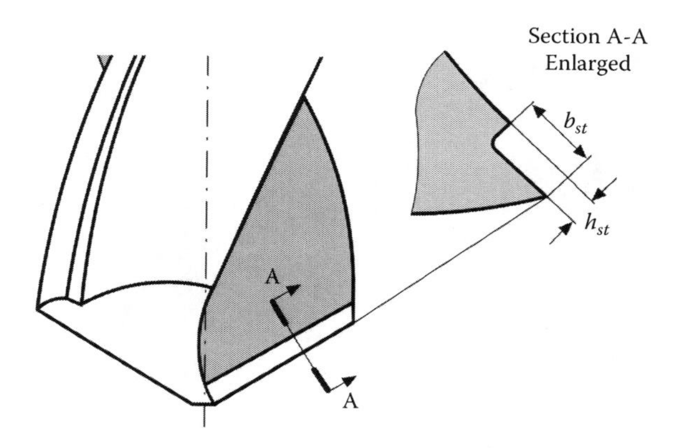

FIGURE 4.88 Chip-breaking step on the drill rake face.

3. Determination of the chip thickness h_C as

$$h_C = \zeta h_D \tag{4.50}$$

4. Determination of the optimal radius of chip curvature $R_{ch\text{-}opt}$. For a wide range of steels,

$$R_{ch\text{-}opt} = 6.5 h_C$$

5. Calculating the depth h_{st} as $h_{st} \geq h_C$
6. Calculating the step width b_{st} as

$$b_{st} = \sqrt{h_{st}(2R_{ch\text{-}opt} - h_{st})} \tag{4.51}$$

The chip breakability can be enhanced if a cutting edge is provided with chip splitting grooves often called nicks. Initially developed for spade drills and evolved into ANSI Standard B94.49-1975, such nicks have been also used for twist drills since the eighteenth century (French and Goodrich 1910 [reprinted in 2001]). Figure 4.89a shows drill design according to US Patent No. 1,383,733 (1921). The drill has the major cutting edges 1 and 2 that are provided with chip splitting grooves 3 and 4 extending over the length of flutes 5 and 6. A number of various drill designs with chip splitting nicks were introduced. A detailed analysis of their effect on drill performance was presented by Nakayama and Ogawa (1985). The design of a twist drill having multiple nicks ground as radial grooves on its flank shown in Figure 4.89b was described as early as 1940 by Veremachuck (Fig. 244 in Veremachuk 1940) who also extensively studied application specifics of this design in its comparison with the standard twist drill. In this drill, the major cutting edges (lips) 7 and 8 are located so that preferable distributions of the rake and flank angles along these edges are achieved as well as the thinning of the chisel edge 9. Flank surfaces 10 and 11 are provided with radial grooves 12 and 13, respectively, that aims to separate the forming chip into rather narrow strips.

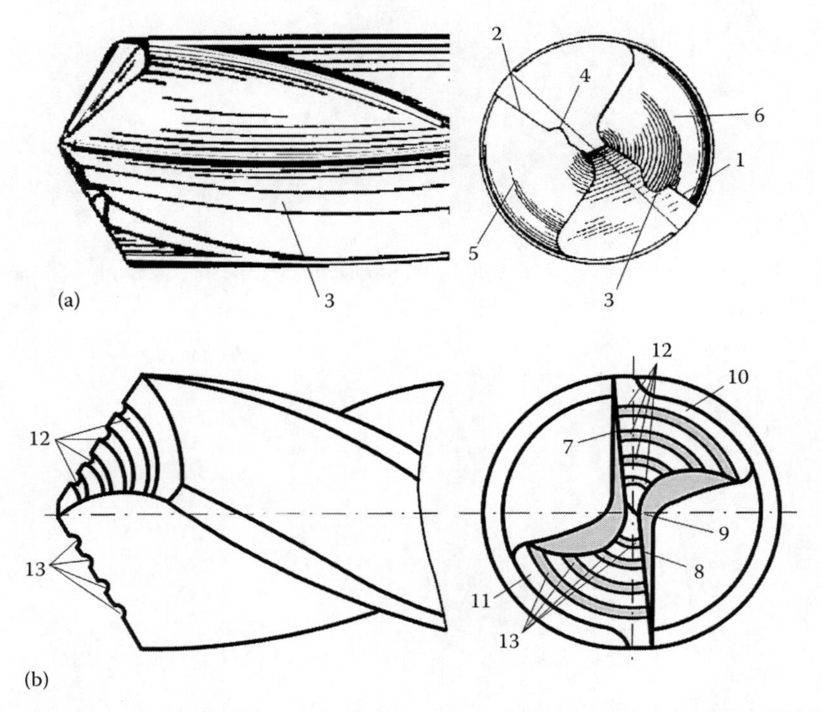

FIGURE 4.89 Drills with radial chip splitting grooves: (a) made on the rake face and (b) made on the flank surface.

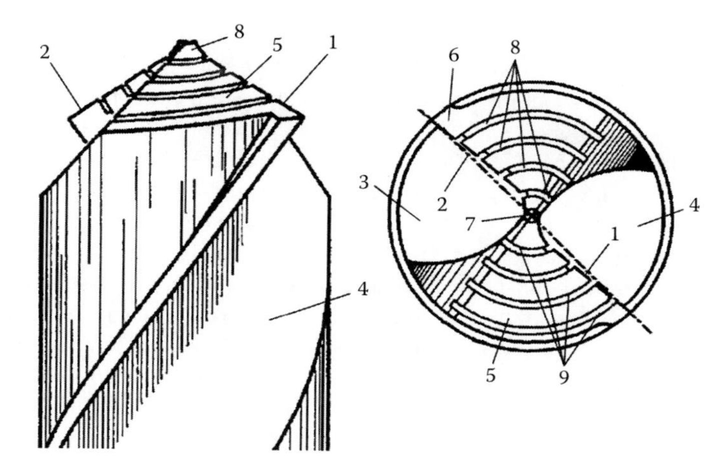

FIGURE 4.90 Drill design according to US Patent No. 5,452,971 (1994).

Figure 4.90 shows a drill design according to US Patent No. 5,452,971 (1994). According to Nevilles, the inventor, it is completely a new concept in twist-drill theory, design, and method of construction that will revolutionize the rotary end cutting tool systems. The drill has two curved major cutting edges 1 and 2 formed as intersection lines of flutes 3 and 4 and flank surfaces 5 and 6. These edges are connected by the chisel edge 7. Two series of offset volumetric grooves 8 and 9 (deep nicks ground as radial grooves) ground on each flank surface as shown in Figure 4.90. The patent claims that such a drill produces precise, near mirrorlike well-finished holes in various work materials. Besides, its penetration rate is nearly four times faster and tool life seven times longer in comparison to a comparable-sized standard twist drill.

A number of concerns about a drill having radial chip splitting grooves can be identified. The major design concern is a difficulty to assure the flank angle on the sides of the grooves to prevent friction of these sides in drilling. According to the design shown in Figure 4.89b, these grooves are rather shallow and have round cross section; thus, severity of the problem is low. In the design shown in Figure 4.90, these grooves are deep and having rectangular cross section; thus, the problem becomes of real concern. To apply the flank angle on the groove sides, their profile should be of a fishtail cross section that is virtually impossible to apply using standard machines and grinding wheels. The application problem is that the radial grooves should be reapplied with relatively high accuracy on each successive re-sharpening. In the author's opinion, these two prime problems prevent practical applications of the discussed designs.

The chip rigidity and thus its enhanced breakability can be achieved if a drill has a means to increase the rigidity of the forming chip. US Patent No. 2,204,030 (1940) offers a drill design where the major cutting edges 1 and 2 are made with ribs (projections) 3 and 4 as shown in Figure 4.91. The formed chip

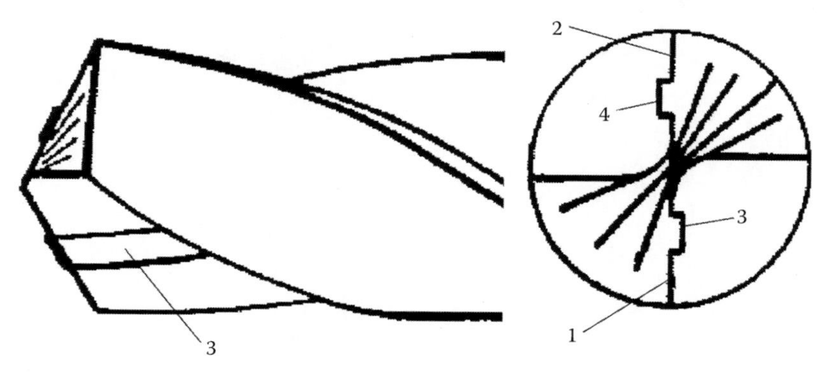

FIGURE 4.91 Enhancing chip rigidity by providing ribs on the drill rake faces.

would have a rib of rigidity so its breakability would be much greater. To achieve the same objective, US Patents No. 867,639 (1907) and 1,404,546 (1922) offer various combinations of the previously described means located asymmetrically about the axis of rotation. Although the described ideas of chip rigidity enhancement are widely used in modern designs of cutting inserts including those for drills, the formation of additional rigidity means that it is always accompanied by increasing the drilling torque and force.

Often the described measures are not sufficient to obtain the desirable chip shape and to avoid significant force due to chip interaction with the side wall of the flute and with the walls of the machined hole. The latter should be particularly avoided as it results in damage to hole quality and significant friction. Therefore, the chip flute profile should be made so that it helps to curl the chip into an easy transportable shape and then transport this chip away from the machining zone out of the hole being drilled.

To understand a need for more sophisticated shapes of the chip flute, consider a simple model of chip flow shown in Figure 4.77. As discussed earlier, the length of the chip produced by the periphery point A (its nearest vicinity to assume some finite chip width) over one drill revolution is much greater that that by point B, which is the inner end of the major cutting edge where this edge meets with the chisel edge. As a result, the forming chip 1 flows along straight part 2 of the flute and then over its curved part 3 as shown in Figure 4.77 reaching the walls of the machined hole.

To deal with the problem, various approaches are used. The first one is to use groove-type and obstruction-type chip breakers as those used in singe-point tool operations (Jawahir and VanLuttervelt 1993; Jawahir and Zhang 1995; Jawahir et al. 1997). Figure 4.92a shows a drill with groove 1 placed on the rake face 2 near the major cutting edge 3 that facilitates chip breaking and prevents chip clogging (Sahu et al. 2003). Figure 4.92b shows a drill according to US Patent No. 2,966,081 (1960). In this drill, the flute profile is of a concave shape 4 and the chip-breaking step (VIEW B) is provided so that the cutting edge 5 is formed.

Although these chip breakers may produce broken chips suitable for evacuation from the machining zone, they are rarely used in drilling because of the following:

- Groove geometry (parameters r_{gr} and h_{gr}) and orientation (angle τ_{gr}) as well as chip step geometry (h_{st}, b_{st}, and r_{st}) are application-specific, that is, they work for rather narrow range of the combination of the work materials properties, feed, speed, MWF, etc. A small departure from one of these parameters may ruin chip breaking.

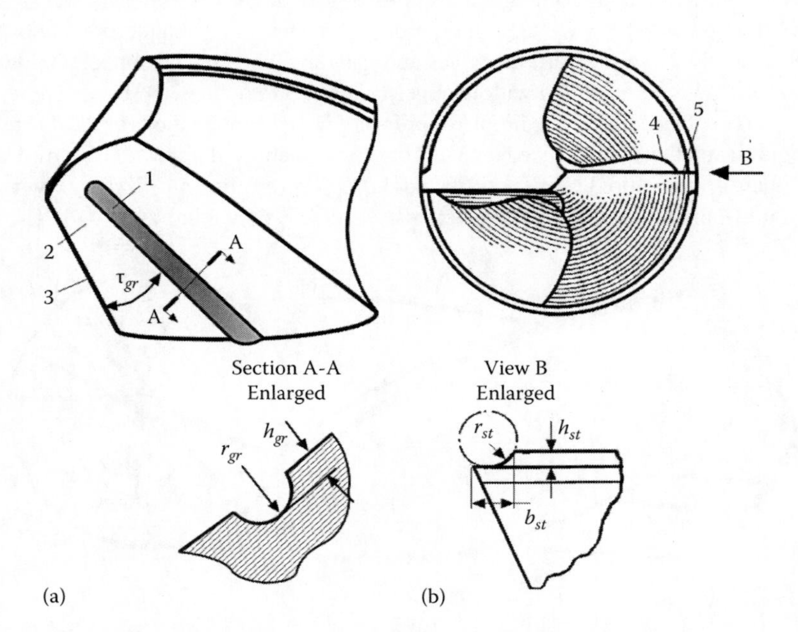

FIGURE 4.92 (a) Groove-type and (b) obstruction-type chip breakers for a drill.

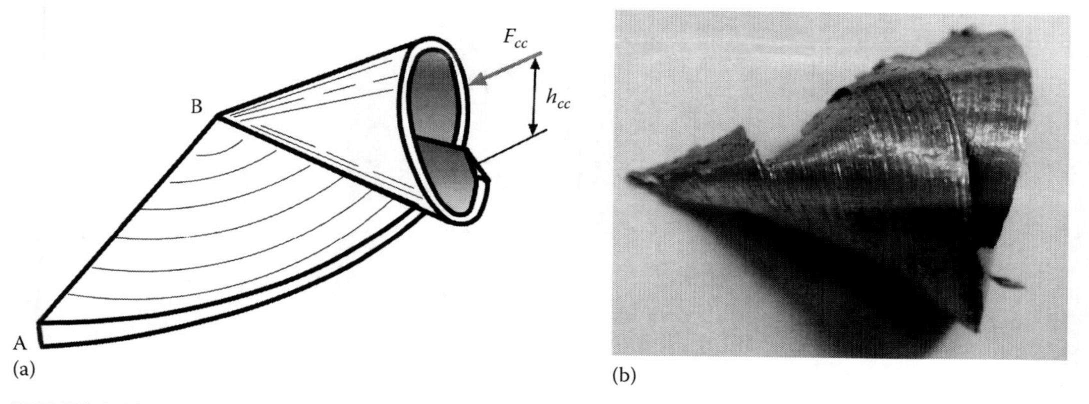

FIGURE 4.93 Model of chip curling: (a) a model, and (b) a real chip fragment.

- Groove-type and obstruction-type chip breakers increase the drilling torque and the axial force.
- The groove and chip step geometry should be reproduced with relatively high accuracy on each drill re-sharpening. Particularly with the groove shown in Figure 4.92a, a great amount of the tool material should be removed to restore the original shape of the drill that significantly lowers the number of possible resharpenings.

More intelligent ways to design the flute profile to achieve a suitable shape of the chip can be understood if one considers a simple model of chip shape formation (chip curling during drilling) shown in Figure 4.93a. In this picture, the letters A and B designate the corresponding ends of the cutting edge AB. To curl such a chip into cone-like shape and then break it when the deformation in its root reaches the strain at fracture or to form long tight curls (Figure 4.93b), a certain force F_{cc} should be applied to the chip at a certain distance h_{cc}. Because the force F_{cc} is the reaction force from some obstacle made on the flute, it depends on the drilling process parameters while the direction of this force and the distance h_{cc} can be varied by flute profile parameters. This is the principle that outlines many patented drill designs that differ only by particular values of these two parameters.

Figure 4.94a shows flute profile according to US Patent No. 5,622,462 (1997). As seen, each major cutting edge (lip) consists of the outer straight part 1 and inclined inner part 2 having concave portion 3. This apex 4 is formed to apply the force F_{cc} shown in Figure 4.93. The inner part is built-in in the flute profile that assures its consistency over each successive drill regrinding. Figure 4.94b shows a similar flute profile according to US Patent No. 5,931,615 (1999). As can be seen, the major cutting edge 5 is located on the drill transverse axis. It has concave part 6 that ends with apex 7. The location of the major cutting edge on the drill transverse axis improves the distribution of the flank and rake angles over the straight part of the cutting edge and decreases the plastic deformation of the chip that reduces the drilling torque and force. However, complicated concave part 6 should be ground on each successive drill regrind.

There are two obvious disadvantages of the profiles shown in Figure 4.94. The first one is that the cutting geometry over the radiused parts and apexes is not favorable so that one should expect an increase in the drilling torque and force. The second is that the arm (h_{cc} in Figure 4.93) is small, which may not create the sufficient bending moment at the chip root to break the forming chip into pieces while force F_{cc} is excessive because the radius of the radiused part is too small. However, it has been found that the breaking up of the forming chip into small pieces is not always desirable. Rather, the efficiency of drilling is improved, and less down time is needed if the drill flute profile is constructed so that the chip coming out from the cutting edge is formed into long sustained curls as shown in Figure 4.81.

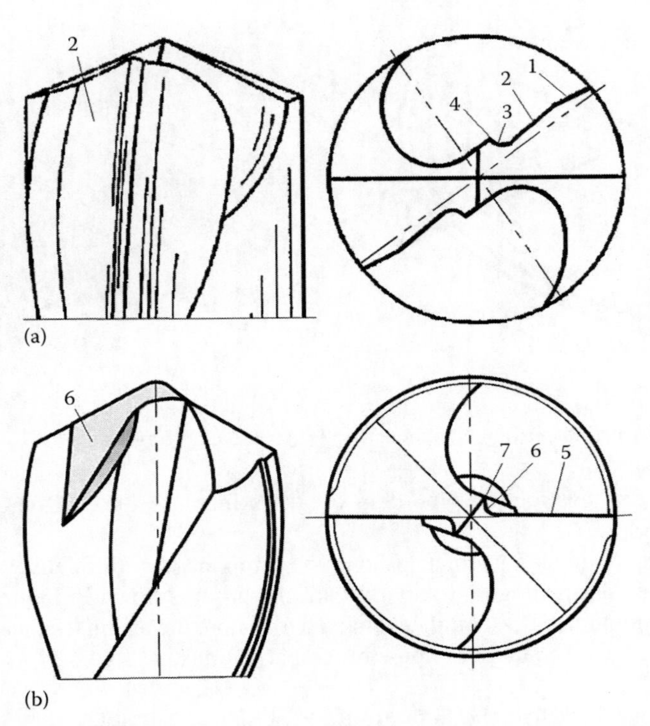

FIGURE 4.94 Two-flute profiles with a mean for chip curling as a part of the major cutting edge: (a) according to US Patent No. 5,622,462 (1997), and (b) according to US Patent No. 5,931,615 (1999).

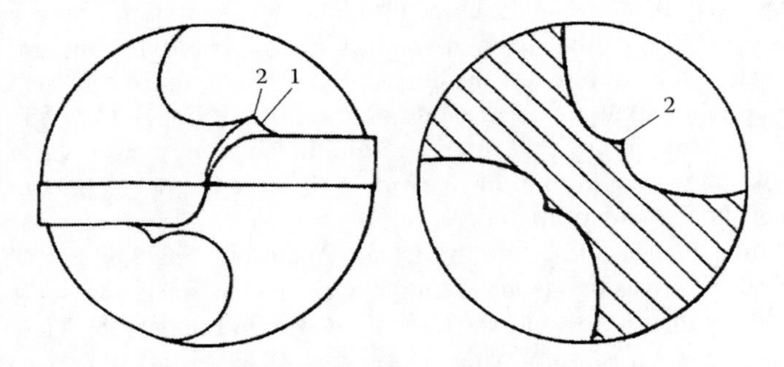

FIGURE 4.95 Drill design according to US Patent No. 4,222,690 (1980).

Figure 4.95 shows the flute profile according to US Patent No. 4,222,690 (1980) where curved part 1 and apex 2 belong to the flute. Many other known flute profiles originated from the same idea. The difference is in the location of apex and the radius of the curved part. For example, Figure 4.96 shows a twist drill in which a concavely shaped surface 1 extends from the inner end 2 of the major cutting edge 3 toward the outer periphery of the drill body and apex 4 is located on the side wall of flute 5.

As extreme cases, it is worthwhile to consider some special profiles of the chip flute and thus drill cross sections shown in Figure 4.97. In Figure 4.97a, the flute is formed by the cutting edge portion 1 that becomes concave portion 2 extending to apex 3 located almost at the intersection of the flute and the relieved part of the drill body. According to US Patent No. 4,583,888 (1986), the position of apex 3 is selected so the chip contact with the wall of the machined hole is prevented. It can be accomplished by selecting the web thickness $d_{wb} = (0.25\ldots0.35)d_{dr}$ and the flute-width ratio (FWR) = 0.4…0.8. According to US Patents No. 4,983,079 (1991) and 5,088,863 (1992), a significant reduction in the axial force, drilling torque, and thus drilling power as well as improvement in chip

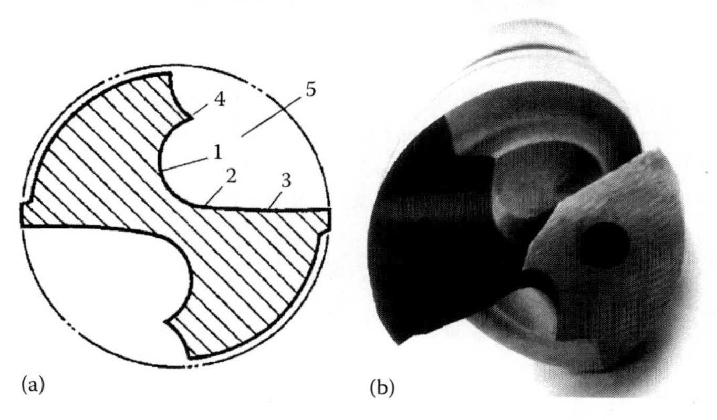

FIGURE 4.96 Drill body and the chip-curling apex located on the side wall of the flute: (a) flute profile, and (b) drill appearance.

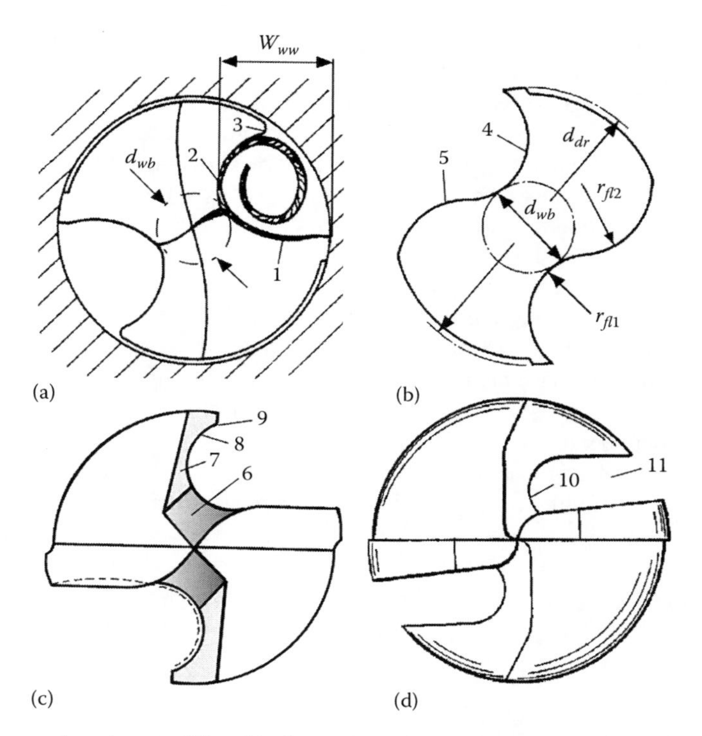

FIGURE 4.97 Some other shapes of the chip flute: (a) profile according to US Patent No. 4,583,888 (1986), (b) double-radius chip flute profile, (c) flute profile according to US Patent No. 5,716,172 (1998), and (d) flute profile for extremely heavy-duty and deep-hole drills.

transportation from the machining zone in heavy-duty drilling operation is achieved when distance W_{ww} is selected properly from the range of $(0.45\ldots0.65)d_{dr}$.

Figure 4.97b shows double-radius chip flute profile. According to US Patent No. 4,744,705 (1988), significant reduction in drilling power and improvement stability of drilling operations are achieved when the web thickness $d_{wb} = (0.25\ldots0.50)d_{dr}$, $r_{fl1} = (0.2\ldots0.3)d_{dr}$, and $r_{fl2} = (0.25\ldots0.40)d_{dr}$. It is understood that the cutting edge formed as the line of intersection of the concave portion 4 of the flute with that flank surface would be concave. Convex part 5 of the flute profile is to curl the forming chip.

Figure 4.97c shows the flute profile according to US Patent No. 5,716,172 (1998) meant for heavy-duty drills. It has the front chip curling means consisting of two flat surfaces 7 and 8 that are made to fold and then curl the chip emerging from the cutting edges into the concave flute 8. Flute 8 is provided with apex 9 to prevent the contact of the chip and the wall of the drilled holes.

Figure 4.97d represents the generic idea of extremely heavy-duty and deep-hole drills with FWR of 0.5…0.02 to increase torsional rigidity to the drill shank (e.g., US Patents No. 4,565,473 [1986] and 4,975,003 [1990]). The concave part 10 of the chip flute 11 has a small radius to bend, curl, and break the forming chip or just direct it into the flute without breaking as a small diameter curl or a string.

4.5.8 Chisel Edge

4.5.8.1 General

The chisel edge length, its location angle known as the chisel angle, and geometry (the rake and flank angles) of this edge define to a large extent the performance of the drill. Moreover, these parameters are of crucial importance in the design of HP drills and thus are pillars of *VPA-balance*d drill design concept. The standard drill nomenclature always presents the chisel edge as a single design component of a drill as shown in Figure 4.14. Moreover, a great number of practical engineers experience difficulties in understanding that the chisel edge is the cutting edge, not an indenter penetrating the workpiece as it is often presented in the professional literature. Moreover, a picture of the chisel edge that acts as an indenter penetrating into the workpiece (Figure 4.98) taken out of context from a classical paper by Galloway (1957) is presented in many manufacturing books, and there is a notion circulating in the professional literature that in drilling this edge smashes or extrudes the work material.

As discussed earlier, a normal or non-modified chisel edge is responsible for 50%–60% of the total axial force. The reduction of the resultant axial force is vitally important when one tries to increase the allowable penetration rate of the drill. As the chisel edge is the major contributor to this axial force, one should (1) reduce the length of this edge and/or (2) improve the geometry of this edge. Therefore, any further increase in the drill penetration rate should be correlated with the modification of the chisel edge geometry. To apply such a modification, the geometry of the chisel edge should be clearly understood.

In reality, if the chisel edge passes through the axis of rotation, then there are two chisel edges—each one starts from the inner end of one of the major cutting edges (lips) and extends to the center of rotation (edges 2-3 and 4-3 in Figure 4.99a). Each chisel edge has its rake and flank faces with the corresponding rake and clearance angles. To maintain drill symmetry, the lengths of the chisel edges and their angles should be the same. Figure 4.99a helps to visualize the rake and flank faces of these two edges as well as the directions of the chip flow. Figure 4.99b shows

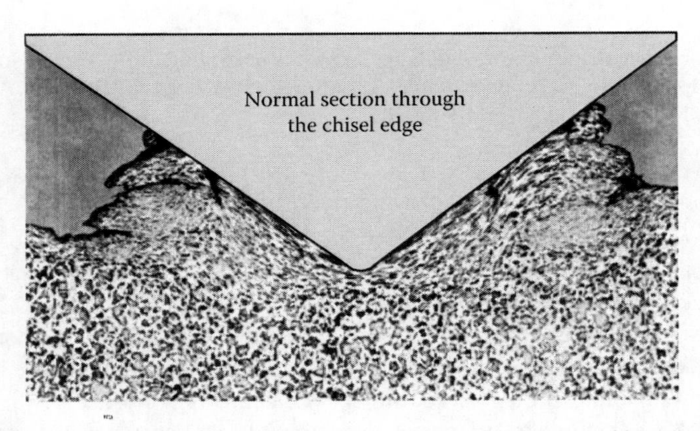

FIGURE 4.98 Chisel edge penetration into the workpiece commonly presented by manufacturing books.

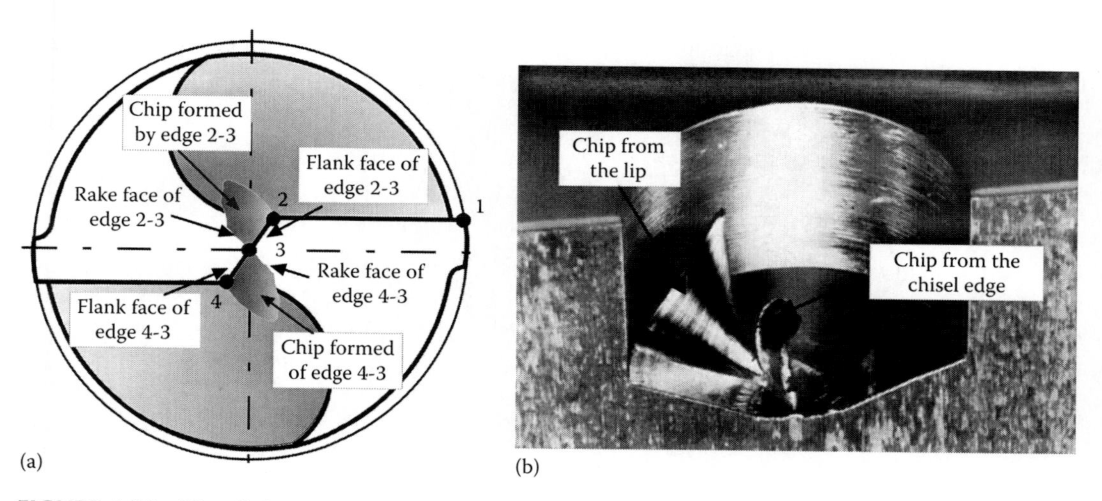

FIGURE 4.99 Visualizing the chisel edge: (a) rake and flank faces of two chisel edges 2-3 and 4-3 and (b) partially formed chips obtained using a quick-stop device.

the partially formed chips obtained using a quick-stop device where the chips formed by the major cutting edge (lip) and that formed by the chisel edge are clearly distinguishable (Galloway 1957).

4.5.8.2 Case 1: The Primary Flank Is Planar and Its Width Is Equal to or Greater than $2a_o$

Figure 4.100 shows a model to determine the geometry of the chisel edge for the simplest yet common case when the flank faces of the major cutting edges (lips) 1-2 and 4-5 have the same T-hand-S clearance angles $\alpha_{n1} = \alpha_{n2}$ and the width of the primary flank is equal to or greater than $2a_o$. This figure shows the chisel edge 2-4 (its two parts 2-3 and 4-3) as formed by two flank planes 1 and 2 having T-hand-S clearance angles α_{n1} and α_{n2}. It is obvious that in the considered case, the chisel edge is a line of intersection of two flank planes.

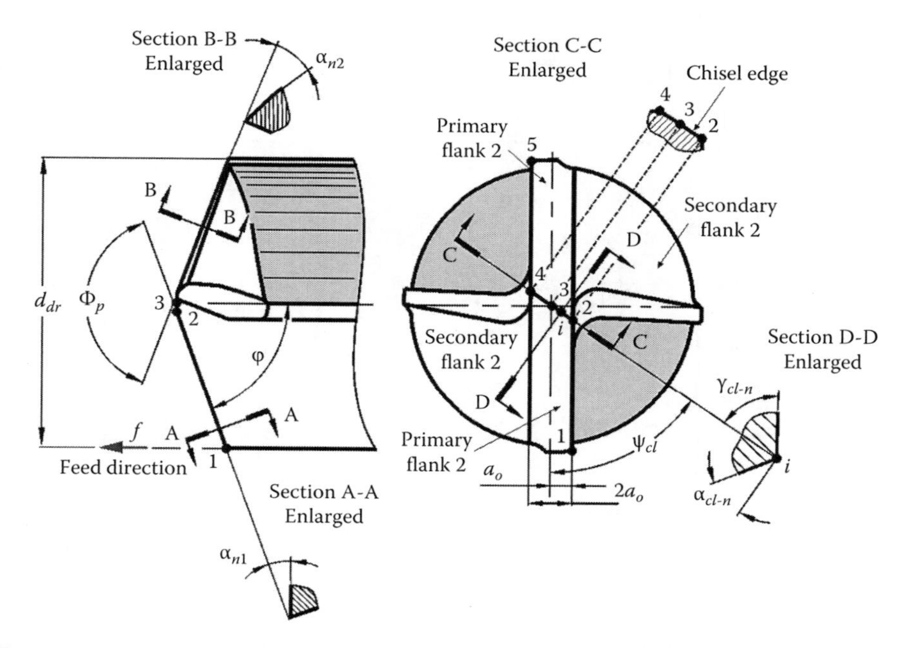

FIGURE 4.100 Model for the chisel edge formed by two flank planes having the same T-hand-S clearance angles.

An important inspection equation for the considered case was obtained by the author earlier (Astakhov 2010). It relates the T-hand-S normal clearance angles $\alpha_{n1} = \alpha_{n2} = \alpha_n$ and the chisel edge angle ψ_{cl} as

$$\psi_{cl} = \arctan\frac{\cos\varphi}{\tan\alpha_n} = \arctan\frac{\cos(\Phi_p/2)}{\tan\alpha_n} \tag{4.52}$$

Using this equation, one can inspect the clearance angle of the major cutting edge. To do that, the actual point angle, Φ_p, and the chisel edge angle, ψ_{cl}, are measured in a simple manner with, for example, a toolmaker's microscope PG1000, which is a common piece of equipment in many machine shops. Then, the T-hand-S clearance angle $\alpha_{n1} = \alpha_{n2} = \alpha_n$ is calculated as

$$\alpha_n = \arctan\frac{\cos(\Phi_p/2)}{\tan\psi_{cl}} \tag{4.53}$$

and compared with that shown in the tool drawing.

For example, for a drill having $\Phi_p = 140°$ and measured $\psi_{cl} = 65°$, the T-hand-S clearance angle $\alpha_n = \arctan(\cos(\Phi_p/2)/\tan\psi_{cl}) = \arctan(\cos 70°/\tan 65°) = 9.1°$. This value is compared with that shown in the drill drawing, for example, as shown in Figure 4.50. Figure 4.101 shows how the chisel edge angle varies with the point angle for four common T-hand-S clearance angles of the major cutting edge (8°, 12°, 15°, and 18°). As can be seen, this angle decreases with the point angle and with the T-hand-S clearance angle.

The total length of the chisel edge $l_{2\text{-}4}$ (distance 2-4 in Figure 4.100), which actually is the sum of the lengths of two chisel edges, that is, $l_{2\text{-}3} + l_{4\text{-}3}$, is calculated as

$$l_{cl} = l_{2\text{-}4} = \frac{2a_0}{\sin\psi_{cl}} = \frac{2a_0}{\sin(\arctan(\cos\varphi/\tan\alpha_n))} \tag{4.54}$$

Figure 4.102 shows how the total chisel edge length varies with the point angle for four common T-hand-S clearance angles of the major cutting edge (8°, 12°, 15°, and 18°). As can be seen, this length increases with the point angle and with the T-hand-S clearance angle.

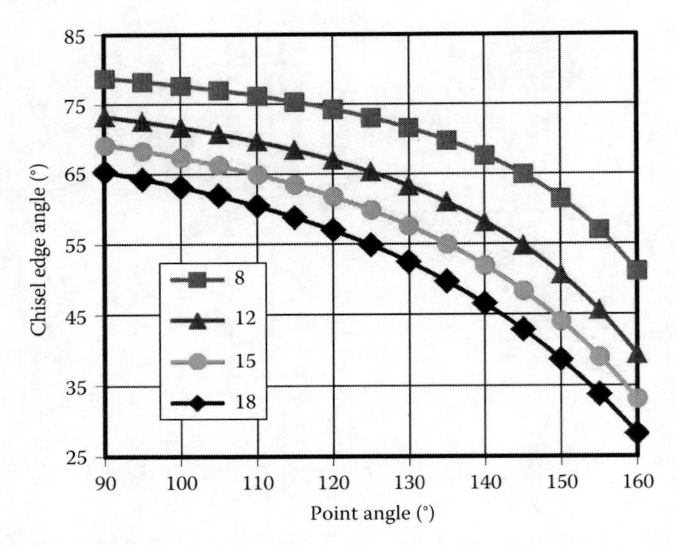

FIGURE 4.101 The chisel edge angle as a function of the point angle for four common T-hand-S clearance angles of the major cutting edge (8°, 12°, 15°, and 18°).

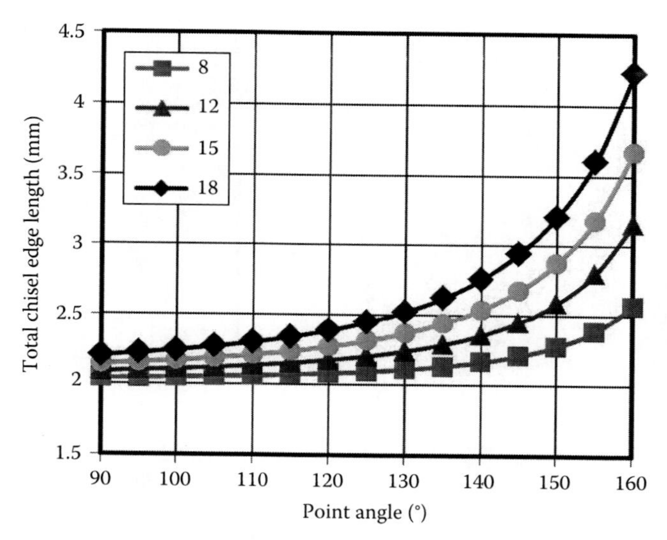

FIGURE 4.102 The total length of the chisel edge as a function of the point angle for four common T-hand-S clearance angles of the major cutting edge (8°, 12°, 15°, and 18°). A drill of 14.48 mm dia., $a_o = 0.6$ mm.

As the chisel edge passes the axis of rotation (at least, theoretically), the rake and clearance angles of this edge are the same in T-hand-S and T-mach-S, that is, $\gamma_{cl\text{-}1} = \gamma_{cl\text{-}n2} = \gamma_{cl\text{-}n} = \gamma_{cl\text{-}nm}$ and $\alpha_{cl\text{-}1} = \alpha_{cl\text{-}n2} = \alpha_{cl\text{-}n} = \alpha_{cl\text{-}nm}$. Therefore, the further analysis is made for only one-half of the chisel edge, that is, subscripts 1 and 2 are omitted.

The rake and clearance angles of the chisel edge in T-use-S are considered accounting for the fact that the point angle of this edge is zero, that is, the coincident T-hand-S, T-mach-S, and T-use-S working planes are simultaneously the normal planes for the chisel edge. In Figure 4.100, they are represented by a section plane D-D. As a result, the T-hand-S and T-mach-S rake and clearance angles ($\gamma_{cl\text{-}n}$ and $\alpha_{cl\text{-}n}$) do not vary along the chisel edge. This is not the case for these angles in T-use-S ($\gamma_{cl\text{-}ne}$ and $\alpha_{cl\text{-}ne}$) as a given point i of the cutting edge; these angles are calculated as

$$\gamma_{cl\text{-}ne\text{-}i} = \gamma_{cl\text{-}n} + \eta_{e\text{-}i} \tag{4.55}$$

$$\alpha_{cl\text{-}ne\text{-}i} = \alpha_{cl\text{-}n} - \eta_{e\text{-}i} \tag{4.56}$$

where $\eta_{e\text{-}i}$ is the resultant cutting speed angle calculated using Equation 4.26.

The author showed earlier (Astakhov 2010) that the normal T-hand-S (T-mach-S) clearance angle of the chisel edge $\alpha_{cl\text{-}n}$ is calculated as

$$\alpha_{cl\text{-}n} = \arctan\frac{1}{\tan\varphi\sin\psi_{cl}} = \arctan\frac{1}{\tan(\Phi_p/2)\sin\psi_{cl}} = \arctan\frac{1}{\tan(\Phi_p/2)\sin(\arctan(\cos\varphi/\tan\alpha_n))} \tag{4.57}$$

and the normal T-hand-S (T-mach-S) rake angle of the chisel edge, $\gamma_{cl\text{-}n}$

$$\gamma_{cl\text{-}n} = \alpha_{cl\text{-}n} - 90° \tag{4.58}$$

Equation 4.57 shows that for a given point angle Φ_p, the normal T-hand-S (T-mach-S) clearance angle of the chisel edge $\alpha_{cl\text{-}n}$ is determined by the T-hand-S normal clearance angle of the major cutting edge α_n. Because the normal T-hand-S (T-mach-S) rake angle of the chisel edge, $\gamma_{cl\text{-}ni}$, is fully determined by $\alpha_{cl\text{-}n}$ (Equation 4.58), this angle is also determined by α_n. In other words, the geometry of the chisel edge

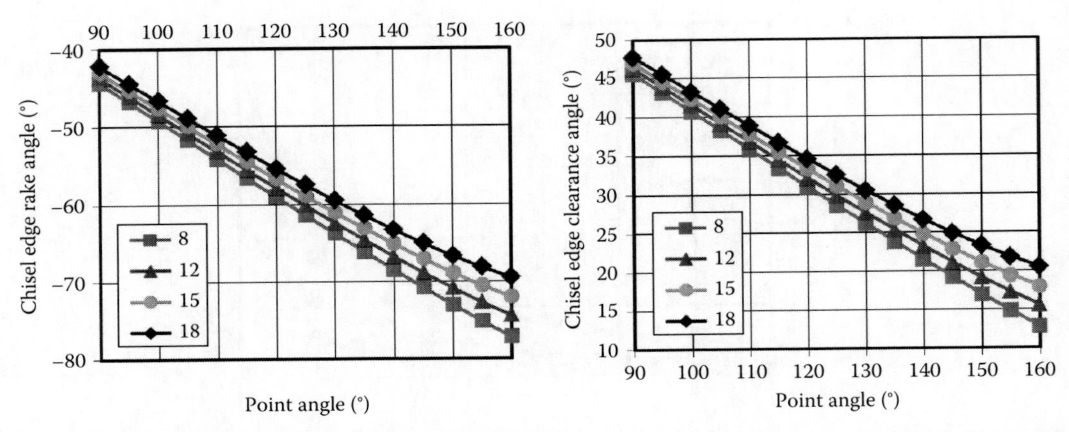

FIGURE 4.103 Variations of the normal T-hand-S (T-mach-S) rake and clearance angles of the chisel edge with the point angle for four common T-hand-S clearance angles of the major cutting edge (8°, 12°, 15°, and 18°).

(for a given point angle Φ_p) is fully determined by the T-hand-S normal clearance angle of the major cutting edge α_n. This angle also determines the total length of the chisel edge as follows from Equation 4.54. Therefore, this angle is one of the major design parameters of the drill, and thus, (1) this angle should be clearly indicated on the tool drawing, (2) tight tolerance should be assigned on this angle, (3) this angle should be the same for the major cutting edges to assure drill symmetry and the proper shape of the chisel edge, (4) this angle should be included in the tool inspection procedure and inspection report. The latter, however, with extremely rare exceptions, is not a common practice in the tool industry, and (5) this angle should not be less than 10° for steels and 14° for aluminum alloys work materials for HP drills.

Figure 4.103 shows variations of the normal T-hand-S (T-mach-S) rake angle of the chisel edge with the point angle for four common T-hand-S clearance angles of the major cutting edge (8°, 12°, 15°, and 18°). As can be seen, $\gamma_{cl\text{-}n}$ is highly negative, for example, for $\Phi_p = 140°$ and $\alpha_n = 12°$, it is $\gamma_{cl\text{-}n} = -68°$ while $\alpha_{cl\text{-}n} = 23°$ for the same parameters.

The second term in the equations for the determination of both the chisel edge normal rake and clearance angles in T-use-S ($\gamma_{cl\text{-}ne}$ and $\alpha_{cl\text{-}ne}$) (Equations 4.55 and 4.56) is the resultant cutting speed angle η_e defined by Equation 4.26. It was discussed in Section 4.5.6 that this angle is small for points on the major cutting edge. This is, however, not nearly the case for the chisel edge where this angle must be taken into consideration in any analysis of the chisel edge geometry. Figure 4.104 shows

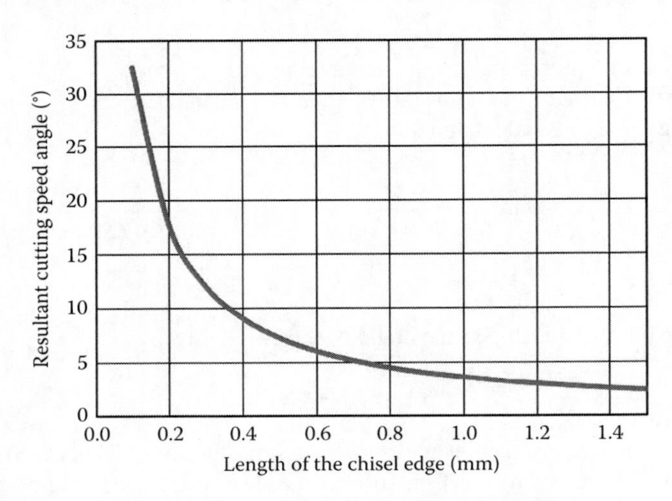

FIGURE 4.104 Variation of the resultant cutting speed angle over the length of the chisel edge (the feed per revolution $f = 0.4$ mm/rev and the length of the chisel edge 1.5 mm).

an example of variation of the resultant cutting speed angle over the length of the chisel edge for a common drill design. As can be seen, angle η_e becomes significant in the vicinity of the axis of rotation. As a result, the chisel edge normal rake and clearance angles in T-use-S ($\gamma_{cl\text{-}ne}$ and $\alpha_{cl\text{-}ne}$) defined by Equations 4.55 and 4.56, respectively, deviate from their T-hand-S (T-mach-S) values.

Figures 4.105 and 4.106 show examples for a common drill design. As can be seen, the T-use-S rake angle $\gamma_{cl\text{-}ne}$ becomes much less negative in the vicinity of the axis of rotation compared to its T-hand-S (T-mach-S) values (Figure 4.103) that improves chip formation. The problem, however, appears to be with the T-use-S clearance angle $\alpha_{cl\text{-}ne}$. On one hand, this angle decreases toward the axis of rotation so that the chisel edge does not brake as it should be with high clearance angles. On the other hand, the interference of the chisel edge flank face and the bottom of the hole being drilled takes place as the T-use-S clearance angle becomes very small and then negative in the vicinity of the axis of rotation as clearly seen in Figure 4.106. As follows from Equations 4.26 and 4.56, the higher the feed per revolution, the greater the problem. This is because such interference leads to a significant increase of the axial force in drilling and thus limits the allowable penetration

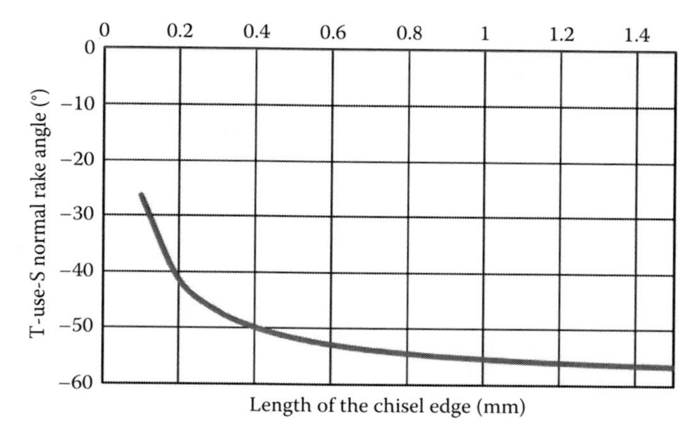

FIGURE 4.105 Variation of the T-use-S rake angle over the length of the chisel edge (the T-hand-S normal clearance angle $\alpha_n = 8°$, point angle $\Phi_p = 120°$, feed per revolution $f = 0.4$ mm/rev, and the length of the chisel edge 1.5 mm).

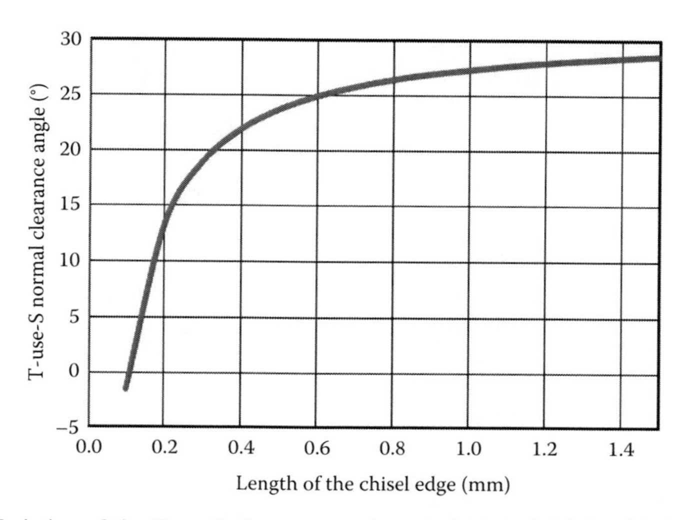

FIGURE 4.106 Variation of the T-use-S clearance angle over the length of the chisel edge (the T-hand-S normal clearance angle $\alpha_n = 8°$, point angle $\Phi_p = 120°$, feed per revolution $f = 0.4$ mm/rev, and the length of the chisel edge 1.5 mm).

rate—the major objective of HP drills. Being important for high-penetration rate drills, this problem, however, is not always of prime concern for such drills as another much more significant yet completely abandoned in the literature on drill design and geometry problem arises when one tries to increase the drill penetration rate.

As discussed in Section 4.5.6, the bottom of the hole being drilled is not a conical surface as commonly depicted in the literature. It is actually a hyperboloid (Figure 4.69) described by Equation 4.31. In practical calculations of the geometry of the major cutting edges and on the part drawing, such a hyperboloid is approximated by a cone. Being a reasonable approximation in the consideration of the geometry of the major cutting edges, such an approximation, however, is totally unacceptable in considerations of working conditions of the chisel edge. The following explains the reason.

Figure 4.107 shows a model of cutting by the chisel edge (Vinogradov 1985). Particular interest in the further considerations presents section A-A in this figure. The intersection lines *ab* and *cd* of the bottom of the hole being drilled by the major cutting edges and the axial cross section (section A-A in Figure 4.107) are branches of hyperbolas having asymptotes k_1 and k_2. Distance *ac* is equal to the length of the chisel edge, that is, $ac = 2a_o/\sin\psi_{cl}$. A right-hand coordinate system is set as is shown in Figure 4.107. Its origin is at the intersection of the asymptotes k_1 and k_2 with the drill longitudinal axis. The z_c-axis is along the drill longitudinal axis, while x_c- and y_c-axes are set as shown in Figure 4.107 to form the right-hand Cartesian coordinate system.

In this coordinate system, an equation of hyperbolas *ab* and *cd* is obtained from Equation 4.31 as

$$\frac{x^2}{a^2} - \frac{z^2}{(a_o/\tan\varphi)^2} = 1 \tag{4.59}$$

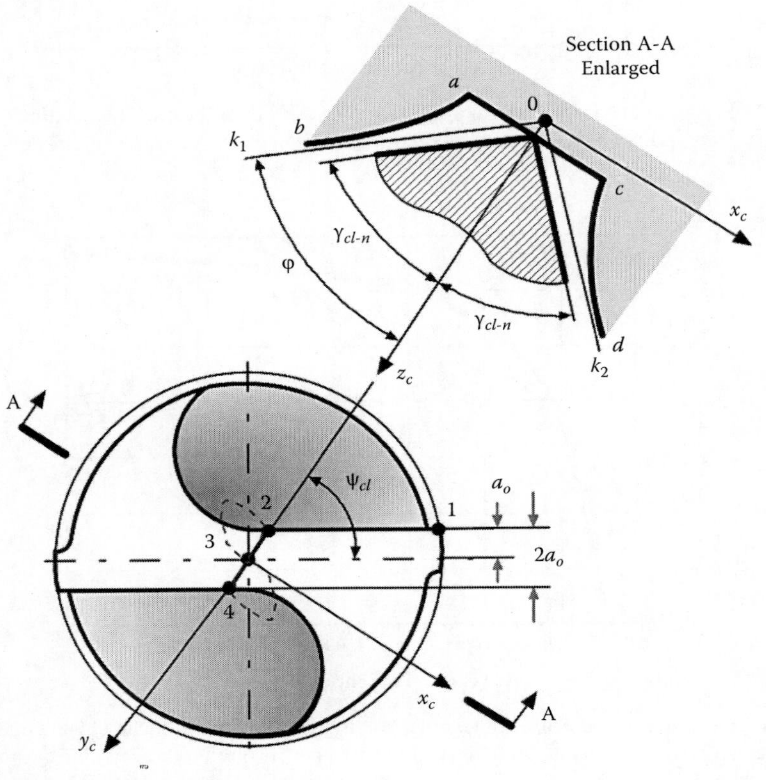

FIGURE 4.107 Model of cutting by the chisel edge.

The discussed problem is in a small gap between the chisel edge rake faces and the bottom of the hole being drilled represented by branches *ab* and *cd* of hyperbola. As shown in Figure 4.105, the rake angle of the chisel edge is highly negative so the uncut (undeformed) chip thickness in its transformation into the chip undergoes significant plastic deformation. As discussed in Appendix A, this deformation is fully characterized by the chip compression ratio ζ defined by the ratio of the chip thickness h_C and the uncut chip thickness h_D. As discussed in Chapter 1, the uncut chip thickness is calculated as $h_D = f_z \sin\varphi$, so if the cutting feed in drilling is $f = 0.4$ mm/rev and the drill has two teeth and point angle 140° (half-point angle $\varphi = 70°$), then $f_z = 0.2$ and $h_D = 0.2\cdot\sin 70° = 0.19$ mm. In machining of an aluminum alloy at the discussed highly negative rake angle, the chip compression ratio reaches 8 (Astakhov 2010), while that in machining of a high-manganese steel, it is 6 (Vinogradov 1985). As such, the chip thickness $h_C = 1.52$ mm in machining of aluminum and $h_C = 1.14$ mm in machining of high-manganese steel.

For normal drilling to take place, the chip formed by both halves of the chisel edge should pass through a narrow gap between the chisel edge rake face and the bottom of the hole being drilled that may create a problem as shown in Figure 4.108. To understand the problem, let's consider a practical example: drill of 20 mm dia., $2\varphi = 120°$, $\alpha_n = 15°$, $a_o = 1.5$ mm, and chisel edge length $ac = 1.7$ mm.

Due to a significant length of the chisel edge, first consider the gaps between the rake face of this edge and the bottom 0c. These gaps are measured over the x_c-axis in the direction parallel to the z_c-axis as

$$\delta_{cl\text{-}x} = x\tan(90° - \gamma_{cl\text{-}n}) \tag{4.60}$$

For the considered case, they are

x (mm)	0.2	0.4	0.6	0.8	1.0	1.7
$\delta_{cl\text{-}x}$ (mm)	0.13	0.26	0.39	0.52	0.65	1.13

The kinematic gaps will be greater than these calculated by *f*/4.

In its further flow, the chip must pass the gap between the rake face of the chisel edge and the walls of the bottom of the hole being drilled created by the major cutting edges. The gap thickness varies over the z-direction as

$$\delta_{cl\text{-}z} = \left[x\tan(90° - \gamma_{cl\text{-}n}) + a_o\tan^{-1}\psi_{cl}\tan^{-1}\varphi\right] - \frac{\sqrt{x^2-(2a_o)^2}}{\tan\varphi} \tag{4.61}$$

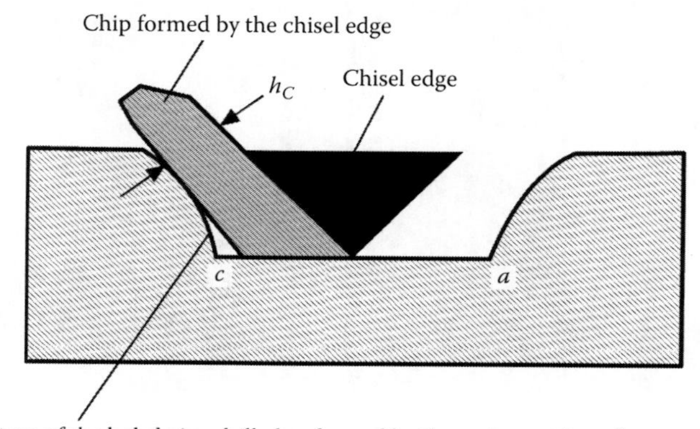

FIGURE 4.108 Schematic representation of the problem.

For the considered case, they are

x (mm)	1.7	2.0	2.5	3.0	5.0
$\delta_{cl\text{-}z}$ (mm)	1.10	0.93	0.97	0.96	1.0

The obtained results show that the gaps are small compared to the chip thickness. As such, when the chip formed by the chisel edge attempts to squeeze through the gap smaller than its thickness, the following might happen:

1. When the chip thickness is not much greater than the gap width, further chip plastic deformation takes place. As such, the chip is further deformed by the chisel edge rake face and by non-strain hardened work material. This explains why the axial force due to chisel edge is high (see Figure 4.25).
2. When the chip thickness is much greater than the gap width (a common case for high-penetration rate drills), then the chip spreads over the entire rake and flank faces of the chisel edge as shown in Figure 4.109. Any attempt to increase the drill penetration rate leads to harmful consequences as shown in Figures 4.110 and 4.111.

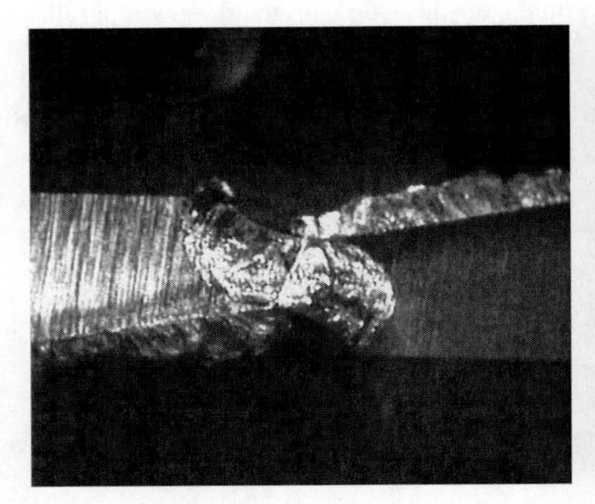
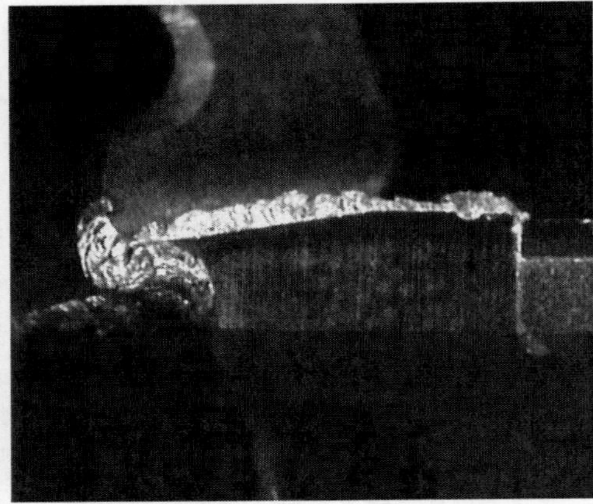

FIGURE 4.109 Chisel edge covered by the aluminum chip/built-up edge of a PCD-tipped drill.

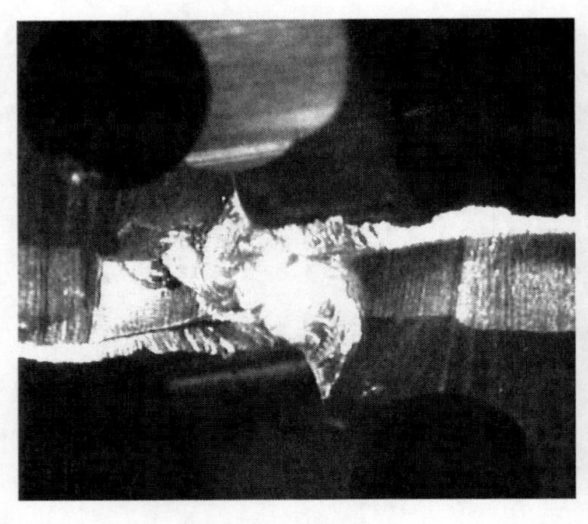
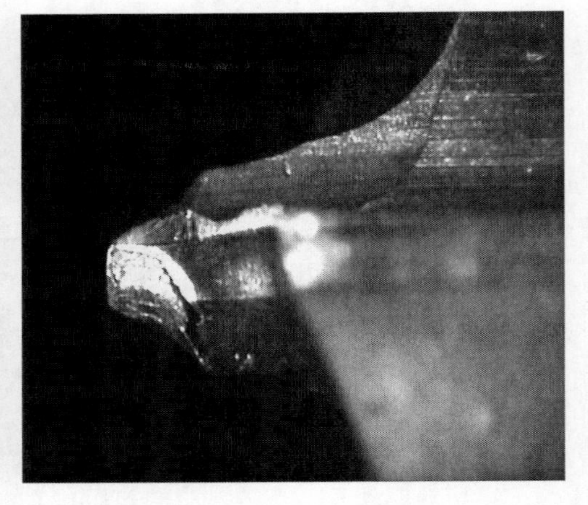

FIGURE 4.110 Consequences of an attempt to increase the drill penetration rate in machining of aluminum alloy A380—a cemented carbide straight-flute drill.

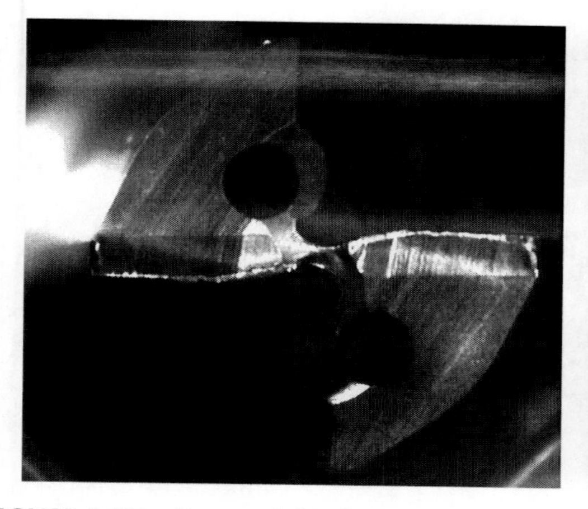
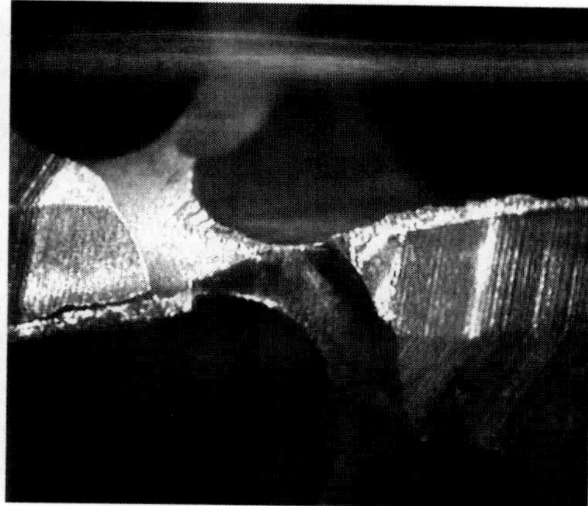

FIGURE 4.111 Fractured chisel edge region.

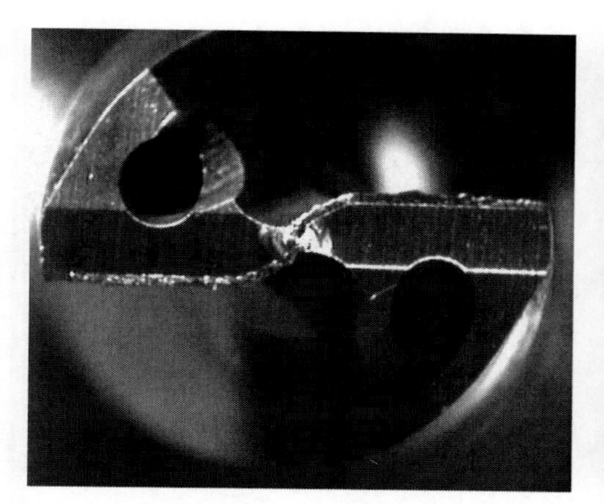
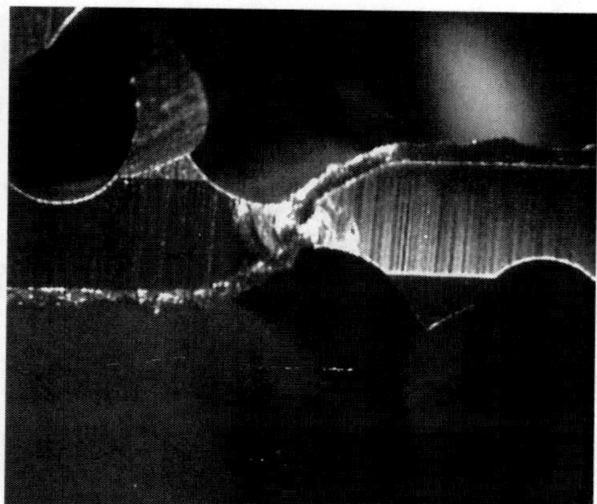

FIGURE 4.112 The chisel edge covered by the aluminum chip/built-up edge of a twist drill having a plane primary flank.

3. Attempts to apply deep gashes and use of a twist drill instead of a straight-flute drill do not solve the problems as shown in Figure 4.112, which remains in the same severity even if the length of the chisel edge is reduced by application of the web thinning. The observations of failures of thousands of drills of different designs and geometries and made of various tool materials allow to conclude that the known drill transverse (radial) vibration commonly attributed to the lack of drill symmetry occurs due to the described problem with the flow of the chip formed by the chisel edge. Figure 4.113 shows side margins with evidences of transverse (radial) vibrations. Such vibrations not only affect the machined hole shape and tolerance (important in drilling precision holes, e.g., in the automotive industry) but also reduce tool life. The harder the tool material, the greater the harmful effect of this vibration.
4. Attempts to change the planer flank face used in the drills shown in Figures 4.109 through 4.112 into conical or helical do not solve the problem. Figure 4.114 shows the consequences of an attempt to increase the drill penetration rate in machining of aluminum alloy A380 using a high-performance cemented carbide twist drill having a helical point grind.

FIGURE 4.113 Evidences of the drill transverse (radial) vibrations.

FIGURE 4.114 Consequences of an attempt to increase the drill penetration rate in machining of aluminum alloy A380 with a high-performance cemented carbide twist drill having a helical point grind.

4.5.8.3 Case 2: The Primary Flank Is Planar and the Width of the Primary Flank Is Equal to a_o

Significant improvement in drill self-centering and in the geometry of the chisel edge is achieved when the width of the primary flank is equal to a_o, while the secondary flank plane is applied with the normal clearance angle greater than that of the primary clearance angle. A model that visualizes the essential geometry parameters of such a drill is shown in Figure 4.115. As can be seen, the flank surface of each major cutting edge (lip) consists of two planes. The so-called primary flank plane is applied to the cutting edge 1-2 with the normal T-hand-S clearance angle $\alpha_{n1\text{-}1}$ and extends from this cutting edge to the drill transverse axis so that this plane is the flank plane for the chisel edge 2-3. The secondary flank plane is then applied with the normal T-hand-S clearance angle $\alpha_{n1\text{-}2}$, so that this plane serves as the rake plane for the chisel edge 4-3. Symmetrically, the primary and secondary flank planes are applied to the major cutting edge 4-5.

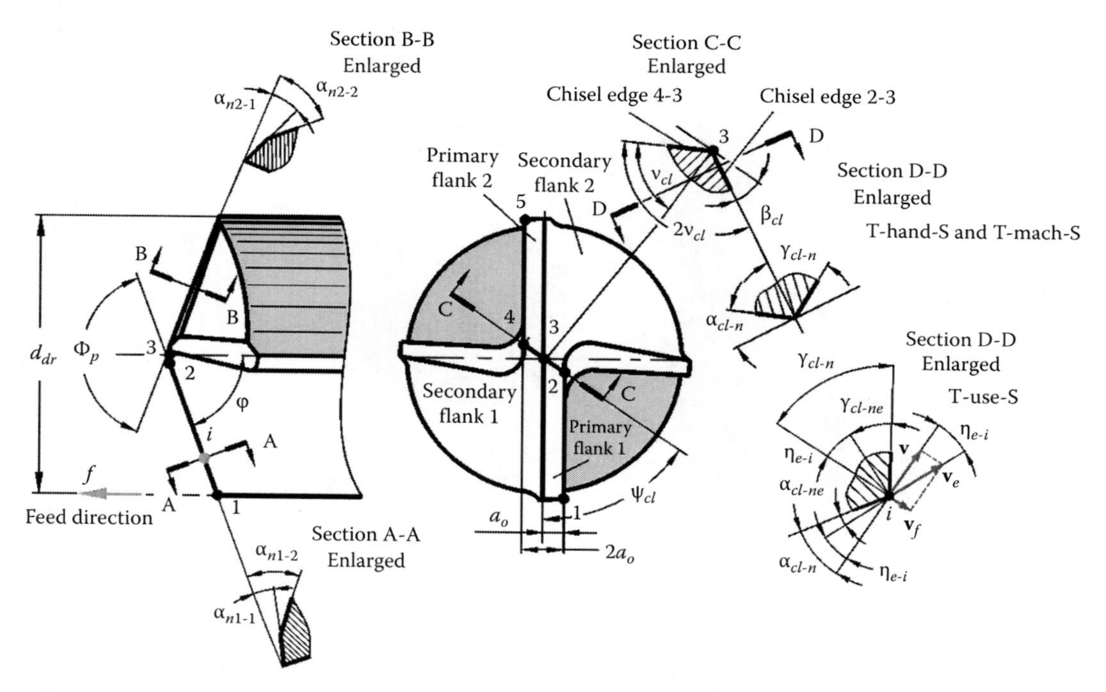

FIGURE 4.115 Chisel edge parameters visualization when formed by two flank planes having different clearance angles.

In the case considered, the two chisel edges 2-3 and 3-4 are no longer lines belonging in the tool back plane (Figure 4.65) as was in the previous case. Rather, each chisel edge makes an angle β_{cl} with this plane (section C-C in Figure 4.115). This angle is calculated as

$$\tan\beta_{cl} = \frac{(\tan\alpha_{n2\text{-}2} - \tan\alpha_{n1\text{-}1})\sin\psi_{cl}}{2\sin(\Phi_p/2)} \tag{4.62}$$

where the chisel edge angle ψ_{cl} is the angle measured in the tool back plane as the angle between the projection of the chisel edges 2-3 and 3-4 into this plane and the tool transverse axis as shown in Figure 4.115. The chisel edge angle in this case is calculated as

$$\psi_{cl} = \arctan\frac{2\cos(\Phi_p/2)}{\tan\alpha_{n2\text{-}2} + \tan\alpha_{n1\text{-}1}} \tag{4.63}$$

In tool drawings, the half chisel wedge angle ν_{cl} or chisel wedge angle $2\nu_{cl}$ is normally indicated as shown in Figure 4.115. The chisel wedge angle $2\nu_{cl}$ is calculated as

$$2\nu_{cl} = 180° - 2\arctan\frac{(\tan\alpha_{n2\text{-}2} - \tan\alpha_{n1\text{-}1})\sin\psi_{cl}}{2\sin(\Phi_p/2)} \tag{4.64}$$

The apex 3 formed at the intersection of chisel edges 2-3 and 4-3 (Figure 4.115) can be regarded as the centering point of the drill. As this apex first touches the workpiece at the beginning of drilling,

FIGURE 4.116 Chisel edge shape when it is formed by two flank planes having different flank angles.

it helps to reduce drill wandering and thus reduces drill transverse vibrations at the hole entrance, that is, a drill with such a point gains some self-centering ability. It was also found that this shape of the chisel edge makes the chisel wedge stronger and less susceptible to chipping. Figure 4.116 shows the discussed feature of the chisel edge.

As the chisel edge passes the axis of rotation (at least, theoretically), the rake and clearance angles of this edge are the same in T-hand-S and T-mach-S, that is, $\gamma_{cl\text{-}n1} = \gamma_{cl\text{-}n2} = \gamma_{cl\text{-}n} = \gamma_{cl\text{-}nm}$ and $\alpha_{cl\text{-}1} = \alpha_{cl\text{-}n2} = \alpha_{cl\text{-}n} = \alpha_{cl\text{-}nm}$. Therefore, the further analysis is made for only one-half of the chisel edge, that is, subscripts 1 and 2 are omitted.

The normal clearance angle of the chisel edge is calculated as

$$\alpha_{cl\text{-}n} = \arctan\left(\frac{1}{\tan(\Phi_p/2)\sin\psi_{cl}}\cos\beta_{cl}\right) \tag{4.65}$$

and the normal rake angle is calculated using Equation 4.58.

The length of each chisel edge $l_{2\text{-}3}$ (distance 2-3) that is equal to $l_{4\text{-}3}$ (distance 4-3) calculates accounting for β_{cl} as

$$l_{2\text{-}3} = l_{4\text{-}3} = \frac{2a_o}{2\sin\psi_{cl}\cos\beta_{cl}} \tag{4.66}$$

The rake and clearance angles of the chisel edge in T-use-S are considered accounting for the fact that the half-point angle of this edge is $\nu_{cl.}$ As such, the T-use-S rake and clearance angles considered point i of the chisel edge are calculated as

$$\gamma_{cl\text{-}ne\text{-}i} = \gamma_{cl\text{-}nm} + \upsilon_{e\text{-}i} \tag{4.67}$$

$$\alpha_{cl\text{-}ne\text{-}i} = \alpha_{cl\text{-}nm} - \upsilon_{e\text{-}i} \tag{4.68}$$

where $\upsilon_{e\text{-}i} = \arctan(\tan\eta_{e\text{-}i}\sin\nu_{cl})$ where η_e is determined according to Equation 4.26.

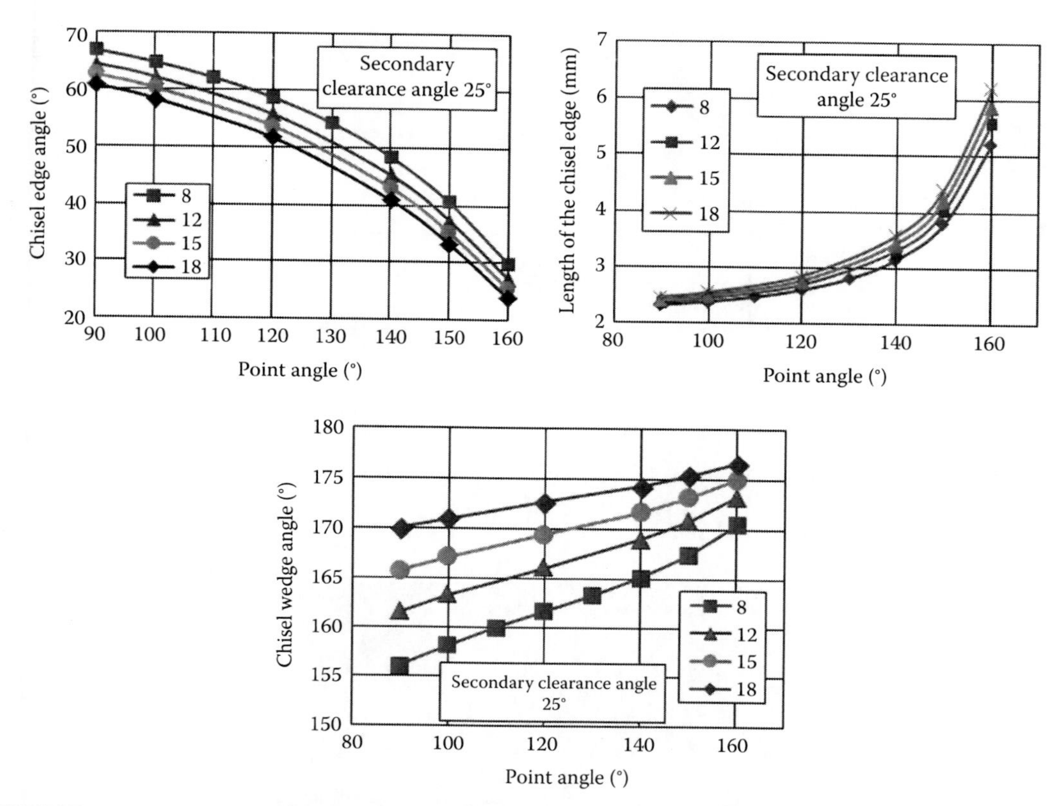

FIGURE 4.117 Chisel edge geometry parameters for a drill of 14.48 mm dia., $a_o = 0.6$ mm when the flank surface is formed by two planes as shown in Figure 4.115.

Figure 4.117 gives an example of the chisel edge parameters for the considered case. As can be seen, if the flank face is formed by two planes as shown in Figure 4.115, then the angle and thus the length of the chisel edge both increase. The chisel wedge angle $2\nu_{cl}$ and thus drill self-centering increase with the reduction of the point angle.

Figures 4.118 and 4.119 show examples of the kinematic rake and clearance angles distribution over the chisel edge. Although these are better compare to the previously-considered case,

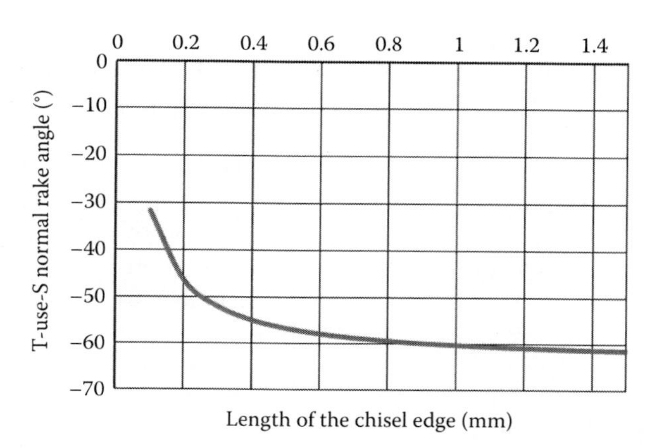

FIGURE 4.118 Variation of the T-use-S rake angle over the length of the chisel edge (the T-hand-S normal clearance angles: $\alpha_{n1} = 10°$ and $\alpha_{n2} = 25°$, point angle $\Phi_p = 140°$, feed per revolution $f = 0.4$ mm/rev, and the length of the chisel edge 1.5 mm).

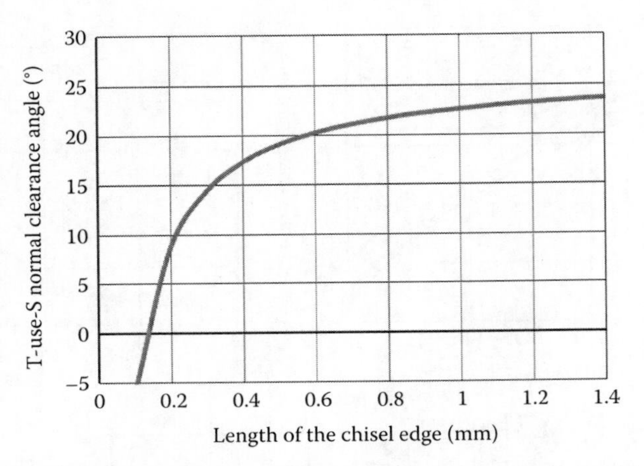

FIGURE 4.119 Variation of the T-use-S clearance angle over the length of the chisel edge (the T-hand-S normal clearance angles: $\alpha_{n1} = 10°$ and $\alpha_{n2} = 25°$, point angle $\Phi_p = 140°$, feed per revolution $f = 0.4$ mm/rev, and the length of the chisel edge 1.5 mm).

FIGURE 4.120 A common point grinding flaw—the second flank planes do not form the intended chisel edge.

the interference of the chisel flank adjacent to the axis of rotation exists for this point geometry. The following should be taken into consideration:

- To gain the self-centering advantage with less negative rake angle and thus greater allowable penetration rate as the gap between the chisel edge rake face and the wall of the hole being drilled (Figure 4.108) becomes greater, the point grinding machine setup should be very accurate, particularly for small drill diameters (less than 10 mm). Figure 4.120 shows a common point grinding flaw—the second flank planes do not form the intended chisel edge as the lines of their intersection with the primary flank faces do not pass through the axis of rotation. The net result is a non-straight chisel edge consisting of three segments.
- Experience shows that the gains due to the self-centering point and less negative rake angle are balanced by the increased length of the chisel edge so that both considered cases are practically the same in terms of the contribution of the chisel edge to the axial force of drilling.

4.5.8.4 Drill Flank Is Formed by Two Surfaces (Generalization: Tertiary Flank Plane and Split Point)

To improve the chisel edge geometry, mainly to reduce its negative rake angle, the second flank surface can be angularly located with respect to the primary flank surface. Figure 4.121 shows one of the earlier designs of such a flank face where the primary and secondary flank surfaces are not planes. This drill

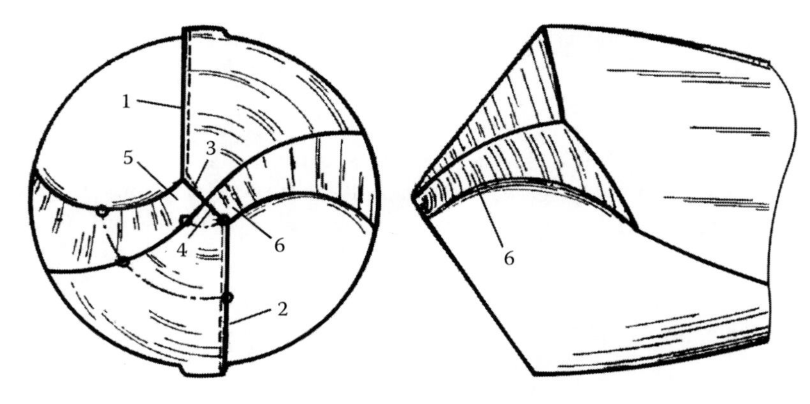

FIGURE 4.121 Drill geometry according to US Patent No. 3,564,947 (1964).

has two major cutting edges 1 and 2 and two chisel edges 3 and 4. The chisel edges are provided with fully developed lands 5 and 6 that extend to the heels. Such a location of the secondary flanks improves the geometry of the chisel edges and provide better conditions for the chip formed by the chisel edge to pass the gar between the bottom of the hole being drilled and the rake face of the chisel edge.

In modern designs, the primary and secondary flank surfaces are normally planes. A general objective in forming two-plane flank can be thought of as the finding of the location of flank planes *R* and *F* (Figure 4.122) to assure the desired chisel edge angle, ψ_{cl} (to maintain a certain length of the chisel edge). The location of plane *R* is uniquely defined by the selected point angle Φ_p and the desired T-hand-S clearance angle $\alpha_{n\text{-}R}$ at point 1. Therefore, the problem reduces to finding the

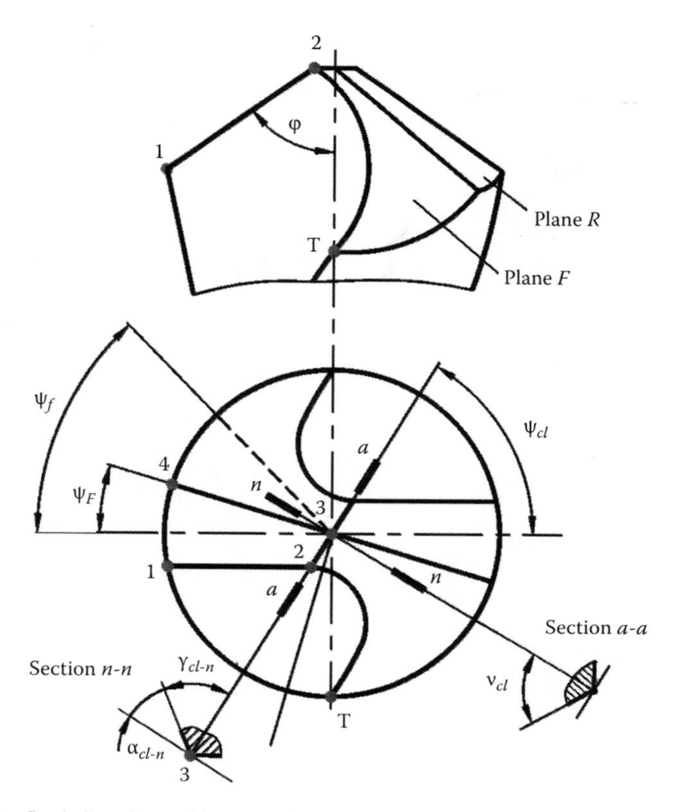

FIGURE 4.122 The flank face formed by two planes—general case.

location of plane F. The solution of such an ambitious problem is presented earlier (Rodin 1971; Astakhov 2010). In this section, a much less ambitious but practical evaluation of the geometry parameters is considered.

In practice, plane R having T-hand-S clearance angle $\alpha_{n\text{-}R}$ is first ground. Then, the grinding wheel is set with angle $\alpha_{n\text{-}F}$ and the drill is revolved by a certain angle ψ_f and the second flank plane F is ground. Angles $\alpha_{n\text{-}F}$ and ψ_f together with the half-point angle φ define the geometry of the chisel edge, namely, its inclination angle ψ_{cl}, half-wedge angle ν_{cl}, T-hand-S (T-mach-S) rake $\gamma_{cl\text{-}n}$, and clearance $\alpha_{cl\text{-}n}$ angles.

Angle ψ_F is determined as (Vinogradov 1985)

$$\psi_F = \arccos\left(\frac{A}{\sqrt{B+C+D+E}}\right) \tag{4.69}$$

where

$$A = \cos\psi_f\,(\sin\alpha_{f\text{-}F} + \tan\psi_f\,\tan^{-1}\varphi) - \cos\alpha_{f\text{-}F}\,\tan\alpha_{f\text{-}R}$$

$$B = \sin^2\psi_f\left(\frac{1}{\tan^2\varphi} + \sin^2\alpha_{f\text{-}F}\right)$$

$$C = \cos^2\alpha_{f\text{-}F}\,(\tan\alpha_{f\text{-}R} - \cos\psi_f\,\tan\psi_f)^2$$

$$D = \frac{1}{\tan^2\varphi}\left[\left(\frac{\cos\alpha_{f\text{-}F}}{\cos\alpha_{f\text{-}R}}\right) - \cos\psi_f\right]^2$$

$$E = 2\sin\psi_f\,\tan^{-1}\varphi\cos\alpha_{f\text{-}F}\left[\left(\frac{\cos\alpha_{f\text{-}F}}{\cos\alpha_{f\text{-}R}}\right) - \tan\alpha_{f\text{-}R}\right]$$

where the T-hand-S clearance angles of flank planes F and R in the working plane are

$$\alpha_{f\text{-}F} = \arctan\left(\frac{\tan\alpha_{n\text{-}F}}{\sin\varphi}\right)$$

$$\alpha_{f\text{-}R} = \arctan\left(\frac{\tan\alpha_{n\text{-}R}}{\sin\varphi}\right)$$

The chisel edge angle is calculated as

$$\psi_{cl} = \frac{\sin(\varphi_{o1}+\varphi)}{\tan\alpha_{o1}\sin\varphi_{o1}\sin\varphi - \tan^{-1}\psi_R\,(\sin\varphi\cos\varphi_{o1} - \sin(\varphi_{o1}+\varphi))} \tag{4.70}$$

where

$$\varphi_{o1} = \arctan\left[\frac{\cos\psi_F}{(\sin\psi_F\,\tan\alpha_{o1} - \tan\varphi_o)}\right]$$

where φ_o is the angle between planes R and F measured in the axial plane perpendicular to the intersection rib 3-4 (Figure 4.122):

$$\varphi_o = \arctan(\tan^{-1}\varphi\cos\psi_F + \sin\psi_F\tan\alpha_{o1})$$

$$\alpha_{o1} = \arctan(\cos\psi_F\tan\alpha_{oN} + \sin\psi_F\tan\varphi_o)$$

$$\alpha_{oN} = \arctan\left[\frac{\left(\tan\alpha_{f-F} - \sin(\psi_F - \psi_f)\tan^{-1}\varphi_o\right)}{\cos(\psi_F - \psi_f)}\right]$$

$$\psi_R = \arctan\left(\frac{\tan\alpha_R}{\cos\varphi}\right)$$

The chisel wedge angle $2\nu_{cl}$ is calculated as

$$2\nu_{cl} = 180° - 2\arctan\left(\sin\psi_{cl}\tan\alpha_{o1} - \cos\psi_{cl}\tan^{-1}\varphi_{o1}\right) \quad (4.71)$$

The T-hand-S (T-mach-S) normal rake angle of the chisel edge is calculated as

$$\gamma_{cl\text{-}n} = -\arctan\left[\frac{\sin\psi_{cl}}{\left(\cos\psi_{cl}\tan\nu_{cl} + \tan^{-1}\varphi_{o1}\right)}\right] \quad (4.72)$$

The T-hand-S (T-mach-S) normal clearance angle of the chisel edge is calculated as

$$\alpha_{cl\text{-}n} = \arctan\left[\frac{\left(\tan^{-1}\varphi - \cos\psi_{cl}\right)\tan\nu_{cl}}{\sin\psi_{cl}}\right] \quad (4.73)$$

Analysis of Equations 4.69 through 4.73 reveals the following features of the discussed geometry of the flank face:

- It allows to gain the self-centering advantage with less negative rake angle and thus greater allowable penetration rate as the gap between the chisel edge rake face and the wall of the hole being drilled (Figure 4.108) becomes greater. The problem with the chip flow over the rake face of the drill can be solved permanently.
- The rake angle of the chisel edge is mainly determined by the T-hand-S working angle $\alpha_{f\text{-}F}$ and is less dependent on ψ_F.
- The clearance angle of the chisel edge depends entirely on the chisel edge angle as this secondary plane F does not affect the primary flank face.
- When the T-hand-S normal clearance angles of the primary R and secondary F planes are chosen to be small, then interference on top of the secondary flank heel (point T in Figure 4.122) can occur. The fulfillment of the following conditions prevents such interference:

$$0.5\left[d_{dr}\left(\tan\alpha_{o1} - \tan^{-1}\varphi\right) - f/4\right] \geq 0.5\ldots2.0 \text{ mm} \quad (4.74)$$

The proper use of the discussed geometry of the flank face allows to reduce the negative T-hand-S (T-mach-S) rake angle to –50° that significantly improves chip formation on the chisel edge. The price to pay is an increased length of the chisel edge. For example, when $\psi_f = 15°$ (as such $\psi_F = 16.3°$), $\alpha_{n\text{-}R} = 15°$,

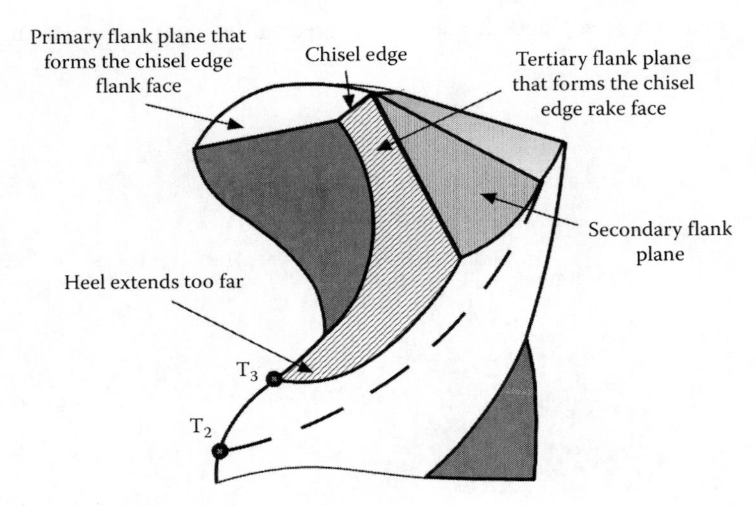

FIGURE 4.123 The flank face consisting of three flank planes.

$\alpha_{n-F} = 25°$, and $2\varphi = 120°$, the chisel edge angle $\psi_{cl} = 43.8°$. For comparison, when the width of the primary flank is equal to or greater than $2a_o$ (case 1 considered in Section 4.5.8.2), this angle is equal 61.8°.

The implementation practice of such drills showed that when the T-hand-S (T-mach-S) rake angle is reduced to −50°, the drill terminal end becomes weak as the top of the heel point *T* locates too far from the cutting edge (Vinogradov 1985). It may cause drill vibration at the hole entrance and undermines the use of the additional stabilizing margins.

To ease the problem with the location of point *T* and significantly improve the rake face geometry of the chisel edge, the flank face consisting of three flank planes can be used. A drill with such a flank face is shown in Figure 4.123. When the tertiary flank plane is applied, the axial clearance angle α_{f-F} and the location angle ψ_F of this plane shouldn't be great as the secondary flank is meant to *win* some space so that the location of point *T* should be much closer to the cutting edge. Figure 4.123 qualitatively compares locations of this point when two (point T_2) and three (point T_3) are applied.

According to the author's experience, good results can be achieved when the first two flank planes are ground in the manner discussed in Section 4.5.8.3 (Case 2) and the tertiary plane is applied just to improve the rake angle of the chisel edge while keeping its length unchanged. Any violation of this condition can ruin the drill. Figure 4.124 shows an example of improper location of the tertiary flank plane

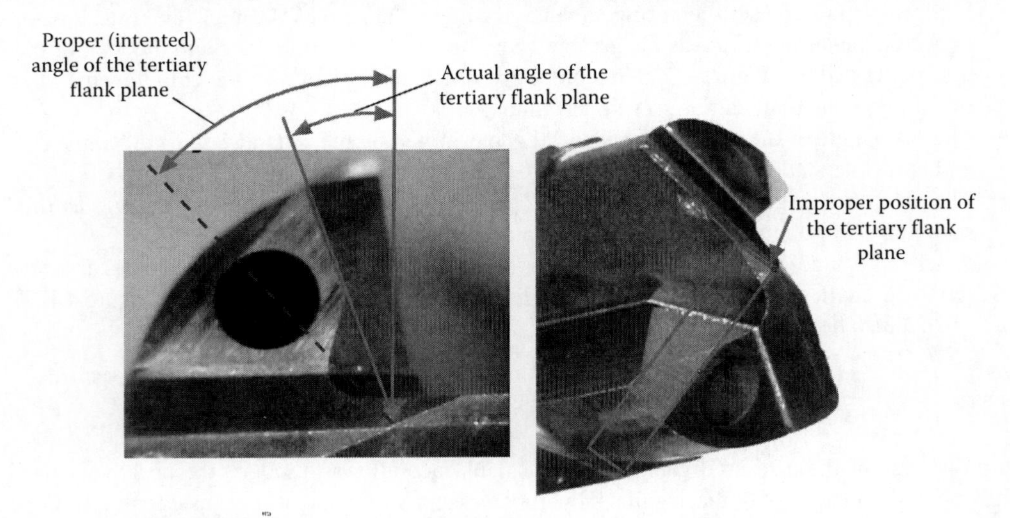

FIGURE 4.124 Improper location of the tertiary flank plane.

FIGURE 4.125 Consequences of the improper location of the tertiary flank plane: (a) interparty applied ternary plane and (b) chipping of the drill periphery.

on a full (cross) PCD drill meant for high-penetration rate drilling. Figure 4.125 shows that formed by this plane, the chisel edge is too long while its rake face geometry (assures less negative rake angle and condition of chip free motion over the rake face as shown in Figure 4.108) is not improved. These resulted in drill transverse vibrations and thus chipping of the drill corners of the major cutting edges on the first drilled hole. Traces of the chip flow can clearly be seen on the *non-working* flank faces of both parts of the chisel edge where no chip should be found provided that the chip has sufficient room to flow in the gap between the chisel edge rake face and the wall of the bottom of the hole being drilled.

The foregoing analysis reveals that trying to improve the rake face geometry (both the rake angle and the gap between this face and the bottom of the hole being drilled), one unavoidably runs into the problem with weakening the drill (the location of the top of the heel) no matter how many flank faces are used. To solve such a problem, a local modification of the rake face geometry can be used. Such a modification is known as the split-point flank geometry.

As early as in 1923, Oliver patented (US Patent No. 1,467,491) (1923) a very distinctive drill with split point shown in Figure 4.126. The drill has two major cutting edges 1 and 2 and two chisel edges

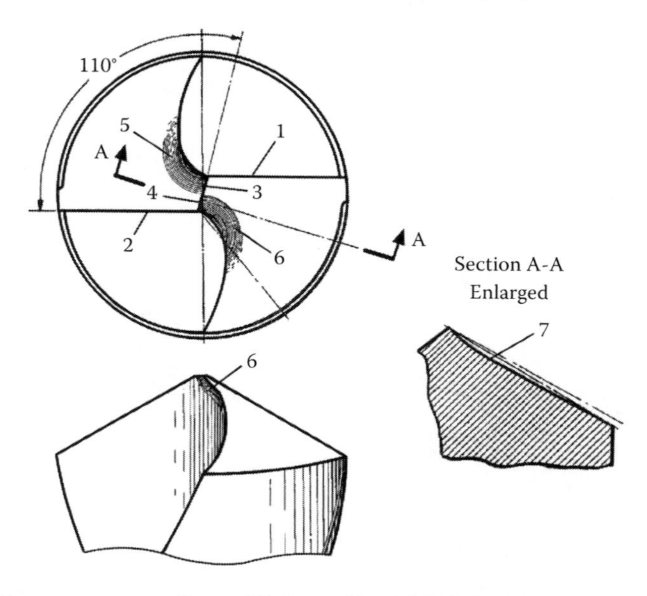

FIGURE 4.126 Drill geometry according to US Patent No. 1,467,491 (1923).

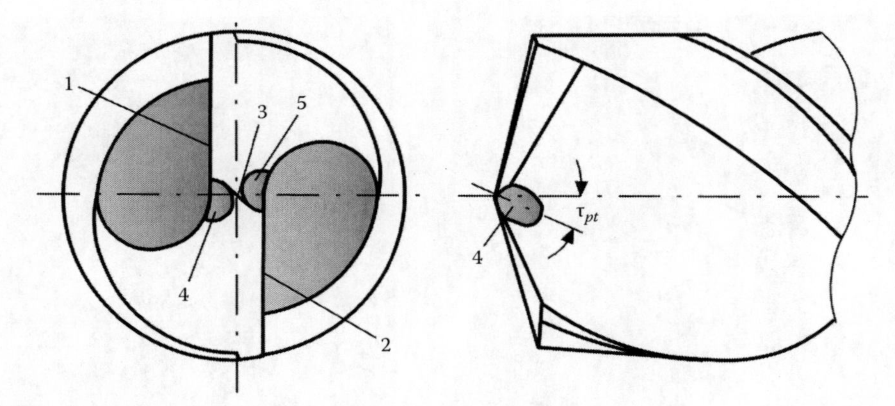

FIGURE 4.127 Drill geometry according to US Patent No. 5,590,987 (1997).

3 and 4. The rake faces of these edges are provided with two depressions 5 and 6. As a result, the rake face 7 obtains the rake angle, which can be varied according to particular work material. As claimed in the patent description, such geometry allows at least 50% higher penetration rate compared to a drill with the standard geometry. Although it is a great design, it was, in the author's opinion, much ahead of the time as grinding fixtures available in that time did not allow reproducing such a flank geometry with any reasonable accuracy that is crucial for its proper implementation. Moreover, the available drilling systems were not capable of significant increase in the drill penetration rate due to the lack of strength of the feeding mechanisms and feed accuracy. As a result, this great flank geometry was not requested for more than 50 years.

Figure 4.127 shows a drill having the simplest partial split geometry. As seen, the drill has two major cutting edges 1 and 2 and the chisel edge 3. Two pits 4 and 5 are ground as shown to provide rake faces to the corresponding parts of the chisel edge. Angle τ_{pt} is adjusted to provide the intended rake angle to the parts of the chisel edge and facilitate chip removal (Bhattacharyya et al. 1971). Still, this geometry is less efficient compared to that shown in Figure 4.126.

The full realization of the potential of the split-point geometry is offered by US Patents No. 4,556,347 (1985) and 4,898,503 (1990). Such geometry is shown in Figure 4.128. According to these patents, the chisel edge 1 is provided with the rake face (notch) 2 having rake angle of between 5° and 10°, while the angle of the notch to the drill axis is selected to be between 32⁰ and 38°. As claimed by US Patent No. 4,556,347 (1985), the comparison of the performance of this drill with a commercially available precision twist drill conforming to NAS 907 standard showed significant improvement in tool life when drilling difficult-to-machine materials. For example, a 5.8 times increase in tool life was achieved in drilling Inconel 718 of 44 HRC.

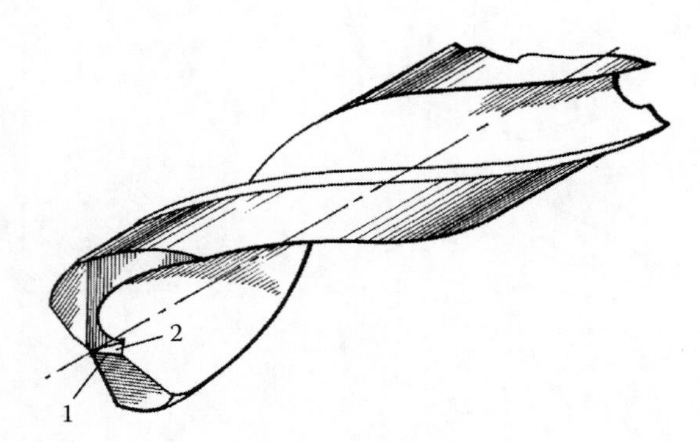

FIGURE 4.128 Drill geometry according to US Patents No. 4,556,347 (1985) and 4,898,503 (1990).

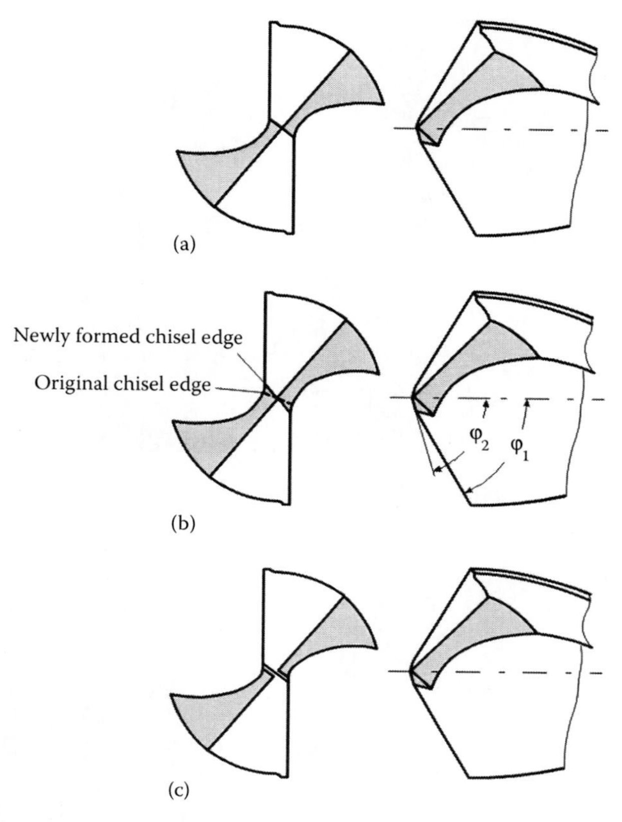

FIGURE 4.129 Split-point geometries recommended by ACTI. (From *Metal Cutting Tool Handbook*, 7th edn., The Metal Cutting Tool Institute by Industrial Press, New York, 1989.)

American Cutting Tool Institute recommends (*Metal Cutting Tool Handbook* 1989) the following split-point geometries shown in Figure 4.129. Figure 4.129a shows a split point with notches ground right along and coinciding with the chisel edge. Figure 4.129b shows the split point similar to Figure 4.129a, except that the notches are at *an angle* to the original chisel edge. This type of the split point enables the drill to center itself as the apex of the drill point contacts the surface of the workpiece first. Note that the half-point angle of the apex, φ_2, is greater than that of the drill (φ_1). Figure 4.129c shows a split point with the original chisel edge intact. The notches are parallel but separated. While the highly negative rake at the chisel edge remains, it does so only on the part of the contact length of the rake face. Moreover, such geometry provides a means of egress for the chip formed by the chisel edge.

The best application results with the discussed split-point geometries are achieved when the rake angles of the chisel edge are selected properly, the rake faces of the chisel edges are ground symmetrically, and the proper grade of the tool material combined with rigid systems and internal MWF supply are used. Figure 4.130 shows WSTAR solid-carbide drills for machining of aluminum developed and manufactured by Mitsubishi Materials Corporation.

Further improvements in split-point drills are achieved when the sharp corner between the chisel edge and the major cutting edge is rounded as shown in Figure 4.131. As can be seen in this figure, the chisel edge becomes *S* shaped due to a special grind of the tertiary flank surface that is obviously not a plane. A reduction of the axial force by 20% and smooth chip flow from the chisel edge is achieved with such a design of the chisel edge and its rake face. It is clear that *S*-shaped split point must be ground on special CNC grinders to assure symmetry of both parts of the chisel edge that is a crucial requirement to achieve the intended performance of such drills.

FIGURE 4.130 WSTAR drill by Mitsubishi Materials Corporation.

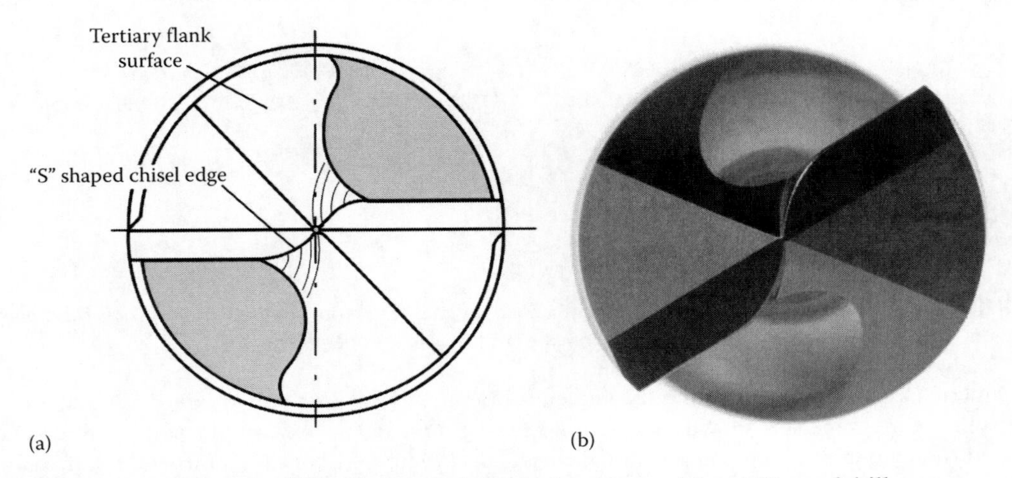

FIGURE 4.131 A split-point drill with *S*-shaped chisel edge: (a) a model, and (b) a real drill.

Although the split-point geometry seems to solve all the problems with the rake face geometry (a positive rake angle of this edge can be applied) and the flow of the chip formed by the chisel edge (the chip formed by the chisel edge apparently has plenty of room between the rake face of the chisel edge and the walls of the hole being drilled), it is not quite so. Another significant problem remains with the most known designs of the split-point chisel edge. As this problem is not clear for many drill designers, its clarification is needed as this should be solved in any *VPA-balanced* drill design.

Figure 4.132 shows the length l_{sp} of the rake face of the split-point chisel edge and the transition radius r_{sp} between this face and the third flank plane. According to the traditional split-point geometry, length l_{sp} is at maximum at point 2 and then decreases toward point 3 becoming virtually zero at this point.

To comprehend the problem, one should compare this length with the tool–chip contact length known as the length of the tool–chip interface discussed in Section C.3.2.1. According to Equations C.26 and C.27, this length is calculated knowing the uncut chip thickness h_D and the chip compression ratio ζ as

$$l_c = h_D \zeta^{k_r} \tag{4.75}$$

where $k_r = 1.5$ when $\zeta < 4$ and $k_r = 1.3$ when $\zeta \geq 4$.

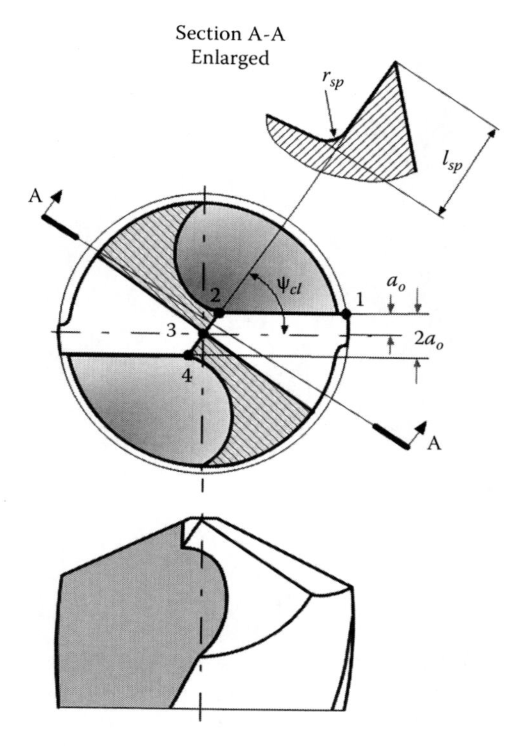

FIGURE 4.132 Representation of the length of the rake face of the split-point chisel edge.

As discussed earlier (Section 4.5.8.2), the chip compression ratio for the chip formed by the chisel edge in drilling of aluminum alloys may be as high as 8 while that in drilling steels can be as high as 6; hence, $k_r = 1.3$ in Equation 4.75 should be considered. As such, the tool–chip contact length for the example considered in Section 4.5.8.2 is in drilling an aluminum alloy $l_c = 2.84$ mm and in drilling of a high-manganese steel $l_c = 1.95$ mm. The condition of the chip free flow can be thought of as $l_{sp} > l_c$ so that the forming chip does not have an obstacle in its way to disturb its flow. Moreover, the transition radius r_{sp} should be rather generous exceeding at least the chip thickness h_C. Unfortunately, these two conditions are not met in the traditional split-point geometry that restricts its use in high-penetration rate drills.

Figure 4.133 shows a drill design according to US Patent No. 6,071,046 (2000) that combines several previously described modifications to the chisel edge geometry. It has *S* helical relieved point grind, partial split geometry, and the further modification of the remaining sections of the chisel edge that allows greater tool–chip contact length on the rake face of the chisel edge. The drill has two major cutting edges 1 and 2 and two portions 3 and 4 of the chisel edge provided with the rake surface 5 and 6. The remaining part 7 of this chisel edge having a concave shape is provided with recesses 8 and 9 formed in the front flank faces, each recess providing one chisel edge cutting section with the rake surface and the other with the flank surface. The portion of each recess that constitutes the rake surface merges with an adjacent web-thinning recess surface. Thereby, improved chip evacuation from the major cutting edges and, particularly, from the chisel edge cutting section may be effectively obtained as claimed by the patent. Self-centering ability is another advantage of this geometry. A complicated point grind and its metrology as well as sensitivity to the drill runout (sets the middle of the chisel edge out of the center of rotation) are shortcomings of this geometry.

4.5.8.5 Modifications of the Chisel Edge

The first question to ask before reading this section is rather simple: "Why does one need to modify the chisel edge even further if the previously discussed 'latest and greatest' designs apparently

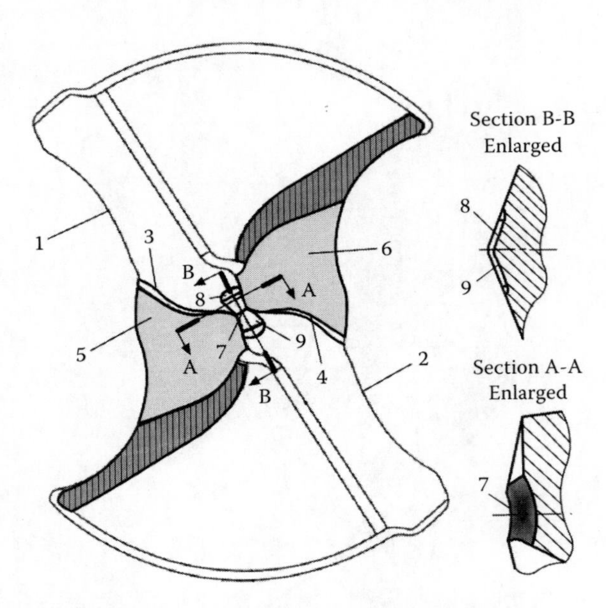

FIGURE 4.133 Drill design according to US Patent No. 6,071,046 (2000).

'cure' the problems of this edge?" A simple answer to this question is also: "They don't." Although one may argue that the split-point design can be further improved to account for the real tool–chip contact length assuring the proper direction of the chip flow from the chisel edge, the inherent problem with interference of this edge (see Figures 4.106 and 4.119) cannot be solved using this or other previously discussed designs. Having noticed many problems in drill performance due to the chisel edge, its modifications were attempted.

There are two principal directions in modifications of the chisel edge. The first one is its complete elimination (a radical approach *let's first eliminate then think*), while the second is altering its location defining it as based on the kinematics and mechanics of metal cutting (an intelligent approach *cure where it hurts* imbedded in the *VPA-balanced* drill design concept).

4.5.8.5.1 Eliminating the Chisel Edge

The problems created by the chisel edge were always in the center of attention in the drill design (Thornley et al. 1987b). The most radical solution to these problems is the total elimination of this edge from the drill point design. As early as 1911, Mather patented a drill with no chisel edge (US Patent No. 989,379 [1911]). In the proposed drill shown in Figure 4.134, a common twist drill 1 is

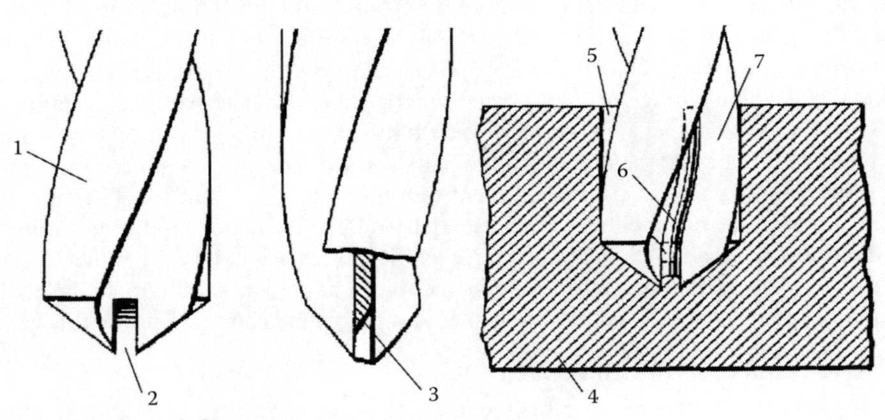

FIGURE 4.134 Drill design according to US Patent No. 989,379 (1911).

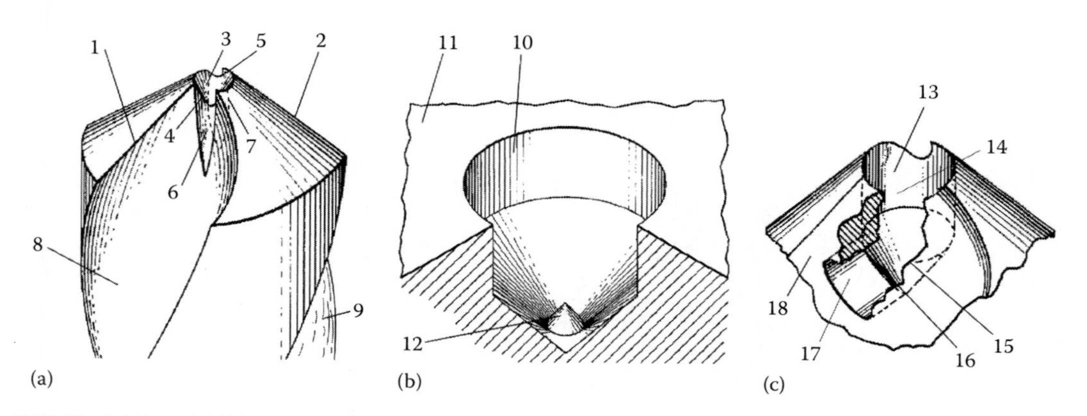

FIGURE 4.135 Drill design according to US Patent No. 3,028,773 (1962): (a) the basic design, (b) the shape of the bottom of the hole being drilled, and (c) a modification of the basic design.

provided at the apex of the web with the slot 2. This slot having the width equal to the web diameter extends upward into the drill body and terminates at the upper end in the inclined surface 3. When drill 1 penetrated into workpiece 4 drilling hole 5, core 6 forms. This core either breaks due to bending when its front end comes into contact with inclined surface 3 or, when the work material is ductile, bends into the chip flute 7 and thus is removed at the end of drilling.

The next basic solution of the chisel edge problems shown in Figure 4.135a is described in US Patent No. 3,028,773 (1962). As can be seen, a drill has two major cutting edges 1 and 2. Conical surface 3 is formed instead of the chisel edge. This surface 3 has cutting edges 4 and 5 and two secondary flutes 6 and 7, which communicate with the flutes 8 and 9 of the major cutting edges 1 and 2. When such a drill works, that is, drills a hole 10 in workpiece 11, protuberance 12 of conical shape (Figure 4.135b) is formed by the cutting edges 4 and 5. As claimed by the inventor, protuberance 12 acts as a journal for the drill as it progresses through the workpiece. This journal and bearing arrangement supports the drill at a point adjacent to the drill center thereby assuring that the hole being drilled and the drill remain concentric.

Figure 4.135c shows a portion of another embodiment of the twist drill shown in Figure 4.135a. As can be seen, the opening 13, concentric with the drill axis, comprises of a cylindrical portion 14 and conical portion 15. According to the patent, the major portion of the bearing load is sustained by surface 14. Cutting edge 16 that maintains the maximum length of the protuberance acts along conical surface 15 that has hole 17. Hole 17 is in communication with the front part 18 of flute 8.

Although the idea of self-support of the discussed drill is valuable and can be successfully implemented, obvious drawbacks of the proposed design and geometry should be noted:

- It is rather difficult to grind the proposed drill as a specially designed and dressed grinding wheel together with grinding fixture and metrological equipment are needed. Re-sharpening of this drill presents even a greater challenge.
- The internal cutting edges, that is, 4 and 5, actually shave the work material so that they should be made sharp to perform the shaving action. Smallest built-up due to low speed and high contact pressure can easily disable these edges.
- The bending of the protuberance into hole 17 may actually disturb rather than improve self-support of the drill.

The subsequent developments, for example, described in US Patents No. 4,143,723 (1979), 4,342,368 (1982) and well summarized in US Patent No. 4,373,839 (1983) did not offer new ideas. Rather, various design applications of the ideas described earlier particular to various drill configurations were attempted.

A special drill point design with split-point geometry shown in Figure 4.136 was developed by Hallden (US Patent No. 2,334,089 [1939]) that, in the author's opinion, opened a new line of truly

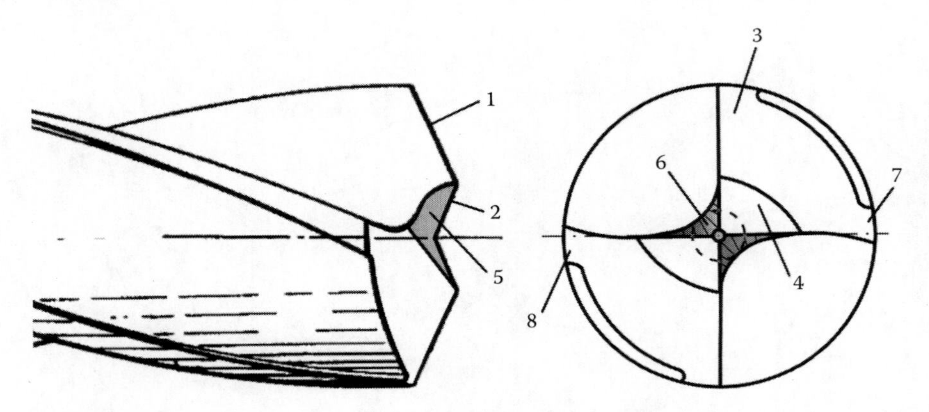

FIGURE 4.136 Drill geometry according to US Patent No. 2,334,089 (1939).

self-stabilizing drills. These can be referred to as double-apex drills (widely known as W-point drills). Each major cutting edge of this drill has two portions, namely, the outer 1 and the inner 2 cutting edges and corresponding flank surfaces 3 and 4. Flank surface 4 is shaped as an invert conical surface as shown in Figure 4.136. Rake face 5 of the region of inner cutting edge 2 adjacent to the drill center is formed by providing (grinding) gash 6.

There are a number of advantages of the proposed design not realized by the inventor and apparently by the subsequent specialists who did not notice such a great leap ahead in the design of drills:

- Because the cutting edges locate on the axis of the drill, there are no variations of the rake and clearance angles along these edges over the drill radius as in the traditional drill design.
- While drilling, inner cutting edge 2 forms a conical surface on the drill bottom that definitely stabilizes the drill preventing its wandering and thus improving drilling stability and quality of the hole being drilled. This is due to the presence of stabilizing cone 1 formed at the bottom of the hole being drilled as shown in Figure 4.137. In the presence of cone, any unbalanced radial force does not act on the walls 2 of the hole being drilled through the side margins thus preventing drill radial vibration and deterioration of the drilled hole. Rather it acts on the cone forcing one of the inner cutting edges to cut deeper to compensate this unbalanced force stabilizing in this manner drill performance. Additional supporting lands 7 and 8 (Figure 4.136) shown in the patent would interfere with drill stability as any redundant supports in mechanical systems.

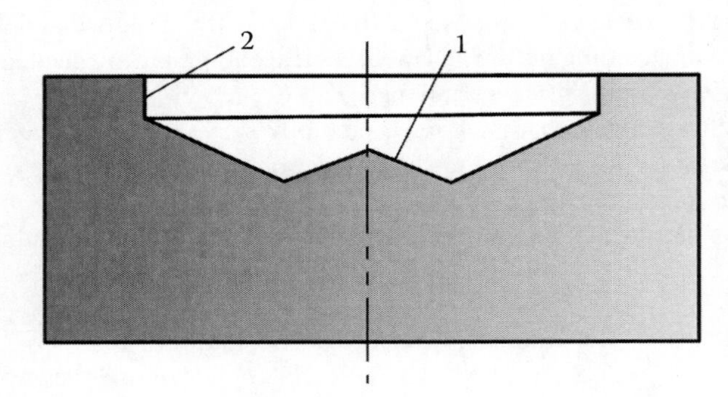

FIGURE 4.137 The bottom of the hole being drilled by the drill shown in Figure 4.136.

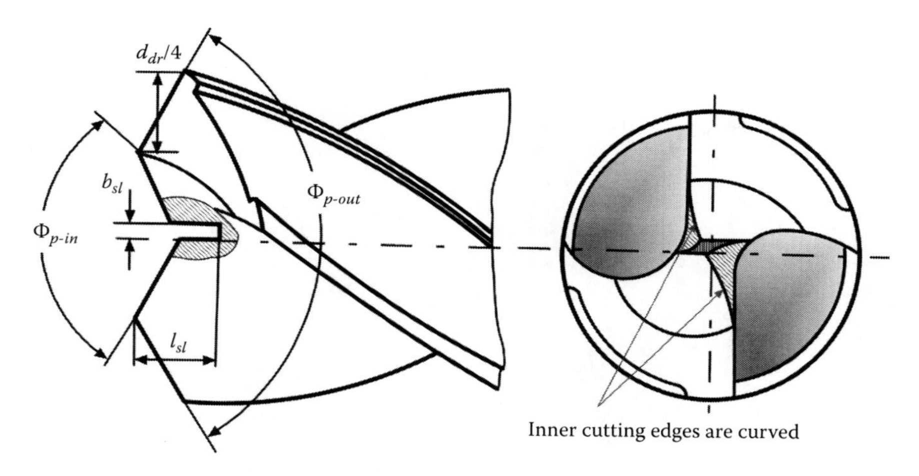

FIGURE 4.138 One of the modern designs of double-point drills.

The drill design shown in Figure 4.136 has a significant drawback: the inner cutting edges must meet exactly at the axis of rotation. Otherwise, a pin forms at the bottom of the hole being drilled. The diameter of this pin is equal to double the shift of the point of intersection of the inner cutting edge and the axis of rotation.

One of the modern designs of a double-apex drill is shown in Figure 4.138. A slot having length l_{sl} and width b_{sl} at the intersection of inner cutting edges is to compensate the manufacturing and installation inaccuracies. As shown, the inner cutting edges are not straight. Their curvature depends on the lead of the helical flute. A number of tests of such a drill on steels and on the automotive aluminum alloy revealed that the roundness of hole drilled is almost two times better than that with the best standard drills. It allowed recommending such a drill as rougher for cored holes (spool holes) in the valve bodies of the automatic transmissions to assure proper working conditions of the semifinished and finished reamers. The design of this drill requires further optimization for application-specific conditions to gain its maximum potential.

4.5.8.5.2 Altering the Location of the Chisel Edge

A discussed earlier, the inherent problem of the chisel edge passing through the axis of rotation is interference of this edge with the bottom of the hole being drilled as shown in Figures 4.106 and 4.119. One may argue, however, that the discovered interference is of *theoretical significance* as it is not commonly observed in practice. The simple answer to this *undisputable argument* is that it is not observed due to drill runout. To exemplify the answer, let's consider the test results of a study where the drill was deliberately installed with runout and tool life was measured as a function of this runout.

Figure 4.139 shows the influence of drill runout on tool life measured in the total length of drilled holes (drilling titanium alloy VT5). As can be seen in this figure, drilling was virtually impossible by drills of length > $6d_{dr}$ (60 mm) having no runout due to severe vibrations set since the beginning of drill engagement. When runout was 0.1 mm, the total length of drilled holes was 45 mm till the onset of vibrations. When this runout was 0.48 mm, the total length of drilled holes increased to 470 mm. As such, the sti*cking out* drill corner had prevailing wear. It was concluded that an increase of drill runout up to 0.5 mm resulted in an almost 50-fold increase in tool life. A less profound result (up to 25-fold tool life increase) was observed when the same drill having the length of 85 mm was tested (curve 2 in Figure 4.139).

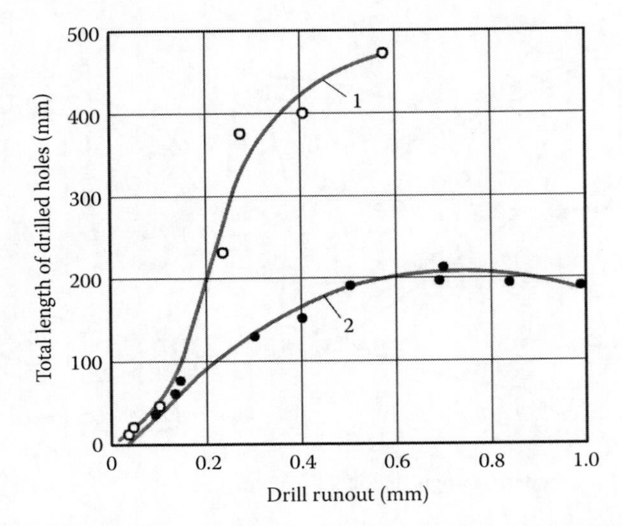

FIGURE 4.139 Influence of drill runout on tool life measured in the total length of drilled holes: (1) drill length 60 mm and (2) drill length 85 mm. Work material—titanium alloy VT2 (85.455–91.35% Ti, 5.5–7% Al, 2–3% Mo, 0.8–2.3 %Cr), tensile strength, ultimate 1000–1250 MPa, hardness HB 270; drill—material HSS M2, diameter 9 mm, point angle $\Phi_p = 118°$, T-hand-S normal clearance angle $\alpha_n = 16°$, cutting speed $v = 32$ m/min, cutting feed $f = 0.13$ mm/rev. Holes of 30 mm length were drilled.

The drilling torque was measured in the tests. It was found that this torque reduced from 22.6 to 13.7 N-m when drill runout increased from zero to 0.25 mm. Such a decrease (and thus corresponding increase in tool life) was explained by a reduction of friction forces on the drill margins.

Naturally, drill runout causes the drilled hole to be oversize. Its amount as shown in Figure 4.140 depends on the properties of the work material. The study points out that hole oversize is normally within drilled hole tolerance, while for greater runouts, it proposes a simple but practical way to deal with this issue, namely, to decrease the drill diameter accordingly.

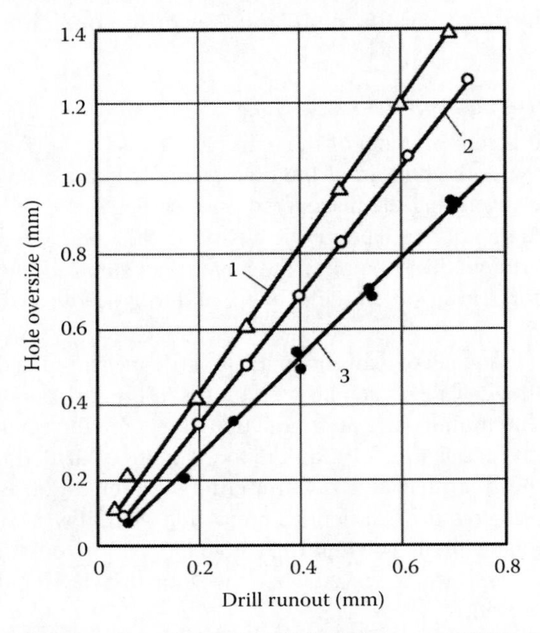

FIGURE 4.140 Influence of drill runout on oversized holes. The test conditions and the drill were the same as in Figure 4.139: (1) ANSI stainless steel 303, (2) ANSI steel 4140, and (3) titanium alloy VT5.

In the author's opinion, all the obtained results can be explained using the result presented in this chapter as follows:

- The drilling torque minimum is achieved when runout increases to 0.25 mm. Any further increase in this runout does not affect the drilling torque while tool life continues to increase further. Therefore, the drilling torque, and thus a reduction of friction on the *inside* margin, is not the root cause of tool life improvements
- Titanium alloys as a whole have unique mechanical properties compared to many steels. Unfortunately, the differences in properties are not considered in the analysis of machinability of titanium alloys, while emphasis is always placed on the low thermoconductivity of titanium alloys compared to steels. In the considered case, ANSI steel 4140 has the following key mechanical properties related to drilling: tensile strength, ultimate $s_{ul} = 655$ MPa, hardness HB 180, and modulus of elasticity $E = 200$ GPa, while titanium alloy VT5 has $s_{ul} = 1120$ MPa, hardness HB 285, and $E = 115$ GPa. As can be seen, VT5 has much higher strength and hardness while much lower E compared to steel 4140
- Relatively small elasticity modulus of titanium alloy VT5 causes significant springback (see Section 4.5.9.5), a.k.a. elastic recovery of the work material. It definitely affects the friction condition on the drill margins gripping them firmly compared to steel 4140 if the standard back taper is applied on the drill. It explains why the drilling torque reduces with increasing drill runout up to 0.25 mm. Any further increase in runout does not affect the drilling torque as runout becomes greater than springback of the wall of the hole being drilled
- Relatively small elasticity modulus and high strength of titanium alloy VT5 cause the major problems when the standard design of the drill chisel edge is used. The problems occur both on the rake and flank faces of this edge. On the rake face, when formed normally, the heavily serrated chip of high strength cannot pass through the gap between the rake face of the chisel edge and the walls of the bottom of the hole being drilled (see Section 4.5.8.2), while normally *light* interference of the chisel edge (Figures 4.106 and 4.119) becomes a significant factor. These cause radial vibration of the drill noticed in the tests.

For years, the discussed problems were not profound as drills, drill holders, machine spindles, etc., resulted in drill significant runout that *automatically* solved the discussed problems. Improvements in the accuracy of the components of the drilling system achieved for the last 10–15 years significantly reduced the total runout of drills so that the discussed problems became noticeable.

The discussed example signifies a need to understand drill design and geometry in modern-age drilling.

As an example of designs that combine the advantages of the shifted chisel edge revealed in the previously discussed study while not causing problems with oversize holes, consider the geometry of a drill with a non-central chisel edge developed, studied, and successfully implemented for drilling difficult-to-machine materials (Vinogradov 1985). The basic geometry of such a design is shown in Figure 4.141a. As can be seen, the drill has two major cutting edges 1 and 2 and the chisel edge 3. This chisel edge is shifted by distance Oc from the axis of rotation 0 by grinding an additional flank surface 4 adjacent to the inner part of the major cutting edge 2. This additional surface alters the cutting edge 2 in a way that the drill force balance disturbed by the shift of the chisel edge is restored.

The results of geometry analysis, multiple tests, and implementation results proved that the drill geometry shown in Figure 4.141a has the following advantages:

1. No one point of the chisel edge has zero cutting speed.
2. Self-centering ability without compromising the strength of the drill tip, which is important in machining difficult-to-machine heat-treated work materials. This self-centering ability is due to the cone formed at the center of the bottom of the hole being drilled as shown in Figure 4.141b. As seen, point *a* of the chisel edge is the first point of the drill that

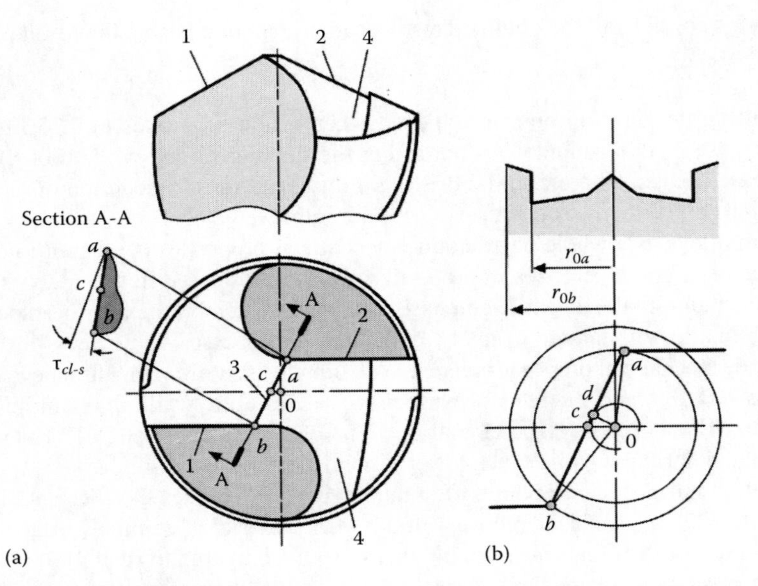

FIGURE 4.141 The drill with shifted chisel edge: (a) model of the drill geometry and (b) geometry of the bottom of the hole being drilled.

touches the workpiece due to angle τ_{cl-s}, which is formed as a result of the discussed shift. As shown in Figure 4.141b, the chisel edge forms the complicated shape of the bottom of the hole being drilled with stabilizing cylinder of r_{0a} radius and conical surface having the radius of the base equal to distance 0*d*.

3. Positive rake angles (higher than normally achieved even with the best split-point geometry) over the chisel edge with the maximum at point *d* reduce the cutting force and drilling torque.
4. Extremely high inclination angles over the chisel edge with the maximum of 90° at point *d*. Because the built-up edge does not form under any cutting conditions when the cutting edge inclination angle is high, the stability of drilling, quality of machined holes, and tool life of drills having the discussed geometry are much superior over the known drill point geometries.

Naturally, many particular designs and drill geometries can be developed using the idea of the shifting of the chisel edge from the drill axis. However, knowledge of tool geometry, particularly T-use-S geometry combined with mechanics and tribology of metal cutting, is a prerequisite for such developments.

4.5.9 Point Angle and Margin

In the author's opinion, the point angle and parameters of the margin (its width and back taper) should be considered together as they have multiple mutual correlations considered in this section.

As discussed in Appendix C, the clearance angle α distinguishes the cutting tool among other separating tools, while the tool cutting edge angle κ_r is the most important geometrical parameter of this tool. In drilling tool, the point angle Φ_p (or the half-point angle φ) is considered instead of κ_r for convenience of measurements. These angles are related as

$$\frac{\Phi_p}{2} = \varphi = \kappa_r \tag{4.76}$$

Therefore, the point angle Φ_p (or the half-point angle φ) should be regarded at the same level as κ_r. Unfortunately, this is not nearly the case in the practice of drill design, selection, and applications where the real influence of this angle is underestimated.

The 118° point angle is the *standard* angle for most drills used in industry although the 140° point angle recently became *standard* for cemented carbide drills. As discussed earlier in this chapter, since the beginning of the twenty-first century, 118° point angle is considered to be a good compromise for general-purpose drills for a variety of different work metals. Although carbon-steel drills were replaced by HSS drills and now cemented carbide drills have taken over, and although the grinding technique has changed from hand grinding to simple fixture grinding and then to specialized drill fixture grinding, and eventually to CNC drill grinding, the *standard* 118° point angle has not been changed. It mysteriously suits many users who do not want to deal with the application-specific drill geometry thinking that it is up to the drill manufacturers to suggest the optimum drill (including drill material and its coating) for a given application.

The author's many-year experience, however, shows otherwise. Drill manufacturers always try to sell the so-called *on-the-shelf* products, which they can produce in mass quantity at low manufacturing cost and with decent quality. With such drills, the lead time (the time between a purchase order and actual tool delivery) is minimal and the tool is relatively inexpensive. These real-world conveniences often overshadow the potential gains in efficiency (tool life, productivity, etc.) that can be achieved with application-specific drills. The logic of many production practitioners is simple: "We just buy more drills."

Experienced practitioners in the field who care about the system efficiency and the quality of the machined parts normally pay more attention to application-specific point angles. Table 4.5 shows the application-specific values for the point angle. The use of these application-specific point angles increases productivity and the quality of the drilling operation that, considered together, results in higher efficiency of drilling operations.

TABLE 4.5
Recommended Point Angles for Drills

Work Material	Tensile Strength (MPa)	Hardness (HB)	Point Angle (±3°)
Soft steels	<400	<120	110–120
Structural steels. Ordinary carbon steels with low to medium carbon content (<0, 5% C)	<550	<200	118–125
Carbon steels with high carbon content (>0, 5% C)			
Ordinary low alloy steels. Ferritic and martensitic stainless steels	500–700	180–260	130–135
Tool steels and difficult-to-machine alloys	>700	240–320	140
Cast irons		140–200	90–100
		200–240	110–115
		>240	118–125
Brass			110–129
Copper			120–130
Aluminum alloys			
Low or no silicon (long chip type)			130–140
High silicon (short chip type)			115–120
Magnesium alloys			130–140
Nickel			118–125
Zink alloys			100–115
Molded plastics			118–125
Laminated plastics			125–135
Carbon			80–90

The influence of the point angle, and thus the rationales behind the selection of its optimal, for a given machining condition, can be understood easily if one considers what happens when this angle deviates from its *standard* (118°) value (hereafter the words *increase* and *decrease* refer to this value). As such, *the optimal* should be clearly defined by the corresponding objective of optimization. Although the point angle affects almost every facet of drill performance, the further brief analysis of the drill point *optimality* concerns with the following: the axial/radial force ratio, uncut chip thickness, chip flow direction, and cycle time as these are of concern in the design and implementation of HP drills.

4.5.9.1 Axial/Radial Force Ratio

As discussed in Section 4.4, when drill works, the radial and axial components of the cutting force act on the major cutting edge as shown in Figure 4.142. As discussed in Section C.3.3.1 and as following from Figure 4.142, the axial and radial components of the drilling force are related through the half-point angle as

$$\frac{F_A}{F_R} = \tan\varphi \tag{4.77}$$

As follows from this equation, the radial component increases and the axial component decreases when the point angle decreases. This can be a valuable means at tool/process designer disposal to increase the allowable drill penetration rate normally restricted by the axial force particularly for long drills. This works, however, when the drill is symmetrical, that is, if the force balance discussed in Section 4.4 is maintained. This is because when the radial component increases, a small violation of drill symmetry may result in a significant unbalanced radial force that lowers tool life and ruins the quality of drilled holes. Because modern CNC drill grinders significantly improve drill symmetry, tool designers/process planers should revisit this powerful means to be used to improve drilling productivity.

4.5.9.2 Uncut Chip Thickness (Chip Load) and Chip Flow

As discussed in Chapter 1, the nominal thickness of cut known in the literature as the uncut (undeformed) chip thickness or chip load and the nominal width of the cut known in the literature as the uncut (deformed) chip width in drilling are calculated as

$$h_D = f_z \sin\varphi \tag{4.78}$$

$$b_D = \frac{(d_{dr}/2)}{\sin\varphi} \tag{4.79}$$

where f_z is the feed per tooth (mm/rev).

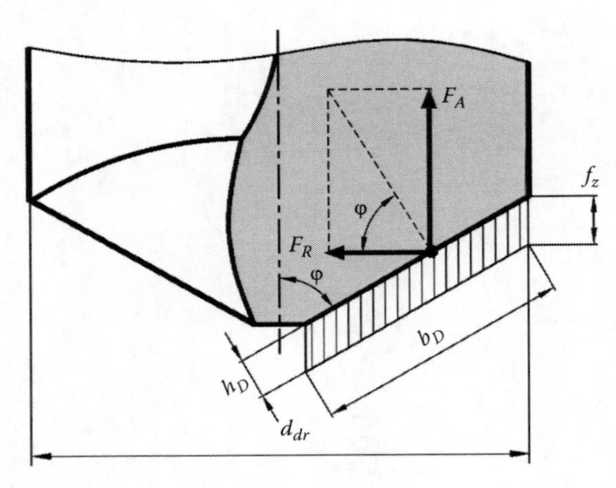

FIGURE 4.142 Axial and radial components.

It follows from Equations 4.78 and 4.79 that when the point angle decreases, the uncut (undeformed) chip thickness decreases that sets less pressure on the unit length of the cutting edge and thus may result in greater tool life or, alternatively, may allow a higher feed per tooth and thus higher drill penetration rate. On the other hand, the proportionally wider chip is formed by the cutting edge that requires special means of its handling (both breaking and transportation over the chip flute). Therefore, there is an optimal point angle for a given work material and drill parameters (material and geometry). The latter can be adjusted in a rather wide range to achieve higher drilling efficiency.

Small point angles not only cause wider chips but also significantly change the direction of chip flow. Special shapes of the chip flute as those shown in Figures 4.96 and 4.97 should be used to handle the formed chip. If the point angle becomes even smaller, another significant problem arises that restricts any further decrease in this angle. This problem is associated with a small radius of curvature of the machined surface as a part of the hyperboloid approximated as a cone (Figure 4.70c). This happens when the radius of the chip free surface r_{fs} (shown in Figure 4.84) reaches a certain critical value that depends on the work material. When it happens, the chip width h_C is greater than the clearance between the chip free surface and the machined surface (Figure 4.84) so that normal chip formation becomes virtually impossible.

4.5.9.3 Exit Burr and Delamination

A burr is defined as a rough projection left on a workpiece after drilling or cutting. Burr formation in drilling is one of the serious problems in precision engineering and mass production where the productivity of advanced manufacturing systems is often reduced due to additional deburring operations. Therefore, understanding the drilling burr formation and its dominant parameters is essential for controlling the burr size.

The burr forms as the drill approaches the exit; the workpiece deflects and then plastically deforms due to the thrust (axial) force created by the drill. Once this deformation is initiated, it increases as the drill cuts more of the workpiece material. When the drill exits the workpiece, the remaining material deflects creating an exit burr. Two burr types are normally distinguished as described in Figure 4.143 (Min et al. 2001; Sokołowski 2010). When there is no subsequent deburring operation used, the uniform burr type (Figure 4.143a) is preferable as it is smaller. When there is subsequent deburring, the crown burr type (Figure 4.143b) is preferred as it is much easier for deburring.

Studies have shown that the height, H_b, and thickness, T_b, of the drilling burr increase when the point angle increases. As an example, Figure 4.144 shows the influence of the drill point angle on the burr parameters in drilling ANSI 304L steel (Kim and Dornfeld 2002). As can be seen, smaller

(a)

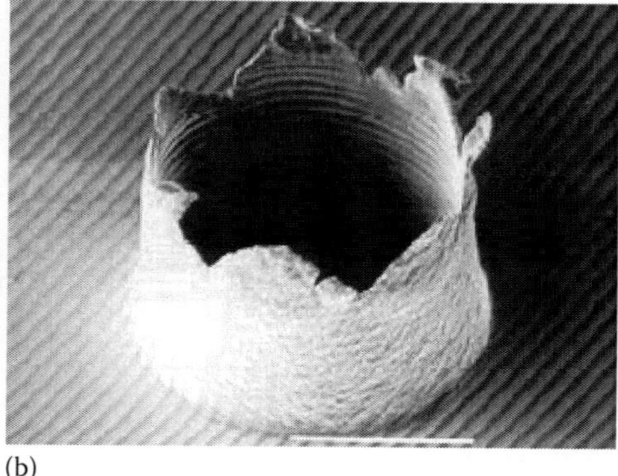

(b)

FIGURE 4.143 Examples of drilling burr shapes: (a) uniform and (b) crown burrs.

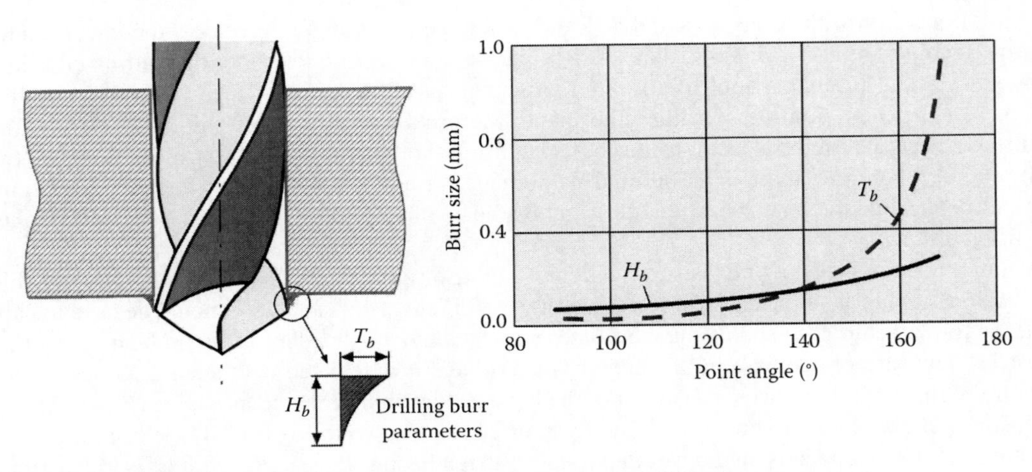

FIGURE 4.144 Showing the influence of the point angles on the drilling burr parameters.

point angles are beneficial in terms of reduction of the exit burr parameters. The simplest modification of the standard point grind of a carbide drill having primary point angle of 140° and secondary point angle of 70°–75° to minimize the existing burr parameters is shown in Figure 4.145. Its implementation increases tool life and reduces drilling exit burr when drilling a wide variety of engineering materials (Pilny et al. 2012).

The selection of the optimal point angle tends to be the real challenge of machining composite parts. A drill cutting through a metal part simply has to remove the material and clear the hole. By contrast, a drill cutting through a layered composite structure is likely to push the layers ahead of it, producing unacceptable delamination on the exit side (Khashaba 2012). The problem is that there are a great number of composite and plastic materials with considerably irreverent *drillability* that require a special drill geometry, primarily point and rake angles for a given application (Davim and Reis 2003; Davim 2004a,b).

The quality of the drilled holes such as roughness/waviness of its wall surface, roundness, and axial straightness of the hole section causes a number of problems in service life of composite parts. As pointed out by Khashaba (2012), bolting and riveting are extensively used as a primary method

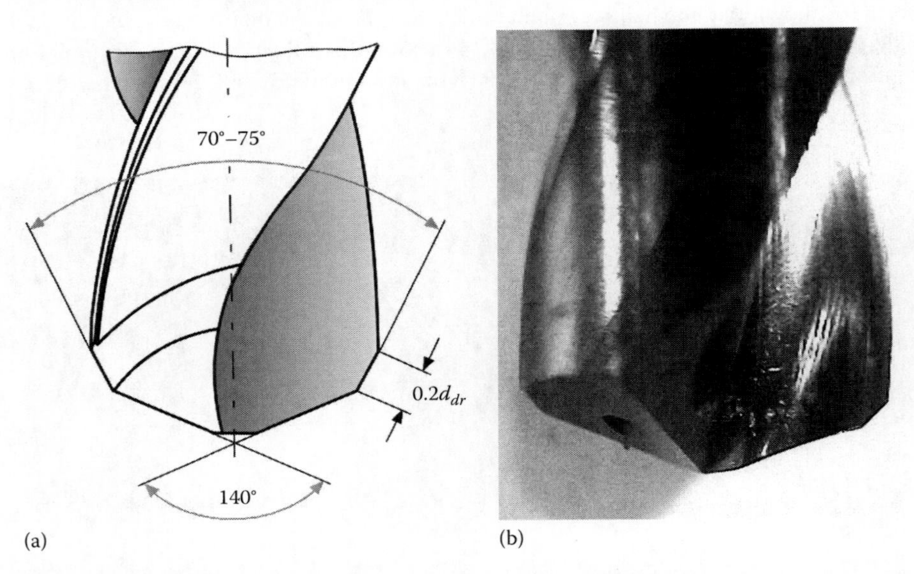

FIGURE 4.145 A double-point drill: (a) the point geometry, and (b) a real drill.

of forming structural joints of composite parts in the aerospace and automobile industries. High stresses on the rivet, for example, is often found as the root cause of its failure. Microcracking and delamination left after drilling significantly reduce the composite strength. Therefore, the quality of the drilled holes is found to be critical to the life of the riveted joints. Reduced mechanical properties of fiber-reinforced plastics (known as FRP), due to the stress concentration caused by the softening and re-solidification of the matrix material that has different thermal properties than the fiber, is another problem that occurrs in drilling of FRP. The low thermal conductivity and sharp temperature gradients in FRP lead to thermal damages and burning the matrix.

Damages associated with drilling FRP composites were observed at the entrance as well as at the exit of the drilled hole, in the form of peel-up and push-out delaminations, respectively (Khashaba 2012). Peel-up delamination occurs as the drill enters the laminate and is shown schematically in Figure 4.146a. After the cutting edge of the drill makes contact with the laminate, the cutting force at the drill corners is the driving force for peel-up delamination. It generates a peeling force in the axial direction through the slope of the drill flute. The flute tends to pull away the upper laminas and the material spirals up before it is machined completely. This action results in separating the upper laminas from the uncut portion held by the downward acting thrust force and forming a peel-up delamination zone at the top surface of the laminate. The peeling force is a function of tool geometry (the drill point angle) and friction conditions at the chip flute defined by both the surface conditions of the drill rake face and flute profile.

Push-out delamination occurs as the drill reaches the exit side of the material as is shown schematically in Figure 4.146b. As the drill approaches the end, the wall thickness of the material becomes smaller so that its resistance to deformation due to the axial (thrust) force imposed by the drill reduces. At some point, the thrust force exceeds the interlaminar bond strength causing an exit delamination zone as the tool pierces through the exit side. This happens before the laminate is completely penetrated by the drill. In practice, it has been found that the push-out delamination is more severe than that of peel-up (Khashaba 2012). Large point angles are beneficial at the drill entrance, while small ones cause little damage at the drill exit (Heisela and Pfeifroth 2012).

An intelligent selection of the proper parameters of drill geometry and suitable cutting conditions can significantly reduce the push-out delamination by lowering the axial force (Abrão et al. 2008; Ahmad 2009). Unfortunately, the optimum geometrical and machining conditions found for one FRP may not be suitable for a seemingly similar FRP. For example, the standard point angle of 118° was found to be optimal in drilling AS4/PEEK CFRP, while the effect of the point angle was found to be marginal in drilling T300/5208 CFRP (Ahmad 2009). In the author's opinion, however, the influence of the chisel edge on the axial force was not properly accounted for that led to the obtained result. In any case of drilling, the high rake angle (and thus fast helix angle) combined with high clearance angle of the major cutting edge seem to be beneficial in drilling of carbon FRPs. A small point angle tends to produce better exit quality. However, too small a point angle can give the tool poor strength. For CFRP, the optimal compromise seems to be a point angle of 90°.

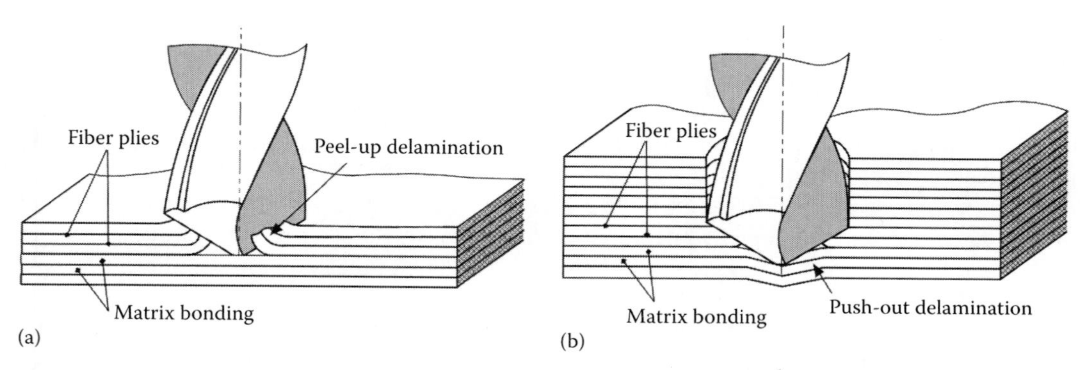

FIGURE 4.146 Delamination in FRP: (a) peel-up at the drill entrance and (b) push-out at the exit.

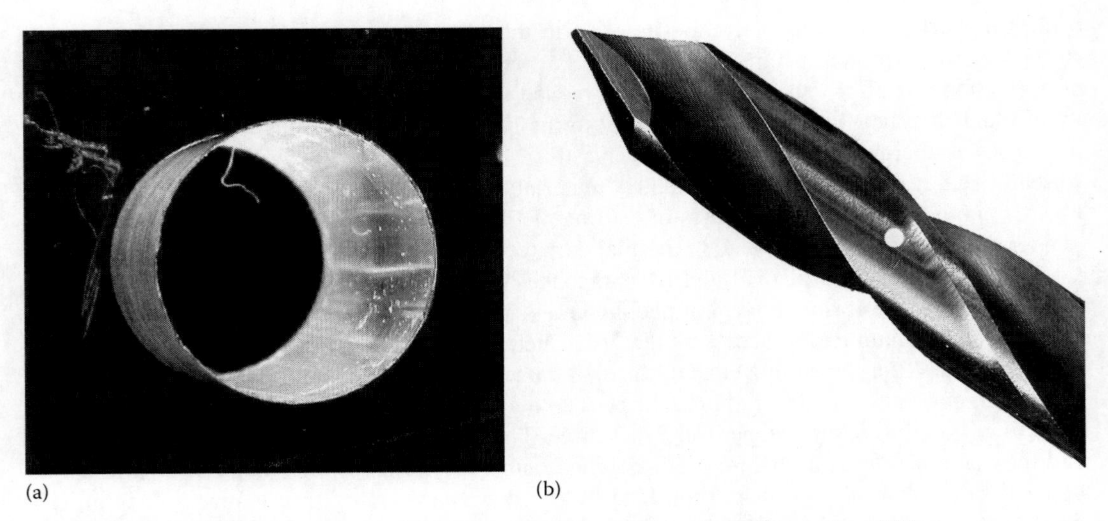

(a) (b)

FIGURE 4.147 Drilling of acryl: (a) achievable quality and (b) the optimal geometry drill.

Although a small point combined with aggressive rake angle of the major cutting edge is beneficial for FRPs, it may not be suitable for other plastics. For example, according to Basic Car Audio Electronics Co., the best results in drilling acryl (transparent hole wall with no feed marks; no chipping on the drill entrance and exit) as shown in Figure 4.147 are achieved when a small point drill having a zero T-hand-S rake angle over the entire major cutting edge and standard helix angle (30°) is used.

4.5.9.4 Cycle Time

The total drill travel necessary for making a given hole increases when the point angle increases. Moreover, in high-penetration rate drilling, the feed is decreased at the drill entrance ramping to the full feed after the drill margins are fully engaged. It may not be important when one hole is drilled, but in high-volume automated production where a number of holes are drilled and a drilling operation is critical in terms of the cycle time (e.g., tap holes in a valve body in the automotive industry), the point angle becomes a factor significantly affecting the cycle time. It might be one of the reasons why the standard point angle for such application was increased from the traditional 120°–140°.

4.5.9.5 Back Taper

Two other areas of drill design that impact its performance are back taper and margin widths. The purpose of back taper is to reduce the heat due to friction while the tool is engaged in the workpiece. Conventional drill designs have back taper values that correspond to standards established within the industry (0.03–0.12 mm/100 mm). Most HP drills, on the other hand, have back taper values that are virtually double those preestablished guidelines. These higher values create more relief while the drill is in the workpiece, minimizing heat.

In conjunction, the margin widths of HP drills are typically narrower than those of conventional drills. The margin guides the drill through the inner hole wall. For HP drills, a thinner margin width is required to reduce rubbing on the wall.

As discussed in Section C.3.3.2, the tool's minor cutting edge angle is an important parameter of the cutting tool geometry. In drills, the minor cutting edge is defined as the line of intersection of the flute face and the margin. Figure 4.148 shows the definition of the tool minor cutting edge angle for a straight-flute drill (in the picture, φ_1 is shown significantly exaggerated for clarity). It is defined according to the standard definition provided by ISO 3002-1 standard (see Appendix C) as the acute angle between the projection of the minor (side) cutting edge into the reference plane (a plane that contains the drill longitudinal axis) and the direction set by the vector of the cutting feed.

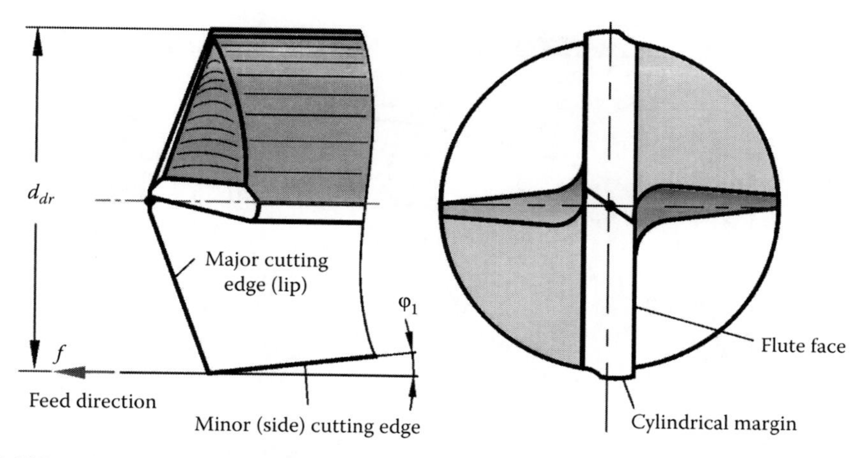

FIGURE 4.148 Visualization of the drill minor cutting edge angle.

The drill minor cutting edge angle φ_1 is not indicated on the tool drawing. Rather, it is *hidden* in the so-called back taper. The sense of this back taper is as follows. Normally, the drill (reamer) margin is made with back taper toward the tool shank so that the side cutting edge formed as the intersection of this margin and the flute has minor cutting edge angle φ_1. According to the accepted definition (*Metal Cutting Tool Handbook* 1989), the back taper is a slight decrease in diameter from the front to back in the body of the drill. For twist drills, the back taper Δ_{bt} (included) is assigned on the drawing as a diameter decrease $(d_{dr} - d_{dr1})$ per working length L_{wr} (as shown in Figure 4.149). As such, the tool minor edge angle is calculated as

$$\varphi_1 = \arctan \frac{\Delta_{bt}}{2L_{wr}} \tag{4.80}$$

The common perception of the role and importance of the back taper stems from the experience with twist drills. As mentioned earlier, according to the prevailing notion in twist drilling, the purpose of back taper is to reduce the heat due to friction while the tool is engaged in the workpiece thus to prevent binding of a drill in the hole being drilled. Although such binding is one of the most common failure modes of twist drills, its cause is not well understood. Therefore, a need is felt to clarify this important issue.

To do so, let's consider a simple tensile testing and its strain–stress diagram. During tensile testing of a material specimen, the stress–strain curve is a graphical representation of the relationship between stress, derived from measuring the load applied on the specimen, and strain, derived from measuring its deformation, that is, elongation, compression, or distortion. The slope of the

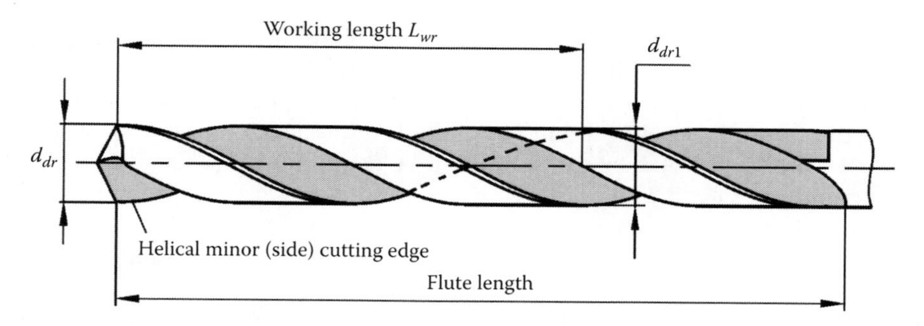

FIGURE 4.149 Back taper applied over the working length.

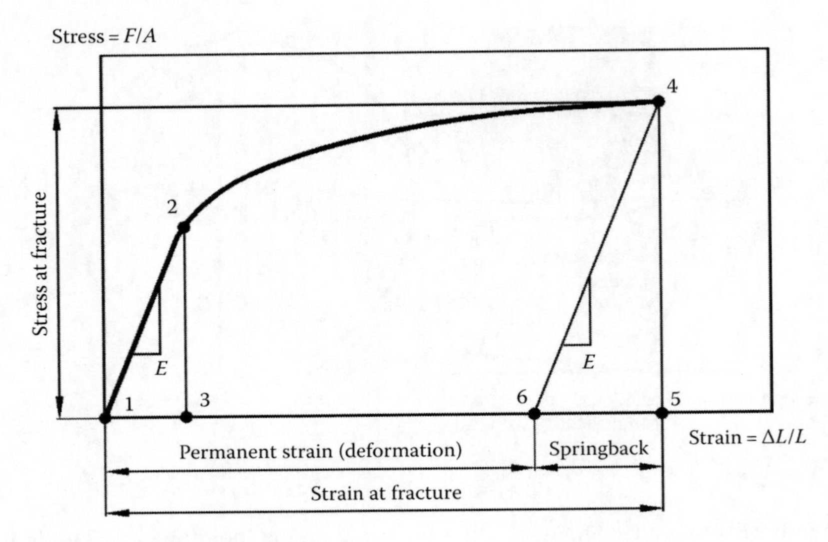

FIGURE 4.150 A simple strain–stress diagram.

stress–strain curve at any point is called the tangent modulus; the slope of the elastic (linear) portion of the curve (line 1–2 in Figure 4.150) is a property used to characterize materials and is known as the Young's modulus E.

Young's modulus, E, can be calculated by dividing the tensile stress by the tensile strain in the elastic (initial, linear) portion (lines 1–2 in Figure 4.150) of the stress–strain curve:

$$E \equiv \frac{\text{Tensile stess}}{\text{Tensile strain}} = \frac{F/A}{\Delta L/L} = \frac{FL}{A\Delta L} \tag{4.81}$$

where

F is the force exerted on an object under tension
A is the original cross-sectional area through which the force is applied
ΔL is the amount by which the length of the object changes
L is the original length of the object

In the diagram shown in Figure 4.150, point 1 represents unstressed material. When forces F are applied, the work material deforms first elastically up to point 2 on the diagram. This point represents the so-called elastic limit. Within this limit, the work material is subjected to only elastic deformation, so if the applied stress is released, the material regains its initial size. Distance 1-3 on the strain axis represents the maximum elastic deformation. If the applied stress exceeds the elastic limit, the material exhibits a combination of the elastic and plastic deformations. The applied stress can grow further up to point 4 on the diagram where fracture occurs. The strain corresponding to point 4 is known as the strain at fracture. In Figure 4.150, it is represented by distance 1-5 on the strain axis. After fracture, however, the applied stress is released and the permanent strain found in the work material (represented by distance 1-6 in Figure 4.150) is less than that at fracture by the elastic strain represented by distance 6-5 in the stress–strain diagram. As such, the location of point 6 is readily found by drawing a line from point 4 parallel to line 1-2. As such, distance 6-5 in the stress–strain diagram is known as elastic recovery in materials testing or springback in materials processing.

The next step is to apply the concept of elastic recovery to drilling. As discussed by the author earlier (Astakhov 2006), when the work material is being cut, the material is deliberately overstressed

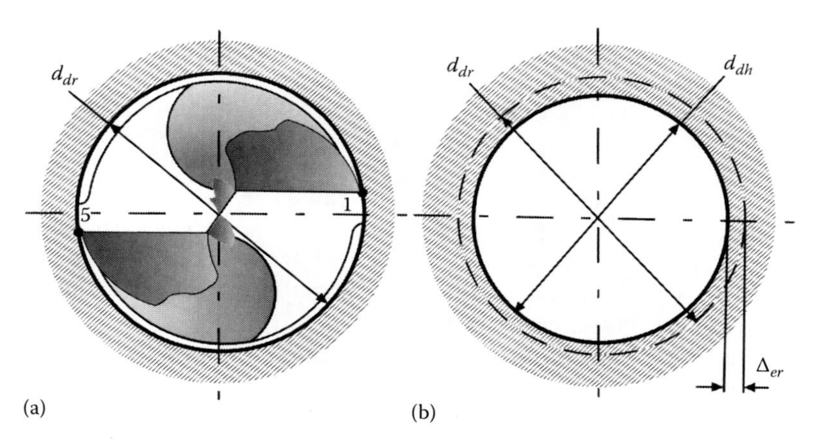

FIGURE 4.151 Model of elastic recovery in drilling: (a) the diameter of the hole being drilled at the point of cut, and (b) elastic recovery behind this point.

beyond the elastic limit in order to induce a permanent deformation and then separation of the stock to be removed. As such, the fracture of the chip from the wall of the hole being drilled occurs so the strain, and stress at fracture are achieved at any point of separation of the work material. Figure 4.151 shows a simple model where no drill runout is considered for the sake of simplifying further considerations. In Figure 4.151a, the drill corners 1 and 5 separate the chip from the rest of the workpiece forming the wall of the hole being drilled. The diameter of this wall is always equal to that of the drill, that is, d_{dr}.

After the corners 1 and 5 of the drill pass a certain part of the machined hole, the load due to cutting is removed so that the applied stress returns to zero. As a result, the machined hole shrinks due to springback (elastic recovery) because elastic deformation is recoverable deformation. Note that this recovery is not only due to mechanical action of corners 1 and 5 but also due to heating and subsequent cooling of the work material adjacent to the hole being drilled. In other words, the thermal energy due to plastic deformation of the work material and friction between the tool and the workpiece in their relative motion causes thermal expansion of the work material around the drill terminal end. When the corners 1 and 5 advance further, the work material contracts due to cooling by MWF. As a result of the mentioned mechanical and thermal factors, the diameter of the hole being machined d_{dh} becomes smaller than that of the diameter of the drill, d_{dr} by Δ_{er} as shown in Figure 4.151b. Therefore, if no back taper is applied to the drill's margins, drill will be gripped by the contracted wall of the machined hole.

Several variables influence the amount of springback. Among others, the stress at fracture (defines the height of the starting point of the unloading line represented by point 4 in Figure 4.150) and the modulus of elasticity (defines the slope of the unloading line represented by line 4-6 in Figure 4.150) are of importance. It is obvious from the diagram shown in Figure 4.150 that the higher the strength of the work material, the greater the springback; the lower elasticity modulus, the higher springback. Therefore, the tool cutting edge angle of the minor cutting edge should be made work material-specific as, on one hand, it is desirable that the diameter of drill does not change significantly with the number of re-sharpenings but, on the other hand, the said binding should not occur either. Unfortunately, drill manufacturers do not pay much attention to this important issue so that the back taper applied to drills varies significantly from one drill manufacturer to the next.

Consider a few practical examples. To the first approximation, the said springback can be determined as the ratio of the ultimate strength of the work material, σ_{UTS}, and its elasticity modulus, E, that is, springback = σ_{UTS}/E. As the modulus of elasticity is almost the same for wide group of steels ($E = 200$ GPa), the springback is determined by the strength of the steel. For cold-drawn steel

AISI 1012, having σ_{UTS} = 270 MPa, springback = 0.00135, while for annealed steel AISI 1095, having σ_{UTS} = 650 MPa, springback = 0.00325. Unfortunately, the same twist drill is used for drilling these two work materials, that is, no account is taken for the difference in springbacks of these two materials. The matter gets worse when titanium or aluminum alloys are drilled. For commonly used annealed titanium alloy Ti-Al6-4V (Grade 5), σ_{UTS} = 880 MPa, E = 113.8 GPa, and springback = 0.00772, that is, four times greater than that of medium-carbon steels. This explains known difficulties with drill binding in machining of titanium alloys. For common in the aerospace industry aluminum alloy 6061-T6 σ_{UTS} = 310, MPa E = 68.9 MPa, and springback = 0.00450, which is, greater than that of steels. Therefore, for successful drilling of titanium and aluminum alloys, the back taper applied to drills should be adjusted correspondingly.

As previously discussed, another factor that affects the possibility of drill binding in the hole being drilled is thermal expansion. The drill margin(s) made cylindrical, i.e. with no clearance always rubs against the wall of the hole being drilled. The higher springback, the higher thermal energy released due to this rubbing, the higher temperature rises. When back taper is sufficient, the discussed rubbing occurs only over the small portion of the margin adjacent to the drill periphery corner. If, however, the back taper is insufficient, the discussed rubbing occurs over a great portion of the margin causing significant contact temperatures. As the contact temperature rises, the drill and the hole being drilled expand. This expansion depends on many factors primarily on the thermal characteristic of the work material as its temperature expansion coefficient (CTE), thermal conductivity, mass, etc. One should realize that for the same contact temperature and CTE for the workpiece and the drill, the drill expands greater as its mass is much smaller so it heats up faster.

Consider a few practical examples. For tool materials, HSS M4 (common tool materials for HSS drills), CTE = 12.24 (μm/m)/°C and for WC6Co carbide, CTE = 5.4 (μm/m)/°C. For work materials, steel AISI 1045 CTE = 13 (μm/m)/°C, steel AISI 1095 CTE = 12.4 (μm/m)/°C, and Ti alloy CTE = 9.2 (μm/m)/°C. As follows from these data, (1) HSS drill expands much greater than carbide drills, which should be accounted for in assigning the proper back taper for HSS drills and (2) Ti alloy has much smaller CTE than HSS. As a result, back taper on HSS drills should be made greater in drilling of titanium alloys.

When the back taper is insufficient, the result depends on how small is actual back taper and many other particularities of the hole-making operations. The simplest outcome of insufficient back taper is the so-called scoring (Astakhov 2010). When scoring happens in drilling, the surface finish of machined holes deteriorates significantly. When the back taper is made even smaller than that which caused simple scoring, the drill transverse vibrations and the increased wear on the margins occur. Figure 4.152 shows the appearance of chatter marks and wear on the drill margin.

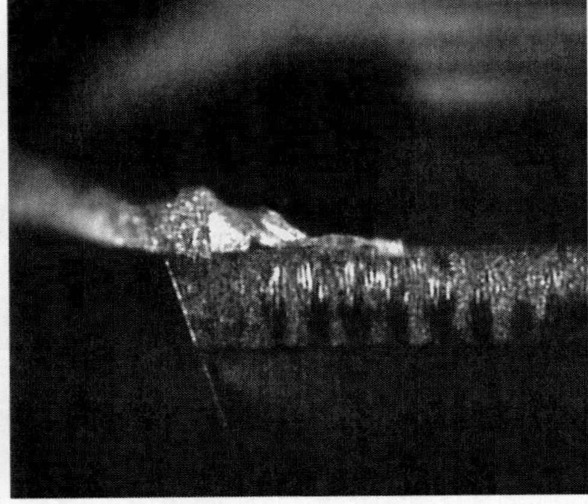

FIGURE 4.152 Chatter marks and wear on the drill margin.

(a)

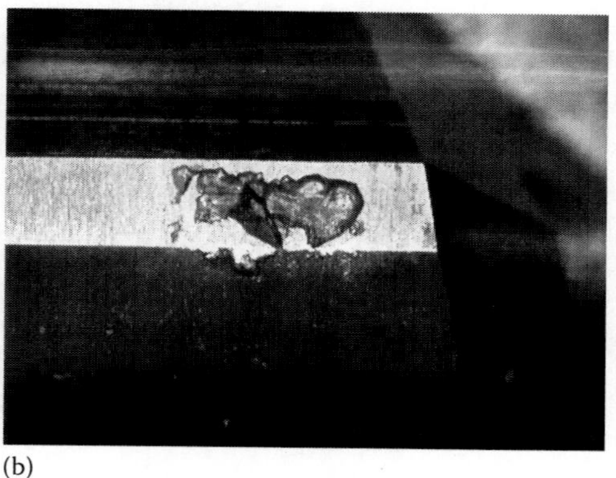
(b)

FIGURE 4.153 Wear of the drill major (a) and trailing (b) margins due to high contact pressure caused by the lack of back taper.

When the back taper is made even smaller, the tool condition deteriorates rapidly. Figure 4.153 shows the common wear pattern that occurs on drill margins due to high contact stresses caused by insufficient back taper. As seen, the margin adjacent to the major cutting edge is ruined due to a high contact pressure, and the built-up edge is formed on the auxiliary margin. When this contact pressure becomes high enough, particularly for multi-edge axial tools such as reamers, the tool breaks by an excessive torque (Astakhov 2010). Often, such a breakage is wrongly attributed to a misalignment problem not paying attention that the tool is gripped in the hole. Even after the tool breaks, its removal presents a significant problem. When such a failure happens, the part autopsy (see Chapter 2) should be used, that is, the part should be carefully sectioned to carry out the root cause analysis of this failure.

Unfortunately, this issue is not understood in the automotive industry. There are two major reasons for that. First is that the role of back taper became significant only recently when other imperfections of the machining system (misalignment, parameters of MWF (coolant) in terms of its clearness and content, spindle runout, workholding fixture accuracy and rigidity, etc.) were improved. US Patent No. 6,054,304 (2000) claims that the great tool life and high-penetration rate in gundrilling are direct results of increasing the back taper to 0.3 mm/100 mm instead of the old notion of back taper 0.08 mm/100 mm used for years. The second issue is that PCD-tipped drills and reamers used for finishing tight-tolerance operations have relatively short PCD tips. As such, the measuring of the actual back taper is virtually impossible even if the advanced tool presetting machine as, for example, Kelch or Zoller, is used because the difference in the tool diameter due to back taper over the actual length of an insert is beyond the recognition range of such machines.

4.5.9.6 Margin and Minor Cutting Edge

The role of the minor cutting edge in turning is discussed in Section C.3.3.2 where a hypothetical cutting tool (Figure C.38a) is used to facilitate understanding of this role. To understand the real significance of the minor cutting edge in drilling, consider a hypothetical drill shown in Figure 4.154. As can be seem, this drill has a single (major) cutting edge and no minor (side) cutting edge. Figure 4.155 shows the axial cross section of the hole being drilled by this tool. For clarity, the feed per revolution is significantly exaggerated. Although this profile looks odd, Figure 4.156 explains this result. Figure 4.156a shows two successive positions of the discussed drill. As can be seen, because there is no side cutting edge provided, a part of the work material represented by triangle ABC forms at each drill revolution that results in the hole profile shown in Figure 4.155. Moreover, if the half-point angle φ increases as shown in Figure 4.156b, the interference of the workpiece and the drill shank takes place. When $\varphi \to \pi/2$ (an extreme case for the considered hypothetical drill), contour ABB′C representing the materials uncut by the major cutting edge changes assuming a rectangular shape having the maximum cross-section area.

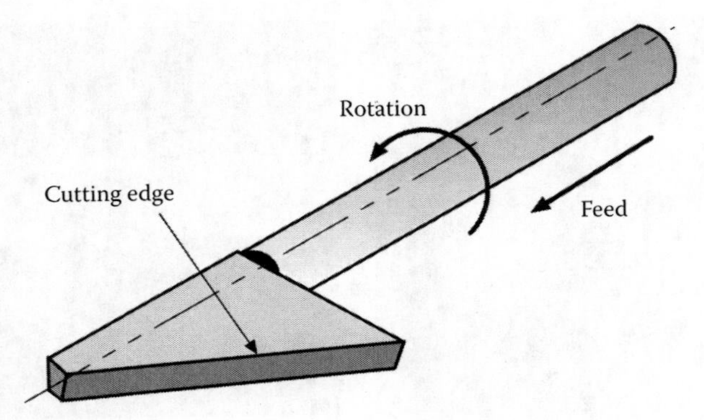

FIGURE 4.154 A hypothetical cutting tool used to explain the role of the minor cutting edge in drilling.

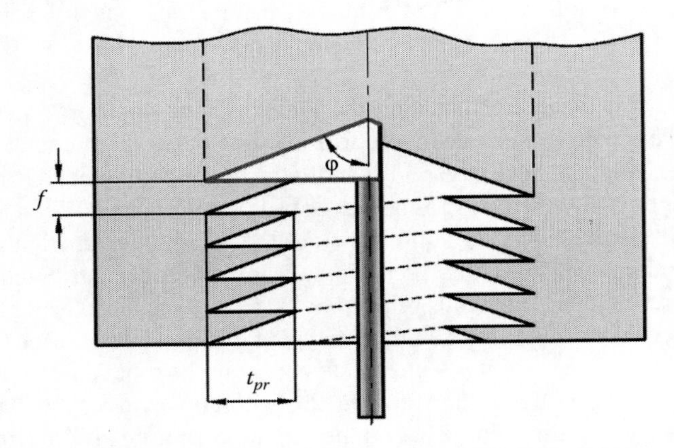

FIGURE 4.155 Profile of the hole drilled by the hypothetical drill shown in Figure 4.154.

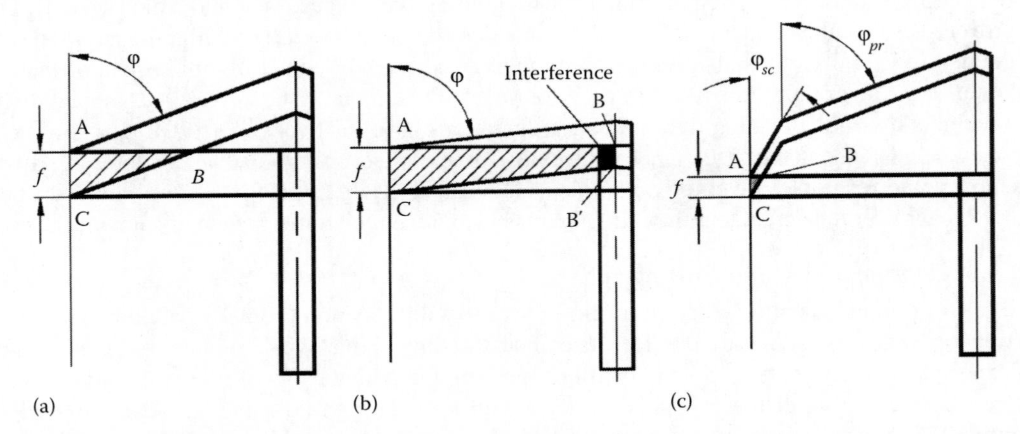

FIGURE 4.156 Explanation of the profile shown in Figure 4.155 and influence of φ on the uncut chip thickness cut by the minor (side margin) cutting edge: (a) for a normal point angle, (b) for large point angle, and (c) for double-point angle construction.

A real drill is made with the minor (side) cutting edge that cuts the material left by the major cutting edge (represented by area ABC in Figure 4.156). As such, the following is true: the greater the half-point angle φ, the more uncut material is left for the minor (side) cutting edge. Unfortunately, this edge is not meant for cutting, and thus it does not have the clearance angle, which makes drilling unstable and causes premature wear of the drill periphery corners. Figure 4.157 shows a margin

FIGURE 4.157 Aluminum deposit of the margin.

of a straight-flute drill for high-speed machining of high-silicon aluminum alloy with a high feed. As can be seen, the aluminum deposit formed on this margin as it is the minor cutting edge adjacent to the drill corner and cuts the work material.

The foregoing analysis explains significant improvement in tool life and drilled hole quality in drilling difficult-to-machine materials when a double-point drill shown in Figure 4.145 is used. Figure 4.156c provides the physical explanation. As can be seen, when the major cutting edge of the discussed hypothetical drill is modified by the second point angle having a much smaller half-point angle φ_{sc}, the area of triangle ABC that represented the work material to be cut by the minor cutting edge is substantially reduced and that significantly improves cutting conditions on the minor cutting edge.

The number of point angles applied to a drill and their angles φ's is application-specific. It is clear that the harder and stronger the work material, the greater the pressure applied to the part of the drill margin adjacent to the drill corner, the smaller the second/third point angle should be to minimize the amount of the work material cut by the minor cutting edge. Figure 4.158 shows a triple-point drill for machining cast irons. Its implementation not just reduces radial vibrations due to improved conditions at the margins but also results in two- to three-fold reduction in the axial force that, in turn, at least doubles the allowable cutting feed compared to a drill with the standard point grind. The drawback of

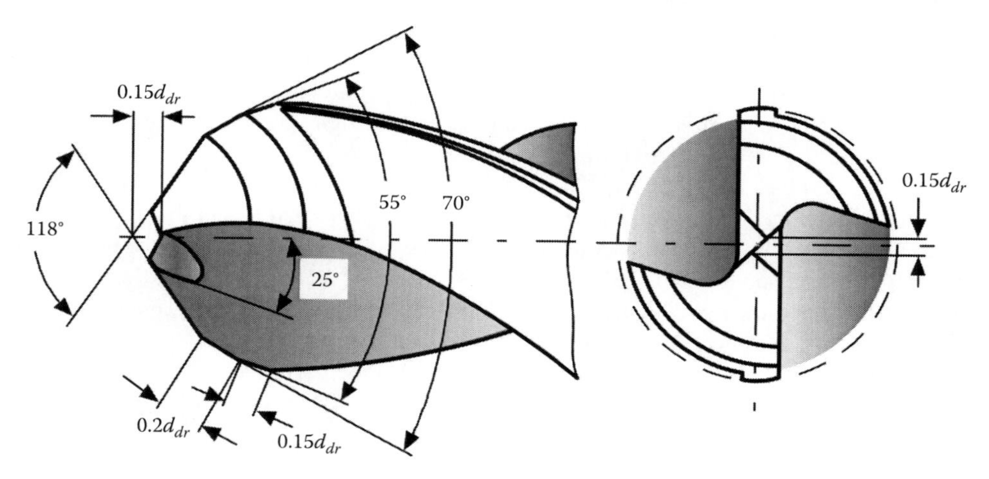

FIGURE 4.158 Triple-point drill for machining cast irons.

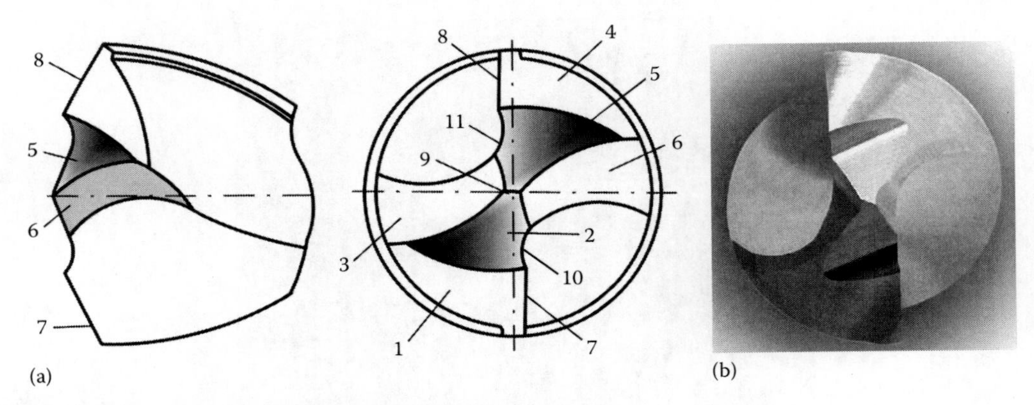

FIGURE 4.159 (a) Generic MFD and (b) real MFD.

this design is that the cutting edge becomes longer with any additional point angle applied to a drill so that the margins are located too far back from the drill point. Moreover, the cycle time increases when this drill design is used due to significant increase in the length of its cutting part.

The latter problem can be solved with the use of the so-called multifacet drills (MFDs). As discussed by Wu and Shen (1983), around 1953, a new type of drill point with multifacet flank shapes was successfully developed experimentally to drill a special alloy steel. Since then, a great variety of such drills has been developed for different work materials. This new drill became known as an MFD for short. A typical MFD is shown in Figure 4.159a. This drill has six facets 1, 2, 3, 4, 5, and 6 in a symmetric form although not all MFDs are symmetrical. Flanks 1 and 4 are ground in the same way as for a standard twist drill so that straight portions 7 and 8 of the major cutting edge are formed. Flanks 3 and 6 are ground to form other cylindrical surfaces near the center of the drill to thin the chisel edge 9, while flanks 2 and 5 are ground to form curved parts 10 and 11 of the major cutting edges. Figure 4.159b shows one practical implementation of such a design.

When properly designed and manufactured, an MFD has the following advantages over a standard twist drill: (1) the axial force (thrust) is reduced up to 70%, (2) self-centering ability, (3) reduced exit burr, and (4) the chip is separated into easy-to-transport pieces and shapes. In spite of these advantages being proven experimentally, these drills did not attract much attention from drill manufacturers and users because of the complexity of their geometry that should be applied on each regrind almost manually. Although some simplified versions of MFD were patented (US Patents No. 5,011,342 [1991] and 5,422,979 [1995]), it did not help in any noticeable increase in the use of such designs.

Modern development of MFD originated at the Xi'an Petroleum Institute (Wang et al. 1993). Figure 4.160 shows some developed geometries known as the four-margin design. This design, which can be of symmetrical or asymmetrical forms, offers a number of beneficial features: (1) the major cutting edges have positive rake angles and preferable angle distribution along their length, (2) thick web that allows high-penetration rate, and (3) four margins increase drill stability and provide separate channels for MWF supply into the machining zone and flutes for chip removal.

Although the developed and tested MFDs offer these advantages and thus it makes good sense to produce and use such drills, drill manufacturers and users are rather reluctant to produce and implement them in industry due to some design and manufacturing concerns (Astakhov 2010). Regardless of the concerns, MFD design is the right direction for development in the author's opinion. Figure 4.161 shows an MFD for drilling in thin steel sheets. Using such a drill, one can drill a round hole with no exit burr.

The next consideration is the margin itself. One may wonder why the cylindrical margins are applied to the major cutting edge in drills and many drilling tools while no margins are applied to plunging endmills. The simple answer is because any plunging endmill is meant to make a relatively short plunge resulting in the machined hole length that hardly exits the mill diameter. On the contrary, drilling tools machine much longer holes so the margins are needed to improve drill

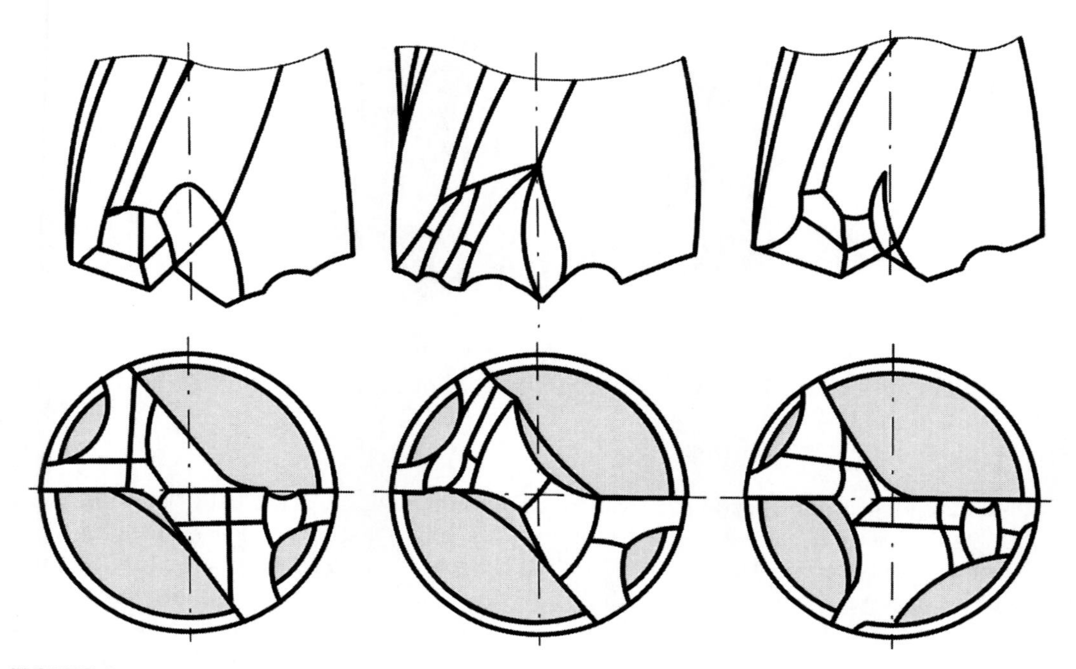

FIGURE 4.160 MFD developed at the Xi'an Petroleum Institute.

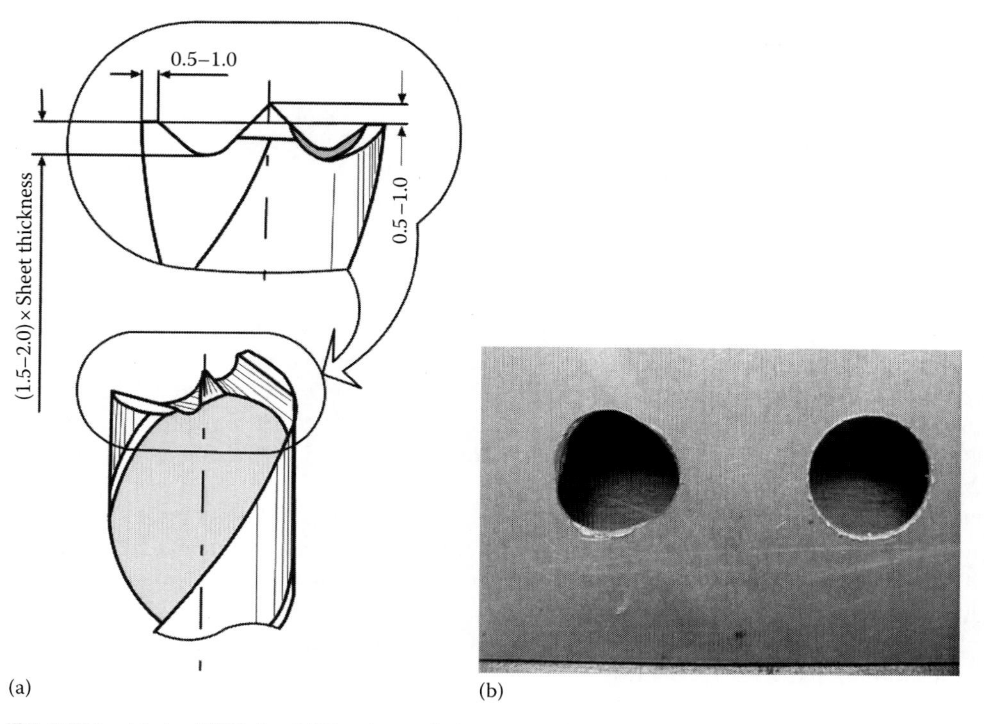

FIGURE 4.161 (a) An MFD for drilling in steel sheets and (b) comparison of the results of drilling with a 12.5 mm standard and an MFD in thin (1.4 mm) aluminum (6160-T6) sheet.

FIGURE 4.162 MAPAL-MEGA-Drill-Reamer.

stability preventing chatter and radial vibrations and thus improving hole accuracy. This is because the margin prevents *side cutting* by the drill corners and minor cutting edges when the drill force balance discussed in Section 4.4 is violated due to imperfections in the drilling system (e.g., unequal length of the drill major cutting edges, drill runout, variation of the properties of the workpiece in the radial direction, cutting through cross holes, etc.). When a drill has only two margins, these margins are capable of stabilizing this drill under the action of the unbalanced radial force $F_{y\text{-}ub}$. However, as discussed in Section 4.4.2, there are some other additional unbalanced force factors in any real drill acting in different directions that may cause drill improper position in the hole being drilled. To minimize the effect of these additional force factors, and thus stabilize a drill in the hole being drilled, many straight-flute and twist drills are manufactured with four margins as shown in Figures 4.5 and 4.18. Advanced drill designs may include a number of additional margins as, for example, MAPAL-MEGA-Drill-Reamer shown in Figure 4.162 that, as thought of, may improve drilled hole shape and its surface roughness.

The role and thus the design of drill margins should be re-thought for high-speed high-penetration rate drilling using modern drilling systems of high accuracy. The major line of thought is to reduce friction between the margins and the wall of the hole being drilled. The traditional margins are cylindrical as shown in Figure 4.163a. Its width a_m is selected in the range of 0.3–0.6 mm depending on the drill diameter. This range was selected experimentally in the nineteenth century and hardly changed since then. As discussed previously, an aggressive back taper is one of the solutions for high-speed drilling. Much better results, however, are achieved if this margin is modified in the manner shown in Figure 4.163b. The narrow margin of width $a_{m1} = 0.10\ldots0.15$ mm is made over the axial distance l_{m1} equal to approximately three times greater than the feed per revolution of the drill, and the clearance angle $\alpha_{m1} = 5°\ldots7°$ is applied. Such a design significantly improves the cutting conditions of the minor cutting edge, while the narrow margin still maintains drill stability. If the drilling system is accurate and rigid, this design can be modified further to that shown in Figure 4.163c where the narrow margin with clearance angle is extended over the whole working part of the drill.

The margin design shown in Figure 4.163d is the most aggressive design meant for drilling relatively short hole in working materials of low elasticity module as, for example, titanium alloys and aluminums, that is, where the expected springback is great. Such a margin is made with no cylindrical portion but with two clearance angles: $\alpha_{m1} = 3°\ldots6°$ and $\alpha_{m2} = 10°\ldots12°$. The modification of this design for longer drill is shown in Figure 4.163e where the margin is CAM relieved (instead of cylindrically ground) with $\alpha_{m1} = 5°\ldots6°$.

The discussed problem with the traditional margin design/location can be solved if the functions of the minor cutting edge and the margin are separated so that each of these drill components can perform its intended function without affecting the performance of the other. Figure 4.164

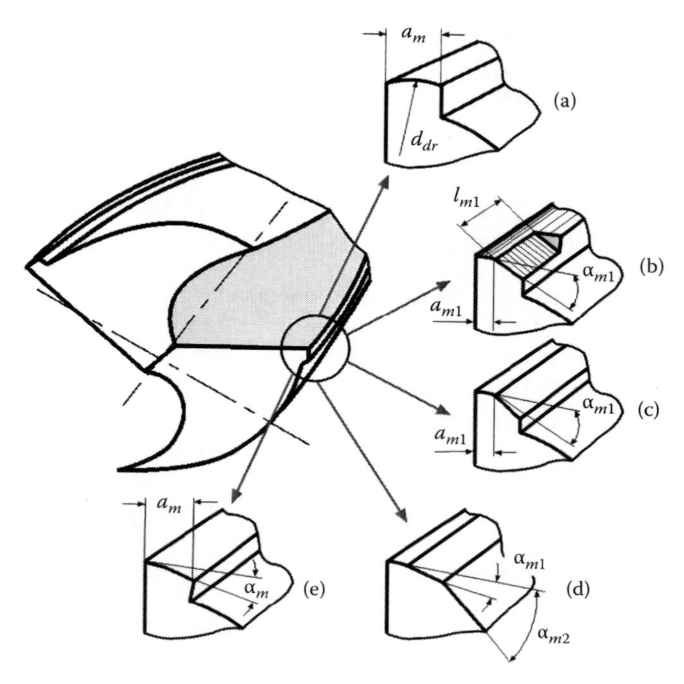

FIGURE 4.163 Traditional and advanced design/geometry of drill margins: (a) standard margin, (b) narrowed margin with the clearance angle applied over the short axial distance, (c) narrowed margin with the clearance angle applied over the whole working part, (d) no cylindrical margin, and (e) CAM relived margin.

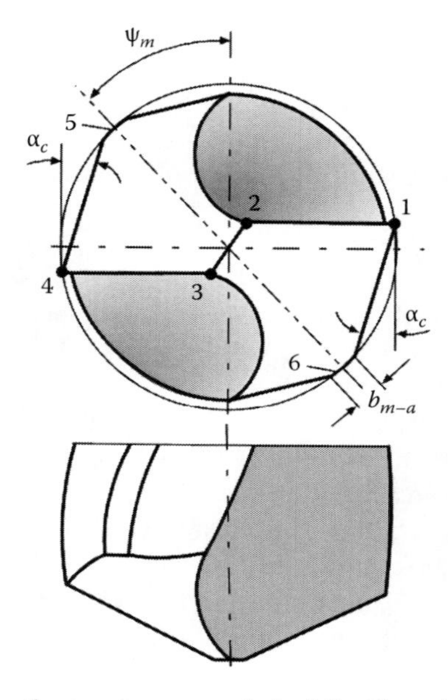

FIGURE 4.164 A drill design where the margins are angularly shifted from the corners of the major cutting edges.

shows a drill design where the margins are angularly shifted from the corners of the major cutting edges. As can be seen, the drill has the major cutting edges (lips 1-2 and 3-4). The minor cutting edges of such a drill that start from corners 1 and 4 are provided with the performance-optimal clearance angles α_c. The drill margins 5 and 6 having width b_{m-a} are shifted angularly from the corners 1 and 4 to achieve maximum drill stability defined by the location angle ψ_m.

Moreover, the diameter of the margins can be made slightly greater than that of corners 1 and 4. When the drill is made so, it yields a finishing tool combining cutting by the cutting edge and burnishing of the machined surface by margins 5 and 6. The roundness and surface roughness of machined holes are improved dramatically.

4.6 DRILL DESIGN OPTIMIZATION BASED ON THE LOAD OVER THE DRILL CUTTING EDGE

The degree of optimality of the geometry of any cutting tool can be assessed through optimality of the geometry of the cutting edge, on one hand, and by uniformity of the load over this edge. The latter is particularly true for cutting tools having long cutting edges over which the cutting conditions change substantially as in drills. For given work material and tool design, the load at point i of the cutting edge is characterized by the cutting speed, v_i, and uncut (undeformed) chip thickness, $h_{D\text{-}i}$. This load determines the tool wear rate (Astakhov 2004) and thus tool life.

Often for tools with long cutting edges, the cutting speed and uncut chip thickness may vary over the cutting edge. Therefore, accounting for these variations, one should establish a criterion of the load on a given part of the cutting edge to be able to determine the most loaded portions of the cutting edge. As pointed out by Rodin (1971), one of the prime reserves in cutting tool improvements is to make the load evenly distributed. As a criterion of optimality, the wear rate (in the sense as it was introduced by the author [Astakhov 2004]) or tool life (if the tool life of the considered portion of the cutting edge is greater compared to other parts) can be used.

In the simplest case, the following empirical formula is used to correlate the cutting speed with tool life and cutting parameters:

$$v = \frac{C_T C_w C_{cc}}{T^{m_v} h_D^{m_t} b_D^{m_b}} \tag{4.82}$$

where

C_T, C_w, and C_{cc} are constant depending upon the tool material, work material, and cutting conditions, respectively
T is tool life
h_D and b_D are the uncut chip thickness and its width, respectively
m_v, m_t, and m_b are powers to be determined experimentally using cutting tests (e.g., as discussed by Astakhov (Chapter 5 in Astakhov [2006])).

To analyze the distribution of the load over a given cutting edge, a point i on this edge is selected to be the base point. For this point, the cutting speed is v_i, uncut chip thickness is $h_{D\text{-}i}$, and tool life is T_i. For any other point of the cutting edge, for example, point p, tool life T_P can be calculated using Equation 4.82. The load coefficient at this point can be calculated as $k_{N\text{-}p} = T_i/T_p$. If $k_{N\text{-}p} > 1$, then this point p is loaded more than the base point i.

The load coefficient k_N can also be determined as the ratio of the uncut (undeformed) chip thickness $h_{D\text{-}p}$ at the considered points to that $h_{D\text{-}i}$ corresponding to tool life T_i determined using Equation 4.82, that is,

$$k_{N\text{-}p} = \frac{h_{D\text{-}p}}{h_{D\text{-}i}} \tag{4.83}$$

If k_N's for various point of the cutting edge are known, then the uniformity of the load over this edge can be assessed.

4.6.1 Uncut Chip Thickness in Drilling

Although the uncut (undeformed) chip thickness can be easily determined using the vector analysis (Astakhov 2010), it was found instructive to visualize this important characteristic of the cutting process graphically to help a tool designer/optimizer to develop its material sense and only then to derive an equation for its calculation. Figure 4.165 shows a model for visualization of the uncut chip thickness. In this model, a drill having two lips is shown. There are two surfaces of the cut in the axial (the y_0z_0 plane) section apart from each other by $f/2$.

In the model shown in Figure 4.165, the uncut chip thickness h_D is determined for point i of the cutting edge as follows. The normal to the surface of cut lays in the plane n-n, which is normal to the cutting edge at point i. As this plane crosses the z_0-axis (the longitudinal axis of the drill) at point b, line ib is the normal to the surface of cut at point i. The normal section plane n-n crosses the two consecutive surfaces of cut located in the axial direction by distance $f/2$, so curves $Ar_{b\text{-}r}$ and $Ar_{b\text{-}a}$ are intersection lines. For the sake of simplicity, these curves can be replaced by circular arcs having point b as their center. Segment id of the normal to the point of cut is between point i and point d formed at the intersection of arc $Ar_{b\text{-}a}$ with this normal (line ib). This segment id is a graphical interpretation of the uncut (undeformed) chip thickness.

To determine the uncut chip thickness analytically, consider a vector **F** along the direction of the cutting feed f and the normal to the surface of cut $\mathbf{N}_p$. If the angle between these two vectors is designated as ε_{NF}, then the uncut chip thickness can be represented as

$$h_D = \frac{1}{2} f \cos \varepsilon_{NF} \tag{4.84}$$

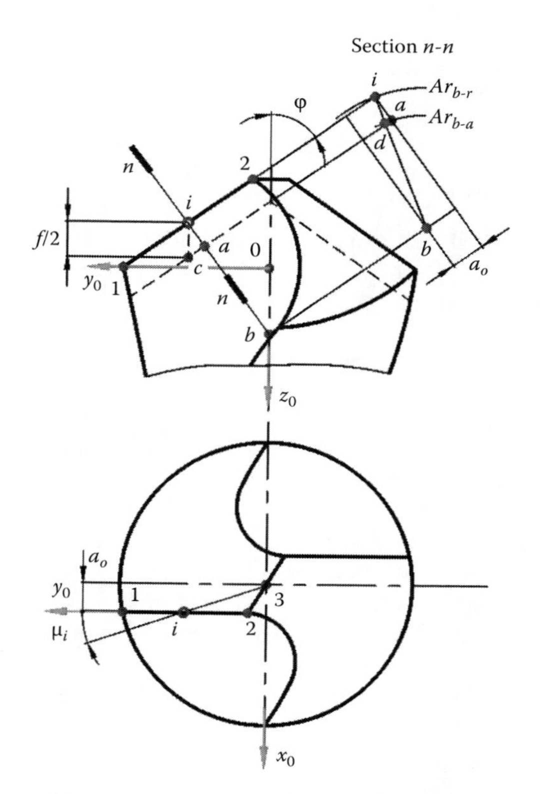

FIGURE 4.165 Model for graphical determination of the uncut chip thickness.

The angle between two vectors **F** along the direction of the cutting feed f and the normal to the surface of cut $\mathbf{N}_p$ is calculated as

$$\cos \varepsilon_{NF} = \frac{\mathbf{N}_p \cdot \mathbf{F}}{\|\mathbf{N}_p\| \|\mathbf{F}\|} \tag{4.85}$$

A unit vector of the normal to the surface of cut in the $x_0y_0z_0$ coordinate system is

$$\mathbf{N}_p = -\mathbf{i}\sin\mu_i\cos\varphi + \mathbf{j}\cos\mu_i\cos\varphi - \mathbf{k}\cos\mu_i\sin\kappa \tag{4.86}$$

so its modulus is

$$\|\mathbf{N}_p\| = \sqrt{\sin^2\mu_i\cos^2\varphi + \cos^2\mu_i\cos^2\varphi + \cos^2\mu_i\sin^2\varphi} = \sqrt{\cos^2\mu_i + \sin^2\mu_i\cos^2\varphi} \tag{4.87}$$

Unit vector **F** in the feed direction is

$$\mathbf{F} = -\mathbf{k}f \tag{4.88}$$

Substituting Equations 4.86 through 4.88 into Equation 4.85, one can obtain

$$h_D = \frac{f}{2}\frac{1}{\sqrt{1+(\cot\varphi/\cos\mu_i)^2}} = \frac{f}{2}\frac{1}{\sqrt{1+(\cot\varphi/\cos(\arcsin(a_o/r_i)))^2}} \tag{4.89}$$

This equation can be used to determine the uncut (undeformed) chip thickness for any point of the cutting edge 1-2.

4.6.2 Load Distribution over the Cutting Edge

The load distribution over the drill cutting edge is determined for a drill that rotates with the angular velocity ω and fed with the feed f. As discussed in Chapter 1, the cutting speed at point i of the cutting edge located at radius r_i is calculated as

$$v_i = r_i\omega \tag{4.90}$$

The cutting speed changes over the cutting edge 1-2 in proportion to the radius from the inner point 2 to the periphery point 1. It follows from Equation 4.89 that the uncut chip thickness also varies over cutting edge 1-2 as angle μ_i depends on the location of a considered point on this edge.

Calculations showed (Rodin 1971) that for standard twist drills with the web diameter d_{wb} = $0.15d_{dr}$ having the point angle $\Phi_p = 120°$ and chisel edge angle $\psi_{cl} = 55°$, the uncut chip thickness h_D changes from $0.43f$ at the periphery point 1 to $0.35f$ at point 2 where the major cutting edge intersects the chisel edge. Thus, h_D at point 2 is 18% lower than that at point 1.

When only the influences of the uncut chip thickness and cutting speed on tool life are considered, the condition when tool life is constant (T = Const) can be represented as

$$v = \frac{C_{vt}}{h_D^{m_{vt}}} \tag{4.91}$$

Substituting Equation 4.90 into Equation 4.91 and accounting for the fact that the angular velocity is constant, one can obtain

$$r_i = \frac{c_{vt-\omega}}{h_{D-i}^{m_{vt}}} \tag{4.92}$$

where $c_{vt\text{-}\omega} = c_{vt}/\omega$.

If the uncut chip thickness corresponding to constant tool life for point i is $h_{D\text{-}i}$, then that point p on this edge can be represented as

$$\frac{r_p}{r_i} = \left(\frac{h_{D\text{-}i}}{h_{D\text{-}p}}\right)^{m_{vt}} \tag{4.93}$$

Experiments showed (Rodin 1971) that for medium-carbon steels, $m_{vt} \cong 0.5$. If it is assumed that the uncut chip thickness is equal to 1 at the periphery point 1, then the theoretical uncut chip thickness assuring the equality of tool life for other points of the cutting edge 1-2 can be determined by the following formula:

$$h_{D\text{-}i} = \left(\frac{r_{dr}}{r_i}\right)^2 \tag{4.94}$$

Table 4.6 shows theoretical and real (for a drill having point angle $\Phi_p = 120°$ and the web diameter $d_{wb} = 0.15d_{dr}$) uncut chip thickness calculated for various points of the cutting edge 1-2. The load coefficients k_N for the same points are also shown. As can be seen, the load coefficient for the periphery point 1 is many folds greater than that for the central part of the cutting edge 1-2. This is the prime cause for drill nonuniform wear observed in practice. Observations showed that tool wear at the drill corner is normally much greater than that of central parts of the cutting edge.

4.6.3 Drills with Curved Cutting Edges

The twist drill used in industry for more than 140 years for most widely employed metalworking operation (drilling) can be regarded as an unperfected cutting tool. As discussed in this chapter, severe problems associated with this cutting tool are great variations of the load coefficient and tool geometry parameters over the cutting edge. For example, the normal rake angle varies from +30° at drill periphery part of the cutting edge (point 1 in Figure 4.165) (point 1) to −30° at the inner end of this edge (point 2 in Figure 4.165); the load coefficient varies from 1 at point 1 to 0.035 at point 2. These are the greatest variations among general-purpose cutting tools used in industry.

TABLE 4.6
Theoretical and Real Uncut Chip Thickness and Load Coefficients for Points of Cutting Edge 1-2

r_i/r_{dr}	Theoretical Uncut Chip Thickness	Real Uncut Chip Thickness	Load Coefficient
1	1	0.43f	1
0.6	2.8	0.43f	0.36
0.2	25	0.38f	0.035

Although researchers, tool engineers, and professionals in the field are well aware about the importance of the region of the major cutting edge adjacent to periphery point 1 in tool life consideration, the influence of other parts of the major cutting edge (lip) as well as the chisel edge on tool life is not well understood. For example, improving conditions of the chisel edge (web thinning or splitting) cause improvement not only in self-centering ability but also in tool life defined by the wear of the drill corners (point 1). This influence also follows from the results of the tests carried out with standard drill of 30 mm dia. used for enlarging pre-drilled holes in gray cast iron (Rodin 1971). Series of tests with pre-drilled holes of 26, 17.5, and 11.5 mm were carried out. It was found that tool life in machining of pre-drilled hole of 26 mm dia. was 29.5 min while that in machining pre-drilled holes of 11.5 mm dia. was three-fold lower. This result exemplifies the inter-influence of various parts of the cutting edge on tool life. Unfortunately, this inter-influence was not a subject of extensive research studies.

For simplicity, consider a drill with the so-called diametral cutting edges, that is, the major cutting edge (lip) is along the drill radius as in the drill design shown in Figure 4.136. For such a drill, $\mu_i = 0$ for any point i of the major cutting edge so that, as follows from Equation 4.89, the uncut (undeformed) chip thickness is constant along this cutting edge. The cutting speed changes significantly along the cutting edge causing nonuniform wear of the drill along this edge. The objective of the further considerations is to alter the shape of the cutting edge to achieve uniform drill wear.

According to Equation 4.91, the cutting speed and the uncut chip thickness under T = Const for any point i of the cutting edge 1-2 correlate as

$$v_i = \frac{C_{vt}}{h_{D\text{-}i}^{m_{vt}}} \tag{4.95}$$

For diametral cutting edge, the uncut chip thickness for point i of this edge is calculated as

$$h_{D\text{-}i} = \frac{f}{2}\sin\varphi_i \tag{4.96}$$

where φ_i is the half-point angle for point i of the major cutting edge as in general this angle can vary along the major cutting edge.

As discussed in Chapter 1, the cutting speed at point i is calculated as

$$v_i = \frac{2\pi n}{1000} r_i \tag{4.97}$$

where

$\pi = 3.141$

r_i is in millimeters

n is the rotational speed in rpm or rev/min.

Substituting Equations 4.96 and 4.97 into Equation 4.95, and after some rearrangements, one can obtain

$$\tan\varphi_i = \frac{(A_i/r_i)^{1/m_{vt}}}{\sqrt{1-(A_i/r_i)^{2/m_{vt}}}} = \frac{dr_i}{dz_0} \tag{4.98}$$

Differential equation (4.98) defines the shape of the uniformly loaded cutting edge in the y_0z_0 plane. Its numerical differentiation, however, shows that the length of the drill point under the accepted conditions is too long for practical applications. The closest known shape of the cutting edge to the obtained result is the drill design with the ellipsoidal point shown in Figure 4.60d. The obtained result shows why even though complicated and difficult to grind as the grinding accessories and programs are not fully developed, this drill point is still in use showing remarkable results when applied properly. The problem is in grinding of such complicated profile. The calculation of a suitable grinding wheel profile used to generate the clearance surface of a twist drill involves complicated numerical methods, and this is of a major concern. The available methods for modeling grinding wheel profile are based on the theory of enveloping or theory of conjugations and computer simulation methods. So far, only particular solutions to the problem are found, for example, the toroidal grinding method for curved cutting edges of twist drills (Fetecau et al. 2009). Although Ivanov et al. (1998) proposed a generalized analytical method for profiling of all types of rotation tools for forming helical surfaces, it was not noticed by the developers of CNC drill grinders.

Another feasible possibility is varying angle μ_i, that is, the cutting edge can be curved not only in the y_0z_0 plane but also in the x_0y_0 plane. The advantages of such geometry were found out by the trial and error method as early as at the beginning of the twentieth century. For example, US Patent No. 1,309,706 (1917) describes a drill design shown in Figure 4.166 where the major cutting edges (lips) 1 and 2 are curved in both the y_0z_0 and x_0y_0 planes. The advantages as a remarkable increase of tool life and drilling *smoothness* are explained using an intuitive but very precise perception as "By this constriction the work of removing metal, that is, the cutting, is distributed in such a manner that a unit of length of edge at the periphery does no more work than a unit of length of edge nearer the axis and therefore that the amount of heat produced in removing the metal is more nearly uniform for each unit length of the cutting edge ..." If this heat results in the uniform temperatures equal to the optimal cutting temperature (Section 2.6.9.2) along this edge, then the maximum tool life and uniform tool wear along the cutting edge can be achieved.

The model of optimization of the cutting edge shape for the considered case was discussed by the author earlier (Astakhov 2010). It was shown that because the uncut chip thickness varies over the cutting edge due to variation of angle μ_i, the distribution of this angle should be as follows:

$$\cos\mu_i = \frac{f}{2}\frac{\cot\varphi_i}{\sqrt{(f^2/4r_i^2)-1}} \tag{4.99}$$

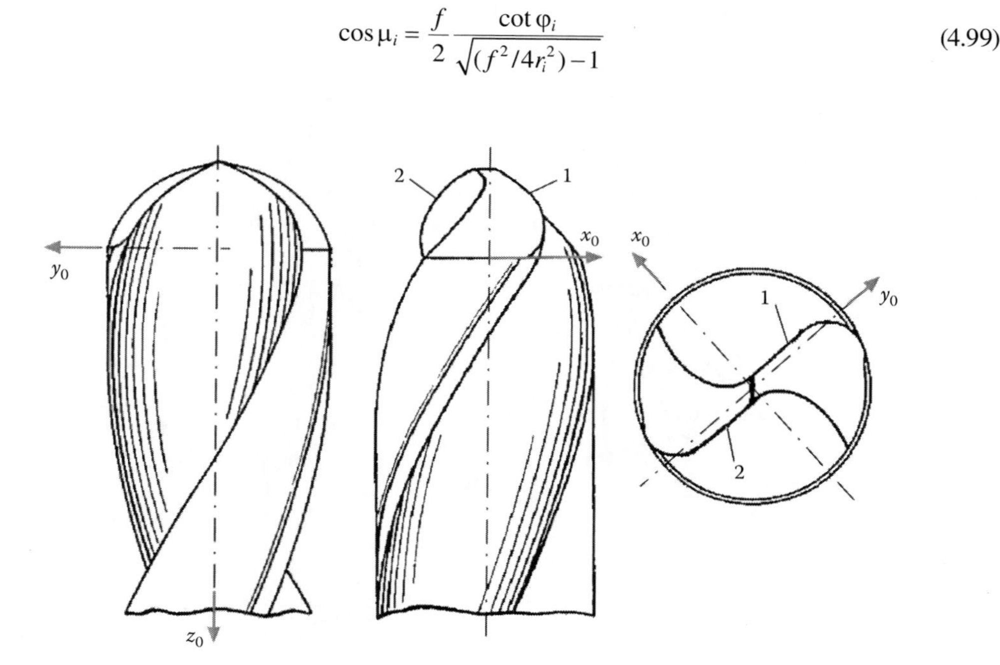

FIGURE 4.166 Drill geometry according to US Patent No. 1,309,706 (1917).

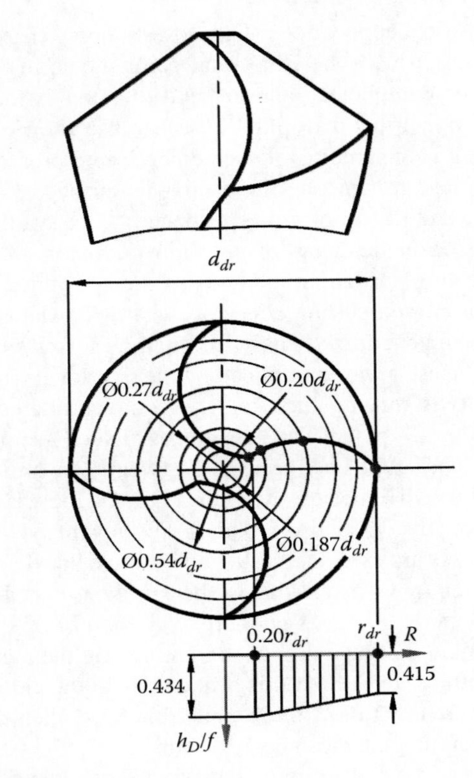

FIGURE 4.167 Calculated curvilinear cutting edge.

Figure 4.167 shows a drill geometry where the cutting edge is constructed using angles μ_i calculated by Equation 4.99. As can be seen, the uncut chip thickness varies along the cutting edge following a linear fashion with the minimum at the drill periphery. The obtained results constitute the background for drills with curved cutting edges.

Another way to improve performance of a twist drill is to achieve the constant rake angle over the entire length of the major cutting edges (lips). The major benefit of such a goal is *the balanced drill* where the unit chip flows from the unit length of the cutting edge do not cross each other. It follows from Figure 4.78 that the suitable distribution of the normal rake angle long the cutting edge can be achieved by continuous variation of the tool cutting edge angle (the half-point angle) φ along this cutting edge. Schematically, such a point grind of a twist drill is shown in Figure 4.168.

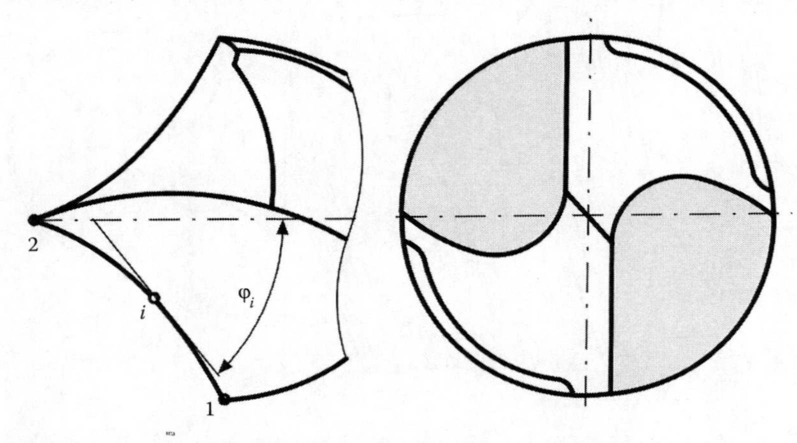

FIGURE 4.168 A point grind of a twist drill with the variable half-point angle.

TABLE 4.7
Tool Cutting Tool Angles for the Curved Cutting Edge 1-2

r_i/r_{dr}	0.40	0.69	0.84	0.95	1.00
φ	15°	30°	45°	60°	90°

Each point of the cutting edge, for example, point i, is characterized by its own point angle φ_i, which is the angle between the tangent to the cutting edge at i and the drill longitudinal axis.

To determine the shape of the cutting edge 1-2 that results in a constant rake angle along this edge, the graph similar to that shown in Figure 4.78 is used. For a given rake angle, say 29°, one may draw a horizontal line in Figure 4.78 that corresponds to this angle. Points of intersection of this line with corresponding curves define point angles Φ_p (=2φ). Table 4.7 shows the calculated half-point angles for the considered case.

A drill shown in Figure 4.169a is described in US Patent No. 5,273,380 (1993). As seen, it is similar to that shown in Figure 4.168. According to this patent, the drill is provided having a novel drill point that includes the concave cutting edges. As claimed, this drill is characterized by much greater tool life due to improved heat dissipation along the concave cutting edge.

A great disadvantage of the drill shown in Figure 4.169a is that the uncut chip thickness is the greatest drill periphery that lowers tool life. This issue can be easily resolved if a second point grind similar to that shown in and discussed in Section 4.5.9.6 is applied as shown in Figure 4.169b. ICS Cutting Tools (Casco, WI,) manufactures such a drill calling it the Perfect Point drill (Figure 4.170). As claimed by ICS Cutting Tools, this point geometry offers advantages not found with other conventional or specialty point styles. Most notably, these include an improved cutting action that generates superior surface finishes with excellent drilling precision and accuracy. This design also provides excellent stability while substantially reducing the chatter and vibration so common with other drill point shapes. The drill combines the best features of metal and wood cutting drills and is well suited

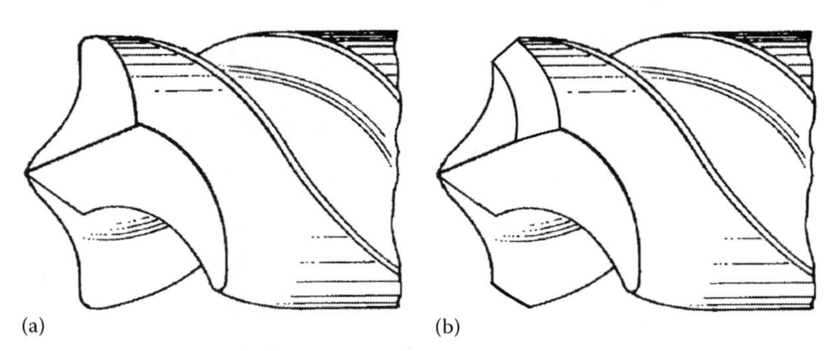

FIGURE 4.169 Practical realization of the concept geometry: (a) drill design according to US Patent No. 5,273,380 (1993) and (b) its proposed modification.

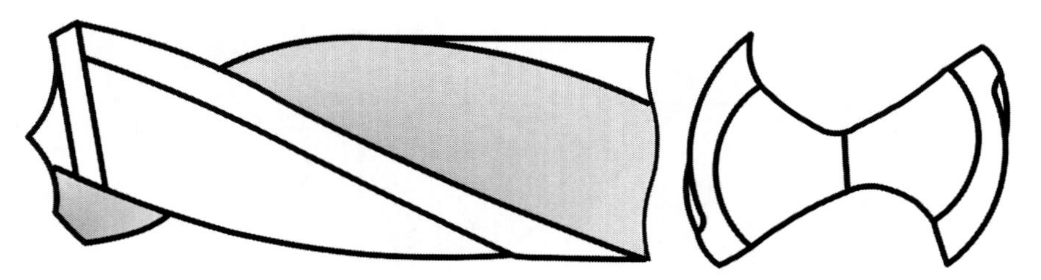

FIGURE 4.170 The Perfect Point drill.

for drilling holes in a variety of materials. It may be used for drilling ferrous and nonferrous metals, natural and composite wood products, and a range of plastics and similar manufactured materials. Unlike other point designs, the Perfect Point geometry consistently produces perfectly round and uniform holes with no metal burr breakthrough or wood grain tears and splinters in virtually any material thickness.

Another possibility to achieve the constant rake angle over the entire length of the major cutting edges (lips) and thus to design *the balanced drill* can be though of as a drill geometry where the cutting edge is constructed using an alternation of μ_i through the distance a_o. In other words, the major cutting edge of a drill is curved in the x_0y_0 plane so that each point of this edge has its own a_o. As discussed earlier, the distance a_o directly affects the rake angle. Therefore, the curvature of the cutting edge can be selected so that a uniform rake angle along the cutting edge 1-2 is achieved. If the tool designer knows what the rake angle is to be achieved, then he can draw a horizontal line on corresponding graph similar to those shown in Figure 4.79. The intersection points of this line with corresponding curves indicate a_o's along the cutting edge are used to achieve the objective.

For example, for a twist drill having point angle Φp = 120° and helix angle of the drill flute ω_d = 30° when a 20° invariable rake angle is to be achieved along the major cutting edge, a_o's for points of the major cutting edge as shown in Table 4.8. Figure 4.171 shows a graphical representation of these results. Experimental study of a drill with this geometry showed that tool life increases while the drilling torque decreases.

It is interesting to mention that the similar profile shown in Figure 4.42 was developed as early as 1882 empirically. It is not a surprise then that Sandvik Coromant Co. has developed a similar drill

TABLE 4.8
Distances a_o's along the Cutting Edge to Achieve a 20° Invariable Rake Angle

r_i/r_{dr}	0.43	0.50	0.73	0.85	0.95
a_o	$-0.07d_{dr}$	0	$0.15d_{dr}$	$0.30d_{dr}$	$0.50d_{dr}$

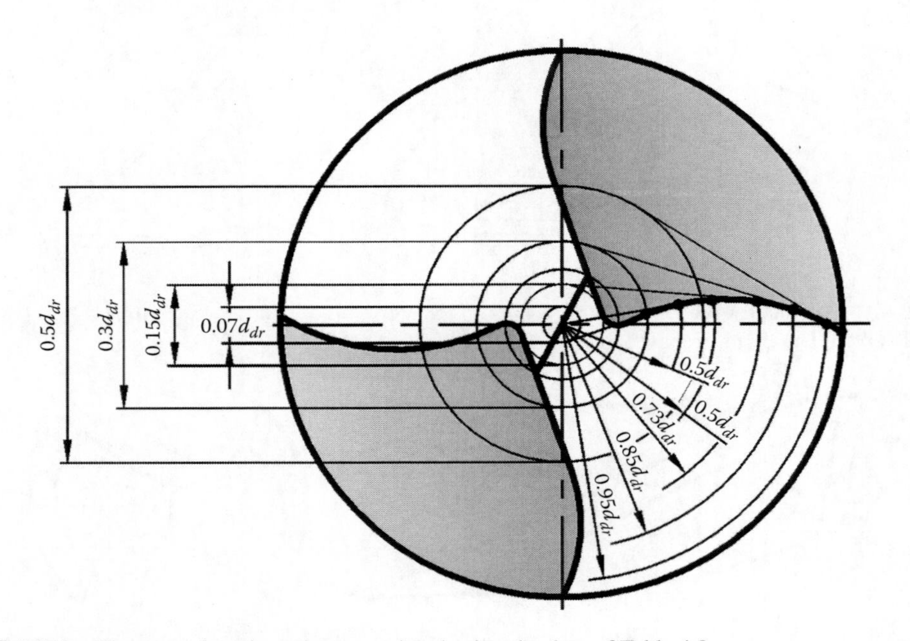

FIGURE 4.171 The curved cutting edge constricted using the data of Table 4.8.

CoroDrill Delta-C R846, which fully resembles the discussed drill geometry. Reportedly, CoroDrill Delta-C R846 shows great results in drilling of Ni/Co-based heat resistant alloys as well as titanium alloys and stainless steels.

4.6.4 Generalization

Reading this chapter, one may ask a logical question: "Why do we need to know particularities of the drill geometry and the means of its optimization as drills are standard on-shelf items that can be readily selected/bought using colorful catalogs of numerous drill manufacturers?" In fact, drill diameters, lengths, shank types, and sizes are standardized by many ISO and DIN standards, for example, ISO 235:1980 parallel shank jobber and stub series drills and Morse taper shank drills; ISO 494:2009 cylindrical shank twist drill (long series); ISO 2306: 1972 drills for use prior to tapping screw threads; and *DIN 6539:1991* continuous parallel shank solid hardmetal twist drills and dimensions. Therefore, a need is felt to provide a clear answer to this question.

According to various estimates, 75% or more of the tools used in certain aerospace manufacturing operations are specials. In the automotive industry, the use of special drills exceeds 80%. What exactly determines if a tool is a special is a matter for discussion. Although even when produced in large quantities, custom tools lack the economy of scale characteristic of standard tools, there are a number of advantages in the use of custom drills in these industries so that part makers' willingness to pay a premium and take advantage of special drills can be readily explained. Standard drills are cheaper, but there is value in the more expensive custom HP drills; they offer more capability. The cost of machining time is way greater than that of the drill so that even if a special HP tool cost twice compared to a standard, its productivity saves 10 times more due to saving the machining time. If a shop uses the custom HP drills properly, their productivity can go way up and their profits can go way up too. With many operations reluctant to make large capital purchases, a custom HP tool may enable a shop to increase productivity without purchasing new equipment.

The other advantage of custom HP drills is that they are application-specific and can be designed and manufactured according to the recent trends in the development of new tool materials, coatings, manufacturing, control, and inspection equipment (both in drill manufacturing and its use). Unfortunately, many standards on drills lag far behind these trends.

The matter discussed in this chapter provides a clear way for optimization of drill parameters and explains the rationale behind the known high-performance drills. It argues that a special HP drill can be *assembled* out of design/geometrical components needed to achieve the goal of optimization. Such an assembly is significantly facilitated when a circular diagram shown in Figure 4.172 is constructed. This diagram visualizes the requirement to drilling operation and thus makes it easy to select the design/geometry parameters of the HP drill.

It should be clear that

- High-penetration rate, high-efficient drilling operation cannot be achieved by just doubling the penetration rate with the standard drills as they are not designed and made for such conditions. Moreover, the tolerances on the important (for HP drills) design/geometrical parameters of standard drills, for example, on runout, lip-height variation, and back taper, are too wide and thus are not acceptable for HP drills. A common mistake in the designing of HP drills is to assign the same tolerances as those for standard drills on the *secondary* features, as, for example, on the back taper. This chapter explains why such a practice should not be used
- The optimization of the drill design/geometry parameters makes sense if and only if the drilling system supports such an optimization (see Section 1.2.2.2)
- No one particular drill design/geometry can maximize all the parameters shown in Figure 4.172 simultaneously

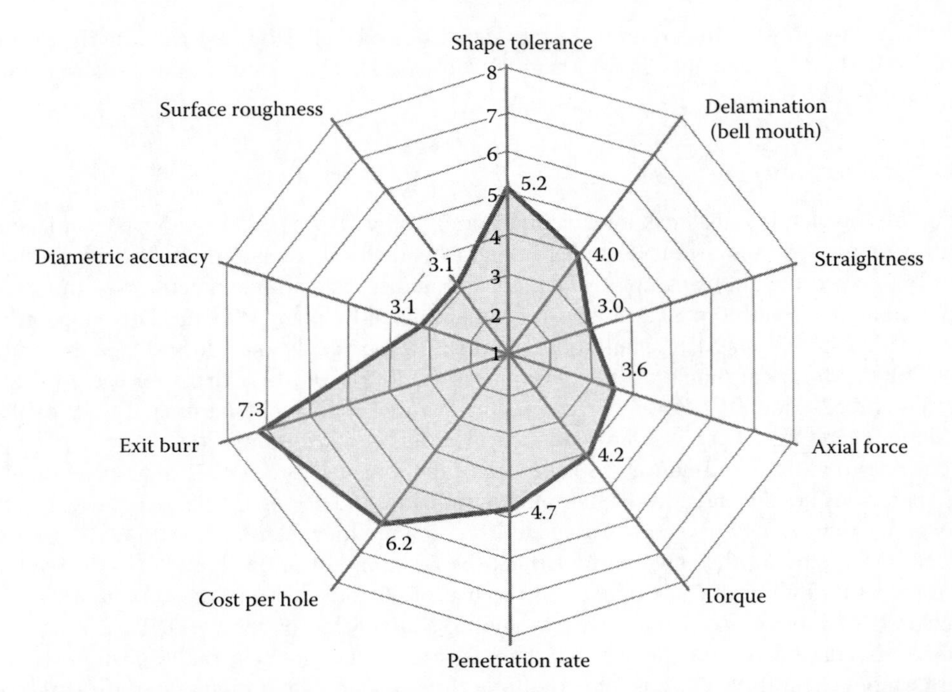

FIGURE 4.172 Circular diagram showing relative importance (scale 1–8) of the parameters of a drilling operation.

- Adding any additional parameter of the tool geometry may increase the cost of the tool as additional programming, grinding wheel, grinding time, etc., may be required
- The geometry/design parameters that affect the goal of prime importance should be selected first and then adjusted by going down through other parameters. In the example shown in Figure 4.172, the point angle that results in the tolerable exit burr (Figure 4.144) is selected. Then, as the cost per drilling hole is the second important objective, the double-point drill should be selected to minimize the cycle time and improve quality of machining holes.
- The design/geometry parameters optimized for one tool material or even tool material grade may not be suitable for the other as the friction/adhesion, achievable contact surface conditions, sharpness of the cutting edge, temperature-affected properties, and many other parameters hidden from the view of manufacturing specialists may change the tool performance dramatically.

REFERENCES

Abele, E. and Fujara, M. 2010. Simulation-based twist drill design and geometry optimization. *CIRP Annals—Manufacturing Technology* 59:145–150.

Abrão, A.M., Campos Rubio, J.C., Faria, P.E., and Davim, J.P. 2008. The effect of cutting tool geometry on thrust force and delamination when drilling glass fibre reinforced plastic composite. *Materials & Design* 29:508–513.

Agapiou, J.S. 1993a. Design characteristics of new types of drill and evaluation of their performance drilling cast iron. I. Drills with four major cutting edges. *International Journal of Machine Tools and Manufacture* 33:321–341.

Agapiou, J.S. 1993b. Design characteristics of new types of drill and evaluation of their performance drilling cast iron. - II. Drills with three major cutting edges. *International Journal of Machine Tools and Manufacture* 33:343–365.

Ahmad, J. 2009. *Machining of Polymer Composites*. Boston, MA: Springer-Verlag.

Armarego, E.J.A. and Cheng, C.Y. 1972a. Drilling with flat rake face and conventional twist drills—Part I: Theoretical investigation. *International Journal of Machine Tool Design and Research* 12:7–35.

Armarego, E.J.A. and Cheng, C.Y. 1972b. Drilling with flat rake face and conventional twist drills. Part II: Experimental investigation. *International Journal of Machine Tool Design and Research* 12:37–54.

Astakhov, V.P. 2004. The assessment of cutting tool wear. *International Journal of Machine Tools and Manufacture* 44:637–647.

Astakhov, V.P. 2006. *Tribology of Metal Cutting*. London, U.K.: Elsevier.

Astakhov, V.P. 2010. *Geometry of Single-Point Turning Tools and Drills. Fundamentals and Practical Applications*. London, U.K.: Springer.

Astakhov, V.P. 2011a. Authentication of FEM in metal cutting, Chapter 1. In *Finite Element Method in Manufacturing Processes*, Davim, J.P. (ed.). New York: Wiley.

Astakhov, V.P. 2011b. Drilling. In *Modern Machining Technology*, Davim, J.P. (ed.). Cambridge, U.K.: Woodhead Publishing. pp. 79–212.

Astakhov, V.P. and Shvets, S. 2004. The assessment of plastic deformation in metal cutting. *Journal of Materials Processing Technology* 146:193–202.

Astakhov, V.P. and Xiao, X. 2008. A methodology for practical cutting force evaluation based on the energy spent in the cutting system. *Machining Science and Technology* 12:325–347.

Beyer, W.H. 1987. *CRC Standard Mathematical Tables*, 28th edn. Boca Raton, FL: CRC Press.

Bhattacharyya, A., Bhattacharyya, A., Chatterjee, A.B., and Ham, I. 1971. Modification of drill point reducing thrust. *ASME Journal of Engineering for Industry* 93:1073–1078.

Bhushan, B. (Editor-in-Chief). 2001. *Modern Tribology Handbook*, Two Volume Set. Boca Raton, FL: CRC Press.

Chen, W.-C. 1997. Applying the finite element method to drill design based on drill deformations. *Finite Elements in Analysis and Design* 26:57–81.

Chen, Y.R. and Ni, J. 1999. Analysis and optimization of drill cross-sectional geometry. *SME Paper MR99-162*.

Davim, J.P. and Reis, P. 2003. Study of delamination in drilling carbon fiber reinforced plastics (CFRP) using design of experiments. *Composite Structures* 59:481–487.

Davim, J.P., Reis, P., and Antonio, C.C. 2004a. Drilling fiber reinforced plastics (FRPs) manufactured by hand lay-up: influence of matrix (Viapal VUP 9731 and ATLAC 382-05). *Journal of Materials Processing Technology* 155–156:1828–1833.

Davim, J.P., Reis, P., and Antonio, C.C. 2004b. Experimental study of drilling glass fiber reinforced plastics (GFRP) manufactured by hand lay-up. *Composites Science and Technology* 64:289–297.

De Beer, C. 1970. The web thickness of twist drills. *Annals of the CIRP* 18:81–85.

Fetecau, C., Stan, F., and Oancea, N. 2009. Toroidal grinding method for curved cutting edge twist drills. *Journal of Materials Processing Technology* 209:3460–3468.

Fiesselmann, F. 1993. An option for doubling drilling productivity. *The Fabricator* 23 (4):36–38.

French, L.G. and Goodrich, C.I. 1910 (reprinted in 2001). Principles of deep hole drilling. In *Deep Hole Drilling. Machinery's Reference Series No. 25*. Bradley, IL: Lindsay Publication Inc. pp. 3–19.

Fugelso, M.A. 1990. Conical flank twist drill points. *International Journal of Machine Tools and Manufacture* 30 (2):291–295.

Fuh, K.-H. and Chen, W.-C. 1995. Cutting performance of thick web drills with curved primary cutting edges. *International Journal of Machine Tools and Manufacturing* 35 (7):975–991.

Fujii, S., DeVries, M.F., and Wu, S.M. 1970. An analysis of drill geometry for optimum drill design by computer-I: Drill geometry analysis *ASME Journal of Engineering for Industry* 92:647–656.

Galloway, D.F. 1957. Some experiments on the influence of various factors on drill performance. *ASME Transactions* 79 (191–231).

Granovsky, G.E. and Granovsky, V.G. 1985. *Metal Cutting (in Russian)*. Moscow, Russia: Vishaya Shkola.

Heisela, U. and Pfeifroth, T. 2012. Influence of point angle on drill hole quality and machining forces when drilling CFRP. *Procedia CIRP* 1:471–476.

Hibbeler, R.C. 2010. *Mechanics of Materials*, 8th edn. Boston, MA: Prentice Hall.

Hsieh, J.-F. 2005. Mathematical model for helical drill point. *International Journal of Machine Tools and Manufacture* 45 (7–8):967–977.

Ivanov, V., Nankov, G., and Kirov, V. 1998. CAD-orientated mathematical model for determination of profile helical surfaces. *International Journal of Machine Tools and Manufacture* 38:1001–1015.

Jawahir, I.S., Balaji, A.K., Stevenson, R., and van Luttervelt, C.A. 1997. Towards predictive modeling and optimization of machining operations. Paper read at *Manufacturing Science and Engineering. Proceedings 1997 ASME International Mechanical Engineering Congress and Exposition*, November 16–21, Dallas, TX.

Jawahir, I.S. and Van Luttervelt, C.A. 1993. Recent developments in chip control research and applications. *Annals of the CIRP* 42 (2):659–693.

Jawahir, I.S. and Zhang, J.P. 1995. An analysis of chip curl development, chip deformation and chip breaking in orthogonal machining. *Transactions of NAMRI/SME* XXIII:109–114.

Kennedy, B. 2006. The straight story. *Cutting Tool Engineering* 58 (1):58–64.

Khashaba, U.A. 2012. Drilling of polymer matrix composites: A review. *Journal of Composite Materials* 47:1817–1832. http://jcm.sagepub.com/content/early/2012/07/09/0021998312451609.

Kim, J. and Dornfeld, D. 2002. Development of an analytical model for drilling burr formation in ductile materials. *ASME Journal of Engineering Materials and Technology* 124 (2):192–198.

Lorenz, G. 1979. Helix angle and drill performance. *Annals of the CIRP* 28:83–86.

Metal Cutting Tool Handbook, 7th Ed. 1989. New York: The Metal Cutting Tool Institute by Industrial Press.

Min, S., Kim, J., and Dornfeld, D.A. 2001. Development of a drilling burr control chart for low alloy steel AISI 4118. *Journal of Materials Processing Technology* 113 (1–3):4–9.

Mollin, R.A.1995. *Quadrics*. Boca Raton, FL: CRC Press.

Nakayama, K. and Ogawa, M. 1985. Effect of chip splitting nicks in drilling. *Annals of the CIRP* 34 (1):101–104.

Narasimha, K., Osman, M.O.M., Chandrashekhar, S., and Frazao, J. 1987. An investigation into the influence of helix angle on the torque-thrust coupling effect in twist drills. *The International Journal of Advanced Manufacturing Technology* 2 (4):91–105.

Olson, W.W., Batzer, S.A., and Sutherland, J.W. May 1998. Modeling of chip dynamics in drilling. Paper read at *Proceedings of CIRP International Workshop on Modeling of Machining Operations*, Atlanta, GA.

Oxford, Jr., C.J. 1955. On the drilling of metals—I. Basic mechanics of the process. *Transactions of the ASME* 77:103–114.

Pilny, L., De Chiffre, L., Pıska, M., and Villumsen, M.F. 2012. Hole quality and burr reduction in drilling aluminium sheets. *CIRP Journal of Manufacturing Science and Technology* 5:102–107.

Radzevich, S.P. 2010. *Kinematic Geometry of Surface Machining*. Boca Raton, FL: CRC Press.

Rodin, P.R. 1971. *Cutting Geometry of Twist Drills (in Russian)*. Kyiv, Ukraine: Technika.

Sahu, S.K., Ozdoganlar, O.B., DeVor, R.E., and Kapoor, S.G. 2003. Effect of groove-type chip breakers on twist drill performance. *International Journal of Machine Tools and Manufacture* 43:617–627.

Sokołowski, A. 2010. On burr height estimation based on axial drilling force. *Journal of Achievements in Materials and Manufacturing Engineering* 43 (2):734–742.

Spur, G. and Masuha, J.R. 1981. Drilling with twist drills of different cross section profiles. *CIRP Annals—Manufacturing Technology* 30:31–35.

Stabler, G.V. 1964. The chip flow law and its consequences. Paper read at *Proceedings of the Fifth International MTDR Conference*. Birmingham, UK, p. 243–251.

Takashi, M. and Jürgen, L. 2010. Simulation of drilling process for control of burr formation. *Journal of Advanced Mechanical Design, Systems, and Manufacturing* 4 (5):966–975.

Thornley, R.H., Wahab, A.B.I., and Maiden, J.D. 1987a. A new approach to eliminate twist drill chisel edge. Part 1. Asymmetrical configuration. *International Journal of Production Research* 25 (4):589–602.

Thornley, R.H., Wahab, A.B.I.El., and Maiden, J.D. 1987b. Some aspects of twist drill design. *International Journal of Machine Tools and Manufacture* 27:393–397.

Tsai, W.D. and Wu, S.M. 1979. Computer analysis of drill point geometry. *International Journal of Machine Tool Design and Research* 19:95–108.

Vable, M. 2012. *Mechanics of Materials*. Houghton, MI: Michigan Technological University. http://www.me.mtu.edu/~mavable/MoM2nd.htm.

Veremachuk, E.S. 1940. *Deep Hole Drilling (in Russian)*. Moscow, Russia: State Publishing House of Defence Industry.

Vinogradov, A.A. 1985. *Physical Foundation of the Drilling of Difficult-to-Machine Materials with Carbide Drills (in Russian)*. Kyiv, Ukraine: Naukova Dumka.

Wang, J. and Zhang, Q. 2008. A study of high-performance plane rake faced twist drills. Part I: Geometrical analysis and experimental investigation. *International Journal of Machine Tools and Manufacture* 48:1276–1285.

Wang, S., Shu, L., and Xiao, Z. 1993. A new twist drill. *Cutting Tool Engineering* (2):32–35.

Wright, J.D. and Armarego, E.J.A. 1983. An analysis of conical drill point grinding—The generation process and effects of setting errors. *Annals of the CIRP* 32 (2):1–5.

Wu, S.M. and Shen, J.M. 1983. Mathematical model for multifacet drills. *ASME Journal of Engineering for Industry* 105:177–182.

5 PCD and Deep-Hole Drills

Don't design for everyone. It's impossible. All you end up doing is designing something that makes everyone unhappy.

Leisa Reichelt (United Kingdom), a user experience consultant (Freelance)

5.1 PCD DRILLING TOOLS

PCD drilling tools were developed to meet the challenges of both new work materials and HSM primary in the automotive and aerospace industries.

5.1.1 Challenges of Work Materials

There are two distinctive groups of work materials: metal and polymer based.

5.1.1.1 Metal-Matrix Composites

MMCs offer high strength-to-weight ratio, high stiffness, and good damage resistance over a wide range of operating conditions, making them an attractive option in replacing conventional material for many engineering applications. Typically, MMCs are aluminum, titanium, copper, and magnesium alloys, while the reinforcement materials are silicon carbide, aluminum oxide, boron carbide, graphite, etc., in the form of fibers, whiskers, and particles. Probably the single most important difference between fiber-reinforced and particulate composites or conventional metallic materials is the anisotropy or directionality of properties, that is, particulate composites and conventional metallic materials are isotropic, while the fiber-reinforced composites are generally anisotropic. Particulate-reinforced composites offer higher ductility. Their isotropic nature as compared to fiber-reinforced composites makes them an attractive alternative (Shin and Dandekar 2012). Although there are a great number of MMCs, high-silicon aluminum-matrix MMCs have the highest usage due to high volume of car manufacturing.

In the context of a global competition, manufacturing companies are compelled to improve productivities through the optimizations of their production operations including machining. In the automotive industry, transmission and engine components are made of high-silicon aluminum alloys, which have a high strength-to-weight ratio. Aluminum is cast at a temperature of 650°C (1200°F). It is alloyed with silicon (9%) and copper (3.5%) to form the Aluminum Association 380 alloy (UNS A03800). Silicon increases the melt fluidity and reduces machinability. Copper increases hardness and reduces ductility. By greatly reducing the amount of copper (less than 0.6%), the chemical resistance is improved making AA 360 (UNS A03600) well suited for use in marine environments and in automotive transmissions (valve bodies, case, and torque converter housings). High-silicon alloys are used in automotive transmissions (pump cover) and engines (for cylinder castings). One of the most common alloys is AA 390 (UNS A03900) with 17% silicon for high wear resistance.

Machining of such alloys presents a great challenge due to their unique properties, namely, the combination of a soft easy-to-adhere Al-matrix and highly abrasive particles including silicon and sludge (Tomac and Tonnessen 1992; Hung et al. 1996; El-Gallab and Sklad 1998; Andrewesa et al. 2000). Figure 5.1a shows a SEM micrograph of 380 Al alloy as it is reported to the customer and appears in manufacturing books. In reality, however, real die aluminum castings supplied to automotive plants contain clusters of sludge as shown in Figure 5.1b. The presence of this sludge and

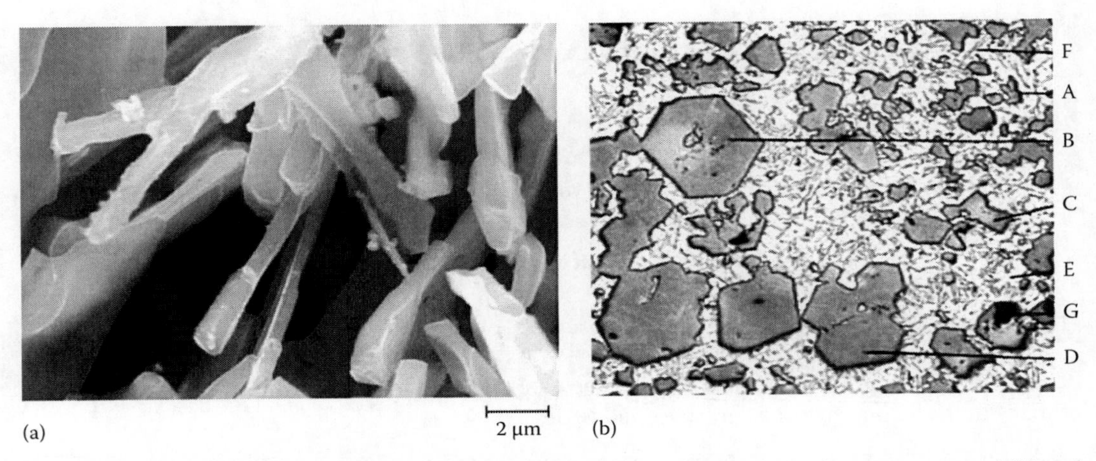

FIGURE 5.1 Micrographs of 380 Al alloy: (a) SEM image commonly presented to customers and (b) real microstructure that shows coarse crystals of sludge (A, B, C, and D). The remainder of the structure consists of aluminum matrix (E), eutectic silicon (F), and Al2O3 (G).

silicon particles makes Al alloys highly abrasive that causes premature tool wear and significant heat generation during machining. The latter causes thermal distortions of the machined parts, resulting in location and diametric errors in machining.

The soft easy-to-adhere Al matrix is also the root cause of one of the major failure modes of cutting tools because it adheres to the tool cutting edge forming the BUE (Astakhov 2010) that causes rapid degradation of the cutting ability of the carbide cutting tools as shown in Figure 5.2. As discussed in Chapter 2, the presence of significant BUE completely undermines the advanced geometry of the chisel edge. It dulls the tool so it can no longer cut through the material. This is the principal problem encountered in the machining of high-silicon Al alloys. As a result of this issue, tool life is significantly shortened causing high tooling cost per unit (a machined part). Even when the most advanced nanocoatings are used, the silicon carbide fibers quickly dull the cutting edge of the tungsten carbide tooling. Therefore, tungsten carbide tools are used only for roughing operations.

Over the last few decades, unattended machining centers and manufacturing cells have been rapidly developed for factory use in the automotive industry. As well understood by specialists in the automotive industry, the principal difficulty preventing automated, unattended, and around-the-clock operations is the cutting tool. Short tool life, a great scatter in tool life, lack of reliable data, and the lack of effective sensors to monitor the unmanned production systems are major contributors to the problem. Worn or fractured tools result in the manufacture of products outside the specifications and significant scrap of almost finished parts. Moreover, breakage of the cutting tool can cause damage to the machine tool itself. This issue is particularly problematic to high-cost ceramic-bearing spindles. These factors invariably lead to increases in the manufacturing costs, loss of manufacturing capacity, and unnecessary use of energy, materials, and labor.

Among the five basic types of tool materials used in the modern automotive industry, PCD tools have rapidly advanced in the machining of aluminum alloy die castings in the automotive industry (Coelho et al. 1995). Today, practically all finishing *hole* operations and all flat surfaces of the Al alloy engine parts and parts of transmissions are machined using these tools. PCDs allow high-speed aluminum marching that increases productivity, efficiency, and quality of the machined part. The high efficiency of these tools comes with a high cost. The costs of PCD tools are three to five times higher than that of carbide tools for similar applications in the automotive industry, and PCD tools are highly sensitive to any inaccuracy (runout, alignments, etc.) in machine tools; thus, special machine tools are required for their effective use. However, when a PCD tool is made and run properly, its tool life is up to tenfold higher than that of the best (the best carbide grade and

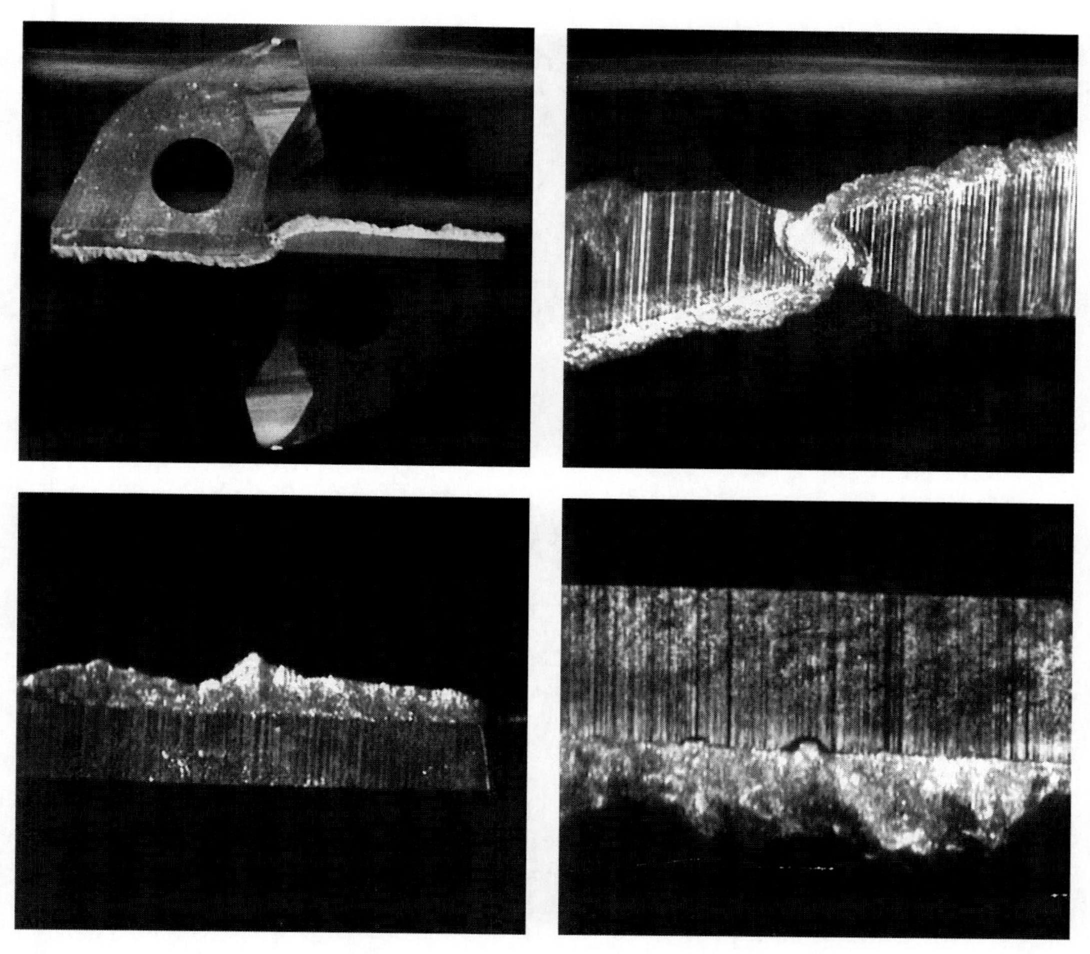

FIGURE 5.2 BUE on a carbide drill and its harmful consequences.

optimized tool design) carbide not to mention mirror-shining surface finish of the machined holes and ability to maintain tight tolerances over the whole tool lifetime. The latter cannot be achieved with cemented carbide tools even in principle. In modern setting of HSM, the difference in the performance of cemented carbide and PCD tools is even greater. As a result, PCD tools are widely used in the automotive industry.

5.1.1.2 Polymer-Based Composite Materials

Fiber-reinforced polymer (FRP) composites are a class of material that offer numerous advantages over monolithic metals and other homogeneous materials. CFRPs consist of a wide range of composite materials with different fiber types, fiber orientation, fiber content, and matrix materials. Due to their greater strength-to-weight ratio, the composites are widely used in various structures and components. The aerospace industry is making a major effort to incorporate an increasing number of composite materials into various components and structures. Each Boeing 787 Dreamliner, for example, is made up of (by weight) 50% composite, 20% aluminum, 15% titanium, 10% steel, and 5% other materials. By volume, the aircraft is 80% composite. Each 787 contains approx. 35 tons of CFRP, a good portion of it stacked in conjunction with aluminum or titanium alloys (Meguid and Sun 2005). Aircraft designers and manufacturers like stack materials because they combine the high strength of metals with the low weight and corrosion resistance of composites. The variety of stack

FIGURE 5.3 An example of a composite/metal stack combination.

materials is increasing nearly as fast as the applications, replacing aluminum honeycomb materials, which consist of honeycomb paper sandwiched between layers of aluminum. Often, stacks have various layers of composite, or a composite/metal stack combination (Figure 5.3), or foam or other core materials, and are then wrapped with composites. Some stack materials may also incorporate a thin copper mesh designed to protect against lightning strikes. Overall stack thicknesses can vary from less than 6 to 60 mm (Destefani 2011).

As the use of such materials has been steadily increasing in many industries, there is an increasing need for cost-effective method of producing high-quality holes in such materials with dimensions that are within narrow tolerances. Delamination generally represents the main concern when drilling composites (see Chapter 4), because of the lowering of fatigue strength as well as a poor assembly tolerance. Figure 5.4 shows the appearance of delamination. Thrust force has been widely

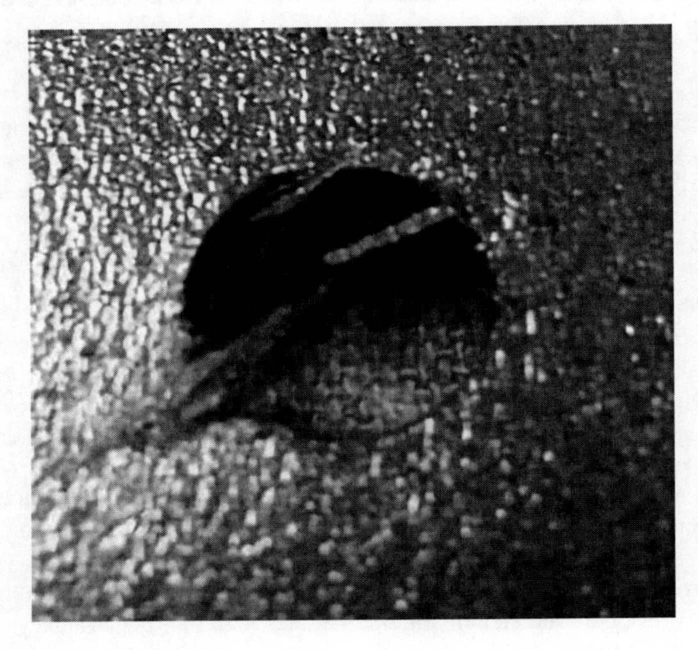

FIGURE 5.4 Appearance of delamination.

cited by various authors as being responsible for delamination; thus, the drill geometry should be developed to reduce the axial (thrust) force, thus reducing delamination. Drill design and application engineers rarely see anything but test panels and for proprietary reasons may have only a generic idea of the materials making up the stack so that the old-fashioned trial-and-error method combined with experience is used mostly in the development of drills for such applications. Various drill geometries and tool materials have been tested for the drilling of composite materials. PCD drills found wide application in drilling various FRPs.

5.1.1.3 Similarity and Differences

The use of the PCD tool material is beneficial in drilling both MMC and FRP as almost pure abrasion wear was found to be the case in drilling such materials. This, however, is the only similarity, while there are a number of differences. The major differences are as follows:

- The chip in drilling FRP is powderlike so that its removal does not present a problem. In drilling MMC, a great amount of the chip is formed, which should be transported over the chip flutes out of the hole being drilled.
- Aggressive drill geometry is normally used for FRP to reduce the drilling torque and axial force to reduce delamination. MMCs are generally much stronger so that the drilling torque and axial force are much higher, which imposes restrictions on the drill design.
- The drilling temperature is normally much greater in drilling MMC as the formed chip is heavily deformed due to highly ductile matrix material and due to friction of this formed chip in its sliding over the tool–chip interface.
- MMCs possess the low elasticity modulus and high thermal conductivity that causes great springback of the wall of the hole being drilled (Chapter 4) and thus friction on the drill margins.
- The need to cool down the machining zone and facilitate chip transportation call for the use of high-pressure internal MWF (coolant) supply in drilling MMCs, while MWF is not normally used in drilling FRPs.
- The tolerances on holes (diameter, shape, location, etc.) are normally much tighter in drilling MMCs.

The listed differences clearly indicate that a PCD drill used with great success for drilling FRP cannot be *automatically* applied for drilling MMCs although a number of such attempts are known. They are discussed later in this chapter.

5.1.2 PCD-Tipped Drilling Tools

Since PCD blanks became commercially available, multiple drilling tools for machining nonferrous materials were designed and implemented. Figure 5.5 shows PCD-tipped drilling tools by MAPAL Co. Figure 5.6 shows a two-stage PCD-tipped reamer with internal high-pressure MWF supply through properly designed nozzles. Because the cost of earlier PCD blanks was high, initial applications of PCD were mainly for expensive finishing tools, for example, reamers where both surface finish and diametric accuracy were of prime concern. Those years, specialist learned how to braze PCD tips and how to trim and finish grind PCD-tipped tools through the painful trial-and-error method.

Eventually, with the further development of the technology and lowering cost of the PCD tool material, PCD-tipped drills found their way to the shop floor. A typical PCD-tipped straight-flute drill with internal high-pressure MWF supply is shown in Figure 5.7. It has a cemented carbide body and the two PCD tips brazed in the pockets made (normally EDMed) on the periphery of its terminal end. As can be seen, such a drill has two combined major cutting edges (a.k.a. the lips), each consisting of the PCD and carbide portions. The idea behind this design was crystal

FIGURE 5.5 Various PCD-tipped drilling tools by MAPAL Co.

FIGURE 5.6 Two-stage PCD-tipped reamer.

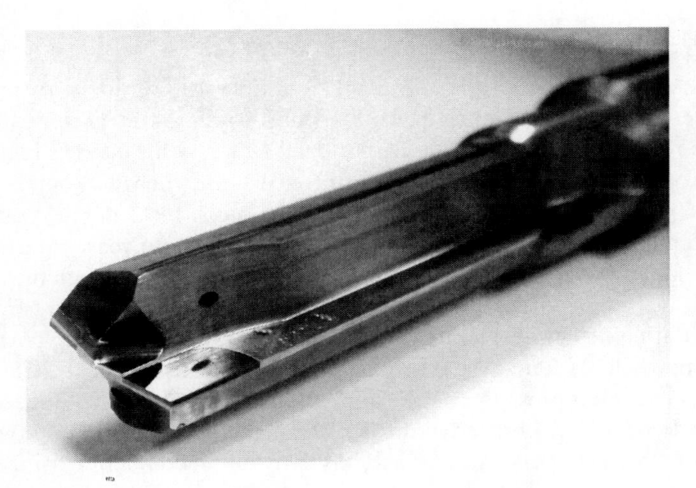

FIGURE 5.7 PCD-tipped straight-flute drill.

clear—as maximum cutting speed and thus tool wear (under proper drill design and normal drilling conditions) occur at the drill periphery regions, the placement of a PCD tip in this region of the major cutting edge should solve the problem, that is, it should help to increase tool life while enjoying the high quality of PCD machining. Soon, however, it was found that this is not always the case.

When PCD-tipped drills were used at relatively low (for the PCD tool material) cutting speeds and moderated penetration rates, their performance was satisfactory. The advantages due to switching to the PCD tool material were marginal as the gain in tool life was overshadowed by the increased cost of the tool. Moreover, it was soon found that such drills are more sensitive to the conditions of the drilling system, particularly to the accuracy of the tool holders and machine spindles, imperfections in castings (e.g., inclusions), and the accuracy of the tool presetting. The next generations of the drilling systems were designed and made to support the performance of the PCD tool materials in hope to gain the maximum advantages of PCD-tipped drills. Rigid and powerful high-speed spindles, internal high-pressure through-tool MWF supply, advanced machine controllers, high-accuracy tool holders, laser presetting machines, etc., were introduced to support high penetration rate and thus high efficiency of these drills. Soon, however, it was found that PCD-tipped drills have two inherent problems that, in the author's opinion, cannot be resolved even in principle.

The first problem is that the tool life of the carbide portion of the drill often defines tool life. An example is shown in Figure 5.8 where a PCD-tipped drill for machining a high-silicon automotive aluminum alloy is shown at the end of its tool life. As can be seen, the carbide portion of the drill is excessively worn, which caused radial tool vibrations that, in turn, caused the chipping of PCD tips. Moreover, on tool re-sharpening, a great amount of the carbide portion should be ground off to restore the normal working condition of this portion. This is not feasible for the PCD inserts as they are much harder and can easily be damaged by the heat generated in grinding. Although there are a few ways to deal with the problem, for example, un-brazing the PCD inserts, even more problems occur with the subsequent EDM trimming and finish grinding. As an example, Figure 5.9 shows the drill reconditioned by EDG (see Section B3 in Appendix B) with no subsequent grinding/polishing. Although it may appear the simplest and thus the fasted method of drill reconditioning, the performance of such a drill is inferior to that finished with the grinding wheel.

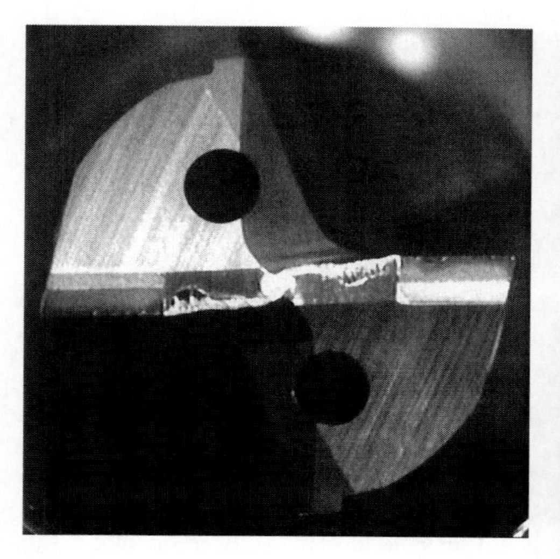

FIGURE 5.8 Showing that the carbide portion often defines tool life of PCD-tipped drills.

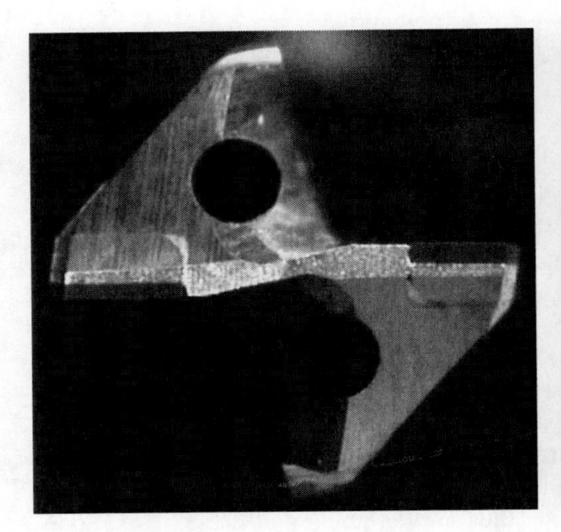

FIGURE 5.9 Showing drill reconditioning using EDG.

The second problem with PCD-tipped drills is probably the oldest, yet the most urgent, problem that requires solution in high-efficiency (H-E) drilling. They are the gaps between the PCD inserts and the carbide body. These gaps are caused by the following:

- When a PCD insert is brazed into a tool body using a common brazing technology (e.g. the induction brazing), the brazing filler material (BFM) does not adhere to the PCD layer, that is, the only carbide substrate is actually brazed to the drill body.
- When a tool is re-tipped, it is difficult to remove the remaining BFM from the pocket of the carbide body so that when new PCD inserts are brazed, their position may not be as intended. One may wonder, however, why not to apply more heat and thus to melt the remaining of BFM. The problem is that the amount of heat in PCD brazing is always limited to avoid overheating of the PCD tool material (discussed in Appendix B in great details). When a tool with excessive gaps between the PCD tips and the drill body (similar to that shown in Figure 5.10) is used in high-speed drilling of highly abrasive work material, the work material goes into the discussed gaps, grinds down BFM, and creates a splitting wedge that may cause chipping of the PCD layer as illustrated in Figure 5.11.

One may argue, however, that the gaps shown in Figures 5.10 and 5.11 are much too excessive, and thus a properly manufactured new PCD-tipped drill may not have such enormous gaps. Therefore, the problem is in the quality of drill manufacturing rather than in its design. The author's experience shows otherwise. Figure 5.12 shows the very beginning of the tearing of BFM by the highly abrasive work material after only a few hundred drilling cycles on a new tool made with minimum possible gaps between the PCD inserts and the carbide body. This drill was made after painstaking efforts to improve manufacturing quality of PCD-tipped drills. Although tool life of this drill is significantly improved, the problem remains particularly when the drill is re-tipped. Figure 5.13 clearly shows that the work material flows even into the smallest gap between the carbide body and the PCD insert causing chipping of the PCD layer.

Another inherent problem with PCD-tipped drilling tools is variation of the thickness of PCD inserts from one manufacturing batch of PCD blanks to the next. Because not much can be done to increase this thickness H_{db} (see Figure 3.79) and it is not feasible to grind each insert when this thickness is greater than the depth of the pocket made in the tool body, steps on the rake face are

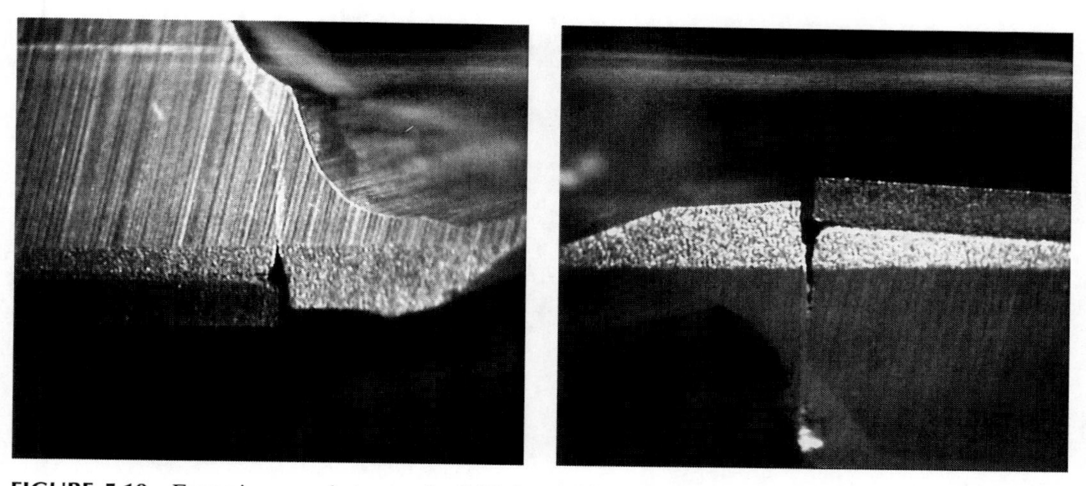

FIGURE 5.10 Excessive gaps between the PCD-brazed inserts and the carbide body in a reconditioned PCD-tipped drill.

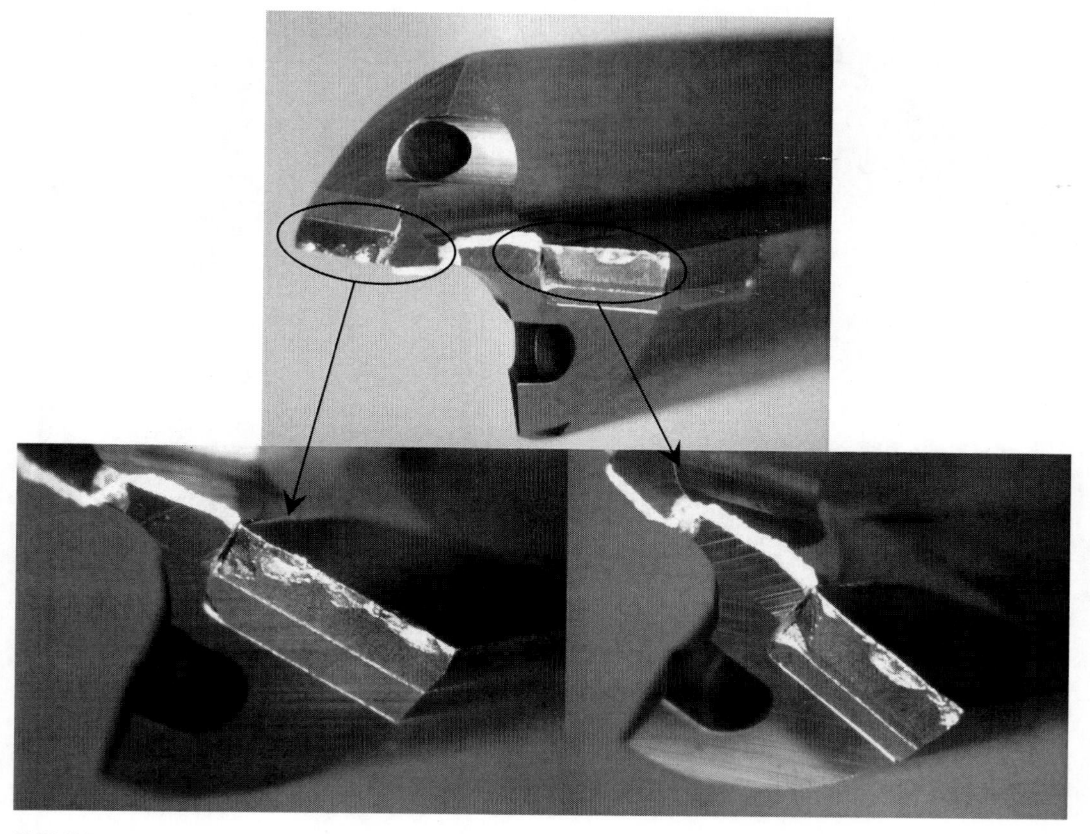

FIGURE 5.11 Chipped PCD portions of PCD inserts as a result of work material flow into the gaps between the PCD inserts and the carbide body.

FIGURE 5.12 Showing that the problems begin to occur after few drilling cycles even for a new PCD-tipped drill with minimal gaps between the PCD-brazed inserts and carbide body.

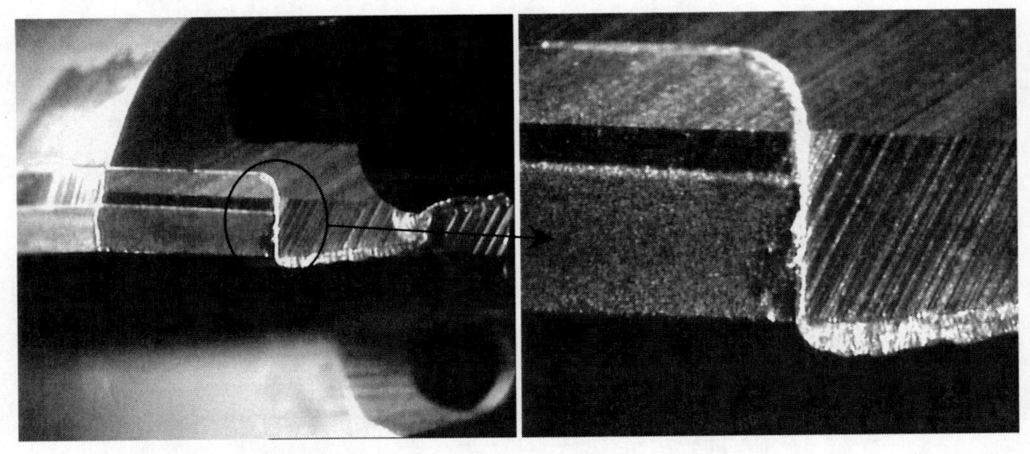

FIGURE 5.13 Showing that the work material flows even into the smallest gap between the carbide body and the PCD insert causing chipping of the PCD layer.

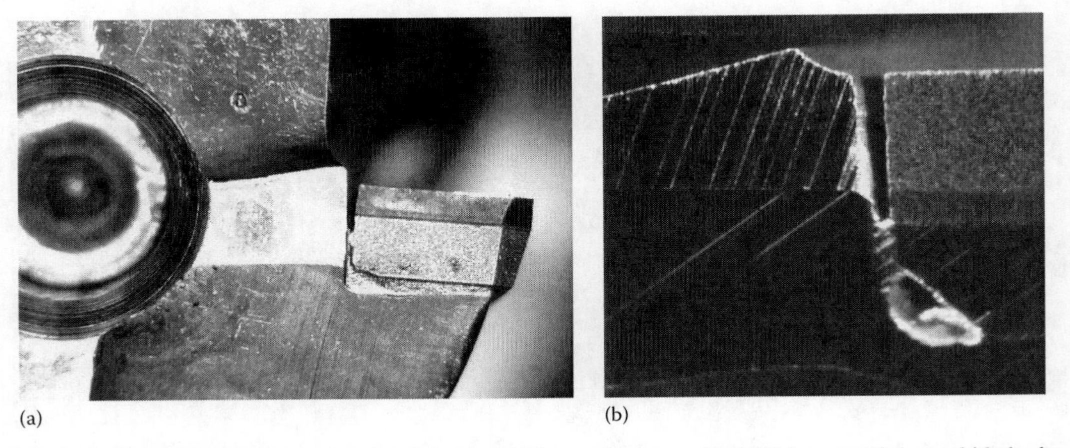

(a) (b)

FIGURE 5.14 Step on the rake face combined with the gap between the PCD insert and the carbide body: (a) in a reamer, and (b) in a drill.

FIGURE 5.15 Steps in the flute.

often found on the PCD-tipped tool. Such steps can be seen in Figures 5.10 and 5.11. Figure 5.14 shows a step on the rake face combined with the gap between the PCD insert and the carbide body. Figure 5.15 shows steps in the flute on a helical multistage drill. Such steps disturb normal chip transportation over the flute causing chip clogging in high-penetration-rate drilling.

5.1.3 Full-Face (Cross) PCD Drills

The discussed disadvantages of PCD-tipped drills were known for a long time, practically since the beginning of their practical use. Therefore, a number of attempts (with some great successes) were made in the development of the so-called full-face (cross-PCD) drills. These can be broadly divided into two major design (technological) approaches: (1) PCD is sintered in a part of the drill body and (2) a fully sintered PCD segment(s) is brazed into the drill body.

5.1.3.1 First Approach: PCD Is Sintered in a Part of the Drill Body

The idea of this type of drill is that its PCD cutting part is not sintered only on one side with a cemented carbide substrate as discussed in Chapter 3. Rather, cemented carbide embraces three sides of PCD on sintering and then becomes a part of the drill terminal end. The essence of multiple design variations of the so-called veined drill idea pioneered by Precorp Co. (Spanish Fork, UT) is in sintering a PCD layer inside a carbide blank to form the nib. The simplest nib design is shown in Figure 5.16 initially developed for printed circuit board PCD drills in the range of 0.15–3.2 mm diameter according to US Patent 4,713,286 (1987).

Precorp Co. uses a seven-step process to produce its veined drills. These are shown in Figure 5.17. The process begins with a solid-carbide blank, which is then slotted. Diamond powder is

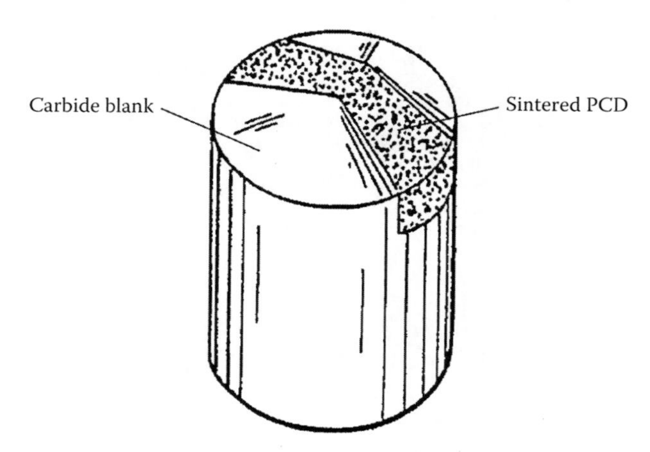

FIGURE 5.16 Design of the nib according to US Patent 4,713,286 (1987).

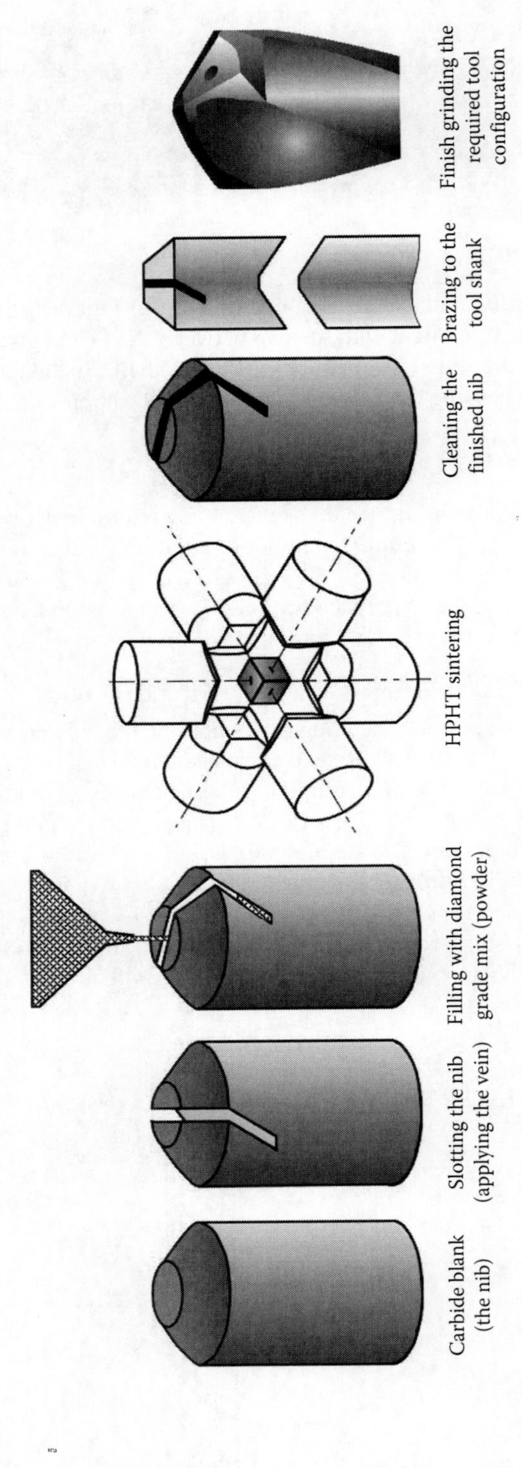

FIGURE 5.17 Steps in manufacturing a modern veined drill used by Precorp Co.

inserted into the slot of the blank. The blank is then placed in the company's press and subjected to HPHT sintering process. The finished PCD-embedded blank, referred to as a nib, is then brazed to a carbide blank a good distance from the cutting edge, depending on the drill diameter. This creates the specified length of the tool to be completed. The particularities of the drill geometry are then ground to produce the finished PCD-veined drill.

The spacing of the braze interface keeps it away from the heat generated in the cutting zone, thereby aiding the integrity of the bond. This strategy is a key since the Achilles heel of a diamond tool is the integrity of the braze as discussed in Chapter 3 and later in this chapter. This feature alone can be reason enough to explore the application of a PCD-veined drill against a more common PCD-tipped variety.

As pointed out previously, the PCD-veined drill was developed for drills of 0.15–3.2 mm diameter. These drills were originally designed for the electronics and aerospace industries, but further application testing has brought this technology to the attention of other industries to drill green carbide, aluminum MMCs, reinforced ceramic composites, and various grades of aluminum. Its application success, however, led to further development of such a technology for larger drill diameters. The problem was that if a nib having design as in Figure 5.17 is used for drill of large diameter, then a significant amount of PCD material should be ground off after the sintering to achieve the desired drill shape.

To address this issue, new nib designs shown in Figure 5.18 were proposed (US Patent 4,762,455, 1988). Figure 5.18a shows the design that is particularly applicable for drills with webs of up to approx. 0.8 mm wide. In this design, narrow elongated veins of PCD are imbedded in the margins of the web at the major edges to form the actual cutting surfaces or lips. The veins extend inwardly from the circumference to the respective lands along chords that are equally spaced from the drill diameter and overlap at the midpoint of the web. The widths of the veins are selected so that veins overlap laterally and axially at the midpoint to form a short vein section across the full width of the web at the midpoint.

Figure 5.18b shows an alternative nib design applicable for larger diameter drills. In this design, narrow elongated veins of PCD are imbedded in the margins of the web to form the major edges of the drill. The cemented carbide cylindrical blank is made with a central bore, and a pair of oppositely directed slots or grooves is formed along the spaced parallel chords of the blank extending inwardly from the periphery and tangent to the central bore. The slots are approx. 0.25–0.5 mm wide, and the slots and bore are approx. 0.8–1.3 mm deep. After sintering, fluting of the drill is significantly simplified because a very small amount of PCD of the vein is removed.

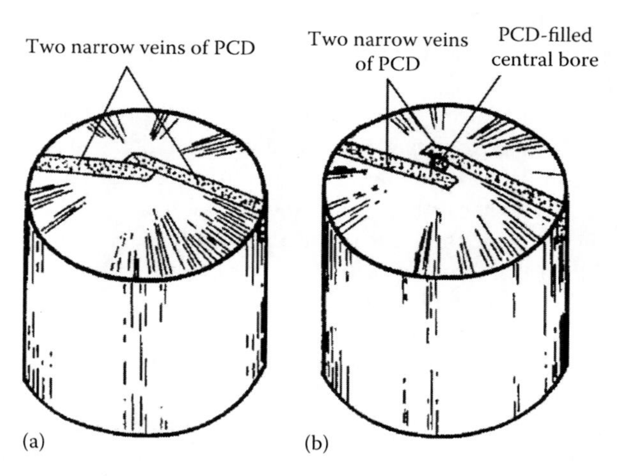

FIGURE 5.18 Nib designs for larger diameters of PCD-veined drills: (a) for small drill diameters and (b) for large drill diameters.

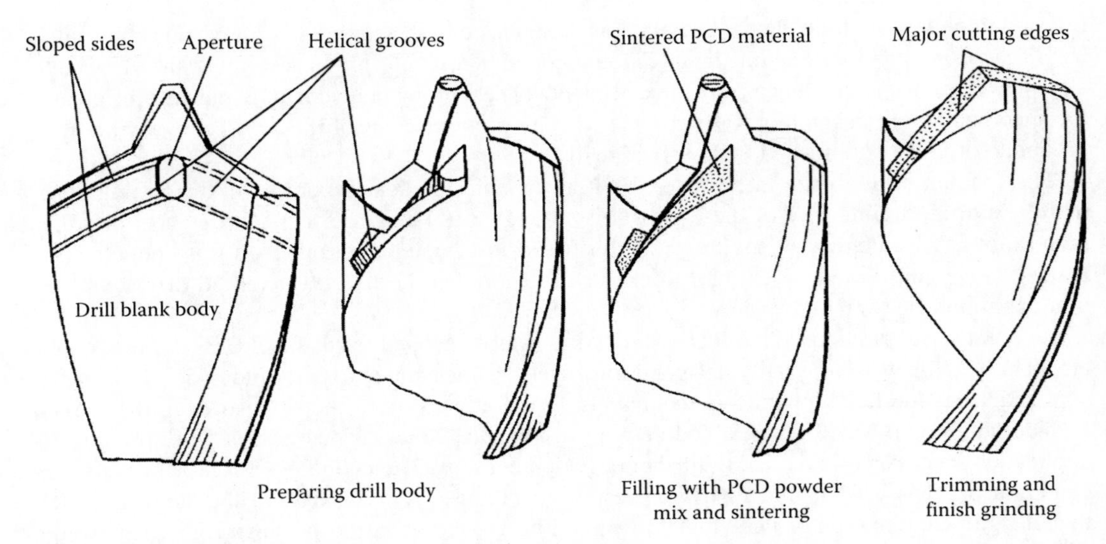

FIGURE 5.19 PCD drill design and the principal steps in its manufacturing developed by Smith International. (Courtesy of Megadiamond Industries, Provo, UT.)

Smith International (Megadiamond Industries) developed another version of the considered drill type that could be referred to as the intersintered drill. The idea of this design and the principal steps in its manufacturing are shown in Figure 5.19 (US Patent No. 4,991,467 (1991)). Referring to this figure, a twist drill blank body is made of cemented carbide. An aperture and two helically formed grooves are made at the end of the body. The sides of the grooves are sloped to assure that the PCD powder is packed in the groove with no voids. After HPHT sintering, the drill is cleaned, trimmed, and finish ground to expose the diamond cutting edges.

Figure 5.20 shows the process steps in manufacturing of the intersintered drill. The process begins with a carbide blank having two flutes. The blank diameter is made a bit greater than that of the finished drill diameter. This blank made of cemented carbide is formed with an aperture and helical grooves. PCD grade powder is prepared. Cobalt (6%–15%) as the catalyst material is added to the grade mixture. The percentage of cobalt is preferred to be 13%. The PCD powder having a size range from 3 to 60 μm is used. The gap filling diamond powder is also used and is preferred to be from 1 to 3 μm. The grade powder containing PCD and cobalt mixture is then mixed with a small percentage of wax binder. The percentage of wax binder is about 1%. The mixture of grade powder with wax is then trawled into the helical grooves and through the aperture of the blank. These activities are shown by steps 1 and 2 in Figure 5.20.

A *getter* is used in the manufacturing process to form better diamond bonds. The *getter* is a reactive metal that reacts with contaminates and oxides to facilitate better diamond bonding. The diamond getters disable the impurities in the mix that may affect the strength of diamond bonds. A typical getter material is selected in the group consisting of zirconium, columbium, tantalum, and hafnium. In the considered case (step 3 in Figure 5.20), a wire of the columbium getter is laid over the powder pressed in the helical grooves. The blank is then placed into a deep draw can that (a can is a vessel) be made of the reactive metals such as columbium (step 4) to minimize contaminates in the grain powder and thus to gain better strength of diamond bonds. The can containing the drill body is then run through a die to make the powder more compact (step 5). While still in the can, the drill body is then run through a *dewax* cycle in a vacuum furnace (step 6). During this cycle, the temperature is slowly ramped to 400°C for 2 h and then increased to 620°C for approx. 1.5 h, and then the can is allowed to cool for an additional 4 h. The process takes about 12 h with a 4–5 h slow cool down from the maximum temperature (620°C). After the can is taken from the vacuum furnace, another open-ended can made

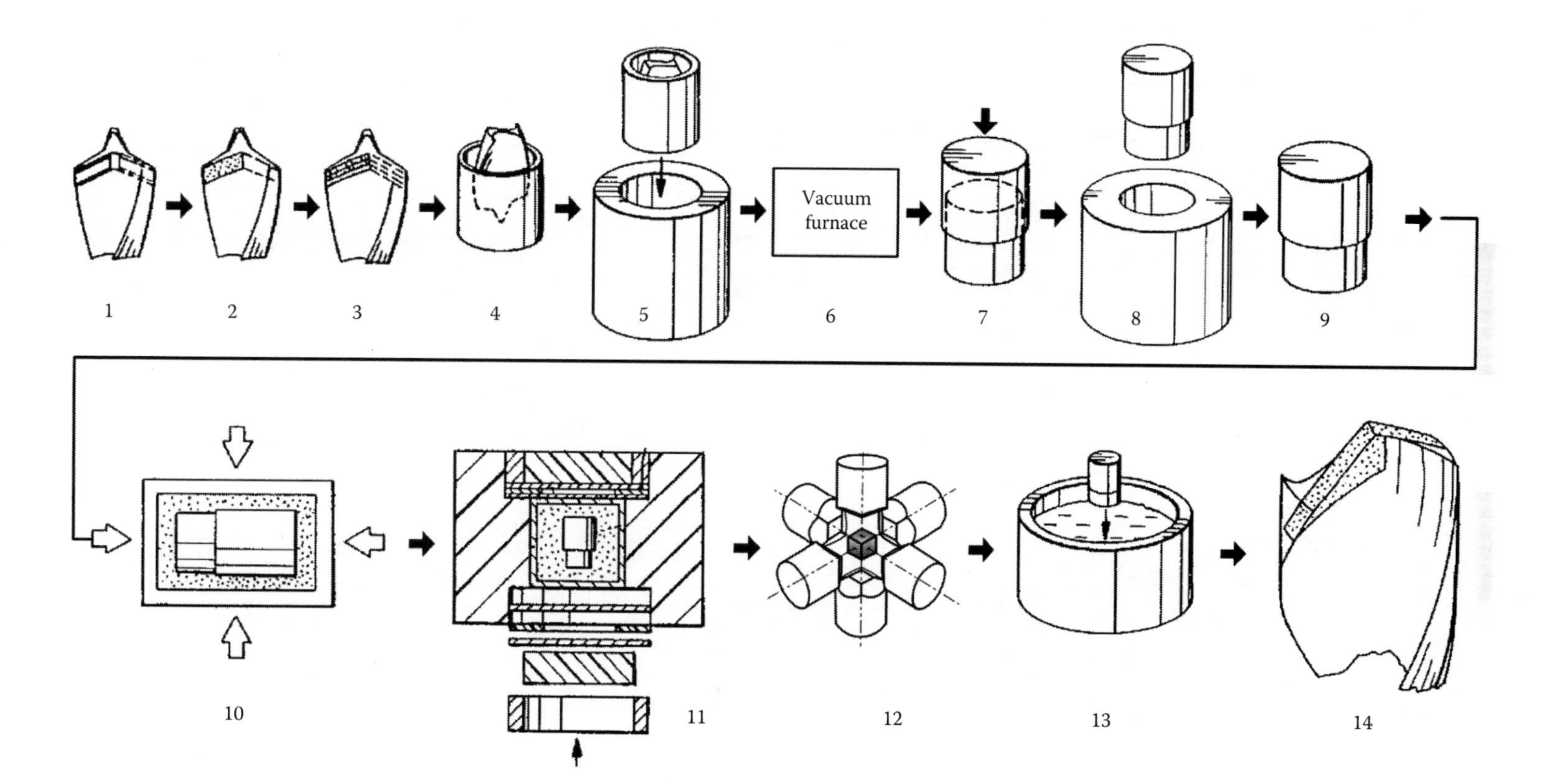

FIGURE 5.20 Detailed process steps in manufacturing of the intersintered drill.

of columbium is placed over the compressed can (step 7) and then run through a die (step 8) to obtain the completely sealed assembly (step 9). This assembly is then run through a precompact stage. It is first surrounded by salt and then put in a precompact press to further compact the assembly (step 10).

The compacted assembly is loaded into a pyrophyllite cube (step 11) and is then subjected to HPHT sintering in a cubic press (step 12). The time of the press is approx. 10 min. The temperature is ramped up to 815°C for about 4 min; the cube is held at HPHT for about 1 min and then allowed to cool down for approx. 5 min to reduce residual stress within the finished twist drill.

The sintered can is subsequently broken out of the pyrophyllite cube and then is dipped into a bath of fused sodium hydroxide to remove columbium cover and wire wound around the helix groove filled with the sintered diamond (step 13). The finished blank known as the nib is then brazed to a drill shank to finish the product (step 14).

As mentioned at the beginning of this section, PCD drills of this type, that is, when PCD is sintered in a part of the drill body, were initially developed for printed circuit board, and then this drill type has brought to other applications to drill green carbide, aluminum MMCs, reinforced ceramic composites, and various grades of aluminum. This type of full-face PCD drills has some limitation in its application to high-strength materials.

The sandwich-type PCD drill technology can be considered as the development of the previously discussed intersintered drill technology. This approach was pioneered by the development of the sandwich PCD drill design in the early 1980s (US Patent No. 4,627,503, 1986). The drill includes a sintered PCD wafer having a PCD layer sintered together with two carbide substrates as shown in Figure 5.21. Such blanks are readily available in the market nowadays. As the wafer has two cemented carbide substrates, it can be assembled with the drill body having an axial slot and then brazed to this body. After the brazing, the drill is finish trimmed and ground.

The major drawback of this approach is a need to grind a substantial amount of carbide and then PCD to achieve the desired drill geometry. Such grinding was a huge problem at the time when this design was developed. The second problem is that a significant amount of cobalt should be placed in the mix (grade powder) to assure sintering as the two carbide substrates act as the shields limiting the pressure and temperature needed to fully sinter a *standard* (low cobalt) grade powder. The resultant product (PCD) is not as strong as that *normally* sintered. Thirdly, relatively thin carbide

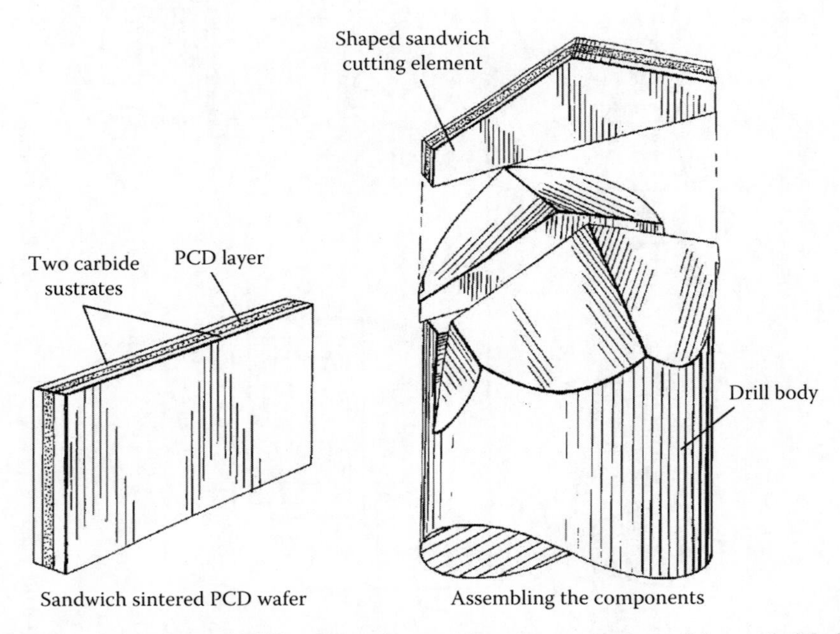

FIGURE 5.21 Sandwich sintered PCD wafer and assembling the components of the sandwich drill.

FIGURE 5.22 PCD blanks for manufacturing of sandwich drills.

substrates did not provide sufficient protection to the PCD layer on brazing, so it often happens that the PCD layer is overheated (this problem is to be discussed later). Note that the modern technologies (sintering, EDMing, grinding) reduce the severity of these problems.

Figure 5.22 shows typical blanks for manufacturing such drills that are now commercially available from various PCD manufacturers. The quality of these blanks may differ significantly, which affects drill performance in HP applications. To exemplify this statement, let's consider the results of applications of the two types of drills (SW1 and SW2, for clarity of the further considerations) made using the sandwich technology. These drills were used for a highly demanding application, namely, drilling of high-silicon aluminum alloy (A390 having 17% of Si) at a high penetration rate. The PCD sandwiches for these drills were made by two well-known companies: the first company uses the proper technology while the other uses the veined technology.

Figure 5.23 shows appearance of two drills (SW1) after completing tool life. Note that the geometry of these drills was not optimized according to the VPA-Balanced© design concept. As a result, some aluminum deposit can be seen on the web-thinning part of the right drill shown in Figure 5.23 as the gash ground on the rake face did not provide sufficient room for the flow of the chip made by this part of the cutting edge. However, this minor drawback did not affect drill performance and machined hole accuracy.

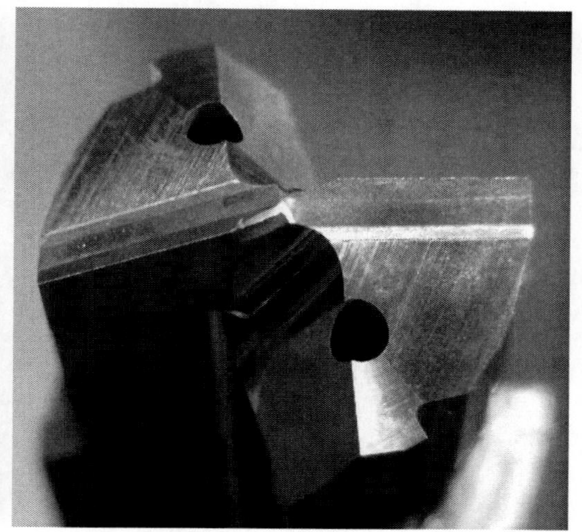

FIGURE 5.23 Appearance of two drills (SW1) after completing tool life.

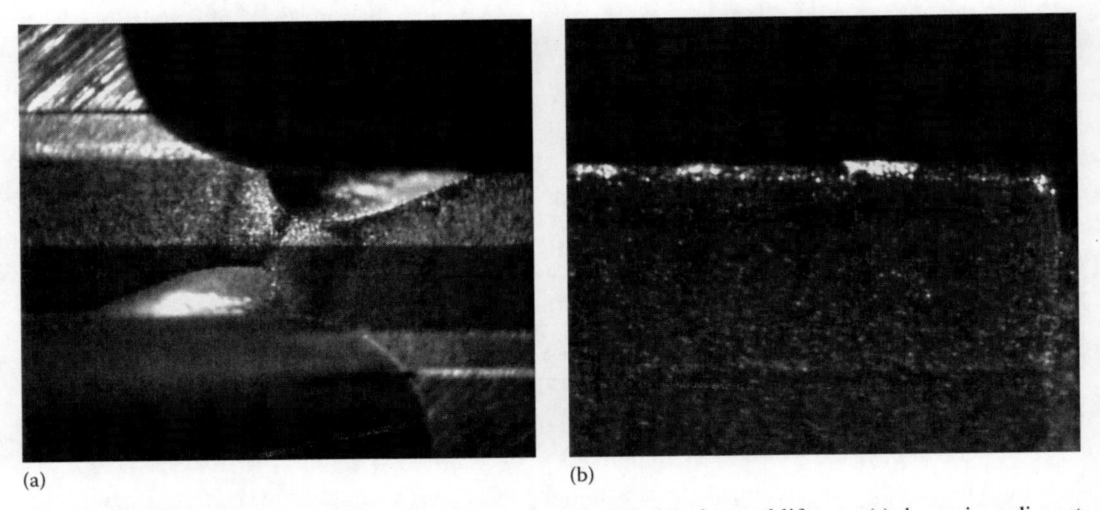

(a) (b)

FIGURE 5.24 Particularities of the wear pattern of the SW1 drill after tool life run: (a) the region adjacent to the chisel edge and (b) a part of the major cutting edges adjacent to the drill corner.

Figure 5.24a and b show particularities of the wear pattern of the SW1 drill after tool life run. The following can be concluded from these patterns:

- The region adjacent to the chisel edge shows impressions made by the chip formed by the chisel edge. Because both the rake faces of this edge were not designed properly (see Section 4.5.8), the chip formed by this edge did not have enough room to escape in HP drilling. As a result, a great force was imposed on the central part of the drill. Nevertheless, the drill survived and made the complete tool life with no problems.
- The major cutting edges adjacent to the drill corners (Figure 5.24b) are rounded by wear, so the micro-BUE in traces can be seen on this rounded part. The drill corner that actually determined the drill ability to cut proper holes is not worn, so the tool life of this drill can be extended by at least 30%.

Figure 5.25 shows appearance of the rake face of the SW1 drill. As can be seen, the *aluminum painting* of this face is minimal as original grinding marks can be seen. It means that the

FIGURE 5.25 Appearance of the rake face (drill SW1) after tool life run.

FIGURE 5.26 Unmelted cobalt particles on the rake face of SW2 drill.

amount of cobalt that is unavoidably greater for this kind of PCD technology did not cause BUE as expected. Moreover, the color of the PCD is black (not dark black of *usual* PCD though). No cobalt lakes or other signs of its presets were observed under a detailed microscopic examination of the rake face of a new SW1 drill, that is, when the carbide layer is removed and the rake face is polished.

The analyses of performance of SW2 drills (made using pure veined technology) showed that two important issues cannot be resolved in principle in manufacturing of these drill types. First, a significant amount of carbide and PCD are ground off to bring the veined blank to the final shape of the drill. This grinding damages PCD tool material developing a network of microcracks. Second, the sintering of PCD requires high temperature and isostatic pressure as discussed in Chapter 3 in great detail. However, when the sintering is carried out with a carbide nib (veined drills) or between two carbide plates (sandwich drills), these pressures and temperatures are much lower than those in normal PCD sintering. This forces PCD blank manufacturers to use an extensive amount of cobalt (up to five times more than in normally sintered PCDs. Pools (a.k.a. lakes) of cobalt can often be seen on PCD surfaces after grinding. Figure 5.26 shows unmelted cobalt particles on the rake face of drill SW2.

A great amount of cobalt and unmelted cobalt particles used in manufacturing of SW2 drill create all the known problems with such drills. These problems can be briefly summarized as follows:

1. BUE still persists on this drill type although in different form. When such drills are used, the ground rake face is covered by a thin layer of aluminum that eventually develops BUE with all previously listed consequences. Figure 5.27 shows BUE on the rake face of a sandwich drill. The rake face of this is covered by a thin layer of aluminum that creates significant friction force in chip formation at the tool–chip interface. The test results show that the development of the BUE does not correlate with tool wear. Rather, it begins to develop at the first drilled holes as shown in Figure 5.28.
2. Because different parts of PCD do not have the same exposition to sintering temperature and pressure, the properties of the PCD vein layer vary along its length deteriorating to the drill's center. As a result, when used for high-strength aluminum alloys, the following happens:
    a. Aluminum forms BUE as a thin layer on the rake face that significantly increases contact stresses.
    b. The sharp cutting edge collapses due to high contact stresses becoming round, which increases the cutting force even further.

FIGURE 5.27 BUE formed on the rake face of SW2 drill. Note that the rake face is covered by a thin layer of aluminum.

FIGURE 5.28 Showing that BUE begins to develop at the first drilled hole with SW2 drill.

3. The relatively low strength of PCD sintered in this way combined with high forces causes drill breakage in the center as shown in Figure 5.29. All the attempts to improve manufacturing quality of these drills by better grinding of the rake face, improving the drill geometry, etc., were of little help.

The distinguished appearance of SW2 drill can be characterized as follows:

- The color of the rake face of SW2 drills is dark brown instead of as with SW1 drills.
- Cobalt lakes and spots can be seen under small magnification.
- The rake face cannot be polished due to high cobalt content.
- The major cutting edges are not as sharp as their should be for PCD drilling tools.

5.1.3.2 Second Approach: PCD Segment(s) Is Brazed into the Drill Body

Figure 5.30 shows modifications to the design concept shown in Figure 5.21 to reduce the grinding stock on PCD (US Patent No. 4,527,643, 1995). As can be seen, a PCD layer is sintered on the substrate top forming five distinctive cutting edges (two major cutting edges, one chisel, and two side margins). This design concept, however, was not used widely in practice for a variety of reasons.

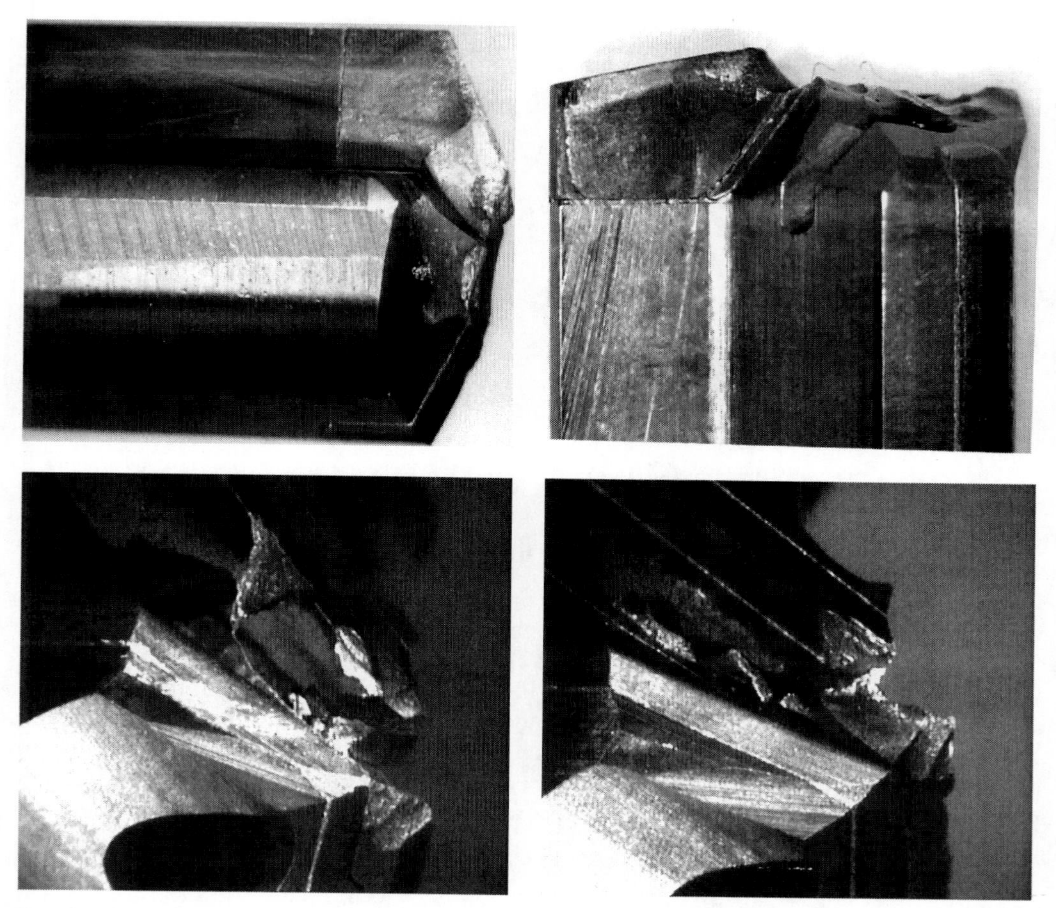

FIGURE 5.29 Typical pattern of breakage of sandwich PCD drill in machining of high-silicon aluminum alloys.

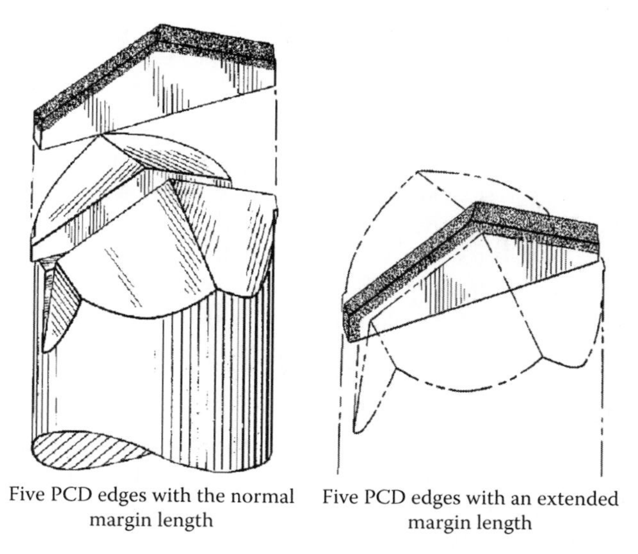

FIGURE 5.30 Modifications to the design concept shown in Figure 5.21.

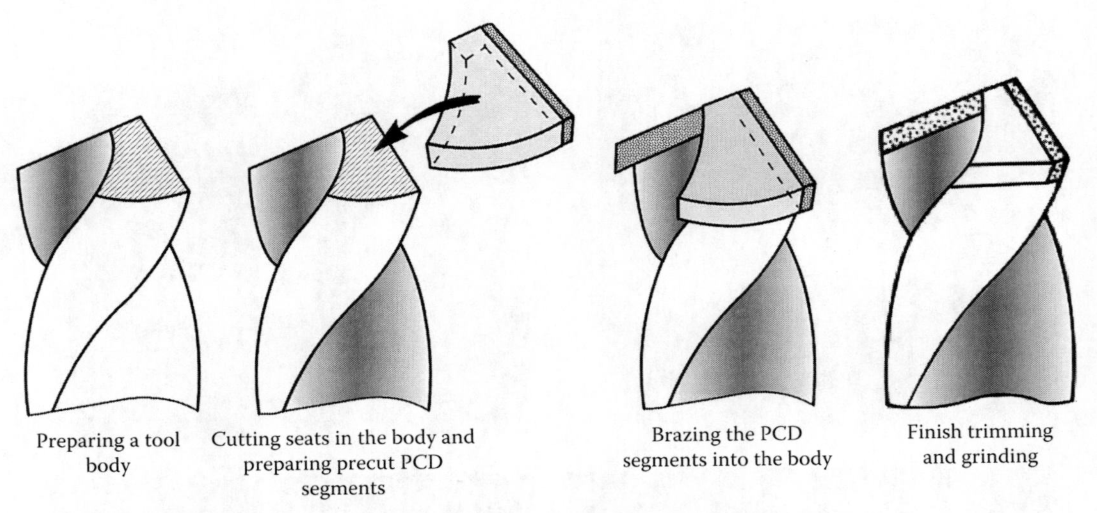

FIGURE 5.31 Manufacturing steps in realizing the first design concept of PCD drill.

To reduce the amount of ground PCD and to promote its SYNDITE PCD products for drilling application, De Beers Industrial Diamond Division developed several design concepts at De Beers Technical Service Centre (Ascot, United Kingdom) (Sani and Clark 1992).

Figure 5.31 shows the manufacturing steps in making a drill according to the first simplest design concept of PCD drill. A standard drill body was ground with two flat seats. Two precut (EDMed) PCD segments are then brazed on these flats seats. After brazing and cleaning, the drill was finish trimmed and ground to its final shape. An analysis of the manufacturing procedures and the results of the use of the drill made according to this concept showed the obvious drawbacks of this concept:

1. It was difficult to locate the segments on the flat seats on brazing so that an extra stock should be allowed for trimming and grinding. This significantly added cost in manufacturing such drills.
2. On drilling, the cutting force (the drilling torque) is acting directly on the braze joints on both segments in their most vulnerable shearing direction. This led to the disengagement of the segments from the drill body particularly when the drill is partially worn so that the cutting force (the drilling torque) increases.

To address these drawbacks, the second design concept was proposed by De Beers. In this concept, the drill body is made with two locating shoulders. Figure 5.32 shows the manufacturing steps in making a drill according to the second design concept of PCD drills. As can be seen, once the segments were brazed onto the drill, the PCD was automatically in line with the flute line. While drilling, the cutting force (the drilling torque) is acting on the locating shoulders that relieve the load on the brazing joints. It was anticipated that this would reduce production costs to a great extent. However, matching of the two adjoining segments was found to be a critical operation and required compound grinding of the segments. This is because when this concept was developed, no modern PCD grinding and EDMing techniques were developed.

In general, the previous drill design performed well when drilling abrasive materials of low toughness. However, weakness at the web became apparent when tough materials were drilled. This design was finally abandoned on the basis of (1) difficulty in matching the two butting PCD segments and (2) occasional failure at the brazed point and at the web.

To eliminate the complicated grinding operation to butt the segments, it was decided that a new concept, shown in Figure 5.33, should be tested (the third design concept). In this concept, a small

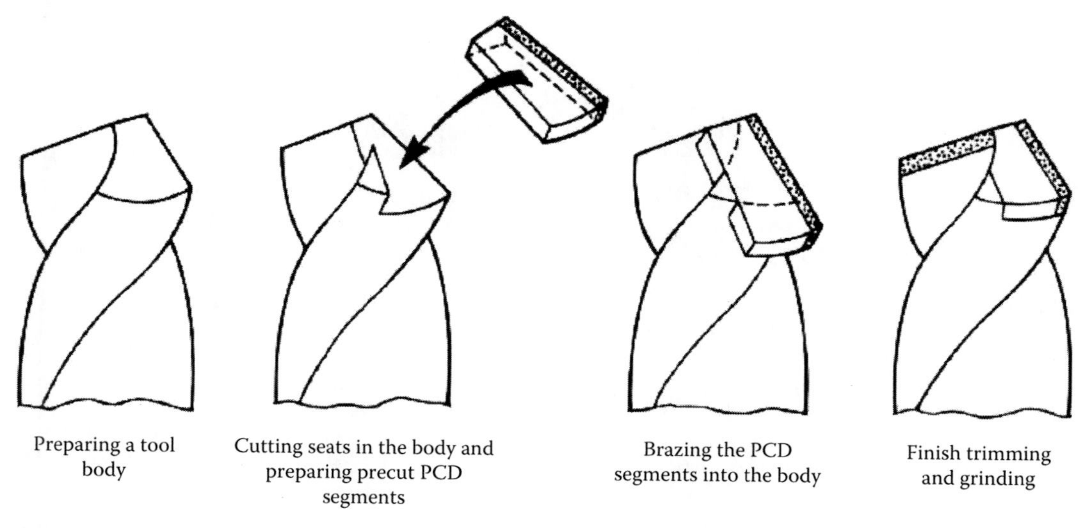

FIGURE 5.32 Manufacturing steps in realizing the second design concept of PCD drill.

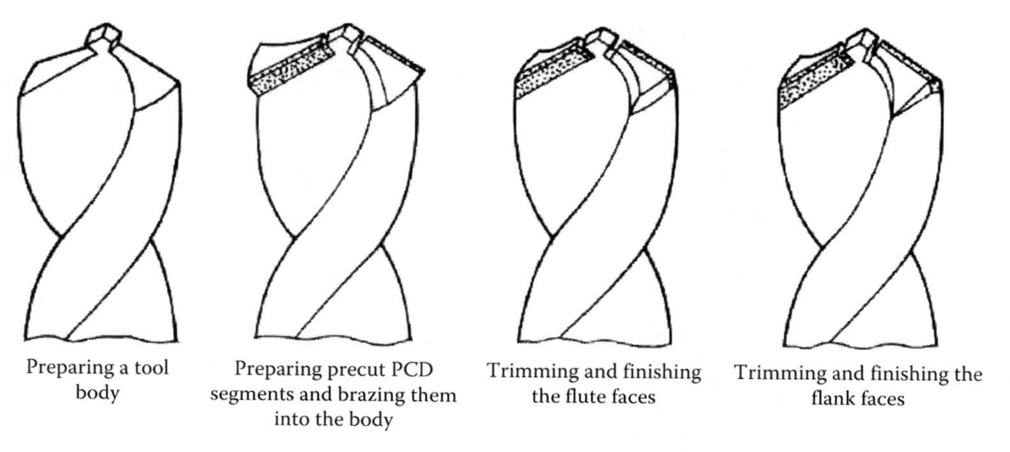

FIGURE 5.33 Manufacturing steps in realizing the third design concept of PCD drill.

region adjacent to the chisel edge is left in the center of the drill. This not only simplifies segment location but also assists in butting the two PCD segments. However, the implementation of this concept, it revealed the following drawbacks:

1. Occasional breakage of the web during trials.
2. Difficulties to maintain the web thickness because as the thickness of the PCD increased, an unacceptably large non-PCD cutting edge resulted. This not only affected the tool performance but also spoiled the cosmetic appearance of the finished product.

To overcome the limitation on the PCD length and to adhere to the original cost objective, which dictated little or no flute grinding, the following forth concept was tested. The drill cutting face was partially removed perpendicular to the profile of the flute and replaced by PCD segments, as shown in Figure 5.34. Using this design approach, one can achieve an increase in the thickness of the PCD in the flute area to a great extent without deviating from the flute line. Another positive aspect of this design was that the braze line was no longer in line with the web. The braze line formed an angle close to 20° to the web that significantly improved the web strength compared to the previous designs.

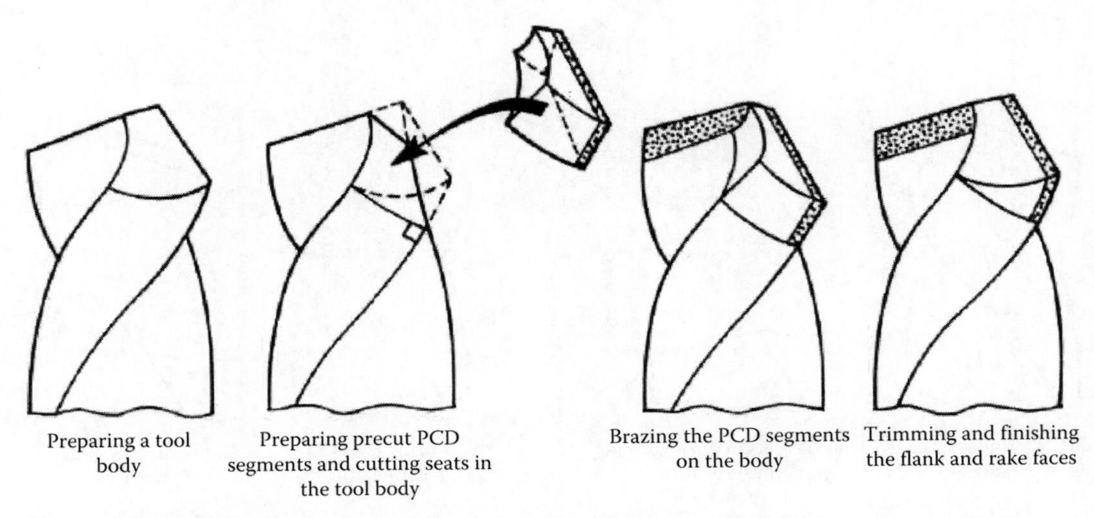

FIGURE 5.34 Manufacturing steps in realizing the forth design concept of PCD drill.

However, major drawbacks to this design also became apparent:

1. Difficulties in positioning and in compound grinding to produce an acceptable match between the two PCD segments.
2. The thickness of the web was found to be directly related to the height of the PCD. As the PCD height increased beyond a certain limit, the thickness of the web increased to such an extent that it could no longer be accepted from a cosmetic and performance points of view.

Based on the experience gained in the manufacturing and implementation of the previously discussed design concepts of PCD drills, the following successful approach (the fifth design concept) was finally taken, which overcame (to a certain extent, at least) most of the listed limitations. The basic manufacturing steps are illustrated and described in Figure 5.35 (US Patent No. 5,195,403, 1993). A great advantage of this design is significantly reduced grinding stock in the flute area, while the flank faces are formed in the stage of PCD segment preparation. A typical finished drill is shown in Figure 5.36.

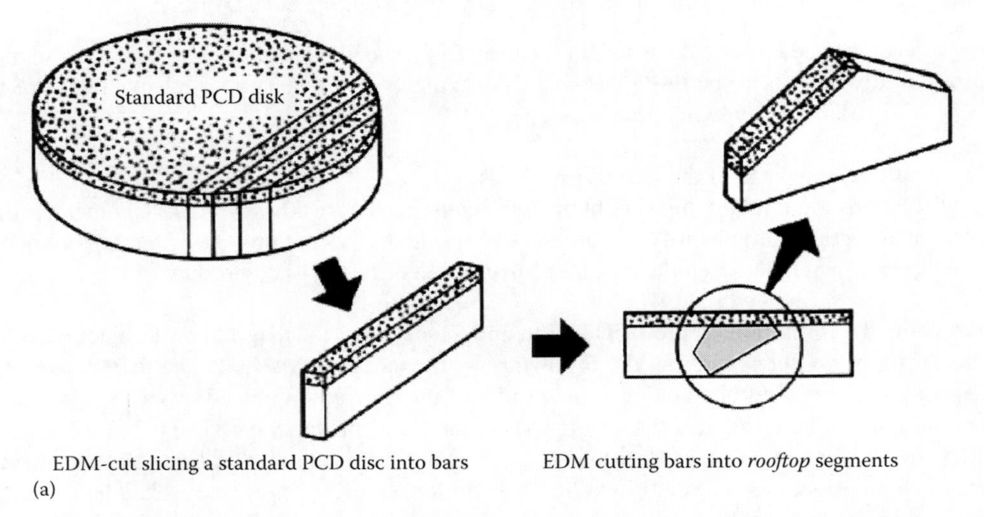

FIGURE 5.35 The basic manufacturing steps in realizing the fifth design concept of PCD drill: (a) slicing PCD segment from a PCD disc.

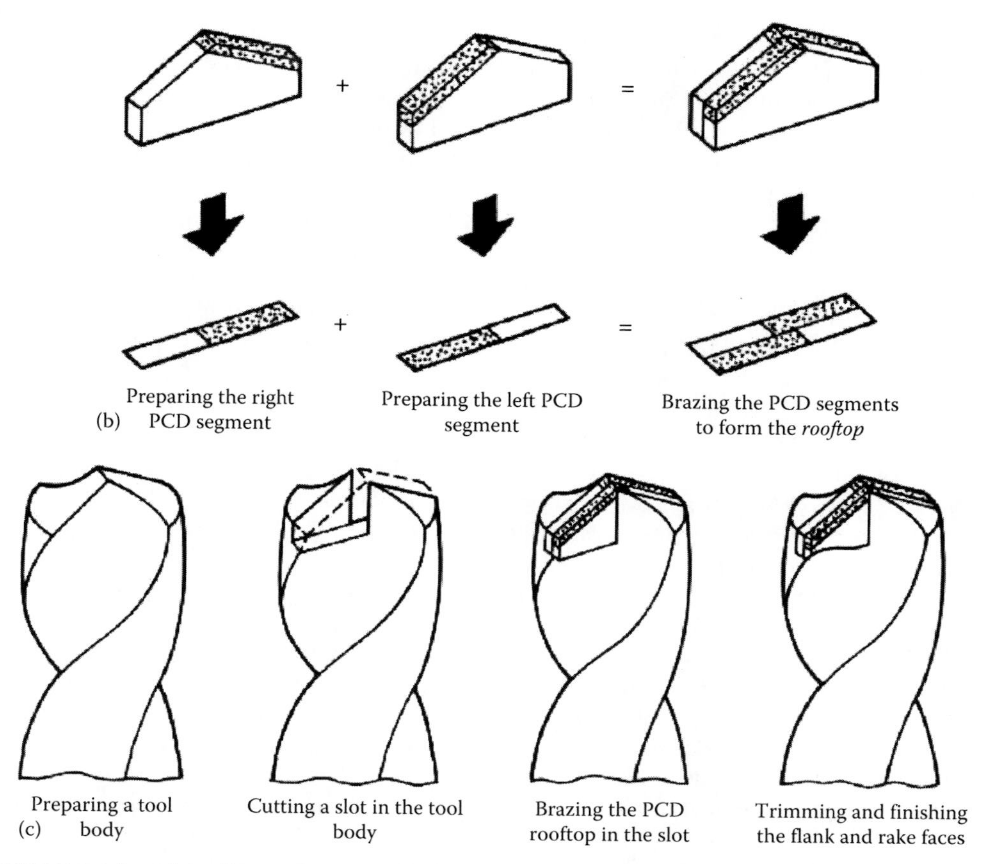

FIGURE 5.35 (continued) The basic manufacturing steps in realizing the fifth design concept of PCD drill: (b) brazing two PCD segments, and (c) steps in finishing the drill.

FIGURE 5.36 A typical finished PCD-tipped twist drill made according to the fifth design concept.

Twist- and straight-flute drills made to this fourth concept were used by PCD drill manufacturers for a long time for the drilling of nonferrous and composite materials at relative low cutting speeds. For example, the performance of such drills was assessed using comparative tests that were carried out using 8.5 mm dia. De Beers PCD-tipped, standard HSS, and WC drills. The drilling parameters used throughout the tests were spindle rotational speed, 2200 rpm; feed rate, 120 m/min; and depth of hole, 25 mm. In machining of silicon flour–filled epoxy resin (sifter) blocks, HSS and WC drills were capable to drill only one hole while De Beers PCD-tipped drill produced more than 120 holes. After drilling 120 holes, the PCD-tipped drill was examined and no sign of wear was observed at the cutting edges or close to the plane of the braze. The high-abrasion-resistance characteristics of PCD compared to other tool materials are not something new, and the test was primarily carried out to ensure that the plane of braze is not exposed to the flow of passing swarf.

To evaluate the abrasion and toughness characteristics of the PCD drill, drilling trials in Al–SiC, a highly abrasive MMC, were carried out. The SiC inclusions in the metal matrix induced a severe interrupted cutting action. This MMC was found to be an ideal media to evaluate the toughness and the stability of the brazed joint. The following drilling results were obtained: HSS drill was capable to drill only 2 holes, WC 4 holes, and PCD-tipped drill produced more than 150 holes. The PCD drill still exhibited no sign of wear or damage after 150 holes.

Despite multiple successful implementations, frequent failures of such drills were reported. These failures occurred due to far from perfect EDMing, brazing, and grinding of PCD as many modern technologies to perform these operations were not yet fully developed at the time. EDM and grinding machine tool builders, brazing professionals, and tool application specialists learned on the go. The major obstacle in such learning was a rather limited market for PCD drills in the automotive and aerospace industries. This was combined with relatively high cost of PCD drills. A limited number of tool manufacturing companies were capable of producing such drills efficiently maintaining their quality and consistency.

Another variation of PCD drill design is shown in Figure 5.37 (US Patent No. 5,195,404, 1993). The manufacturing steps in making such a drill include cutting an elongated strip from a standard PCD disk along parallel planes that are inclined at an angle of 30° to the imaginary plane normal to the upper surface of the disk. The resultant strip is then cut along the dashed lines so that a number of prismatic cutting inserts were obtained. One of the inserts is shown in Figure 5.37. The drill body is prepared by cutting slots for the cutting inserts. Two cutting inserts are brazed into these slots and then drill finish trimmed and ground.

According to the inventors, this design has the following advantages:

- Minimum amount of PCD material is needed to make a drill.
- Minimum amount of PCD material can be trimmed/ground to produce the final shape of the drill.
- The design of insert pockets and the location of the brazing areas assure the best resistance to the cutting force while drilling.

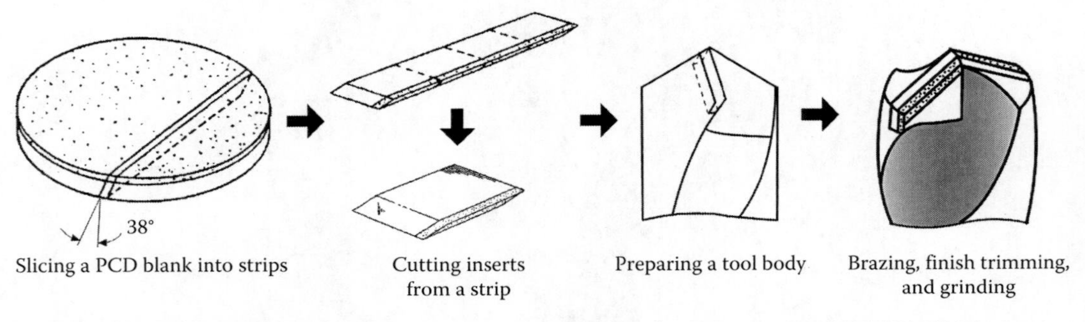

FIGURE 5.37 Manufacturing steps in making a drill according to US Patent No. 5,195,404 (1993).

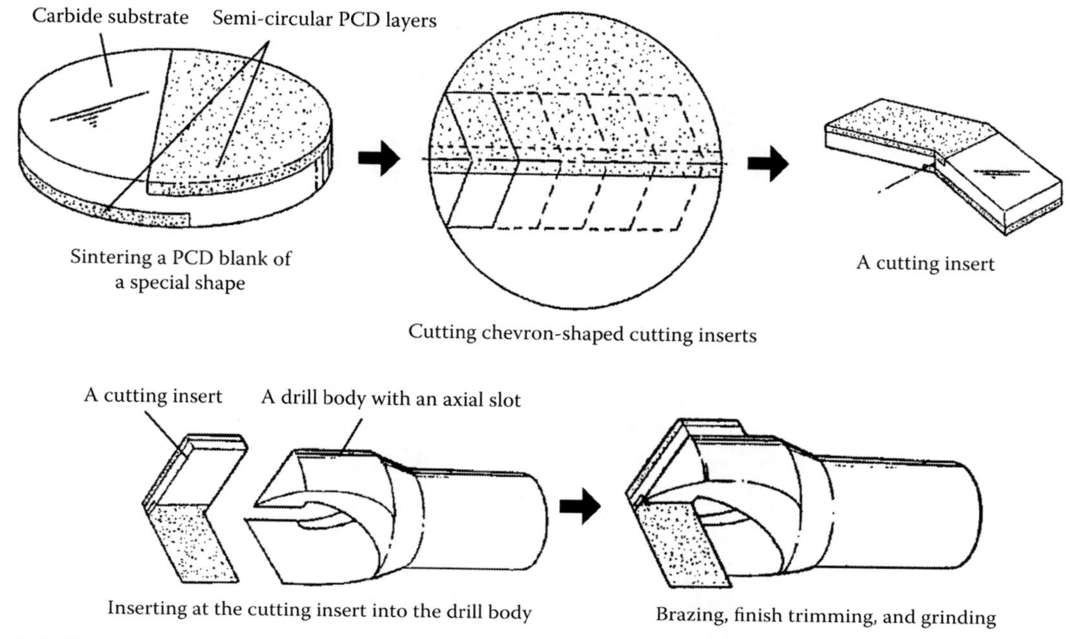

FIGURE 5.38 Manufacturing steps in making a drill according to US Patent No. 5,232,320 (1993).

The use of a composed PCD blank in the manufacturing of a PCD drill design is described in US Patent No. 5,232,320 (1993). Manufacturing steps in making this drill are shown in Figure 5.38. As can be seen, a composite PCD blank has a semicircular PCD layer on both of its parallel faces. This blank is cut in a number of chevron-shaped cutting inserts according to the cutting pattern illustrated in Figure 5.38. A single cutting insert has two diamond layers on opposite wings of the chevron that overlap across the centerline of the inserts.

A boring head shown in Figure 5.38 has a pair of arms that define opposed, staggered mounting faces. These faces lie on either side of the diameter of the body. The cutting insert is inserted into the tool body and then brazed into position to these faces. After brazing, the cutting insert is trimmed and then ground to the final tool geometry. The resultant chisel edge of the drill has two portions of PCD on either side of a central carbide portion. This cutting insert has the particular advantage to the previously discussed twin inserts as the chisel edge is formed as a single unit and thus better resists cutting forces on the rake faces of the two portions of the chisel edge. Note also that there is no brazing filler layer passing through the center of the chisel edge where the contact force is higher compared to the other portions of the chisel and other cutting edges.

It is understood that the cutting insert of this design can be fitted to a variety of drilling tools, for example, twist- and straight-flute drills.

As a variation of this design concept, Figure 5.39 shows the drill concept (design) and its manufacturing stages according to US Patent No. 5.443,337 (1995). The manufacturing of this drill starts with a stub made of cemented carbide having flat parallel upper and lower surfaces for ease of fabrication. An array of rectangular cross-sectional slots are made in this stub by EDMing as shown in Figure 5.39a. Then the slots are packed with fine mesh powder of PCD (the grade powder) as shown in Figure 5.39b. Then the loaded stub is subjected to HPHT sintering. After sintering and cleaning, the stub is cut (along the dashed lines in Figure 5.39b) into a *chip* containing elongated rectangular *veins* of PCD material fused into the surface of a cemented carbide matrix. The resultant shape of the *chip* is shown in Figure 5.39c. As can be seen, a half of the side carbide walls from each side are removed to expose PCD material. This *chip* is cut in half as shown in Figure 5.39d and one side

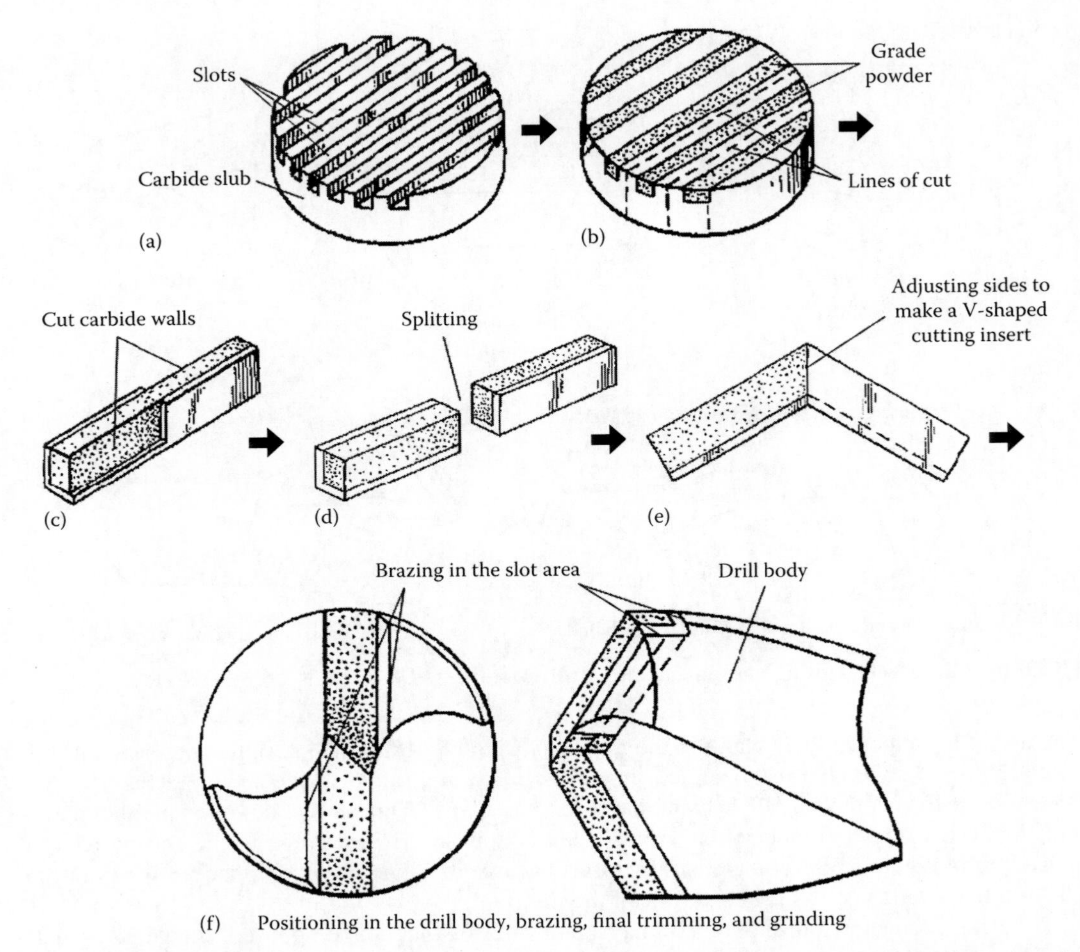

FIGURE 5.39 Manufacturing steps in making a drill according to US Patent No. 5.443,337: (a) the design of a stub made of cemented carbide with an array of rectangular slots, (b) the slots packed with fine mesh powder of PCD, (c) the resultant shape of the chip, (d) cutting chip into two parts, (e) a resultant V-shaped combined insert made using these parts, and (f) positioning the insert in the drill body, brazing, final trimming, and grinding.

of each half is shaped so that a resultant V-shaped combined insert shown in Figure 5.39e can be assembled. This assembly is then inserted in the tool body where a rectangular slot having parallel, longitudinally disposed outer walls is cut inward. The assembly is brazed into the drill body, and then the tool is cleaned, finish trimmed, and ground to its final shape/dimensions (Figure 5.39f).

The attempt to use a solid PCD block with no substrates on the drill face started in the 1980s. One of the earlier drill designs is shown in Figure 5.40 (US Patent No. 4.697,971, 1987). In this simplest design, a PCD nib having a rather thick layer of PCD and rather long thick substrate is brazed on the end of a correspondingly shaped drill body. After cleaning, the brazed blanks are fluted and then pointed to obtain the final desired drill geometry. Note that the patent describes the brazing procedure and fixtures in great details as a chance to overheat the diamond layer with the brazing technology of the 1980s was great.

Enjoying the full PCD face as a lucrative advantage, this drill design, however, requires a substantial removal of the PCD material that is both wasteful and ineffective as grinding of a great amount of PCD is expensive and time-consuming. Moreover, the heat generated on grinding can be equally (to brazing) harmful for a PCD layer that may reduce the quality of the drill.

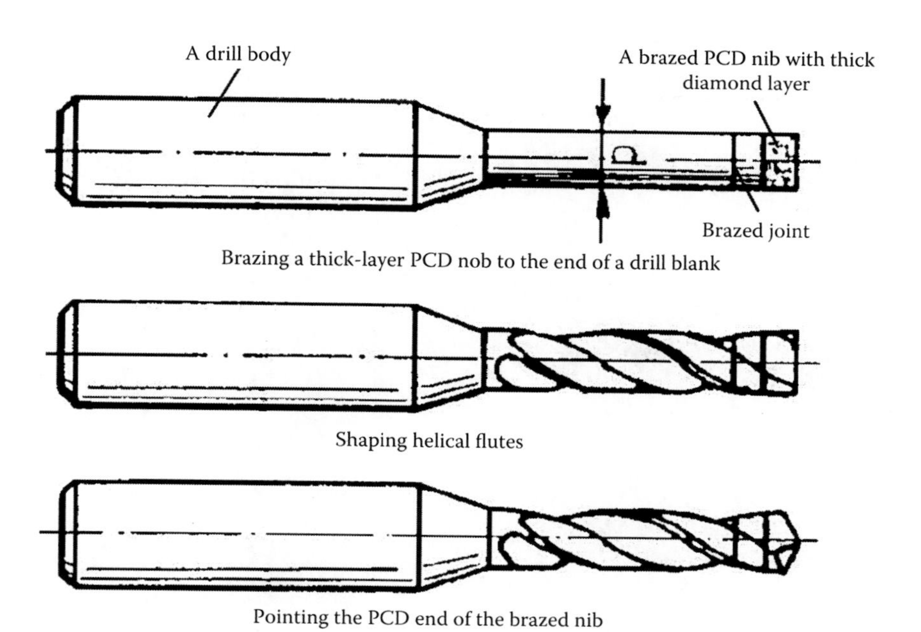

FIGURE 5.40 Earlier drill design with the full PCD face.

The introduction of new, high-strength, highly abrasive composite materials such as MMCs and fiber-reinforced plastics and the wider application of materials such as aluminum with a high silicon content led to a significant increase in the demand for PCD cutting tools. Although PCD products are used extensively in machining applications, their use in drilling has been much more limited. The high cost of the finished tool, due to the complicated manufacturing route, was considered to be the main contributing factor.

The De Beers Technical Service Centre pioneered the development of PCD drills with freestanding PCD and worked out a method for their manufacturing (Sani 1994). This method of manufacture became possible due to the development of a brazing procedure of freestanding PCD layers (i.e., without an integral carbide backing) onto carbide and steel. In the course of this work, freestanding PCD was successfully brazed, in the presence of argon, using a titanium-activated braze alloy.

To manufacture PCD-tipped twist drills, standard HSS and carbide drills are modified by machining a pocket at the tip of the drill and then brazing an unbacked (freestanding) PCD segment into the pocket. The pocket can be produced by either grinding or EDMing. Unlike most previous designs, the pocket is not cut square to the axis of the drill but parallel to the cutting edges. This produces a pocket in the form of an inverted *V* as shown in Figure 5.41.

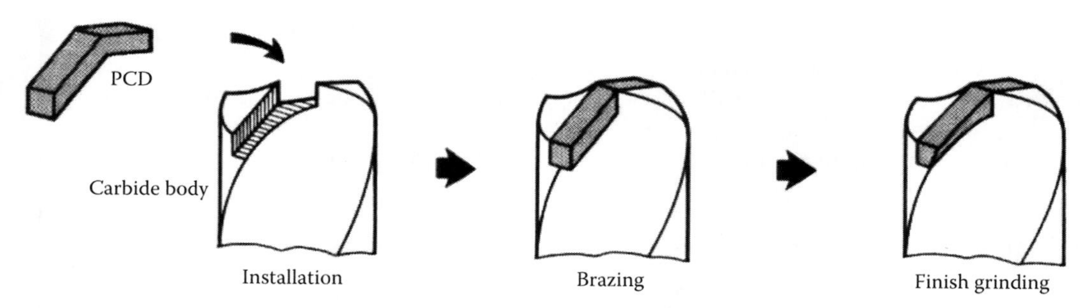

FIGURE 5.41 PCD-tipped drill manufacture using freestanding PCD.

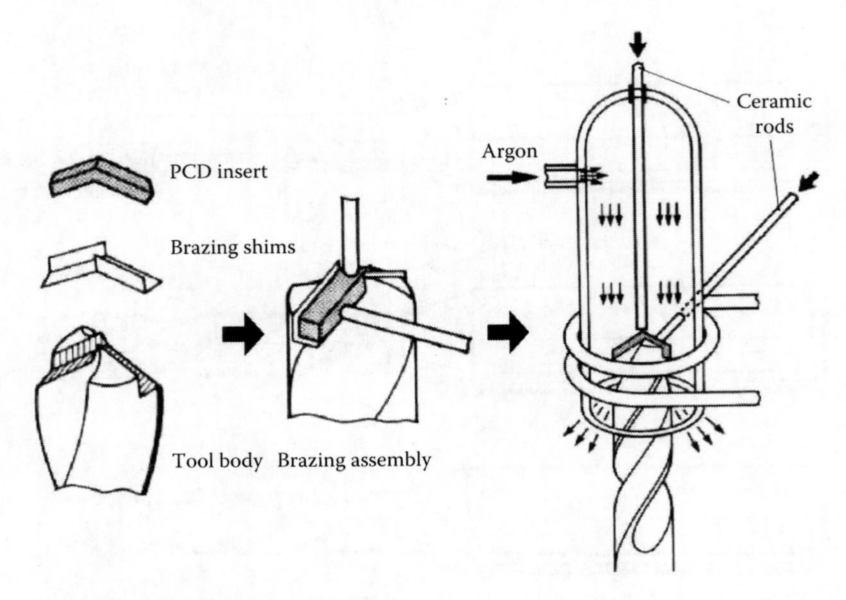

FIGURE 5.42 Steps involved in the brazing procedure.

The reason for machining the pocket in this form is twofold:

1. To braze the PCD tip, the *V*-shaped segment is simply placed into the pocket that, by virtue of its matching geometry, causes the PCD tip to be automatically centered. Brazing is carried out by pressing the segment against the side of the pocket, while maintaining gentle pressure on the apex. Using this arrangement, no special setting-up operation is needed. In fact, once the segment and the pocket are cut to the correct geometries, a perfect braze can be guaranteed each time (Figure 5.42).
2. Apart from locating the segment, the *V*-shaped pocket has a dramatic effect on the amount of material that needs to be removed from the flute area. By eliminating the PCD from certain areas, standard grinding instead of sophisticated flute grinding can be used to produce the necessary angles in the flute area.

Figure 5.43 shows a drill made according to this method after the brazing (Figure 5.43a) and finished product (Figure 5.43b). This drill design and its manufacturing steps were much ahead of the time so they have not attracted much attention from PCD drill users and manufacturers because unbacked PCDs (with no substrate) were not yet common and the brazing procedure was cumbersome for most of the tool manufacturers. A great potential of this drill is not recognized until now. Therefore, it is instructive to list major advantages of this design concept:

1. Any grade of PCD, including leached and thermally stable as well as CVD grades, can be used.
2. The force balance is preferable compared to other drill designs.
3. Practically all known drill point geometry/designs can be then applied to make application-specific drills.
4. Minimum trimming/grinding of the brazed PCD insert is required.

The author's experience in the design and implementation of this kind of drill allows to state the following:

1. The manufacturing of this drill design is mainly limited due to the definite lack of understanding of their advantages from the users and tool manufacturers. Moreover, many tool manufacturers are reluctant to invest in vacuum furnaces required to manufacture such tools.

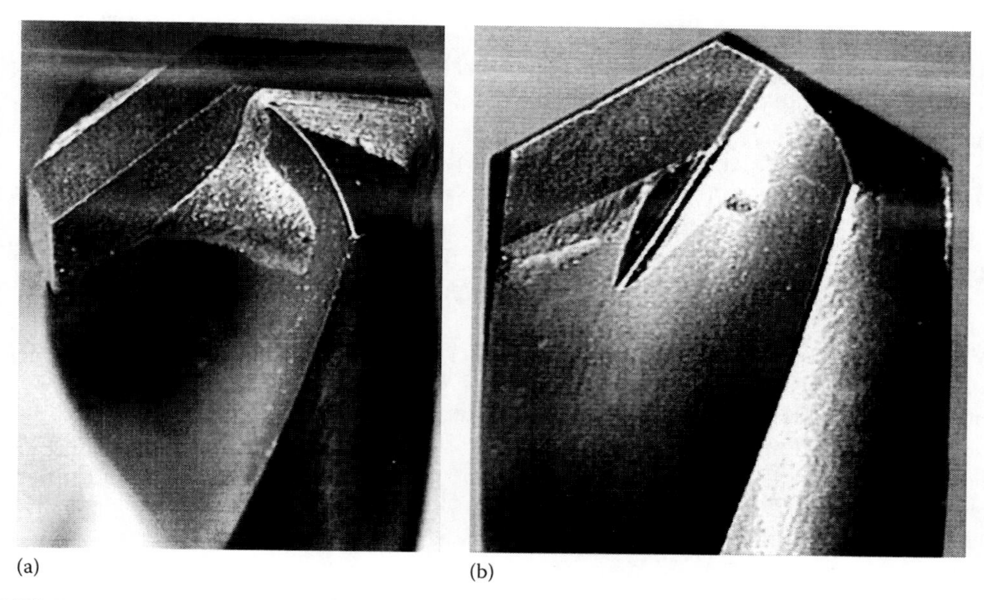

(a) (b)

FIGURE 5.43 Showing a drill after the (a) brazing and (b) finished product.

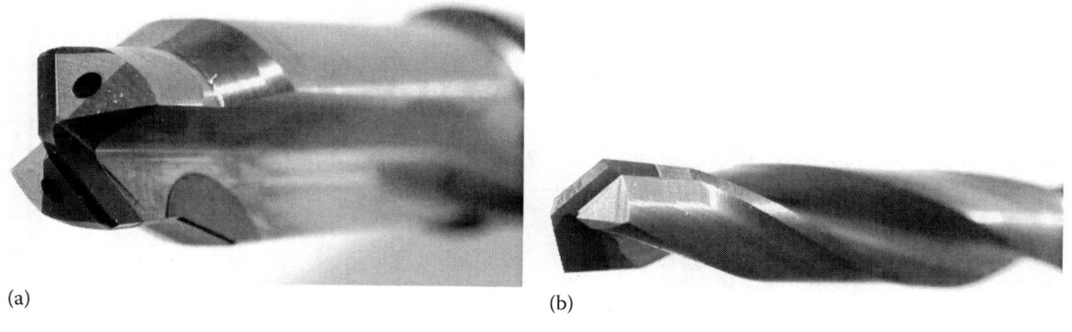

(a) (b)

FIGURE 5.44 Full-face (cross) PCD straight-flute (a) and twist (b) drills. (Courtesy of Fullerton Co., Saginaw, MI.)

2. The performance of such drills is very sensitive to their proper application-specific design and accuracy of manufacturing. In other words, such drills are not forgiving compared to other kinds of PCD drill designs.
3. When properly designed and made, this kind of drill outperforms any other types of full-face (cross) PCD drills allowing much higher drilling productivity (the penetration rate).

Figure 5.44 shows full-face (cross) PCD straight-flute and twist drills by Fullerton Co. Optimized versions of these tools were tested and then implemented for HSM of high-silicon aluminum alloys in a power train plant of the automotive industry.

5.2 DEEP-HOLE DRILLS

5.2.1 Introduction

The term deep-hole machining (DHM) relates to a machining operation of a hole with an excessive depth-to-diameter ratio (known as *L/D*). For many years, it was considered that DHM range begins when $L/D > 5$. With the introduction of some modern machinery and tooling including high-pressure solid-carbide straight-flute and twist drills, *L/D* greater than 15 can nowadays be accepted as a reasonable indicator. The use of the *L*/*D* criterion, however, is not nearly sufficient to make the proper decision about the tool when high machined hole specifications, for example, diametric accuracy, hole straightness and shape, and surface roughness, are required. This is because when properly used, DHM tools produce superior holes with close tolerances when a group of special tools, called self-piloting tools (hereafter SPTs), is used.

The techniques of self-piloted drilling began developing in the late eighteenth century with the growing need for more accurate bores in rifle barrels and cannons. While SPDs are still used for this purpose, their use has been extended to an increasingly wider variety of applications. Today SPDs are the most efficient tools, if not the only tools, for producing extremely deep holes, regardless of the precision required.

While most applications involve hole depths varying from 10 to 30 diameters, it is common to encounter drills with depth-to-diameter ratios of 100 to 1. Furthermore, holes with depth-to-diameter ratios of 300 to 1 have been successfully drilled (Swinehart 1967). While such features may be accomplished with a twist drill, the extra problems involved in getting MWF (the coolant) in and the chips out of the hole make it quite difficult to drill beyond a *L*/*D* about 20 to 1 and completely impractical for *L*/*D* beyond 50 to 1.

Besides high *L*/*D*s, SPDs are also capable of drilling very straight deep holes. The hole that the tool has machined is actually a continuation to the guide bushing so that the tool will continue to machine a straight hole along the same initial direction. Although the SPTs were developed primarily for producing deep holes, their ability to produce holes of good surface finish and close diametric tolerance is often attractive to engineers requiring close tolerance holes of much shorter lengths. SPDs can even be economical for holes as short as one diameter, under certain conditions.

Usually in order to produce a finished hole, two to five operations are required: drilling, boring, rough and finish reaming, and finally honing. Not only are these operations costly by themselves, but each carries many other hidden costs that may not be so obvious. These include the cost of multiple handling and transporting of the parts between several machines, making extra setups, and carrying more inventory, and, under critical conditions, they can require extra inspections that involve the procuring and maintaining of gages and allied equipment.

SPDs can eliminate such subsequent operations if one or more of the following conditions exist: (1) the precision requirements for the drilled hole (either size, finish, straightness, location, or all four) are such that they are difficult to attain by the more conventional methods, (2) the material is one in which an SPD can readily produce the degree of precision required, (3) the configuration of the part is such that it would be difficult to index from one station to another, (4) the location of the hole must be held accurately with respect to other holes or surfaces, and (5) the size of the hole or the configuration of the part and/or machine would require special tooling and/or fixtures.

However, SPDs also have their limitations in deep-hole applications. A tool as long as 100 diameters or more is rather flimsy and obviously cannot withstand as much torque or thrust as a shorter tool. Not only will the torque cause torsional deflection or twisting, but the axial force can cause axial deflection or buckling of the tool. Due to this fact, SPDs are very sensitive to even small changes in their geometrical parameters and require both the proper design and the skillful regrinding procedure. The drilling process outcomes depend to a great extent on the conditions of the machine, for example, rigidity, alignment and accuracy of feed motion, MWF maintenance, and intelligent process control.

5.2.2 Common Classification of Deep-Hole Machining Operations

In general, deep-hole machining operations are classified by the method of MWF supply and swarf transportation from the machining zone. It can be gundrill (gun)-type, STS-type (known also as Boring and Trepanning Association (BTA)-type), and ejector-type machining.

The principle of gundrilling is shown in Figure 5.45. MWF (the coolant) is supplied under high pressure to the tool holder and then to the drill shank. Then it flows through a kidney-shaped internal passage made in the shank and then through the coolant passage made in the tip. After cooling

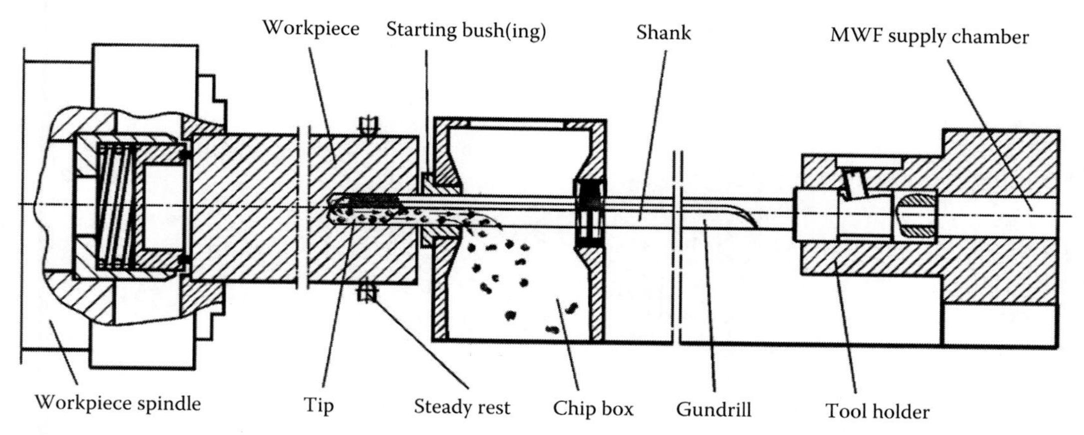

FIGURE 5.45 Gundrilling principle.

and lubricating the machining zone, MWF carries away the formed chips as the mixture (often referred to as the swarf) over the external V-shaped flute made on the shank. The swarf ends up in the chip box from where it passes through the chip separator and flows into the MWF tank. After cleaning and cooling down, MWF is pumped back to the gundrill.

Figure 5.46 shows the principle of single tube system (STS) drilling. The STS tool assembly consists of a boring bar and a single- or a multiedged drill head secured at its terminal end. MWF is supplied under high pressure through the inlet of the pressure head and flows through the annular channel between the boring bar and the bore wall toward the drill head. After cooling and lubricating the machining zone, MWF carries away swarf through the interior of the drill head and boring bar. Contrary to gundrilling, the returning chips do not come in contact with the bore wall; thus, better surface finish can be achieved.

The tubular cross section of the boring bar possesses greater buckling stability compared to the gundrill's shank; thus, greater feed rates can be achieved in STS drilling. As such, the annular chip removal channel is of much greater cross section compared to the V-flute of gundrill; thus, more chip per unit time can be transported without clogging. Moreover, even when chip clogging occurs due to poor chip breaking, the inlet pressure increases helping to push forward the chip cluster that clogs the chip removal channel. A requirement for reliable sealing between the face of the workpiece and the pressure head is a price to pay for this advantage.

Figure 5.47 shows the principle of ejector drilling. MWF is supplied to the inlet of the collet chuck. Then this fluid separates into two portions. The first portion flows in the annular channel

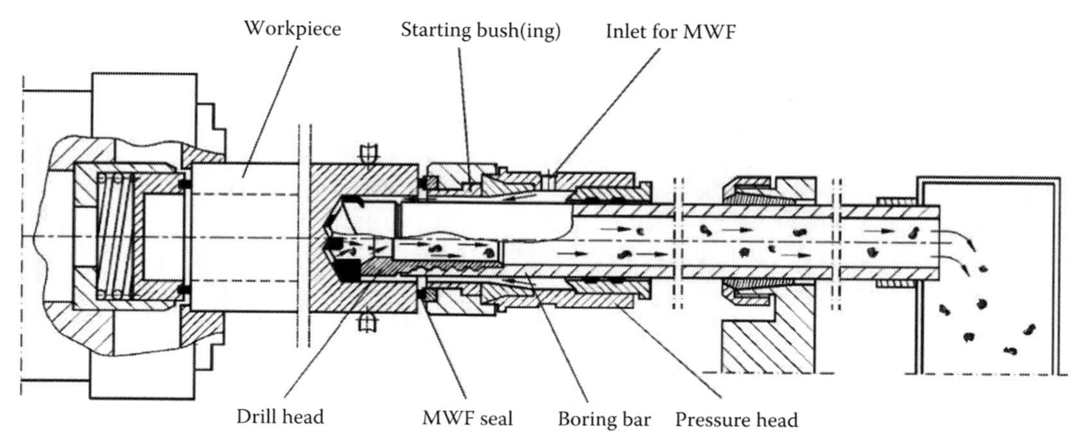

FIGURE 5.46 STS drilling principle.

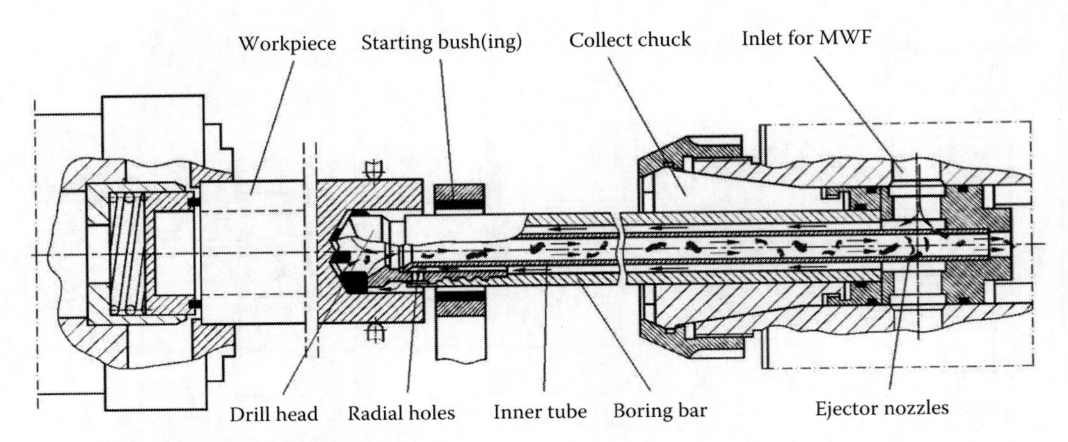

FIGURE 5.47 Ejector drilling principle.

formed by the boring bar, called the outer tube, and the inner tube. In the drill head, MWF flows outside through the radial holes made in the drill head; thus, this part reaches the machining zone where it cools and lubricates the cutting elements and the bearing areas. Then the MWF–chip mixture (swarf) goes into the inner tube. The second portion flows through the ejector nozzle(s) made on the inner tube. As a result, this flow creates a partial vacuum (the ejector effect) in the inner tube that sucks the swarf into the inner tube as a vacuum cleaner.

The great advantages of ejector drilling are as follows: (1) there is no need to have a nonreliable seal between the face of the workpiece and the starting bushing and (2) much lower inlet MWF pressure is needed. As a result, ejector drilling does not always require a special drilling machine so that this type of drilling can be used in many general-purpose machines or even machining centers as one of the common drilling operations.

However, this versatility comes at certain costs. First, it is not available for hole diameters below 20 mm. Second, it is suitable only for the work materials that generate easy-to-control chips. That eliminates most nickel-based and many nonferrous alloys. This is because if the chip even slightly clogs the chip removal channels including the inner tube, there is no pressure available to push the formed chip cluster through as the maximum pressure created by the ejector is much less than 0.1 MPa. As a result, MWF leaves the radial coolant holes and then flows between the boring bar and the bore outside the drilling coolant circuit. When this happens, the drill normally breaks.

5.2.2.1 Force Balance and the Meaning of the Term *Self-Piloting Tool*

5.2.2.1.1 Theoretical (Intended) Force Balance

Although there are a great number of designs of SPTs, two important features, namely, the theoretical force balance and the locating principle, are the same. Therefore, they should be considered before particular designs of SPTs are analyzed further in the subsequent sections.

To comprehend the concept of SPTs, one should first consider the force balance of such a tool. Following methodology of constructing the force balance diagram used in Chapter 4 (Section 4.4), one can construct a force balance diagram for the tool shown in Figure 5.48. For convenience and simplification of further considerations, the right-hand $x_0y_0z_0$ coordinate system, illustrated in Figure 5.48, is set as follows:

1. The z_0-axis along the longitudinal axis of the drill, with sense as shown in Figure 5.48, toward the drill holder.
2. The y_0-axis passes through periphery corner and is perpendicular to the z_0-axis. The intersection of these axes constitutes the coordinate origin as shown in Figure 5.48.
3. The x_0-axis is perpendicular to the y_0 and z_0 axes as shown in Figure 5.48.

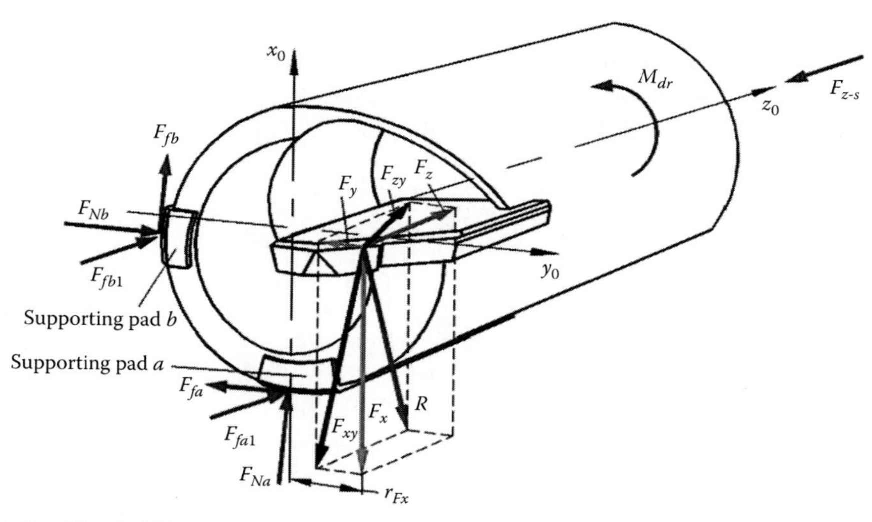

FIGURE 5.48 Simple SPD head force balance.

Figure 5.48 shows the force balance in the T-mach-S $x_0y_0z_0$ coordinate system (Astakhov and Osman, 1996a and 1996b). When an SPT works, the cutting force generated is due to the resistance of the workpiece material to cutting. This force is a 3D vector that can be thought of as applied at a certain point of the drill rake face. The cutting force R (or the resultant cutting force for multi-edge tools) can be resolved into three components, namely, the power (tangential) F_x, axial F_z, and radial F_y components, respectively. These components are commonly referred to as the tangential, radial, and axial forces (Sakuma et al. 1980). The axial force is balanced (equal in magnitude and oppositely directed) by the axial force $F_{z\text{-}s}$ of the feed mechanism of the deep-hole machine, while the tangential and radial forces sum to create force F_{xy} (acts in the x_0y_0 plane) that (in contrast to other axial tools as twist drills, reamers, and milling tools) generally is not balanced, regardless of the number of the cutting edges used. To prevent drill bending due to this unbalanced force, some special measures are taken. The term *deep-hole drilling* has grown to mean that the unbalanced cutting force F_{xy} generated in the cutting process is balanced by the equal and opposite force due to supporting pads, which bear against the wall of the hole being drilled.

Due to the action of the mentioned forces on the supporting pads, these pads ensure balancing of the tool in the x_0y_0 plane while simultaneously providing for a unique additional machining operation known as burnishing. The fact that the pads bear against the wall of the hole being drilled behind the side cutting edge effectively means that the tool is provided with its own guide. The concept of self-piloting (sometimes referred to as self-guidance), meaning the tool guiding or steering itself along the bore, has been recognized as the major underlying principle of the design of SPTs (Griffiths 1993; 1995a).

Nowadays, multiedge and multisupporting pad SPTs are used; thus, the described self-piloting feature may not be obvious. As discussed by the author earlier (Astakhov 2010), no matter how many cutting edges an SPT has, as long as there is an unbalanced radial force acting on the supporting pads of the tool, it remains self-piloting. Alternatively, no matter how many supporting pads a hole-making tool has, it should not be regarded as self-piloting if there is no (at least, theoretically) unbalanced radial force to cause self-guiding.

Not only does the self-piloting design make it possible to machine deep holes, but it also provides stable cutting conditions for the cutting insert or cutting portion of the tool through eliminating the radial vibrations that often cause drill failure in ordinary drills (Sakuma et al. 1981). When machining conditions are selected properly, a machining operation with SPT is very stable and consistently produces holes of high quality. Although the use of SPTs requires a number of additional accessories as starting bushings, high-pressure coolant system, etc., the benefits gained with the use of these tools are greater as SPTs proved capable of maintaining close size control and producing holes of good surface finish that met the output requirements of honing. The elimination of the whole sequence of standard hole-making operations makes the use of SPTs appealing even for machining shallow holes. Thus, the use of SPTs is considered whenever one or more of the following conditions exist:

- The $L/D > 15$
- High-precision requirements difficult to attain by conventional bore machining operations
- The tight position tolerance on the longitudinal axis of the machined hole
- The tight shape tolerance on the machined hole.

5.2.3 Additional Force Factors in Real Tools

The additional bending moments similar to those discussed in Chapter 4 (Section 4.4.2) are considered in this section. One important feature of the force balance shown in Figure 5.48 is that the tangential and radial forces (or their resultant F_{xy}) are normally fully balanced by the normal (F_{Na} and F_{Nb}) and tangential (F_{fa} and F_{fb}) reactions acting on the supporting pads (Griffiths 1993; Richardson and Bhatti 2001). In the design of BTA and ejector heads, the arms $a_{zx\text{-}ub}$ and $a_{zy\text{-}ub}$ (Figure 5.49) (the distances from the point of total force application A to the corresponding apexes B and C on the supporting pads) are small so that the unbalanced moments due to F_x and F_y forces are insignificant. The axial force F_z is balance by the $F_{z\text{-}s}$. The problem, however, is that the arm $e_{zy\text{-}ub}$ is significant so the additional bending moment due to the axial force presents a problem.

The equilibrium condition in the z_0y_0 plane is given in the following equation:

$$F_z a_{zy\text{-}ub} = F_{fa1}\left(\frac{d_{dr}}{2}\right) + F_{fb1}\left(\frac{d_{dr}}{2}\right) \tag{5.1}$$

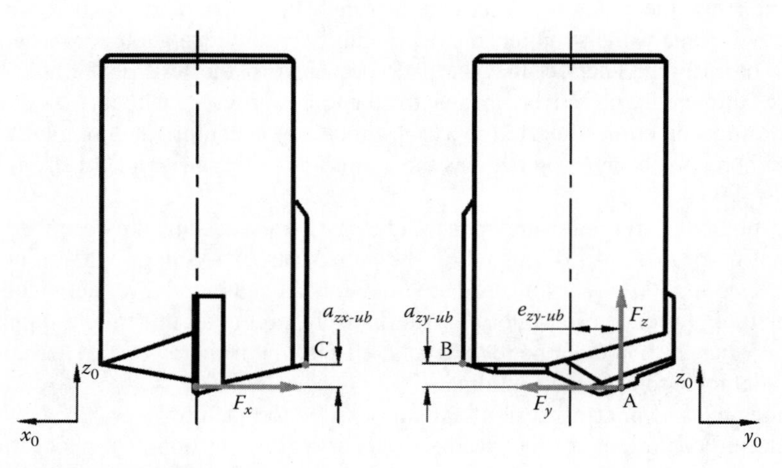

FIGURE 5.49 System of unbalanced loads.

Because the friction forces F_{fa1} and F_{fb1} are small compared to the axial force F_z, the additional bending moment

$$M_{b\text{-}Fz} = F_z a_{zy\text{-}ub} - \left(F_{fa1}\left(\frac{d_{dr}}{2}\right) + F_{fb1}\left(\frac{d_{dr}}{2}\right)\right) \tag{5.2}$$

tends to bend the drill in the counterclockwise direction in the z_0y_0 plane.

A number of measures have been undertaken to reduce harmful consequence of this moment. The most common is to use very shallow feed rates, thus reducing the axial force. This measure, however, results in low productivity in SPT machining. Another common measure is to introduce additional array of the supporting pads at the rear end of the drill head.

The problem with the mentioned force balance was solved when SPTs with the partitioned cutting edge were introduced. In such drills, the cutting inserts are located on both sides of the x_0 axis. Figure 5.50a shows a traditional drill head and Figure 5.50b shows a head with the partitioned cutting edge. A simplified force model in the y_0z_0 plane for a drill with the partitioned cutting edge is shown in Figure 5.51. Using this figure, one can obtain a simple condition of equilibrium, where there is no bending moment in this plane:

$$F_{z1}r_{in1} + F_{z2}r_{in2} = F_{z3}r_{in3} + F_{pa2}r_{dr} \tag{5.3}$$

Although the design shown in Figure 5.50b was apparently introduced to improve the chip removal as the chip flow is separated into two portions that flow in two chip mouths, it was soon found that such drills allow much greater feeds then those found in conventional twist drilling. Nowadays, practically all STS and ejector drills having diameter greater than 18 mm use the partitioned cutting edge design.

There are some attempts to use the same idea for gundrills. Figure 5.52 shows an example of a gundrill having the chip removal flute (the main discharge port 1) that opens to a distal end face of the drilling head. The secondary (auxiliary) discharge port 2 opens to the distal end face at a substantially opposite position in the radial direction with respect to the main discharge port 1. A bypass flow pass port 3 connects secondary (auxiliary) discharge port 2 and the chip removal

(a)

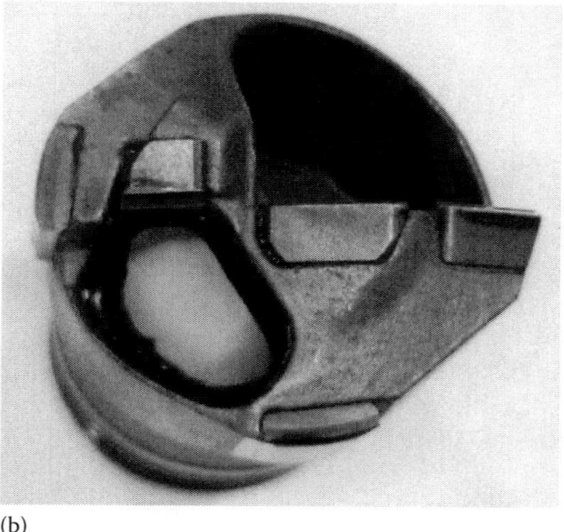

(b)

FIGURE 5.50 Drill designs: (a) with single cutting insert and (b) with partitioned cutting edge made by three cutting inserts.

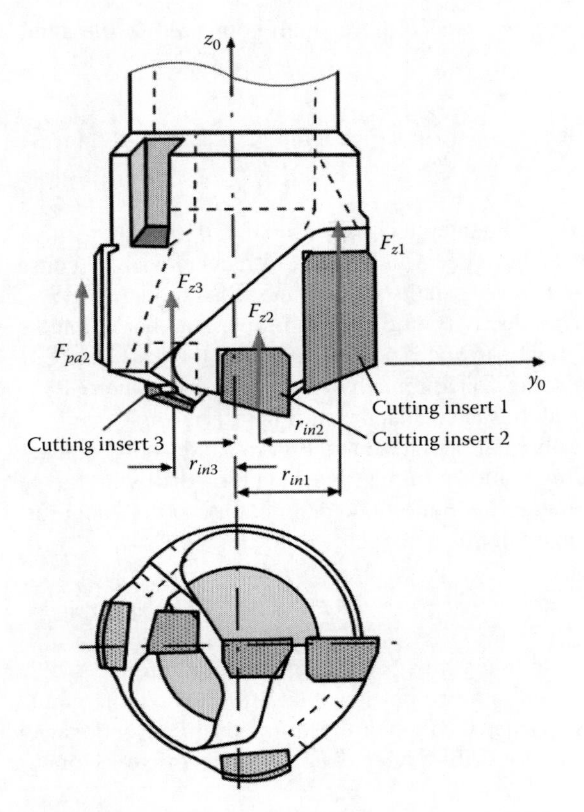

FIGURE 5.51 Simplified force balance in the y_0z_0 plane.

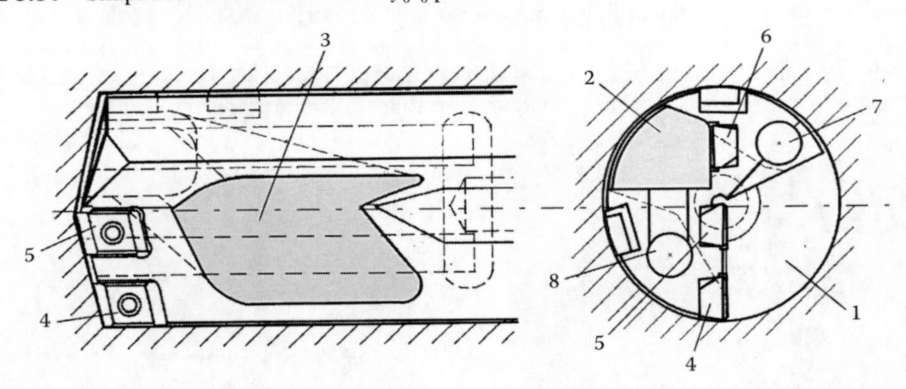

FIGURE 5.52 Gundrill with the partitioned cutting edge.

flute 1. The cutting edge is partitioned so that two cutting inserts 4 and 5 are located in the front of the chip flute 1, while the cutting insert 6 is located in front of the secondary discharge port 2. Two coolant holes 7 and 8 assure proper MWF distribution, so each discharge port has sufficient MWF flow rate to transport the formed chip.

5.2.4 Common Feature of SPTs: Supporting Pads

5.2.4.1 Locating Principle of SPTs

As discussed in the previous section, the term *SPT* has grown to mean that the intentionally unbalanced cutting force generated in the cutting process due to a special tool design is balanced by the equal and opposite force due to the supporting pads, which bear against the wall of the hole being drilled. As such, SPT guides itself initially in the starting bushing and then in the hole being drilled so that it can be considered as self-piloting (Swinehart 1967; Sakuma et al. 1978, 1981; Griffiths and Grieve 1985; Astakhov and Osman 1995). However, the guiding and burnishing actions of the supporting pads depend on their design/geometry and location that until now presented some difficulties in understanding the rationale behind the proper selection of these important features.

To exemplify this statement, Figure 5.53 shows the recommendations by ISCAR Co. that summarizes practically all designs of the supporting (bearing) area discussed in the literature over the last 200 years. As can be seen, a great variety of possible designs starting with a single supporting continuum and finishing with three supporting pads are offered with vague explanations on their implementation conditions. No explanation why a particular design of the supporting area (pads) is suitable for the recommended conditions is provided. Therefore, a need is felt to clarify the issue and thus to help tool/process designer/planner to make an intelligent selection of the proper design/location of the bearing area of SPT for a given application.

In the author's opinion, the previously discussed universally accepted definition of self-piloting refers to stable drilling conditions and thus is restricted only to the case where all forces acting on an SPT are completely balanced. Unfortunately, these ideal conditions can hardly exist in the practice of deep-hole machining where additional forces due to a number of real world imperfections (inaccuracies in real deep-hole machining system including alignments, clearances, drill design, and manufacturing inaccuracies) affect process stability.

Stability is defined as a complex issue consisting of the following subitems: entrance, static, and dynamic stabilities (Astakhov 2002a). Experience shows that depending upon a particular combination of the parameters of a drilling system, one or another type of stability plays the chief role. Entrance stability, however, always plays an important role because it is not possible in any practical situation to achieve zero clearance in the starting bushing because of a number of different factors. Among them, the clearance in the starting bushing that changes with each tool regrind due to tool back taper and wear of the starting bushing is of prime concern.

The author's analysis of gundrill failures in drilling aluminum alloy engine heads resulted in the conclusion that the entrance instability is the prime cause for the formation of the so-called bell mouth, which is essentially the heavily deformed tapered part of the machined hole at the beginning of the drilled hole. Although such phenomena are observed in everyday practice of gundrilling, it has never been regarded as an essential factor affecting drill performance because the duration of entrance instability is very short (0.3–0.8 s), so it can hardly be noticed in the production cycle, particularly when the existing control systems are used (Astakhov 2002a,b).

Figure 5.54 shows a typical example of variation of components of the cutting force and coolant pressure during a drilling cycle. As can be seen in the figure, the axial force does not change significantly at the entrance, while the tangential and radial forces do. It was found that the radial and power force fluctuations are 10- to 20-fold larger compared to the rest of the cycle. The duration of entrance instability is approx. 0.28 s so it can hardly be noticed while drilling. Based on a number of the experimental results, the following can be concluded:

1. The clearance between the drill and the starting bushing is the most critical factor in entrance stability. When this clearance is zero (in the experiments, it was achieved by the selective combination of a gundrill and a starting bushing), the instability was at its minimum depending on the gundrill design.
2. Among tool parameters, the design and location of the supporting pads (supporting continuum) are critical.
3. Experimental comparison of gundrills with the supporting continuum (Profile G in Figure 5.53) and two supporting pads (Profile A in Figure 5.53) shows that the force fluctuations are 30%–50% smaller when the clearance between the gundrill and the starting bushing was kept at minimum. The opposite was true, however, when this clearance was increased.

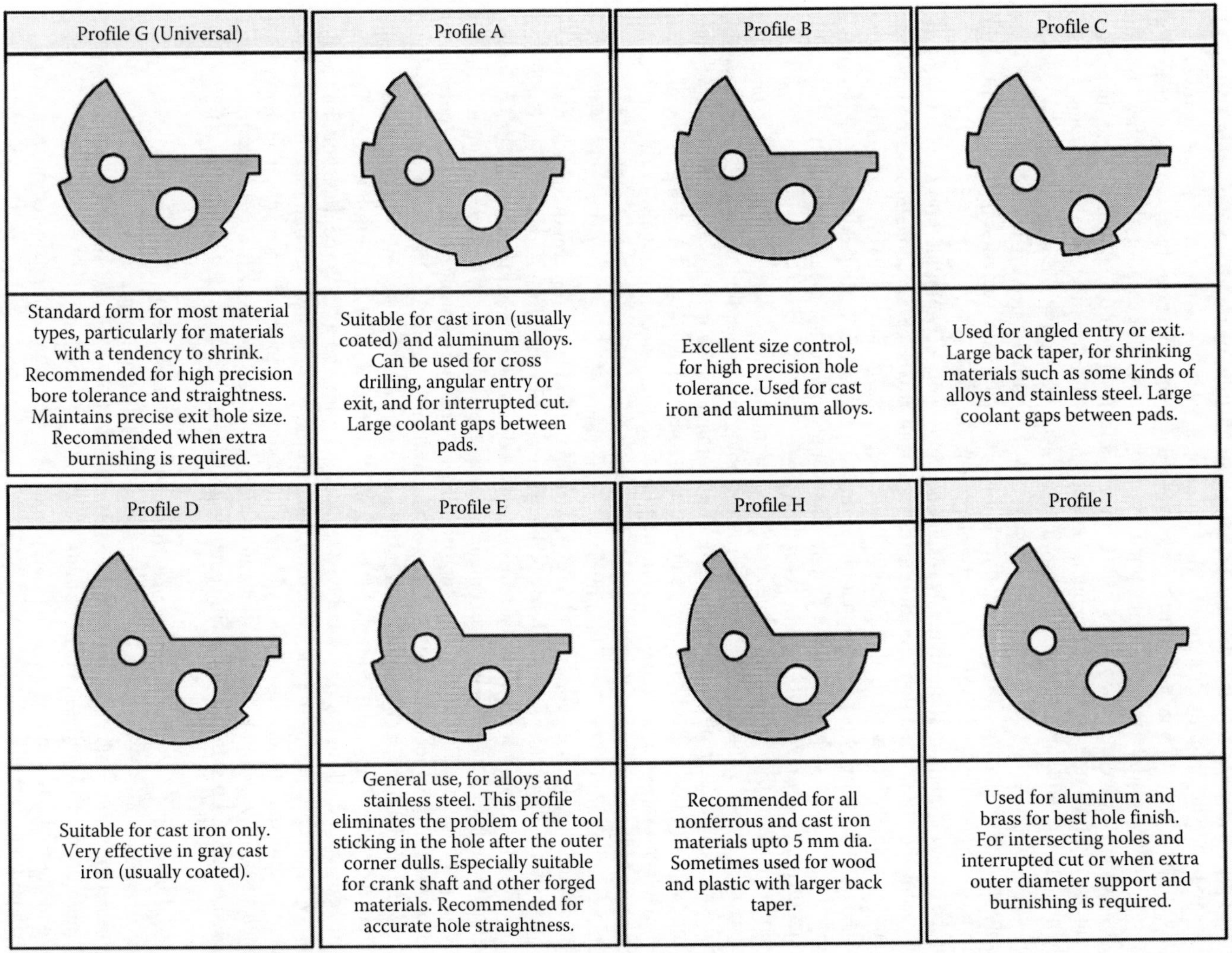

Profile G (Universal)	Profile A	Profile B	Profile C
Standard form for most material types, particularly for materials with a tendency to shrink. Recommended for high precision bore tolerance and straightness. Maintains precise exit hole size. Recommended when extra burnishing is required.	Suitable for cast iron (usually coated) and aluminum alloys. Can be used for cross drilling, angular entry or exit, and for interrupted cut. Large coolant gaps between pads.	Excellent size control, for high precision hole tolerance. Used for cast iron and aluminum alloys.	Used for angled entry or exit. Large back taper, for shrinking materials such as some kinds of alloys and stainless steel. Large coolant gaps between pads.

Profile D	Profile E	Profile H	Profile I
Suitable for cast iron only. Very effective in gray cast iron (usually coated).	General use, for alloys and stainless steel. This profile eliminates the problem of the tool sticking in the hole after the outer corner dulls. Especially suitable for crank shaft and other forged materials. Recommended for accurate hole straightness.	Recommended for all nonferrous and cast iron materials upto 5 mm dia. Sometimes used for wood and plastic with larger back taper.	Used for aluminum and brass for best hole finish. For intersecting holes and interrupted cut or when extra outer diameter support and burnishing is required.

FIGURE 5.53 Design and location of the supporting areas as recommended by ISCAR Co.

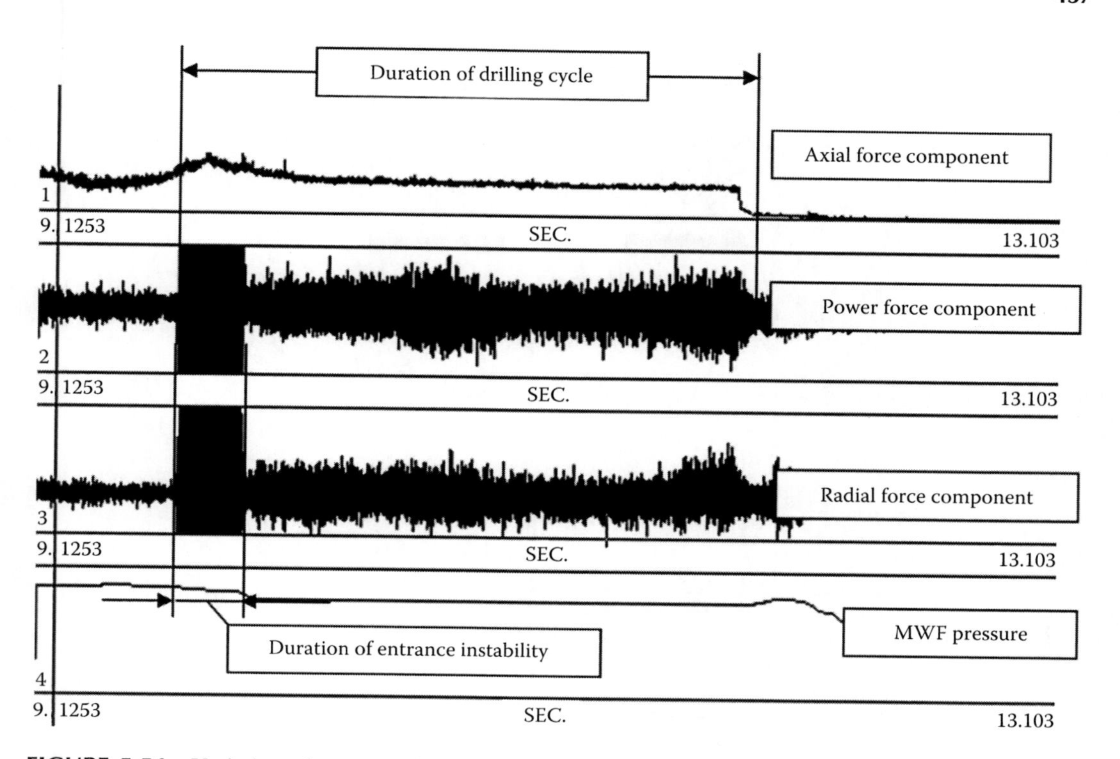

FIGURE 5.54 Variation of components of the cutting force and MWF pressure during a drilling cycle. The cutting regime: drill rotational speed 8000 rpm; feed rate 1118 mm/min; pressure of MWF 7 MPa; the approach angles, the outer cutting edge 45° and the inner cutting edge 30°; the normal flank (relief) angles, the outer cutting edge 12° and the inner cutting edge 20°; the radial clearance in the starting bushing 10 μm; and the distance between the face of the starting bushing and that of the workpiece 0.3 mm. The same scale for cutting forces is used. Time duration is shown from the beginning of recording.

The entrance of an SPT into the workpiece consists of two distinctive stages (Astakhov 2002b). The first stage is the entrance of the SPT's cutting part. At this stage, the cutting force and its components change (in terms of their magnitude and direction) steadily with the corresponding increase of the length of the cutting edges engaged in cutting. It is important to note that at this first stage, the supporting (burnishing) elements of SPT remain in the starting bushing. At this stage, the location of the supporting (burnishing) elements of the SPT plays the chief role. The second stage, which completes the entrance, is the entrance of the supporting (burnishing) SPT elements into the workpiece. At this stage, the design (i.e., the front chamfer length and angle) of the supporting (burnishing) elements of SPT is important as this defines the parameters of the bell mouth.

Figure 5.55 shows the initial relative location of a gundrill and the starting bushing. For clarity, the initial location of the gundrill is considered to be coaxial with the starting bushing, that is, the origin of the gundrill's original coordinate system O_1 is assumed to be coincident with the center O of the starting bushing having diameter d_{sb}. Although in practice this is rarely the case, it is shown (Astakhov 2002b) that the initial location of the gundrill does not affect the drill's behavior at the entrance unless severe additional forces due to the misalignment of the gundrill and the starting bushing are high enough to affect the proper entrance and working conditions of the gundrill.

Figure 5.55a illustrates the initial position of a gundrill having the supporting continuum *ab*. In this figure, d_{dr} is the drill's nominal diameter, d_{sb} is the diameter of the starting bushing, and Δ_{sb} is the nominal clearance between the drill and the starting bushing. It is clear that $\Delta_{sb} = (d_{sb} - d_{dr})/2$. Figure 5.55b illustrates the initial position of a gundrill having two supporting pads, *a* and *b*. The body clearance having the diameter d_{bc} is ground between these pads as shown. All other notations are the same as in Figure 5.55a.

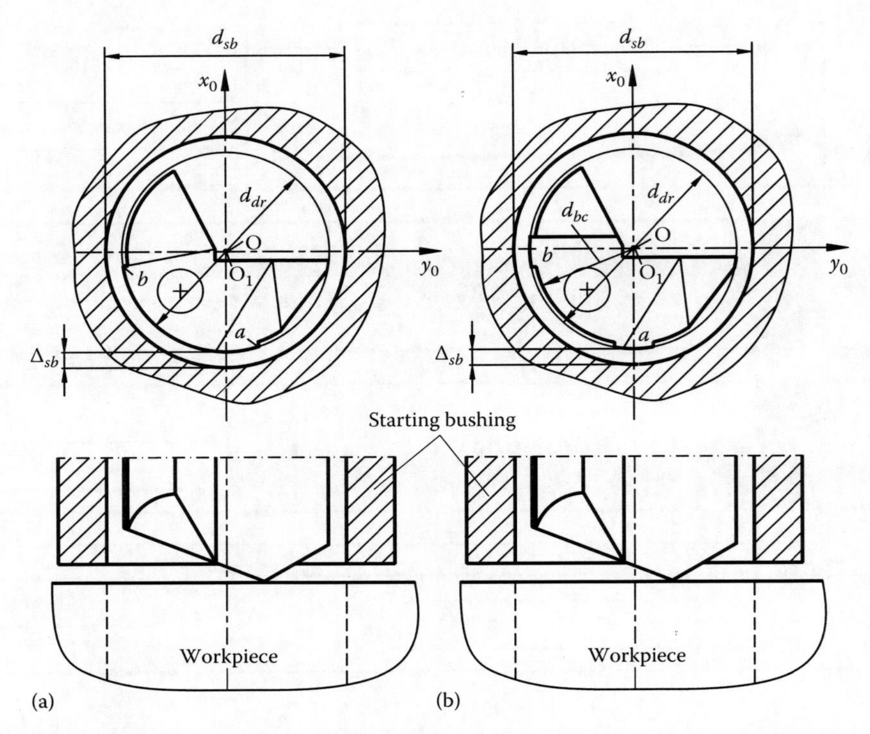

FIGURE 5.55 Initial position of the gundrill in the starting bushing: (a) gundrill having supporting continuum *ab* and (b) gundrill having two supporting pads *a* and *b*.

Figure 5.56 shows the comparison of locations of the gundrills shown in Figure 5.55 after the first stage of drills entrance is over developed using the developed model of drill entrance (Astakhov 2002a). The comparison of these locations allows drawing of the following conclusions.

The shift OO_1 after the first stage of the drill entrance that defines the extent of the bell mouth (defined as a tapered part at the drilled hole entrance) is smaller for the gundrill design with supporting continuum. This is explained as follows. As shown in Figure 5.56a, the force in the x_0y_0 plane, F_{xy}, shifts the drill in the direction of its action defined by the angle ε_1. The drill moves until its highest (in this direction) point d_1 touches the surface of the starting bushing. This defines the final location of the drill with the support continuous in the starting bushing. Although the force F_{xy} creates an additional tilting moment acting clockwise with respect to point d_1,

$$M_s = F_{xy}m_1 \tag{5.4}$$

this moment cannot tilt the drill as the cutting edges of the drill are already fully engaged with the workpiece (shown by the dashed area around side margin in Figure 5.56a).

On the contrary, the shift OO_1 after the first stage of the drill entrance that defines the extent of the bell mouth is greater for the gundrill design with two supporting pads (*a* and *b* in Figure 5.56b). This is explained as follows. As shown in Figure 5.56b, the force in the x_0y_0 plane, F_{xy}, shifts the drill in the direction of its action. The drill moves until the edge d_a of the leading supporting pad touches the surface of the starting bushing. When it happens, a tilting moment M_a having the arm l_{da} ($M_d = F_{xy}l_{da}$) causes the drill rotation until the edge d_b of the trailing supporting pad touches the surface of the starting bushing. This happens because nothing prevents drill to rotate counterclockwise with respect to the edge d_a. As a result, the shift OO_1 is greater than that in Figure 5.56a. The greater the angle between the supporting pads, the greater the shift OO_1 and thus the extent of the bell mouth.

While drilling, the gundrill design with two supporting pads has definite advantages in terms of drill proper guiding as the proper circle is defined by three points (supporting pads [*a* and *b*] and side margin

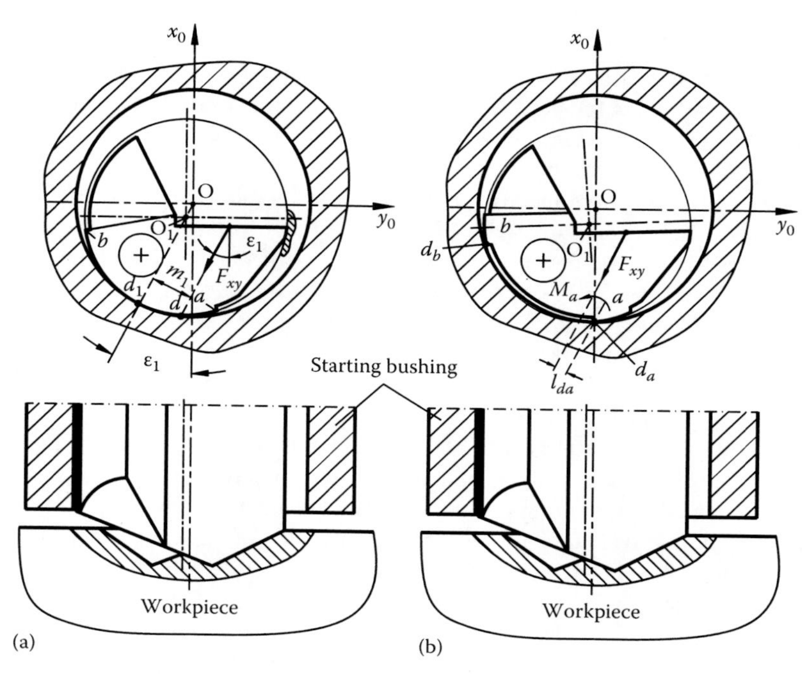

FIGURE 5.56 Final locations of the gundrills shown in Figure 5.55 after the first stage of drill entrance: (a) gundrill with the sporting continuum and (b) gundrill with two supporting pads.

of the outer cutting edge) (Astakhov et al. 1995; Astakhov and Galitsky 2005) and optimum normal stress to enhance the burnishing action by these pads (Sakuma et al. 1980; Griffiths and Grieve 1985). Moreover, great variety of work materials and tool geometries (the approach angles of the outer an inner cutting edges) can be used as long as the direction of F_{xy} is still within the angle between supporting pads.

On the contrary, the gundrill design with supporting continuum has a great bearing area, so it is difficult to assure the proper burnishing action as the normal contact stress is relatively low. As the drill has no well-defined three points for its location (guidance), these are selected by the drill itself within the continuum *ab* depending upon the variation of angle ε_1 when drilling. This affects the consistency of the quality of machined holes including their roughness, straightness, and diametric accuracy. Moreover, unnecessary friction over a great bearing area causes higher drilling torque that is highly undesirable particularly for long drills.

Depending upon the properties of the work material and drill geometry, the direction of F_{xy} may vary in a great range. This combined with great variation in the clearances in the starting bushing, the distance between the face of the bushing and that of the workpiece, the front chamfer in the starting bushing, etc., explain a number of configurations of drill designs with supporting continuum shown in Figure 5.53.

In the author's opinion, the variety of drill designs with supporting continuum shown in Figure 5.53 originates in the following:

1. Some gundrill users use standard ACME twist drill bushings on their machines that give excessive clearance in the starting bushing even for a new bushing.
2. Often, starting bushing is excessively worn because normally there is no procedure/control plan item related to their validation and thus replacement. Often, it is left to the machine operator's discretion. Moreover, it is rather difficult to access to the starting bushing in many gundrilling machines used in the automotive industry so that its inspection/replacement is difficult and time consuming.

These open a wide room for the use of drill designs with supporting continuum as it covers to a certain extent the lack of proper selection of starting bushing and machine maintenance.

5.2.4.2 Optimal Location of the Supporting Pads

As the location of the supporting pads of SPT defines its stability, it is important to determine the optimum pad location as one of the most important design parameters as it determines the quality of the drilled holes. In further consideration, the so-called static stability is considered as the criterion of such optimality. This parameter has been defined by many researches as being the most important (Pfleghar 1975, 1977; Stockert and Weber 1977a,b; Stockert and Thei 1978; Sakuma et al. 1980; Astakhov et al. 1995; Richardson and Bhatti 2001).

As discussed earlier, when SPT works, the cutting force is generated due to the resistance of the workpiece material. As shown in Figure 5.48, this force is a 3D vector applied at a certain point of the cutting edge. Its tangential and radial components (forces), F_x and F_y, sum to create force F_{xy} (acts in the x_0y_0 plane) that (in contrast to other axial tools as twist drills, reamers, end milling tools) generally is not balanced, regardless of the number of the cutting edges used. The supporting pads provide the means to balance this force.

Figure 5.57a shows the forces in the x_0y_0 plane for a gundrill. In this figure, angles ε_1 and ε_{xy} are location angles of F_{xy} with respect to the y_0 and x_0 axes, respectively. This figure also shows the location angles ψ_a and ψ_b of the supporting pad a and b with respect to the x_0-axis and the central angle between these pads, ψ_{ab}. If one compares Figure 5.48 and Figure 5.57a, one may notice that in the latter the forces in the x_0y_0 plane are applied at the coordinate origin, while in Figure 5.48, the point of their application is located at certain distance r_{Fx} from the x_0-axis. The difference, which is simply parallel displacement of the force's origin, is taken care of by the application of a moment equal to moment M_s defined by Equation 5.4. It can be shown (Astakhov 2002b) that this moment is the resistant moment having the magnitude equal to and the direction opposite to the drilling torque.

The optimum location of the supporting pads is achieved when their normal reaction forces F_{Na} and F_{Nb} (due to action of the resultant force F_{xy}) become equal. The analysis of the force system shown in Figure 5.57b revealed (Astakhov 2002b) that it is impossible to achieve equal normal forces on the supporting pads by keeping their symmetrical location relative to the action angle ε_{xy} of the resultant force F_{xy} as follows in Figure 5.58.

The problem is that the friction coefficient μ_{sp} is stable as are the friction forces F_{fa} and F_{fb} only when the supporting pads bear against the starting bushing. At the entrance, μ_{sp} changes over each step of drill entrance because the amount of plastic deformation done by the supporting pads changes continuously. When μ_{sp} increases as it happens when the leading supporting pad (pad a) enters the workpiece, the normal reactions on the supporting pads change in different ways as illustrated in Figure 5.59. The reaction on the trailing supporting pad (pad b)

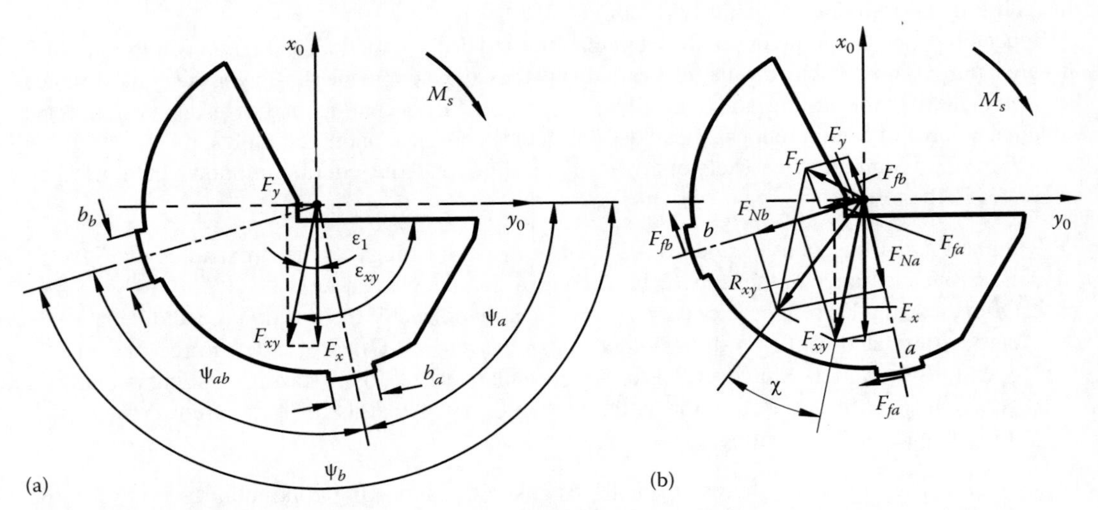

FIGURE 5.57 Model of forces acting on the gundrill: (a) action forces and (b) action and reaction forces.

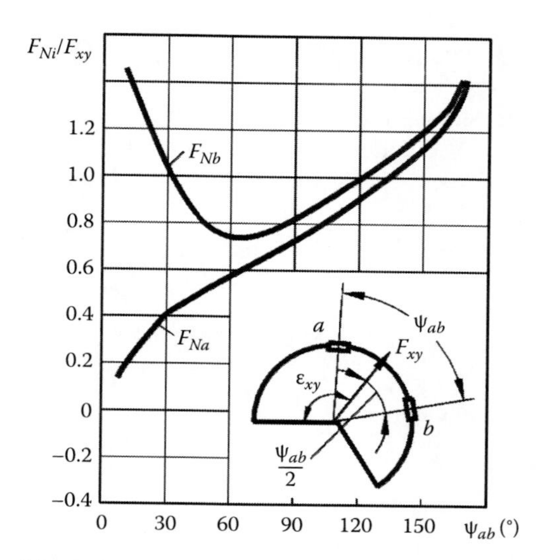

FIGURE 5.58 The relationship $F_{Ni}/F_{xy} = f(\psi_{ab})$ under symmetrical location of the supporting pads relative to the resultant force F_{xy}.

increases under any relative position of the pads, but the rate of this increase is higher when the angle between the supporting pads ψ_{ab} is larger.

The normal reaction of the leading pad (pad *a*) depends to a large degree on the location of the trailing pad (pad *b*). The increase of μ_{sp} may lead to a significant increase of F_{Na} (when $\psi_b \approx \varepsilon_{xy}$) as well as to its significant decrease (when $\psi_b \ll \xi_{xy}$). It follows from Figure 5.59 that under such conditions, the loss of stability can occur due to separation of the leading supporting pad from the bore surface, that is,

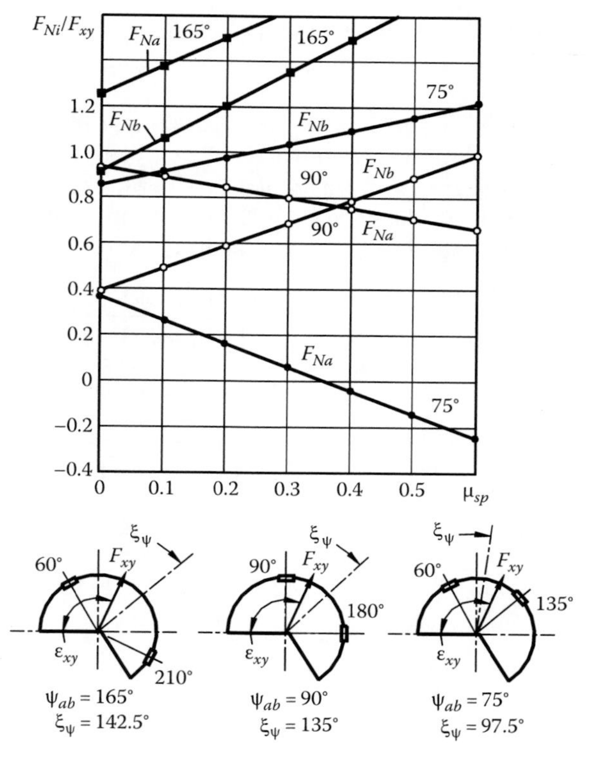

FIGURE 5.59 The relationship $F_{Ni}/F_{xy} = f(\mu_{sp})$ for different locations of the supporting pads.

when $F_{Na}/F_{xy} < 0$ even though the direction of the resultant force F_{xy} is located in the limits of the included angle ψ_{ab}, that is, the drill should be stable according to the existent stability concept.

Analysis of the force model shown in Figure 5.57b revealed that when $\mu_{sp} \neq 0$, the condition of the pads' equal load can be achieved under their unsymmetrical location relative to the direction of F_{xy}. In this case, the system becomes self-stabilizing when the optimum location angles are

$$\psi_{a-opt} = \frac{\pi}{2} + \varepsilon_1 - \arccos q_{na} + \chi \tag{5.5}$$

$$\psi_{b-opt} = \frac{\pi}{2} + \varepsilon_1 + \arccos q_{nb} + \chi \tag{5.6}$$

where q_{na} and q_{nb} are pads' parameters

$$q_{na} = \frac{F_{Na}}{F_{xy}} \quad \text{and} \quad q_{nb} = \frac{F_{Nb}}{F_{xy}} \tag{5.7}$$

When a gundrill is self-stabilizing, the condition of equal pads load is

$$q_{na} = q_{nb} = \frac{0.5}{u_{sp}} \tag{5.8}$$

and the parameter u_{sp} depends upon the central angle between the pads, ψ_{ab}, as shown in Figure 5.60. χ is the angle of action of the resultant force R_{xy} (Figure 5.57b):

$$\chi = \arcsin\left[\frac{F_{Nb}}{F_x}\mu_{sp}\cos\varepsilon_1 \sin\psi_{ab}\right] \tag{5.9}$$

Figure 5.61 shows the example of determination of the optimum location angles of the supporting pads under given design conditions. For comparison, the experimental points are also shown in this figure.

The proposed methodology can be further simplified for some particular cases. One of these happens when it is desired that $\psi_b = 180°$ for the so-called *micable* drill, that is, the diameter of which can be measured with a micrometer over the margin and supporting pad *b*. As soon as the location of one of the supporting pads is fixed, the force model is significantly simplified. For the considered case shown in Figure 5.62, the location angle of the supporting pad *a* is then determined as

$$\psi_{1a} = 2\arctan\frac{(1-\mu_{sp}) - k_F(1 - m_d/r_{dr})(1+\mu_{sp})}{(1+\mu_{sp}) - k_F(1 - m_d/r_{dr})(1-\mu_{sp})} \tag{5.10}$$

where $k_F = F_y/F_x$.

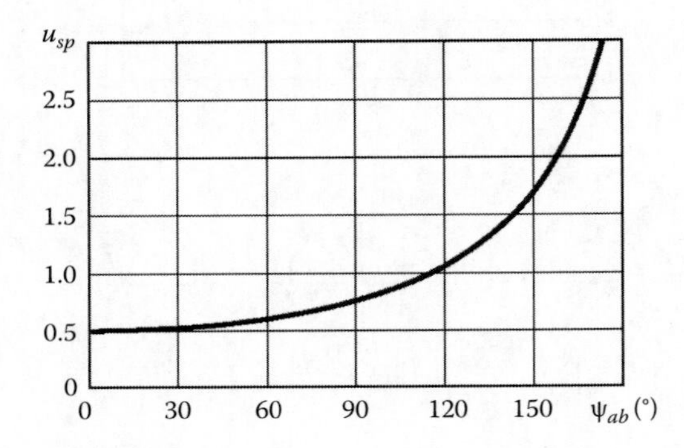

FIGURE 5.60 The relationship $u_{sp} = f(\psi_{ab})$.

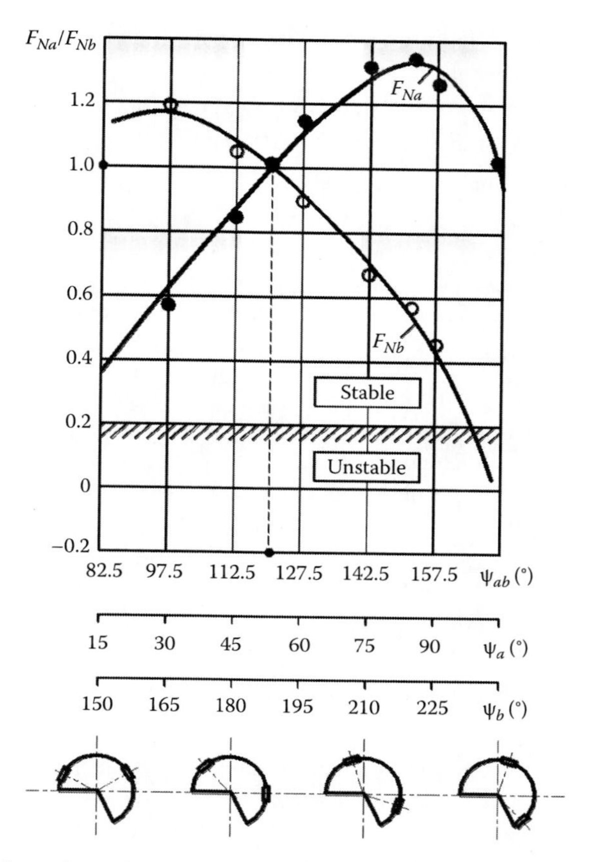

FIGURE 5.61 Finding the optimum location angles of the supporting pads under given $\psi_{ab} = 135°$, $\mu_{sp} = 0.15$. Experimental points are shown for the 8 mm dia. ASTVIK 15-03-02 design gundrill. The gundrilling regime: spindle rotational speed of 6800 rpm, feed rate of 850 mm/min, and MWF flow rate of 12 L/min.

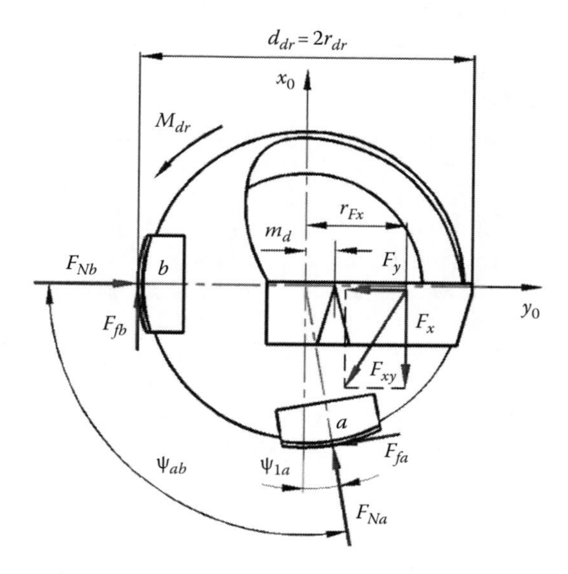

FIGURE 5.62 A model to determine the location angle of supporting pad *a*.

TABLE 5.1
Results of the Pad Location Angle Calculation Using Equation 5.7

m_d (mm)	ψ_{1a}	ψ_{ab}
$0.10d_{dr}$	29°40′–27°30′	119°40′–117°30′
$0.15d_{dr}$	16°30′–15°20′	122°–120°40′
$0.20d_{dr}$	18°–17°30′	126°–125°

As follows from Equation 5.10, the location angle ψ_{1a} increases with friction coefficient μ_{sp} and the ratio k_F becomes smaller when the distance m_d increases.

Table 5.1 shows the results of calculations of the pad location angle using Equation 5.10 with an STS drill of 26 mm diameter. For the case of drilling AISI steel 1040 (Bescrovny 1984), $k_F = 0.10 - 0.33$ and $\mu_{sp} = 0.3$ were adopted using the experimental data. To verify the obtained results experimentally, a series of the tool life tests was carried out with an STS 26 mm diameter drill having $m_d = 0.10d_{dr}$ and $\psi_{ab} = 118°30'$. The work material was AISI 1040 steel (HB170) and the depth of drilling was 1650 mm. The experimental results showed that the wear of the front ends of supporting pads was uniform amounting 1.4–1.7 mm, which is close to that of the side margin.

5.2.4.3 Location Accuracy of the Supporting Pads

The location accuracy of the supporting pad is defined by two sets of factors, namely, external and internal. The internal factors include design features and manufacturing inaccuracies in tool manufacturing. The external factors include all other factors in the deep-hole machining system that can affect the accuracy of location. The problem is complicated so only its application side is considered in this section.

Regardless of a particular type of SPT, the length of a supporting pad, l_{sp}; its width, b_{sp}; back taper; and the diameter, d_{sp}, of the cylinder on which these pads locates define the apparent contact area between the pad and the supporting surface. The actual contact area depends also on the starting bushing diameter, d_{sb} (or the pilot hole diameter when used instead of the starting bushing). The diameters d_{sp} and d_{sb} determine the theoretical clearance, $\delta_{sb} = (d_{sb} - d_{sp})/2$, in the starting bushing as shown in Figure 5.63a.

5.2.4.3.1 Transverse Direction (x_0y_0 Plane)

While the drill is in the starting bushing and no force is applied yet, the theoretical drill location in the starting bushing (or in the pilot hole prepared earlier in the previous operation) is as shown in Figure 5.63a. When drilling starts, the cutting force F_{xy} (Figure 5.62) causes the drill head (tip) radial shift until it touches the piloting surface by its supporting pads (Astakhov 2002b). As such, the theoretical contact area becomes a line because the diameter of the starting bush is always greater than that of the supporting pad as seen in Figure 5.63b. The actual contact area, the maximum contact stresses, $\sigma_{a\text{-}max}$ and $\sigma_{b\text{-}max}$, and their distributions depend on the clearance in the starting bushing, Δ_{sb}, the angle between the supporting pads, ψ_{ab}, and their nominal widths, b_a and b_b.

Analysis of the model of relative location of the supporting pads and the starting bushing shown in Figure 5.63b allows introducing a compound criterion of location accuracy of SPTs in the starting bushing. This criterion is termed as the contact angle, α_{ct}. Its meaning directly follows from Figure 5.63b. As can be seen,

$$\alpha_{ct} = 0.5(\nu_1 - \nu_2) \tag{5.11}$$

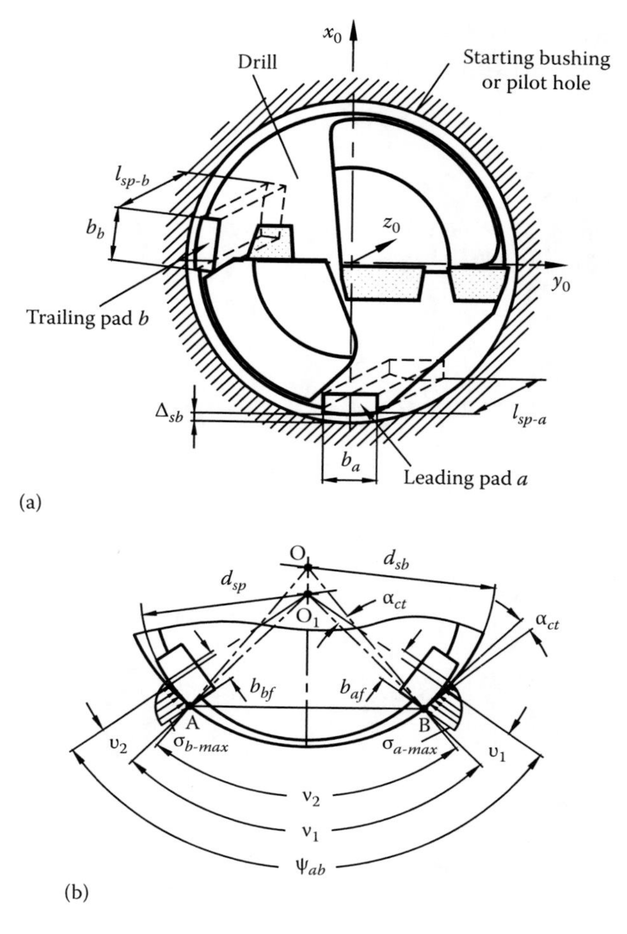

FIGURE 5.63 Model of the SPT tool location in the starting bushing: (a) initial location and (b) model for calculating the pressure angle.

The sense of angles ν_1 and ν_2 can be seen in the model shown in Figure 5.63b. It follows from this model that angle ν_1 can be represented as

$$\nu_1 = \psi_{ab} - (\upsilon_1 + \upsilon_2) \tag{5.12}$$

where

$$\upsilon_1 = \arcsin \frac{b_a}{d_{sp}} \tag{5.13}$$

and

$$\upsilon_2 = \arcsin \frac{b_b}{d_{sp}} \tag{5.14}$$

Normally, the leading and trailing supporting pads are made of the same width, that is, $b_a = b_b = b_{sp}$. When this is the case, Equation 5.12 becomes

$$\nu_1 = \psi_{ab} - 2\arcsin\frac{b_{sp}}{d_{sp}} \tag{5.15}$$

As can be seen in Figure 5.63b, chord AB is sheared by two triangles, namely, ABO_1 and ABO. From triangle ABO_1, it follows that

$$AB = d_{sp}\sin\left(\frac{\psi_{ab}}{2} - \arcsin\frac{b_{sp}}{d_{sp}}\right) \tag{5.16}$$

From triangle ABO, it follows that

$$AB = d_{sb}\frac{\sin\nu_2}{2} \tag{5.17}$$

Combining Equations 5.16 and 5.17, one can obtain

$$\nu_2 = 2\arcsin\left(\frac{d_{sp}}{d_{sb}}\sin\left(\frac{\psi_{ab}}{2} - \arcsin\frac{b_{sp}}{d_{sp}}\right)\right) \tag{5.18}$$

Substituting Equations 5.15 and 5.18 into Equation 5.11, one can obtain

$$\alpha_c = 0.5\left(\psi_{ab} - 2\arcsin\frac{b_{sp}}{d_{sp}} - 2\arcsin\left[\frac{d_{sp}}{d_{sb}}\sin\left(\frac{\psi_{ab}}{2} - \arcsin\frac{b_{sp}}{d_{sp}}\right)\right]\right) \tag{5.19}$$

The relative influence of the parameters of Equation 5.19 on the location accuracy was determined by the relative change of the contact angle, $\Delta\alpha_{ct-i}$ with incremental changes $\Delta\psi_{ab-k}$, Δb_{sp-j}, and $(\Delta(d_{sp}/d_{sb}))_l$ as

$$\Delta\alpha_{ct\text{-}i} = \left|\frac{\alpha_{ct-i} - \alpha_{ct-1}}{\alpha_{ct-1}}\right| \cdot 100\% \tag{5.20}$$

$$\Delta\psi_{ab\text{-}k} = \left|\frac{\psi_{ab\text{-}k} - \psi_{ab\text{-}1}}{\psi_{ab\text{-}1}}\right| \cdot 100\% \tag{5.21}$$

$$\Delta b_{sp\text{-}j} = \left|\frac{b_{sp\text{-}j} - b_{sp\text{-}1}}{b_{sp\text{-}1}}\right| \cdot 100\% \tag{5.22}$$

$$\Delta\left(\frac{d_{sp}}{d_{sb}}\right)_l = \left|\frac{(d_{sp}/d_{sb})_l - (d_{sp}/d_{sb})_1}{(d_{sp}/d_{sb})_1}\right| \cdot 100\% \tag{5.23}$$

where $\Delta\alpha_{c\text{-}i}$, $\Delta\psi_{12\text{-}k}$, $\Delta b_{sp\text{-}j}$, and $(d_{sp}/d_{sb})_l$ are current i, k, j, and l values of the corresponding parameters and $\alpha_{c\text{-}1}$, $\psi_{12\text{-}1}$, $b_{sp\text{-}1}$, and $(d_{sp}/d_{sb})_1$ are certain initial values of these parameters.

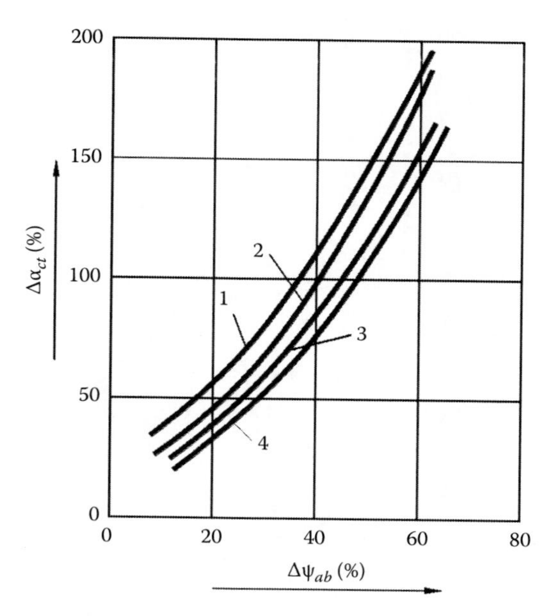

FIGURE 5.64 Effect of the angle between the supporting pads, ψ_{ab}, on the pressure angle, α_{ct}: 1—b_{sp} = 2 mm, drill/starting bushing diameters ∅8H6/g6; 2—b_{sp} = 4 mm, drill/starting bushing diameters ∅30H6/g6; 3—b_{sp} = 4 mm, drill/starting bushing diameters ∅40H6/g6; and 4—b_{sp} = 4mm, drill/starting bushing diameters ∅60H6/g6.

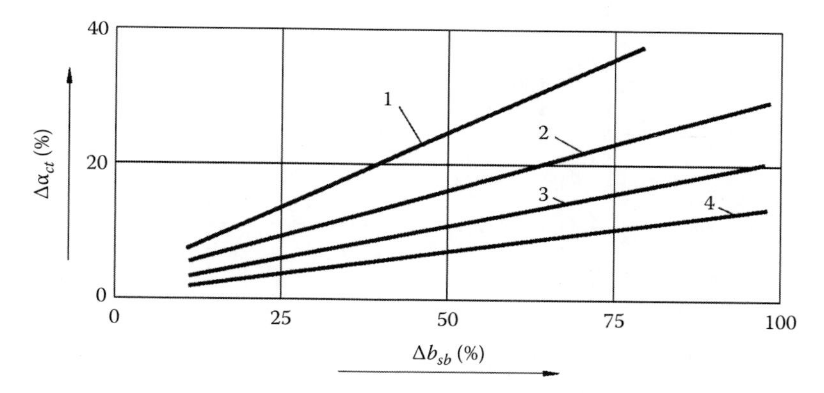

FIGURE 5.65 Effect of the width of the supporting pads, b_{sp}, on the pressure angle, α_{ct}: 1—b_{sp} = 2 mm, ψ_{ab} = 110°, drill/starting bushing diameters ∅8H6/g6; 2—b_{sp} = 4 mm, ψ_{ab} = 110°, drill/starting bushing diameters ∅30H6/g6; 3—b_{sp} = 4 mm, ψ_{ab} = 110°, drill/starting bushing diameters ∅40H6/g6; and 4—b_{sp} = 4 mm, ψ_{ab} = 110°, drill/starting bushing diameters ∅60H6/g6.

Correlations $\Delta\alpha_{ct} = f(\Delta\psi_{ab})$, $\Delta\alpha_{ct} = f(\Delta b_{sp})$, and $\Delta\alpha_{ct} = f(\Delta(d_{sp}/d_{sb}))$ are shown in Figures 5.64 through 5.66, respectively. Their analysis shows that the chief influence on the location accuracy of the drill in the starting bushing (pilot hole) has the ratio (d_{sp}/d_{sb}), that is, the clearance in the starting bushing (pilot hole). The angle between the supporting pads and of their width has lesser influence. This conclusion found its full experimental confirmation in the analyses of the entrance stability (Astakhov 2002a,b) and mechanism of hole deviation (Katsuki et al. 1987).

5.2.4.3.2 Axial Direction (z_0y_0 Plane)

The location accuracy in the z_0y_0 plane defines the actual contact length of the supporting pad. The axis of the drill tip may have one of two possible angular deviations from the axis of the starting bushing or from the theoretical axis of the hole being drilled. This deviation is defined as the

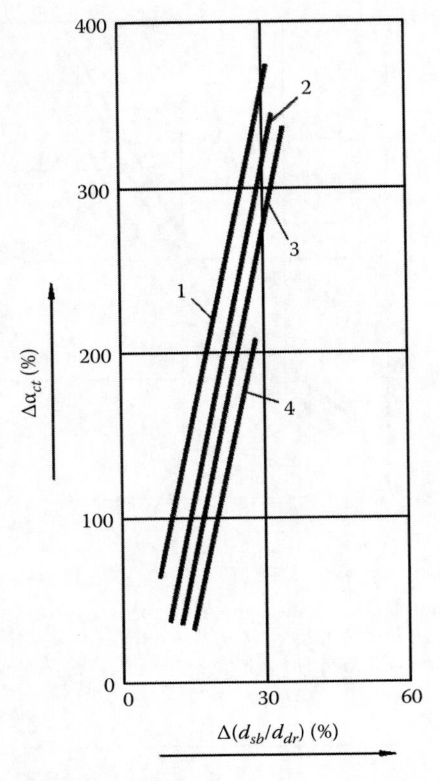

FIGURE 5.66 Effect of $\Delta(d_{sb}/d_{dr})$ on the pressure angle, α_{ct}: 1—b_{sp} = 4 mm, ψ_{ab} = 50°; 2—b_{sp} = 4mm, ψ_{ab} = 80°; 3—b_{sp} = 4 mm, ψ_{ab} = 100°; and 4—b_{sp} = 4 mm, ψ_{ab} = 120°.

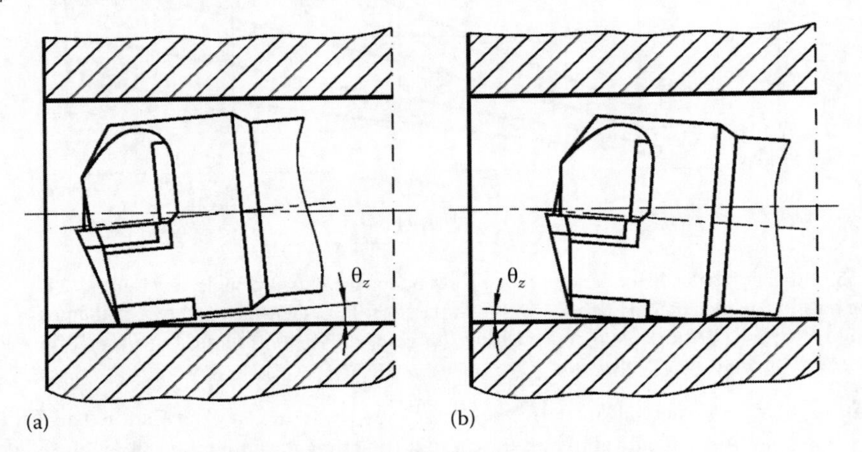

FIGURE 5.67 The (a) positive and (b) negative longitudinal contact angles.

longitudinal contact angle θ_z as shown in Figure 5.67. This angular deviation is an unavoidable factor in any SPT drilling because of the following:

- Two additional bending moments discussed in Section 5.2.3 are always the case for any tool design. These moments cause drill bending
- Linear and/or angular misalignment between the axis of the drill tip and that of the starting bushing that is always the case in real deep-hole machining systems
- Installation error for detachable drill heads.

The angle of the axial deviation, θ_z, due to the listed causes can be represented as the sum

$$\theta_z = \theta_1 + \theta_2 + \theta_3 \tag{5.24}$$

where

- θ_1 is the angular displacement due to the additional bending moments caused by the cutting forces
- θ_2 is the angular displacement due to the overall misalignment (including its radial and angular components)
- θ_3 is the angular displacement due to clamping when a detachable drill head is used.

When the longitudinal contact angle θ_z is positive, the excessive wear of the front ends of the supporting pad can be observed. When this angle is negative, the excessive axial force and the vibration of the gundrilling system are the results. In practice, however, due to the action of the addition bending moments shown in Figure 5.49 and due to the back taper, the contact angle θ_z is often positive.

The derivation of the complete model to determine the longitudinal contact angle θ_z is rather bulky and, what is more important, impractical as too many particularities of drill design and accuracy of the components of the drilling system are involved. Having noticed nonuniform pad wear in the x_0y_0 and z_0y_0 plane, a number of solutions were proposed to reduce this kind of pad wear. They are discussed next.

5.2.4.3.2.1 Supporting Pad Design Geometry to Reduce Their Nonuniform Wear

Improving the design/geometry of the SPT supporting pads aims both to reduce their nonuniform wear (and thus increase the tool life) and to improve machined surface integrity. There are a number of ways to achieve these objectives.

In the author's opinion, the simplest and the most effective way to prevent edge contact shown in Figure 5.63b is offered by US Patent No. 4,596,498 (1986). According to this patent (Figure 5.68), the first and the second supporting pads are provided with cylindrical outer surfaces having radii of curvature, ρ_{sp1} and ρ_{sp2}, less than that of the bore. Further, the location and radii of curve of the supporting pads are arranged so that the contact with the bore wall is established in the central portion of the circumferential surface rather than at either the leading or trailing edges of the supporting pads. Moreover, besides reducing wear of the supporting pads and improving surface integrity of the machined bores, such a design and arrangement allows to use various coatings on the supporting pads without danger that coating will break off or wear rapidly due to edge contact.

US Patent No. 3,751,177 provides an improvement in pad location by means of a supporting (guide) pad mounted on a pivot so that it can rock and adapt itself to the surface of the hole, thereby

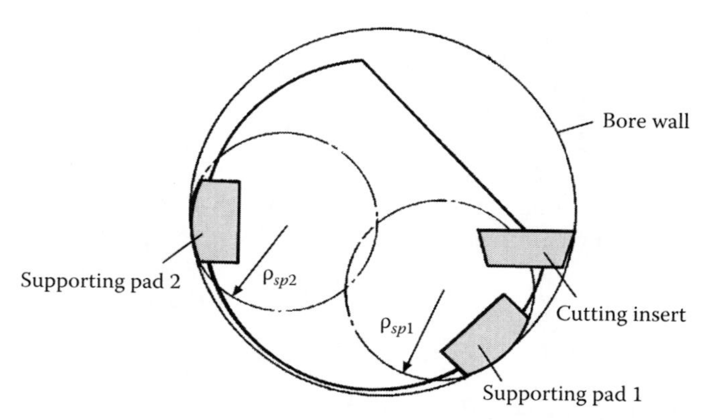

FIGURE 5.68 Pad design to prevent edge contact according to US Patent No. 4,596,498 (1986).

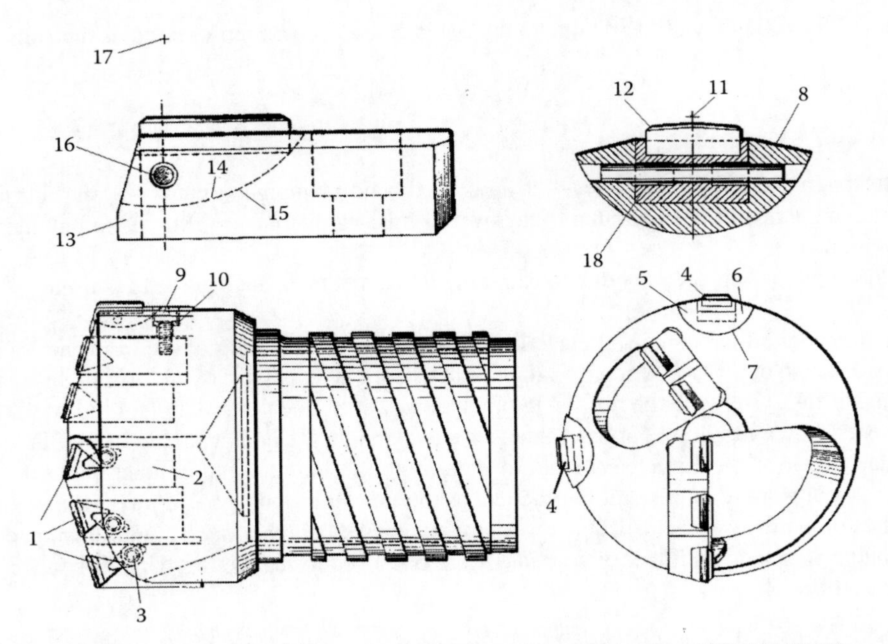

FIGURE 5.69 Pad design to prevent edge contact according to US Patent No. 3,751,177 (1973).

securing a full surface contact. The drill shown in Figure 5.69 has a number of cutting inserts (1) locked in supports (2) by clamps (3). The supports (2) are attached to the drill in a suitable way, for example, by screws or by brazing. It is provided with two supporting (guide) pads (4) for supporting and guiding it in the hole. The pads are mounted on a pivot so that they can rock around both a longitudinal (the z_0y_0 plane) and a transverse (the x_0y_0 plane) axes. The pivoting around the longitudinal axis is obtained by means of a support (5) with a cylindrical sliding surface (6), which rests on the matching surface (7) in a depression made in the drill body. The periphery surface (8) of support (5) is approximately of the same diameter as the drill body. The support is fastened with a screw (9) and a resilient washer (10) permitting rocking of the support.

The pivot axis (11) lies outside the pad that prevents the leading edge (12) of the pad from contacting the bore wall while drilling. This lessens the friction and facilitates the entering of the lubricant between the pad and the bore walls. By suitable position of the pivot axis (11), the contact pressure can be properly distributed.

The pivoting on the transverse axis is obtained by a support (13) having a cylindrical surface (14) sliding along a matching surface (15) in the support (5). The pad is held by a tubular pin (16) within the bore through both supports (5 and 13). As the axis of the pin lies outside the pivot center (17), the rocking requires bending of the pin. This is made possible by an annular space (18) extending along a part of the pin (16). The pivot mounting around two perpendicular axes provided by the two described cylindrical sliding joints may be obtained by a single spherical joint.

Figure 5.70 shows the pad shaping according to US Patent Application 2010/0158623 (2010). As can be seen, the pad shape is arranged to its best performance at the entrance of the hole being drilled (Astakhov 2002b) providing generous front smooth chamfer. When such a tool drills, this chamfer assures proper plastic deformation of the hole walls (Griffiths and Grieve 1985).

5.2.5 Gundrills

5.2.5.1 History

Present-day deep gundrilling has its origin in the firearm industry of the eighteenth century. During the period between years 1500 and 1750, the town of Suhl in Germany was known as the center of

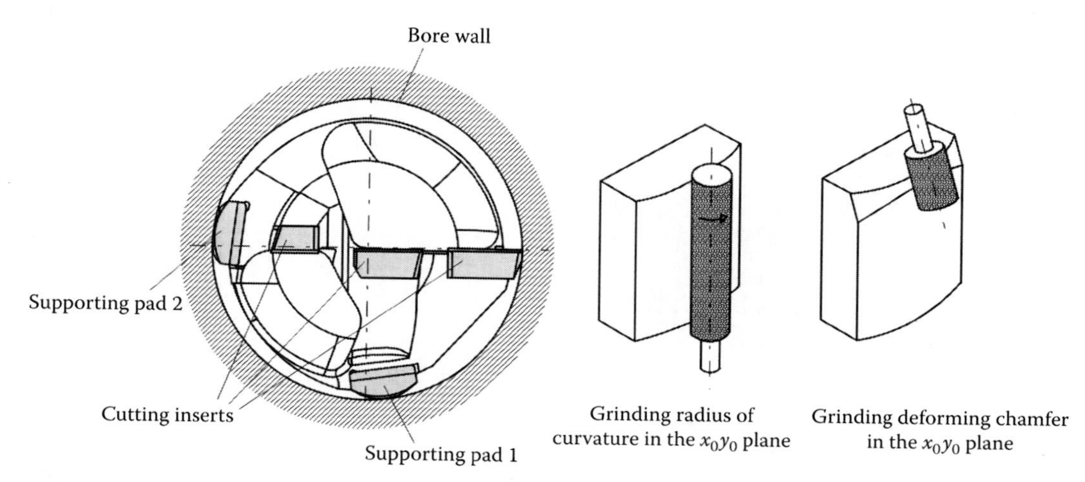

FIGURE 5.70 Supporting pad shaping according to US Patent Application 2010/0158623 (2010).

deep-hole drilling. In such a drilling, a water mill was used as the machine tool and a spade drill bit was the cutting tool. Two barrels could be drilled simultaneously by two parallel boring spindles. The feed and the thrust were provided by the operator actuating a lever, which supplied certain force amplification. Jean Maritz developed an advanced horizontal boring machine in 1754. A drilling head had been mounted on the end of a boring bar that was rotated by animal power and a downward feed motion was applied to the gun barrel. The first boring machine in which the workpiece was rotated and feed motion was given to the drilling tool appears to have been used about 1758 by J. Verbruggen in Germany.

A major survey of the methods and equipment available for making guns was published in 1794 by G. Monge's in his book, *L'Art de Fabriquer les Canons* at the investigation of the Committee of Public Safety of the French Revolution. This survey shows that these earlier-date SPTs were in wide use, albeit primarily for finishing gun barrels.

An article published in 1886 by Landis describing gun barrel manufacture shows that by this time, not only was there a variety of SPTs available but the refinement of pressurized cutting oil systems was employed (Landis 1886). SPTs illustrated in this paper are remarkably similar in design to present-day tools. He even showed a tool with replaceable supporting pads and an indexable cutting edge. This very detailed and comprehensive article shows that the advantages of SPTs had been recognized, and they were employed to improve the efficiency of gun manufacturing for centuries. Although a number of improvements have been made since then, for example, the application of tungsten carbide and specially blended MWF (coolants), it remains a fact that the foundation of modern deep-hole machining was laid by the turn of the twentieth century. Should patent experts have read Landis' paper, more than a half of the existing patents on 'new' designs of gundrills would not have been granted in the author's opinion.

5.2.5.2 Basic Design and Geometry

The basic design and geometry parameters of a commonly used gundrill are shown in Figure 5.71. The gundrill consists of a drill body having a shank (1) and a tip (2). The tip is made up of a hard wear-resistant material such as tungsten carbide. The other end of the shank incorporates an enlarged driver (3) having the machine-specific design. The shank is of tubular shape having an elongated passage (4) extending over its entire length and connecting to the MWF supply passage (5) in the driver. The shank has a V-shaped flute (6) on its surface that serves as the chip removal passage. The shank length depends mainly on the depth of the drilled hole as well as on the lengths of the bushing and its holder, chip box, etc., so it is determined from the tool layout.

The tip is larger in diameter than the shank that prevents the shank from coming into contact with the walls of the hole being drilled. Flute (7) on the tip, which is similar in shape to flute (6),

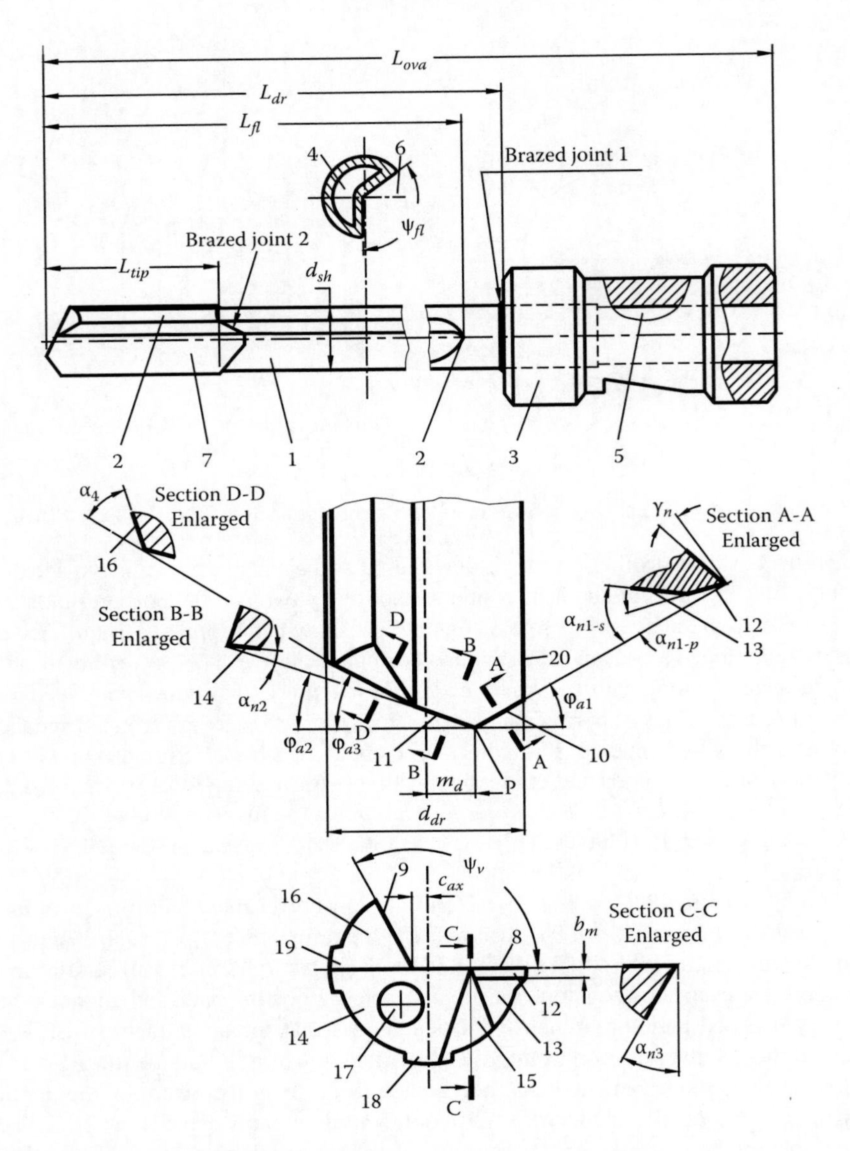

FIGURE 5.71 Geometry of a common gundrill.

extends along the full length of the tip. This flute is bounded by side faces (8 and 9) known as the cutting face and side face, respectively. The depth of this flute is such that the cutting face (8) extends past the axis (distance c_{ax}) of the tip, which is also the axis of the drill body. The angle ψ_v between the side and cutting faces is known as the profile angle of the tip, which is usually equal or close to the V-flute profile of the shank.

The terminal end of the tip is formed with the approach cutting edge angles φ_{a1} and φ_{a2} of the outer (10) and inner (11) cutting edges, respectively. These cutting edges meet at the drill point. The location of the drill point (defined by the distance m_d in Figure 5.71) can be varied for optimum performance depending on the work material and the finished hole specifications. One common point grind calls for the outer angle, (φ_{a1}), to be 30° and the inner angle, (φ_{a2}), to be 20°. The geometry of the terminal end largely determines the shape of the chips and the effectiveness of the cutting fluid, the lubrication of the tool, and removal of the chips. The process of chip formation is also governed by other cutting parameters such as the cutting speed, feed rate, and work material.

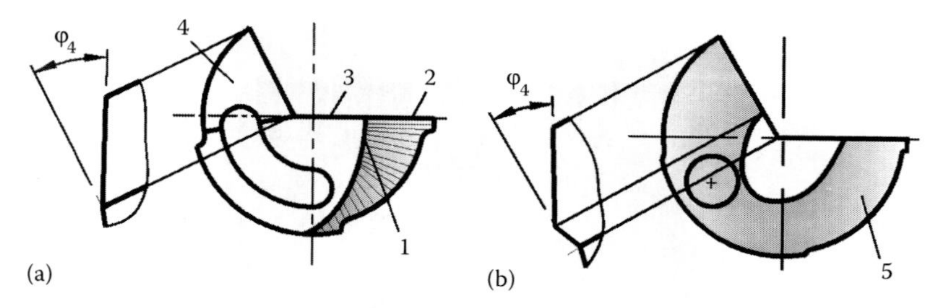

FIGURE 5.72 Alternative grinds of the flank surfaces: (a) combined helical and planar and (b) helical.

The prime flank surface (12) having normal primary flank angle $\alpha_{n1\text{-}p}$ of 7° to 10° is applied to the other cutting edge (10). To assure drill-free penetration, the secondary flank surface (13) having normal flank angle $\alpha_{n1\text{-}s}$ of 12° to 20° is applied as shown in Figure 5.71. Flank surface (14) having normal flank angle α_{n2} of 8° to 12° is applied to the inner cutting edge (11). To assure drill-free penetration, that is, to prevent the interference of the drill's flanks with the bottom of the hole being drilled, the auxiliary flank (15) (normal flank angle α_{n3}) and shoulder dub-off (16) (flank angle φ_4) are provided. Their location and geometry are uniquely defined for a given gundrill.

Another common shape of the flank surface is a helical surface rather than a planar surface. The helical flank surface is normally applied to the flank of the outer cutting edge, while other flanks are planar. Different manufacturers have different standards on the lead and generating diameter of the helical flank surface, depending upon drill diameter and design of the grinding fixture.

Modern designs use great lead and generating diameters so that the flank surface of the outer cutting edge does not affect the shape of the flank plane of the inner cutting edge or the shoulder dub-off as shown in Figure 5.72a (Zhang et al. 2004). The lead and generating diameter of this surface is relatively large, so the line, or rib, of intersection (1) of outer (2) and inner (3) flank surfaces does not extend too far from the vertical axis. Therefore, shoulder dub-off (4) is not affected by the helical surface. In older designs, for example, US Patent No. 2,325,535 (1943), which is still in use, a relatively small lead and generating diameter of the helix surface is used so that this surface passes through the outer flank and shoulder dub-off as shown in Figure 5.72b. In this figure, a helical convex surface (5) is applied to the outer cutting edge. This helical convex surface has a small lead and generating diameter so that it passes through a flank surface of the inner cutting edge.

5.2.5.3 Chip Breaking

Among deep-hole drills, gundrills are the only drills without chip-breaking steps ground on the rake face. This makes re-sharpening of gundrills much simpler and faster using standard grinding wheels and fixtures often even by a gundrilling machine operator. This simplicity, however, comes at some cost. The rake face with no chip-breaking step is possible because the chip formed at the inner cutting edge should impinge on the chip formed by the outer cutting edge and thus should serve as an obstacle chip breaker. In other words, the collision of these two chip flows should result in the formation of the so-called backbone at their interface which, colliding with the rotating bottom of the hole being drilled (or, at worst, the sidewalls of the hole being drilled), causes the breakage of such a combined chip. This point is illustrated in Figure 5.73.

To achieve reliable chip breaking, the following parameters can be varied: distance m_d (Figure 5.71) that defines the width of the chips formed by the outer and inner cutting edges; the approach cutting edge angles φ_{a1} and φ_{a2} (Figure 5.71) that define the chip thickness (under a given cutting feed), the interaction angles of the chips formed by the outer and inner cutting edges; the cutting feed (feed per revolution) that defines the chip thickness; the cutting speed that defines the temperature of the formed chip; and the shoulder dub-off angle φ_4 that defines the interaction conditions of the chips and MWF (Astakhov 2010). In the author's experience, reliable chip breaking

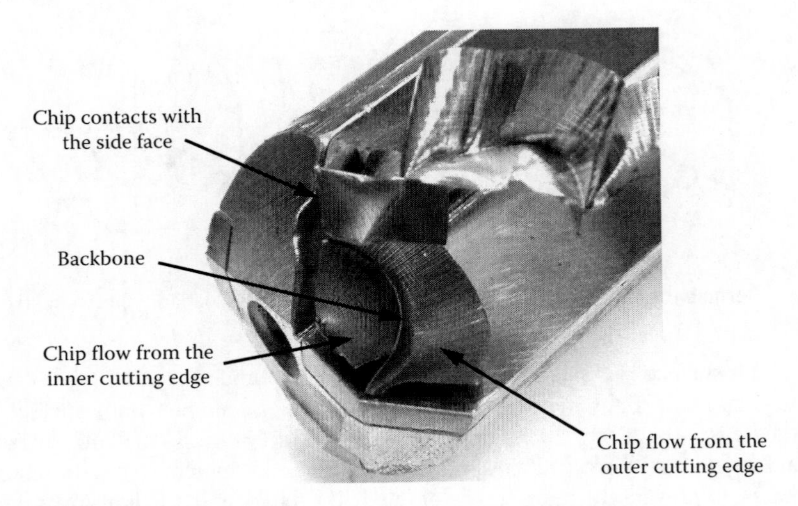

FIGURE 5.73 Interaction of the chip flows from the outer and inner cutting edges.

FIGURE 5.74 Improved tool geometry.

can be practically always achieved (besides few exceptions) if one understands the interinfluence of the listed parameters and thus varies them accordingly. For example, marginal chip breakage is achieved with the drill shown in Figure 5.73 because the chip flow from the outer cutting edge is much stronger so that the intersection rib collides with the side of the chip removal flute. To solve the problem, distance m_d was increased so that reliable chip breaking was achieved due to the interaction of the discussed rib with the bottom of the hole being drilled as shown in Figure 5.74.

Unfortunately, this is not always the case so some means enhancing chip breaking/separation are added to the basic gundrill design. These can be broadly divided in two groups. The chip-breaking/separation features in the design and geometry of drills of the first group are applied on drill manufacturing so that there is no need to reproduce them on drill re-sharpening. Figure 5.75 shows an example of the drill design developed in the 1930th with the chip-separating steps applied to the rake face of the outer cutting edge. All the design/geometries of the drills of the first group are just variations of this principle.

For example, Figure 5.76 shows the so-called nicked gundrill according to Japanese Patent Laid-Open No. 2007-50477. As can be seen, the drill has the driver (1) and the shank (2). This nicked drill is configured so that each cutting edge has a plurality of concave groove-like nicks (3, 4, and 5 as shown in Figure 5.76) arranged, so on drilling they divide the chip in the width direction. The disadvantages of such a design, for example, possible undesirable interaction of chip flows from nicks, were listed in the description of US Patent Application Publ.

FIGURE 5.75 Chip-separating steps applied to the rake face of the outer cutting edge.

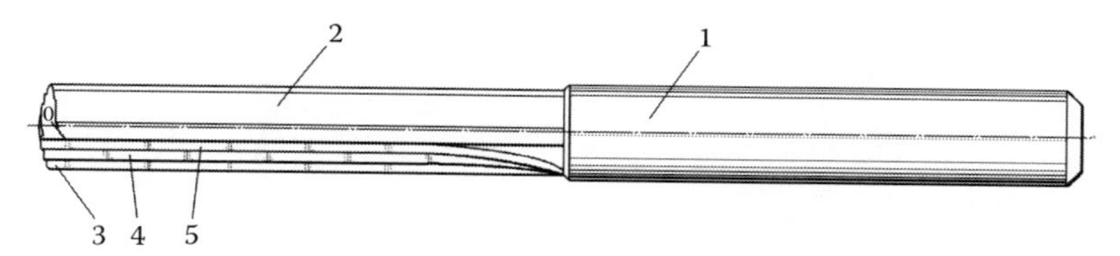

FIGURE 5.76 Design of gundrill according to Japanese Patent Laid-Open N 2007-50477.

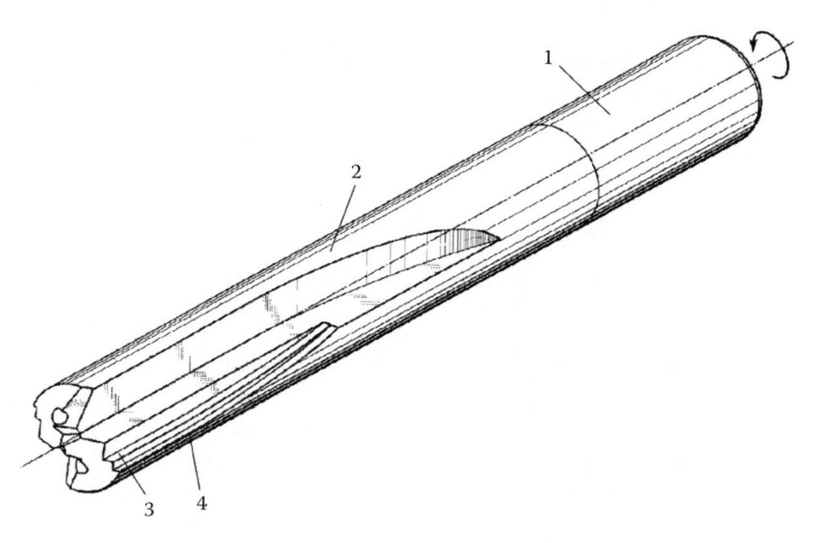

FIGURE 5.77 Drill design according to US Patent Application Publ. No. US 2012/0063860.

No. US 2012/0063860, which offers a modified design shown in Figure 5.77. The gundrill in this figure has the driver (1) and shank (2). The tool has a plurality of rake faces (e.g., 3 and 4 shown in Figure 5.77) arranged in a steplike manner and spaced in the front and rear direction of the tool rotation direction. The inventors believe that the chip is segmented into strips having narrow width that improve chip controllability.

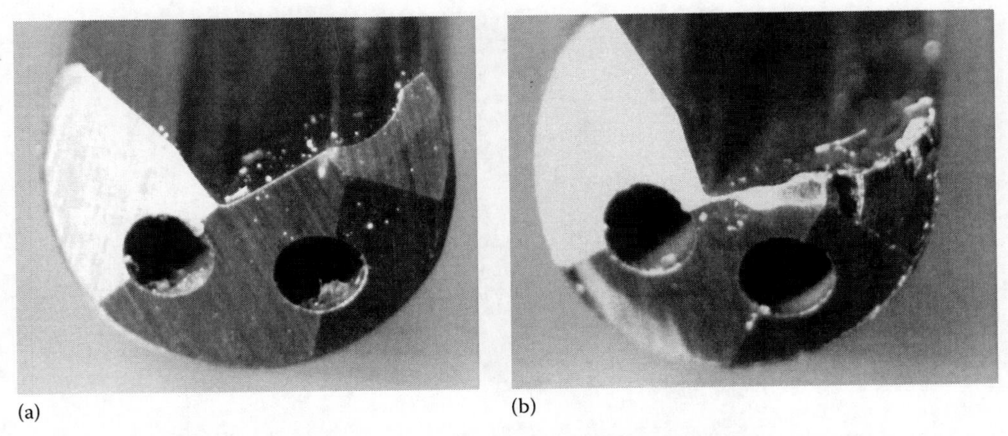

FIGURE 5.78 Chip-separating groove on the gundrill's rake face: (a) new drill and (b) wear pattern.

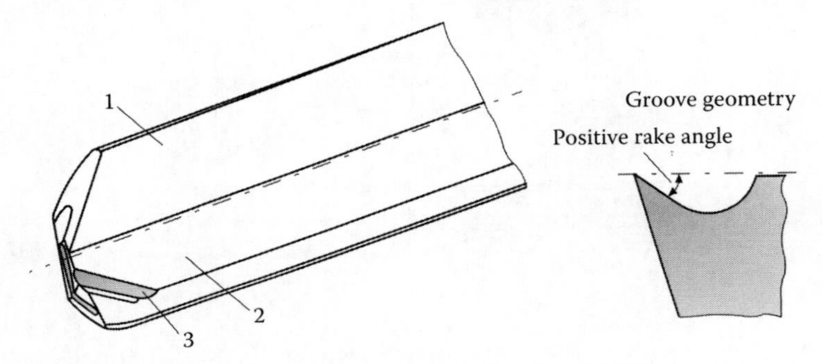

FIGURE 5.79 Gundrill tip geometry according to US Patent No. 7,753,627 (2010).

According to the author's experience, however, application of the grooves on the gundrill's rake face is not that effective as such grooves only separate the formed chip in a number of flows providing no means for its breakage. Moreover, in machining of difficult-to-machine work materials, grooves on the rake face lower tool life. As an example, Figure 5.78 shows chip-separating groove on the gundrill's rake face. As can be seen in Figure 5.78b, deep wear notches form at the edges of the groove lowering tool life.

The chip-breaking/separation features in the design and geometry of drills of the second group are applied on drill manufacturing as well as on each re-sharpening so that the gundrill losses its re-sharpening simplicity. As an example, Figure 5.79 shows the gundrill tip geometry according to US Patent No. 7,753,627 (2010) where the tip (1) having the rake face (2) is provided with chip-breaking groove (3). The advantages of such gundrills are reliable chip breaking and cutting with positive rake angle. The latter enhances tool life by reducing the cutting force/power and also makes it suitable for NDM machining (discussed in Chapter 6). The price to pay is the necessity to restore the grove geometry with high accuracy on each successive tool regrind so that a special grinding fixture and wheel are needed. This is not a problem, however, when a CNC point grinding machine is used (which actually should be the case with HP gundrills). Another drawback is that much greater amount of the tool material has to be removed on regrinding to restore the chip-breaking flute, which lowers the total number of regrinds. This should be compensated with increased tool life between regrinds.

5.2.5.4 Common Recommendations on the Selection of Tool Geometry/Design Parameters

Gundrills with the planar flank faces (Figure 5.71) became the most popular in modern manufacturing for drilling holes of 3–30 mm diameter. The recommended parameters of the geometry of

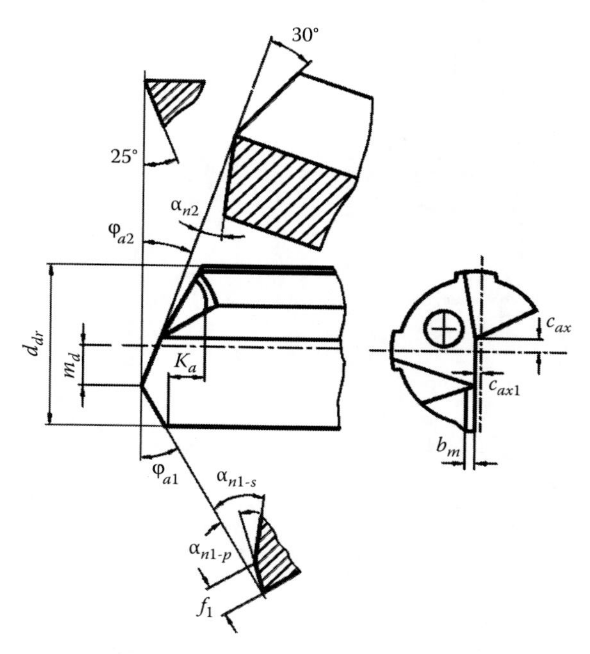

FIGURE 5.80 Parameters of a gundrill with planar flanks.

TABLE 5.2
Standard Parameters of Gundrills with Planar Flanks

Work Material	Hardness HB	m_d (mm)	φ_{a1} (°)	φ_{a2} (°)	$\alpha_{n1\text{-}p}$ (°)	$\alpha_{n1\text{-}s}$ (°)	α_{n2} (°)
Steel	To 240	$0.20d_{dr}$	45	20	7	15	20
	240–320	$0.25d_{dr}$	35	25	7	15	20
Cast iron	120–300	$0.25d_{dr}$	30	30	6	8	15

such drills are shown in Figure 5.80 and Table 5.2. Other parameters are as follows: $c_{ax1} = 0.02d_{dr}$, $c_{ax} = 0.05d_{dr}$, $b_m = (0.02\ldots0.04)d_{dr}$, $f_1 = (0.03\ldots0.05)d_{dr}$, and $K_a \geq 0.5$ mm.

Figure 5.81 shows the geometry of a gundrill for drilling low-surface-roughness (Ra = 0.63 – 0.32) holes. Tables 5.3 and 5.4 list the values of the parameters shown in this figure.

5.2.5.5 Particularities of the Rake and Flank Geometries

The rake face shown Figure 5.80 is used in approx. 99% of gundrills due to its apparent simplicity. As such, the T-hand-S rake angle of any point of a gundrill is considered to be zero, and it does not change with re-sharpening. Moreover, because the shift of the outer and inner cutting edges with respect to the centerline (CL) (distance c_{ax1}) is small, these angles are considered to be diametral, that is, the rake angles in T-hand-S and T-mach-S are the same because the assumed, as well as the actual, vector of the cutting speed is approximately perpendicular to the cutting edge. This simplicity, however, is only apparent because there is a not well-understood issue in the region near the axis of rotation.

Although the gundrill does not have the chisel edge, an *Achilles' heel* of straight-flute and twist drills, the problem with cutting in the region close to the axis of rotation does not magically disappear. In basic designs (similar to that shown in Figure 5.71) presented in many literature sources, the cutting edge is shown as passing through the axis of rotation, so there should be no problem with cutting in the region close to the axis of rotation. One should realize that a certain tolerance on the cutting edge location with respect to CL should be allowed.

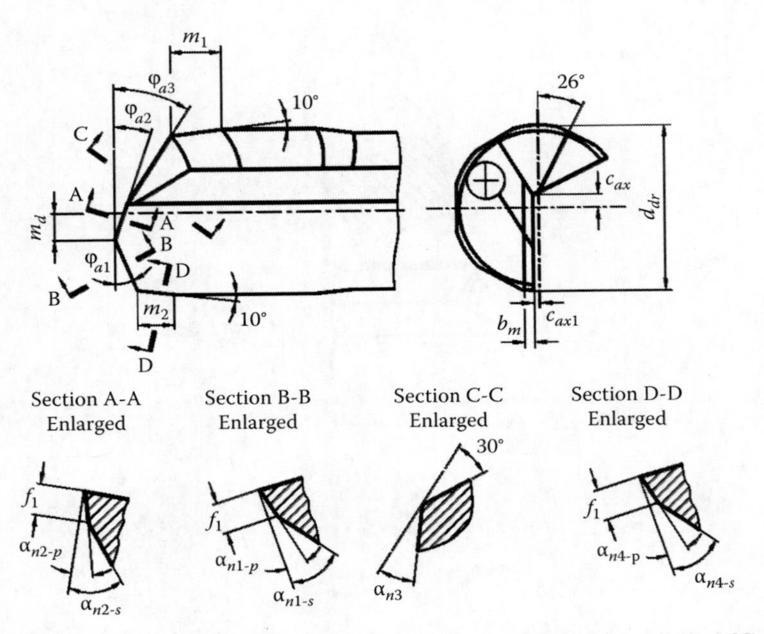

FIGURE 5.81 Parameters of a gundrill for drilling low-surface-roughness (Ra = 0.63–0.32) holes.

TABLE 5.3
Dimensional Parameters of Gundrills with Planar Flanks for Holes of Ra = 0.63–0.32 μm

Drill Diameter	m_d (mm)	m_1 (mm)	m_2 (mm)	c_{ax} (mm)	c_{ax1} (mm)	b_m (mm)	f_1 (mm)
From 3 to 5	0.7	0.5	—	0.4	0.08	0.12	—
Over 5 to 8	1.2	1.0	1.0	0.6	0.10	0.15	0.3
Over 8 to 12	2.0	1.3	1.3	0.7	0.15	0.20	0.5
Over 12 to 15	2.4	1.6	1.6	0.8	0.20	0.25	0.5
Over 15 to 20	2.8	2.0	2.0	1.0	0.25	0.30	0.6

TABLE 5.4
Geometrical Parameters of Gundrills with Planar Flanks for Holes of Ra = 0.63–0.32 μm

Work Material	Hardness HB	φ_{a1} (°)	φ_{a2} (°)	$\alpha_{n1\text{-}p}$, $\alpha_{n2\text{-}p}$, $\alpha_{n4\text{-}p}$ (°)	$\alpha_{n1\text{-}s}$, $\alpha_{n2\text{-}s}$, $\alpha_{n4\text{-}s}$ (°)	α_{n3} (°)
Steel	To 240	35	15	12	18	20
	240–320	30	20	7	16	20
Cast iron	120–300	30	20	6	8	15

Figure 5.82 shows the possible locations of the cutting edge with respect to CL (the y_0-axis). Figure 5.82a shows the theoretical location of the cutting edge where its projection into the x_0y_0 plane coincides with the projection of the y_0-axis, that is, the cutting edge passes through the center of rotation. In reality, a certain shift from its ideal location is the case. The rake face can be made so that the location of the cutting edge is as shown in Figure 5.82b. This is

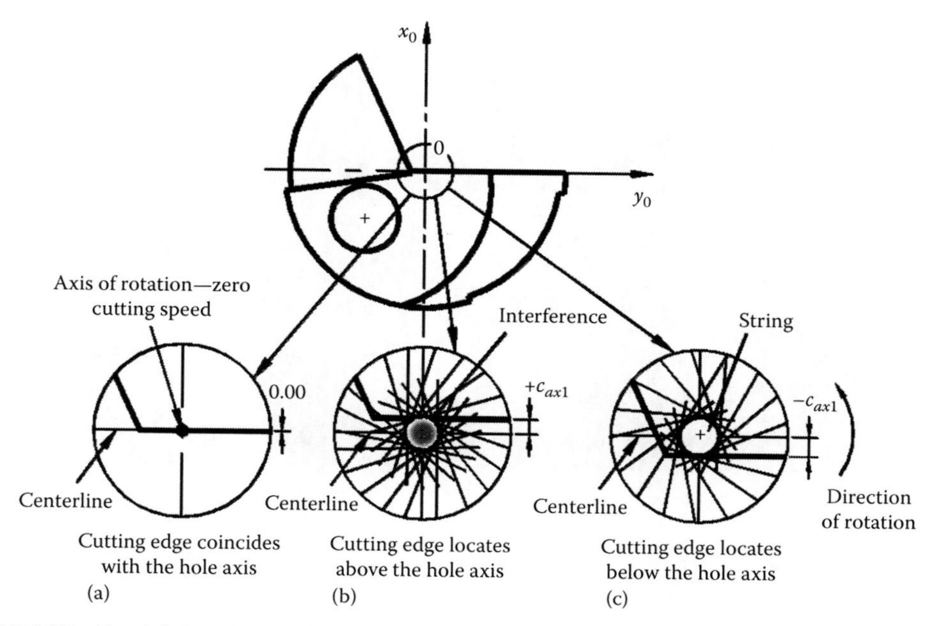

FIGURE 5.82 Possible locations of the cutting edge with respect to the CL (the y_0-axis).

the worst-case scenario because of interference between the flank surface of the drill and the bottom of the hole being drilled as clearly seen in this figure. In other words, when the cutting edge is shifted by a certain distance c_{ax1} up with respect to the y_0-axis, the drill has no means to remove the cylindrical core having the radius equal to c_{ax1} as shown in Figure 5.82b. When this distance is relatively small, the drill bends to compensate for this core. When c_{ax1} exceeds a certain threshold (depending upon drill particular drill design and parameters of the gundrilling system), the tip simply breaks.

To avoid the interference, the rake face should be always located so that the projection of the cutting edge into the x_0y_0 plane occupies a position right on or slightly below CL (the y_0-axis) as shown in Figure 5.82c and in the common recommendation discussed in the previous section (see Figures 5.80 and 5.81). When the cutting edge locates below CL, a string is formed as the result of such a location. The theoretical string diameter would be equal to $2(-c_{ax1})$. The string will be attached to the bottom of the hole being drilled as shown in Figure 5.83. Figure 5.84 shows examples of the strings. It is clear that this string is undesirable particularly when blind holes are drilled.

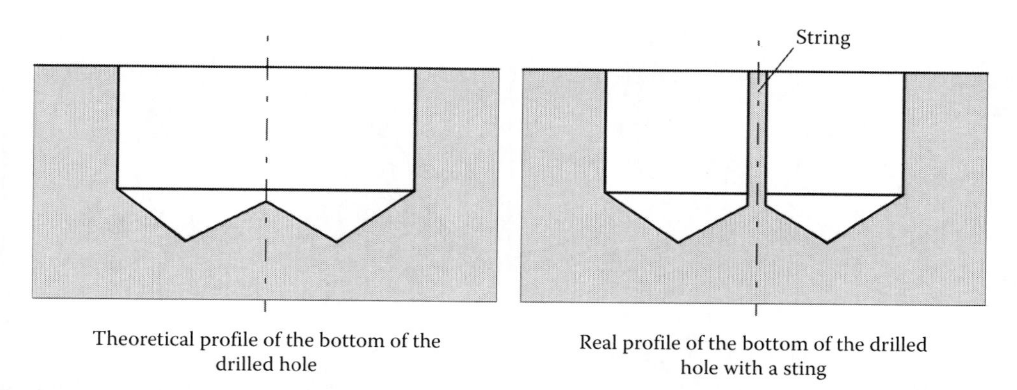

FIGURE 5.83 Theoretical and real profile of the bottom of the drilled hole.

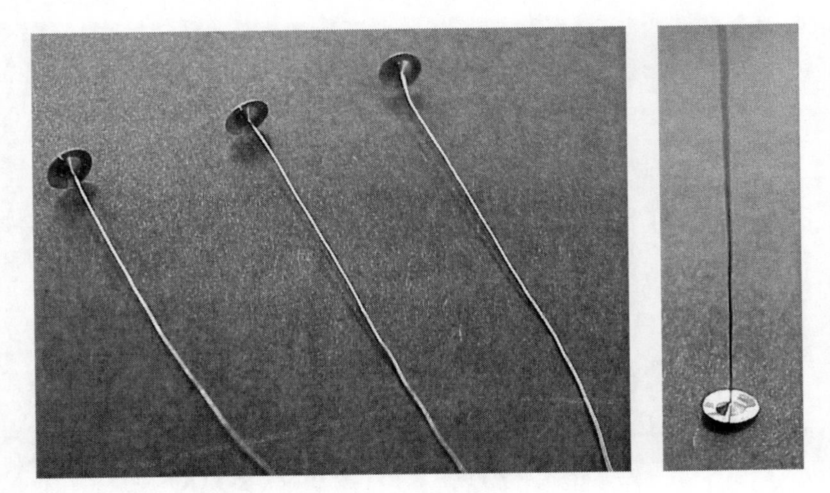

FIGURE 5.84 Strings.

It is also worthwhile to explain the role of c_{ax1} at the entrance of the hole being drilled. Figure 5.85a shows the ideal location of a drill tip in a starting bushing (no misalignment). In this figure, Δ_{sb} is the radial clearance between the tip and the bushing and c_{ax1} is the location distance of the cutting edge behind CL. It is clear that the intended (theoretical) sting diameter is calculated as $d_{st} = 2c_{ax1}$. When drilling begins, the cutting force applied to the cutting edge shifts the tip toward the bushing walls as explained earlier in Section 5.2.4. As such, when the drill rotates, the string changes it location at the entrance with respect to the center of the drilled hole (Figure 5.85b). Eventually, when the supporting part enters the hole, the drill occupies a new location and its longitudinal axis coincides with the axis of rotation. The string, however, does not change its diameter. It shifts together with the tip after the full tip entrance into the hole being drilled. A small additional force due to this shift does not normally present any problem. This is not the case, however, when the workpiece rotates and the gundrill is stationary. This case is shown in Figure 5.85c. As such, the maximum diameter of the string increases becoming equal to $d_{st} = 2(c_{ax1} + \Delta_{sb})$. If Δ_{sb} is significant (due to improper diameter of the starting bushing or due to bushing wear), the large additional force acts on the rake face that often causes tip breakage. Besides, the interference between the string and the sidewall of the V-flute may take place that affects the position of the gundrill with respect to the axis of rotation. To avoid this, the distance c_{ax} (Figure 5.85a) should not be more than $c_{ax} = 0.05d_{dr}$.

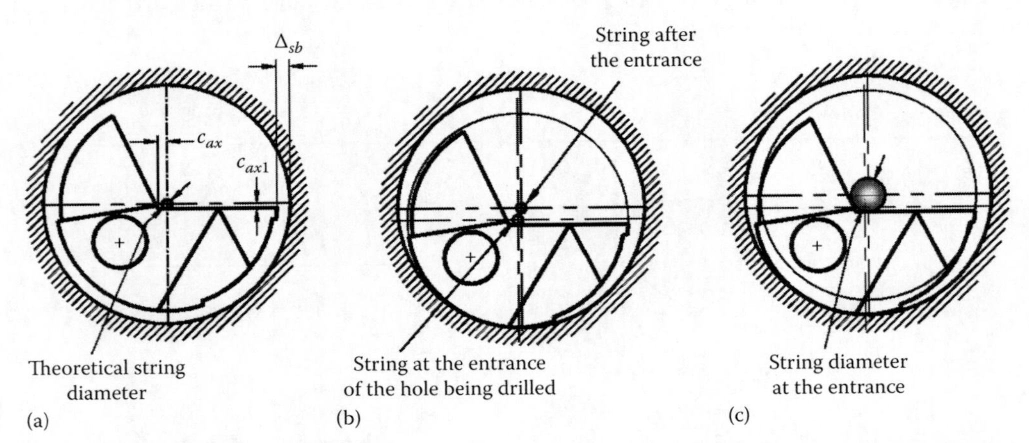

FIGURE 5.85 (a) Theoretical string diameter, (b) string diameter the hole entrance when drill rotates, and (c) string diameter the hole entrance when workpiece rotates.

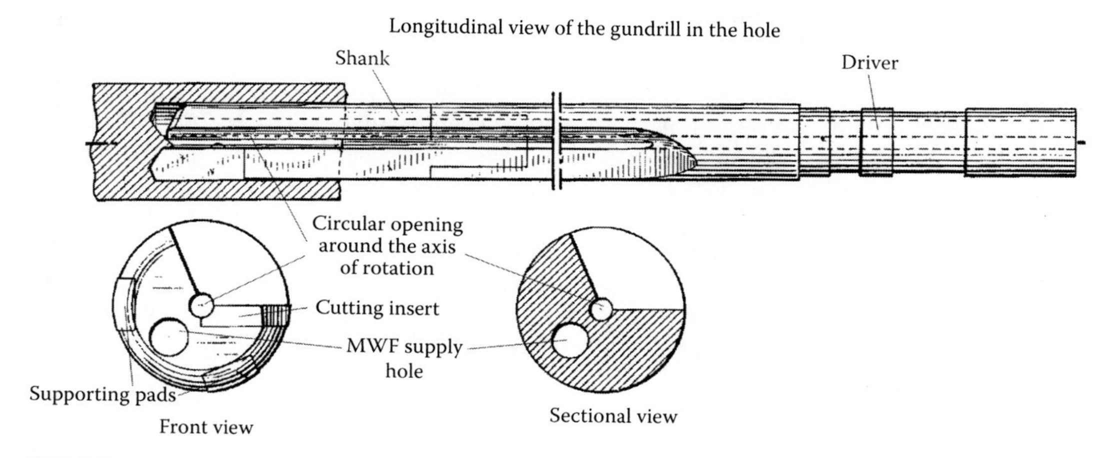

FIGURE 5.86 Gundrill design according to US Patent No. 2,418,021 (1947).

In practice, however, the string presents some problems only when blind holes are being drilled in special high-alloy steels. Otherwise, the sting usually breaks into relatively small portions and does not create any problems. Moreover, multiple testings of gundrills have shown that proper selection of distance c_{ax1} results in lower cutting forces and greater drill stability. The author's application experience allows to suggest that this distance should be assigned within the range of 0 to $-c_{ax1} = 0.02d_{dr}$. This should be included in an inspection report and clearly indicated on any gundrill drawing.

The discussed string can provide some useful benefits in gundrilling if its diameter is deliberately increased to a rod (Figure 5.86, a gundrill design according to US Patent No. 2,418,021). As can be seen, the gundrill has the same components as the usual gundrill, namely, the driver, shank supporting pads, and MWF supply passage made in the shank. The difference is in the circular opening that extends from the front of the gundrill to its opposite end passing through the shank as a hole. In the operation of the drill, the presence of the circular opening causes the formation of a rod. This rod extends from the bottom of the hole being drilled throughout the length of the workpiece, thus forming a kind of support around which the drill rotates. The rod is removed after the drilling operation is over. Technically, the discussed drill is a trepanning tool as it does not drill the entire work material. Such a drill can be used only for through holes as it was intended for rifle barrels, cam shafts, etc.

The discussed design pioneered a number of variations of gundrill designs using the same or similar idea adjusting it to the needs of improved gundrill manufacturing quality and requirement to make more versatile tools. Figure 5.87 shows a gundrill design according to US Patent

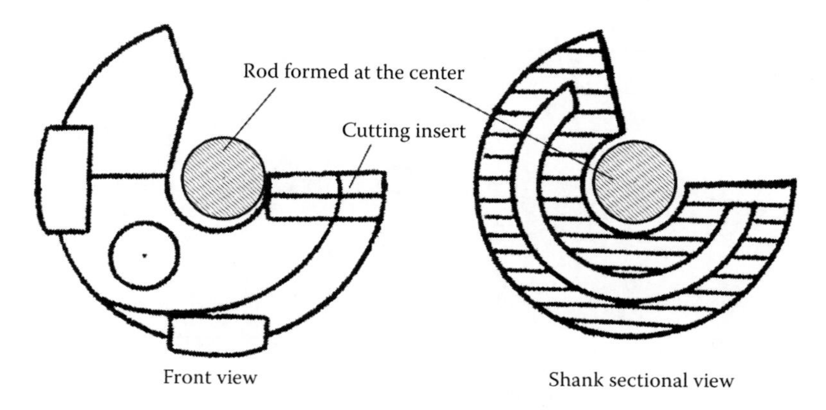

FIGURE 5.87 Gundrill design according to US Patent No. 3.089,359 (1963).

No. 3.089,359 (1963). As claimed, the significant opening is made for the sake of using replicable cutting inserts in which location accuracy on the shank does not affect the conditions of cutting in the center as the rod diameter is great, so only small variation of this diameter can be caused by the differences in the cutting insert position. The circular opening on the tip requires a special cross-sectional configuration of a tubular shank.

In modern gundrills of small-diameter ground on CNC grinders, the location of the cutting edge is precisely controlled so that this distance is small, and thus, the resultant small-diameter string does not present problems as it breaks itself into small pieces. In modern gundrill with indexable inserts, a universal mean to deal with a partially formed string is shown in Figure 5.88, where a general sting deflector is a part of the insert design (e.g., US Patent No. 4,565,471, 1986, and US Patent Application, 2011/0033255). As can be seen, when a partially formed string contacts the deflector, it bends and then fractures from the bottom of the hole being drilled. The design and location of the chip deflector depends on the type and design of a particular drill and thus can be used with any type of drill when one tries to solve the problems that unavoidably occur for any drill in the region adjacent to the axis of rotation.

The geometry of major and auxiliary flank faces of standard gundrills having planar flank faces is simple as the major cutting edges are located practically on CL. As a result, the clearance angles in T-hand-S and T-mach-S are the same because the assumed, as well as the actual, vector of the cutting speed is approximately perpendicular to the cutting edge. Therefore, the formulae correlating clearance angle in different standard reference planes are the same as for straight-flute drills in T-hand-S considered in Chapter 4. Particularities of more complicated and rarely used flank geometries are considered by the author earlier (Astakhov 2010).

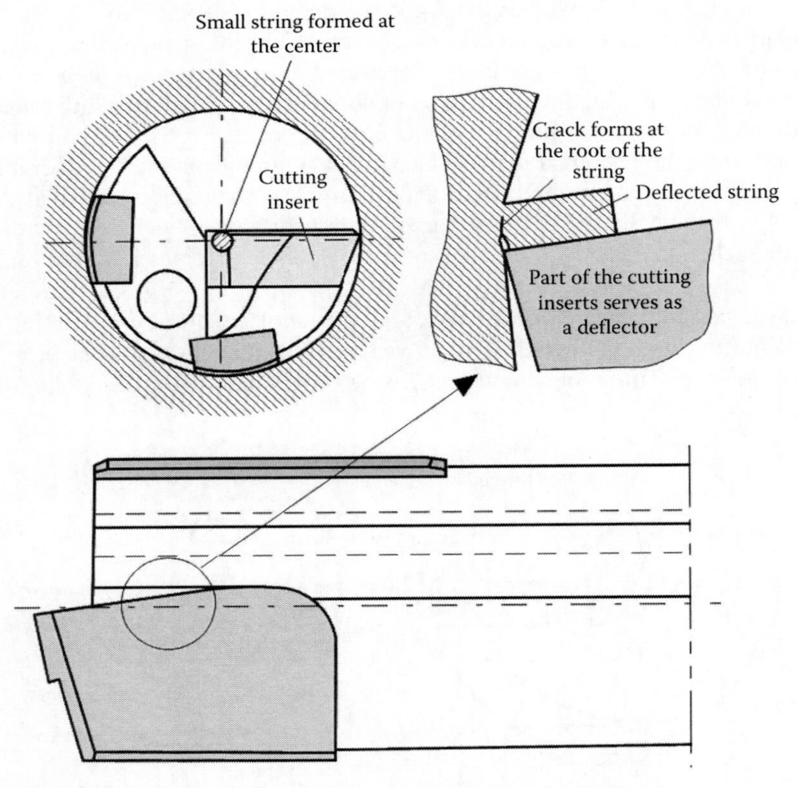

FIGURE 5.88 Dealing with the string in modern gundrills with indexable inserts.

5.2.5.6 Shank

Shanks consist of tubing with a flute rolled or swaged to the required shape (Figure 5.89). The most common flute form is a *V* shape that matches the flute on the tip although other shapes are still in use for some specific applications as each of the listed shapes has its own advantages. These may include simplicity of manufacturing, chip removal space, torsional and buckling stiffness, and MWF supply cross-sectional area. Although the shank is as critical as the tip and driver, practically no information on the shank design, material, and manufacturing requirements is available in the published literature.

The major dilemma in the shank design is the proper selection of its external diameter and wall thickness (or the internal diameter). On one hand, the MWF (coolant) channel through the length of the shank must be large enough to fully supply MWF to the tip. This leaves a limited wall thickness for the shank and explains why the shank is probably the most critical component of the gundrill. On the other hand, the wall thickness should be large enough to withstand the axial force and drilling torque. This wall thickness combined with the shank length is a major constraint on the allowable feed. The better the shank design and its material, the higher the allowable feed per revolution and thus penetration rate of gundrilling.

The shank length L_{sh} is another critical parameter of the shank. Needless to say, this length should be as short as possible. As discussed in Chapter 1, the proper way to determine this length is

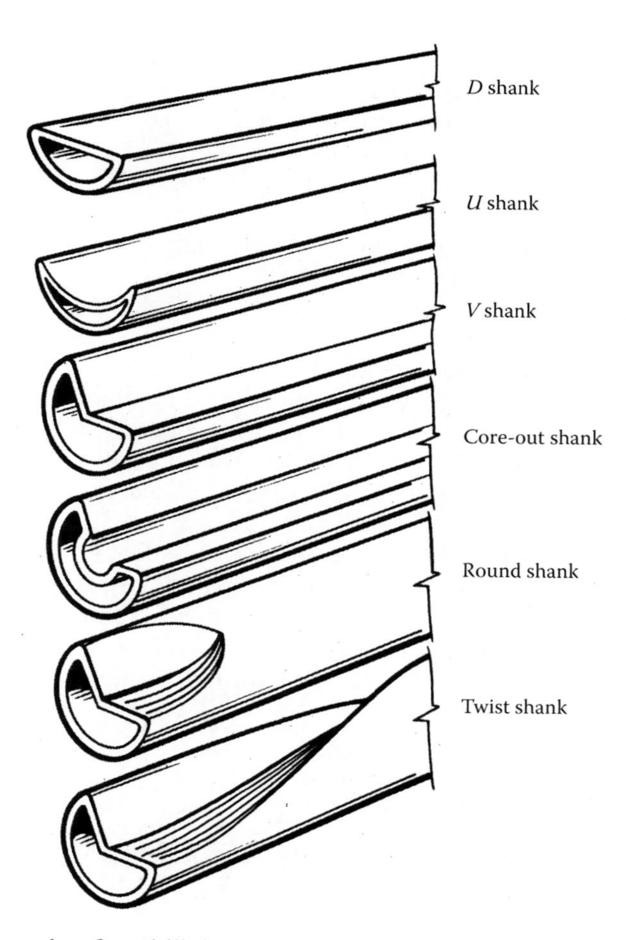

FIGURE 5.89 Various styles of gundrill shanks.

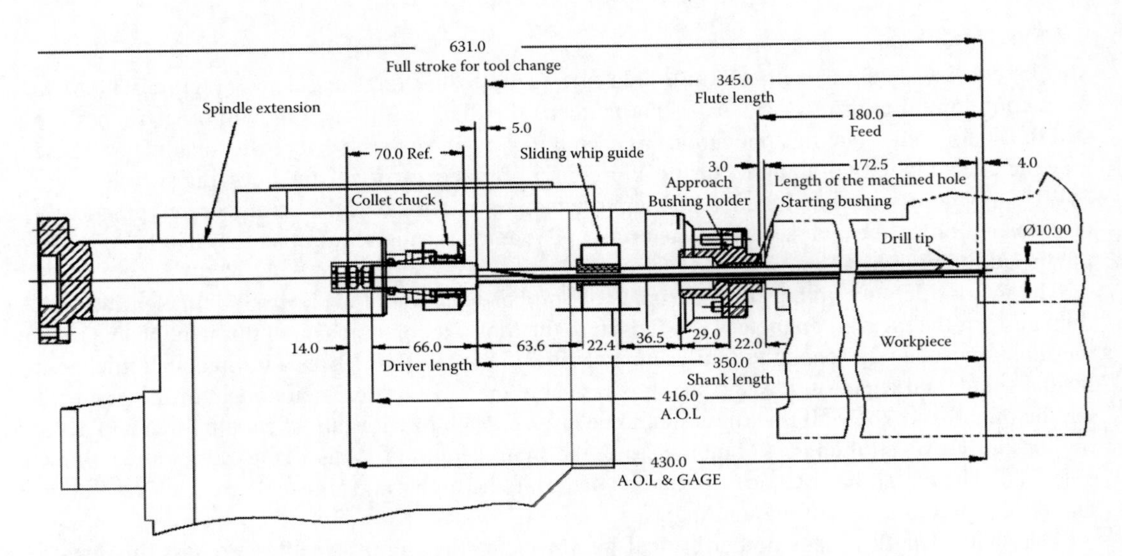

FIGURE 5.90 A typical tool layout to determine the length of the shank, flute, and drill.

to design the tool layout for a given gundrilling operation. An example of the tool layout is shown in Figure 5.90. The following sequence is recommended in determining the shank length: start with the length of the machined hole then add the approached and overshoot distances and then the length of the bushing including bushing holder and chip box.

Once the shank length is determined, an important decision is to be made about the use of steady rest(s) often called whip guides and their number to restrict whipping of the shank. Although whipping is mentioned in many gundrill-/gundrilling-related technical and trade literature sources, its nature is not revealed. A need is felt to clarify the issue.

Because a gundrill is a drill, the resultant force factors the shank must withstand are the same as discussed in Chapter 4, Section 4.4.3 (Figure 4.19), that is, the axial force F_{ax} and the drilling torque M_{dr} as shown in Figure 5.91. The drilling torque causes drill twisting, while the axial force causes its buckling due to extensive drill length. It may be logically assumed that whipping (shown in Figure 5.91 as $w(z_0)$) is a result of shank buckling due to the action of the axial force. The problem, however, is that whipping is observed with no drilling when a gundrill is just installed into the starting bushing and rotates. It was noticed that the higher the rotation speed, the greater the whipping.

Our understanding of gundrill whipping based on observation of many gundrilling systems allows to represent whipping as consisting of two components, namely, forced and *natural*. The forced whipping occurs when a gundrill rotates, and it is due to the action of the axial force as considered in Chapter 4. When the drill is stationary, some static bending of the shank is observed due to the action of the axial force. Natural whipping is caused by a distributed load

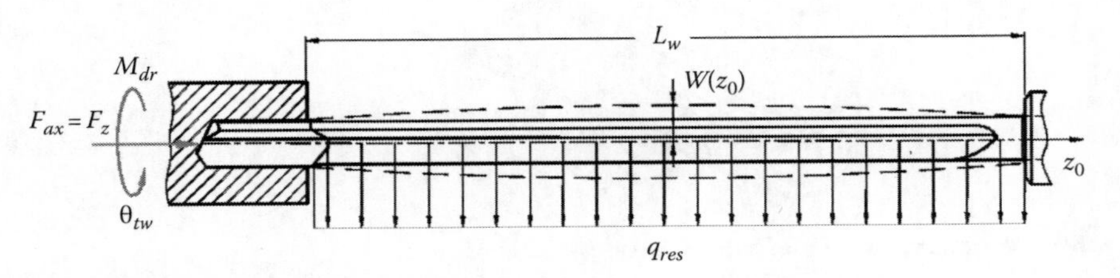

FIGURE 5.91 Force factors acting on a gundrill.

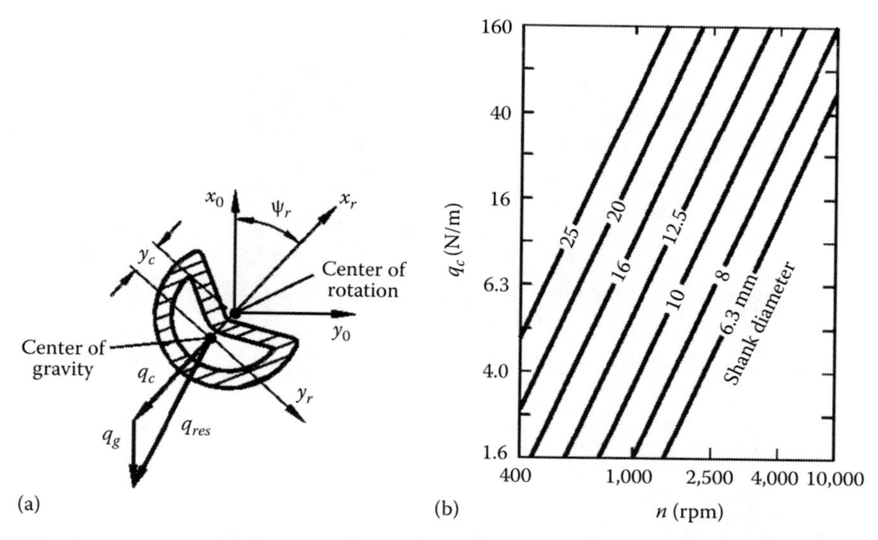

FIGURE 5.92 Showing the nature of (a) the distributed load on a gundrill and (b) its centrifugal component.

$\mathbf{q}_{res}$ acting on the gundrill shank as shown in Figure 5.92a. This distributed load is vectorial load consisting of two components, namely,

$$q_{res} = q_c + q_g \tag{5.25}$$

where

- $\mathbf{q}_c$ is the centrifugal component arising because the center of rotation of the shank does not coincide with its center of gravity
- $\mathbf{q}_g$ is the load due to shank weight

The magnitudes of these distributed loads are calculated as

$$q_c = A_{sh}\rho_{sh}y_c\omega_{sh}^2 \ \text{(N/m)} \tag{5.26}$$

where

- A_{sh} is the shank cross-sectional area (m^2)
- ρ_{sh} is the density of the shank material (kg/m^3)
- y_c is the distance between the center of shank rotation and its center of gravity as shown in Figure 5.92a
- ω_{sh} is the shank angular velocity (s^{-1}).

Figure 5.92b shows an example of q_c for some common shank diameters (for shank having flute angle $\psi_c = 110°$ and wall thickness $\delta_{sh} = 0.12d_{sh}$).

The magnitude of the load due to shank weight is calculated as

$$q_g = A_{sh}\rho_{sh}g \ \text{(N/m)} \tag{5.27}$$

where g is the acceleration due to gravity (m/s^2). When gundrilling on the Earth, $g = 9.81$ m/s^2.

The rotating coordinate system $x_r y_r$ is introduced to obtain the final equation for distributed load q_{res} as shown in Figure 5.92a. As such, the angle between the stationary and the rotating axes is a function of time, t as $\psi_r = \omega_{sh}t$. Therefore,

$$q_{res} = \sqrt{q_c^2 + q_g^2 - 2q_cq_g\cos\omega_h t} \tag{5.28}$$

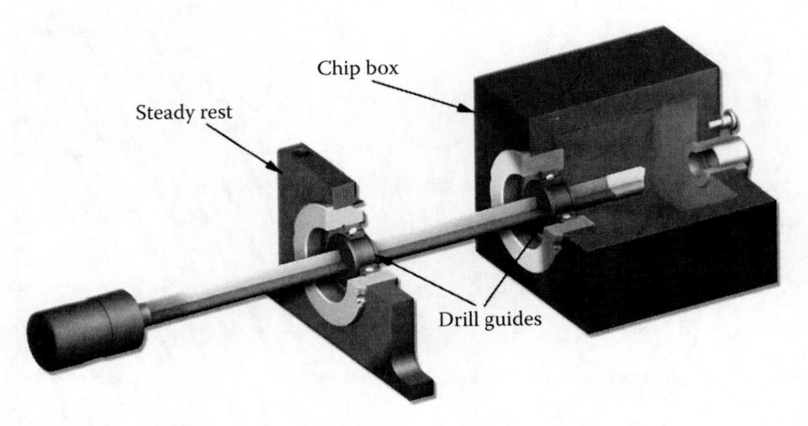

FIGURE 5.93 Steady rest for a gundrill by International Drill Guide Co.

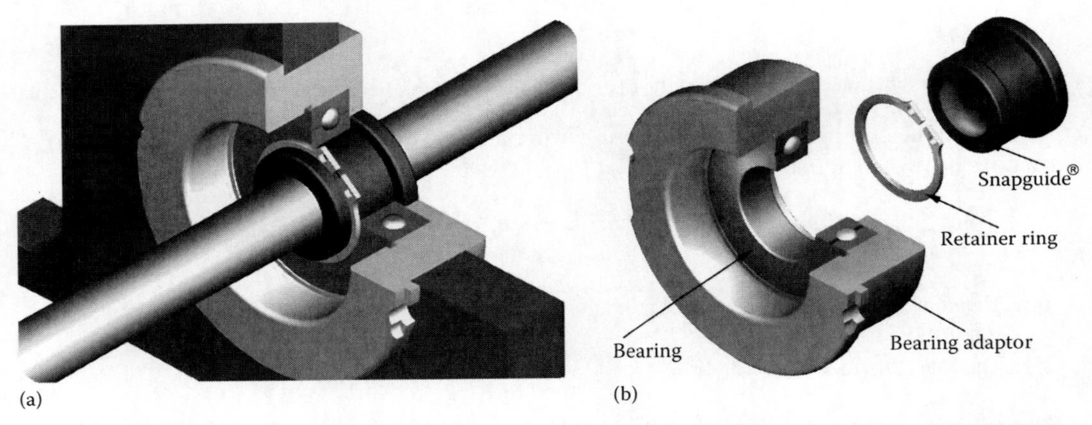

FIGURE 5.94 Snapguide® steady rest bushing. (Courtesy of International Drill Guide Co., Willoughby, OH): (a) assembly, and (b) its components.

As can be seen from Equation 5.28, the resultant distributed load q_{res} changes with the rotation angle ψ_r, and its maximum is achieved when vectors $\mathbf{q}_c$ and $\mathbf{q}_g$ have the same direction.

A steady rest is installed on the gundrilling machine as shown in Figure 5.93 to limit whipping (both its components—forced and natural). Figure 5.94 shows a simple and reliable design by International Drill Guide Co. that provides significant help in damping vibration and drill whipping. These flexible plastic bushings support the shank with a slight compression fit and are mounted in a bearing on the steady rest.

Figure 5.95 helps to make the decision about the use of steady rests. For example, if the shank length of an 8 mm drill rotating at 3000 rpm is 400 mm, then a whip guide is needed because the maximum allowable distance between supports for these conditions is 350 mm. Information similar to that shown in Figure 5.95 should be always requested from gundrill manufacturers for the design of the layout for a given gundrilling operation.

The foregoing analysis suggests that same as with twist and straight-flute drills, a gundrill shank should withstand buckling and torsion (see Section 4.4). The buckling stability is enhanced by the proper location of steady rest according to Figure 5.95 so that resistance to torsion is of prime concern in the selection of shank material and cross-sectional shape. Figure 5.96 shows how the torsion rigidity GJ (G is the shear modulus of the shank material, and J is the polar moments of inertia—see Figure 4.22) varies with the shank diameter for the annular and V-shaped cross sections. As can be seen, the torsion rigidity is always significantly greater for the annular cross section. Heat treatment of the shank improves the torsion rigidity but such an improvement is rather marginal (Nikolov 1986).

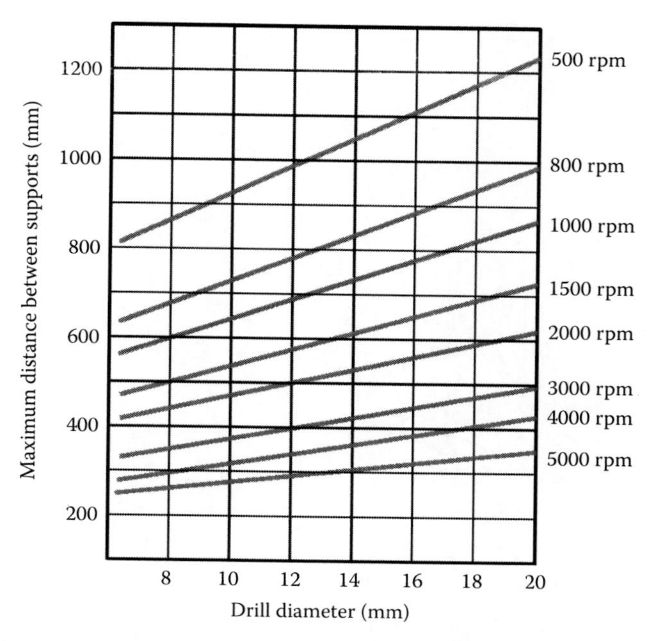

FIGURE 5.95 Maximum allowable distance between supports for high-penetration-rate gundrills.

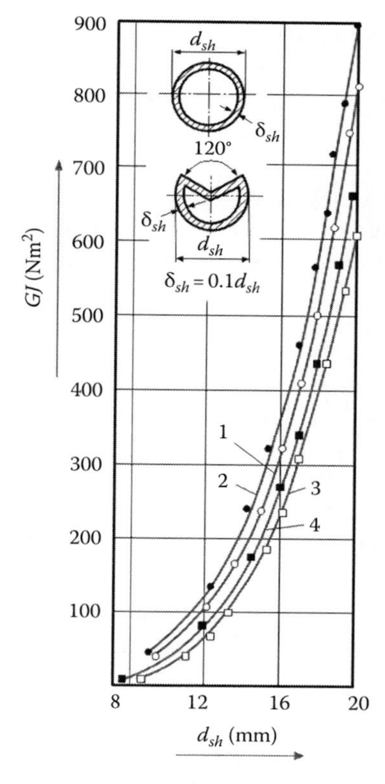

FIGURE 5.96 Torsion rigidity GJ versus shank diameter for the annular and V-shaped cross sections: The data shown for shanks made of AISI steel 1040 (G = 0.8 MPa): (1 and 2) annular cross section, non-heat-treated and heat-treated, respectively, and (3 and 4) V-shaped cross section, non-heat-treated and heat-treated, respectively.

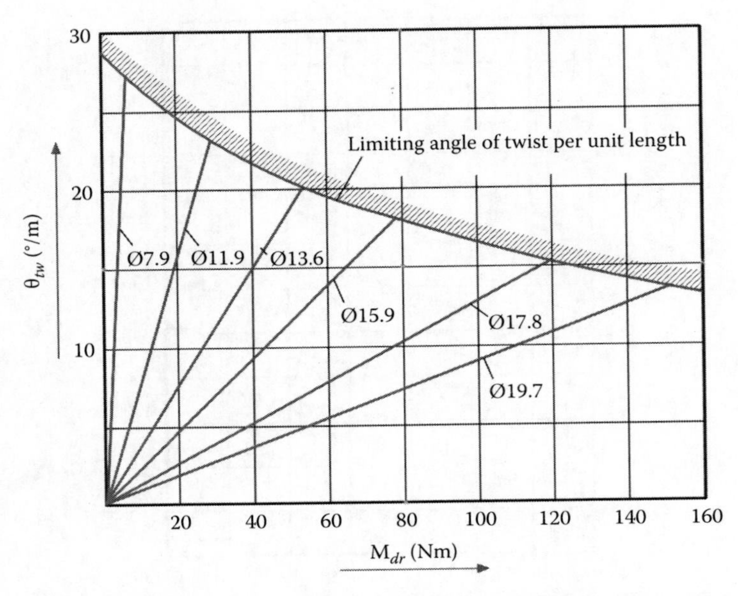

FIGURE 5.97 The limiting angle of twist as a function of the drill diameter and applied torque.

As discussed in Chapter 4, Section 4.4, torsion deformation defined by the limiting angle of twist is the limiting factor. This angle is calculated as

$$\theta_{tw} = \frac{M_{dr} \cdot 180°}{GJ} \quad (5.29)$$

Heat treatment of shanks improves the limiting angle of twist (Figure 5.97) allowing greater torques and thus penetration rates. The use of twist shanks results in a favorable distribution of the load q_c over the shank length and due to the axial force (thrust)–torque coupling (see Section 4.4.3.5) allows greater penetration rates.

5.2.5.7 Drivers

Drivers are mounted at the end of the shank. There are a number of various styles of drivers. The basics styles are shown in Figure 5.98. Drivers 1–4 are plane drivers that are most common. The plane cylindrical driver (driver 1) is mostly used in HP drills as it is installed into a precision shrink-fit or hydraulic holder. The standard tolerance h6 is used for its diameter. Plane drivers 2, 3, and 4 have means to accommodate a setscrew to transmit the drilling torque: driver 2 has flats, driver 3 has tapered undercut, and driver 4 has tapered flat. They are used in normal precision gundrilling.

Drivers 5 and 6 are preset drivers as they have the adjustments for the depth of presetting. Driver 5 has the front adjustment, while driver 6 has the rear adjustment by means of a locking hollow screw. Some drivers have been made with key, pin, or tang drive or with locking thread (driver 8). Locking into the socket can be accomplished by using a threaded cap over the front of the socket or a collet closure in the socket.

Normally MWF (the coolant) is fed through the center of the driver from its rear end. This is possible when the spindle has a central hole through it to supply MWF. When the spindle has no hole through it, a special rotary union can be mounted in front of the spindle or around the driver where crossholes shown in driver 7 carry MWF at right angle to the central coolant of the gundrill.

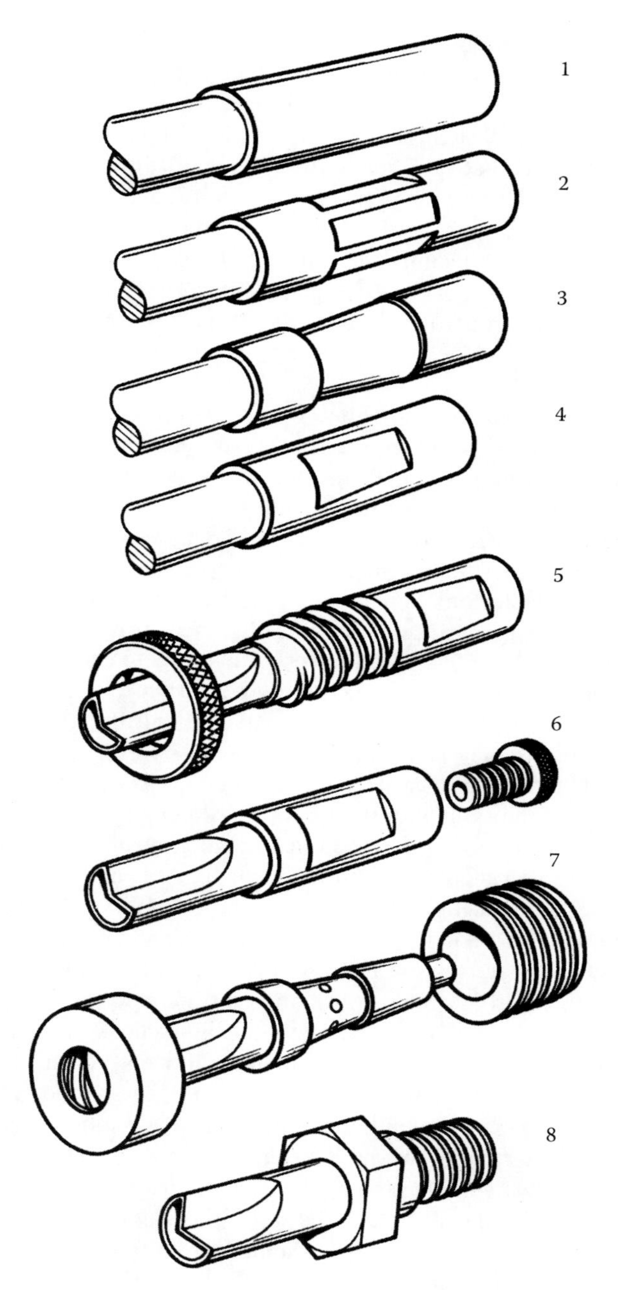

FIGURE 5.98 Types of drivers for gundrilling.

5.2.5.8 HP Gundrills and Gundrilling

HP gundrilling has emerged during the last 10 years as the process allows the penetration rate to be more than 900 mm/min (approx. 35 ipm) for aluminum alloys, more than 250 mm/min (10 ipm) for cast irons, and more than 180 mm/min (7 ipm) for alloy steels (Astakhov 2004). It became possible due to significant improvements in the manufacturing quality of gundrills including the quality of their components, implementation of better gundrilling machines equipped with advanced controllers as well as their proper maintenance, application of better MWFs, better training of engineers and operators, and many other gundrilling system-related factors.

As discussed in Chapter 1, drilling efficiency is a drilling system property. Although it is so, there should be a certain gage to align the properties of the components of such a system. In the author's opinion, an HP gundrill should be considered as this gage so that the properties/characteristics of other components of the drilling system should be selected to fully support the performance of such a gundrill. As discussed in Chapter 1, no matter how good a drill is, it will not show its best performance if the components of the drilling system do not fully support this performance. However, the opposite is also true, that is, no matter how good the drilling system is, which may include the most advanced (and thus expensive) components, high drilling efficiency cannot be achieved if the gundrill used in this system is not capable of HP gundrilling. Therefore, this section considers both principles of HP gundrills and properties of supporting gundrilling systems.

5.2.5.8.1 HP Gundrilling System

The most common method of gundrilling is one where the gundrill rotates and the workpiece is stationary. This method imposes special requirements on the accuracy of gundrilling machines and their components. The alignment of the gundrill components should be next to perfect when drilling holes with diameters less than 10 mm (0.4 in.) in light materials such as aluminum alloys, that is, when the rotational speed (6,000–15,000 rpm) and the feed rate (600–1000 mm/min [24–40 in/min]) are high. The clearance in the starting bushing, drill holder–starting bushing and whip guide alignments, and the accuracy of the feed motion are key factors in using this method.

Although awareness of the importance of machine alignment grows in industry, there are at least three important issues that remain. First, there is no simple way to check the discussed alignment in many gundrilling machines built in production lines. Normally, it takes many hours to clean up the space for such an inspection. Second, the discussed gundrilling machines do not have any means to correct alignment when needed. Commonly, shims are used to adjust this alignment that reduces the machines dynamic stability. Third, the alignment is normally checked between the starting bushing holder and the spindle of the machine. Although it is an important parameter, it is not sufficient. It should be clearly understood that the alignment in the system *actual gundrill holder–actual starting bushing* should be examined although it is not that easy accounting for the current method and accessories used for misalignment inspection. For HP gundrilling application, the discussed misalignment should not exceed 0.004 mm. It should be checked between the actual gundrill holder and the actual starting bushing. The use of modern laser alignment systems with digital targets significantly simplifies the verification of this parameter.

Yet another important issue in the consideration of alignment is the steady rest (whip guide) (Figure 5.93). Unfortunately, there is not much data on the influence of this alignment on the drill performance. An extremely important and unknown fact to end users follows from the comparison of data presented in Figure 5.99: the whip guide alignment affects the deviation of the hole axis much more than that of the starting bushing. Unfortunately, there is no simple way to check and to correct the alignment of the whip guide(s) on many machines used in industry.

The clearance between the gundrill tip and the starting bushing is another important but often ignored system parameter. Ideally, this clearance should be next to zero. However, it is not possible in any practical situation due to a number of different factors (drill-free rotation and penetration, tip back taper, wear of the tool and starting bushing, etc.). The excessive clearance in the starting bushing is the prime cause for entrance instability in gundrilling when the rotational speed and feed rate are high. This instability causes the formation of a bell-shaped part at the hole entrance known as the bell mouth. According to the author's experience, excessive clearance in the starting bushing caused by the use of nonspecialized for gundrilling or excessively worn starting bushing is the prime cause for the so-called unpredicted drills' failures (Astakhov 2001). Therefore, when one uses HP gundrilling, only specialized gundrilling bushings should be used and gundrills should not be re-sharpened beyond approx. 1/3 of the original tip length.

High-pressure MWF delivery is necessary to cool the workpiece and the tool, to provide lubrication between tool and workpiece, as well as to carry away chips from the cutting area along the flute

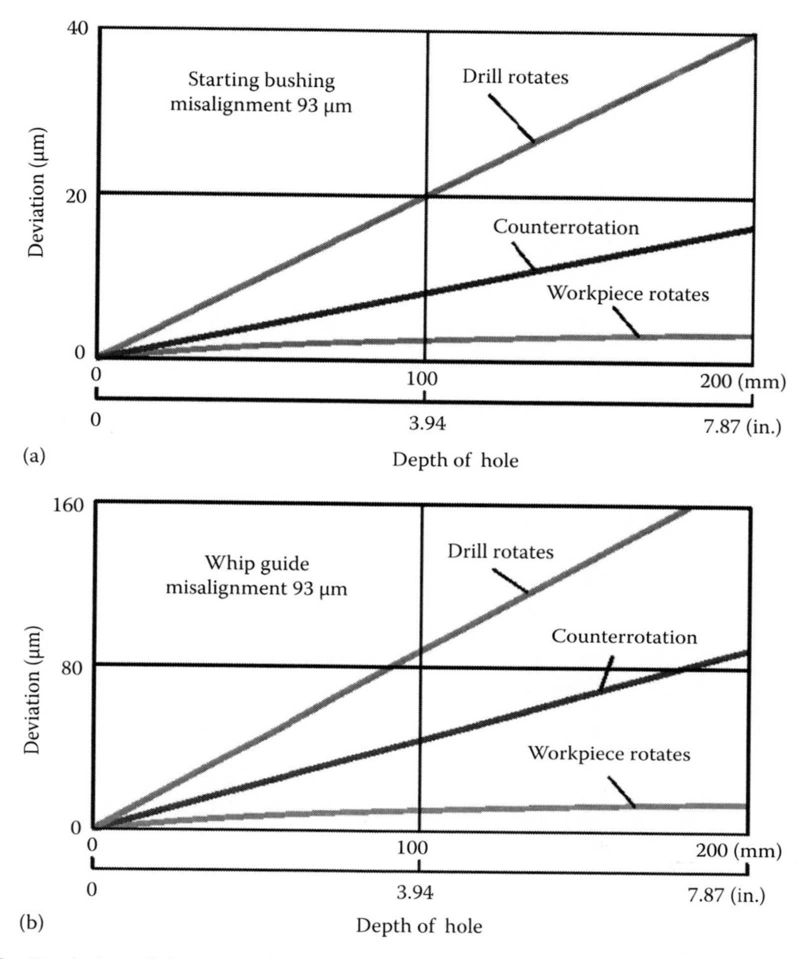

FIGURE 5.99 Deviation of the hole axis (position error) for different gundrilling methods: (a) influence of the misalignment of the starting bushing and (b) influence of the misalignment of the whip guide. Gundrill 10 mm (0.394 in.) diameter, $n = 1600$ rpm, $f = 0.04$ mm/rev (0.0016 in./rev).

to the chip box. Cooling action dissipates both the external heat of friction and the internal heat of plastic deformation due to cutting and burnishing. Lubrication between the workpiece and the drill contact areas reduces contact stresses and amount of the thermal energy generated on these areas, so it reduces adhesion and/or diffusion wear of gundrills. To effectively carry chips away, MWF should have a sufficient combination of viscosity and velocity. Improper selection of this combination causes chip clogging in the flute that lead to an increase in torque and probable drill breakage.

In simple terms, the MWF flow rate needed for reliable chip removal in gundrilling can be calculated using the assessment of the work needed to be done by the MWF flow. This work can, for horizontally located gundrill, be represented as

$$W_{fl\text{-}req} = (F_{r\text{-}a} + F_{r\text{-}t})L_{dr} \tag{5.30}$$

where

$F_{r\text{-}a}$ is the resistance force due to atmospheric pressure (N)
$F_{r\text{-}t}$ is the resistance force due to chip transportation (N)
L_{dr} is the length of drilling (m).

The resistance force due to atmospheric pressure is calculated as

$$F_{r\text{-}a} = p_{atm}A_{v\text{-}fl} \tag{5.31}$$

where
p_{atm} is the atmospheric pressure
$p_{atm} \approx 0.1$ MPa (see Section 6.3.2)
$A_{v\text{-}fl}$ is the cross-sectional area of the chip flute (V-flute) (m²).

The resistance force due to chip transportation is calculated as (Astakhov and Joksch 2012)

$$F_{r\text{-}t} = K_{ch}G_{ch} \tag{5.32}$$

where
$K_{ch} = 0.15$ is the proportionality coefficient
G_{ch} is the weight of the chip produced by the drill (N)

$$G_{ch} = 0.785 d_{dr}^2 f n \gamma_w \tag{5.33}$$

where
d_{dr} is the drill diameter (m)
f is the feed per revolution (m/rev)
n is the spindle rotational speed (rev/s)
γ_w is the specific weight of the work material (N/m³)

$$\gamma_w = \rho_w g \tag{5.34}$$

where
ρ_w is the density of the work material (kg/m³)
g is acceleration due to gravity (m/s²).

The work done by the MWF flow is calculated as (Astakhov and Joksch 2012)

$$W_{fl} = \gamma_{fl}\frac{Q_{fl}^3}{A_{v\text{-}fl}^2 2g} \tag{5.35}$$

where
Q_{fl} is the flow rate of MWF (m³/s)
γ_{fl} is the specific weight of MWF (N/m³).

The condition of the reliable chip transportation is

$$W_{fl\text{-}req} \geq W_{fl} \tag{5.36}$$

which, accounting for Equations 5.30 and 5.36, yields an expression for minimum required MWF flow rate (m^3/s)

$$Q_{fl\text{-}min} \geq \sqrt[3]{\frac{W_{fl\text{-}reg} A_{v\text{-}fl}^2 2g}{\gamma_{fl}}} \tag{5.37}$$

There are three basic types of coolants used in gundrilling:

1. Oil-based MWFs known as *straight oils* are generally used for gundrilling of alloy steels on stand-alone machines having their own MWF supply system. Compared to water-soluble MWFs, oil-based MWFs significantly reduce tool wear, yield better surface finishes, and generally improve the accuracy of drilling. However, it happens only when such MWFs contain extreme pressure additives of sulfur (2.5%–3.5%) for high-alloy steels and heat-treated cast irons, chlorine (3.5%–5%) for light ferrous materials, and 10%–14% of fat. Low coolant viscosity aids in good heat dissipation and good load-carrying capacity. It also reduces the risk of pump starving when cold starting, improves the efficiency of filters, and reduces the amount of oil carried off with the chips. The kinematic viscosity should generally not exceed 20–30 cSt/20°C. In exceptional cases, 45 cSt/20°C can be used.
2. Water-based MWFs known as *water-soluble oils* are used for machining aluminum alloys on in-line gundrilling machines where MWF is supplied from a central coolant pump station. These MWFs are generally more economical and can be used for nonferrous metals and high machinability steels under light cutting conditions. Extreme pressure additives contained in water-soluble oils prevent work material adhesion to the tip. These MWFs also contain film strength enhancers (animal and vegetable fats) to reduce friction and wear. Dilution 8 to 1 is normally used, while dilution greater than 10 to 1 significantly reduces film strength and creates vapor pockets at high tool load areas in the gundrilling zone that reduces tool life dramatically.
3. Synthetics are water-based MWFs that are easier on the environment but harder on the gundrill. It is used, however, in gundrilling cast irons with great success.

MWF filtration is essential to system performance. Because MWF collects and circulates considerable quantities of both coarse and fine chips, it must be carefully purified in the interests of both tool life and hole quality. Poor filtration leads to increased MWF temperatures and rapid failure of the coolant pump. It also causes premature failure of solenoid valves, leaking servo valves, and bearing failure in the rotating coupling. Cartridge filters of size in the range of 5–10 μm for high-precision holes should be used. Filtration of particles of 15–20 μm for precision holes and in the range of 20–30 μm for normal holes should be guaranteed. Drilling cast iron requires rough filters, magnetic drums, or rolled media, followed by a bag-type or woven media polishing filter.

The MWF temperature defines to a large extent its cooling, lubricating, and transportation abilities. This is particularly important with oil-based MWFs. About 40°C–50°C (100°F–120°F) is generally recommended as the maximum temperature of MWF. It can often be maintained by circulation through a heat exchanger or even by installing a fan to blow across the surface of the coolant reservoir. When precise holes are to be drilled, refrigeration systems may be necessary.

MWF pump(s) plays a more important role than one may think. Unfortunately, most of the gundrilling coolant supply systems have the inferior type of pumps, called variable-displacement pumps. A variable-displacement pump is designed to maintain *set* pressure. If an obstruction is encountered by the MWF flow, the *set* pressure (the pressure seen on the gage by the operator) will be maintained but the flow rate supplied to the tool will actually decrease because MWF will be diverted through the pump's internal relief valve. As a result, the obstruction (in the case of chip clogging in the flute of the tool) can, in fact, be worsened and quickly lead to drill failure. Experienced practitioners in industry change these pumps with fixed-displacement pumps.

5.2.5.8.2 HP Gundrills

It is worthwhile to point out that gundrills can be also classified according to drill length-to-diameter ratio. When this ratio is less than 10, gundrills are called short; when it is between 11 and 50, gundrills are called normal; when it is between 51 and 100, gundrills are called long; and when this ratio exceeds 100, gundrills are called extra-long. As discussed previously, a gundrill consists of three major parts, namely, the tip, shank, and driver. The design of the listed components of a gundrill depends on this ratio. For example, nowadays short gundrills are made of a solid-carbide rod so that the distinction between the components becomes only functional.

Tip material. The design of an HP gundrill begins with the tip as the tip performance for short and normal gundrills and even long defines the overall performance of the drill. The design particularities include the grade of the tool material (normally sintered carbide) and tip macro- and microgeometry.

Hundreds of different carbide grades are used in metal cutting depending on the work material, MWF used, cutting operations, and required quality of the machined part and so on. Surprisingly, only a very few carbide grades are used to make gundrill tips. There was a time not long ago when only two grades of carbide were used to produce gundrills, namely, C2 and C3. It is still believed that C2 is a *forgiving* carbide and thus can be used on any drilling machine with much less than perfect working conditions. A *small* price to pay includes relatively short tool life and poor surface finish. The other is C3 proving to be much harder and thus more wear resistant. Unfortunately, it is also more brittle and therefore cannot be used in gundrilling systems having excessive misalignment and runout. Recent advancements in the development of carbide materials and their technology resulted in the appearance of new micro- and submicrograin carbides. The use of such carbides for gundrill tips significantly enhances the tip strength and its wear resistance. Another problem, however, has emerged: how to select the proper grade for a given application from the great variety of available grades. Experience shows that the improper selection of even submicrograin good-quality carbide leads to premature tool failure. In other words, the margin for errors in the carbide selection becomes significantly smaller with the discussed variety of different carbide grades.

It seems that there should not be any problem for a tool/process designer in the selection of the suitable carbide grade if the corresponding correlations between the properties of carbide (carbide grades) and drilling conditions (including work material, its metallurgical state, machine conditions, and coolant) are known. Unfortunately, this is not the case in practice. As discussed in Chapter 3, the properties of carbide grades presented in the carbide manufacturers' data (brochures, catalogs, etc.) have little to do with gundrill working conditions though some of these tool materials were developed especially for gundrilling applications. For example, the hardness that is usually listed as the prime parameter is given at room temperature, and no dependence of this parameter upon the temperature in the region of cutting temperatures (which reaches 600°C–800°C) is ever provided. TRS (see Chapter 3 for details) is another example of nonrelevant parameters accounting for the method of its determination (ASTM B528-99 standard). The most important parameters as fracture toughness, homogeneity, purity, and wear resistance (both adhesion and abrasion) are not normally available.

Unfortunately, there is no relevant information or reliable data on the behavior of multiple coatings used for gundrills. Because gundrills are re-sharpened by removing a certain layer of the tool material from their flank surfaces, the use of coatings is only justified if prime tool wear takes place on the rake face as the crater wear and/or on the supporting pads.

Fortunately for gundrill designers/users, there are a few companies, for example, CERATIZIT, which provide the following:

- Application-specific carbide grades
- Close tolerances on the bolt and coolant hole diameters of their carbide rods
- Complete metallurgical report/structure of their product upon request
- High consistency of cemented carbide quality.

The tips of HP gundrill should be made from such carbide rods.

The shape, dimensions, and location of the coolant passage in the gundrilling tip differ from one carbide manufacturer to another. In terms of shape, gundrill manufacturers have adopted various shapes for this passage: one or two circular holes or a single kidney-shaped hole (Astakhov 2010). It is interesting to mention here that the shape, dimensions, and location of this passage are determined by carbide manufacturers having very limited knowledge in gundrilling. Moreover, there is an opinion among gundrilling practitioners that the kidney-shaped hole makes the tip weaker causing its breakage.

In any consideration of the discussed parameters of the coolant passage, one has to realize that these parameters are of hydraulic nature (Astakhov et al. 1994). Delivering the sufficient MWF flow rate with minimal pressure losses should be considered as the objective of all coolant channels and passages in a gundrill. As discussed in Chapter 6, these pressure losses determine the inlet MWF pressure for a given flow rate or, vice versa, determine the flow rate through the gundrill under a fixed inlet MWF pressure (as often used in practice). The total pressure loss is the sum of (1) the frictional losses in the tubular shank, (2) pressure losses due to flow sudden contraction from the shank passage into the coolant passage in the tip, (3) the frictional losses in the coolant passage in the tip, and (4) pressure losses due to coolant flow interaction with the bottom of the hole being drilled. Experiments have shown (Astakhov et al. 1994; Astakhov 2010) that for short and normal length gundrills, losses (2)–(4) are an order greater than (1), while for long and extra-long gundrills, these losses are of the same order. The minimum pressure losses occur when the tips having a kidney-shaped coolant passage are used. As such, the kidney-shape hole does not make the tip weaker. In other words, it does not cause tip breakage as shown by FEM analysis and by direct testing. Rather, tip fracture is often caused by improper brazing and insufficient MWF flow rate. Therefore, the kidney-shape coolant passage in the tip should be unconditionally adopted for HP gundrills because these tools require a high coolant flow rate that normally cannot be achieved with one- and two-hole coolant passages under the inlet pressure allowed by the many existing gundrilling machines used in the automotive industry.

The design and location of the supporting pads. These two parameters greatly affect entrance stability, deviations of the holes axis, and diametric accuracy. As discussed in Section 5.2.4, the results of multiple studies have conclusively proven that gundrills with the supporting continuum, which are now common in the automotive industry, are always inferior to that with the supporting pads. The use of the supporting continuum does not have any advantages in gundrill performance. Although it only simplifies drill's periphery grinding in drill production, the supporting continuum results in unstable drill locating in the starting bushing and in the hole being drilled. Gundrills with two and three (when crosshole gundrilling is the case) optimally located supporting pads should become common in HP gundrills. Their use results in the reduction of force fluctuations by 30%–50% when proper angle between the supporting pads and sufficient radial relief are used.

Tip geometry. Although the understanding of the rake and flank geometries and their adjustments for a particular work material are important stages in the design of any drill including gundrills, it is only the tip of the iceberg. Ultimately, any serious tool designer pursuing the best tool design should consider a reasonable balance in justifying a number of objective functions of the tool design. The following objective functions must be considered in the gundrill design:

- Meeting hole quality requirements
- Achieving maximum tool life under the desirable productivity (the penetration rate)
- Producing chip shapes suitable for reliable removal
- Assuring stability at the entrance to the hole being drilled (preventing an excessive bell mouth).

Although these objective functions seem to be independent, it was shown (Astakhov 2010) that they are just different facets of the proper tip design/geometry as the tip does the actual cutting and burnishing. In other words, one can start the design with practically any feature and then develop others using

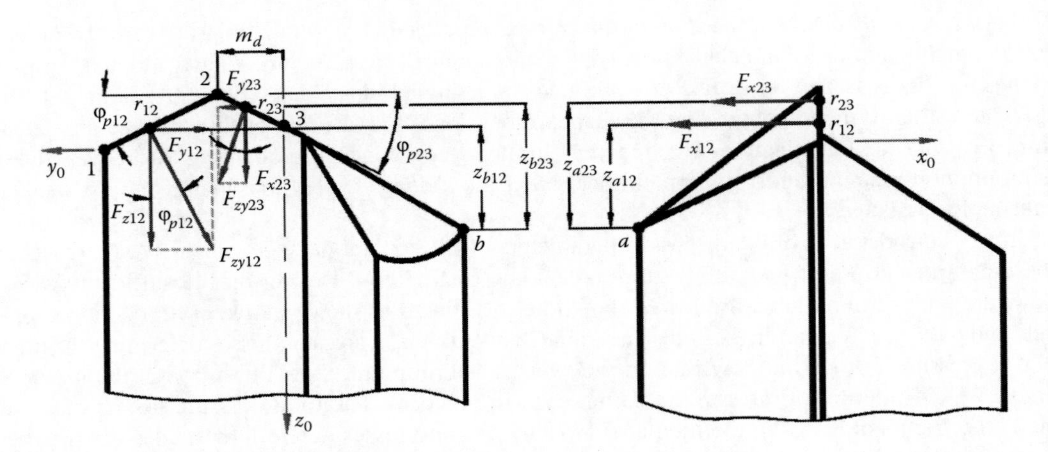

FIGURE 5.100 Components of the cutting force causing additional bending moments.

iteration procedure to balance the final design. In the author's opinion, however, it is easy and more logical to start with the consideration of the influence of the cutting force components.

Following the introduced cutting force and its components in SPT (Figure 5.48), one can represent the components of the cutting force as the projections of this force on the axes of the original (T-mach-S) coordinate system for cutting edges 1–2 and 2–3 as shown in Figure 5.100. Besides the previously discussed bending moments due to the axial components of the cutting force, the following bending moments act in the y_0z_0 plane

$$M_{y_0z_0} = F_{y12}z_{b12} - F_{y23}z_{b23} \tag{5.38}$$

and in the x_0z_0 plane

$$M_{x_0z_0} = F_{x12}z_{a12} + F_{x23}z_{a23} \tag{5.39}$$

Because the power component of the cutting force, F_x, are normally (besides some special cases) much greater than the radial component F_y and because the bending moment in the x_0z_0 plane calculates at the sum of the moments due to F_{x12} and F_{x23}, the bending of the gundrill in this plane by angle θ_{xz} (Figure 5.101) is normally of prime concern. The greater the angle θ_{xz}, the

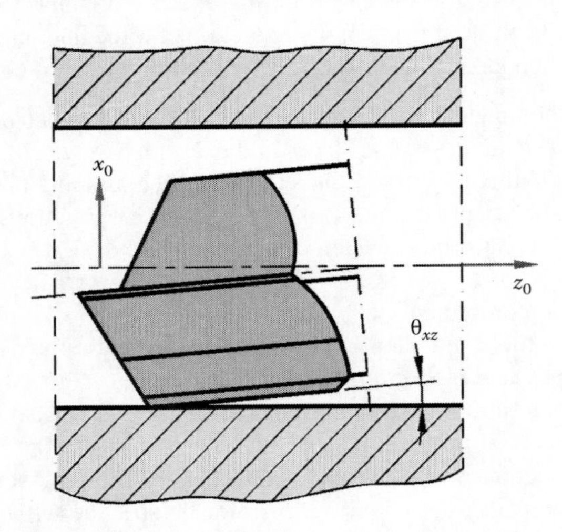

FIGURE 5.101 Bending of the gundrill in the x_0z_0 plane.

lower the quality (surface finish and diametric accuracy) of the machined holes. Although as discussed earlier in Section 5.2.3 the bending moment in the x_0z_0 plane is the *Achilles' heel* of any SPT, it is more profound in gundrilling due to small drill diameters and thus weak shanks. The problem is greatest in the facet point grind shown in Figure 5.71. To assure drill-free penetration, that is, to prevent the interference of the drill's flanks with the bottom of the hole being drilled, the auxiliary flank (15) (normal flank angle α_{n3}) is applied that pushes the location of point "a" far behind the cutting edge so that the arms z_{a12} and z_{a23} (Figure 5.100) become significant. This explains why the facet point grind often shows no advantages over the CAM point grind although the former is much superior in terms of drilling itself. The mentioned arms are much smaller when the CAM point grind is used. Therefore, the tool geometry parameters should be optimized to minimize the bending moment in the x_0z_0 plane—this constitutes *Rule No. 1* in the HP drill tip geometry/design.

It follows from Figure 5.100 that the approach angles of the outer, φ_{p12}, and inner, φ_{p23}, cutting edges determine the ratio of axial F_z and radial F_y forces as parts of F_{zy}. It follows from Equation 5.38 and Figure 5.100 that the additional bending moment in the y_0z_0 plane can be eliminated if the approach angles and distance m_d are selected so that

$$F_{y12}z_{b12} = F_{y23}z_{b23} \tag{5.40}$$

However, this scenario is highly undesirable by any means. First, the trailing supporting pad has to be firmly pressed to the wall of the hole being drilled to assure the static stability in drilling and to take its share in burnishing of the hole being drilled. Second, if the condition defined by Equation 5.40 is justified, then the length of the outer cutting edge and its angle φ_{p12} should be rather small compared to that of the inner cutting edge. If it is the case, then the formed chip is directed into machined surface ruining its surface finish. Third, as discussed later, there will be a number of problems with the direction of MWF flow in the machining zone. Therefore, force component $F_y = F_{y12} - F_{y23}$ should be significant.

A problem in HP gundrilling is that the arm of this component (with respect to point "b" in Figure 5.100) can also be significant.

Figure 5.102 shows a typical example when the so-called stepped-slash point grind common in the automotive industry is used. Obviously, the bending moment in the y_0z_0 plane affects the drill deflection, particularly for small drill diameters and for long gundrills where the shank is not rigid enough to withstand the additional bending moment in this plane (Hasegawa and Horiuchi 1975). Therefore, the tool geometry parameters should be optimized to minimize the bending moment in the y_0z_0 plane—this constitutes *Rule No. 2* in the HP drill tip geometry/design.

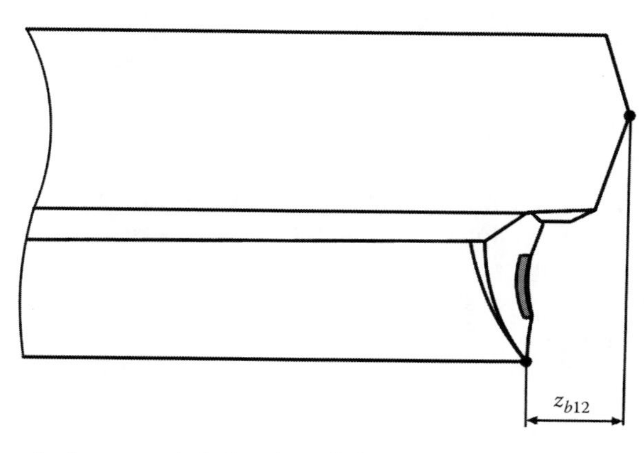

FIGURE 5.102 Arm z_{b12} in the stepped-slash point grind.

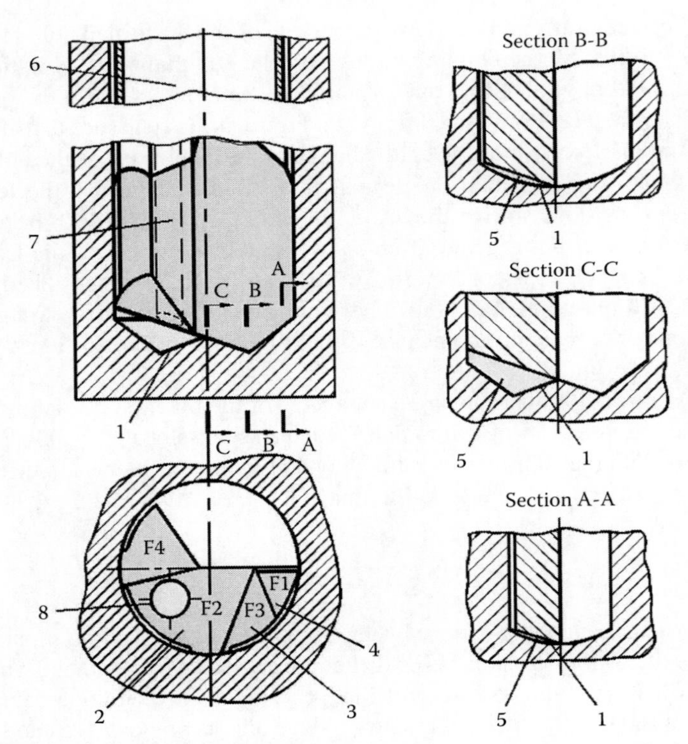

FIGURE 5.103 Concept of the bottom clearance space.

While drilling, the gundrill geometry results in the formation of the sculptured surface 1 known as the bottom of the hole being drilled (Figure 5.103). This bottom, from one side, and the drill's flanks (2, 3, and 4), from the other side, form a limited space (5) named as the bottom clearance space. The topology of the bottom clearance space can be appreciated in different cross sections as shown in Figure 5.103.

MWF (the coolant) is supplied into the bottom clearance space (5) under pressure through the internal passage (6) of the shank and coolant passage (8) made in the tip (7). The MWF pressure in the bottom clearance space directly affects tool life as it has a major influence on the cooling and lubrication conditions on the tool flank (of the inner, outer, and side cutting edges) surfaces as well as on the supporting pads. The higher the MWF pressure, the higher the tool life. This constitutes *Rule No. 3* in the HP drill tip geometry/design. High MWF pressure in the bottom clearance space provides a better penetration of the MWF into the extremely narrow passages (see Section A-A in Figure 5.103) between the tool flanks and the bottom of the hole being drilled, that is, better conditions for lubricating and cooling of the flank contact areas. This is particularly important for MWF penetration in the regions adjacent to the drill periphery point where this MWF is mostly needed. Unfortunately, it is not achieved in standard gundrills and thus the tool wear in this zone is much higher than that in any other.

The topology of the bottom clearance is rather complicated as it is formed by two sculptured surfaces: the bottom of the hole being drilled and the flank surfaces of the gundrill. The bottom of the hole being drilled in general case is formed by two adjacent hyperboloids of revolution. If the cutting edges are not straight, this bottom has even more complicated shape. In particular case, however, when the cutting edges are straight, horizontal (in T-mach-S), and located on the y_0 axes, the bottom (1) (Figure 5.103) consists of two adjacent conical surfaces (one is reverse due to the outer cutting edge and the other is direct as formed by the inner cutting edge). A section of the bottom of the hole being drilled is shown in Figure 5.104.

The complicated shapes of the drill flanks and the bottom make the shape of the bottom clearance space very irregular and complex. One may ask a logical question: why is it important to know

FIGURE 5.104 Section of the bottom of the hole being drilled.

the topology of this space? There are two prime reasons for that. First is to assure drill-free penetration into the hole being drilled with no interference. Second is to assure the preferable MWF flows in this space to reduce tool wear.

The condition of the drill-free penetration into the hole being drilled is justified if there is no interference of the drill flanks and the bottom of the hole being drilled. The term *interference* is commonly associated with the interaction of coherent waves. In tool design, this term is to describe the hypothetic overlapping or even contact of any surfaces or points of the cutting tools, which is not supposed to overlap or to be in contact with the workpiece. It should be clear that when interference takes place, free penetration is theoretically impossible. In reality, it is not always so because a gundrill is not a rigid body (shank), so when *light* interference occurs, the drill usually bends (of course, to a certain extent) so that its cutting edges come into contact and the surfaces of interference just rub against the workpiece. Interference leads to additional forces on the tip and elevated temperatures due to rubbing and to drill vibration. It ruins the quality of drilled holes and reduces tool life. Therefore, any interference in any kind of drilling should be prevented—this constitutes *Rule No. 4* in the HP drill tip geometry/design.

Because the bottom clearance space has complicated shape, it is very difficult to assure the absence of interference in the 2D design by taking cross sections as shown in Figure 5.103. A possible way to analyze interference is 3D modeling where a section of the hole being drilled is considered in its contact with the cutting edges. Rotating this tool with respect to the workpiece, one can visualize the clearances and interferences in the bottom clearance space (Astakhov 2010). Practically any commercial CAD software is fully capable of such a modeling. This method, however, is not yet common in the gundrill design because it requires real understanding of tool geometry and profile generating mathematics. As a result, interference in the design gundrill is assured *by eye*. At best, it is verified by a few 2D cross sections. The tip point grind design of many modern gundrills remains the same over the last 50+ years when CAD systems were not available, so the tip was designed so that interference is avoided *for sure* by grinding tool secondary flanks. As such, other parameters as drill bending (Rule Nos. 1 and 2) and the MWF pressure in the bottom clearance space (Rule No. 3) are neglected.

To understand these conditions, one should ask a logical question: how many flank planes are really needed for free penetration of a gundrill? Normally, four flank surfaces (planes) are provided. Figure 5.103 shows F1, F2, F3, and F4 flank planes. Planes F1 and F2 are flank planes of the outer and inner cutting edges, while planes F3 and F4 are meant to assure drill-free penetration into the hole being drilled. Because there are no recommendations on the selection of the location of planes F3 and F4, most of gundrill manufacturers and users follow the common pattern established more than 50 years ago (Figure 5.71). It is shown by the author earlier that the introduced rules of the HP

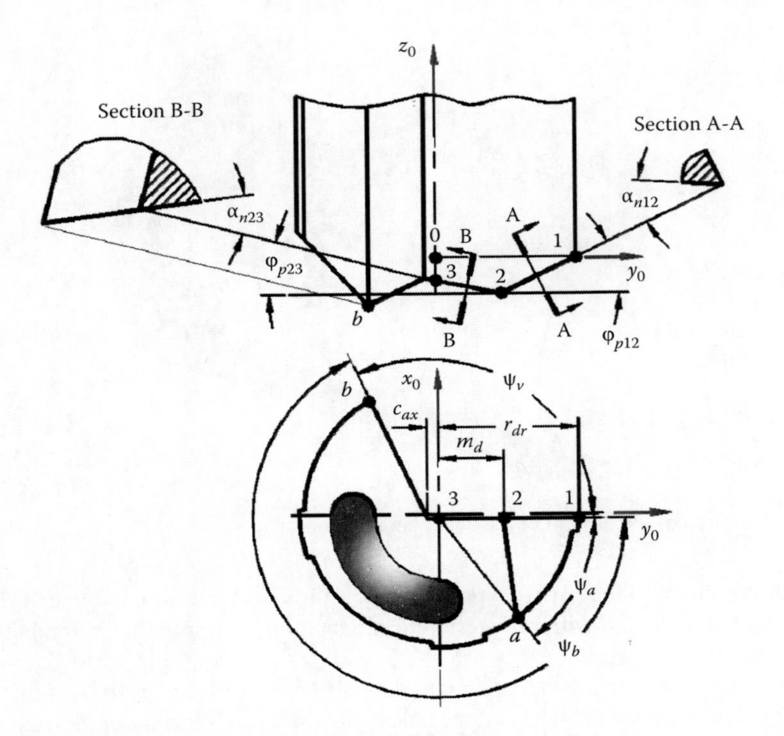

FIGURE 5.105 A gundrill is made with just two major flanks.

gundrill design can be justified simultaneously following *VPA-Balanced* drill design concept where parameters of planes F3 and F4 are determined analytically (Astakhov 2010).

To explain this concept, Figure 5.105 shows a gundrill made with just two flank planes. The free penetration of the drill's flank can be verified by the set of the clearances δ_{cl} between the flank points and the bottom of the hole being drilled (Astakhov 2010). The drill's flank touches the bottom when the value of this clearance is equal to zero. The value of δ_{cl} varies continuously along the flank and the graphical analysis shows that the most *dangerous* points of the drill flank (the possibility of interference with the bottom) are located at the periphery points *a* and *b* shown in Figure 5.105. Figure 5.106 shows visualization of these points on a solid model of a gundrill. Figure 5.107 shows visualization of interference of point *b*.

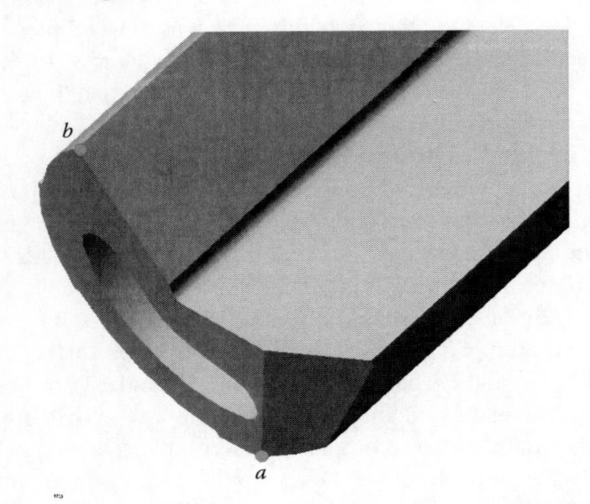

FIGURE 5.106 Visualization of the most *dangerous* points *a* and *b* of the drill flank.

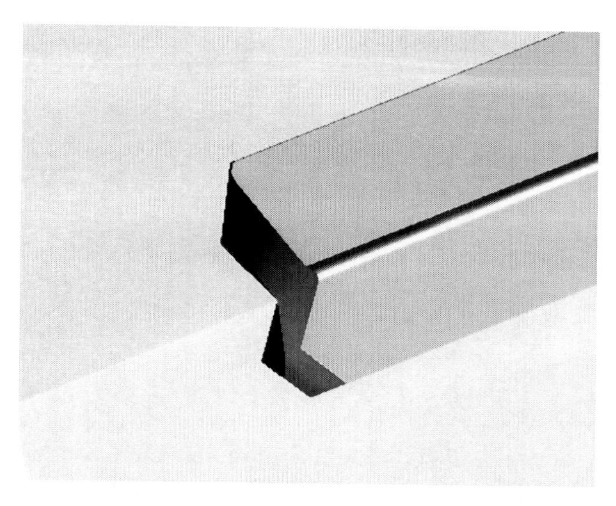

FIGURE 5.107 Visualization of interference of point "b."

It was shown by the author earlier (Astakhov et al. 1995a and b) that the conditions of the gundrill-free penetration into the hole (no interference with the bottom of the hole being billed) are as follows:

for point "a,"

$$\xi_{3a} \geq \xi_L \tag{5.41}$$

for point "b,"

$$\xi_{4b} \geq \xi_L \tag{5.42}$$

where

$$\xi_L = \arctan\left(\tan\varphi_{p12} - \frac{m_d}{r_{dr}}(\tan\varphi_{p12} - \tan\varphi_{p23})\right) \tag{5.43}$$

$$\xi_{3a} = \arctan\left[\left(\cos\psi_a + \tan\alpha_{2n}\frac{\sin\psi_a}{\sin\varphi_{p23}}\right)\tan\varphi_{p23}\right] \tag{5.44}$$

$$\xi_{4b} = \arctan\left[\left(\cos\psi_b + \tan\alpha_{2n}\frac{\sin\psi_b}{\sin\varphi_{p23}}\right)\tan\varphi_{p23}\right] \tag{5.45}$$

where

$$\psi_b = \psi_{1b} + \arcsin\left(\frac{c_{ax}}{r_{dr}}\sin\psi_{1b}\right) \tag{5.46}$$

$$\psi_a = 180 - \alpha_a - \arcsin\left(\frac{m_d}{r_{dr}}\sin\alpha_a\right) \tag{5.47}$$

where ψ_{1b} (=360° − ψ_v) as before is the angle of the sector of the x_0y_0 plane corresponding to the drill body and

$$\alpha_a = \frac{\sin(\varphi_{p12} - \varphi_{p23})}{\tan\alpha_{n1}\cos\varphi_{p23} - \tan\alpha_{n2}\cos\varphi_{p12}} \tag{5.48}$$

If $\alpha_a > 0$, then its value calculated using Equation 5.48 is substituted into Equation 5.47; otherwise,

$$\psi_a = \alpha_a - \arcsin\left(\frac{m_d}{r_{dr}}\sin(180 - \alpha_a)\right) \tag{5.49}$$

When conditions set by Equations 5.41 and 5.42 are justified, there is no need to make flank surfaces F3 and F4 to assure drill-free penetration; thus, the drill can be as that shown in Figure 5.105. However, for practical parameters of the tool geometry, flank surfaces F3 and/or F4 are often needed when one of these or both conditions are not valid. Therefore, the next question to be answered is about what the geometry of these flanks F3 and F4 should be.

Figure 5.108 shows the limiting positions of flank F3 and F4 when these flanks touch the bottom of the hole being drilled, that is, when the bending moments in the x_0z_0 and y_0z_0 planes are at the absolute minimum. There are four geometry parameters needed to make these flanks, namely, the location angle, the approach angle, the depth, and the flank angle. These can be determined graphically or analytically using the model shown in Figure 5.108.

Analytical determination of the introduced geometrical parameters of flanks F3 and F4 resulted in the following. It is obvious as it directly follows from Figure 5.108 that $\varphi_{pF3} = \varphi_{pF4} = \varphi_{pF1}$ ($\varphi_3 = \varphi_4 = \varphi_1$) as the considered part of the bottom of the hole being drilled is formed by the outer cutting edges 1–2.

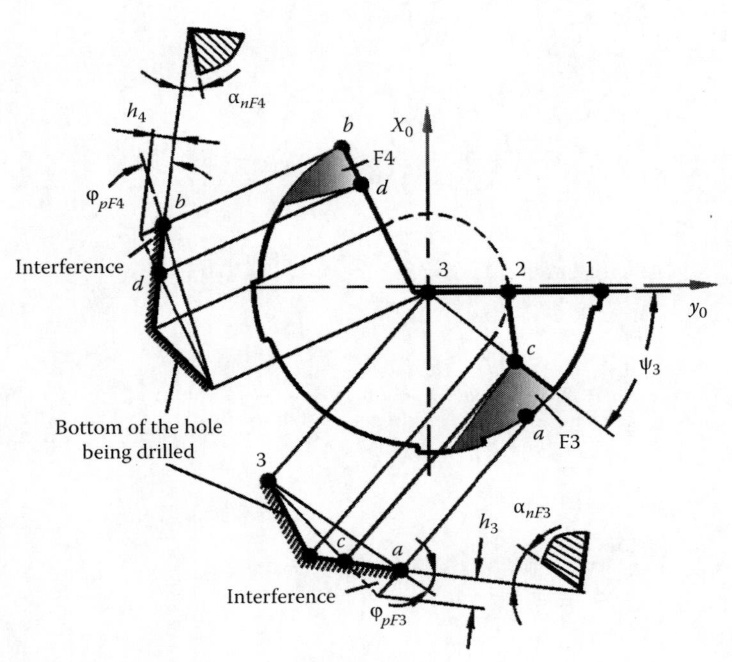

FIGURE 5.108 Model for flank surfaces F3 and F4.

If the conditions set by Equation 5.41 are not satisfied, then the flank F3 should be applied. Its geometrical parameters are calculated as follows.

The location angle ψ_3 is calculated as (Astakhov et al. 1995a and b)

$$\psi_3 = 2\arctan\frac{\tan\alpha_{n1}}{\sin\varphi_{p23}} \tag{5.50}$$

The clearance angle of flank F3 is calculated as

$$\alpha_{n3} = \arctan\left(\sin\varphi_{pF3}\tan(\psi_m - \psi_3)\right) \tag{5.51}$$

where

$$\psi_m = \arcsin\frac{\cos\psi_3 + \sin(\tau_{13} + \alpha_{p2})}{2\sin(c_m/2)} \tag{5.52}$$

$$c_m = 90 + \tau_{13} + \alpha_{p2} - \psi_3 \tag{5.53}$$

$$\tau_{13} = -\arcsin\left(\frac{(r_{dr} - m_d)\tan\varphi_{p12} + m_d\tan\varphi_{p12}}{r_{dr}}\right)\frac{\cos\alpha_{p2}}{\tan\varphi_{p23}} \tag{5.54}$$

$$\alpha_{p2} = \arctan\frac{\tan\alpha_{2n}}{\sin\varphi_{p23}} \tag{5.55}$$

and the depth h_3 is calculated as

$$h_3 = r_{dr}\sin\theta_{\psi 2}\tan(\alpha_{n3} + \theta_{\psi 3}) + \frac{f}{360}(90 + \tau_{13} + \theta_{2b})\frac{\cos\alpha_{n3}}{\cos\varphi_{pF3}} \tag{5.56}$$

where

$$\theta_{\psi 3} = \arctan\frac{\tan\alpha_{n1}\cos\psi_3}{\tan\varphi_{pF3}\sin\psi_3 + 1} \tag{5.57}$$

$$\theta_{\psi 2} = \arccos\left(\frac{m_d}{r_{dr}}\frac{1-(\tan\varphi_{p12}/\tan\varphi_{p23})}{1-(\tan\varphi_{p12}/\tan\varphi_{p23}) - 2\sin^2\alpha_{p1}(1+\tan\alpha_{p1}\tan(\tan\alpha_{p2}/\tan\alpha_{p1}))}\right) \tag{5.58}$$

$$\alpha_{p1} = \arctan\frac{\tan\alpha_{n1}}{\sin\varphi_{p12}} \tag{5.59}$$

If the conditions set by Equation 5.42 are not satisfied, then the flank F4 should be applied. Its geometrical parameters are calculated as follows.

The clearance angle of flank F4 is calculated as

$$\alpha_{n4} = \arcsin \frac{1-\cos(\theta_{\psi 4}-\psi_v)}{r_{dr}\sin(\theta_{\psi 4}-\psi_v)-c_{ax}\sin\psi_v} r_{dr}\sin\varphi_{p12} \tag{5.60}$$

where

$$\theta_{\psi 4} = \arccos\left(\frac{\tan\xi_L \tan\xi_L \sin\varphi_{p23} - \tan\alpha_{2n}\sqrt{\tan^2\alpha_{n2}+\sin^2\varphi_{p23}-\tan^2\xi_L\tan^2\xi_L}}{\sin^2\varphi_{p23}+\tan^2\alpha_{2n}}\right) \tag{5.61}$$

and the depth h_4 is calculated as

$$h_4 = \cos\varphi_{p12}\left[\begin{array}{l} r_{dr}(\tan\varphi_{p12}-\tan\xi_{4b})-c_{ax}\cos\psi_a(\tan\varphi_{p23}-\tan\xi_{4b}) \\ -m_d(\tan\varphi_{p12}-\tan\varphi_{p23}) \end{array}\right]+\frac{f\psi_v}{360}\frac{\cos\alpha_{n4}}{\cos\varphi_{pF4}} \tag{5.62}$$

As an example, let us consider a gundrill having the parameters shown in Table 5.5. For these parameters, one needs to determine using the introduced model if flank planes F3 and F4 are needed. Table 5.6 shows the results of the calculations. Because $\xi_{3a} \geq \xi_L$ (the condition set by Equation 5.41), flank plane F3 is not needed. On the contrary, as $\xi_{4a} < \xi_L$, flank F4 is needed. Its geometry is calculated and shown in Table 5.6. Taking into account the grinding, measurements, and installation

TABLE 5.5
Parameters of Gundrill 1

Drill diameter	$2r_{dr}$	31.75 mm
Approach cutting edge angle of the outer cutting edge	$\varphi_1(\varphi_{p12})$	25°
Main cutting angle of the inner cutting edge	φ_2 (φ_{p23})	−15°
Normal flank angle of the outer cutting edge	α_{n1}	20°
Normal flank angle of the inner cutting edge	α_{n2}	8°
Size of V-shaped flute extension	c_{ax}	1.8 mm
Angle of the sector corresponds to the V-shaped flute	ψ_v	116°
Angle of the sector corresponds to the drill body	ψ_c	244°
Offset of the point P of the cutting edge	m_d	7.94 mm
Maximum cutting feed	f	0.2 mm/rev

TABLE 5.6
Grinding Parameters of Gundrill 1

ξ_L	ξ_a	ξ_b	φ_4	α_{n4}	h_4 (mm)
5.65°	5.84°	1.01°	25°	6.37°	1.065

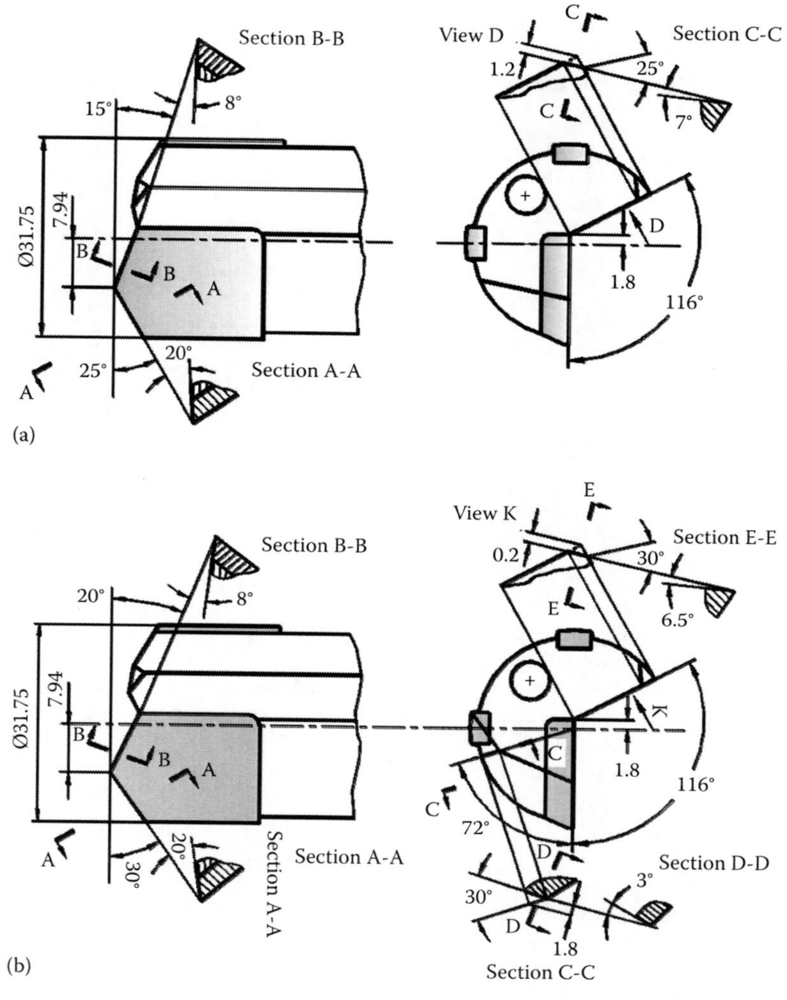

FIGURE 5.109 Gundrills with the calculated parameters: (a) gundrill 1 and (b) gundrill 2. (US Patents US Patents 7,147,411 (2006) and 7,195,428 (2007)).

tolerances, the parameters $\alpha_{n4} = 7°$ and $h_4 = 1.2$ mm were accepted. A gundrill with these parameters is shown in Figure 5.109a.

Let us gundrill 2 has the same parameters as shown in Table 5.5 except the approach angles of the outer and inner cutting edges which are now $\varphi_1 = 30°$ and $\varphi_2 = -20°$.

For these parameters, one needs to determine using the introduced model if flank planes F3 and F4 are needed. Table 5.7 shows the results of the calculations. As seen in this table, both conditions of free penetration (Equations 5.41 and 5.42) are not justified. Therefore, to assure gundrill-free penetration into the workpiece, flanks F3 and F4 are needed.

TABLE 5.7
Grinding Parameters of Gundrill 2

ξ_L	ξ_3	ξ_9	φ_3	ψ_3	α_{n3}	h_3 (mm)	φ_4	α_{n4}	h_4 (mm)
6.09°	3.78°	3.71°	30°	72.1°	2.45°	1.43	30°	6.15°	0.44

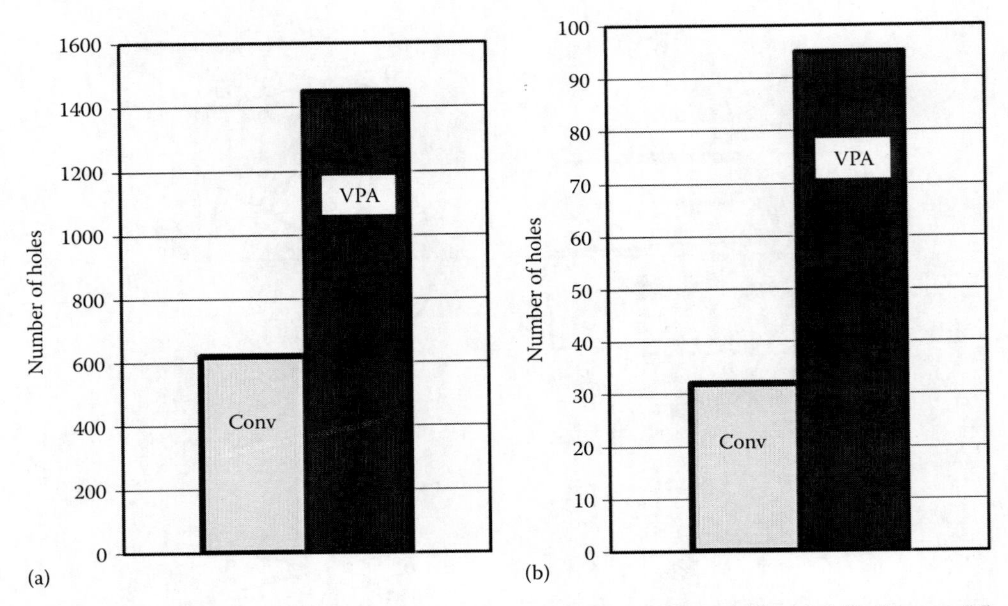

FIGURE 5.110 Tool life comparison of the conventional with the VPA gundrills: (a) drill diameter 5.5 mm, length 200 mm; drilling regime: rotational speed 5100 rpm, feed rate 155 mm/min; machine, Excello NC; work material, nodular cast iron of HB240 hardness and (b) drill diameter 5 mm, length 900 mm; drilling regime: rotational speed, 2800 rpm, feed rate 32.5 mm/min; machine, Technidrill; work material, SS15-15LC Mod. of HB300 hardness HB300.

Based on the calculated results shown in Table 5.7, parameters $\psi_3 = 72°$, $\alpha_{n3} = 3°$, $h_3 = 1.8$ mm, $\alpha_{n4} = 6.5°$, and $h_4 = 0.7$ mm were accepted. A drill with these parameters is shown in Figure 5.109b.

Figure 5.110 shows some results of the comparison of the best conventional gundrills used in the automotive industry (light bars) with the *VPA-Balanced* gundrills (dark bars).

Shank and driver. The shank must be designed and made properly. Although there are a number of issues that affect shank performance, the excessive corner radii and shank material-related considerations are of prime concern in HP gundrilling (Astakhov 2004).

Gundrill shanks must be made of a high-yield-strength material and properly heat treated. Unfortunately, these issues are not always followed by gundrill manufacturers. First, high-yield-strength materials present problems (such as excessive warping, wrinkling, and cracking) when the V-flute is formed (crimped or even swaged) using old tube crimping technology. So tubular products made of 4130 and 34Cr6Mo steels having moderate strength are common in the gundrilling industry. Second, very few gundrill manufacturers understand the proper heat treatment procedure for shanks and thus the fact that it must include a thermomechanical rather than pure thermal relief of the stresses formed on producing the V-flute. The best structure of the shank for short gundrills is a tempered martensitic structure, while for normal and long gundrills, the upper bainitic structure (see Chapter 3) is the best choice. This is the only structure that possesses a very unique combination of high hardness, increased toughness, and great wear resistance suitable for gundrill shanks. Unfortunately, no one shank of gundrills produced today has this structure.

When the shank is brazed to the tip and to the driver (brazed joints 1 and 2 in Figure 5.71), the excessive heat from this brazing often ruins the results of the heat treatment at the brazed joints. Often this heat causes high residual thermal stresses hidden in the tip. When an increased drilling torque occurs due to, for example, chip clogging or tool wear, the tip fails (Astakhov 2004). Therefore, the use of low-temperature, high-strength BFMs combined with infrared in-process temperature control followed by a 100% torque test is mandatory for gundrill brazing operations.

Experience shows that when the shank is made of high-yield-strength material, properly heat treated to achieve small grain size binate structure, and properly connected to the driver (using a low-temperature brazing filler metal), the increase in the gundrill penetration rate can be as high as twice compared to gundrills commonly used today.

While a gundrill is re-sharpened and then eventually re-tipped many times, the shank and the driver are still the same. Therefore, the driver should be made of tool steel, hardened, and ground.

Not fully realized in the tooling industry, there is a huge difference in the design shank and drivers for rotating HP gundrills compared to nonrotating. Although everybody talks about the importance of the alignment in the system *tool holder–starting bushing* asking for 4 μm, the alignment inside the gundrill is completely ignored. Often, the OD of the shank is not ground; the hole for the shank in the driver is just drilled, so when such a shank is installed in the driver and then brazed, the resultant misalignments between the shank and driver may easily exceed 20 μm. Often, the tip final configuration is completely ground before it is brazed into the shank, which not only adds to the misalignment but also ruins the whole idea of the application-specific back taper applied to this tip. In the author's opinion, such a practice should not be used for HP gundrills.

Figure 5.111 shows an example of the driver design for HP rotating gundrill. The following should be noted:

1. The driver is made of tool steel and heat treated to high hardness.
2. The proper datum (datum A) is specified.
3. The hole for the shank is made with H6 tolerance. The requirements on its surface roughness and runout with respect to the datum are properly set.
4. The step on the rear end of the hole for the shank is properly dimensioned to prevent edge contact of the shank in the final assembly.

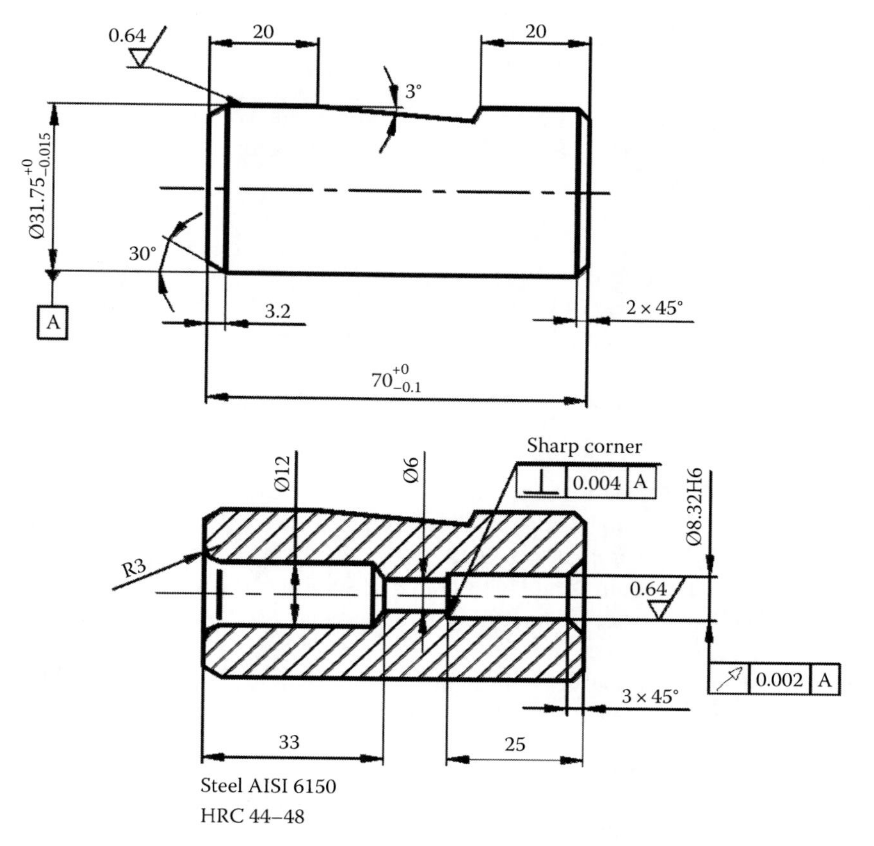

FIGURE 5.111 Example of the driver design for HP rotating gundrill.

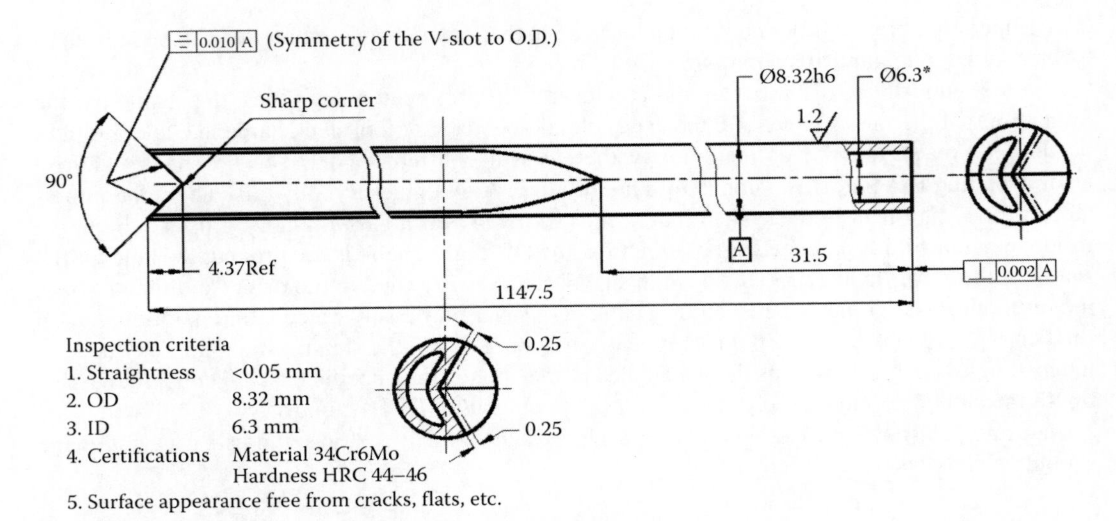

FIGURE 5.112 Example of the shank design for HP rotating gundrill.

Figure 5.112 shows an example of the shank design for HP rotating gundrill. The following should be noted:

1. The shank is made of a high-quality steel and heat treated to high hardness.
2. The proper datum (datum A) is specified.
3. The shank is ground to meet the requirements on its OD and surface roughness.
4. The V-groove for the tip is properly dimensioned with the corresponding form tolerance.

Figure 5.113 shows an example of the tip ground on the tube for HP rotating gundrill that should always be the case for HP gundrills. The following should be noted:

1. Even small features/details of the tip profile to be ground are clearly indicated with tolerances.
2. High requirements to surface finish of the tip.
3. Clear indication of the place of brazing and reference on the corresponding work instruction where the procedure, materials, and operating regime are clearly specified.

Figure 5.114 shows the final assembly requirements for HP rotating gundrill. The following should be noted:

1. Clear indication of the place of brazing and reference on the corresponding work instruction where the procedure, materials, and operating regime are clearly specified. The contact of the shank face and the shoulder in the driver are assured by applying the axial force on brazing.
2. Torque and flow rate requirements with reference on the corresponding work instruction where the procedures and equipments used are specified.

Figure 5.115 shows an example for the VPA tip grind according to US Patents No. 7,147,411(2006) and 7.195,248 (2007). The following should be noted:

1. High requirements to the surface roughness of the flank faces of the major cutting edges
2. The location of the shoulder dub-off and the auxiliary flanks
3. High normal clearance angle of the outer cutting edge
4. A flat for MWF passage adjacent to the side margin.

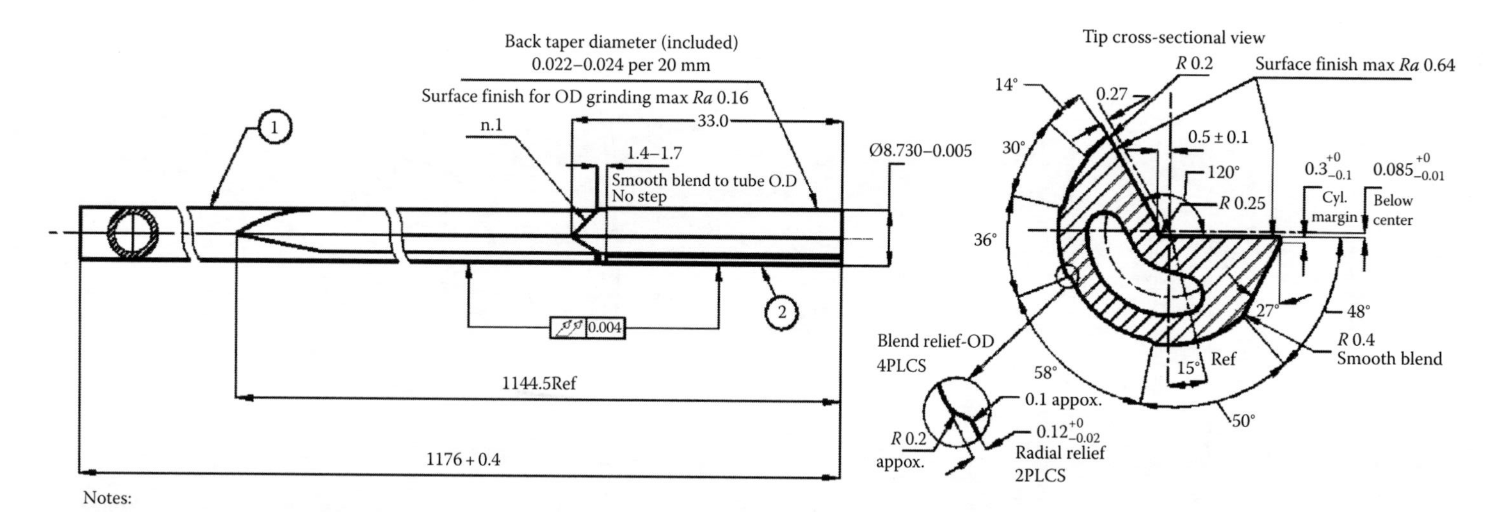

Notes:
1. Brazing according to work instruction No.XXX-2012a.
2. After OD grinding, tip should be handled with a plastic cup on it.
3. Brazing joint should be clean inside and outside.
4. Coating of the tip after brazing: $TiAl_2CN/VCN$ nanoscale mutilayer.

FIGURE 5.113 Example of the tip ground on the tube for H-E rotating gundrill.

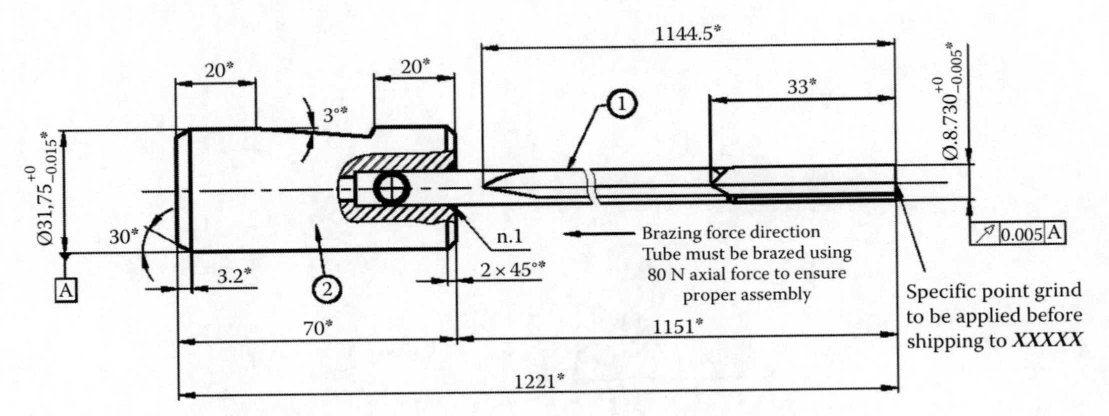

Notes:

1. Brazing according to work instruction No. XXX-2010c.
2. Flow check 100% according to work instruction No. XXX-2012c: min should be not worse. that 55 L/min at 7 MPa as tested using a straight oil of min 25 mm/s viscosity.
4. Torque test 100% according to work instruction No. XXX-2012c[2]: min torque should not be less that 3.2 Nm.
5. Brazing joint should be clean inside and outside.
6. *Reference dimensions.

FIGURE 5.114 Final assembly requirements for HP rotating gundrill.

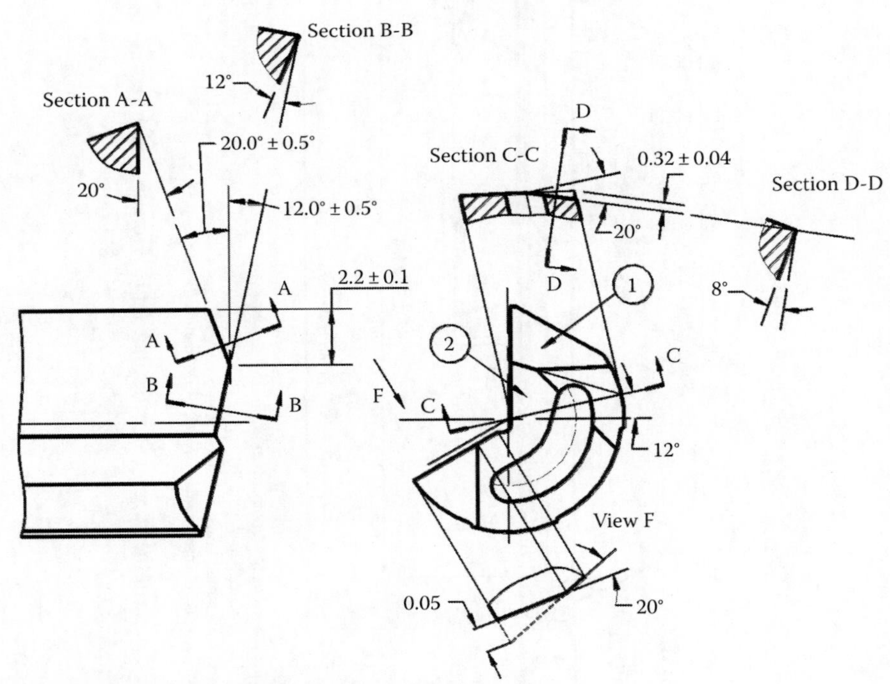

1. Surface finish on flanks 1 and 2 not worse than *Ra* 0.32.
2. Use grinding instruction VPA 15-03-2012 to follow the proper sequence.
3. Ground surface should not have any visible burns or discolorations.

FIGURE 5.115 Example for the VPA tip grind according to US Patents No. 7,147,411 (2006) and 7.195,248 (2007).

FIGURE 5.115 (continued) Example for the VPA tip grind according to US Patents No. 7,147,411 (2006) and 7.195,248 (2007).

5.2.6 STS Drills

5.2.6.1 History

Many literature sources including German books (e.g., König and Klocke 2008) name STS drills as BTA drills and maintain that the process was developed in Germany in the late 1930s. According to an archive search attempted by the author, it is not exactly so. This method of MWF supply through the annular gap between the boring bar and the bore wall using the pressure head and swarf removal through the interior of the drill head and boring bar was used as an industrial technology as early as in the mid-1930s. Moreover, drilling heads with carbide inserts secured on the boring bar using rectangular threads were used. The cutting speed used was as high as 112 m/min. A two-spindle Fritz Werner horizontal deep-hole drilling/boring machine was used in the production of gun and artillery barrels (Veremachuk 1940). Figure 5.116 shows examples of HSS and

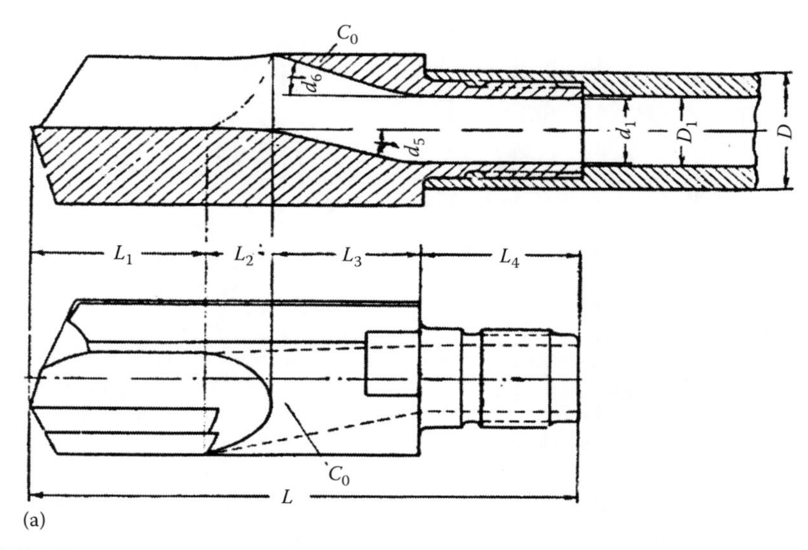

FIGURE 5.116 Examples of HSS and carbide drilling heads used in the production of gun and artillery in earlier 1930th: (a) HSS drill head.

(continued)

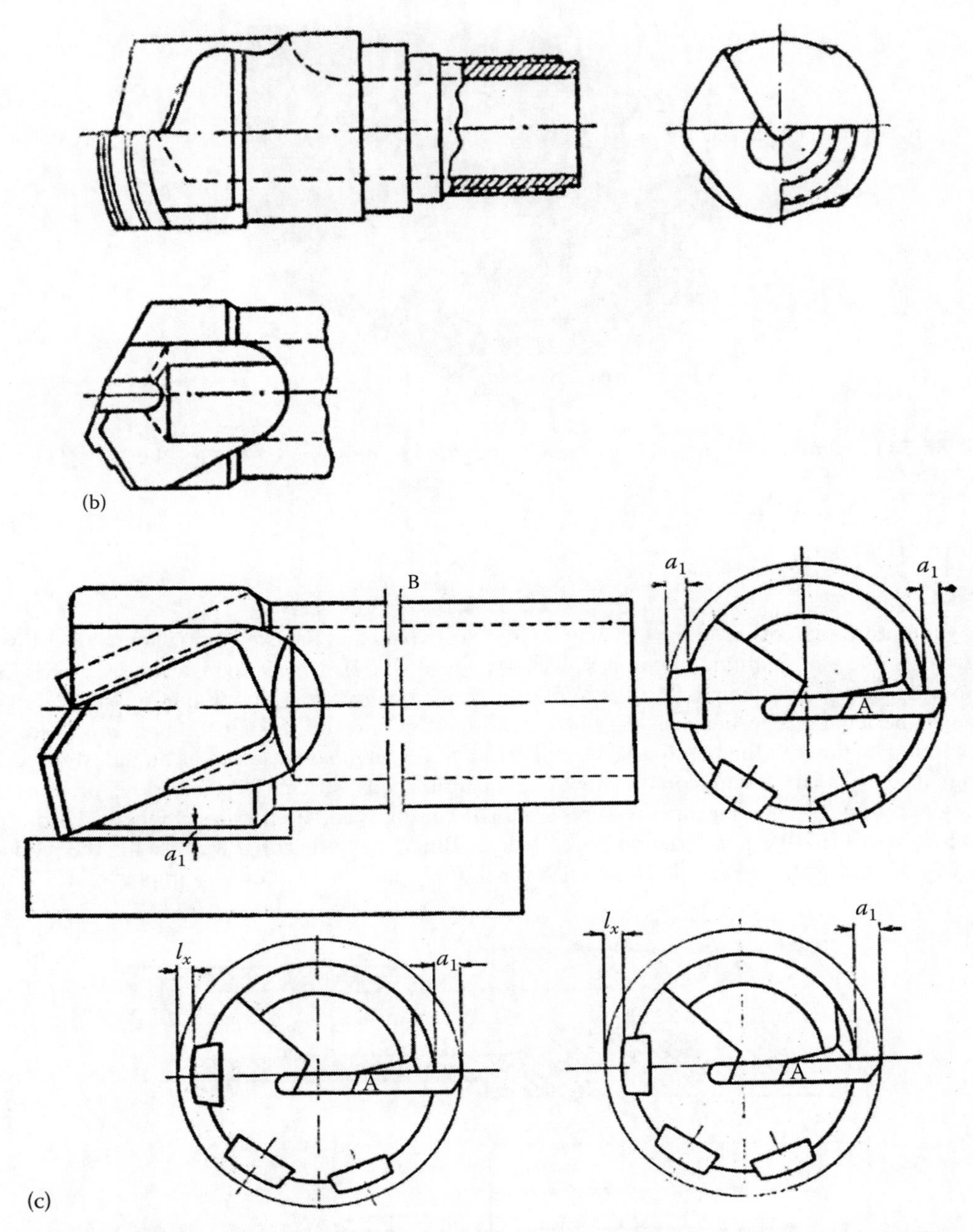

FIGURE 5.116 (continued) Examples of HSS and carbide drilling heads used in the production of gun and artillery in earlier 1930th: (b) drill head with brazed carbide cutting inserts and (c) drill head with replaceable carbide cutting inserts and supporting pads. (From Veremachuk, E.S., *Deep Hole Drilling* (in Russian), State Publishing House of Defence Industry, Moscow, Russia, 1940.)

carbide drilling heads. As can be seen, these closely resemble the designs of modern drilling heads. Moreover, the method of mounting these drilling heads on the boring bar with a square multi-start thread and two centering shoulders is the same as used today.

In 1942, Beisner made a better design of the drilling head. This design is shown in Figure 5.117. The major improvements were bringing the chip mouth closer to the cutting edges so that more efficient cooling and lubrication of the contact areas can be achieved by increasing MWF velocity over

FIGURE 5.117 Beisner's design of the drilling head.

these areas. Moreover, better conditions for chip removal over the chip mouth can be achieved under conditions of relatively small (compared to modern drilling conditions) MWF flow rates. The additional stabilizing pad is added behind the cutting edge in the axial direction to limit drill vibrations.

The Boring Trepanning Association (BTA) was founded in 1945 to commercialize the made-in-Germany *BTA* deep-hole products abroad, particularly in France, the United Kingdom, and the United States. Gebruder Heller in Bremen undertook the task, which resulted in the formation of the BTA group founded in the US and UK subsidiaries. A US federal trademark registration was filed for BTA by Gebruder Heller GmbH (Bremen–Mahndorf) on December 02, 1959, and USPTO has given the BTA trademark serial number of 72087104. The BTA trademark was filed in the category of Musical Instrument Products. The description provided to the USPTO for BTA is Industrial Oils and Greases. It was cancelled on September 21, 2001. Currently, the American Heller Corporation owns this trademark. The description of their goods and services is "Boring Tools and Machines for Solid Boring, Trepanning Counter Boring and Finish Boring."

5.2.6.2 Basic Operations

Although the STS method was initially developed as drilling deep-hole technology, its implementation in modern manufacturing significantly broadened over the years since its introduction. Today it includes the following basic operations (Figure 5.118):

- STS solid drilling
- STS counter boring
- Trepanning.

Figure 5.119 shows STS drills with single cutting insert and with three cutting inserts brazed in the drill body. These drills are used for small hole diameters starting with 11 mm. Figure 5.120 shows the drilling heads with indexable cutting inserts and supporting pads. The minimum diameter of the drill with two cutting inserts is 24 mm (Figure 5.120a), while that with multiple cutting inserts are used in the diameter range of 150–350 mm.

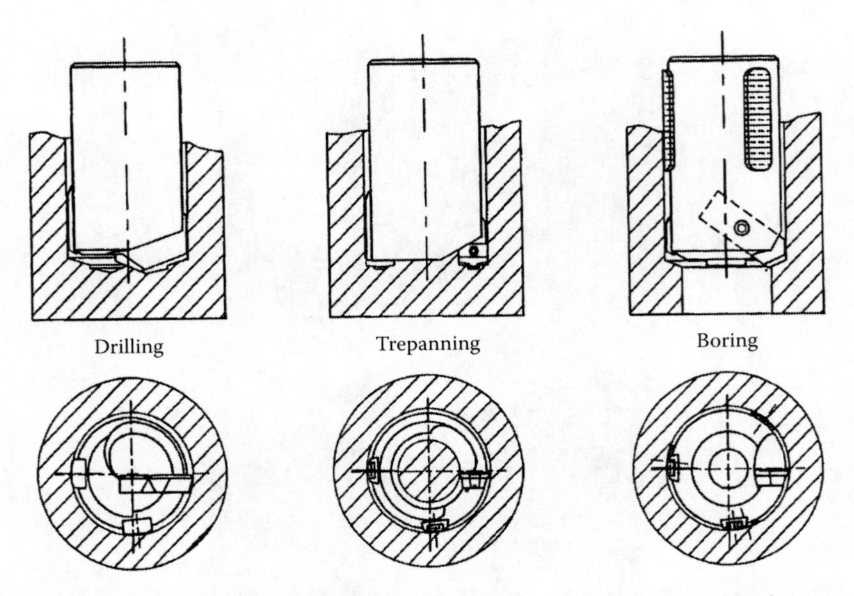

FIGURE 5.118 Basic STS hole-manufacturing operations: drilling, trepanning, and boring.

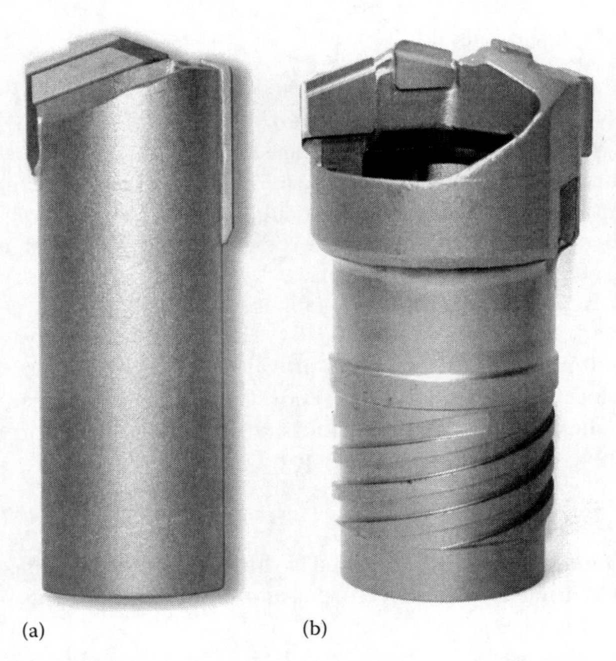

FIGURE 5.119 STS drills with single cutting insert (a) and with three cutting inserts (b) brazed in the drill body by BTA Heller Inc.

Instead of making a hole by cutting all the metal into chips, trepanning (pronounced TREE-panning or treh-PAN-ing) removes a solid core of material by cutting around it. This is an advantage when cutting expensive alloys, as the solid core can be used to make other parts, or, if it is recycled, is more valuable than chips. The trepanning head is completely hollow as can be seen in Figure 5.121, and the cutting process is similar to the solid STS drilling, but it requires less spindle power, as it cuts less material at each revolution. Normally, the indexable trepanning tools cover a range of diameters from 90 to 600 mm and provided with either an internal or external fast lead thread.

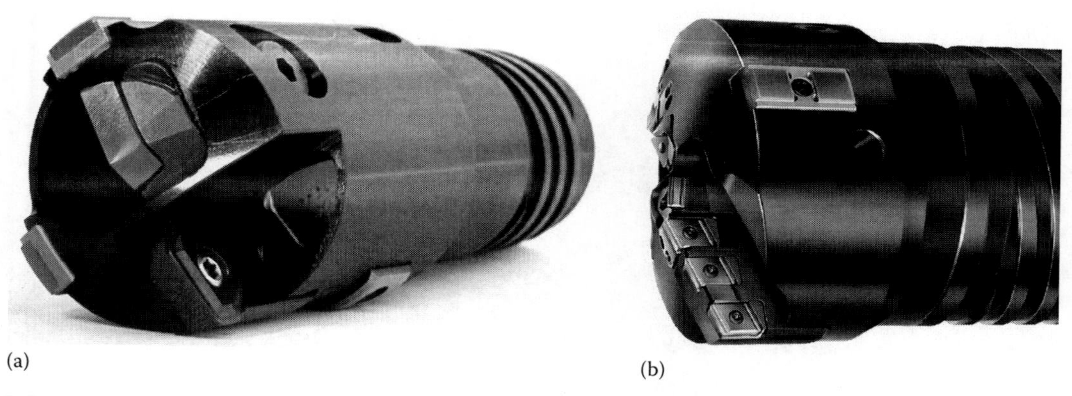

(a) (b)

FIGURE 5.120 STS drill with indexable cutting inserts and supporting pads: (a) drill for smaller hole diameters (by BTA Heller Inc.) and (b) drill for large drill diameters (type 43 drill head by Botek Co.).

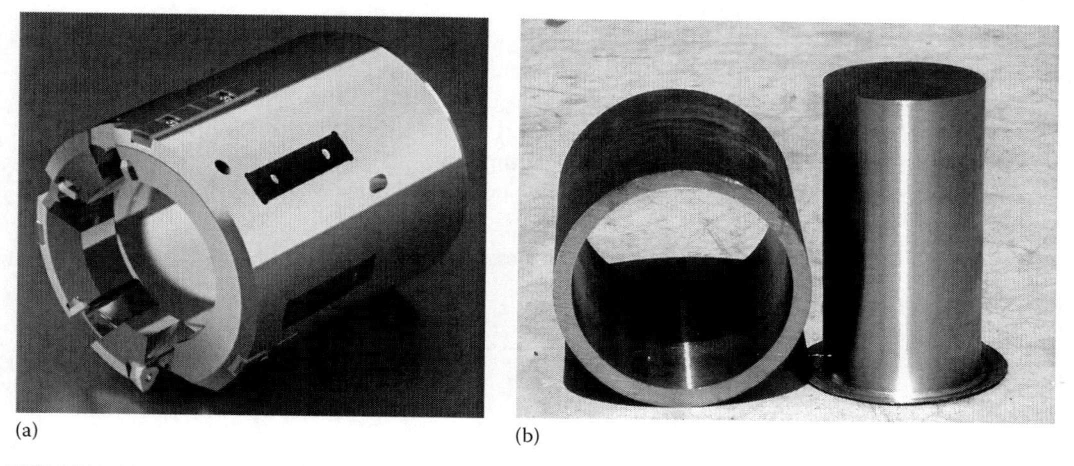

(a) (b)

FIGURE 5.121 Trepanning tool (a) and the trepanned workpiece and its core (b).

The carbide inserts of multiple carbide grades, coatings, and chip breakers can be used although precision ground inserts provide improved chip control when trepanning specialty materials.

Skiving and roller burnishing are performed when close diameter, roundness, and surface finish tolerances are required, often for hydraulic cylinder applications. For a hydraulic cylinder to operate effectively, the cylinder's diameter must be precisely round and have a mirrorlike surface finish to ensure a tight seal between it and the mating internal piston. This is commonly achieved through skiving and subsequent roller burnishing inside a tubular workpiece. Skiving uses a set of carbide blades positioned around the diameter of a tool to slice away chips and create a geometrically round bore. Roller burnishing, a cold-working process, uses multiple rollers to compress the peaks of material left behind after skiving to generate an extremely smooth surface finish. Burnishing also introduces a residual stress layer into the cylinder wall, which improves cylinder fatigue life.

A skiving tool is essentially a modified floating reamer, using multiple knives in a rapid stock removal process. This utilizes high penetration rates with low radial engagements. Roller burnishing uses one or more rollers to cold work the surface of the bore. Rollers are pressed against the bore, plastically deforming the top layer of the metal, compressing peaks, and filling in valleys. Roller burnishing can produce surface finishes of 1 μm Rz. Skiving and roller burnishing are often combined into one tool to complete both operations in a single pass in these deep-hole machining operations. Figure 5.122 shows a combined three-stage boring, skiving, and roller-burnishing tool.

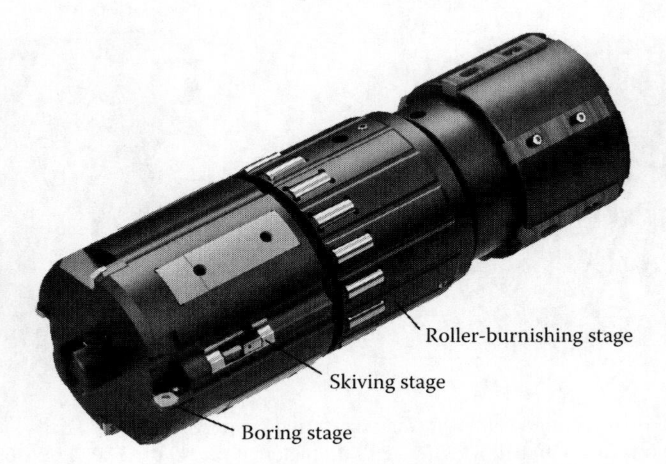

FIGURE 5.122 Combined three-stage boring, skiving, and roller-burnishing tool.

5.2.6.3 Basic Geometry of STS Drills

Figure 5.123 shows the particularities of the tool geometry of the simplest BTA drill. Although one may think that visually it is different from that for a gundrill (Figure 5.71), it is practically the same in reality. The shown drill has a carbide cutting insert (1) brazed into the tool body (2). The cutting insert includes the outer (3) and the inner (4) cutting edges. The cutting insert has side margin (5) ground as a circular land as in any other drills. The leading (6) and trailing (7) supporting pads

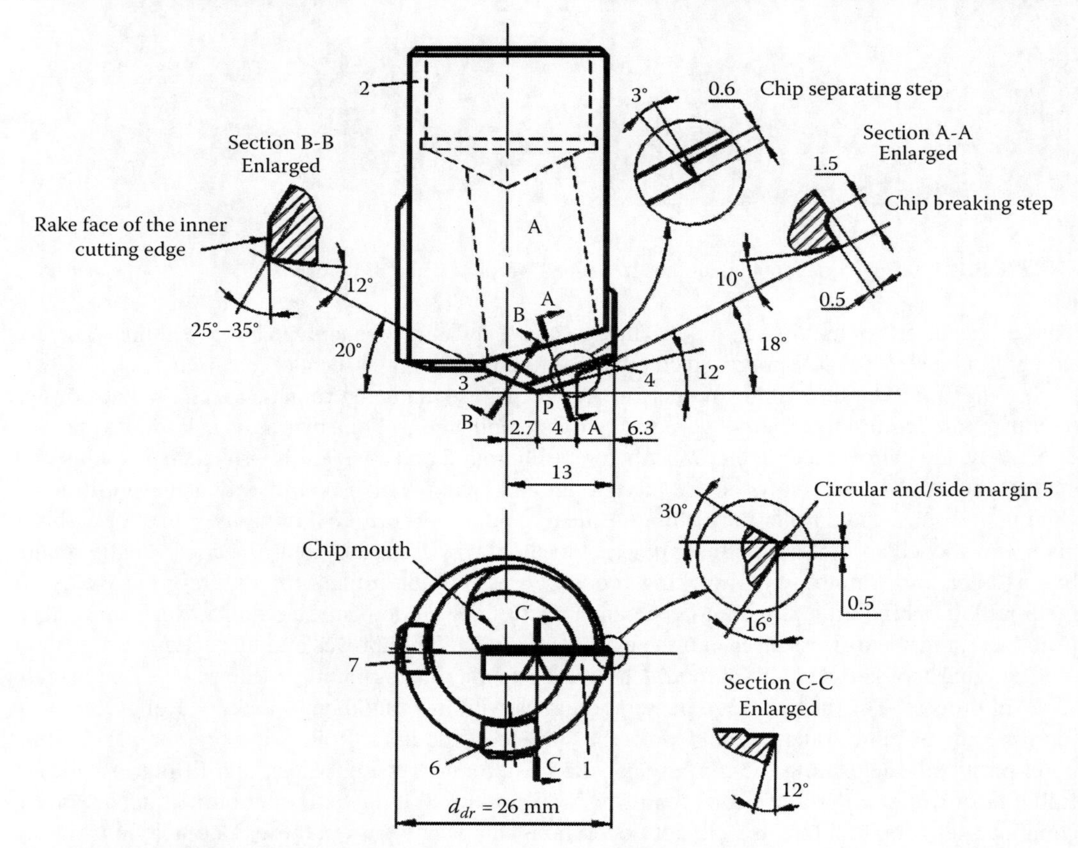

FIGURE 5.123 Particularities of the tool geometry of the simplest STS (BTA) drill.

are mounted on the drill body to balance the drill. However, this drill has the following particularities compared to the gundrill shown in Figure 5.71:

1. The geometry of the outer cutting edge:
   a. The chip-breaking step is provided.
   b. The chip-separating step of 0.6 mm divides the outer cutting edge into two parts with different approach angles (12° and 18°).
2. The geometry of the inner cutting edge:
   a. The offset of the drill point P equal to 2.7 mm is much smaller compared to the gundrill geometry. This offset is approx. 15% of the drill radius, while in gundrill, it is normally 50%. It is compensated, however, by a relatively large approach angle (20°) of the inner cutting edge compared to the outer.
   b. The inner edge rake face is ground with high negative rake angle 25°–35° (the reason will be explained later in this chapter).
3. The drill body has the chip mouth.

The presence of the chip-breaking and chip-separating steps as well as the different approach angles of the portions of the outer cutting edge is explained by the necessity of reliable chip breaking. The chip should be broken into small pieces to be able to pass through the chip mouth and then transported without clogging as swarf through the interior of the drill head and boring bar.

Figure 5.124 shows the parameters of the cutting insert. Table 5.8 shows a fragment of recommended parameters for drills 40–60 mm diameter.

Figure 5.125 shows particularities of the tool geometry of a three-insert STS drill with indexable inserts and supporting pads. Although such a drill is more complicated and thus expensive compared to that shown in Figure 5.123, its major advantages (tested by the author) are as follows:

- Up to 30% higher allowable feed per revolution.
- The design with indexable cutting inserts and replaceable supporting pads allows restoring a worn drill in short time without sending the drill for re-sharpening.

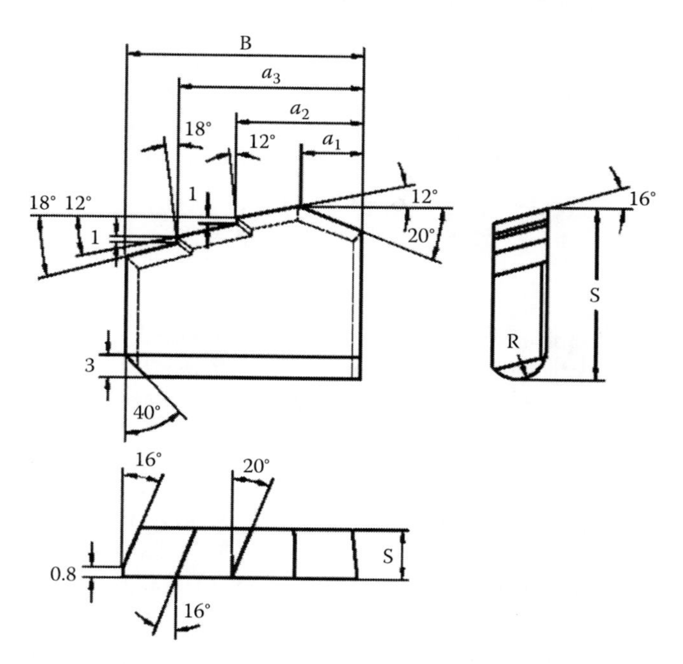

FIGURE 5.124 Parameters of the cutting insert.

TABLE 5.8
Standard Parameters of the Cutting Insert

Drill Diameter (mm)	*B* (mm)	*L* (mm)	a_1 (mm)	a_2 (mm)	a_3 (mm)	*R* (mm)
40–46.99	26	24	7	13.5	20	2.3
47–51.99	28	24	7	14	21	3.4
52–56.99	31	24	8	16.5	24	3.4
57–60.99	33	25	8.4	16.6	24.8	3.4

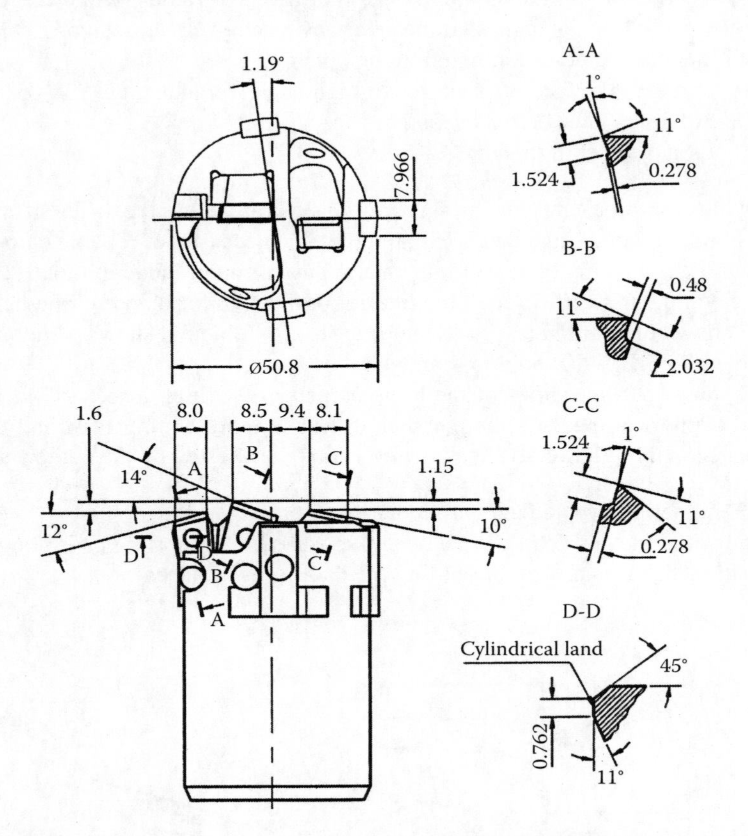

FIGURE 5.125 Particularities of the tool geometry of a three-insert STS drill with indexable inserts and supporting pads.

- Different grades of carbide/coating can be used on each insert.
- Different grades of carbide of the cutting edges and different chip breaker parameters can be used on the same drill body that allows using the same body for drilling various work materials.

5.2.6.4 Power and Force

Understanding the partition of the total power supplied to an STS drill allows the proper assessment of drilling efficiency and the optimization of the drilling operation including the drill design optimization. The method of assessment of the energy partition was proposed by Stockert and Weber 1977b. According to this method, the initial data on the drilling torque can be obtained by analyzing this torque on the drill entrance as shown in Figure 5.126. It can be seen in this figure that steady growth of the drilling torque is interrupted by two steps corresponding to two chip separation steps ground

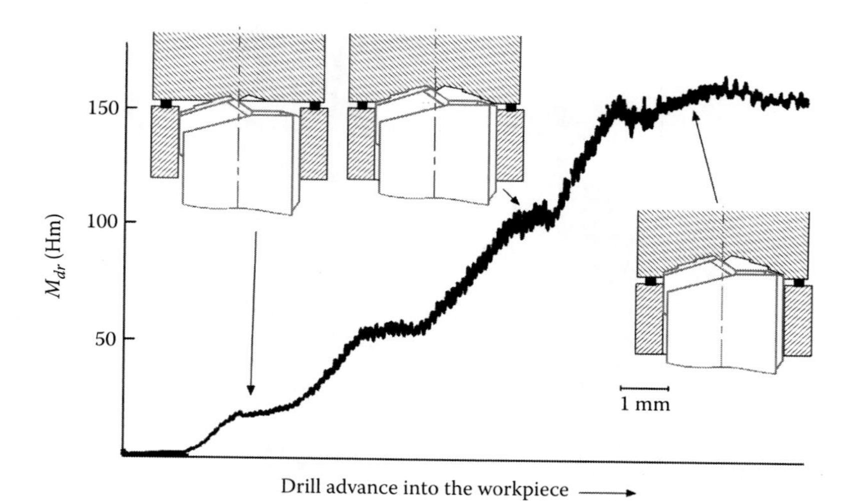

FIGURE 5.126 Variation of the drilling torque at drill entrance into the workpiece. A solid STS drill of 500 mm diameter; cutting speed $v = 90$ m/min, cutting feed $f = 0.14$ mm/rev; and work material AISI steel 1060.

on the cutting edge. As the length of the cutting edge does not increase while the step enters the workpiece, the drilling torque does not increase during this time period. The third step in the torque curve (Figure 5.126) approximately corresponds to the full load of the cutting edge as the supporting pads remain in the starting bushing. A further increase in the drilling torque occurs when these pads enter the workpiece. After this, the curve of the drilling torque slightly fluctuates around the constant value known as the total (resultant) torque in drilling the full length of the hole.

Knowing the rotational speed and torque records, one can assess the power spent in drilling and its partition. Figure 5.127 shows the power balance. In this balance, the total power spent by the supporting pads is divided into two parts. The first part is the power spent in pure friction of the supporting pad against the workpiece, and the second part is the energy spent in burnishing of the drilled surface by these pads. This became possible as a series of tests with pre-burnished holes was

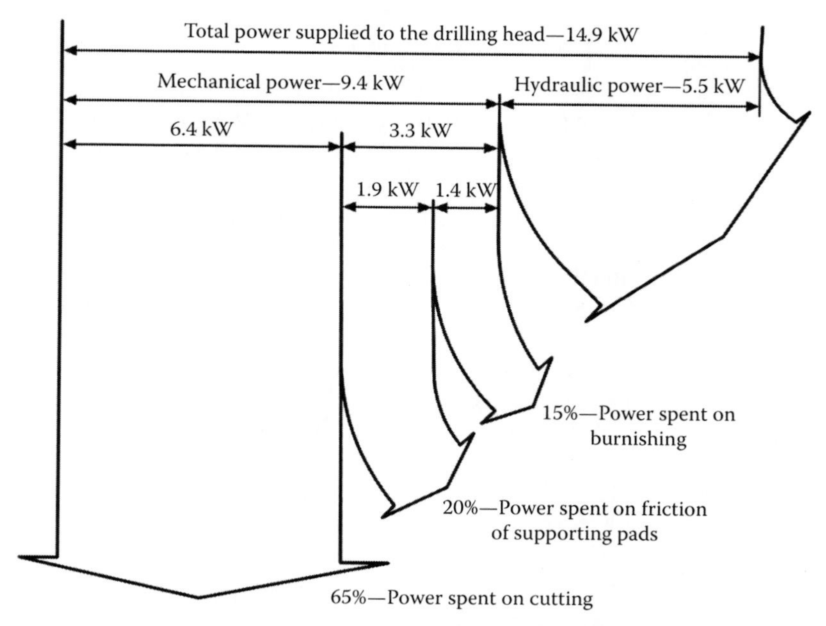

FIGURE 5.127 Power balance for the conditions indicated in Figure 5.126. MWF flow rate $Q_{fl} = 227$ L/min.

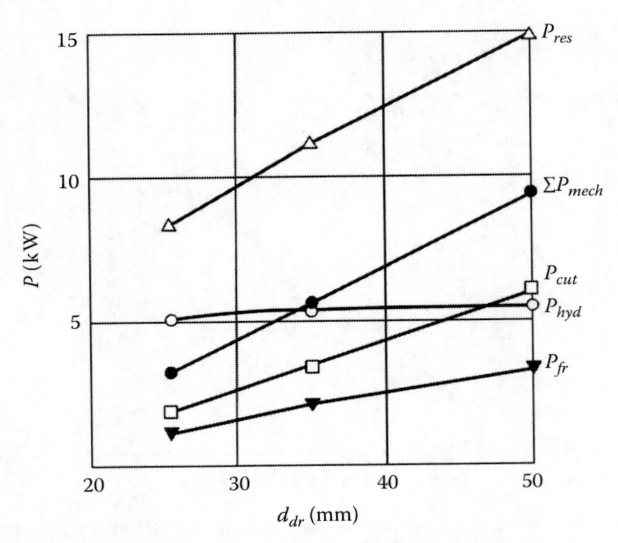

FIGURE 5.128 The total (resultant) power (P_{res}) and its components: ΣP_{mech}, total mechanical power; P_{cut}, cutting power; P_{fr}, total power spent on friction and burnishing of the supporting pads; and P_{hyd}, hydraulic power.

used to assess the power spent in pure friction. The diameter of these holes was adjusted so that no change in the surface roughness occurred, so there was no burnishing.

To verify that the obtained result on the power balance is not drill diameter/design specific, three other series of tests were carried out with STS drill having 25.5, 35, and 50 mm diameter. The same work material AISI steel 1060 and the cutting speed v = 90 m/min were used in all the tests. For the test were carried out with drill of 25.5 mm dia. drill 25.5 mm, the cutting feed f = 0.08 mm/rev; MWF flow rate Q_{fl} = 108 L/min. For drill of 35 mm dia., the cutting feed f = 0.10 mm/rev; MWF flow rate Q_{fl} = 154 L/min. For drill of 50 mm dia., the cutting feed f = 0.14 mm/rev; MWF flow rate Q_{fl} = 227 L/min. The obtained results are shown in Figure 5.128. As can be seen, the power partition is practically the same as presented in Figure 5.127.

The analysis of the power balance reveals the following:

1. The highest portion of the power (approx. 65%) is spent on chip formation, so it should be *considered* first in the drilling optimization. Such optimization may include the use of the application-specific chip breakers (both geometry and dimensions) and optimal clearance angles. In the author's opinion, however, the cutting conditions near the axis of rotation where the rake angle of –30° is used (Figure 5.123) should be optimized first.
2. The energy spent on friction is greater than that on burnishing. It should be clear that burnishing is a useful function of the tool as it improves integrity of the drilled surface while energy spent on friction is wasted. Although it cannot be avoided, it certainly can be minimized by the optimum pad location and pad geometry and by the proper selection of the combination of the pad and work materials. Pad coating and polishing can also help.
3. The hydraulic power constitutes about 37% of the total power spent in drilling. In reality of modern machine, the portion of the hydraulic power is much greater as the power of secondary equipment as, for example, chillers (for maintaining MWF temperature), automated chip crushing, and centrifuge system should also be taken into consideration.

Optimization of the cutting tool geometry/design and the proper selection of the drilling regime may require the determination of the cutting force components and the location of the point of application of the resultant force. There are two principal ways of determining these parameters, namely, experimental and theoretical (Astakhov and Osman 1996a, Astakhov and Galitsky 2005, 2006).

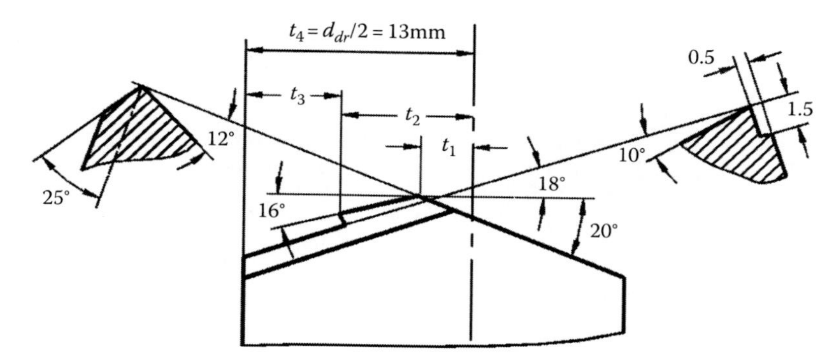

FIGURE 5.129 Geometry parameters used in the tests.

The former is more accurate but valid only for the test conditions. The latter is much less accurate including too many unknown variables. However, in the author's opinion, a tool designer/process planner/manufacturing engineer should be aware of both, particularly in the way of representing the results obtained using each method.

As an example of experimental determination of the cutting forces, consider an STS tool shown in Figure 5.123 having the geometry shown in Figure 5.129. In order to obtain detailed information on the cutting force components and drilling torque, the tests were carried out for each part of the cutting insert (shown as t_1, t_2, t_3, and t_4 stages in Figure 5.129) (Bescrovny 1984). In the tests, the drilling head was located in the starting bushing mounted in a dynamometer. The dynamometer design is discussed in Astakhov (2001). The methodology of the tests was as described Astakhov, V.P. and Shvets, S.V. 2001.

Briefly, the test conditions were as follows:

- Machine—a special CNC deep-hole drilling machine with a high-pressure MWF supply system capable of delivering a flow of up to 220 L/min and generating a pressure of 4.5 MPa was used
- The stationary workpiece–rotating tool working method was used in the experiments.
- Workpiece material was AISI 1040, having ultimate tensile strength 580 MPa. Test bar of 150 mm length and 60 mm diameter was used
- Cutting tool—specially designed SPT heads of 26 mm dia. with internal chip removal were used. The tool material was carbide ISO P10. As shown in Figure 5.129, the cutting edge is divided into three sections: the outer, middle, and inner cutting edges. The outer and middle edges have 18° and 12° approach angle, respectively. These edges are separated by a step, which are at 18° and 12°, respectively. The inner cutting edge is reversed. The offset is approx. 15% of the drill radius. The rake angle for the inner cutting edge is –25° in comparison to 0° for the outer and middle cutting edges. Tolerances for all angles were ±0.5°. The roughness Ra of the face and flank of the drills was less than 0.45 μm. Each cutting edge was examined at magnification of 15× for visual defects such as chips or cracks.
- Parameters of experiments—the following parameters were selected as the cutting conditions: $t_1 = 2.7$–6.5 mm; $t_2 = 1.8$–6.3 mm; $t_3 = 3.0$–6.3 mm; $t_4 = d_{dr}/2 = 13$ mm; $f = 0.062$–0.125 mm/rev; and $v = 40$–90 m/min.
- Statistical analysis of the results—the 22 factorial, complete block type of the design of experiments was used to establish the experimental force relationship (Astakhov 2012a). The mathematical models obtained as regression equations have been statistically analyzed. Such an analysis included the examinations of variance homogeneity, die significance of the model coefficients, and the model adequateness (Astakhov 2012b).

The experimental results are shown in Table 5.9.

TABLE 5.9
Experimental Results

Cutting Edge Number (Figure 5.129)	Statistical Relationships for Cutting Force Components		
	F_x (N)	F_y (N)	F_z (N)
1	$1497t^{0.98}f^{0.81}$	$450t^{1}f^{0.81}$	$594t^{0.94}f^{0.61}$
2	$1560t^{0.99}f^{0.78}$	$585t^{1.07}f^{0.96}$	$636t^{0.93}f^{0.66}$
3	$1620t^{0.94}f^{0.77}$	$770t^{0.92}f^{0.90}$	$728t^{0.93}f^{0.63}$

The experimental results showed the following:

1. The width of the stage (the width of cut) t_i and the cutting feed f have the prime influence on the force components, while the influence of the cutting speed is negligibly small.
2. The data obtained allow to find out not only the force magnitudes but the point of the resultant force application. It was proven in the test that the force components can be considered as applied in the middle of the corresponding stages of the cutting edges.
3. For practical calculation, components F_x and F_y can be determined as

$$F_z = (0.60...0.65)F_x \tag{5.63}$$

and

$$F_y = (0.30...0.33)F_x \tag{5.64}$$

Similar to gundrills, the problem with cutting in the region close to the axis exists in BTA drills.

5.2.6.5 Problem with the Core

It is discussed in Section 5.2.5.5 that although the gundrill does not have the chisel edge, the problem with cutting in the region close to the axis of rotation does not magically disappear. To avoid interference with the bottom of the hole being drilled, the rake face of gundrills is located below CL allowing formation of a string. A number of ways to deal with this string were also discussed.

It should be not a surprise that the problem with cutting in the region close to the axis of rotation exits in STS and ejector drilling. Traditionally, the stage of the cutting edge adjacent to the center of rotation is ground with a highly negative rake angle (25°–30°, see Figure 5.119), so the formed core at the center of the hole being drilled is bent and thus fractured by this rake face. In the author's opinion, this problem is not normally discussed in the literature on the subject because when the basic geometry of STS drills was introduced, the cutting feed used for drilling was relatively low so that the core fracture contribution was not profound. Moreover, other problems, for example, chip breakage, spindle-starting bushing alignment, quality and accuracy of shanks (tubes), and supplying of the sufficient flow rate of MWF, greatly overshadowed the problem with the core fracturing. When the listed problems were eventually fixed and when higher feeds were attempted, the problem with core fracturing became important.

One can understand the problems easily if he or she analyzes the data presented in Table 5.9. As can be seen, the contribution to the total cutting force of stage 1 of the cutting edge is practically the same as that of stage 3 although stage 3 removes approximately five times greater volume of the work material per one revolution. It means that the cutting by stage 1 of the cutting edge requires fivefold more energy per unit volume of the work material removed.

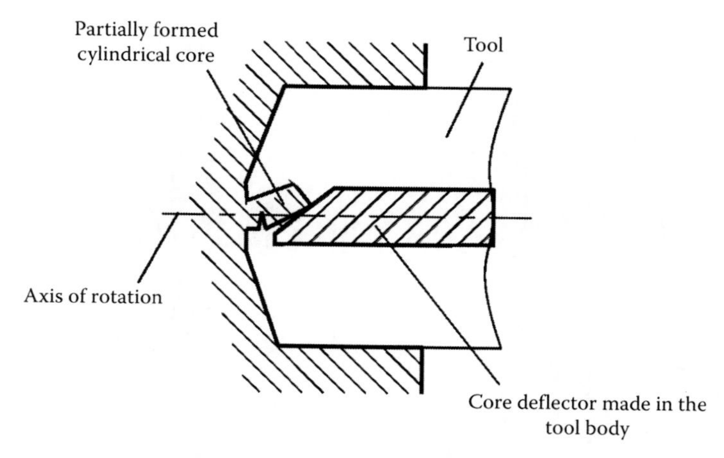

FIGURE 5.130 Core deflector as a part of the drill body (US Patent No. 4,565,471 (1986)).

A universal approach to deal with a partially formed core is shown in Figure 5.130. In this design, the core deflector is a part of the drill body (US Patent No. 4,565,471 (1986)). As can be seen, when a partially formed string contacts the deflector, it bends and then fractures from the bottom of the hole being drilled. The design and location of the chip deflector depends on the type and design of a particular drill and thus can be used with any type of drills when one tries to solve the problems that unavoidably occur for any drill in the region adjacent to the axis of rotation.

Figure 5.131 shows the design according to US Patent No. 4,616,964 (1986) where the side of the cutting insert deflects the partially formed core. As can be seen, the area around the axis of rotation is made a nonmachining zone by shifting the cutting inserts from this axis. As a result, the cutting edge is deliberately shifted from the zone of very low cutting speeds reducing the axial force and thus allowing greater penetration rates. The distance between the axis of rotation and the cutting edge of the insert defines the radius of the forming core. To deflect the core, this distance is made larger than that at the rear end of the insert in the axial direction. As a result, the core is deflected and then fractured.

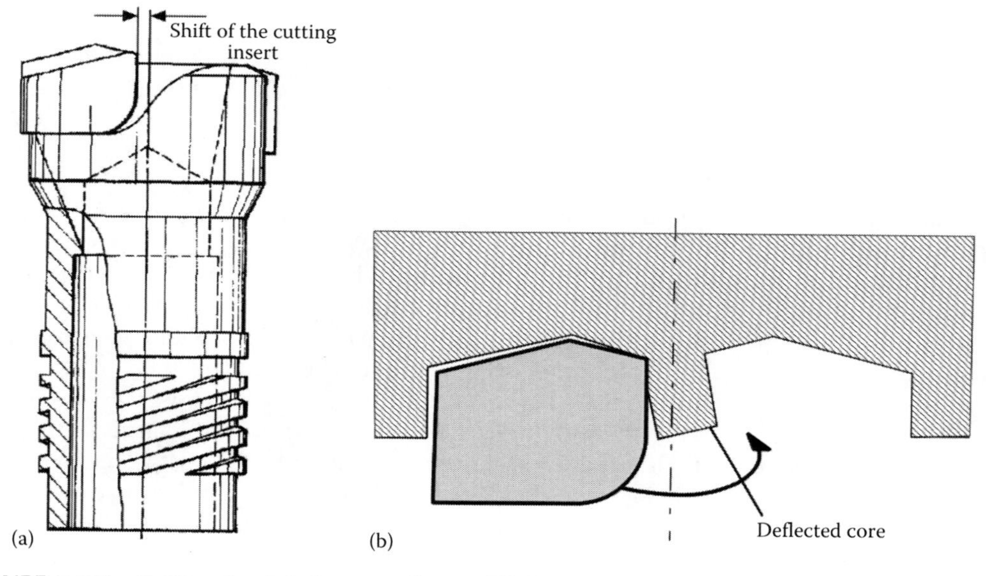

FIGURE 5.131 Drilling head design according to US Patent No. 4,616,964 (1986): (a) drill design, and (b) core deflection in dilling.

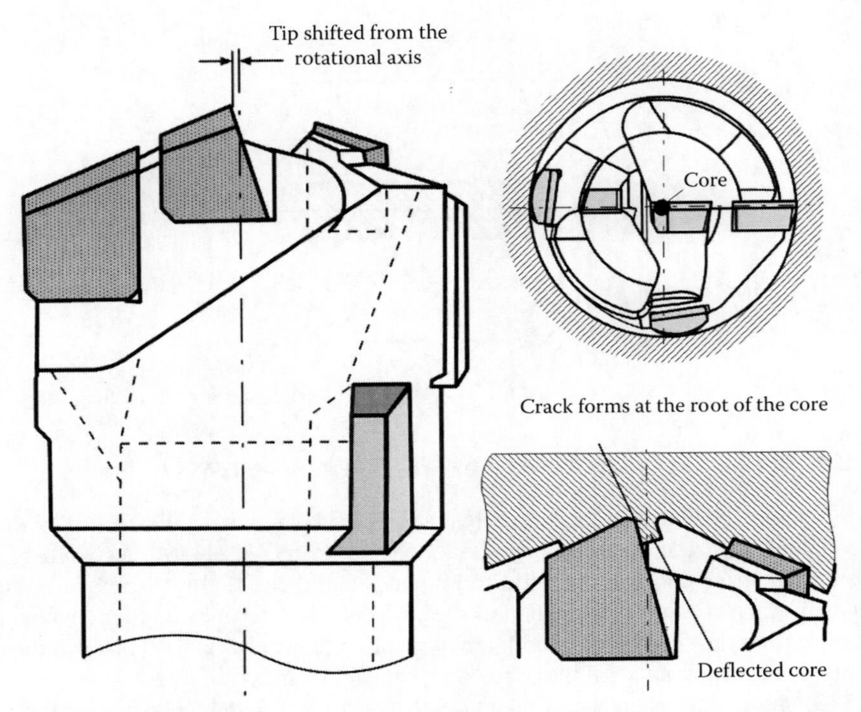

FIGURE 5.132 Drilling head design according to US Patent Application No. 2011/0067927 (2011).

Figure 5.132 shows the modification of the previously described idea for the drilling head design according to US Patent Application No. 2011/0067927 (2011). This drilling head includes screw-clamped carbide inserts. As can be seen, the central cutting insert forms an angle in a specific range with respect to the insert tip (or the axis of the drill). As a result, lateral displacement of the partially formed core takes place by the side surface of the cutting insert. To protect the edge of the insert from chipping, the central cutting insert locates well above CL so only the side of this insert participates in the core deflection and fracture.

In the author's opinion, practically all the known designs utilize the same idea—deflection and the fracture of the partially formed core. This still results in increase cutting force and in destabilization of drill smooth performance.

5.2.6.6 Problem with the Pressure Distribution

The architecture of the drilling head developed by Beisner addressed the conditions of MWF delivery parameters (relatively low flow rate) assuring smooth MWF flow in the machining zone and proper removal of the chips generated by the cutting insert. The introduction of the indexable cutting inserts and replicable supporting pads gave drilling heads some new *squared* style, while not much care about the MWF conditions around the supporting pad and in machining zone was taken as the modified Bernoulli equation (see Section 6.3.3.3) seems to be forgotten.

The modified Bernoulli equation relates the static pressure in the MWF flow with the velocity of this flow. The higher the velocity, the lower the static pressure. Because the diameter of the drilling head is made rather close to that of the hole being drilled and because relative high flow rate is to be delivered into the machining zone of STS drills, the velocity of MWF should be high so that the static pressure of the flow around the supporting pads and the side of the cutting insert can be rather low. Having noticed the problem with insufficiently lubricated supporting pads, the tool manufacturers tried to alter the pad design instead of addressing the problem using the modified Bernoulli equation.

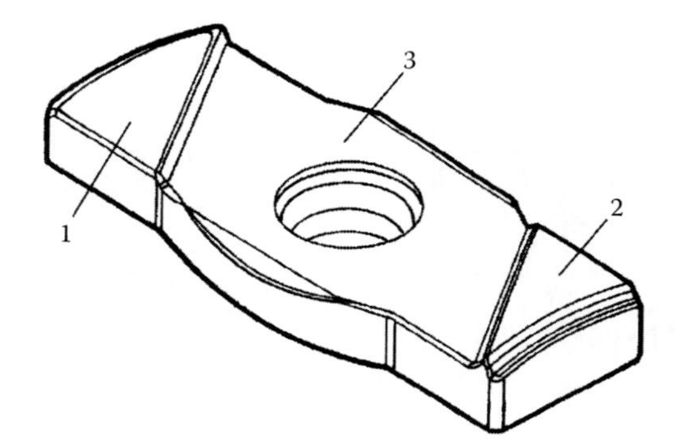

FIGURE 5.133 The improved pad design according to US Patent No. 6,602,028 (2003).

Figure 5.133 shows an example of the modified pad design according to US Patent No. 6,602,028 (2003). The full contact surface of such a pad is reduced to two triangular areas (1 and 2 in Figure 5.133), while a countersink (3) is formed between these two contact areas in the hope to improve pad's cooling and lubricating by the flow of MWF.

The special geometry of the drill head results in the formation of a special shape of the bottom of the hole during drilling. The space enclosed between this bottom from one side and the flanks of the drill head from the other is called the bottom clearance space as was discussed earlier in the considerations of gundrilling. The MWF pressure in the bottom clearance space has a major influence on the cooling and lubricating conditions of the flank and rake contact areas. Increasing the cutting fluid pressure in the bottom clearance space provides better penetration of the oil-based cutting fluid to the narrow passages between the tool flanks and the bottom, that is, better conditions for lubrication and cooling of the flank contact areas. Therefore, the pressure in the regions close to the cutting edge(s) should be as high as possible under a given flow rate. This leads to a considerable reduction in flank wear and therefore increases the tool life.

Unfortunately little attention has been paid to the influence of the drill design parameters on the MWF pressure distribution in the bottom clearance. The *smooth* design of the STS drilling head with a single cutting insert (Figure 5.119) was gradually replaced with body designs of STS drilling head with partitioned brazed and indexable cutting inserts that should have changed considerably the architecture of the flank faces (e.g., as shown in Figure 5.120). No adjustments of this architecture of the flanks were made. To comprehend the difference in the MWF fluid pressure distribution in the bottom clearance space between the STS drilling heads with a single and with indexable cutting inserts, a series of tests were carried out for STS drill having 50.8 mm diameter. The static (the drill just brought to close contact with the bottom of the hole) and dynamic (the drill was rotating) pressure distributions were measured Astakhov, Subramanya, and Osman 1995; Astakhov, Abi-Karam and Osman 1998). Figures 5.134 and 5.135 show examples of the experimental results. A subsequent analysis of these results revealed the following:

1. The static and the dynamic pressure distributions in the bottom clearance space were not uniform, contrary to the currently held view. Both distributions change significantly in the bottom clearance, and there exist regions with zero pressure and even negative pressure under static and dynamic conditions.
2. The rotation of the tool leads to a significant change in the pressure distribution in the bottom clearance space. The rotation makes this distribution more uniform in the case of the STS drill with a single cutting insert and less uniform in the case of the STS drill with indexable inserts. This seems to be due to the swirling effect produced by the 'rough' features of the drill head.
3. In deep-hole drilling, the tool life is defined by the flank wear of the cutting edge(s). It is known that flank wear is influenced by the rate of penetration of MWF to the contact areas between the bottom of the hole being drilled and the flank(s). This rate depends

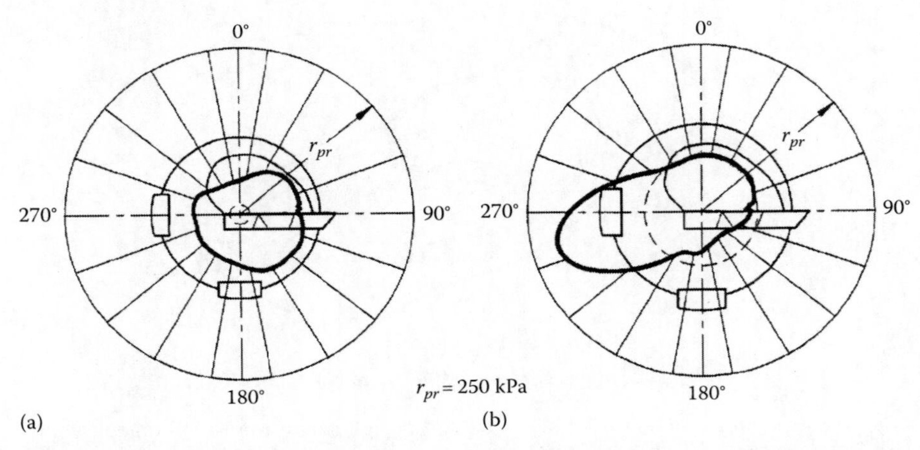

FIGURE 5.134 Examples of the dynamic (the drill rotates) MWF pressure distribution in the bottom clearance space for the STS drill with a single cutting insert: (a) position of the pressure transducer corresponds to the radius r = 3.1 mm and (b) position of the pressure transducer corresponds to the radius r = 13.1 mm.

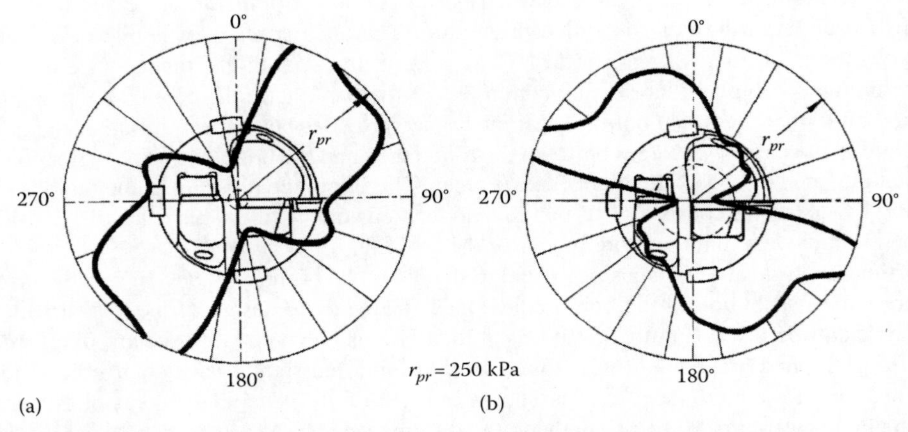

FIGURE 5.135 Examples of the dynamic (the drill rotates) MWF pressure distribution in the bottom clearance space for the STS drill with indexable inserts: (a) position of the pressure transducer corresponds to the radius r = 3.1 mm and (b) position of the pressure transducer corresponds to the radius r = 13.1 mm.

upon both the properties of the cutting fluid and its static pressure in the region adjacent to the flank contact areas. The experiment results show that the STS drill with indexable inserts has a higher cutting fluid pressure in these areas. For the STS drill with a single cutting insert, the zone of maximum pressure is shifted to the direction of the second supporting pad and plays little role in improving the cooling and lubrication of the cutting edge. However, a number of zones with negative (compared to atmospheric) pressure and with recirculating stagnating were found for the STS drill with indexable inserts.

4. The architecture of the STS drill with a single cutting insert provided better conditions for the chip removal process due to much more uniform pressure distribution at the chip mouth of the drilling head.
5. The results also showed that the architecture of the drill head can be changed to achieve a better pressure distribution, avoid negative pressure regions in the bottom clearance space, and provide better conditions for cooling, lubrication, and the chip removal.

5.2.6.7 Addressing the Problems

Figure 5.136 shows an improved architecture of the drilling head according to US Patent No. 6,682,275 (2004). As can be seen, additional MWF channels 1, 2, and 3 are made to provide better MWF delivery to the supporting pads and to the flanks of the cutting inserts. The major difference, however, is in the location of the chip mouth (4). As can be seen, it is located much further down in the axial direction from the cutting inserts that definitely improves chip removal and MWF flows in the bottom clearance space. Still intuitive and not complete, this architecture is a step forward in the right direction.

Figure 5.137 shows an example of optimized STS drilling head for a range of bore diameters from 20 to 60 mm (Vinogradov 1985). The distinctive feature of this drill is the presence of the

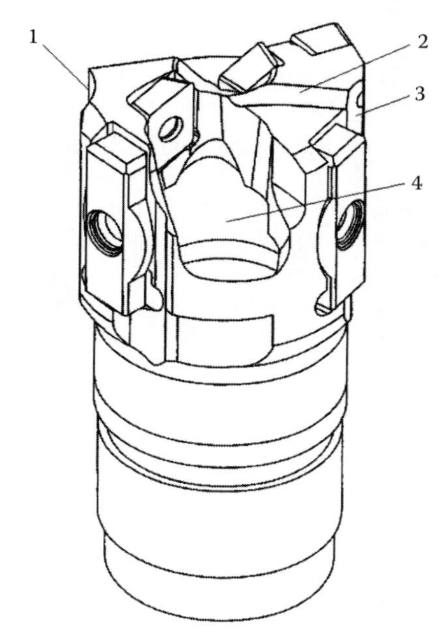

FIGURE 5.136 Improved architecture of the drilling head according to US Patent No. 6,682,275 (2004).

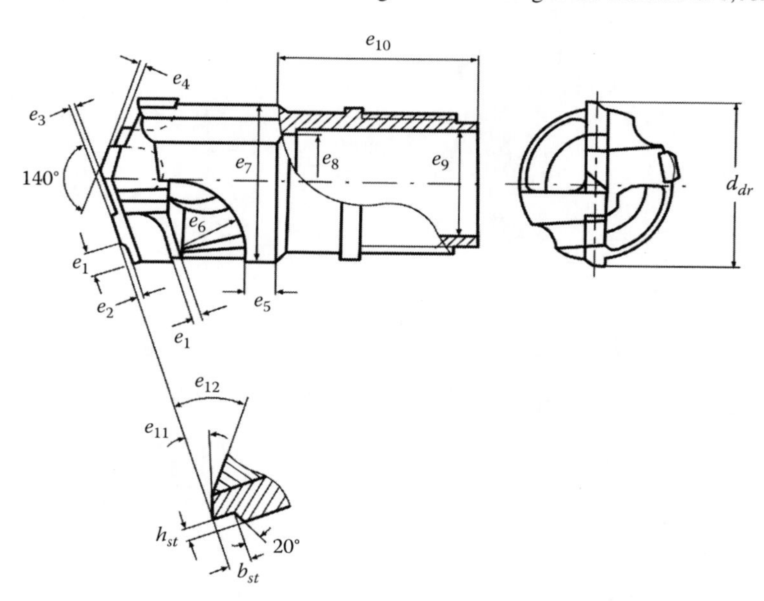

FIGURE 5.137 An example of the optimized design of the STS drilling head.

chisel edge. This chisel edge is shifted from the center of rotation and is similar to that discussed in Chapter 4 (Section 4.5.8.5, Figure 4.141) where advantages of such a chisel edge are listed. Among them, two are most relevant: (1) no one point of the chisel edge has zero cutting speed and (2) self-centering ability without compromising the strength of the drill tip. The problem with the core is permanently solved. Moreover, the architecture of the drill includes the optimized MWF flow passes that assure proper cooling and lubrication of the contact areas.

The proper accounting for the force balance allowed using one supporting pad, which should be the target for any STS and ejector drilling head designs. The following particularities of the drill designs should also be noticed:

1. Diameter e_8 is selected to be equal to d_{dr} – (1–3) mm.
2. The bore of diameter e_8 is made so that the distance between its left shoulder and the end of the cutting insert does not exceed 3 mm.
3. The radius of the z_0y_0 projection of the chip mouth e_6 = (0.4–0.6) e_9.
4. Size e_5 = (4–8) mm.
5. Size e_{10} and thread parameters are standard for STS drills.
6. The point angle is 140° for wide range of difficult-to-machine materials.
7. Size e_3 depends on the drill diameter and the cutting feed. It is 0.2–0.5 mm.
8. The step on the outer cutting insert having size e_2 is ground for chip separation, and it should be approximately $2f$ but should not be greater than $0.2b_{st}$.
9. The step size $e_4 \geq e_2$.
10. The rake angle of all the cutting inserts is zero. However, it can be increased to 5° in drilling of difficult-to-machine materials.
11. The primary clearance angle e_{11} = 10°–12°, while secondary clearance angle e_{12} = 25°–20°.

5.2.7 Ejector Drills

The principle of ejector drilling was invented by Kurt Faber in 1963. US Patent No. 3,304,815 (1967) fully describes this principle. The prime objective of this invention was "to provide a separate rearward action on the return flow of the flushing medium thereby lessening the pressure of this medium at the front of the drill and also lessening the tendency of breakage." As a result, ejector drilling does not always require a special drilling machine as it can be used in many general-purpose machines or even machining centers as one of common drilling operations. To make it possible, Sandvik Coromant Co. offers a variety of ejector adapters (examples are shown in Figure 5.138) designed for different machines and their spindle particularities.

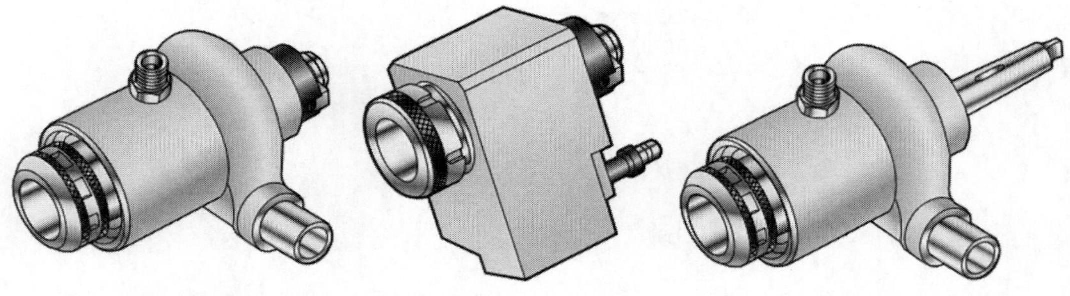

FIGURE 5.138 Adapters for ejector drills offered by Sandvik Coromant Co.

There are three basic myths about ejector drilling circulated in the trade (and unfortunately sometimes in scientific) literature and textbooks:

1. This type of deep-hole drilling is an alternative to STS drilling.
2. The pressure and flow rate needed for ejector drilling are much lower than those for STS drilling.
3. Ejector drilling is suitable only for a group of relatively easy-to-machine work materials.

None of these myths are true.

As discussed in Chapter 6, in practice of MWF systems, machine tools, and cutting tool design and implementation, the pressure of MWF is always considered to be of prime concern, while its flow rate is totally neglected. In the authors' opinion, this is one of the most severe misconceptions in MWF applications as it affects various facets of machining efficiency, productivity, and quality. The MWF flow rate is of prime concern, while its pressure is only a means to assure the flow rate needed to assure that MWF can perform its intended actions in machining (Astakhov and Joksch 2012). As such, four basic actions of MWFs are commonly considered: (1) cooling, (2) lubricating, (3) chip transportation from the machining zone (or even from the working zone), and sometimes (4) chip control. All of these actions are directly defined by the MWF flow rate.

Although all of these actions are important, reliable chip removal is one of the first and foremost requirements to any deep-hole drilling application. It is particularly true for STS and ejector drilling. The MWF (coolant) is supplied to a deep-hole drill and then, after performing its cooling and lubricating actions in the machining zone, it carries away the chips in the form of swarf through the interior of the drill head and boring bar or inner tube. As a result, after the machining zone, a two-phase flow, that is, the chip–coolant mixture, should be considered. When the MWF flow rate is insufficient to transport the formed chip, the chip clogs the chip removal passages. Therefore, it is of prime concern to determine the sufficient flow rate.

The flow of the chip–coolant mixture through the chip removal passage may occur in different transportation modes. Figure 5.139 shows the influence of the MWF velocity on the chip transportation and the velocity profiles of the mixture in a tube located horizontally. When the MWF velocity is low (zone 1 in Figure 5.139), the MWF does not have any effect on the chip accumulated in the tube. As such, the velocity profile shows that the MWF moves, while the chip does not. Increasing the MWF velocity leads to its interaction with the chip (zone 2 in Figure 5.139). When it happens, a part of the chip layer moves with the MWF, whereas the other part forms a slow-moving (gliding) layer at the bottom. This transportation mode is called heterogeneous with a gliding layer. Because

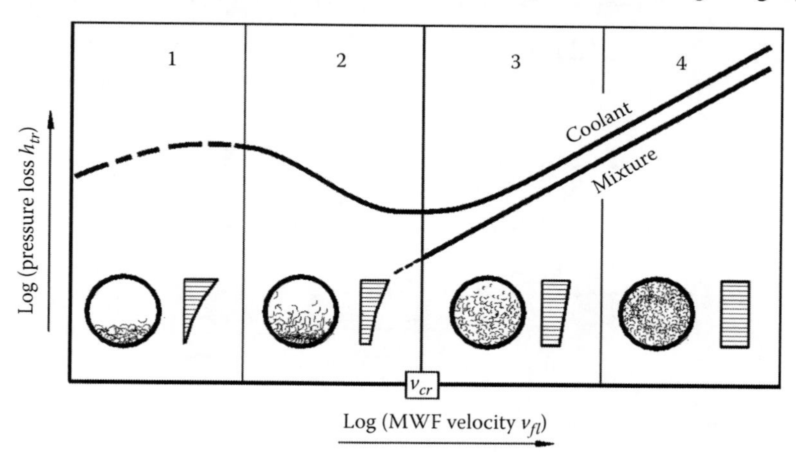

FIGURE 5.139 Flow modes and velocity profiles in a tube with chips.

a lot of chips concentrate at the tube bottom, there is a possibility of formation of chip clogging. Such a mode, however, is considered to be acceptable in gundrilling where achieving the middle of the second zone is desirable. This is because high MWF pressure is needed in gundrillig due to small cross-sectional areas of the coolant channels to deliver the MWF flow rate corresponding to this mode.

Even further increase in the MWF velocity leads to the heterogeneous transportation mode where the coolant–chip mixture forms (zone 3 in Figure 5.139). When the MWF velocity becomes sufficient to reach this transportation mode, all the chip fragments flow in the mixture although the concentration profile is still not fully uniform as shown in Figure 5.139. The coolant–chip mixture velocity corresponding to the beginning of the heterogeneous mode is referred to as the critical velocity, v_{cr}. It is clear that the flow rate corresponding to this critical velocity called the critical MWF flow rate is calculated through the tube diameters and v_{cr} as discussed in Chapter 6.

Although the further increase in the MWF flow rate leads to the pseudohomogeneous flow mode (zone 4 in Figure 5.139), the pressure losses in swarf transportation become significant. Moreover, achieving this regime requires a great coolant flow rate that is not always feasible in deep-hole drilling.

The foregoing analysis implies that no matter what type of drill, STS or ejector, is used, at least the minimum MWF flow rate should be delivered to achieve the critical velocity, v_{cr}, in the boring bar or inner tube to assure reliable chip transportation.

Figure 5.140 presents for the first time a side-by-side comparison of the STS and ejector drilling principles in terms of MWF supply parameters provided that the designs of the drilling heads, work materials, and machining parameters (the cutting speed and feed) used are the same. In other words, the shape and amounts of chip generated and thus to be transported out are the same for both drills. Being reasonable and practical, these assumptions simplify the analysis of the models shown in Figure 5.140.

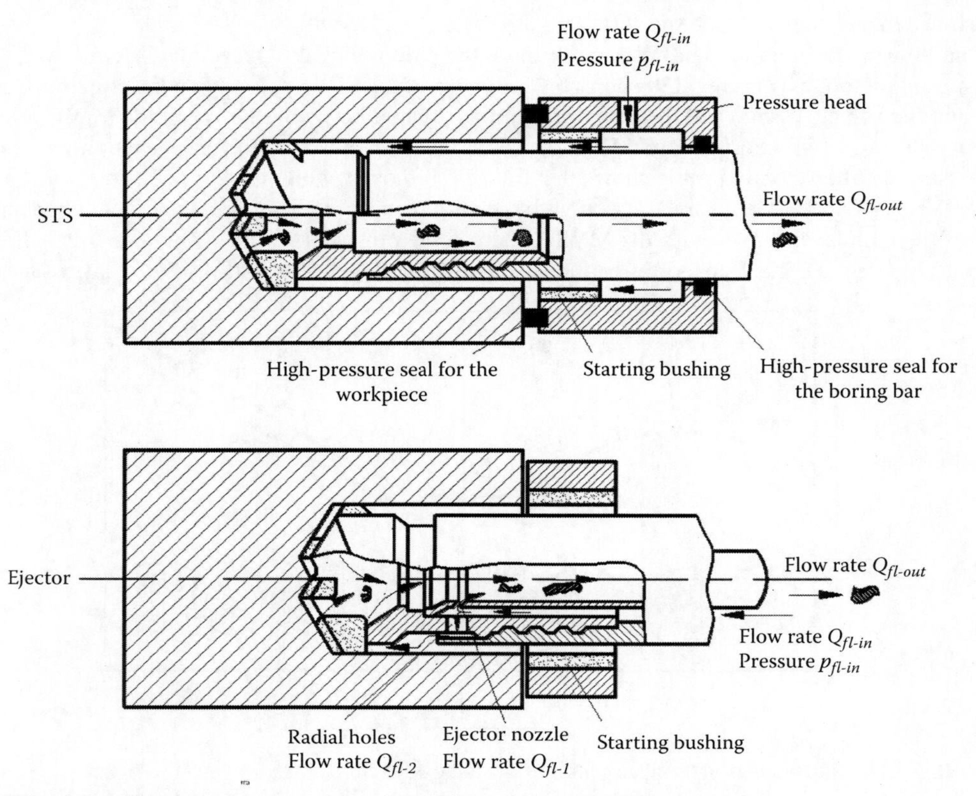

FIGURE 5.140 Side-by-side comparison of the STS and ejector drilling principles.

The pressure and flow balance for the STS drill shown in Figure 5.140, namely, the outlet MWF flow rate $Q_{fl\text{-}out}$, are almost equal (except for possible minor leaks) to the inlet MWF flow rate $Q_{fl\text{-}in}$. As MWF from the boring bar flows into a tank with atmospheric pressure, the outlet pressure is equal to the atmospheric pressure and the inlet pressure is equal to the sum of the major and minor pressure losses calculated from the inlet hole of the pressure head to the exit of the rear end of the boring bar as discussed in Chapter 6 in detail. The experimental results and calculations show that the MWF pressure losses in the annular gap between the boring bar and the wall of the hole being drilled is normally the highest out of major pressure losses (Astakhov, Subramanya, and Osman 1995). Moreover, this portion of the total pressure loss increases gradually as the drill goes deeper. Because in high-penetration-rate STS drilling a great amount of chip is generated, the pressure losses in the mixture in its flow inside the boring bar are significant.

A device called the pressure head is used to deliver the MWF flow into the annular clearance between the boring bar and starting bushing and then the wall of the hole being drilled as drilling progresses. Although in the literature it is often depicted in the manner shown Figure 5.140, the functions and the design of the pressure head are wider and more complicated.

The pressure head is a unit of the drilling machine. This unit

- Directs MWF into the drill
- Mounts drill bushing
- Contains locating bore and pilot for work holding components to provide support for the workpiece
- Contains precision rotating spindle with high thrust and radial load capacity (for rotating workpieces)
- Contains boring bar seal packing glands.

Figure 5.141 shows an example of the pressure head design capable of delivering 10 MPa MWF pressure (Shertladse et al. 2006). Rotating spindle (1) is installed on the bearings mounted in the head housing (2). The workpiece locator (3) mounted on the front end of the spindle has a tapered surface to locate the workpiece (4). The taper contact between the workpiece and the locator assures reliable seal under the action of surface springs (5) and due to MWF pressure acting on the face of

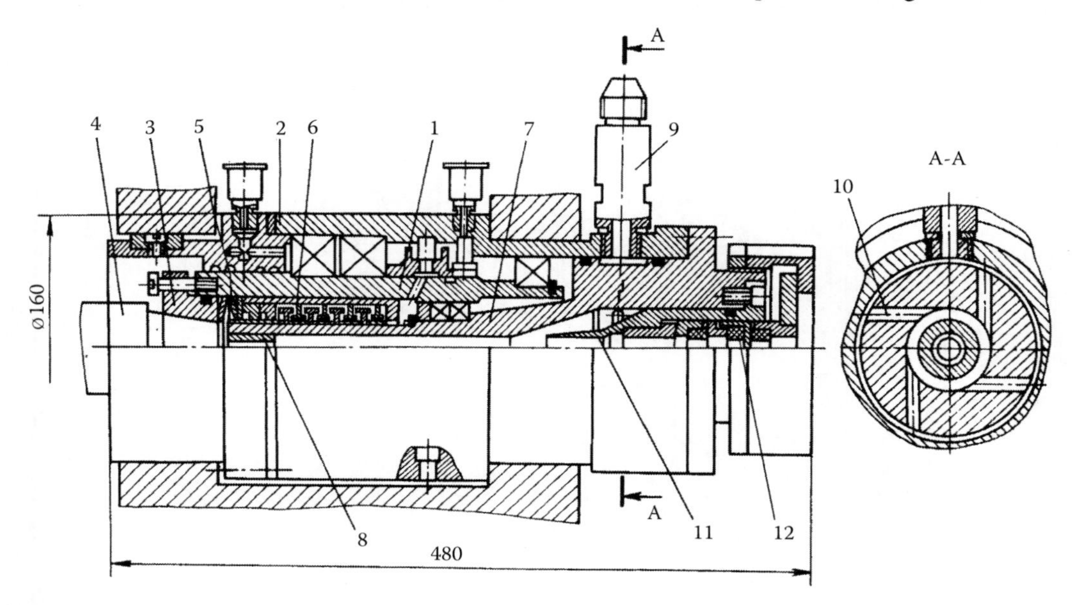

FIGURE 5.141 Pressure head.

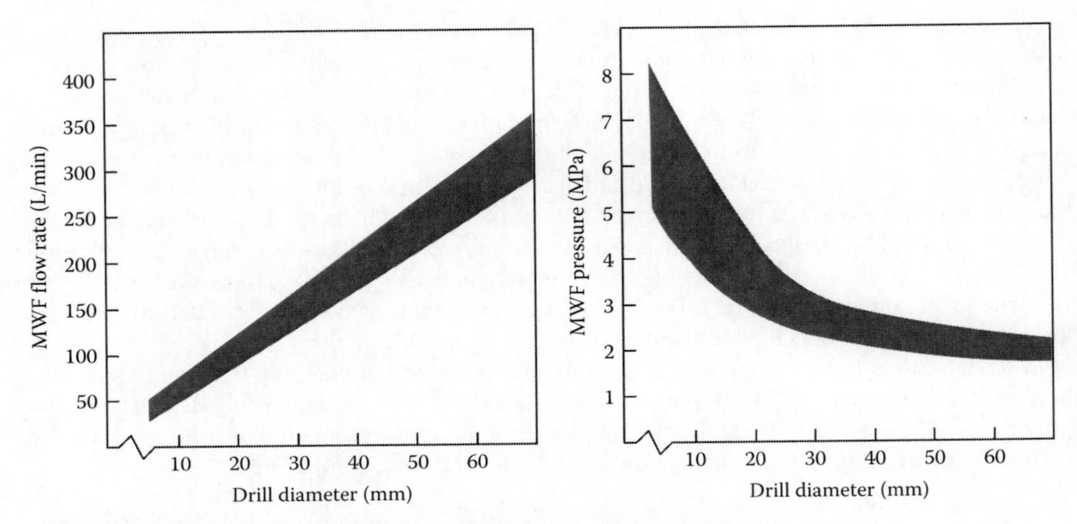

FIGURE 5.142 Pressure and flow rate of MWF as functions of STS drill diameter.

the locator. The labyrinth seal package (6) prevents MWF leaks between the spindle and the stationary unit (7) having the starting bushing (8) in it. MWF is supplied through the connector (9) and then through tangential holes (10). The directing flange (11) is to unify the coolant flow and thus prevent any MWF flow fluctuations. The seal (12) is for the boring bar.

Figure 5.142 shows the required MWF pressure and flow rate depending upon BTA (STS) drill diameter. Because of significant flow rates of MWF involved in STS drilling and high requirement to MWF clearness, a typical MWF supply system shown in Figure 5.143 includes a high-volume tank having the volume equal to the maximum MWF minute flow rate per 10 min (e.g., if the maximum flow rate is 200 L/min, then the tank volume should not be less than 2000 L); array of MWF filters that commonly includes magnetic, centrifugal, and paper-cartridge filters; refrigerant chiller to stabilize the MWF temperature; and high-pressure pumps. As a result, the shop-floor area occupied by such a system often is the same or even larger than that occupied by the STS deep-hole machine itself (Astakhov and Joksch 2012).

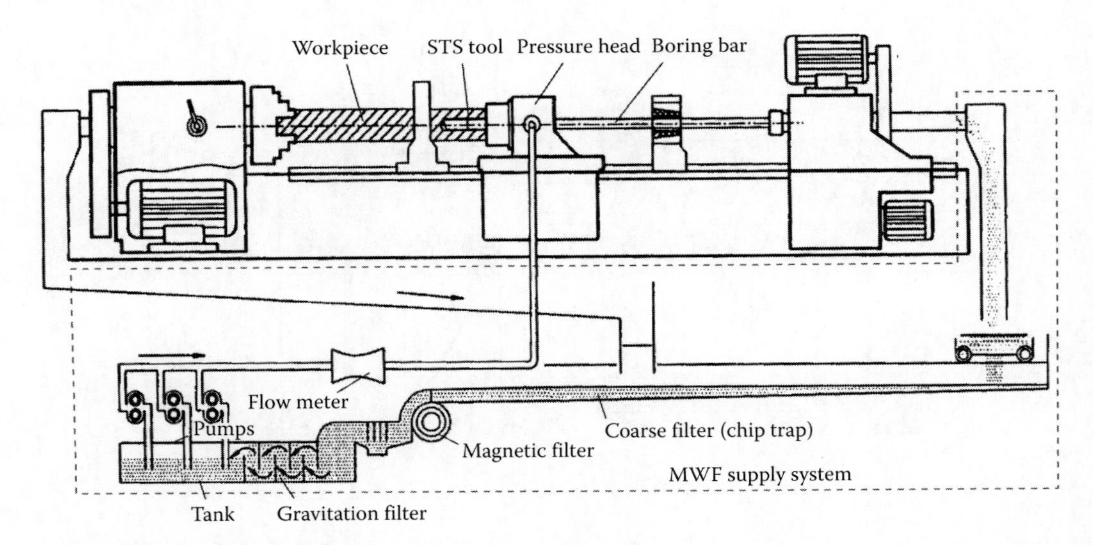

FIGURE 5.143 Schematic of MWF supply system for an STS deep-hole machine.

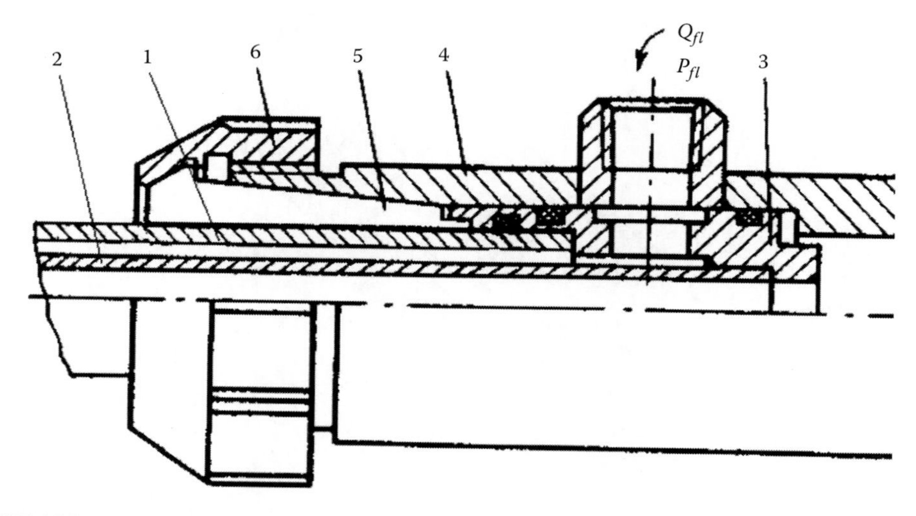

FIGURE 5.144 Basic design of the adapter for nonrotating ejector drills.

The pressure and flow balance for the ejector drill are shown in Figure 5.140. Similar to the STS system, the outlet MWF flow rate $Q_{fl\text{-}out}$ is almost equal (except for possible minor leaks) to the inlet MWF flow rate $Q_{fl\text{-}in}$. This flow rate needed for chip transportation as well as for cooling and lubrication of the cutting and bearing contact areas is passed to the drill head by means of a two-tube system using an adapter (Figure 5.138).

Figure 5.144 shows the basic design of the adapter for nonrotating ejector drills. As can be seen, the outer (1) and inner (2) tubes are mounted into the sealing sleeve (3) installed into the adapter body (4). The precision collet (5) is tightened by the nut (5) to secure the outer tube in the adapter. The design is simple and reliable and the whole unit is not expensive.

The outer tube referred to as the boring bar takes up the drilling torque and axial force, while the inner tube is rather thin. The inner tube is divided by two parts. The front part installed in the drilling head is the flange and the rest is the tube. The adjacent ends of these two parts are slightly conical forming an annular slit termed as the ejector nozzle. In the case shown in Figure 5.140, the ejector is the annular nozzle located in the drilling head as was in the original patent by Kurt Faber (US Patent No. 3,304,815 (1967)).

Once the flow of MWF supplied in the annular clearance between the boring bar and the inner tube reaches the drilling head, it is divided into two branches as indicated by arrows in Figure 5.140. The first flow termed as the ejector flow having the flow rate $Q_{fl\text{-}1}$ passes through the ejector nozzle, and thus, it is directed rearwardly by this nozzle. The second flow having the flow rate $Q_{fl\text{-}2}$ passes through the radial holes in the drilling head and then goes to the machining zone as indicated by arrows in Figure 5.140. The MWF flow around the ejector drill head as thought of by Sandvik Coromant Co. is shown in Figure 5.145. As can be seen, this second flow then returns with the formed chips rearwardly from the machining zone inside the drilling head and joins the flow ejected through the ejector nozzle. This happens because the first flow passing through the ejector nozzle lessens the pressure (compared to the atmospheric pressure) in the inner tube in front of the nozzle. After these two flows join again, they transport the chip over the inner tube similar to that as in STS drilling.

The lessening of the pressure by an ejector is known as the ejector effect (Astakhov, Subramanya, and Osman 1995). When it is sufficient, no seal is needed between the face of the workpiece and the starting bushing. As a result, with simple adapters, the ejector drilling can be used on most standard machine tools and even on CNC machines as a hole-making operation, and moreover, it can potentially substitute STS drilling as was widely advertised when ejector drills became available. Although it is partially true, the practice of ejector drilling showed its limitations and the proper

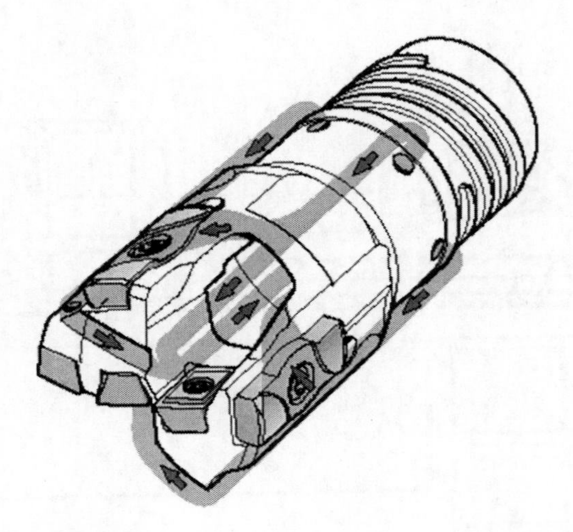

FIGURE 5.145 MWF flows around the ejector drill head.

application range/conditions were gradually established in industry. To understand the limitations of ejector drilling and thus to establish its proper implementation conditions, one needs to understand the basic particularities of the ejector as a hydraulic apparatus.

As discussed by the author earlier (Astakhov et al. 1991, 1995c and d; Astakhov et al. 1996; Astakhov, Subramanya, and Osman 1995; Astakhov, Subramanya, and Osman 1995c)), the ejector is characterized by three similarity numbers:

Relative flow rate

$$q_e = \frac{Q_{fl\text{-}2}}{Q_{fl\text{-}1}} \tag{5.65}$$

As explained previously, the flow rate $Q_{fl\text{-}1}$ passes through the ejector nozzle and the flow rate $Q_{fl\text{-}2}$ passes through the radial holes in the drilling head flowing to the machining zone and then sucked into the inner tube due to the ejector effect.

Relative hydraulic head

$$h_e = \frac{H_{fl\text{-}1}}{H_{fl\text{-}2}} \tag{5.66}$$

where

$H_{fl\text{-}1}$ is the hydraulic head (see Chapter 6 for the definition) of the flow that passes through the ejector nozzle

$H_{fl\text{-}1}$ is the hydraulic head of the flow just in front of the ejector nozzle(s)

Ejector modulus

$$m_e = \frac{A_{en}}{A_{mc}} \tag{5.67}$$

where

A_{en} is the cross-sectional area of the ejector nozzle(s)

A_{mc} is the cross-sectional area of the mixing chamber where two flows are mixed

In the considered case, it is the cross-sectional area of the inner tube.

The three similarity numbers define the hydraulic efficiency of the ejector as

$$\eta_e = q\frac{h}{1-h} \tag{5.68}$$

The experimental result shows (Astakhov et al. 1991; Astakhov et al. 1996) that the partial vacuum (the ejector effect) is reliably formed in the machining zone when the MWF pressure in front of the ejector nozzle is at least 0.3–0.4 MPa. As such, the maximum vacuum even with the optimal ejector can be as high as −0.06 MPa. Therefore, the inlet MWF pressure should be high enough to achieve this target accounting for the hydraulic resistance of the MWF circuit from the adapter inlet to the ejector nozzle. Moreover, the ejector modulus defines the following relationship:

$$q_e = 0.205m_e - 0.007m_e^2 - 0.041 \tag{5.69}$$

that is, knowing m_e (which is the design parameter—see Equation 5.67) and flow rate $Q_{fl\text{-}1}$ passed through the ejector nozzle, the flow rate $Q_{fl\text{-}2}$ can be determined. If the separation of Q_{fl} into $Q_{fl\text{-}1}$ and $Q_{fl\text{-}2}$ in the drilling head does not correspond to q_e according to Equation 5.69, the following may happen:

- The flow rate $Q_{fl\text{-}2}$ is greater than that determined by q_e according to Equation 5.69. In this case, the excess of MWF would flow outside as the gap between the starting bushing and face of the workpiece is not sealed. As far as the flow rate that is sucked by the ejector is sufficient to transport the chip from the machining zone to the ejector nozzle region, there is no problem with drilling besides the waste of energy
- The flow rate $Q_{fl\text{-}2}$ is smaller than that determined by q_e according to Equation 5.69. In this case, the air will be sucked into machining through the same gap to compensate the difference. Note that the flow rate of the sucked air is much greater than the difference because the density of the air is much smaller than that of MWF.

The normal working regime of an ejector drill should be considered when *light* suction of the air takes place.

Figure 5.146 shows the efficiency of ejectors with annular nozzles as a function of the relative flow rate q_e and modulus m_e. As can be seen, the maximum efficiency is about 20% so as the conservation law implies no energy gain is made by installing the ejector in the drill. This efficiency

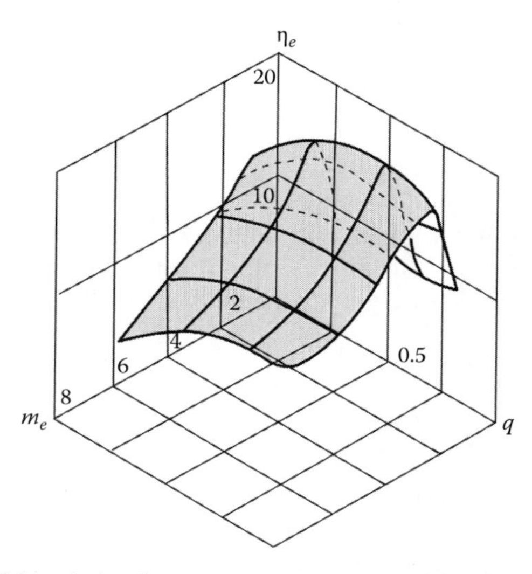

FIGURE 5.146 Influence of the relative flow rate q_e and modulus m_e on the energy efficiency of the ejector.

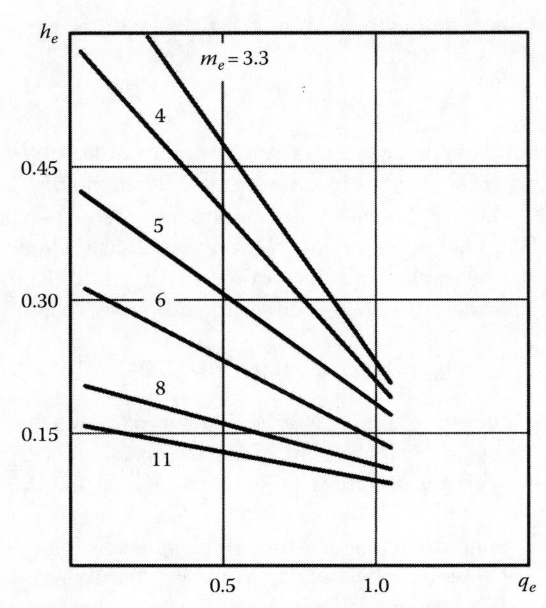

FIGURE 5.147 Generalized energy characteristics of ejectors with annular nozzles under maximum energy efficiency.

is the price to pay for advertised advantages of ejector drills. It can be seen in Figure 5.146 that maximum efficiency is achieved when m_e = 3.5–4.0, while for commercial drills, $m_e > 10$.

Based upon maximum efficiency, the general characteristic $h_e - f(q_e, m_e)$ is obtained as shown in Figure 5.147. As can be seen, the greater the modulus m_e, the less hydraulic head h_e can be generated by the ejector. For a given ejector drill, that is, for the given hydraulic head before the ejector nozzle, H_{ej} and its m_e, one can determine the hydraulic head H_{mx} of the chip–MWF mixture in the inner tube just after the ejector nozzle as

$$H_{ej} = hH_{mx} \tag{5.70}$$

or in terms of more convenient manometric pressure

$$\frac{p_{mx}}{\rho_{fl}} = h\frac{p_{ej}}{\rho_{mx}} \tag{5.71}$$

where ρ_{fl} and ρ_{mx} are the densities of MWF and MWF–chip mixture, respectively.

Knowing the hydraulic head H_{mx}, one can determine the maximum length of the inner tube under the condition of reliable chip transportation in a given ejector drill based upon the following energy condition:

$$H_{mx} \geq \pm\Delta h_{ver} + \Delta h_L + \Delta h_m \tag{5.72}$$

where Δh_{ver} is the difference in elevations between the outlet of the inner tube and the ejector nozzle (see Figure 6.16) accounted for only in vertical or inclined drills. As such, the sign "–" is used when the ejector nozzle is located below the outlet of the inner tube (the common case) and the sign "+"

when the opposite is true; Δh_L is the major hydraulic loss as discussed in Chapter 6. One can calculate this loss using Equation 6.26 in Chapter 6 as

$$h_L = \lambda \frac{l_{in\text{-}ej}}{d_{in}} \frac{Q_{mx}^2}{2gA_{in}^2} \tag{5.73}$$

where

$l_{in\text{-}ej}$ is the length of the inner tube from the ejector nozzle to its outlet
d_{in} is the diameter of the inner tube
Q_{mx} is the volumetric flow rate of the MWF–chip mixture (can be thought of as equal to the total flow rate of MWF supplied to the ejector drill, that is, $Q_{mx} = Q_{mx}$ can be accepted to the first approximation)
A_{in} is the cross-sectional area of the inner tube
λ is the D'Arcy coefficient often referred to as the friction factor (see Equation 6.26). The values of this coefficient depending upon the flow conditions were presented by the author earlier (Astakhov et al. 1996):

Δh_m is the major hydraulic loss as discussed in Chapter 6. One can calculate this loss using Equation 6.32 in Chapter 6 as

$$h_m = K_{fr} \frac{Q_{mx}^2}{2gA_{in}^2} \tag{5.74}$$

where K_{fr} is the flow resistance coefficient accounted for only in rotating ejector drills. The values of this coefficient depending upon the flow conditions were presented by the author earlier (Astakhov et al. 1996).

The hydraulic calculation has shown that the length $l_{in\text{-}ej}$ cannot exceed approx. 2 m when the ejector nozzle is located in the drilling head as shown in Figure 5.140. This limits the allowable length of ejector drill to an approx. 1.8 m, while STS drills do not have this limitation. Therefore, the notion that the ejector drill can be considered as an alternative to STS drills does not have any ground.

The MWF flow rate delivered to the ejector drill Q_{fl} should be greater than that for the STS drill for normal functioning of the ejector and reliable chip removal from the machining zone to the ejector nozzle. Because the cross-sectional area of the annular clearance between the inner and outer tubes used for MWF delivery into the drilling head is much smaller than the cross-sectional area of the annular clearance between the outer tube (the boring bar) and the wall of the hole being drilled, the pressure needed to deliver even the same as for STS drill flow rate to the ejector drill is higher for the same drill length when the ejector nozzle is located in the drilling head.

The notion that the ejector drill required less inlet pressure stems from the alternative (to that shown in Figure 5.140 and originally patented by Faber) ejector nozzle(s) location in some designs of modern ejector drills, for example, by Sandvik Coromant Co. This location in the rear end of the inner tube in standard ejector drills is shown in Figure 5.148. Figure 5.149 shows such an ejector

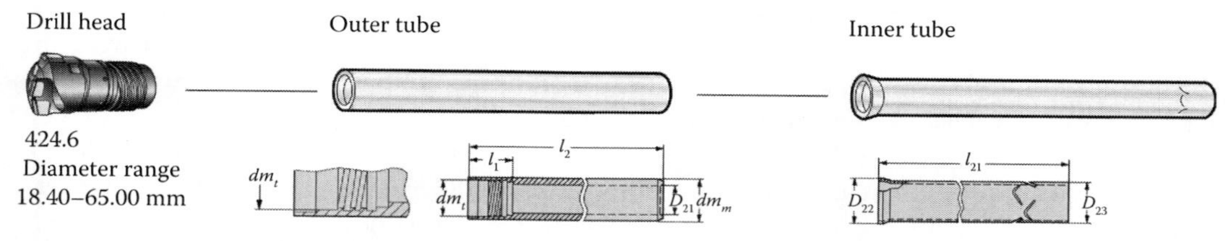

FIGURE 5.148 Ejector nozzles location in the rear end of the inner tube in standard ejector drills by Sandvik Coromant Co.

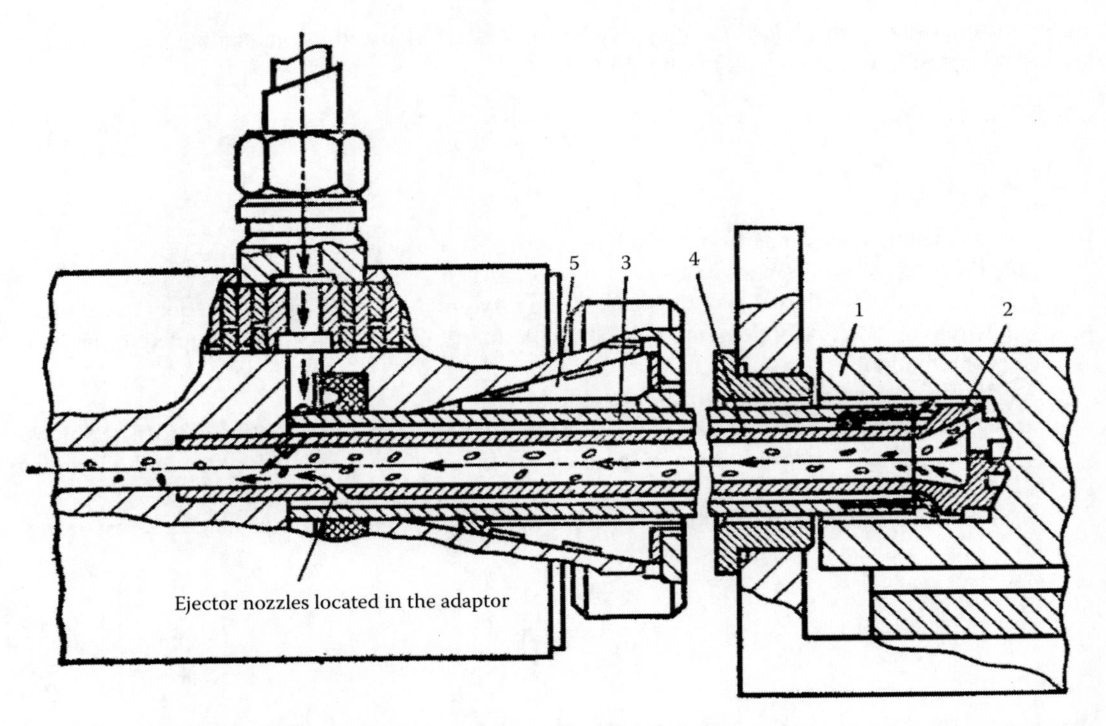

FIGURE 5.149 The ejector nozzles are located in the adapter according to the design of the inner tube shown in Figure 5.148.

drill in operation. The same as discussed previously, the workpiece (1) is drilled by the drilling head (2) located at the end of the outer tube (3). The inner (4) tubes are installed in the drilling head and in the connector (5) to form the annular MWF channel. The difference is, however, that the total flow rate Q_{fl} is divided into two, not in the drilling head but in the adapter. The first flow termed as the ejector flow having the flow rate $Q_{fl\text{-}1}$ passes through the ejector nozzles located at the rear end of the inner tube, and thus, it is directed rearwardly by these nozzles as indicated by the arrows on the figure. The second flow having the flow rate $Q_{fl\text{-}2}$ passes through annular clearance between the outer and inner tubes to the drilling head, then its passes through the radial holes in the drilling head and then goes to the machining zone as indicated by arrows in Figure 5.149. Due to the ejector effect created by the ejector nozzle in the inner tube, this flow rate is sucked from the machining zone into the inner tube providing the transportation of the chip as shown in the picture.

The obvious advertised advantage of such nozzle location is that only small portion $Q_{fl\text{-}2}$ of the total flow rate should pass through really narrow clearance between the outer and inner tubes that require a much smaller inlet pressure for the normal operation of ejector drills. Moreover, as the inlet pressure is not a concern, the relative flow rate q_e can be made optimal so the hydraulic head produced by the ejector can be maximized according to Figure 5.147. However, this great hydraulic head is not needed anymore after the adapter as the inner tube ends so that there is no more hydraulic resistance to overcome.

The second concern with such a design is far more important. As discussed previously, the maximum vacuum even with the optimal ejector can be as high as −0.06 MPa. In other words, only 0.06 MPa times the density of MWF–chip mixture is available to overcome the pressure loss in the inner tube over the distance from the machining zone to the ejector nozzles. This significantly restricts the maximum allowable length of the ejector drills of such a design. This also explains why this length does not exceed one meter for the standard ejector drills.

In the description of the original patent on the ejector drill (US Patent No. 3,304,815 (1967)), Kurt Faber, the inventor, suggested that "for improving the ejector effect, especially in long drills, one or more extra ejector devised may be provided along the drill." The foregoing analysis, however, suggests that as far as the flow rate for an additional ejector is taken from that supplied to the ejector drill, no gains can be achieved in terms of increasing the ejector effect and thus increasing the allowable length of the ejector drill. As discussed by the author earlier, some improvement in the chip removal from the ejector drills is achieved when an additional ejector with independent MWF is used (Astakhov, Subramanya, and Osman 1995). Such an additional ejector can be imbedded in the design of the adapter for ejectors drill or can be made as an attachment to the standard adapters. Figure 5.150a shows one of the possible designs of such adapters for nonrotating ejector drills, and Figure 5.150b shows the design of the attachment for rotating drills. As can be seen, an additional ejector is provided by both designs. An independent MWF line sullying additional flow rate $Q_{e\text{-}adt}$ is provided. As the design of the ejector nozzle can be made optimal and, moreover, adjustable, the achieved improvements can be so great that the ejector on the inner tube is no longer needed.

The last myth that ejector drilling is suitable only for a group of relatively easy-to-machine work materials stems from the use of the standard chip breakers initially developed for STS drilling

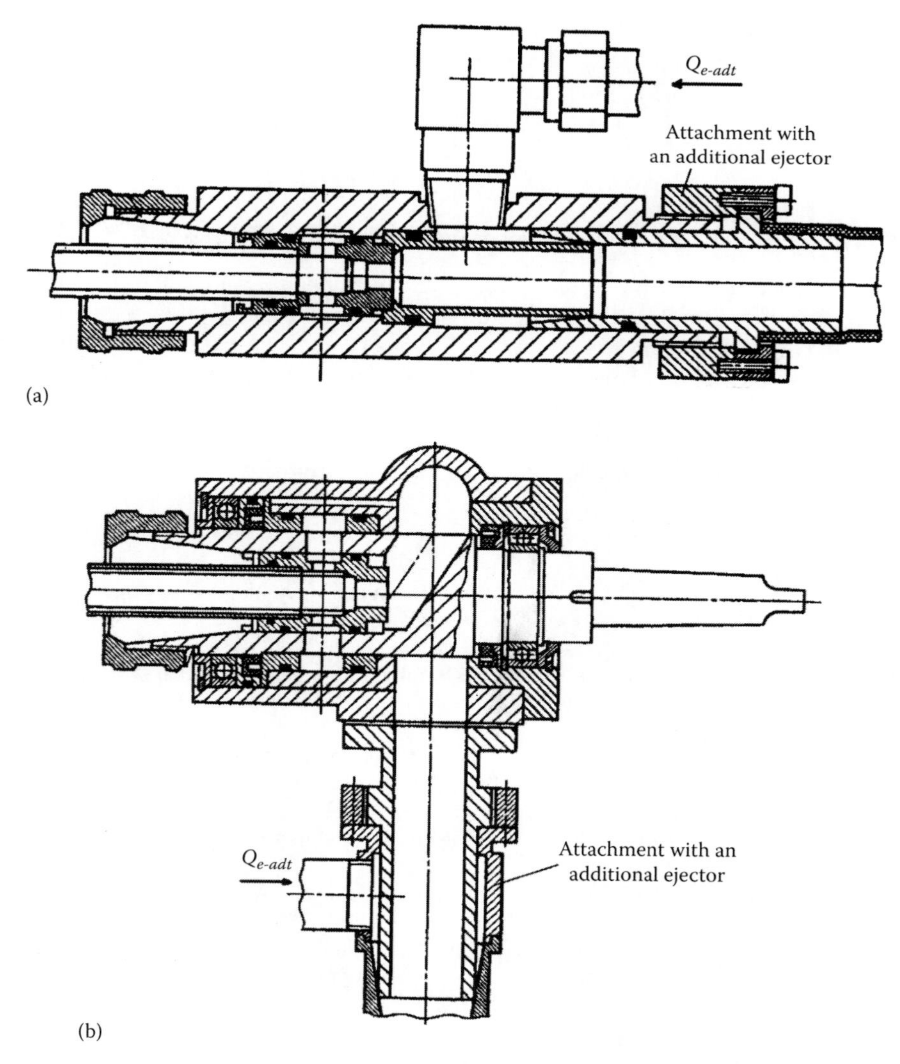

FIGURE 5.150 Adapters for ejector drill with attachments having additional ejectors: (a) for nonrotating drills and (b) for rotating drills.

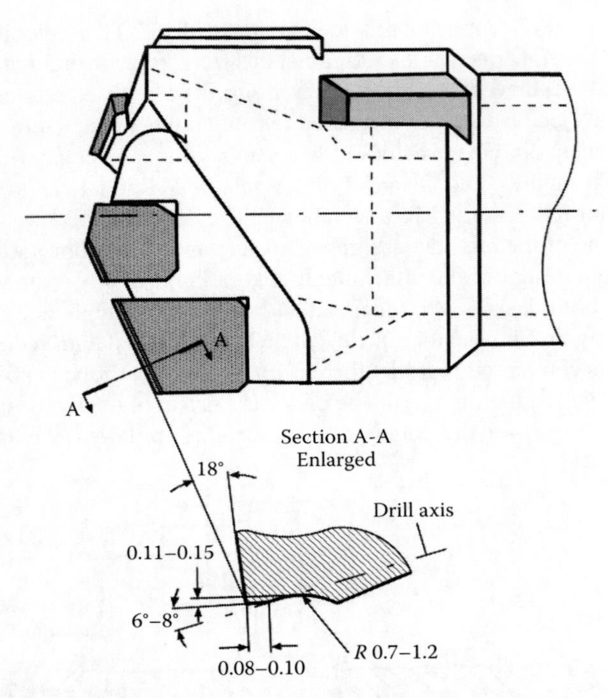

FIGURE 5.151 Chip breaker for ejector drill for drilling difficult-to-machine materials.

(shown as chip-breaking steps in Figures 5.123, 5.125, 5.129, and 5.137). Such a chip breaker works well for STS drill where the MWF flow rate through the machining zone is great, and if, by any chance, a small chip clogging occurs say in the chip mouth, the external pressure increase to push the clogged chip through. As it is not the case in ejector drilling, much more intelligent handling of chip breaking is needed for ejector drills. The author has experience with successful productive ejector drilling (20 and 40 mm hole diameter and 1.2 m length) of soft austenitic stainless steel (AISI 321) using the chip breaker design shown in Figure 5.151.

REFERENCES

Andrewesa, C.J.E., Feng, H.-Y., and Laub, W.M. 2000. Machining of an aluminum/SiC composite using diamond inserts. *Journal of Materials Processing Technology* 102(1–3):25–29.

Astakhov, V.P. 2001. Gundrilling know how. *Cutting Tool Engineering* 52:34–38.

Astakhov, V.P. 2002a. The mechanisms of bell mouth formation in gundrilling when the drill rotates and the workpiece is stationary. Part 1: The first stage of drill entrance. *International Journal of Machine Tools and Manufacture* 42:1135–1144.

Astakhov, V.P. 2002b. The mechanisms of bell mouth formation in gundrilling when the drill rotates and the workpiece is stationary. Part 2: The second stage of drill entrance. *International Journal of Machine Tools and Manufacture* 42:145–152.

Astakhov, V.P. 2004. High-penetration rate gundrilling for the automotive industry: System outlook. SME Paper TPO4PUB249, pp. 1–20.

Astakhov, V.P. 2010. *Geometry of Single-Point Turning Tools and Drills: Fundamentals and Practical Applications*. London, U.K.: Springer.

Astakhov, V.P. 2012a. Chapter 1: Design of experiment methods in manufacturing: Basics and practical applications. In *Statistical and Computational Techniques in Manufacturing*, Davim, J.P. (ed.). London, U.K.: Springer.

Astakhov, V.P. 2012b. Statistical and computational techniques in manufacturing. In *Statistical and Computational Techniques in Manufacturing*, Davim, J.P. (ed.). London, U.K.: Springer.

Astakhov, V.P., Abi Karam, S., and Osman, M.O.M. 1998. The static and dynamic pressure distribution in the machining zone of BTA drills. *Journal of Manufacturing Science and Engineering* 120(4):820–822.

Astakhov, V.P., Al-Ata, M., and Osman, M.O.M. 1997. Statistical design of experiments in metal cutting. Part 2: Application. *Journal of Testing and Evaluation, JTEVA* 25(3):328–336.

Astakhov, V.P., Frazao, J., and Osman, M.O.M. 1991. On the design of deep-hole drills with non-traditional ejectors. *International Journal of Production Research* 28(11):2297–2311

Astakhov, V.P., Farazao, J., and Osman, M.O.M. 1994. Effective tool geometry for uniform pressure distribution in single edge gundrilling. *ASME Journal of Engineering for Industry* 116(4):449–456.

Astakhov, V.P. and Galitsky, V. 2005. Tool life testing in gundrilling: An application of the group method of data handling (GMDH). *International Journal of Machine Tools and Manufacture* 45:509–517.

Astakhov, V.P. and Galitsky, V.V. 2006. The combined influence of various design and process parameters of gundrilling on tool life: Experimental analysis and optimization. *The International Journal of Advanced Manufacturing Technology* 36(9–10):852–864.

Astakhov, V.P., Galitsky, V.V., and Osman, M.O.M. 1995a. An investigation of the static stability in self piloting drilling. *International Journal of Production Research* 33(6):1617–1634.

Astakhov, V.P., Galitsky, V.V., and Osman, M.O.M. 1995b. A novel approach to the design of self-piloting drills. Part 1. Geometry of the cutting tip and grinding process. *ASME Journal of Engineering for Industry* 117:453–463.

Astakhov, V.P. and Joksch, S. 2012. *Metal Working Fluids: Fundamentals and Recent Advances*. Cambridge, U.K.: Woodhead Publishing Limited.

Astakhov, V.P. and Osman, M.O.M. 1996a. An analytical evaluation of the cutting forces in self-piloting drilling using the model of shear zone with parallel boundaries. Part 1: Theory. *International Journal of Machine Tools and Manufacture* 36(11):1187–1200.

Astakhov, V.P. and Osman, M.O.M. 1996b. An analytical evaluation of the cutting forces in self-piloting drilling using the model of shear zone with parallel boundaries. Part 2: Application. *International Journal of Machine Tools and Manufacture* 36(12):1335–1345.

Astakhov, V.P., Osman, M.O.M., and Al-Ata, M. 1997. Statistical design of experiments in metal cutting—Part 1: Methodology. *Journal of Testing and Evaluation* 25(3):322–327.

Astakhov, V.P. and Shvets, S.V. 2001. A novel approach to operating force evaluation in high strain rate metal-deforming technological processes. *Journal of Materials Processing Technology* 117(1–2):226–237.

Astakhov, V.P., Subramanya, P.S., and Osman, M.O.M. 1995c. Theoretical and experimental investigations of coolant flow in inlet channels of the BTA and ejector drills. *Journal of Engineering Manufacture, Part B, Proceeding of the I.Mech.E.* 209:211–220.

Astakhov, V.P., Subramanya, P.S., and Osman, M.O.M. 1995d. An investigation of the cutting fluid flow in self-piloting drills. *International Journal of Machine Tools and Manufacture* 35(4):547–563.

Astakhov, V.P., Subramanya, P.S., and Osman, M.O.M. 1996. On the design of deep hole drills with ejectors. *International Journal of Machine Tools and Manufacture* 36(2):155–171.

Bescrovny, A.M. 1984. On the location of the supporting pads in deep-hole tools (in Russian). *Rezanie i Instrument* 31:99–103.

Coelho, R.T., Yamade, S., Aspinwall, D.K., and Wise, W.L.H. 1995. The application of polycrystalline diamond (PCD) tool materials when drilling and reaming aluminium based alloys including MMC. *International Journal of Machine Tools and Manufacturing* 35(5):761–774.

Destefani, J. 2011. Hole in four… or more. *Cutting Tool Engineering* 63(1):32–36.

El-Gallab, M. and Sklad, M. 1998. Machining of Al/SiC particulate metal matrix composites. Part I. Tool performance. *Journal of Materials Processing Technology* 83(2):151–158.

Griffiths, B.J. 1993. Modeling complex force system, part 1: The cutting and pad forces in deep drilling. *ASME Transactions, Journal of Engineering for Industry* 115:169–176.

Griffiths, B.J. and Grieve, R.J. 1985. The role of the burnishing pads in the mechanics of the deep drilling process. *International Journal of Production Research* 23:647–655.

Hasegawa, X. and Horiuchi, G. 1975. On the motion of drill tip and accuracy of the hole in gundrilling. *Annals of the CIRP* 24:53–58.

Hung, N.P., Boey, F.Y.C., Khor, K.A., Phua, Y.S., and Lee, H.F. 1996. Machinability of aluminum alloys reinforced with silicon carbide particulates. *Journal of Materials Processing Technology* 56(1–4):966–977.

Katsuki, A., Sakuma, K., Tabuchi, K., Onikura, H., Akiyoshi, H., and Nakamuta, Y. 1987. The influence of the tool geometry on axial deviation in deep drilling: Comparison of single- and multi-edge tools. *JSME International Journal* 30:1167–1174.

König, W. and Klocke, F. 2008. *Fertigungsverfahren Band 1: Drehen, Fräsen, Bohren*. Berlin, Germany: Springer.

Landis, A.B. 1886. Deep holes by continuous boring. *American Machinist* 27:4.

Meguid, S.A. and Sun, Y. 2005. Intelligent condition monitoring of aerospace composites: Part I—Nano reinforced surfaces & interfaces. *International Journal of Mechanics and Materials in Design* 2(3–4):37–52.

Nikolov, S. 1986. Improving quality with deep-hole tools (in Bulgarian). *Mashinostroenie* (4):23–27.

Pfleghar, F. 1975. Bestimmung der Reibungszahl an den Fűhrungsleisten einscheidiger Bohrwerkzeuge. *Industrie-Anzeiger* 97:1997–1998.

Pfleghar, F. 1977. Aspekte zur konstruktiven Gestaltung von Tiefbohrwerkzeugen. *WT-Zeitschrift fűr Industrielle Fartigung* 67:211–218.

Richardson, R. and Bhatti, R. 2001. A review of research into the role of guide pads in bta deep-hole machining. *Journal of Materials Processing Technology* 110(1):61–69.

Sakuma, K., Taguchi, K., and Katsuki, A. 1980. Study on deep-hole-drilling with solid-boring tool: The burnishing action of guide pads and their influence on hole accuracies. *Bulletin of the JSME* 23(185):1921–1928.

Sakuma, K., Taguchi, K., Katsuki, A., and Takeyama, H. 1981. Self-guiding action of deep hole drilling tools. *Annals of the CIRP* 30(1):311–315.

Sakuma, K., Taguchi, K., and Kinjo, S. 1978. Study on deep-hole drilling with solid-boring tools—The effect of tool materials on the cutting performance. *Bulletin of the JSME* 21(153):532–539.

Sani, M.N. 1994. Further developments in PCD-tipped drills. *Industrial Diamond Review* 194:6–7.

Sani, M.N. and Clark, I.E. 1992. PCD-tipped twist drills—A new design concept. *Industrial Diamond Review* 52(5):233–237.

Shertladse, A.G., Ivanov, V.I., and Kareev, V.N. 2006. *Hydraulic and Pneumatic Systems* (in Russian). Moskow, Russia: Vishyja Shkola.

Shin, Y.C. and Dandekar, C. 2012. Mechanics and modeling of chip formation in machining of MMC. In *Machining of Metal Matrix Composites*, Davim, J.P. (ed.). London, U.K.: Springer.

Stockert, R. and Thei, T.P. 1978. Einfluβ von Gewichts-und Fliehkräften auf die Auslegung von Tiehbohrwerkzeuge. *Industrie-Anzeiger* 100:56–57.

Stockert, R. and Weber, U. 1977a. Reitrag zűr konstruktiven Auslegung mehrschneidiger Tiefbohrwerkzeuge. *Industrie-Anzeiger* 99:390–391.

Stockert, R. and Weber, U. 1977b. Untersuchung der energie–verhdltnisse beim tiefbohren mit einschneidigen BTA-Vollbohropfen. *Industrie-Anzeiger* 99(26):39–40.

Swinehart, H.J., ed. 1967. *Gundrilling, Trepanning, and Deep Hole Machining*. Dearborn, MI: SME.

Tomac, N. and Tonnessen, K. 1992. Machinability of particulate aluminum matrix composites. *CIRP Annals—Manufacturing Technology* 41:55–58.

Veremachuk, E.S. 1940. *Deep Hole Drilling* (in Russian). Moscow, Russia: State Publishing House of Defence Industry.

Vinogradov, A.A. 1985. *Physical Foundation of the Drilling of Difficult-to-Machine Materials with Carbide Drills* (in Russian). Kyiv, Ukraine: Naukova Dumka.

Zhang, W., He, F., and Xiong, D. 2004. Gundrill life improvement for deep-hole drilling on manganese steel. *International Journal of Machine Tools and Manufacture* 44(2–3):237–331.

6 Metalworking Fluid in Drilling

Making the simple complicated is commonplace; making the complicated simple, awesomely simple, that's creativity.

Charles Mingus Jr. (April 22, 1922–January 5, 1979), Highly influential American jazz double bassist, composer, and bandleader

6.1 INTRODUCTION

There are three equally important pillars of the successful metalworking fluid (hereafter MWF) application: (1) selection of the proper MWF, (2) delivery of this MWF into point of application, and (3) MWF maintenance. The proper physical delivery of MWFs to the machining zone is one of the most important of successful MWF application (Springborn 1967) because, unless MWF is delivered with the parameters needed and carefully placed, it cannot perform its functions. Unfortunately, this is still one of the most neglected aspects of proper MWF application in many practical manufacturing operations and particularly in drilling operations. This is because the modern books on the MWF (Byers 1994; Byers 2006) as well as the known books on drilling tool design/applications do not consider this aspect. This chapter aims to provide the basic guidelines for the proper MWF delivery and application in drilling with emphasis on HP drilling while essential parameters of MWF to be maintained as important to drills performance are also addressed.

There are two aspects of the proper MWF delivery:

1. Delivery of MWF with proper parameters, that is, velocity and temperature. This includes the complete hydraulic circuit of the MWF delivery system with all apparatus (pumps, valves, filters, control systems, piping, rotating units, internal MWF channels in spindles and in tool bodies, etc.)
2. Point of allocation of MWF that includes the location and pointing direction of the MWF nozzles and exit orifices with respect to the tool cutting edge(s) or other reference components of the cutting tool.

It should be clear that when MWF is selected and maintained properly, the two listed aspects are independently equal assuring the success in MWF application.

This chapter considers the practical aspects of MWF supply in drilling operations including gundrilling (some important aspects of the MWF application STS and Ejector drills are considered in Chapter 5). Among the common methods of MWF supply, high-pressure through-tool MWF application, as used in HP drilling operations, and near-dry machining (NDM), as gaining its popularity, are considered in detail to provide manufacturing engineers and process and tool designers with knowledge needed to make proper choices in the design/selection/optimization of drilling systems and HP drills.

6.2 MWF APPLICATION METHODS

Figure 6.1 shows the basic methods of MWF supply into the machining zone. Selection of a particular method is a matter of multifacet consideration, which includes developing a business case. This selection is primarily based on the process economy that blends the system efficiency, past experience available, yearly program, quality requirements, machine and MWF system available, etc.

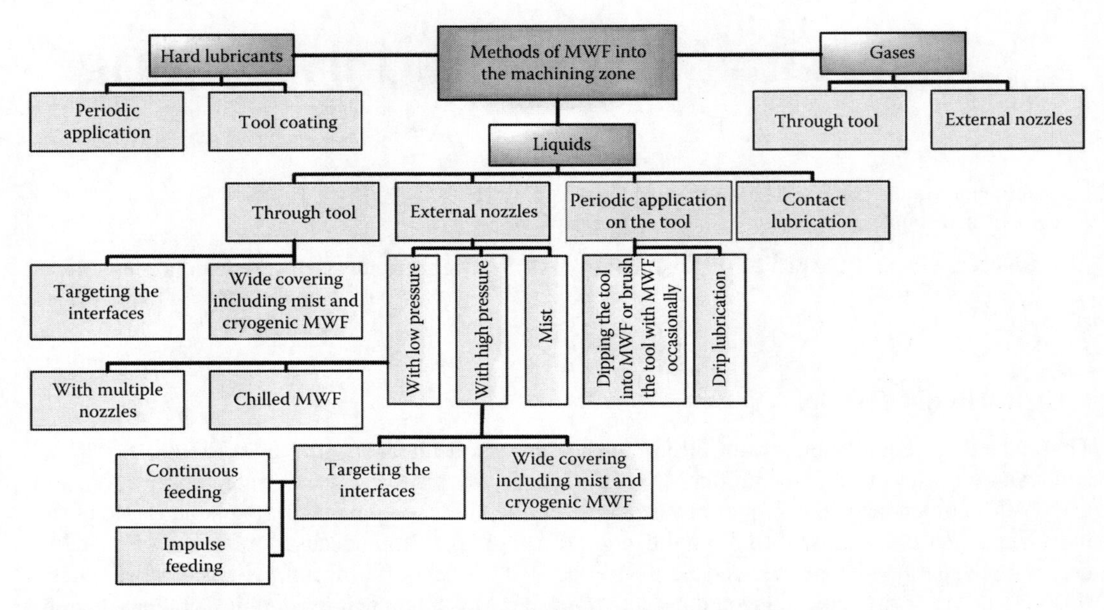

FIGURE 6.1 Methods of MWF supply into the machining zone.

Among the methods presented in Figure 6.1, MWF in the form of liquid covers more than 95% of applications. Within this category, MWF application using external nozzles (also known as flood application) and through tool are in wide use.

6.2.1 Flood Application

Manufacturing engineers and machinists have long relied on the use of MWF to extend tool life and machine to tight tolerances consistently and produce fine finish of machined surfaces. Since its introduction, traditional low-pressure *flood* MWF application has been used till today.

An example of modern flood MWF application in drilling is shown in Figure 6.2. MWF limits heat-related damage to the cutting tool, improves tribological conditions of the cutting tool, and flushes away chips. If a flow of MWF over the tool and part fails to produce the desired results, and the expected improvements do not materialize, the answer, typically, is to use a better MWF brand, increase its concentration, improve filtration, and, which is the most common, supply even more MWF by increasing the MWF flow rate.

FIGURE 6.2 Flood application of MWF in drilling.

For flood MWF application, a great number of various designs of MWF supply units as well as acid resistant polyester flexible tubes and nozzles are commercially available that can suite a particular operation and/or machine.

The following advantages of flood MWF supply can be listed as follows:

- Great versatility and simplicity of adjustment of the nozzle (nozzles) location/direction
- Good chip flushing ability from the machining zone for *open zone* operations such as milling, turning, external broaching, and grinding
- Simple adjustment of MWF flow rate
- Simple maintenance of MWF supply systems.

The following drawbacks of flood MWF supply are as follows:

- Relatively low MWF velocity often wrongly referred to as low pressure that limits MWF access in *closed zone* machining operations such as drilling, taping, reaming, and boring. As a result, this method is not suitable for machining holes with $L/D > 3$ unless periodic withdraws of the tool from the hole being machined for cleaning from chips are used including peck drilling
- The location/direction of the nozzle (nozzles) depends on the machine operator experience and on the convenience to perform/observe a particular operation. In modern CNC machining, however, this problem is often solved by locating the MWF nozzles on the tool holder or in the machine
- A *messy* environment and time-consuming cleanup for gaging. In mass production, special power washing operations are used for parts between processes and final gaging and inspection.

In CNC machines, the location/direction of the nozzle (nozzles) problem with flood MWF application can be resolved by programmable coolant nozzles. Such a nozzle is controlled automatically by the part program to direct MWF precisely at the cutting area for any particular tool used in the machine, eliminating constant adjustments by the operator. The programmable nozzle can be a part of the machine (e.g., HAAS) or can be bought and installed on a CNC machine (e.g., SpiderCool shown in Figure 6.3). It is understood that such a solution is feasible when a number of repeated parts are to be machined and *open zone* operations are prevalent.

FIGURE 6.3 Programmable nozzles for CNC machines by SpiderCool Co.

6.2.2 Through-Tool MWF Application

When higher requirements to productivity and quality are the case, many machine shops, plants and manufacturing companies turn to through-tool MWF application. It has not always been possible for shops to use high-velocity through-MWF tooling. Usage was limited when shops were predominantly equipped with manual machines. Manually operated machine tools require the operator to remain near the actual machining process. Today, however, CNC machine tools are in full force almost everywhere. With the movement of the cutting tool completely programmed, an operator no longer has to remain near the machine to monitor the progress of jobs. As a result, it become possible to limit the exposure of operators to MWF by the sheet-metal or high-strength-plastic enclosures that surround the most modern CNC machine tools. These enclosures keep chips and MWF within the work area, allowing the operator to stay safe and dry while applying a heavier and more constant volume of MWF. Continuously bathed with MWF, the tool can make chips at peak productivity and efficiency.

To use a through-MWF system with drilling tools (drills, reamers, taps, etc.) requires installation of a special MWF delivery means. In simplest cases, tool holders with integrated extension arms that fit over the port fitting of various designs are used. Such tool holder is widely available from major drilling tool suppliers. Figure 6.4 shows an example of a tool holder by Big Kaiser Co. When a tool holder is placed into the machine spindle, the extension arm completes the connection, and MWF can pass freely from the spindle port to the internal coolant ducts within the tool itself. These types of tool holders are usually compatible with most machining centers' tool magazines and changers. A great disadvantage of this type of tool holders is relatively to low allowable MWF pressure (normally it does not exceed 1.8 MPa). It limits the application range of these tool holders.

A wide spread in internal high-pressure MWF supply started when machines with spindles equipped for internal MWF supply similar to that shown in Figure 6.5 became available. A rotary union is used to introduce the flow of MWF into the rotating spindle. When a through-MWF tool is connected to a properly equipped spindle and tool holder, MWF flows through the tool's internal coolant channels (often referred to as ducts) and out of the tool's nozzles.

With a through-MWF system, MWF is usually pumped through the tool and then through the outlet MWF orifice (nozzle) to the machining zone. There are three principle objectives in the use

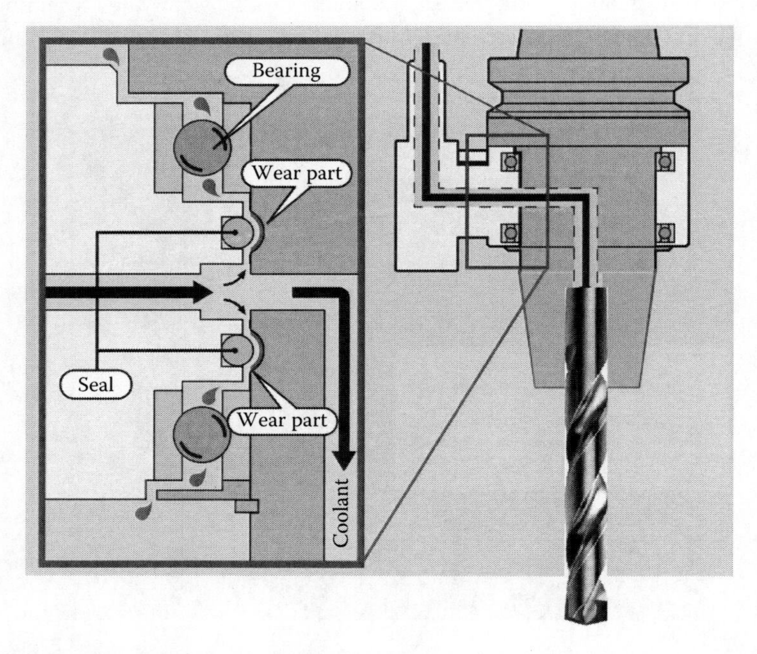

FIGURE 6.4 Tool holder with internal MWF supply.

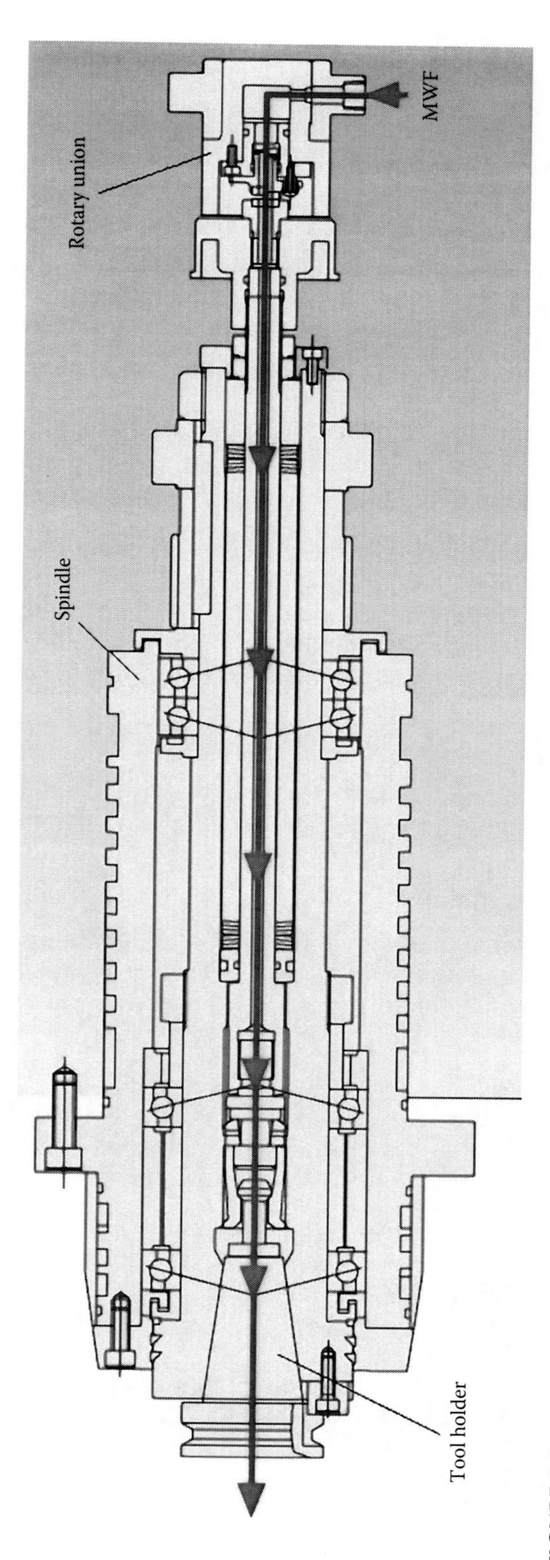

FIGURE 6.5 A spindle with internal MWF supply.

of this technology: (1) increase tool life, (2) facilitate chip breakage and chip removal, and (3) assure dimensional stability of the machined surface (particularly important for finish reamers, e.g., in the automotive industry). Depending upon which one of these three is more important for a particular application, a specific method of MWF-through supply is selected.

Although this technology is normally considered as the whole, there are a number of particularities in the through-MWF system. They can be classified as follows:

By MWF pressure used

- Low-pressure applications where the inlet MWF pressure is under 2 MPa (300 psi)
- Medium-pressure applications where the inlet MWF pressure is in the range of 2 MPa (300 psi)—6.9 MPa (1000 psi)
- High-pressure applications where the inlet MWF pressure is in the range of 6.9 MPa (1000 psi)—20 MPa (3000 psi)
- Ultrahigh-pressure applications where the inlet MWF pressure is over 20 MPa.

By point of MWF application

- The outlet MWF nozzle is directed into the bottom of the hole being drilled. As such, the outlet orifices (nozzles) of the coolant channels are made in the tool flank faces as shown in Figure 6.6. This is common for drills and drilling tools for blind holes.
- The outlet MWF nozzle is directed into the rake face (Figure 6.7). This is used when through holes are to be machined (e.g., cored or previously-drilled holes).

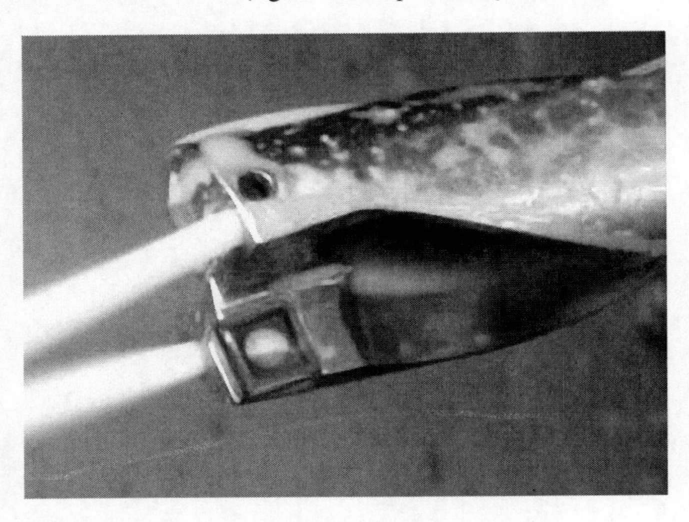

FIGURE 6.6 Coolant directed into the bottom of the hole being drilled.

FIGURE 6.7 Coolant directed at the rake face.

6.2.3 Near-Dry (Minimum Quantity Lubricant) Application

The costs of maintaining and eventually disposing of MWF, combined with health and safety concerns, have led to a heightened interest in either eliminating MWF altogether or limiting the amount of MWF applied. The former process is known as dry machining while the latter is referred to as NDM or MQL machining. Generally speaking, NDM is machining with the supply of very small quantities of lubricant to the machining zone. It was developed as an alternative to flood and internal high-pressure MWF supply to reduce MWFs consumption. In NDM, MWF is supplied as a mixture of air and an oil in the form of aerosol (often referred to as mist).

6.2.4 Application of CMWFs

This involves the application of a cryogenic fluid in the machining process, primarily as an MWF. This fluid is applied as an external spray through a nozzle to perform both conductive and convective cooling of the machining zone or could be applied indirectly to cool the cutting tool through conduction alone.

6.3 HIGH-PRESSURE MWF SUPPLY: THEORY, APPARATUS, AND PARTICULARITIES OF TOOL DESIGN

In practice of MWF systems, machine tools, and cutting tool design and implementation, the pressure of MWF is always considered to be of prime concern while its flow rate is totally neglected. In the author's opinion, this is one of the most severe misconceptions in MWF applications as it affects various facets of machining efficiency, productivity, and quality. This section aims to explain and thus resolve the long-standing issue showing that the MWF flow rate is of prime concern while its pressure is only a mean to deliver the flow rate needed to assure that MWF can perform its intended actions in machining. There are, however, few exceptions where the combination of the flow rate with high MWF static pressure plays an important role.

Four basic actions of MWFs in drilling operations are commonly considered: (1) cooling, (2) lubricating, (3) chip transportation from the machining zone (or even from the working zone), and sometimes (4) chip control. This section considers these actions as related to the MWF flow rate.

6.3.1 Flow Rate

The volumetric flow rate (also known as the volume flow rate or rate of fluid flow) is the volume of fluid that passes through a given surface per unit time (e.g., cubic meters per second [m^3/s] in SI units or cubic feet per second [ft^3/s] in the US customary units or imperial units). It is usually represented by the symbol Q_v. Figure 6.8 shows the model of the fluid flow through a pipe. In this model, A_{ch} is

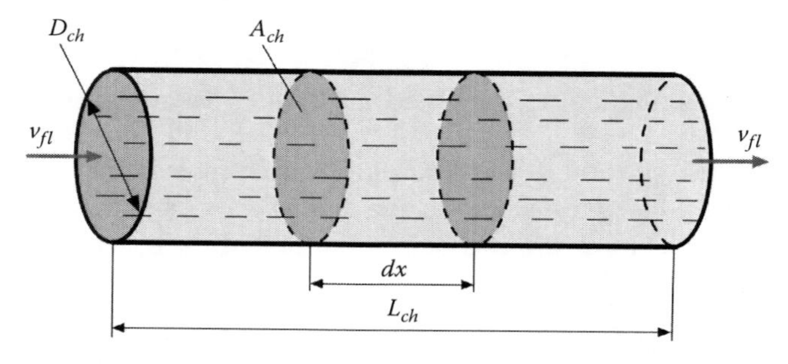

FIGURE 6.8 Model to derive the equation of the volumetric flow rate in a pipe.

the cross-sectional area of the pipe and v_{fl} is the fluid velocity. Considering a unit volume $A_{ch} \times dx$ that passes a reference point during time dt, one can obtain the following equation to calculate the volumetric flow rate as

$$Q_v = \frac{d(A_{ch}x)}{dt} = A_{ch}\frac{dx}{dt} = A_{ch}v_{fl} \tag{6.1}$$

It is obvious that for a round tube,

$$A_{ch} = \frac{\pi D_{ch}^2}{4} \tag{6.2}$$

where D_{ch} is the internal pipe diameter as shown in Figure 6.8.

Other common units for Q_v are as follows: L/s = 10^3 cm^3/s = 10^{-3} m^3/s; L/min = 0.17×10^{-4} m^3/s; gal/s = 3.788 L/s = 0.003788 m^3/s; ft^3/min = 4.719×10^{-4} m^3/s.

If the fluid is incompressible, that is, its density does not change as the substance flows along, then the volume flow rate follows a conservation of volume—the volume that flows in must flow out. This is the case for most MWFs used in metalworking industry.

The mass flow rate is the mass of substance that passes through a given surface per unit time. Its unit is mass divided by time, so kilogram per second in SI units and slug per second or pound per second in the US customary units. It is usually represented by the symbol Q_m.

The mass flow rate for fluids (including practically all MWFs) can be calculated from the density of the fluid, ρ_{fl}, the cross-sectional area through which the fluid is flowing, A_{ch}, and its velocity, v_{fl}, as

$$Q_m = \rho_{fl}v_{fl}A_{ch} \tag{6.3}$$

Comparing Equations 6.1 and 6.3, one can express the mass flow rate through the volumetric flow rate as

$$Q_m = \rho_{fl}Q_v \tag{6.4}$$

6.3.2 Pressure

6.3.2.1 Definition

Pressure is defined as force per unit area. It is usually more convenient to use pressure rather than force to describe the influences upon fluid behavior. The standard unit for pressure is the pascal, which is a newton per square meter. The symbol of pressure is p. Mathematically,

$$p = \frac{F_{fl}}{A_n} \tag{6.5}$$

where

p is the pressure
F_{fl} is the normal force
A_n is the area

Manufacturing engineers are most often interested in pressures that are caused by a fluid because:

- If fluid is flowing through a horizontal pipe and a leak develops, a force F_n must be applied over the area of the hole, A_h, that causes the leak. This pressure, p_{fl}, is called the fluid pressure (the force must be applied divided by the area of the hole). This is schematically shown in Figure 6.9a.
- If a vertical container contains a fluid, the mass of the fluid will exert a force on the base of the container. This pressure is called the hydrostatic pressure. Hydrostatic pressure is the pressure caused by the mass of a fluid. This is shown schematically in Figure 6.9b.

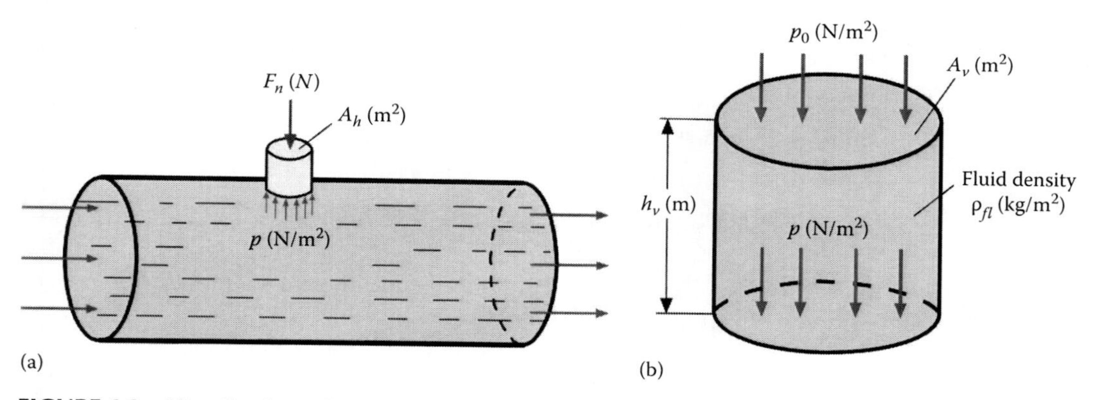

FIGURE 6.9 Visualizations of (a) the fluid pressure and (b) the hydrostatic pressure notions.

Because the mass of the fluid is $A_v \times h_h \times \rho_{lf}$, the hydrostatic pressure is calculated as

$$p_h = p_0 + \frac{A_v h_v \rho_{fl}}{A_v} = p_0 + h_v \rho_{fl} \tag{6.6}$$

where p_0 is the external pressure. If the container is open and does not move, then this pressure is equal to the atmospheric pressure p_{atm}.

Table 6.1 presents useful pressure conversions for the units commonly used in manufacturing engineering.

The following types of pressure are normally distinguished:

- Atmospheric pressure, p_{atm}, is the pressure caused by the weight of the earth's atmosphere. Often atmospheric pressure is called barometric pressure.
- Absolute pressure, p_{abs}, is the total pressure. An absolute pressure of 0 is a perfect vacuum.
- Gage pressure, p_{gauge}, is pressure relative to atmospheric pressure.
- Vacuum is a gage pressure below atmospheric pressure. It is used so that a positive number can be reported.

Absolute pressure, gage pressure, and the atmospheric pressure are related as

$$p_{abs} = p_{atm} + p_{gauge} \tag{6.7}$$

TABLE 6.1
Pressure Units

	Pascal (Pa)	Bar (bar)	Technical Atmosphere (at)	Atmosphere (atm)	Pound-Force per Square Inch (psi)
1 Pa	≡1 N/m²	10^{-5}	1.0197×10^{-5}	9.8692×10^{-6}	145.04×10^{-6}
1 bar	100,000	≡1 bar	1.0197	0.98692	14.5037744
1 atm	98,066.5	0.980665	≡1 kgf/cm²	0.96784	14.223
1 atm	101,325	1.01325	1.0332	≡1 atm	14.696
1 psi	6.894×10^3	68.948×10^{-3}	70.307×10^{-3}	68.046×10^{-3}	≡1 lbf/in.²

Example reading: 1 Pa = 1 N/m² = 10^{-5} bar = 10.197×10^{-6} at = 9.8692×10^{-6} atm, etc.

Note: Often in modern manufacturing, the megapascal is used as the pressure unit. It is clear that 1 MPa = 1×10^6 Pa.

The standard atmosphere is defined as the pressure equivalent to 760 mm of Hg at sea level and at 0°C. The unit for the standard pressure is the atmosphere, atm. Pressure equivalents to the standard atmosphere (atm) are

$$\begin{aligned} 1\,\text{atm} &= 760\,\text{mmHg} = 76\,\text{cm Hg} \\ &= 1.013\times 10^5\,\text{N/m}^2\,(\text{or Pa}) = 101.3\,\text{kPa} = 1.013\,\text{bar} \\ &= 14.696\,\text{psi}\,(\text{or lbf/in.}^2) = 29.92\,\text{in. Hg} = 33.91\,\text{ft H}_2\text{O} \end{aligned} \tag{6.8}$$

The units, psi and atm, often carry a trailing *a* or *g* to indicate that the pressure is absolute or gage pressure. This is the meaning of atma or atmg if nothing is noted otherwise.

6.3.2.2 Pressure Measurement

Although there are a number of various methods for pressure measurements, three basic techniques essential for further considerations are considered.

The piezometer tube manometer. The simplest manometer is a tube, open at the top, which is attached to the top of a vessel containing liquid at a pressure (higher than atmospheric) to be measured. An example can be seen in Figure 6.10a. This simple device is known as a piezometer tube. As the tube is open to the atmosphere, the pressure measured is relative to atmospheric, so it is gage pressure. The pressure at point A is the pressure due to column of fluid above A, that is,

$$p_A = \rho_{fl} g h_A \tag{6.9}$$

where g is acceleration due to gravity. In SI units, $g = 9.81$ m/s². In Equation 6.9, h_A is called hydraulic head as it represents the height to which a given pressure can elevate a column of a liquid.

The U-tube manometer. Using a U-tube enables the pressure of both liquids and gases to be measured with the same instrument. The U-tube is connected as shown in Figure 6.10b and filled with a fluid called the manometric fluid (density $\rho_{fl\text{-}m}$). The fluid whose pressure is being measured should have a mass density less than that of the manometric fluid, and the two fluids should not be able to mix readily—that is, they must be immiscible.

Because pressure in a continuous static fluid is the same at any horizontal level, so pressure at point B is the same as at point C, that is, $p_B = p_C$. For the left-hand arm, pressure at B = pressure at A + pressure due to height h_{A-B}, that is,

$$p_B = p_A + \rho_{fl} g h_{A-B} \tag{6.10}$$

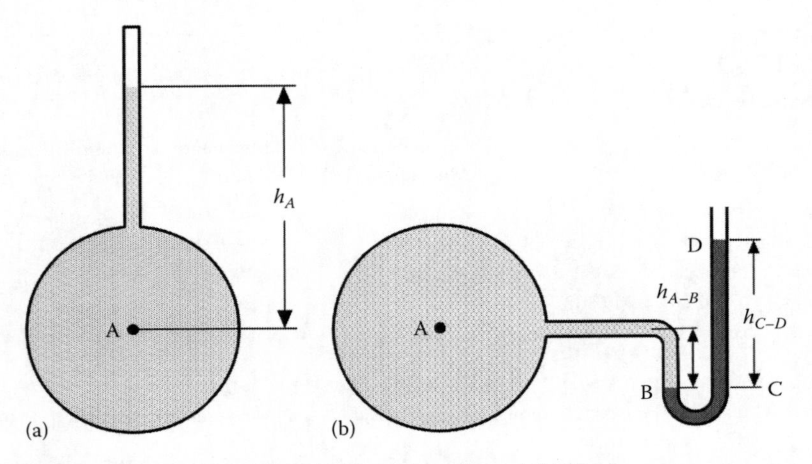

FIGURE 6.10 Pressure measurements by (a) piezometer tube and (b) U-tube manometers.

For the right-hand arm, pressure at C = pressure at D + pressure due to height h_{C-D}, that is,

$$p_C = p_{atm} + \rho_{fl\text{-}m} g h_{C-D} \tag{6.11}$$

Because gage pressure is measured, atmospheric pressure can be subtracted, that is, $p_B = p_C$. Then pressure at point A is calculated as

$$p_a = \rho_{fl\text{-}m} g h_{C-D} - \rho_{fl} g h_{A-B} \tag{6.12}$$

The U-tube manometer is very useful to indicate the difference between two pressures (differential pressure), or between a single pressure and atmosphere (gage pressure), when one side is open to atmosphere. With both ends of the tube open, the liquid is at the same height in each leg as shown in Figure 6.11a. When positive pressure is applied to one leg, the liquid is forced down in that leg and up in the other. The difference in height, h, which is the sum of the readings above and below zero, indicates the pressure (Figure 6.11b). When a vacuum is applied to one leg, the liquid rises in that leg and falls in the other. The difference in height, h, which is the sum of the readings above and below zero, indicates the amount of vacuum (Figure 6.11c). A closed-end manometer gives absolute pressure. When a manometer, which has one leg, sealed, and the other leg open, measures atmospheric pressure, it is called a barometer.

A Bourdon gage is one of the most useful pressure gages in a manufacturing environment. Its design includes a C-shaped tube closed at one end, called the Bourdon tube (Figure 6.12). When the

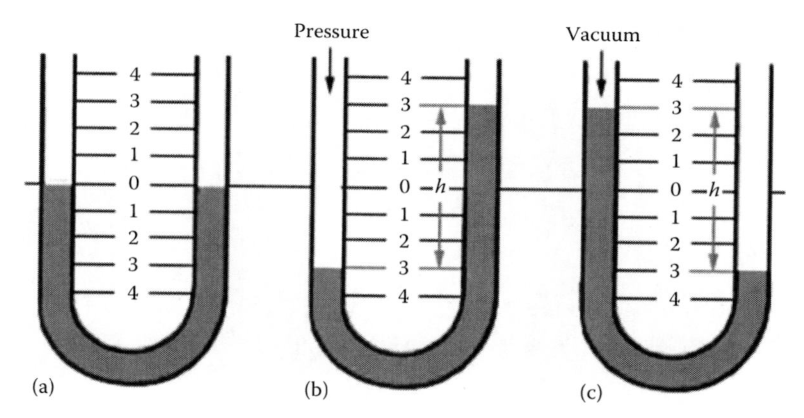

FIGURE 6.11 Measuring pressure with the U-tube manometer.

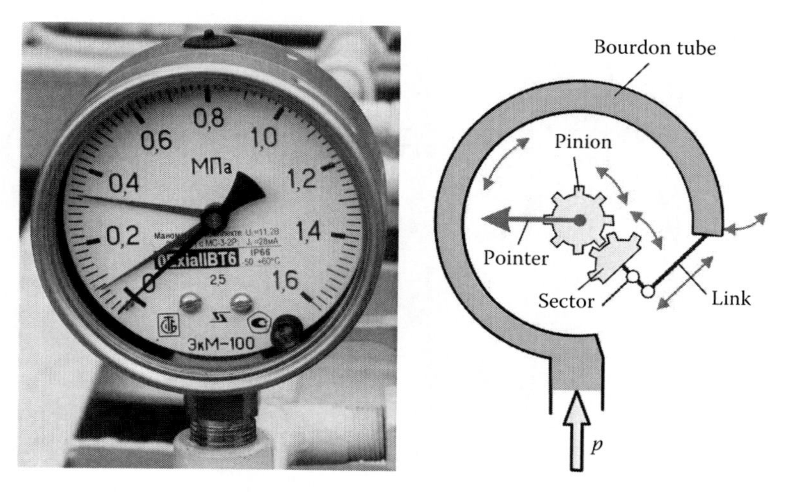

FIGURE 6.12 Bourdon gage.

pressure inside the tube increases, the tube uncurls slightly, causing a small movement at its closed end. A system of levers and gears magnifies this movement and turns a pointer, which indicates the pressure on a circular scale.

6.3.3 Pressure Loss in MWF Supply to the Machining Zone

6.3.3.1 Simple Tests to Understand Phenomenology of Pressure Losses

Experience shows that the pressure of a liquid moving along a pipe decreases with the distance from the beginning of the tube. A simple experiment with a plastic tube connected to a water tap and having three holes located over its length as shown in Figure 6.13 shows that the height of the *fountain* from a hole decreases with proportion to the distance from the tap because the pressure of the water decreases with the tube length.

To carry out a more intelligent test, one may use a setup shown in Figure 6.14. In this setup, piezometer tube manometers are installed on the tube at points 1, 2, and 3 located as shown to measure pressure drop. A measuring cup (measuring volume, V_{mc}) and a stopwatch (to measure the time needed to fill the measuring volume, t_{mc}) are used to calculate the volumetric flow rate

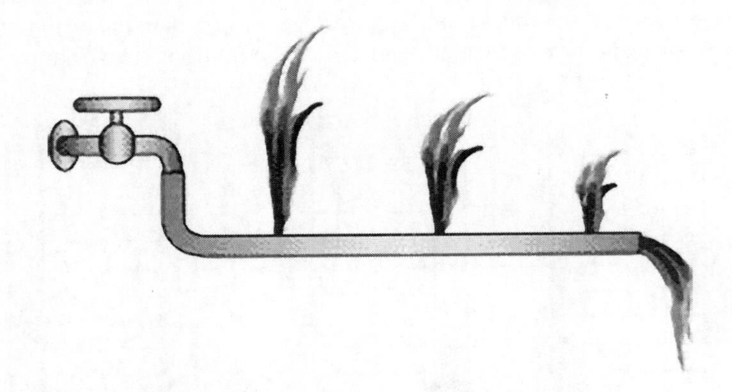

FIGURE 6.13 Simple test showing pressure losses over the length of a tube.

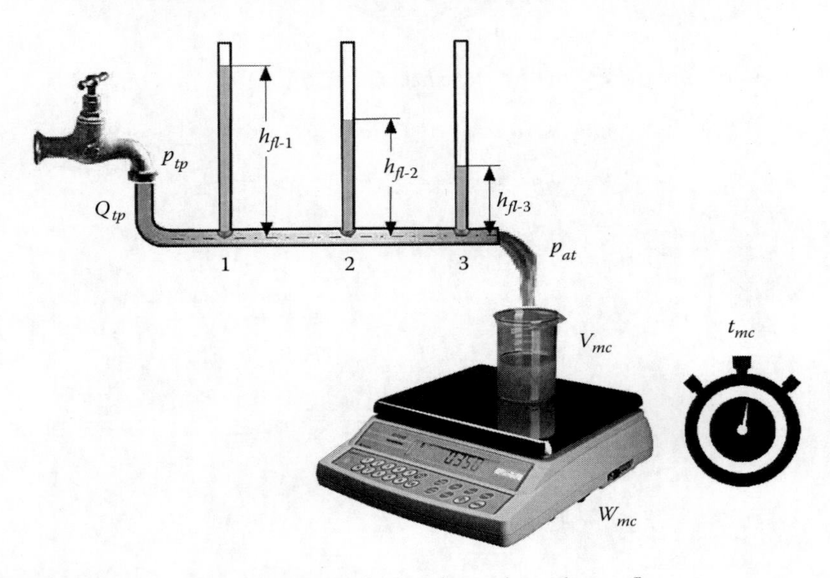

FIGURE 6.14 Simple setup to access pressure drop, volumetric, and mass flow rates.

as $Q_v = V_{mc}/t_{mc}$ while a digital balance is used to measure the mass (weight) of liquid, w_{mc}, over time t_{mc} to calculate the mass flow rate as $Q_m = w_{mc}/t_{mc}$.

Carrying out a series of simple tests, one should arrive to the following conclusions essential to the design of the MWF channels in cutting tools:

- The pressure drop along the length of the tube is in direct proportion to its length, that is, $h_{fl\text{-}3} < h_{fl\text{-}2} < h_{fl\text{-}1}$. Piezometer tube manometers are used in the setup to visualize pressure drops (in the form of hydraulic heads as per Equation 6.9). If Bourdon gages are used instead of piezometer tube manometers, then according to Equation 6.9,

$$p_{tp} > p_1\left(=\frac{h_{fl\text{-}1}}{\rho_{fl}g}\right) > p_2\left(=\frac{h_{fl\text{-}2}}{\rho_{fl}g}\right) > p_3\left(=\frac{h_{fl\text{-}1}}{\rho_{fl}g}\right) \geq 0 \quad \text{i.e. } p_{atm} \tag{6.13}$$

as a Bourdon gage measures the gage pressure.
- If the tap pressure, p_{tp}, is fixed, then the flow rate decreases as (1) the length of the tube increases and (2) the diameter of the tube decreases. As such, the tube diameter affects the flow rate in much greater proportion than its length.
- If an obstacle to the fluid flow is placed in the tube in the form of a diaphragm, additional tap or the tube is just squeezed that changes its configuration and/or reduces its cross-sectional area, the flow rate (under fixed p_{tp}) also decreases.
- If the free end of the tube is elevated with respect to the horizontal position of the tap then the flow rate decreases (under fixed p_{tp}). The opposite is true if this end is lowered.

6.3.3.2 Electrical Analogy to Comprehend the Relationship between the Flow Rate and Pressure

The electronic–hydraulic analogy (derisively referred to as the drain-pipe theory by Oliver Heaviside (Nahin 2002)) is the most widely used analogy for *electron fluid* in a metal conductor. Since electrical current is invisible and the processes at play in electronics are often difficult to demonstrate, the various electronic components are represented by hydraulic equivalents. Electricity (as well as heat) was originally understood to be a kind of fluid, and the names of certain electric quantities (such as current) are derived from hydraulic equivalents. Like all analogies, it demands an intuitive and competent understanding of the baseline paradigms (electronics and hydraulics). Nowadays, the opposite is true. A wide availability of inexpensive volt and ampere-meters as well as the various electronic components made as child development experimental sets to be used as fun science projects made understanding of electrical current flow much easier than that of fluid flow. So, reverse analogy became more instructive.

There are two basic paradigms:

1. Version with pressure induced by gravity. Large tanks of water that are held up high, and the potential energy of the water head is the pressure source. This is reminiscent of electrical diagrams with an up arrow pointing to +*V*, grounded pins that otherwise are not shown connecting to anything, and so on.
2. Completely enclosed version with pumps providing pressure only; no gravity. This is reminiscent of a circuit diagram with a voltage source shown and the wires actually completing a circuit.

Flow and pressure variables can be calculated in a fluid flow network with the use of the hydraulic ohm analogy. The method can be applied to both steady and transient flow situations.

Although analogies of practically all electric (electronic) components are considered in the literature, the following will be used in the further consideration:

1. A pipe completely filled with water is equivalent to a piece of wire.
2. Electric potential is equivalent to hydraulic pressure.
3. Voltage (electrical tension) is also called *potential difference* measured in volts. It is equivalent to a difference in hydraulic pressure between two points.
4. Electrical current measured in amperes is equivalent to a hydraulic volumetric flow rate, that is, the volumetric quantity of flowing fluid over time.
5. Ideal voltage source is equivalent to a dynamic pump. A pressure meter on both sides shows that when this kind of pump is driven at a constant speed, the difference in pressure stays constant.
6. Ideal current source is equivalent to a positive displacement pump. A current meter (little paddle wheel) shows that when this kind of pump is driven at a constant speed, it maintains a constant speed of the little paddle wheel.
7. The electrical resistance of an object (or material) is a measure of the degree to which the object opposes an electrical current passing through it. Normally, electrical resistance is due to connecting wires and due to an electrical component called resistor are considered. In the hydraulic analogy, the former is equivalent to the resistance due to friction in pipes while the latter is due to a constriction in the bore of the pipe, which requires more pressure to pass the same amount of fluid. All pipes have some resistance to flow, just like all wires, and have some resistance to current.

Figure 6.15 shows an electrical analogy of a simple hydraulic circuit. For electrical circuit, Ohm's law is valid, that is,

$$I = \frac{V}{R} \tag{6.14}$$

where
I is the electrical current in amperes
V is the voltage in volts
R is the total electrical resistance in ohms

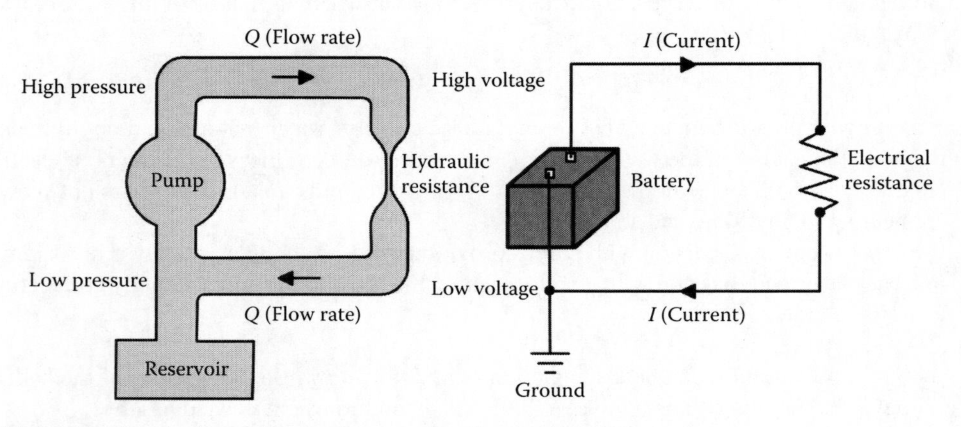

FIGURE 6.15 Electrical analogy of a simple hydraulic circuit.

For hydraulic circuit, the Poiseuille's law is valid, that is,

$$Q = \frac{\Delta p}{R} \tag{6.15}$$

where

Q is the flow rate, in m^3/s
Δp is the pressure difference in Pa
R is hydraulic resistance in $N/(m^4 \cdot s)$

The discussed analogy is useful for the following:

- Understanding a simple fact that for a given pressure (voltage), the MWF flow rate (current) decreases as hydraulic (electrical) resistance increases. If the flow rate optimal for tool performance is known, then the pump pressure needed to deliver this flow rate can be calculated for a given hydraulic resistance of the hydraulic circuit. Because in hydraulic circuits of machining tools, the maximum hydraulic resistance is associated with the tool MWF delivery channels, and because the pump pressure of many practical MWF delivery systems is rather limited, the only way to assure the optimal flow rate is to reduce the hydraulic resistance of the tool. This is discussed in the sections to follow
- Calculating the flow rates through multiple channels made in multiple-stage cutting tools. This is discussed later in this chapter.

6.3.3.3 Modified Bernoulli Equation

The experimental facts described in Section 6.3.3.1 and the idea of hydraulic resistance can be explained if the modified Bernoulli equation is applied. This equation relates the parameters of fluid flow between two cross sections (1-1 and 2-2) of a tube shown in Figure 6.16 in the following form (Fox and McDonald 1985):

$$\frac{p_1}{\rho_{fl} g} + \frac{v_1^2}{2g} + h_1 = \frac{p_2}{\rho_{fl} g} + \frac{v_2^3}{2g} + h_2 + h_{ls} \tag{6.16}$$

where

subscripts 1 and 2 refer to the cross sections 1-1 and 2-2
p_1 and p_2 are the static pressure at these sections
v_1 and v_2 are the velocities of the liquid through the considered sections
h_1 and h_2 are the elevations of sections 1-1 and 2-2 with respect to the reference level, respectively
h_{ls} is the overall hydraulic head (pressure) loss between the considered sections

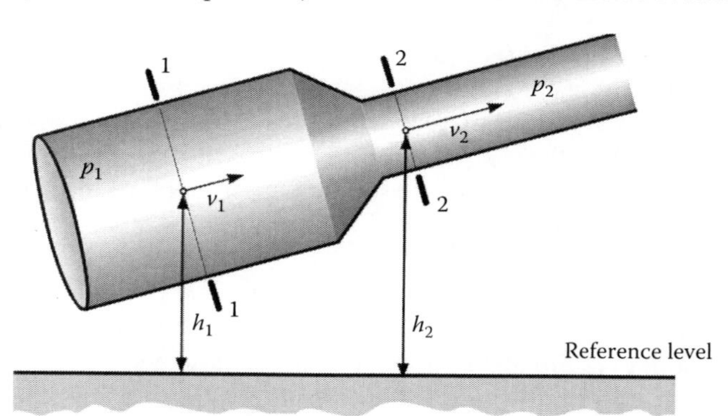

FIGURE 6.16 Model of fluid flow in a tube.

Expressing Equation 6.16 in terms of head loss, one can write

$$h_{ls} = \frac{p_1 - p_2}{\rho_{fl} g} + \frac{v_1^2 - v_2^2}{2g} + (h_1 - h_2) \quad (6.17)$$

Equation 6.17 provides a method to determine h_{ls} experimentally by measuring static pressures and velocities at sections 1-1 and 2-2. In many practical applications of cutting tools, sections 1-1 and 2-2 locate at the same vertical distance from the reference level (i.e., $h_1 = h_2$). Then Equation 6.17 becomes

$$h_{ls} = \frac{p_1 - p_2}{\rho_{fl} g} + \frac{v_1^2 - v_2^2}{2g} \quad (6.18)$$

Overall head loss, h_{ls}, is regarded as the sum of major losses, h_l, due to frictional effects in fully developed flow in constant-area tubes, and minor losses, h_m, due to entrances, fittings, tube area changes, and so on:

$$h_{ls} = h_l + h_m \quad (6.19)$$

The terms *major* and *minor* are conditional. They originated from traditional pipe hydraulics where the frictional losses are much greater than those due to local resistances. In electrical analogy, the major losses are those in the connecting wires while the minor losses are in resistors. When the whole electrical grid is considered starting with a generator at the power station, then the losses in transmission wires are significant. If a local circuit with a battery is considered then the minor losses in resistors play the major role. In full analogy, it often happens in hydraulic conduits of cutting tools that minor losses are much greater than major.

6.3.3.4 Types of Flow

There are two types fluids flow (called regimes): (1) laminar flow and (2) turbulent flow. There is a wide gradation between these two types that is called transitional flow. These flow regimes are characterized as follows:

- *Laminar flow.* In laminar flow, fluid molecules move in straight, parallel lines down current. If a dye is added to a fluid that is in the laminar flow regime, the dye would not mix into the fluid; it would streak out in an approximately straight line. Laminar flow is characteristic of slow-moving fluids and flows where the fluid is very viscous, like some mineral oils.
- *Turbulent flow.* In contrast, turbulent flow is characterized by complex motion of fluid molecules. Molecules move in all directions in bursts of upward, downward, and forward motion and even some backward movement. There is abundant mixing in the flow, and an added dye would mix into the water very quickly.
- *Transitional flow.* Transitional flows have some characteristics of laminar flow and some of turbulent flow. For example, dye may take some time to mix into the flow, but it does mix.

6.3.3.5 Viscosity

Viscosity is a measure of the resistance of a material to flow, that is, how *thick* and easily deformed it is. Viscosity is sort of like the amount of friction within a substance. Walking through air is easy, because there is not much friction between air molecules. Air has a low viscosity. Swimming is more difficult because the water drags on the body. This is due to the *friction* between adjacent water molecules, that is, higher viscosity.

On the most basic level, as it applies to MWFs, viscosity can be regarded as a measure of a fluid's resistance to deformation under shear stress or how resistant a fluid is to pouring. A product expressed as *thick* is slow to flow or pour versus one that flows easily being *thin*. In the application and management of MWFs, viscosity comes directly into play in the following: (1) dealing with *straight oils*. Viscosity is probably the single most important physical characteristic in describing this type of MWFs. How *thick* or *thin* an oil is largely determines how much lubrication (normally due to extreme-pressure additives) the oil can provide. Thicker oils have longer *heavy-duty* molecules not readily squeezed from the contact interfaces under high contact stresses than thin oils, while thin oils tend to *wet* work surfaces much more rapidly and completely, which is important when the cutting speed increases. (2) Selecting pumps to move working solution or concentrate. The viscosity of the fluid has a major effect on the size and type of pump required. (3) Designing bulk storage and delivery systems for straight oils, coolant concentrates, and washing compounds. Water-soluble MWFs have very low viscosity. In fact, they are so close to the viscosity of water that they normally are not measured.

Viscosity of straight oils is temperature dependent. Typically, the higher the temperature of an MWF, the lower its viscosity. The viscosity of MWFs typically is measured at 38°C (100°F), a convenient number because it is the temperature used to rate *standard* oils and is close to the *typical* temperature at which MWFs are used.

Two types of viscosity are normally distinguished: (1) dynamic or true viscosity and (2) kinematic viscosity. To understand the physical difference between these two types, one should refer to the method by which they are measured. In measuring kinematic viscosity using the most common Zahn cup test, a cup with a calibrated hole in the bottom is filled to a reference line. The kinematic viscosity is determined by the time it takes for a known volume of liquid to run out of the hole. If a liquid of a similar measured dynamic viscosity but greater density is tested in the same apparatus, it will show a lower kinematic viscosity, because its greater mass per unit volume (known as density) pushes the liquid through the orifice at a greater rate. Therefore, dynamic viscosity is often referred to as true viscosity as it represents the true *fluidity* of a fluid.

Dynamic viscosity is represented by the Greek letter μ; when reported in SI units, they are pascal-seconds ($Pa \cdot s$ or $N \cdot s/m^2$), which is defined as 1 kg/m/s. The centimeter–gram–second system (*CGS*) physical unit is the poise (P) or more commonly, particularly in ASTM standards, as centipoise (cP). Centipoise is the preferred unit because at 20°C, water has a viscosity of 1.0020 cP. 1 poise = 100 centipoise = 1 g/cm/s = 0.1 $Pa \cdot s$ or 1 centipoise = 1 $mPa \cdot s$. Kinematic viscosity is represented by the Greek letter ν; the unit of physical measure in SI units is m^2/s and in *CGS* is stokes or sometimes centistokes (cS or cSt). 1 stoke = 100 centistokes = 1 cm^2/s = 0.0001 m^2/s and 1 cSt = 10^{-6} $m^2 \cdot s^2$.

Conversion between kinematic and dynamic viscosities for a given fluid (subscript *fl*) is possible given the formula: $\nu_{fl}\rho_{fl} = \mu_{fl}$. In CGS system, kinematic viscosity (cSt) × density (g/mL) = dynamic viscosity (cP) or dynamic viscosity (cP)/density (g/mL) = kinematic viscosity (cSt).

SSU or Seconds Saybolt Universal is an older method of experimentally deriving the viscosity of a fluid. This test measures the amount of time in seconds required for 60 mL of liquid to flow through a calibrated orifice under controlled conditions as per ASTM D 88. This system is a kinematic viscosity rating, and 1.0 SUS = 0.2158 cSt.

6.3.3.6 Reynolds Number

Reynolds number (*Re*) predicts the extent of turbulence in a fluid based on how fast the fluid is flowing, the geometry of the flow (how deep and wide it is, etc.), and the density and viscosity of the fluid. The Reynolds number is a nondimensional parameter defined by the ratio of (1) dynamic pressure $\left(\rho_{fl} \cdot v_{fl}^2\right)$ and (2) shearing stress $((\mu_{fl} \cdot v_{fl})/L_{cr})$ and can be expressed as

$$Re = \frac{\rho_{fl} v_{fl}^2}{(\mu_{fl} v_{fl})/L_{cr}} = \frac{\rho_{fl} v_{fl} L_{cr}}{\mu_{fl}} = \frac{v_{fl} L_{cr}}{\nu_{fl}} \tag{6.20}$$

where L_{cr} is the characteristic length (m).

For a round tube, the characteristic length is the hydraulic diameter, d_h; thus,

$$Re = \frac{\rho_{fl} v_{fl} d_h}{\mu_{fl}} = \frac{v_{fl} d_h}{\nu_{fl}} \tag{6.21}$$

The hydraulic diameter is not the same as the geometrical diameter in a noncircular tube and can be calculated using the generic equation

$$d_h = \frac{4A_{tb}}{p_{w\text{-}t}} \tag{6.22}$$

where

A_{tb} is the cross-sectional area of the tube (m^2)
$p_{w\text{-}t}$ is wetted perimeter of the tube (m)

Based on Equation 6.22, the hydraulic diameter of a circular tube of radius r_{tb} (Figure 6.17a) can be expressed as

$$d_h = \frac{4\pi r_{tb}^2}{2\pi r_{tb}} = 2r_{tb} \tag{6.23}$$

that is, the hydraulic diameter of a standard circular tube is two times its radius. In other words, it coincides with the tube diameter.

The hydraulic diameter of an annular tube having radii $r_{tb\text{-}1}$ and $r_{tb\text{-}2}$ as shown in Figure 6.17b is calculated as

$$d_h = \frac{4\pi\left(r_{tb\text{-}1}^2 - r_{tb\text{-}2}^2\right)}{2\pi\left(r_{tb\text{-}1} - r_{tb\text{-}2}\right)} = 2(r_{tb\text{-}1} - r_{tb\text{-}2}) \tag{6.24}$$

The hydraulic diameter of a rectangular tube having sides a and b as shown in Figure 6.17c can be calculated as

$$d_h = \frac{4ab}{2(a+b)} = \frac{2ab}{a+b} \tag{6.25}$$

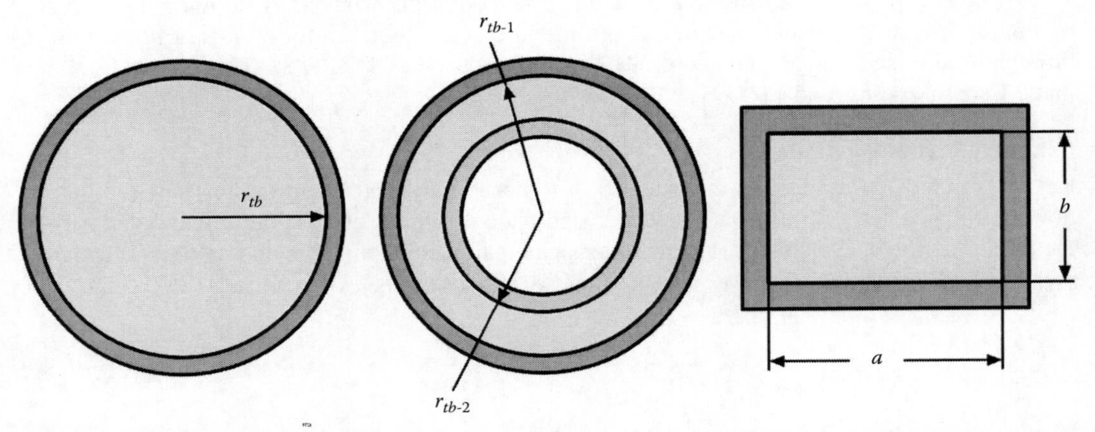

FIGURE 6.17 Simple models to calculate the hydraulic diameter for common cross sections of tubes.

The Reynolds number is used to determine if the flow is laminar, transient, or turbulent. The flow is

- *Laminar* when $Re < 2300$
- *Transient* when $2300 < Re < 4000$
- *Turbulent* when $Re > 4000$.

Example 6.1

Problem

A water-soluble MWF having viscosity 0.658×10^{-6} m²/s at 40°C is flowing at 6 L/min in a 1.8 mm dia. coolant channel made in a tool. Is the flow in the channel turbulent, transient, or laminar?

Solution

The volumetric flow rate of the MWF in the coolant channel in SI units is

$$Q_v = 6\text{L/min} \times 0.17 \times 10^{-4} = 1 \times 10^{-4}\ \text{m}^3/\text{s}$$

The cross-sectional area of the coolant channel is

$$A_{ch} = \frac{3.14 \times (1.8 \times 10^{-3})^2}{4} = 2.54 \times 10^{-6}\ \text{m}^2$$

The velocity of MWF in the coolant channel is calculated as

$$v_{fl} = \frac{Q_v}{A_{ch}} = \frac{1 \times 10^{-4}}{2.54 \times 10^{-6}} = 39.32\ \text{m/s}$$

The Reynolds number is calculated as

$$Re = \frac{39.32 \times 1.8 \times 10^{-3}}{0.658 \times 10^{-6}} = 107{,}562$$

Because $Re \gg 4{,}000$, the fully developed turbulent flow is the case.

6.3.3.7 Major Pressure Losses: Friction Factor

The friction losses in a tube are usually calculated using the Darcy–Weisbach (sometimes referred to as the Fanning equation) formula (Fox and McDonald 1985):

$$h_l = \lambda \frac{l_{tb}}{d_{tb}} \frac{v_{fl}^2}{2g} \tag{6.26}$$

where

λ is Darcy coefficient often referred to as the friction factor
l_{tb} and d_{tb} are the length and diameter of the tube, respectively
v_{fl} is the velocity of the fluid

It must be emphasized that the friction factor λ is empirical and can only be determined through experiments. It depends primarily upon flow type and pipe surface roughness.

6.3.3.7.1 Laminar Flow

The underlying assumption of laminar flow is the condition of uniform viscosity across the diameter of the tube, that is, each fluid molecule within the pipe is exerting a similar force against its immediate neighbor toward the periphery. This yields the following:

- Fluid adjacent to the conduit wall is motionless, that is, its velocity $v_{fl\text{-}w} = 0$
- Maximum fluid velocity ($v_{fl\text{-}c}$) is in the center of the tube
- Fluid velocity is related to the distance from the tube's center by a parabolic function. This is why laminar flow sometimes is termed parabolic flow.

For fully developed laminar flow, the roughness of the tube or pipe can be neglected. The friction factor depends only on the Reynolds number, *Re*, and can be expressed using the Poiseuille's equation as

$$\lambda = \frac{64}{Re} \tag{6.27}$$

Although Equation 6.27 allows to calculate λ and, using Equation 6.26, pressure (hydraulic head) losses in a tube, a greater visualization of the influence of various parameters of laminar flow on the flow rate through a round tube can be achieved if this equation is applied to calculate the volumetric flow rate, Q_v, through a round tube of radius r_{tb} having pressure p_1 at the entrance and p_2 at the end as shown in Figure 6.18. The length of the tube is l_{tb} and the dynamic viscosity of the fluid is μ_{fl}. For these parameters, the volumetric flow rate is calculated as

$$Q_v = \frac{\pi(p_1 - p_2)r_{tb}^4}{8\mu_{fl} l_{tb}} \tag{6.28}$$

Several important properties of laminar flow follow from Equation 6.28. Suppose the original flow rate is 6 L/min. The effect of changes of the parameters of Equation 6.28 is as follows: (1) If the length of the tube is doubled, then Q_v = 3 L/min; (2) if the viscosity of the fluid is doubled, then Q_v = 3 L/min; (3) if the pressure difference is doubled, then Q_v = 12 L/min; (4) if the tube radius, r_{tb}, is doubled, then Q_v = 96 L/min; and (5) a 19% increase in the tube radius, t_{tb}, doubles the flow rate.

6.3.3.7.2 Turbulent Flow

The simplest way to calculate the friction factor for turbulent flow is to use the Blasius equation, which applies to smooth pipes in the range 3,000 < *Re* < 100,000:

$$\lambda = \frac{0.3164}{Re^{0.25}} \tag{6.29}$$

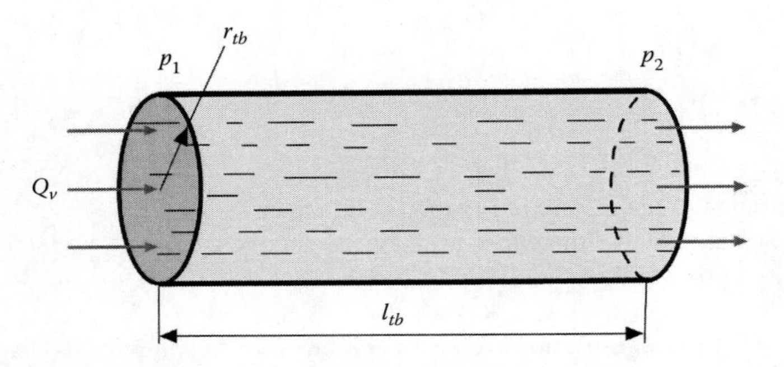

FIGURE 6.18 Parameters to calculate pressure drop in laminar flow.

The real tubes and coolant channels are not smooth, so the friction factor should be determined accounting for the so-called relative roughness of the tube (channel), k_{tb}, which is the ratio of the average absolve roughness, R_z, to the tube diameter, d_{tb}. The friction factor is then calculated using the Colebrook equation:

$$\frac{1}{\lambda^{1/2}} = -2\log\left(\frac{2.51}{Re\lambda^{1/2}}\right) + \frac{k_{tb}}{3.72} \quad (6.30)$$

Because the friction factor is involved on both sides of Equation 6.30, this equation must be solved by iteration.

A graphical representation of the Colebrook equation is the Moody diagram shown in Figure 6.19.

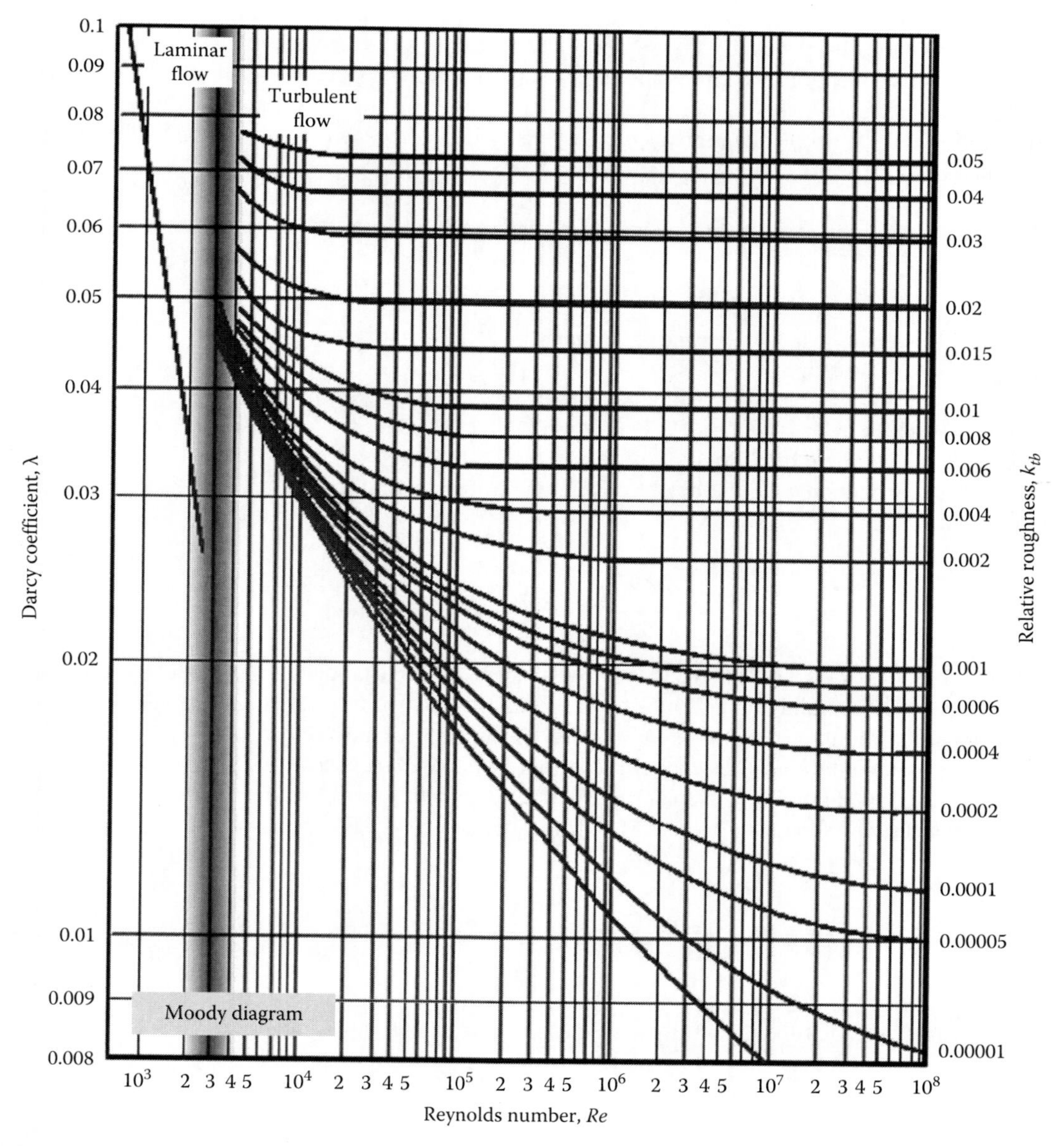

FIGURE 6.19 Moody diagram.

6.3.3.7.3 *Transient Flow*

If the flow is transient, $2300 < Re < 4000$, the flow varies between laminar and turbulent flow, and the friction coefficient should be determined experimentally for a given set of conditions (Kwon 2008). Tests indicate (Stenphenson 1981) that the pressure (head) losses during transient flow are higher than those predicted using the friction factor applicable for other flow conditions. The energy is absorbed during flow reversals when the velocity is low and the friction factor consequently relatively high. Therefore, whenever it is possible, this type of flow should be avoided in the design of MWF supply systems of drilling tools. Unfortunately, this is not always possible.

6.3.3.8 Minor Losses (Losses Due to Form Resistance)

Although they normally account for a major portion of the total pressure loss in hydraulic circuits of drilling tools due to entries and exits, fittings and contractions/expansions and so on, they are traditionally referred to as minor losses. These losses represent additional energy dissipation in the flow, usually caused by secondary flows induced by curvature or recirculation. Therefore, the understanding of the concept of minor pressure losses is paramount for a drilling system/tool designer.

The minor pressure losses are characterized by the flow resistance coefficient K_{fr}, which, for the uniform distribution of the static pressure and density in a considered cross section of the flow, is calculated as

$$K_{fr} = \frac{\Delta p}{\left(1/2\rho_{fl} v_{fl\text{-}b}^2\right)} \tag{6.31}$$

where

Δp is the pressure loss due to minor resistance
$v_{fl\text{-}b}$ is the fluid velocity over the so-called base area

Base areas for all flow resistances, K_{fr}, is normally given based on the smallest cross-sectional area of the component where the velocity if highest.

In terms of Equation 6.19, the head loss due to a minor hydraulic resistance is readily obtained from Equation 6.31 as

$$h_m = K_{fr} \frac{v_{fl\text{-}b}^2}{2g} \tag{6.32}$$

Although the values of the flow resistance coefficients are well tabulated in practically all books on the subject (Bhave and Gupta 2006), it was found instructive to present and explain here some important cases commonly found in drilling tool MWF supply channels.

The first group of minor losses considered in this section includes those due to expansion and contraction of the flow (Figure 6.20). An abrupt contraction of a tube cross-sectional area (Figure 6.20a) gives rise to the so-called shock losses. The base area is area A_2. For this case, the flow resistance coefficient K_{fr} is calculated as

$$K_{fr\text{-}Cont} = 0.5\left(1 - \frac{A_2}{A_1}\right)^{0.75} \tag{6.33}$$

It is obvious that if the tube is round, then Equation 6.33 can be represented as

$$K_{fr\text{-}Cont} = 0.5\left(1 - \frac{d_{tb\text{-}2}^2}{d_{tb\text{-}1}^2}\right)^{0.75} \tag{6.34}$$

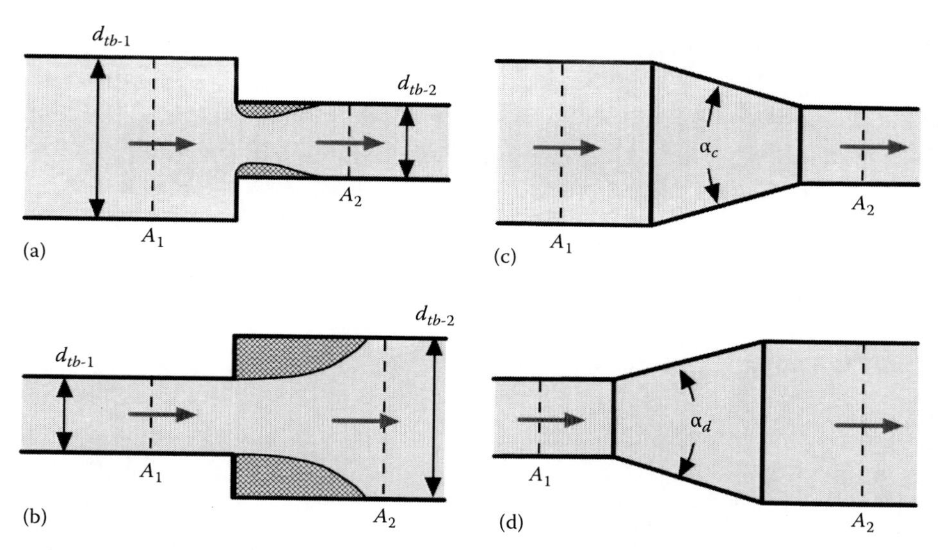

FIGURE 6.20 Expansion and contraction of the flow: (a) abrupt contraction, (b) abrupt expansion, (c) conical contraction, and (d) conical expansion.

When the cross section abruptly expands (Figure 6.20b), the phenomenon is basically similar to that observed when shock losses occur during abrupt contraction. The base area is area A_1. For this case, the flow resistance coefficient K_{fr} is calculated as

$$K_{fr\text{-}Cont} = \left(1 - \frac{A_1}{A_2}\right)^{0.75} \tag{6.35}$$

It is obvious that if the tube is round, then Equation 6.35 can be represented as

$$K_{fr\text{-}Cont} = \left(1 - \frac{d_{tb\text{-}1}^2}{d_{tb\text{-}2}^2}\right)^{0.75} \tag{6.36}$$

A comparison of Equations 6.33 and 6.35 shows that the flow resistance coefficient K_{fr} is twice greater for abrupt expansion than contraction under the same flow conditions.

The flow resistance coefficient K_{fr} for conical contraction (Figure 6.20c) depends on angle of confuser, α_c. It is calculated as follows:

for $\alpha_c < 45°$

$$K_{fr\text{-}ConCont} = 0.8 \sin\left(\frac{\alpha_c}{2}\right)\left(1 - \frac{A_2}{A_1}\right) \tag{6.37}$$

for $\alpha_c > 45°$

$$K_{fr\text{-}ConCont} = 0.5\sqrt{\sin\left(\frac{\alpha_c}{2}\right)}\left(1 - \frac{A_2}{A_1}\right) \tag{6.38}$$

The flow resistance coefficient K_{fr} for conical expansion (Figure 6.20d) depends on angle of diffuser, α_d. It is calculated as follows:

for $\alpha_d < 45°$

$$K_{fr\text{-}Con\,Exp} = 2.6\sin\left(\frac{\alpha_d}{2}\right)\left(1-\frac{A_1}{A_2}\right)^2 \tag{6.39}$$

for $\alpha_d > 45°$

$$K_{fr\text{-}Con\,Exp} = \left(1-\frac{A_1}{A_2}\right)^2 \tag{6.40}$$

The next group of minor losses considered in this section includes those due to an orifice in a straight tube (Figure 6.21).

For thick edge orifice shown in Figure 6.21a, the base area is area A_2. For this case, the flow resistance coefficient K_{fr} is calculated as

$$K_{fr\text{-}Thick\,Edge} = \left[0.5\left(1-\frac{A_2}{A_1}\right)^{0.75} + a_{te}\left(1-\frac{A_2}{A_1}\right)^{1.375} + \left(1-\frac{A_2}{A_1}\right)^2 + \frac{\lambda l_{tb\text{-}2}}{d_{tb\text{-}2}}\right] \tag{6.41}$$

where

$$a_{te} = (2.4 - l_{tb\text{-}2}) \times 10^{\left(0.25+\left(0.535 l_{tb\text{-}2}^8 / 0.050 + l_{tb\text{-}2}^8\right)\right)} \tag{6.42}$$

where λ is the friction factor (Darcy coefficient) for the tube having length $l_{tb\text{-}2}$.

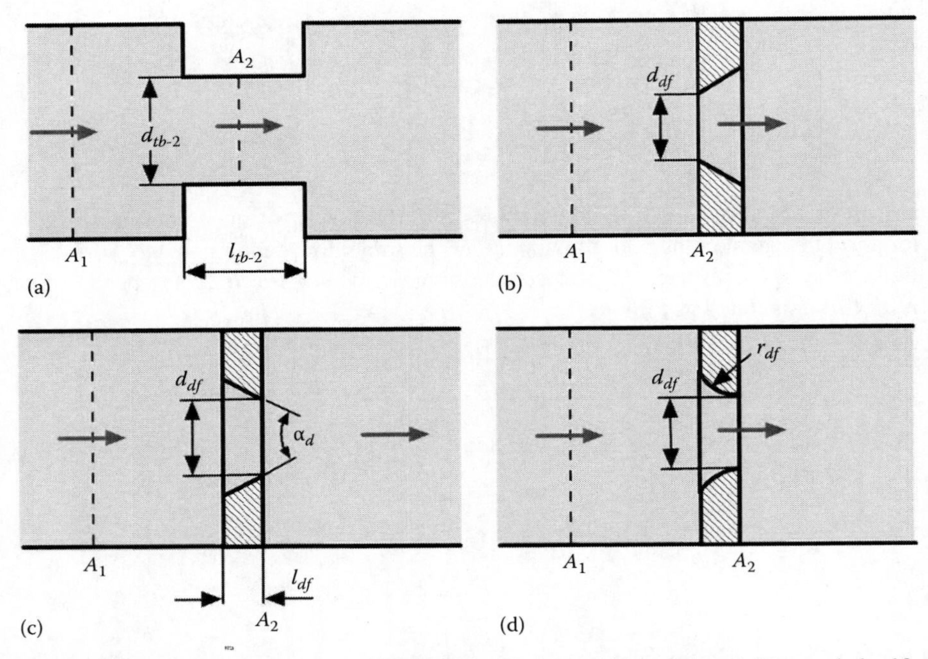

FIGURE 6.21 Orifice in a straight tube: (a) thick edge, (b) sharp, (c) beveled, and (d) rounded orifices.

For sharp orifice (often referred to as a diaphragm) shown in Figure 6.21b, the base area is area A_2. For this case, the flow resistance coefficient K_{fr} is calculated as

$$K_{fr\text{-}Sharp\ Orifice} = \left[\left(1-\frac{A_2}{A_1}\right)+0.707\left(1-\frac{A_2}{A_1}\right)^{0.375}\right]^2\left(\frac{A_2}{A_1}\right)^2 \quad (6.43)$$

For beveled orifice shown in Figure 6.21c, the base area is area A_2. For this case, the flow resistance coefficient K_{fr} is calculated as

$$K_{fr\text{-}Beveled\ Orifice} = \left[\left(1-\frac{A_2}{A_1}\right)+\sqrt{b_{bo}}\left(1-\frac{A_2}{A_1}\right)^{0.375}\right]^2 \quad (6.44)$$

where

$$b_{bo} = 0.13+0.34\times10^{-\left[3.4\left(l_{df}/d_{tb\text{-}2}\right)+88.4\left(l_{df}/d_{tb\text{-}2}\right)^{2.3}\right]} \quad (6.45)$$

For rounded orifice shown in Figure 6.21d, the base area is area A_2. For this case, the flow resistance coefficient K_{fr} is calculated as

$$K_{fr\text{-}Rounded\ Orifice} = \left[\left(1-\frac{A_2}{A_1}\right)+\sqrt{b_{ro}}\left(1-\frac{A_2}{A_1}\right)^{0.75}\right]^2 \quad (6.46)$$

where

$$b_{ro} = 0.03+0.47\times10^{-\left(r_{df}/d_{tb\text{-}2}\right)} \quad (6.47)$$

The next group of minor losses considered in this section represents those due to flow resistances of elbows (Figure 6.22). The flow resistance coefficient $K_{fr\text{-}el1}$ of a 90° smooth elbow (Figure 6.22a) is a function of the turbulent factor f_t, which, in turn, depends on radius r_3 as presented in Table 6.2. Table 6.3 tabulates the flow resistance coefficients.

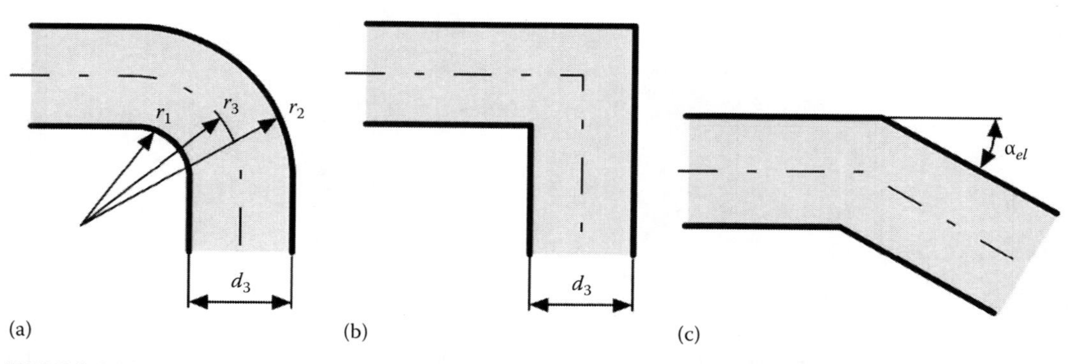

FIGURE 6.22 Elbows: (a) rounded, (b) 90° miter bend, and (c) non-90° miter bend.

TABLE 6.2
Turbulence Coefficient, f_t

$2r_3$ (mm)	8	10	12	19	25	32	38	50	75	100
f_t	0.031	0.029	0.027	0.025	0.023	0.022	0.021	0.019	0.018	0.017

TABLE 6.3
Flow Resistance Coefficient for Smooth Elbows

$2r_3/d_3$	1	1.5	2	3	4	6	8	10	12	14	16	20
$K_{fr\text{-}el1}/f_t$	20	14	12	13	15	17	24	30	34	38	42	50

TABLE 6.4
Flow Resistance Coefficient of a Non-90° Miter Bend Elbow

α_{el} (°)	15	30	45	60	75	90
$K_{fr\text{-}el3}/f_t$	4	8	15	25	40	60

The flow resistance coefficient K_{fr} of a 90° miter bend elbow (Figure 6.22b) is $K_{fr\text{-}el2} = 60f_t$. The flow resistance coefficient $K_{fr\text{-}el3}$ of a non-90° miter bend elbow (Figure 6.22c) depends on the so-called miter angle α_{el} as represented in Table 6.4.

The foregoing considerations suggest that a tool/carbide perform designer should try to select the miter angle as small as is possible and use smooth elbows wherever is possible.

The last useful group of minor losses considered in this section represents losses due to flow resistances that occur on a fluid discharge from a straight tube (Figure 6.23). In the case of uniform velocity distribution (Figure 6.23a), $K_{fr\text{-}ds1} = 1$. When the axis of a coolant channel does

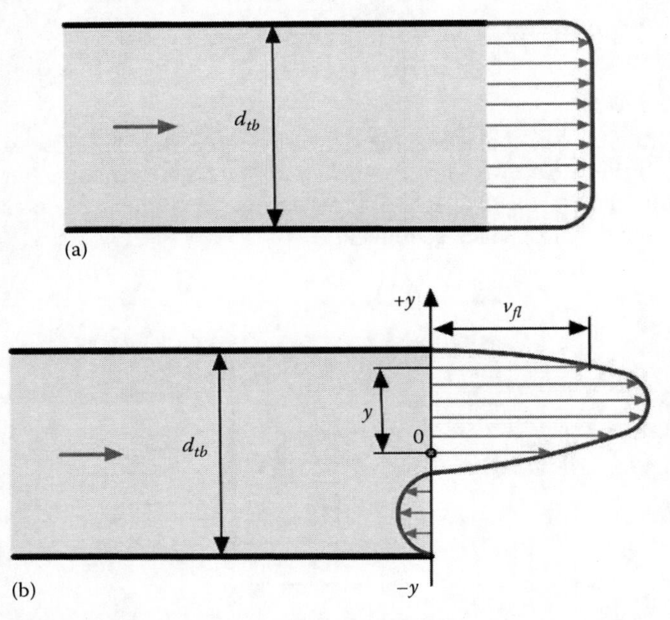

FIGURE 6.23 Free discharge from a straight tube: (a) uniform velocity distribution and (b) asymmetrical velocity distribution.

not coincide with the axis of rotation and when rotational speed is high (as, e.g., in HP drilling/reaming), the fluid velocity in this coolant channel is not distributed uniformly due to the centrifugal force. For a particular yet common case, the fluid velocity distributes over the tube plane section as (Figure 6.23b)

$$\frac{v_{fl}}{v_{fl\text{-}0}} = 0.585 + 164\sin\left(0.2 + 3.9\frac{y}{d_{tb}}\right) \tag{6.48}$$

The flow resistance coefficient $K_{fr\text{-}ds1}$ is in the range of 0.60–3.67 depending upon the rotational speed and the distance of the coolant channel from the axis of rotation. At high rotational speeds (approx. 20,000–25,000 rpm) and for small tools (drills and reamers of up to 12 mm diameter), $K_{fr\text{-}ds1} = 0.7–0.9$.

Example 6.2

The simplest case: A drill is not in the hole being drilled and does not rotate.

Problem

Find the MWF pressure needed to deliver $Q_v = 12$ L/min flow rate for the drill shown in Figure 6.24. Use the following parameters in calculations: MWF—a water-soluble MWF having viscosity 0.658×10^{-6} m²/s at 40°C and density $\rho_{fl} = 985$ kg/m³; diameter of the tool holder (tool shank) $d_h = 12$ mm; drill has two coolant holes of $d_{ch} = 1.3$ mm; the length of the coolant holes $L_{ch} = 80$ mm; the coolant hole is EDM'd with surface finish $R_z = 10$ µm.

Solution

The volumetric flow rate of the MWF in the coolant channel in SI units is

$$Q_v = 12(\text{L/min}) \times 0.17 \times 10^{-4} = 2.04 \times 10^{-4}\ \text{m}^3/\text{s}$$

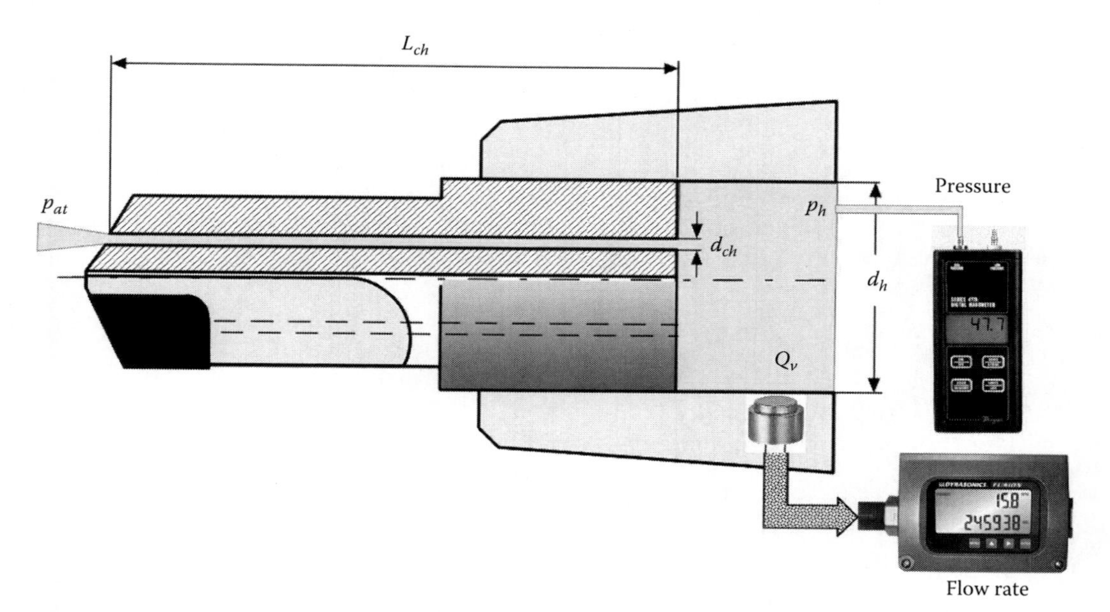

FIGURE 6.24 Model for Example 6.2.

The cross-sectional areas of the tool holder and the coolant hole

$$A_h = \frac{3.14 \times (12 \times 10^{-3})^2}{4} = 113.04 \times 10^{-6}\ \text{m}^2$$

$$A_{ch} = \frac{3.14 \times (1.3 \times 10^{-3})^2}{4} = 1.33 \times 10^{-6}\ \text{m}^2$$

The velocities of the MWF in the holder and in the coolant holes are calculated as

$$v_h = \frac{Q_v}{A_h} = \frac{2.04 \times 10^{-4}}{113.04 \times 10^{-6}} = 1.8\ \text{m/s}$$

$$v_{ch} = \frac{Q_v/2}{A_{ch}} = \frac{(2.04 \times 10^{-4}/2)}{1.33 \times 10^{-6}} = 76.7\ \text{m/s}$$

Note that in the last formula, Q_v is divided by 2 as there are two coolant holes in the drill.

The Reynolds numbers in the holder and in the coolant holes are calculated as

$$Re_h = \frac{v_h d_h}{\nu_{fl}} = \frac{1.8 \times 12 \times 10^{-3}}{0.658 \times 10^{-6}} \approx 32{,}827$$

$$Re_h = \frac{v_{ch} d_{ch}}{\nu_{fl}} = \frac{76.7 \times 1.3 \times 10^{-3}}{0.658 \times 10^{-6}} \approx 151{,}525$$

As in the tool holder and in the coolant holes $Re \gg 2000$, the fully developed turbulent flow takes place in these MWF passages.

The relative roughness of the tube (channel), k_{tb}, which is the ratio of the average absolve roughness, R_z, to the tube diameter, d_{tb}, is calculated for the considered case as $k_{tb} = 0.010/1.3 = 0.0077$. Using this value of k_{tb}, one finds the value of the friction factor (Darcy coefficient) λ in the Moody diagram (Figure 6.19) to be 0.037.

Major hydraulic head losses are calculated only for a coolant hole using Equation 6.26 as

$$h_{ch} = \lambda \frac{L_{ch}}{d_{dh}} \frac{v_{ch}^2}{2g} = 0.037 \frac{80}{1.3} \frac{76.7^2}{2 \times 9.81} = 739.61\ \text{m}$$

In the considered case, when the drill is stationary and the discharge head losses of the MWF flow from the coolant channel are neglected, the only minor head loss is due to flow contraction from the tool holder into the coolant hole(s). For this case, the flow resistance coefficient K_{fr} is calculated using Equation 6.33 as

$$K_{fr\text{-}Cont} = 0.5\left(1 - \frac{A_{ch}}{A_h}\right)^{0.75} = 0.5\left(1 - \frac{1.33}{113.04}\right)^{0.75} = 0.496$$

Minor head losses are calculated using Equation 6.32 as

$$h_m = K_{fr\text{-}Cont} \frac{v_{ch}^2}{2g} = 0.496 \frac{76.7^2}{2 \times 9.81} = 148.72\ \text{m}$$

The total head losses are

$$h_\Sigma = h_{ch} + h_m = 739.61 + 148.72 = 888.33\ \text{m}$$

The required pressure in the tool holder is calculated as

$$p_h = h_{\Sigma}\rho_{fl}g = 888.33\times 985\times 9.81 = 8,583,799.5\,\text{Pa} \approx 8.58\,\text{MPa}(\approx 1244\,\text{psi})$$

Unfortunately, modern machines and manufacturing cells are not normally equipped with MWF systems capable to generate such high MWF pressure. Therefore, some improvements of the design parameters are needed to reduce the required MWF pressure. As the total hydraulic head losses consist of two parts, namely, major and minor losses, some obvious improvements were suggested to decrease these losses.

To reduce major losses due to MWF friction, the coolant hole after EDM was polished to achieve surface finish of R_z = 1.2 μm. As such, the relative roughness of the tube (channel) k_{tb} = 0.0012/1.3 = 0.00092. Using this value of k_{tb}, one finds the value of the friction factor (Darcy coefficient) λ in the Moody diagram (Figure 6.19) to be 0.02. Major hydraulic head losses are calculated using Equation 6.26 as

$$h_{ch} = 0.02\frac{80}{1.3}\frac{76.7^2}{2\times 9.81} = 369\text{ m}$$

To reduce minor head losses, confusors with generous edge radii were made at the entrance of each coolant hole as shown in Figure 6.25. It was found experimentally that K_{fr} = 0.025. Minor head losses are calculated using Equation 6.32 as

$$h_m = 0.25\frac{76.7^2}{2\times 9.81} = 74.97\text{ m}$$

The total head losses are

$$h_{\Sigma} = h_{ch} + h_m = 369 + 74.97 = 443.97\text{ m}$$

The required pressure in the tool holder is calculated as

$$p_h = h_{\Sigma}\rho_{fl}g = 443.97\times 985\times 9.81 = 4,290,015.5\,\text{Pa} \approx 4.3\,\text{MPa}(\approx 623.5\,\text{psi})$$

This pressure is normally within the range of MWF pressures available in modern machines.

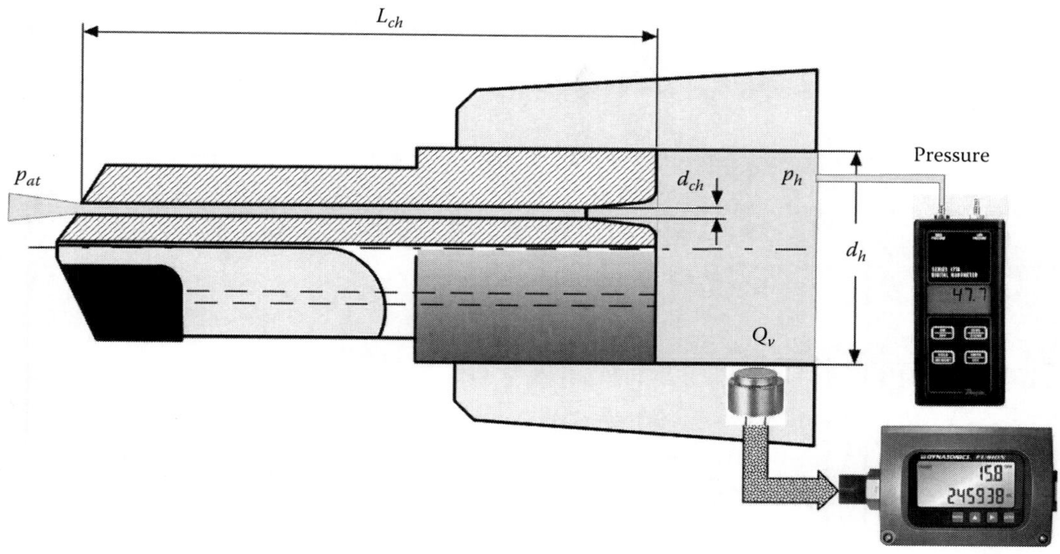

FIGURE 6.25 Model of the modified tool.

6.3.3.9 Solution of the Reverse Problem

Often in practice, the solution of the reverse problem, namely, finding the flow rate under a given MWF pressure in the tool holder, p_h, is required. The exact solution of this problem is feasible only for relatively simple models as that shown in Figure 6.24. For this model,

$$\frac{p_h}{\rho_{fl} g} = \lambda \frac{L_{ch}}{d_{dh}} \frac{v_{ch}^2}{2g} + K_{fr} \frac{v_{ch}^2}{2g} \tag{6.49}$$

For the considered case, $v_{ch} = ((Q_v / n_{ch})/A_h)$ where n_{ch} is the number of coolant holes in the tool. Substituting this equation into Equation 6.49, one can obtain

$$\frac{p_h}{\rho_{fl} g} = \lambda \frac{L_{ch}}{d_{dh}} \frac{(Q_v / n_{ch})^2}{A_{ch}^2 2g} + K_{fr} \frac{(Q_v / n_{ch})^2}{A_{ch}^2 2g} \tag{6.50}$$

or

$$Q_v = n_{ch} \sqrt{\frac{2 p_h A_{ch}^2}{\rho_{fl}(\lambda(L_{ch}/d_{dh}) + K_{fr})}} \tag{6.51}$$

6.3.3.10 Practical Coolant Channel Configurations

For end round tools (drills and reamers), tungsten carbide rods are used. These rods are manufactured as standards with no holes (solid) (Figure 6.26); with one (Figure 6.27), two (Figure 6.28), or

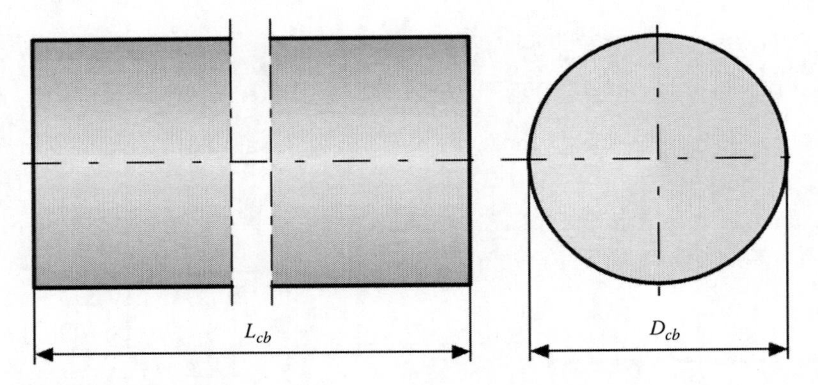

FIGURE 6.26 Solid carbide rod.

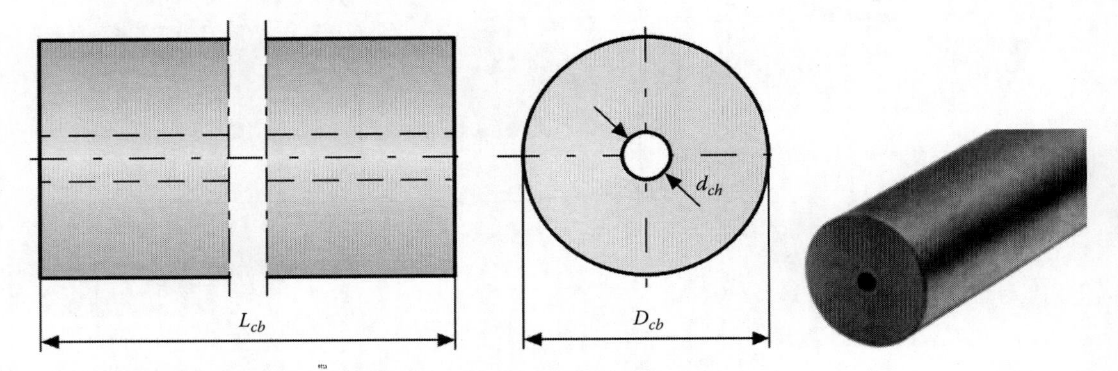

FIGURE 6.27 Carbide rod with one straight central coolant hole.

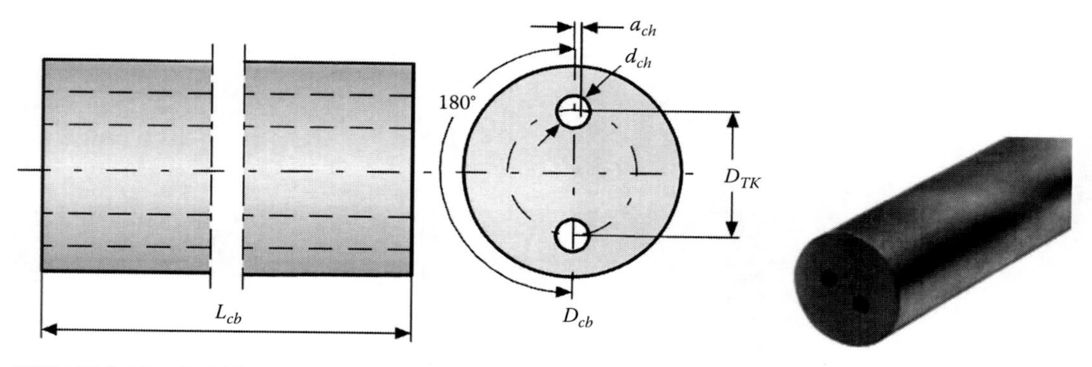

FIGURE 6.28 Carbide rod with two straight parallel coolant holes.

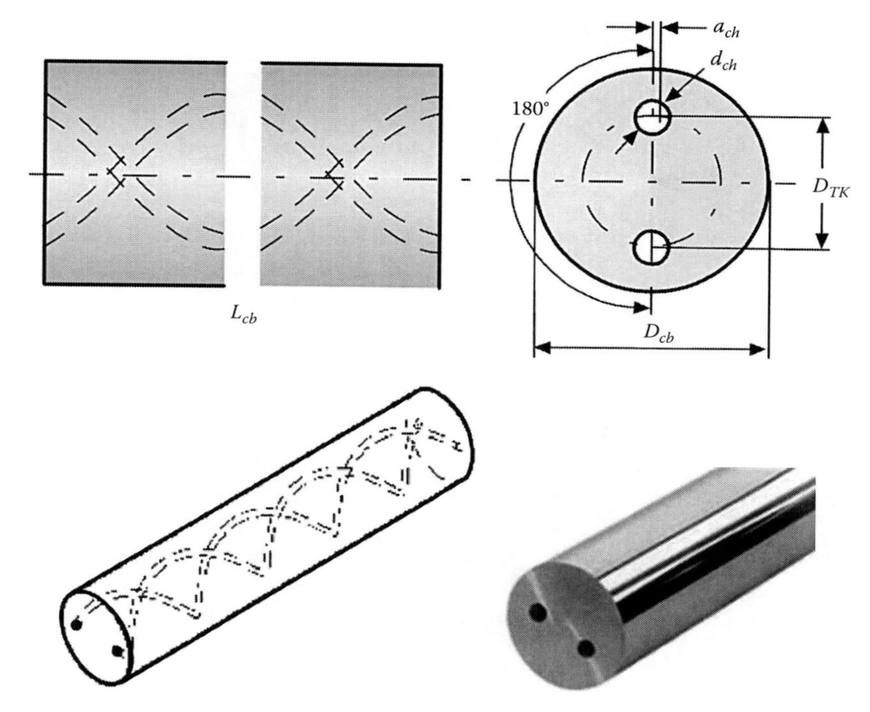

FIGURE 6.29 Carbide rod with two helical coolant holes.

three straight holes; and with two (Figure 6.29) or three (Figure 6.30) twisted to 30° or 40° helical holes. All rods are normally available in standard lengths, L_{cb}, of 310 and 330 mm. Hole(s) diameter, d_{ch}, and diameter of their location (D_{TK} known as the bolt diameter) as well as manufacturing tolerances (e.g., the tolerance on hole location, a_{ch}) are rod manufacturer specific, and these data are available for each standard configuration in the corresponding carbide manufacturers' catalogs. Note that these parameters are not standardized and thus vary from one manufacturer to the next. Custom lengths, hole, and location dimensions are available upon request.

When carbide rods with no holes (solid) (Figure 6.26) are used for tools with internal MWF supply, the coolant holes are made by EDMing a hole(s) to the tool designer specification. Although it looks like an attractive option because (1) these rods are on-shelf products and (2) tool designer has wide possibility to assign any desired location of coolant holes, in reality only relatively short straight coolant holes with rough internal surface that cause great pressure losses can be manufactured (see Example 6.2).

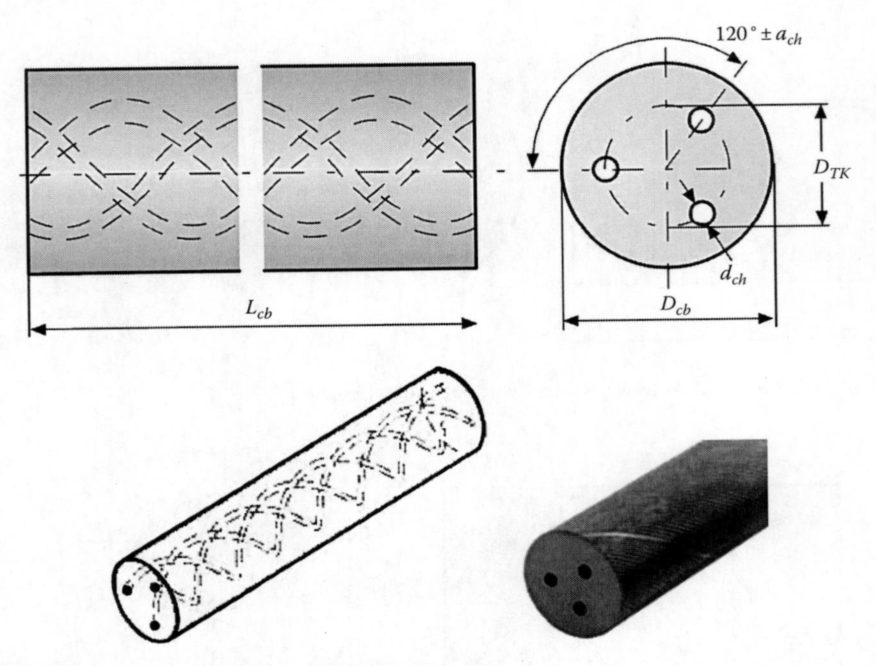

FIGURE 6.30 Carbide rod with three helical coolant holes.

Carbide rods with one straight hole (Figure 6.27) also fall into a category of relatively inexpensive on-shelf products. It seems to be an attractive option for many straight flute and helical reamers regardless of the number of teeth. However, not many specialists in the field realize that its use is limited to reamers with shallow flutes for blind holes. While the former is self-obvious because the bending and torsional strengths of the reamer are limited by distance s_{min} (Figure 6.31), the latter requires more detailed explanation due to its high importance in high-speed and MQL machining.

It is obvious that if a carbide rod with one straight hole is used as a blank for a reamer that cuts through holes, the whole flow rate of MWF will flow forward through the preexisting hole without cooling and lubricating the cutting edges. To prevent this from happening, a central coolant hole is normally plugged, and a series of radial holes that connect flutes and the central hole are made. The location of the outlets of these radial holes presents a problem.

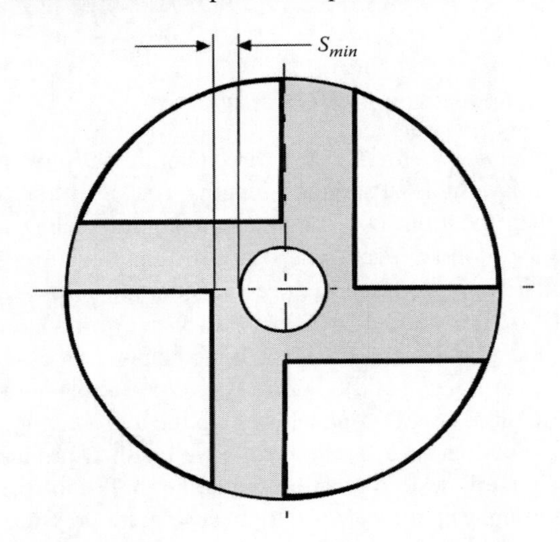

FIGURE 6.31 Cross section of a reamer.

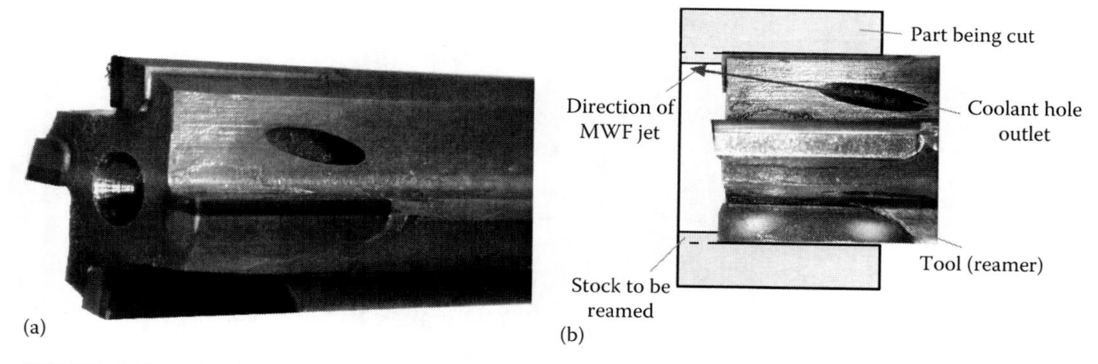

FIGURE 6.32 Showing: (a) improper location of the coolant hole outlet and (b) model of MWF jet flow to explain the issue.

Figure 6.32 helps to clarify the issue that is poorly understood by many tool manufacturers. As can be seen, the central coolant hole is plugged from the reamer face by an epoxy compound and an array of coolant holes were EDMed to reach each cutting tooth (Figure 6.32a). However, as these coolant holes originate from the central coolant hole, the outlet of the coolant holes is directed upwards with respect to the cutting edges so that near-dry reaming takes place because MWF does not reach the cutting part of the tool as shown in the simple model presented in Figure 6.32b.

The discussed issue is much more serious as one might think. In high-speed machining of high-silicon aluminum alloys in the automotive industry, the location and orientation of the outlets of the coolant holes are of vital importance. Such alloys in high-speed machining expand rapidly due to heat generated in machining. Then the machined (drilled or reamed) holes contract so that the diameter of the machined hole becomes smaller (according to the author's experience, up to 12 μm) if reamers having the orientation of the outlets of the coolant holes shown in Figure 6.33 are used. The problem turned many reaming operations into the least reliable and most costly in the whole manufacturing of automatic transmissions. Moreover, spiraling (*tiger strips*) was often observed on the surface of the reamed holes. An example of spiraling is shown in Figure 6.34.

Forced to solve the problem, the tool manufacturer tried to select the different grade of PCD tool material, increase the back taper, change of the cutting geometry, and so on with no success. As a last resort, the diameter of reamers was increased to compensate for holes shrinkage. As this shrinkage changes from one batch of die casting to the next, the whole *salvation* project turned into a nightmare.

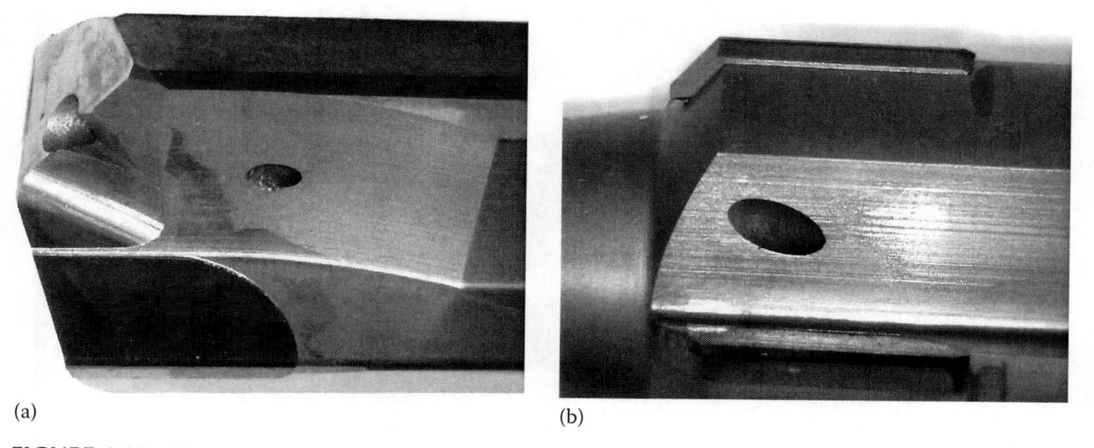

FIGURE 6.33 Improper orientation of the outlet of the coolant hole made in PCD drill-reamer: (a) on the drilling stage, and (b) on the reaming stage.

FIGURE 6.34 Example of spiraling observed in the machined hole.

Only when the proper coolant holes location and orientation of their outlets developed by a team of tool engineers led by the author were implemented, the problem was completely solved. The diameter of the reamed hole becomes equal to that of the reamer. Spiraling disappeared completely. The surface finish, cylindricity, and position deviation of the machined holes were improved dramatically. Threefold improvement in tool life and the scatter in tool life within the scatter of die casting properties variation were also achieved. A negative side of the proposed solution was that all the very expensive multistaged PCD reamers deliberately made oversize to compensate for machined holes construction became obsolete so that they were subjected to re-tipping to their normal diameter.

Figure 6.35 shows another example of improper location of the coolant holes on a reamer. As can be seen, this hole is located too far from the cutting edges. As a lot of chips form in HP reaming, these chips completely block MWF from reaching the contact surfaces where this MWF is mostly needed.

Figure 6.36 is an example of proper location and orientation of a coolant hole. As can be seen, it is directed toward the cutting edge. An angle of 50°–35° between the axis of such a hole and the

FIGURE 6.35 Improper location of the outlet of the coolant hole made in a PCD reamer.

FIGURE 6.36 Proper location and orientation of a coolant hole.

rake face was found to be a good choice. Naturally, such an angle cannot be achieved if one central coolant hole is used. Therefore, a carbide rod with two straight parallel coolant holes should be selected as a blank to make such a tool.

Carbide rods with helical coolant holes (Figures 6.29 and 6.30) are used for drills and reamers with helical flutes. MWF flow in curved, in general, and in helical, in particular, channels is much more complicated than that in straight holes. This flow attracted much attention from researchers because of enormous engineering applications in the heat exchanger networks, heating or cooling coils, and many others.

When fluid flows through a curved pipe, the presence of curvature generates a centrifugal force that acts at the right angle to the main flow direction. Thus, the flow becomes different from that in straight tube. Its distortion due to a secondary flow is a continuous function of the helical (coil) geometry (Das 1993). Because the flow in the small bending radius helical pipe is strongly influenced by centrifugal force, the critical Reynolds number, Re_{cr}, that indicates the onset of turbulent flow is no longer a constant value but a function of $De = Re \times (d_{ch}/D_{TK})^{1/2}$ where De is the Dean criterion characterizing effect of centrifugal force, d_{ch} is the tube diameter, and D_{TK} is the helical diameter (the diameter of the coil) (Ju et al. 2001). Although there are a number of different equations to calculate the critical Reynolds number (Ju et al. 2001), that given by Ito is most widely used:

$$Re_{cr} = 2 \times 10^4 \left(\frac{d_{ch}}{D_{TK}} \right)^{0.32} \tag{6.52}$$

A practical friction diagram of helically coiled tube that accounts for the effect of diameter ratios was presented by Grundman (1985). Later, Hart et al. (1988) presented a tube friction chart for laminar and turbulent flow for single-phase and two-phase flow through helically coiled tubes. Normally, the friction factor for helical coolant channel, λ_{hl}, is greater than that for straight tube, λ. Based on the experimental data presented by Ju et al. (2001), the following practical formulae to calculate the friction factor for helical coolant channel, λ_{hl} can be used:

when $De < 11.6$, it is laminar flow:

$$\lambda = \frac{64}{Re}, \quad \frac{\lambda_{hl}}{\lambda} = 1 \tag{6.53}$$

when $De > 11.6$, $Re < Re_{cr}$, it is laminar flow with big vortex:

$$\lambda = \frac{64}{Re}, \quad \frac{\lambda_{hl}}{\lambda} = 1 + 0.015 Re^{0.75} \left(\frac{d_{tb}}{D_{hl}} \right)^{0.4} \tag{6.54}$$

when $De > 11.6$, $Re > Re_{cr}$, it is turbulent flow:

$$\frac{\lambda_{hl}}{\lambda} = 1 + 0.11 Re^{0.23} \left(\frac{d_{tb}}{D_{hl}} \right)^{0.14} \tag{6.55}$$

where λ is calculated using Equation 6.30 or it is selected from the Moody diagram (Figure 6.19).

Rotation of a tool made with a central coolant channel increases the critical Reynolds number, Re_{cr}, that indicates the onset of turbulent flow because the flow is strongly influenced by the centrifugal force. This increase is a complicated function of rotating speed, coolant channel diameter, and roughness of its surface. However, the influence of rotation diminishes when the Reynolds number increases. As a result, the friction factor becomes a constant for transitional type of flow and then slightly decreases when turbulent type of flow takes place. This decrease, however, becomes insignificant for high Reynolds numbers (Harvey et al. 1971) normally found in MWF flow in end cutting tools.

Rotation of a tool made with two coolant holes parallel to the rotational axis causes additional internal pressure due to the normal acceleration of MWF in these coolant channels. This additional pressure can be estimated provided that the bolt diameter (D_{TK} [mm]), MWF density (ρ_{fl} [kg/m^3]), coolant holes diameter (d_{ch} [mm]), and their length (L_{ch} [mm]), rotational speed (n [rpm]) are known:

Linear speed of the center of the coolant channel is

$$v_{ch} = \frac{\pi D_{TK} n}{60 \times 1000} \text{(m/s)} \tag{6.56}$$

Mass of MWF in a channel is

$$m_{fl\text{-}ch} = \frac{\pi\left(d_{ch} \times 10^{-3}\right)^2}{4}\left(L_{ch} \times 10^{-3}\right)\rho_{fl}\,\text{(kg)} \tag{6.57}$$

Normal acceleration due to rotation is calculated as

$$a_{n\text{-}fl} = \frac{v_{ch}^2}{(D_{TK}/2) \times 10^{-3}}\,(\text{m/s}^2) \tag{6.58}$$

Normal force due to this acceleration is calculated as

$$F_{n\text{-}fl} = m_{cl\text{-}ch} a_{n\text{-}fl}\,\text{(N)} \tag{6.59}$$

Additional pressure loss in MWF in the coolant channel is calculated as

$$p_{n\text{-}rot} = \frac{F_{n\text{-}fl}}{0.5\pi\left(d_{ch} \times 10^{-3}\right)\left(\left(L_{ch} \times 10^{-3}\right)\right)}\,\text{(Pa)} \tag{6.60}$$

Practical calculations, however, showed that this additional pressure loss is normally insignificant. For example, when $D_{TK} = 10$ mm, $\rho_{fl} = 985$ kg/m^3, $d_{ch} = 1.3$ mm, $L_{ch} = 120$ mm, $n = 25{,}000$ rpm, the additional pressure loss is $p_{n\text{-}rot} = 0.018$ MPa.

6.3.3.11 Pressure Loss in the Machining Zone

The concept of the machining zone and the importance of MWF flows in this zone were discussed in Chapter 5 for SPDs. The same basic ideas are fully applicable for *normal* straight flute and twist drills. In full analogy with deep-hole drilling (see Section 5.2.5.8, Figure 5.103), the concept of the bottom clearance space is considered. Figure 6.37 shows that while drilling, the drill geometry results in the formation of the conical surface 1 known as the bottom of the hole being drilled. This bottom, from one side, with the drill's primary 2, 3 and secondary 3, 4 flanks, from the other side, forms two limited spaces named as the bottom clearance spaces 6. MWF (the coolant) is supplied into the bottom clearance space 6 under pressure through the internal passages 7 having outlets 8 and 9 on the flank faces. Changes in the direction and cross-sectional areas in the MWF flow from the orifices 8 and 9 into chip flutes 10 and 11 cause pressure losses.

Because the diameter of the coolant channels and their location on the flank surface as well as the shape of the flank surface vary from one tool to the next, the pressure losses should be determined for a particular drill design experimentally. To do this, the pressure needed to deliver a given flow rate is measured for the drill assembly similar to that shown in Figure 6.24. Then, the workpiece is applied to create the bottom clearance space as shown in Figure 6.37. If the flow rate is kept the same, the measured pressure difference allows to calculate the flow resistance coefficient K_{bc} of the bottom clearance space for a particular drill design.

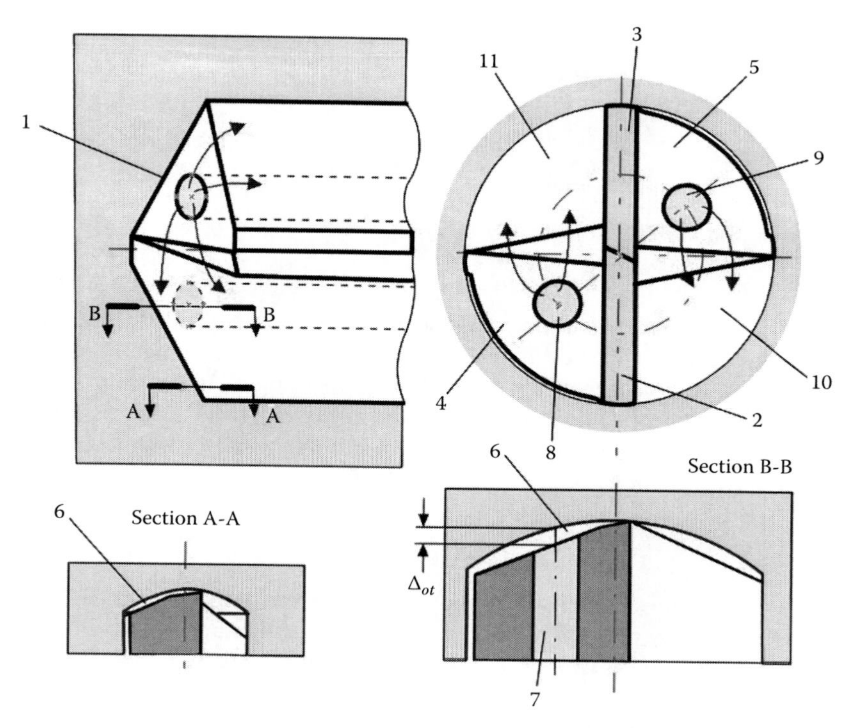

FIGURE 6.37 Concept of the bottom clearance space (the machining zone).

Example 6.3

Determination of the flow resistance coefficient K_{bc} of the bottom clearance space

Problem

Determine the flow resistance coefficient K_{bc} of the bottom clearance space for a tool having the same parameters as in Example 6.2 if the pressure difference was $p_{h2} - p_{h1} = 1.2$ MPa where p_{h1} is the pressure determined experimentally for the setup shown in Figure 6.24, and p_{h2} is the pressure determined experimentally for the setup shown in Figure 6.38 keeping the same volumetric flow rate of the MWF $Q_v = 12(\text{L/min}) \times 0.17 \times 10^{-4} = 2.04 \times 10^{-4}$ m³/s.

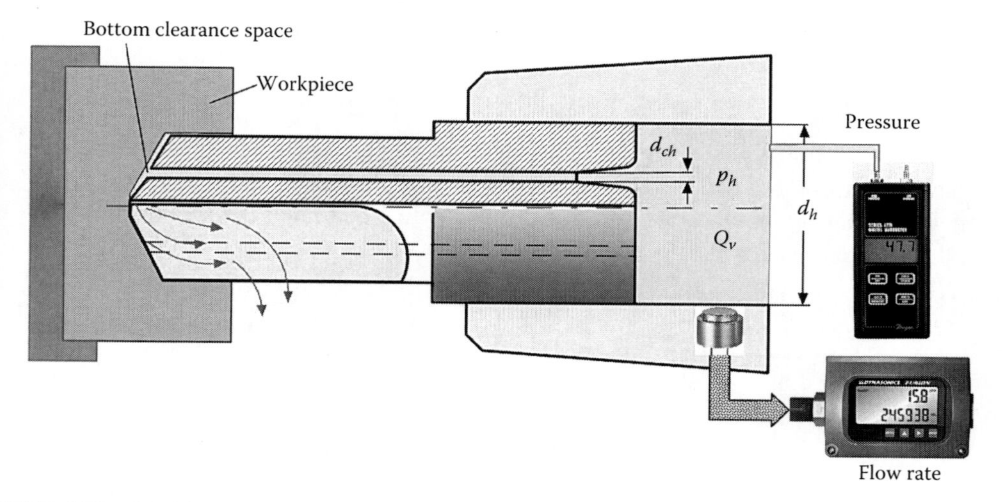

FIGURE 6.38 Model to determine pressure losses in the bottom clearance space.

Solution

To determine the flow resistance coefficient K_{bc}, one can use Equation 6.31, that is,

$$K_{fr} = \frac{\Delta p}{\left(1/2\rho_{fl}v_{fl-b}^2\right)} = \frac{1.2\times10^6}{(1/2\times985\times9.81)} = 0.208$$

where $v_{fl\text{-}b} = v_{ch}$ in Example 6.2, that is, $= 76.7$ m/s.

6.3.3.12 Total Pressure Loss and the Pressure Needed to Deliver the Desirable Flow Rate

The foregoing considerations allow introducing the model of the total pressure loss in drilling based on electrical analogy discussed in Section 6.3.3.2. Figure 6.39 shows a model. As can be seen, the total pressure loss consists of three electrical (hydraulic) resistances, namely, the pressure loss due to flow contraction from the tool holder into the coolant hole(s), R_1; major pressure losses due to friction in the coolant holes, R_2; and pressure loss in the machining zone, R_3. As these three resistances are located in series, the total resistance, R, is calculated as

$$R = R_1 + R_2 + R_3 \tag{6.61}$$

This total resistance R defines the pressure needed to deliver the desired flow rate into the machining zone. For example, using the data of Examples 6.2 and 6.3, the total pressure needed to deliver the volumetric flow rate of the MWF $Q_v = 12(\text{L/min}) \times 0.17 \times 10^{-4} = 2.04 \times 10^{-4}$ m³/s is calculated as

$$p = p_m + p_{ch} + p_{bc} = h_m\rho_{fl}g + h_{ch}\rho_{fl}g + \Delta p_{bc} \tag{6.62}$$

Substituting the data obtained in Example 6.2, $h_\Sigma = h_{ch} + h_m = 369 + 74.97 = 443.97$ m, and in Example 6.3, $\Delta p_{bc} = 1.2$ MPa, into Equation 6.62, one can obtain

$$p = h_\Sigma\rho_{fl}g + \Delta p_{bc} = 888.33\times985\times9.81 + 1.2\times10^6 = 4.3\times10^6 + 1.2\times10^6 = 5.5\times10^6\ \text{Pa} = 5.5\,\text{MPa} \tag{6.63}$$

Note that the model shown in Figure 6.39 is basic so that any other resistance can be added/subtracted for a particular case of drill design. Moreover, the foregoing analysis suggests that any of the resistances included in the consideration can be optimized to satisfy the design and flow requirements.

6.3.4 MWF Flows Management in the Bottom Clearance Space

6.3.4.1 Two-Flute Drills

Let's consider one of these two bottom clearance spaces as they are normally absolutely the same. The topology of the bottom clearance space 6 formed by the bottom of the hole being drilled and

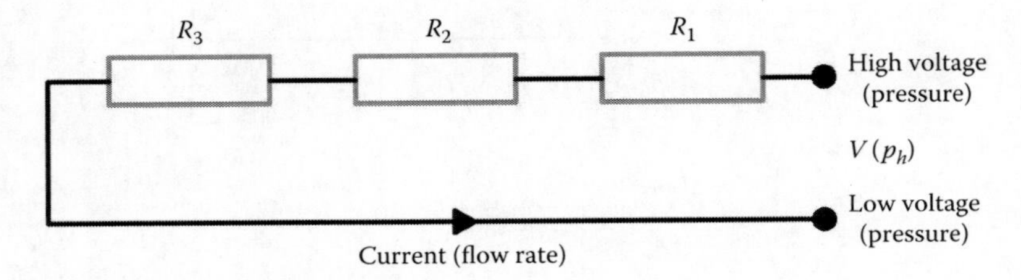

FIGURE 6.39 Model to calculate the total pressure loss.

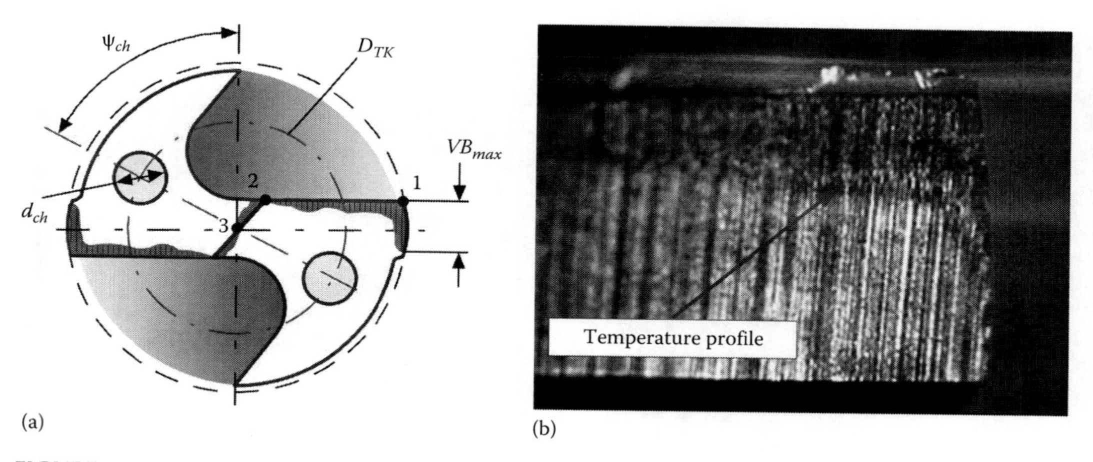

FIGURE 6.40 Flank wear: (a) typical flank wear pattern and (b) temperature profile with the maximum at the drill periphery point.

flanks, primary 2 and secondary 4, can be appreciated in different cross sections parallel to the drill longitudinal axis. Two of such sections, A-A and B-B, are shown in Figure 6.37. MWF (the coolant) is supplied into the bottom clearance space 6 under pressure through the internal passage 7.

Tool life and quality of the machined hole are defined by tool wear while the reliable chip transportation is determined by the MWF properties and flow rate (considered further). Therefore, to meet the first two chief objectives, MWF supplied into the machined zone should reduce tool wear while, to meet the third objective, the flow rate of MWF should be sufficient for reliable chip transportation over the chip removal channels (flutes).

As discussed in Chapter 2, drill wear appears on the flanks of the tool below the cutting edge(s) forming a wear land. Figure 6.40a shows the *normal* wear pattern for a drill. As discussed earlier in Chapter 2, the maximum wear (VB_{max} in Figure 6.40a) is normally at the region of the flank face adjacent to the drill's periphery corners. The three prime reasons for such a location of maximum wear are as follows:

1. The maximum cutting speed is found in the region adjacent to periphery point 1 of the cutting edge (Figure 6.40). As a result, the maximum temperature is commonly found in this region.
2. The amount of material removed by the part of the cutting edge adjacent to the drill periphery point 1 is much greater than that by other regions of the major cutting 1–2 and chisel edge 2–3.
3. The distance passed by the periphery point 1 per one drill revolution is greater than that by other points of the cutting edge. As the wear, particularly abrasive, is directly proportional to the length of the path traveled, the maximum wear is found in the region adjacent to this point 1.

As a result, the wear pattern on the flank faces is not uniform with the maximum at the periphery point 1 as shown in Figure 6.40a. Figure 6.40b shows discoloration of the carbide tool material beneath the cutting edge at some early stage of tool life. The profile of this discoloration resembles temperature distribution over the tool flank face. The flank wear pattern follows this profile (see Figures 2.48 and 2.50).

The foregoing consideration suggests that the region of the flank face adjacent to drill periphery point 1 should be provided with the greatest amount of MWF flow to improve cooling and lubrication in this region, i.e. where these two prime actions of MWF are mostly needed. Unfortunately, reality proves otherwise, that is, the discussed region is provided with the least portion of the total MWF flow rate.

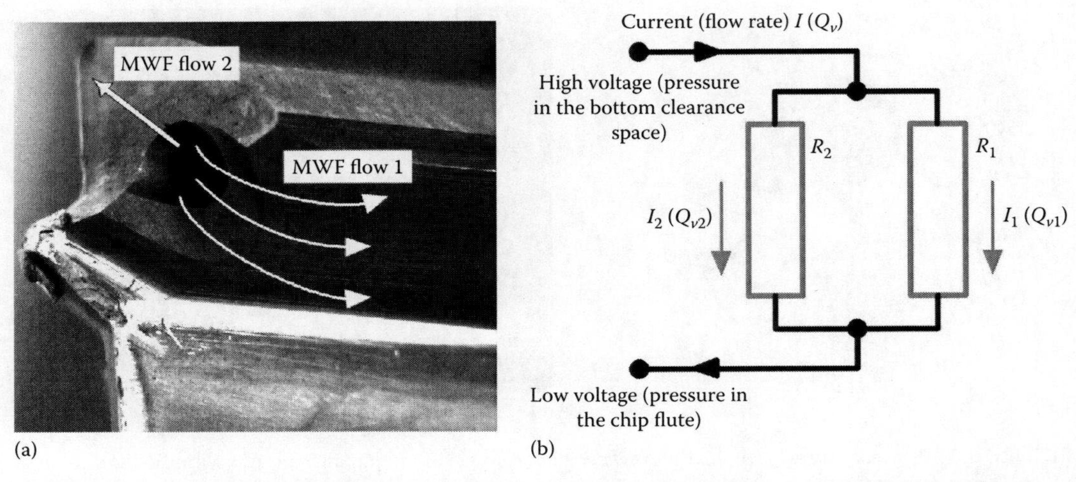

FIGURE 6.41 Showing: (a) model of MWF flows in the bottom clearance space and (b) its electrical analogy.

To understand the problem, one should consider the model of MWF flows in the bottom clearance space shown in Figure 6.41a and its electrical analogy shown in Figure 6.41b. As can be seen, the MWF flow (having flow rate Q_v) in the bottom clearance space separates into two principal flows. The first having flow rate Q_{v1} flows directly into the chip flute and thus does not participate in cooling and lubrication of the region of the tool adjacent to the drill periphery corner. The second one having flow rate Q_{v2} flows and thus provides the cooling and lubrication of this region. Then the second flows through the very narrow gap between the tool body and the wall of the hole being drilled eventually joining the chip flute due to tool back taper.

According to the electrical analogy shown in Figure 6.41b, flow rates Q_{v1} and Q_{v2} are in reverse proportion to the hydraulic resistances R_1 and R_2 of the corresponding flow paths. As clearly seen in Figure 6.41a, the first flow has almost no resistance while the second flow goes over an array of narrow passages starting with that shown in Section A-A in Figure 6.37. In many practical tool designs, $R_1/R_2 = 10–20$, so that only 5%–10% of the total flow rate supplied into the bottom clearance space actually participates in cooling and lubrication of the drill periphery regions.

A logical question about importance of the MWF flow rate in providing cooling action to the region adjacent to the drill periphery should be answered. Earlier (Astakhov 2006), the author introduced a parameter called the cooling intensity K_h useful in accessing the MWF cooling ability

$$K_h = v_{MWF}^{0.65} k_{MWF}^{0.67} C_{P\text{-}MWF}^{0.33} \gamma_{MWF}^{0.33} \nu_{MWF}^{-0.32} \tag{6.64}$$

where

v_{MWF} is the velocity of MWF
k_{MWF} is the thermal conductivity of MWF
$C_{P\text{-}MWF}$ is the specific heat of MWF
γ_{MWF} is the specific weight of MWF

As can be seen, the velocity of MWF v_{MWF} affects its cooling ability almost as much as its thermal conductivity k_{MWF} and much more than its specific heat $C_{P\text{-}MWF}$. Therefore, to intensify the cooling of the region adjacent to the drill periphery, thus to increase tool life, a higher flow rate (defines v_{MWF}) should be assured. According to the electrical analogy of the bottom clearance space shown in Figure 6.41b, it can be accomplished either by increasing the pressure in the bottom clearance (increasing voltage) or by decreasing the hydraulic (electrical) resistance R_2, that is, providing more room for MWF to flow.

Experience shows that significant intensification of the cooling action of MWF is achieved by increasing its velocity, which has a dual result. First, it increases the convection heat transfer coefficient and second, high-velocity jets of MWF blow a boundary layer formed on high-temperature surfaces. This explains the efficiency of the high-pressure MWF supply, which increases the velocity of MWF. Although it is often claimed that the high-pressure supply of MWF increases its penetration ability into the tool–chip and tool–workpiece interfaces (Mazurkiewicz et al. 1989; Li 1996), in reality it is not so because the contact stresses are much higher than the maximum pressure of MWF. It was conclusively proven that tool life (tool wear) does not significantly increase in cutting under very high static pressure (Entelis and Berlinder 1986). In reality, high pressure of MWF increases its velocity, which, in turn, significantly improves the cooling action of this fluid. Moreover, it does not affect the cutting forces (Crafoord et al. 1999).

As introduced in Chapter 5 as *Rule No. 3* in HP drill tip geometry/design (Section 5.2.5.8.2), one of the two ways to increase the velocity of the MWF flow that goes into the drill periphery region is to increase MWF pressure in the bottom clearance space. One of the feasible solutions was proposed by Colvin in his design of drill flanks. According to this design (US Patents 6,056,486 [2000] and 6,270,290 [2001]) called the *pressure tip tool* (CTE Staff Members 2001), which is shown in Figure 6.42, the flank surfaces 1 and 2 of the drill containing the outlet orifices 3 and 4 of the internal coolant passages are modified. The proposed modification can be seen in cylindrical cross section A-A, where the contour lines have been added and accentuated to show the topography of the flank surfaces. Each flank surface has three distinctive regions. Starting from the cutting edge 5, the first region 6 of the flank surface inclines at normal clearance angles over the major cutting edge (lip) 5. The second region is a recessed surface 7 where the internal coolant passage intersects the flank surface creating its outlet orifice 8. In drilling, this recessed surface serves at the MWF reservoir together with the bottom of the hole being drilled. The third region is a trailing dam 9.

Figure 6.43a, Section A-A, shows the visualization of the MWF flow in the bottom clearance space in a standard drill having curved or flat flank surfaces 1 and 2. The MWF flow supplied to this space through the internal coolant passage 3 separates into a number of elementary flows. The directions of these flows are shown by streamlines. Most of the MWF deflects from bottom of the hole being drilled 4 and flows into flute 5 as the flow goes in the direction of the least hydraulic resistance. Due to low MWF pressure in the bottom clearance space, MWF does not flow to the narrow passage between flank 6 and the bottom of the hole being drilled 4 to the tool flank–workpiece contact interface 7 and thus does not provide cooling and lubrication to this important interface particularly in the regions adjacent to the drill periphery point 8 where these functions are mostly needed.

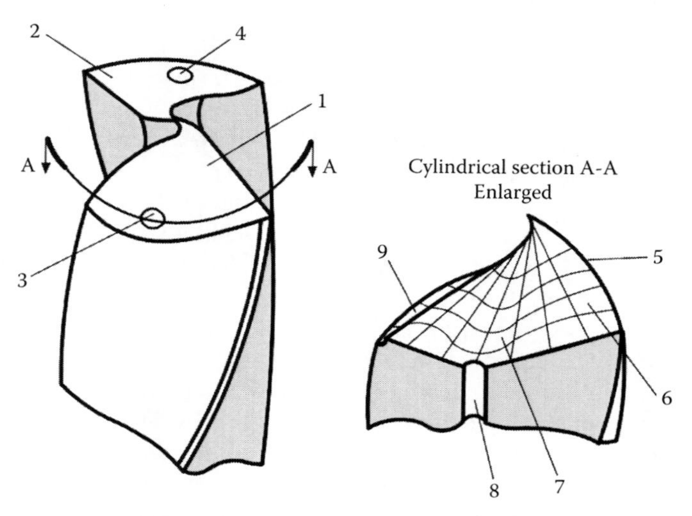

FIGURE 6.42 The pressure tip tool according to US Patent Nos. 6,056,486 (2000) and 6,270,298 (2001).

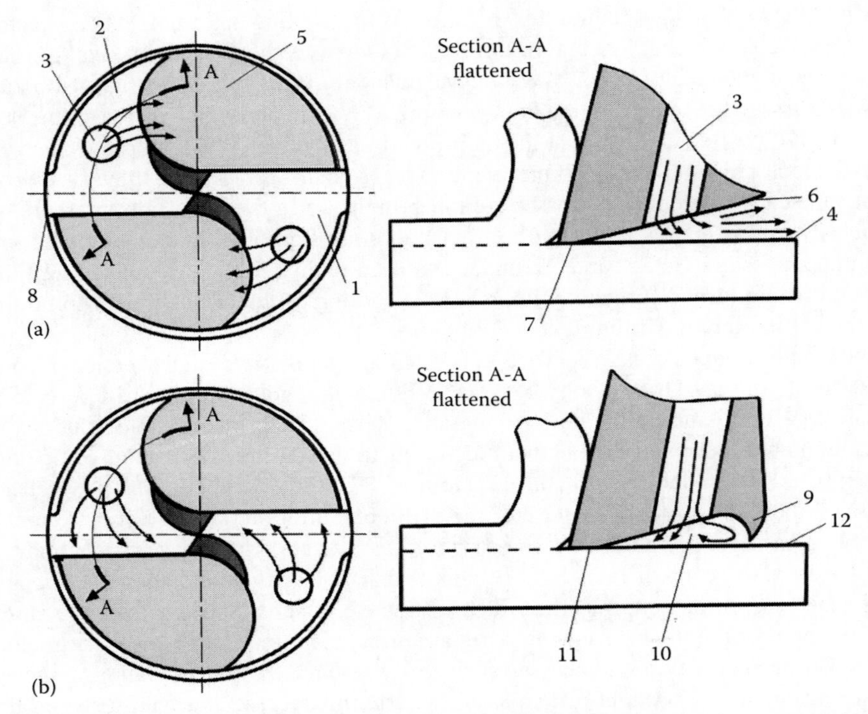

FIGURE 6.43 Visualization of the MWF flow in the bottom clearance space for the common (a) and for the pressure tip (b) drills.

Figure 6.43b shows the MWF flow model in the bottom clearance space in the pressure tip drill. As seen, a trailing dam 9 is added to the flank surface that increases hydraulic resistance for the MWF flow that goes into the chip flute (R_1 in Figure 6.41b). As a result, MWF pressure in the bottom clearance space 10 increases so thus the penetration ability of MWF into the tool flank–workpiece interface 11 also increases.

Although a minimum sixfold increase in tool life with this tool is reported (CTE Staff Members 2001), such an improvement does not fall into the category *plug-and-play solution*. The major issue is the necessity of the accurate drill geometry (flank model) as the drill performance, determined by MWF pressure in the bottom clearance space 10 is very sensitive to the gap between the trailing dam 9 and the bottom of the hole being drilled 12. If this gap is negative, the interference of the tool flank with the bottom of the hole being drilled takes place that ruins drill performance. If this gap is excessive, MWF pressure in the bottom clearance space 10 and is insufficient to improve drill performance. The difference between these two extremes is of a tenth of a millimeter order. Therefore, a mathematical model of the bottom of the hole being drilled (accounting for a particular cutting feed) should be combined with that of the drill flank in order develop the model of the bottom clearance space and then to develop a subroutine for a CNC grinding machine to provide drill flanks with the proper geometry. The second issue is a requirement to the drilling machine to be used to deliver a high pressure at the MWF flow rate needed for the drill. It is understood that other components of the drilling system should also support this drill design.

An example of practical realization of the second way, that is, decreasing the hydraulic (electrical) resistance R_2, and thus, providing more room for MWF to flow, is the design of the so-called G-drills, which became popular in industry. The principle of such a design is shown in Figure 6.44. The distinguishing feature of G-drills is a great body relief, that is, gaps 1 and 2 between the drill body and the wall of the hole being drilled. These gaps are much greater than that in usual drills. This allows MWF from the coolant holes 3 and 4 to separate into approximately equal flows

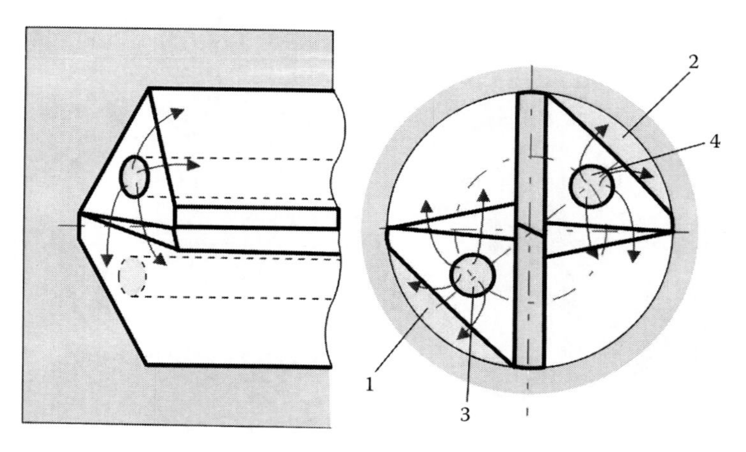

FIGURE 6.44 Principle of the G-drill design.

(flows 1 and 2 in Figure 6.41a) due to decreased hydraulic resistance R_2. As a result, a significant increase in tool life is achieved when the parameters of these two flows are favorable. It happens if the MWF supply system is capable of supplying sufficient flow rate and when the diameter and location of coolant holes 3 and 4 are suitable for the proper distribution of the supplied flow rate.

6.3.4.2 Gundrills

As discussed in the previous section and in Chapter 5 (Section 5.2.5), the MWF pressure in the bottom clearance space defines tool life, drill stability, and the quality of drilled holes so it should be maximized. As discussed by Astakhov et al. (1994b, 1995; Astakhov and Osman 1996), this pressure is determined by hydraulic resistance of the outlet section of the shoulder dub-off (shown as *F*4 in Figure 5.103). To acquire full control over this pressure, one needs to know the answers to the following questions: (1) What is the shoulder dub-off? (2) How can the contour of the outlet cross section be arranged in order to achieve maximum efficiency? (3) Are there limitations on reducing the length of the shoulder dub-off? (4) How does this length correlate with the actual cross-sectional area of the outlet section of the side passage? This section aims to provide the answers to these questions showing how *Rule No.* 3 in HP drill tip geometry/design (Section 5.2.5.8.2) can be realized in the gundrill design.

Figure 6.45 shows the real outlet cross-sectional area of the side passage of the standard gundrill geometry. It appears in the cross-sectional plane Q–Q, which is perpendicular to the x_0y_0 plane and inclined at an angle ψ_v (the profile angle of the V-flute) to the y_0 axis. In the cross-sectional plane Q–Q, this area is enclosed by the polygon $C_1C_2C_3C_4C_5$. Figure 6.46 presents a 3D visualization of the outlet cross section of the side passage.

As shown in Figure 6.45, side C_1C_5 of the polygon $C_1C_2C_3C_4C_5$ belongs to the drill. It is a line of intersection the sidewall of the V-flute and the shoulder dub-off. The sidewall of the V-flute is always a plane—otherwise it is not possible to keep the same profile of the V-flute having profile angle ψ_v. Therefore, side C_1C_5 is always a flat (2D) line, that is, always lies in plane Q–Q. This fact significantly simplifies the analysis of the outlet cross-sectional area of the side passage.

The flank surfaces should assure drill-free penetration into the workpiece (see Section 5.2.5). In other words, there should be no interference between the drill's flanks and the bottom of the hole being drilled. Because there is no model for the flank geometry available, these flanks are ground down with great angles to avoid interference. Although shoulder dub-off surface (in most known designs, it is a plane as shown in Figure 5.103) does not participate directly in cutting, it is one of the most important design components of a gundrill in terms of MWF management in the machining zone. Unfortunately, specialists in the field do not yet understand its role and thus it was thought of as a design component that can be of any shape convenient in grinding.

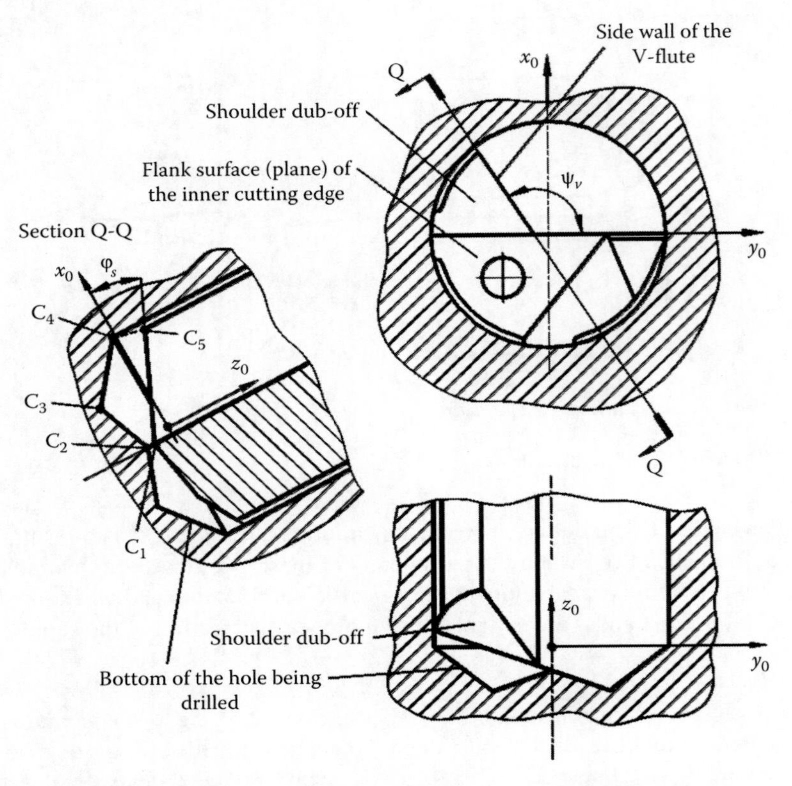

FIGURE 6.45 Representation of the real outlet cross-sectional area of the side passage.

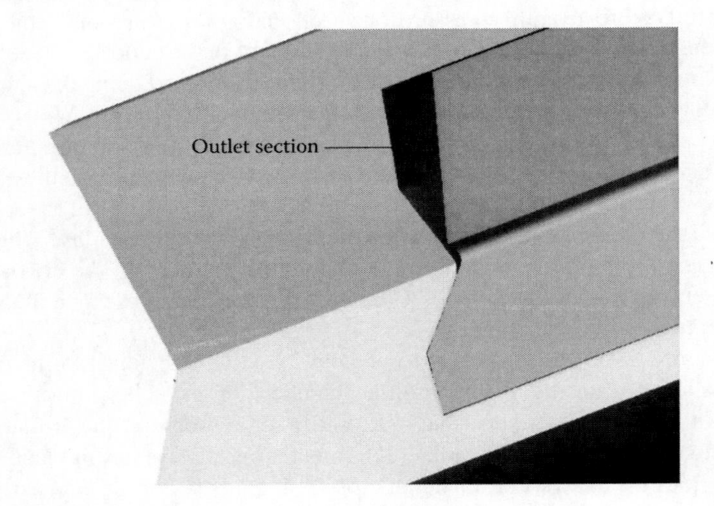

FIGURE 6.46 Visualization of the outlet section of the side passage.

There are several widely recognized designs of the shoulder dub-off surfaces (Figure 6.47).

The slash or general purpose shown in Figure 6.47a is the most common for European made drills (Botek, TBT, Gildemaster, Heller) although some Japanese (Toshiba) and North American (Drill Masters) companies also use such a design. In the author's opinion, this is still the best design although it was used for more than a hundred years (with small modifications). When gundrills were used on old or retrofitted machines, which were not capable to deliver high MWF (coolant) pressure, a notch was added on the shoulder dub-off surface as shown in Figure 6.47b to increase the MWF

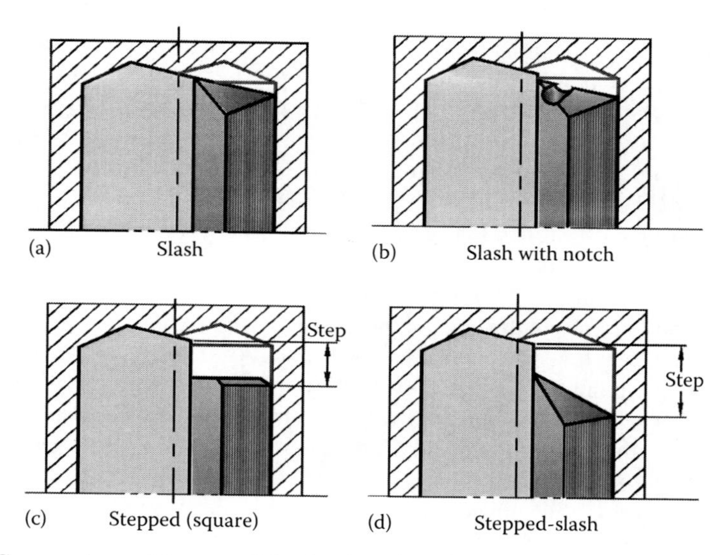

FIGURE 6.47 Commonly used designs of the shoulder dub-off.

flow rate and thus to assure reliable chip removal. A drawback of this design is that the notch should be ground every time when the drill is re-sharpened. Because the notch is ground manually, it was found difficult to keep its dimensions the same from one regrind to another. To solve this problem and to achieve even greater flow rate under a give pressure, the stepped (square) shown in Figure 6.47c and the stepped-slash shown in Figure 6.47d designs (known as point grinds as discussed in Chapter 5) were introduced. Although modern gundrilling machines are capable to deliver high coolant pressure, the stepped-slash design is still in wide use today particularly in the automotive industry where a number of old gundrilling machines are widely used.

In the author's opinion, the stepped square and the stepped-slash designs were introduced by the gundrill manufacturers that had shallow knowledge of gundrilling. Although the intent behind these designs was to increase the MWF flow rate through gundrills and thus improve chip removal, the application of these designs improves only the apparent flow rate (measured by a flow meter installed on the machine) and thus creates more problems than solves.

The known designs shown in Figure 6.47 differ significantly in the outlet cross-sectional area of the side passage. Table 6.5 presents an example of the comparison of the various tip designs for an

TABLE 6.5
Comparison of the Cross-Sectional Area of the Outlet Passage of the Side Passages and MWF Pressure in the Bottom Clearance Space for the Known Designs Shown in Figure 6.47

Design	Cross-Sectional Area of the Outlet Section of the Side Passage, A_{ot} (mm²)	Axial Cross-Sectional Area of the Bottom Clearance Space, A_{BS} (mm²)	Bottom Clearance Space Ratio, A_{ot}/A_{BS} (%)	Static Drilling Fluid Pressure in the Bottom Clearance Space (MPa)
Slash	6.32	5.64	112	0.15
Slash with a notch	8.82	5.64	156	0.10
Stepped	14.82	5.64	263	0.04
Stepped-slash	24.06	5.64	423	0.01

Gundrill parameters: diameter $D = 8$ mm, $r_{dr} = 5$ mm, $m_d = 2$ mm, $c_{ax} = 0.5$ mm, $\varphi_1 = 30°$, $\varphi_2 = 20°$, $\alpha_{n2} = 12°$, $\psi_v = 120°$, $\varphi_4 = 30°$. For slash design with a notch: $A_n = 2.5$ mm². For the stepped design: $S_t = 3$ mm, $\varphi_4 = 0°$. For the stepped-slash design: $S_t = 3$ mm, $\varphi_4 = 30°$. Drilling fluid flow rate is 18.74 L/min.

8 mm diameter gundrill. The various designs are ranked by the cross-sectional area of the outlet section in an effort to develop design parameters that are useful. Table 6.5 also presents a comparison of static MWF pressure in the bottom clearance space for the discussed designs. Controlling the area of the outlet passage by careful tip design enables one to greatly improve chip removal and tip cooling. Traditionally, large outlet passage areas are used in practice as it is believed that they allowed for a greater MWF flow rate for a given inlet MWF pressure. However, it has been determined that the apparent MWF flow rate is unimportant if this MWF does not adequately cool and lubricate the tip or if stagnation pockets form in the region of the drill point.

Another drawback of the discussed designs is not readily apparent to gundrill manufacturers and users. The chip transportation from the machining zone can be thought of as consisting of two stages, namely, chip pickup and chip transportation along the V-flute. As discussed in Chapter 5 (Section 5.2.7, Figure 5.139) and by Astakhov (2010a), Astakhov (2010b), the latter requires at least critical MWF velocity and thus its flow rate. The former, however, has to be explained. The author proposed to define the chip pickup as the initial MWF—chip interaction that should result in picking up the chip (formed by the cutting edges) by the MWF flow from the outlet cross section of the side passage. As such, the MWF is considered as just left the outlet cross section of the side passage, and chip is considered as just formed on the rake face adjacent to the outer and inner cutting edges. As a result of proper pickup, the chip should be found in the MWF flow so that its normal transportation over the V-flute is possible.

The author showed (Astakhov et al. 1995) that the chip pickup depends on the angle β_{cp}, which the MWF flow (entering the V-flute) makes with the z_0 axis (Figure 6.48). This angle depends on practically all parameters of gundrill geometry as well as the location of the outlet of the coolant passage in the tip. A rule of thumb here is *the greater* β_{cp}, *the better chip pickup*. It was also observed that when this angle is less than 40°, a stagnation zone forms on the tool rake face and the chips experience difficulties to join the MWF flow. For the standard stepped-slash design, angle β_{cp} is less than 40°, which explains why chip clogging happens in the V-flute in the proximity of the cutting edges, although the overall MWF flow rate is very high. As a result, drill breakages are often observed due to chip clogging at the drill tip.

The length of the shoulder dub-off is the length that the shoulder dub-off makes in its intersection with the side wall of the V-flute. For a common gundrill, this length C_1C_5 coincides with that of the side wall of the V-flute (see Figure 4.65). The sense of the length of the shoulder dub-off for HP drills is shown in Figure 6.49 by the length C_6C_5. Note that the counter of the outlet cross section of the side passage is flat (2D). Moreover, it should always be considered in the section plane Q–Q.

As follows from the gundrill geometry, side C_1C_2 entirely depends on c_{ax}, which is a design parameter of a gundrill. Side C_2C_3 is made by the inner cutting edge so that its inclination and length are uniquely defined by the geometry of the inner cutting edge. Side C_3C_4 is made by the

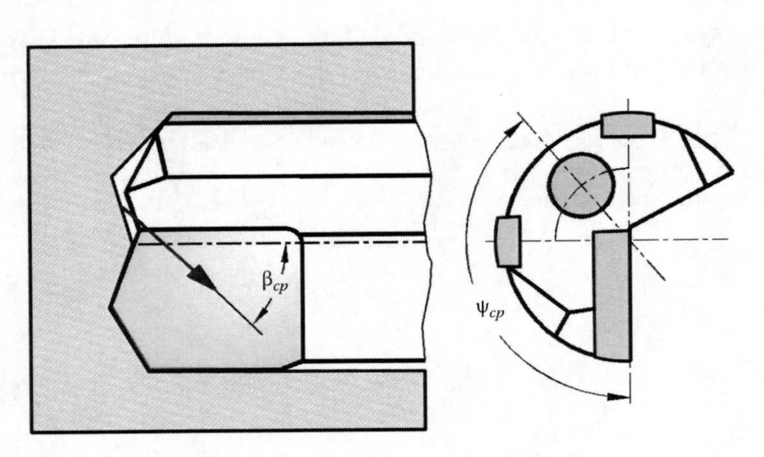

FIGURE 6.48 Sense of angle β_{cp}.

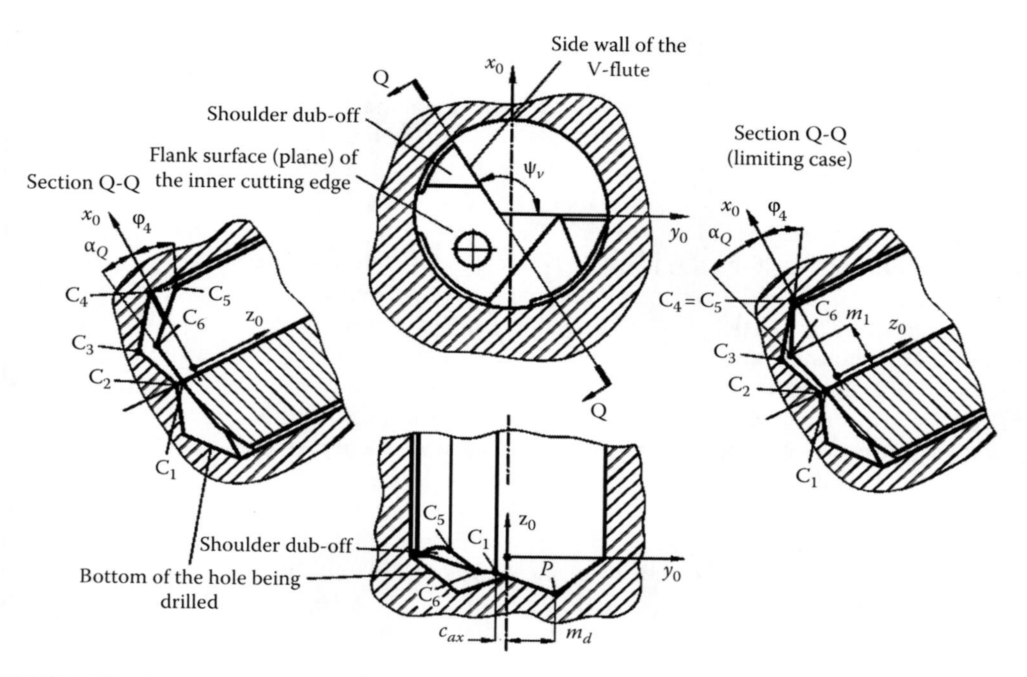

FIGURE 6.49 Graphical representation of the length of the shoulder dub-off.

outer cutting edge so that its inclination and length are uniquely defined by the geometry of the outer cutting edge. Side C_4C_5 is uniquely defined by the location of point C_5 with respect to the bottom of the hole being drilled. Therefore, as follows from Figure 6.49, the cross-sectional area of the outlet section of the side passage can only be altered by varying the length and inclination of sides C_5C_6 and C_6C_1. Consider the position and the length of these segments separately.

Side C_6C_1 is uniquely defined by its inclination angle α_Q and position of point C_6. Figure 6.49 also shows the limiting case where the cross-sectional area of the outlet section of the side passage is at minimum. To achieve this minimum, a tool design is required to meet three conditions: (1) $m_1 = m_d$ (m_d is the location distance of point P), (2) sides C_1C_6 and C_2C_3 should be parallel, and (3) side C_6C_5 and side C_3C_4 should be also parallel.

In general, the inclination angle α_Q calculates as

$$\alpha_Q = \arctan\left(\frac{\tan\alpha_{n2}}{\cos\varphi_2}\sin\psi_b + \tan\varphi_2\cos\psi_b\right) \tag{6.65}$$

where ψ_b is the location angle of point C_5 in the x_0y_0 plane (Astakhov 2010a). Because the inclination angle of side C_2C_3 is equal to φ_2, then sides C_6C_1 and C_2C_3 are parallel when

$$\alpha_Q = \varphi_2 \tag{6.66}$$

Substituting Equation 6.66 into Equation 6.65, one obtains the condition of parallelism of sides C_1C_6 and C_2C_3 as

$$\alpha_{n2} = \arctan\left[\frac{\sin\varphi_2(1-\cos\psi_b)}{\tan\psi_v}\right] \tag{6.67}$$

As it follows from Figure 6.49, sides C_6C_5 and C_3C_4 are parallel when $\varphi_4 = \varphi_1$. If $\varphi_4 > \varphi_1$, then an excessive gap between the shoulder dub-off and the bottom of the hole being drilled is the case. If $\varphi_4 < \varphi_1$, then interference between the shoulder dub-off and the bottom takes place.

In the limiting case when $c_{ax} = 0$ thus $\psi_b = \psi_v$, the cross-sectional area of the outlet section of the side passage is equal to zero. This is the limiting case where the pressure in the bottom clearance space could be made very high. The maximum pressure, however, may be restricted by the maximum MWF pressure available in the machine. Therefore, in real-world situations, the cross-sectional area of the outlet section of the side passage is selected using this and other restrictions. Figure 6.50 shows an example of an optimized cross-sectional area of the outlet section of the side passage.

It was discussed in Section 6.3.4.1 (Equation 6.64) that the velocity of the MWF affects its cooling ability almost as much as its thermal conductivity and much more than its specific heat (Astakhov 2006, 2008). Therefore, it can be concluded that high MWF static pressures in the bottom clearance space allow MWF penetration in the narrow coolant passages in the bottom clearance, the sufficient cooling ability of MWF flow is achieved only when the MWF flow has high velocity around the places where the cooling action is needed.

Figure 6.51 shows the MWF flow model in the bottom clearance space in a standard gundrill having the stepped-slash design (Figure 6.47d). The MWF flow supplied to this space through kidney-shaped orifice 1 separates into a number of elementary flows. The directions of these flows are shown by streamlines. Most of the MWF deflects from bottom of the hole being drilled 2 and flows into chip removal passage 3. The streamlines of the rest MWF flow shown in Figure 6.51 indicate that MWF

FIGURE 6.50 An example of the optimized cross-sectional area of the outlet section of the side passage.

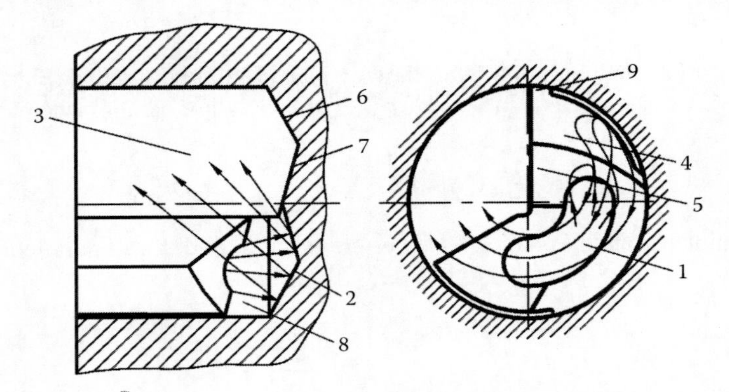

FIGURE 6.51 Model of MWF elementary flows in the bottom clearance space.

elementary flows directed toward the flanks 4 and 5 of the outer 6 and inner 7 cutting edges, respectively. As can be seen, these elementary flows should make loops to come back as these should pass through to the side passage 8 in order to exit the bottom clearance space. These loops result in the formation of multiple vortexes in the bottom clearance space (Astakhov et al. 1994a,b) that significantly reduce the MWF velocity and thus its cooling ability. Moreover, MWF does not flow to the narrow passage between flank 4 and the bottom of the hole being drilled and thus does not provide cooling and lubrication to the region adjacent to the drill periphery point 9 where they are mostly needed.

To solve the problem by assuring an MWF flow around the periphery region and the working part of the side cutting edge, a gundrill design should be modified in the manner shown in Figure 6.52 (US Patent No. 7,195,428 [2007]). In this modified drill, a part of the periphery of the tip 1 adjacent to the side cutting edge 2 is removed to form an auxiliary side 3 MWF passage extending longitudinally over the whole length of the tip and over a certain length $L_{sh\text{-}ax}$ of the shank. The outlet of this passage is formed on the shank connecting the auxiliary side passage with the chip removal flute of the shank. For small gundrills, a radial relief having radius $r_{sh\text{-}ax}$ can be made to increase MWF flow rate as shown in Figure 6.52 View A-2.

When such a gundrill works, the MWF is supplied through the outlet orifice 1 of the tip coolant passage into the bottom clearance space 2 as shown in Figure 6.53. In this space, the MWF flow separates into a number of elementary flows. The directions of these flows are shown by streamlines. A greater part of MWF deflects from bottom of the hole being drilled 3, cools and lubricates the region adjacent to the inner cutting edge 4 and the drill point P, and then flows through the side passage 5 formed between the shoulder dub-off surface 6 and the bottom of the hole being drilled 3 into chip removal passage 7 where it cools and lubricates the rake faces of the inner 4 and outer 8 cutting edges. This flow picks up the formed chip for transportation along the chip removal passages

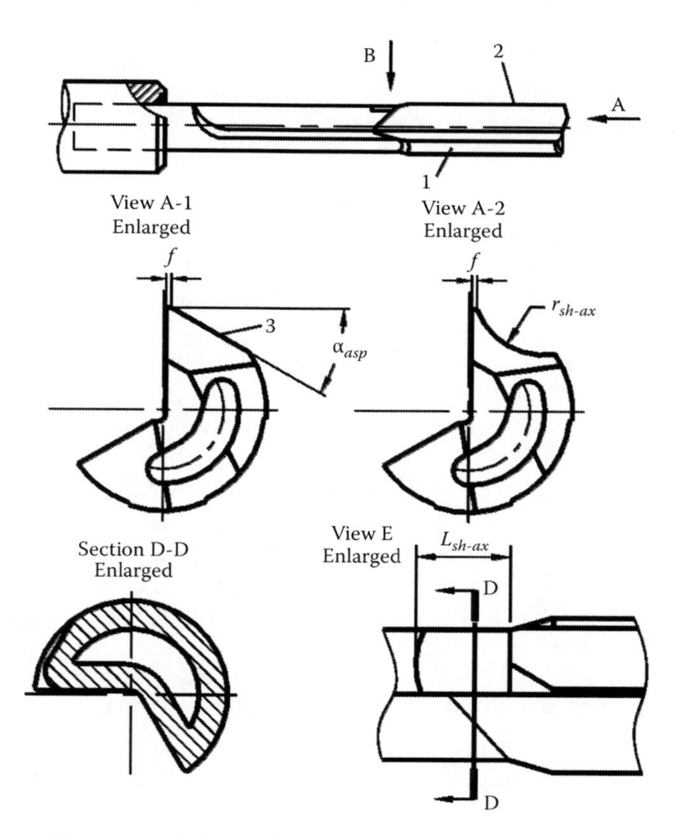

FIGURE 6.52 A gundrill design with the auxiliary side passage.

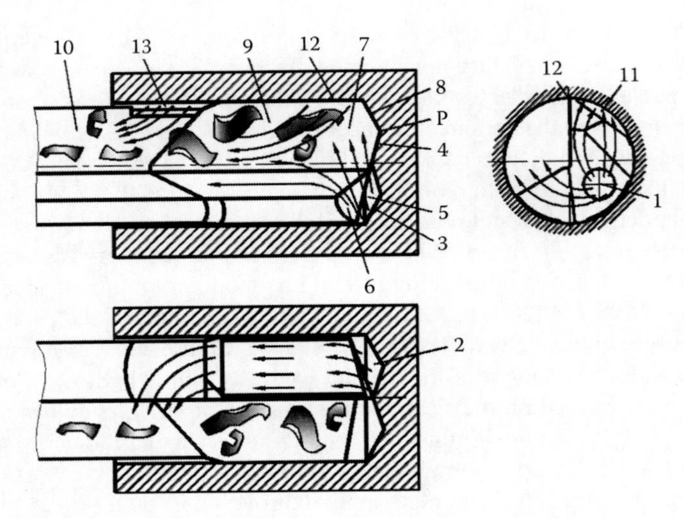

FIGURE 6.53 Model of MWF flow in a gundrill with the auxiliary side passage.

(V-flutes) 9 and 10 made of the tip and that of the shank, respectively, as shown in Figure 6.53 by the streamlines of MWF flow.

Because the part of the tip periphery surface is removed, an auxiliary passage 11 is formed between the periphery surface of the tip portion adjacent to the side cutting edge 12 and the walls of the hole being drilled. This passage continues over the entire length of the tip and then eventually meets with the chip removal passage 10 of the shank through passage 13 made on the shank. As a result, a part of the MWF supplied to the bottom clearance space 2, flows through passages 11 and 13 and then joins the rest of the MWF flow with the chip as show in Figure 6.53. The separation on the MWF supplied into the bottom clearance space 2 into two flows prevents the formation of vortexes as shown by the authors earlier (Astakhov et al. 1995).

6.3.5 Coolant Channels Network: Ejector Effect

Modern tools are usually made with several stages, that is, designed to perform several operations such as drilling, reaming, chamfering, etc. As such, a network of the coolant channels and passages is made in the tool with internal MWF supply to deliver MWF to multiple cutting edges and supporting pads. Figure 6.54 shows an example. The design challenge is usually to assure that the desired flow rate through each channel in proportion to generated heat, formed chip volume, and other considerations of a particular stage.

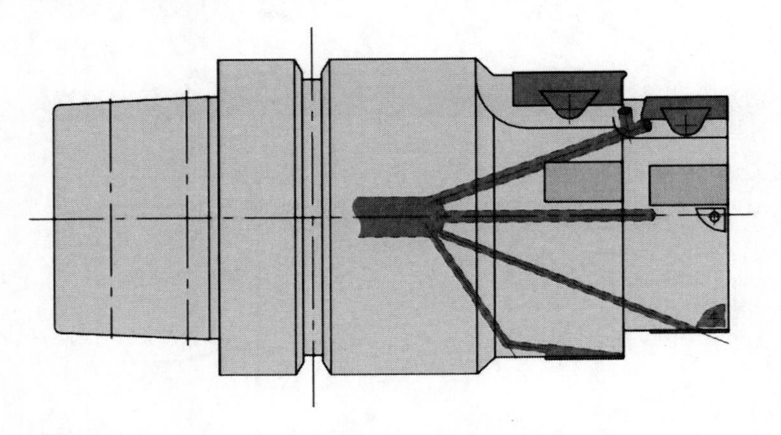

FIGURE 6.54 Example of a network of the coolant channels in a tool.

This problem is easily solved when electrical analogy of the hydraulic circuit is considered as discussed in Section 6.3.3.2. As such, the voltage is considered as the pressure while electrical current is considered as the flow rate. This allows one to apply Kirchoff's laws, developed for electrical circuits, for and for hydraulic circuits. Adopted for hydraulic circuits, these laws are as follows:

1. For any node of the circuit, $\Sigma_{in}Q_v = \Sigma_{out}Q_v$.
2. For any closed circuit, the overall pressure is equal to the algebraic sum of pressure losses, that is, $p_h/\rho_{fl}g = \Sigma h/\rho_{fl}g$.

As an example, consider a three-stage reamer having three coolant holes to deliver a certain flow rate to each stage. A circuit that is an electrohydraulic model for this reamer is shown in Figure 6.55. As in the previous considerations, p_h is the MWF pressure in the tool holder, and Q_v is the volumetric flow rate supplied to the tool holder. Node a, therefore, corresponds to the tool holder while node b corresponds to the chip removal flute. As the static pressure in this flute is small, this pressure is considered to be equal to atmospheric pressure, that is, the gage pressure is equal to zero. Therefore, the total pressure drop from nodes a to b is equal to p_h.

The considerations in Section 6.3.3.2 allow to conclude that the electrical current (and thus the flow rate) in each path (channel) is reversely proportional to the electric (hydraulic) resistance of this path (channel). Such an analogy enhance understanding and simplifies further consideration. The voltage between point a and b is the same for all three paths. Using Ohm's law (Equation 6.14), one can write

$$V_{ab} = I_1R_1 = I_2R_2 = I_3R_3 \tag{6.68}$$

As the voltage is analogous to the pressure drop and the pressure drop is the same no matter which channel forms the path, the following is valid (see Equation 6.50):

$$\frac{p_h}{\rho_{fl}g} = \left(\lambda_1 \frac{L_{ch1}}{d_{dh1}} \frac{1}{A_{ch1}^2 2g} + K_{fr1} \frac{1}{A_{ch1}^2 2g}\right) Q_{v1}^2 = \left(\lambda_2 \frac{L_{ch2}}{d_{dh2}} \frac{1}{A_{ch2}^2 2g} + K_{fr2} \frac{1}{A_{ch2}^2 2g}\right) Q_{v2}^2$$
$$= \left(\lambda_3 \frac{L_{ch3}}{d_{dh3}} \frac{1}{A_{ch3}^2 2g} + K_{fr3} \frac{1}{A_{ch3}^2 2g}\right) Q_{v3}^2 \tag{6.69}$$

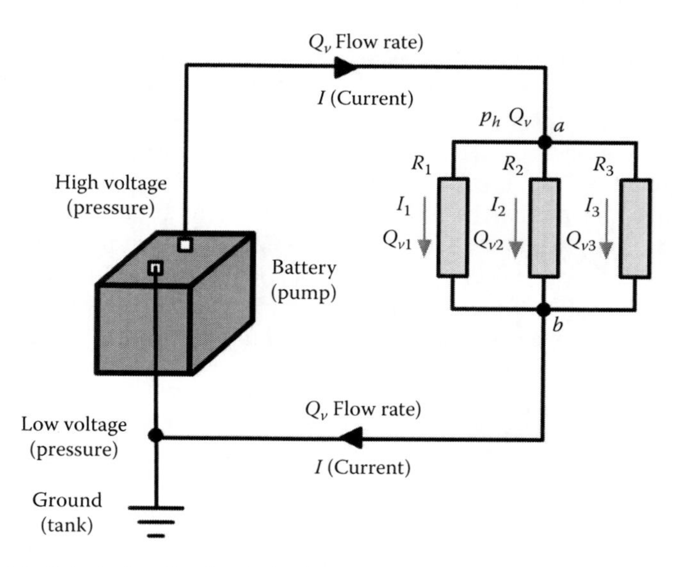

FIGURE 6.55 Model of the hydraulic circuit for three-stage reamer.

If the channels are round holes having diameters d_1, d_2, and d_3, respectively, and $g = 9.81$ m/s^2, then Equation 6.69 can be rewritten in terms of these diameters as

$$\frac{p_h}{\rho_{fl} g} = \left(\lambda_1 \frac{L_{ch1}}{12.09 d_1^5} + K_{fr1} \frac{1}{12.09 d_1^4}\right) Q_{v1}^2 = \left(\lambda_2 \frac{L_{ch2}}{12.09 d_2^5} + K_{fr2} \frac{1}{12.09 d_2^4}\right) Q_{v2}^2$$

$$= \left(\lambda_3 \frac{L_{ch3}}{12.09 d_3^5} + K_{fr3} \frac{1}{12.09 d_3^4}\right) Q_{v3}^2 \tag{6.70}$$

As such

$$\frac{Q_{v1}}{Q_{v2}} = \sqrt{\frac{\left(\lambda_2 \left(L_{ch2}/12.09 d_2^5\right) + K_{fr1}\left(1/12.09 d_2^4\right)\right)}{\left(\lambda_1 \left(L_{ch1}/12.09 d_1^5\right) + K_{fr1}\left(1/12.09 d_1^4\right)\right)}} \tag{6.71}$$

$$\frac{Q_{v1}}{Q_{v3}} = \sqrt{\frac{\left(\lambda_3 \left(L_{ch3}/12.09 d_3^5\right) + K_{fr3}\left(1/12.09 d_3^4\right)\right)}{\left(\lambda_1 \left(L_{ch1}/12.09 d_1^5\right) + K_{fr1}\left(1/12.09 d_1^4\right)\right)}} \tag{6.72}$$

According to the Kirchoff's laws known as the continuity condition in hydraulics

$$Q_v = Q_{v1} + Q_{v3} + Q_{v3} \tag{6.73}$$

It follows from Equations 6.70 through 6.72 that for round channels, the diameter of the channel is the most powerful means at the tool designer's disposal to distribute the total flow rate as needed because the other parameters involved in these equations do not differ significantly from one coolant channel to the next.

Although the foregoing consideration is correct in many practical cases, and, moreover, one intuitively feels that the flow rate between different branches distributes in the reverse proportion to their hydraulic resistance, the electrical analogy used is not always strict because apart from the electrical current, a MWF flow has mass and thus inertia. When the inertial term $(v_{fl}^2/2g)$ in the Bernoulli equation (Equation 6.16) becomes significant, the flow rates distribution as it was presented previously may not be valid. Moreover, *negative* hydraulic resistance in terms of the electrical analogy can be the case, that is, a reverse flow, or suction can take place.

To exemplify the last statement, let us consider a perfume atomizer used in perfume bottles and its model shown in Figure 6.56. In these pictures, the reservoir is the large portion of the perfume bottle that holds the perfume. The perfume is pulled out of the reservoir using a vacuum created due to the Venturi effect which is another name for the ejector effect discussed in Chapter 5 (Section 5.2.7). Vacuum pressure in a classic atomizer is created when the flexible bulb at the end of a flexible tube is squeezed. This pushes a high-velocity flow of the air along a horizontal tube in what is called the Venturi effect—air speeds up, and its static pressure becomes below that of atmospheric that pushes the liquid in the reservoir into the flow of air.

To explain the issue in mathematical and physical terms, let us consider a model of MWF flow shown in Figure 6.57 where Q_v is the flow rate supplied. Considering cross sections 1-1 and 2-2, one can conclude that the flow rates are the same for these sections. The MWF velocity through section 1-1 is

$$v_{1\text{-}1} = \frac{Q_v}{\pi\left(d_{1\text{-}1}^2/4\right)} \tag{6.74}$$

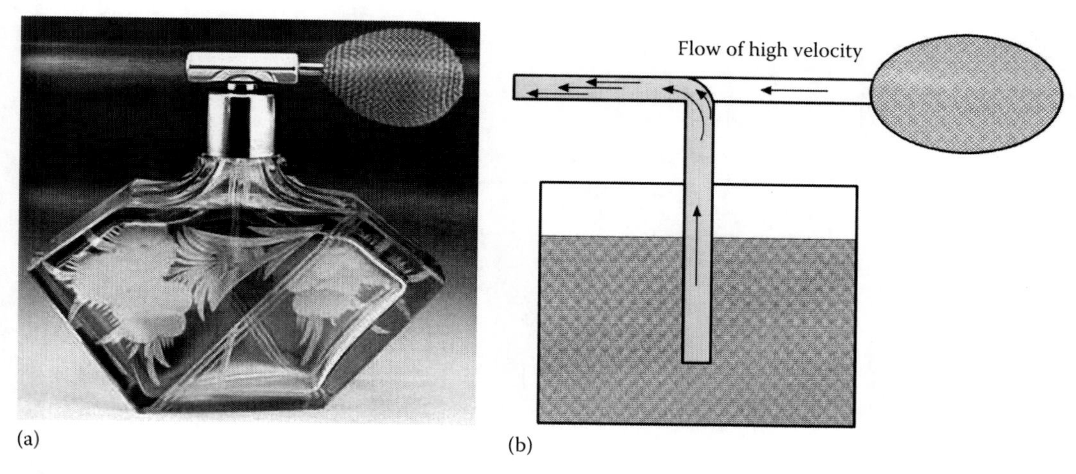

FIGURE 6.56 Spray bottle (a) and its simplified model (b).

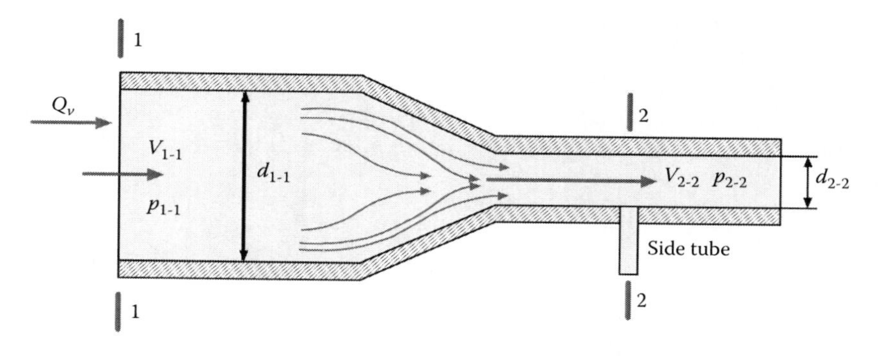

FIGURE 6.57 Model of MWF flow.

The MWF velocity through section 2-2 is

$$v_{2\text{-}2} = \frac{Q_v}{\pi\left(d_{2\text{-}2}^2/4\right)} \tag{6.75}$$

The Bernoulli equation (Equation 6.16) for the MWF flow between these sections is

$$\frac{p_{1\text{-}1}}{\rho_{fl}g} + \frac{v_1^2}{2g_1} = \frac{p_{2\text{-}2}}{\rho_{fl}g} + \frac{v_2^2}{2g} + h_{1\text{-}2} \tag{6.76}$$

where

- subscripts 1 and 2 refer to the cross sections 1-1 and 2-2
- $p_{1\text{-}1}$ and $p_{2\text{-}2}$ are the static pressure at these sections
- v_1 and v_2 are the velocities of MWF through the considered sections
- $h_{1\text{-}2}$ is the overall hydraulic head (pressure) loss between the considered sections

From Equation 6.76, one can obtain

$$p_{2\text{-}2} = p_{1\text{-}1} + \frac{1}{2}\rho_{fl}v_{1\text{-}1}^2 - \frac{1}{2}\rho_{fl}v_{2\text{-}2}^2 - h_{1\text{-}2}\rho_{fl}g \tag{6.77}$$

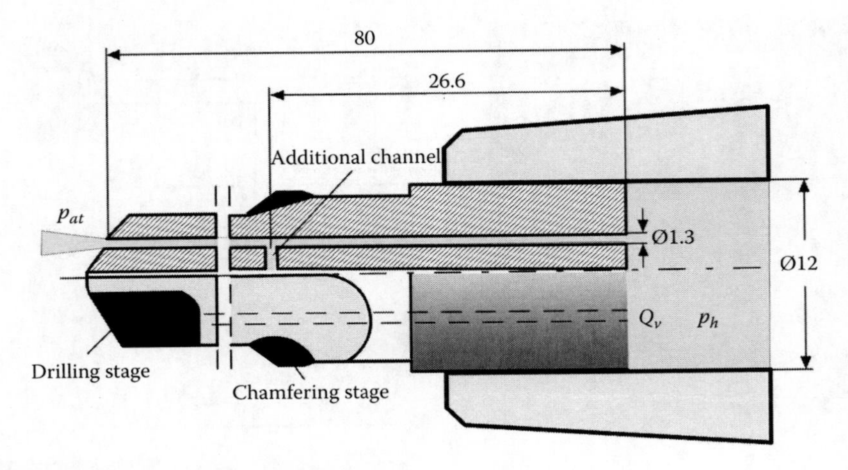

FIGURE 6.58 PCD-tipped drill with a chamfering stage.

Note that pressure $p_{2\text{-}2}$ in this equation is the gage pressure. If, for a certain combination of the parameters of the right side of Equation 6.77, pressure $p_{2\text{-}2}$ becomes less than zero (i.e., less than atmospheric pressure), then the hole made in section 2-2 as shown in Figure 6.58 would suck whatever is in the chip flute rather than supply MWF through this hole. This is called the ejector (Venturi) effect (sometimes wrongly referred to as the Bernoulli effect as it was discovered many years after Bernoulli death) (Astakhov et al. 1991, 1996).

Example 6.4

Problem

A tool having the same parameters as in Example 6.2 was modified compared to that shown in Figure 6.24 by adding a chamfering stage as shown in Figure 6.58. In order to introduce MWF to this stage, two additional coolant channels (one for each cutting edge of the chamfering stage) were added to the tool. Check if any MWF will come out from these additional channels.

Solution

According to the data and result of Example 6.2, $p_{1\text{-}1} = p_h = 8{,}583{,}799.5$ Pa, $v_{1\text{-}1} = v_h = 1.8$ m/s, $v_{2\text{-}2} = 76.7$ m/s, $\rho_{fl} = 985$ kg/m³.

The head loss from the holder to the additional channel is the sum of the major and minor losses. Major hydraulic head losses are calculated only for a coolant hole using Equation 6.26 as

$$h_{ch} = \lambda \frac{L_{ch}}{d_{dh}} \frac{v_{ch}^2}{2g} = 0.037 \times \frac{26.6}{1.3} \times \frac{76.7^2}{2 \times 9.81} = 227.0\ \text{m}$$

Minor head loss is due to flow contraction (from the tool holder into the coolant hole(s)). Therefore, they are the same as in Example 6.2, that is, $h_m = 148.72$ m

The total head losses are

$$h_{\Sigma 1\text{-}2} = h_{ch} + h_m = 227.00 + 148.72 = 375.72\ \text{m}$$

The static MWF pressure at the entrance of the additional coolant channel is calculated using Equation 6.77 as

$$p_{2\text{-}2} = 8{,}034{,}080.0 + \frac{1}{2} \times 985 \times 1.8^2 - \frac{1}{2} \times 985 \times 76.7^2 - 375.72 \times 985 \times 9.81$$

$$= 1{,}506{,}230.68\ \text{Pa} \approx 1.51\ \text{MPa}$$

Therefore, Equations 6.71 and 6.72 are valid for this case.

Consider the situation with the improved MWF supply channels as discussed in Example 6.2. Major hydraulic head losses are calculated using Equation 6.26 as

$$h_{ch} = 0.02 \times \frac{26.6}{1.3} \times \frac{76.7^2}{2 \times 9.81} = 122.7 \text{ m}$$

Minor head loss is due to flow contraction (from the tool holder into the coolant hole(s)). Therefore, they are the same as in Example 6.2, that is, $h_m = 74.96$ m

The total head losses are

$$h_\Sigma = h_{ch} + h_m = 122.7 + 74.96 = 197.96 \text{ m}$$

The static MWF pressure at the entrance of the additional coolant channel is calculated using Equation 6.77 as

$$p_{2\text{-}2} = 4{,}299{,}918.89 + \frac{1}{2} \times 985 \times 1.8^2 - \frac{1}{2} \times 985 \times 76.7^2 - 197.96 \times 985 \times 9.81 = -508{,}666.6 \text{ Pa}$$

As the gage pressure $p_{2\text{-}2} < 0$, then suction into the additional coolant channel takes place rather than supply of MWF through this channel.

6.4 NEAR-DRY (MINIMUM QUANTITY LUBRICANT) DRILLING OPERATIONS: THEORY, APPARATUS, AND PARTICULARITIES OF TOOL DESIGN

6.4.1 Challenges with MWF

MWF have undergone intense regulatory scrutiny during the last 20 years. The United Auto Workers petitioned the Occupational Safety and Health Administration (OSHA) to lower the permissible exposure limit for MWF from 5.0 to 0.5 mg/m^3. In response, OSHA established the Metalworking Fluid Standards Advisory Committee (MWFSAC) in 1997 to develop standards or guidelines related to metalworking fluids. In its final report in 1999, MWFSAC recommended that the exposure limit be 0.5 mg/m^3 and that medical surveillance, exposure monitoring, system management, workplace monitoring, and employee training are necessary to monitor worker exposure to metalworking fluids. Eventually, this recommendation became a mandatory requirement in the automotive industry.

In the current, competitive manufacturing environment, end users of MWF are looking to reduce costs and improve productivity. As a result, a closer look at the cost of MWF was taken. Surprisingly, it was found that MWF represent a significant part of the manufacturing costs. Just two decades ago, MWF accounted for less than 3% of the cost of most machining processes. These MWFs were so cheap that few machine shops gave them much thought. Times have changed and today MWFs account for up to 15% of a shop production cost (Graham 2000), while some European automotive companies reported 16.9% (Brinksmeier et al. 1999). The costs of purchase, maintenance, and disposal of MWF are more than twofold higher than the tool-related costs (in some European countries with strict pollution control regulations), although the main attention of researchers, engineers, and managers has been focused on the reduction of the cutting tools-related costs. Moreover, MWFs, especially those containing oil, have become a huge liability. Not only does the Environmental Protection Agency (EPA) regulate the disposal of such mixtures, but many states and localities also have classified them as hazardous wastes.

It is estimated that MWF consumption is more than 100 million gallons per year in the United States (Feng and Hattori 2000). A typical large automobile metal processing facility utilizes more than 600,000 gal (>2.28 million liters) of MWF concentrates per year and more than 300,000 gal (>1.14 million liters) of straight oil per year. In Germany, MWF consumption is about 75,500 tons/year. In Japan, MWF consumption is 100,000 kL of water-immiscible MWF (disposal cost ¥35–50 per liter), 50,000 kL of water-soluble MWF without chlorine (disposal cost ¥300 per liter), and 10,000 kL of water-soluble MWF with chlorine (disposal cost ¥ 2,250 per liter) (Feng and Hattori 2000). The costs of maintaining and eventually disposing of MWF, combined with the aforementioned health and safety concerns, have led to a heightened interest in either eliminating MWF altogether or limiting the amount of MWF applied. The former process is known as dry machining whereas the latter is referred to as near-dry machining (NDM) or minimum quantity lubricant (MQL) machining. However, non-systemic considerations from the management standpoint and misunderstanding of physics of MWF actions in the cutting process significantly slowdown implementations of these seemingly attractive and cost-effective technologies.

Generally speaking, NDM (MQL) is machining with the supply of very small quantities of lubricant to the machining zone. It was developed as an alternative to flood and internal high-pressure coolant supply to reduce MWFs consumption. In NDM, the cooling media is supplied as a mixture of air and an oil in the form of aerosol (often referred to as mist). An aerosol is a gaseous suspension (hanging) into air of solid or liquid particles. In NDM, aerosols are oil droplets dispersed in a jet of air. Aerosols are generated using the process called atomization, which is the conversion of a bulk liquid into a spray or mist (i.e., a collection of tiny droplets), often by passing the liquid through a nozzle.

An atomizer is an atomization apparatus; carburetors, airbrushes, misters, and spray bottles (Figure 6.56) are only a few examples of atomizers used ubiquitously. In internal combustion engines, fine-grained fuel atomization is instrumental to efficient combustion. Despite the name, it does not usually imply that the particles are reduced to atomic sizes. Rather, droplets of 1–5 μm are generated. Because MWF cannot be seen in the working zone and because the chips look and feel dry, this application of MQL is called NDM.

6.4.2 Understanding the Subject

Any project on NDM should start with understanding the subject. To begin, one should ask oneself a few simple questions: Why does NDM work at all? Why does it reportedly show somehow better results than flood or even high-pressure MWF supply? In other words, how is it possible *physically* because flood or high-pressure MWF definitely removes much more thermal energy from the machining zone? Why does aerosol-containing oil and air mixture have (according to all NDM research and promotional papers) greater cooling ability than that of water-soluble MWFs although the heat capacity of water is 10 times greater than that of oil and much greater than that of air keeping in mind that heat removal due to oil droplets evaporation is negligibly small (Astakhov 2006) because of inherent oil properties and tiny oil flow rate (30–60 mL/h)? Through finding physically justifiable answers to these simple questions, one easily finds that the aerosol used in NDM cannot physically provide the same cooling action as flood and especially high-pressure MWF supply techniques regardless of ungrounded claims used in many sales and promotional materials, which unfortunately found their way even to many professional and even some research publications.

The author's analysis of the implementation practice of NDM reveals that the most common mistake made by many professionals and practitioners in the field is to use the existing machines, tools, tooling, controllers, and part designs for NDM without understanding the physics of machining and without system considerations of the whole machining process. This has been resulting in multiply failures of this seemingly attractive technology. Although the blame should be equally shared by academia and industrial scientists, the major impact in the discrediting of this technology belongs to NDM apparatus and accessories manufacturers

FIGURE 6.59 Idealized image of NDM.

because they used the full power of their sales force to promote their products without much care about the end result. In their sales presentations, they show

1. Idealized pictures of NDM principle similar to that shown in Figure 6.59 where oil droplets mysteriously fly only to the cutting tip of the tool and the tool contact areas happen to be readily exposed to these droplets thus to be cooled and lubricated sufficiently.
2. Pie charts, where the MFW application cost triples that of the tooling cost, which is not always the case.
3. OSHA and EPA set requirements on MWF exposure limits and MWF disposal conveniently forgetting that the exposure to the aerosol (mist) in NDM can be much more hazardous if no special precautions are taken.

Normally, it does not take much time to convince an inexperienced end user that his/her company can be *lean and green* in short time and at reasonable cost. In reality, however, it happens rather rarely.

Experience of implementing of NDM shows that this technology can be successful if and only if the whole NDM system of components is considered and special attention is paid to each and every component individually while maintaining the coherence of the system. Figure 6.60 shows such a system. Therefore, a 360° analysis of all components listed in Figure 6.60 is a prerequisite before any consideration on practically implementing NDM. Unfortunately, this is not the case in practice.

The success of NDM can be achieved if all components of machining or manufacturing system are suitable for this technology and if there is clear understanding of the essence, particularities, and limitation of this technology. This understanding should stem not from sales pitches by NDM equipment manufacturers and reports by environmentally concerned enthusiastic amateur groups but rather from a detailed analysis of all components of the machining system in terms of their suitability for NDM. To do this, certain qualification and system thinking of a responsible manager/specialist are prerequisites. One should clearly realize the truth: at the present stage of development, NDM can substitute flood and high-pressure MWF supply only in particular special cases.

Ideally, a part to be machined, a machine tool (manufacturing cell, production line, etc.) and cutting tools should be designed specifically for NDM. Although some machine tools have been retrofitted for NDM, retrofitting an existing machine does not appear to be an attractive option in mass production manufacturing while in the mold- and die-making industry a number of successive implementations of NDM have been reported particularly in hard milling.

When it is feasible, the implementation of NDM starts from the part design, which should make NDM easier in terms of chip removal and evacuation. For example, deep blind holes are to be avoided giving preference to cored holes; threads should be designed to make them suitable for form

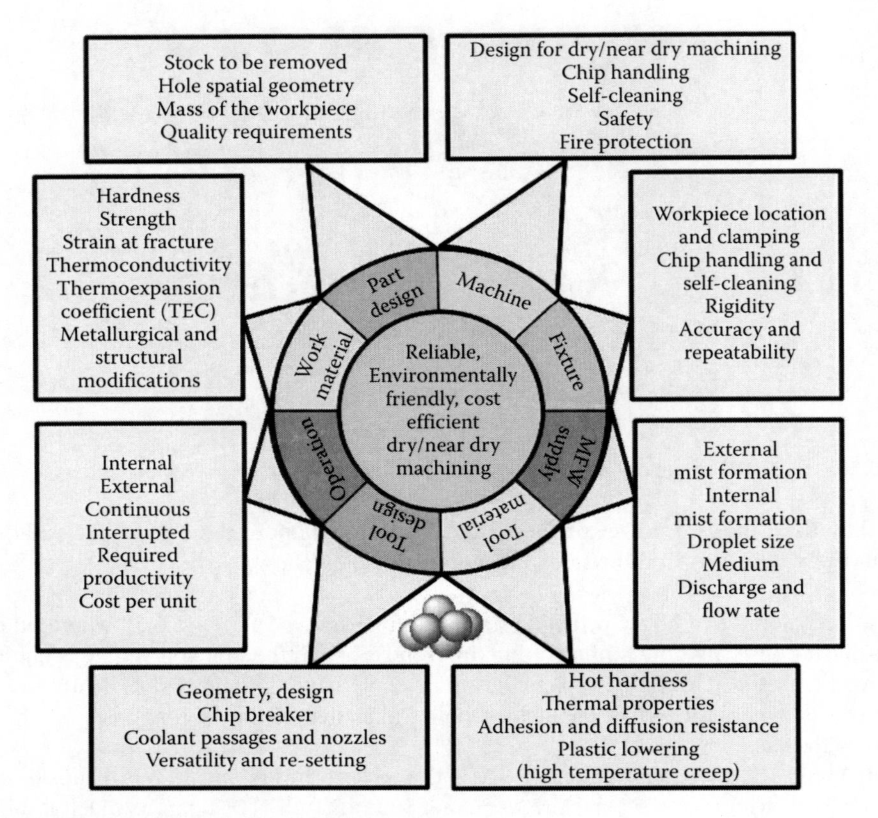

FIGURE 6.60 Basic components of the NDM system.

rather than cutting threading taps; preference should be given to open flat faces suitable for face and interpolating milling; and parts (particularly aluminum parts in the automotive industry) should be designed to be less susceptible to thermal distortion due to heat generated in machining with the aid of FEM thermal distortion analysis. The choice of the part material and its metallurgical state should also be made, accounting for machinability and susceptibility to thermal distortion.

The cutting tools should be designed and the cutting tool material should be selected specially for NDM. Tool geometry and chip-breaking problems should be one of the prime concerns as, if this problem is not resolved, it can easily turn the whole NDM operation into a production nightmare. The design of the aerosol supply channels should be suitable, but also the tool geometry, tool materials, and design of the tool body (back tapers, reliefs, undercuts, and supporting elements) should be optimized for NDM.

Additional procedures in tool setting and maintenance as well as additional equipment for aerosol verification must be implemented and followed.

Therefore, to make NDM technology more user-friendly and to reduce the failure rate in applying this technology, a system analysis of its applicability should precede any implementation commitment. Regardless of the scope of a particular NDM project, this analysis should always include two principal stages:

1. Machining process stage where the work material, parameters of the machining regime, cutting tool design and material, workpiece design, and tolerances (including spatial) are considered together in order to assess feasibility of NDM in principle on one hand and to understand what has to be changed (optimized, re-designed, etc.) to implement NDM on the other hand. The chief objective of the first stage is to reduce the amount of material to be cut, that is, the volume of the chip produced. The details of the first stage have been discussed by the author earlier (Astakhov 2010b).

2. Practical NDM setup and apparatus. At this stage, a particular NDM type, the aerosol components and parameters should be selected and tested in terms of their suitability for the considered project (Astakhov 2008). NDM stability should be verified, and the data for the efficiency analysis should be collected for critical machining operations, tools, and work materials to be involved with NDM.

It should be also remembered that besides *cutting functions*, MWFs also perform two important technological functions: (1) Cooling the workpiece so that it does not distort excessively and does not get excessively hot. As a result, proper in-process and postprocess gaging as well as interoperation handling (loading/unloading and transportation) should be considered. Note that these do not present problems in high-pressure MWF application. This is not nearly the case in NDM. (2) Flushing away chips so the tool, the workpiece, the fixture, and the machine are not damaged. Therefore, additional means to substitute these function should be developed and implemented in NDM.

6.4.3 Implementation Aspects

Regardless of the scope of a particular NDM project, one needs to be prepared to deal with the following aspects in the course of its execution:

1. *Aerosol generation*. This includes the selection of oil and suitable aerosol concentration as well as the design of the atomizer.
2. *Aerosol delivery into the machining zone*. This is a combined aspect that includes the choice of NDM type and assurance that the selected parameters of aerosol do not change with tool changing and exchanging, start/stop machining cycle, change of a particular machining operation, etc.
3. *Chip removal from the machining zone*. This aspect is critical for hole-making operations as the chip formed in the machining zone should be continuously removed to prevent chip clogging, re-cutting, additional heating of the part being machined, etc.
4. *Part thermal distortion*. This aspect plays an important role when thin-wall parts made of a highly thermoconductive work material are machined on a production line with short cycle time using a number of successive cutting tools as, for example, in the automotive industry.
5. *Chip removal from the working zone of the machine*. As chips accumulate in the working zone, it must be continuously removed from all the surfaces of the machine tool surrounded this zone.

The author's experience shows that any failure to consider the listed aspects normally leads to failure of the whole NDM project or enormously increases its cost and implementation. The following sections discuss some important issues in practical implementation of the NDM technology in hope that the knowledge gained can reduce the severity of NDM implementation painstaking efforts.

6.4.4 Aerosol (Mist)

6.4.4.1 How Aerosol Is Generated

As mentioned previously, in NDM, the cooling medium is supplied as a mixture of air and an oil in the form of an aerosol (often referred to as mist). An aerosol is a gaseous suspension (hanging) into air of solid or liquid particles. Aerosols are generated using a process called atomization. An atomizer is an ejector in which the energy of compressed gas, usually air taken from the plant supply, is used to atomize oil. Oil is then conveyed by the air in a low-pressure distribution system to the machining zone.

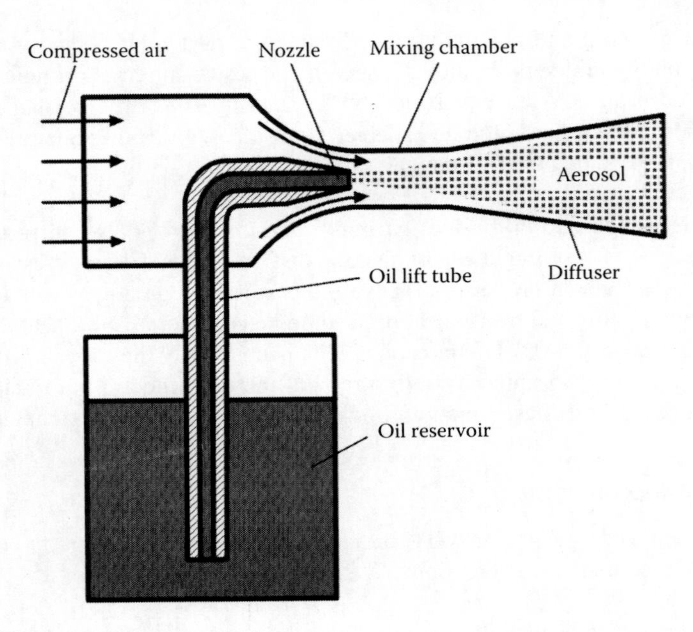

FIGURE 6.61 Schematic of a simple atomizer.

The principle of the atomizer is shown in Figure 6.61. As the compressed air flows through the Venturi path, the narrow throat around the discharge nozzle creates a Venturi effect in the mixing chamber, that is, a zone where the static pressure is below the atmospheric pressure (often referred to as a partial vacuum) (Astakhov et al. 1991, 1996). This partial vacuum draws the oil up from the oil reservoir where the oil is maintained under a constant hydraulic head. The air rushing through the mixing chamber atomizes the oil stream into an aerosol of micron-sized particles.

The design of the atomizer is critical in NDM as it determines the concentration of the aerosol and the size of droplets. Unfortunately, this is one of the most neglected aspects in casual application of NDM. Several machine tool companies have patented their designs (e.g., US patents No. 6,923,604 [2005] and No. 6,602,031 [2003]). In such designs, the position of the discharge nozzle is controlled by the machine controller so the parameters of the aerosol are changed depending upon the machining conditions.

6.4.4.2 Aerosol Composition

In the simplest cases of NDM, the aerosol is an air–oil mixture. The discharge of the oil in this mixture is selected to be in the range of 30–600 mL/h depending upon the design of the NDM system, the nature of the machining operation, the work material, and many other factors. Unfortunately, not many recommendations of the mixture composition are available. Advanced NDM (ANDM) uses aerosol that includes not only oil but also some other components, for example, oil-on-water droplet (OoW NDM).

OoW NDM includes the supply of water droplets covered with a thin oil film (Itoigawa et al. 2006; Yoshimura et al. 2005, 2006). As claimed, this method possesses both great cooling and lubricating abilities. The former is due to water properties (high specific heat capacity, density and thermal conductivity compared to air) and its evaporation. The latter is due to the specific droplet configuration.

The concept of OoW NDM is shown in Figure 6.62, which shows an ideal OoW droplet moving toward a hot surface. When the droplet reaches the tool or hot workpiece surface, the lubricant oil spreads over the surface in advance of water spreading. The water droplets are expected to perform three tasks: carrying the lubricant, spreading the lubricant effectively over the surface due to inertia and cooling the surface due to its high specific heat and evaporation. To make this concept practical, that is, to generate OoW droplets, a specially designed discharge nozzle is needed.

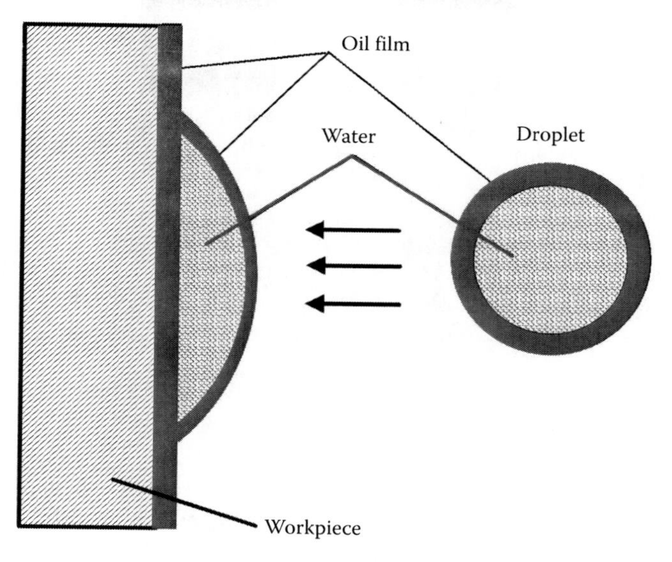

FIGURE 6.62 The concept of the oil-on-water.

6.4.4.3 Aerosol Parameters Control

Modern high-pressure MWF supply systems used in production lines and manufacturing cells are well equipped for strict control of MWF parameters. Digital flow meters, pressure gages, and temperature sensors assure stable trouble-free MWF supply to the cutting tool. Should any MWF-related problem occur, it is normally handled intelligently by machine controllers. This is not the case with NDM.

Aerosol parameters control is probably the weakest link in NDM because there are no simple ways to control aerosol (mist) concentration and droplet size supplied the tool. Although these two parameters are of crucial importance for stability, reliability, and efficiency of NDM, little attention has been paid in the literature (both research and technical) to address this issue or at least to point out its existence and actuality. A detailed description of the experimental methodology to control the parameters of aerosol was presented by Dasch and Kurgin (2010). They developed a new simple method to measure the mist concentration in the aerosol that leaves tool coolant holes: an oil filter such as used in a lawn mower was fitted with a bushing to provide a snug fit to the tool but still allow rotation. Air passed through the filter, but mist was collected. The filter was weighed before and after a timed test with the mist flowing to determine the amount of mist exiting the tool. The method allows to measure the mist concentration actually exiting the tool during tool rotation. Therefore, losses in the spindle, tool holder, and tool were accounted for even when the tool was rotating.

In their test (Dasch and Kurgin 2010), all instrumental measurements of particle concentration and size were made from a 25 cm diameter stainless steel duct leading from the machining enclosure to a mist controller/blower unit. Air was pulled through the duct at a flow rate of 14 m^3/min. Mist was sampled from the duct using isokinetic sampling. Continuous measurements of mist concentration were made with a DataRAM continuous particle monitor (MIE, Inc.), which measures particle levels using a scattered light beam. Mist droplet size distributions were made using a MOUDI cascade impactor (Applied Physics, Inc.), on which particles were collected on 13 stages covering a particle diameter range of 0.03–18 μm. The filters on each stage were weighed to provide a mist size distribution and a mass median particle diameter.

It is understood that such sophisticated equipment is not available on the shop floor, so simple and used-friendly apparatuses to control aerosol (mist) concentration and droplet size at the tool must be developed and made commercially available. In the author's opinion, this is a vital issue for successful implementation of NDM.

6.4.5 Classification of NDM

Unfortunately, there are no accepted classifications of NDM so it is very difficult for a practical engineer or plant manager to make the proper choice about the regimes of NDM and equipment needed. The simplest way to classify NDM is by the method of aerosol supply into the machining zone (Astakhov 2008):

1. NDM 1: NDM with external aerosol supply. In NDM 1, the aerosol is supplied by an external nozzle placed in the machine similar to a nozzle for flood MWF supply.
2. NDM 2: NDM with internal (through-tool) aerosol supply. In NDM 2, the aerosol is supplied through the tool similar to the high-pressure method of internal MWF supply.

As the name implies, NDM with an external aerosol supply (NDM 1) includes the external nozzle that supplies the aerosol. There are two options in NDM 1:

1. *NDM 1.1 with an ejector nozzle.* The oil and the compressed air are supplied to the ejector nozzle, and the aerosol is formed just after the nozzle, as shown in Figure 6.63. In other words, the nozzle itself serves as the atomizer; thus, certain means to control air–oil concentration ratio are provided by the nozzle design.
2. *NDM 1.2 with a conventional nozzle.* The aerosol is prepared in an external atomizer and then supplied to a conventional nozzle, as shown in Figure 6.63. The nozzle design is similar to that used in flood MWF supply.

NDM 1.1 is probably the cheapest and simplest method. For example, the Spra-Kool Midget unit shown in Figure 6.64 is advertised as an economical method of applying an MWF spray for machining. The Spra-Kool unit works on an air pressure of 0.2–1.0 MPa, which should be adjusted on the compressor. Attaching the ball-check fitting to the air supply, dropping the suction tube into an oil container, and locating the nozzle by means of the spring-wire attaching clip, one can get NDM for a cost of US $35. The soft wire in the nose can be bent to direct the spray to the work. It is designed for easy transfer from one machine to another.

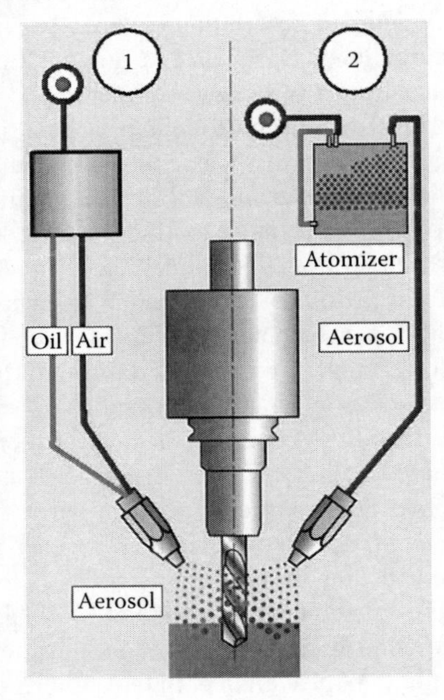

FIGURE 6.63 The principles of (1) NDM 1.1 and (2) NDM 1.2.

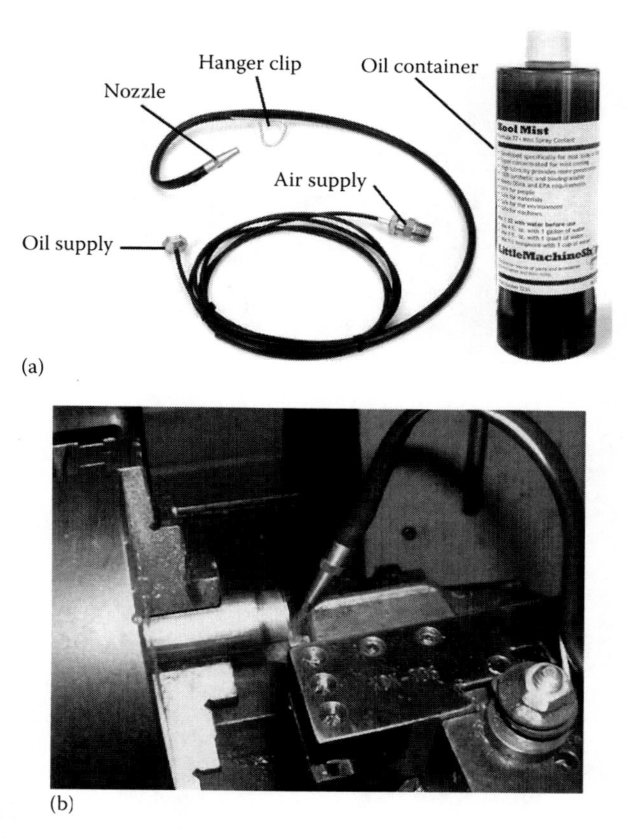

FIGURE 6.64 Spra-Kool Midget (a) and its mounting on a lather (b).

In reality, however, adjustments are not that simple. If no special precautions are taken, the unit generates a dense mist that covers everything in the shop, including the operator's lungs. To prevent this from happening and to gain the full control on the parameters of aerosol, one needs to have (design or buy) a hydraulic unit similar to that shown in Figure 6.65. When such a unit is used, the parameters of the aerosol can be adjusted in a wide range in terms of droplet size and oil flow rate by setting appropriate air and oil flow rates and by adjusting the pressure in the oil reservoir. Moreover, such a device prevents oil spills as it shuts down the oil supply line when the air supply is not available.

NDM 1.2 is probably simpler in terms of adjustment of aerosol parameters. An external atomizer required is an off-the-shelf product. Moreover, the same nozzle as that used for flood MWF can be used, so that NDM 1.2 can be used with the same sets of nozzles installed in the machine.

NDM 1 has the following advantages:

- Inexpensive and simple retrofitting of the existing machines.
- The same cutting tools used for flood MWF will work.
- Easy to use and maintain equipment.
- The NDM equipment can be moved from one machine to another.
- Relative flexibility of NDM 1.2 as the position of nozzle relative to the machining zone can be adjusted for the convenience of operator. As such, the parameters of the aerosol do not depend on the particular nozzle location. There are restrictions, however, on the length and minimal radius of curvature of the line that connects the nozzle with the atomizer.
- Various standard and special nozzle designs are available with NDM 1.2 to suite most common metal machining operations and tool designs. It is proven to be particularly effective in face milling and sawing.

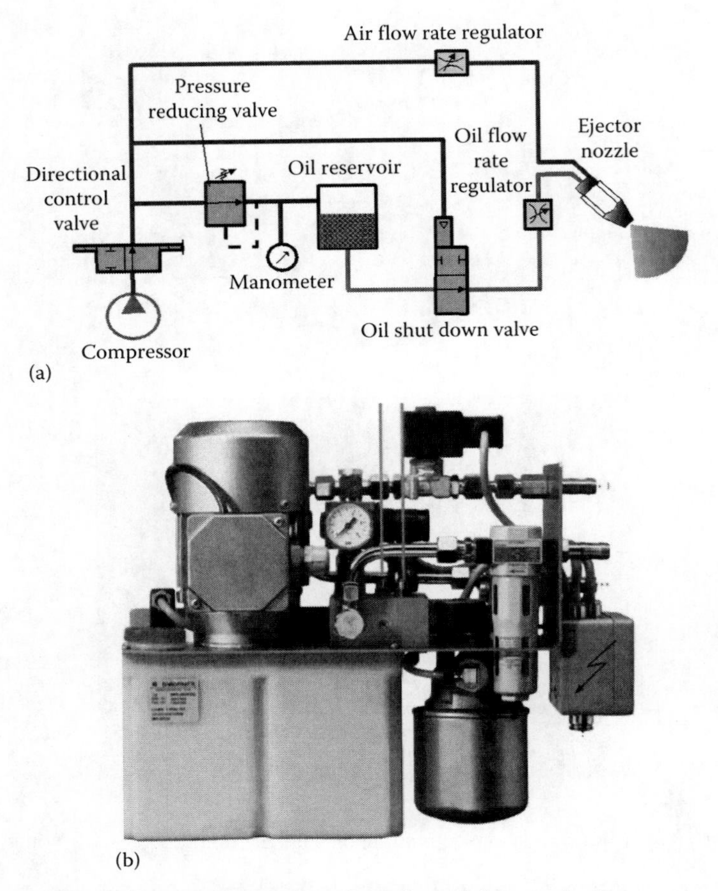

FIGURE 6.65 Aerosol control unit (a) hydraulic schematic, and (b) actual unit.

NDM 1 has the following disadvantages:

- Both NDM 1.1 and 1.2 do not work well with drills and boring tools as an aerosol cannot penetrate into the hole being machined and thus cannot provide any help in the cooling and lubrication of the machining zone and chip removal from the hole being machined.
- A critical aspect for NDM 1.1 and an important aspect for NDM 1.2 is the location of the nozzle relative to the working part of the tool. For both methods, this location must be fixed, that is, should not change as the tool moves. Note that this issue is not that important in flood MWF supply where gravity and the energy of MWF help to cover a much wider range of possible nozzle locations. Moreover, in NDM 1.1, the parameters of the aerosol depend on the distance from the nozzle to the cutting tool so the nozzle location is even more critical and not certain. It makes it difficult for the operator who has to re-adjust this nozzle even within the same operation when the tool, for example, an end mill, goes from slot-milling to side-milling as different parts of the tool are involved in cutting.
- The parameters of the aerosol should be adjusted for each particular metal machining operation and work material.

As the name implies, NDM with internal aerosol supply (NDM 2) includes internal passages for aerosol supply. There are two options in NDM 2: NDM 2.1 with an external atomizer and NDM 2.2 with an internal atomizer located in the spindle of the machine.

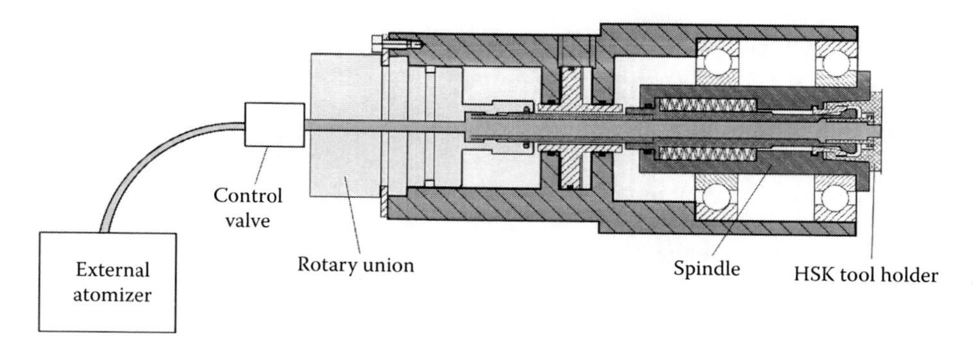

FIGURE 6.66 Principle of NDM 2.1.

In NDM 2.1 (a principle is shown in Figure 6.66), the aerosol is prepared in an external atomizer and then supplied through the spindle and the internal channels made in the tool. When NDM 2.1 is used on machining centers or manufacturing cells, the aerosol supply unit has to react to the frequent tool changes that nowadays take only 1 or 2 s, setting the proper aerosol parameters for each given tool/operation. If the aerosol unit is shut down every time a tool change takes place, then it requires some time to fill the whole system with aerosol again. VOGER, an NDM equipment supplier, has developed the bypass principle illustrated in Figure 6.67.

The aerosol is produced continuously and supplied to the directional control valve, which allows aerosol into the spindle as soon as a tool change is over.

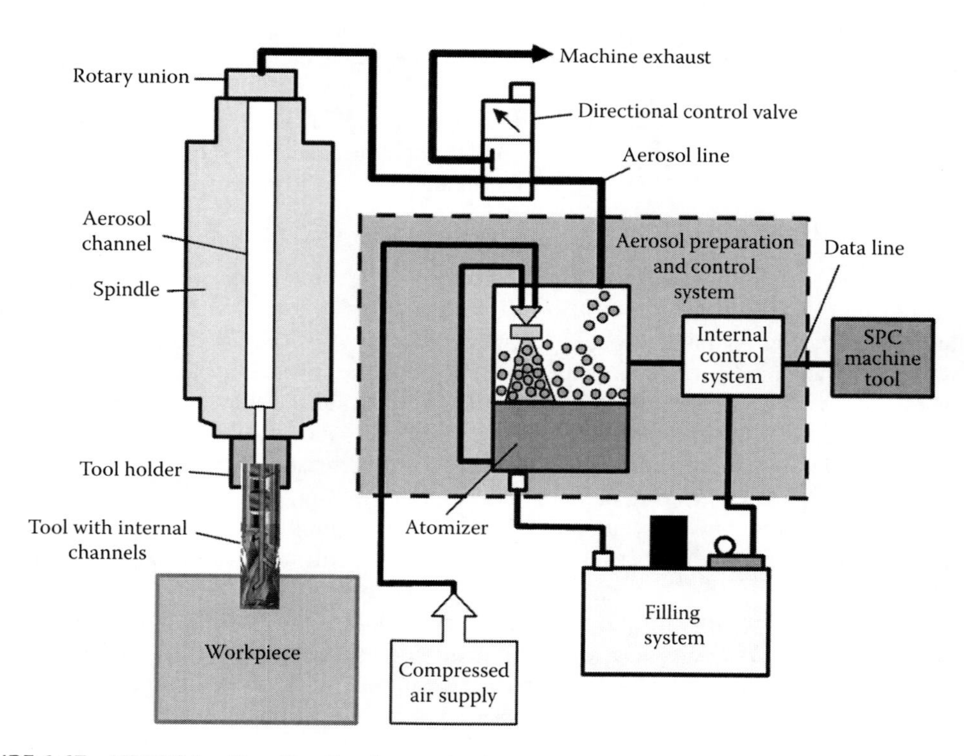

FIGURE 6.67 NDM 2.1 with a directional control valve.

NDM 2.1 has the following advantages:

- Low initial cost
- The possibility of keeping two MWF supply systems on the machine: flood (or high-pressure through the tool) and NDM. Although it may appear that keeping two coolant supply systems on the same machine increases manufacturing costs and setup time, it serves rather a psychological cause providing some assurance of a proven backup should something go wrong with NDM
- Relatively simple installation and control
- Accurate control of the aerosol parameters so that they can be easily adjusted by the machine controller for a given tool or even operation with the same tool.

NDM 2.1 has the following drawbacks:

- Spindle rotation creates a centrifugal force field that coats the wall of the aerosol delivery channel with oil that must be removed periodically. For a high-volume production manufacturing factory (plant, shop, line, cell), this downtime may be intolerable and costly. This additional cost can easily offset the savings on NDM
- Special care should be taken on the position of the flexible line that connects the external aerosol unit with the spindle. Firstly, it should be as close to the machine as possible. Secondly, the radius of curvature of this line should not be less that 200 mm to prevent aerosol decomposition. Often, there is no easy way to meet these two requirements particularly for large-size machines and when shop-floor space is crowded. To solve the problem, UNIST Inc. (Grand Rapids, MI) developed a simple unit where compressed air and oil are supplied to the spindle of a machine separately through a coaxial hose and the aerosol is formed in the inlet of the spindle (Figure 6.68). As such, the location of a MQL unit is not critical as the coaxial hose can be as long as needed for convenient location of the unit
- The parameters of the aerosol that enters the machining zone can be different than those controlled in the outside unit depending upon the spindle speed and the conditions (including design) of the supply channels. As the latter is a critical yet not well-developed and understood issue, it requires additional explanations.

Although the first listed drawback is normally mentioned in some publication, its severity is not fully appreciated. As discussed, aerosol formed in an external atomizer has to pass through the gauntlet of the spindle. As pointed out by Dasch and Kurgin (2010), the gauntlet becomes more impassable at higher spindle rotations as the lubricant (oil in the considered case) droplets in aerosol are centrifugally thrown to the sides of the spindle. As a result, the flow rate of oil decreases as shown in Figure 6.69. Whereas 35 mL/h of oil in aerosol could pass through the machine at 0 rpm, this volume deteriorated rapidly as the rotational speed was increased. Even at a fairly low speed of 2500 rpm, only 3 mL/h was exiting the tool—that is, almost 90% of the mist has been lost in the spindle. The oil that was thrown to the sides of the spindle accumulated and was randomly ejected during the testing, leading to high variability in the readings. This issue has also been documented by Aoyama (2002).

To overcome the problem with the reduction of the mixture flow rate in the rotating spindle, another design classified as NDM 2.2 conducts the oil through the spindle via a central tube within the surrounding air duct as shown in Figure 6.70. The air and oil are mixed to form the aerosol in a mixing chamber located close to the tool. Because the air–oil aerosol is influenced only by the spindle rotation for a short distance, the aerosol flow rate is not significantly affected by spindle rotation and thus not changed with time as with MDM 2.1. Moreover, the discharge response from the tool tip is said to be significantly improved.

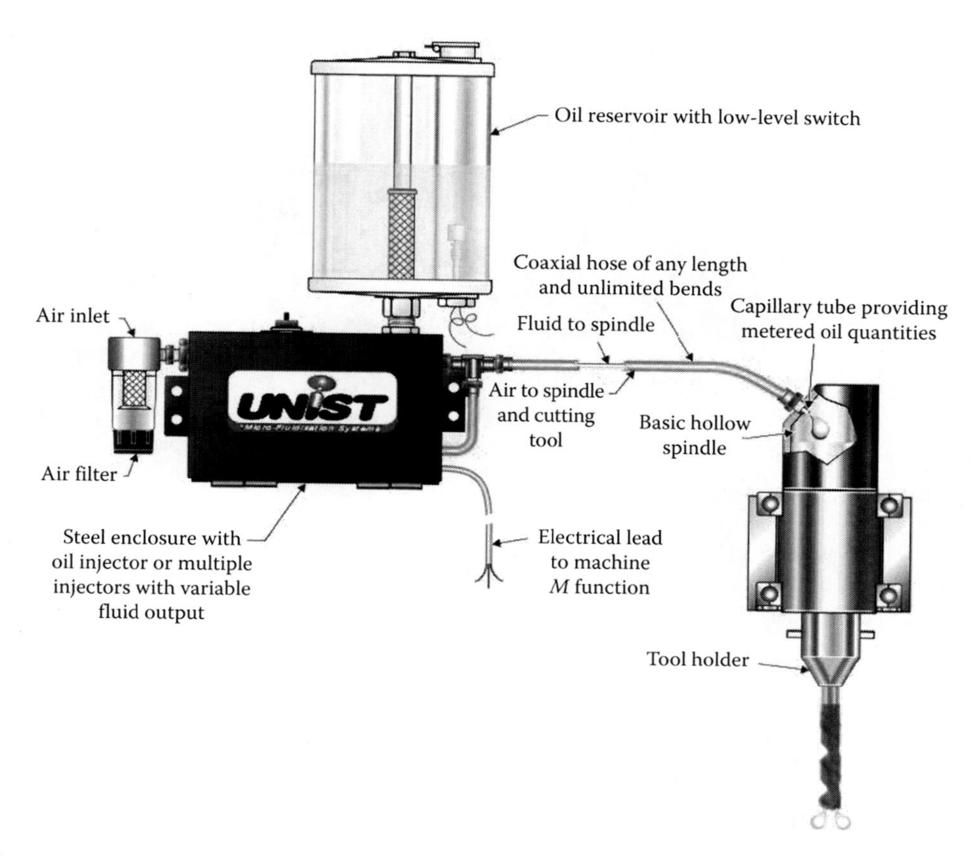

FIGURE 6.68 NDM 2.1 type unit by UNIST Inc.

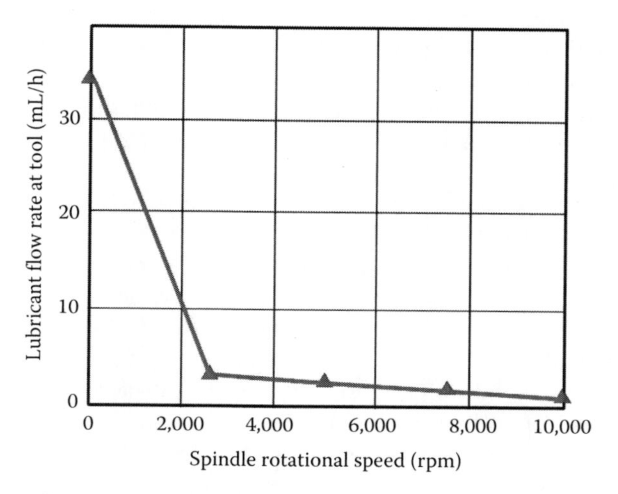

FIGURE 6.69 Lubricant flow rate as affected by the spindle rpm in NDM 2.1.

There is, however, the price to pay for the discussed improvements. In NDM 2.1, the aerosol is prepared in an external atomizer where its parameters can be precisely controlled/adjusted. This is not the case with NDM 2.2, where the atomizer is located inside the spindle or tool holder and designed as the reverse ejector with rather poor characteristics (Astakhov et al. 1991). As a result, there is no way to make any adjustment in operation or even when the tool is installed in the spindle.

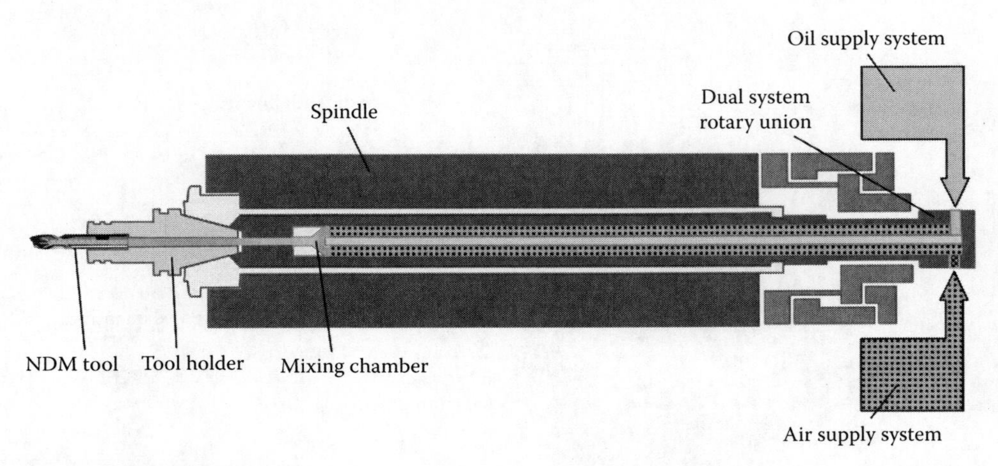

FIGURE 6.70 Principle of NDM 2.2.

Even if one finds a *good* adjustment suitable for a particular tool (its coolant channels), this adjustment may not be suitable for the next tool that may have a considerably different design of these channels. In contrast, in NDM 2.1, the aerosol parameters can be easily programmed for each and every tool to be used in a given machine.

For both NDM 2.1 and NDM 2.2, a critical issue is the introduction of the aerosol into the tool internal coolant channels. The major requirements are as follows: (1) there should be no gaps in the tool holder, that is, the aerosol line should be perfectly sealed; (2) as the tool coolant channels are not normally coaxial with the aerosol supply tube, some means to assure smooth transition from this tube to the tool channels should be a part of the holder or tool design. As pointed out by Stoll and Furness (2006), any cavity in the transport path will disturb the smooth and steady flow of the aerosol to the cutting zone due to the centrifugal forces acting on the oil droplets. This disruption in flow could potentially lead to catastrophic tool failure. To meet these requirements, a number of special designs of tool holders have been developed (Astakhov and Joksch 2012).

As pointed out by Stoll and Furness (2006), although there are several interface designs currently available in the marketplace, none of the available concepts have a *fool-proof* design that eliminates the potential for cavities in the flow path. The author's experience shows that there are two problems: (1) the *adjustment screw-tool shank* interface is inside the tool holder so that it is hidden, and thus there is no easy way to verify the presence or absence of the gap, and (2) in modern industries, tools are set using tool presetting machines (e.g., by Zoller or Kelch) having a length adjustment device. As such, a string passing through a tool holder is used to set the tool length. With discussed changes to the design of HSK holders, it is rather difficult to use such a device. Consequently, caution must be exercised when assembling these tools to ensure that there are no disruptions to the aerosol flow.

Figure 6.71 shows the MAPAL design of the rear end of the tool shank to assure smooth aerosol flow from the supply tube into the tool coolant holes. This picture also shows the design of the adjustment screw mechanism with a spring to assure tight contact between the shank and adjustment screw in the tool holder on tool setting. As such, four basic designs of the adjustment screw are commercially available to fit any design of the rear end of the shank in terms of number and location of the coolant holes.

The design of the mixing chamber and ejector nozzle as well as the controlling of the oil discharge are critical for the application and are normally patented by the machine tool manufacturers. Being integrated into a machine tool, such systems are much more complicated and expensive. Figure 6.72 shows the Bielomatik MQL two-channel system where compressed air and oil are supplied in coaxial channels made in the spindle and thus remain separated until immediately prior the tool holder where the mixing chamber is formed to produce aerosol, which is then supplied into the

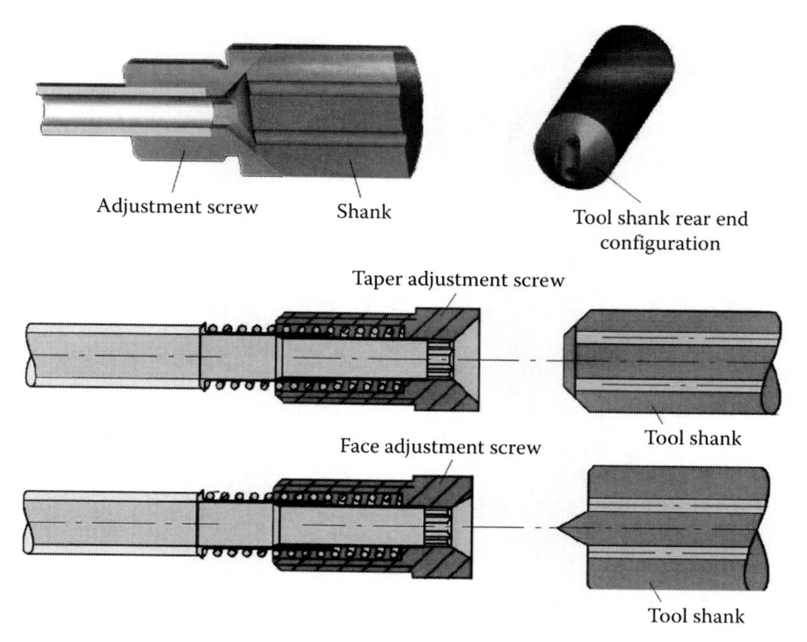

FIGURE 6.71 Tool shank rear end configuration and design of the adjustment screw mechanism by MAPAL. (Courtesy of Mapal USA Co., Port Huron, MI.)

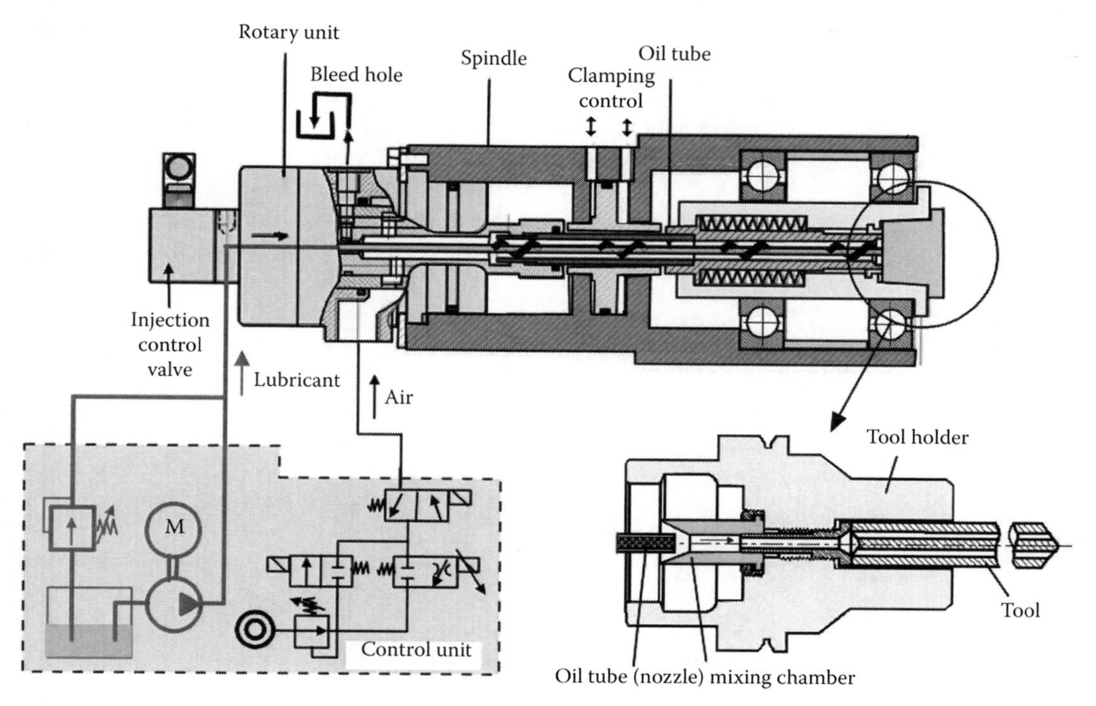

FIGURE 6.72 Bielomatik MQL two-channel system.

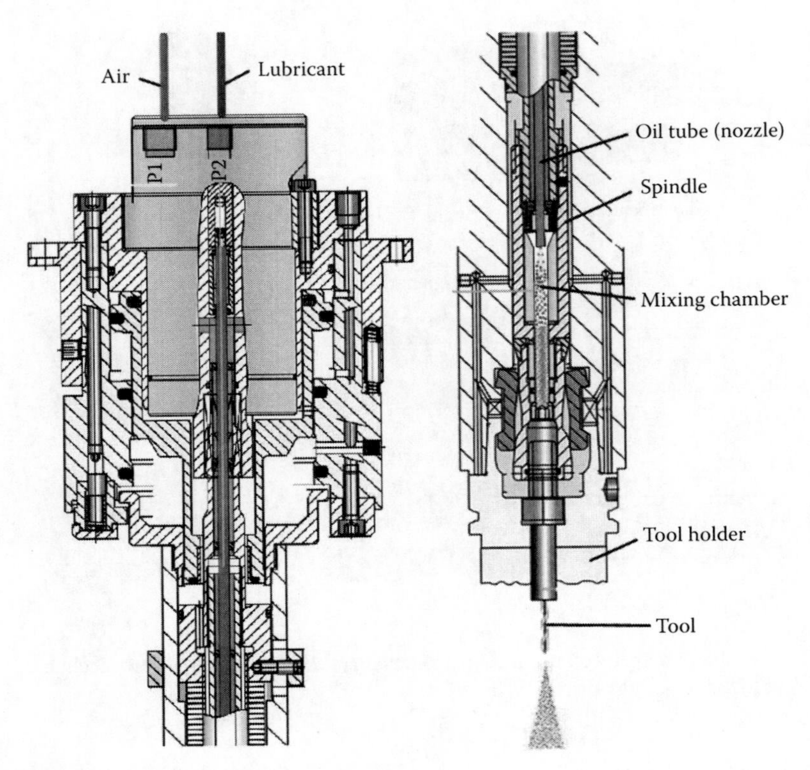

FIGURE 6.73 GAT MQL two-channel system.

cutting tool. Figure 6.73 shows GAT MQL two-channel system. The distinguishing feature of this system is that the mixing chamber is located in the spindle rather than in the tool holder is in the Bielomatik (Figure 6.72).

Dasch and Kurgin (2010) proved the advantage of NDM 2.2 using a Bielomatik system similar to that shown in Figure 6.72. The lubricant (oil) flow rate was calibrated using their special calibration tool in which the oil flow was monitored during a specific time period. Three control flow rates—10, 74, and 117 mL/h—were used in the tests. The actual lubricant flow rate in aerosol was measured at the tool tip using a weighed filter. The results are shown in Figure 6.74. At 0 rpm, these values were 8.7, 59, and 87 mL/h, respectively, somewhat lower than the Bielomatik calibration, indicating

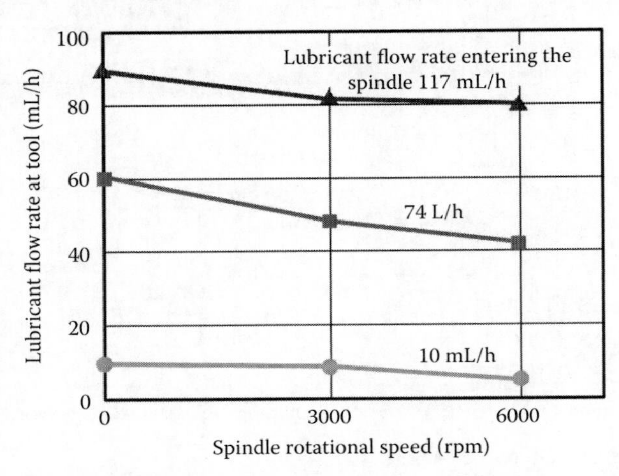

FIGURE 6.74 Lubricant flow rate as affected by the spindle rpm in NDM 2.2.

possible losses in the spindle, tool holder, or tool. There was some decrease in mist flow rate at higher rpm's, but not nearly to the extent found with the single channel system.

In implementing this NDM 2.2 approach, some important issues need to be addressed:

- Two nested rotary couplings (sometimes referred to as unions) must be provided for the air and oil connection, which raises questions regarding reliability and durability particularly when an attempt to retrofit an existing machine is made
- The system plumbing must be prevented from shifting and changing the spindle's residual unbalance
- The forces due to spindle rotation at high-speed machining may affect the formation of aerosol.

As the spindle speeds increase, these issues become more of a problem. To gain maximum advantage of NDM 2.2, a special computer-controlled unit must be installed on the machine and networked with the machine controller to adjust the parameters of the aerosol for each particular tool/operation. All of these factors make retrofitting of an existing machine with NDM 2.2 rather cumbersome and the economic gain becomes uncertain.

6.4.6 Cutting Tool

The next important issue to be considered is the cutting tool. The implementation of NDM requires specially designed cutting tools if the real advantages of NDM are to be achieved. Unfortunately, a common notion is that with external NDM (NDM 1.1 and NDM 1.2), the same drilling tools used for flood MWF applications will work (Woods 2006). Although this is true in some particular cases, the maximum advantage of NDM is achieved if the cutting tool is specially designed for NDM, including specific tool geometry and tool material (including coating).

Cutting tools and tool holders for NDM 2 should be designed or modified especially for NDM. The major changes for a cutting tool meant to work with internal high-pressure MWF supply are (1) tool geometry, (2) design of internal aerosol supply (often referred to as coolant) channels and their outlets (both shape and location with respect to the cutting edge(s)), and (3) design of chip flutes.

6.4.6.1 Modification to the Tool Geometry

Unfortunately, the modifications of the tool geometry suitable for NDM have never been considered in literature and in practical applications although, in the author's opinion, this issue is of prime importance. The importance stems from the known fact that the tool geometry, particularly for end tools such as drills and reamers, determines to a large extent the amount of heat generated in machining (Astakhov 2010a). The ignorance of this issue is readily explained by the lack of knowledge on the subject reflected in practical tool designs (Astakhov 2010a).

The following modification should be implemented in practical tool design for NDM:

1. The clearance angle should be increased compared to *normal* tools discussed in Chapter 4 to reduce friction between the tool flank face and workpiece. To prevent generation of unnecessary heat in NDM, the clearance angle should be increased by 30%–40% compared to its optimal value (Astakhov 2010a) established for machining with high-pressure MWF supply. Moreover, the clearance between tool body and the walls of the hole being drilled (the so-called body clearance) in drills and reamers should be increased at least two times to allow aerosol free flow around the tool body. G-drills (Figure 6.44) body clearances can be useful.
2. Back taper should be increased. As discussed in Chapter 4 (Section 4.5.9.5), back taper is one of the most important parameters of the tool geometry, which directly affects the tool life and quality of the machined surface particularly in finishing reaming. Moreover, a significant amount of heat is generated due to insufficient back taper in the improperly

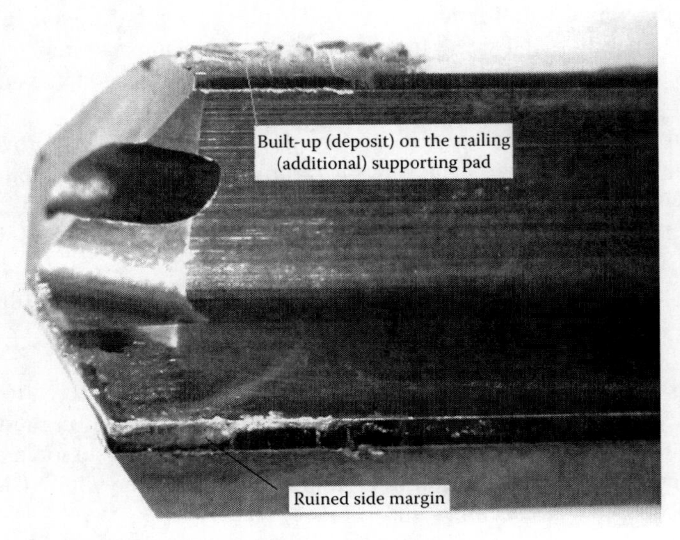

FIGURE 6.75 Consequences of lack of back taper in NDM.

designed drilling tools particularly in reamers used in the automotive industry. When high-pressure MWF supply tools are used, the heat generated due to insufficient back taper is removed by the MWF. Obviously, this is not the case in NDM where there is not enough mass of a cooling media to remove the excessive heat. This heat causes rapid expansion and then contraction of the work material that normally results in high contact pressures on tool margins. As a result, the side margin is ruined due to a high contact pressure and the buildup is formed on the additional supporting pad as shown in Figure 6.75. Therefore, the standard back taper used in drills and reamers for high-pressure MWF supply should be increased two to three times for NDM applications. To prevent tool transverse instability, four to six flute reamers should be used instead of traditional two-flute reamers.

Measures should be implemented to reduce adhesive friction between tool and work materials to its possible minimum. Thermally induced adhesion between the tool and work material (discussed in Section 2.6.8) is the principle issue in NDM particularly in high-speed machining. It causes premature tool wear and chip jamming in the machined hole as it adheres to the contact surface.

Among many directions to reduce the severity of adhesion friction, the following three should be used in the tool design for NDM:

- Selection of application-specific tool material that has minimum adhesion with the work material
- Application of a coating that reduces adhesion
- High polish of all contact and chip-removing surfaces on the tool.

Unfortunately, none of the listed measures are fully implemented in the automotive industry by one or another reason. To exemplify this statement, consider drills used in the automotive industry for drilling holes in high-silicon aluminum alloys. In such application, coarse grind of the tool contact surfaces combined with non-application-specific carbide grade are two prime issues. Coarse grind of the drill rake (Figure 6.76a) and flank (Figure 6.76b) surfaces causes aluminum buildup on these surfaces that lowers tool life (see Figure 2.60).

Surprisingly for many, an improper tool material has the same effect. The problem is in popular submicrograin carbide grades. As discussed in Chapters 2 and 3, although these grades possess a great combination of hardness and toughness, they contain 10%–12% (or even more) of cobalt as

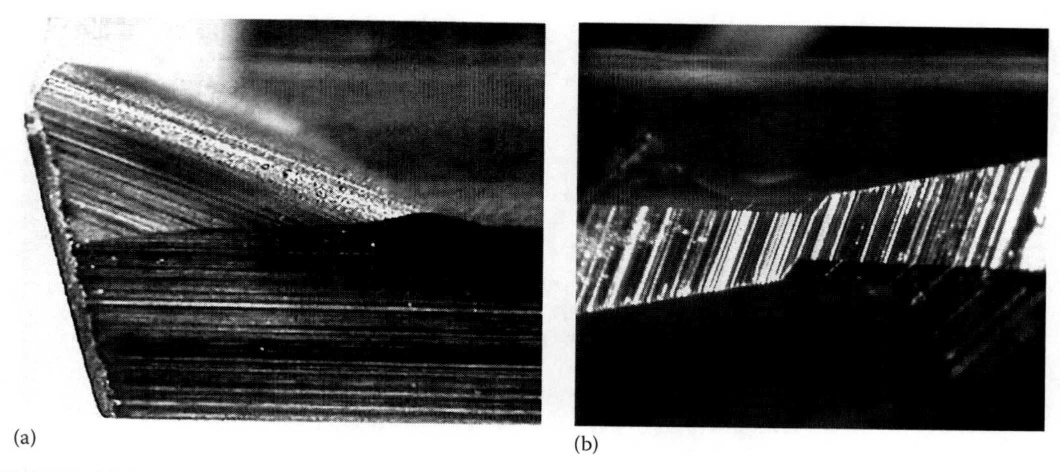

FIGURE 6.76 Coarse grind of the tool rake (a) and flank (b) surfaces.

the matrix material (binder) that holds carbide particles together. In machining of common steels, it does not present any problems so the users enjoy the mentioned advantage of such carbide grades. In NDM of automotive aluminum alloys, however, great cobalt content causes buildup of aluminum on the contact surface because the aluminum matrix adheres to cobalt not to carbide particles. It often causes the whole chip to stick to the tool rake face as shown in Figure 2.60 (Section 2.6.8) causing rapid tool failure as the chip clogs the chip removal channel. The higher the cutting speed, the greater the adhesion. Tool material selection and tool coating selection are the two primary techniques used by tool designers to reduce the occurrence of the built-up edge.

6.4.6.2 Modification to Design of Internal Aerosol Supply Channels and Their Outlets (Both Shape and Location)

End cutting tools such as drills and reamers should be designed with special channels having the maximum cross-sectional area allowable by the tool strength and extremely smooth configurations with no dead ends. A dedicated line should go to every contact (cutting, burnishing, locating, etc.) area. Void and trapped areas in the aerosol flow path should be eliminated. Exit holes and nozzles must be positioned to provide aerosol directly to the chip-forming zones, and special consideration should be given to changing position of these holes with tool re-sharpenings. In multistage reamers, the cross-sectional area of these holes should be selected to provide balanced aerosol supply to all stages or, better, balanced in proportion to the amount of work material cut by various stages. An example of channel design is shown in Figure 6.77.

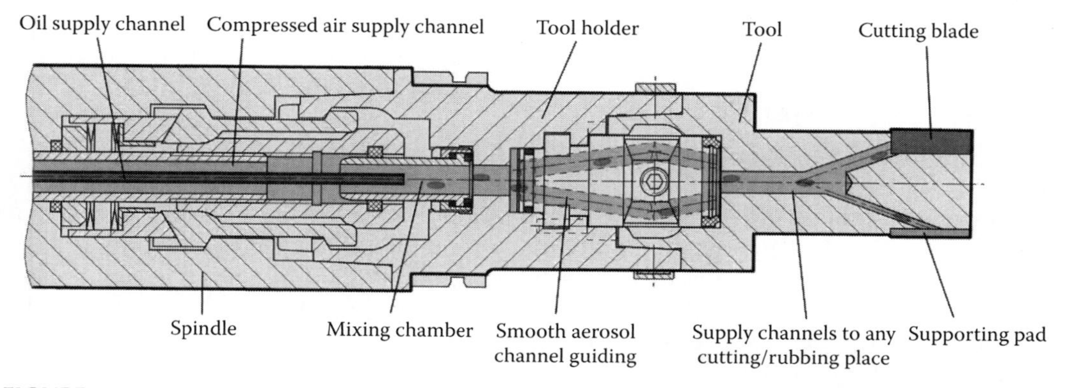

FIGURE 6.77 Reamer coolant channels designed for NDM. (Courtesy of Mapal USA Co., Port Huron, MI.)

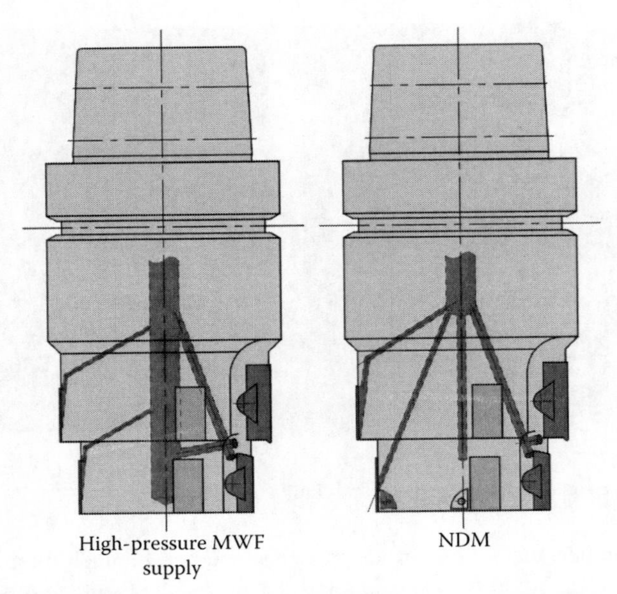

FIGURE 6.78 Comparison of internal channel designs for a two-stage high-pressure MWF supply and NDM reamers. (Courtesy of Mapal USA Co., Port Huron, MI.)

Figure 6.78 shows comparison of internal channel designs for a two-stage high-pressure MWF supply and NDM reamers. Although it seems to be clear that the coolant holes in NDM should go directly to the cutting edges or rubbing parts, such a 2D representation is not sufficient; thus, a lot of improper location of the outlets of the coolant holes in NDM are found in many cutting tools. All flaws discussed in Section 6.3.3.10 should not be tolerated because such design/manufacturing mistakes can easily ruin the whole idea of NDM.

The foregoing consideration implies that standard inexpensive tools cannot be used in NDM, as tools for NDM should be substantially re-designed or, at least, modified, which makes them special. Therefore, one should be prepared to bear greater initial tooling costs, and these costs should be a part of the analysis made prior to considering an NDM application. As these new tools are not off-the-shelf products stored by manufacturers and distributors, the lead time to deliver, modify, and repair this tool is much longer than with standard tools.

New procedures in tool presetting should also be introduced. One of the most critical of these is the verification of the aerosol flow behavior after each setting or re-setting of the cutting tool (Stoll and Furness 2006). To do this, special equipment that matches the production machinery (the same aerosol system, oil, oil flow rate, operating parameters, etc.) should be available for spray pattern testing of the tool–tool holder assembly. Naturally, this adds time and thus cost that should also be included in the cost analysis.

6.4.7 Chip Management

MWFs perform two important technological functions: (1) cooling the workpiece so that it does not get excessively hot so it not distort excessively—as a result, its proper in-process and post-process gaging as well as interoperation handling (loading/unloading and transportation) do not present problems—and (2) flushing away chips so the tool, the workpiece, the fixture, and the machine are not damaged. The challenge for NDM is to provide substitutes for these critical functions. Although since the late 1990s a number of research programs and pilot projects have been conducted to establish and verify the fundamentals of dry cutting (Klocke and Eisenblaetter 1997), it is not yet obvious whether such substitutions can be achieved economically in industrial applications.

Smooth, continuous chip evacuation is a key feature in any high-productivity machining operation, particularly when high-penetration-rate tools, that is, in the automotive industry in machining of aluminum alloys, are considered. High-pressure MWF is effective in ejecting chips from the drills, reamers, taps, etc. In NDM, air replaces the liquid, which presents a problem as air cannot produce the ejection force that a liquid at a high flow rate can exert. A feasible solution should be sought in the location of aerosol exit holes, increased and optimized chip flutes and adjusted speeds and feeds.

Even if the chips are removed from the hole being machined, they must still be removed from the tool and fixtures into a collection hopper to prevent damage to the tool, workpiece, and machine tool. To do this, special vacuum hoods have to be designed and installed. Such a hood can be a part of the machine or the tool holder. It is obvious that these hoods make tool change, tool maintenance, and setting more difficult and time-consuming. Vacuum chip-collecting hoods can also be a part of the tool assembly. Often, a turbine wheel is placed in such a hood to increase air pressure and turbulence thus to increase chip removal capability. Each tool that produces great amount of chips should be equipped with such a hood, which increases the cost of the tool and complicates its setting.

The worktable mechanism offers one approach to moving chips to the collection mechanism. In some designs, the table mechanism flips the workpiece from an upright position for loading to an upside-down position for machining. The flipping actions throws off the stray chips collected on the fixture and worktable, and the majority of these chips fall down onto a chip conveyor belt built into the base of the machine.

Another design necessity for reducing machine downtime for cleanup is to design the walls of all interior surfaces surrounding the working area to be steep enough that chips will slide down and drop onto the chip conveyor belt. As this is not always possible, some machine designs offer blowers that can blast loose chips that have adhered to the walls. Properly located blowers enable the machine to be cleaned automatically without opening the machine hood, avoiding significant downtime.

6.5 INCREASING TOOL LIFE WITH CMWF

6.5.1 Where Does It Hurt?

One of three objective of the use of MWF is tool life increase. Regardless of the selected criterion (criteria) of tool life, namely, tool wear/breakage, dimensional accuracy of machined parts, surface finish, and efficiency of the process (Astakhov 2006), tool wear is solely and specifically what causes a tool to fail. As discussed in Chapter 2, in practical machining, abrasion and adhesion types of tool wear are most common in spite of many researches talks about diffusion (Arsecularatne et al. 2006). In machining of difficult-to-machine material (e.g., titanium alloys and Inconels), the plastic lowering of the cutting edge is of prime concern in tool wear considerations (Astakhov 2004a) as discussed in Chapter 2. Therefore, it is important to determine what types of wear/failure take place in a given operation in order to select a suitable MWF that helps to reduce the wear rate.

Unfortunately, such a consideration is not of concern to specialists who select MWFs and their applications parameters (e.g., concentration, temperature, method of application). In the authors' opinion, this is the prime *sin* in the whole business of MWF. When one has a medical problem and goes to see a doctor, the doctor is expected to ask the following first professional question: *where does it hurt?* rather than starting the visit with writing a prescription. Translating this situation into the *tooling language*, an MWF application specialist/manufacturer/supplier and so on, when asked to provide a suggestion for a new MWF for a given operation or for an MWF that is going to replace the existing one to improve tool life, should ask: *what is the issue/problem the cutting tool has now—what is its current common failure mode?* If the answer involves tool wear, then the tool wear type/mechanism should be clearly identified and a suitable composition, concentration, flow rate, temperature, clearness, etc., of MWF that reduce the wear rate of a given type should be subjected and thus justified.

As discussed in Chapter 2 (Section 2.6.8) the common mechanisms of tool wear are pure abrasion and adhesion and abrasion-adhesive wear combined with the plastic lowering of the cutting edge (Astakhov 2006). As abrasion-adhesive wear combined with the plastic lowering of the cutting edge are targeting mechanisms of tool wear in HP drilling, one should realize that they are temperature dependent (Astakhov 2006). Highest tool temperatures occur at the tool–chip interface. When the rake face is heated to temperature of 900°C–1200°C, a plastic flow begins in volumes adjacent to this face. This flow takes place due to adhesion bonding between the rake face and the chip. In case of carbide tool materials, plastic deformation is greater in cobalt matrix. This plastic deformation results in tearing-off of carbide (WC and TiC) grains from *soft* cobalt layers that undergo severe plastic deformation, *plowing* this *soft* layer by inclusions contacting in the work material, and *spreading* of the tool material on the chip and workpiece contact surfaces. If the temperature increases further, a liquid layer forms between tool and workpiece due to diffusion leading to the formation of low-melting-point compound Fe_2W having melting temperature $T_m = 1130°C$. This layer is quickly removed in cutting (Makarow 1976). When plastic lowering occurs, it changes the clearance and flank cutting tool angles that leads to the further increase in the tool wear rate. Chipping and then breakage of the cutting wedge occur at the final stage. The proper application of MWF can significantly slow down plastic lowering. As such, the cooling action of MWF is the most important. Therefore, the proper selection and application of MWF for HP drilling should provide means to control this temperature.

6.5.2 How CMWF Can Reduce Cutting Temperature

As discussed by the author earlier (Astakhov 2006), MWF does not penetrate to the tool–chip interface and thus cannot provide direct cooling and lubrication action at this interface to reduce the maximum temperature at the tool–chip contact. However, the use of MWF can reduce this temperature by affecting the power spent in plastic deformation of the work material, which can be as high as 60%–80% of the total power required by the cutting system for its existence. This power per unit volume of the work material being removed can be, to the first approximation, estimated as (Astakhov 1998/1999)

$$\text{Power per unit volume} = \frac{1}{2}\sigma_{UTS}\varepsilon_f \tag{6.78}$$

where

σ_{UTS} is the ultimate strength
ε_f is the strain at fracture of the work material

For austenitic stainless steels, σ_{UTS} is normally lower than that of medium and high carbon plain and alloy steels while ε_f is significantly greater. This explains much higher temperatures and lower tool life in machining of austenitic stainless steels as more power is required for their plastic deformation in machining. For a large group of difficult-to-machine materials, both σ_{UTS} and ε_f are greater than those of medium and high carbon plain and alloy steels, which make them difficult-to-machine as more power is required for their plastic deformation in transforming the layer being removed into the chip. To reduce power required for plastic deformation of the work material, one can reduce σ_{UTS} or ε_f or both.

The reduction of the ultimate strength σ_{UTS} is effective for high-strength work materials, for example, Inconel 718. To achieve this, the so-called thermally enhanced machining (TEM) that includes plasma-enhanced machining (PEM) is used (Sun et al. 2010). TEM uses external heat sources to heat and softens the workpiece locally in front of the cutting tool. As temperature rises, the strength and work hardening of the work material reduce while the strain at fracture increases not to the same proportion. The net effect is the reduction of the work of plastic deformation and thus improved tool life.

The reduction of strain at fracture ε_f is normally achieved with the use of MWF due to the so-called Rebinder effect (Astakhov 2006). Our current understanding of the Rebinder effect is that the

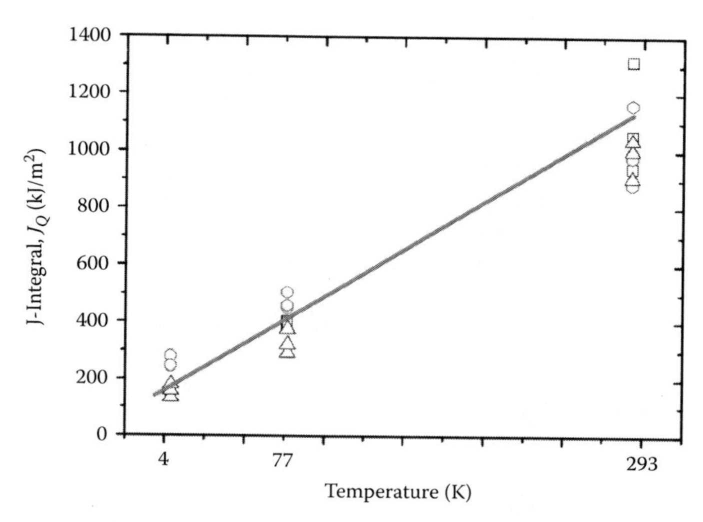

FIGURE 6.79 Fracture toughness of AISI steel 316L as a function of its temperature.

alternation of the mechanical and physical properties of materials is due to the influence of various physiochemical processes on the surface energy (Rebinder 1979). The surface energy reduces due to the penetration of the foreign atoms (from MWF decomposition), then the energy required for fracture reduces.

6.5.3 Why It Works

Despite the discussed shortcomings, the use of LN2 in metal cutting is beneficial in certain applications (Yildiz and Nalbant 2008), while it does not affect the process or make it worthwhile for others. There are a number of particular machining situations that should be considered on a case-by-case basis.

LN2 application as MWF was found beneficial in machining of some grades of stainless steel. Consider steel ANSI 316L. The ultimate strength σ_{UTS} for this steel at 273 K (0°C) is 653 MPa, and its yield strength at this temperature is 329 MPa. At 77 K (the temperature of LN2), the ultimate strength σ_{UTS} becomes 1532 MPa, and the yield strength becomes 726 MPa (Sa et al. 2005). According to the prevailing metal cutting theory where the so-called flow shear strength is considered as the only work material property (Astakhov 2010a), the cutting force should increase at least two-folds if such a steel is cut at 77 K compared to cutting at 273 K. As a result, tool life should be lower, the opposite is actually the case in reality.

The prime reason for that is that the fracture toughness of this work material reduces markedly at 77 K as shown in Figure 6.79. Therefore, much less energy should be spent for its purposeful fracture in cutting (Astakhov 2010a) that increases tool life. The reason for the reduction of fracture toughness is explained (Sa et al. 2005) as to be due to a solid phase transformation in this steel at low temperature, that is, austenite having great fracture toughness transforms into martensite having much lower fracture toughness.

6.6 APPLICATION OF CMWFs

6.6.1 Basics

As pointed out by Ghosh (2006), cryogenic MWFs (CMWFs) have been used in industrial quenching and cold treating applications for quite some time, but their application in the machining industry as coolants is relatively new. However, the history of cryogenic machining research traces back more than 50 years. One of the earliest known applications of cryogenic machining involved the application of liquid carbon dioxide in machining in the early 1950s. Use of liquid nitrogen as a coolant in machining was first documented in the mid-1960s. However, issues related to economical machining with cryogenic fluids, variable material behavior at cryogenic temperatures and lack of efficient delivery systems prevented industrial adoption and caused the level of research to drop off significantly in the seventies and early eighties. Since then, the need for more efficient machining of difficult-to-machine materials as well as the desire to eliminate the use of oil or water-based cutting fluids has prompted renewed interest from several research groups and industries.

Cryogenic machining involves the application of a cryogenic fluid in the machining process, primarily as MWF. This fluid is applied as an external spray through a nozzle to perform both conductive and convective cooling of the machining zone or could be applied indirectly to cool the cutting tool through conduction alone. Generally the term *cryogenic* refers to fluids that have boiling point lower than –150°C (–238°F). By that definition, cryogenic fluids include liquefied gases of air, nitrogen, argon, oxygen, hydrogen, and helium. Liquid nitrogen (commonly referred to as LN2 in the literature) has been the most widely used cryogenic coolant in machining operations, mainly because of its inert behavior and due to its wide availability, established handling procedures, and relatively low cost. However, there have been other efforts involving liquid carbon dioxide, cold gases, and solid/gas mixtures as coolants or cutting medium that are not technically cryogenic but considered so in metal cutting publications for the sake of simplicity.

There is a principle problem in the evaluation of the results of using cryogenic liquids as MWF in metal cutting because these liquids affect not only the machining process but also the properties of the work and tool materials as discussed in Chapter 3. So it is difficult to estimate the net effect of a cryogenic fluid used as MWF.

6.6.1.1 Work Material

As pointed out in Chapter 3, the thermal treatment of metals must certainly be regarded as one of the most important developments of the industrial age. After more than a century, research continues into making metallic components stronger and more wear-resistant. One of the more modern processes being used to treat metals (as well as other materials) is cryogenic tempering.

It is known that cryogenic treatment provides for three documented transformations in metals. First, in heat-treated steels, retained austenite is transformed to martensite, creating a more uniform grain structure and homogeneous steel. This provides for a tougher and more durable material as the voids and weaknesses of an irregular grain (or crystal) structure are eliminated. This mechanism also provides for better thermal properties and better heat dissipation in cryogenically treated steels. Additionally, it is this mechanism that leads to friction reducing qualities in metals, especially when final machining, polishing, grinding, or honing are done after deep cryogenic treatment. It is also why cryogenically treated steels show more uniform hardness than nontreated steels.

The second mechanism relates to modification in the carbon microstructure of cryogenically treated steels. Before and after treatment micrographs show the formation of carbides within the steels. The technical description of this is called *the precipitation of eta-carbides.*

The third mechanism relates to stress relief. It is based on Einstein's (and the German physicist Bose's) observation that matter is at its most relaxed state when it has the least amount of molecular activity or kinetic energy. Freezing a component/workpiece/part to cryogenic temperatures removes heat and thus reduces the molecular activity in the metal. This *relaxes* the metal and reduces residual stresses in it. These hidden stresses propagate when the part is placed into service and cause failures due to fatigue. Hence, by reducing residual stresses, one greatly reduces or eliminates failures due to cracking or what people term *metal fatigue.*

6.6.1.2 Tool Material

The particularities of cryogenic heat treatment of tool materials are discussed in Chapter 3 (Sections 3.3.7.3 and 3.4.8) and in Chapter 2 (Section 2.6.13.2). It generally improves tool life of various tool materials although the mechanism of such an improvement is different for each type of tool materials. Moreover, the result varies from one batch of a given tool material to the next due to variations of manufacturing parameters of this material.

The simplest mechanism is for HSSs. When a tool made of this steel is exposed to cryogenic tempering, the retained austenite is transformed to martensite. As a result, the hardness and wear resistance of the tool increase. Considerably different mechanism takes place when composite tool materials are treated. Such tool materials include distinctive solid phases in their structure. The higher material performance, the greater the difference in physical and mechanical properties of these phases. This is particularly true for tool materials where extremely hard nonmetallic particles (WC, diamond, boron nitride Al_2O_3, etc.) are held together by the matrix material (e.g., cobalt). As these materials are shaped, great deformation and temperatures are applied. As there are great differences in the behavior of the metallic and nonmetallic solid phases under these severe conditions, interfacial defects such as microcracks, residual stresses, and even defects of the atomic structure occur (Komarovsky and Astakhov 2002). These defects lower the strength, wear resistance, and other useful mechanical and physical properties of tool materials. Moreover, as the population of these defects is of random nature, the great scatter in the performance data is observed in the practice. For example, even when the cutting tool is of high quality and the machining system fully supports its performance, the scatter of the tool life can reach 40% due to microdefects in the tool material (Astakhov 2006).

The essence of cryogenic tempering is the healing microdefects and strengthening of interfacial bounds that enhances the physical and mechanical properties of materials. This healing is achieved by introducing a high-density energy flux (the rate of transfer of energy through a unit area [$J \cdot m^2/s$]) through the entire cross section of the material.

6.6.2 Commonly Accepted Rationale behind the Use of CMWF

The common perception and rationale provided in promotion of CMWF supply is rather naïve: high temperature in metal cutting is *logically* chilled by CMWF having extremely low temperature. As such, not much attention is paid to physics, that is, heat capacity of CMWF, thermoconductivity, and other *relevant* physical characteristics. The temperature difference between metal cutting and CMWF does its *magic* blinding many specialists in the field. The setup for cryomachining demonstration looks so *cool* that spectators intuitively feel that they observe *the cutting edge* technology of the future or are even present in a launch of a space craft to far away constellations.

In reality, however, LN2 is not a particularly good coolant. It has a low temperature, but it also has a low heat capacity, low thermoconductivity, and low heat of vaporization. Nitrogen gas is also known to be a good thermoinsulator.

Specific heat capacity (often shortened to specific heat) is the amount of heat or thermal energy required to increase the temperature of a certain quantity of a substance by one unit of measure. For example, at a temperature of 15°C, the heat required to raise the temperature of a water sample by 1°C is 4.186 J/g. In other words, water has heat capacity of 4.186 kJ/(kg · K). LN2, however, has heat capacity of 2.04 kJ/(kg · K), that is, two times smaller. Moreover, gas nitrogen has heat capacity of 1.04 kJ/(kg · K), that is, four times smaller than that of water. Because in machining, LN2 evaporates, the overall heat capacity of mixture LN2+ nitrogen gas is approximately threefold lower than that of water, that is, to remove the same amount of heat, the mass flow rate of mixture LN2+ nitrogen gas should be three times greater than that with water. Obviously, this is not the case in metal cutting.

Thermoconductivity is the amount of heat a particular substance can carry through it in unit time. Usually expressed in W/(m · K), the units represent how many Watts of heat can be conducted through a 1 m thickness of said material with a 1 K temperature difference between the two ends. The thermoconductivity of water is 0.67 W/(m · K) while that of LN2 is 0.14 W/(m · K), that is, more than four times lower.

The amount of energy released or absorbed by any substance during a phase transition is called the latent heat. If heat is continuously added, changes of phase from solid to liquid and then from

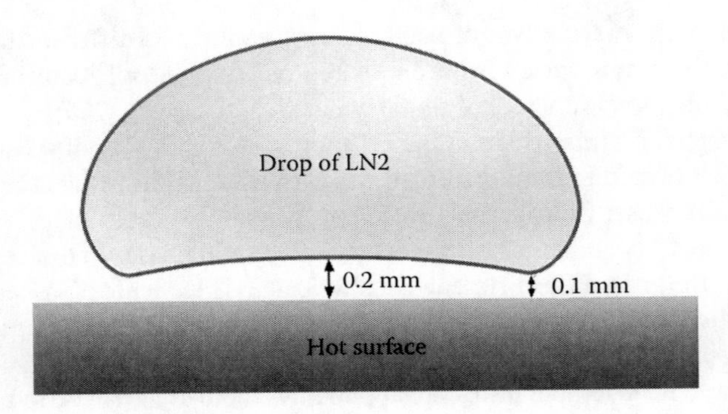

FIGURE 6.80 Visualization of the Leidenfrost effect.

liquid to vapor occur. These changes are called phase transitions. The latent heat absorbed during the liquid–vapor transition is called the latent heat of vaporization. This energy overcomes the intermolecular forces inside the liquid. The latent heat of vaporization of LN2 is around 200 kJ/kg, that is, very low.

As a common example used to demonstrate poor cooling ability of LN2, consider what happens if one tries to chill 10 L of magma with 10 L of LN2. Accounting for specific gravity, one finds that 10 L of LN2 has a mass of around 8 kg. The latent heat of vaporization of LN2 is around 200 kJ/kg. So the vaporization of 10 L of LN2 would take 1600 kJ of energy. The heat capacity of magma is 1.7 kJ/(kg·K). As 10 L of magma would have a mass of 25 kg, there is around 40 kJ/K in 10 L of magma. If it is assumed that the magma starts out at 1500 K, it would take a 40 K cooling for the magma to deliver 1600 kJ of energy to the LN2, leaving the magma at toasty 1460 K, which is a very small change in the overall temperature.

The LN2 would evaporate in a huge cloud, and the magma would be relatively untouched. The reason is that the latent heat of vaporization of LN2 is small. One would be much better off if he or she uses water instead of LN2 because water has a much higher latent heat of vaporization, over an order of magnitude larger than LN2.

The use of LN2 as MWF in metal cutting invariably results in a chilling rate less than the theoretical potential discussed in the previously considered example. The shortfall is attributed to the well-known Leidenfrost effect. The Leidenfrost effect is a phenomenon in which a liquid, in near contact with a mass significantly hotter than the liquid's boiling point, produces an insulating vapor layer, which keeps that liquid from boiling rapidly (Figure 6.80). This is most commonly seen when cooking; one sprinkles drops of water in a skillet to gage its temperature—if the skillet's temperature is at or above the Leidenfrost point, the water skitters across the metal and takes longer to evaporate than it would in a skillet that is above boiling temperature but below the temperature of the Leidenfrost point.

6.7 MWF ESSENTIAL PARAMETERS TO BE MAINTAINED IN HP DRILLING

For companies to realize the full potential of HP drilling, a strict maintenance of MWF is needed as the MWF parameters deterioration can easy turn successful HP drilling operations into a disaster in terms of efficiency/productivity. In other words, the current practice of ignoring even the basic aspects of the proper MWF maintenance in many machine shops cannot be tolerated in HP drilling operations. The problem is that many machine shop specialists do not clearly understand the requirements to MWF in HP drilling operations including drilling, reaming, boring, and tapping. Even in large plants, for example, in the automotive industry where the process efficiency is closely monitored, the MWF aspect in this efficiency is almost totally ignored while the cutting tool supplier/cutting tool manager companies are held responsible for the process efficiency. Commonly, the MWF maintenance business is outsourced so the objective to the outside MWF commodity company responsible for MWF maintenance is to reduce the MWF supply/maintenance costs while not correlating this objective with the efficiency of machining operations. Moreover, among the parameters of the MWF/coolant maintenance data supplied to the plant management by this company, the MWF concentration (as the target set by the MWF supplier or set by the corporate internal standard) and bacteria (mold) as set by safety requirements are of concern while other parameters may not be controlled or requested for consideration. In the author's opinion, it is one of the major hurdles in the implementation of HP drilling and other unattended machining operations in modern manufacturing world.

This section does not aim to cover the subject of MWF maintenance thoroughly as it is out of the scope of the book. Rather, this section aims to raise awareness of the problem pointing out the major MWF parameters to be controlled in HP drilling and their proper reporting. It argues that even the basic maintenance practices can greatly improve an MWF's longevity, the shop environment, and the overall level of part quality.

6.7.1 Concentration

All MWFs are formulated to work within a specific concentration range. These working concentration limits are determined and recommended by the manufacturer for optimal fluid performance. All fluids have a *minimum concentration* level, which must be adhered to maintain bio-stability, good corrosion protection, and cutting performance. If this concentration level is allowed to drop below the minimum level for any given length of time, various problems can arise. Bio-instability and biomass development, low pH, corrosion of ferrous alloys (red rust), emulsion splitting, poor cutting performance, and ultimately the metal removal fluid disposal are some problems that can occur.

MWF concentration should be set by the MWF control plan and checked on a daily basis by means of an optical device called a refractometer. There are various types and styles of refractometers available for purchase, ranging from basic handheld optical refractometers to more sophisticated digital units. The best type of refractometer is one that is used regularly in the shop, as concentration is the most critical factor for optimal tool life. The readings (the daily log) should be posted on the quality board for each department. Once per week, the refractometer reading should be verified by the more accurate laboratory analysis.

6.7.2 Water Quality

Good water chemistry is essential for the long life and proper performance of MWFs. Sufficient volumes of good quality water must be available on demand for charging systems and for restoring the makeup of a solution. It may be necessary to install storage tanks to maintain adequate water quantities to meet all production requirements. Tap water is readily available, but it comes from a variety of sources. It is often used as the initial source for MWF systems. Tap water chemistry can vary widely depending upon its original source, geographical location, and pretreatment conducted on site or by the local water authority.

Aqueous chemical and physical properties such as pH, conductivity, alkalinity, total hardness (calcium and magnesium levels), other ion and elemental levels, surface tension, turbidity, ECA (electrokinetic charge), foam characteristics, and microbiological levels (bacteria, fungi, yeast, and algae) may all influence MWF performance. These properties can affect corrosion protection of MWF, residue properties, foam, emulsion stability for semisynthetics and soluble oils, susceptibility to microbiological attack, charge density of both true solutions and emulsions, filtering properties, and wetting. Ion analysis inductively coupled plasma-atomic emission spectroscopy [ICP-AES] should be used to monitor process water regularly to establish a baseline because water quality can change dramatically during the year as seasonal precipitation patterns change. The parameters that should be evaluated continuously are pH and hardness.

The pH of water indicates whether it is acidic (pH of 0.0–7.0), neutral (pH of 7.0), or basic or alkaline (pH of 7.0–14.0). Most water used commercially in the United States exhibits a pH range from 6.4 to 8.5, depending upon the original source and type of pretreatment conducted by the local water authority. Water used for MWFs should exhibit an optimal pH range from 7.0 to 8.5. It should be checked on a daily basis using a pH tape, and the results should be posted on the quality board.

Total hardness indicates the presence of dissolved minerals and their salts in water. Predominant ions are calcium and magnesium. Other ions contributing to hardness include iron, zinc, aluminum, potassium, and silicon. Total hardness is reported in parts per million (ppm) of calcium carbonate ($CaCO_3$). It can also be reported in units called grains. One grain of hardness is equivalent to 17 ppm of calcium carbonate. Water hardness is typically defined using the following scale: 0–49 ppm is very soft, 50–124 ppm is soft, 125–249 ppm is medium, 250–369 ppm is hard, and 370 ppm and above is very hard.

Hardness can readily affect MWF performance. Soft water may degrade the performance of all fluids by promoting the formation of foam. This condition is especially likely when using synthetics in grinding operations and semisynthetics and soluble oils in both machining and grinding applications. Foam can drastically impair fluid performance by contributing to poor wetting and coverage properties, which diminish lubricating, cooling, and proper film coverage for in-process corrosion protection. Foam can also hamper fluid detergency, making it more difficult to handle and filter swarf.

When dense foam forms, it can lessen the filtering capabilities of a system by interfering with indexing mechanisms and by creating poor filter beds. Foam can also suspend tramp oils, preventing skimmers and other mechanical devices from removing them effectively. Tramp oils can act as a matrix, becoming finely suspended on a dense bed of foam. This development further intensifies a dense foam layer. Excessive foam can also lead to housekeeping issues by causing system barges and return lines, such as floor troughs, to overflow. Foam can also cause pump cavitation, creating excessive wear and premature mechanical failure.

As hardness increases, it can adversely affect the stability of semisynthetic and soluble oil emulsions. The formation of hard water soaps from calcium and magnesium ions and anionic components (typically fatty acid-based emulsifiers) can radically alter emulsion particle size. This development will rapidly lead to scum formation and lose emulsions in which cream and free oil are present. Both semisynthetics and soluble oils are harmed by calcium soaps. Semisynthetics are especially hampered by high magnesium levels.

Hardness can build up in water stored for use as well as in metalworking fluid systems as a result of evaporation. Many systems can loose from 5% to 25% of their water on a daily basis, depending upon system size, openness to plant environment, time of year, geographical location, plant conditions (air temperature and circulation patterns), and MWF temperature as the water circulates. Adding hard water to make up for loss resulting from evaporation will cause levels of hardness in the system to rise rapidly. As a result, emulsion instability in semisynthetic and soluble oils will lead to corrosion problems, susceptibility to emulsification of tramp oils and microbiological attack, poor tool life, inferior surface finish, foam, and filtering problems. The optimal hardness level is in the range of 125–200 ppm for HP drilling. It must be maintained by using the appropriate blend of water types for proper MWF performance.

6.7.3 MWF Filtration

According to the author's experience, filtration of MWF is the least understood and thus the most neglected aspect of MWF implementation. Although there is always a great price to pay for such ignorance, in high-speed high-efficiency machining, as a whole, and in HP drilling, in particular, this price is enormous. Surprisingly, there is no drive to resolve the long-standing problem once and forever at national level; levels of a particular industry, that is, the automotive industry; or the manufacturing plant level. Manufacturing specialists completely rely on the MWF suppliers' expertise often hiring them as their MWF commodity managers in hope that they know what they are doing and thus is acting to the best interest of the high-efficiency production.

The whole consideration of filtration begins with answering the following question: why is good MWF filtration needed? There are many benefits to proper filtration of MWFs. Metal particles removed during the grinding or machining process can quickly build up in the fluid causing costly damage to the parts and the machine tool. The common term for this dirt is *swarf*, and it must be removed quickly from MWF by proper filtration. The proper application of MWF (1) improves product finish and reduces scrap, (2) increases tool life, (3) increases the life of MWF reducing disposal cost, (4) reduces machine tool wear and downtime for repairs, and (5) improves the health and safety of the operator.

The second important question to be answered is as follows: what level of MWF filtration do I need? The answer to this question depends whom do you talk to. The general rule established by MWF providers for selecting the level of filtration or particle size in microns can be represented as follows (MicroTech Filtration): machining operations such as turning, milling, and drilling require 30–40 μm filtration. Semifinish grinding requires 15–20 μm filtration. Finish grinding with close tolerances and high requirement to surface roughness requires 5–10 μm filtration. Ultra-precision components machining such as grinding superhard materials like carbide, glass, and ceramic or honing, EDMing, and superfinishing of hardened materials requires 1–5 μm filtration.

Intelligent MWF users have completely different opinion on the choice of filtration level gained through extensive many-year experience. For example, Carbide Processor Inc. suggests that MWF should be filtered to a particle level that is 10% of the tightest desired tolerance (Carbide Processor Inc.). The rationale behind this suggestion is simple as shown in Figure 6.81. Particles in the coolant get between the tool and the machined surface during machining. The tool material (e.g., sintered carbide) is hard, but it chips. As the tool edge rotates into the work, it traps a particle. The hard particle then causes micro-fracturing of the edge of the tool. This is like chipping obsidian with deer antlers to make a knife or like cutting diamonds. A large amount of force is concentrated in a very small area and the tool material fractures. A chipped edge requires more energy to cut so its tool life reduces and such an edge makes rougher cuts. A significant part of the reason carbide becomes dull is micro-fracturing and microchipping. Therefore, for successful HP drilling, MWF should be filtered to the size of the little circle (it reality it should be 2–8 μm). The second smallest particle (10 μm) is more than enough to deflect the tool.

The major problem not realized by many shop specialists in drilling is actually in the swarf composition. When metal is being removed, the swarf consisting of the chip and particles of micron and submicron metal and nonmetal fines are produced. Commonly, books on MWF address the filtration of the chip formed (e.g., Byers 2006) while the analysis of fines in terms of their origin, size, and properties including hardness is not considered. Although such books include chapters on metal cutting process, operations, and tool wear, the covered subjects are too far from reality so that the influence of MWF clearness on tool performance is not mentioned.

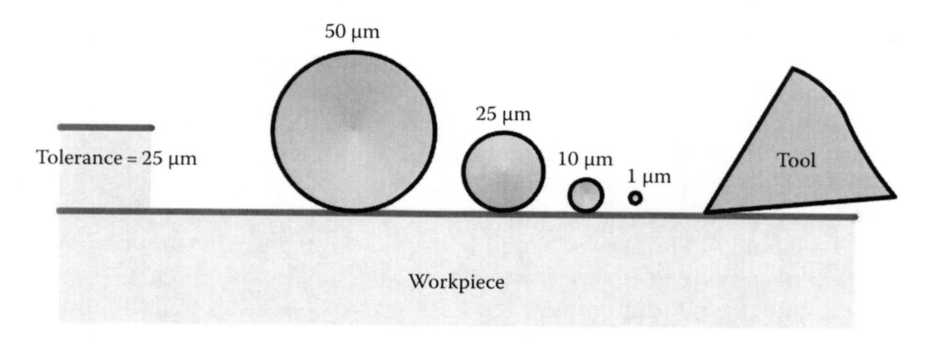

FIGURE 6.81 Particle size versus tolerance.

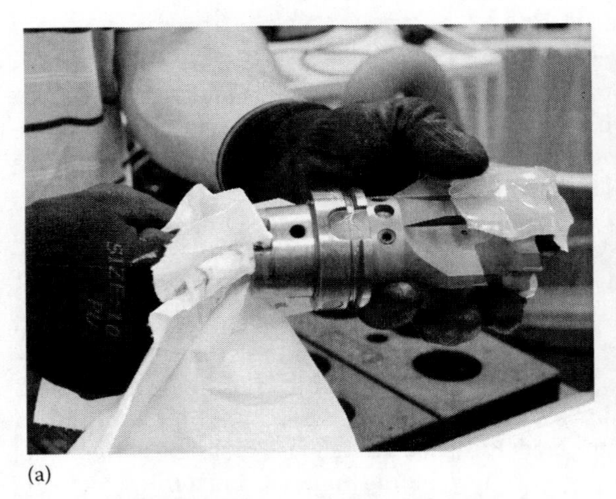

(a) (b)

FIGURE 6.82 Showing a paper towel after hand cleaning of a PCD reamer removed from the machine (a) and (b) the paper towel after cleaning.

The discussed fines are actually micron and submicron particles of the tool material. The grain size of modern carbide hardly exceeds 1 μm so very hard particles of tungsten carbide (WC) of this size are found in MWF as a result of tool wear (see Chapters 2 and 3). The grain size of modern PCD for hole-making operations is in the range of 2–10 μm so very hard particles of diamond of this size are found in MWF as a result of tool wear (see Chapters 2 and 3). Figure 6.82 shows a paper towel after hand cleaning of a PCD reamer removed from the machine. As can be seen, the tool is actually *coated* by diamond fines. If these hard fines are not addressed by means of filtration, they will be continually picked up by the coolant pumps and reintroduced onto the cutting zone, thus reducing overall tool life and resulting in the deterioration of surface roughness. The problem is that 20–30 μm filters are common even in the advanced manufacturing facilities. The argument that MWF commodity management companies put forward in any meeting on the matter is as simple as this: finer filters would require often change of clogged filter cartridges so it may increase the machine downtime and overall cost of the MWF management program. Surprisingly, no one from manufacturing asks two simple questions:

1. What is the law of the universe that forbids installing two parallel filters so that no interruption in production would take place when one filter is serviced?
2. What is the ratio of the saving on filter cartridges and losses due to poor part quality, low tool life, wear of expensive machines, etc.?

These small fines can take hours to settle out of the fluid. When they eventually do settle out, they produce a very dense mud on the bottom of the coolant sump. This mud tends to accumulate in stagnant areas, such as in the corners of the coolant sump and dead zones. This mud can build up to such a high degree that the bulk fluid and its additives (microbicides) are sealed off to the bottom layers. This, in turn, allows anaerobic bacteria and biofilms to form and create an anoxic (no oxygen) environment in the sump.

Good filtration equals longer MWF life. In the past, the high cost of filtration media and equipment was prohibitive for most machine shops to realize a cost–benefit. Now because of the high prices of MWFs and the even higher costs of their disposal, all shops should be implementing the proper MWF filtration. The level of filtration depends on the machining operation, the part tolerance, and the required level of surface roughness. Filtration as low as 5 μm may be required in order to carry out HP drilling efficiently. Mechanical filtration removes contaminants by forcing MWF through a filter medium with holes smaller than the contaminants. Mechanical filters with fine filtration media can remove particles as small as 1 μm, but filtration under 5 μm is not recommended because many of the MWF additives will be removed as they are not fine blended. A typical mechanical filter for MWF used in HP drilling would use a 6–10 μm filter cartridge in the so-called absolute rating. Absolute rating means that no particles greater than a certain size will pass through the filter and is based on the maximum pore size of the filtering medium.

The clearance of MWF is characterized not only by the size of the particles in it but primarily by the number of particles of a given size in the unit of MWF volume. This clearance is reported as the required cleanliness rating in accordance with ISO standards. ISO 4406 is an internationally recognized standard that expresses the level of particulate contamination of MWF. The standard is also used to specify the required cleanliness level for MWF systems. ISO 4406 cleanliness rating system is based on a number of contamination particles larger than 2, 5, and 15 μm in a 1 mL MWF sample. Once the number and size of the particles are determined, the points are plotted on a standardized chart of ISO range numbers to convert the particle counts into an ISO 4406 rating. The ISO 4406 rating provides three range numbers that are separated by a slash, such as 16/14/12. In this rating example, the first number 16 corresponds to the number of particles greater than 2 μm in size; the second number 14 corresponds to the number of particles greater than 5 μm in size; and the third number 12 corresponds to the number of particles greater than 15 μm in size. All three values for applicable range numbers can be determined through the use of the ISO 4406 standardized chart based on the actual number of particles counted within the 1 mL sample for each size category (>2, >5, >15 μm). For example, if a 1 mL sample contained 6000 2 μm particles, 140 5 μm particles, and 28 15 μm particles, the fluid would have a cleanliness rating of 20/14/12. The number of 2 μm particles (6,000) falls in the range greater than 5,000 but less than 10,000, which results in an ISO 4406 range number of 20. The number of 5 μm particles (140) falls in the range greater than 80 but less than 160, which results in an ISO 4406 range number of 14. The number of 15 μm particles (28) falls in the range greater than 20 but less than 40, which results in an ISO 4406 range number of 12.

ISO Contaminations System Code 19/16/13 is the goal in HP high-speed drilling while 20/18/14 should be considered as the goal for current MWF supply systems. Portable particle counters are available for in-house testing. The advantage is that measurement results can be obtained quickly, as opposed to the 1–2-week waiting period typical for laboratory testing when a sample of MWF is sent to an external company (lab) for the analysis.

In order for companies to realize the most return out of their machine tools and the increasingly expensive cutting tools they use, they must change their thinking about MWFs in general. MWFs account for around 8%–12% of the direct manufacturing cost of the part being manufactured but can impact the overall manufacturing costs of the part by as much as 95%. Instead of seeing MWF as a necessary evil, it should be viewed as an asset. MWFs are a liquid tool and should be treated as such. Taking proper care of them will increase a company's productivity and profitability. Choose to neglect them, and it can cost the business dearly. Therefore, the decision on establishing the proper MWF maintenance program should not be difficult to make.

REFERENCES

Aoyama, T. 2002. Development of a mixture supply system for machining with minimal quantity lubrication. *Annals of the CIRP* 51(1):289–292.

Arsecularatne, J.A., Shang, L.S., and Montross, C. 2006. Wear and tool life of tungsten carbide, PCBN and PCD cutting tools. *International Journal of Machine Tools and Manufacture* 45:482–491.

Astakhov, V.P. 1998/1999. *Metal Cutting Mechanics*. Boca Raton, FL: CRC Press.

Astakhov, V.P. 2004. The assessment of cutting tool wear. *International Journal of Machine Tools and Manufacture* 44:637–647.

Astakhov, V.P. 2006. *Tribology of Metal Cutting*. London, U.K.: Elsevier.

Astakhov, V.P. 2008. Ecological machining: Near-dry machining. In *Machining Fundamentals and Recent Advances*, Davim, P.J. (ed.). London, U.K.: Springer, pp. 195–223.

Astakhov, V.P. 2010a. *Geometry of Single-Point Turning Tools and Drills: Fundamentals and Practical Applications*. London, U.K.: Springer.

Astakhov, V.P. 2010b. Metal cutting theory foundations of near-dry (MQL) machining. *International Journal of Machining and Machinability of Materials* 7(1/2):1–16.

Astakhov, V.P., Frazao, J., and Osman, M.O.M. 1991. On the design of deep-hole drills with non-traditional ejectors. *International Journal of Production Research* 28(11):2297–2311.

Astakhov, V.P., Farazao, J., and Osman, M.O.M. 1994a. Effective tool geometry for uniform pressure distribution in single edge gundrilling. *ASME Journal of Engineering for Industry* 116(4):449–456.

Astakhov, V.P., Frazao, J., and Osman M.O.M. 1994b. On the experimental optimization of tool geometry for uniform pressure distribution in single edge gundrilling. *ASME Journal of Engineering for Industry* 118:449–456.

Astakhov, V.P., Galitsky, V.V., and Osman, M.O.M. 1995. A novel approach to the design of self-piloting drills with external chip removal, Part 2: Bottom clearance topology and experimental results. *ASME Journal of Engineering for Industry* 117:464–474.

Astakhov, V.P. and Joksch, S. 2012. *Metal Working Fluids for Cutting and Grinding: Fundamentals and Recent Advances*. London, U.K.: Woodhead.

Astakhov, V.P. and Osman, M.O.M. 1996. An the improvement of tool life in self-piloting drilling with external chip removal. *Journal of Engineering Manufacture, Part B, Proceeding of the IMechE* 210:243–250.

Astakhov, V.P., Subramanya, P.S., and Osman, M.O.M. 1996. On the design of ejectors for deep hole machining. *International Journal of Machine Tools and Manufacture* 36(2):155–171.

Bhave, P.R. and Gupta, R. 2006. *Analysis of Water Distribution Networks*. Oxford, U.K.: Alpha Science.

Brinksmeier, E., Walter, A., Janssen, R., and Diersen, P. 1999. Aspects of cooling application reduction in machining advanced materials. *Proceedings of the Institution of Mechanical Engineers* 213:769–778.

Byers, J.P. 1994. *Metalworking Fluids*. New York: Marcel Dekker.

Byers, J.P. 2006. *Metalworking Fluids*, 2nd edn. Boca Raton, FL: CRC Press.

Carbide Processor Inc. Available from http://www.carbideprocessors.com/pages/machine-coolant/how-fine-to-filter-coolant.html.

Crafoord, R., Kaminski, J., Lagerberg, S., Ljungkrona, O., and Wretland, A. 1999. Chip control in tube turning using a high-pressure water jet. *Proceedings of the Institution of Mechanical Engineers, Part B* 213:761–767.

CTE Staff Members. 2001. Farewell to BUE. *Cutting Tool Engineering* 53(2):43–47.

Das, S.K. 1993. Water flow through helical coils in turbulent condition. *The Canadian Journal of Chemical Engineering* 71:971–973.

Dasch, J.M. and Kurgin, S.K. 2010. A characterisation of mist generated from minimum quantity lubrication (MQL) compared to wet machining. *International Journal of Machining and Machinability of Materials* 7(1/2):82–95.

Entelis, S.G. and Berlinder, E.M. 1986. *Cooling-Lubricating Media for Metal Cutting* (in Russian). Moscow, Russia: Machinostroenie.

Feng, F.S. and Hattori, M. 2000. Process cost and information modeling for dry machining. NIST Publication. http://www.mel.nist.gov/msidlibrary/doc/cost_process.pdf. (Accessed May 17, 2008.)

Fox, R.W. and McDonald, A.T. 1985. *Introduction to Fluid Mechanics*. New York: John Wiley & Sons.

Ghosh, R. 2006. Technology assessment on current advanced research projects in cryogenic machining. McLean, VA: The Association for Manufacturing Technology.

Graham, D. 2000. Dry out, *Cutting Tool Engineering* 52:1–8.

Grundman, R. 1985. Friction diagram of the helically coiled tube. *Chemical Engineering and Processing* 19:113–115.

Hart, J., Ellenberger, J., and Hamersma, P.J. 1988. Single-and two-phase flow through helically coiled tubes. *Chemical Engineering Science* 43:775–783.

Harvey, P.H., Nandapurkar, S.S., and Holland, F.A. 1971. Friction factors for a tube rotating about its own axis. *The Canadian Journal of Chemical Engineering* 49:207–209.

Itoigawa, F., Childs, T.H.C., Nakamura, T., and Belluco, W. 2006. Effects and mechanisms in minimal quantity lubrication machining of an aluminum alloy. *Wear* 260(3):339–344.

Ju, H., Huang, Z., Xu, Y., Duan, B., and Yu, Y. 2001. Hydraulic performance of small bending radius helical coil-pipe. *Journal of Nuclear Science and Technology* 38(10):826–831.

Klocke, F. and Eisenblaetter, G. 1997. Dry cutting. *Annals of the CIRP* 46(2):519–526.

Komarovsky, A.A. and Astakhov, V.P. 2002. *Physics of Strength and Fracture Control: Fundamentals of the Adaptation of Engineering Materials and Structures*. Boca Raton, FL: CRC Press.

Kwon, H.J. 2008. Head loss coefficient regarding backflow preventer for transient flow. *KSCE Journal of Civil Engineering* 12(3):205–211.

Li, X. 1996. Study of the jet-flow rate of cooling in machining. Part 1. Theoretical analysis. *Journal of Materials Processing Technology* 62:149–156.

Makarow, A.D. 1976. *Optimization of Cutting Processes* (in Russian). Moscow, Russia: Mashinostroenie.

Mazurkiewicz, M., Kubala, Z., and Chow, J. 1989. Metal machining with high-pressure water-jet cooling assistance—A new possibility. *ASME Journal of Engineering for Industry* 111:7–12.
MicroTech Filtration. Available from http://www.microtechfiltration.com/Questions.html.
Nahin, P.G. 2002. *Oliver Heaviside: The Life, Work, and Times of an Electrical Genius of the Victorian Age*. Baltimore, MD: John Hopkins University Press.
Rebinder, P.A. 1979. *Surface Effects in Dispersion Systems* (in Russian). Moscow, Russia: Nauka.
Sa, J.W., Kim, H.K., Choi, C.H., Kim, H.T., Hong, K.H., Park, H.K., Bak, J.S., Lee, K.W., and Ha, E.T. 2005. Mechanical characteristics of austenitic stainless steel 316ln weldments at cryogenic temperature. In *Fusion Engineering 2005, Twenty-First IEEE/NPS Symposium*, Knoxville, TN.
Springborn, R.K. 1967. *Cutting and Grinding Fluids: Selection and Application*. Dearborn, MI: SME.
Stenphenson, D. 1981. *Pipeline Design for Water Engineers*, 2nd edn. Amsterdam, the Netherlands: Elsevier.
Stoll, A. and Furness, R. 2006. Near dry machining (MQL) is a key technology for driving the paradigm shift in machining operations. SME Paper TP07PUB22:1–16.
Sun, S., Brandt, M., and Dargusch, M.S. 2010. Thermally enhanced machining of hard-to-machine materials—A review. *International Journal of Machine Tools and Manufacture* 50(8):663–680.
Woods, S. 2006. Near dry. *Cutting Tool Engineering* 58(3):16–17.
Yildiz, Y. and Nalbant, M. 2008. A review of cryogenic cooling in machining process. *International Journal of Machine Tool Design and Research* 48:947–964.
Yoshimura, H., Itoigawa, F., Nakamura, T., and Niwa, K. 2005. Development of nozzle system for oil-on-water droplet metalworking fluid and its application to practical production line. *Transactions of JSME, Series C* 48(4):723–729.
Yoshimura, H., Itoigawa, F., Nakamura, T., and Niwa, K. 2006. Study on stabilization of formation of oil film on water droplet cutting fluid. *Transactions of the Japan Society of Mechanical Engineers C* 72:941–946.

7 Metrology of Drilling Operations and Drills

The thing I hate about an argument is that it always interrupts a discussion.

G. K. Chesterton **English writer (1874–1936)**

Metrology of a drilling operation consists of two closely related parts. The first part is the primary part that relates to the various tolerances on the hole being drilled. Therefore, the first logical part of this chapter is devoted to these tolerances, their standard designations in the part drawing, and proper interpretation in part manufacturing. The latter is vital in the selection of the measuring equipment, design of part fixtures and tool layouts, and selection of the proper tool and machining regime.

The second logical part of this chapter is entirely based on its first part and even overlaps the discussed notions/definitions of this part to some extents. However, it relates to the drill metrology considered for the first time in this book. It presents the author's unique vision of the subject explaining the meaning of this term, significant drill metrological parameters, their proper definitions, inspection/measurements using various measuring equipment, and their proper assigning by the drill drawing.

7.1 INTRODUCTION

Holes, also known as bores, in general are one of the most common features of mechanical components. In precision applications, surface finish, eccentricity, relative position, shape, and diameter of the holes directly relate to the performance and even service life of the component. Besides diametric error, the general term *poor surface quality* can be further categorized into several different types of problems, including (1) out-of-roundness, (2) out-of-perpendicular, (3) nailheading (bellmouthing), and (4) off-location. Deviation from the permissible tolerances on these features can impede operation and result in premature failure of the system. Therefore, it is important to inspect and validate how the accuracy parameters of machined holes conform to those assigned by the part drawing/manufacturing drawing for the operation. To do this, the following applies:

1. The quality parameters assigned by the drawing should be understood, and thus, appropriate hole-making operation should be selected.
2. The results of bore machining should be inspected using industrial metrology fundamentals, that is, appropriate gages and measuring methodology.

Metrology of drilling operations is based upon geometrical product specification (GPS). The idea of GPS is to give assurance in obtaining the following essential properties of a product:

- Functionality
- Safety
- Dependability
- Interchangeability.

GPS is implemented through a series of standards. GPS standards have been divided into four groups:

1. Fundamental GPS standards (fundamental rules for dimensioning and tolerancing).
2. Global GPS standards (e.g., ISO 1 on the standard reference temperature).
3. General GPS standards.
4. Complementary GPS standards (e.g., technical rules for drawing indications).

At the first design stages of a component, it is drawn to be an ideal, perfectly manufactured object, that is, having perfect form, texture, and size. Manufacturing processes involved in the production of this designed component cause various deviations of the finished component from its ideal conditions. For example, there can be variations in dimension, form, and surface texture. Because these variations can have a great effect on its service performance, it is, therefore, critical that the definitions of geometry are standardized and understood, so that the variation that is inherent in manufacturing processes can be taken into account. To be able to understand the geometrical variations within component parts, a set of requirements have been produced. These requirements are the part of GPS standards that cover small-scale features (surface texture) and large-scale features (sizes and dimensions, geometrical tolerances, and geometrical properties of surfaces) (Figure 7.1).

GPS is an internationally accepted concept covering all of the different requirements indicated in a technical drawing relating to the part geometry, that is, size, distance, radius, angle, form, orientation, location, runout, surface roughness, surface waviness, surface defects, and edges, as well as all related verification principles, measuring instruments, and their calibration. In other words, this system covers specification of the requirements for the micro- and macrogeometry of a part with associated requirements for verification and calibration of related measuring instruments.

As explained in the introduction of ISO 14660-1:1999, geometrical features exist in three *worlds*:

1. The world of specification, where several representations of the future workpiece are imagined by the designer.
2. The world of the part, the physical world.
3. The world of inspection, where a representation of a given workpiece is used through sampling of the workpiece by measuring instruments.

For a manufacturing/process engineer/planner, it is important to understand the relationship between these three worlds to assure that the manufacturing process/operation meets the quality requirements assigned by the part designer. In other words, proper understanding of the quality requirements assigned

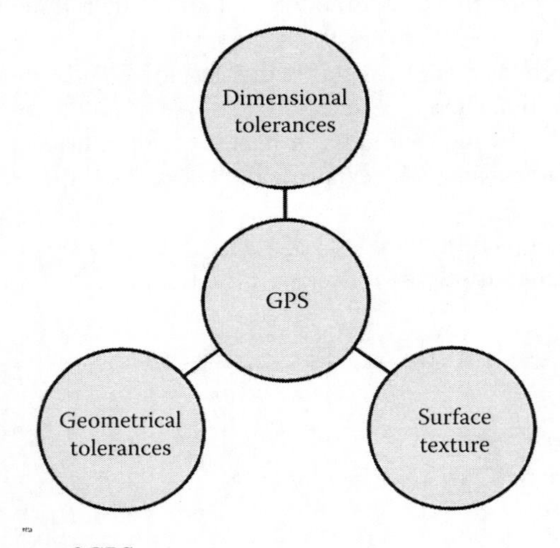

FIGURE 7.1 General structure of GPS.

by the part designer allows him or her to deploy proper machining operation(s) and gages (including measuring equipment and methodologies as well as a suitable quality assurance system).

ISO 14660 defines standardized terminology for geometrical features in each world as well as standardized terminology for communicating the relationship between each world. It is, therefore, of importance that all persons involved in design, manufacturing, and metrology are equipped with knowledge of the requirements of GPS. It is critical that the communication between the relevant engineering departments involved in GPS is clear, precise, and accurate and that each department understands the product design requirements.

As defined by ASME Y14.5M-1994, geometrical dimensioning and tolerancing known as GD&T is an international language used in drawings to accurately describe a part. The language consists of a well-defined set of symbols, rules, definitions, and conventions that can be used to describe the size, form, orientation, and location tolerances of part features. Moreover, GD&T is an exact language that enables designers to *say what they mean* in a drawing, thus improving product designs. Production uses the language to interpret the design intent, and inspection looks to the language to determine the inspection methodology. By providing uniformity in drawing specifications and interpretation, GD&T reduces controversy, guesswork, and assumptions throughout the manufacturing and inspection process. A manufacturing engineer/process designer/planner, part quality inspector, and many other specialists involved in part production must possess some basic working knowledge of GD&T to truly understand the designer's intent in terms of part accuracy.

7.2 STANDARD REFERENCE TEMPERATURE

Temperature variation is one of the most significant sources of gaging error. As manufacturing tolerances get tighter and the margin for gaging error gets smaller, it becomes an issue that must be addressed. In practice, ambient temperatures fluctuate from hour to hour and month to month in shop floor environments, and workpieces vary in temperature as the result of operations and/or seasonal ambient variations. Temperature-induced drift is a subtle effect that is often overlooked or ignored.

The most important GPS standard is ISO 1:2002, Geometrical Product Specifications (GPS)—Standard reference temperature for geometrical product specification and verification. This standard defines the standard reference temperature for all dimensional measurements, and all other GPS standards refer to it. GPS standard ISO 1:2002 states: "The standard temperature for geometrical product specification and verification is fixed at 20°C." What does it mean for manufacturing? First, during manufacture, the component must be measured close to 20°C; how close to 20°C depends on the materials and the tolerances involved. Second, the final inspection should either be made

- In a temperature-controlled room at 20°C
- By comparison with known artifacts of similar materials at a temperature close to 20°C
- By a measuring gage/machine with thermal compensation
- By the operator with manual corrections.

7.3 SMALL-SCALE FEATURES

The actual cross section of the machined surface viewed at high magnification is far different from the ideal flat, cylindrical, or curvilinear surface indicated in a drawing. Geometrically, the surface is seen to have a large number of minute irregularities (peaks and valleys) superimposed on more widely spaced undulations (waviness). Metallurgical examination of hardened steel components will show the transition from an amorphous layer through re-hardened and tempered zones, which finally merge with the original structure obtained by heat treatment. Analysis by sophisticated methods will reveal the presence of tiny cracks and changing residual stress at different depths below the surface (Astakhov 2010b).

All these features of an engineering surface (made on the finished part) constitute the so-called surface integrity that determines to a great extent the behavior of the part in service (Astakhov 2010b).

Thus, for example, surface texture has a significant effect on the frictional and lubricant-retention properties of the surface. Waviness determines whether mating will be accurate so that leakage can be avoided in sealed joints, metallurgical damage affects the wear resistance of the part, and residual stresses influence the fatigue strength of critical parts besides causing harmful deformations. Therefore, it is evident that *surface quality* is of critical importance and its proper evaluation is of utmost interest to the shop engineer/process planner.

The irregularity of a machined surface is a result of the machining process including complex influence of many parameters of the machining system such as the choice of cutting tool (geometry and material), feed and speed, part configuration and fixture, machine tool, coolant, and environmental conditions. This irregularity consists of high and low spots machined into a surface by the cutting tool. The standard term for these peaks and valleys is the surface texture. Because surface texture defines to a great extend the performance of the surface (e.g., its wear resistance), it is quantified by the part drawing in certain standard ways.

The term surface finish is well known but the concept is understood more in qualitative terms than in quantitative terms. This is evident from the fact that many industries continue to specify finish as rough, good, smooth, glossy, mirror, etc. None of these terms are sufficiently accurate and besides, they tend to convey different meanings to different people. It is in the common interest to adopt standard quantitative designation of surface finish known as its roughness and standard methods of evaluation with appropriate inspection techniques that will eliminate the variable subjective factor.

7.3.1 Definition of Surface Profile, Cutoff (Sampling) Length, and Centerline

The basic term involved is shown in Figure 7.2. The surface profile is usually measured in a direction perpendicular to the lay of the surface—that is, the predominant direction of the scratch marks. As can be seen in Figure 7.2, surface profile is composed of three distinct types of irregularities. The first type of irregularity is a form of error that is usually of a magnitude that can easily be detected by conventional measuring methods. The second type of irregularity consists of waviness with fairly regular spacing that can be attributed to vibrations of the machine. The third type of irregularity consists of closely spaced peaks and valleys superimposed on the first two types of irregularities. These peaks and valleys can be correlated with the shape, size, and motion of the cutting tool.

The international consensus of opinion has been to relate surface texture (commonly referred to as roughness or surface finish) to the height of the closely spaced irregularities over a short length. This length, which is called the cutoff (sampling) length, is specified so that variations due to waviness or form errors are excluded. Since the peaks and valleys have different heights, it is evident that some kind of average height should be determined. To do this, a surface profile should be analyzed. The profile shown in Figure 7.3 is a highly magnified sectional view of the surface. Within the selected cutoff length, it is possible to draw a CL

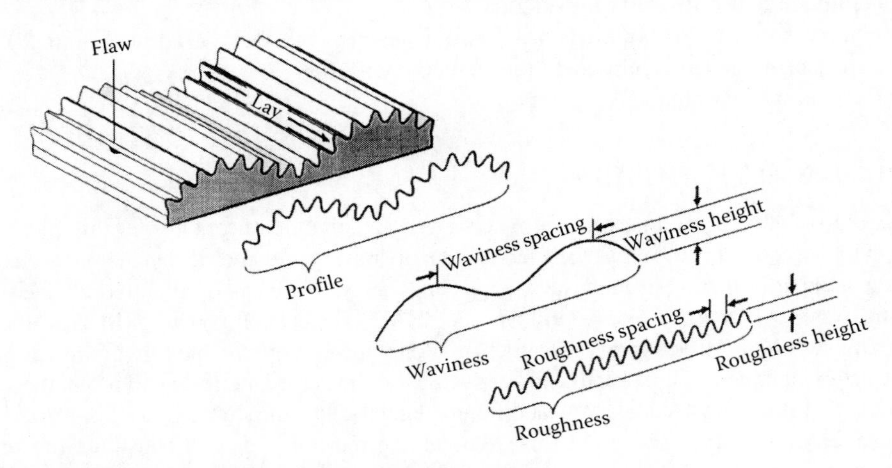

FIGURE 7.2 Visualization of the basic surface microfeatures.

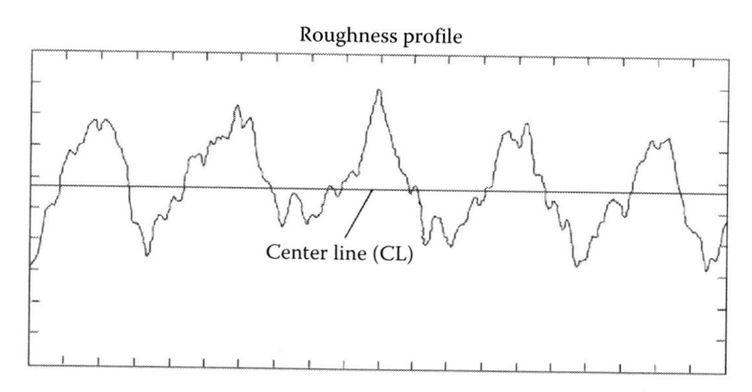

FIGURE 7.3 Surface profile and the CL.

in such a way that the areas of the peaks above the line are equal to the areas of the valleys below the line. This CL serves as the baseline for determination of many surface roughness characteristics.

7.3.2 Common Characteristics of Surface Texture (Roughness) Used in Drilling Operations

There are a great number of ISO standards for characterization and measuring surface texture, for example, ISO 1302 Indication of surface texture in technical product documentation; ISO 3274 Surface texture: Profile method—Nominal characteristics of contact (stylus) instruments; ISO 4287 Surface texture: Profile method—Terms, definitions, and surface texture parameters; ISO 4288 Surface texture: Profile method—Rules and procedures for the assessment of surface texture; and ISO 8785 Surface imperfections—Terms, definitions, and parameters.

There are several different methods of surface texture (roughness) measurement in use today. Some of the commonly used methods in drilling operations are shown in Figure 7.4. The method used on any given part depends largely on where in the world the part is manufactured and the measurement parameters the manufacturer and the customer prefer to use. It is not uncommon for different parties involved in the production to use different methods for surface texture (roughness) measurements.

In North America, the most common parameter for surface texture is average roughness (*Ra*). *Ra* is calculated by an algorithm that measures the average length between the peaks and valleys and the deviation from the mean line on the entire surface within the sampling length (see Figure 7.4). *Ra* averages all peaks and valleys of the roughness profile and then neutralizes the few outlying points so that the extreme points have no significant impact on the final results. It's a simple and effective method for monitoring surface texture and ensuring consistency in measurement of multiple surfaces.

The advantages of using *Ra* are as follows:

- The most commonly used parameter to monitor a production process
- Default parameter in a drawing if not otherwise specified
- Available even in the least sophisticated measuring instruments
- Statistically a very stable, repeatable parameter
- Good for regular (cutting) and for random (grinding) type surfaces
- A good parameter where a machining operation is under control and where the conditions are always the same, for example, cutting tool, speeds, feeds, MWF (lubricant).

The disadvantages of using *Ra* are as follows:

- Not a good discriminator for different types of surfaces (no distinction is made between peaks and valleys)
- Not a good measure of sealed surfaces.

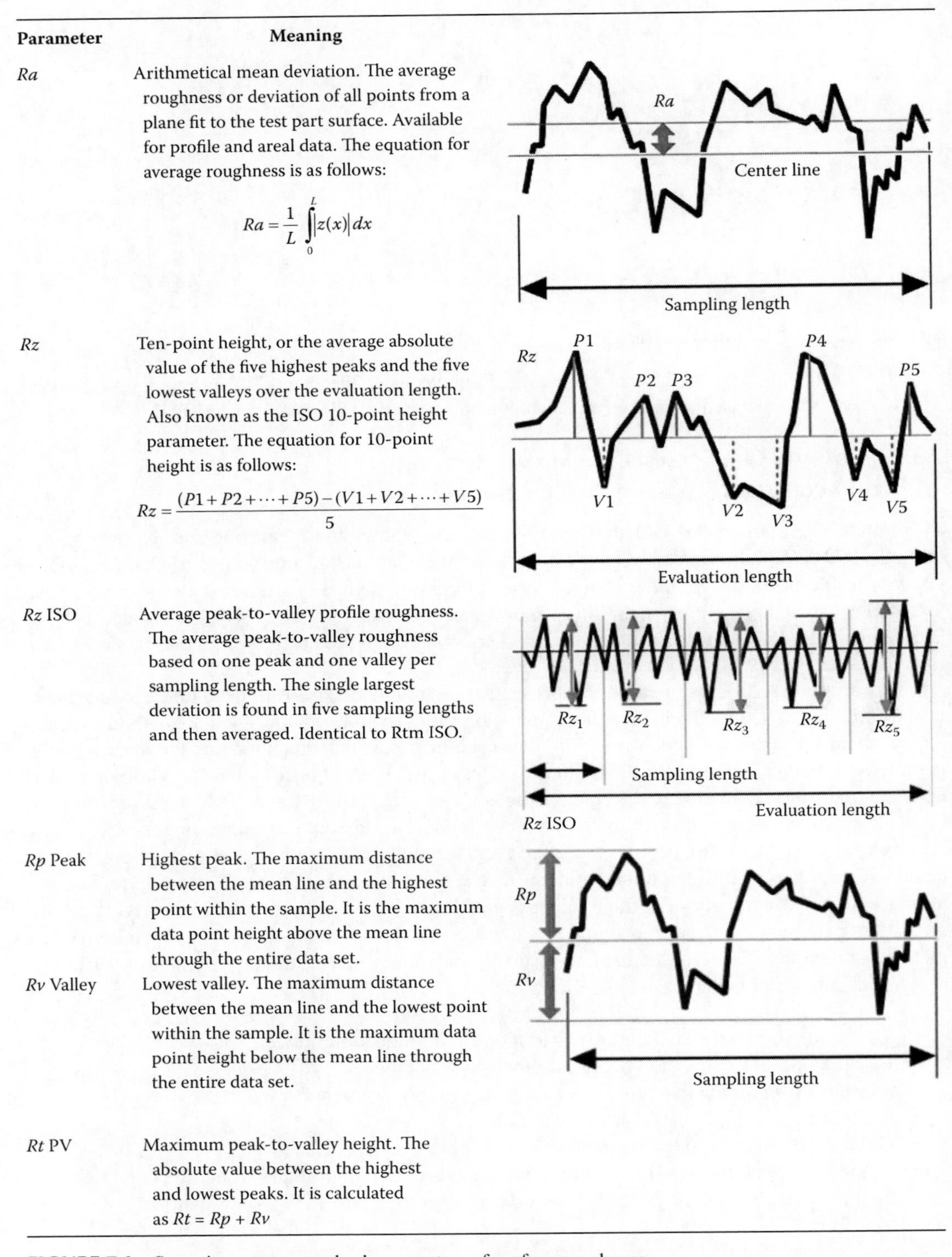

Parameter	Meaning
Ra	Arithmetical mean deviation. The average roughness or deviation of all points from a plane fit to the test part surface. Available for profile and areal data. The equation for average roughness is as follows: $$Ra = \frac{1}{L}\int_0^L \lvert z(x)\rvert\, dx$$
Rz	Ten-point height, or the average absolute value of the five highest peaks and the five lowest valleys over the evaluation length. Also known as the ISO 10-point height parameter. The equation for 10-point height is as follows: $$Rz = \frac{(P1+P2+\cdots+P5)-(V1+V2+\cdots+V5)}{5}$$
Rz ISO	Average peak-to-valley profile roughness. The average peak-to-valley roughness based on one peak and one valley per sampling length. The single largest deviation is found in five sampling lengths and then averaged. Identical to Rtm ISO.
Rp Peak	Highest peak. The maximum distance between the mean line and the highest point within the sample. It is the maximum data point height above the mean line through the entire data set.
Rv Valley	Lowest valley. The maximum distance between the mean line and the lowest point within the sample. It is the maximum data point height below the mean line through the entire data set.
Rt PV	Maximum peak-to-valley height. The absolute value between the highest and lowest peaks. It is calculated as $Rt = Rp + Rv$

FIGURE 7.4 Some important standard parameters of surface roughness.

Parameter	Meaning
Rmax ISO	Maximum peak-to-valley profile height. The greatest peak-to-valley distance within any one sampling length.

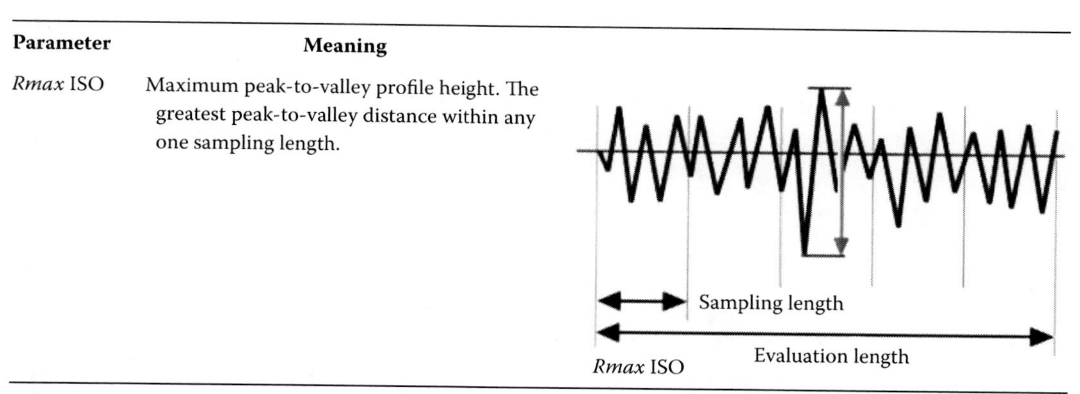

FIGURE 7.4 (continued) Some important standard parameters of surface roughness.

In Europe, the most common parameter for surface texture is mean roughness depth (*Rz*). *Rz* is calculated by measuring the vertical distance from the highest peak to the lowest valley within five sampling lengths and then averaging these distances. *Rz* averages only the five highest peaks and the five deepest valleys; therefore, extremes have a much greater influence on the final value. Over the years, the method of calculating *Rz* has changed but the symbol *Rz* has not. As a result, there are three different *Rz* calculations still in use and it is very important to know which calculation is being defined before making the measurement. Sometimes, *Rmax*—maximum peak-to-valley height—is also specified. The applications of these parameters are as follows:

- *Rz* is more sensitive than *Ra* to changes in surface finish as maximum profile heights and not averages are being examined
- *Rmax* is useful for surfaces where a single defect is not permissible, for example, a seal with a single scratch
- *Rz* and *Rmax* are used together to monitor the variations of surface finish in a production process. Similar values of *Rz* and *Rmax* indicate a consistent surface finish, while a significant difference indicates a surface defect in an otherwise consistent surface.

Rz to *Ra* conversion is recommended as follows: Standard BS 1134/1-1972: $Rz = (4–7)Ra$; Siemens Co. $Rz = (4–10)Ra$. Actual ratio depends upon the shape of the profile. An exact conversion of the peak-to-valley height *Rz* and the CL average height *Ra* can neither be theoretically justified nor empirically proved. For surfaces that are generated by manufacturing methods of the group *metal cutting*, a diagram for the conversion from *Ra* to *Rz* and vice versa is shown in supplement 1 to standard DIN 4768 part 1, based on comparison measurements. Table 7.1 presents some common values according to this standard, parameters *Rp* and *Rpm* are also often used. *Rp*, per ISO 4287, is the max height of any peak to the mean line within one sampling length. *Rpm*, the mean leveling depth—per rules of ISO 4288—is an averaging of *Rp* over 5 cutoffs, according to ASME B46.1-2002. When *Rp* is calculated over the evaluation length, it is designated as *Rpm*. Note that many surface gages measure *Rpm* but report the result as *Rp*.

Applications of these parameters are as follows:

- *Rpm* is useful in predicting bearing characteristics of a surface
- A low value of *Rpm* and large value of *Rz* indicates a plateau surface
- The ratio *Rpm*/*Rz* quantifies the asymmetry of profile
- *Rpm* is recommended for bearing and sliding surfaces and surface substrates prior to coating
- *Rp* is a good parameter to control coating quality.

TABLE 7.1
Comparison of Roughness Values

DIN ISO 1302	Roughness Values, *Ra*	(μm)	0.025	0.05	0.1	0.2	0.4	0.8	1.6	3.2	6.3	12.5	25	50
		(min)	1	2	4	8	16	32	63	125	250	500	1000	2000
	Roughness grade number		N1	N2	N3	N4	N5	N6	N7	N8	N9	N10	N11	N12
Suppl. 1 to DIN 4768/1	Roughness Values, *Rz* in μm	From	0.1	0.25	0.4	0.8	1.6	3.15	6.3	12.5	25	40	80	160
		To	0.8	1.6	2.5	4	6.3	12.5	20	31.5	63	100	160	250

The surface texture requirement in the drawing applies to the evaluation length. Certain parameters are defined on the basis of the sampling length and others on the basis of the evaluation length (see ISO 4287, ISO 12085, 13565-2, and ISO 13565-3). When the parameter is defined on the basis of the sampling length, the number of sampling lengths constituting the evaluation length is of decisive importance. Therefore, surface finish is related to the closely spaced irregularities only, and other defects like waviness and form error should be disregarded. The visualization of the lengths involved in surface roughness measurements is shown in Figure 7.5.

In surface roughness measurements, the cutoff wavelength (commonly designated as λ_c) of a profile filter determines which wavelengths belong to roughness and which ones to waviness. Note that this length is selected at the beginning of surface roughness measurement according to Table 7.2, which represents current industrial practice of ignoring the *profile filter* step and using a sampling length equal to the cutoff wavelength. The sampling length is the reference for roughness evaluation. This length is equal to the cutoff length. Traversing length is the overall length traveled by the stylus of a surface measuring gage known as profilometer. Evaluation length is a part of the traversing length (Figure 7.5) from where the values of the surface parameters are determined.

The cutoff wavelength λ_c has to be chosen in such a way as to include a sufficient number of primary irregularities for the purpose of averaging. Evidently, a large cutoff (sampling) length is necessary for drilled surfaces where individual toolmarks are widely spaced, compared to ground or lapped surfaces produced by the overlapping trajectories of tiny abrasive grains.

The foregoing considerations on surface roughness measurements imply that the actual methodology of surface is as shown in Figure 7.5 as an example of *Rz* measurement. As can be seen,

$$Rz = \frac{1}{5(Rz_1 + Rz_2 + Rz_3 + Rz_4 + Rz_5)} \tag{7.1}$$

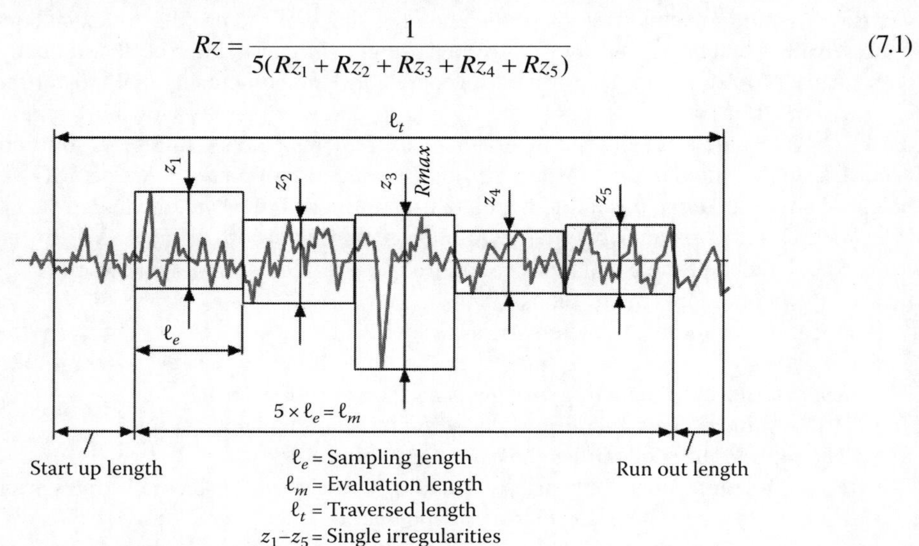

FIGURE 7.5 Meaning of the traversed, evaluation, and sampling lengths.

TABLE 7.2
Selection of Cutoff Wavelength λ_c in Surface Roughness Measurements

Roughness Parameter		Cutoff	Sampling/Evaluation Length
Rz (μm)	*Ra* (μm)	λ_c (mm)	l_e/l_m
Up to 0.1	Up to 0.02	0.8	0.08/0.4
Over 0.1	Over 0.02	0.25	0.25/1.25
Up to 0.5	Up to 0.1		
Over 0.5	Over 0.1	0.8	0.8/4
Up to 10	Up to 2		
Over 10	Over 2	2.5	2.5/12.5
Up to 50	Up to 10		
Over 50	Over 10	8	8/40
Up to 200	Up to 80		

The maximum roughness depth *Rmax* is the largest of the five roughness depths considered in a sense shown in Figure 7.5.

7.3.3 Designation of Surface Texture (Roughness) Parameters

Standard ISO 1302 (2002) is rather complicated document so only its part relevant to drilling operation is covered in this section.

7.3.3.1 Basic Symbols

The basic graphical symbols and their explanations are presented in Table 7.3. When complementary requirements for surface texture characteristics have to be indicated, a line shall be added to the longer arm of any of the graphical symbols illustrated as shown in Figure 7.6. For use in the written text of, for example, reports or contracts, the textual indication for (1) is APA, for (2) MRR, and for (3) NMR. These abbreviators stand for APA, any process allowed; MRR, material removal required; and NMR, no material removed.

When the same surface texture is required on all surfaces around a workpiece outline (integral features), represented in the drawing by a closed outline of the workpiece, a circle shall be added to

TABLE 7.3
Graphical Symbols without Inscription

Symbol	Meaning
	Basic graphical symbol: may only be used in isolation when its meaning is *the surface under consideration* or when explained by a note.
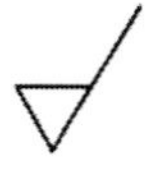	Expanded graphical symbol: machining surface with no indication of other details; in isolation, this expanded graphical symbol may only be used when its meaning is *a surface to be machined*.
	Expanded graphical symbol: surface from which removal of material is prohibited; this expanded graphical symbol may also be used in a drawing relating to a manufacturing process to indicate that a surface is to be left in the state resulting from a preceding manufacturing process, regardless of whether this state was achieved by removal of material or otherwise.

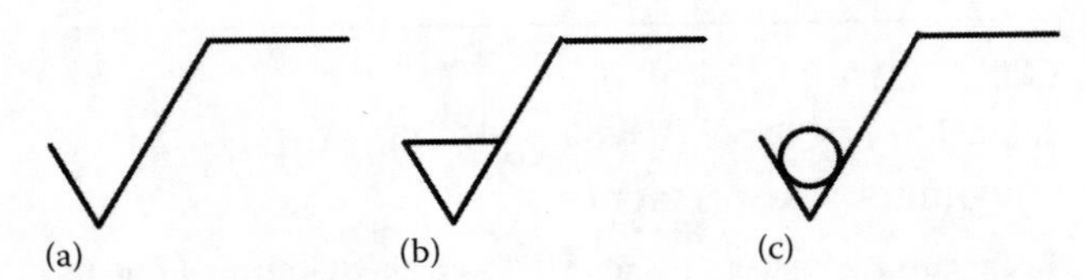

FIGURE 7.6 Complete graphical symbol: (a) for APA, for (b) MRR, and for (c) NMR.

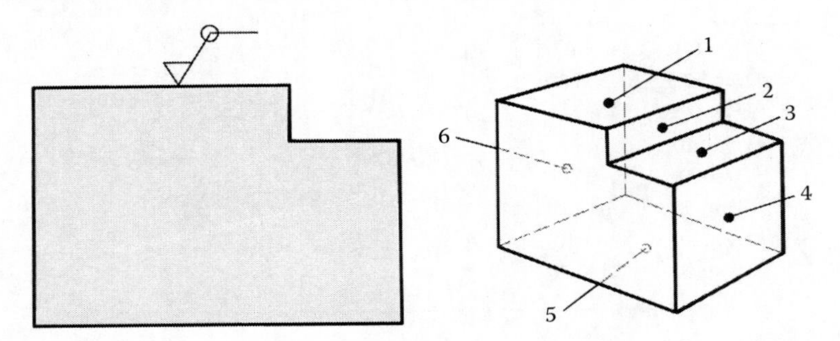

FIGURE 7.7 The outline in the drawing represents the six surfaces shown on the 3D representation of the workpiece (the front and rear surfaces not included).

the complete graphical symbol illustrated in Figure 7.6, as shown in Figure 7.7. Surfaces should be indicated independently if any ambiguity may arise from the all-around indication.

7.3.3.2 Composition of Complete Graphical Symbol

In order to ensure that a surface texture requirement is unambiguous, it may be necessary, in addition to the indication of both a surface texture parameter and its numerical value, to specify additional requirements (e.g., transmission band or sampling length, manufacturing process, surface lay and its orientation, and possible machining allowances). It may be necessary to set up requirements for several different surface texture parameters in order that the surface requirements ensure unambiguous functional properties of the surface.

The mandatory positions of the various surface texture requirements in the complete graphical symbol are shown in Figure 7.8.

The complementary surface texture requirements in the form of surface texture parameters, numerical values, and transmission band/sampling length should be located at the specific positions in the complete graphical symbol in accordance with the following:

1. *Position a—single surface texture requirement.* Indicates the surface texture parameter designation, the numerical limit value, and the transmission band/sampling length (Table 7.2). To avoid misinterpretation, a double space (double blank) should be inserted between the parameter designation and the limit value.
2. *Positions a and b—two or more surface texture requirements.* Indicate the first surface texture requirement at position *a*. Indicate the second surface texture requirement at position *b*. If a third requirement or more is to be indicated, the graphical symbol is to be enlarged

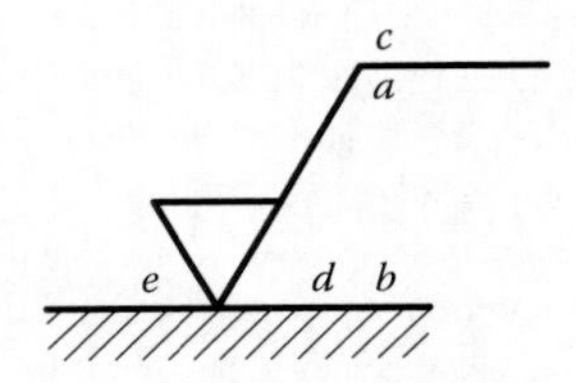

FIGURE 7.8 Positions (*a*–*e*) for location of complementary requirements.

accordingly in the vertical direction, to make room for more lines. The positions *a* and *b* are to be moved upward when the symbol is enlarged.

3. *Position c—manufacturing method.* Indicates the manufacturing method, treatment, coatings, or other requirements for the manufacturing process to produce the surface, for example, drilled, turned, ground, and plated.
4. *Position d—surface lay and orientation.* Indicates the symbol of the required surface lay and the orientation, if any, of the surface lay, for example, "=," "X," and "M" (see Figure 7.9).
5. *Position e—machining allowance.* Indicates the required machining allowance, if any, as a numerical value given in millimeters.

Figure 7.10 shows the control elements in indication of surface texture requirements in engineering drawings. In this designation, the surface texture requirement shall be indicated as a unilateral or bilateral tolerance. When the parameter designation, the parameter value, and the transmission band are indicated, they are understood as a unilateral upper tolerance limit of the parameter in question (*16%-rule* or *max-rule* limit). If the parameter designation, the parameter value, and the transmission band indicated are to be interpreted as a unilateral lower tolerance limit of the parameter in question (16% or max limit), then the parameter designation shall be preceded by the letter *L*, for example, *L Ra* 0.32.

Symbol	Meaning	Direction of tool marks
R	Radial lay orientation relative to the center of the surface displaying the surface texture symbol.	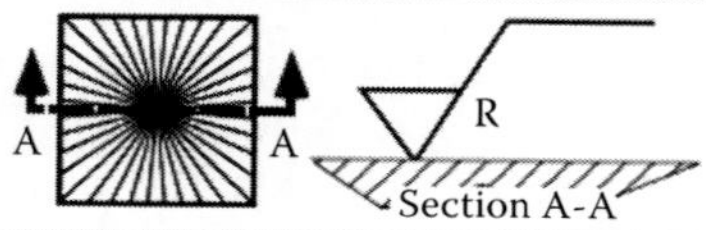
⊥	Perpendicular orientation relative to the surface in the view displaying the surface texture symbol.	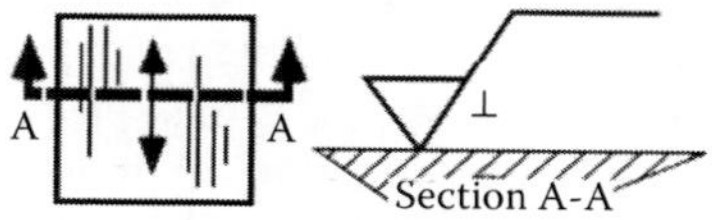
X	Angular lay orientation in both directions relative to the surface in the view displaying the surface texture symbol.	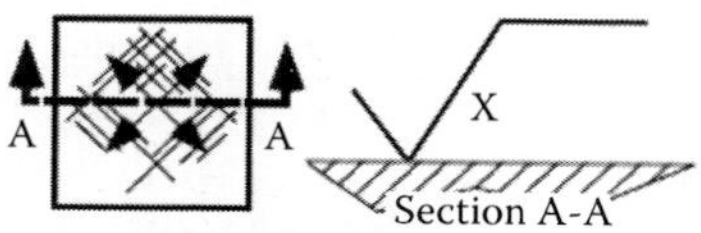
M	Multidirectional lay orientation relative to the surface in the view displaying the surface texture symbol.	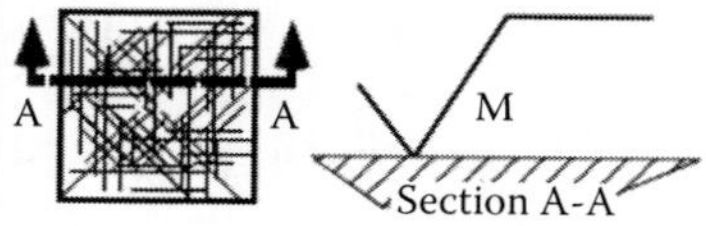
C	Circular lay orientation relative to the center of the surface displaying the surface texture symbol.	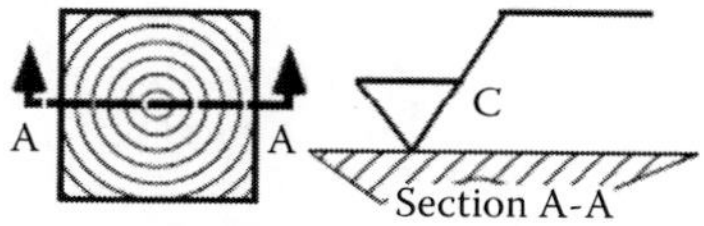
=	Parallel lay orientation relative to the surface in the view displaying the surface texture symbol.	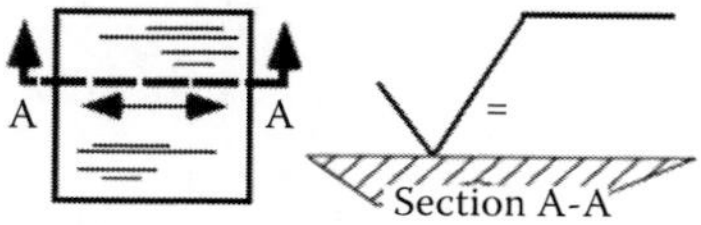
P	Particulate, non-directional, or protuberant lay orientation relative to the surface displaying the surface texture symbol.	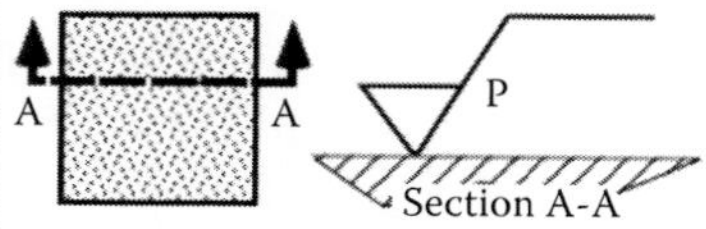

FIGURE 7.9 Lay symbols and their meaning.

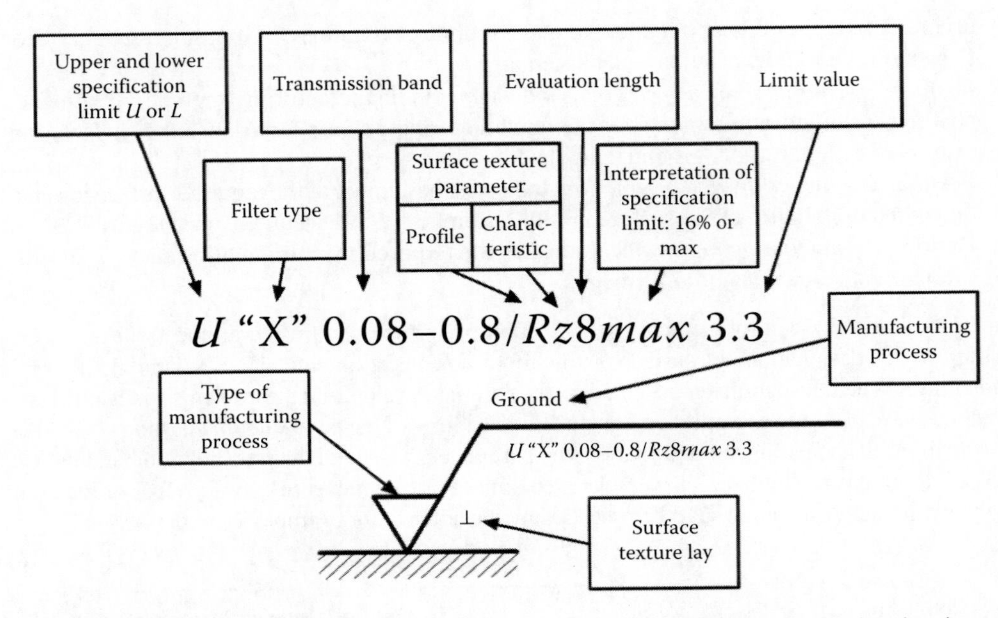

FIGURE 7.10 Control elements in indication of surface texture requirements in engineering drawings.

A bilateral tolerance is indicated in the complete symbol by placing the requirement for the two tolerance limits above each other, the upper specification limit (*16%-rule* or *max-rule* limit) preceded by *U* being indicated over the lower specification limit preceded by *L*. Where the upper and lower limits are expressed by the same parameter with different limit values, the *U* and *L* may be omitted provided the omission does not leave any doubt. The upper and lower specification limits, however, are not necessarily expressed by means of the same parameter designation and transmission band. Figure 7.11 shows an example.

There are two different ways of indicating and interpreting the specification limits of surface texture: (1) the *16% rule* and (2) the *max rule* defined by standard ISO 4288:1996. The *16% rule* is defined as the default rule for all indications of surface texture requirements. This means that the *16% rule* applies to a surface texture requirement when a parameter designation is applied (see Figure 7.12). If the *max rule* is to be applied to a surface texture requirement, *max* shall be added to the parameter designation.

Where no transmission band is indicated in connection with the parameter designation, the default transmission band applies to the surface texture requirement (see Figure 7.12 for no transmission band indicated). Certain surface texture parameters do not have a defined default transmission band,

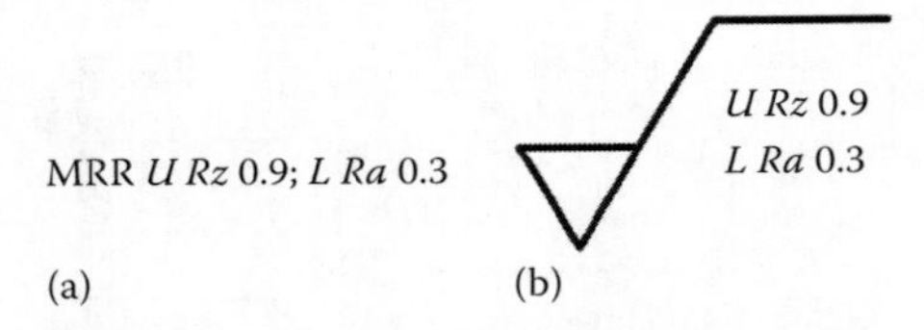

FIGURE 7.11 Bilateral surface specification: (a) in text and (b) on drawing.

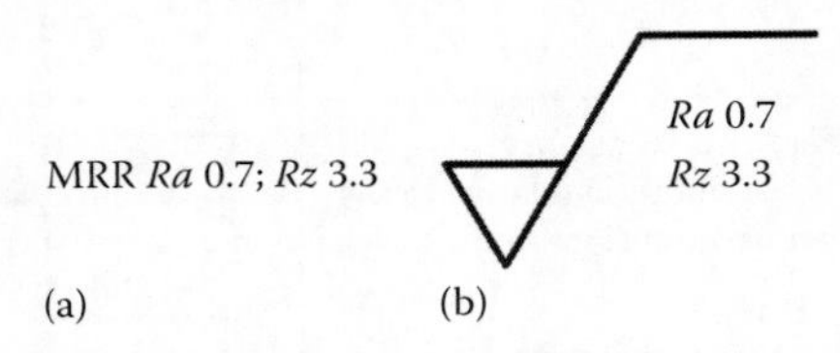

FIGURE 7.12 Parameter indication where *16% rule* applies (default transmission band): (a) in text and (b) on drawing.

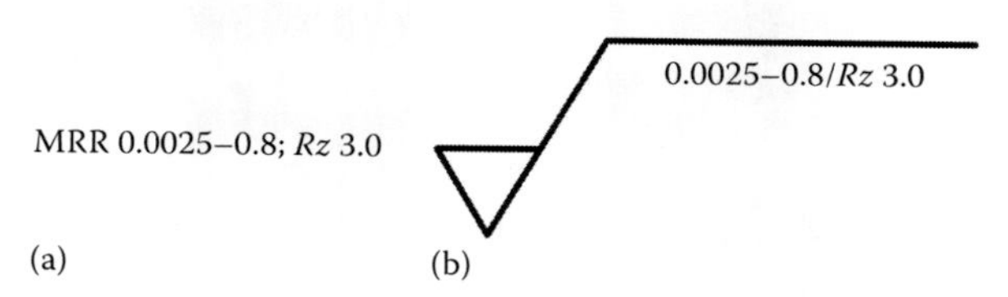

FIGURE 7.13 Indication of transmission band in connection with surface texture requirement: (a) in text and (b) on drawing.

a default shortwave filter, or a default sampling length (long-wave filter). Consequently, the surface texture indication shall specify transmission band, shortwave filter or long-wave filter, to ensure that the surface texture requirement is unambiguous.

To provide assurance that the surface is controlled unambiguously by the surface texture requirement, the transmission band shall be indicated in front of the parameter designation separated from it by an oblique stroke (/). The transmission band shall be indicated by the inclusion of the cutoff values of the filters (in millimeters), separated by a hyphen ("-"), the shortwave filter indicated first, and the long-wave filter second (Figure 7.13).

7.3.3.3 Practical Designation on Tool Drawings

Standard ISO 1302 (2002) allows simplified designation of the surface texture parameter(s) in a part drawing. Figure 7.14 shows an example. Although such a designation is simple and commonly found in the vast majority of drawings, one should understand the requirements set by the designation shown in this Figure. They are as follows:

1. Surface roughness on all surfaces except one:
   a. One single, unilateral/upper specification limit.
   b. $Rz = 6.1$ μm.
   c. *16% rule*, default (ISO 4288).
   d. Default transmission band (ISO 4288 and ISO 3274).
   e. Default evaluation length ($5 \times \lambda_c$) (ISO 4288).
   f. Surface lay, no requirement.
   g. Manufacturing process involves material removal.
2. The hole with a different requirement has a surface roughness:
   a. One single, unilateral/upper specification limit.
   b. $Ra = 0.7$ μm.
   c. *16% rule*, default.
   d. Default transmission band (ISO 4288 and ISO 3274).
   e. Default evaluation length ($5 \times \lambda_c$) (ISO 4288).
   f. Surface lay, no requirement.
   g. Manufacturing process involves material removal.

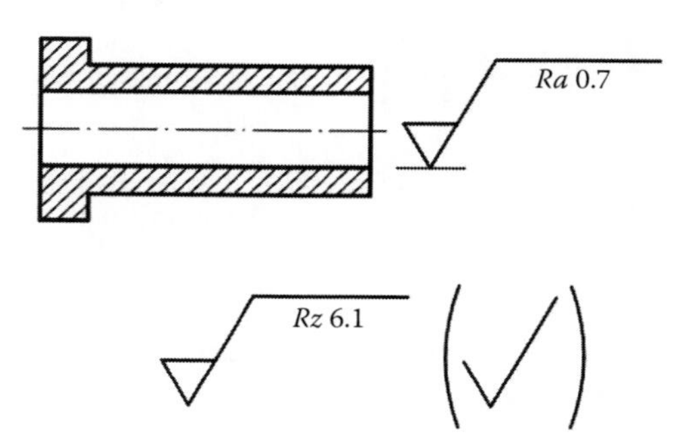

FIGURE 7.14 Example of simplified surface texture designation.

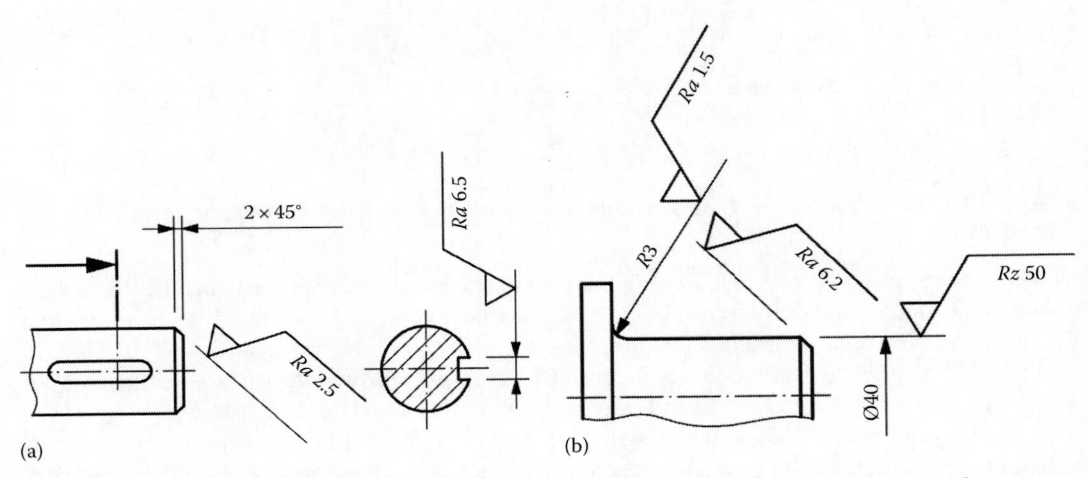

FIGURE 7.15 Two examples of simplified surface texture designation: (a) indication of surface texture and dimensioning are combined using the same dimension line and (b) surface texture and dimensioning are indicated together on an extended dimension line or separated on the respective projection line and dimension line.

Figure 7.15 shows two other examples. The same as in the previous example, the following parameters are set: a single, unilateral upper specification limit, *16% rule*, default (ISO 4288), default evaluation length (5 × λ_c) (ISO 3274), default transmission band (ISO 4288 and ISO 3274), surface lay, no requirement, and manufacturing process involves material removal.

Designation of surface texture according to American standard ASME/ANSI Y14.5M-2002 is similar to that set by ISO 1302 (2002) with few exceptions. Figure 7.16 shows positions for complementary requirements that are different from those shown in Figure 7.8. As can be seen, the major difference is that if the surface texture is expressed in *Ra* (as recommended), then it is shown outside the basic symbols with no symbol *Ra* shown. Figure 7.17 shows examples.

7.3.3.4 Preferred Surface Roughness

Typical surface roughness values vary widely depending on the processes employed (Figure 7.18). Even for given process, roughness values depend on a number of factors (Figures 7.19 and 7.20). For instance, drilling shown in Figure 7.21 indicates a range of anywhere from 12.5 to 0.2 μm. This is because the surface texture and its roughness are system parameters that depend on the design and conditions of the components of the drilling system. Major factors influencing the

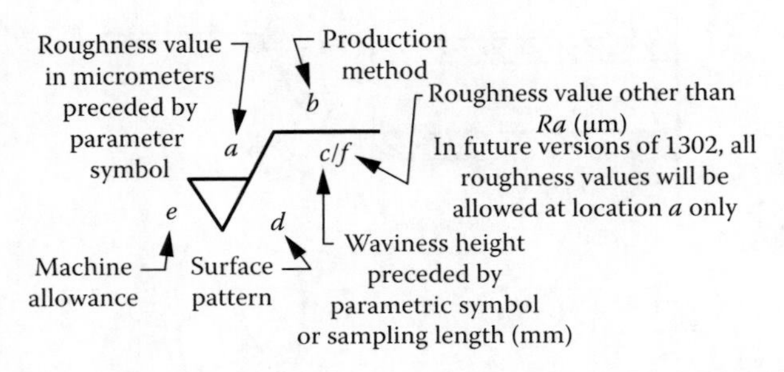

FIGURE 7.16 Positions for location of complementary requirements set by ASME/ANSI Y14.5M-2002.

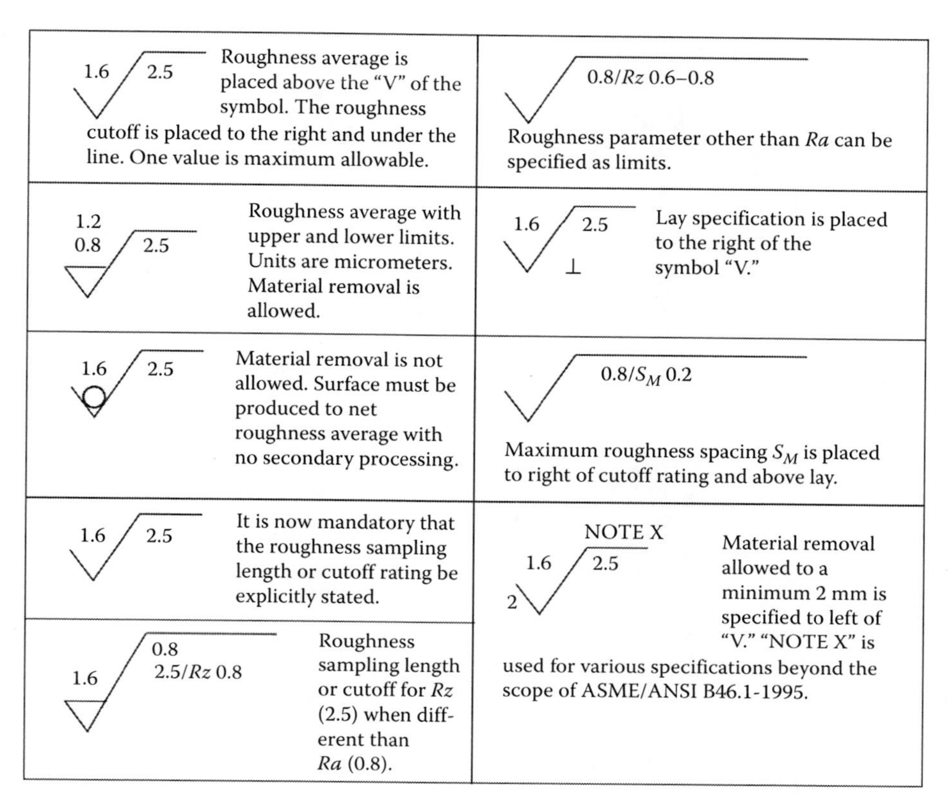

FIGURE 7.17 Samples of surface texture applications.

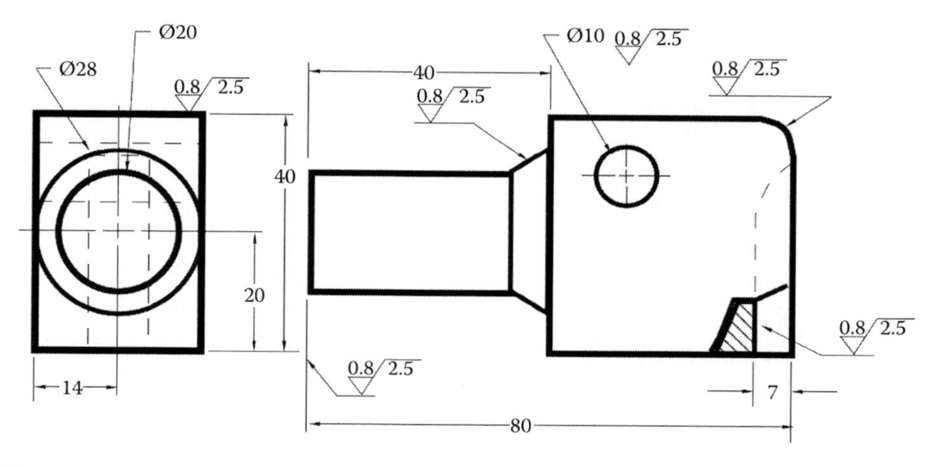

FIGURE 7.18 Examples of locations for surface texture symbols.

ultimate roughness value achievable include the following: the mechanical properties of the material itself, cutting speed and feed, drilling tool design and geometry, MWF characteristics, and tool holder. A minor change in any factor may have a profound effect on the roughness of the surface produced. As previously discussed, the process *optimality* as a system characteristic can be judged by the difference of the theoretical surface finish and that obtained in a given application.

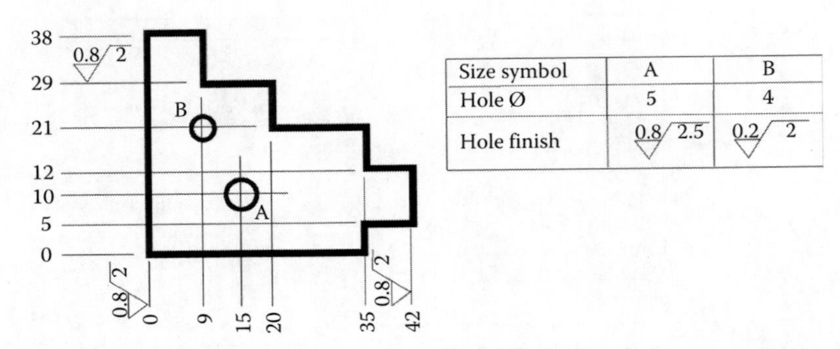

Size symbol	A	B
Hole Ø	5	4
Hole finish	0.8/2.5	0.2/2

FIGURE 7.19 Example of a drawing with rectangular coordinate dimensioning without dimension lines.

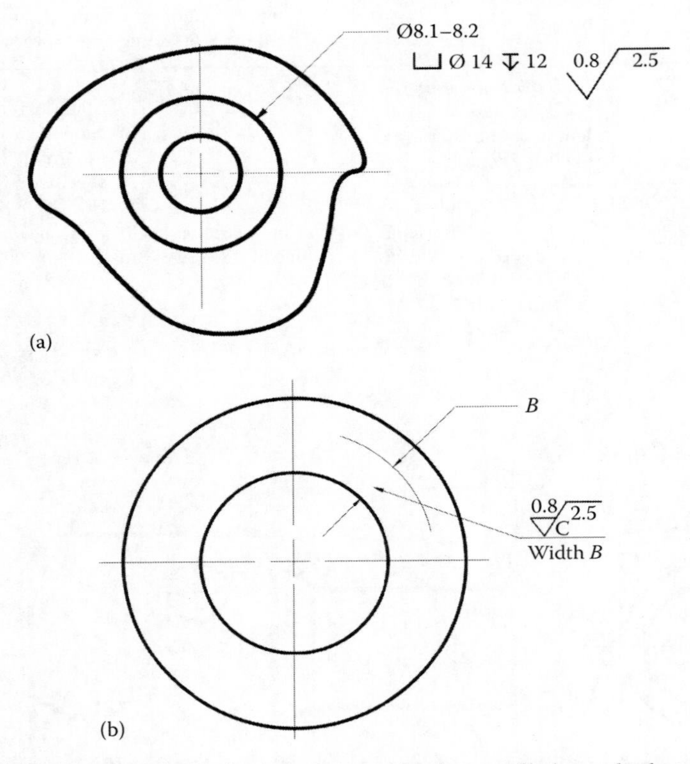

FIGURE 7.20 Examples where the surface texture specifications applied to only the second of a two-step process (a) and limiting a surface texture specification to only part of a surface precision (b).

7.3.3.5 Different Methods for Designating Surface Texture

Although many years have passed since the acceptance of the two basic standards for surface texture designation, namely, ISO 1302 (2002) and ASME/ANSI Y14.5M-2002, many machine shops, manufacturing companies, and even countries still use *traditional* designation adopted many years ago to avoid confusion in their new and old drawings as these old standards are widely accepted and seem to be well understood within a company (country). Table 7.4 shows some common examples. Therefore, it is of importance for any drilling manufacturing specialist to analyze a part drawing before designing a drilling operation in order to understand the real requirements to the surface texture assigned by this drawing as these real requirements define to a large extent the selection/design of the components of the drilling operation, its efficiency, and gaging needed.

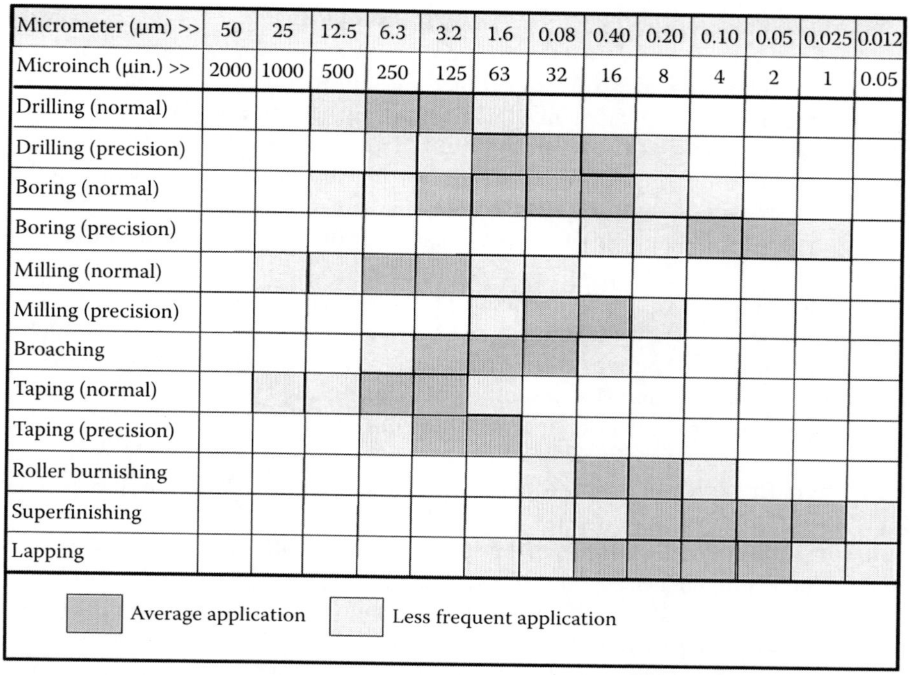

Micrometer (μm) >>	50	25	12.5	6.3	3.2	1.6	0.08	0.40	0.20	0.10	0.05	0.025	0.012
Microinch (μin.) >>	2000	1000	500	250	125	63	32	16	8	4	2	1	0.05
Drilling (normal)													
Drilling (precision)													
Boring (normal)													
Boring (precision)													
Milling (normal)													
Milling (precision)													
Broaching													
Taping (normal)													
Taping (precision)													
Roller burnishing													
Superfinishing													
Lapping													

Average application   Less frequent application

FIGURE 7.21 Surface roughness produced by common hole-making operations.

TABLE 7.4
Different Methods for Designation *Ra* 0.8

Symbol	Standard
32 AA	ANSI B46.1-1962
32 CLA	BS 1134-1961
RNE 0.8	Old US MIL Specifications
0.8a	JIS B0601-1976
▽▽▽	JIS B0601-1976 DIN3141-1960
0.8 ▽▽▽	JIS B0601-1976
N6	ISO 1302-1978
0.8 √	ANSI B46.1-2002
√ *Ra* 0.8	ISO 1302-2002
32 √	Common US Designation

7.4 LARGE-SCALE FEATURES: BORE TOLERANCING

Dimensions are a part of the total specification assigned to parts designed by engineering. However, the engineer in industry is constantly faced with the fact that no two objects in the material world can ever be made exactly the same. The small variations that occur in repetitive production must be considered in the design. To inform manufacturing specialists how much variation from exact size is permissible, the designer uses a tolerance or limit dimension technique. A *tolerance* is defined as the total permissible variation of size or the difference between the limits of size. *Limit dimensions* are the maximum and minimum permissible dimensions. Proper tolerancing practice ensures that the finished product functions in its intended manner and operates for its expected life.

All bore dimensions applied to the drawing, except those specifically labeled as basic, gage, reference, maximum, or minimum, will have an exact tolerance, either applied directly to the dimension or indicated by means of general tolerance notes. For any directly tolerated decimal dimension, the tolerance has the same number of decimal places as the decimal portion of the dimension (Gooldy 1998).

Engineering tolerances may broadly be divided into three groups: (1) *size tolerances* assigned to dimensions such as length, diameter, and angle; (2) *geometrical tolerances* used to control a hole shape in the longitudinal and transverse directions; and (3) *positional tolerances* used to control the relative position of mating features. Interested readers may refer to Gooldy (1998) and Smith (2002).

Some basic standard definitions useful for further considerations are as follows:

- *Size*—a number expressing, in a particular unit, the numerical value of a linear dimension (ISO 286-1:1988).
- *Basic size*—the nominal diameter of the shaft (or bolt) and the hole. This is, in general, the same for both components.
- *Size tolerance*—the difference between the maximum and minimum limits of size (Note: Size tolerance is an absolute value without sign) (ISO 286-1:1988).
- *Tolerance zone*—the zone contained between two lines representing the maximum and minimum limits of size, defined by the magnitude of the tolerance and its position relative to the zero line (ISO 286-1:1988).
- *Lower deviation*—the difference between the basic size and the minimum possible component size.
- *Upper deviation*—the difference between the basic size and the maximum possible component size.

In a graphical representation of tolerance limits, a straight horizontal line that represents the basic (nominal) diameter of the bore and to which the deviations are of zero deviation is known as the zero line. The positive deviations from the basic size (diameter) are shown above this line and the negative deviations below it. Figure 7.22 shows a graphical representation of the previous definitions.

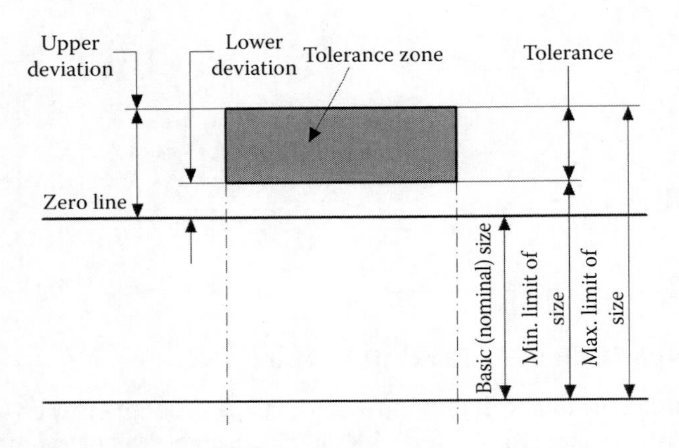

FIGURE 7.22 Graphical representation of the basic size deviation and tolerance.

The ISO system of limits and fits (ISO 286-2-2010 *GPS—ISO code system for tolerances on linear sizes—Part 2: Tables of standard tolerance classes and limit deviations for holes and shafts*) covers standard tolerances and deviations for sizes up to 3150 mm. The system is based on a series of tolerances graded to suit all classes of work from the finest to the coarsest, along with different types of fits that range from coarse clearance to heavy interference. Here, *fit* is the general term used to signify the range of tightness that may result from the application of a specific combination of tolerances in the design of mating parts.

The system ISO defines 28 classes of basic deviations for bores. These classes are marked by capital letters (A, B, C,..., ZC). The tolerance zone for the specified dimensions is prescribed in the drawing by a tolerance mark, which consists of a letter marking the basic deviation and a numerical marking of the tolerance grade (e.g., H7, H8, D5). Table 7.5 presents tolerances commonly used in drilling operations. Figure 7.23 shows the graphical representation of common bore tolerances for a bore of 30 mm dia. Among the shown tolerances, H6–H11 are most common for drilling operations.

TABLE 7.5
Bore Tolerances Commonly Used in Drilling Operations

	Nominal Sizes								
Over	3	6	10	18	30	40	50	65	80
Incl.	6	10	18	30	40	50	65	80	100
Tolerance	**Micrometers**								
G6	12	+14	+17	+20	+25	+29	+34		
	+4	+5	+6	+7	+9	+10	+12		
G7	+16	+20	+24	+28	+34	+40	+47		
	+4	+5	+6	+7	+9	+10	+12		
G8	+22	+27	+33	+40	+48	+56	+66		
	+4	+5	+6	+7	+9	+10	+12		
H6	+8	+9	+11	+13	+16	+19	+22		
	0	0	0	0	0	0	0		
H7	+12	+15	+18	+21	+25	+30	+35		
	0	0	0	0	0	0	0		
H8	+18	+22	+27	+33	+39	+46	+54		
	0	0	0	0	0	0	0		
H9	+30	+36	+43	+52	+62	+74	+87		
	0	0	0	0	0	0	0		
H10	+48	+58	+70	+84	+100	+120	+140		
	0	0	0	0	0	0	0		
H11	+75	+90	+110	+130	+160	+190	+220		
	0	0	0	0	0	0	0		
J6	+5	+5	+6	+8	+10	+13	+16		
	−3	−4	−5	−5	−6	−6	−6		
J7	+6	+8	+10	+12	+14	+18	+22		
	−6	−7	−8	−9	−11	−12	−13		
J8	+10	+12	+15	+20	+24	+28	+34		
	−8	−10	−12	−13	−15	−18	−20		

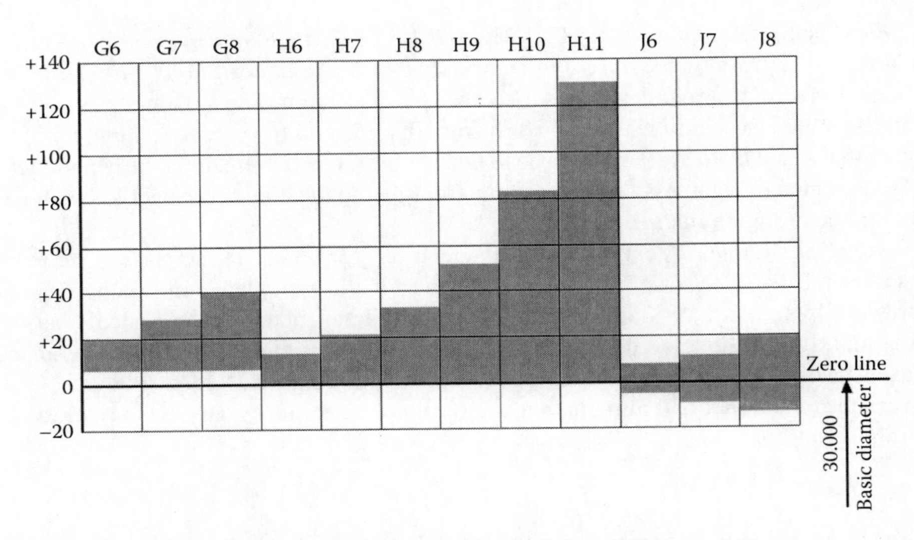

FIGURE 7.23 Graphical representation of common bore tolerances for a bore of 30 mm dia.

7.5 LARGE-SCALE FEATURES: GEOMETRICAL TOLERANCES

7.5.1 Concept and Standard Symbols

The purpose of geometrical tolerancing is to describe the allowable deviations from the ideal part geometry of products. A universal language of symbols is set by various standards for geometrical tolerancing, much like the international system of road signs that advise drivers how to navigate the roads. Geometrical tolerancing symbols allow a design engineer to precisely and logically describe part features in a way they can be manufactured and inspected maintaining accuracy needed for part intended performance in service.

When utilizing geometrical tolerancing and dealing with the component's real geometrical surface, the deviations from the nominal shape, orientation, and location can be either a single or related to a datum feature of geometrical tolerance type (Figure 7.24). The single classification

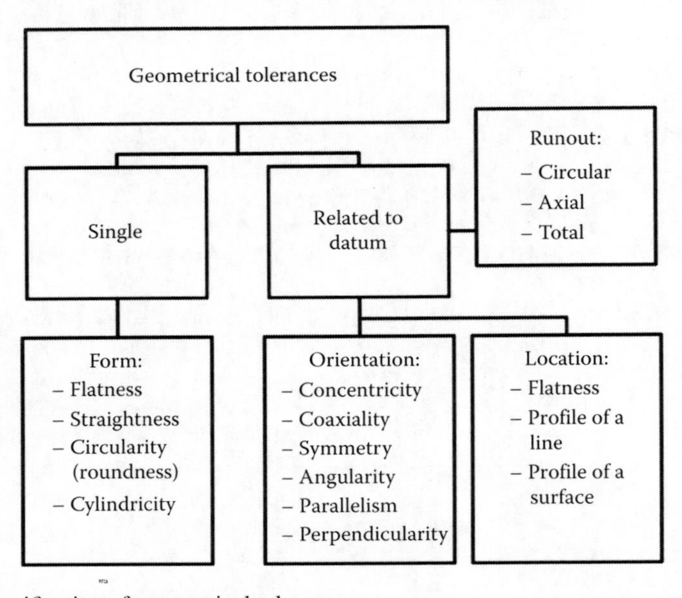

FIGURE 7.24 Classification of geometrical tolerances.

relates to form tolerances that are not normally related to a datum, for example, roundness. It is possible to look at the profile tolerance types as form tolerances as they do not always relate to a datum.

There are sets of standards for geometrical tolerancing: GPS set by ISO standards and GD&T set by American standards. The objective of these standards is to define requirements and rules for describing the geometrical requirements for part and assembly geometry. Proper application of these standards ensures that the allowable part and assembly geometry defined in the drawing leads to parts that have the desired form and fit (within limits) and function as intended.

The utilization of geometrical tolerancing in drawings is by

- Geometrical references (datum features)
- Feature control frames
- Geometrical characteristics (symbols)
- Tolerance shapes
- Tolerance zones (values).

The symbols for the drawing indications of geometrical tolerances according to ISO 1101, ISO 5459, ISO 286, and ISO 10579 are shown in Figure 7.25, and Table 7.6 shows the symbols for the geometrical tolerance characteristics. Table 7.7 presents a geometrical tolerance reference chart relevant to drilling operations. With the symbols, all kinds of geometrical tolerances can be expressed.

7.5.2 Definitions of Basic Terms

There are some important terms used in geometrical tolerancing, which have to be understood thoroughly in order to comply with the requirements of corresponding standards. A *feature* is a general term applied to a physical portion of a part, such as a surface, hole, or slot. An easy way to remember this term is to think of a feature as a part surface (Krulikowski 1998). The part shown in Figure 7.26 contains seven features: the top and bottom, the left and right sides, the front and back, and the hole surface.

A *feature of size* (*FOS*) is one cylindrical or spherical surface, or a set of two opposed elements or opposed parallel surfaces, associated with a size dimension. A key part of the FOS definition is that the surfaces or elements *must be opposed* (Krulikowski 1998). An axis, median plane, or centerpoint can be derived from a FOS. FOS dimension is a dimension that is associated with a FOS. Figure 7.27 shows some important examples of features of size and features of size dimensions (Henzold 2006).

Tolerance zone is space limited by one or several geometrically perfect lines or surfaces and characterized by a linear dimension, called a tolerance. A geometrical tolerance applied to a feature defines the tolerance zone within which that feature shall be contained. Unless a more restrictive indication is required, for example, by an explanatory note, the toleranced feature may be of any form or orientation within this tolerance zone.

7.5.3 Definitions of Geometrical Tolerance-Related Terms

Figure 7.28 shows the structure of geometrical tolerances.

Form deviation is the deviation of a feature (geometrical element, surface, or line) from its nominal form (Figure 7.29). If not otherwise specified, form deviations are assessed over (or along) the entire feature. Form deviations are originated, for example, by the looseness or error in guidances and bearings of the machine tool, deflections of the machine tool or the workpiece, error in the

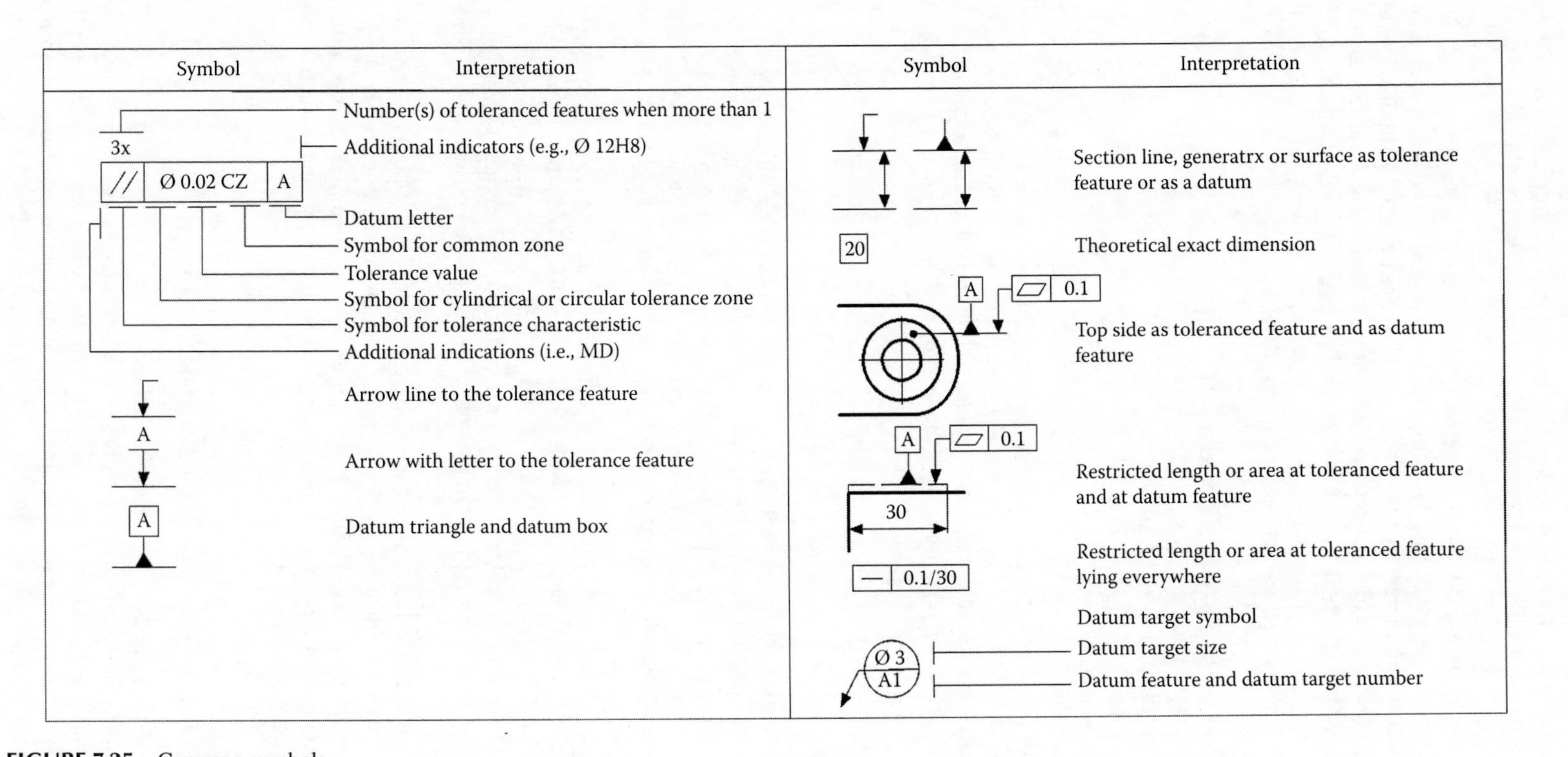

FIGURE 7.25 Common symbols.

TABLE 7.6
Modifiers

Symbol	Interpretation (Modifier)
Ⓕ	Free state
Ⓛ	Least material condition (LMC)
Ⓜ	Maximum material condition (MMC)
Ⓢ	Regardless of feature size (RFS) (ASME Y14.5)
Ⓤ	Unsymmetrical tolerance zone (ASME Y14.5)
Ⓔ	Envelope requirement
Ⓡ	Reciprocity requirement
Ⓣ	Target method (ASME Y14.5)
LE	Allies to line element
NC	Not convex
ACS	Any cross section

fixture of the workpiece, hardness deflection, or wear. The ratio between width and depth of local form deviations is in general more than 1000:1.

Form tolerance (Figure 7.24) is the permitted maximum value of the form deviation. According to ISO 1101, there are defined form tolerance zones within which all points of the feature must be contained. Within this zone, the feature may have any form, if not otherwise specified. The tolerance value defines the width of this zone (Figure 7.29). Form tolerances limit the deviations of a feature from its geometrical ideal line or surface form. Special cases of line forms with special symbols are straightness and roundness (circularity) (Figure 7.30). Special cases of surface forms with special symbols are flatness (planarity) and cylindricity. Figure 7.31 shows an example of cylindricity tolerance—the surface shall be contained between two coaxial cylinders with a radial distance 0.05 (Henzold 2006).

Orientational deviation is the deviation of a feature from its nominal form and orientation. The orientation is related to one or more (other) datum feature(s). The orientational deviation includes the form deviation (Figure 7.29). If not otherwise specified, orientational deviations are assessed over the entire feature. Orientational deviations are originated similarly as form deviations. They are originated also by erroneous fixture of the workpiece after remounting on the machine tool.

Orientation tolerance is the permitted maximum value of the orientation deviation. According to ISO 1101, there are defined orientation tolerance zones within which all points of the feature must be contained. The orientation tolerance zone is in the geometrical ideal orientation with respect to the datum(s). The tolerance value defines the width of this zone (Figure 7.29) (Henzold 2006).

TABLE 7.7
Geometrical Tolerancing Reference Chart

Type of Tolerance	Geometrical Characteristics	Symbol	Can Be Applied to a Feature	Can Be Applied to a Feature Size	Can Affect Virtual Condition?	Datum Reference Used?	Can Use Ⓜ Modifier?	Can Use Ⓢ Modifier?	Standard
Form	Roundness	○	Yes	No	No	No	No	No (note 5)	ISO 12181
Form	Cylindricity	⌭	Yes	No	No	No	No	No (note 5)	ISO 12180
Form	Straightness	—	Yes	Yes	Yes (note 1)	No	Yes (note 1)	No (note 5)	ISO 12780
Location	Symmetry	⌯	No	Yes	Yes	Yes	No	No (note 5)	ISO 2768
Location	Concentricity	◎	No	Yes	Yes	Yes	No	No (note 5)	ISO 2768
Location	Position tolerance	⌖	No	Yes	Yes	Yes	Yes	Yes	ISO 1101
Orientation	Perpendicularity	⊥	Yes	Yes	Yes (note 1)	Yes	Yes (note 1)	No (note 5)	ISO 2768
Orientation	Parallelism	//	Yes	Yes	Yes (note 1)	Yes	Yes (note 1)	Yes (note 4)	ISO 2768
Runout	Circular runout	↗	Yes	Yes	Yes (note 1)	Yes	No	No (note 5)	ISO 1101 ISO 2768
Runout	Total runout	⌰	Yes	Yes	Yes (note 1)	Yes	No	No (note 5)	ISO 1101 ISO 2768

Notes: (1) When applied to a FOS. (2) Can also be used as a form control without a datum reference. (3) When a datum FOS is referenced with the MMC modifier. (4) When an MMC modifier is used. (5) Automatic per rule #3.

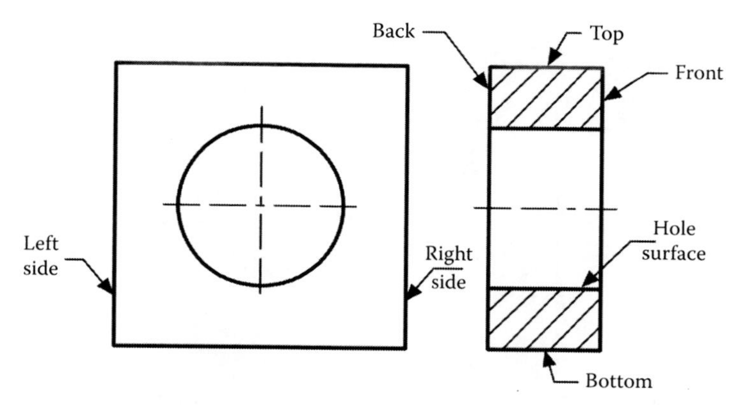

FIGURE 7.26 Examples of features.

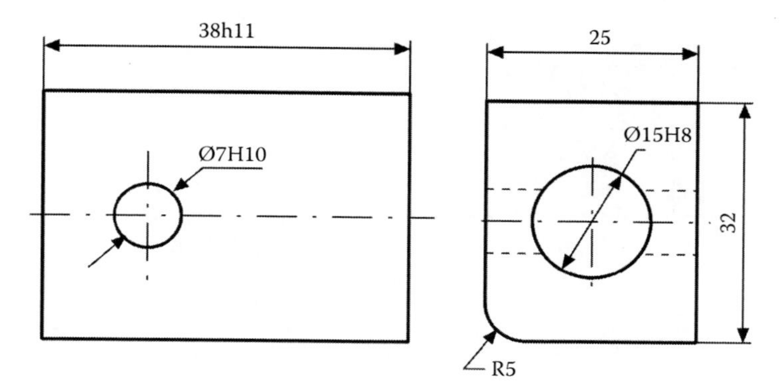

FIGURE 7.27 Examples of features of size and features of size dimensions.

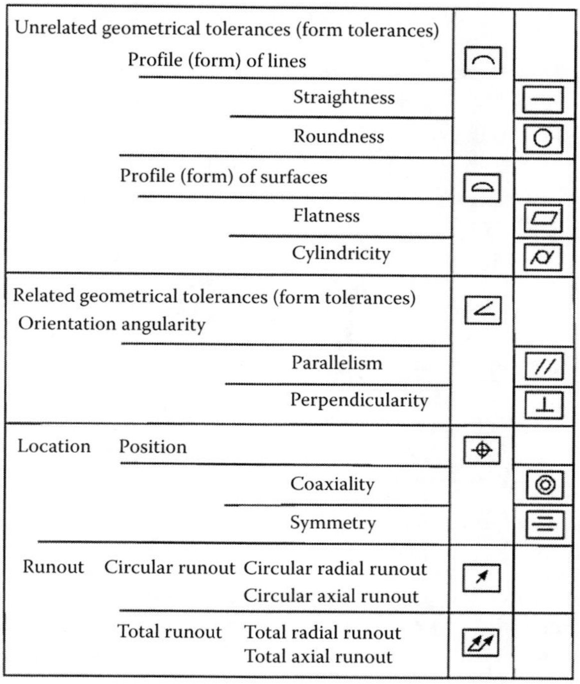

Geometrical tolerance	Symbol	Symbol
Unrelated geometrical tolerances (form tolerances)		
Profile (form) of lines	⌒	
Straightness		—
Roundness		○
Profile (form) of surfaces	⌓	
Flatness		▱
Cylindricity		⌭
Related geometrical tolerances (form tolerances) Orientation angularity	∠	
Parallelism		//
Perpendicularity		⊥
Location Position	⌖	
Coaxiality		◎
Symmetry		⌯
Runout Circular runout Circular radial runout Circular axial runout	↗	
Total runout Total radial runout Total axial runout	⌰	

FIGURE 7.28 Structure of geometrical tolerances.

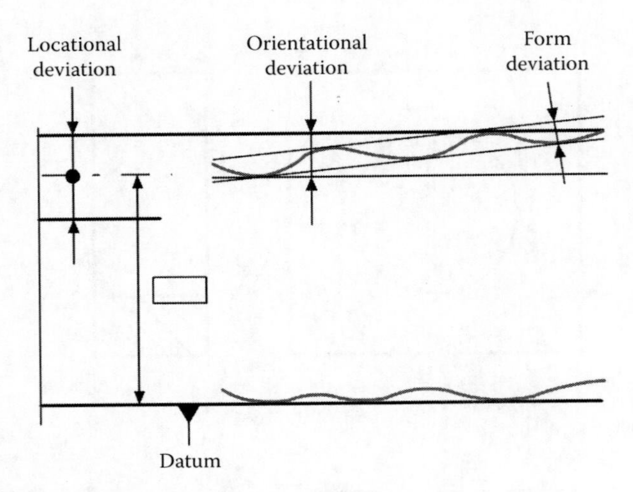

FIGURE 7.29 Form, orientational, and locational deviation.

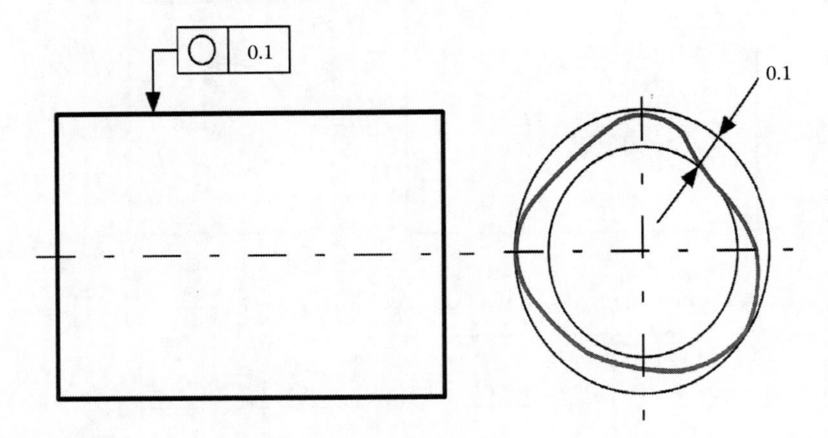

FIGURE 7.30 Form deviation: roundness (circularity).

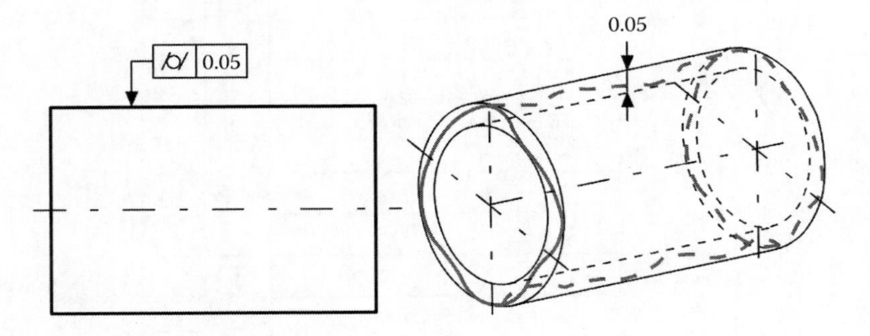

FIGURE 7.31 Form deviation: cylindricity.

Orientation tolerances limit the deviations of a feature from its geometrical ideal orientation with respect to the datum(s). Special cases of orientation with special symbols are parallelism (0°) and perpendicularity (90°) (Figure 7.32). The orientation tolerance also limits the form deviation of the toleranced feature, but not of the datum feature(s). If necessary, a form tolerance of the datum feature(s) must be specified (Henzold 2006).

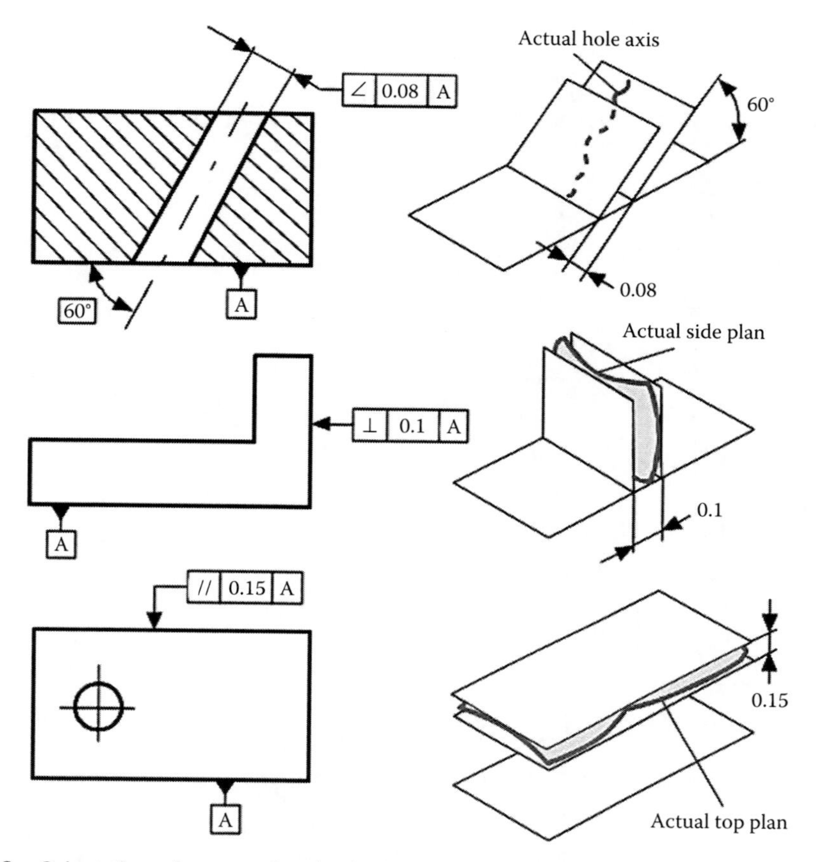

FIGURE 7.32 Orientation tolerances: drawing indications and tolerance zones.

Locational deviation is the deviation of a feature (surface, line, point) from its nominal location. The location is related to one or more (other) datum feature(s). The locational deviation includes also the form deviation and the orientational deviation (of the surface, axis, or median face) (Figure 7.29). If not otherwise specified, locational deviations are assessed over the entire feature. Locational deviations are originated similarly as size, form, and orientational deviations (Henzold 2006).

Location tolerance is twice the permitted maximum value of the locational deviation. According to ISO 1101, there are defined location tolerance zones within which all points of the feature must be contained. The location tolerance zone is in the geometrical ideal orientation and location with respect to the datum(s). The tolerance value defines the width of this zone (Figure 7.29). Location tolerances limit the deviations of a feature from its geometrical ideal location (orientation and distance) with respect to the datum(s). Special cases of location with special symbols are coaxiality (when toleranced feature and datum feature are cylindrical) and symmetry (when at least one of the features concerned is prismatic) where the nominal distance between the axis or median plane of the toleranced feature and the axis or median plane of the datum feature is zero (Figure 7.33). The location tolerance also limits the orientation deviation and the form deviation of the toleranced feature (plane surface or axis or median face), but not the form deviation of the datum feature(s). If necessary, a form tolerance for the datum feature(s) must be specified (Henzold 2006).

Runout tolerances are partly orientation tolerances (axial circular runout tolerance, axial total runout tolerance) and partly location tolerances (radial circular runout tolerance, radial total runout tolerance). However, according to ISO 1101, they are considered as separate tolerances with separate symbols because of their special measuring method.

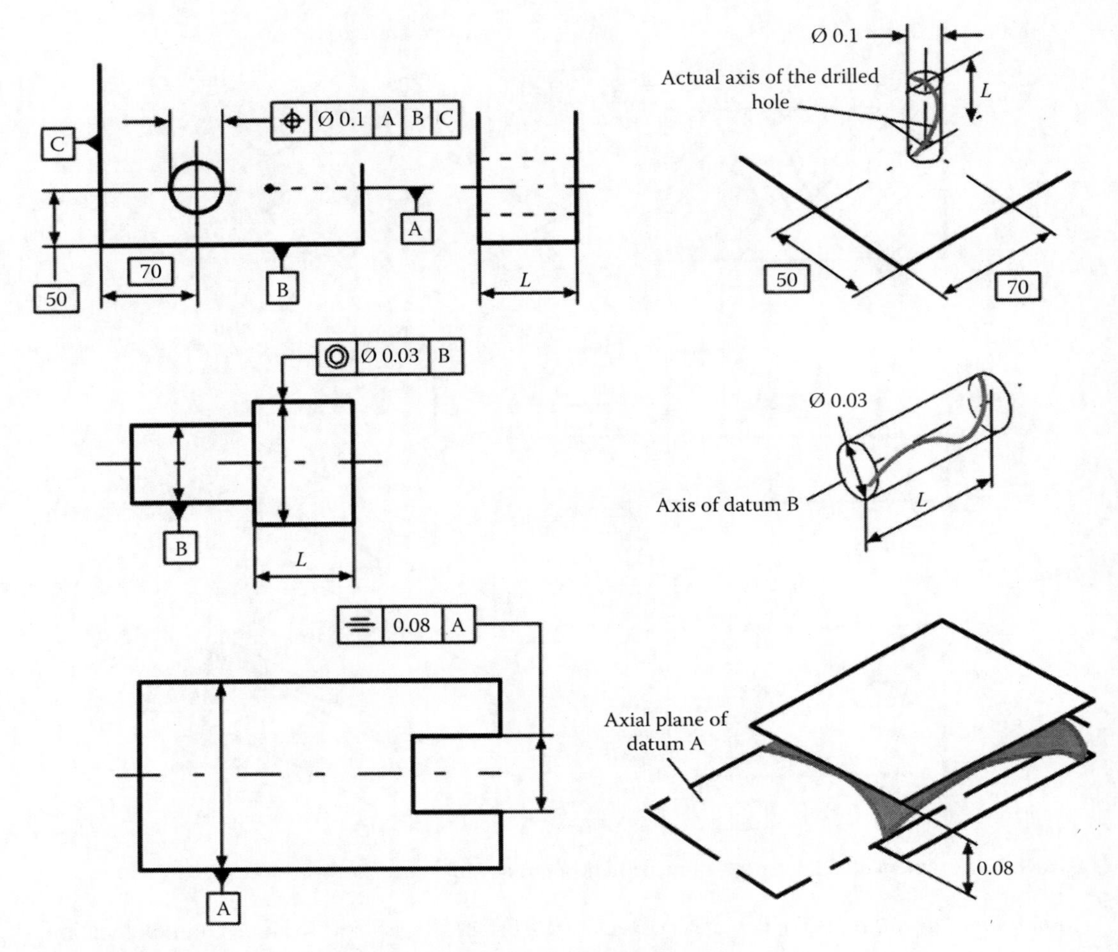

FIGURE 7.33 Location tolerances: drawing indications and tolerance zones.

Circular radial runout (Figure 7.34a). In each plane perpendicular to the common datum axis, the profile (circumference) shall be contained between two circles concentric with the datum axis and with a radial distance of 0.05 mm.

Total radial runout tolerance (Figure 7.34b). The surface shall be contained between two cylinders coaxial with the datum axis and with a radial distance of 0.05 mm. During checking of the circular radial runout deviation, the positions of the dial indicator are independent of each other. However, during checking of the total radial runout deviation, the positions of the dial indicator are along a guiding (straight) line parallel to the datum axis. Therefore, the straightness deviations and the parallelism deviations of the generator lines of the toleranced cylindrical surface are limited by the total radial runout tolerance, but not by the circular radial runout tolerance (Henzold 2006).

7.5.4 Datum Features

In the author's opinion, the datum specification distinguishes a drawing from a picture. This is particularly true for any tool drawing that *must* begin with the assignment of the proper datum.

Datums are theoretically perfect points, lines, axes, surfaces, or planes used for referencing features of an object. They are established by the physical datum features that are identified in the drawing. Identification of datum features is done by using a datum feature symbol defined by

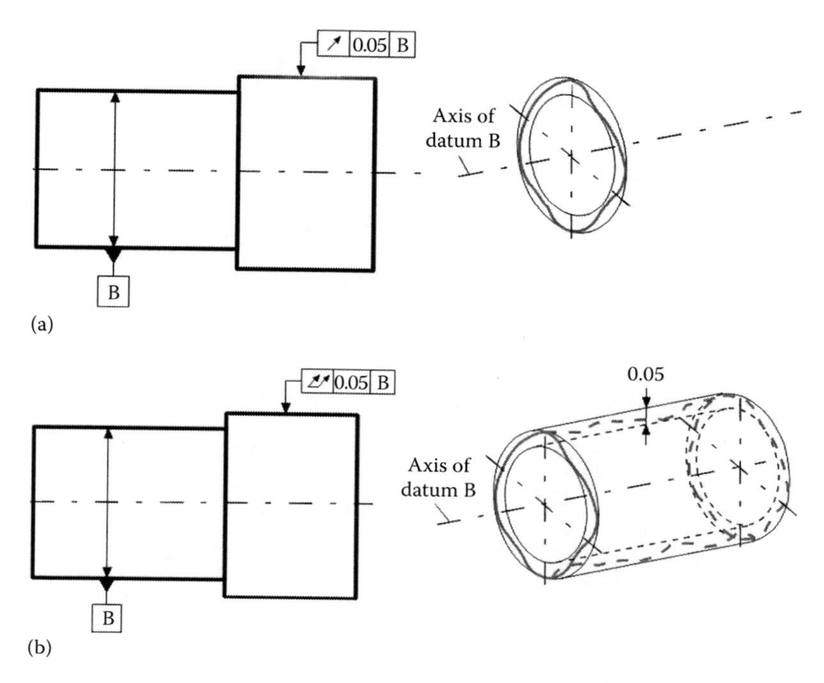

FIGURE 7.34 (a) Circular radial runout tolerance and (b) total radial runout tolerance: drawing indications and tolerance zones.

standard ISO 5459:2011 *Geometrical product specifications (GPS)—Geometrical tolerancing—Datums and datum systems*. This symbol consists of a capital letter enclosed in a square frame. A leader line extends from the frame to the selected feature. A triangle is attached to the end of the leader and is applied in the appropriate way to indicate a datum feature (Figure 7.35).

The datum may be established by the following:

- One single datum feature (e.g., Figure 7.34)
- Two or more datum features of the same priority as a common datum (e.g., a common axis, as shown in Figure 7.36), indicated by a hyphen between the datum letters in the tolerance frame
- Two or more datum features with different but not specified priorities (indicated by a sequence of datum letters in the tolerance frame without separation) (Figure 7.37). Because such an indication is not unequivocal and may lead to different measuring results for the same workpiece, this indication has therefore been eliminated from ISO 1101 (now labeled as former practice)
- Datum features of different priorities (e.g., three-plane datum system according to ISO5459) (Figure 7.38).

FIGURE 7.35 Datum symbol.

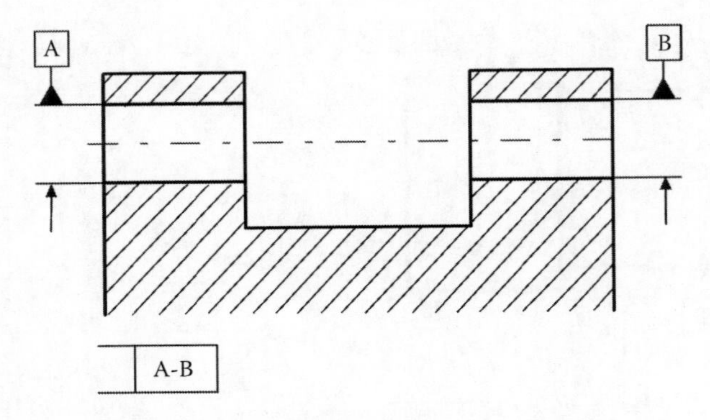

FIGURE 7.36 Datums of same priority as common datums.

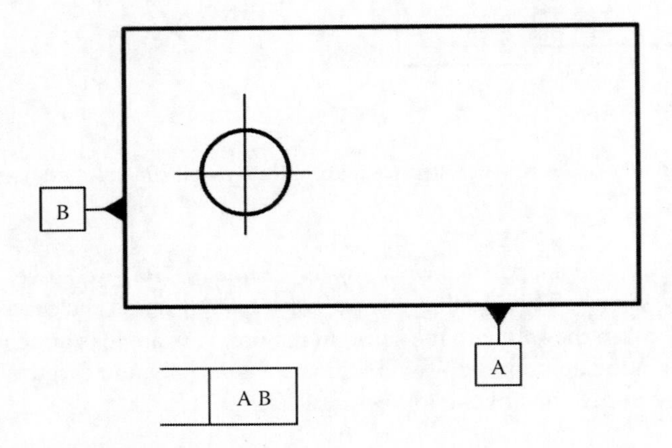

FIGURE 7.37 Datums with priorities not specified.

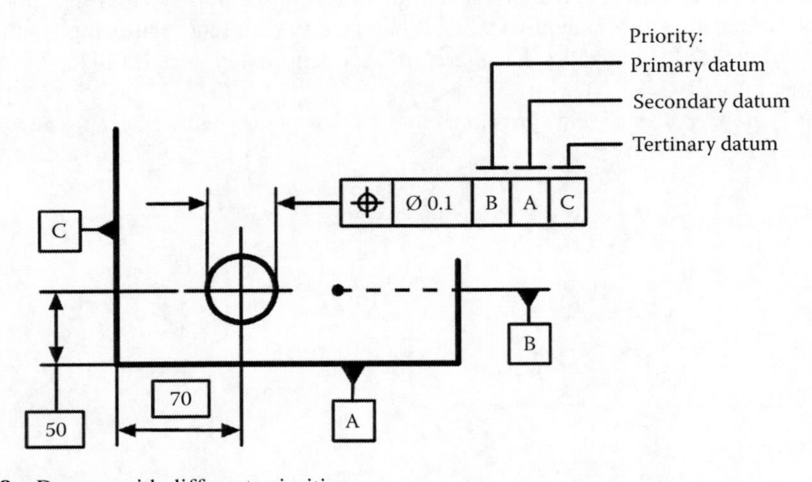

FIGURE 7.38 Datums with different priorities.

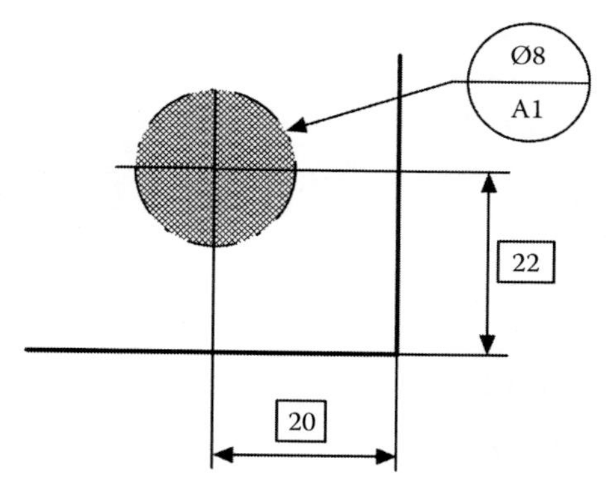

FIGURE 7.39 Datum target frames.

Datums with different priorities (order of precedence) are classified as follows (Henzold 2006):

1. *The primary datum*—datum feature orientated to the minimum requirement relative to the simulated datum feature.
2. *The secondary datum*—datum feature orientated without tilting relative to the primary simulated datum feature (only by translation and rotation) according to minimum requirement relative to the secondary simulated datum feature. The secondary simulated datum feature is perpendicular to the primary simulated feature.
3. *The tertiary datum*—this is defined as a feature or features used to complete the coordinate system in relation to the primary and secondary datums.

Where a component is rough machined, datum target points are used to establish a datum. Datum target frames are used to define the points, lines, or areas where the manufacturing locations or measurement points should be defined from, to create the coordinate system. Figure 7.39 shows an example of a datum target frame. This example shows the location zone for manufacture or measurement to be anywhere within the circular area of a diameter of 8 mm with a theoretical center being at 20 mm by 22 mm from the corner. A1 is the reference name (datum feature and datum target number). The target is an area so it is hatched.

7.6 BORE GAGING

Bore gaging, already a significant factor in manufacturing, has become increasingly important in metrology because of the growing concern for total quality assurance (TQA). Compared with OD measurement, bore gaging creates more engineering challenges by the very nature of its special role in QA. This is especially true when measuring difficult-to-assess internal features such as splines, threads, and deep bores.

As machine capability increases, so does the demand for dimensional gaging to cope with tighter limits on tolerances and greater complexity of component parts. Manufacturers are concerned not only with making the component to a specific tolerance band but also with size variation from component to component because it gives them better control of their process. To meet this demand, the gage they use must be able to discriminate size variation better than before. Moreover, many manufacturers now require that any in-process or post process gage records and sends the measured size of a component to a statistical process control (SPC) or data-collection system. This section aims to familiarize manufacturing professionals with the basic concepts of bore gaging.

7.6.1 Bore Gage Classification and Specification

To measure the bore tolerances, modern manufacturing requires the use of gages. A *gage* is defined as a device for investigating the dimensional fitness of a part for specific function.

Gaging is defined by ANSI as a process of measuring manufactured materials to assure the specified uniformity of size and contour required by industries. Gaging thereby assures the proper functioning and interchangeability of parts; that is, one part will fit in the same place as any similar part and perform the same function, whether the part is for the original assembly or replacement in service (Nee 2010).

Bore gages may be classified as follows:

1. Master gages.
2. Inspection gages.
3. Manufacturer's gages.
4. Gages that control dimensions.
5. Gages that control various parameters of bore geometry.
6. Fixed limit working gages.
7. Variable indicating gages.
8. Post process gages.
9. In-process gages.

Master gages are made to their basic dimensions as accurately as possible and are used for reference, such as for checking or setting inspection of manufacturer's gages. *Inspection gages* are used by inspectors to check the manufactured products. *Manufacturer's gages* are used for inspection of parts during production.

Post process gages are used for inspecting parts after being manufactured. Basically, this kind of gage accomplishes two things: (1) it controls the dimensions of a product within the prescribed limitations and (2) it segregates or rejects products that are outside these limits. Post process gaging with feedback is a technique to improve part accuracy by using the results of part inspection to compensate for repeatable errors in the bore manufacturing operations. The process is normally applied to CNC machines using inspection data to modify the part program and on tracer machines using the same data to modify the part template.

In-process gages are used for inspecting parts during the machining cycle. In today's manufacturing strategy, in-process gages and data-collection software provide faster feedback on quality. Indeed, the data-collection and distribution aspect of 100% inspection has become as important as the gaging technology itself. Software, specifically designed to capture information from multiple gages, measure dozens of product types and sizes, and make it available to both roving inspectors and supervising quality personnel as needed, is quickly becoming part of quality control strategies as an integral part of the measurement systems analysis (MSA). In conjunction with CNC units, in-process gaging can automatically compensate for workpiece misalignment, tool length variations, and errors due to tool wear.

7.6.2 Components of Gage Accuracy

The accuracy of a gage is determined by three factors:

1. *Resolution*. For a gage to accurately determine whether or not a component is in tolerance, the general rule is that the gage resolution should be near 10% of the tolerance
2. *Reproducibility and Repeatability* (*R&R*)
3. *Linear accuracy*. This is the value of maximum deviation from the true size a gage will be capable of measuring across its entire working range. As this Figure is intrinsically linked to the range of the gage, it means nothing unless considering both the value of deviation and the range together. Many manufacturers mistakenly use the 10% rule that establishes an appropriate resolution and apply it to linear accuracy. This is incorrect because it does not take into account the range of the gage.

7.6.3 Bore Gage Types

In choosing a bore gage, one should first eliminate those types of bore gages that are least appropriate for the application. For example, for a production environment, noncontact measurement techniques, including optical and laser methods, tend to be bulky, relatively expensive, and inflexible when measuring special internal features. Such gaging requires the part to be taken to the gage. Not all types of noncontact systems are suitable for a production environment, and many entail high maintenance costs.

Gages that control dimensions are used to control bore diameter (Nee 2010). These gages can be either post process or in-process gages. Further, these gages can be either *fixed limit gages* or *variable indicating gages*. These cages are classified as (1) plug/pin gages, (2) needle gages, (3) chamfer gages and countersink gages, (4) dial caliper gages, (5) dial bore gages, (6) electronic bore gages, (7) hole micrometers/ingages, and (8) telescopic gages.

The detailed description of bore gages and their selection and tolerances was presented by the author earlier (Astakhov 2014).

7.7 DRILL METROLOGY

For centuries of drilling history, drill metrology was not considered seriously as the tolerances on drilled holes were wide open. Primitive hand gages and eyeballing measurements relying more on common sense and experience than on results of accurate measurements were considered common practice in drill metrology. Nowadays, however, with widening use of HP drills and modern drilling systems, the tolerances on drilled holes became the same as the were for reamed or even for ground holes not long ago. The dimensional parameters of drills to make such holes as well as parameters of drill geometry became of prime concern. Therefore, this section for the first time in the drilling literature discusses and explains drill metrology, its parameters, and their importance, defines the proper terms, presents the measuring methodologies, and provides suggestions on the tolerances and their proper specification for common and HP drills.

The drill metrological features can be thought of as the dimensional and geometrical non-datum features, for example, drill diameter and length, point and chisel angles, and the datum features, for example, drill runout.

7.7.1 Drilling Tool Diameter

Although it may appear that the drill diameter and its tolerance are the simplest part of drilling metrology as its selection is based upon the tool layout (see Chapter 1), it is not always the case, as a number of important system considerations not normally considered in the literature in the field should be involved in the proper selection of drill diameter tolerance. This section aims to explain some critical issues in the selection of the proper drill diameter and assigning its tolerance limits.

The obvious answer to the question *what should be the drill diameter* is *it should be equal to the diameter of the hole to be drilled with this drill*. Being perfectly correct under the ideal conditions, this answer, however, does not account for the tolerances on the hole, manufacturing tolerance on the drill, shrinkage of the hole after drilling, and the back taper of the drill, that is, that the drill diameter reduces with each successive re-sharpening (see Section 4.5.9.5).

7.7.1.1 Existing Tolerances

Drilling tool diameter tolerances are considered using practical examples. An unprepared user may think that the drill tolerances should correspond to the hole (bore) tolerances shown in Table 7.5 and depicted in Figure 7.23. As bores of H8–H11 tolerances are mostly drilled, the drill tolerances should primarily conform to these tolerances being somewhere within the corresponding tolerance zones. Reality, however, proves otherwise.

Tables 7.8 through 7.10 show tolerances on drills according to most commonly used sources of such tolerances. As can be seen, being a bit courser, the tolerances according to standard ASME B94.11M-1993 for general-purpose HSS twist drills are in the same line as those by standard DIN 338, while tolerances used by one of the major drill manufacturers depend on the drill material, that is, the tolerances on HSS drills are not nearly the same as those on carbide drills.

To understand the problem with drill tolerances, consider a graphical representation of drill tolerances for the most common H10 bore tolerance. Figure 7.40 shows an example for a bore of

TABLE 7.8
Drill Diameter Tolerance according to Standard ASME B94.11M-1993 for General-Purpose HSS Twist Drills

Diameter of Drill		Tolerance			
		Inches		Millimeters	
Inches	Millimeters	Plus (+)	Minus (−)	Plus (+)	Minus (−)
From #97 to #81	From 0.15 to 0.33	0.0002	0.0002	0.005	0.005
Over #81–1/8	Over 0.33–3.18	0.0000	0.0005	0.000	0.013
Over 1/8–¼	Over 3.18–6.35	0.0000	0.0007	0.000	0.018
Over ¼–½	Over 6.35–12.70	0.0000	0.0010	0.000	0.025
Over ½–1	Over 12.70–25.40	0.0005	0.0012	0.000	0.030
Over 1–2	Over 25.40–50.80	0.0000	0.0015	0.000	0.038
Over 2–3½	Over 50.80–88.98	0.0000	0.0020	0.000	0.051

TABLE 7.9
HSS Drill Diameter Tolerances according to DIN 338 Used by Leading Tool Manufacturers (h8 ISO)

Drill Diameter (mm)	Tolerance (μm)
Over 3–6	−0/−18
Over 6–10	−0/−22
Over 10–18	−0/−27
Over 18–30	−0/−33
Over 30–50	−0/−39
Over 50–65	−0/−46

TABLE 7.10
Common Tolerances on Carbide Drills by Leading Tool Manufacturers (m7 ISO)

Drill Diameter (mm)	Tolerance (μm)
Over 3–6	+16/+4
Over 6–10	+21/+6
Over 10–18	+25/+7
Over 18–30	+29/+8
Over 30–50	+34/+9
Over 50–65	+41/+11

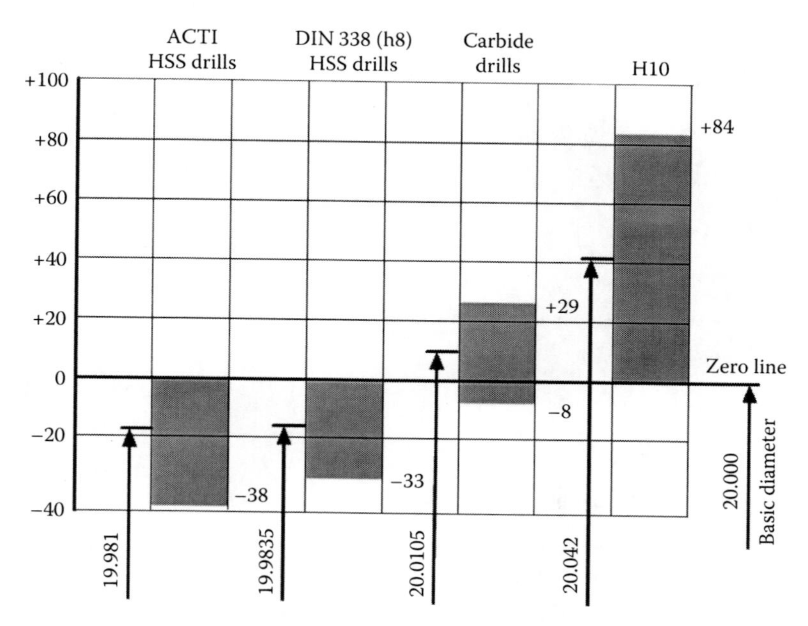

FIGURE 7.40 Example of drill diameter tolerances for a hole of 20H10 dia.

20 mm nominal diameter. As can be clearly seen, if a drill is made according to standard ASME B94.11M-1993, then the system runout should be 0.061 mm to achieve the middle of the tolerance zone of the bore; if a drill is made to DIN standard, then the system runout should be 0.0585 mm to achieve the middle of the tolerance zone of the bore; and if a modern carbide drill is used, then the system runout should be 0.0315 mm to achieve the middle of the tolerance zone of the bore. The latter is more realistic for new drilling systems where rigid spindles and precision shrink-fit and hydraulic tool holders are used, while the former two are allocable for older drilling systems. In other words, the standards and recommendations were adopted a long time ago, and thus, they do not consider the recent changes in the accuracy of components of the drilling system, while the tool manufacturing companies are much more adaptive to these rapidly changing conditions.

The drilling tool diameter and its tolerance are the system-dependent parameters. For certain particular cases, one of which is considered further, the issue of the selection of the proper drilling tool diameter and its tolerances becomes critical. Therefore, a methodology for calculating the proper drilling tool diameter and its tolerance zone should be available. Such a methodology is considered in the next section.

7.7.1.2 Methodology to Calculate the Drilling Tool Diameter and Its Tolerance Zone

The selection of the proper drilling tool diameter is important when a tight-tolerance hole is to be machined or when the tool that follows the drill is sensitive to the drilled hole size. As an example, this section discusses a methodology of the selection of the proper diameter and tolerances on a tap drill (used to drill a hole for further taping with a threading tap). Figure 7.41 shows the model to calculate tap drill diameter, its tolerance, and minimum allowable diameter after regrind for metric threads with H tolerance position. In the model shown in Figure 7.41, the following applies:

D_{1min} is the minimum minor diameter of internal thread (Table 7.11).

D_{1max} is the maximum minor diameter of internal thread (Table 7.11).

A_{max}, A_{min} are the maximum and minimum growth (elastic recovery) of the minor diameter on tapping. These depend on the properties of the work material. Table 7.12 shows A_{max} for 8%Si aluminum alloys (e.g., A380), while A_{min} is approximately equal to $A_{max}/2$ (GOST 19257-73).

$d_{1\text{-}max}$ and $d_{1\text{-}min}$ are maximum and minimum diameters of the tap hole.

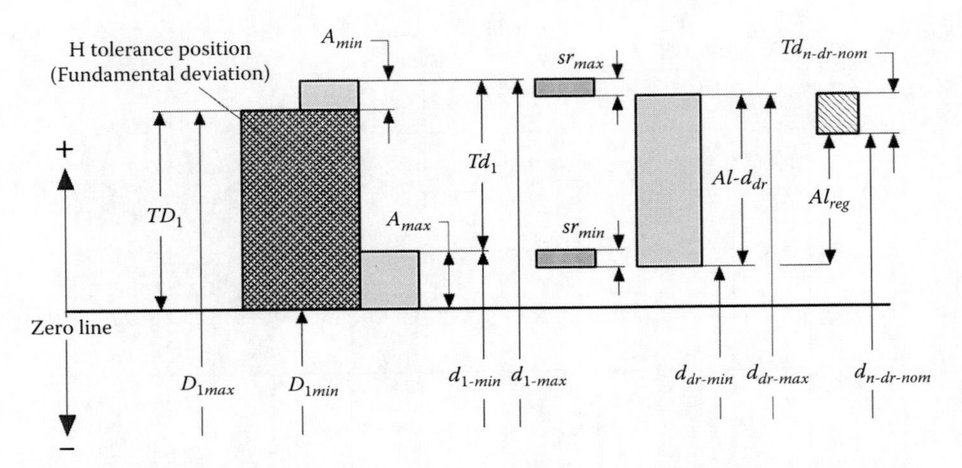

FIGURE 7.41 The model including metric tolerance system for screw threads with H tolerance position.

TABLE 7.11
Internal Metric Thread—M Profile Limiting Dimensions for M6, M8 and M10 threads

Basic Thread Designation	Tolerance Class	Minor Diameter, D_1 (mm) Min.	Max.
M5 × 0.75	6H	4.134	4.334
M6 × 1	6H	4.917	5.153
M8 × 1.12	6H	6.647	6.912
M10 × 1.5	6H	8.376	8.676

TABLE 7.12
Maximum and Minimum Growth (Recovery) of the Minor Diameter on Tapping for 8%Si Aluminum Alloys

Pitch (mm)	0.8	1	1.25	1.5
A_{max} (mm)	0.064	0.080	0.100	0.120

$Td_1 = d_{1\text{-}max} - d_{1\text{-}min}$ is the tolerance on the tap hole.
sr_{max} is the maximum drill setting runout.
sr_{min} is the minimum drill setting runout.
$d_{dr\text{-}min} = d_{1\text{-}min} - sr_{min}$ is the minimum diameter of the tap drill (cutoff diameter for regrinds) (to be shown in the drawing).
$d_{dr\text{-}max} = d_{1\text{-}max} - sr_{max}$ is the maximum diameter of the tap drill.
$Al\text{-}d_{dr} = d_{1\text{-}max} - d_{1\text{-}min}$ is the total allowance for the tap drill diameter variation.
$d_{n\text{-}dr\text{-}nom}$ is the chosen nominal diameter of new tap drill (to be shown in the drawing).
$Td_{n\text{-}dr\text{-}nom} = d_{dr\text{-}max} - d_{n\text{-}dr\text{-}nom}$ is the tolerance on a new drill (to be shown in the drawing).
Al_{reg} is the allowance for regrinding.

The methodology allows determining (1) the diameter of a new drill and its tolerance and (2) the minimum diameter of the tap drill (cutoff diameter for regrinds). It includes the following simple steps:

1. *Determination of* D_{1min} and D_{1max} using the tolerance zone assigned by the part drawing. For a given thread size, these are determined from Table 7.8. For example, for M6 × 1 thread, $D_{1min} = 4.917$ mm and $D_{1max} = 5.153$ mm. Graphical representation of the obtained diameters shown in Figure 7.42 (similar to that shown in Figure 7.41) makes this and further steps much more transparent helping to avoid numerical errors.
2. Determination of the maximum and minimum diameter of the machined hole prior recovery. As follows from Figure 7.41, $d_{1\text{-}max} = D_{1max} + A_{max}/2$ and $d_{1\text{-}min} = D_{1min} + A_{max}$. For example, for M6 × 1 thread, $A_{max} = 0.08$ mm (Table 7.8); therefore, $d_{1\text{-}max} = 5.153 + 0.08/2 = 5.193$ mm and $d_{1\text{-}min} = 4.917 + 0.08 = 4.997$ mm as can be seen in Figure 7.42.
3. Establishing the drilling system runout (see Section 7.7.5.5) depending upon the particular holder and setting practice used. For example, for the tap drill meant for M6 × 1 thread, $sr_{max} = 20$ μm = 0.02 mm for a standard drill setting.
4. *Calculating the minimum tool diameter.* In the considered example, the minimum diameter of the tap drill (cutoff diameter for regrinds) is $d_{dr\text{-}min} = \mathrm{d}_{1\text{-}min} - \mathrm{sr}_{min}$. The minimum runout can be zero in the most conservative case. For M6 × 1 thread, the minimum diameter is then calculated as $d_{dr\text{-}min} = 4.997 - 0.000 = 4.997$ mm as indicated in Figure 7.42.
5. *Calculating the maximum tool diameter.* In the considered example, the maximum diameter of the tap drill is calculated as $d_{dr\text{-}max} = \mathrm{d}_{1\text{-}max} - \mathrm{sr}_{max}$ as follows from Figure 7.41. In the considered example for M6 × 1 thread, $d_{dr\text{-}max} = 5.193 - 0.02 = 5.173$ mm as indicated in Figure 7.42.
6. *Calculating the tolerance on a new tool.* For the tap drill, it is calculated as $Td_{n\text{-}dr\text{-}nom} = d_{dr\text{-}max} - d_{n\text{-}dr\text{-}nom}$ (Figure 7.41). In the considered example for M6 × 1 thread, $Td_{n\text{-}dr\text{-}nom} = 5.173$ mm $- 4.997 = 0.176$ mm as indicated in Figure 7.42.
7. *Selecting the nominal tool diameter.* For the tap drill, it should be selected in the range of $d_{dr\text{-}min} - d_{dr\text{-}max}$ (Figure 7.41). In the considered example for M6 × 1 thread, special and standard tools are considered. The nominal drill diameter of 5.15 mm is selected for special tool while that of 5.10 mm—for standard.
8. *Assigning tolerance zone and deviations for the drilling tool.* For the considered example for M6 × 1 thread, the tolerance zone of the diameter of the special drill is assigned to be 0.026 mm as shown in Figure 7.42.

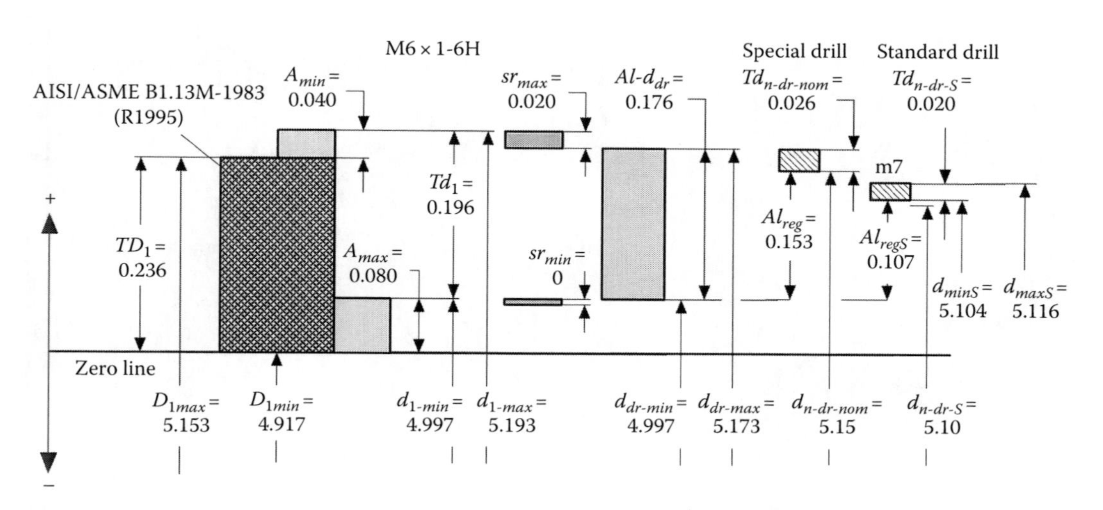

FIGURE 7.42 Graphical representation of tap drill diameter calculations for M6x1-6H tap drill.

Therefore, for a new special HP drill, its diameter should be indicated in the drawing as ø5.15+0.026/−0 and the minimum diameter of the tap drill (cutoff diameter for regrinds) as ø4.977 mm. Therefore, the allowance for drill regrinding is Al_{reg} is 0.153 mm as shown in Figure 7.42. In case of the standard drill, ISO tolerance m7 is indicated in catalogs of leading tool manufacturers (e.g., Sandvik Coromant and MAPAL)—for drill diameter of 5.10 mm, it is +0.016/+0.004 as shown in Figure 7.42. It means than the maximum diameter of the selected standard drill is 5.160 mm, while its minimum diameter is 5.104 mm. As such, as clearly seen in Figure 7.42, the allowance for drill regrinding Al_{regS} is 0.107 mm.

7.7.1.3 Assessment of the Results

Suggested tap drill diameter and tolerance are ø5.15+0.026/−0 mm and the minimum diameter of the tap drill (cutoff diameter for regrinds) is ø4.977 mm. According to the old rule, established by manufacturing practice long time ego, the tap drill diameter is selected as the tap size minus the pitch. This rule somehow made its way into ISO 2306-72 (revision date February 28, 2008), according to which the drill diameter for M6 × 1 is 5 mm with no tolerances assigned. Accounting for this standard, all reference books on tool selection and some tool manufacturers' catalogs recommend this diameter. If one uses a standard 5 mm drill with the standard tolerance and if this drill is made of a carbide tool material, then according to Table 7.10, the tolerance limits are +0.016/+0.04, that is, at the low end of the acceptable limit according to Figure 7.42. Due to back taper, such a drill will be out of the acceptable tolerance after few regrinds. If, however, the drill is made of HSS, its tolerance according to Table 7.9 is h8, that is, 0/−0.018 mm. As can be seen, the drill diameter can be well below the acceptable minimum according to Figure 7.42. Moreover, when such a drill is re-sharpened, its diameter becomes even smaller.

Figure 7.43 shows what happens on tapping when a standard carbide tap drill subjected to five re-sharpenings is used. As can be seen, the tap flute is full of chips. Not only cutting threads (two to three first relieved threads) but also practically all the threads participate in cutting, which creates high cutting torque and leads to tap breakage. Even when the tap can manage to complete the full thread, it breaks on retraction (Figure 7.44) as the hole becomes smaller due to bore springback after tapping. Moreover, the chip left in the tap flute causes its recutting and severely damages the thread as can be clearly seen in Figure 7.44.

The problems with tap breakage and poor quality of the threads were solved in the setting of the powertrain plants of one on the world's largest automotive company. The tap drills having the diameters calculated using the proposed methodology were used for threading holes M5, M6, M8, M10, and M12. A great reduction in the tap breakage, part scrap, and significant improving in thread quality were achieved.

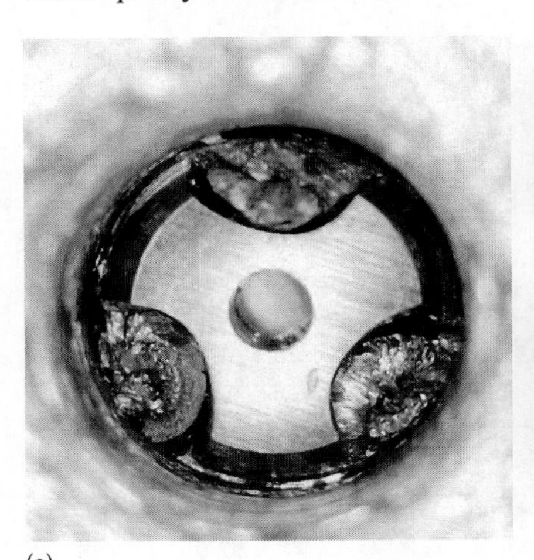

(a)

(b)

FIGURE 7.43 Tap can break due to significant amount of chips accumulated in the flute (a) and due to a high tapping torque as practically all threads are involved in cutting (b).

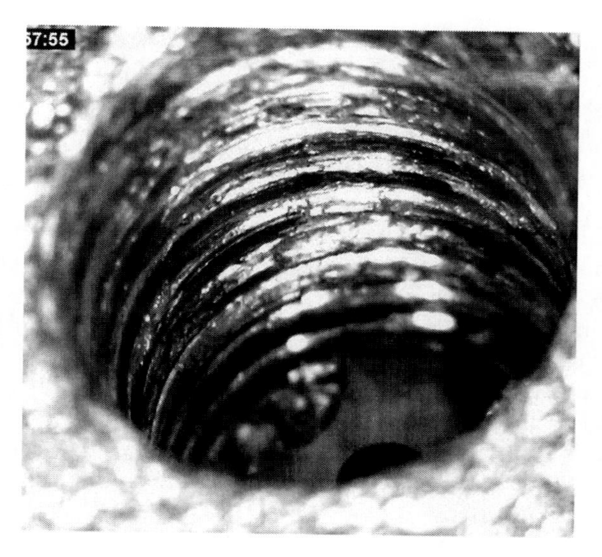

FIGURE 7.44 Showing poor quality of the machined thread and the tap broken on its retraction.

The foregoing considerations suggest that the tolerance on the drill should be calculated carefully rather than relying on multiple tables and recommendations found in various literature reference sources. Many of these recommendations and even standards were developed a long time ago (although recently revised) so that they do not reflect the changes that have occurred in industry over the past decade. In the considered example, the tap drill of 5 mm served its objective when tap drills were used in old machining system characterized by a great system runout (spindle runout + tool-holder runout). As such, the system runout can reach 0.15 mm that overlaps the lack of tap drill diameters. In modern manufacturing with high-speed spindles and precision shrink-fit and hydraulic holders, the system runout normally does not exceed 5–10 μm (0.005–0.010 mm) that creates the problems discussed previously with the diameter of tap drills. Figure 7.45 shows the quality of thread cut when the calculated diameter of the tap drill was used.

FIGURE 7.45 Thread cut when the calculated diameter of the tap drill was used.

7.7.1.4 Drill Diameters in the Tool Drawing

Figure 7.46 shows how the diameters with tolerances of the steps of a drill are indicated in the drawing. Note that the drill diameters are absolute, that is, a non-datum feature as measured between the drill corners regardless of a particular location of these corners with respect to other features.

7.7.2 Shank Diameter

As discussed in Chapters 1 and 5, drills can have a great number of shank designs/configurations. For HP drills, however, cylindrical shanks are recommended because they assure the highest accuracy of drill location in the holder. In the case when the shank is cylindrical, the shank diameter (d_{sh} in Figure 7.46) is the diameter of the cylindrical shank.

Table 7.13 shows the shank diameters according to standard ASME B94.11M-1993 for general-purpose HSS twist drills. They are a way too coarse and, what is more important, are not suitable for modern hydraulic and shrink-fit drill holders. Table 7.14 shows the tolerances of the shank diameter (straight shank drills) used by leading drill manufacturers for modern HP drills.

As the shank is used as the datum feature and as it is installed with high precision in a tool holder, no scribing/etching (e.g., tool number, manufacturer's logo, order numbers/letters) that compromises these properties should be placed on the shank for HP drills as when such a drill is installed in the high-precision holder for its measurement or presetting, this scribing/etching affects the accuracy of drill location in the drill holder. According to the author's experience, tool inspectors/presenters often use a diamond file to *smooth-out* the scribing/etching on the drill shank (e.g., as those shown in Figure 7.47) to achieve the desirable accuracy of inspection/presetting. Figure 7.48 shows an

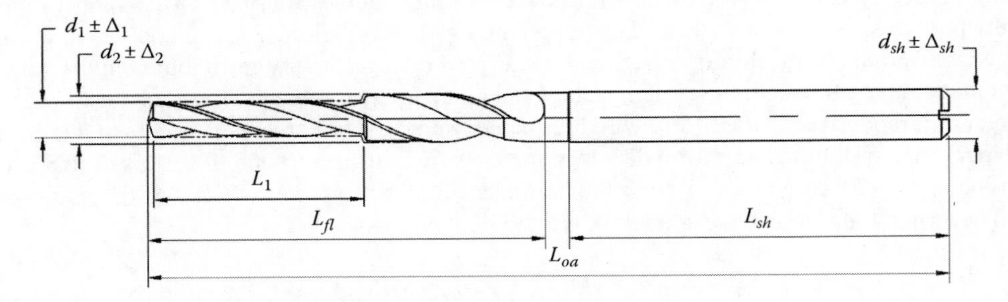

FIGURE 7.46 Drill diameter and length in a drawing of a drill.

TABLE 7.13
Shank Diameter Tolerances (Straight Shank Drills) according to Standard ASME B94.11M-1993 for General-Purpose HSS Twist Drills

Diameter of Drill		Tolerance			
		Inches		Millimeters	
Inches	Millimeters	Plus (+)	Minus (−)	Plus (+)	Minus (−)
From #97 to #81	From 0.15 to 0.33	0.0002	0.0002	0.005	0.005
Over #81–1/8	Over 0.33–3.18	0.0000	0.0025	0.000	0.064
Over 1/8–¼	Over 3.18–6.35	0.0005	0.0030	0.013	0.076
Over ¼–½	Over 6.35–12.70	0.0005	0.0045	0.013	0.114
Over ½–2	Over 12.70–50.08	0.0005	0.0030	0.013	0.076

Source: Reprinted from ASME B94. 11M-1993 by permission of The American Society of Mechanical Engineers.

TABLE 7.14
Shank Diameter Tolerance (Straight Shank Drills) Commonly Used by Leading Drill Manufacturers

Diameter of Drill (mm)	Tolerance (mm) Plus (+)	Tolerance (mm) Minus (−)
Over 0–3	0.000	0.006
Over 3–6	0.000	0.008
Over 6–10	0.000	0.009
Over 10–18	0.000	0.011
Over 18–30	0.000	0.013

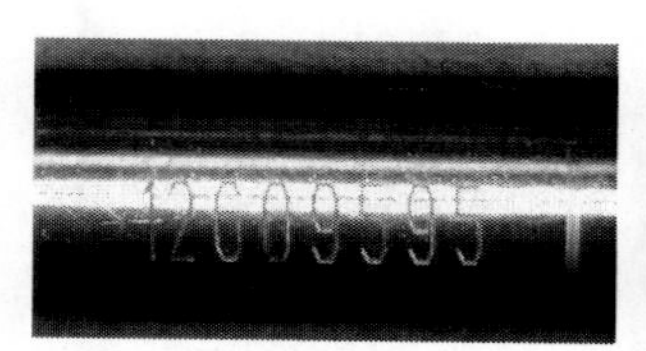

FIGURE 7.47 Scribing/etching placed on the tool shank.

FIGURE 7.48 An example of the proper placement of the tool information.

example of the proper placement of the tool information. For straight-flute drills, the tool information can also be placed on a side of the chip flute at its end.

7.7.3 Overall Length/Flute Length/Shank Length

Overall length (size L_{oa} in Figure 7.46) is the length from the drill point to the end of its shank. Although it may appear to be the simplest part of the drill parameter selection based upon the tool layout (see Chapter 1), it is not always so as a number of important system considerations should be involved in such a selection that is not normally considered in the literature in the field. This section aims to explain some critical issues in the selection of the proper drill length.

The obvious answer to the question "what should be the drill length?" is "it should be equal to the length of the hole to be drilled with this drill + the length of the transition part between the end of the chip flute and the shank + the length of the drill shank as it follows from the tool layout." Being perfectly correct, this answer, however, does not account for drill re-sharpening as the length of the flute (size L_{fl} in Figure 7.46) becomes shorter while the shank length (size L_{sh} in Figure 7.46) remains the same. Moreover, due to back taper, the drill diameter decreases with each re-sharpening so it may happen that this diameter can be undersized (below the low limit allowed by the hole tolerance) in precision drilling. Therefore, the selection of the initial overall length of a drill should account for these two factors, bearing in mind the number of re-sharpenings as this defines the total tool life. When the minimum allowable drill diameter is the major restrictive factor on drill length, for example, for tap drills, this diameter should be included in the drill drawing, and a note in this drawing should clearly state that the drill cannot be used any further if its diameter is less than the indicated minimum allowable drill diameter. While some drill manufacturers tried to include the minimum allowable length, arguing that the minimum diameter is a function of the back taper so that this length can be calculated and checked even with a simple metallic ruler, it is not correct. This is because the amount of back taper can vary significantly so that for the same minimum length, the corresponding minimum diameter may significantly differ.

The second issue in the consideration of the drill length is the minimum allowable flute length. As discussed in Chapter 4 (Figure 4.20), the length $L_{dr\text{-}1}$ is called the setting drill length (gage length), and it is measured from the drill corner to a certain reference point (e.g., from the rear datum face of HSK holder). As the drill becomes shorter with each re-sharpening, its shank portion held in the tool holder is decreased as the drill is pushed out in the tool presetting to keep the same gage length. As such, however, there is no way to compensate for the corresponding decrease in the flute length (size L_{fl} in Figure 7.46). Eventually, after a certain number of re-sharpenings, there is always a danger that the flute length L_{fl} can become shorter than the depth of the hole being drilled.

Figure 7.49 shows the consequences of using a drill that has the flute length L_{fl} shorter than the depth of the hole being drilled. Figure 7.49a shows that the part (although it is made of high-strength

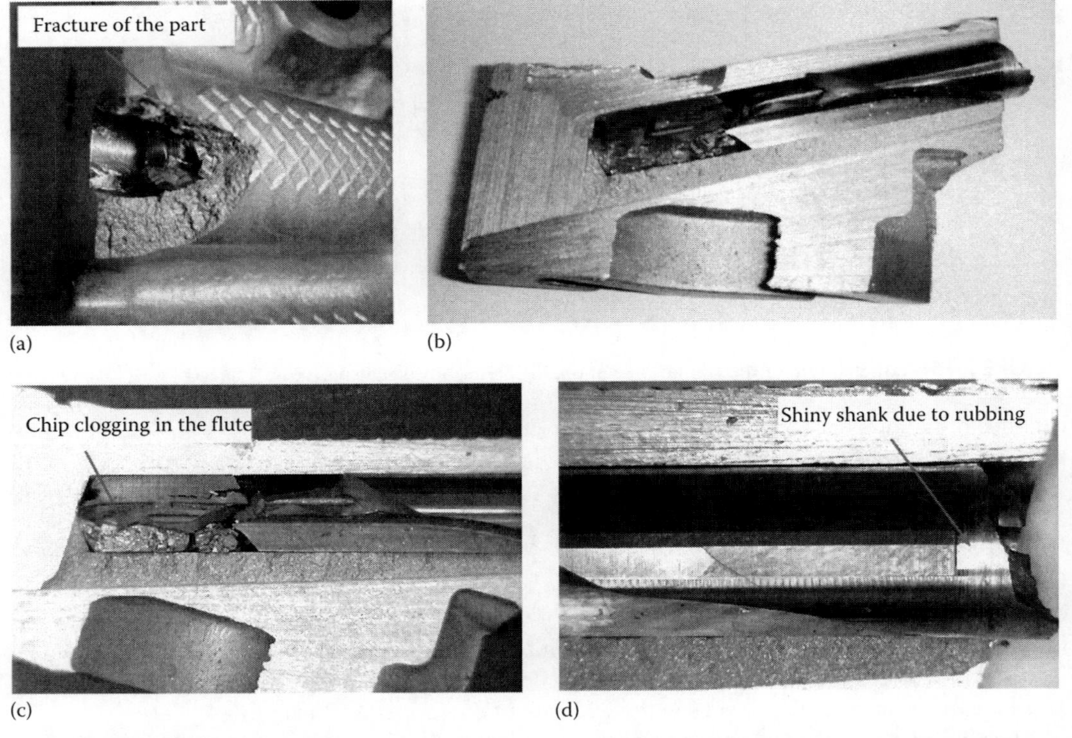

FIGURE 7.49 Consequences of the insufficient chip flute length: (a) fractured part, (b) broken tool, (c) chip clogging the flute, and (d) shiny part of the shank beyond the drill body relief due to rubbing against the wall of the hole being drilled.

aluminum alloy A390) was actually fractured as the formed chip had no room to escape. Tool breakage (Figure 7.49b) was also a result. Figure 7.49c shows that the elements of the formed chip look as welded because chip clogging creates high compressive force. Figure 7.49d shows a shiny part on the tool shank that indicates that the drill flute was completely buried in the hole so the portion of the shank behind the flute having no body relief rubs against the wall of the hole being drilled.

To prevent this from happening, the flute length should always be longer than the depth of the hole being drilled. A question is how much longer. In the author's opinion, the flute length (size L_{fl} in Figure 7.46) as it is commonly shown in the tool drawings makes sense only in tool manufacturing, while in use, it is rather a misleading feature. This is because it is not a functional length that can assure reliable chip removal because its depth and profile change toward the end of the flute. Instead, the active flute length $L_{fl\text{-}a}$, which is shorter than the flute length L_{fl}, should be considered. For the twist drill (Figure 7.50a), this length is from the drill point to the place of the flute where it begins to change its depth, while for the straight-flute drill (Figure 7.50b), this length is measured from the drill point to the beginning of the flute recess (washout as it is commonly referred to in tool drawing and the corresponding radius is given r_{ws}). The minimum allowable active flute length $L_{fl\text{-}a\text{-}min}$ should be equal to the maximum depth of the drilled hole plus a length equal to the drill diameter d_{dr}. This flute length should be clearly indicated in the drill drawing so that drill is not to be re-sharpened beyond this minimum length.

The next question is about the length of the body relief L_{br} (Figure 7.50). Figure 7.51 shows what happens if the body relief is shorter than the depth of the hole being drilled. As can be seen, severe

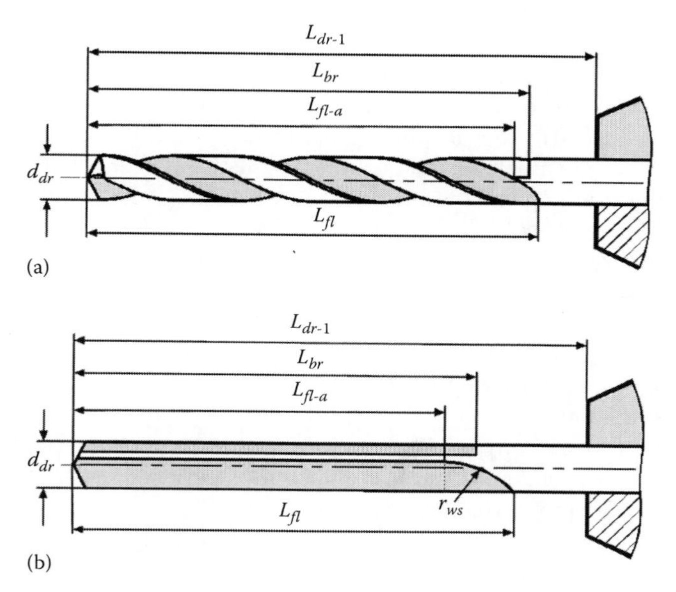

FIGURE 7.50 Additional parameters to standard drill length to be considered: (a) in a twist drill and (b) in a straight-flute drill.

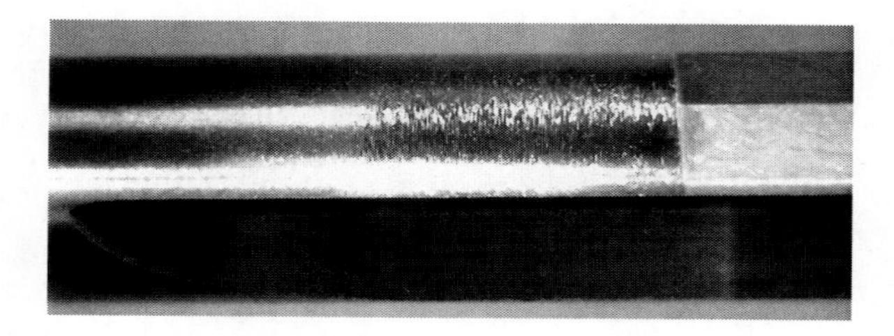

FIGURE 7.51 Showing what happens if the body relief is shorter that the depth of the hole being drilled.

FIGURE 7.52 Showing the consequences of location at the end of the chip flute too close to the face of the drill holder.

rubbing of the cylindrical part of the shank took place that ruined the quality of the drilled hole. Therefore, the minimum body relief length $L_{br\text{-}min}$ should always be greater than the maximum depth of the drilled hole. The rule of thumb is to make this length equal to the active flute length $L_{fl\text{-}a}$. This is because the minimum allowable active flute length $L_{fl\text{-}a\text{-}min}$ cannot be readily measured directly on the drill, while the body relief length L_{br} can. As a result, the body relief length L_{br} can be measured before each successive drill regrind so that the drill is not reground beyond the minimum allowable active flute length $L_{fl\text{-}a\text{-}min}$.

Another important issue that came out recently with applications of HP drills concerns the actual drill length as installed in the holder, $L_{dr\text{-}1}$ (Figure 7.50). As discussed in Chapter 4 (Section 4.4.3, Figure 4.20), this length should be kept to its possible minimum to assure drill resistance to the high axial force and drilling torque arising in drilling. It is considered by the tool layout designers of HP drilling operations that this length can be almost the same as the flute length so that the end of the flute is located near the drill holder face. The author's experience shows that this is not a good idea. Figure 7.52 shows the problem. As can be seen, deep impressions that damage the tool holder are made by the great amount of the chip particular to HP drilling operations. The matter becomes worse when an abrasive work material, for example, a high-silicon aluminum alloy, is drilled. To prevent this from happening, length $L_{dr\text{-}1}$ should be greater than the flute length L_{fl} by at least 0.8 of drill diameter.

Standards DIN ISO 2768-1 and ASME B94.11M-1993 (Tables 7.15 and 7.16) establish tolerances on the overall length and on the length of the flute. Note that these tolerances are rather coarse as the modern drill grinding machines are capable to maintain approx. 1/10 of these tolerances. As a result, leading drill manufacturers established their own tighter tolerances for HP drills.

7.7.4 Datum

The definition and meaning of the datum is discussed in Section 7.5.4. The datums are positioned in the technical drawing differently depending on the specific requirements and functionality of

TABLE 7.15
Tolerances on the Overall Drill Length according to Standard ASME B94.11M-1993 for General-Purpose HSS Twist Drills

Diameter of Drill		Tolerance			
		Inches		Millimeters	
Inches	Millimeters	Plus (+)	Minus (−)	Plus (+)	Minus (−)
From #97 to #81	From 0.15 to 0.33	1/32	1/32	0.8	0.8
Over #81–1/8	Over 0.33–3.18	1/8	1/8	3.2	3.2
Over 1/8–½	Over 3.18–12.70	1/8	1/8	3.2	3.2
Over ½–1	Over 12.70–25.40	1/4	1/4	6.4	6.4
Over 1–2	Over 25.40–50.80	1/4	1/4	6.4	6.4
Over 2–3½	Over 50.80–88.90	3/8	3/8	9.5	9.5

Source: Reprinted from ASME B94. 11M-1993 by permission of The American Society of Mechanical Engineers.

TABLE 7.16
Tolerances on the Flute Length according to Standard ASME B94.11M-1993 for General-Purpose HSS Twist Drills

Diameter of Drill		Tolerance			
		Inches		Millimeters	
Inches	Millimeters	Plus (+)	Minus (−)	Plus (+)	Minus (−)
From #97 to #81	From 0.15 to 0.33	1/64	1/64	0.4	0.4
Over #81–1/8	Over 0.33–3.18	1/8	1/16	3.2	1.6
Over 1/8–½	Over 3.18–12.70	1/8	1/8	3.2	3.2
Over ½–1	Over 12.70–25.40	1/4	1/8	6.4	3.2
Over 1–2	Over 25.40–50.80	1/4	1/4	6.4	6.4
Over 2–3½	Over 50.80–88.90	3/8	3/8	9.5	9.5

Source: Reprinted from ASME B94. 11M-1993 by permission of The American Society of Mechanical Engineers.

the feature or features. It is important to identify these requirements so as not to make fundamental errors when manufacturing or measuring the component. For example, care must be taken whether an axis, surface/feature extension, or target datum is chosen as the datum. Figure 7.53 shows a typical example of datum identification in a drawing of drilling tools. In Figure 7.53a, the datum is the axis of the component, while in Figure 7.53b, the datum is a feature extension (surface). As such, considerably different metrologies involved in the inspection of part deviations with respect to the datum are used. If, for example, the datum is a reamer shank axis (the most common datum for drilling tools), then the tool can be located in a precision collet chuck or a hydraulic tool holder, and the runout of the working end of the tool is measured by rotating the tool in a tool setting machine and measuring the maximum and minimum deviation. If, however, the datum is indicated as shown in Figure 7.53b, the surface of the shank is the datum. As such, the shank must be positioned in a precision V-block and rotated while measuring the maximum and minimum deviation of the tool working end. Besides a very few special cases, the latter is incorrect as the tool working datum is

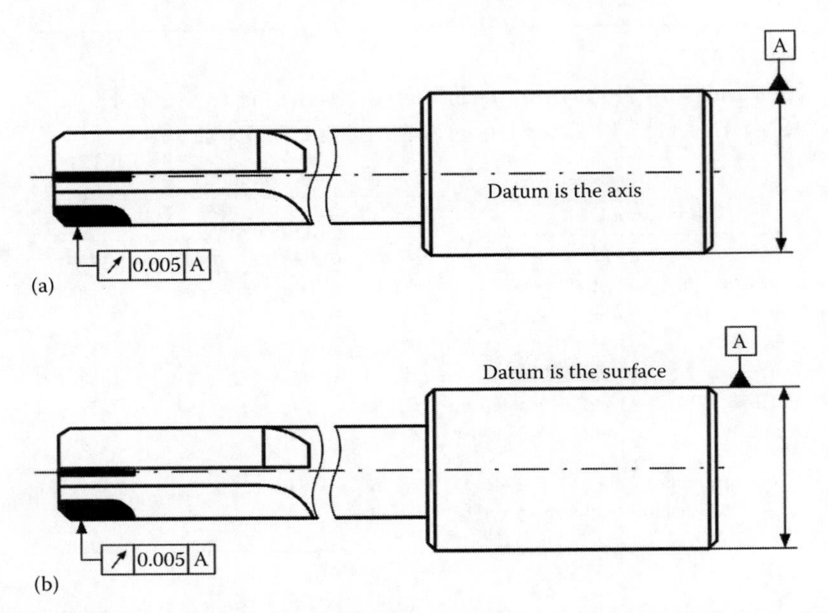

FIGURE 7.53 Datum designation position changes the datum feature: (a) datum is the axis of the shank and (b) datum is the surface of the shank.

the axis of the shank as most modern tool holders use this datum. Unfortunately, many tools in drawings used in the automotive industry have the datum indicated as shown in Figure 7.53b.

7.7.5 Runout (Straightness)

7.7.5.1 Concept

Although the general concept of runout is discussed in Section 7.5.3, a need is felt to clarify its particular definition for drilling tools. In drilling tool measurements, runout is often abbreviated as T.I.R. that should be understood as total indicated reading not total indicated runout as commonly thought of in many machine shops and even in the trade and professional reference literature.

To comprehend the concept, one should first consider the starting point, which is the axis of the datum as shown in Figure 7.54a. In this figure, the datum surface is represented by a circle centered on two perpendicular lines. If the circle is rotated around the intersection of these lines in the measuring machine spindle, it would appear to not move as the axis of the datum is assumed to have

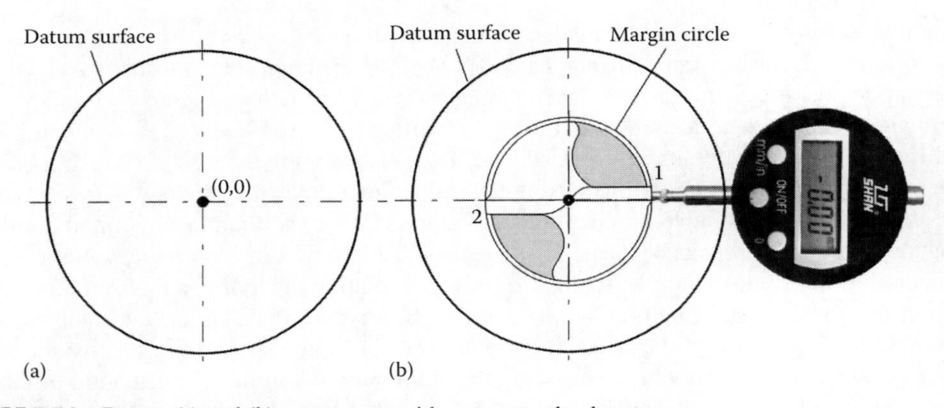

FIGURE 7.54 Datum (a) and (b) zero runout with respect to the datum.

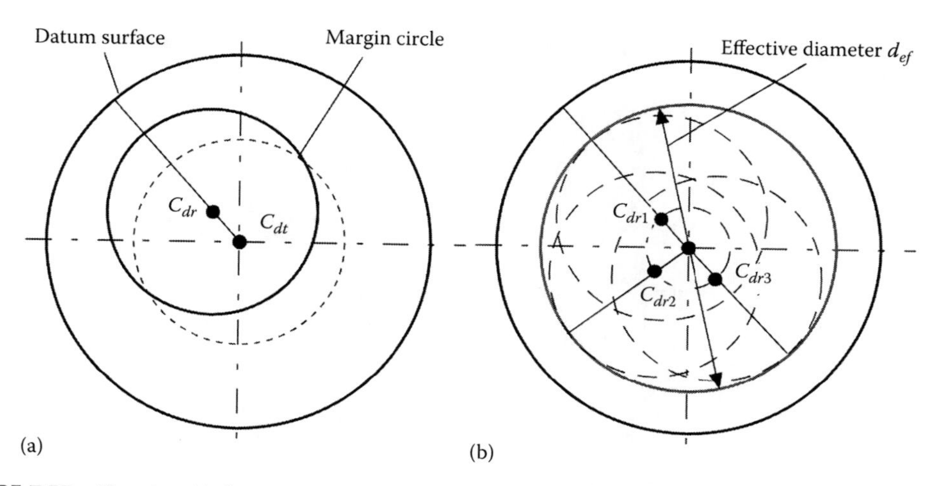

FIGURE 7.55 Showing (a) the center of the margin circle that does not coincide with that of the datum and (b) the sense of the effective diameter.

a zero runout. Next, a drill is added to the drawing as shown in Figure 7.54b. As shown, the drill shares its center with that of the datum so its runout is zero. It implies that if dial test indicators are used and set to zero at margin 1 and then the drill is turned so that margin 2 is found to be in the place of margin 1, the reading of the indicator is still zero. In other words, a hypothetical cylinder called the margin cylinder formed in drill rotation and represented by the margin circle at the drill periphery corners 1 and 2 as shown in Figure 7.54b shares the same axis with the datum. When one obtains this result in real (less than ideal) manufacturing world, it usually means the indicator is broken, hitting a stop, or not touching the margin as there is always some runout in real drills.

Let us consider what happens if the margin cylinder is not aligned with that of the datum. This situation is shown in Figure 7.55a. The dashed circle is the previous position of the margin circle, that is, its ideal location, and the solid-line circle is the (new) actual location of this circle. The center of the actual margin circle is marked as C_{dr}, while that of the datum is C_{dt}. A line from C_{dt} through C_{dr} can be always drawn and this line will always be a radius.

Figure 7.55b shows what happens when the datum surface turns in the measuring machine spindle. As can be seen, the center of the margin circle traces out a small circle. This circle goes to zero diameter when the margin circle's center is aligned with that of the datum. Note that as the spindle turns, the outer side of the drill will define a circle known as the effective diameter d_{ef}. The drill runout Δ_{dr} then can be calculated as $(d_{ef} - d_{dr})/2 = \Delta_{dr}$.

Figure 7.56 shows the original meaning of the term T.I.R. The test indicator set on margin 1 at drill periphery corner and its reading is marked as shown in Figure 7.56a. Then the drill is rotated by 180° (for two-flute drill) and a new reading of the indicator is marked as shown in Figure 7.56b. The absolute difference in indicator readings is called T.I.R. Although nowadays the drill runout is rarely measured as shown in Figure 7.56 as noncontact digital measuring machines are widely available for such a purpose, the term T.I.R. is still in common use to indicate the original meaning of runout.

In all previous considerations of the runout measurement, it was assumed that the runout of the datum is zero. To assure that this is the case, modern drill measuring and presetting machines used to inspect/preset modern high-efficiency drills are equipped with high-precision spindles and tool-holding means.

7.7.5.2 Importance

For HP drills, runout is the single most important factor affecting tool life as it has the critical effect on machining accuracy and tool life. To quantify this statement, Big Kaiser Co. tested 3 mm dia. HSS and carbide drills (Shin and Dandekar 2012). Each drill was tested under the same conditions,

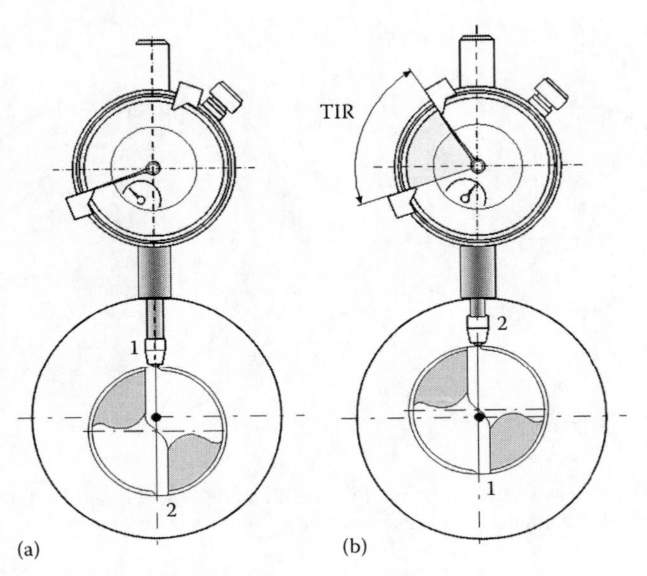

FIGURE 7.56 Original meaning of TIR: (a) the indicator point is place on margin 1 and (b) the indicator point is placed on margin 2.

TABLE 7.17
Test Conditions

Work Material	Speed, Carbide Drill (m/min)	Speed, HSS Drill (m/min)	Feed (mm/rev)	Hole Depth, $3d_{dr}$ (mm)	Hole Depth, $5d_{dr}$ (mm)
Steel ANSI 1055	76	28	0.1	9	15

with only runout changed. Table 7.17 lists test conditions. The wear land of 0.2 mm was selected as the tool life criterion. Figure 7.57 presents the test results which can be summarized as follows:

- Carbide drills have the highest sensitivity to diminished tool life due to runout. Improving runout from 15 to 2 μm tripled tool life of the solid-carbide drill
- HSS tools were slightly less sensitive than their solid-carbide counterparts to diminished life. Improving runout from 15 to 2 μm produced a 230% improvement in tool life. Through-coolant HSS tools were even less sensitive to diminished tool life, producing only a 160% improvement in tool life.

7.7.5.3 Method of Measurement and Tolerancing According to Standards

The drill runout tolerance is assigned the maximum allowable runout (MAR) by various standards. The MARs according to the known standards are rather coarse as they were developed a long time ago when the drill runout itself was a small fraction of the total drilling system runout.

Standard DIN 1414-2 provides a method of measuring drill runout as shown in Figure 7.58. As can be seen, the actual surface of the shank, not its axis, is selected as the datum that, as discussed previously, undermines the actual drill location in any real holder. For drills greater than 2 mm diameter, MAR is calculated as

$$\text{Runout} = 0.03 + 0.01\frac{L_{oa}}{d_{dr}} \tag{7.2}$$

Table 7.18 shows some examples.

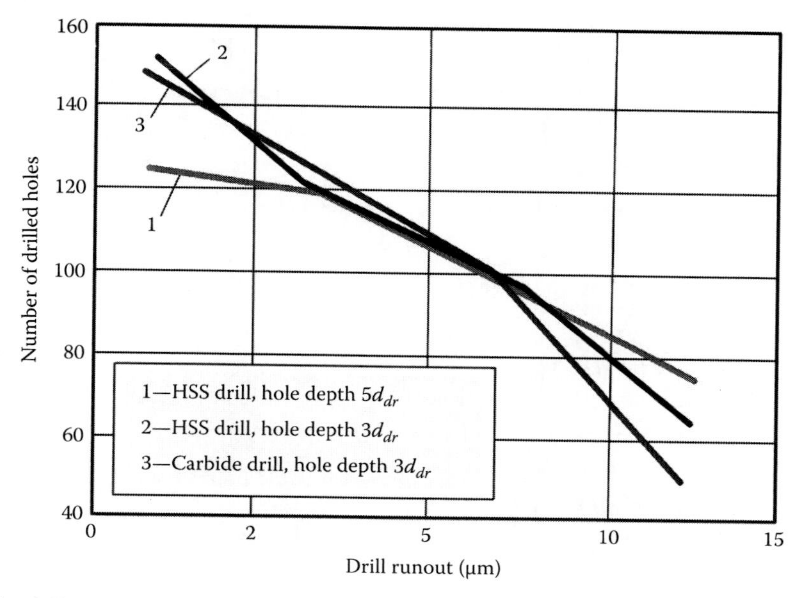

FIGURE 7.57 Influence of drill runout on tool life.

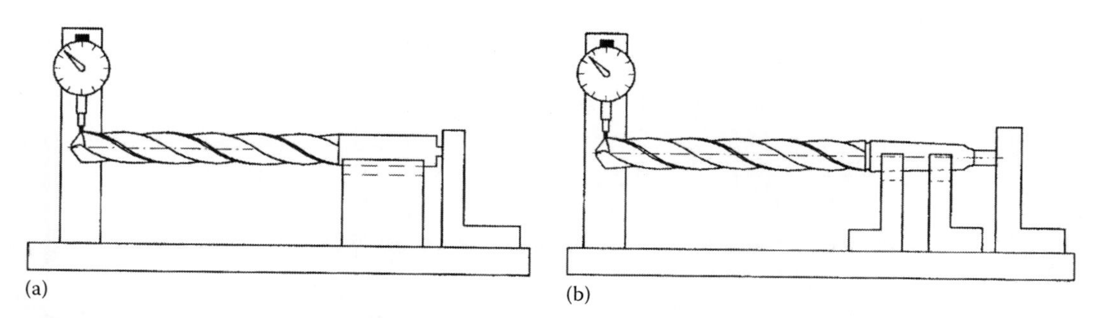

FIGURE 7.58 Measuring runout according to standard DIN 1414-2: (a) for straight shank drills and (b) for taper shank drill.

Standard ASME B94.11M-1993 recommends the selection of MAR (Jobbers length HSS two-flute drills) as follows:

- For drills less than 0.03125 in. (0.8 mm), MAR can be selected according to the manufacturer's discretion.
- For drills 0.03125 in. (0.8 mm) and larger, MAR can be calculated per the following formula:

$$\text{Maximum TIR} = \left(0.0001316 \times \frac{\text{overall length}}{\text{drill diameter}}\right) + 0.00368 \text{ in.} \quad (7.3)$$

Table 7.19 shows some examples of MAR calculated using Equation 7.3.

Analysis of the data given by Equations 7.2 and 7.3 and Tables 7.18 and 7.19 shows that MARs of the drill runout are excessive. Moreover, MAR increases when the drill diameter decreases, which is in direct contradiction with common drill manufacturing practice.

TABLE 7.18
Examples of MAR Assigned by Equation 7.2 (Jobbers Length Drill General Purpose HSS Twist Drills)

Diameter × Length (mm)	Maximum Runout (mm)
3 × 70	0.26
4 × 83	0.24
6 × 102	0.20
8 × 114	0.17
10 × 130	0.16
15 × 181	0.15

TABLE 7.19
Examples of MAR Calculated Using by Equation 7.3

Diameter	Maximum TIR
1/16 in. (approx. 1.6 mm)	0.008 in. (0.203 mm)
1/8 in. (approx. 3.2 mm)	0.007 in. (0.178 mm)
1/4 in. (approx. 6.4 mm)	0.006 in. (0.152 mm)
3/8 in. (approx. 9.5 mm)	0.005 in. (0.127 mm)
1/2 in. (approx. 12.5 mm)	0.005 in. (0.127 mm)

TABLE 7.20
MAR (Tolerance on the Radial Runout) for HSS Drills (GOST 20698-75)

	Radial Runout Tolerance (mm)			
	Cylindrical Shank		Taper Shank	
Drill Diameter (mm)	Precision	Normal	Precision	Normal
>3–6	0.03	0.05	0.04	0.07
>6–10	0.04	0.06	0.05	0.08
>10–20	0.05	0.07	0.06	0.10

Tables 7.20 and 7.21 show MARs according to Russian standards for HSS and carbide drills. Indian standard ICS 25.100.30 (2007) states that MAR shall be within 0.02 mm. In the author's opinion, many existing standards and recommendations where MAR (tolerance on the radial runout) is assigned are severely outdated and thus cannot be used as guidelines for assigning MAR on HP drills. To the first approximation, MAR for Jobbers length carbide drills should not be greater than 0.005 mm for drills from 3 to 10 mm and 0.008 mm for drills over 10–20 mm dia. The problem, however, is often not with drill runout itself as the drilling system runout affects the efficiency of the drilling operation and tool life. Drill runout is only a part of the total system runout.

7.7.5.4 Assigning in Drill Drawings

Figure 7.59 shows the proper assignment of MAR in drill drawings. As previously discussed, it consists of the GD&T runout symbol, runout tolerance, and datum information. In the case shown,

TABLE 7.21
MAR (Tolerance on the Radial Runout) for Carbide Drills (GOST 17277-71)

Drill Diameter (mm)	Precision	Radial Runout Tolerance (mm)
From 1 to 2	High	0.02
Over 2–3		
Over 3–6		0.03
Over 6–12		
From 1 to 2	Normal	0.04
Over 2–3		
Over 3–6		0.06
Over 6–12		

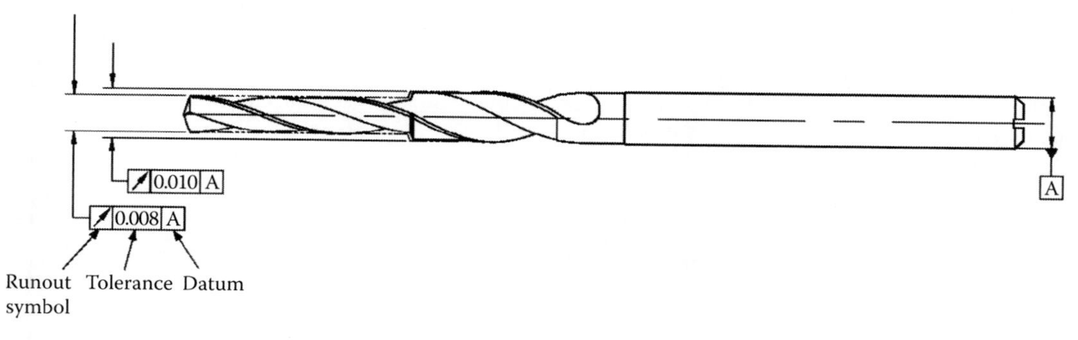

FIGURE 7.59 Assigning runout tolerance as MAR in the drill drawing.

the datum selected is the common datum for drilling tools, that is, the axis of the shank. In case of multistage drills, the runout tolerance should be assigned for each stage.

7.7.5.5 System Runout

7.7.5.5.1 The Essence

To achieve truly HP drilling, the system (total) runout should not exceed 15 μm. This runout, measured after the drill has been installed into the tool holder and then with the tool holder into the spindle, is the sum of the runouts in four components of the drilling system: the machine spindle + the spindle/tool holder interface + the tool holder/tool interface + the tool itself (Figure 7.60). If the shop uses an existing machine for HP drilling and has no plans to replace/refurbish the spindle, then the

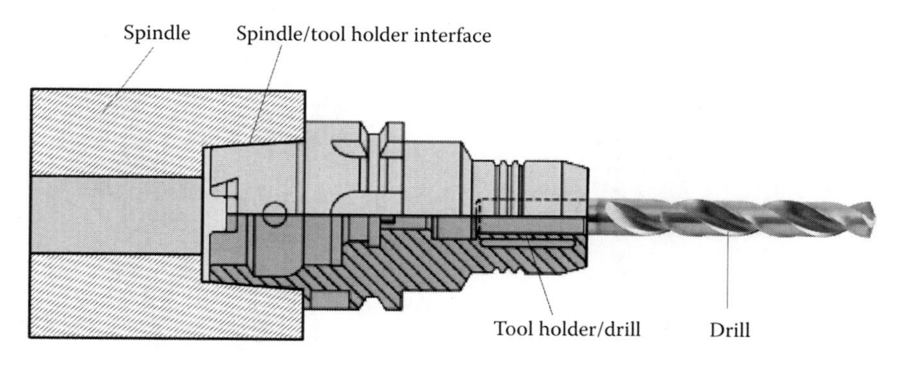

FIGURE 7.60 Components of the drilling system that contribute to a total runout.

first two contributors of the error are beyond the control of the process designer. However, on new machines designed for HP drilling, either of the two is likely to be the most significant contributor to the system runout. The spindles used in most of machine tools today typically offer concentricity and tool holder location tolerances sufficient to leave ample room in this 20 μm margin, while those meant for HP drilling are 5–10 μm or even better for automotive applications. A drill designed and made for HP drilling applications is not likely to contribute much to the system runout as modern HP drills are manufactured to tight runout tolerances.

That leaves the tool holder clamping interface as the most significant contributor to the system runout. In fact, a wrong tool holder alone may account for more than 20 μm of error. Many shops use tool holders, such as 3-jaw drill chucks, that allow drill runout to exceed 25 μm (0.001″). Extrapolating from the test data shown in Figure 7.57, a solid-carbide drill with runout of 25 μm (0.001″) would produce fewer than 25 holes. A higher-quality chuck, though more expensive, could improve tool life dramatically.

Savings can be measured in cost per hole as it is common practice in the automotive industry. An average price for the 3 mm dia. carbide drills used in the tests (Shin 2012) is $40. With runout of 2 μm (0.00008″), this drill can produce 148 holes, at $0.27 per hole. With runout of 15 μm (0.0006″), the cost per hole nearly triples to $0.80 per hole. As a result, manufacturers willing to accept 0.0006″ runout are passing up an opportunity to cut drilling costs by 66%. Note that this simple cost calculation does not include improvements in surface quality and diametric and shape accuracy (including their repeatability) of machined holes. In the automotive industry, the latter can nearly double the calculated cost saving allowing to abolish semi-finished drilling operations.

Static TIR (runout) is a combination of angular (azimuthal) TIR and radial (offset) TIR (Figure 7.61). Angular TIR is the result of a misalignment (skew) between the rotational axis of the cutting tool and the central axis of the collet/spindle system. For example, if a collet tool holder is used, the causes of this type of runout include improper use of set screws in a two-point collet, poorly aligned central collet bore,

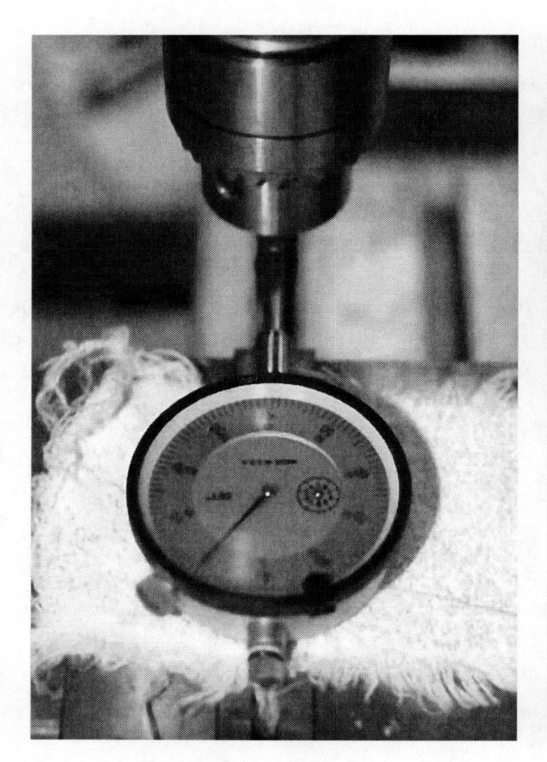

FIGURE 7.61 Measuring total runout.

worn spindle taper, and debris between the collet and spindle bore tapers. Radial runout is the result of a lateral (parallel) offset between the rotational axis of the tool and the central axis of the collet/spindle system. The most common causes are an offset collet bore and mounting a tool with a shank smaller than the minimum diameter of the collet gripping range. Naturally, measurements of static TIR always include a combination of both angular and radial runouts.

The only measurement of runout that truly represents the real situation in actual drills is taken while the spindle is turning at operating speed as static dial indicator measurements have very little correlation with spindle performance while in operation. The runout measured this way is known as dynamic runout (often referred to as dynamic TIR) might also be a result of dimensional infidelity but can include other factors such as anisotropic (uneven) material density (rotational imbalance), worn-out spindle bearings, dynamic properties of the tool holder/spindle coupling, and rotational resonances. While a static runout of less than 5 μm TIR is achievable on high-precision drilling machine spindles, this information does not provide proper indication of the more relevant runout measurement at operating speeds. Therefore, the dynamic runout should be measured as part of a TQA program. To do this, a noncontact measuring system should be used. One of such systems is the Lion Precision (Minneapolis, MN) Dynamic Runout System. The capacitance-based, noncontact Lion system can carry out dynamic measurements of spindles at speeds in excess of 80,000 rpm.

7.7.5.5.2 Measurement

There are two basic types of total runout in drilling tools: static and dynamic. Static total runout (often referred to as static TIR) is a result of problems with the physical dimensions, or arrangement, of the components of the spindle/tool holder/tool system. While drill runout (TIR) can easily been measured using one of many high-accuracy tool inspection machines available in the market, static total TIR should be measured after the drill has been installed into the tool holder and then into the spindle. In the simplest case, it can be measured similar to its definitions given in Figure 7.56, as shown in Figure 7.62. As the drill has known (and small) runout, a pin gage having the same diameter as the drill is often installed in the tool holder to simplify the measurements. Such a measurement yields a much more accurate result, as the real total TIR is measured instead of that based on two random points represented by the drill margins.

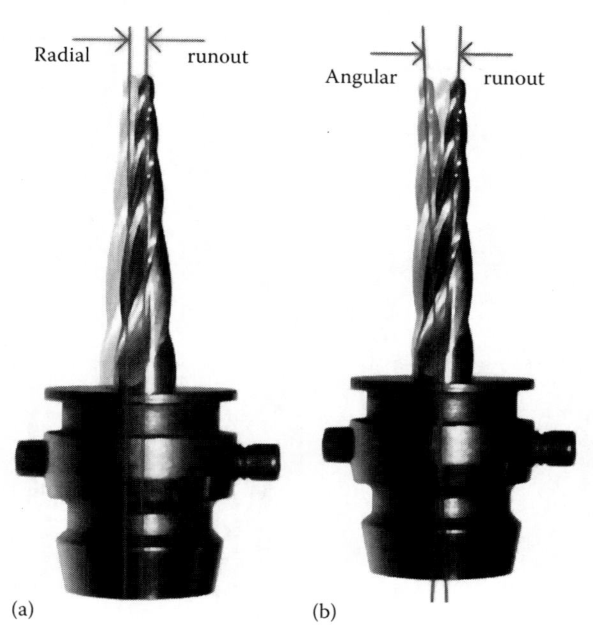

FIGURE 7.62 Visualization of radial TIR (a) and angular TIR (b).

7.7.5.5.3 Adjustment

Even a precision tool holder has a runout accuracy of 3–5 μm, and thus, after the drilling tool has been installed into the tool holder and spindle, the total runout is at best 10–12 μm. While acceptable for many drills of normal accuracy, it is excessive for HP drills and combined drills (e.g., those used to replace three-pass operations into one in the automotive industry). Therefore, various compensation procedures are used to reduce this total runout. As the saying goes, accuracy is the sum total of your compensating mistakes.

When the most popular high-precision shrink-fit tool holders are used, there are no means available for compensation of the total runout—it is what it is as they say. Some compensation becomes available when a hydraulic holder is used and the tool/holder assembly is preset in the spindle of a modern presetting machine (e.g., Zoller). The total runout is measured and the tool is rotated to the position where the direction of the maximum runout of the spindle/holder assembly is opposite to that of the tool. Then the tool is locked in the holder in this position. However, such a compensation method has two pitfalls:

1. It is time-consuming. It is a factor in high-throughput production environment where the *load* on each presetting machine is up to 40 tools per shift.
2. Modern drilling tools are made with small runouts so that this compensation may not be sufficient.

Therefore, a tool holder capable of correcting the total runout can be beneficial.

Compensating tool holders appeared only recently. Figure 7.63 shows the basic idea of adjustable tool holder as seen by ISCAR. This company has introduced a new tool holder designed to easily adjust for radial and angular misalignment. The new GYRO is an adjustable tool holder that can be used on drilling, tapping, and reaming applications. The design of this tool holder allows for a smooth and easy adjustment to account for any radial or angular tool assembly misalignment. It can adjust angularly by as much as 1°, radially by 0.08 in. (approx. 2 mm).

There are a great number of adjustable tool holders appearing almost monthly from the major drilling tool and tool-holder manufacturers. The most recent have the modular design that allows one to interchange the tapers and clamping portion of the tool minimizing the amount of inventory.

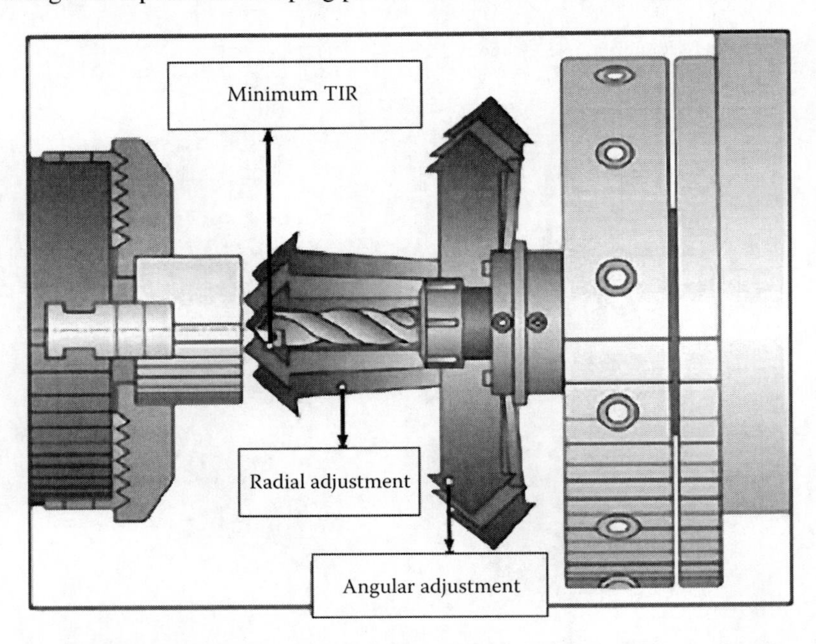

FIGURE 7.63 The basic idea of adjustable tool holder. (Courtesy of ISCAR Co., Galilee, Israel.)

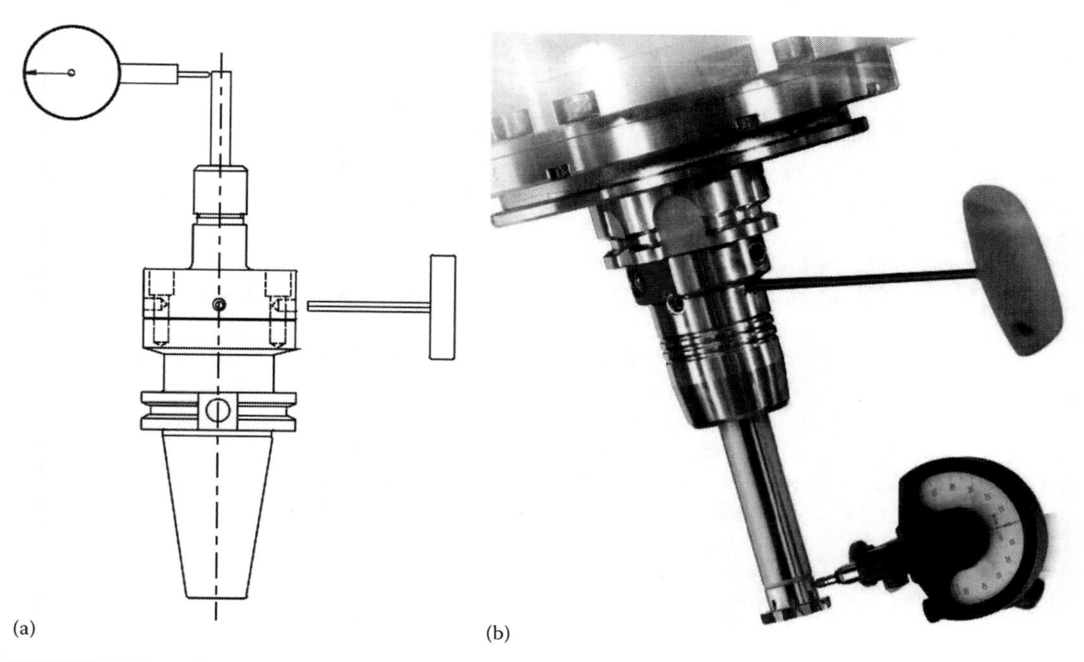

FIGURE 7.64 Adjustment to the minimum runout: (a) schematic, and (b) real tool holder with the tool.

These tool holders normally have adjustment screws to minimize TIR by turning them using a key in the manner shown in Figure 7.64. Normally, laser-sensor presetting machines are used instead of dial indicators as shown in Figure 7.64 for clarity. The differences between the available adjustable tool holders are in precision, amount of the adjustments, repeatability, sustainability, balancing, etc. More significant issue, however, is the ability of an adjustable tool holder to maintain the set compensation during the whole duration of tool operation in the machine. Note that it can be measured in months for PCD tools.

7.7.6 Point Angle and Lip Height

7.7.6.1 Point Angle Tolerances

Standard DIN 1414-2 assigns tolerance ±3° for the point angle (for drills H having $\Phi_p = 118°$ and drills W having $\Phi_p = 130°$). Standard ASME B94.11M-1993 provides the point angles and their tolerances as shown in Table 7.22.

TABLE 7.22
Point Angles and Their Tolerances according to Standard ASME B94.11M-1993

Drill Diameter			
Inches	**Millimeters**	**Included Angle (°)**	**Tolerance (°)**
From 1/16 to ½	From 1.59 to 12.70	118	±5
Over ½–1½	Over 12.70–38.10	118	±3
Over 1½–3½	Over 38.10–88.90	118	±2

Source: Reprinted from ASME B94. 11M-1993 by permission of The American Society of Mechanical Engineers. All rights reserved.

Based on the author's experience and capabilities of modern drill point grinding machines, the point angle should be application specific and its tolerance should be within ±1° for HP drills.

7.7.6.2 Lip Height Tolerances

The idea of drill symmetry is simple: the cutting edges of a drill (the drill point) should be symmetrical with respect to the axis of rotation. It implies that regardless of a particular point angle, the two half-point angles φ_1 and φ_2 (Figure 7.65) must be equal. Similarly, the length of the cutting edges (lips) L_1 and L_2 should be equal (Figure 7.65). As every feature has a tolerance, the tolerances on the drill's corresponding features should be assigned to achieve these conditions with the desired accuracy.

To assign the proper tolerance of the drill features to assure the symmetry of the major cutting edges (lips), one needs to understand what happens when these edges are not symmetrical. Figure 7.66a shows a drill having the major cutting edges (lips) of equal length but unequal half-point angles. One cutting edge does most of the cutting. The consequences are oversized machined holes and shortened tool life due to excessive wear of the corner of the *loaded* cutting edge. Similar results are obtained when the cutting edges are not of the same length as shown in Figure 7.66b. In this case, however, greater (than

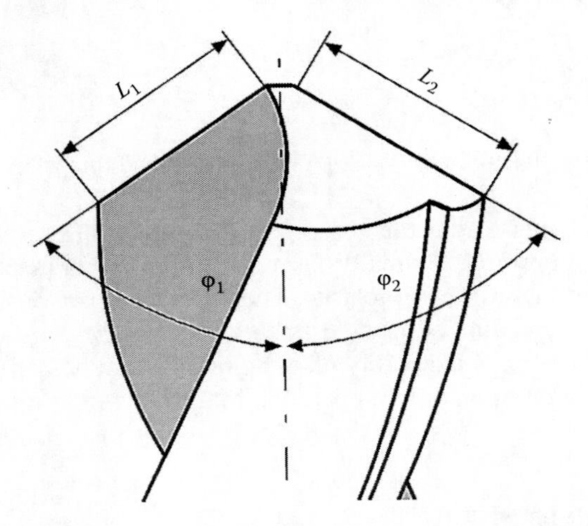

FIGURE 7.65 Half-point angles and length of the major cutting edges (lips) must be equal.

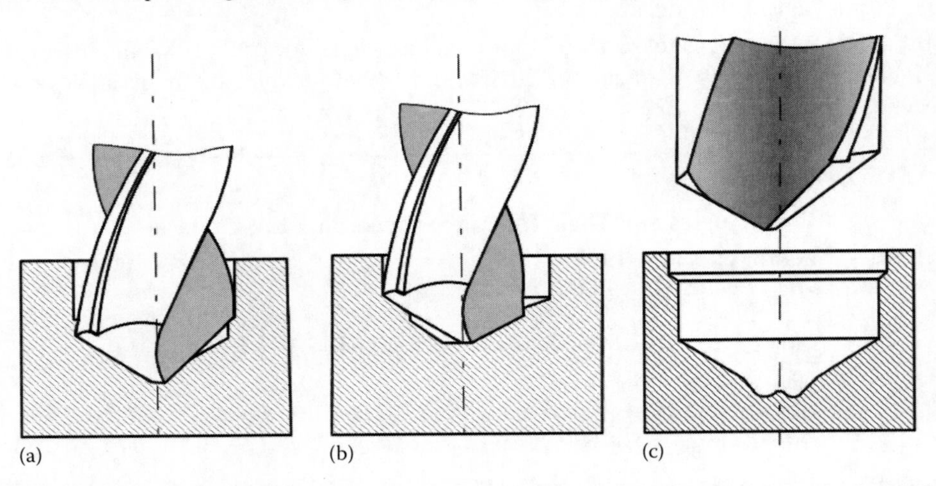

FIGURE 7.66 Showing what happens when the major cutting edges (lips) are not symmetrical: (a) half-point angles are not equal, (b) the edges are not of the same length, and (c) half-point angles and cutting edges lengths are not equal.

FIGURE 7.67 Visual consequences of a severe lip height variation—unequal chip appearance.

in the previous case) hole oversize and wear of one drill corner are normally found. Figure 7.66c shows consequence of a combination of both unequal lengths and half-point angles. Figure 7.67 shows visual consequence of a severe lip height variation. As can be seen, the chips flowing from the flutes are of different appearance.

Standards DIN 1414-1:2006 and ASME B94.11M-1993 do not assign any tolerance on the half-point angle. Rather, to assure symmetry of the major cutting edges (lips), these standards assign tolerances on the so-called relative lip height (widely known as the lip height variation), understood as the difference in indicator reading on the cutting lips at a given distance from the drill axis measured as shown in Figure 7.68. DIN standard 1414-2: 98-06 specifies the point of the measurement to be at the center of the major cutting edge. Both standards provide the same methodology of the measurement of the relative lip height: Rotate the drill in a V-block against a back (front by standard DIN 1414-2) end stop. Measure the cutting lip height variation on a comparator or with an indicator set at a location approximately 75% of the distance from the center to the periphery of the drill (in the middle of the major cutting edges according to DIN standard 1414-2: 98-06). Table 7.23 and Figure 7.69 show the tolerances on the lip height according to ASME B94.11M-1993 and DIN 1414-1:2006, respectively (The Metal Cutting Tool Institute 1989).

According to Russian standards GOST 20698-75 and GOST 17277-71, the lip height variation is assigned through the axial runout of the cutting edges. The tolerances are shown in Tables 7.24 and 7.25. Although the tolerances set by these standards are tighter compared to those shown in Table 7.23 and Figure 7.69, they are still too coarse for HP drills. Having realized this issue, the leading drill manufacturers significantly tightened lip height variation tolerance as modern drill point grinding machines allow

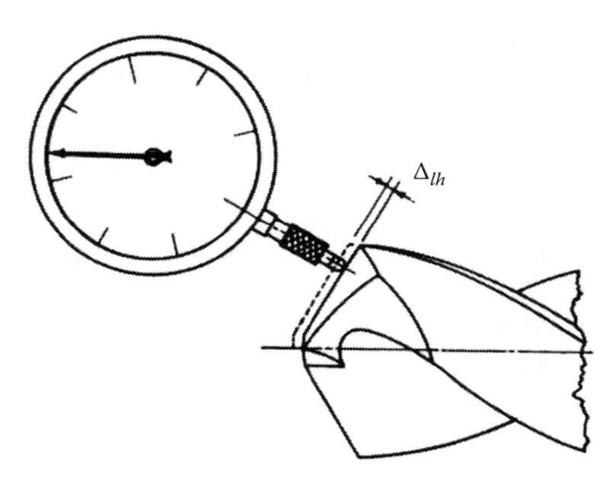

FIGURE 7.68 Relative lip height Δ_{lh} measurement.

TABLE 7.23
Tolerances on the Lip Height for General-Purpose Two-Flute HSS Drills

Drill Diameter Range	Tolerance (Total Indicator Variation)
1/16–1/8 in. inclusive	0.0020 in.
1.59–3.18 mm inclusive	0.051 mm
Over 1/8–1/4 in. inclusive	0.0030 in.
Over 3.18–6.35 mm inclusive	0.076 mm
Over 1/4–½ in. inclusive	0.0040 in.
Over 6.35–12.70 mm inclusive	0.102 mm
Over ½–1 in. inclusive	0.0050 in.
Over 12.70–25.4 mm inclusive	0.127 mm
Over 1–3½ in. inclusive	0.0060 in.
Over 25.4–88.90 mm inclusive	0.152 mm

Source: Reprinted from ASME B94. 11M-1993 by permission of The American Society of Mechanical Engineers.

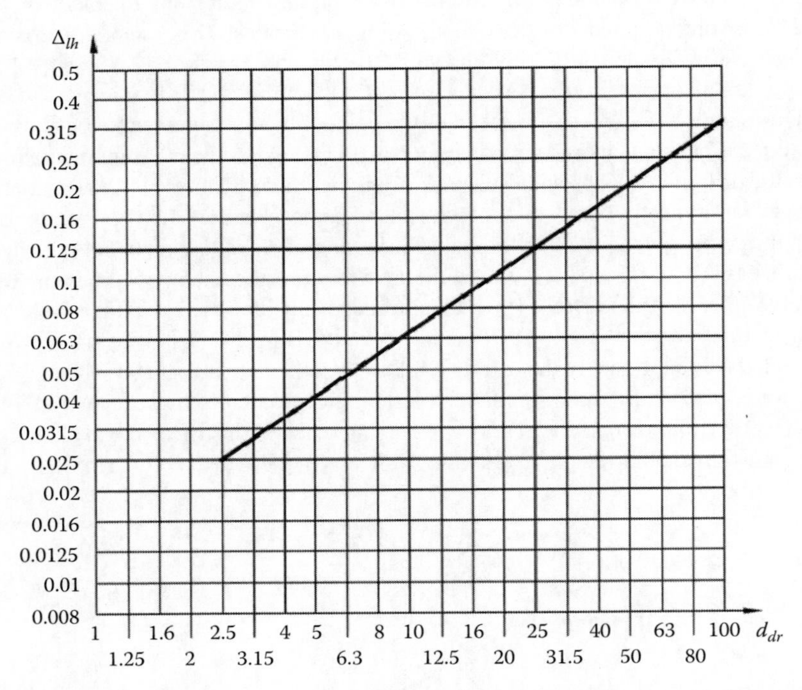

FIGURE 7.69 Tolerances on the lip height for general-purpose two-flute HSS drills (standard DIN 1414-1).

this with no additional cost. Unfortunately, there are a number of drill manufacturers that just follow the data shown in Table 7.23 and Figure 7.69.

To appreciate the tolerance on the lip height set by various standards, one needs to understand its effect on the cutting conditions of a drill. Figure 7.70 defines the involved terms. This figure shows by solid lines the theoretical profile of the major cutting edges in the reference plane. This profile is perfectly symmetrical with respect to the axis of the datum so that the cutting extensions of the

TABLE 7.24
Axial Runout Tolerances for HSS Drills (GOST 20698-75)

Drill Diameter (mm)	Axial Radial Runout Tolerance (mm)	
	Precision	Normal
>3–6	0.02	0.03
>6–10	0.03	0.04
>10–20	0.04	0.05

TABLE 7.25
Axial Runout Tolerances for Carbide Drills (GOST 17277-71)

Drill Diameter (mm)	Precision	Axial Runout Tolerance (mm)
From 1 to 2	High	0.02
Over 2–3		
Over 3–6		0.03
Over 6–12		
From 1 to 2	Normal	0.04
Over 2–3		
Over 3–6		0.06
Over 6–12		

projections of the major cutting edges intersect at point O on the axis of the datum and the major cutting edges OA and OB are of the same length. The theoretical uncut chip thickness cross section is also shown. This uncut chip thickness cross section is the same for both cutting edges. The profile of a drill having the lip height variation Δ_{lh} (as presented by standards ASME B94.11M-1993 and DIN 1414-1:2006) or the axial runout $\Delta_{lh\text{-}ax}$ (as defined by standards GOST 20698-75 and GOST 17277-71) is shown by the dashed line. As can be seen, the major cutting edges O_1A_1 and O_1B_1 are not of the same length so that extensions of their projection in the reference plane intersect at point O_1, which does not lie on the axis of the datum.

To facilitate understanding of the lip height variation, a model that correlates the uncut (undeformed) chip thickness and lip height variation was developed (Kobayashi 1967). Its modernized version is shown in Figure 7.71. When lip heights of a drill are the same, the cutting edges OA_0 and OB_0 of this hypothetical drill are the same and so their working conditions represented by the uncut (undeformed) chip thickness are absolutely the same. When this hypothetical drill is rotated by a half revolution (180°) while the cutting feed is applied, its theoretical point designated by O moves into a new position O_n located in the datum (rotational) axis. The major cutting edges OA_0 and OB_0 become OA_n and OB_n as shown in Figure 7.71.

Because a two-flute drill is considered, it is clear (see Chapter 1) that the distance OO_n is equal to the feed per tooth f_z and the uncut chip thickness (the chip load):

$$h_{D-O} = f_z \sin\varphi \tag{7.4}$$

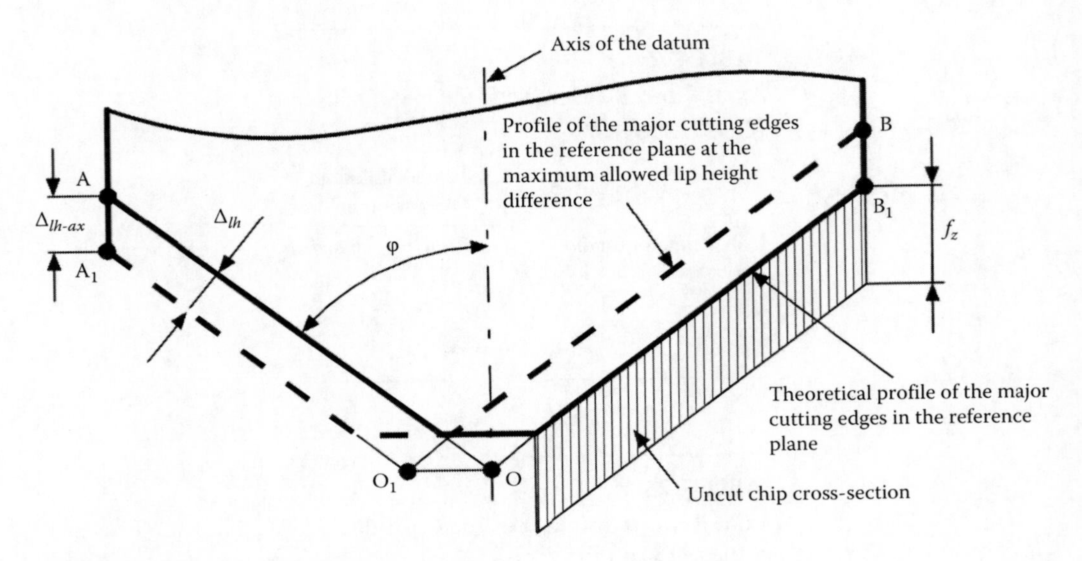

FIGURE 7.70 Defining the terms involved.

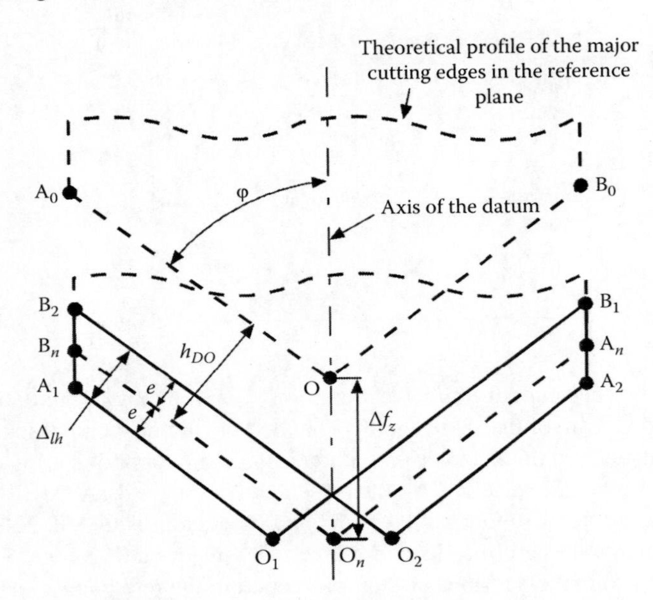

FIGURE 7.71 Model to determine the influence of lip height variation.

Any real drill is manufactured with the lip height difference Δ_{lh} that results in uneven cutting edges, that is, the length of the cutting edge O_1A_1 is not equal to that of O_1B_1. According to standards ASME B94.11M-1993 and DIN 1414-1:2006, the lip height difference Δ_{lh} is assessed by the rotation of the drill by a half revolution (180°) with no feed (against an end stop). After this rotation, the center O_1 is found in the location O_2 so that the cutting edges O_1A_1 and of O_1B_1 become O_2A_2 and of O_2B_2, respectively. As follows from the model shown in Figure 7.71, the distance A_1B_2 is equal to the axial runout, $\Delta_{ih\text{-}ax}$ (as defined by standards GOST 20698-75 and GOST 17277-71), and the lip height variation defined by standards ASME B94.11M-1993 and DIN 1414-1:2006 is calculated as

$$\Delta_{lh} = \frac{A_1B_2}{\sin\varphi} = \frac{\Delta_{lh\text{-}ax}}{\sin\varphi} \quad (7.5)$$

In the model shown in Figure 7.71, the distance e designates the variation of the uncut chip thickness by each cutting edge compared to the hypothetical drill when the real drill is rotated by a half revolution (180°). As follows from this model,

$$e = 0.5\Delta_{lh\text{-}ax} \sin\varphi = 0.5\Delta_{lh} \tag{7.6}$$

If h_{D-1} designates the uncut chip thickness for the cutting edge O_1A_1 and h_{D-2} designates the uncut chip thickness for the cutting edge O_1B_1, then these can be calculated accounting for Equations 7.4 and 7.6 as

$$h_{D-1} = h_{D-O} + \Delta f_z = f_z \sin\varphi + 0.5\Delta_{lh\text{-}ax} \sin\varphi = f_z \sin\varphi + 0.5\Delta_{lh} \tag{7.7}$$

and

$$h_{D-2} = h_{D-O} - \Delta f_z = f_z \sin\varphi - 0.5\Delta_{lh\text{-}ax} \sin\varphi = f_z \sin\varphi - 0.5\Delta_{lh} \tag{7.8}$$

The variation of the uncut chip thickness in one drill revolution is calculated as the difference between h_{D-1} and h_{D-2}, that is,

$$\Delta h_{D1-2} = h_{D-1} - h_{D-2} = \Delta_{lh\text{-}ax} \sin\varphi = \Delta_{lh} \tag{7.9}$$

As stated previously, in the author's opinion, the tolerances on the lip height are too coarse for HP drills. The developed model allows to substantiate this statement showing that these tolerances are unacceptable even for usual drills. As an example, let us consider an 18 mm diameter, HSS drill having standard $\Phi_p = 118°$ meant to drill a medium carbon steel AISI 1045 (HB170). According to recommendations (Astakhov 2011), the cutting feed is selected to be $f = 0.30$ mm/rev so that the feed per tooth is $f_z = 0.15$ mm/rev. According to DIN 1414-1:2006 (Figure 7.69), the maximum lip height difference for this drill can be $\Delta_{lh\text{-}max} = 0.10$ mm. The uncut chip thicknesses for the major cutting edges are calculated using Equations 7.7 and 7.8 as

$$h_{D-1} = f_z \sin\varphi + 0.5\Delta_{lh\text{-}max} = 0.15 \sin 59° + 0.5 \cdot 0.1 \approx 0.18 \text{ mm} \tag{7.10}$$

$$h_{D-2} = f_z \sin\varphi - 0.5\Delta_{lh\text{-}max} = 0.15 \sin 59° - 0.5 \cdot 0.1 \approx 0.08 \text{ mm} \tag{7.11}$$

As can be seen, when a drill is made with the maximum allowed lip height difference, the desired 50%/50% chip load (force) balance (discussed in Chapter 4) becomes 70%/30%, which provides evidence to the author's statement about the coarse tolerances on the lip height assigned by the discussed standards.

Table 7.26 presents the recommended axial runout tolerances for HP drills based on the author's experience and capabilities of modern drill point grinding machines. The implementation practice of HP drills used in advanced powertrain plants in the automotive industry shows the feasibility and cost-efficiency of these tolerances.

TABLE 7.26
Recommended Axial Runout Tolerances for High-Efficiency Drills

Drill Diameter (mm)	Axial Runout Tolerance (mm)
From 1 to 3	0.003
Over 6–20	0.005

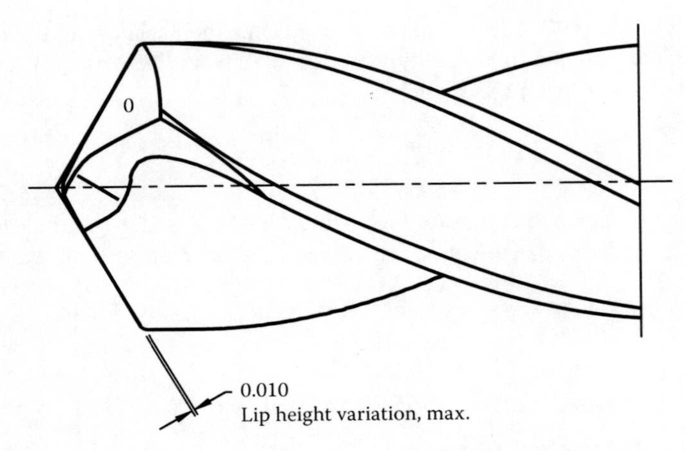

FIGURE 7.72 Common designation of the lip height variation tolerance by leading drill manufacturers.

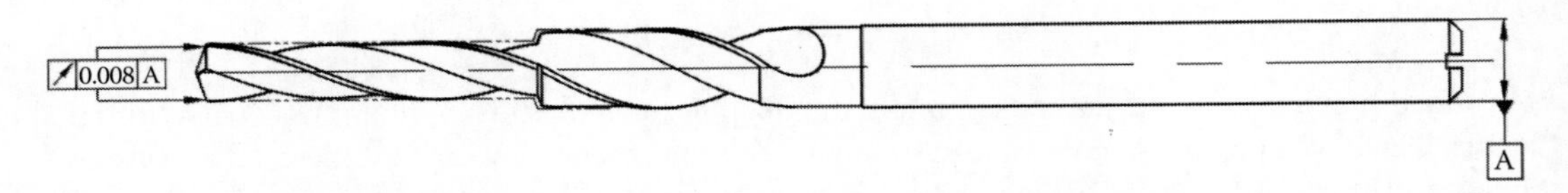

FIGURE 7.73 Proper representation of the lip height variation as the axial runout in tool drawings.

7.7.6.3 Assigning in Drill Drawings

When specially requested by a demanding customer, leading drill manufacturers reluctantly set the lip height variation tolerance in a tool drawing in the manner shown in Figure 7.72. Such a representation is incorrect as no datum for measuring this variation is indicated. In the previously discussed methodology for the lip height variation set by standard DIN 1414-2, the surface of the tool shank is used as the datum so it must be clearly indicated in tool drawings.

In the author's opinion, the axial runout of the major cutting edges (lips) is the best way to control lip height variation and to assign its tolerance in drill drawings. All modern drill inspection and presetting machines actually measure this runout in drill inspection/presetting with respect to the axis of the drill shank. Figure 7.73 shows the proper representation of the lip height variation as the axial runout in tool drawings.

7.7.7 Web Thickness, Centrality (Symmetry) of the Web, and Flute Spacing

7.7.7.1 Web Thickness

As discussed in Chapter 4 (Section 4.4.3.3), the web diameter known as the web thickness is selected to be $d_{wb} = (0.125\ldots0.145)d_{dr}$. However, the full web thickness is rarely used at the drill point in modern drills. Rather, web thinning is applied to reduce the web thickness at the drill point. This is achieved by grinding the gashes on the rake faces as discussed in Chapter 4 (Figures 4.74 through 4.76). Therefore, the web thickness is rather a design parameter that should be left to the discretion of drill designers. As a result, many standards including ASME B94.11M-1993 and GOST 20698-75 do not set any requirements/tolerances on this drill feature.

Surprisingly, standards DIN 1414-1 and DIN 1411-2 define the minimum web thickness as shown in Figure 7.74. DIN 1414-1 represents the margin width as a function of the drill diameter as shown in Figure 7.75. Standard DIN 1414-2 discusses the method of its measurement, suggesting to use a caliper, a stylus instrument, or a micrometer.

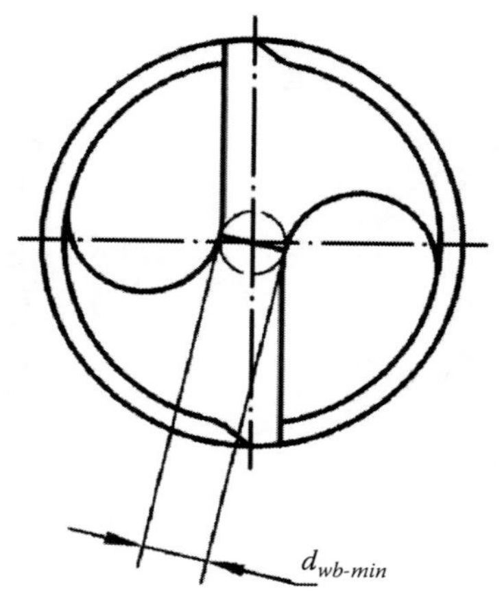

FIGURE 7.74 Minimal web thickness.

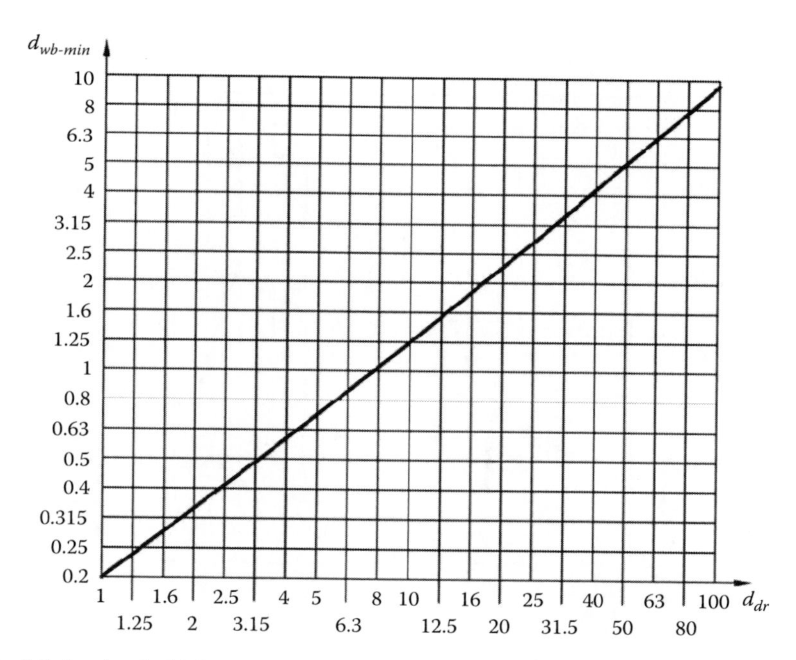

FIGURE 7.75 Minimal web thickness as a function of the drill diameter according to standard DIN 1414-2.

7.7.7.2 Centrality (Symmetry) of the Web

The web symmetry notion according to standard DIN 1414-1 is shown in Figure 7.76a. According to this designation, it is the radial shift of the web axis with respect to that of the datum, which is, according to the designation provided by this standard, the axis of the shank. Figure 7.76b shows the method of its measurement according to standard DIN 1414-2. As can be seen, the surface of the shank is used as the datum not its axis as per standard DIN 1414-1 (Figure 7.76a). One may wonder why symmetry tolerance shown in Figure 7.76a is used to set the interlocation of two cylindrical surfaces instead of runout or coaxiallity. This is because two points of the web circle as measured by an indicator (Figure 7.76b) do not define a circle or cylinder. Figure 7.77 shows the tolerances on the web symmetry according to standard DIN 1414-1.

Standard ASME B94.11M-1993 provides another term and method of measurement for this feature. It is referred to as centrality of the web. The method of measurement is as follows:

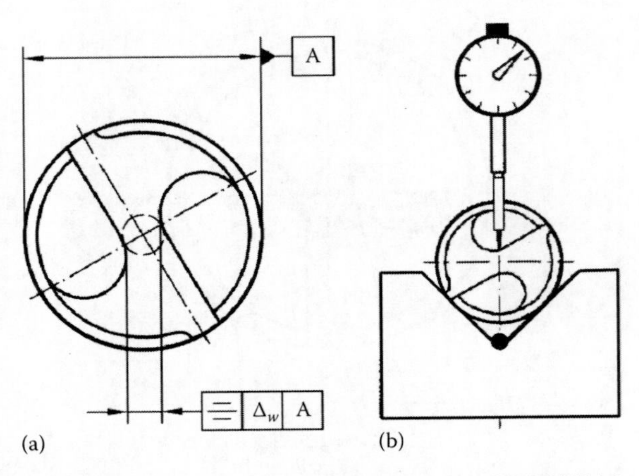

FIGURE 7.76 Web symmetry according to standards DIN 1414-1 and DIN 1414-2: (a) idea and designation in drawings and (b) method of measurement.

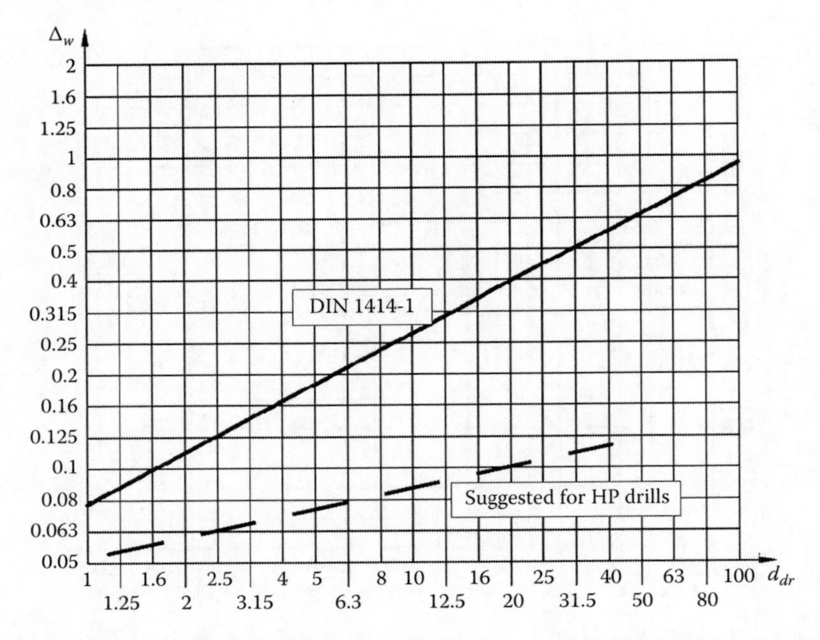

FIGURE 7.77 Tolerances on the web symmetry according to standard DIN 1414-1 and suggested for HP drills.

Rotate the drill in a close fitting bushing. Record the difference in indicator reading of the web at the point as the drill is indexed 180°. Table 7.27 provides tolerances on centrality of web for general-purpose two-flute HSS drills according to standard ASME B94.11M-1993. Table 7.28 shows the tolerances on the web symmetry according to Russian standard 20698-75 for precision HSS drills.

A simple comparison of the tolerances presented in Figure 7.77 and Tables 7.32 and 7.33 shows that the tolerances on centrality of the web set by standard ASME B94.11M-1993 are much tighter than those set by other listed standards. However, the method of measurement of the tolerance on this feature provided by ASME B94.11M-1993 is highly questionable as the incorrect datum is selected and the suggested *close fitting bushing* is not a well-defined term particularly in the presence of back taper applied to drills.

TABLE 7.27
Tolerances on Centrality of Web for General-Purpose Two-Flute HSS Drills

Drill Diameter Range	Tolerance (Total Indicator Variation)
1/16–1/8 in. inclusive	0.0030 in.
1.59–3.18 mm inclusive	0.076 mm
Over 1/8–¼ in. inclusive	0.0040 in.
Over 3.18–6.35 mm inclusive	0.102 mm
Over ¼–½ in. inclusive	0.0050 in.
Over 6.35–12.70 mm inclusive	0.127 mm
Over ½–1 in. inclusive	0.0070 in.
Over 12.70–25.4 mm inclusive	0.178 mm
Over 1–3½ in. inclusive	0.0100 in.
Over 25.4–88.90 mm inclusive	0.254 mm

Source: Reprinted from ASME B94. 11M-1993 by permission of The American Society of Mechanical Engineers.

TABLE 7.28
Tolerances on the Web Symmetry according to Russian Standard GOST 20698-75 for Precision HSS Drills

Drill Diameter Range (mm)	Maximum Asymmetry of the Web (mm)
From 3 to 6 inclusive	0.10
Over 6–10 inclusive	0.15
Over 10–20 inclusive	0.20

In the author's opinion, the standard tolerances on web symmetry are not sufficiently tight for HP drills. The recommended tolerances on web symmetry for HP drills based on the author's experience and capabilities of modern drill point grinding machines are shown in Figure 7.77. The implementation practice of HP drills used in advanced powertrain plants in the automotive industry shows the feasibility and cost-efficiency of these tolerances.

7.7.7.3 Flute Spacing

Another closely related feature is the flute spacing. Its meaning is shown in Figure 7.78a. According to standard DIN 1414-1, the flute spacing is measured on the face of the drill, as close as possible to the outer corners using a V-block and a dial gage as shown in Figure 7.78b. The methodology is as follows: Place the drill on the V-block so that one major cutting edge touches the gage pin as shown in Figure 7.78b. Set the gage to zero and rotate the drill through 180° so that the other cutting edge is in contact with the gage pin, and take the reading, which is taken as the flute spacing A_{sp}. The deviation is then calculated as $\Delta_{fs} = A_{sp}/2$. Figure 7.79 shows the tolerances on the flute spacing according to standard DIN 1414-1.

The method for measuring the flute spacing provided by standard ASME B94.11M-1993 is absolutely the same. Table 7.29 provides tolerances on the flute spacing for general-purpose two-flute HSS drills according to standard ASME B94.11M-1993.

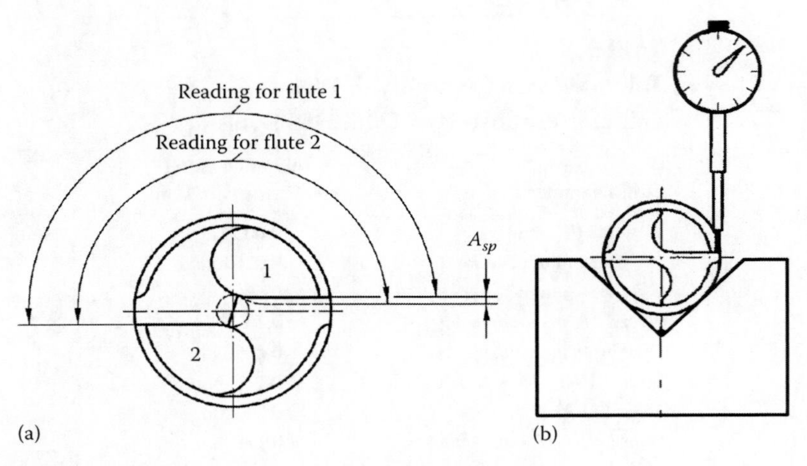

FIGURE 7.78 The flute spacing: (a) meaning and (b) measuring.

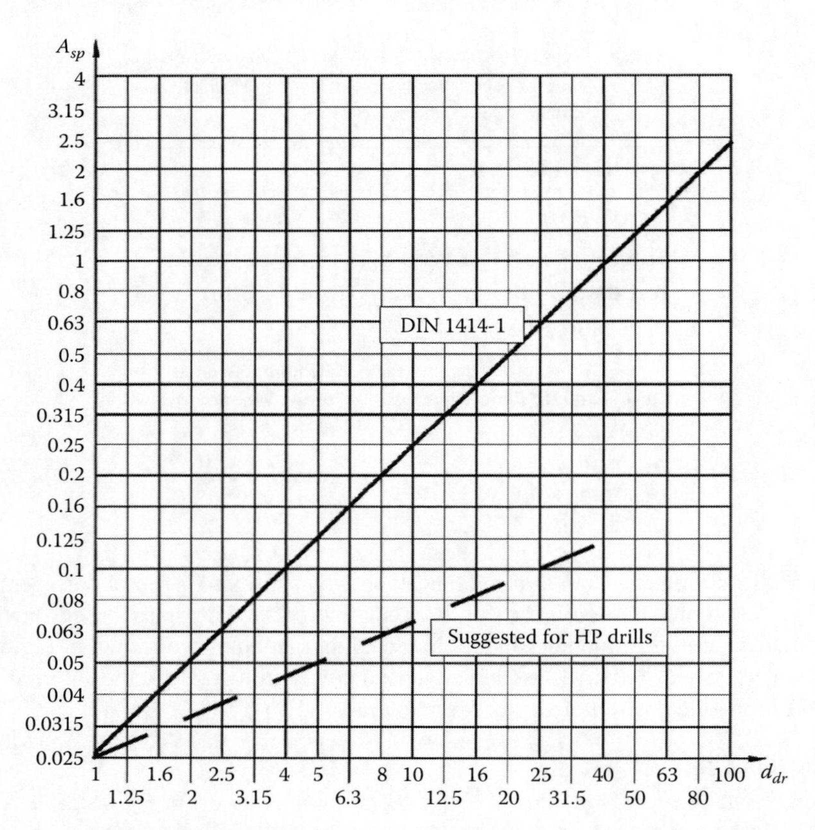

FIGURE 7.79 Flute spacing tolerance according to standard DIN 1414-1 and suggested for HP drills.

Table 7.30 presents the tolerances on the flute spacing according to Russian standard 20698-75 for precision HSS drills.

In the author's opinion, the standard tolerances on the flute spacing (Figure 7.79, Tables 7.29 and 7.30) are not sufficiently tight for HP drills. The problem with this error is that there is no adequate metrology support for the detection of this error. The standard measuring methodologies include the location of a drill in a V-block, so the drill real axis of rotation is not properly defined. Common tool presetting machines allow focusing only on one lip and then rotating the drill to focus on the other.

TABLE 7.29
Tolerances on Centrality of Web for General-Purpose Two-Flute HSS Drills

Drill Diameter Range	Tolerance	
	Total Indicator Variation	Actual Deviation
1/16–1/8 in. inclusive	0.0030 in.	0.0015 in.
1.59–3.18 mm inclusive	0.076 mm	0.038 mm
Over 1/8–1/4 in. inclusive	0.0060 in.	0.0030 in.
Over 3.18–6.35 mm inclusive	0.152 mm	0.076 mm
Over ¼–½ in. inclusive	0.0100 in.	0.0050 in.
Over 6.35–12.70 mm inclusive	0.254 mm	0.127 mm
Over ½–1 in.″ inclusive	0.0140 in.	0.0070 in.
Over 12.70–25.4 mm inclusive	0.356 mm	0.178 mm
Over 1–3½ in. inclusive	0.0260 in.	0.0130 in.
Over 25.4–88.90 mm inclusive	0.660 mm	0.330 mm

Source: Reprinted from ASME B94. 11M-1993 by permission of The American Society of Mechanical Engineers.

TABLE 7.30
Tolerances on the Flute Spacing according to Russian Standard GOST 20698-75 for Precision HSS Drills

Drill Diameter Range (mm)	Maximum Flute Spacing Error (mm)
From 3 to 6 inclusive	0.06
Over 6–10 inclusive	0.08
Over 10–20 inclusive	0.10

As such, the flute spacing error known also as the web eccentricity cannot be detected. Existing tool geometry measurement machines do not include this feature in their basic programs.

When machining aluminum and its alloys, the web eccentricity, however, can be observed after a short time of cutting. Figure 7.80 shows an example. As can be seen, the two parts of the chisel edge are not equal that causes hole size and location problems. When machining steels, web eccentricity causes the so-called chisel edge *walking* hindering precise hole location at the tool entrance. Figure 7.81 shows three hole entrances drilled by tools having (1) excessive web eccentricity, (2) allowable (by the standards) web eccentricity, and (3) less than 1° web eccentricity. In practice of CNC machining, to prevent chisel edge *walking*, a spot (Figure 4.7) or center (Figure 4.8) drill is first used to make hole starts, thus to assure their accurate locations and improve entrance conditions for the drill. In practice of the automotive industry, a short end mill is widely used to start the hole (cut pilot holes) when an inclined hole entrance or uneven/out of location core hole is to be drilled.

The recommended tolerances on web symmetry for HP drills based on the author's experience and capabilities of modern drill point grinding machines are shown in Figure 7.79.

7.7.8 Chisel Edge Centrality

As discussed in Chapter 4, the direction of the chisel edge is fully defined by its angle ψ_{cl}, which is an outcome of many drill geometrical parameters, for example, the clearance angle of the major

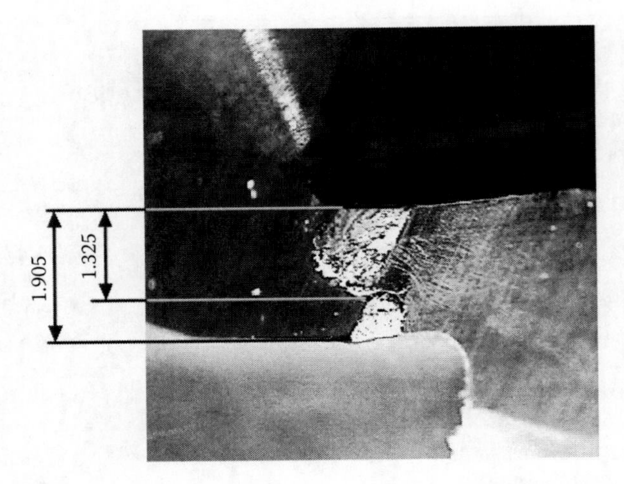

FIGURE 7.80 Web eccentricity/lip height error as marked by aluminum deposit on the rake faces of the chisel edge in drilling of aluminum alloy.

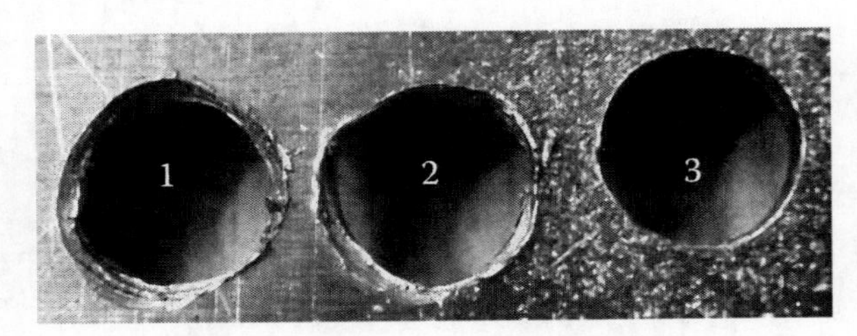

FIGURE 7.81 Hole entrances made by a drill with (1) excessive web eccentricity, (2) maximum allowable web eccentricity, and (3) with web eccentricity less than 1°.

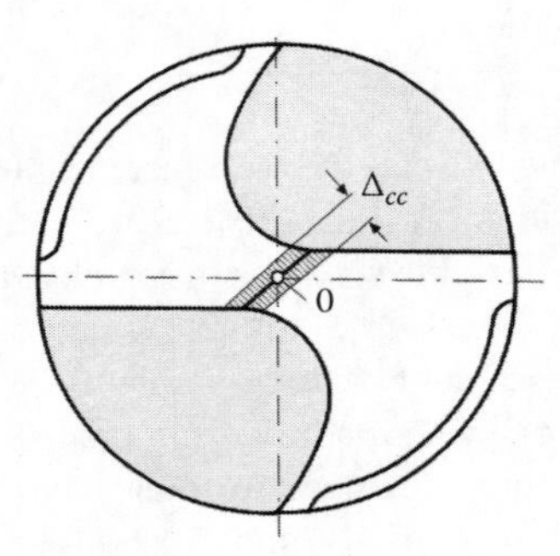

FIGURE 7.82 Chisel edge centrality tolerance.

cutting edges (lips), type of flank surfaces, and point angle. Ideally, this edge should pass through the center of the drill designated by point 0 in Figure 7.82, which is the point on the axis of the datum feature. In reality, however, it happens rather rarely so actual chisel edges on real drills do not pass through the center 0. In the author's opinion, a tolerance on the chisel edge centrality Δ_{cs} in the sense shown in Figure 7.82 should be introduced for HP drills as this feature has almost the same effect as the web eccentricity.

Unfortunately, the existing standards, including DIN 1414-1 and ASME B94.11M-1993, do not define such a feature and thus its tolerance. Table 7.31 presents the recommended tolerances on chisel edge centrality for HP drills.

TABLE 7.31
Recommended Tolerances on Chisel Edge Centrality for High-Efficiency Drills

Drill Diameter Range (mm)	Chisel Edge Centrality (mm)
From 3 to 6 inclusive	0.010
Over 6–10 inclusive	0.014
Over 10–20 inclusive	0.018

7.7.9 Back Taper

The definition and importance of application-specific back taper for HP drills are discussed in Chapter 4 (Section 4.5.9.5). It was pointed out that the amount of the back taper for HP drills should be much greater than that for common drills. Moreover, the higher the yield strength of the work material and the lower its modulus of elasticity, the greater back taper should be used.

Standard DIN 1414-1 recommends the back taper to be in the range of 0.02–0.08 mm/100 mm of drill length. Standard DIN 1414-2 suggests measuring it from the outer corners toward the shank of the drill using a micrometer or an indicator. No tolerance is assigned on the back taper.

Standard ASME B94.11M-1993 provides the tolerances per unit of flute length shown in Table 7.32. This *unit length* is not specified by the standard and the tolerances are too coarse. Probably, the standard assigns the back taper of the drill per 1 in. (25.4 mm) of its length as in this case, the provided data are in good agreement with those set by standard DIN 1414-1.

Table 7.33 shows back taper for drills for machining difficult-to-machine alloys provided by Russian standard GOST 20698-75 for precision HSS drills. As can be seen, the values of the back taper are much greater than those according to the DIN and ASME standards.

The drawbacks of the values of the back taper and its tolerance provided by various standards are as follows:

1. The length of the flute over which the back taper should be applied is not specified.
2. Besides Russian standard GOST 20698-75, the back taper is assigned in a rather wide range with no reference to the properties of the work material, so one may wonder which particular value of the back taper should be used in a particular case of drilling.

TABLE 7.32
Tolerances on Back Taper for General-Purpose Two-Flute HSS Drills

Diameter of Drill		Tolerance per Unit of Flute Length	
Inches	Millimeters	Inches	Millimeters
From #97 to #81	From 0.15 to 0.33	None	None
Over #81–1/8	Over 0.33–3.18	0.0000–0.00008	0.000–0.020
Over 1/8–¼	Over 0.33–6.35	0.0002–0.00008	0.005–0.020
Over ¼–½	Over 3.18–12.7	0.0002–0.00009	0.005–0.022
Over ½–1½	Over 12.7–25.4	0.0003–0.00011	0.007–0.028
Over 1–3½	Over 25.4–88.90	0.0004–0.00015	0.010–0.038

Source: Reprinted from ASME B94. 11M-1993 by permission of The American Society of Mechanical Engineers.

TABLE 7.33
Back Tapers according to Russian Standard GOST 20698-75 for Precision HSS Drills

Drill Diameter Range (mm)	Back Taper per 100 mm (mm)
From 3 to 6 inclusive	0.08–0.10
Over 6–10 inclusive	0.10–0.12
Over 10–20 inclusive	0.12–0.15

3. Only ASME B94.11M-1993 assigns a tolerance on the back taper. However, the assigned tolerances are too coarse.
4. The measuring of the back taper on the HP PCD (CVD) drills and reamers cannot be accomplished as recommended by standard DIN 1414-2 or even on the modern drill presetting machines because the length of the cutting part of the tool is too short to distinguish the difference in diameters in the presence of even smallest runout. On the other hand, these tools are extremely sensitive to values of back taper. Modern tool geometry inspection machines can handle such measurements provided they are properly programmed and used for this geometrical feature.

According to the author's experience, the back taper used for HP drills should be 0.08–0.15 mm/100 mm of drill length for machining of carbon and alloyed steels and cast irons, while for work materials of low elasticity modulus and/or high yield strength (e.g., high-silicon aluminum alloys, titanium alloys, heat-resistance alloys), it can be increased up to 0.4 mm/100 mm of drill length. In the author's opinion, the tolerance on the back taper should be approximately 1/8 of the back taper value selected for a given case of drilling and clearly indicated in the tool drawing.

7.7.10 Margin Width

As discussed in Chapter 4 (Section 4.4.3.3), the margin width, b_m, for common drills is selected using the following formula:

$$b_m = (0.2\ldots0.5)\sqrt[3]{d_{dr}} \tag{7.12}$$

As it is a purely design parameter of the drill, many national and international standards do not set any requirement on this design feature. Standard 1414-1:2006-11, however, set such requirements. Figure 7.83 shows the margin width as represented by this standard. Such a representation is

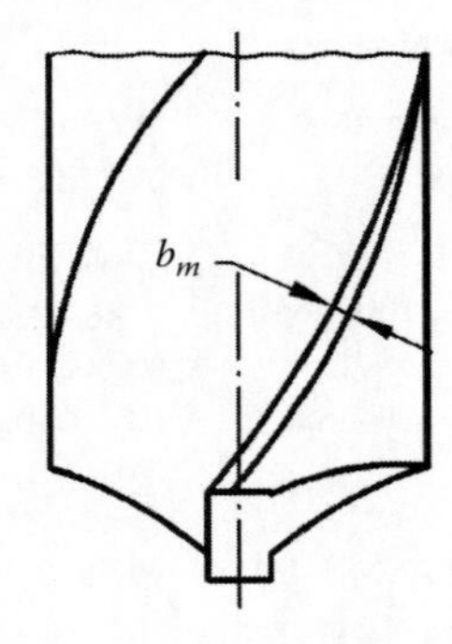

FIGURE 7.83 Margin width representation according to standard 1414-1:2006-11.

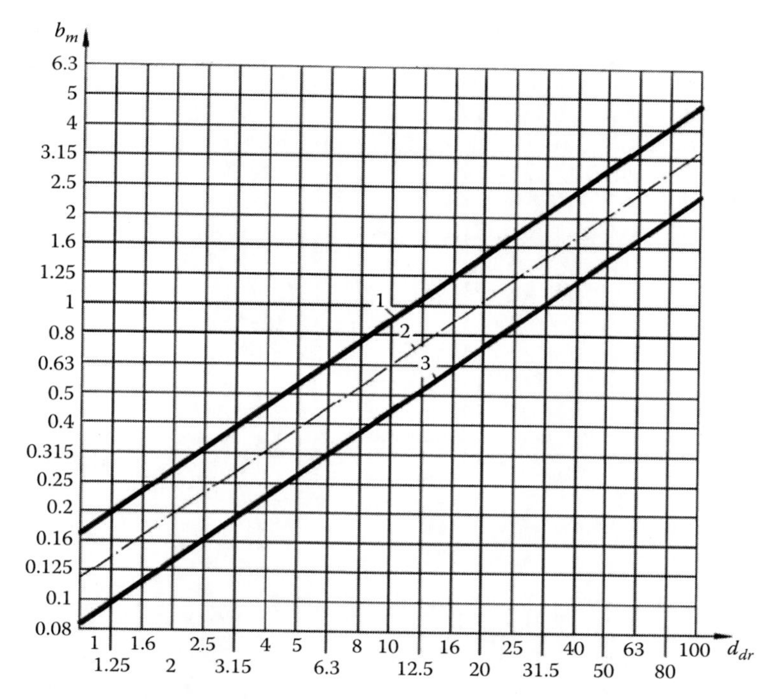

FIGURE 7.84 Margin width and its tolerance as a function of the drill diameter according to standard 1414-1:2006-11: (1) the upper limit, (2) the nominal value, and (3) the lower limit.

misleading as the actual width of the margin can only be viewed in a cross section perpendicular to the drill axis as shown in Chapter 4, Figure 4.37.

Figure 7.84 shows margin width and its tolerance (represented by the upper and lower limits) as a function of the drill diameter according to standard 1414-1:2006-11. A simple comparison of these data with those given by Equation 7.12 shows that they are almost the same. As discussed in Chapter 4, such values of the margin width are excessive for HP drills even if the lower limit given by Figure 7.84 is considered. This issue is discussed in Chapter 4 (Section 4.5.9.6).

7.7.11 Angle of Helix

The definition of the helix angle for twist drills is discussed in Chapter 4 (Section 4.5.3, Figure 4.48). Standard 1414-1:2006-11 is probably the only standard that establishes the requirements on this angle. Standard 1414-2 provides a measuring procedure for this angle as follows: the helix angle, ω_d (see Figure 7.85), shall be measured at both minor cutting edges (helical margins, *auth.*) using a TMM, whose reticle is to be aligned with the minor cutting edge, as shown in Figure 7.85, and the reading of the angle is to be taken. Figure 7.86 shows the lead of helix versus the drill diameter for low helix angle (*H*), standard helix angle (*N*), and high helix angle (*W*) drills and their tolerances.

In the author's opinion, these data are redundant and obsolete. It is redundant as the lead of helix is calculated (see Equation 4.20) as

$$p_{hl} = \frac{\pi d_{dr}}{\tan \omega_d} \tag{7.13}$$

It is obsolete because modern drills are ground using CNC machines having virtually zero helix angle error.

Moreover, the measuring procedure for the helix angle is provided by standard 1414-2 as it is not clear how to *align* the TMM reticle (straight line) and the projection of the helix on the reference

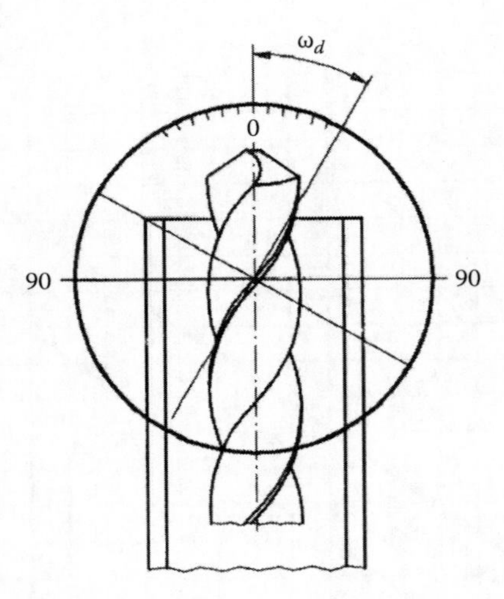

FIGURE 7.85 Measurement of helix angle.

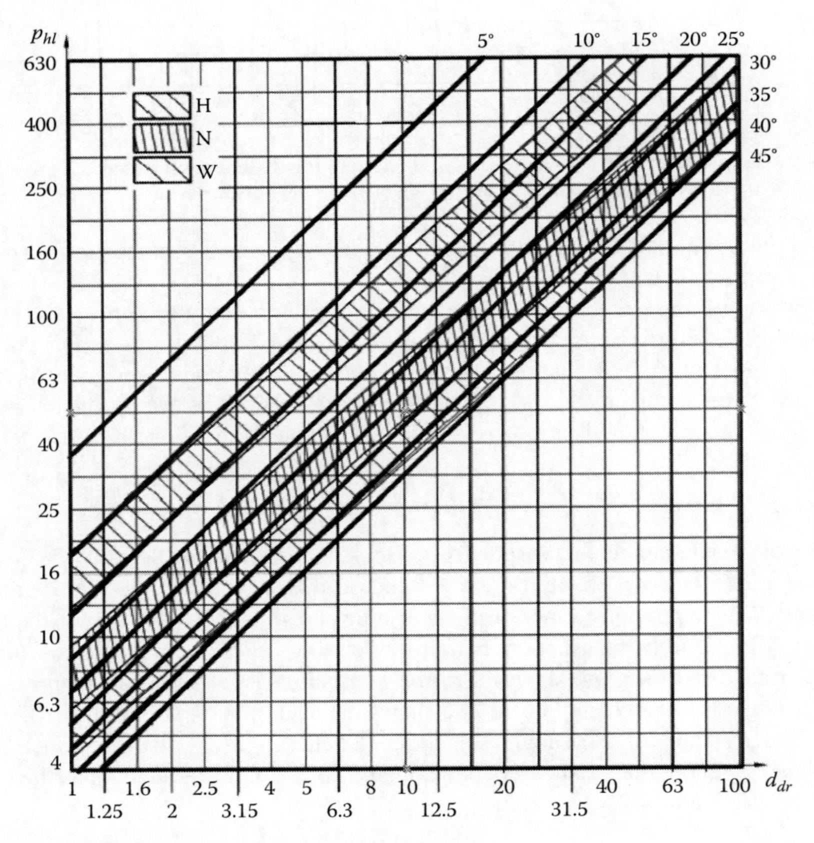

FIGURE 7.86 The lead of helix versus the drill diameter.

plane as shown in Figure 7.85. The proper way to measure the helix angle at the drill margin is to set the TMM reticle as the tangent to the projection of the helical surface into the reference plane at the point where this projection intersects the datum axis as shown in Figure 7.85. The problem is that this axis is not visible in measuring with such a microscope so the helical angle can be measured only approximately.

7.7.12 Clearance Angle

Russian standard GOST 2034-80 recommends the tolerance on the clearance angle to be ±4° for $d_{dr} < 3$ mm dia., while for $d_{dr} \geq 3$ mm diameter, this tolerance is ±3° for HSS twist drills. There is no indication in which plane in T-had-S and how it is measured.

Standard DIN 1414-1:2006-11 does not set any tolerance on the clearance angle. Rather, it provides just its definition in the manner shown in Figure 7.87. Such a definition is incorrect in many respects:

1. The shown angle is not *flat* as the boundary lines are located in different planes. While in the upper line (boundary), the clearance angle is drawn from the drill corner, the other (lower) line (boundary) is drawn from the top of the chisel edge that undermines the definition of the clearance angle according to any tool geometry standard.
2. The cutting edge is not parallel to the drill transverse axis. Rather, the cutting edge is inclined so that the drill corner lies on this axis. The side angle at the drill corner measured this way can be regarded as the T-mach-S side clearance angle (Chapter 4). However, as discussed in Chapter 4 (Figure 4.66), this is not the clearance angle shown in tool drawings where the T-hand-S side clearance angle is indicated. For proper measuring of the side clearance angle at drill corner, this corner should be above the drill axis by distance a_o and both boundaries defining this angle should lie in the same working plane through drill corner.

Standard DIN 1414-2 provides methodology for measuring the clearance angle as follows: The side clearance angle α_f shall be measured at the edge of one of the lands using a V-block with stop and a TMM. Place the drill in the V-block so that it is in contact with the stop. Rotate the drill until its axis lies approximately at the land edge (the major cutting edge will appear as a straight line parallel to the datum plane). Fix the drill in this position. Adjust the reticle so that it is over the land edge and take the reading. Rotate the drill through 180° and repeat this procedure.

It can be seen that the definition and tool location according to standard DIN 1414-2 are not the same as those defined by its *twin* standard DIN 1414-1:2006-11 in terms of the location of the major cutting edge and the boundaries of the clearance angle. Moreover, there are a number of problems with this methodology:

1. No particular system of drill geometry measurement is indicated although only the clearance angle in T-hand-S can be measured with a TMM.
2. In T-hand-S, the side clearance angle should be represented as shown in Figure 4.55, that is, the drill corner should be above the drill axis by distance a_o (Figure 4.66).
3. The term *datum plane* is not defined by the standard. The T-hand-S reference plane is actually meant.

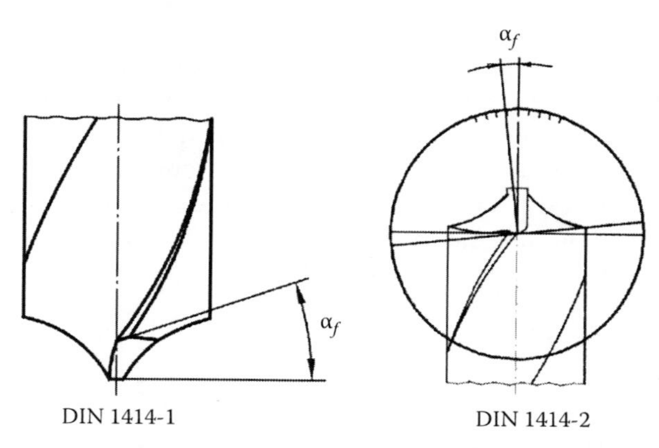

FIGURE 7.87 Clearance angle interpretations according standards DIN 1414-1 and DIN 1414-2.

FIGURE 7.88 The clearance angle as depicted by USCTI book. (From The Metal Cutting Tool Institute, *Metal Cutting Tool Handbook*, 7th edn., Industrial Press, New York, 1989.)

Standard ASME B94.11M-1993 does not provide neither the methodology of the clearance angle measurement nor the tolerances on this angle. However, in USCTI book (The Metal Cutting Tool Institute 1989), which is the explanations to this standard, the clearance angle is depicted in the manner shown in Figure 7.88 and defined as *the angle, measured across the margin, at the periphery of the drill*. The position of the cutting edge shown in Figure 7.88 is not defined and it is not the same as defined by standard DIN 1414-1 and DIN 1414-2.

In the author's opinion, the definitions of the side clearance angle by both standards are misleading. For example, the drill shown in Figure 7.87 (as represented in according to DIN 1414-2) has *negative* clearance angle as the top of its heels is ahead of the drill corners in the axial direction so that these tops will make the first contact with the workpiece (instead of drill corners) causing interference of the flank faces. In Figure 7.88, the lower boundary that defines this angle should be drawn from the drill corners not from the end of the heel as shown in the picture because the line of intersection of the flank face with the drill body is not a straight line. The upper boundary that defines this angle in Figure 7.88 is drawn from the top of the chisel edge, which is in direct contradiction with the standard (ISO 3002) definition of the clearance angle. Moreover, the plane of this angle measurement is unanswered as two boundaries shown in Figure 7.79 do not belong to the same plane.

To understand the major issue when defining and measuring the clearance angle, one should realize that any angle in the tool geometry consideration is measured in a certain reference plane as discussed in Appendix C and Chapter 4, that is, the boundaries of any angle should be in the same plane of measurement. Moreover, any inspection is a comparison of what was actually made to that set by the drawing. As the tool drawing sets the clearance angle at drill corner in T-hand-S (see example in Chapter 4, Figure 4.54), then it should be measured in the same system to make a meaningful comparison. Figure 7.89 shows the proper tool location in a TMM for the side clearance angle measurement at the drill corner A. As can be seen, the major cutting edge AB is set parallel to the drill longitudinal axis and shifted by distance a_o (see Chapter 4). Line 1 passing through drill corner A is the first boundary of the clearance angle that is set at 90° to the drill longitudinal axis. It represents the tangent to the assumed machining surface at point A. Line 2 passing through drill corner A is the second boundary of the clearance angle that is tangent to the flank face at point A. Note that both boundaries defining the side clearance angle at the drill corner lie in the same working plane through this corner. Under this setting, the side clearance angle at the drill corner is measured between the normals to line 1 and line 2 as shown in Figure 7.89. Although such setting for clearance angle determination at point A is commonly recommended by various standards and manuals, the problem with proper determining of the tangent to the curved flank at point A, particularly for small clearance angles and small drill diameters, is important though not clearly understood. Therefore, a need is felt to clarify this long-standing problem.

When the primary flank face is flat, there is no problem with defining the position of line 2 as the line of intersection of this flank face and the drill margin or cylindrical body is almost a

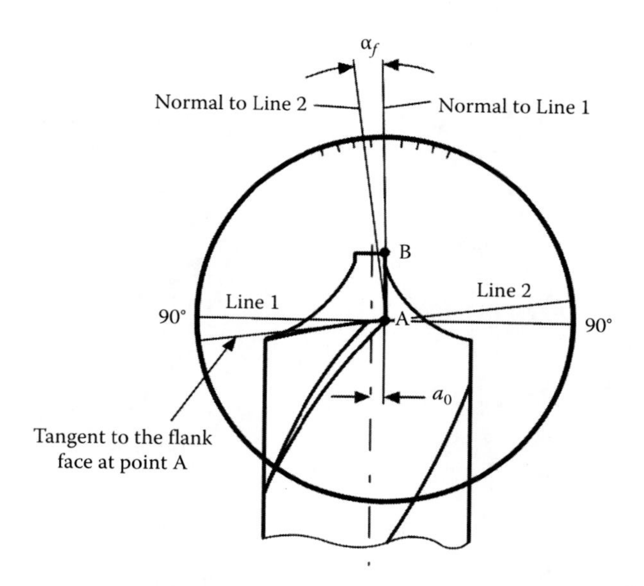

FIGURE 7.89 Showing the proper tool location in a TMM for the side clearance angle measurement at the drill corner.

straight line when viewed in the manner shown in Figure 7.89. In this simple yet becoming common case, line 2 is just set along this line of intersection so that the side clearance angle α_f can be measured accurately. The problem is that this angle is not a part of the tool drawing for drills with the primary flat flanks because the normal clearance angle is indicated in the tool drawing for such drills as discussed in Chapter 4.

When the primary flank is not flat, the curve of intersection of the flank face and the drill margin and/or cylindrical body should be considered and characterized properly. In general, a curve is characterized by the curvature, radius of curvature, and center of curvature (Quadrini et al. 2007). They are a measure of *how much* or *how quickly* a line curves compared to a straight line. Figure 7.90 visualizes the basic terms involved. In the x–y coordinate system shown in this figure, any point on this curve is characterized by its slope or tangent shown for point A in Figure 7.90. The rate of change of this tangent inclination with respect to the x-axis along the curve is called the curvature.

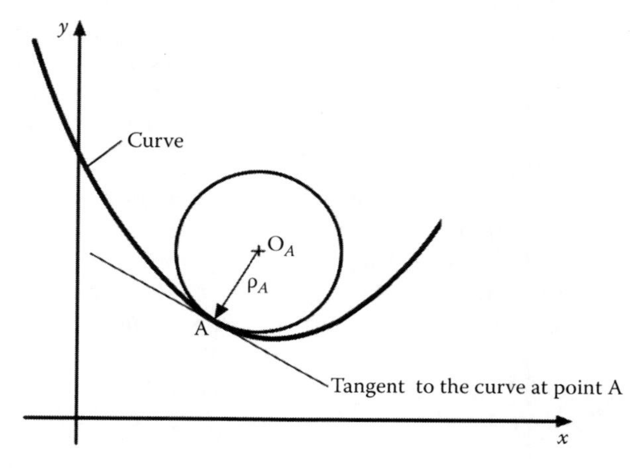

FIGURE 7.90 Curve parameters at point A.

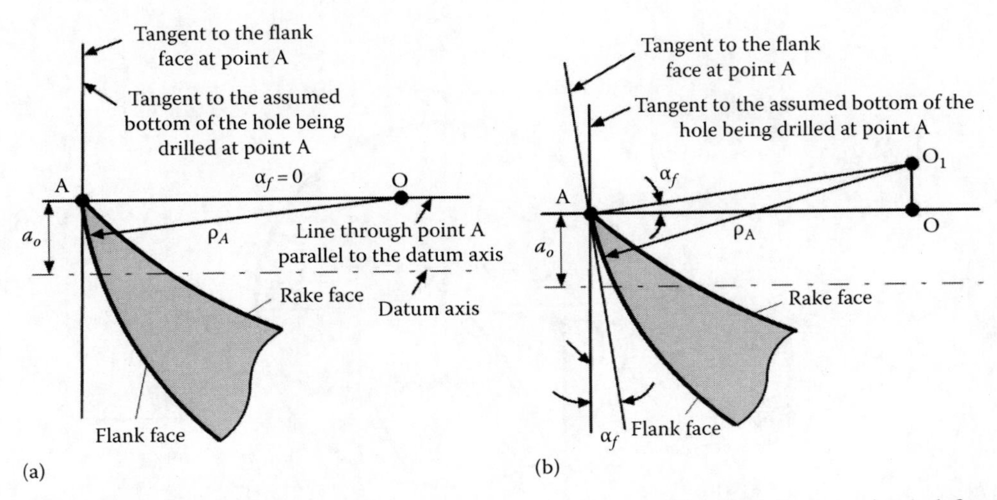

FIGURE 7.91 Showing how the location of the center of curvature of the flank face at point A defines the side clearance angle α_f: (a) $\alpha_f = 0$ and (b) $\alpha_f > 0$.

The center of curvature at considered point A is O_A. It is defined as the center of the circle whose center lies on the concave side of a curve on the normal to the tangent to a given point of the curve and whose radius ρ_A is equal to the radius of curvature at that point.

The next step is to apply these properties of a curve to measuring the clearance angle. The proper determination of the clearance angles in various section planes for a drill with quadratic (Figure 4.61) and helical (known as CAM-relieved in the industry) point drills requires the assessment of the parameters of the curves formed as the intersection line of the sculptured flank face and the corresponding reference plane of the clearance angle measurement in T-hand-S. To determine the side clearance angle α_f (the clearance angle in the working plane in T-hand-S) at a point A of the major cutting edge, the working plane P_f through point A should first be defined. As shown in Figure 7.91, it is defined by two lines: one is the tangent to the assumed bottom of the hole being drilled—it is the vertical line through point A as discussed in Chapter 4. Another is a line through point A drawn parallel to the axis of the datum. It is actually the normal to the tangent to the assumed bottom of the hole being drilled at point A.

The intersection line of the flank surface and the defined working plane is a curve for which parameters must be evaluated to measure the side clearance angle α_f. According to the definition given in Chapter 4, the side clearance angle is the angle between two tangents: the tangent to the assumed bottom of the hole being drilled and the tangent to the curve representing the flank surface in the working plane. Figure 7.91a shows the case where the center of curvature O of this curve lies on the line through point A parallel to the datum axis. As such, the tangent to this curve at point A coincides with that to the assumed bottom of the hole being drilled. By definition, the side clearance angle in this case is equal to zero.

The case shown in Figure 7.91a presents a significant problem in tool inspection with *any,* even the most advanced, measuring equipment. For example, when one tries to follow the procedure presented in standard DIN 1414-2 (Figure 7.89) to measure a small side clearance angle (6°–8°), it is difficult to establish the position of line 2 (Figure 7.89) particularly for small drill diameters even when a modern TMM is used. Even the most advanced tool inspection machines measure this angle by measuring the distances between the tool flank curve and the tangent to the assumed bottom of the hole being drilled represented by a straight line through point A so that rather misleading results are common in such measurements. Moreover, visually, the flank surface appears as the *proper* grind as per Figure 7.88.

Figure 7.91b shows the proper location of the center of curvature O_1 of the curve representing the flank surface in the working plane. It has to be above the normal of the tangent to the assumed bottom of the hole being drilled at point A. As such, the side clearance angle at point A is determined as

$$\alpha_{f\text{-}A} = \frac{\arcsin OO_1}{\rho_A} \tag{7.14}$$

where ρ_A is the radius of curvature of the flank face at point A.

No matter what type of nonflat frank surface grinding is used, the meaning of distance OO_1 is the same and it is always determined by Equation (7.14).

7.7.13 Surface Roughness

Standard ASME B94.11M-1993 does not specify surface roughness on drills.

Standard GOST 2034-80 specifies the following parameters of drill surfaces roughness:

- Flank faces: for high-precision drills A1 and for drills B1, $Ra = Rz = 3.2$ μm; for drills B, $Rz = 6.3$ μm.
- For margins: for high-precision drill A1, $Ra = Rz = 3.2$ μm; for drills B1 and B, $Rz = 6.3$ μm.
- Flute face: at the drill point, $Rz = 3.2$ μm and then $Rz = 6.3$ μm.
- Shank: for high-precision drills, $Ra = 0.63$ μm; for drills of normal precision, $Ra = 1.25$ μm.

Standard DIN 1414-1:2006-11 specifies surface roughness for various drill elements as shown in Table 7.34. Standard DIN 1414-2 specifies the following places for measuring surface roughness of the drills:

The arithmetical mean deviation of the profile Ra shall be measured using a suitable stylus instrument at the following points:

a. In the flute, at right angles to the helix angle, $\omega_{d,}$ at a distance equal to the drill diameter from the point of the drill (measurement 1), and at half the flute length (measurement 2)
b. At the point, across the grinding direction close to the major cutting edges
c. Across the land, between the flute and the end of the relief
d. On the Morse taper, close to point A as in DIN 228-1 (the gage line diameter)

When measuring in the flute (item a), the length of measurement shall extend to the minor cutting edge (the margin). Measurement across the land (item c) shall be made along the drill axis in either direction.

In the author's opinion, the roughness of drill's surfaces according to standards GOST 2034-80 and DIN 1414-1:2006-11 is course for HP drills. Figure 7.92 shows the recommended surface

TABLE 7.34
Surface Roughness of the Elements of a Twist Drill

Drill Element	Drill Diameter ≤15 mm	Drill Diameter >15 mm
Land (margin)	$Ra = 0.8$ μm	$Ra = 0.8$ μm
Shank	$Ra = 0.8$ μm	$Ra = 0.8$ μm
Flute	$Ra = 0.8$ μm	$Ra = 1.6$ μm
Flute at drill point	$Ra = 0.8$ μm	$Ra = 0.8$ μm

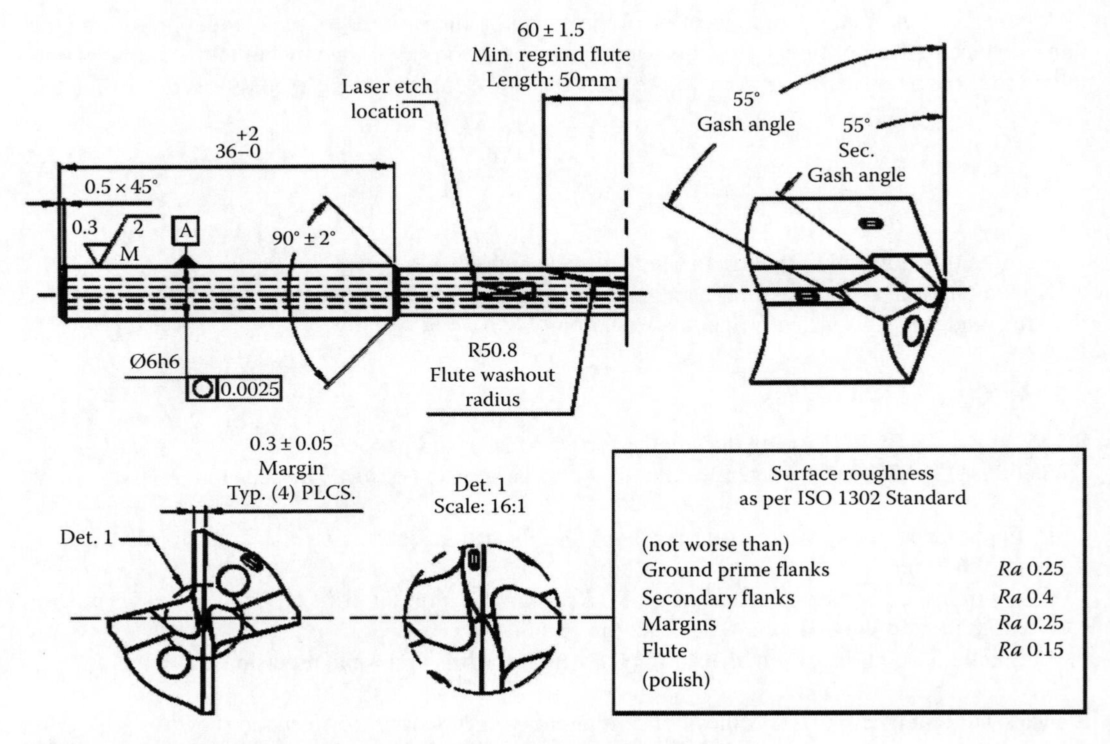

FIGURE 7.92 Recommended surface roughness on the drill components.

roughness on the components of HP carbide drills based on the author's experience and capabilities of modern drill grinding machines. It is represented as a fragment of a drill drawing. Note the following:

- Surface roughness (not worse than) is assigned for any component of the drill point and the flute. The standard for its assessment is clearly indicated
- The direction of the surface roughness measurement is clearly indicated on each drill component
- The surface roughness on the should be always assigned. The axis of the shank is the datum so the shank should be made almost perfect in order to establish this axis properly by a high-precision tool holder. To do this, (a) the shank is ground to mirror-shining surface finish $Ra = 0.3$ μm, and the mixed direction of grinding marks is assigned. The latter is achieved when most modern grinding machines having the second compensating grinding wheel are used to grind the shank and (b) 2.5 μm roundness tolerance is assigned, that is, the shank roundness should not be worse than 2.5 μm at any diameter along its length. Tolerance h6 is assigned on the shank diameters, which is standard for hydraulic and shrink-fit precision tool holders
- The place for toolmarking is clearly indicated, and it is not on the shank
- The flute washout radius, the flute length, and the minimum regrind flute length are shown.

The author clearly understands that there are only very few drill manufacturers in the world capable of achieving and properly inspecting such characteristics of the drill components as state-of-the art production and inspection equipment in addition to well-trained personnel (operators and engineers) are required. The testing and implementation of the drills made to these requirements proved their high efficiency in highly demanding automotive applications.

7.7.14 Drill Inspection

Any, even the simplest, inspection is a part of a quality program, so the terms involved should be clearly defined. This section defines the objective and basic terms of drill inspection to help both the drill manufacturers and the drill users to achieve HP drilling operations:

Article—an article is defined as a design or geometrical component of a drill. The article exists if and only if a characteristic(s) of this article is (are) assigned by the drill drawing or by other related document(s) properly referenced in the drill drawing.

Characteristic—a dimensional, visual, functional, mechanical, or material feature or property, which describes and constitutes the design of an article and can be measured, inspected, tested, or verified to determine conformance to design requirements. This includes drawing requirements (including requirements set in the drawing title block) and/or specification requirements determined from drawing notes. Note that dimensions/features identified as *Reference* (i.e., noted as *REF* or are in parenthesis in the drawing) are exempt and are not required to be inspected/reported.

First Article Inspection (FAI)—a complete verification that the article being inspected complies with the requirements identified in engineering drawings, specifications, and purchase orders as well as any other applicable design requirement(s). The FAI package includes the FAI form, annotated drawing, and all supporting documentation (i.e., certifications test reports). Note that the FAI can also be referred to as the First Article Inspection Report (FAIR) or the Initial Sample Inspection Report (ISIR).

Defect—a defect is defined as a nonconforming characteristic.

Defective drill—a drill that contains one or more defects.

Lot—a collection of drills bearing identification and treated as a unique entity from which a sample is to be drawn and inspected to determine conformance with the acceptability criteria.

Homogeneous lot—a group of drills manufactured at approximately the same time that are expected to share similar quality levels for selected characteristics.

Lot size—the number of drills in a lot.

Random sample—a sample selected in such a way that each drill of the population has an equal chance of being selected.

Reject—to refuse to accept. Rejection in an acceptance sampling sense means to decide that a lot has not been shown to satisfy the acceptance criteria based on the information obtained from the sample. Rejected lots may be immediately submitted for material review disposition upon completion of sampling inspection or may be screened 100% for any detected defects. In this case, all defective drills should be submitted for material review.

Sample—part of a population selected according to some rule or plan.

Sample size—the number of drills selected as representative of a population.

Sampling plan—the instructions given to personnel responsible for performing sampling inspection. Sample inspection plans (as a part of Standard Operating Practice) should be developed, authorized for use, and used in drill inspection and are defined in attachments to this.

FAI is required whenever any of the following occur:

- The tools are being manufactured for the first time by this manufacturer
- There is a revision to the tool engineering drawing
- There is a change in raw material used (use of an alternate material listed in the drawing)
- There is a change in tooling (new, replacement, or major modification) used in drill manufacturing
- There is a change in the manufacturing process that may affect form, fit, or function of the part

- The drill has not been manufactured in 6 months
- There is a change in the plant of manufacture (either due to a natural or man-made event)
- There is a change in the supplier's sub-tier special process provider (i.e., heat-treat, coating/coating facility)
- There is a customer requirement for FAI as a part of the corresponding QC procedure or when the tools in a particular manufacturing batch do not perform as expected, that is, have lower than usual tool life and break.

Table 7.35 presents an example of the process flow for a hypothetical VPDrills company. The following should be noted in this table:

1. Quality inspection is a serious procedure that must be carried out according to the developed quality plans, procedure, instructions, and manual to assure the high quality and sustainability of manufactured drills.
2. The inspection including FAI is based exclusively on the tool drawing that must include all the information essential to tool performance and manufacturing. In other words, if an article is not properly specified by the tool drawing, it cannot be inspected according to the basic definition of inspection.

The inspection includes the following:

1. *Visual inspection* carried out according to the note(s) in the tool drawing. For example, a note in a tool drawing "3. CUTTING TIP AND ITS EDGES SHOULD NOT HAVE ANY VISIBLE CHIPS, CRACKS OR DEFECTS @ x25" implies that the drill should be cleaned before visual inspection that is carried out under a magnification glass/microscope having magnification ×25. Another example is the location and authenticity of drill marking (e.g., *laser* etch *location* in Figure 7.92). Any visual inspection relies on the experience and judgment of the inspector.
2. *Dimensional inspection* includes inspection of the tool dimensional and geometrical articles. Dimensional inspection planning is an activity to generate specific instructions to inspect manufactured drills based on the design. Properly developed inspection plans ensure consistency of measurement results. Inspection planning activity and data models are necessary to enable inspection planners and product designers to effectively communicate during product design and inspection process planning.
3. *Physical inspection* includes the inspection of physical characteristics of the article assigned by the drawing, for example, hardness, thickness, and consistency of the coating layer, and MWF flow rate through the coolant holes of the tool under a specified inlet MWF pressure. Physical inspection can be either nondestructive or destructive, for example, when the strength of brazing of the cutting insert and the tool body is evaluated.

The visual inspection is carried out according to the note in the tool drawing or according to the company's quality procedure/manual for such an inspection. The physical inspection characteristics and procedures as related to the tool material(s) are discussed in Chapter 3. Therefore, the section to follow concentrates on the dimensional inspection the tool dimensional and geometrical articles as this type of inspection is the most common yet least understood in terms of geometrical articles including their proper definition and tolerances.

7.7.15 Dimensional Inspection (Metrological) System: Flowchart

The dimension inspection is a part of the dimension metrology system. Its flow chart shown in Figure 7.93 includes four principal stages: design, planning, execution, and analysis. According to the author's experience, the first stage presents the major problem in drill inspection.

TABLE 7.35
Process Flow (Company VPDrills)

	Major Steps	Activities
01	FAI sample selection	After all modifications to methods of manufacture have been made, a drill from the first production lot shall be chosen for FAI.
02	Drawing selection	1. Internal FAI: Use the manufacturing shop copy (SC) drawing unless otherwise specified on the inspection plan of VPDrills. 2. Customer FAI: Use the customer print (CP) approved by the customer (if one exists). Otherwise use the customer copy (CC) if a CP is not available or SC as specified on the inspection plan.
03	Characteristic identification	1. Determine the FAI form from the database of VPDrills to be used to complete the FAI. Refer to the quality plan for customer-specific requirements for completing the FAI. 2. Obtain a hard copy of the engineering drawing and manually place consecutively numbered circles (balloons) adjacent to each characteristic in the drawing. As an alternate, a CAD program may (e.g., Inventor) be used to identify the Inventor characteristics in the drawing. 3. Record each characteristic in the appropriate location on the FAI form corresponding to the numbered circle in the drawing. For example, the first circled characteristic in the drawing would be recorded in the row with item no. 1 identified. Subsequent rows are added consecutively until all characteristics are identified. Reference characteristics may be omitted in the FAI form.
04	First Article Inspection	1. Unless otherwise approved by VPDrills quality, all inspection methods must be in accordance with VPDrills specification VPA-1203-2012 *Standard Measuring Methods* and stated for each characteristic on the form. Units of measure are those that are specified in the VPDrills engineering drawing. 2. Record the variable measurement of all characteristics in the *Actual Measure* block (or equivalent) of the row corresponding to the item number noted in the drawing: a. All dimensional, hardness, and test requirements must be reported with a measurement value. b. Dimensions stating *TYP* or multiple locations (e.g., 2×, 4×) must report measurement values for each feature. c. All basic dimensions must be reported (unless exclusion is approved by VPDrills quality). d. Geometrical features (i.e., true position) for multiple features must report the condition for each feature. 3. For characteristics that are recorded on material and/or test report certifications, type *CONFORMS* in the *Actual Measure* block (or equivalent), reference the certification number in the *Comment* block (or equivalent), and include a copy of the certification(s) with the FAI submittal. 4. For characteristics that require visual inspection, type *CONFORMS* in the *Actual Measure* block (or equivalent) and state *VISUAL* in the *Inspection Method* block (or equivalent). 5. For characteristics that are inspected using a functional or go/no-go gage, type *CONFORMS* in the *Actual Measure* block (or equivalent) and state the gage type or gage number in the *Inspection Method* block (or equivalent). 6. Indicate acceptance or rejection by typing *A* or *R* in the block provided. 7. Sign and date the form in the appropriate boxes. *Note*: an electronic signature or typed signature is acceptable.

(*continued*)

TABLE 7.35 (continued)
Process Flow (Company VPDrills)

	Major Steps	Activities
05	FAI report and samples	1. For suppliers (e.g., materials, coating), the following applies: a. For the first production shipment, the FAI form, annotated (ballooned) drawing, and applicable material/process certifications are to be included with the FAI sample. b. The FAI sample(s) must be identified. Attach an identifying tag to the FAI sample(s) using string (do not use tape or wire). Identification shall include the supplier's name, VPDrills purchase order part number and drawing revision, and VPDrills purchase order number, as well as any identification (sintering press number, serial number, etc.) that will link the sample to the source of manufacture. An orange label to the top and one side of the shipping container should be attached. The label shall identify the shipment as containing the FAI parts. 2. For VPDrills, attach an identifying tag to the FAI sample(s) using string (do not use tape or wire).
06	Review and approval of supplier FAI (quality)	1. Review the FAI package for completeness and approve (as necessary). If additional information or changes are needed, contact receiving inspection to resolve with the supplier. 2. For an acceptable FAI and for parts with a unique inspection plan in the business system (VPA-Q-01), do the following: a. Update the FAI status. b. For each characteristic in the inspection plan, verify the inspection method that identifies the same inspection method the supplier used on the FAI form.
07	Review and approval of VPDrills FAI (quality)	1. Review and approve the completed FAI form. 2. File the approved FAI (with all applicable certifications) in the respective part data folder. 3. Update the FAI status in the VPDrills inspection plan (for parts with an FAI inspection sequence). 4. When required, coordinate submittal to the customer.

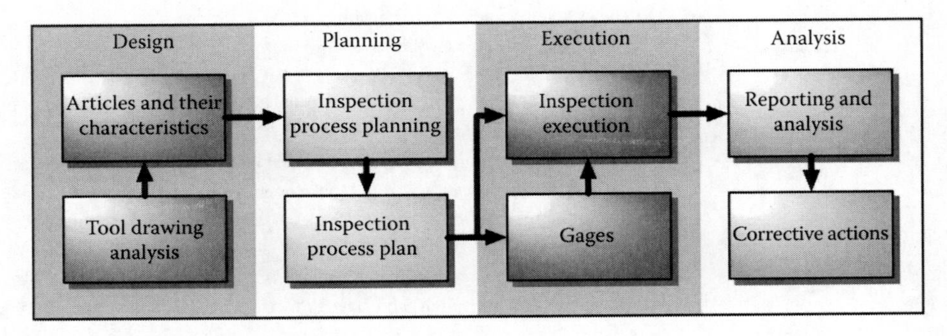

FIGURE 7.93 Dimension metrology system.

While dimensional articles and their characteristics indicated in the proper tool drawing are clear (at least, most of the time), geometrical articles is of prime concern as their definitions, designation in tool drawings, and tolerancing vary from one drill drawing to the next, from one drill manufactures to the other. This is due to coexistence of many contradictive sources of information on the matter including national and international standards as well as books including textbooks and web-based materials. Therefore, the subsequent section aims to provide methodological help to specialist to carry out this stage.

7.7.16 Dimensional Inspection (Metrological) System: Design Stage

As can be seen in Figure 7.93, the dimension inspection system originates from the tool drawing, so it completely relies on the information (and its accuracy) about articles provided by this drawing. Provided that one has a proper drill drawing where all information essential to the tool manufacturing and performance is indicated properly, he or she is fully equipped to go to the second design step (Figure 7.93) that includes compiling a list of articles and their characteristics. Tables 7.36 and 7.37 list articles to be considered and provide their proper definitions with references to the corresponding graphical representations. Although these tables give definitions for the listed articles, some additional explanations are required to the additional terms used in the article definitions, feasibility of their measurement, and adjustments to particularities of various drill designs.

7.7.16.1 Diameter-/Length-Related Articles

Back taper is defined as article 07N in Table 7.36. This article characteristic (designated as Δ_{bt} in Chapter 4 [Section 4.5.9.5]) is assigned in the tool drawing as the difference in the diameters over a specified reference length $L_{bt\text{-}r}$ (see Figure 7.94), that is,

$$\Delta_{bt} = d_{b1} - d_{b2} \tag{7.15}$$

The first diameter d_{b1} is normally the drill diameter and the second d_{b2} is measured according to the standard definition at the reference distance (length) $L_{bt\text{-}r}$ from the first as shown in Figure 7.94. For example, standard DIN 1414-1 recommends the back taper to be in the range of 0.02–0.08 mm/100 mm of drill length, that is, the reference length is 100 mm as discussed in Section 7.7.9. This rather great reference length $L_{bt\text{-}r}$ creates two problems:

- The overall drill length can be shorter than length $L_{bt\text{-}r}$ or the length of the special drill material (e.g., PCD) can be much shorter than this length
- Back taper is normally applied along some distance from the drill point, and thus, the rest of the drill is made cylindrical.

In both cases, back taper cannot be physically measured over the reference length $L_{bt\text{-}r}$.

To solve the problem with back taper assessment, the second diameter $d_{b2\text{-}a}$ is measured at some known feasible distance $L_{bt\text{-}a}$ from the drill as shown in Figure 7.94. Using geometric similarity as

$$\frac{d_1 - d_2}{L_{bt\text{-}r}} = \frac{d_1 - d_{2a}}{L_{bt\text{-}a}} \tag{7.16}$$

one can determine the back taper as

$$\Delta_{bt} = (d_1 - d_{2a})\frac{L_{bt\text{-}r}}{L_{bt\text{-}a}} \tag{7.17}$$

Margin width, b_m, is defined as article 08N; body clearance diameter, d_{cl}, as article 09N; and land width, W_L, as article 010N in Table 7.36. The characteristics of these articles are measured in the back plane as shown in Figure 7.95.

Coolant hole diameter, d_{ch}, is defined as article 14N in Table 7.36. It is measured in the back plane. Figure 7.95 shows that two diameters are measured in the case of two-hole drill, that is, $d_{ch\text{-}1}$ and $d_{ch\text{-}2}$. It is obvious that if there are more than two coolant holes, the diameter of each hole is measured.

The diameter of the auxiliary margins is defined as article 15N in Table 7.36. Figure 7.96 shows the drill designs where the diameter of the auxiliary margins, $d_{m\text{-}a}$, is not the same as that of the drill. Figure 7.96a shows drill design discussed in Chapter 4 (Figure 4.165). Such a design includes burnishing margins, which have a diameter that is slightly greater than that of the drill to improve drill stability and achieve greater quality of the machined surface. Figure 7.96b shows a PCD-tipped drill. The diameter of the auxiliary margins is slightly smaller than that of the drill (normally 6–12 μm).

TABLE 7.36
List and Definitions of the Non-Datum Articles

#	Article	Definition	Graphical Representation
01N	Drill diameter	The diameter over the margins of the drill measured at the periphery corners.	Size d_{dr} in Figure 7.50.
		If a drill has several steps, the diameter of each step of the drill is measured over the margins at the corresponding periphery corners.	Sizes d_1 and d_2 in Figure 7.46.
02N	Overall length	The length from the drill point to the end of its shank.	Size L_{oa} in Figure 7.46.
03N	Flute length	The length from the drill point to the end of the flute.	Size L_{fl} in Figure 7.46.
04N	Active flute length	The length from the drill point to the place of the flute where it begins to change its profile.	Size $L_{fl\text{-}a}$ in Figure 7.50.
05N	Length of the body relief	The length from the drill point to the end of the body relief.	Size L_{br} in Figure 7.50.
06N	Step length	The distance between the corners of consecutive steps.	Size L_1 Figure 7.46.
07N	Back taper	The difference in diameters over the margins of the drill per unit (specified) length of the drill.	See Figure 7.94.
08N	Margin width	Margin is the cylindrical portion of the land that is not cut away to provide clearance. The width of the margin is measured in the back (perpendicular to the drill longitudinal axis) plane.	See Figure 7.95.
09N	Body clearance diameter	The diameter of the body clearance cylinder, that is, the portion of the land that has been cut away to prevent its rubbing against the walls of the hole being drilled. This diameter is measured in the back (perpendicular to the drill longitudinal axis) plane.	See Figure 7.95.
10N	Land width	The distance between the leading edge and the heel of the land measured at right angles to the leading edge.	See Figure 7.95.
11N	Shank diameter (if cylindrical)	Shank is the part of the drill by which it is held and driven.	See Figure 7.53.
12N	Shank out of roundness (circularity)	Out of roundness is defined as a half of the maximum difference in the diameters of two concentric circles within which each circular element of the shank surface is located. It is measured in a plane perpendicular to the shank longitudinal axis.	See definition in Figure 7.30 and designation in a tool drawing in Figure 7.92.
13N	Shank length	The distance between two ends of the shank in the axial direction.	See Figure 7.92.
14N	Coolant hole diameter	A coolant hole is the passage through which the MWF is fed through the drill body (for drills with internal MWF supply). The diameter of each coolant hole is measured and reported. The coolant hole diameters is measured in the back (perpendicular to the drill longitudinal axis) plane.	See Figure 7.95.

TABLE 7.36 (continued)
List and Definitions of the Non-Datum Articles

#	Article	Definition	Graphical Representation
15N	Diameter of the auxiliary margins (if different than that of the primary margins)	The diameter over the auxiliary margins of the drill measured at the heels.	See Figure 7.96.
16N	Surface condition of the rake face	Surface roughness *Ra* or *Rz* measured in the direction specified by the tool drawing.	See Figure 7.92.
17N	Surface roughness of the primary flank face	Surface roughness *Ra* or *Rz* measured in the direction specified by the tool drawing. If not specified, measured in the direction parallel to the cutting edge.	See Figure 7.92.
18N	Surface condition 2 of primary flank face	Direction of grinding marks with respect to the primary cutting edge.	See Figure 7.92.
19N	Surface roughness of secondary flank face	Surface roughness *Ra* or *Rz* measured in the direction specified by the tool drawing.	See Figure 7.92.
20N	Surface roughness of the margin	Surface roughness *Ra* or *Rz* measured in the direction specified by the tool drawing.	See Figure 7.92.
21N	Surface roughness of the flute	Surface roughness *Ra* or *Rz* measured in the direction specified by the tool drawing.	See Figure 7.92.
23N	Surface roughness of shank	Surface roughness *Ra* or *Rz* measured in the direction specified by the tool drawing.	See Figure 7.92.
24N	Flute profile	Profile of the chip flute.	
25N	Helix angle	The angle between the tangent to the projection of the helical surface into the reference plane and the datum axis. Discussed in Section 7.7.11.	See Figure 7.85.

This is because these margins are used only for improving drill entrance stability, particularly in drilling cored holes. When the drill is fully engaged in cutting, these margins should not touch the surface of the machined hole by PCD tips. In both cases, the diameter of the auxiliary margins, d_{m-a}, is of critical importance for drill performance so it should be tightly toleranced and thus inspected.

Flute profile is defined as article 24N in Table 7.36. Although the tool profile can be controlled by measuring the boundary curves and their radii of curvature, it is not feasible for many drills having complicated flute profiles. The easiest way to inspect the accuracy of the actual tool profile is to import DXF overlays (2D CAD outline curves) into the measurement view.

7.7.16.2 Major Reference Plane

The characteristics of many important drill articles are measured in or with respect to the major reference plane that should be first distinguished among infinite numbers of reference planes containing the axis of the datum (commonly, the axis of the shank) as was introduced in Chapter 4 (Section 4.5.5). The major reference plane is defined as the plane containing the axis of the datum and as follows:

1. In the sense of the reference plane discussed in Appendix C, it should be perpendicular to the assumed direction of the cutting speed, which in T-hand-S is always perpendicular to the cutting edge. In geometrical sense, it is parallel to the projections of the major cutting edges into the back plane when the major cutting edges are 2D straight or curved lines as shown in Figure 7.97a.
2. The plane of radial symmetry at 180° for other shapes of the major cutting edges. Figure 7.98 shows an example for CoroDrill 860 drill (Sandvik Coromant Co.).

TABLE 7.37
List and Definitions of the Datum-Related Dimensional Articles

#	Article	Definition	Graphical Representation
01D	Drill runout	The distance between the datum axis and the center of the margin circle measured in the back (perpendicular to the datum axis) plane through the drill corners.	See Figure 7.59.
Articles related to the major cutting edges			
02D	Point angle, Φ_p	Measured in the major reference plane as the angle between the projections of the major cutting edges (lips) into this plane. When the projections of the major cutting edges into the major reference plane are not straight lines, the point angle $\Phi_{p\text{-}i}$ is measured for a point of consideration *i* on this projections as shown in Figure 7.96c. When the drill point is made with more than one point angle (e.g., a triple-point drill shown in Figure 4.158), the point angle is measured for each part.	See Figure 7.97.
03D	Half-point angle, φ	It is measured individually for each projection of the major cutting edges as the angle between the projection of a considered major cutting edge (lip) into the major reference plane and the axis of the datum as shown in Figure 7.96b. In the case of a two-flute drill considered in Figure 7.96b, two half-point angles, φ_1 and φ_2, are measured. When the projections of the major cutting edges into the major reference plane are not straight lines, the half-point angle φ is measured for a point of consideration *i* on this projection of the corresponding major cutting edge. In the case of a two-flute drill shown in Figure 7.96c, two half-point angles, φ_{1-i} and φ_{2-i}, are measured. When the drill point is made with more than one point angle (e.g., a triple-point drill shown in Figure 4.158), the half-point angles are measured for each part of the considered cutting edge.	See Figure 7.97.
04D	Flute spacing, a_o	It is measured in the back plane. It is defined as the distance between the projection of the cutting edge in the back plane and the line of intersection of the major reference plane and the back plane. See explanations in Section 7.7.16.3.	Figures 7.97a and 7.99
05D	Relative lip height (lip height variation), Δ_{lh}	Measured in the major reference plane, the normal (relative lip height) or the axial (parallel to the datum axis and known as the axial runout) distance between the projections of two or more cutting edges into the major reference plane conditionally revolved about the axis of the datum to bring all the edges on the same side of the major reference plane with respect to the axis of the datum. In the case of two-flute drill, one projection of the major cutting edge is revolved by 180°. See explanations in Section 7.7.16.3.	Figure 7.100

TABLE 7.37 (continued)
List and Definitions of the Datum-Related Dimensional Articles

#	Article	Definition	Graphical Representation
06D	Normal rake angle	The angle between rake face and the reference plane P_r measured in the normal plane P_n. If the rake face surface is not planar, then the plane tangent to the curved rake face surface in the considered point on the cutting edge is used instead of the rake plane. See explanations in Section 7.7.16.3.	Figure 7.101
07D	Side rake angle	The angle between rake face and the reference plane P_r measured in the working plane P_f. If the rake face surface is not planar, then the plane tangent to the curved rake face surface in the considered point on the cutting edge is used instead of the rake plane. See explanations in Section 7.7.16.3.	Figure 7.101
08D	Back rake angle	The angle between rake face and the reference plane P_r measured in the back plane P_p. If the rake face surface is not planar, then the plane tangent to the curved rake face surface in the considered point on the cutting edge is used instead of the rake plane.	Figure 7.101
O9D	Normal clearance angle (primary flank)	The angle between the tool cutting edge plane P_s and the tool primary flank plane measured in the normal plane P_n. It is clear that if the primary flank surface is not planar, then the plane tangent to the curved flank surface in the considered point on the cutting edge is used instead of the flank plane. See explanations in Section 7.7.16.3.	Figure 7.101
10D	Side clearance angle (primary flank)	The angle between the tool cutting edge plane P_s and the tool primary flank plane measured in the working plane P_f. It is clear that if the primary flank surface is not planar, then the plane tangent to the curved flank surface in the considered point on the cutting edge is used instead of the flank plane. See explanations in Section 7.7.16.3.	Figure 7.101
11D	Back clearance angle (primary flank)	The angle between the tool cutting edge plane P_s and the tool primary flank plane measured in the back plane P_p. It is clear that if the primary flank surface is not planar, then the plane tangent to the curved flank surface in the considered point on the cutting edge is used instead of the flank plane. See explanations in Section 7.7.16.3.	Figure 7.101
12D	Normal clearance angle (secondary flank)	The angle between the tool cutting edge plane P_s and the tool secondary flank plane measured in the normal plane P_n. It is clear that if the secondary flank surface is not planar, then the plane tangent to the curved flank surface in the considered point on the cutting edge is used instead of the flank plane. See explanations in Section 7.7.16.3.	Figure 7.101

(continued)

TABLE 7.37 (continued)
List and Definitions of the Datum-Related Dimensional Articles

#	Article	Definition	Graphical Representation
13D	Side clearance angle (secondary flank)	The angle between the tool cutting edge plane P_s and the tool secondary flank plane measured in the working plane P_f. It is clear that if the secondary flank surface is not planar, then the plane tangent to the curved flank surface in the considered point on the cutting edge is used instead of the flank plane. See explanations in Section 7.7.16.3.	Figure 7.102
14D	Back clearance angle (secondary flank)	The angle between the tool cutting edge plane P_s and the tool secondary flank plane measured in the back plane P_p. It is clear that if the secondary flank surface is not planar, then the plane tangent to the curved flank surface in the considered point on the cutting edge is used instead of the flank plane. See explanations in Section 7.7.16.3.	Figure 7.101
15D	Width of primary clearance, b_{fn}	Length of the projection of the primary clearance surface into the tool cutting edge plane P_s measured in the normal plane P_n.	Figure 7.101
16D	Edge preparation	A transition curve between the rake and the flank surfaces measured in the T-hand-S normal plane (direction normal to the theoretical cutting edge). The radius of the cutting edge, R_{ce}, that approximates the transition curve between the rake and flank faces is measured in the normal plane. See explanations in Section 7.7.16.3.	Figure 7.101
Articles related to the chisel edge and its region			
17D	Chisel angle, ψ_{cl}	The angle between the projection of the chisel edge in the back plane and the major reference plane measured in the back plane.	Figure 7.103a
18D	Chisel edge length, l_{cl}	The length of the projection of the chisel edge into the back plane.	Figure 7.103a
19D	Chisel edge centrality (symmetry), Δ_{cc}	The distance between the projection of the chisel edge into the back plane and the axis of the datum.	Figure 7.103b
20D	Centrality of the web, Δ_w	The difference of the distances from the axis of the datum and the cutting edges measured along a line drawn through the point of the intersection of the axis of the datum with the back plane perpendicular to the major reference plane. See explanations in Section 7.7.16.4.	Figure 7.103c
21D	Web thickness, d_{wb}	The distance between the points of the major cutting edges measured along a line drawn through the point of the intersection of the axis of the datum with the back plane perpendicular to the major reference plane. See explanations in Section 7.7.16.4.	Figure 7.103c
22D	Radius of the end of the chisel edge, r_{cl}	Only for *S* chisel edge drills. The radius of the circular arc tangent to both the straight portions of the chisel edge and the corresponding major cutting edge. See explanations in Section 7.7.16.4.	Figure 7.104a
23D	Mismatch of the intersection lines of the secondary flank faces with the primary flank faces, m_f	The distance between the points of intersection of the lines of intersection of the primary and secondary flank faces with the chisel edge measured in the back plane. See explanations in Section 7.7.16.4.	Figure 7.104b

TABLE 7.37 (continued)
List and Definitions of the Datum-Related Dimensional Articles

#	Article	Definition	Graphical Representation
24D	Chisel wedge angle, ν_{cl}	Measured in the chisel edge reference plane as the angle between the projections of the rake and flank surfaces of the chisel into this plane.	Figure 7.105
25D	Chisel wedge symmetry, Δ_{clw}	Only for self-centered designs of the chisel edge. The difference in the half chisel wedge angles, $\nu_{cl\text{-}1}$ and $\nu_{cl\text{-}1}$, measured in the chisel edge reference plane.	Figure 7.105b
26D	Chisel edge rake angle, γ_{cl}	The angle between the chisel edge rake face and the chisel edge reference plane measured in the chisel edge normal plane. If the rake face surface of the chisel edge is not planar, then the plane tangent to the curved rake face surface in the considered point on the cutting edge is used instead of the rake plane. See explanations in Section 7.7.16.4.	Figure 7.106
27D	Chisel edge clearance angle, α_{cl}	The angle between the flank face of the chisel edge and the tool chisel edge plane, $P_{s\text{-}cl}$, measured in the chisel edge normal plane. If the flank face surface of the chisel edge is not planar, then the plane tangent to the curved flank face surface in the considered point on the cutting edge is used instead of the flank plane. See explanations in Section 7.7.16.4.	Figure 7.106
28D	Only for generic design of split-point drills. Length of the rake face, l_{sp}	The length of the rake face measured in the chisel edge normal plane. See explanations in Section 7.7.16.4.	Figure 7.107
29D	Only for generic design of split-point drills. Transition radius, r_{sp}	Radius between the rake face of the chisel edge and the chisel edge split gash measured in the chisel edge normal plane.	Figure 7.107a
Gash-related parameters (generic)			
30D	Gash face angle, $\psi_{gh\text{-}b}$	The angle between the sides of the gash measured in the back plane. See explanations in Section 7.7.16.5.	Figure 7.108
31D	Gash inclination angle, $\lambda_{gh\text{-}b}$	The angle between the straight and gashed potions of the major cutting edge measured in the back plane. See explanations in Section 7.7.16.5.	Figure 7.108
32D	Extent of the gash past the CL, a_4	Measured in the back plane, the distance between the end of the straight portion of the gash and the vertical axis of the drill drawn through the axis of the datum. See explanations in Section 7.7.16.5.	Figure 7.108
33D	Radius of the gash, r_{gh}	Radius of the gash measured in the back plane. See explanations in Section 7.7.16.5.	Figure 7.108
34D	Gash plane angle, $\psi_{gh\text{-}r}$	The angle between the gash direction and the axis of the datum measured in the major reference plane. See explanations in Section 7.7.16.5.	Figure 7.108
Articles related to the coolant holes			
34D	Bolt diameter, d_{bd}	Diameter of a circle drawn from the axis of the datum and passing through the centers of the coolant holes. Measured in the back plane.	Figure 7.112
35B	Coolant holes location angle, ψ_{ch}	Angle between the line drawn through the centers of the coolant holes and the major reference plane measured in the back plane.	Figure 7.112

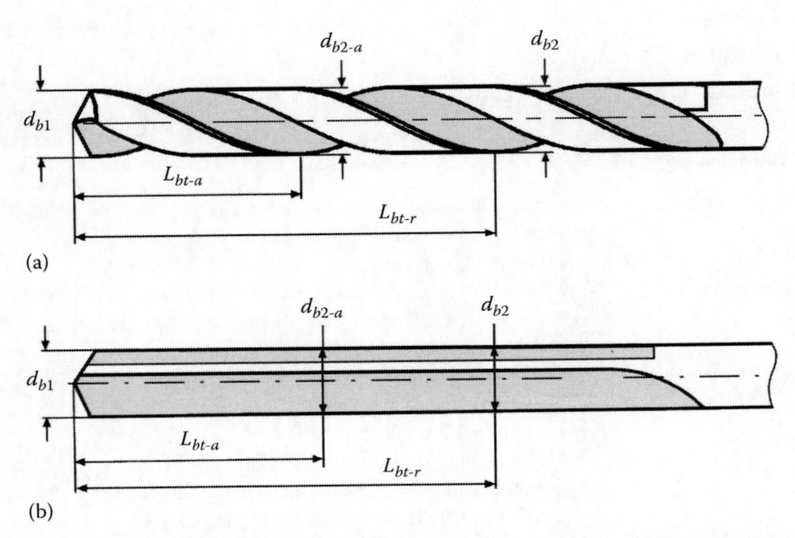

FIGURE 7.94 Back taper particularities: (a) for a twist drill and (b) for a straight-flute drill.

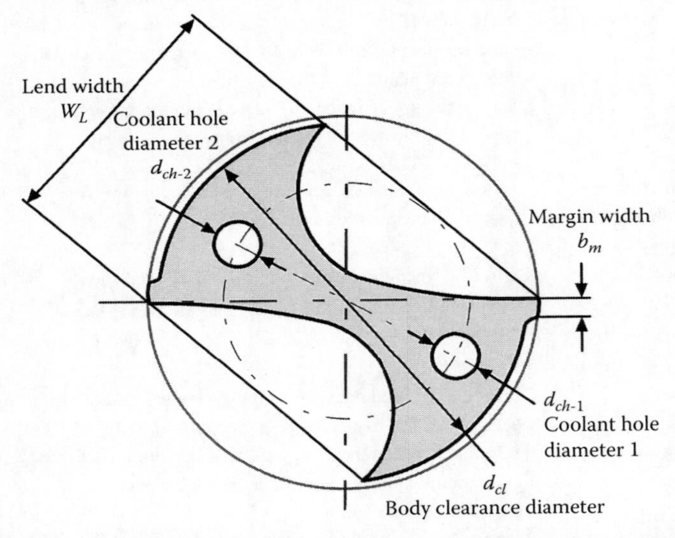

FIGURE 7.95 Showing graphical representations of margin width (article 08), body clearance diameter (article 09), land width (article 10), and coolant hole diameters (article 14) in Table 7.36.

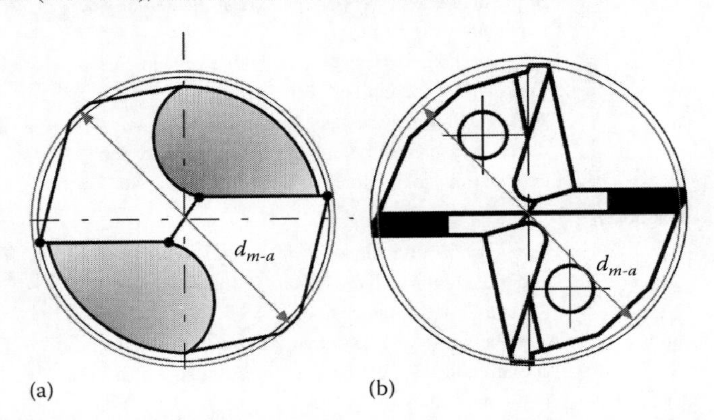

FIGURE 7.96 Diameter of the auxiliary margins is not the same as that of the drill: (a) greater and (b) smaller.

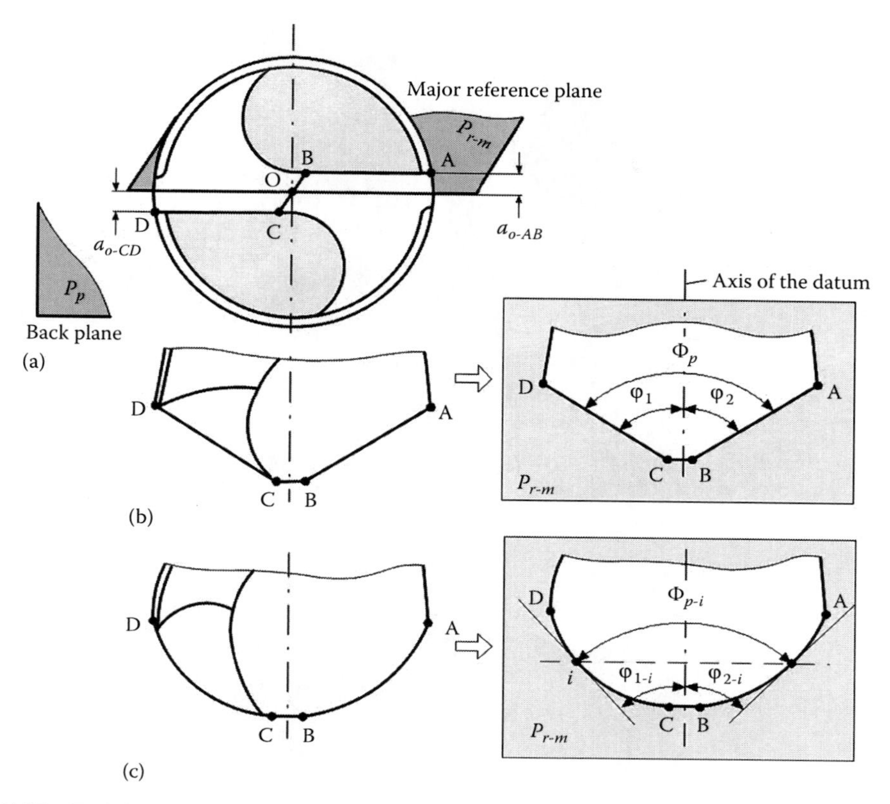

FIGURE 7.97 Definition of the major reference plane (a). Tho cases are considered: (b) the projection of the major cutting edges into the reference plane are straight lines, and (c) these projection are curves.

Once the major reference plane is defined, the drill back plane can also be defined as the plane perpendicular to the major reference plane. The characteristics of the many important articles of the drill geometry are measured in the reference and back planes.

7.7.16.3 Articles Related to the Major Cutting Edges

The point angle, Φ_p, is defined as article 02D in Table 7.37. *The half-point angle*, φ, is defined as article 03D in Table 7.37. *Flute spacing*, a_o, is defined as article 04D in Table 7.37. In the case of two-flute drill considered in Figure 7.97a, two heights, namely, $a_{o\text{-}AB}$ and $a_{o\text{-}CD}$, are measured as the flute spacing for the major cutting edges AB and CD, respectively, so that the absolute value of the difference $a_{o\text{-}AB} - a_{o\text{-}CD}$ is then compared with the tolerance A_s (Section 7.7.7.3, Figure 7.79.)

When the projection of the major cutting edge into the back plane is not a straight line, then an array of flute spacings can be measured for many points of the cutting edge (e.g., to control the shape of the cutting edge of the drill shown in Figure 4.172). Figure 7.99 shows an example of measuring the flute spacing. Two heights, namely, $a_{o\text{-}i1}$ and $a_{o\text{-}i2}$, located on the major cutting edges on the same radius r_i are measured, and then their difference is compared with the tolerance assigned on the cutting edge shape. Note that assessment of a_o's when compared to the profile of the complicated shape cutting edges is the most proper way to compare the shapes of such cutting edges even when the most sophisticated drill measuring machines are used.

Lip height variation, Δ_{lh}, is defined as article 05D in Table 7.37. Figure 7.100 provides a graphical representation of its measuring procedure for a two-flute drill. As can be seen, the projections AB and CD of the major cutting edges are brought into the same side of the axis of the datum by rotating AB about the axis of the datum by 180°. As such, projection AB occupies a new location A_rB_r.

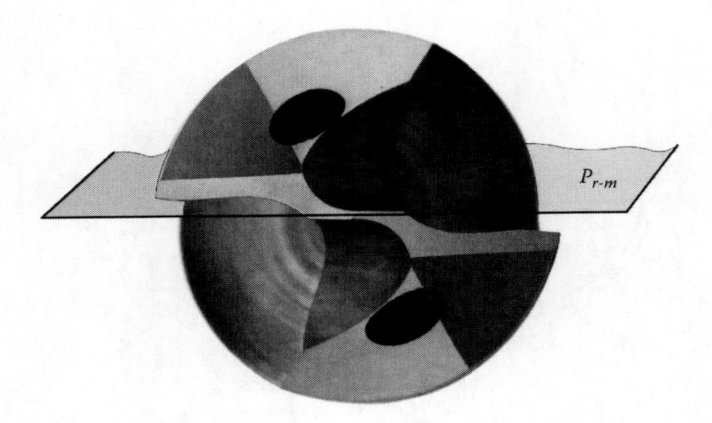

FIGURE 7.98 Showing an example of the definition of the major reference plane as the plane of radial symmetry at 180° for CoroDrill 860 drill. (Courtesy of Sandvik Coromant Co., Fair Lawn, NJ.)

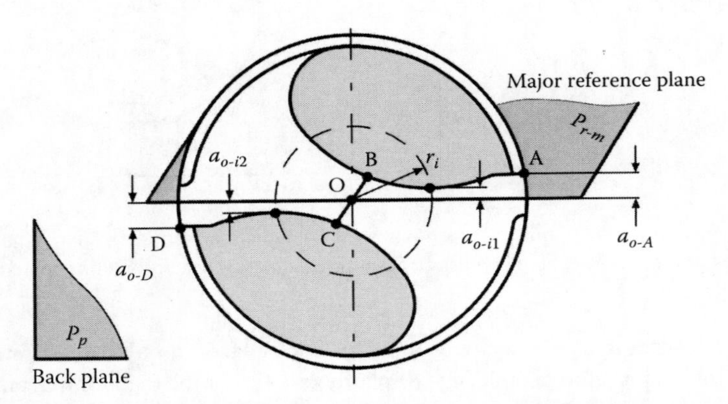

FIGURE 7.99 Measuring the flute spacing for the points of the cutting edge located on the same radius r_i.

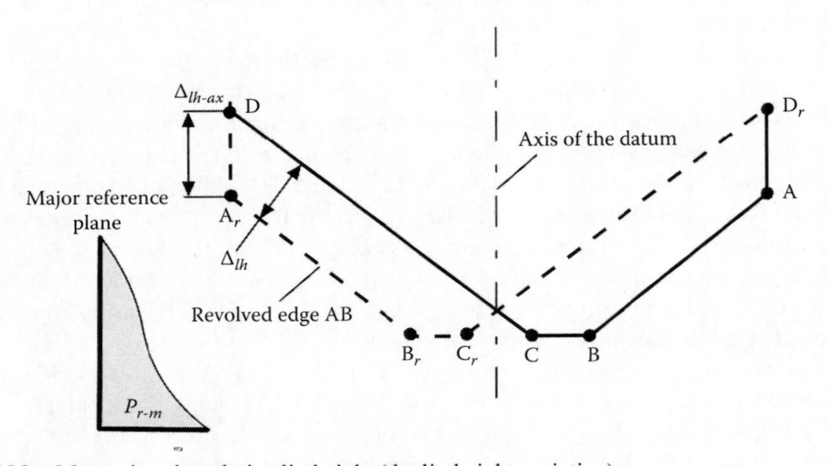

FIGURE 7.100 Measuring the relative lip height (the lip height variation).

The normal Δ_{lh} (the relative lip height) or axial $\Delta_{lh\text{-}ax}$ (the axial runout) can then be measured for any point of the projections A_rB_r and CD. The measured value can be compared to those assigned by the drill drawing and the referenced standards.

Rake and clearance (primary) angles are defined as articles 06D-11D in Table 7.37. The definitions of the planes of measurement and the corresponding articles are given in Chapter 4 (Section 4.5). It is discussed that the set of measuring plane (sometime called reference planes) should be defined for the considered point of the major cutting edge in T-hand-S. Figure 7.101 shows graphical representation of these planes for the point of consideration *i*. They are the normal, working, and back planes represented by sections A-A, B-B, and C-C, respectively (i.e., as should be shown in a tool drawing). These planes are related to the major reference plane as they are defined with respect to this plane as the following:

- The reference plane $P_{r\text{-}i}$ drawn through point *i* is parallel to the major reference plane.
- The normal plane $P_{n\text{-}i}$ drawn through point *i* is perpendicular to the major reference plane and to projection of the major cutting edge on which point *i* is located into the reference plane. Note that if the cutting edge is not straight, then the tangent to the cutting edge at *i* is considered.
- The working plane $P_{f\text{-}i}$ drawn through point *i* parallel to the datum axis is perpendicular to the major reference plane.
- The back plane $P_{p\text{-}i}$ drawn through point *i* perpendicular to the datum axis is perpendicular to the major reference plane.
- The cutting edge plane $P_{s\text{-}i}$ drawn through point *i* is perpendicular to the major reference plane and contains the major cutting edge. Note that if the cutting edge is not straight, then the cutting edge plane that contains the tangent to the cutting edge at *i* is considered instead of the actual cutting edge.

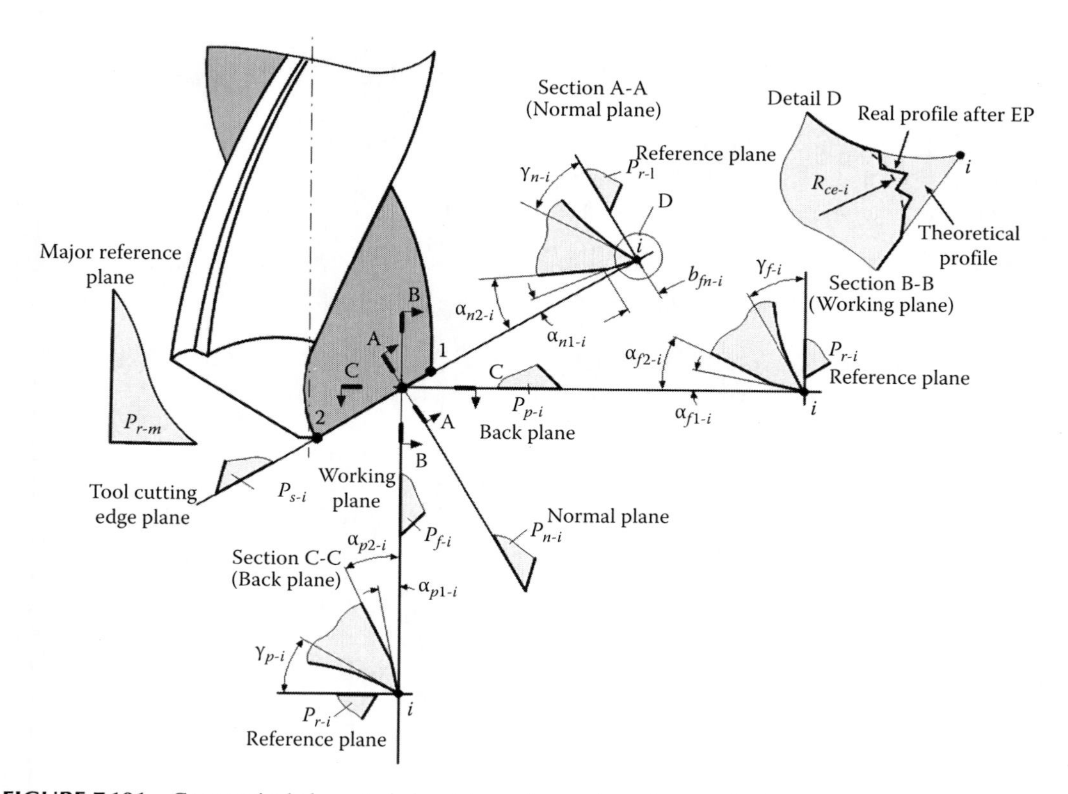

FIGURE 7.101 Geometrical characteristic of the major cutting edge at point *i*.

It should be clear that not all the listed rake and clearance angles are indicated in a tool drawing. As discussed earlier, when the primary flank face is ground flat, then the normal clearance angle is indicated in the tool drawing. As such, this angle is the same (in T-hand-S) for any point of the major cutting edge. If needed, the angles in other reference planes can be calculated as discussed in Chapter 4 (Section 4.5). In the author's opinion, advanced tool inspection machine having powerful on *board* computers should do such a calculation automatically and include the results (at least as an option) in the inspection report.

When the flank face is not flat, then the side clearance angle is indicated at point A, that is, at the drill corned as shown in Figure 4.55 in Chapter 4. In this case, the clearance angles are not the same for different points of the major cutting edge. Moreover, their distribution over this edge depends not only on the selected shape of the flank surface but also on the setting of the grinding fixture to generate the chosen shape of the flank face. The type of a particular surface selected to be a flank surface and setting parameters of the grinding setup (machine) are not indicated in the tool drawing. Therefore, it is important in this case to measure the clearance angle at least at three points of the major cutting edge, for example, at the periphery corner, at the middle, and at its inner end at the point of intersection of this edge with the chisel edge. In the author's opinion, advanced tool inspection machine having powerful on *board* computers should provide distribution of the T-hand-S clearance angle over the major cutting edge. According to the author's experience, when such information is available, the distribution of the clearance angle over the major cutting edge can be optimized to extend tool life and to prevent the interference of the flank face and the bottom of the hole being drilled in HP drilling.

The rake angle is not normally indicated in the tool drawing as discussed in Chapter 4 (Section 4.5.5) because of the following:

- It is assumed to be zero in T-hand-S for straight-flute drills.
- It is uniquely defined by the angle of helix for twist drills (see Equations 4.24 and 4.25).

For HP drills, however, the rake angle is indicated in tool drawings in the following cases:

- For plane rake faced twist drills (e.g., shown in Figure 4.59)
- For straight-flute and twist drills when the rake face is partially (the most common case) or fully modified by gashes (e.g., shown in Figure 4.51).

In the author's opinion, advanced tool inspection machines having powerful on *board* computers should provide distribution of the rake angle over the major cutting edge as a part of the tool inspection report.

Clearance angles of the secondary flank surface are defined as articles 12D–14D in Table 7.37. They are measured in the same reference plane as those for the primary flank as shown in Figure 7.101.

The width of primary clearance, b_{fn}, is defined as article 15D in Table 7.37. It is measured in the normal plane $P_{n\text{-}i}$ as shown in Figure 7.101.

Edge preparation is defined as article 16D in Table 7.37. Figure 7.102 shows edge preparation (EP) types according to standard ANSI B212.4. The same types are defined by standard ISO 1832: 2004/2005. As discussed in Appendix C (Section C.4.2), out of these many standard types, more than 95% of drills and drilling tools receive a radius hone (Figure C.52). In real cutting tools, however, the theoretical profile does not turn into a smooth curve with a defined radius. The real profile of the transitional curve between the rake and the flank faces is similar to that shown in Figure 7.101 (Detail D). Although such a curve is no longer an edge (by the edge definition as to be a line), it is conditionally called the cutting edge. This real profile is approximated by a certain radius R_{ce} and is regarded as the EP radius.

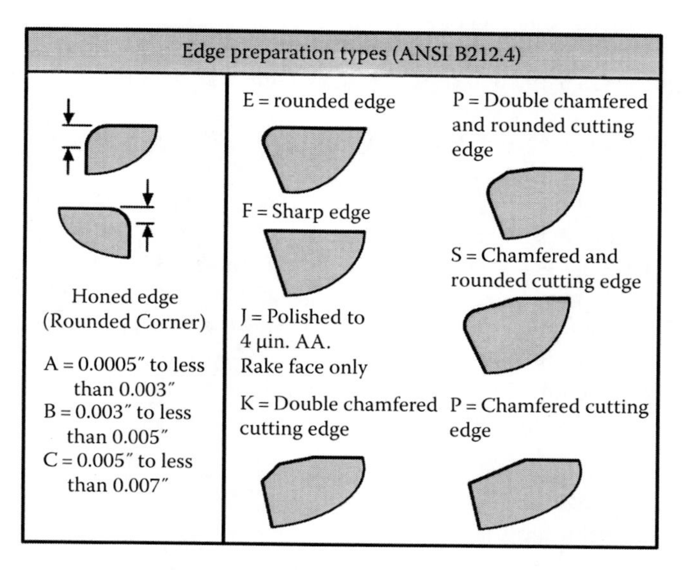

FIGURE 7.102 Edge preparation types according to standard ANSI B212.4.

As discussed in Appendix C (Section C.4.2), high-end EP measuring machines/microscopes do this approximation automatically (Figures C.62 and C.63) or even are capable of selecting the best curve rather than an arc to approximate the real profile. Moreover, the complete transition surface over the entire length of the cutting edge can be visualized digitally (Figure C.65). Unfortunately, no metrology for these advanced parameters is available yet, and thus, these parameters are not parts of the tool drawing. These examples are just another confirmation to the author's point that the level of the available manufacturing and inspection equipment is far ahead of the drill design practice and metrology defined by the international and national standards in various developed countries.

In the author's opinion, the following additional parameters should be included in the drawing of the advanced drills and drilling tools:

1. The radius of the cutting edge with the tolerance so that the real transition surface should be then approximated by the radius. The radii of this surface then can be measured over the entire length of the cutting edge (Figures C.62 and C.64) and compared with the tolerance assigned.
2. The surface roughness of the transition surface that should be better than those on the rake and flank faces as the surface roughness improvement (on the tool contact surfaces) is the leading idea of EP (see Appendix C). Figure C.49 shows a comparison of surface roughness before and after applying EP. Therefore, the inspection of surface roughness of the transition surface should be carried out to assure that the intended effect of EP is actually achieved.

The discussed measurements can be used as the prime parameter in capability studies of various available and emerging EP techniques (Figure C.55). Consistency/processing time/cost/improvement in drill performance sequence can be evaluated to select the optimal parameters of EP and its most efficient technique/technology.

The ability to approximate the real shape of the transition surface shape by the best-fit curve can be used in the future to apply real metal cutting theory to the optimization of the contact conditions.

7.7.16.4 Articles Related to the Chisel Edge and Its Region

Chisel edge angle, ψ_{cl}, is defined as article 17D in Table 7.37. Figure 7.103a shows its graphical representation.

Chisel edge length, l_{cl}, is defined as article 18D in Table 7.37. It is measured in the back plane. Figure 7.103a shows its graphical representation.

Chisel edge centrality (*symmetry*), Δ_{cc}, is defined as article 19D in Table 7.37. Figure 7.103b shows its graphical representation.

Centrality of the web, Δ_w, is defined as article 20D in Table 7.37. It is measured in the back plane. Figure 7.103c shows the graphical representation of the principle of the measurement of this article for a two-flute drill. Points C and B on the projection of the major cutting edges into the back plane are points of intersection of the chisel edge with the major cutting edges, and the line through center O perpendicular to the major reference plane is parallel to line EF. As such, the web centrality is calculated as

$$\Delta_w = |l_{w1} - l_{w2}| \tag{7.18}$$

Alternatively, the web centrality can be represented as

$$\Delta_w = \frac{|d_E - d_F|}{2} \tag{7.19}$$

where

d_E is the diameter of the circle with the center at point O and passing through point E

d_F is the diameter of the circle with the center at point O and passing through point F as shown in Figure 7.103c.

The obtained result can then be compared with the standard and recommended tolerances shown in Figure 7.77.

Web thickness, d_{wb}, is defined as article 21D in Table 7.37. It is measured in the back plane. In Figure 7.103c, it is the distance between points E and F, that is, the web thickness is calculated as

$$d_{wb} = l_{w1} + l_{w2} \tag{7.20}$$

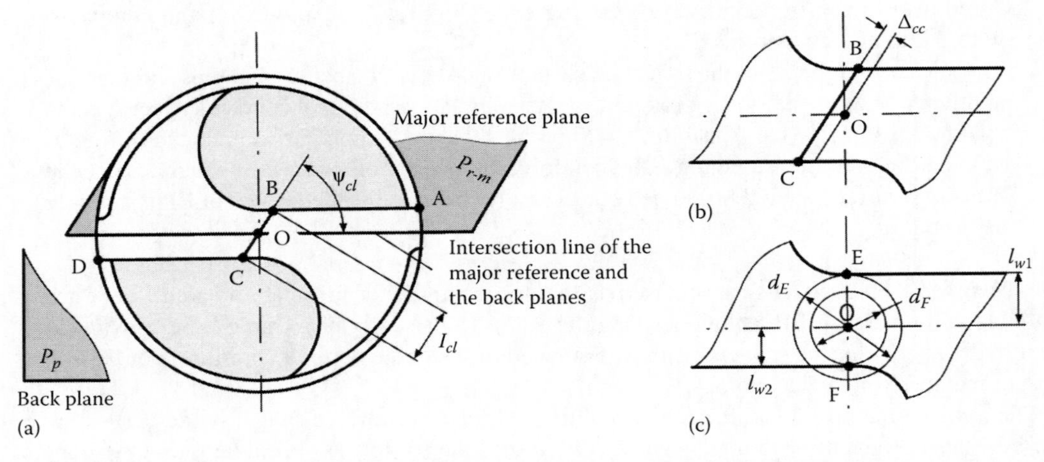

FIGURE 7.103 Characteristics of the chisel edge and the web in the back plane measured with respect to the major reference plane: (a) chisel edge angle, ψ_{cl}, and chisel edge length, l_{cl}, (b) chisel edge centrality (symmetry), Δ_{cc}, and (c) centrality of the web, Δ_w.

Alternatively, the web thickness can be represented as

$$d_{wd} = \frac{(d_E + d_F)}{2} \tag{7.21}$$

Mismatch of the intersection lines of the secondary flank faces with the primary flank faces, m_f, is defined as article 22D in Table 7.37. It is measured in the back plane as shown in Figure 7.104b. Note that individual mismatches m_{f1} and m_{f2} measured in the back plane between points E and F and the major reference plane with the back plane can also be measured to correct a grinding flaw (Figure 4.121).

Chisel wedge angle, ν_{cl}, is defined as article 24D in Table 7.37. It is measured in the chisel edge reference plane that is defined as the reference plane revolved about the axis of the datum by angle ψ_{cl} with respect to the major reference plane as shown in Figure 7.105b.

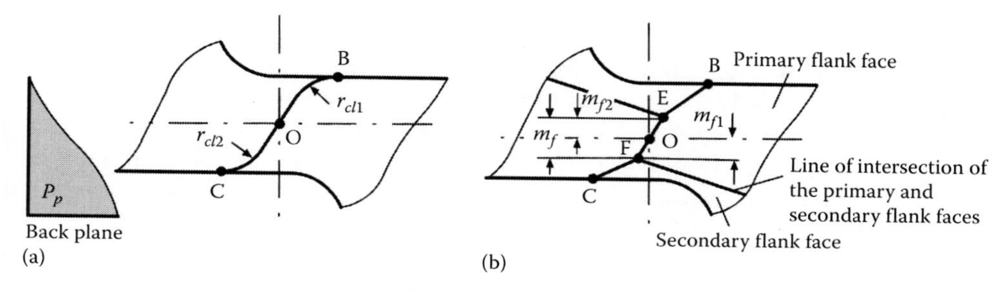

FIGURE 7.104 Graphical representation of (a) the radius of the end of the chisel edge and (b) mismatch of the secondary flank faces.

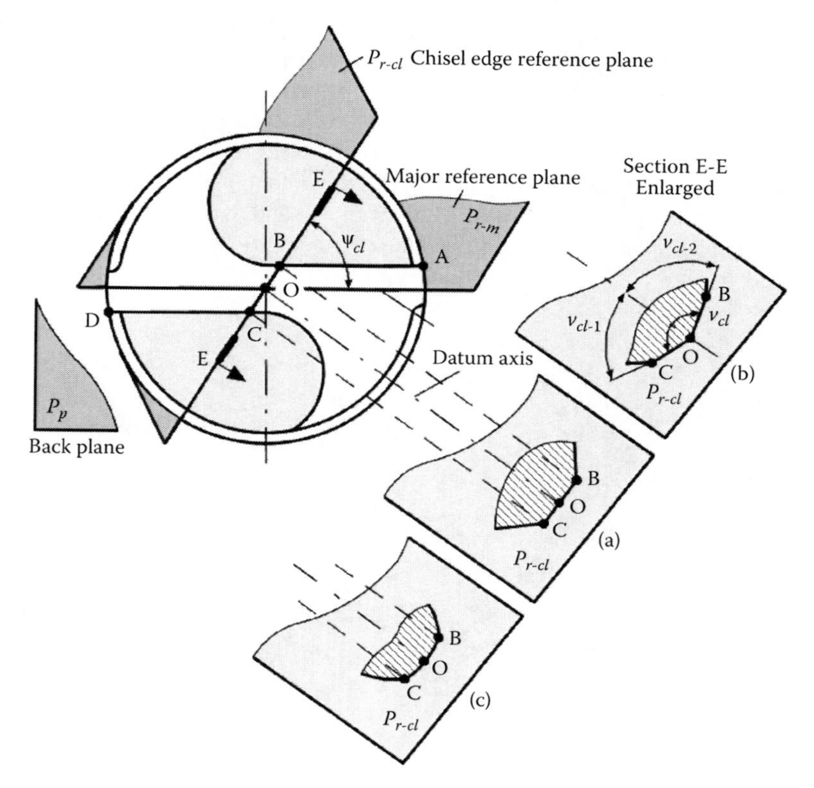

FIGURE 7.105 Chisel edge reference plane and the chisel wedge geometry in this plane: (a) chisel wedge angle is equal to 180°, (b) chisel wedge angle is less than 180°, and (c) curved chisel edge.

As discussed in Chapter 4 (Section 4.5.8.2), when the flank faces of the major cutting edges (lips) have the same T-hand-S clearance angles $\alpha_{n1} = \alpha_{n2}$ and the width of the primary flank is equal or greater than $2a_o$, then the chisel wedge angle is equal to 180° as shown in Figure 7.105a. As such, when drill first touches the surface of the workpiece, it does so (at least, theoretically) by the whole length of the chisel wedge as this edge is perpendicular to the axis of rotation and feed motion (the axis of the datum in the geometry considerations). Such an engagement often presents problems causing poor shape of the machined hole at the entrance. As discussed in Chapter 4, Section 4.5.8.3, significant improvement in drill self-centering and in geometry of the chisel edge is achieved when the width of the primary flank is equal to a_o, while the secondary flank plane is applied with the normal clearance angle greater than that of the primary clearance angle. As such, the chisel wedge angle is less than 180° as shown in Figure 7.105b, so when drill first touches the surface of the workpiece, it does so (at least, theoretically) by the central point connecting two portions of the chisel edge. For this and many other similar designs of drills with some self-centering ability, the chisel wedge angle, ν_{cl}, should be measured. Note that this angle is not a part of the tool drawing as its particular value is an outcome of many other parameters of the tool geometry (Section 4.5.8.3). However, this angle can and should be used to compare different drill geometries in terms of their effect on the drill's self-centering ability. The smaller the chisel wedge angle, the higher the self-centering ability of the drill.

For drills with not planar flank faces, the chisel edge is not a straight line as shown in Figure 7.105c. As can be seen, such drills possess some self-centering ability that, to the first approximation, can be characterized by the radius of curvature of the chisel edge projection into the chisel edge reference plane.

Chisel wedge symmetry, Δ_{clw}, is defined as article 25D in Table 7.37. It is measured in the chisel edge reference plane as shown in Figure 7.105b

Chisel edge rake angle, $\gamma_{cl\text{-}n}$, is defined as article 26D in Table 7.37. It is measured in the chisel edge normal plane, $P_{n\text{-}cl}$, which is defined as a plane perpendicular to the projection of the chisel edge into the chisel edge reference plane. Figure 7.106 shows graphical representation of these planes for point G of the chisel edge when the chisel wedge angle is equal to 180° (Figure 7.106a) and when this angle is less than 180° (Figure 7.106b). The rake angle of the chisel edge is defined as the angle between the chisel edge rake face and the chisel edge reference plane measured in the chisel edge normal plane. When the rake face of the chisel edge is sharply curved as, for example, when formed by pits (Figure 4.128), the tangent to the rake face of the chisel edge at point of consideration (point G in Figure 7.106) is drawn, and the chisel edge rake face is considered as the angle between this tangent and the chisel edge reference plane.

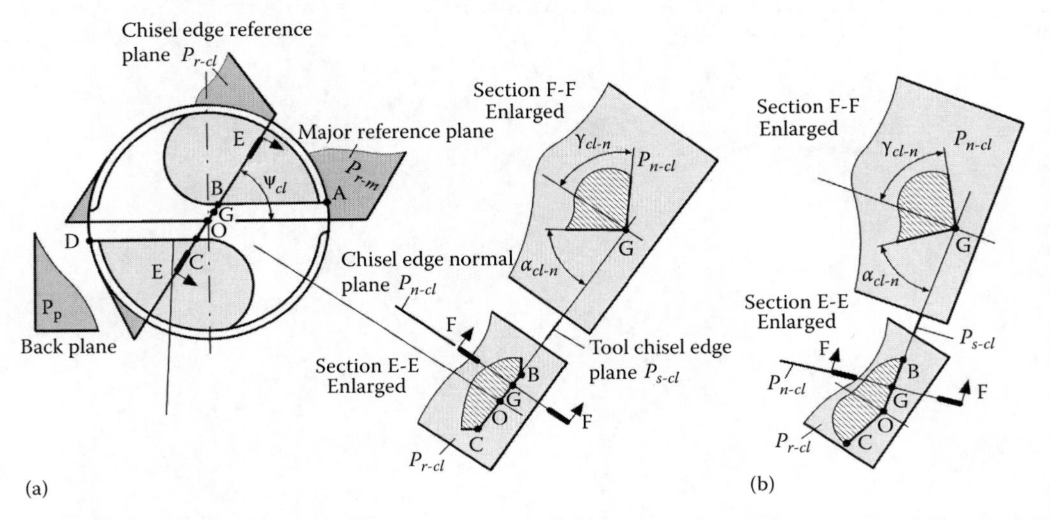

FIGURE 7.106 Chisel edge rake and clearance angle representation: (a) when the chisel wedge angle is equal to 180° (Figure 7.104a) and (b) when the chisel wedge angle is less than 180° (Figure 7.104b).

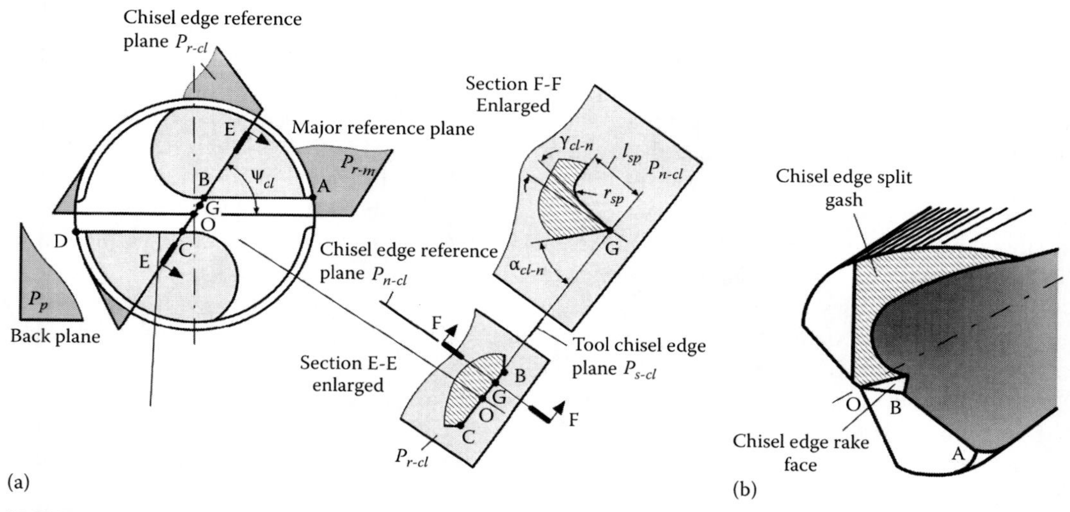

FIGURE 7.107 Defining parameters of the generic split-point drill design: (a) length of the rake face of the chisel edge and the transition radius and (b) generic configuration of the rake face.

Note that, besides very special cases, the rake angle of the chisel edge varies along this edge so that it should be measured at various points of this edge to assess the distribution of the chisel edge rake angle over this edge. This distribution is an important design parameter for HP drills as discussed in Chapter 4 (Section 4.5.8). Moreover, for partially split chisel edges (e.g., as shown in Figure 4.128), the various points of the chisel edge may have considerable different rake angles.

Chisel edge clearance angle, α_{cl-n}, is defined as article 27D in Table 7.37. It is measured in the chisel edge normal plane, P_{n-cl}, as the angle between the flank face of the chisel edge and the tool chisel edge plane, P_{s-cl}, as shown in Figure 7.106. The tool chisel edge plane P_{s-ch} drawn through point G is perpendicular to the chisel edge reference plane and contains the chisel edge.

In some particular cases, the chisel edge is not straight so the cutting edge plane contains the tangent to the cutting edge at G that is considered instead of the actual chisel edge. If the flank face surface of the chisel edge is not planar, then the plane tangent to the curved rake face surface in the considered point on the cutting edge is used instead of the flank plane.

Note that, besides very special cases, the clearance angle of the chisel edge varies along this edge so that it should be measured at various points of this edge to assess the distribution of the chisel edge clearance angle over this edge.

Length of the rake face of the split-point chisel edge, l_{sp} (only for generic design of split-point as one shown in Figure 7.107b), is defined as article 28D in Table 7.37. It is measured in the chisel edge normal plane as shown in Figure 7.107a. Note that l_{sp} varies over the chisel edge as can be seen in Figure 7.107b, so its distribution over this edge should be measured (the details are discussed in Section 4.5.8).

Transition radius between the rake face of the chisel edge and the chisel edge split gash, r_{sp}, is defined as article 29D in Table 7.37. It is measured in the chisel edge normal plane as shown in Figure 7.107a.

7.7.16.5 Articles Related to Gashes

There are a number of ways to dimension gashes applied to the drill point to modify the rake faces of the major cutting edges (called web thinning) and/or to modify the extent and rake geometry of the chisel edge as discussed in Chapter 4. As a gash is normally a 3D article, its proper dimensioning and tolerancing present some problems in the tool drawing and thus in tool inspection.

One of the common ways to dimension gashes is by assigning the gash direction in the major reference plane by the gash reference angle (shown in Figure 4.51 as the web thin heel angle) and by the gash inclination angle in the back plane shown as the web thin notch face angle as shown in Chapter 4, Figure 4.51. The gash parameters are the actual gash opening angle (shown in Figure 4.51 as the web thin notch angle) and the actual radius (shown in Figure 4.51 as the web thin wheel radius). The extent of the gash is controlled by the resultant web thickness (shown in Figure 4.51). Such a representation of the gash parameters is easier for tool manufacturing but is not exactly strict for tool drawing where the representation of the plane perpendicular to the general gash direction to show the actual gash opening angle and radius is rather difficult.

Figure 7.108 shows one of the feasible ways of representation of gashes as, in the author's opinion, the tool designer/developer should put in the drawing the parameters essential to the tool performance and the process development/manufacturing specialist should Figure out how to *translate* these parameters into the position and shape of the grinding wheel. Once being a difficult job attainable only to few high-qualification specialists, this problem nowadays becomes almost routine as drills are designed using a 3D CAD program, so drill digital CAD model can be directly uploaded into a CNC grinding machine so that the grinding wheel shape is selected properly, its radius is dressed correspondingly, and it is positioned in the correct way with respect to the drill while applying the gash.

In Figure 7.108, the direction of the gash in the back plane (the face view of the drill) is defined by the gash face angle, $\psi_{gh\text{-}b}$, and the gash inclination angle, $\lambda_{gh\text{-}b}$, while its extent is determined by the distance a_4 from the CL to the beginning of the gash radius that, as discussed in Chapter 4, is an important drill design parameter. The radius of the gash in the back plane, r_{gh}, is also a part of the tool drawing. In the major reference plane, the gash plane angle, $\psi_{gh\text{-}r}$, is defined. If more than

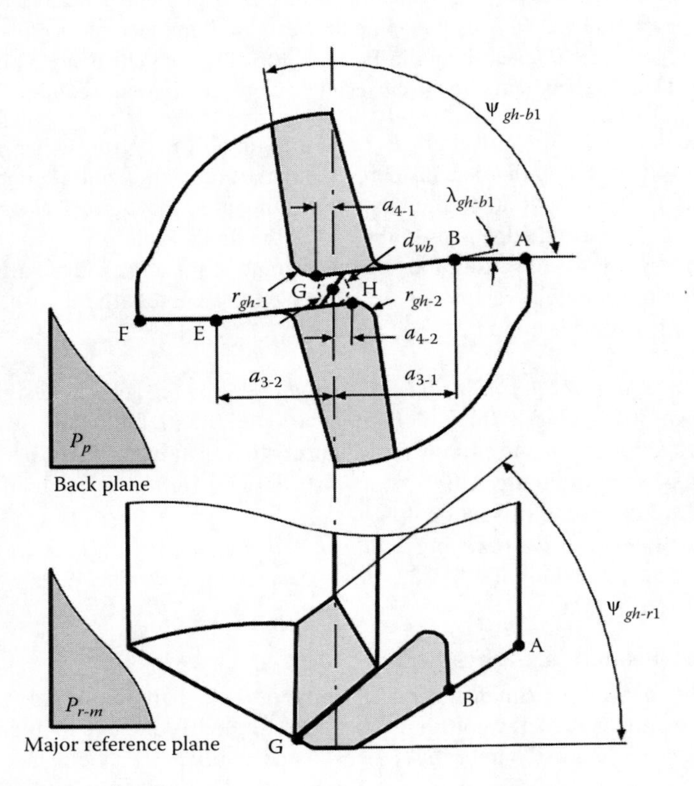

FIGURE 7.108 Parameters of gashes.

one gash is applied (e.g., as shown in Figure 7.92), they should be dimensioned correspondingly. Therefore, these parameters are included in Table 7.37:

- *Gash face angle*, $\psi_{gh\text{-}b}$, is defined as article 30D in Table 7.37. It is measured in the back plane as shown in Figure 7.108.
- *Gash inclination angle*, $\lambda_{gh\text{-}b}$, is defined as article 31D in Table 7.37. It is measured in the back plane as shown in Figure 7.108.
- *Extent of the gash past the CL*, a_4, is defined as article 32D in Table 7.37. It is measured in the back plane as shown in Figure 7.108.
- *Radius of the gash*, r_{gh}, is defined as article 33D in Table 7.37. It is measured in the back plane as shown in Figure 7.108.
- *Gash plane angle*, $\psi_{gh\text{-}r}$, is defined as article 34D in Table 7.37. It is measured in the major reference plane as shown in Figure 7.108.

The listed articles should be measured for each gashed flute. Figure 7.108 shows a two-flute drill where two characteristics of these articles are to be measured and then compared with the tolerances assigned (hopefully) in the tool drawing.

It is common in the tool industry, however, to show other parameters of the gashes also shown for reference in Figure 7.108. The most common are the web diameter, d_{wb}, as determined by gashes and the beginning of the gashes on the tool rake face (distance a_3). The latter is often shown from the drill periphery corner (point A in Figure 7.108). If shown in the considered tool drawing, these parameters can also be included in Table 7.37. In the author's opinion, being important, these parameters are outcomes of gashing. In other words, it is difficult to Figure out the setting of the gash grinding wheel using these parameters.

7.7.16.6 Articles Related to the Coolant Holes

Two important articles related to the coolant holes location are measured in the back plane:

1. Location diameter, d_{bd} (also known as the bolt diameter in the carbide rods, D_{TK}, as discussed in Section 6.3.3.10), is defined as article 35D in Table 7.37. It is measured in the back plane as shown in Figure 7.109.
2. Coolant holes location angle, ψ_{ch}, is defined as article 36D in Table 7.37. It is measured in the back plane as shown in Figure 7.109.

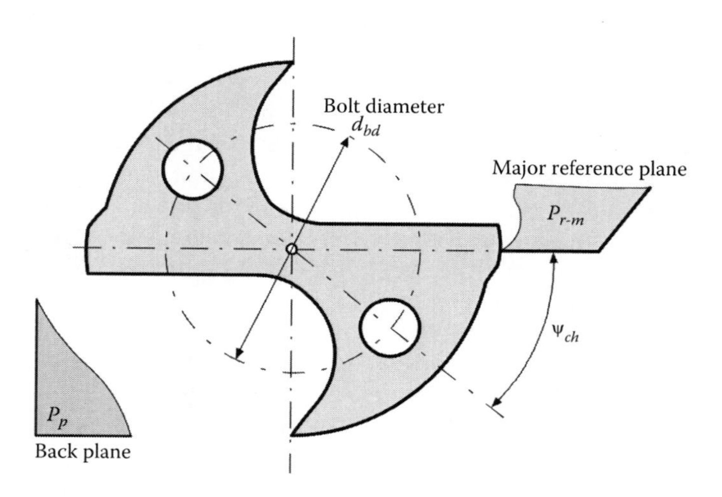

FIGURE 7.109 Parameters defining the location of the coolant holes.

7.7.16.7 Dealing with Nonincluded Articles

It should be clear that not all the articles and their definitions are included in Tables 7.36 and 7.37 as there are can be other design (dimensional and geometrical) articles and their characteristics assigned by the tool drawing for a particular drill. When dealing with such an article, the following sequence is recommended:

1. Define the datum feature(s).
2. Define the standard reference planes (e.g., the major reference plane, the back plane, the working plane) if the article is datum-related.
3. Define the article in the manner similar to definitions provided by Tables 7.36 and 7.37.
4. If needed, provide a simple sketch to facilitate developing its measuring procedure.

7.7.17 Dimensional Inspection (Metrological) System: Planning Stage

This stage has two prime objectives/outcomes:

1. Developing tables similar to Tables 7.36 and 7.37 where the non-datum and datum-related features particular to the drill to be inspected are listed with their proper definition. All the issues related to the listed articles and their characteristics should be resolved together with the drill design development and manufacturing/in-process inspection specialists.
2. Selecting the measuring equipment. Inspection and measuring equipment selected to carry out inspection include all of the tools and devices that are used to verify that a drill's characteristics all conform to the tolerances required by the drill's design/drawing. If needed, special inspection/measuring metrological procedures particular to the selected equipment should be developed.

The details of the first objective are discussed in the previous section where the definitions and graphical representations of the non-datum and datum-related features of the drill to be inspected are explained. Achieving the second objective requires some knowledge of drill gages in terms of their capabilities to achieve the required accuracy of the measurements.

7.7.17.1 Simple Gages

Various handheld gages including calipers and micrometers can be used to measure some geometrical parameters of drills. Such gages are widely available at small costs. Figure 7.110 shows an example of using a simple gage to measure some basic parameters of a drill. Figure 7.111 shows some drill gages available in the market. As can be seen, all these gages are hand gages. Their advantages are low cost and simplicity. Their disadvantage is accuracy of measurements. In the author's opinion, this kind of gage is the best choice if one wants to attempt an ambitious home-improvement project. They are totally unacceptable for inspection of HP drills.

There are a number of simple optical devices for inspection of some features (articles) of the tool geometry. Among them, optical drill analyzers by Titan Co. (model OTWD-1 and OTWD-2) are most universal. The optical drill geometry analyzer shown in Figure 7.112 is an inexpensive optical instrument available for checking twist drill geometry after re-sharpening. Whether drill sharpening is done using a simple manual fixture or on a more expensive drill sharpening machine, it provides a quick and easy way of checking the results. The 5× magnification of the two basic optical units with their built-in scales helps one to insure that all angles are properly ground, web thickness is correct, and the point is properly centered. The use of the unit is simple as the drill to be measured is located into the basic V-block that is part of the unit. This centers it in relation to the built-in scales of the magnifiers. The unit is intended to measure the four components of the

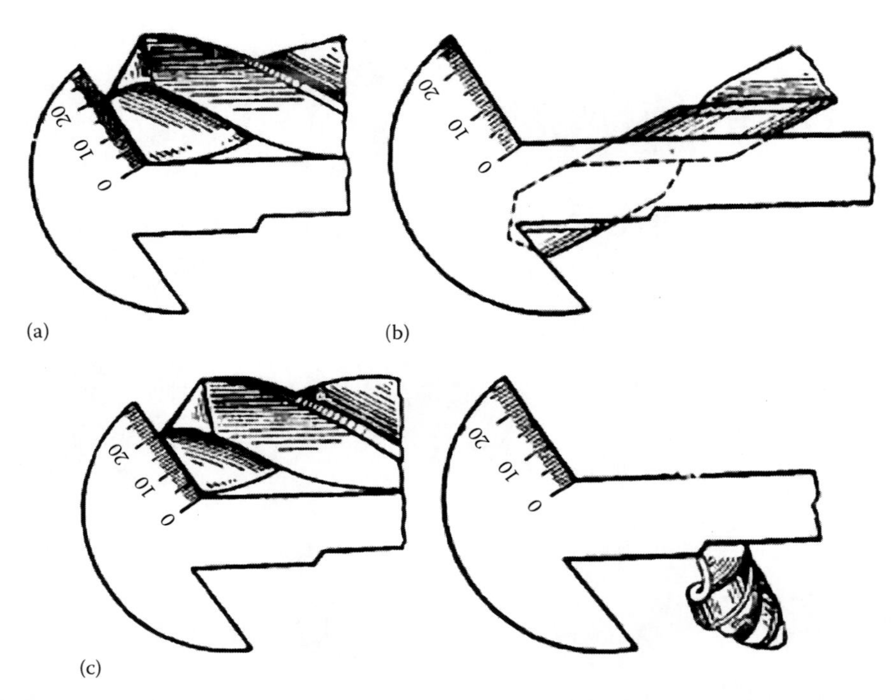

FIGURE 7.110 Measuring drill parameters with a simple gage: (a) measuring the length of the major cutting edges (lips), (b) measuring the angle of helix, and (c) measuring the chisel edge angle (length).

twist drill geometry: the point angle, axial clearance angle at the drill corners, chisel angle, and web centrality. The accuracy of such measurements however is not nearly sufficient (approximately 20 times coarser than required) for HP drills. Being important parameters of the drill geometry for job-shop applications, the listed four parameters are only a small portion of the parameters listed in Tables 7.36 and 7.37.

7.7.17.2 Optical Microscopes/Measuring Systems

Inspection and measurement of HP drills go beyond handheld gages to include CNC drill inspection machines able to execute programmed measurement routines, laser micrometers (in- and post process) and contact and noncontact profilometers that measure surface roughness, and microscopes. The modern inspection systems include optical comparators, toolmaker microscopes, and digital microscopes designed especially for tool geometry inspection. Being different in the design, tool-holding, and measuring principles, these units allow to measure practically all non-datum articles listed in Table 7.36.

One of the major concerns in choosing which optical system is best for a particular shop is the cost. Optical including stereo microscopes are just for inspection, making them less expensive. They range from about $2,000 to $5,000. With a digital camera and some software, the cost increases to $12,000 or $15,000. Measuring microscopes start at about $15,000 and can go as high as $80,000, depending on options. Vision measurement systems start at about $40,000, but they can reach $200,000 or higher. Some pricing crossover exists between the measuring microscope and the vision measurement system categories. For example, depending on the shop's needs, it might be worth looking into spending $40,000 or $50,000 on an automated vision system versus $60,000 on a manual measuring microscope. But with many different options available, the most important consideration is finding the right solution to suit the shop's measurement needs assuring that the measuring system has the needed accuracy, range of drill diameters, and be suitable to inspect critical articles.

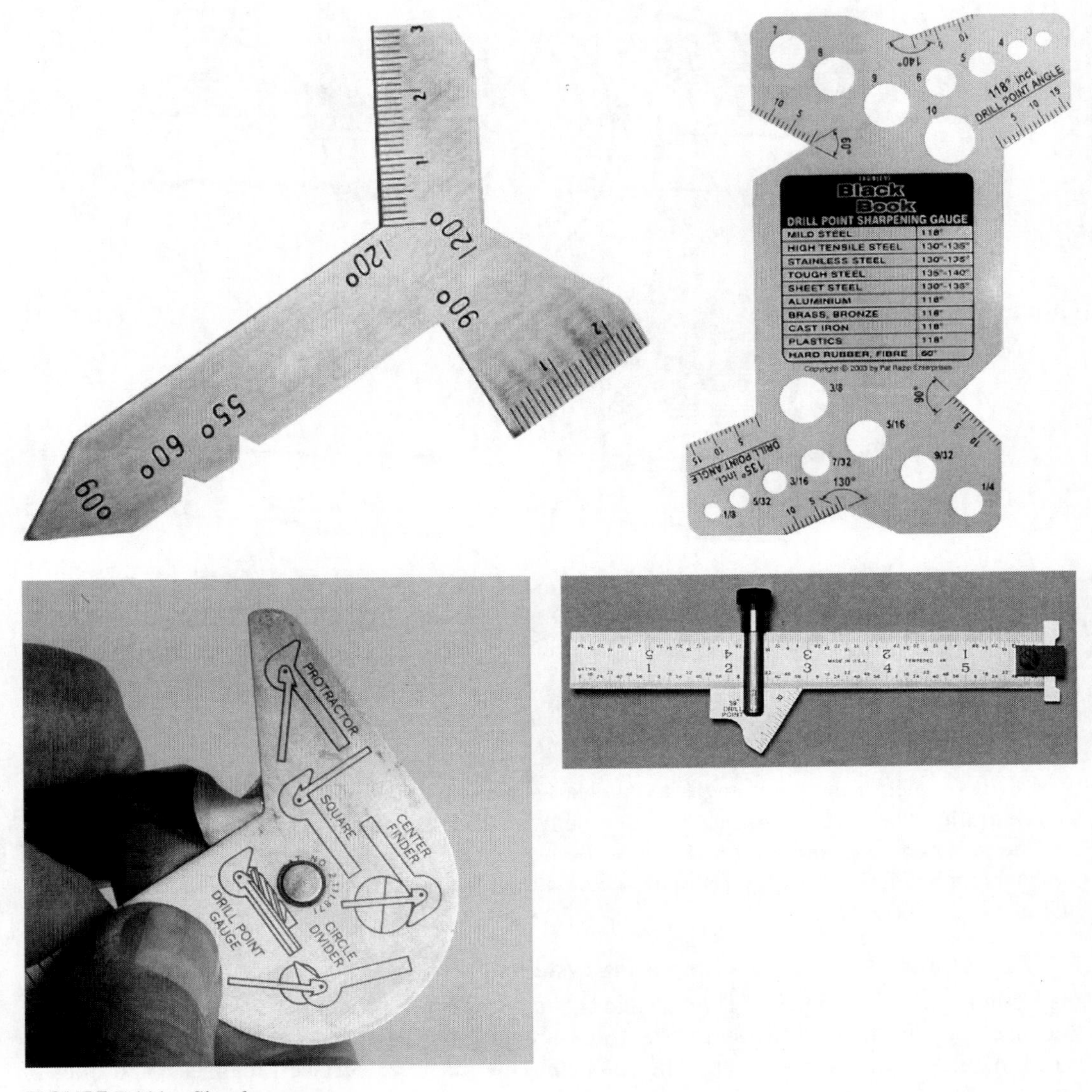

FIGURE 7.111 Simple gages.

For years, the optical comparator was the most commonly used device for tool inspection. The comparator offered reasonably priced technology for checking drills by profile projection. Figure 7.113 shows an optical comparator as a detachable unit of a drill grinding machine. As can be seen, the location of the tool allows measurement in the major and chisel edge reference planes (e.g., shown in Figure 7.107). As such, many important parameters in these reference planes, for example, the point angle, lip height variation and length, and the point angle of the chisel edge, can be measured with reasonable accuracy while the drill is still in the grinding machine.

In modern designs of optical comparators as shown in Figure 7.114, the use of surface illumination such as LED illumination, digital readout and digital cameras, and edge finding further enhanced the measuring capabilities of comparators. Such units are built on a rigid cast base to assure metrological stability. Telecentric optics yield crystal-clear upright and reversed images as 10×–100× magnification lens. Motorized *XY* motion further enhances the accuracy of measurements. When one understands the definitions given in Tables 7.36 and 7.37, he or she can design and make the corresponding tool-holding fixture which enables to inspect of many articles listed in these tables with high accuracy.

FIGURE 7.112 Optical drill analyzer OTWD-2.

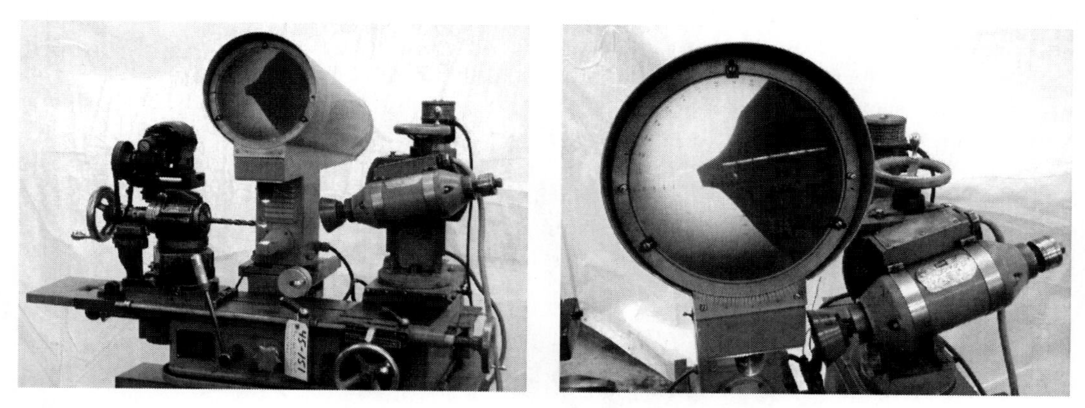

FIGURE 7.113 Optical comparator as a detachable unit of a drill grinding machine.

FIGURE 7.114 Modern computerized comparator with a digital camera.

Tool makers microscope (TMM) is another common device to inspect cutting tools. TMM is a type of a multifunctional device that is primarily used for measuring tools. These microscopes are widely used and commonly seen inside machine and tools manufacturing industries and factories. The main use of TMM is to measure the shape, size, angle, and the position of the small components that fall under the microscope's measuring range. Modern TMMs are outfitted with a CCD camera that has the ability to capture, collect, and store images into specialized computer software.

The glass grading and optics system make TMMs fully functional because, whatever is being viewed under these microscope parts or precision instruments, it is important that the objectives and the eyepiece lenses are made of fine-quality glasses only. These essential parts are what makes the device very durable and gives it the ability to withstand the wear and tear associated with the everyday stress of factory usage.

Some TMMs are equipped with a cross hair reticle on the eyepiece, coupled with a protractor on the tube. High-end TMMs use semiconductor laser devices as directors. Instead of the cross hairs, a red point is virtually marked on the microscope's working surface in order to locate the parts that have to be measured by the microscope. The CCD imaging system can also be used as a measurement system as well. This is another advanced feature of the newer versions of toolmaker's microscope models. A CCD camera that has the ability to measure diameters and distances provides additional conveniences in measurements. The magnification power of TMMs is nothing better than a regular compound microscope. In fact, it has a total magnification power up to 80×. This is due to the fact that these microscopes require good working distances of around 100 mm.

Figure 7.115 shows Mitutoyo 176-808A TMM with digimatic micrometer heads, 30× magnification, although some other TMMs are widely used. Commonly, TMM consists of the cast base, the main lighting unit, the upright with carrying arm, and the sighting microscope. The rigid base is resting on three-foot screws by means of which the equipment can be leveled with reference to the built-in box level. The base carries the coordinate measuring table and consists of two measuring slides: one each for *X* and *Y* directions and rotary circular table provided with the glass plate.

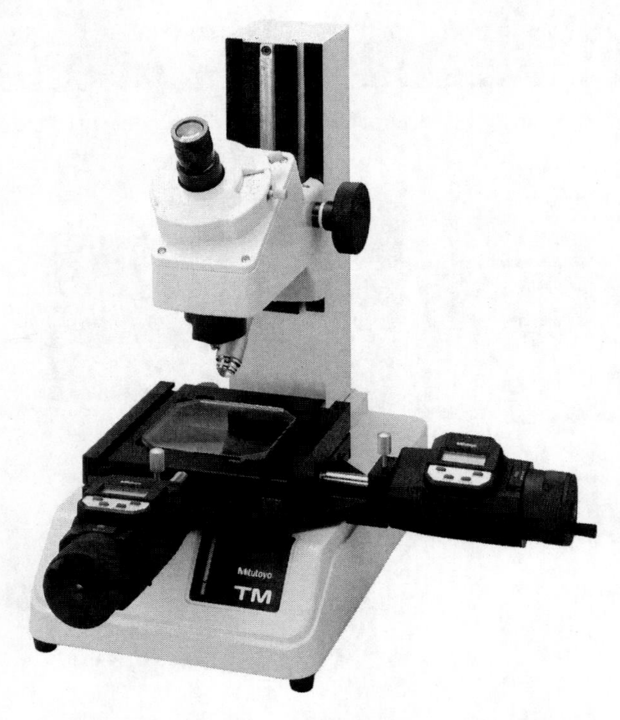

FIGURE 7.115 Mitutoyo 176-808A TMM with digimatic micrometer heads, 30× magnification.

The slides are running on precision balls in hardened guide ways warranting reliable travel. Two micrometer screws with normally digital readouts assure measuring range of 0–25 mm and permit the measuring table to be displaced in the *X* and *Y* directions.

TMM is suitable for the following fields of applications:

- Length measurements in Cartesian and polar coordinates
- Angle measurement of tools, threading tools, punches and gages, templates, etc.
- Thread measurements, that is, profile, major and minor diameters, thread angle, lead, and profile position with respect to the thread axis
- Simple measurements of the parameters of cutting tool geometry
- Tool wear analysis and measurements, including wear pattern parameters.

Stereo microscopes provide a 3D view of the part (a cutting tool) and generally are used for visual inspection, not measurement. A stereo microscope works like the human eye, which means it has good depth of field, that is, one can see a 3D image magnified maybe 20× or 30× on a stereo microscope. It is most common to look through the eyepiece when using stereo microscopes, but they also are available with a digital camera to view the part on a computer screen. Shops that want to document a defect can do so by adding the camera and software.

Stereo microscopes are appropriate for the shop floor as well as the QC lab to inspect a tool after it has been machined. It is popular for users to look for strange artifacts, defects or burrs, and the surface condition of the tool. Another application for stereo microscopes is cutting tool wear inspection. It is used in wear studies including studies of coating suitability, machining regime optimization, and influence of the drill geometry parameters on wear.

While rough measurements can be performed on stereo microscopes, measuring microscopes, or toolmakers' microscopes, produce a 2D image for more critical measurements. While an eyepiece is typically used with stereo microscopes, with measuring microscopes, the image is usually shown on a computer screen using a video camera. The measurement is made in the software using the image from the camera, not using the part itself. It is easier for tool inspectors because they do not have to strain their eyes as they can move the microscope stage in the *X* and *Y* directions and see the tool moving around while looking comfortably at the computer screen. The inspector takes a series of data points, say around a circle, and the software takes those points and calculates the circle that those points make up. He or she can take a number of points and measure diameters, lengths, arcs, and angles. If documented inspection is required, the data can be output in a spreadsheet or SPC program.

7.7.17.3 Specialized Measuring Microscopes

The need for a special microscope to control essential geometrical parameters of the drill was always there. Figure 7.116 shows the pure mechanical measuring fixture according to US Patent No. 3,414,979 (1968). In this inspection fixture, the drill 1 is located on the precision rollers 2 so that the proper datum location of the drill is assured. Dial 3 is to indicate the drill point angle, while adjusted dial indicator 4 is to measure the lip height variation. Dial indicator 5 is connected through a movable carriage 6 (Section A-A) to measure drill runout. The dial indicator 7 is to measure back taper (Section B-B) and other parameters of the drill margins. It is clear that the use of such a fixture requires clear understanding of the features (articles) to be measured/inspected and high skill of the operator for its proper use. It can be considered as a *beta-version* of modern special microscopes. Note, however, that not all modern microscopes are meant for and thus capable of measuring features that the fixture shown in Figure 7.116 can, for example, runout.

In the author's opinion, the HP drill manufacturers and users should have a measuring microscope for proper control of HP drill quality, non-datum article measurements (see Table 7.36), wear studies/control, etc. A few models of such microscopes developed especially for inspection of axial tools are available in the market. Figure 7.117 shows Zoller promBasic measuring microscope as an example. Although the appearance of such microscopes from various manufacturers as EuroTech

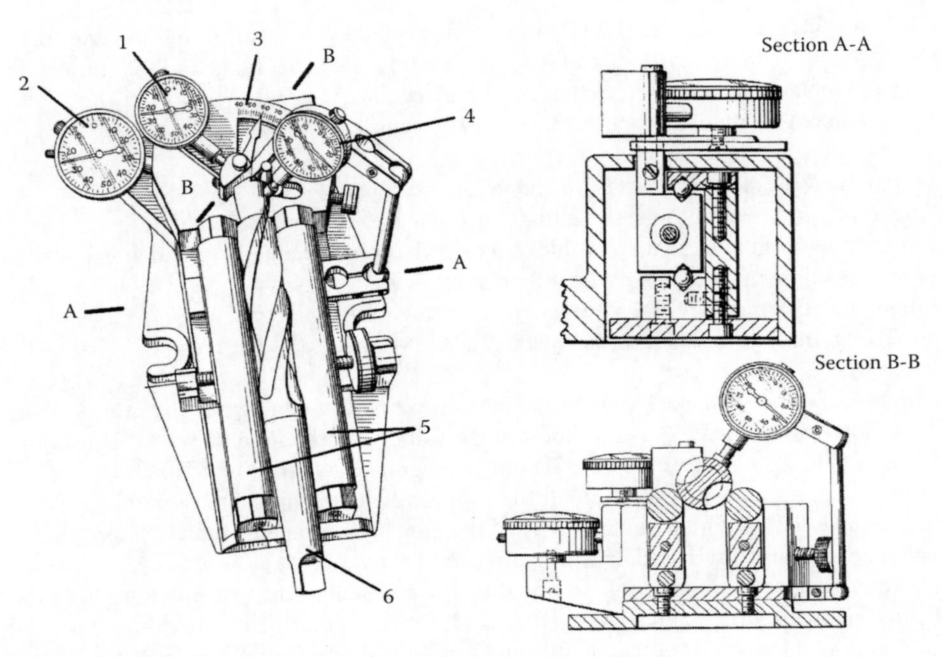

FIGURE 7.116 A measuring fixture according to US Patent No. 3,414,979 (1968).

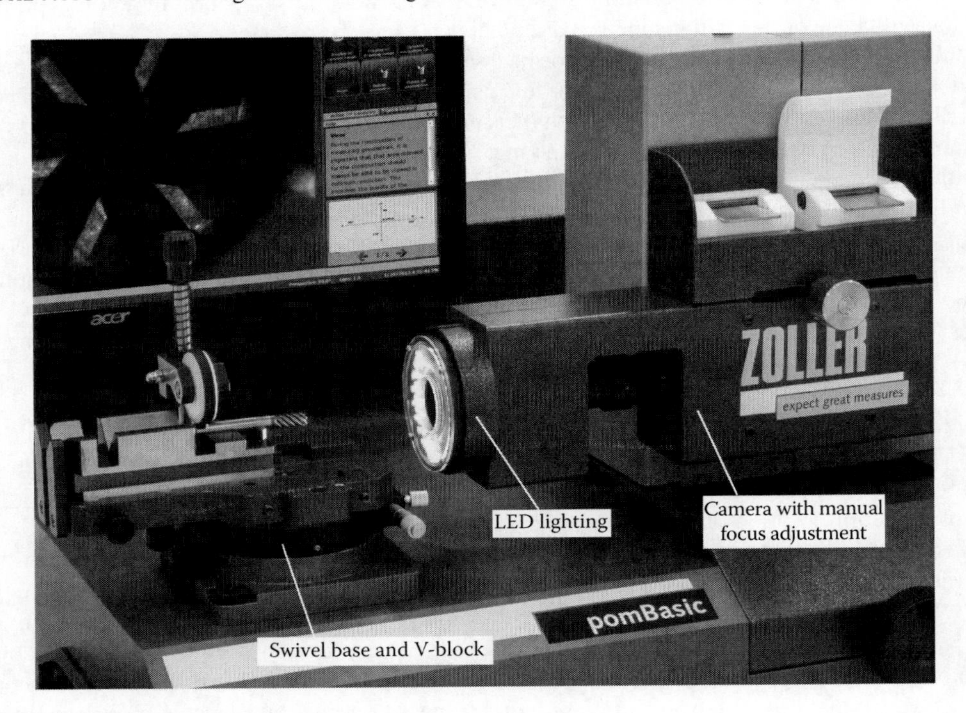

FIGURE 7.117 Zoller promBasic measuring microscope.

(e.g., PG1000), Zoller (see Figure 7.117), Walter, etc., looks almost the same for a casual observer as it includes the swivel base with a V-block, LED lights, camera with the objective, *X* and *Y* (horizontal and vertical) precision motions, and monitor with mouse-click on-screen measurements, their characteristics and measuring abilities are not nearly the same. The only common feature of the microscopes available today is that none of them has a manual containing instructions on what features of a drill

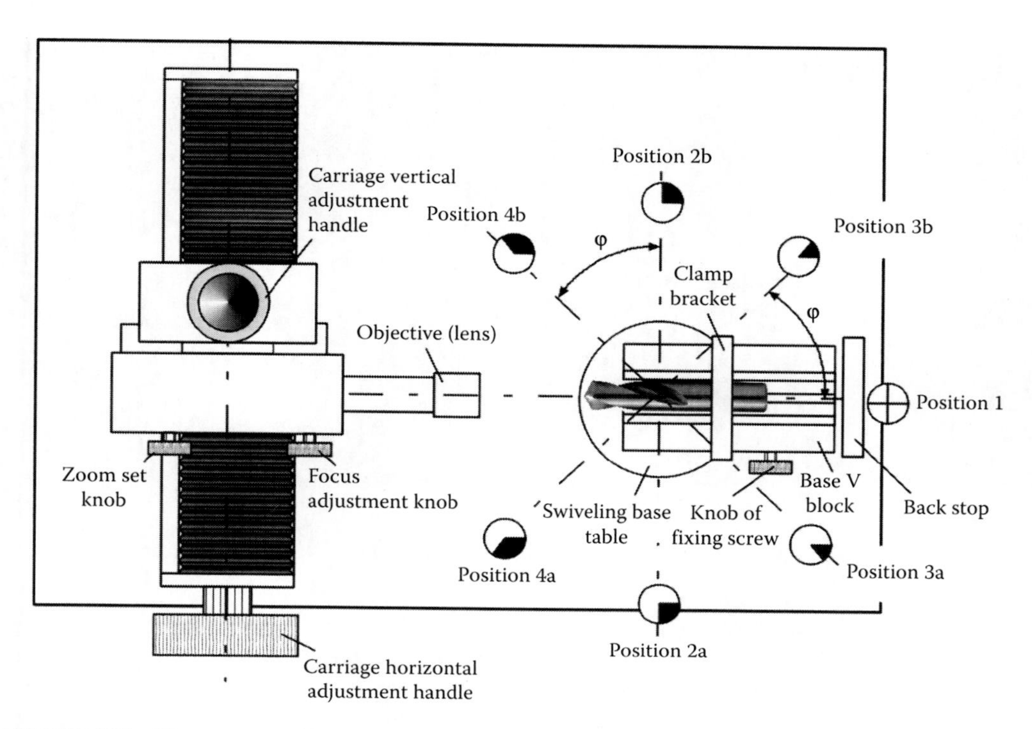

FIGURE 7.118 The proper architecture and basic positions of the tool measuring microscope.

can be measured and corresponding step-by-step procedures. As a result, some end users developed a number of measuring procedures according to their perceptions and understanding of what and how to measure that may or may not be in lines of the definitions provided by Tables 7.36 and 7.37. The rest, however, just use this piece of equipment for visual observation of axial tools in manufacturing and for wear analysis. Therefore, a need is felt to provide explanations on the essential features of such equipment and practical recommendations for their proper use.

Figure 7.118 shows the proper architecture and basic positions of the tool measuring microscope to facilitate the discussion on the procedure of measuring the common parameters (articles) of the drill geometry discussed earlier in this chapter. Each of the positions shown in this figure corresponds to a certain plane of measurement, and thus the parameters (articles) in this particular plane can be properly observed and measured. Note that the procedure discussed in the following text is universal, that is, applicable to any tool measuring microscope available today provided that the swiveling base table of the microscope can be brought to all positions shown in Figure 7.118.

As shown in Figure 7.118, four (for each cutting edge) basic positions of the swiveling base are selected. Position 1 represents the measurements in the back plane, position 2 represents the reference and working planes, position 3 represents the cutting edge plane, and position 4 represents the normal section plane. Corresponding non-datum articles listed and defined in Table 7.36 can be measured in these planes.

The inspection procedure begins with setting the tool in position 1, so it is considered as the starting position. The tool in the V-block and objective are brought together to an appropriate scale where the tool should be at magnification allowed by the monitor size. The tool should be cleaned from small debris using edge cleaning putty. Figure 7.119a shows the view that one should obtain in this position. First, the tool is visually examined for obvious flaws and chipping, excessive asymmetry of features (e.g., the position of the coolant hole with respect to the major cutting edges [lips]), coarse surface finish of the flank faces, etc. Second, the scale is applied (Figure 7.119b) and the following articles listed in Table 7.36 can be inspected: 08N, margin width; 09N, body clearance diameter; and

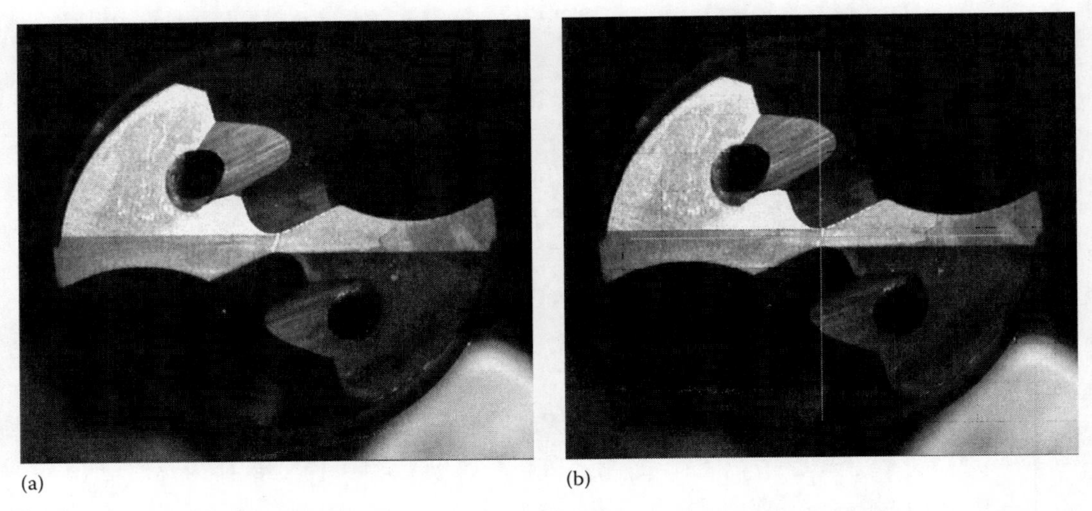

(a) (b)

FIGURE 7.119 View in position 1: (a) view for general assessment and (b) view with coordinates to measure features.

14N, coolant hole diameter. The following articles listed in Table 7.37 can be inspected: 17D, chisel angle ψ_{cl} (approx.); 18D, chisel edge length l_{cl}; 21D, web thickness, d_{wb}; 22D, radius of the end of the chisel edge, r_{cl}; 23D, mismatch of the intersection lines of the secondary flank faces with the primary flank faces, m_f; 30D, gash face angle, $\psi_{gh\text{-}b}$; 31D, gash inclination angle, $\lambda_{gh\text{-}b}$; 32D, extent of the gash past the CL, a_4; 33D, radius of the gash, r_{gh}; 34D, bolt diameter, d_{bd}; and 35D, coolant holes location angle, ψ_{ch} (approx.). Note that the flank and chisel edge wear (see Section 2.6.4) is not measured in the back plane although practically all literature sources depict such wear as viewed in the back plane.

The next step in tool inspection is to bring the swiveling table in position 2a. Normally, the angular scale around the swiveling table is provided as shown in Figure 7.120. This enables the user to set the current position of the table on the scale to any chosen angle with approx. 1° accuracy. Figure 7.121a shows the view that one should obtain in this position. First, the tool is visually examined for obvious flaws in the same way as in position 1. Particularly, surface roughness on the drill margin and chipping of the side cutting edge are assessed.

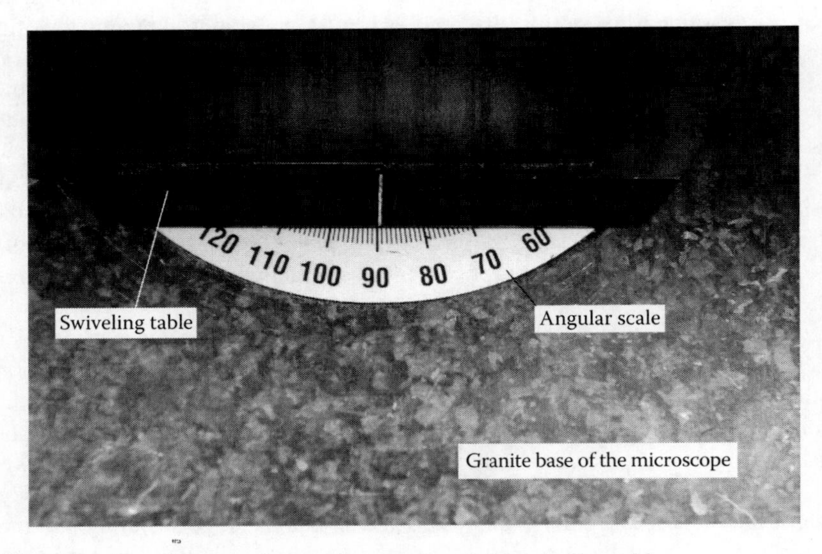

FIGURE 7.120 Angular scale to set the angular position of the swiveling table.

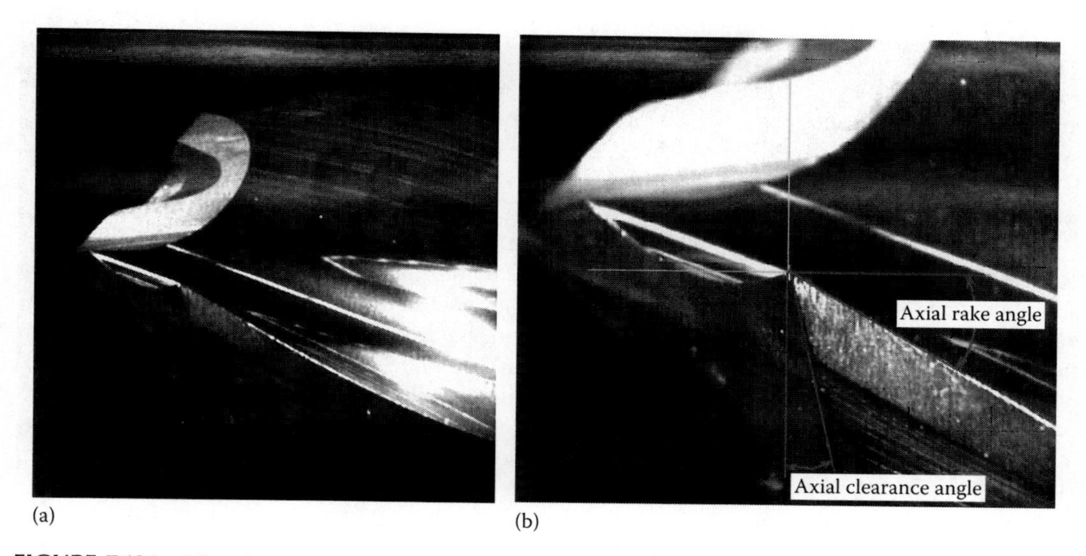

FIGURE 7.121 View in position 2a: (a) view for general assessment and (b) axial clearance angle at drill corner measurement.

The following caharcteristics of the articles can be measured: (1) axial clearance angle at drill corner in the manner discussed in Section 7.7.12 and shown schematically in Figure 7.87 (Figure 7.121b shows actual measurement of this article) and (2) axial rake angle at drill corner. For a straight-flute drill it should be zero if no special measures to modify the rake face geometry, is taken. For a twist drill, this angle should be equal to helix angle, ω_d, (article 25N—in Table 7.36).

The characteristics of other articles listed in Tables 7.36 and 7.37 can be measured: 03N, flute length; 04N, active flute length (approx.); and 05N, length of the body relief. If a drill has steps (stages) with secondary and other cutting edges, then the step distances can be approximately measured. Other useful parameters of articles not listed in Tables 7.36 and 7.37 can also be measured. For example, Figure 7.122 shows re-sharpening notches in front of the second drill step.

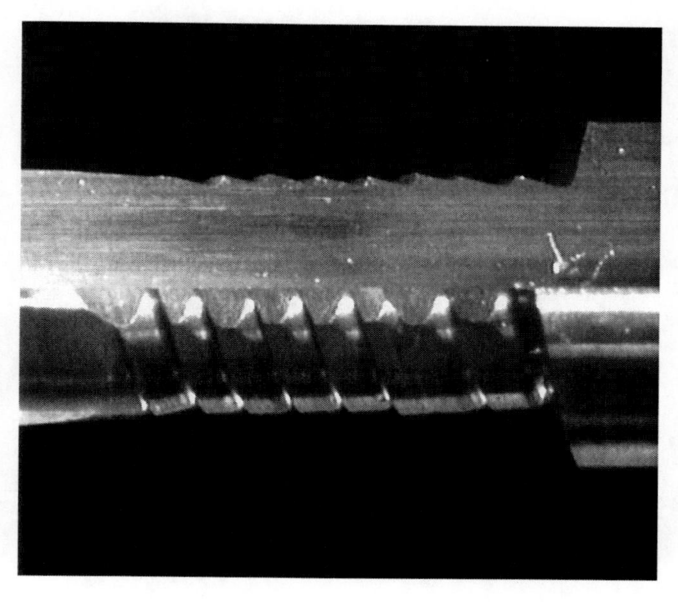

FIGURE 7.122 Showing the re-sharpening notches in front of the second drill step.

Counting these notches, one can conclude that the drill was re-sharpened eight times. If the parameters of the body undercut in front of second, third, etc., cutting edges are important, they can be assessed by increasing the zoom.

The discussed measurement/assessments are related to the right major cutting edge. Rotating the swiveling table by 180°, that is, to position 2b, the same measurement/assessments can be made for the left major cutting edge. When a drill has more than two teeth (the flute), all measurements/assessments were carried out in position 2a by rotating drill by the central angle between teeth (the flute).

Then the drill is rotated by 90° (for two-flute tool) or by 180/z angle (where z is the number of cutting teeth [flutes]) clockwise about the axis of the datum (the axis of the shank). Figure 7.123a shows the view that one should obtain in this position. First, the tool is visually examined for obvious flaws in the same way as in position 1. The following parameters of the corresponding articles in the major reference plane can be measured (see Tables 7.36 and 7.37): 01N, drill diameter (approx.); 02D, point angle Φ_p; 24D, chisel wedge angle, ν_{cl} (approx.); and 34D, gash plane angle, $\psi_{gh\text{-}r}$. Figure 7.123b shows an example of measuring the point. If the drill is made with generic design of split-point drills it is further rotated clockwise by chisel angle ψ_{cl} to measure the length of the rake face, l_{sp} (article 28D in Table 7.38).

The next step in tool inspection is to bring the swiveling table to position 3a. The angle of the swiveling table is set to be a half-point angle φ obtained by dividing the point angle measured at position 2 by two. The achieved drill position with respect to the camera objective corresponds to the cutting edge plane. Figure 7.124a shows the view that one should obtain in this position. First, the tool is visually examined for obvious flaws in the same way as in position 1. Particularly, surface roughness of the drill prime flank surface and microchipping of the cutting edge are at various magnifications. If edge preparation (see Section C.4.2) is applied and when the optical zoom of the microscope allows, the conditions and uniformity of the applied edge preparation can also be assessed.

Drill flank wear is assessed/measured in this (the cutting edge) plane so that this plane is the most important in any assessment of drill performance. If optical zoom of the measuring microscope allows, a detailed analysis of wear topography can be attempted to determine the root cause

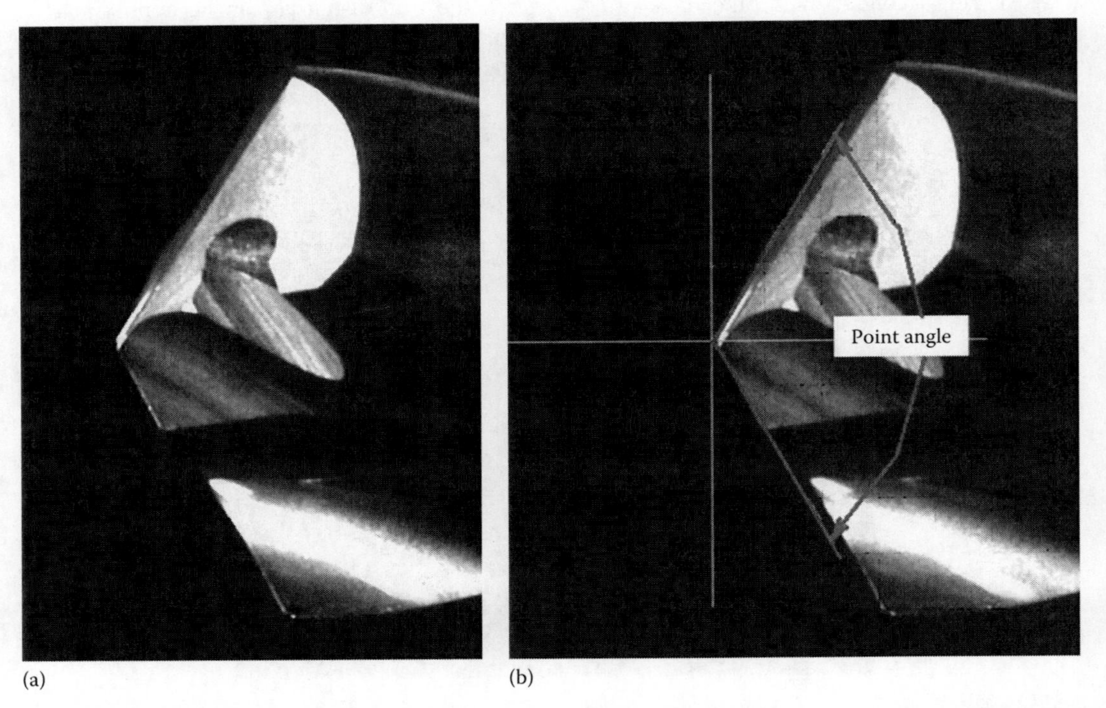

(a) (b)

FIGURE 7.123 View in position 2a at 90°: (a) view for general assessment and (b) point angle measurement.

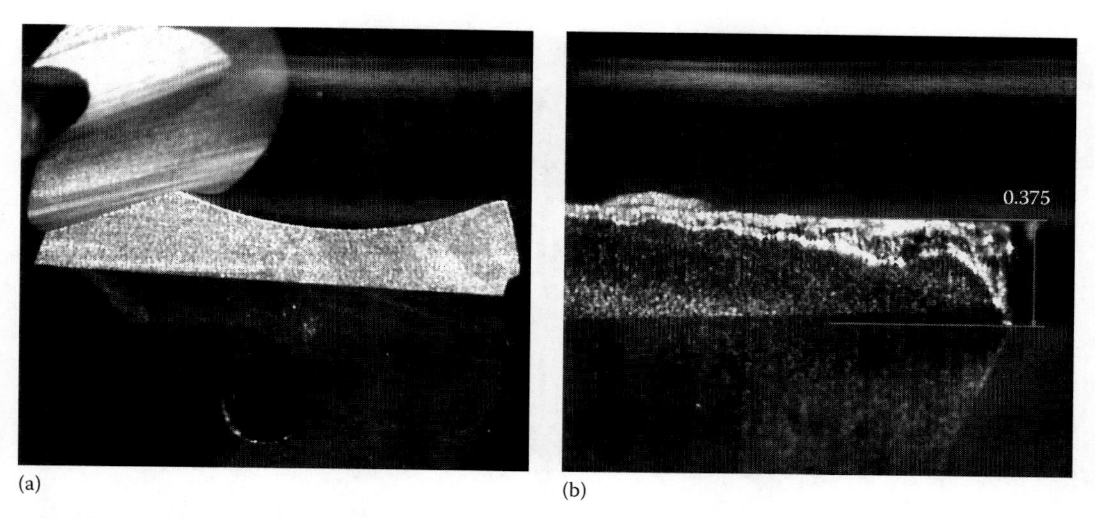

(a) (b)

FIGURE 7.124 View in position 3a: (a) view for general assessment and (b) corner wear measurement.

of flank wear (Chapter 2). Figure 2.51 in Chapter 2 shows various topographies of the drill flank wear in the cutting edge plane. Figure 7.124b shows an example of the measuring of the drill corner wear. Note that HSS and sintered carbide drill wear is much more great than that of PCD drills. So if the wear/damage of PCD tool is to be attempted, much greater optical zoom is required on the measuring microscope.

If a drill has step(s), then the second (third, fourth, etc.) cutting edge and the discussed parameters in the cutting edge plane can be assessed/measured. If a step has the point angle equal to that of the first stage, then the carriage is moved horizontally and zoomed on the cutting edge of the second or other cutting edges. If a stage has the point angle different than the first stage, it is measured in position 2, and the angular position of the swiveling table is set correspondingly by the half-point angle of the considered stage. Figure 7.125 shows the view in position 3 for the cutting edge of the second stage.

The discussed measurement/assessments are related to the right major and second cutting edges. Rotating the swiveling table by 180°, that is, to position 3b, the same measurement/assessments can

FIGURE 7.125 View in position 3 for the cutting edge of the second stage.

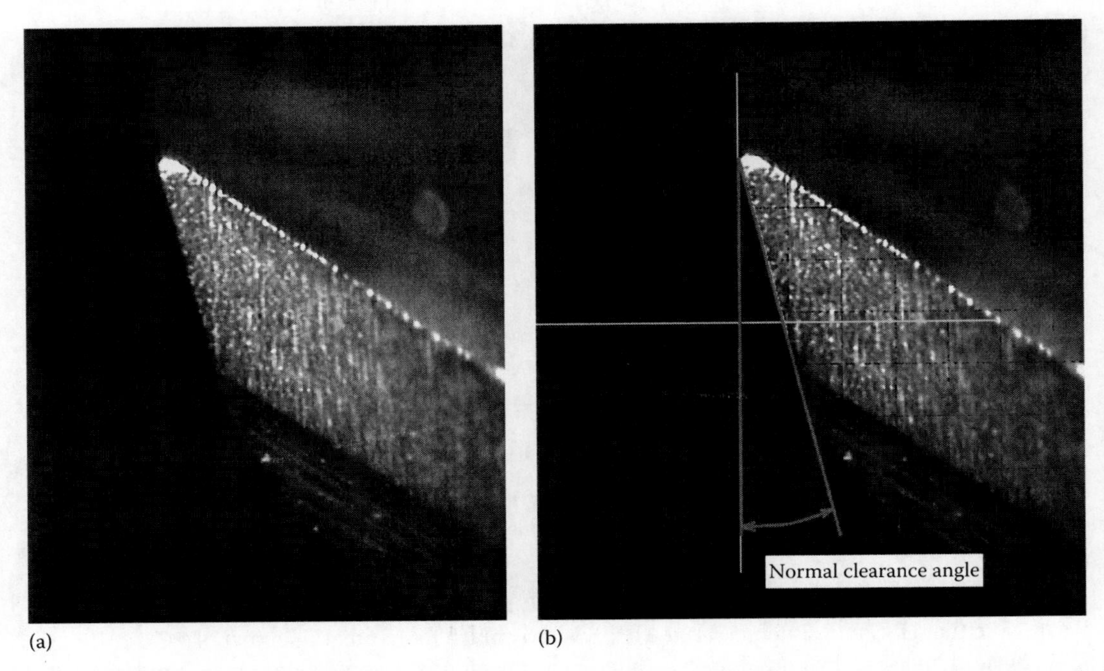

FIGURE 7.126 View in position 4a: (a) view for general assessment and (b) normal clearance angle of the major cutting edge measurement.

be made for the left major and secondary cutting edges. When a drill has more than two teeth flutes, all measurements/assessments are carried out in position 3a by rotating drill by the central angle between teeth flutes. If a drill has step(s), then the second (third, fourth, etc.) cutting edge and the discussed parameters in the cutting edge plane can be assessed/measured.

The next step in tool inspection is to bring the swiveling table to position 4a. The angle of the swiveling table is set to be a half-point angle φ with respect to position 2. Figure 7.126a shows the view that one should obtain in this position at high zoom. Position 4a corresponds to the normal plane for the right major cutting edge (see Figure 7.101) so that the normal clearance angle of the major cutting edge can be measured in the manner shown in Figure 7.126b. This angle is important as it is normally indicated in tool drawings.

The discussed measurement of the normal clearance angle is related to the right major cutting edge. Rotating the swiveling table, to position 4b, the normal clearance angle can be measured for the left major cutting edge. When a drill has more than two teeth flutes, all measurements/assessments were carried out in position 4a by rotating drill by the central angle between teeth flutes. If a drill has step(s), then the second (third, etc.) cutting edge and the discussed parameters in the cutting edge plane can be assessed/measured. If a step has the point angle equal to that of the first stage, then the carriage is moved horizontally and zoomed on the cutting edge of the second or other cutting edges. If a stage has the point angle different than the first stage, it is measured in position 2, and the angular position of the swiveling table is set correspondingly by the half-point angle of the considered stage.

The discussed procedure is related only to some standard articles of a drill. The parameters of other articles that may be particular to a given drill design can be measured in the similar manner knowing the corresponding plane of measurements and adjusting the swiveling table and drill angular position correspondingly. Before measuring/assessing any drill parameters, this parameter and the plane of its measurement should be clearly defined by the tool drawing because one can only inspect parameters indicated in this drawing.

The discussed procedure implies that a measuring microscope should (1) have sufficient space to bring a drill in the positions shown in Figure 7.118 (unfortunately, this is not the case in some modern measuring microscopes where the angle of the swiveling table rotation is restricted by ±90°, that is, it is impossible to bring a drill to position 4. The author advises not to use such microscopes as they severely restrict user's ability to measure parameters of the essential articles listed in Tables 7.36 and 7.37) and (2) have a great optical zoom. This is particularly important for HP drills including PCD drills. While the former is rather obvious, the latter is not well understood so it requires further clarifications.

While TMMs normally have 30× lens (15× eyepiece lens and 2× objective lens) zoom, the measuring microscopes have 6×–15× optical (called—lens) zoom. The manufacturers of the measuring microscopes claim that they have a great digital zoom so that the total magnification should be sufficient for even tiny drill inspection. This common misconception lasts until a blurry image of the drill appears on the monitor no matter how many megapixels the digital camera has.

What is the difference between optical zoom and digital zoom? The short answer is optical zoom = resolution and digital zoom = cropping. Optical zoom is lens-assisted magnification. It is usually measured as the ratio of maximum focal length possible to minimum focal length possible. For example, a 28–110 mm lens will have an optical zoom of 110/28 = 4×. A higher optical zoom is a better choice because a lens can have distortions toward the low or high focal lengths.

Digital zoom is just cropping the image and zooming in (digitally or pixel wise) to the interesting area. It is more like a digital technique and misinforms the customer. This is because digital zoom is not really zoom, in the strictest definition of the term. What digital zoom does is enlarge a portion of the image, thus *simulating* optical zoom. In other words, the camera crops a portion of the image and then enlarges it back to size. In so doing, image quality deteriorates.

What, therefore, is the rule of thumb, when it comes to using zoom? Here it is: always use optical zoom. Whatever you do, buying a microscope or digital camera, choose one that warns you that you are about to use digital zoom or that allows you to disable digital zoom (most do). There is no point in comparing digital zoom with digital zoom or optical zoom with total zoom. Always compare optical zoom with optical zoom.

While considering implementing a measuring microscope, end users should clearly understand the following:

1. Only some of the datum-related articles listed in Table 7.37 can be measured. This is because although the tool shank as the common datum is located in a V-block and tightly pressed against it so that the tool location is correct, the axis of the shank is not visible on the screen, so no measurement of the tool can be taken with respect to this datum.
2. The image on the screen is not digitally correlated with the on-screen measurements. In other words, no edge finder is available for precision measurements as the user using the mouse defines the lines and then measures the distance and angles between these lines. While for many angular measurements such precision is acceptable, linear measurements accuracy may not be sufficient for some non-datum features.

7.7.17.4 CNC Measuring Systems

To evaluate the existing CNC measuring/inspection systems, one needs to understand what would be the ideal measuring/inspection system. The following *wish list* describes the desirable features of such a system:

1. It should be capable to measure/inspect the articles described in Tables 7.36 and 7.37 plus parameters of drill geometry in T-mach-S and T-use-S described in Chapter 4.
2. It should be suitable for fully automated measurements removing the subjectivity of manual and semi-manual measurements. Users are just teaching it basically to go around the drill in relation to the CAD drawing and tell the machine what to measure. The program is just recording the steps one is making in measuring various articles according to the drill drawing. Then this program can be saved and can be run over and over. The program also can be easily edited to make other programs from it.

3. To facilitate the measurements/inspection, the system should include a library of common and standard features (articles) and definition/measuring functions. Corresponding tool geometry standards and nonstandardized parameters should be clearly indicated to help a potential user to make the proper choice of a particular built-in procedure.
4. It should include a combination of very low-distortion optics, precision spindles, rigid frames, and high-resolution video that can be processed by the software to make automatic measurements.

Technically, the ideal system should be able to carry out the following measurement sequence:

1. *Hang* the drill in midair.
2. Scan its fully digital image.
3. Determine the datum according to the drill drawing. For drilling tool, the common datum is the axis of the shank.
4. Measure the datum geometry and store the information. The shank cylindricity and surface roughness are measured.
5. *Idealize* the datum, that is, to make it perfect digitally. The actual shank axis is determined and postulated that this axis is ideal, that is, having no deviation.
6. Digitally correct the scanned image accounting on the corrections made to the datum. As a result, a completely new digital image of the drilling tool is obtained.

The ideal system should allow to define the purpose of measurement. Normally, there should be two options: (a) inspection and (b) research. The sequence of measurements depends upon the chosen option.

The inspection option should allow the following:

1. Load the CAD drawing into the measuring system computer. Note that this drawing should contain well-defined articles and their parameters with tolerances to be inspected. Moreover, they should be clearly specified according to the corresponding tool geometry and tool drawing standards. According to the author's experience, this is the major problem as the articles and their characteristics in many tool drawings are defined using the *free-lens* style, that is, according to the designer's understanding and interpretation of drill features.
2. The measuring system determines articles to be measured from the loaded tool drawing and carries out the corresponding measurement using the corrected digital image of the tool.
3. The inspection protocol is created where the measured and drawing parameters are listed and compared to determine which of the measured parameters are is and which are out of the corresponding tolerance zones assigned by the tool drawing tolerance zone. Such a protocol lists the parameters defined by the drawing versus measured parameters showing the location of the measured parameters with respect to the tolerance zone. For convenience, the out-of-tolerance measured parameters should be highlighted, for example, indicated by a different eye-catching color.

In the research option, the modified digital image of the tool is dealt with. Depending upon the purpose of a particular research, some measurements defined by the tool geometry standards and thus built-in in the measuring system can be attempted as well as the measuring of some parameters in nonstandard reference plane. The system should be able to help the user to determine any section or reference plane easily with respect to the datum. As such, the determination of the geometry parameters in T-mach-S and T-use-S (see Chapter 4) should not present any problem, which helps to optimize the parameters in T-hand-S, that is, those shown in the tool drawing.

Another important aspect of such a system is to measure 3D wear topography and volume of the work tool material through comparing digital images before and after tool run. As discussed by the author earlier (Astakhov 2004), such an assessment is the most objective way to evaluate various tool materials and optimality of the tool design. Collecting and storing the data can also help to evaluate any change in the drilling operation should unexpected tool failure(s) occur due to, for example, problems with castings or the machine spindle. Besides, the machinability of the work materials can be compared objectively.

Moreover, the ideal CNC measuring system should be able to simulate drilling, that is, it should be able to rotate and feed the drill into a digital workpiece. The user should be able to make any desirable cross section of the digitally formed assembly. The following objectives of such simulations are of prime importance in the drill design and optimization:

1. Verification of the absence of interference of the drill flanks with the bottom of the hole being drilled in the manner shown in Figure 5.108.
2. Measuring arms of the unbalanced moments, for example, $a_{x\text{-}ub}$ and $a_{y\text{-}ub}$ shown in Figure 4.18; $a_{zx\text{-}ub}$ and $a_{zy\text{-}ub}$ shown in Figure 5.50; and z_{a12}, z_{a23}, z_{b12}, and z_{b23} shown in Figure 5.101. These are needed for the detailed optimization of the tool geometry as discussed in Chapters 4 and 5.
3. Visualization and quantification of the topology of the bottom clearance space (see Chapters 4 through 6) for optimization of the flank surface design and geometry as well as for improving the MWF flows in this bottom clearance space (see Section 6.3.4).

Unfortunately, among CNC drill measuring systems available today, no one is capable of fulfilling the previously discussed wish list. Therefore, the suitability of the available and emerging systems can be assessed using the previously discussed requirements, that is, by the relative importance, measurement accuracy, measuring complexity, compliance with the drill geometry standards, and the extent to which a particular system complies with the listed requirements.

As not many systems of this caliber are available in the marketplace, such a ranking does not present a problem. What one needs is just to take Tables 7.36 and 7.37 and go over the articles listed in these tables with a sales/technical representative of a perspective measuring machine. In the author's experience, only very few machines can *survive* such an assessment particularly when the datum-related articles are considered. There are a number of reasons for that. The most common are as follows:

1. No one CNC tool measuring machine is supplied with a manual that includes the definitions, drawing, principles, and examples of measurement of drill standard articles listed in Tables 7.36 and 7.37. A short training provided by the machine manufacturers is oriented mainly on the basic functions of the machine. As a result, many measuring machines in the tool manufacturers' and users' possession are used only up to 30% of their capability. Most of them are used as an expensive tool presetting machine as the inspection reports include only the diameters, runouts, and step lengths as discussed in Chapter 2 with examples.
2. Practically no one tool drill geometry measuring machine uses the standard terminology and thus definitions of drill articles listed in Tables 7.36 and 7.37. Rather, a number of own *homemade* terms are used with no proper definitions.

There are two basic CNC tool measuring machines available today in the tool industry, namely, the Zoller Genius 3 Pilot 3 (Figure 7.127a) and a family of the Walter Helicheck machine (Figure 7.127b shows Helicheck Plus L). The capabilities and accuracy of these machines are similar so that an inexperienced user may have a hard time to select which one of these two machines is the best choice for his/her measuring needs. The absence of manuals where the well-defined features of the tool geometry and dialed step-by-step procedures of their measurement using these machines just adds more challenge to the selection challenge. The ignorance of the basic articles and their standard definitions listed in Tables 7.36 and 7.37 from both sides of the fence does not help either.

The author's analysis of the available CNC measuring machines shows that these machines are capable in principle (hardware-wise) of measuring practically all articles listed in Tables 7.36 and 7.37 as well as the articles in T-mach-S and T-use-S. The problem is that just few of them are set as the

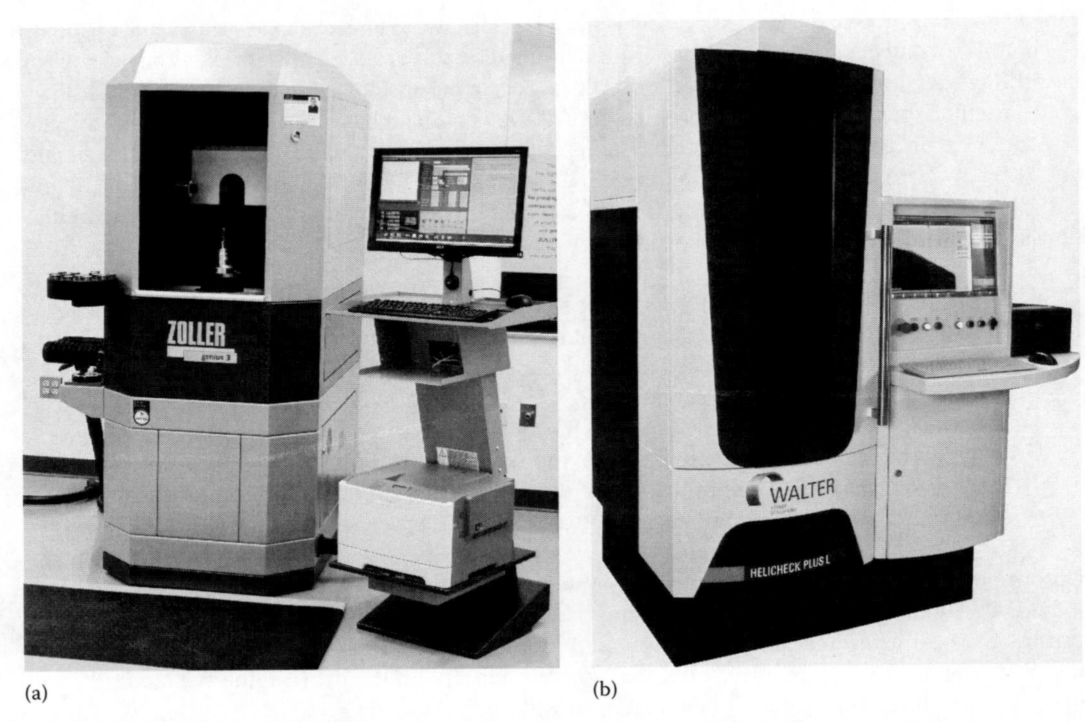

(a) (b)

FIGURE 7.127 CNC tool measuring machines: (a) Zoller Genious 3 and (b) Walters Helicheck Plus.

standard functions, while most require additional programming. Such programming can be attempted using the graphics and definitions presented in this chapter with close collaboration with the specialists of the corresponding companies. The author considers such a situation as normal as CNC measuring machines are in the infant stage of their development and implementation in the cutting tool industry.

The great advantage of CNC measuring machines is proper definition and nullification of the datum. As discussed previously, the datum for any HP drill is the axis of its shank. This datum is considered as to be perfect so that other datum-related parameters of the tool design and geometry are measured with respect to this datum. In reality, however, nothing is perfect because the shank is mounted in the tool holder, which, in turn, is mounted on the measuring machine spindle. As a result, the datum axis of the tool may not be coincident with the axis of spindle rotation. As the measurement of the tool parameters is made with respect to the axis of rotation, the measured results cannot be compared with those assigned by the drawing with respect to the datum.

To resolve the issue, the datum nullification is the first mandatory step in the use of any CNC measuring machine. Figure 7.128 shows its realization on the Zoller Genius 3 Pilot 3 machine. As can be seen, the runout of the shank is actually measured. Then the result of this measurement is used for the corresponding correction of the measurement of other parameters of the drilling tool. Obviously, a drilling tool should be mounted in the tool holder so that a part of its shank can be measured, so the datum can then be nullified. Note that this step is not called properly as *datum nullification*, on the machine. Rather the term *wobbling compensation* is used.

Figure 7.129 shows the two basic locations of the measuring camera. These locations correspond to the positions 2ab and 1 shown in Figure 7.118, correspondingly. Therefore, the non-datum and datum articles can be inspected/measured in the tool reference, back and working planes to inspect practically all features listed in Tables 7.36 and 7.37. As mentioned previously, some of these are built-in features of the machine, while others required additional programming.

Figure 7.130 shows an example of the menu where some of the built-in standard articles are listed. An article or series of articles can be selected by the machine operator so that the machine

FIGURE 7.128 Datum nullification position.

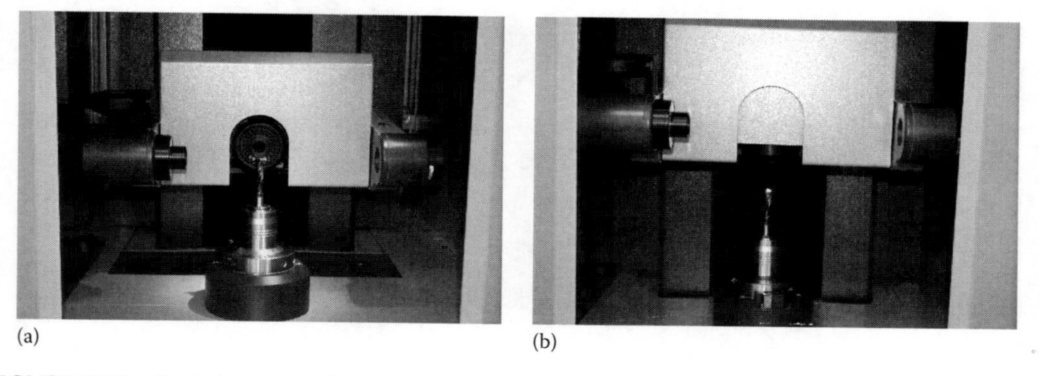

(a) (b)

FIGURE 7.129 Basic locations of the measuring cameras with respect to the drill to be measured: (a) side position and (b) top position.

FIGURE 7.130 Example of the menu where some of the built-in standard features are listed.

can measure the features of the selected articles almost automatically. For example, if one selects to measure the normal T-hand-S clearance angle (wrongly termed as the axial relief angle on the menu), then the next menu for this article with some additional choices and visualization of the measuring flank face(s) appears on the monitor screen as shown in Figure 7.131. Although not obvious, the normal or axial clearance angle in T-hand-S for a selected point of the major cutting edge (lip) can be measured. For example, if one selects option *in relation to cone angle* (which actually stands for the drill point angle) as shown in Figure 7.131 then the normal clearance angle will be measured. Otherwise, the side (axial) clearance angle will be measured if this option is not selected.

Figure 7.132 shows the result of measuring the normal clearance angle at the selected point of the cutting edge. As can be seen, it is 7.8°. The machine, however, offers some additional useful options.

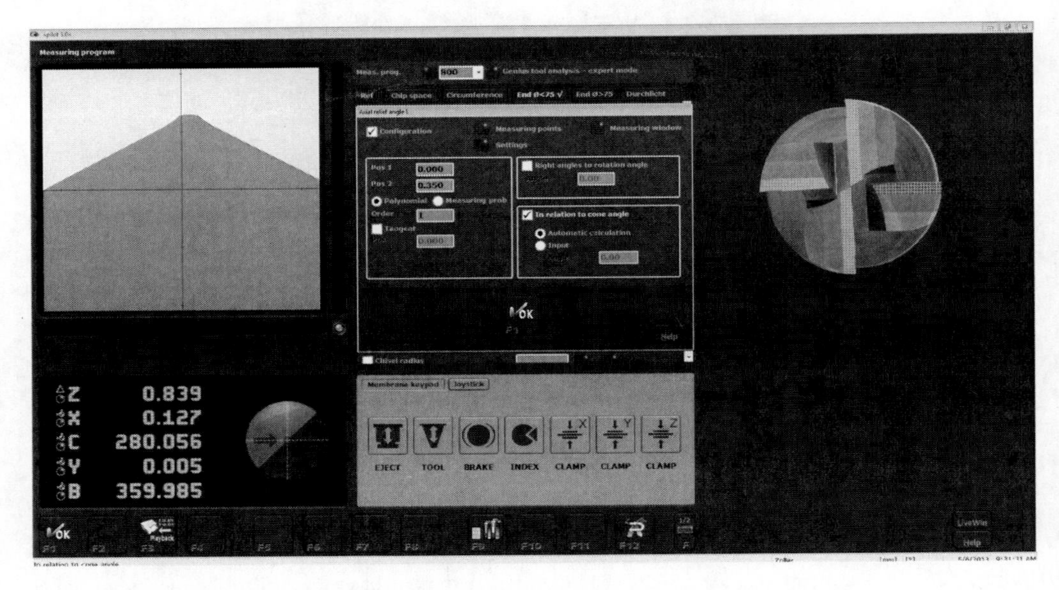

FIGURE 7.131 Menu for measuring the normal clearance angle over the major cutting edge.

FIGURE 7.132 Showing the result of measuring the normal clearance angle in the selected point of the cutting edge.

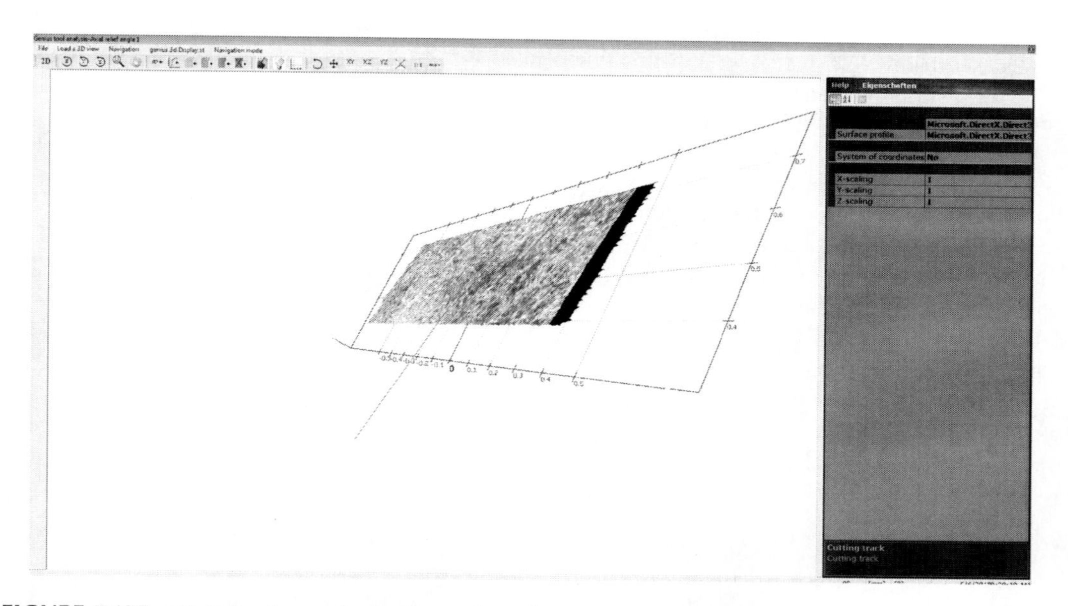

FIGURE 7.133 Digital primary flank plane where the normal clearance angle in the considered point is indicated.

Two of these options are of prime interest. First, the digital image of the primary flank plane with 3D visualization of the clearance angle at the considered point of the cutting edge can be viewed and orbited by the computer mouse as shown in Figure 7.133. As such, one can see the variation (if any) of the normal clearance angle along the major cutting edge. Moreover, the ground texture of the primary flank face can be viewed at high magnification to analyze, for example, grinding marks and micro-defects. Second, the actual shape of the primary flank face in its section by the normal or orthogonal plane through the selected point of the major cutting edge can be viewed as shown in Figure 7.134.

As can be seen, the polynomial of the first order (a straight line) is selected to overlay the actual shape of the section of the primary flank. If the primary flank is not a plane but a 3D surface generated

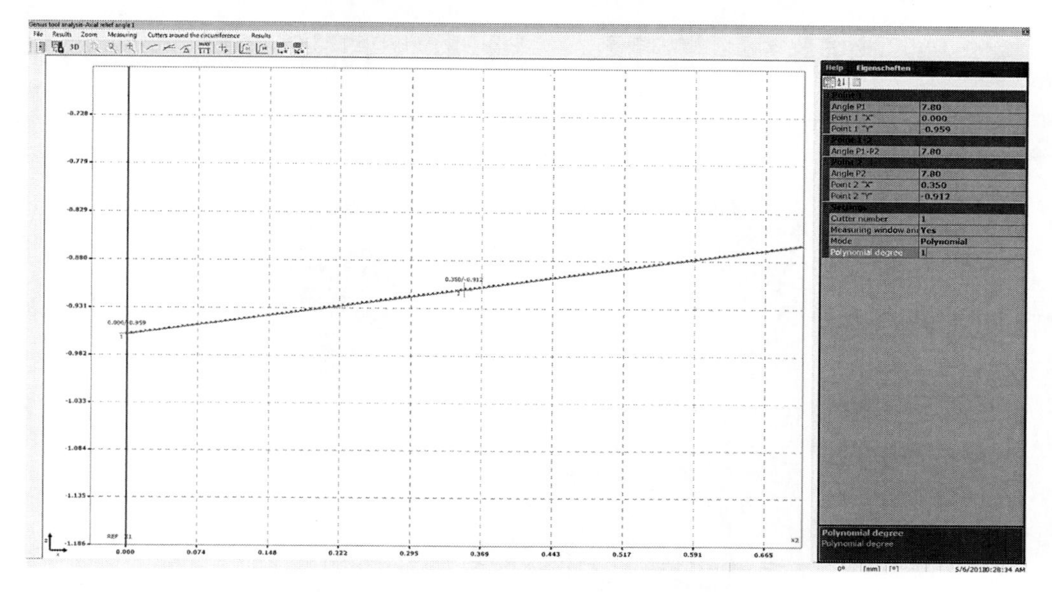

FIGURE 7.134 Showing straightness of the primary flank face as sectioned by the normal or orthogonal plane through the selected point of the major cutting edge.

by a CAM grind (e.g., a part of conical and helical surface as discussed in Chapter 4), then the line of its intersection with the normal or orthogonal plane through the selected point of the major cutting edge is a complicated curve. By selecting a suitable polynomial degree on the screen menu, one can obtain the best fit (approximation) of the flank face so that the normal to the flank face can be drawn and the proper clearance angle in the manner shown in Figure 7.91b can be measured. Such an operation, however, requires additional programming as it is not readily shown in the on-screen menu.

Unfortunately, a great number of articles listed in Tables 7.36 and 7.37 and crucial to HP drill performance are not built-in features that can be measured easily using a series of on-screen menus and options. An example is an attempt to measure the chisel edge centrality (symmetry) Δ_{cc} as listed as article 19D in Table 7.37 and visualized in Figure 7.82. Figure 7.135 shows the screen and menu for the inspection of straight line symmetry that actually stands for flute spacing measurement as defined by article 4D in Table 7.37 and shown in Figure 7.97a. This is selected because it properly visualizes the web centrality. Figure 7.136 shows what should be considered in the centrality of the web measurement (unfortunately not actually highlighted on the monitor in the manner shown in this Figure). The chisel

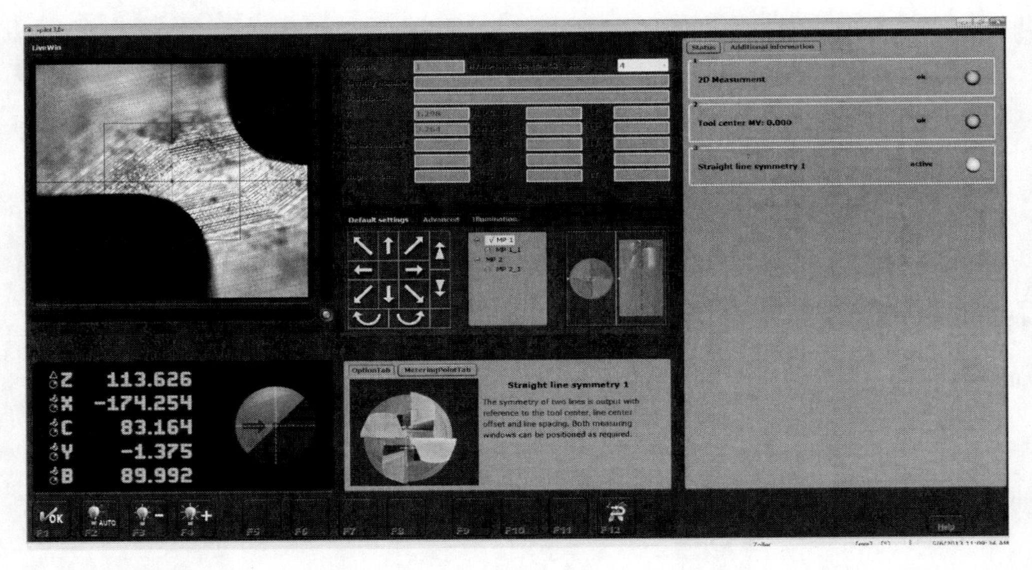

FIGURE 7.135 Showing the screen and menu for the inspection of straight line symmetry used to visualize the web centrality.

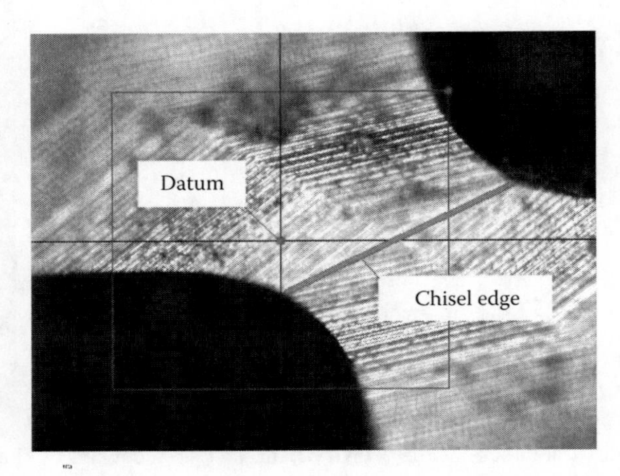

FIGURE 7.136 Showing what should be considered in the centrality of the web measurement.

edge centrality (symmetry) Δ_{cc} is the distance between the datum and the actual chisel edge, that is, as perpendicular from the point designated as the datum in Figure 7.136 to the highlighted chisel edge. If the chisel edge would be highlighted and digitized in the manner shown in Figure 7.136, then the following articles listed in Table 7.37 can be measured in this drill position/screen menu: the chisel edge angle, ψ_{cl} (17D); chisel edge length, l_{cl} (18D); centrality of the web, Δ_w (19D); and web thickness, d_{wb} (21D). The listed parameters are visualized in Figure 7.103.

7.7.17.5 Summary

Inspection and measurement of HP drills go beyond handheld gages to include CNC drill inspection machines able to execute programmed measurement routines. For years, the optical comparator was the most commonly used device for tool inspection. The comparator offered reasonably priced technology for checking drills by profile projection. The use of surface illumination, digital readout, and edge finding further enhanced the device.

However, the comparator lacked the ability of the naked eye or digital camera to see fine details in the geometry of cutting tools. The toolmakers' microscope, with its linear scales and digital readout, was an improvement over the comparator, because tool images could be clearly seen and measured. But like the comparator, the toolmakers' microscope was still a mechanical system limited to hardware upgrades. Like a comparator or microscope, a digital system uses ground glass lenses to magnify features on a tool. However, this type of system also uses a camera to convert the tool image to a digital image.

Early versions of monitor-based video inspection systems used black-and-white cameras, television monitors, and line generators. Measurements were made by linear scales and digital readout or by a pixel count from the monitor screen. In a video system, the number of pixels, or single dots on the monitor screen, determines the display resolution. With pixel count, cutting tools can be easily measured. And because tool images are in a digital format, they can be printed, but they lack supporting documentation such as x-axis, y-axis, and rotational measurements. In the past 10 years, the lower cost and higher sophistication of computers, video imaging cards (frame grabbers), and cameras have boosted the popularity and accuracy of CNC video inspection systems. System manufacturers can use frame grabbers to display live tool images on a computer monitor screen and write custom programs that increase the speed and accuracy of video inspection. The greatest advantage of CNC measuring machines is proper definition and nullification of the datum that sets such machines apart from any other HP drill measuring devices. The introduction of this feature opens a new era in HP drill inspection/measurement that, for the first time in the history of drill measurement, can be carried out exactly according to the proper tool drawing. Modern advanced CNC measuring machines can, in principle, inspect/measure all non-datum and datum articles required for HP drills. Although it is true, three issues remain that slowdown implementation of CNC measuring machines and thus HP drills:

1. *Proper HP drill drawing.* This issue in discussed in Chapters 4 and 5. The articles and their parameters listed in Tables 7.36 and 7.37 should be properly and clearly indicated in drill drawings. Otherwise no one knows what to inspect/measure. The author pointed out earlier (Astakhov 2010a) the basic requirement to tool drawings and presented a number of examples. This chapter includes examples of the proper representation of the datum and non-datum features listed in Tables 7.36 and 7.37 in HP drill drawings.
2. *Outdated standards* (ISO, DIN, etc.) for parameters of the articles listed in Tables 7.36 and 7.37. This chapter provides basic recommendations for the proper selection of such parameters for HP drills.
3. *The limited library of common features/articles to be measured almost automatically.* using CNC measuring machines. In the author's opinion, it is very difficult to justify a \$200,000 nonproduction machine that then is used only to 30% of its capacity and has limited built-in standard articles to be measured automatically, suffers from nonstandard/improper terminology, and has no useful user-friendly manual.

In the author's opinion, the listed issues can be resolved in relatively short time so HP drilling tools can be properly inspected. It should provide a significant boost for wider implementation of HP drills.

REFERENCES

Astakhov, V.P. 2004. The assessment of cutting tool wear. *International Journal of Machine Tools and Manufacture* 44:637–647.

Astakhov, V.P. 2010a. *Geometry of Single-Point Turning Tools and Drills. Fundamentals and Practical Applications*. London, U.K.: Springer.

Astakhov, V.P. 2010b. Surface integrity definitions and importance in functional performance. In *Surface Integrity in Machining*, Davim, J.P. (ed.). London, U.K.: Springer-Verlag, pp. 1–35.

Astakhov, V.P. 2011. Drilling. In *Modern Machining Technology*, Davim, J.P. (ed.). Cambridge, U.K.: Woodhead Publishing, pp. 79–212.

Astakhov, V.P. 2014. Bore-gaging displacement sensors. In *Spatial, Thermal, and Radiation Measurement*, Webster, J. (ed.). Boca Raton, FL: CRC Press, pp. 28-1–28-16.

Gooldy, G. 1998. *Dimensioning, Tolerancing, and Gaging Applied: Ref. ASME Y14.5M-1994 ANSI B4.4 (B89.3.6) AMD ISO 1101*. Upper Saddle River, NJ: Prentice Hall.

Henzold, G. 2006. *Dimensioning and Tolerancing for Design, Manufacturing and Inspection*. Burlington, MA: Butterworth-Heinemann.

Kobayashi, A. 1967. *Machining of Plastics*. New York: McGraw-Hill.

Krulikowski, A. 1998. *Geometric Dimensioning and Tolerancing*. New York: Delmar.

Nee, J.G. 2010. *Fundamentals of Tool Design*. Dearborn, MI: SME.

Quadrini, F., Squeo, E.A., and Tagliaferri, V. 2007. Machining of glass fiber reinforced polyamide. *eXPRESS Polymer Letters* 1(12):810–816.

Shin, Y.C. and Dandekar, C. 2012. Mechanics and modeling of chip formation in machining of MMC. In *Machining of Metal Matrix Composites*, Davim, J.P. (ed.). London, U.K.: Springer, pp. 1–50.

Smith, G.T. 2002. *Industrial Metrology: Surfaces and Roundness*. London, U.K.: Springer.

The Metal Cutting Tool Institute. 1989. *Metal Cutting Tool Handbook*, 7th edn. New York: Industrial Press.

Appendix A: Axial Force, Torque, and Power in Drilling Operations

A.1 COMMON METHODS

Unfortunately, the present stage of development of the metal cutting theory does not allow calculating the axial force and drilling torque (power) even if exact properties of the work material, particularities of the tool geometry, machining regime, and other parameters of the drilling system are known. As a result, many researchers used the so-called mechanistic approach. In the author's opinion, the essence of this approach should be clearly understood.

Armarego and Chang (1972a,b) proposed an approach to predict thrust and torque during drilling for a conventional drill and a modified drill in order to simplify the calculations. The method of calculation used the orthogonal cutting model and the oblique cutting model, and its refined version firstly referred as the mechanistic approach was developed by Wiriyacosol and Armarego (1979). Basically, this method consists of dividing the cutting edges into a limited number of cutting elements. These elements were assumed to be oblique cutting edges on the cutting lip and orthogonal cutting edges on the chisel edge. The calculation used empirical equations established from orthogonal cutting tests. The problem of cutting force calculations was reduced to choosing the number of cutting elements and to determine the empirical equations for some cutting parameters.

Watson (1985a–d) initially used practically the same method, with a more accurate model of the drill geometry particularly related to twist drills and the plasticity-based oblique-machining model of chip formation to develop a model to predict the axial force and drilling torque. He developed a model for the chisel edge and the lip from the orthogonal cutting model and the oblique cutting model, respectively. Watson recognized that the chip flows from the lips and the chisel edges are continuous across their width and that continuity imposes a restriction on the possible variation of the chip flow angle across those edges. Unfortunately, the further researchers who applied the mechanistic approach to drilling did not notice this very important restriction.

The most noticeable from the further researches on the mechanistic approach in drilling is that by Chandrasekharan et al. (1995, 1998). Using the same mechanistic approach, they developed a 3D cutting force model for drilling. Moreover, separate models for axial (thrust) force, drilling torque, and radial forces were suggested.

According to the mechanistic approach, the so-called specific cutting force (normally designated as k_c) for the work material is determined. It is measured in N per square mm and its value has been worked out and tested for most materials and is available in multiple catalogs of drill manufacturers. It is defined as the tangential cutting force needed for a chip with a certain cross section (1 mm^2) or the effective cutting force divided by the theoretical chip area. For example, Table A.1 shows k_c for some common work materials.

TABLE A.1
Specific Cutting Force for Some Work Materials

Work Material	State	k_c New Tool	k_c Worn Tool
Aluminum	Forged annealed cast Si < 13%	500	700
	Forged aged cast Si > 13%	750	1050
FGL250 (DIN 1691) gray cast iron	Cast	1250	1750
FGS500 (ISO 185:1988) cast iron	With spheroidal graphite	1500	2100
Mild steels			
Medium-strength steels		1750	2450
Titanium alloys			
High-strength steels		2000	2800
Austenitic stainless steels			
Nickel-based alloys		3250	4550
Cobalt-based alloys			

Selecting the appropriate value of k_c (from Table A.1 or from many other widely available sources), one can then calculate the following:

The axial thrust F_{ax} (N):

$$F_{ax} = \frac{k' k_c f d_{dr}}{2} \tag{A.1}$$

where

d_{dr} is the drill diameter (mm)
f is the cutting feed (mm/rev)
k' is the tool geometry coefficient that depends on a particular drill geometry. Its *average* value is accepted to be 0.5

Drilling torque M_{dr} (N/m):

$$M_{dr} = \frac{k_c f d_{dr}^2}{8000} \tag{A.2}$$

Cutting power (kW):

$$P_c = \frac{k_c f d_{dr} v}{240{,}000} \tag{A.3}$$

where v is the cutting speed (m/min).

Some leading tool manufacturers, for example, ISCAR (http://www.iscar.com/), provide online calculator of the axial force, torque, and power in drilling. Figure A.1 shows an example of the calculation of these parameters using the ISCAR online calculator. In this picture, κ_r is the tool cutting edge angle (defined and explained later in this book) according to ISO Standard 3002, wrongly termed as the lead angle in the calculator. It is equal to half of the point angle Φ_p (Chapter 1, Figure 1.7). Although it looks simple, an array of questions remains. For example, what is that *effective rake angle*—where do I get it in selecting a drill? What are the metallurgical state, hardness, and other important properties of the work material (in the considered case—steel 1045)?

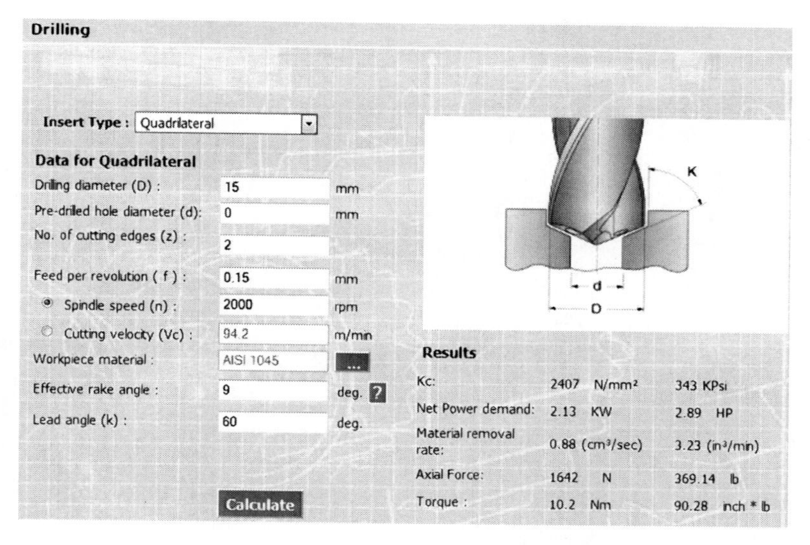

FIGURE A.1 Example of calculations using the online ISCAR calculator.

The leading tool manufacturer Sandvik Coromant offers a methodology that is seemingly better as it accounts for the properties of the work materials. According to this methodology, the specific cutting force for feed per edge, k_{cfz} (N/mm²), is calculated as

$$k_{cfz} = k_{c04}\left(\frac{0.4}{f_z \sin\kappa_r}\right)^{0.29} \tag{A.4}$$

where
k_{c04} is the specific cutting force determined for $f_z = 0.4$ mm/rev
$\kappa_r = 90$

Table A.2 lists this specific cutting force for various work materials.

Knowing k_{cfz}, one can calculate the following:

The axial thrust F_{ax} (N):

$$F_{ax} = 0.5\frac{d_{dr}}{2} f k_{cfz} \sin\kappa_r \tag{A.5}$$

Drilling torque M_{dr} (N/m):

$$M_{dr} = \frac{d_{dr} f k_{cfz} a_p}{2000}\left(1 - \frac{a_p}{d_{dr}}\right) \tag{A.6}$$

where a_p is the depth of cut (Section 1.1.3).

Cutting power (kW):

$$P_c = \frac{d_{dr} f k_{cfz} v}{240{,}000} \tag{A.7}$$

where v is the cutting speed (m/min).

TABLE A.2
Sandvik Coromant Generalized k_{c04}

Material	Condition	Hardness HB	k_{c04} (N/mm²)[a]
Unalloyed steel	C = 0.15%	125	1900
	C = 0.35%	150	2100
	C = 0.60%	200	2250
Low alloy steel	Non-hardened	180	2100
	Hardened and tempered	275	2600
	Hardened and tempered	300	2700
	Hardened and tempered	350	2850
High alloy steel	Annealed	200	2600
	Hardened	325	3900
Stainless steel	Martensitic/ferritic	200	2300
	Austenitic	175	2450
Steel casting	Unalloyed	180	2000
	Low alloyed	200	2500
	High alloyed	225	2700
Hard steel	Hardened steel	55HRC	4500
	Manganese steel 12%	250	3600
Malleable cast iron	Ferritic	130	1100
	Pearlitic	230	1100
Gray cast iron	Low tensile strength	180	1100
	High tensile strength	260	1500
Nodular cast iron	Ferritic	160	1100
	Pearlitic	250	1800
Chilled cast iron		400	3000
Heat resistant alloys	Fe-base, annealed	200	3000
	Fe-base, aged	280	3050
	Ni- or Co-base, annealed	250	3500
	Ni- or Co-base, aged	350	4150
	Ni- or Co-base, cast	320	4150
Aluminum alloys	Non heat treatable	60	500
	Heat treatable	100	800
Aluminum alloys, cast	Non heat treatable	75	750
	Heat treatable	90	900
Copper and copper alloys	Lead alloys, Pb > 1%	110	700
	Brass, red brass	90	750
	Bronze and lead-free copper Including electrolytic copper	100	1750

[a] k_{c04} values are valid for: $f_z = 0.4$ mm/tooth, $\kappa_r = 90°$, $\lambda_{sh} = +6°$.

In the author's opinion, the mechanistic approach is totally outdated so its use should be abolished in metal cutting, in general, and in drilling, in particular research activities. This is because of the following:

1. The approach is based on inadequate model of chip formation (Astakhov 2005).
2. The mechanistic approach is not suitable particularly for drilling. As mentioned earlier, the researchers ignored the Watson warning (Watson 1985a–d) that the chip flows from the lips and the chisel edges are continuous across their width and that continuity imposes a restriction on the possible variation of the chip flow angle across those edges. The division of the cutting edge of the drill in this case does not make any sense as the chip flow

on each element is considered independent from the other, so the chip flow direction is determined by the *theoretical* chip flow direction. This is not nearly the case in drilling as discussed in Chapter 4, Section 4.5.7, which completely ruins the foundation of this approach in drilling.

3. The approach is valid only for the properties of the work material used in the test to determine k_c. A small deviation from these properties can significantly affect the experimentally determined values. Moreover, when blanks come to metal-machining operations, a rather wide tolerance on the part material properties is normally allowed that makes the determined k_c's of no practical significance as this specific cutting force was determined for only some specific set of mechanical properties of the work material.
4. As tool design and geometry become more sophisticated, for example, HP drills having a number of specially designed gashes located at various angles; as the number of the tool materials and coatings increases every month; as more advanced MWF (coolants) are used; etc., k_c determined for some specific conditions becomes useless. It hampers any drill geometry optimization because the change in the drill geometry parameters not reflected in k_c.
5. The determination of k_c for a given set of drilling conditions requires extensive, expensive, and time-consuming testing that is not feasible in modern manufacturing.
6. The determination of k_c for a given set of drilling conditions requires an expensive properly calibrated drilling dynamometer with accessories that nearly double its cost. Besides, as discussed by the author earlier, the cutting force cannot be measured with reasonable accuracy, although this fact has never been honestly admitted by specialists in the field. To appreciate the issue, one should consider the result of the joint program set by College International pour la Recherche en Productique (CIRP) and National Institute of Standards and Technology (NIST) to measure the cutting force in the simplest case of orthogonal cutting (Ivester et al. 2000). The experiments were carefully prepared (the same batches of the workpiece [steel AISI 1045], tools, etc.) under supervision of the NIST and replicated at four different most advanced metal cutting laboratories in the world. Interestingly, although extraordinary care was taken while performing these experiments, there was significant variation (up to 50%) in measured cutting force across these four laboratories. Obviously, such accuracy is not acceptable in calculations of the physical efficiency of the cutting system.

Moreover, many tool and cutting insert manufacturers (not to mention manufacturing companies) do not have a decent dynamometer to measure the cutting force, so it would be rather difficult to determine the cutting force for the discussed calculations. Several dynamometers used in the field are not properly calibrated because the known literature sources did not present the proper experimental calibration for cutting force measurements using piezoelectric dynamometers.

To defend the mechanistic approach, the following must be said. It was developed as a temporary measure to deal with manufacturing culture of the past as discussed in Preface. Drilling using nonrigid drilling machines, inaccurate tool holders, literally dirty MWF (coolant), insufficient machine power and weak feed mechanism, etc., did not require an accurate determination of the force factors in drilling. Bulk, approximate numbers were sufficient to select drilling parameters and design primitive work-holding fixtures. The mechanistic approach was a necessity for metal cutting researchers and tool manufacturers as nothing better had been offered by the metal cutting theory. Having been justified for *bulk* estimation of the force parameters in tool manufacturer catalogs, such an approach should not be used in scientific publications.

Having noticed drawbacks of the specific cutting force-based force parameter calculations, some leading drilling tool manufacturers provide the data for a specific type of drill tool and some work materials. The recent trend, however, is not to include the data on the cutting force and torque in their catalogs as there are too many possibilities in terms of drill design, geometry, materials (both tool and work), coatings, cutting conditions, MWF used, etc.

A.2 DEFORMATION METHOD

As explained in Preface, the time has changed so the *silent revolution* in manufacturing (referred in recent manufacturing publication as the fourth industrial revolution) is taking place. The use of powerful accurate drilling machines with sophisticated controllers capable of monitoring the drilling torque, cutting power, axial force, and many other parameters of drilling is becoming general manufacturing practice. In other words, the force factors of drilling can nowadays be assured with the accuracy that is higher than in any laboratory test using force dynamometers and other measuring equipment. These factors also change their application significance. Before, the maximum axial force and drilling torque were used to check if a machine is capable to carry out a given drilling job. As modern machines have an excess of power, the force factors of drilling can now be used for the optimization of drill design as discussed in Chapter 4. As stated in Chapter 1, the *VPA-Balanced*© design assures the minimum axial force and drilling torque. Two basic questions—why (what for)? and how?—should be answered.

A.2.1 Why Is the Reduction of the Cutting Force/Energy Spent in Cutting Needed

A simple answer is obvious—the reduction of the drilling torque and axial force allows greater penetration rate as these two force factors present the major restriction on this rate as discussed in Chapter 4 (Section 4.4). Although this answer is perfectly correct, it is still superficial as the force factors in drilling are simplifications of the modern general factors, namely, energy spent in cutting. Therefore, the energy required by the cutting system for its existence should be dealt with as this energy determines tool life through the so-called technical resource of the drill (Astakhov 2006). In simple terms, this resource can be thought of as the amount of energy that can be transmitted through the cutting tool (drill) till the criteria of tool life are reached. Therefore, the reduction of the energy required by the cutting system for its existence (i.e., for machining of a given work material at a chosen machining regime) allows increasing both the drill penetration rate (machining cost reduction through increasing productivity) and tool life (cost reduction through the reduction in tool cost and decreasing downtime).

A.2.2 How to Reduce the Cutting Force/Energy Spent in Cutting

A.2.2.1 Ability of the Prevailing Metal Cutting Theory

Although metal cutting, or simply machining, is one of the oldest processes for shaping components in the manufacturing industry and it is widely quoted that 15% of the value of all mechanical components manufactured worldwide is derived from machining operations, machining remains one of the least understood manufacturing operations due to the low predictive ability of machining models (Usui 1982, 1988) despite its obvious economic and technical importance. In the author's opinion, this is due to the common notion that new surfaces in metal cutting are formed simply by *plastic flow around the tool tip* (Shaw 2004). In other words, the metal cutting process is one of the deforming process (metal forming operations that used plasticity of the work material to achieve the final shape of the product) where a single-shear plane model of chip formation constitutes the very core of metal cutting theory, and thus, this process is thought of primarily as a cutting tool deforming a particular part of the workpiece by means of shearing. Although a number of cutting theories and the FEM models have been developed based on this concept, their prediction ability is low so that they are not used in any practical process design and optimization of cutting parameters.

If one considered metal cutting as one of the deforming metal working operation used in industry the a question is why is can be modeled with the same success as other deforming process. To understand that, one can consider state of the art in the closely related deforming process used in industry. Until 10 years ago, the design of metal forming tools was mostly based on knowledge gained through experience, and designing of optimum tools often required a protracted and expensive trial-and-error testing. Today, even in the earlier phases, simulations of the forming process are carried out using FEMs. The most important goals of such simulations are verification of manufacturability of the sheet-metal parts and obtaining vital information on the optimal tool design. As a result, great savings have been achieved due to the introduction of process simulation in metal forming. These savings originate from faster development of tools and from dramatic shortening of trial-and-error testing. In recent years, tool development and production time have been reduced by about 50% due to the use of simulations, and a further 30% reduction over the next few year appears realistic. The simulation of forming tool has already reached the stage where its result can be fed directly into the press tool digital planning and validation process. Thus today, starting from the design model and through practically all process steps as far as the actual design of the press tool, the production of a component can be fully simulated before a first prototype is built (Roll 2008).

Obviously, this is not nearly the case in metal cutting crippled by the inadequate theory. The problem is that the single-shear plane model used as the foundation of this theory does not resemble the reality even to the first approximation (Astakhov 2005).

In a deforming process, the ductility of the work material is the most desired property of the work material, while in metal cutting, the ductility causes useless plastic deformation of the work material in its transformation into the chip. As discussed by the author earlier in the analyses of the energy partition in the cutting system (Astakhov 2006, 2010; Astakhov and Xiao 2008), depending on the ductility of the work material 60–80% of the energy required by the cutting system for its existence is spent for plastic deformation of the layer being removed, which is actually wasted as the deformed chip does not serve any useful purpose. This is the major difference between metal cutting and deforming operations used in industry. Unless it is clearly realized by the researchers and practitioners in the field, no progress in metal cutting modeling can be achieved.

A.2.2.2 Alternative Approach

The earlier-discussed difference between metal cutting and other deforming operations allowed the author to formulate the second metal cutting law named as the deformation law as follows:

> Plastic deformation of the layer being removed in its transformation into the chip is the greatest nuisance in the metal cutting, that is, while it is needed to accomplish the process, it does not add any value to the finished part. Therefore, being by far the greatest part of the total energy required by the cutting system, this energy must be considered as a waste that should be minimized to achieve higher process efficiency.

The lower the energy of plastic deformation, the lower the cutting forces, and the greater the tool life, the better the quality of the machined surface and the process efficiency. Therefore, the prime objective of the cutting process design is to reduce this energy to its possible minimum by the proper selection of the tool geometry, tool material, machining regime, MWF, and other design and process parameters.

In drilling, the reduction of the plastic deformation of the layer being removed in its transformation into the chip can be achieved by altering the state of stress in the deformation zone, that is, the so-called stress triaxiality that is directly correlated with the strain at fracture of the work material, that is, with the amount of plastic deformation needed to form the chip (Astakhov 1998/1999, 2006). As discussed by the author earlier, this stress triaxiality, and thus the amount of plastic deformation, can be varied in a wide range by the tool geometry. This is the essence of significance of the tool geometry—its proper designs/selection for a given drilling tool. As a drilling tool may have parts of the cutting edge of different geometries, the plastic deformation and thus energy spent on each particular part of the cutting edge should be known with high accuracy. To accomplish the clear objective of the metal cutting, that is, to make the introduced deformation law of practical significance, a reliable measure of this energy should be readily available to be used at various levels from a research laboratory to the shop floor.

A.2.3 Chip Compression Ratio as the Most Reliable Measure of Plastic Deformation in Metal Cutting

Although the total cutting force, and thus the cutting force (after transformation energy) on each part of a drilling tool, can be seemingly measured using a dynamometer and a simulation cutting tool contacting only a part of the cutting edge in question, there are some problems as discussed above.

Therefore, to make practical calculations of the energy spent in cutting, another approach has to be used.

There are two characteristics of plastic deformation in metal cutting, namely, CCR and shear strain. Historically, CCR was introduced in the earlier studies on metal cutting as a measure of plastic deformation of the work material in its transformation into the chip (Zorev 1966; Astakhov 1998/1999). A model of chip deformation in the simplest case of cutting (orthogonal cutting) is shown in Figure A.2. A flat section abcd having length L_1 and thickness h_D is distinguished in the layer to be removed by the cutting tool. Once the distinguished section is deformed on its transformation into the chip, the section abcd transforms into section a′b′c′d′. In this transformation, called plastic deformation, the area of the initial section does not change due to conservation of work material volume. However, the dimensions of its sides do change. Length L_1 of side ab becomes length L_2 of side a′b′, while thickness h_D (uncut chip thickness) becomes chip thickness h_C. The CCR represents such a transformation due to plastic deformation as

$$\zeta = \frac{L_1}{L_2} = \frac{h_C}{h_D} \tag{A.8}$$

Although this parameter was widely used in metal cutting tests of the past (Zorev 1966), it was always considered as a secondary parameter to provide only qualitative support to certain

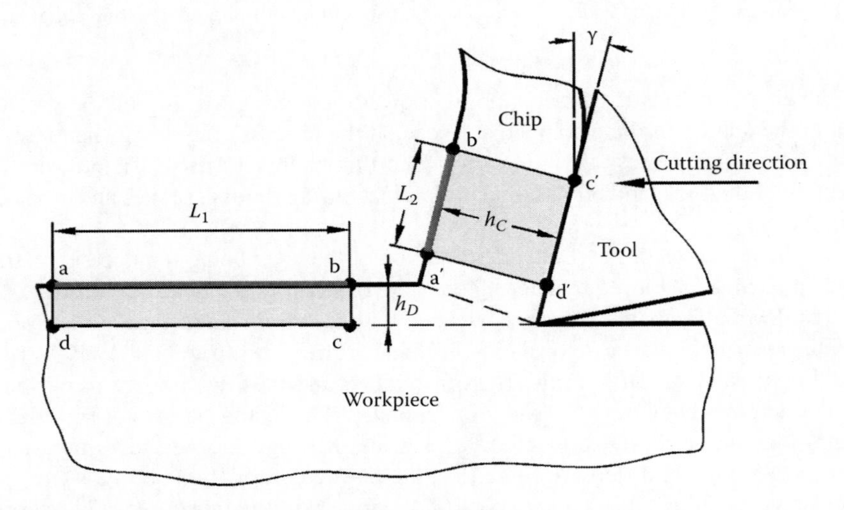

FIGURE A.2 Simple model of the chip plastic deformation in orthogonal cutting.

conclusions. Since the real physical meaning of this parameter has never been revealed, it was gradually abandoned in metal cutting studies because nobody could explain the obtained results. For example, when one obtains $\zeta = 2.5$ in machining of a steel and $\zeta = 4.5$ in machining of a copper alloy at the same cutting speed, he should conclude that the plastic deformation and thus the energy required for this deformation in the latter case are much greater than that in the former. However, the cutting force in machining of the steel is much greater than that in machining of the copper alloy. As the total energy required by the cutting system can be thought of as the product of the cutting force and the cutting speed, then an unexplained contradiction between the values of the cutting force and CCR is obvious. That is why CCR is practically abandoned in modern metal cutting studies. For example, although Shaw in his book (Shaw 2004) dedicated a full chapter to the analysis of plastic deformation in metal cutting, this parameter is not even mentioned. The same can be said about the books by Trent and Trent and Wright (2000), Oxley (1989), and Gorczyca (1987); Altintas (2000) just mentioned its definition in the consideration of the single-shear plane model; Childs et al. (2000) mentioned this parameter as related to the friction coefficient at the tool–chip interface. Not a single modern study on metal cutting correlates this parameter with the amount of plastic deformation in metal cutting.

The shear strain is another characteristic of plastic deformation in metal cutting. It is calculated as

$$\varepsilon = \frac{\cos\gamma}{\cos(\varphi_s - \gamma)\sin\varphi_s} = \frac{1 - 2\zeta\sin\gamma + \zeta^2}{\zeta\cos\gamma} \tag{A.9}$$

where φ_s is the so-called shear angle (Merchant 1945; Zorev 1966; Shaw 2004).

Although Equation A.9 is used in practically all books on metal cutting, there are some obvious problems with these equations in their terms of physical meaning and experimental confirmation (Astakhov 2006). If one calculates shear strain using Equation A.9 (it can be easily accomplished by measuring the actual CCR ζ) and then compares the result with the shear strain at fracture obtained in standard materials tests (tensile or compression), he or she easily finds that the calculated shear strain is much greater (twofold to fivefold) than that obtained in the standard materials tests. Moreover, when the CCR $\zeta = 1$, that is, the uncut chip thickness, is equal to the chip thickness so no plastic deformation occurs in metal cutting (Astakhov and Shvets 2004), the shear strain, calculated by Equation A.9, remains significant with no apparent reason for that. For example, when $\zeta = 1$, the rake angle $\gamma = -10°$, Equation A.9 yields $\varepsilon = 2.38$; when $\zeta = 1$, the rake angle $\gamma = 0°$, then $\varepsilon = 2$; and when $\zeta = 1$, $\gamma = +10°$, then $\varepsilon = 1.68$. As shown by the author earlier (Astakhov 2005), this severe physical contradiction is caused by the incorrect velocity diagram used to derive Equation A.9.

The foregoing analysis suggests that apparently, there is no reliable measure of plastic deformation in metal cutting that can be used in tool and process designs as suggested earlier. However, Astakhov and Shvets (2004) showed that CCR is the only post-process reliable parameter of plastic deformation that objectively reflects the reality. Its proper physical meaning as the most reliable measure of plastic deformation was revealed (Astakhov 2006, 2010) through the direct correlation with work material mechanical properties. It was shown that the elementary work spent over plastic deformation of a unit volume of the work material A_u is calculated as

$$A_u = \frac{K(1.15\ln\zeta)^{n+1}}{n+1} \tag{A.10}$$

so that the total work done by the external force applied to the tool is then calculated as

$$A = A_u v f a_p \tau_{ct} \tag{A.11}$$

where

K, n are parameters of the true stress–strain curve, known as the flow curve (Astakhov 2006)
K is the stress at strain $\varepsilon = 1.0$
n is the strain-hardening coefficient determined as the slope of this curve represented as a log-log plot
τ_{ct} is the time of cutting

The measuring and representation of the cutting speed, v; feed, f; and depth of cut, a_p, are discussed in Chapter 1.

Knowing CCR, one can directly determine the following:

1. Power spent on plastic deformation of the layer being removed, which is the largest portion of the power required by the cutting system and which is the major contributor to the cutting force as

$$P_{pd} = \frac{K(1.15\ln\zeta)^{n+1}}{n+1} v a_p f \tag{A.12}$$

   Dividing this power by the cutting speed, one can obtain the cutting force due to the plastic deformation of the work material.
2. The so-called natural length of tool–chip interface (Equation 4.75).
3. The chip velocity relative to the cutting tool as the cutting speed divided by the CCR.

The complete methodology of determining the cutting forces using the CCR was presented by Astakhov and Xiao (2008). The methodology of the cutting force evaluation on various parts of a drill using CCR was developed by Vinogradov (1985).

A.2.4 Experimental Determination of the Chip Compression Ratio

The simplest method is to measure the chip thickness and then calculate the CCR using Equation A.8 as the ratio h_C/h_D. However, it is not always possible because the chip (a) might have a saw-toothed free surface and (b) be so small and 3D curved. When this is the case, the second method known as the weighting method (Astakhov 2006) can be used. A small (3–7 mm long) straight piece of the chip is separated from the rest of the chip. Then, its length L_c and width b_c are measured. When the piece of the chip selected for the study is not straight, a computer vision system available nowadays in most shops is used to measure its length properly. Then, it is weighed so its weight G_{ch} (N) is determined. The chip thickness is then calculated as

$$h_C = \frac{G_{ch}}{b_c L_c \rho_w g} \tag{A.13}$$

where

ρ_w is the density of the work material (kg/m^3)
$g = 9.81$ m/s^2 is the gravity constant

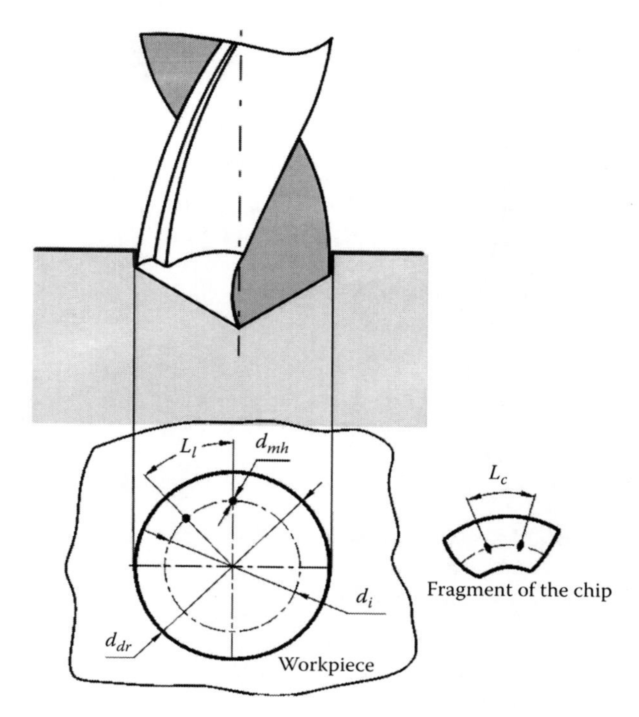

FIGURE A.3 Schematic of the third method.

The third method is the direct method that includes a specially prepared workpiece. The essence of this method is that the workpiece is *marked* before cutting and then the resultant marks on the chip are compared with the original marks. The realization of the discussed method to measure CCR in drilling is shown in Figure A.3. Two small holes of diameter d_{mh} are drilled in the workpiece along the trajectory of the point i of the drill cutting edge (at certain diameters d_i). The arc distance L_l between the centers of these holes is measured. After the test, a chip fragment having marks from two holes is found, and the ark distance L_c between their centers is measured at high magnification using an optical comparator or a digital microscope. CCR is calculated using Equation A.8 as the ratio L_l/L_c.

REFERENCES

Altintas, Y. 2000. *Manufacturing Automation. Metal Cutting Mechanics, Machine Tool Vibrations, and CNC Design*. Cambridge, U.K.: Cambridge University Press.

Armarego, E.J.A. and Cheng, C.Y. 1972a. Drilling with flat rake face and conventional twist drills—I. Theoretical investigation. *Journal of Machine Tools and Manufacture* 12:17–35.

Armarego, E.J.A. and Cheng, C.Y. 1972b. Drilling with flat rake face and conventional twist drills. Part II: Experimental investigation. *International Journal of Machine Tool Design and Research* 12:37–54.

Astakhov, V.P. 1998/1999. *Metal Cutting Mechanics*. Boca Raton, FL: CRC Press.

Astakhov, V.P. 2005. On the inadequacy of the single-shear plane model of chip formation. *International Journal of Mechanical Science* 47:1649–1672.

Astakhov, V.P. 2006. *Tribology of Metal Cutting*. London, U.K.: Elsevier.

Astakhov, V.P. 2010. *Geometry of Single-Point Turning Tools and Drills: Fundamentals and Practical Applications*. London, U.K.: Springer.

Astakhov, V.P. and Shvets, S.V. 2001. A novel approach to operating force evaluation in high strain rate metal-deforming technological processes. *Journal of Materials Processing Technology* 117(1–2):226–237.

Astakhov, V.P. and Shvets, S.V. 2004. The assessment of plastic deformation in metal cutting. *Journal of Materials Processing Technology* 146:193–202.

Astakhov, V.P. and Xiao, X. 2008. A methodology for practical cutting force evaluation based on the energy spent in the cutting system *Machining Science and Technology* 12:325–347.

Chandrasekharan, V., Kapoor, S.G., and DeVor, R.E. 1995. A mechanistic approach to predicting the cutting forces in drilling: With application to fiber-reinforced composite materials. *Journal of Manufacturing Science and Engineering* 117:559–570.

Chandrasekharan, V., Kapoor, S.G., and DeVor, R.E. 1998. A mechanistic model to predict the cutting force system for arbitrary drill point geometry. *Journal of Manufacturing Science and Engineering* 120:563–570.

Childs, T.H.C., Maekawa, K., Obikawa, T., and Yamane, Y. 2000. *Metal Machining: Theory and Application.* London, U.K.: Arnold.

Gorczyca, F.Y. 1987. *Application of Metal Cutting Theory*. New York: Industrial Press.

Ivester, R.W. 2004. Comparison of machining simulations for 1045 steel to experimental measurements. SME Paper TPO4PUB336:1–15.

Ivester, R.W., Kennedy, M., Davies, M., Stevenson, R.J., Thiele, J., Furness, R., and Athavale, S. 2000. Assessment of machining models: Progress report. Paper presented at the *Proceedings of the Third CIRP International Workshop on Modelling of Machining Operations*, August 20, 2000, University of New South Wales, Sydney, New South Wales, Australia.

Merchant, M.E. 1945. Mechanics of the metal cutting process. I. Orthogonal cutting and a type 2 chip. *Journal of Applied Physics* 16:267–275.

Oxley, P.L.B. 1989. *Mechanics of Machining: An Analytical Approach to Assessing Machinability*. New York: John Wiley & Sons.

Roll, K. 2008. Simulation of sheet metal forming—Necessary developments in the future. Keynote Paper NUMISHEET. In *LS-DYNA Anwenderforum*, September 1–5, 2008, Bamberg, Germany, pp. A-1-59–A-1-68.

Shaw, M.C. 2004. *Metal Cutting Principles*, 2nd edn. Oxford, U.K.: Oxford University Press.

Trent, E.M. and Wright P.K. 2000. *Metal Cutting*, 4th edn. Woburn, MA: Butterworth-Heinemann.

Usui, E. 1988. Progress of "predictive" theories in metal cutting. *JSME International Journal* 31:363–369.

Usui, E. and Shirakashi, T. 1982. *Mechanics of Metal Cutting—From "Description" to "Predictive" Theory.* Paper read at On the Art of Cutting Metals—75 Years later, at Phoenix, AZ.

Vinogradov, A.A. 1985. *Physical Foundation of the Drilling of Difficult-to-Machine Materials with Carbide Drills (in Russian)*. Kiev, Ukraine: Naukova Dumka.

Watson, A.R. 1985a. Drilling model for cutting lip and chisel edge and comparison of experimental and predicted results. I—Initial cutting lip model *International Journal of Machine Tool Design and Research* 25:347–365.

Watson, A.R. 1985b. Drilling model for cutting lip and chisel edge and comparison of experimental and predicted results. II—Reversed cutting lip model. *International Journal of Machine Tool Design and Research* 25:367–376.

Watson, A.R. 1985c. Drilling model for cutting lip and chisel edge and comparison of experimental and predicted results. III—Drilling model for chisel edge. *International Journal of Machine Tool Design and Research* 25:377–392.

Watson, A.R. 1985d. Drilling model for cutting lip and chisel edge and comparison of experimental and predicted results. IV—Drilling tests to determine chisel edge contribution to torque and thrust. *International Journal of Machine Tool Design and Research* 25:393–404.

Wiriyacosol, S. and Armarego, E.J.A. 1979. Thrust and torque prediction in drilling from a cutting mechanics approach. *Annals of CIRP* 28:88–91.

Zorev, N.N. (ed.) 1966. *Metal Cutting Mechanics*. Oxford, U.K.: Pergamon Press.

Appendix B: Tool Material Fundamentals

B.1 HIGH-SPEED STEELS

B.1.1 Basics of Metallurgical Notions/Terminology Needed to Understand HSS

To understand HSS, some basic notions and terminology should be introduced.

B.1.1.1 Phase Diagram

This section aims to familiarize readers with the basics of metallurgical notions/terminology used in the whole HSS business. It should help in understanding the text in Chapter 3 and other HSS-related literature and promotional materials. Note that many *fancy* metallurgical terms are used in the latter counting on unfamiliarity of many users with the subject. Thus, this section is written to help many cutting tool specialists to overcome such unfamiliarity equipping them with sufficient knowledge to understand metallurgical particularities of HSS. As such, the information provided is kept to bare-bone minimum based on the author's experience.

The basic metallurgical terms are associated with the so-called phase diagrams. Phase diagrams are maps of the equilibrium phases associated with various combinations of temperature and composition. Binary phase diagrams are two component maps widely used by engineers. They are helpful in predicting phase transformations and the resulting microstructures.

A binary phase diagram shows the phases formed in differing mixtures of two elements over a range of temperatures as shown in Figure B.1. Compositions run from 100% Element A on the left of the diagram, through all possible mixtures, to 100% Element B on the right. The composition of an alloy is given in the form A—x% B. For example, Cu—20% Al is 80% copper and 20% aluminum. Weight percentages are often used to specify the proportions of the alloying elements, but atomic percent may be used. The type of percentage is specified, for example, Cu—20 wt% Al for weight percentages and Cu—20 at% Al for atomic percentages.

Alloys tend to solidify over a temperature range, rather than at a specific temperature like pure elements. By cooling alloys from the liquid state and recording their cooling rates, the temperature at which they start to solidify can be determined and then plotted on the phase diagram. If enough experiments are performed over a range of compositions, a start of solidification curve can be plotted onto the phase diagram. This curve (curve 123 in Figure B.1) will join the three single solidification points and is called the *liquidus* line. The liquidus line separates the all melt phase from the liquid (melt) + crystal phase.

In the same way that sugar dissolves into hot tea (a liquid solution), it is possible for one element to dissolve in another while both remain in the solid state. This is called solid solubility and is characteristically up to a few percent by weight. This solubility limit will normally change with temperature. The extent of the solid solubility region can be plotted onto the phase diagram (curves 145 and 367 in Figure B.1). A solid solution of B in A (i.e., mostly A) is called alpha (α) and a solid solution of A in B (i.e., mostly B) is called beta (β).

Line 46 is referred to as the *solidus* line. It separates the liquid (melt) + crystal phase from the all crystal phase. At alloy compositions and temperatures between the start of solidification and the point at which it becomes fully solid (the eutectic temperature), a mushy mix of either alpha or beta will exist as solid lumps with a liquid mixture of A and B. These partially solid regions are marked

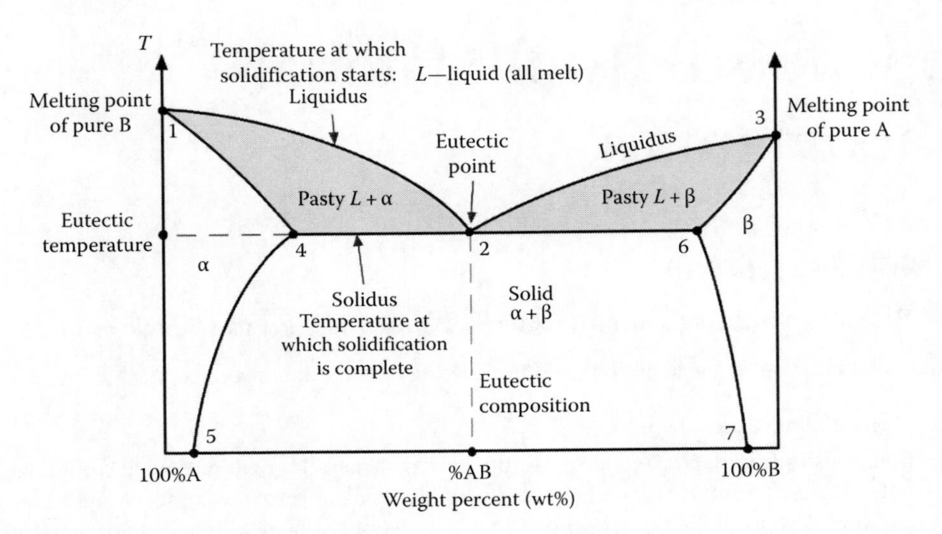

FIGURE B.1 Representation of the phase diagram.

on the phase diagram. The region below the solidus (sometimes referred to as eutectic) line, and outside the solid-solution region, will be a solid mixture of alpha and beta and is labeled to reflect this.

A eutectic system is a mixture of chemical compounds or elements that has a single chemical composition that solidifies at a lower temperature than any other composition. This composition is known as the *eutectic composition* and the temperature is known as the *eutectic temperature*. On a phase diagram, the intersection of the eutectic temperature and the eutectic composition gives the *eutectic point*. Eutectic alloys have two or more materials and have a eutectic composition. When a noneutectic alloy solidifies, its components solidify at different temperatures, exhibiting a plastic melting range. A eutectic alloy solidifies at a single, sharp temperature. The phase transformations that occur while solidifying a given alloy can be understood by drawing a vertical line from the liquid phase to the solid phase on a phase diagram.

B.1.1.2 Structures

Cementite, also known as iron carbide, is a chemical compound of iron and carbon, with the formula Fe_3C (or Fe_2C–Fe). By weight, it is 6.67% carbon and 93.3% iron. It is extremely hard and its hardness increases generally with the proportion of carbon content. The hardness and brittleness of cast iron is believed to be due to the presence of the cementite. It is magnetic below 250°C. Its presence in iron and steel increases hardness but decreases tensile strength. Its composition by weight is 14 parts of iron and 1 part of carbon or 93.45% iron and 6.55% carbon. It occurs in steel that has cooled slowly from a high temperature as a constituent of pearlite. The carbon is almost completely combined with a definite amount of iron to form carbide of iron.

Microstructure is defined as the structure of a prepared surface or thin foil of material as revealed by a microscope above 25× magnification. The microstructure of a material directly and strongly correlates with physical properties such as strength, toughness, ductility, hardness, corrosion resistance, high-/low-temperature behavior, and wear resistance, which in turn govern the application of this material in industrial practice.

Unit cell. The crystal structure of a material or the arrangement of atoms within a given type of crystal structure can be described in terms of its unit cell. The unit cell is a small box containing one or more atoms, a spatial arrangement of atoms. The unit cells stacked in 3D space describe the bulk arrangement of atoms of the crystal. The crystal structure has a 3D shape. The unit cell is given by its lattice parameters, which are the length of the cell edges and the angles between them, while the positions of the atoms inside the unit cell are described by the set of atomic positions measured from a lattice point.

Austenite, also known as gamma phase iron, is a metallic nonmagnetic allotrope of iron or a solid solution of iron, with an alloying element. Austenite exists above the critical eutectoid temperature. It is named after Sir William Chandler Roberts-Austen (1843–1902). Austenite is the solid solution of carbon or iron carbide (Fe_3C) in gamma iron. When carbon steel is heated, particularly no change in the constituents occurs during the heating until a temperature corresponding to the lower critical Ac_1 is reached; this is at about 724°C–727°C. Here, there is a complete change in the nature and structure of the pearlite, and it is known under the general term solid solution usually called *austenite*.

As the temperature is raised from the lower critical Ac_1 to the upper critical Ac_2, which ends at about 852°C–854°C, the last remaining excess of ferrite or cementite depending on whether it is a low- or high-carbon steel will be absorbed by the austenite, so that above the upper critical range Ac_2, the steel is composed entirely of solid solution, the austenite. Austenite is hard and nonmagnetic. Figure B.2 shows the structure of austenite unit cell.

Austenitization means to heat HSS to a temperature at which it changes crystal structure from ferrite to austenite. An incomplete initial austenitization can leave undissolved carbides in the matrix.

Martensite, named after the German metallurgist Adolf Martens (1850–1914), most commonly refers to a very hard form of steel crystalline structure. Martensite is formed by rapid cooling (*quenching*) of austenite that traps carbon atoms that do not have time to diffuse out of the crystal structure. It is the chief constituent of hardened steels. It has fibrous or needle like structure. It is very hard and consists of iron with carbon in varying proportion up to about 2%. When it contains iron and 0.9% carbon, it is termed as hardened site, which corresponds in composition to that of pearlite or martensite with carbon. Martensite is not as tough as austenite.

Martensite, generally regarded as a solid solution of carbon or carbide in alpha ferrite, is the chief and characteristic constituent of hardened steel when cooled rapidly from temperature above the critical range of the transition constituent's austenite to sorbite. Martensite is the hardest and also the most brittle, with little ductility.

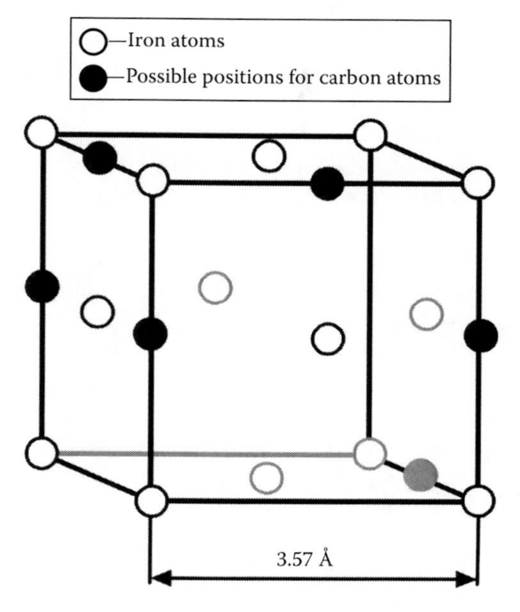

FIGURE B.2 Unit cell in an austenite crystal: FCC.

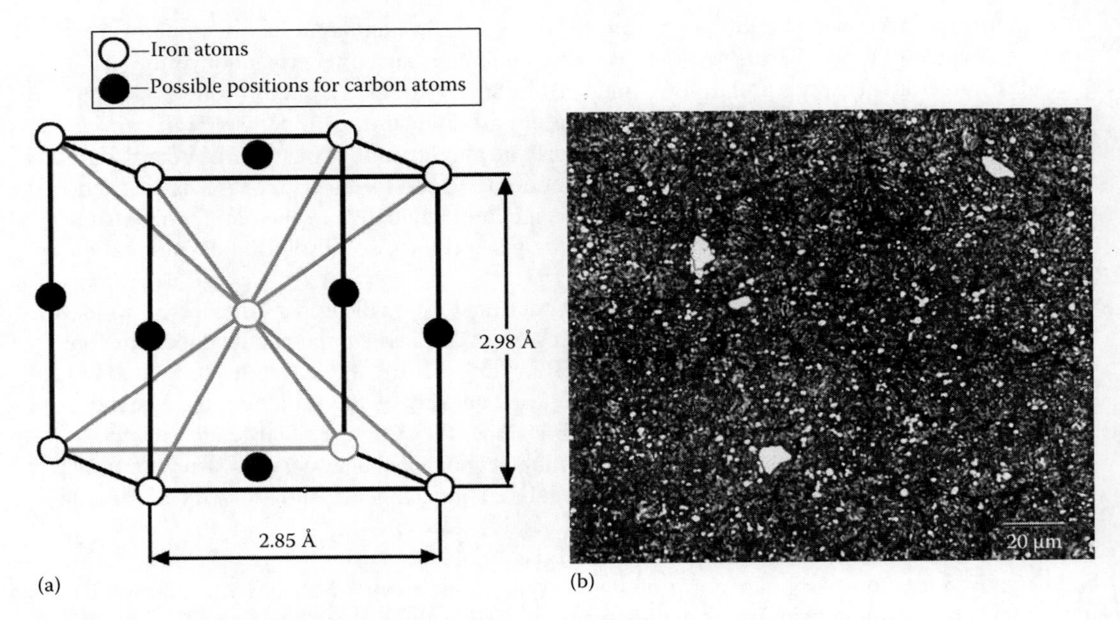

FIGURE B.3 Unit cell in a martensite crystal (a) and microstructure of martensite (b).

Martensitic reaction begins during cooling when the austenite reaches the martensite start temperature (M_s) and the parent austenite becomes mechanically unstable. At a constant temperature below M_s, a fraction of the parent austenite transforms rapidly, and then no further transformation will occur. When the temperature is decreased, more of the austenite transforms to martensite. Finally, when the martensite finish temperature (M_f) is reached, the transformation is complete. Figure B.3 shows the structure of martensite unit cell as well as its typical microstructure (HSS M2).

Ferrite is a body-centered cubic (BCC) form of iron, in which a very small amount (a maximum of 0.02% at 1333°F/723°C) of carbon is dissolved. This is far less carbon than can be dissolved in either austenite or martensite, because the BCC structure has much less interstitial space than the FCC structure. Ferrite is the component that gives steel and cast iron their magnetic properties and is the classic example of a ferromagnetic material. This is also the reason that tool steel becomes nonmagnetic above the hardening temperature—all of the ferrite has been converted to austenite. Most *mild* steels (plain carbon steels with up to about 0.2 wt% C) consist mostly of ferrite, with increasing amounts of cementite as the carbon content is increased, which, together with ferrite, form the mechanical mixture pearlite. Any iron–carbon alloy will contain some amount of ferrite if it is allowed to reach equilibrium at room temperature. Figure B.4 shows the structure of ferrite unit cell as well as its typical microstructure.

Pearlite. It is a mixture of about 87.5% ferrite and 12.5% cementite. The name pearlite is derived from the fact that it shows oblique lighting, under the microscope the rainbow color of the mother of pearl when the etching process has removed a part of surrounding softer ferrite. Soft steels contain ferrite and pearlite. The hardness increases with the proportion of cementite.

Perlite forms during the process of cooling at a slow rate when a red heat and cementite forms a mechanical mixture with ferrite and appears under high magnification as alternate layers of cementite and ferrite. Slow cooling produces coarse pearlite and quicker cooling finer pearlite. The carbon content of pearlite in the plain carbon steel is 0.85%.

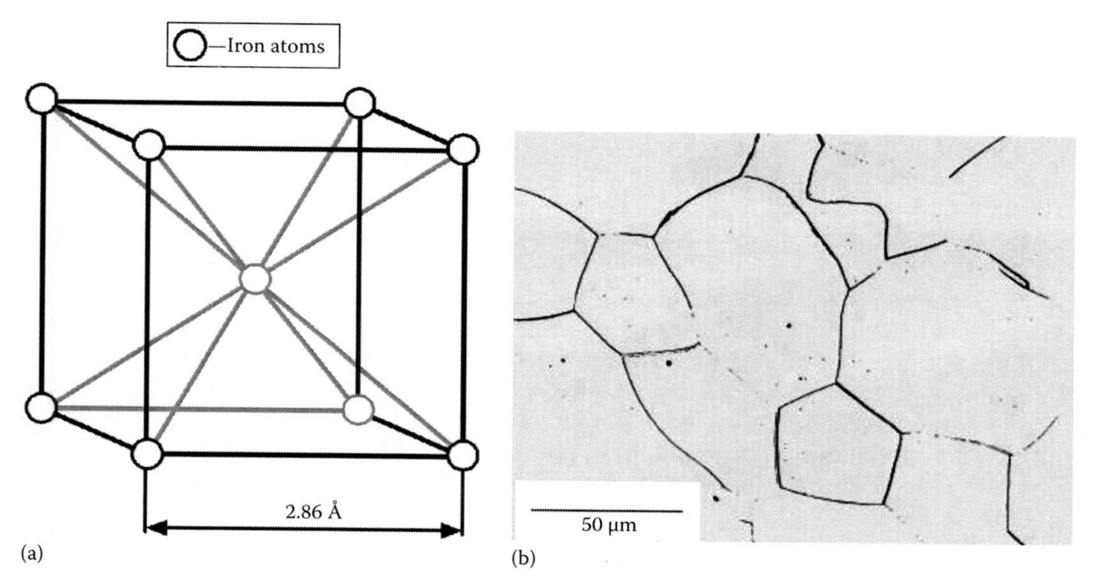

FIGURE B.4 Unit cell in a ferrite crystal BCC (a) and microstructure of ferrite (b).

Bainite is an acicular microstructure (not a phase) that forms in steels at temperatures from approx. 250°C to 550°C (depending on the alloy content). First described by E. S. Davenport and Edgar Bain, it is one of the decomposition products that may form when austenite (the FCC crystal structure of iron) is cooled past a critical temperature of 727°C (about 1340°F). Davenport and Bain originally described the microstructure as being similar in appearance to tempered martensite. A fine nonlamellar structure, bainite, commonly consists of cementite and dislocation-rich ferrite. The high concentration of dislocations in the ferrite present in bainite makes this ferrite harder than it normally would be.

The temperature range for transformation to bainite (250°C–550°C) is between those for pearlite and martensite. When formed during continuous cooling, the cooling rate to form bainite is more rapid than that required to form pearlite, but less rapid than is required to form martensite (in steels of the same composition). Most alloying elements will lower the temperature required for the maximum rate of formation of bainite, though carbon is the most effective in doing so. Bainite is an intermediate of pearlite and martensite in terms of hardness.

The microstructures of martensite and bainite at first seem quite similar; this is a consequence of the two microstructures sharing many aspects of their transformation mechanisms. However, morphological differences do exist that require a TEM to see. Under a simple light microscope, the microstructure of bainite appears darker than martensite due to its low reflectivity.

B.1.1.3 Important Temperatures on Phase Diagrams

Some important boundaries at single-phase fields have been given special names in heat treatment. These include the following:

A_1—the eutectoid temperature, which is the minimum temperature for austenite

A_3—the lower-temperature boundary of the austenite region at low-carbon contents, that is, the gamma/gamma + ferrite boundary

A_{cm}—the counterpart boundary for high-carbon contents, that is, the gamma/gamma + Fe_3C boundary

Sometimes the letters c, e, and r are included in the literature on heat treatment:

Ac_{cm}—in hypereutectoid steel, the temperature at which the solution of cementite in austenite is completed during heating
Ac_1—the temperature at which austenite begins to form during heating, with c being derived from the French *chauffant*
Ac_3—the temperature at which transformation of ferrite to austenite is completed during heating
Ae_{cm}, Ae_1, Ae_3—the temperatures of phase changes at equilibrium
Ar_{cm}—in hypereutectoid steel, the temperature at which precipitation of cementite starts during cooling, with r being derived from the French *refroidissant*
Ar_1—the temperature at which transformation of austenite to ferrite or to ferrite plus cementite is completed during cooling
Ar_3—the temperature at which austenite begins to transform to ferrite during cooling
Ar_4—the temperature at which delta-ferrite transforms to austenite during cooling.

B.1.1.4 Time–Temperature Transformation, aka the Continuous Cooling Transformation Diagram

A time–temperature transformation (TTT) diagram is a plot of temperature versus the logarithm of time for a steel alloy of definite composition. Example of TTT diagram is shown in Figure B.5. It is used to determine when transformations begin and end for an isothermal (constant temperature) heat treatment of a previously austenitized alloy. When austenite is cooled slowly to a temperature below lower critical temperature (LCT), the structure that is formed is pearlite. As the cooling rate increases, the pearlite transformation temperature gets lower. The microstructure of the material is significantly altered as the cooling rate increases. TTT diagram indicates when a specific transformation starts and ends, and it also shows what percentage of transformation of austenite at a particular temperature is achieved.

Cooling rates in the order of increasing severity are achieved by quenching from elevated temperatures as follows: furnace cooling, air cooling, oil quenching, liquid salts, water quenching, and

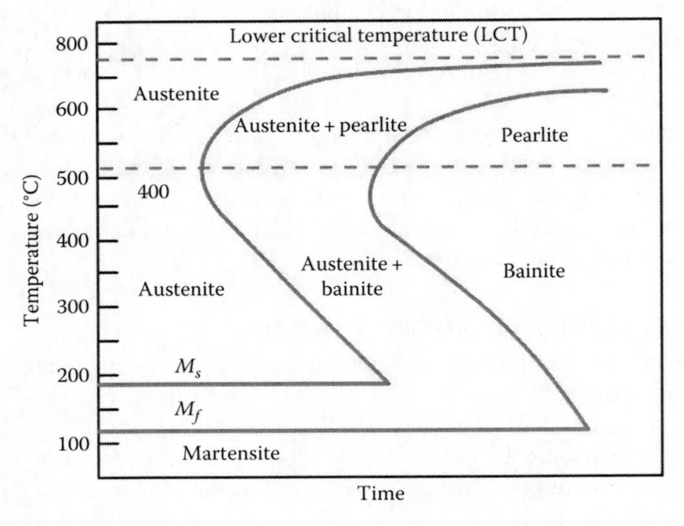

FIGURE B.5 Example of TTT diagram.

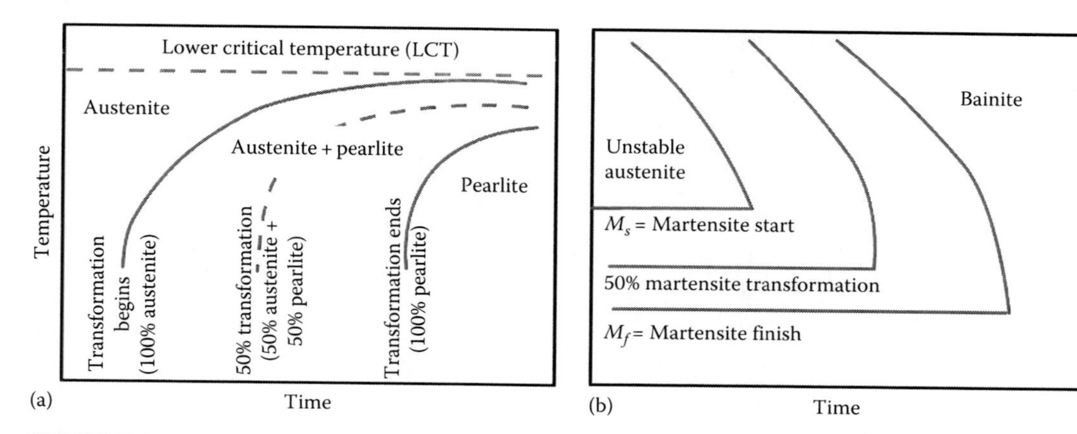

FIGURE B.6 Interpretations of the upper (a) and lower (b) halves of TTT diagram.

brine quenching. If these cooling curves are superimposed on the TTT diagram, the end product structure and the time required to complete the transformation can be found.

In Figure B.5, the area on the left of the transformation curve represents the austenite region. Austenite is stable at temperatures above LCT but unstable below LCT. Left curve indicates the start of a transformation and right curve represents the finish of a transformation. The area between the two curves indicates the transformation of austenite to pearlite, austenite to martensite, or austenite to bainite. In this figure, line M_s represents the highest temperature at which transformation of austenite to martensite starts during rapid cooling and line M_f represents the temperature at which martensite formation finishes during rapid cooling.

Figure B.6a represents the upper half of the TTT diagram. As can be seen, when austenite is cooled to temperatures below LCT, it transforms to other structures due to its unstable nature. A specific cooling rate may be chosen so that the transformation of austenite can be 50%, 100%, etc. If the cooling rate is very slow such as annealing process, the cooling curve passes through the entire transformation area and the end product of this cooling process becomes 100% pearlite. In other words, when slow cooling is applied, all the austenite will transform to pearlite. If the cooling curve passes through the middle of the transformation area, the end product is 50% austenite and 50% pearlite, which means that at certain cooling rates, a part of the austenite can be retained without transforming it into pearlite.

Figure B.6b represents the lower half of the TTT diagram. It shows the types of transformation that takes place at higher cooling rates. If a cooling rate is very high, the cooling curve remains on the left-hand side of the transformation start curve. In this case, all austenite will transform to martensite. If there is no interruption in cooling, the end product will be martensite.

Figure B.7 shows the cooling rates A and B. The end product of both cooling rates is martensite. Cooling rate B is also known as the critical cooling rate, which is represented by a cooling curve which is tangent to the nose of the TTT diagram. Critical cooling rate is defined as the lowest cooling rate that produces 100% martensite while minimizing the internal stresses and distortions. If, however, cooling rate represented by curve A is used, then a higher distortion and higher internal stresses would be the case.

Figure B.8 shows what happens when a rapid quenching process is interrupted (horizontal line of cooling curve C represents the interruption) by immersing the material in a molten salt bath and soaking at a constant temperature followed by another cooling process that passes through bainite region of TTT diagram. The end product is bainite, which is not as hard as martensite.

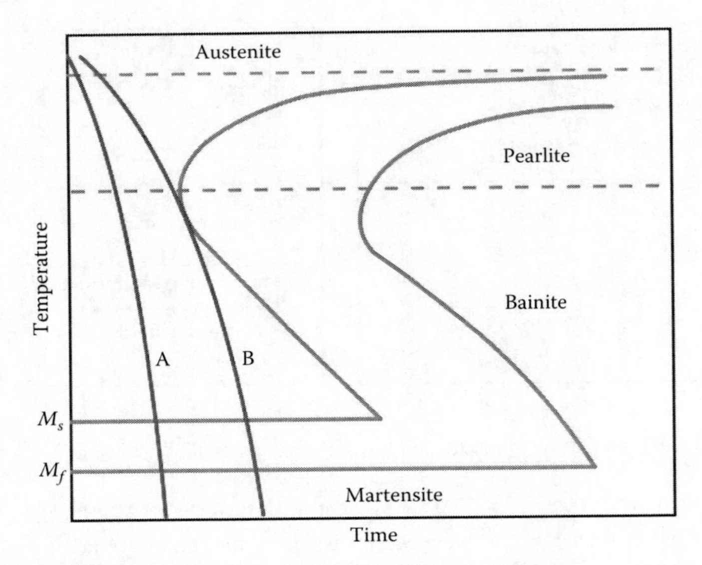

FIGURE B.7 Graphical representation of the critical cooling rate (curve B).

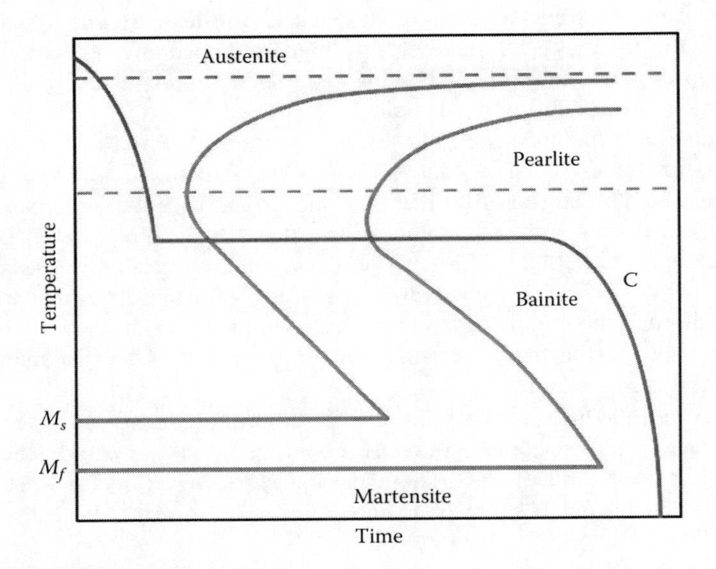

FIGURE B.8 Interrupted quenching.

In Figure B.9, cooling curve D represents a slow cooling process, such as furnace cooling. An example for this type of cooling is annealing process where all the austenite is allowed to transform to pearlite as a result of slow cooling.

Figure B.10 shows a case where the cooling curve E passes through the middle of the austenite–pearlite transformation zone. Such a cooling rate is not high enough to produce 100% martensite so that a structure containing martensite + pearlite is formed. Since the cooling curve E is shown to be tangent to the nose of 50% pearlite, austenite is transformed to 50% pearlite. Since curve E leaves the transformation diagram at the martensite zone, the remaining 50% of the austenite will be transformed to martensite.

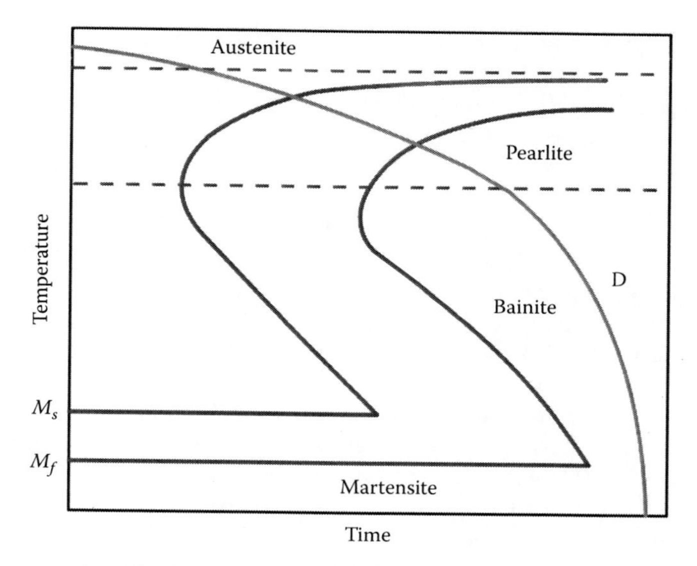

FIGURE B.9 Representation of slow cooling rate (annealing).

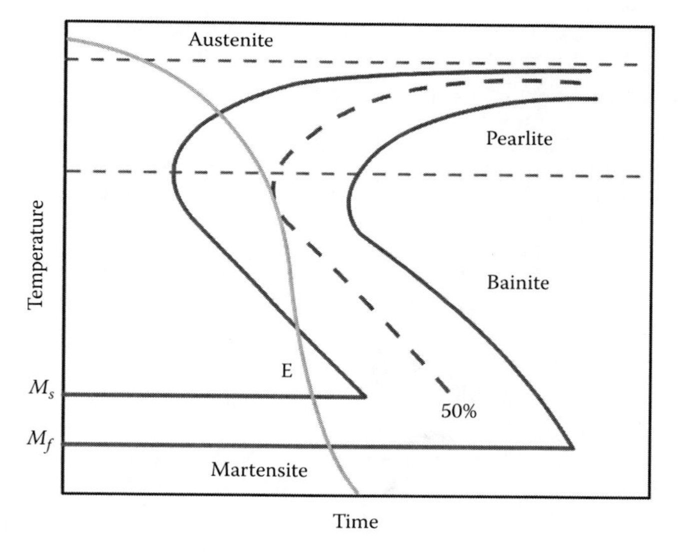

FIGURE B.10 Cooling rate represented by curve *E* permits both pearlite and martensite formation.

B.2 CEMENTED CARBIDES

The manufacturing process of cemented carbides seems to be well known. They are manufactured by a powder metallurgy process consisting of a sequence of steps in which each step must be carefully controlled to obtain a final product with the desired properties, microstructure, and performance. These steps include (Davis 1995) the following:

1. Processing of the ore and the preparation of tungsten carbide powder
2. Preparation of other carbide powders
3. Production (milling) of the grade powders
4. Compacting or powder consolidation
5. Sintering
6. Post-sinter forming

Reading even detailed description of the essence of these steps (Davis 1995), one may wonder why the quality and cost of cemented carbides seemingly of the same composition vary significantly from one manufacturer to the next. These concerns should be addressed properly as the answers are in the carbide manufacturing technology that varies significantly from one manufacturer to the next.

B.2.1 Tungsten Metal Powder

The objective of this stage is to produce tungsten metal powder. The flowchart of this stage is shown in Figure B.11 and the processes involved in this stage are shown in Figure B.12.

Industrial raw materials to process tungsten-bearing minerals are wolframite ([Mn, Fe]WO4), scheelite (CaWO4), ferberite (FeWO4), and huebnerite (MnWO4). The lower limit of workable tungsten concentration ranges from 0.1 to 0.3 wt% WO_3. Concentrations of more than 2 wt% are infrequent. The average WO_3 content is usually about 0.5 wt%. Wolfram Bergbau and Hutten GmbH (WBH) in Austria, a world leading manufacturer of tungsten oxide, tungsten metal, and tungsten carbide, mines about 400,000 tons of scheelite ore with average WO_3 content 0.55 wt%. Depending on the mineral type, different techniques, such as gravity methods and flotation, are used to concentrate the ore up to reduced WO_3 contents of about 40–60 wt%. Tungsten-bearing scrap, which usually has a high tungsten content, is a considerable alternative tungsten source to the natural mineral resources (Neikov et al. 2009).

The main process steps from concentrates or scrap through the formation of high-purity ammonium paratungstate (APT: $3(NH_4)_4O \cdot 7WO_3 \cdot 6H_2O$ and $5(NH_4)_2O \cdot 12WO_3 \cdot 11H_2O$) intermediate to tungsten metal powder are shown schematically in Figure B.11.

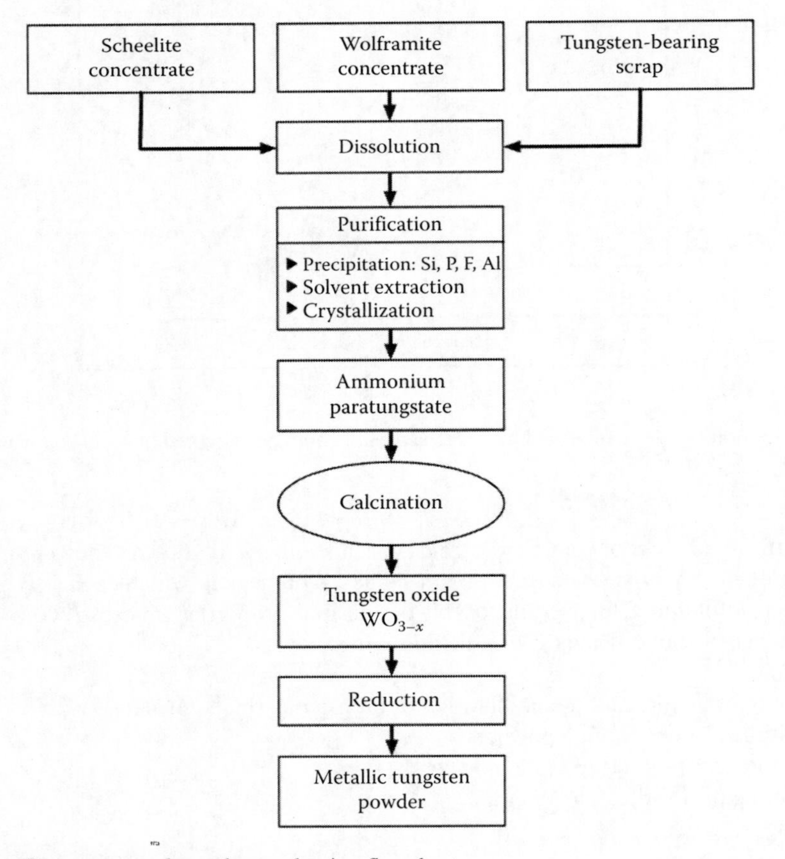

FIGURE B.11 Tungsten metal powder production flowchart.

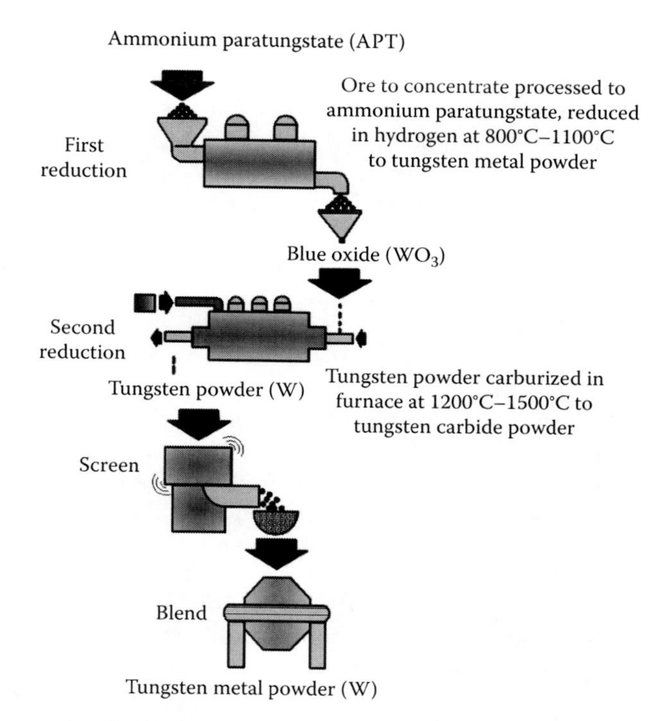

FIGURE B.12 Processes involved in manufacturing of tungsten metal powder.

In the dissolution stage, the tungsten-bearing feed materials (wolframite concentrate and oxidized tungsten scrap) are treated with NaOH or scheelite concentrate is treated with Na_2CO_3 to form water-soluble sodium tungstate Na_2WO_4.

The purification stage includes the following:

- Removal of the aluminum, arsenic, fluorine, phosphorus, and silicon by precipitation.
- Substitution of sodium by NH_4 forms an ammonium solution $(NH_4)_2WO_4$. This is ordinarily done by solvent extraction, which is a liquid ion exchange process, where tungsten is extracted by means of an aliphatic amine dissolved in isodecanol and paraffin. This process is followed by re-extraction into aqueous ammonia solution.
- Crystallization of APT by evaporation of ammonia and water. APT is the intermediate product of the highest purity for metallic tungsten and tungsten carbide powder production.

The calcination stage includes heating the APT to 400°C–900°C, predominantly in rotary furnaces, for ammonia and water elimination. If accomplished in excess of air, the resulting oxide is WO_3 yellow tungsten oxide. In its absence, a slightly lower WO_{3-x} blue tungsten oxide is formed. WO_3 is the most stable a-oxide with homogeneity range WO_3–$WO_{2.96}$; it has three crystalline modifications: monoclinic modification stable at temperatures lower than 727°C, tetragonal modification at temperatures that range from 727°C to 1097°C, and cubic modification at temperatures higher than 1097°C. It sublimes noticeably at 797°C–847°C and exists in the form of polymer molecules in the vapor phase.

Reduction stage with hydrogen reducer: The reduction of the oxide to metal proceeds through several intermediate oxide phases. Generally, two furnace designs for the reduction process are used industrially: the multitube pusher-type and the rotary furnaces. During the reduction process, the sequence in which oxides occur depends on the applied temperature and on the moisture within the powder bed, accordingly, static in pusher-type popular furnaces and dynamic in rotary furnaces.

Although one-step conversion furnaces (more complicated, and thus more expensive) are sometimes used, two-step conversion processes from APT to tungsten are more common (Figure B.12).

The first step is calcination; the second is reduction of the respective oxide to tungsten using hydrogen. Each of these two steps has its own particularities and their regimes vary from one powder manufacturer to the next. The important reduction parameters are temperature, time, hydrogen flow rate, height of oxide powder mass, and apparent density of the oxide. Production of tungsten powder of desired size (0.5–5 μm) is carried out by varying these parameters (Neikov et al. 2009). However, not only grain size but other powder characteristics, for example, grain size distribution, agglomeration, apparent density, and compacting behavior, are also affected by these parameters. It is not always possible to vary these parameters in a controlled manner.

There are a number of other tungsten powder production methods. Among them, the following are frequently used: reduction by solid carbon, precipitation from gaseous phase of tungsten hexafluoride and tungsten hexachloride, amalgam method, and production of powder by a plasma process.

The resultant product, that is, tungsten powder, is characterized by grain size, morphology, and purity. Tungsten has two outstanding properties: the highest melting point of all metals at 3410°C and a very high density of 19.25 g/cm³ at 20°C. All the powder sizes, from ultrafine to hundreds of microns, are produced on an industrial scale, but the bulk of the powder used at present is in the size range of 2–5 μm. Particle sizes ranging from 1 to 10 μm are measured by Fisher subsieve sizer (FSSS) (ASTM B330 and MPIF 32 standards) and correspond to particle sizes observed directly by optical microscopy (OM) (MPIF E 20 and GOST 23402-78 standards) and by SEM. SEM and OM are also the best way to estimate the overall particle morphologies. Particle size distributions can be measured by light scattering techniques (ASTM B 822 standard) or by sedimentation based on x-ray monitoring of gravity sedimentation scattering (SediGraph) (ASTM B 761 standard). Controlling the reduction parameters makes possible the production of both narrow and wide size distributions and provides any specific requirement on tungsten powders for various applications.

Apparent density (ISO 3923 and ASTM B 212 standards), tap density (ISO 3953 and ASTM B 527 standards), and green density and green strength have great importance for determining the behavior of a powder during subsequent consolidation steps, including pressing and sintering.

Although a feature of tungsten powder is a relatively very high purity in comparison with most other metal powders used for powder metallurgical products, trace elements, including molybdenum, aluminum, arsenic, silicon, phosphorus, sodium, potassium, calcium, magnesium, and sulfur, which are removed during APT manufacture, can be found. The contaminants, such as nickel, chromium, iron, and cobalt, characteristic alloying elements of reduction boats or furnace tubes, have to be avoided. The presence of traces of foreign elements has an important effect on the reduction kinetics and, as a consequence, the characteristic of the resulting tungsten powder. Although tungsten powder is only an intermediate in cemented carbide production, its properties influence the quality of the corresponding tungsten carbide powders, thus influencing the sintering behavior and properties of cemented carbides. The origin of such impurities can be in the normal impurity level of the raw materials or in power manufacturing process itself. Although the behavior of different trace impurities at different stages of cemented carbides production has been studied (Uhrenius et al. 1991; Chenguang 1992), only general recommendations on the allowable amount of impurities is available (Upadheaya 1998).

Alternatively, tungsten metal powder can be purchased from a great variety of its manufacturers located mainly in China and India. Domestic manufacturers of tungsten powder provide more authentic information on the quality of their product and SEM images of their powders. For example, Global Tungsten & Powders Corp. (Towanda, PA) supplies 14 types of tungsten powder providing the following information for each type: average particle size—FSSS (ASTM B 330), bulk density (ASTM B 329), standard screen mesh sizes, and LOR (ASTM E 159)—and loss on reduction (H_2O+O). Besides generic SEM images of its products, the company offers an SEM image for a particular supply lot for an analysis fee.

B.2.2 Tungsten Carbide Powder

B.2.2.1 Processes

Conventional manufacturing of tungsten carbide powders accounts worldwide for the bulk of the tungsten carbide powder produced. This technique permits the production of tungsten carbide powders with particle sizes ranging from 0.15 to 50 μm. The major processes involved in such manufacturing are shown in Figure B.13.

Although it is possible to produce tungsten carbide directly from ore, oxide, or APT, or by gaseous carburization, the preferred method is to carburize tungsten metal powder by adding controlled amounts of carbon black. This addition helps control particle size and size distribution, which, together with binder content, determine the resultant properties of the sintered carbide (Davis 1995).

First, *black mix* tungsten powder of the desired particle size and size distribution and high-quality (low ash and sulfur content) carbon black is prepared. Because the two powders differ significantly in density, great care must be taken to ensure uniform distribution of carbon. Mixing is performed in ball mills, attrition mills, or specialized blenders. Ball milling times of 24 h or more may be required for adequate mixing. Attritions mills or specially designed blenders are faster, typically 2–6 h (Neikov et al. 2009). Figure B.14 shows a micrographic appearance of tungsten powder before and after milling process.

The aim of the carburization process is to produce stoichiometric tungsten carbide with 6.13 wt% C or a small excess (0.01–0.02 wt%) of free carbon (Neikov et al. 2009). As discussed in Chapter 3, carbon deficiency results in the formation of brittle eta phase in the final product. The exact amount of the carbon black addition in the black mix is determined by practice as it depends on carburization conditions and equipment used. Finer powders, containing more absorbed oxygen or water vapor, require more carbon black than coarser powders. The senses of the fine and coarse powder are shown in Figure B.15. Gas flow and size (depth) of the black mix charge in the carburization furnace can also influence carbon composition.

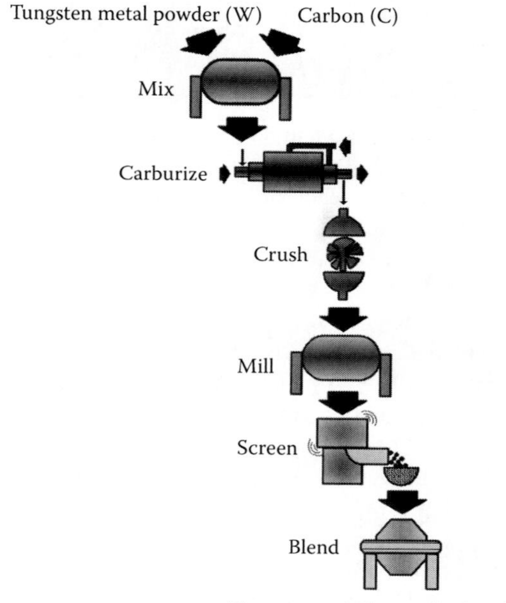

FIGURE B.13 Processes involved in manufacturing of tungsten carbide powder.

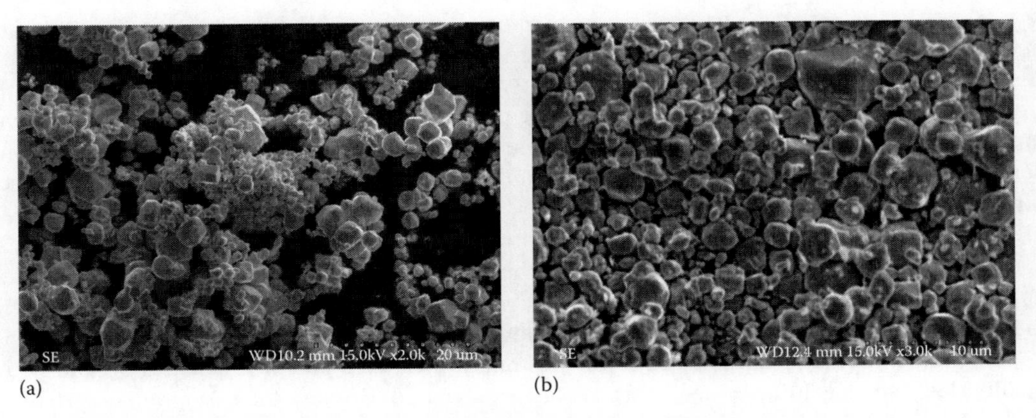

FIGURE B.14 Tungsten powder (a) before milling process and (b) after milling process.

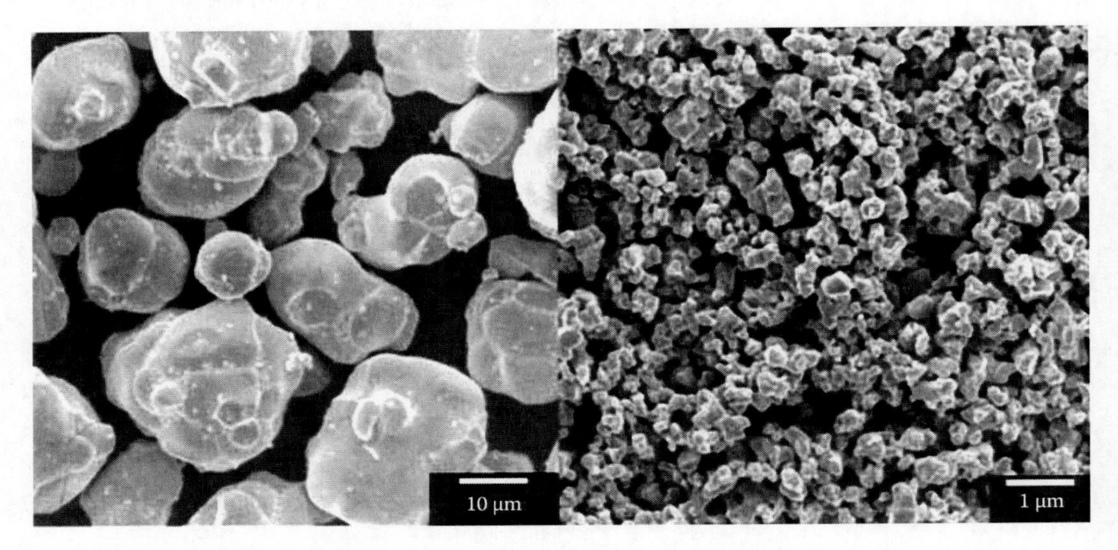

FIGURE B.15 SEM micrographs of coarse and ultrafine tungsten carbide powders.

Carburization is performed in the presence of hydrogen at temperatures ranging from 1400°C to 2650°C. The hydrogen atmosphere reacts with carbon black to form gaseous hydrocarbons (primary methane, CH_4), which then react to form tungsten carbide as $W + CH_4 \rightarrow WC + 2H_2$. The minimum process temperature is 1400°C, while higher temperatures are used for coarser powders. A common arrangement for production of large quantities of tungsten carbide is to charge black mix into covered graphite boats, which are then conveyed as a controlled stocking rate through a pusher-type tube furnace. Smaller batches of tungsten carbide, however, can be produced in an induction-heated graphite crucible where graphite containers are used to maintain the proper carbon content in the powder mass.

Upon exiting the carburizing furnace, the powders are caked or agglomerated. Therefore, carbides normally are subjected to a milling and sieving operation to produce deagglomerated powders with proper size distribution. Typical particle size ranges from 0.5 to 30 µm. Such powders are then ready to be mixed with binder powders for the manufacturing of cemented carbide parts.

Small amounts of vanadium, tantalum, or chromium carbides are frequently added to inhibit grain growth. These additives may be in the form of oxides or metals and are added to tungsten/carbon black mix with the appropriate amount of carbon black. They may be also blended with tungsten powders as pure carbides. Typical additions range from 0.5% to 2%.

There are some other processes, the so-called alternative processes, used to produce carbide powders. One of them is called the conventional calcination–reduction–carburization (CRC) process and offers the potential to manufacture commercial WC ultrafine powders with median grain sizes below 0.5 μm. Both tight process control and process automation are keys to achieve greater uniformity and batch to batch reliability.

Recently, the demands for tungsten carbide powders finer than 1 μm are increasing as material for precision tools such as drills and end mills for metal cutting, printed circuit boards, cutters, and precision pins. Together with the limitation of the conventional technique concerning the efficiency of fine-grained metal powder production, there has been a strong motivation to develop alternative techniques.

Tokyo Tungsten Co. Ltd has developed the direct carburizing process (Neikov et al. 2009), where tungsten trioxide and carbon black react in the following steps: $WO_3 \rightarrow (WO_{2.9}) \rightarrow (WO_{2.72}) \rightarrow WO_2 \rightarrow W \rightarrow W_2C \rightarrow WC$. At the $(WO_{2.72}) \rightarrow WO_2$ step, ultrafine particles are generated and then transformed into tungsten carbide without grain growth. The starting material is prepared in the form of pellets from a mixture of tungsten oxide with carbon black. Subsequent reaction is accomplished in rotary furnaces in two steps: first, the direct reduction of WO_3 to tungsten by carbon under nitrogen, followed by carburization under hydrogen at temperatures up to 1600°C to tungsten carbide powder. This method permits the production of ultrafine powder with particle sizes ranging from 0.1 to 0.7 μm.

Another direct carburization process has been developed by Dow Chemical (Neikov et al. 2009). A mixture of tungsten trioxide and carbon black falls through a vertical reactor at temperatures of approx. 2000°C. Accordingly, the reaction time is a few seconds and results in the formation of an intermediate mixture of W, W_2C, WC, and C. This product has to be subjected to a second more conventional, carburization step. Particle sizes were reduced by controlling the milling time, carbon content, inhibitor addition, and sintering temperature. The characteristic particle size range for this process is between 0.1 and 0.8 μm.

The spray conversion process has been developed for Nanodyne Inc. (Neikov et al. 2009) to produce nanopowders. Aqueous solutions of tungsten and cobalt salts are spray dried, subsequently reduced, and carburized with a gas mixture such as hydrogen and ammonia or carbon dioxide and carbon monoxide in a fluidized bed reactor. The product obtained is relatively large, about 75 μm, hollow WC–Co composite powder particles, which consist of tungsten carbide crystallites with a particle size of about 30 nm. This product is milled in order to obtain a powder suitable for consolidation.

Another technique to produce tungsten carbide nanopowders is a chemical vapor reaction process. This process is based on the gas phase reaction of metal halides with different gas mixtures to produce nanosized powder ranging from 5 to 50 nm.

B.2.2.2 Preparation of Mixed Carbide Crystals

Tungsten–titanium–tantalum (niobium) carbides are used to make steel grades of cemented carbides as such grades have better resistance to cratering and oxidation than the straight grades. These carbides are produced from metal oxides of titanium, tantalum, and niobium. These oxides are mixed with metallic tungsten powder and carbon similar to the preparation of tungsten carbide powder. The mixture is heated under a hydrogen atmosphere or vacuum to reduce the oxides and form solid-solution carbides such as WC–TiC, WC–TiC–TaC, or WC–TiC–(Ta,Nb)C. The Menstruum process is also effective for production of mixed carbide, producing very homogeneous carbide crystals (Lassner and Schubert 1999).

B.2.2.3 Important Properties

The most important chemical side of tungsten carbide powder is its carbon content. Stoichiometric pure tungsten carbide has a total carbon content (C_t) of 6.135 wt%. From the point of view of thermodynamic equilibrium, WC has a very narrow stability range (Lassner and Schubert 1999). Thus, not even a small carbon shortage leads to W_2C subcarbide formation and excess of carbon to the formation of free carbon (C_f). In commercial WC powders, there is always a small portion of free carbon present, about 0.03% at a stoichiometric C_t level. Even when the total carbon is below the stoichiometric level, the coexistence of traces of W_2C and C_f can occur. The carbon balance is important for hard metal manufacturing, because a shortage of carbon leads to the formation of the brittle phases, for instance, Co_2W_3C, and any excess in carbon leads to graphite precipitation. Thus, both cases result in a detrimental effect on the mechanical properties of the final product.

The major important property of tungsten carbide powders, considered as the most relevant to cemented carbide productions, is the average grain size (AGS) sometimes referred to as the average particle size (APS). It is obtained by FSSS measurements and commonly given both as *as supplied* and *lab-milled*, according to ASTM standard procedures (ASTM B330 and 430). Unfortunately, the term *particle size* is not well defined and also depends on the corresponding measurement technique. ASTM B390-92(2000) Standard Practice for Evaluating Apparent Grain Size and Distribution of Cemented Tungsten Carbides was withdrawn in 2010. Nevertheless, there exists an ISO standard (ISO 3252), which defines the difference between particle, grain, and agglomerate. The differences in these terms are clearly defined in Figure B.16. From this figure, it becomes obvious that the often used term *grain size* for WC powders is misleading and should be avoided when talking about measurement values (FSSS, laser diffraction, etc.). Strictly speaking, the term *grain* is only correct in case of single-crystalline particles, which is not the case for most of WC grades.

For most WC powders having APS of 0.5–4 μm, there also exists a close relationship between APS and AGS of the WC sintered cemented carbide structure as long as milling conditions in grade powder preparation and sintering conditions are kept constant. As a result, APS of the WC powder is used as the major steering parameter for cemented carbide structure.

Very fine WC powders (APS < 0.5 μm) are commonly strongly agglomerated. These agglomerates survive any deagglomeration treatment and will finally codetermine the values as obtained by FSSS measurement. In this case, it seems more sensible to assess APS by image analysis obtained by high-resolution SE microscopy.

The particle size distribution of WC is very similar to that of the corresponding W powder. The latter is a consequence of the powder layer height during hydrogen reduction of tungsten oxides and can be somewhat further influenced by the reduction conditions. Moreover, a certain decrease in the coarse fraction can be achieved by heavy milling of the powder either after carburization or during grade powder production.

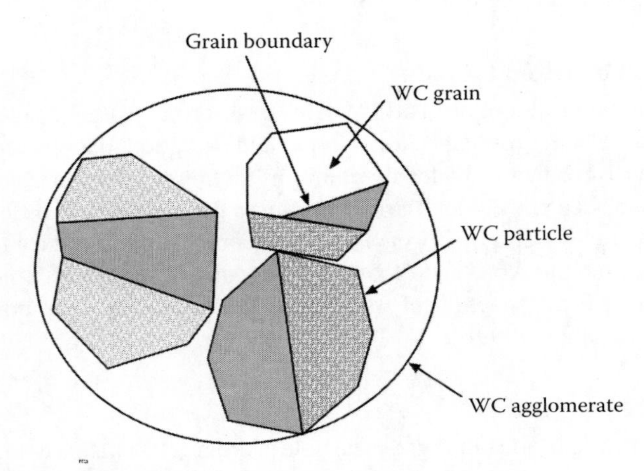

FIGURE B.16 Physical constitution of WC powders (schematic).

B.2.2.4 Commercial WC Powders

Alternatively, similar to tungsten metal powder, WC powder can be purchased from a great variety of its manufacturers located mainly in China and India. Domestic manufacturers of tungsten powder provide more authentic information on the quality of their product and SEM images of their powders. For example, Global Tungsten & Powders Corp. (Towanda, PA) supplies 17 types of WC powder providing the following information for each type: max and min of particle size—FSSS as supplied (ASTM B 330), bulk density (ASTM B 329), coercivity, magnetic saturation, shrinkage, and hardness.

B.2.3 Manufacturing of Grade Powder

Cemented carbide grade powders may consist of WC mixed with a finely divided metallic binder (cobalt, nickel, or iron) or with additions of other cubic carbides, such as TiC, TaC, or NiC, depending on the required properties and application of the cutting tool. The steps in cemented carbide grade powder preparation are shown in Figure B.17.

The necessary amounts of all constituents (WC, other carbides, grain growth inhibitors, additional carbon black, and binder material) and of all further necessary temporary additions (organic solvent, additional organic liquids, lubricant) are weighted or measured precisely and loaded into a suitable blending–milling device. Intensive milling is necessary to break up the initial carbide crystals and to blend the various components such that every carbide particle is coated with binder material. This is accomplished by ball, vibratory, or attrition mills that use carbide balls.

Mill dimensions, rotational speed, ball diameter, loading of milling media, and carbide components must exist in an optimized relationship. In this process, the uptake of iron (due to abrasion) from the mill walls and of cemented carbide from the milling media can be kept at a minimum. The weight ratio of milling media to powder together with the milling time is responsible for the amount of disintegration. The higher the ration, the more pronounced the milling (comminution).

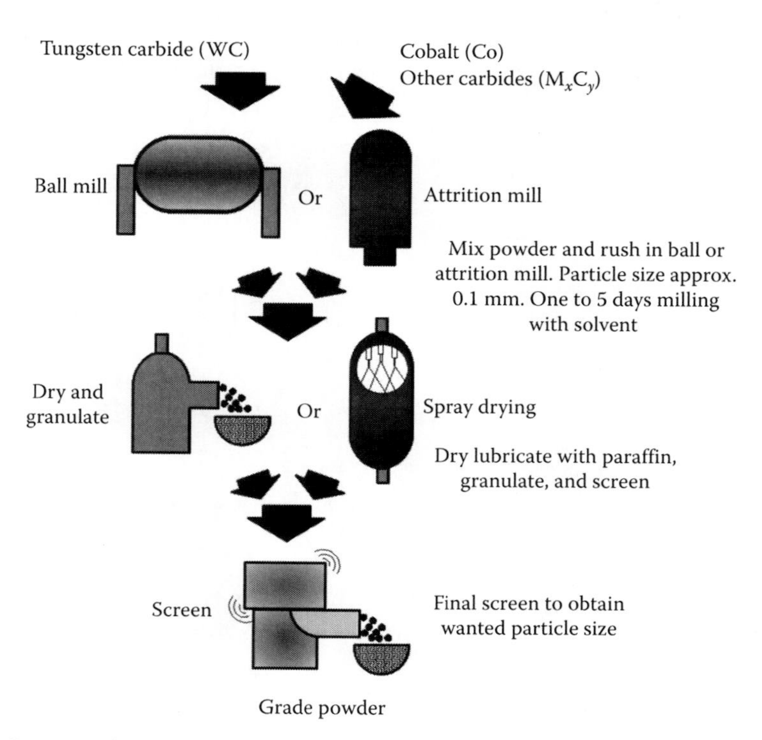

FIGURE B.17 Processes involved in manufacturing of grade powder.

Any particular mill and mill charge combination has a limiting particle size, beyond which no further comminution will occur. For most industrial aggregates, this is in the range of 0.5–1 μm. Only in carefully optimized mills is further comminution possible, finally reaching a size limit of about 100–150 nm. This limit is inherent to the mechanical properties of WC.

During milling, several processes occur simultaneously: breakup of agglomerates, blending, and comminution. In general, the surface energy of the mix is increasing, which enhances the sintering. Sometimes, the carbides are milled alone or with only a small part of the Co (*split milling*) to achieve a more intense milling in a shorter time, while the rest of the binder is added only in the late part of the milling. Cobalt, by its large powder volume (low density compared to WC), decreases the probability for a WC particle to be hit by a ball, thereby prolonging the milling time. In any case, a uniform distribution of the binder is important to avoid increases in porosities in the sintered body, which is particularly important for ultrafine grades. For these grades, longer milling times are therefore mandatory. The use of ultrafine Co powder grades has further contributed to achieving better distributions.

Milling is performed under an organic liquid such as heptane or acetone to minimize heating of the powder and to prevent its oxidation. The liquid is distilled after milling operation. Either before milling or afterward, a solid lubricant (paraffin wax, polyethylene glycol, camphor, etc.) is added to the powder blend in amounts of 1–3 wt%. The lubricant provides a protective coating to the carbide particles and prevents or greatly reduces the oxidation of the powder. The lubricant also imparts strength to the pressed or consolidated powder mix.

After milling, the organic liquid is removed by drying. Grated powder is granulated by either spray drying or mechanical granulation. Granulation is a necessary prerequisite for automated compacting with near-net shaping.

Spray drying is a combination of the distillative separation of the coolant used on milling and the granulation process. It is the most frequently used method for granulation of powder grades on a big scale. The schematic of this process is shown in Figure B.18. In processing, the sludge is forwarded to a spray nozzle (pressure 0.5–1.0 MPa) (Lassner and Schubert 1999). On leaving the nozzle, it is finely divided into droplets (10–50 μm) forming an upward-directed jet. The jet forms a taper, roughly fitting the upper end of the chamber. The forming of granules is rather

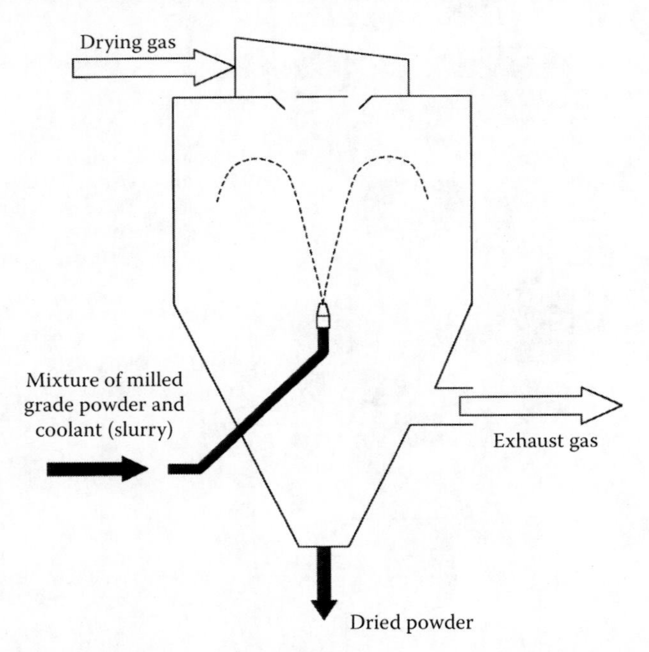

FIGURE B.18 A schematic of a mixed flow drying system using a nozzle atomizer.

FIGURE B.19 Spray-dried WC-Co powder.

complicated mechanism. Hot gas (100°C–130°C) heats the incoming particles and the milling liquid is evaporated. Particles of different speed and moisture grade hit each other, resulting in quite spherical particles of 100–400 μm. Particles big enough settle. The particle size distributing is quite close. To obtain the desired size, the working conditions of the dry spraying chamber must be adjusted.

The granules collected at the bottom are evacuated into a gas-tight container. Critical properties of the granules, besides their size, are also apparent density and hardness. If the hardness is too low, granules can be smashed during handling, while too high a hardness causes granules to partially survive during compacting, yielding an increased residual porosity during sintering. Approximately 5% of the powder remains ungranulated. It is separated by the cyclone and can be used for purposes where granulation is not required. The capacity of a spray dryer is between 120 and 180 kg/h. The size of the spray dryer is about 5 m in height and 2 m in diameter.

Figure B.19 shows an SEM image of spray-dried WC–Co powder. The described granulation method yields a free-flowing powder consisting of spherical particles. It guarantees the highly productive filling of the compacting dies. Spray-dried granules are commonly hollow spheres, so that the filling height during compacting is higher for spray-dried powder as it has a lower apparent density.

B.2.4 Manufacturing of Final Products

The steps in cemented carbide preparation are shown in Figure B.20. They include presintering, sintering, and postsintering operations (processes). The steps are shown for preparing of preforms for drilling tools.

A wide variety of techniques are used to compact the cemented carbide grade powders to the desired shape. Indexable inserts and flat preforms (see Section 3.4.6) are pill processed (pressure applied in one direction) in semiautomatic or automatic presses at pressures 50–150 MPa. Carbide rods are formed by extrusion process (see Section 3.4.6.2).

Unlike most other metal powders, cemented carbide powders do not deform during the compacting process. Generally, they cannot be compressed too much above 65% of the theoretical upper limit for density. Despite this low green density, some carbide manufacturers of drilling preforms have developed various technologies for achieving good dimensional tolerances in the sintered product.

In cold isostatic pressing, the powder is compacted in a tight rubber or plastic container inserted in a liquid under pressure. Besides wear parts and metal-forming tools, mainly bigger pieces are compacted by this method and in most cases need further indirect shaping. The subsequent forming

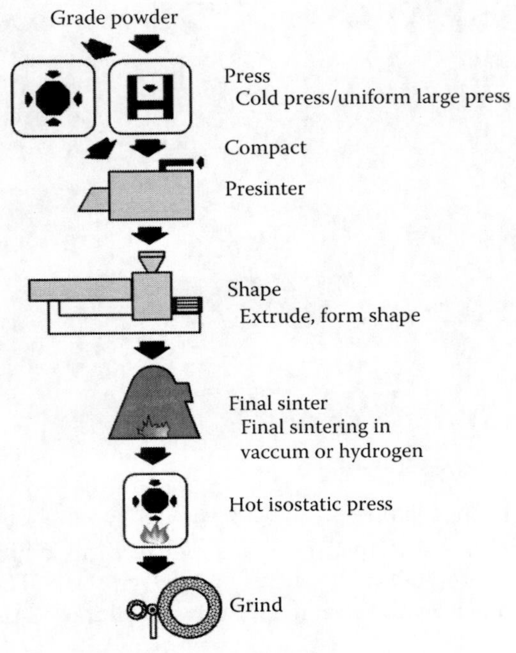

FIGURE B.20 Processes involved in manufacturing of the final product.

can be done either in the *green* state or after presintering. All types of machining and grinding operations can be used. In forming, the sintering shrinkage must also be considered.

Round or profiled rods or tubes can be produced by extrusion. The amount of lubricant must be increased and additional plastifiers added (see Section 3.4.6.2). The corresponding recipes are kept secret.

Injection molding can be regarded as a combination of extrusion and precision die compacting. A highly plastified mixture of the carbide grade powder and lubricants is screwed under pressure into a precision die. The powder-to-plastifier ratio is roughly 90:10 by weight or 50:50 by volume. The procedure is only economic in case of complex shapes and a large number of parts. Dewaxing is carried out very slowly and requires special care, as for extruded material.

A common sintering cycle includes several stages, such as dewaxing, presintering, sintering, and cooling. For pressed-to-shape parts, it is customary to combine dewaxing, presintering, and sintering in one aggregate and heating cycle, thus minimizing energy consumption and handling. It is a prerequisite that the furnace is equipped with a wax condenser to remove the lubricant from the sintering chamber. Only for parts that must be shaped before sintering, a presintering stage is performed at 700°C–1000°C to give the compact sufficient strength to permit forming. This is commonly done in special furnaces.

Sintering is conducted either in hydrogen (for straight grades and low TiC-containing alloys) or, more commonly, in vacuum. For higher TiC- and TaC-alloyed grades, a deep vacuum is mandatory in order to avoid residual gas porosity. In general, there is a tendency today to use vacuum sintering for all grades (Lassner and Schubert 1999).

A typical sintering cycle is shown in Figure B.21 (Lassner and Schubert 1999). At first, the charge is slowly heated up to 400°C–500°C to render proper dewaxing of the compacts. The temperature is then raised gradually to the optimal sintering temperature of 1350°C–1600°C, which depends on the cemented carbide composition and the WC grain size. Several holds at certain temperatures are common during heating to allow outgassing of carbon monoxide (which forms as a result of residual oxygen still present in the densifying compact), unless there is open porosity. Isothermal liquid-phase sintering lasts about 1–1.5 h.

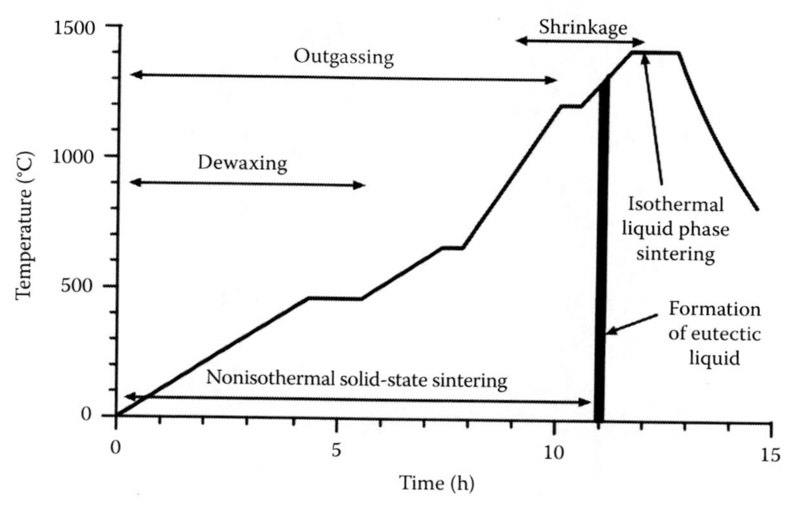

FIGURE B.21 Representative sintering cycle.

Densification starts below the liquidus temperature during heating (nonisothermal sintering). This solid-state shrinkage is especially pronounced in submicron and ultrafine grades. It is due mainly to particle rearrangement processes, which occur as a result of the spreading of cobalt along the WC surfaces, and subsequent contraction under the action of capillary forces. Both processes are activated by the partial dissolution of WC in cobalt.

At about 1300°C–1340°C (binary eutectic temperature WC–Co 1310°C; ternary eutectic temperature W–Co–C 1275°C–1280°C), partial melting occurs, and more WC is dissolved until the eutectic concentration (54% Co and 46% WC) is reached. A further increase in temperature results in additional dissolution of WC and complete melting of the binder phase. In this stage, rapid final densification occurs and the sintered body is practically pore-free (Lassner and Schubert 1999).

A blank being sintered undergoes microstructural changes. At 1400°C, the constitution of the molten phase corresponds to about 50% Co and 50% WC. The liquid phase preferably dissolves the surface layers of the WC particles and grains that have been activated during powder milling, as well as bridges between agglomerated carbides. The morphology of the particles changes from rounded to prismatic (formation of low-energy interfaces). Furthermore, the melt disintegrates large polycrystalline carbide particles into individual grains by penetrating into the large-angle grain boundaries. During isothermal sintering, further adjustment of the microstructure occurs due to solution/reprecipitation and/or coalescence processes, which lead to a more uniform distribution of the cobalt binder and a decrease in carbide contiguity. This is an important aspect for obtaining optimal mechanical properties of the sintered parts. Too high a sintering temperature will promote unwanted grain coarsening.

At the beginning of peak temperature holding, the microstructure is nearly fully dense, and the carbide particles are nearly unchanged in size and shape from the original milled powder stage. They are relatively small, irregularly shaped, and poorly dispersed, tending to agglomerate. The cobalt phase is also poorly dispersed, with many pools or lakes present. A noticeable amount of residual porosity also exists. An example of such a microstructure is shown in Figure B.22.

The main purpose of the final stage of the sintering operation is to develop the microstructure by holding at temperature above the cobalt melting point for a time sufficient to develop a more uniform carbide structure with good cobalt phase dispersion and minimal residual porosity. An example of such a microstructure is shown in Figure B.23. This is usually accomplished by holding for 30–90 min above 1350°C, reaching a peak temperature at about 1425°C and 1500°C

FIGURE B.22 Microstructure of WC-6Co under-sintered cemented carbide.

FIGURE B.23 Microstructure of WC-6Co sintered cemented carbide at the end of final sintering stage.

(Davis 1995). During this period, the cobalt phase, driven by capillarity forces, disperses more evenly. This process also improves carbide particle distribution.

Carbide distribution is further improved during holding by the dissolution of small particles into the liquid phase, with subsequent precipitation onto the larger particles during cooling. This results in a gradual increase in average particle size. Increasing the sintering temperature has a similar influence on grain growth.

One undesirable consequence of grain growth by the solution and precipitation process is the tendency of large tungsten grains to grow at a disproportionally high rate. This discontinuous grain growth occurs more readily at lower cobalt contents (3–6 wt%) with finer average particle size mixes. It is most pronounced when tungsten carbide is the only carbide phase present. Some additions (0.1%–0.5%) of group of VB carbides (vanadium carbide, niobium carbide, and tantalum carbide) inhibit significant grain growth in fine-grained carbide/cobalt compositions. Titanium carbide is also a strong grain growth inhibitor (Davis 1995).

During vacuum sintering, cobalt loss by evaporation should be controlled (Davis 1995). If uncontrolled, as much as 10%–20% of the cobalt content may be lost, thus resulting in a loss of mechanical strength and the formation of rough, coarse-grained surface structure caused by the precipitation of tungsten carbide from the evaporated cobalt. Cobalt evaporation can be minimized by completely enclosing the pressed parts in graphite fixturing, the wall of which is maintained at the sintering temperature.

On cooling, the dissolved WC is precipitated onto already present WC grains. Certain amounts of W and C remain in solid solution, their amount depending on the gross carbon of the respective alloy, and the cooling rate. The W content in solid solution at room temperature varies between 18 and 20 wt% W in substoichiometric alloys and 1–2 wt% W in stoichiometric alloys.

The solutes stabilize the cubic modification of cobalt down to room temperature. Due to the different coefficients of expansion of the carbide(s) and cobalt, the binder phase is under tensile stresses and the carbides under compressive stresses. Uneven macroscopic stresses can lead to distortion of the sintered body.

Shrinkage during sintering corresponds to 17%–25% on a linear scale or 45%–60% by volume. The amount of shrinkage depends on WC powder particle size; the milling, granulation, and compacting conditions; and the grade composition.

As discussed in Chapter 3, the carbon balance of cemented carbide is a very important parameter. The W/C ratio should be close to 1:1. If the carbon content is too low (strongly substoichiometric), η-phases of the M_6C type will be formed and lead to embrittlement of the alloys. If the W/C ratio is higher than stoichiometric, graphite is precipitated (C porosity), which also leads to product deterioration. Thus, only a small gap exists in terms of the carbon in which only WC and Co are present. This gap is smaller in the lower cobalt content carbide grades. Most cemented carbides are slightly substoichiometric to obtain an optimal property relationship and to avoid precipitates. Tungsten in solid solution improves the strength of the binder and makes the material less prone to plastic deformation during machining (solid-solution strengthening).

The carbon balance is severely affected by any traces of oxygen present in the grade powder but also by the sintering atmosphere. The latter can be both decarburizing and carburizing. The ability to maintain the right carbon balance during sintering comes from experience and requires very strict reproducibility in all the foregoing steps. It is most difficult in very low Co grades, where the gap between graphite and η-precipitation is the smallest.

Sometimes, cemented carbides exhibit residual porosity (A and B type), which can occur for several reasons. Since the 1970s, pressure sintering (HIPing) has been applied to remove this porosity. For that purpose, the sintered specimens are loaded in a HIPing device and heated again under Ar or He pressure of 50–150 MPa (temperatures are commonly 25°C–50°C below the vacuum sintering temperature). Hence, any residual porosity can be removed. Exceptions are gas porosities, which can occur either due to incomplete outgassing of the carbon monoxide or due to the presence of impurities.

In recent years, this rather expensive two-stage treatment (vacuum sintering and subsequent HIPing) was substituted by the sinter-HIP process, which was developed in the early 1980s. Sinter-HIP represents low-pressure isostatic pressing up to approx. 7 MPa combined with normal vacuum sintering. At the end of the common sinter heating cycle, the pressure is applied as long as the binder is still liquid. Void-free specimens can be produced.

Hot pressing combines the two steps of compacting and sintering into one. The graphite mold is filled with cemented carbide powder. The mold itself is the resistance heating element. Pressure is applied throughout the heating period. Hot pressing includes high labor costs and is generally only applied for special parts and/or alloys, which are not easy to sinter to full density otherwise.

A large number of cemented carbide products are shaped after sintering because of surface roughness, tolerance, and geometry requirements. For carbide rods used in drilling tools, manufacturing such a shaping removes warpage and *skin*. This forming operation, however, is both time-consuming and expensive (Davis 1995). The sintered products are formed with metal-bonded diamond or silicon carbide wheels, turned with single-point diamond tools, or lapped with diamond-containing slurries.

B.3 DIAMOND

B.3.1 Word Origin and Early History

The name *diamond* is derived from the ancient Greek word αδάμας (*adámas*), meaning *proper, unalterable, unbreakable*, and *untamed*, and from ἀ- (a-), *un-* + δαμάω (*damáō*), meaning *I overpower* and *I tame*. Naturally occurring diamond is a relatively rare polymorph of carbon characterized by a 3D arrangement of tetrahedrally coordinated carbon atoms. On the Earth's surface, diamonds occur in several major kinds of deposits: primary and secondary (both alluvial and littoral). In primary deposits, they are enclosed in host rocks of kimberlite or lamproite that form *pipes* that are downward-tapering, cone-shaped structures of igneous (volcanic) origin. Subsequent erosion of these pipes and fluvial transport of the diamonds lead to the formation of alluvial deposits in river beds and river terraces. The final resting place of diamonds is in littoral or ocean floor deposits.

Separation of diamonds from their enclosing or associated rocks includes crushing, screening, and sieving, use of grease belts, suspension in heavy or dense media, and sorters that use the luminescence of diamonds when they are exposed to x-rays. Diamond crystals are sorted into many categories depending on size, color, clarity, and shape for valuation purposes. Both natural and synthesized diamonds have the highest hardness of all known materials, the highest thermal conductivity at room temperature, a high refractive index and optical dispersion, a low thermal expansion, and a relatively high inertness to chemical attack. This unique combination of properties permits diamond to be foremost in certain applications: as a highly prized gemstone; industrially as an important abrasive unsurpassed in certain cutting, drilling, sawing, machining, grinding, and polishing operations for many materials; and in electronic and optical applications as a heat sink and window material, respectively.

Diamonds were first discovered in ancient times in India and Borneo and later in Brazil in the early 1700s, in alluvial deposits where flowing water had sorted minerals on the basis of density and toughness. These alluvial concentrations of diamonds are derived from primary source rocks by erosion and fluvial transport, where the tumbling action in the transported sediments often concentrates the better-quality diamond crystals by preferentially destroying flawed ones. These processes have occurred over long periods of geologic time in all rivers draining primary sources, and particularly good-quality alluvial diamonds are found in rivers in Angola, Brazil, and India and in areas of southern and western Africa. The Orange River system in South Africa has transported diamonds from inland primary sources to the Atlantic Ocean, where ocean currents have distributed the diamonds along both the littoral zone and the ocean floor off of South Africa and Namibia. Exploration of alluvial deposits can be done by panning of stream sediments or by drilling, pitting, and trenching of streambed and terrace deposits, in conjunction with a search for the heavy mineral assemblages that accompany diamond. Recovery of diamonds from the ocean floor requires the use of sophisticated underwater equipment to collect the diamonds and return them to a processing ship. Alluvial deposits account for approx. 20% by weight (but 41% by value) of the world's annual production of diamonds.

In 1772, Antoine Lavoisier, the great French chemist, pooled resources with other chemists to buy a diamond, which they placed in a closed glass jar. They focused the sun's rays on the diamond with a remarkable magnifying glass and saw the diamond burn and disappear. Lavoisier noted the overall weight of the jar was unchanged and that when it burned, the diamond had combined with oxygen to form carbon dioxide (Krebs 2006). He concluded that diamond and charcoal were made of the same element—carbon. In 1779, Carl (Karl) Scheele showed that graphite burned to form carbon dioxide and so must be another form of carbon (Partington 1962). In 1796, the English chemist Smithson Tennant established that diamond was pure carbon and not a compound of carbon; it burned to form only carbon dioxide. Tennant also proved that when equal weights of charcoal and diamonds were burned, they produced the same amount of carbon dioxide (Barnard 2000). W. Henry and W. Lawrence found carbon to embody three separate structures: cubic, hexagonal, and amorphous. The composition of natural diamond actually consists of approx. 99% carbon 12 and 1% carbon 13 (Davis 1993).

Whether society's infatuation with diamonds rests in its inherent beauty, its list of useful industrial properties, or because of the love, excellence, and purity it has come to represent, our society undeniably values these diamonds with great worth. However, in this technologically driven age, our values come into question as scientists introduce synthetic diamonds into the market. In other words, diamonds like bronze and steel, which were once a scarce commodity, have now become ordinary, reproducible materials. While the news of such may dismay some jewelers and gem connoisseurs, contrasting methods for creating diamonds may also encourage innovation from other industries as they can begin utilizing diamond's material properties. Regardless of whether or not we positively welcome their arrival, synthetic diamonds already have, are, and will invariably shape our world and thereby require further examination.

In order to replicate a diamond, the properties of the diamond itself must be examined. Diamonds are in part so valuable because of the sheer number of useful properties they possess. Spears and Dismukes comment on their superior properties saying, "Choose virtually any characteristic property of a material—structural, electrical or optical—and the value associated with diamond will almost always represent an extremist position among all materials considered for that property." (Spear and Dismukes 1994). Simply stated, in addition to the aesthetic appeal of diamonds, an array of practical uses also exist. Transparent and lustrous, they catch the eye with its unique appearance and magnificent beauty. This may mislead the viewer to also believe they are delicate; however, diamonds, an allotrope of carbon, are also the hardest natural material on Earth.

B.3.2 Important Properties

Figure B.24 shows a simplified phase diagram of carbon. It shows that diamond is a stable form of carbon at very high pressures and temperatures. Under ordinary conditions for temperature and pressure, near 0.1 MPa and room temperature, diamond may be considered a metastable

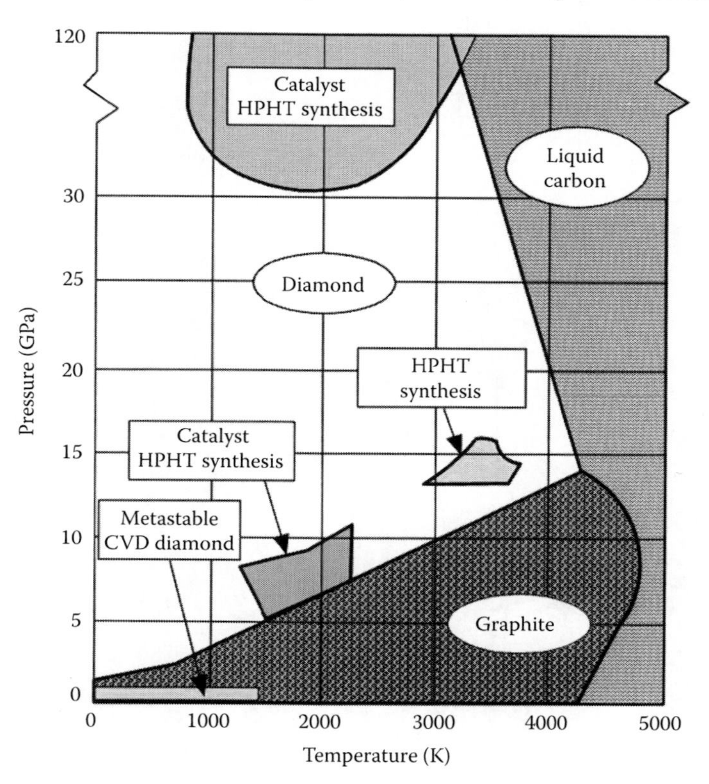

FIGURE B.24 Carbon phase diagram.

form of carbon. Even though it is not the minimum energy state, it does not spontaneously convert to graphite (Asmussen and Reinhard 2002).

A carbon atom has four valence electrons and four vacancies in its outer shell with a ground state electron configuration of $1s^2 2s^2 2p^2$. Such a configuration allows the formation of several allotropic forms: graphite, diamond, lonsdaleite (a form of carbon found in meteorites), and recently discovered 60-atom (or more) carbon spherical molecule, the so-called fullerenes (Burstein 2003).

The structure of graphite is shown in Figure B.25. In the graphite lattice structure, the carbon atoms form continuous hexagons in stack basal planes. Within each plane, the carbon atom is strongly bonded to its three neighbors with a covalent bond having bond strength of 524 kJ/mol. This atomic bonding is threefold coordinated and is known as sp^2. The spacing between basal planes (0.335 nm) is larger than the spacing between atoms in plane (0.142 nm). Such a configuration results in a large anisotropy in the crystal (Pierson 1999; Purniawan 2008).

Diamond, on the other hand, is one form of carbon allotropy with tetrahedral structure, as illustrated in Figure B.26. The orbitals of diamond are bonded to the orbital of four other carbon atoms with a strong covalent bond (i.e., the atoms share pairs of electrons) to form a regular tetrahedron

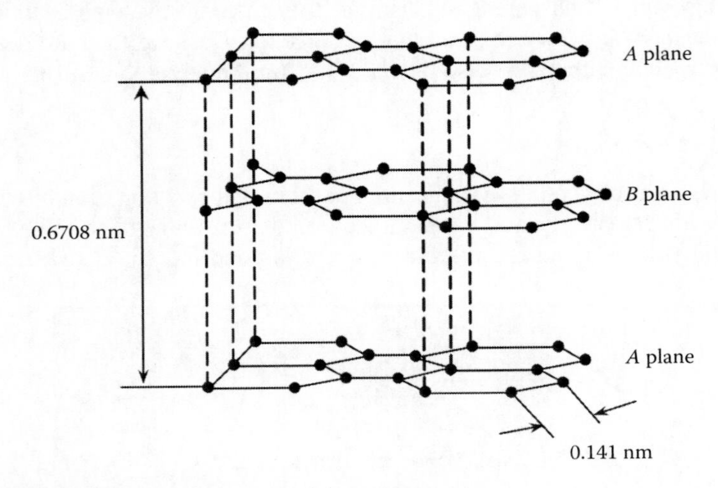

FIGURE B.25 Schematic of graphite hexagonal crystal structures.

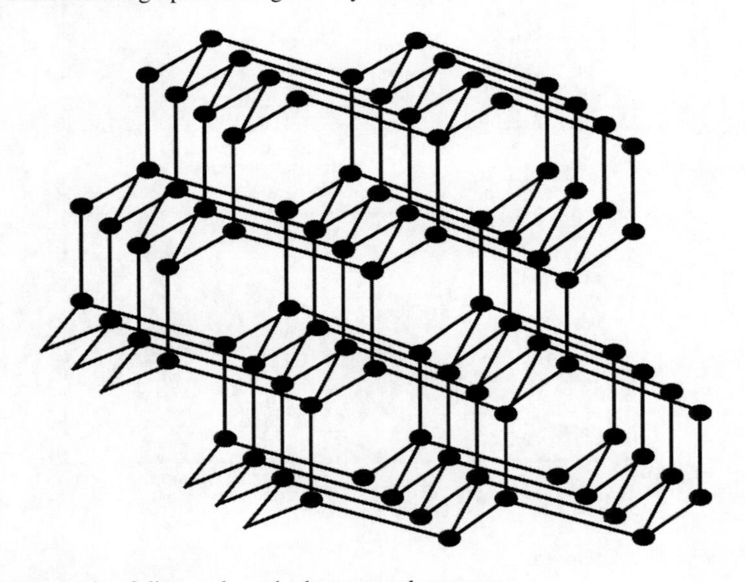

FIGURE B.26 Schematic of diamond octahedron crystal structures.

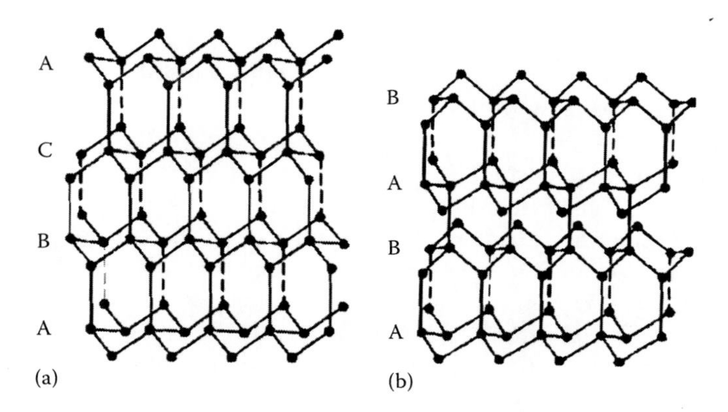

FIGURE B.27 Crystal structures of (a) cubic diamond and (b) hexagonal lonsdaleite. A, B, and C indicate the stacking sequence of sheets of carbon atoms.

with equal angles (109°28″). Unlike graphite, the structure of diamond is isotropic. Each diamond tetrahedron combines with four other tetrahedrons to form a strongly bonded, 3D, and entirely covalent crystalline structure, with a small bond length of 0.154 nm and a high bond energy of 711 kJ/mol or 170 kcal (Pierson 1999; Purniawan 2008).

The lattice constant of the unit cell of the FCC form of diamond is 0.3567 nm with a density of 3.52 g/cm^3. For the hexagonal form lonsdaleite, a = 0.252 nm and c = 0.412 nm with the same density as the cubic form (Figure B.27). In this allotropic form, the stacking of successive layers of carbon atoms is ABABAB as compared to the ABCABC type of stacking of the normal cubic form. Diamond twins readily during growth on the (111) plane resulting in both interpenetration and flat twins called macles (other more complex twin relationships are also possible [Wilks and Wilks 1991]). The growth morphology of natural diamond is typically octahedral, but dodecahedral crystals are common, perhaps because of subsequent solution after growth. Many crystals exhibit evidence of uneven development of crystal faces leading to less regular shapes and of modification by etching and dissolution to produce rounded surfaces and edges. A few diamond crystals exhibit the approximate form of a cube, but the nonplanar irregular surfaces are the result of curved, noncrystallographic forms of growth.

Coated diamonds are dull and opaque due to the presence of an outer layer (formed by a fibrous, columnar growth) that contains a high level of impurities (Wilks and Wilks 1991). Only very small grains (<30 μm) of the hexagonal lonsdaleite have ever been seen, so not much can be said of their morphology.

Besides the single crystals usually implied in a discussion of diamond, it also occurs in nature in the form of polycrystalline aggregates (Wilks and Wilks 1991). These can be classified into two types based on origin: carbonado and ballas. The former is a dark-colored aggregate of previously formed, submicron-sized grains that have become cemented to each other to varying degrees of bonding. Some varieties such as framesite show evidence of mechanical deformation during the aggregation process. Ballas refers to a gray-colored polycrystalline aggregate of diamond grains that grew simultaneously as opposed to an assemblage of previously grown crystals. Specifically the term refers to a ball-shaped PCD with a radial structure (like some hailstones). Contrary to carbonado, ballas is free of secondary phases. Most synthesized compacts have a structure that resembles carbonado rather than ballas. Bort is a general term used to describe low-grade industrial diamond suitable only for use in a fragmented form.

Diamond is nominally pure carbon with a C^{12}/C^{13} ratio of approx. 99:1. Although other elements are reported in chemical analyses, most are considered to be present in oxide, silicate, and sulfide mineral phases as inclusions in the diamond. Only boron and nitrogen are considered to be truly substitutional in the lattice. Hydrogen and oxygen, possibly as OH, may also be important structural contaminants though they may also be present as second-phase gases along with CO, CO_2, H_2O,

CH_4, and other species (Melton 1985). These impurity elements can be present in concentrations from approx. 1 to more than 1000 ppm; a variety of additional elements can occur at even lower concentrations (Wilks and Wilks 1991; Field 1992).

The classification of diamond into four principal types is based on the presence of nitrogen (type Ia and Ib) and boron (type IIb) and the effects of these impurities on the infrared (ir) and ultraviolet (uv) transmission (Wilks and Wilks 1991; Field 1992). Nominally pure diamond without detectable amounts of these impurities is referred to as type IIa. It has the highest thermal conductivity and is transparent out to 10.6 μm. The concentration of boron in natural type IIb diamonds is approx. 0.25 ppm and as much as 270 ppm in synthesized crystals; a crystal with approx. 10 ppm boron is deep blue. The most famous natural type IIb diamond is the blue 45.52-carat Hope diamond. Dissolved nitrogen can range up to 2500 ppm in type Ia natural diamonds and to approx. 500 ppm for synthesized type Ib. Most natural diamonds are type Ia, which is characterized by the presence of platelets, made visible by transmission electron microscopy on the (100) planes. The exact nature of these platelets is still controversial, but they are often called nitrogen platelets (possibly N–N or C–N clusters). An increase in the high-temperature bending strength of type Ia over type IIa diamonds has been attributed to the platelets interfering with dislocation motion (Evans and Wild 1965). When the nitrogen is in this aggregated atomic form, the crystals can be essentially colorless to light yellow, in contrast to the darker yellow type Ib, which has nitrogen as isolated atoms in substitutional positions in the lattice. Most synthesized diamond is type Ib unless special precautions are taken to remove nitrogen from the growth environment. A type Ib diamond can be converted to a type Ia and vice versa by annealing at high pressures and temperatures (Chrenko et al. 1977). This was first established in the laboratory and is now presumed to be the way nature made type Ia diamonds. If this conversion could be carried out in a practical manner, it would be of some interest in improving the value of slightly yellow gemstones (to date, there is no evidence that such treated type Ia diamonds exist in the jewelry marketplace). The latest development in the jewelry trade is treatment of brown type IIa diamonds to lighten or remove their brown color to make them colorless by annealing at very high temperatures and pressures (Smith et al. 2000).

The main inclusions in natural diamond are pyroxene, garnet, spinel, and olivine, and the understanding of the conditions of their formation from laboratory studies is the basis for the determination of the P–T conditions where diamond was formed (Boyd and Gurney 1986). The compounds CO, CO_2, H_2, and H_2O are also found in diamond, and it is possible that diamond nucleated and grew in a C–H–O system in equilibrium with the silicate matrix. Graphite is also a common inclusion in natural diamond. Additional minerals identified as inclusions include oxides (chromite and rutile), sulfides (pyrrhotite), and minerals such as zircon, coesite, kyanite, and metallic iron.

Diamond is thermodynamically unstable at ambient conditions, but unless heated to 650°C in air (oxidation to CO_2) or 1700°C in vacuum or an inert atmosphere (graphitization), it will remain as diamond indefinitely. To convert graphite directly to diamond requires very high pressures and temperatures (see Figure B.24). For synthesis of industrial diamond, these conditions can be lowered by the presence of metal solvents/catalysts. More recent work on the thermodynamically stable region for diamond indicates a positive slope for the melting curve instead of the negative slope deduced previously on the basis of analogy to the P–T diagrams for Si and Ge (Bundy 1989).

Diamond is chemically inert to inorganic acids but can be etched in oxidizing molten salts such as KNO_3 at 600°C. Carbon as diamond or graphite is soluble in several metals, particularly Fe, Ni, Co, and other Group VIII, the rightmost group in the Periodic Table, elements. This imposes a serious limitation on the use of diamond for machining alloys based on these materials in spite of the high hardness of diamond. At the temperatures developed at a cutting tool tip, diamond reacts with these metals. However, similar reactions facilitate the bonding of diamond to metal in diamond tool making because of the wetting of diamond by carbide-forming metals such as W, Ta, Ti, and Zr (Wilks and Wilks 1991; Field 1992).

Diamond is the hardest material known because of its combination of a 3D arrangement of tetrahedrally coordinated C–C bonds with a bond distance of 0.154 nm. The Knoop hardness (K) is in

TABLE B.1
Mechanical Properties of Diamond and Other Hard Materials (GPa)

	Diamond	Cubic BN	Cemented WC	SiC
Bulk modulus	440	370		211
Compressive strength	8.69		4–6	
Modulus of elasticity, E	950–1100	890	460–675	

the range of 68.7–98.1 kN/m^2 (7,000–10,000 kgf/mm^2). Harder materials have been postulated on the basis of bond distance but so far have not been found or made in a useful form (Liu and Cohen 1989). Diamond is twice as hard as cubic BN (K is approx. 44 kN/m^2), which in turn is twice as hard as SiC (K is approx. 23.5 kN/m^2). The hardness of diamond varies as a function of crystallographic orientation; this is important when mounting diamonds for cutting tools or in polishing gemstones (Wilks and Wilks 1991). The *table*, the largest facet of a brilliant cut diamond, is generally the (001) face because it is easier to saw and polish than the (111) face using traditional techniques. The use of lasers to cut diamond diminishes the importance of orientation dependence of hardness in some industrial operations.

Although hard, diamond is also very brittle and cleaves readily on the (111) plane and also on other planes under certain conditions. The ability of diamond to fracture is useful in cutting and grinding applications because the crystals are self-sharpening. For other procedures such as sawing rocks or concrete, tougher diamonds are desirable (Wilks and Wilks 1991). The uncontrolled fracture of a gemstone is disastrous, but in the cutting and polishing industry, large diamonds are first reduced to desired sizes by controlled cleavage or laser cutting to avoid the time-consuming process of sawing (Bruton 1970).

Measurement of the mechanical properties of diamond is complicated, and references should be consulted for the various qualifications (Bruton 1970; Wilks and Wilks 1991). Table B.1 compares the theoretical and experimental bulk modulus of diamond to that for cubic BN and for SiC and compares the compressive strength of diamond to that for cemented WC and the values for the modulus of elasticity E to those for cemented WC and cubic BN.

The thermal conductivity value of 2000 W/(mK) at room temperature for type IIa natural diamond is about five times that of Cu, and recent data on 99.9% isotopically pure C^{12} type IIa synthesized crystals are in the range of 3300–3500 W/(mK) (Field 1992). This property combined with the high electrical resistance makes diamond an attractive material for heat sinks for electronic devices. More impure forms of diamond (type Ia) have lower thermal conductivities (600–1000 W/(mK)).

The averaged value of the coefficient of linear thermal expansion of diamond is 1.18×10^{-6} (m/m K). The relatively low expansion combined with the low reactivity of diamonds, except for carbide formation, leads to some challenges in making strong bonds between diamond and other materials.

Diamond is an electrical insulator (approx. 1016 Ω/cm) unless doped with boron when it becomes a p-type semiconductor with a resistivity in the range of 10 to 100 Ω/cm. n-Type doping has been claimed but is less certainly established. The dielectric constant of diamond is 5.58.

B.3.3 Applications

The first extensive use of diamond in industry was as diamond powder mixed with olive oil. This abrasive slurry was placed on the horizontal face of a spinning cast iron plate to polish rough diamonds (held under pressure against the face) into faceted gems. The beginnings of this art are lost to history but probably started about AD 1500. A major use of industrial diamond today is in the grinding and finishing of various cemented carbide parts and cutting tools. As the cemented carbide industry has expanded (growth has been remarkable since its introduction), the growth of the diamond wheel market has correspondingly increased. In 1935, only about half of all diamonds mined were used in industry. Now, this figure has increased to about 80%.

The selected applications of diamond are based on its high hardness, wear resistance, ability to self-sharpen as it cleaves, thermal conductivity, and chemical inertness. Loose abrasive grain is used in lapping and polishing operations, for example, on the scaife used for faceting gem crystals and polishing rock materials for monuments and buildings. Single-point tools for engraving, turning, or cutting applications are made by mounting an individual diamond in metal holders (Grodzinski 1953; Smith 1974; Wilks and Wilks 1991). Other uses for single diamonds are as phonograph needles, bearings, surgical knives, and wire dies. The differential hardness of diamond is the cause of the nonuniform wear, or loss of roundness, of a wire die. If sintered (polycrystalline) diamond is used for wire dies, the anisotropic wear problem disappears. For many applications such as grinding, cutting, drilling, and sawing, abrasive grains are mounted in resin, metallic, or vitreous bonding materials on wheels, disks, or saw blade sectors. Metallic bonding can be accomplished by electroplating as well. The grooving of concrete runways and roads is a well-known example of the use of impregnated diamond saw blades. An interesting new development is as beads, which are strung on long wires for sawing blocks of stone from quarries. About 90% of industrial diamond is nowadays synthesized.

B.3.4 Types

There are two types of diamond: natural and synthetic diamond. Both can be either single-crystal diamond or polycrystalline diamond (known as PCD). Natural PCD abrasives have long been used for cutting and drilling very hard materials, and scientists continue to explore methods of replicating the properties of carbonado, one of the more industrially important varieties of PCD. Carbonado occurs as black, irregularly shaped, polycrystalline aggregates with grain sizes ranging from less than 1 to several hundred microns (Jeynes 1978; Dismukes et al. 1988). Unlike nonporous microcrystalline diamond aggregates found in kimberlites (e.g., stewartite and framesite), carbonado is recovered only from alluvial deposits in Brazil and the Central African Republic (CAR). Carbonado exhibits the extreme hardness and high thermal conductivity of single-crystal diamond, but it is much less vulnerable to catastrophic cleavage. The random crystallographic orientation that is typical of carbonado and other natural and synthetic PCD combines with a complex defect microstructure to minimize crack propagation (Lammer 1988). In addition, grinding materials composed of PCD tend to wear more uniformly than single-crystal diamond, whose hardness varies with crystallographic orientation. The tight diamond–diamond bonding in carbonado suggests that it is a naturally *sintered* diamond (Hall 1970).

B.3.5 Synthetic Diamond History

The 1797 discovery that diamond was pure carbon catalyzed many attempts to make artificial diamond. If the common carbon of commerce could be turned into diamond, a millionfold increase in the value of the starting material would be obtained. There was, in addition, the interest of the scientific achievement. The earliest claim of success (C. Cagniard de la Tour) was made in the year 1823. The diamond synthesis problem has attracted the interest of thousands. Those pursuing the problem have ranged from charlatans through rank amateurs to the world's greatest scientists. Included among the great are Boyle, Bragg, Bridgman, Crookes, Davey, Despretz, Friedel, Liebig, Ludwig, Moisson, Parsons, Taman, and Wohler.

The literature on diamond synthesis is not very extensive. Most of the work has gone unpublished. Many of the world's large industrial organizations have considered the problem and have spent millions of dollars. So many years passed without success that those working on the problem felt embarrassed to admit that they were so engaged. Another aspect of the problem was the chicanery and fraud. Quite a number have claimed to possess a procedure for converting graphite into diamond, invited the unwary to invest their money, and then vanished.

The earliest successes were reported by James Ballantyne Hannay in 1879 (Hannay 1879) and by Ferdinand Frédéric Henri Moissan in 1893 (Royère 1999). Their method involved heating charcoal at up to 3500°C with iron inside a carbon crucible in a furnace. Whereas Hannay used a flame-heated tube, Moissan applied his newly developed electric arc furnace, in which an electric arc was struck between carbon rods inside blocks of lime. The molten iron was then rapidly cooled by immersion in water. The contraction generated by the cooling supposedly produced the high pressure required to transform graphite into diamond. Moissan published his work in a series of articles in the 1890s (Moissan 1894).

Despite these early successes, many following scientists experienced difficulty in reproducing such results. Spending nearly 30 years of his life and over £30,000, British scientist Charles Algernon Parsons went through great lengths in order to reproduce a high-pressure high-temperature diamond. While claiming to have produced high-pressure high-temperature diamonds, over time he reluctantly admitted that no scientist, past or present, could create synthetic diamonds and that the best they could create were spinels, or a simple class of minerals crystallized in an octahedral form (Davis 1993).

Despite past failures and inspired by the vast industrial uses of diamonds especially during wartime, GE resumed the synthetic diamond project in 1951. Schenectady Laboratories of GE and a high-pressure diamond group were formed. The first authenticated synthesis of diamond was made on December 16, 1954, by H. Tracy Hall. It should be noted that this synthesis was unique in several aspects when compared to prior claims:

1. The diamonds were grown in a very short time, 15–120 s.
2. The yield was high; that is, a significant portion of the graphite starting material was converted so that the diamonds formed could be seen by the naked eye without being separated from the starting material. Also, the diamonds, though tiny (up to 350 μm across) were obtained in sufficient quantity to be felt and held in the hand and could be heard to *scratch* as they were drawn across a piece of glass. These perceptions of the diamonds by the unaided physical senses were remarkably satisfying.
3. Hundreds of crystals were grown, many of which were intergrown with each other and some were twinned. This result contrasts sharply with the way diamonds are found in nature.
4. The discoverer readily duplicated his successful experiment a dozen times, and a short time later the experiment was duplicated by others.
5. The process was immediately recognized as being commercially feasible; indeed, less than 3 years after discovery, diamond grit was being sold on the commercial market.

B.3.6 Synthetic Diamond Technology: High-Pressure Equipment

Early in the 1930s, devices known as piezometers were developed. They include steel thick-walled cylinders equipped with plungers. Pressures up to 0.3 GPa were created using such devises. The great American scientist and Nobel prize winner Percy Bridgman pioneered the development of high-pressure creating devices known as *Bridgman anvil* (Prikhna 2008). Bridgman placed a piezometer into the cavity of a great hydraulic cylinder and obtained a two-stage apparatus, in the internal cavity of which a pressure up to 3 GPa was attained. The hardening of many pressure transmitting liquids at higher pressure became an obstacle to further increase the pressure in apparatuses of this type. Therefore, Bridgman had to use hard minerals like pipestone or pyrophyllite as a pressure medium. It is remarkable that the higher pressures were possible to generate in such a simple device.

Howard Tracy Hall first realized previous high-pressure creating devices failed in two main areas: the pistons and the cylinders. Without finding a solution to this problem, the device would not be able to create an environment suitable for creating the pressure needed. He solved the problem by eliminating the bottom of the bore and using two tops *back to back*. Hall also realized that the steel currently available could not withstand the pressure that would be applied.

Instead, he used cemented carbide, which was necessary to create a device strong enough to withstand the pressure. Hall called his device a *belt* apparatus (Figure B.28) (US Patent 2,947608, 1960; failed August 29, 1955—diamond synthesis) (Hall 1960). Subsequently, three types of equipment are the inventions of H. Tracy Hall. They are the belt, the tetrahedral press, and the cubic press (Hall 1982).

In the belt (Figure B.28), the central portion of the apparatus consists of a tungsten carbide die with a tapered entrance hole on the top and bottom surfaces. This die accepts a cell that is made of pyrophyllite. Pyrophyllite is a hydrous, naturally occurring aluminum silicate. The better grades of this material are mined in South Africa. It is gray in color and can easily be machined with standard machine tools. The material is resistant to heat and has a low thermoconductivity compared to metals. It also serves to transmit pressure and has appropriate frictional and extrusion characteristics to form a gasket between the die cavity and two conical punches, which are thrust into the top and bottom portions of the die. The punches conform to the internal shape of the pyrophyllite cell within the die. The opposing punches are pushed toward each other by a hydraulic ram. The die and the punches are surrounded by compound steel binding rings that exert an inward thrust on the tungsten carbide. This is necessary to prevent the failure of the tungsten carbide by the tensile loads that develop as a result of the compressive loads.

Located within the pyrophyllite cell is a heater-sample tube. The sample tube may be of graphite or a high-melting metal such as tantalum and is connected through a tantalum disk at each end to a current disk. The current disk makes contact with the tip of the carbide punch on

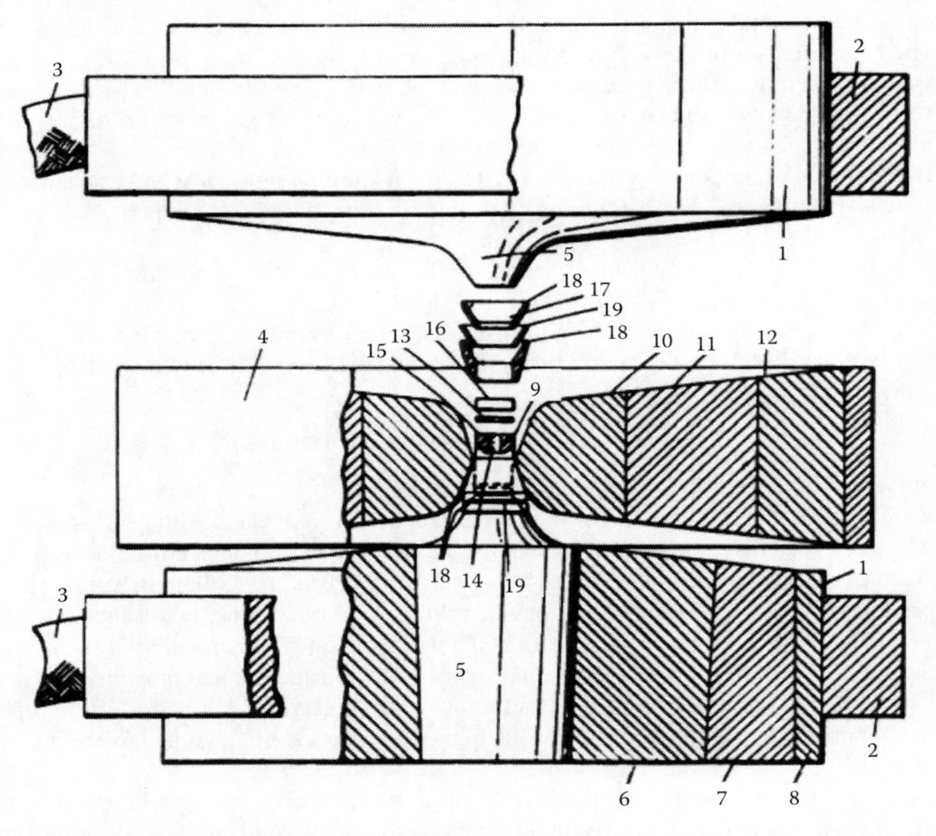

FIGURE B.28 The belt (schematics): ready-assembled punch (1), conductor (2, 3), ready-assembled cylinder (4), punch (5), block of rings that support the punch (6, 7), safety collar of the punch block (8), reaction vessel (9), insert of the cylinder (10), block of rings supporting the cylinder (11, 12), and a set of conductive and sealing parts (13–19).

each end of the cell arrangement. High current electricity is brought through the current rings, the current disk, and the sample-heater tube to heat the contents of the sample-heater tube. The advance of the punches causes extrusion and compression of that portion of the pyrophyllite cell that is caught between the tapering sides of the punch and the tapering internal shape of the die. This compression and extrusion device admits the generation of pressures beyond 100,000 atm in the belt device.

The sample-heater tube usually occupies no more than 25% of the total volume of the pyrophyllite cell assembly. In a commercial machine for diamond production, the circular diameter of the punch tips is approx. 1 in. The distance between the punch tips is about 1½ in. Up to 25 ct of friable diamond grit can be made in an apparatus of this kind in a 5 min cycle. When the belt is used in continuous duty, the punches and the die must be cooled. This is accomplished by placing water-cooling jackets around the binding rings that surround the punches and the die. The hydraulic ram that drives the punches in this commercial-size machine should have a thrust of about 1000 tons.

In the tetrahedral anvil apparatus shown in Figure B.29, four anvils, each having a triangular face and sloping shoulders that make an angle of 35.3° with respect to the plane of the triangular face, impinge in a symmetrical manner along tetrahedral axes upon a tetrahedral cell. The faces of the tetrahedral cell are triangular in shape and are larger in area than the corresponding anvil faces. After the anvils advance and contact the faces of the tetrahedral cell (made of pyrophyllite), additional anvil advancement causes pyrophyllite to extrude into the spaces between parallel sloping shoulders of the anvils. This automatically formed gasket, with continued advance of the anvils, compresses, yields, flows, and extrudes, achieving a balance between frictional pressure and other forces to form a seal and also provide pressure buildup within the cell.

Maximum pressures achieved in tetrahedral anvil devices are of the order of 140,000 atm. Routine, everyday pressures in commercial use are about 70,000 atm. A routine temperature simultaneously held with the 70,000 atm pressure is 1,800°C. Both conditions can be simultaneously held for days. A tetrahedral cell is prepared from a solid pyrophyllite tetrahedron that is cut apart and drilled to provide cavities for the sample, electrical heaters, thermocouples, etc., as required. The assembled cell is held together by a small amount of white glue. All four tetrahedral anvils are electrically insulated from each other for making four independent electrical connections into the cell. As in the case of the belt, heating the sample tube is accomplished electrically (usually high current, low voltage).

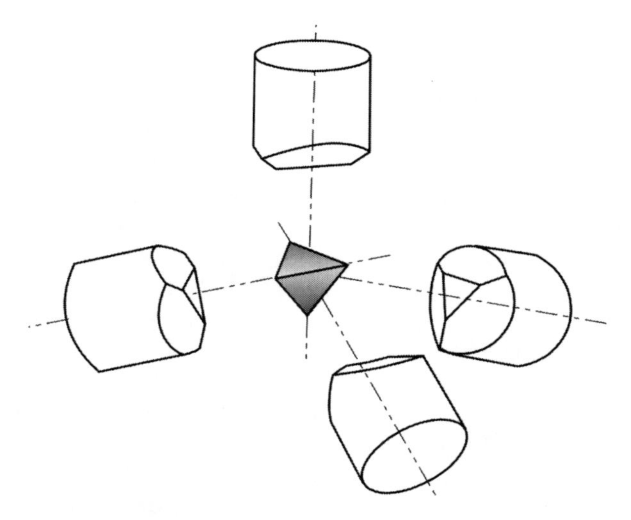

FIGURE B.29 Tetrahedral anvils and cell (schematics): US Patent No. 2,918,699. Four tungsten carbide anvils press together, creating tetrahedral shape with an enormous amount of pressure.

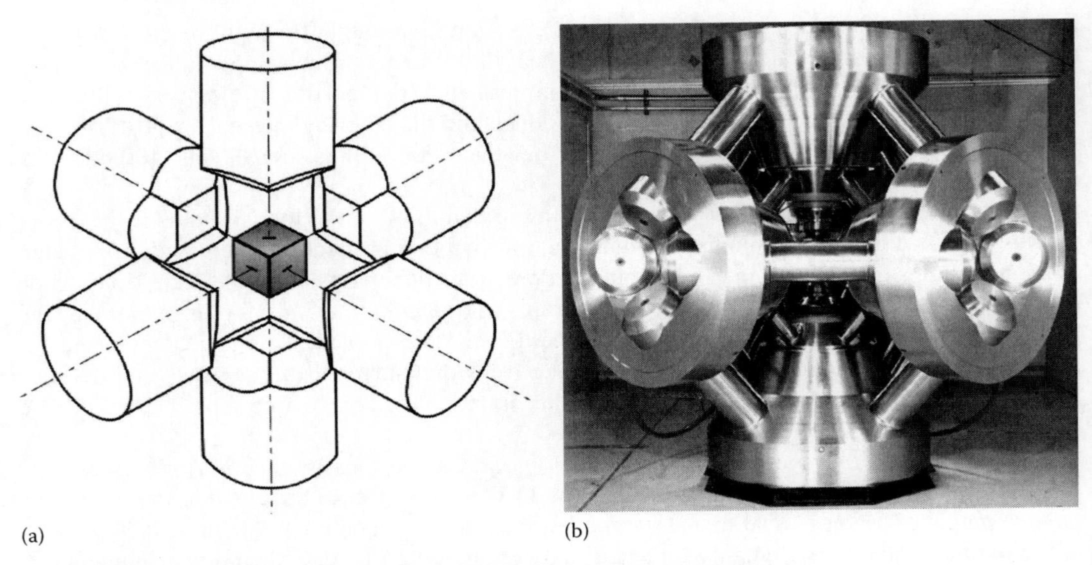

FIGURE B.30 (a) Cubic anvils and cell (schematics) and (b) its modern design.

The cubic press, Figure B.30, is an extension of the tetrahedral anvil press idea. In a cubic press, six cubic anvils are directed along the axes of the Cartesian coordinate system to impinge on a cube of pyrophyllite. As in the case of the tetrahedral cell, the cubic cell is drilled and machined to accept appropriate components for bringing in heating current, for containing the sample, etc. The cubic anvil has a square face whose area is less than the area of the corresponding face of the cubic cell. The sloping shoulders of the cubic anvil make an angle of 45° with respect to the plane of the anvil face. In high-pressure machines of the tetrahedral and cubic type, a hardened steel binding ring is interference fit to the cylindrical surface of the tungsten carbide anvils. These binding rings provide an inward thrust toward the axis of the anvils and increase their life. The six electrically isolated cubic anvils provide six independent electrical connections to the interior of the cell.

In all three devices (belt, tetrahedral press, and cubic press), the cell is used only once. The maximum pressure for routine runs with good anvil lifetime in the cubic press is somewhat lower than in the tetrahedral press, namely, 65,000 versus 70,000 atm. However, the cubic cell is somewhat more convenient in its geometry than the tetrahedral cell. It is obvious that the multianvil press concept (tetrahedral and cubic presses) can be extended to include octahedral presses, prismatic presses, and higher-order geometrical configurations involving a greater number of anvils than six.

An apparatus, which is able to create a pressure up to 8 GPa, was developed by L.F. Vereshchagin with coworkers. The same authors were the first in the USSR to synthesize diamond using this apparatus (Prikhna 2008). A few years earlier, diamond was synthesized in Sweden by von Platen in a two-stage apparatus (von Platen 1962). This apparatus had a six-punch solid-state inner block with a cubic pressure cavity, which was activated by a hydraulic pressure of 0.6 GPa through a plastic shell.

Once the commercial production of synthetic diamonds was organized, the industrial demand grew tremendously for diamond tools for sharpening carbide tools, grinding hard natural stones, and machining semiconducting materials. The transition from laboratory-type to industrial-type HPA was made. The mixture compositions and the technological parameters of synthesis of larger high-strength crystals were developed. To synthesize such crystals, it was necessary to arrange conditions at which a lower quantity of nuclei formed in the mixture grew more slowly. This, however, increased the crystal costs. Influencing this process was possible by a considerable increase of the HPA cavity volume. Because of this, a family of multianvil apparatuses with cavities of cubic, octahedral, and cubo-octahedral shapes driven by single plunger presses was developed (Prikhna 2008).

B.3.7 Synthetic Diamond Technology: The Synthesis

The cubic cell of a 200-ton cubic press similar to that depicted in Figure B.30 is shown in Figure B.31. The body of the cell is made of pyrophyllite, and the edge of the cube is 16 mm, which is 25% longer than the edge of the corresponding square anvil face. The exterior surface of the cell is painted with a suspension of red iron oxide in water and is dried in an oven at 120°C for 0.5 h before use (Hall 1982).

The central hole in the pyrophyllite is approx. 6 mm in diameter and the counterbores on each end are 11 mm in diameter by 4.2 mm deep. The pyrophyllite thermal insulation plug located within the steel current ring. The molybdenum disks locate at the bottom of each counterbore. The central hole is completely filled with alternating disks of graphite and pure nickel.

When the anvils are in a retracted position, the cell is centered on the lower anvil through which heating current will pass. One current ring should touch this anvil and the opposite current ring will touch the opposing anvil when the anvils are closed on the sample. The cell is held in place by placing a piece of adhesive tape across the paper tab onto the flat shoulder of the anvil's binding ring. Oil is then admitted to the hydraulic rams until the anvils contact the cell faces. Oil pressure is allowed to build up to about 60,000 atm within the cell. During anvil advance (after contacting the cell), the 12 edges of the oversize cell (oversize with respect to the cube volume that would be enclosed if the anvils were allowed to touch with no cell present) will extrude between the sloping shoulders of the advancing anvils to automatically form a compressible gasket.

Once the cell is pressurized, heating current is applied. This current is supplied by a stacked core-type welding transformer for which the output voltage is controlled by a variable autotransformer connected to the primary winding. The electrical characteristics of such systems vary, so some experimentation is required to select the correct autotransformer setting to obtain the proper temperature. For an initial trial, it may be set for a current flow of 800 A through the sample. If the setting is correct, the current will start to decrease after about 20 s of heating. As the current falls, voltage will rise. These combined signals indicate that diamond is being made. The temperature inside the cell is near 1500°C, the nickel is molten, and graphite is dissolving in it and crystallizing out of solution as diamond. Pressure is providing the thermodynamic driving force, and liquid nickel is acting as the solvent–catalyst.

The current falls as diamonds form because diamond is an electrical insulator, while graphite is an electrical conductor (actually, diamond will be somewhat of a conductor at 1500°C). The current will probably level out near 200 A as diamond is formed. Heating power is disconnected after about 60 s. The sample will cool to room temperature for about 4 s and pressure is released after an additional 20 s, at which time adjustments will take place in the cell and

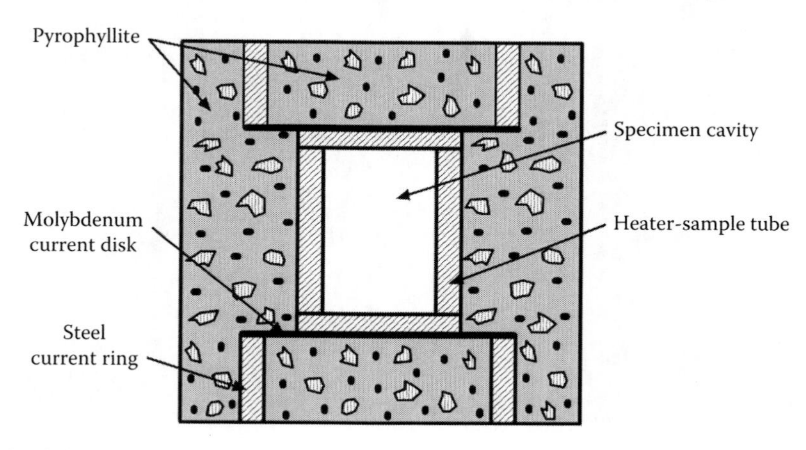

FIGURE B.31 Cubic cell (schematic).

anvils that will lengthen the anvil's life. The cell is now removed from the press and broken open with a hammer. The contents of the central hole can usually be retrieved as a cylinder consisting of a *welded together* mass of nickel, graphite, and diamond crystals. The pyrophyllite surrounding this cylinder will have been transformed into a white, hard ceramic containing kyanite and coesite. The diamond crystals can be retrieved from the cylinder by selective chemical oxidation.

The first treatment consists of concentrated sulfuric acid (H_2SO_4) to which about 10% by weight of sodium or potassium nitrate is added. This mixture and the sample are heated strongly on a hot plate within a shielded fume hood for about an hour. When an appropriate temperature is reached, *clouds* of graphite particles will be streaming upward from the sample. Eventually most of the graphite is oxidized and disappears. At this point, the liquid is carefully decanted and discarded. Following this, concentrated hydrochloric acid (HCl) is added to the sample and they are heated lightly on the hot plate. HCl is then decanted and the residue washed with water. The diamonds will be clean enough at this point to be observed under 10 power magnification. Typical diamond morphological features will be evident. For additional cleaning, the H_2SO_4/$NaNo_3$ and HCl treatments should be repeated until all the graphite and nickel are gone. There will still be some residue from the pyrophyllite. This can be removed by prolonged treatment with concentrated HF solution or by fusion with NH_4F.

In industrial applications, there are a number of quality procedures to obtain final product—diamond powder similar to that shown in Figure B.32. Diamond powders can be of different quality. Some grades can be used for diamond grinding wheels, while others are subjected to further HPHT process to make PCD disks to be used in PCD cutting tools as discussed in Chapter 3.

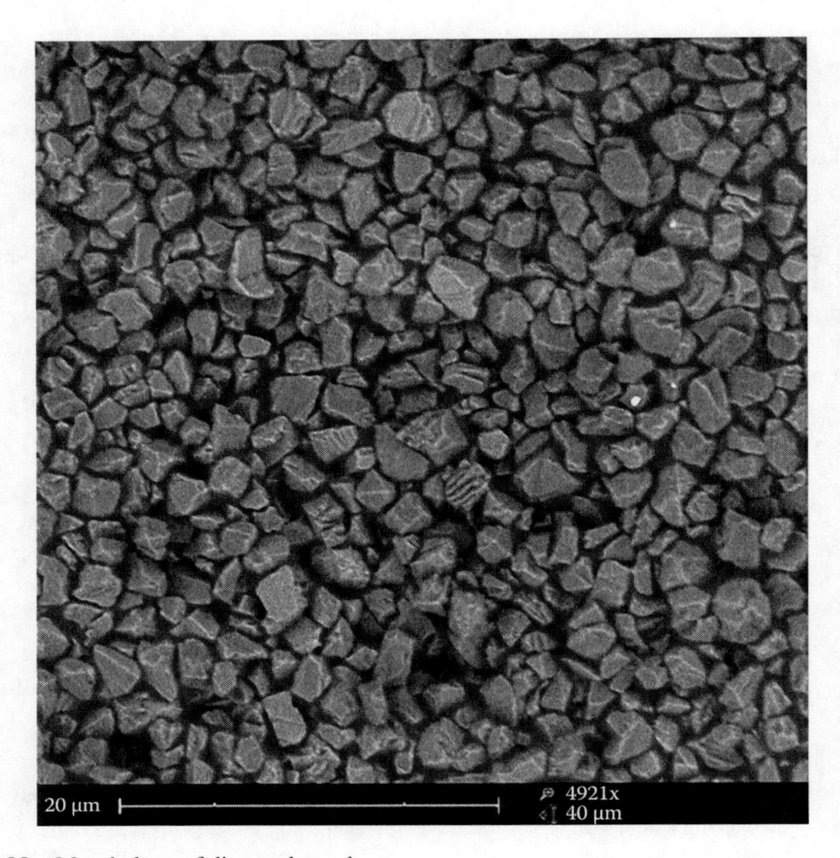

FIGURE B.32 Morphology of diamond powder.

B.3.8 Shaping of PCD and CVD Diamond

It is widely accepted that the real obstacle for the growth of the market for diamond-based products and applications is not only a relatively high cost of the material itself, but foremost the costly process of shaping the manufactured objects. The latter is due to the need to remove relatively large volumes of the diamond material until the final shape is reached. This shaping is very slow, requires the use of expensive equipment, and makes the manufacturing process dependent on expensive specialists with rare skills.

Shaping of PCD and CVD diamond tool blanks is challenging due to their high hardness and low toughness (brittleness). Precision grinding and nontraditional machining processes including electrical discharge machining (EDM), both wire cut EDM (WEDM) and electrical discharge grinding (EDG), various combinations of grinding and EDMing, and laser machining are used to shape PCDs and CVD diamonds (CVD-D).

B.3.8.1 Grinding

The cutting edge geometries and final dimensions of cutting tools are typically produced on CNC tool and cutter grinders. Tools with PCD cutting edges, however, pose a challenge. Grinding PCD pits the hardest known substance (diamond) against an equally hard grinding wheel. The resulting virtual standoff consumes excessive time and wheels. Low *G* ratios, high cutting forces and thus temperatures, high wheel costs, and limited capability of grinding complex tool edge configurations are some of the common problems associated with diamond grinding of PCD. However, the surface finish achieved in grinding is the best compared to other shaping operations.

In grinding PCD, the *G* ratio, that is, the ratio of the volume of PCD workpiece removed to the volume of wheel lost, is very low due to the high hardness of PCD. When diamond wheels are used to grind tungsten carbide, they give typical *G* ratios in the order of 30–500. However, *G* ratios achieved in the grinding of PCD are in the rate of 0.001–0.021, that is, the wheel wears away 500–1000 times faster than the tool being ground. A particular value of the *G* ratio depends on the type of the grinding wheel bond. The resin bond has the *G* ratio as low as 0.001. Metal and vitrified bond diamond wheels have a *G* ratio 20 times larger than that of the resin bond. It is not difficult to imagine that harder bond wheels like metal and vitrified bonds have a higher capacity to hold the diamond grits while grinding. Therefore, the diamond grits have a higher resistance against the grinding force and can remove more materials before a rupture. However, if the holding capacity is too high, the self-sharpening ability of the diamond wheels decreases. In this case, the wheels cannot grind the PCD efficiently and this results in a longer grinding time. A typical stock removal rate with a 6A2 wheel grinding small PCD tools is approximately 0.5 mm^3/min (0.33 $in.^3$/min).

B.3.8.2 Electro-Discharge Machining

One way to limit expense is removing PCD via electrical erosion, in the form of either electrical discharge grinding (EDG) with a copper-alloy disk or wire EDMing (WEDM). The cobalt binder in PCD acts as a conductor for the electrical energy. Because the amount of cobalt is low, electroconductivity of PCD is poorer than that of most metals. This presents a problem with WEDMing coarse PCD grades (>20 μm), because the wire does not cut through a large-size grain. Rather it "jumps" to the nearest cobalt pool, causing rough edges. EDMing has some distinctive advantages. Compared with grinding, the cutting force is much smaller, which does not cause PCD blank deformation. Compared with grinding and FIB, EDMing process efficiency is higher so that the manufacturing cost is lower.

Wire electrical discharge machining (WEDM) is a nontraditional machining process that uses electricity to cut any conductive material precisely and accurately with a thin, electrically charged copper or brass wire as an electrode. During the WEDM process, the wire carries one side of an electrical charge and the workpiece carries the other side of the charge. When the wire gets close to the part, the attraction of electrical charges creates a controlled spark, melting and vaporizing microscopic particles of material. The spark also removes a miniscule chunk of the wire, so after the wire travels through the workpiece one time, the machine discards the used wire and automatically advances new wire. The process takes place quickly—hundreds of thousands of sparks per second—but the wire never touches the workpiece. The spark erosion occurs along the entire length of the wire adjacent to the workpiece, so the result is a part with good surface finish and no burrs, regardless of how large or small the cut. WEDM machines use a dielectric solution of deionized water to continuously cool and flush the machining area, while EDMing is taking place. In many cases, the entire part is submerged in the dielectric fluid, while high-pressure upper and lower flushing nozzles clear out microscopic debris from the surrounding area of the wire during the cutting process. The fluid also acts as a nonconductive barrier, preventing the formation of electrically conductive channels in the machining area. When the wire gets close to the part, the intensity of the electric field overcomes the barrier and dielectric breakdown occurs, allowing current to flow between the wire and the workpiece, resulting in an electrical spark. The material-removal rate can be as high as 6–8 mm^3/min, depending on the particular grade of PCD. The accuracy can be as high as 0.005–0.007 mm.

Electrical discharge grinding (EDG) is a nontraditional thermal process for machining PCDs. EDG has been developed by replacing the stationary electrode used in electrical discharge machining (EDM) with rotating electrode. In the EDG process, material is removed by melting and vaporization, the same as EDM process. In EDG process, an electrically conductive wheel is used as a tool electrode instead of the stationary tool electrode used in EDM. There is no contact with workpiece and tool electrode (rotating wheel) except during electric discharge. Due to the rotational motion of the wheel electrode, the peripheral speed of the wheel transmitted to the stationary dielectric into the gap between workpiece and wheel resulting in the flushing efficiency of the process being enhanced. Therefore, the molten material is effectively ejected from the gap and no debris accumulation takes place, while in EDM, debris accumulation is a major problem which adversely affects the performance of the process. Due to the enhancement in flushing, higher material removal and better surface finish is obtained as compared to the conventional EDM process. At the same machining condition, EDG gives better performances than EDM and machines PCD much faster (two to three times) as compared to conventional grinding.

EDMing requires tight control. Poorly controlled EDM pulses can easily overheat PCD. A recast layer (known as the white layer) on the EDMed surface and dimples due to pulled cobalt from this surface are just a few examples. As discussed in Chapter 2, the thermally damaged PCD contact surface would break down in use. In order to optimize the WEDM process for PCD tool manufacturing, the critical machining requirements must be identified. The optimization of WEDM process includes productivity during roughing operation and quality during finishing operation. In both operations, the electrical parameters including voltage pulse width, peak voltage, peak current, and pulse interval are changed to identify the influence on surface quality and machining rate of PCD. Today's electrical erosion spark generators employ solid-state circuitry that controls spark on-time duration and frequency within nanoseconds.

WEDM can offer an advantage compared to EDG, because it can create small inside radii and other complex tool features that a rotary disk cannot. On the other hand, the rotary electrode is a thick, stiff piece of copper alloy that, unlike a wire, does not offer any deflection so that, all things being equal, a faster material-removal rate can be achieved. The advantage lessens, however, when removing more than 0.3–0.4 mm of excess PCD stock. In this case, WEDMing becomes easier. Moreover, modern WEDM machines are designed so that the wire is presented to the part horizontally rather than in the common vertical alignment. This arrangement enables the wire to travel in a larger range of motion than a vertical wire when shaping a tool clamped in the machine's C-axis.

Having noticed rough cutting edges and the presence of the white layer after EMDing (both WEDM and EDG), posterosion grinding was introduced. The main benefit of posterosion grinding is extended tool life. The decision to follow erosion with grinding to generate a finer finish is usually driven by efficiency. One should decide if he or she is going to put grinding time into a particular

PCD tool, depending on how much more tool life he or she is going to gain in the application. In the author's opinion, HP drilling tools must be posterosion ground, because such tools benefit from a finer finish and contact surface quality than possible with electrical erosion.

Grinding and EDM machine builders have developed units that combine traditional grinding with EDG. Vollmer (Vollmer of America Corp., Carnegie, PA), for example, builds machines that combine conventional grinding with rotary EDG as well as multiaxis wire EDMs for shaping PCD tools. Walter Grinders (United Grinding Technologies Inc., Fredericksburg, VA) also offers machines with EDG and conventional grinding capabilities.

Two basic Helitronic Diamond platforms by Walter Grinders are available for these applications. In a 2-spindle machine, one spindle is dedicated to grinding, and the other can either be for grinding or erosion. The other Helitronic Diamond configuration involves a single-spindle machine with a wheel changer that switches both the grinding or EDG wheel and its cooling manifold. A 12-position carousel can hold a mixture of grinding wheel and/or erosion wheel setups, which can be quickly exchanged to produce different tool features. The Ewamatic line by United Grinding Technologies includes a number of grinding machines with an eroding option.

Another emerging process to enhance the efficiency of shaping PCDs is abrasive electrical discharge grinding (AEDG). In AEDG, the synergistic interactive effect of a combination of electrical discharge machining and grinding is employed to increase machining productivity. In the AEDG process, metallic or graphite electrode used in electrical discharge grinding (EDG) process has been replaced with metallic bond grinding wheel. Therefore, material is removed by the combined effect of electro-erosion and microcutting process (mechanical effect of abrasives). An increase in performance measures of the AEDG process becomes evident when machining special PCD grades.

B.3.8.3 Laser Machining

Lasers applications are increasing in diamond cutting tool manufacturing. The process is based on rapid heating of a workpiece by short pulses of photons which being absorbed in a thin surface layer cause a local vaporization (sublimation) of the work material. The laser beam can be focused into a spot as small as a few microns in diameter. The laser machining is a contactless technique, the hardness of the work material is not a significant factor, but optical absorption and thermal conductivity are important factors, which determine the cutting rate at given laser parameters.

Laser cutting is especially effective for processing CVD-D and PCD. A variety of special lasers emitting in the spectral range from UV to IR are available now for cutting purposes. The use of Nd: YAG and fiber lasers is most common. The high peak performance of common lasers permits a maximum cut depth of up to 10 mm. As the laser beam can be focused on a very small diameter for high precision, fine cuts are possible with a minimum cut width of up to 15 μm (0.0006 in.). In addition, the heat-affected zone along the cut is very small (up to 2 μm). This means that deformations of the parts to be processed can be avoided.

The high energy depth in the focus point of the laser beam causes the material to melt and evaporate. By using an active or neutral process gas, for example, oxygen, nitrogen, or argon, the melted material is blown out. If the workpiece or laser beam is now moved, a cut is created. The smallest possible cut width is dependent on both the beam characteristics and material strength. When cutting fine contours, the precision and dynamics of the cutting machine are of extreme importance.

The cavity size is a function of the size of the laser and the light's focus diameter. A higher wattage laser creates a larger cavity than one with less wattage, and a larger cavity means a higher material-removal rate. For example, the Spectra 820 laser machining center (Wendt GmbH, Meerbusch, Germany) uses a stationary diode-pumped, solid-state laser and oscillates the workpiece to the light source to achieve material-removal rates and impart surface finishes similar to finish grinding. Unlike grinding, where parameters such as wheel diameter or wheel speed are continually changing and interacting with each other, the laser machine's parameters are predictable and do not interact with each other.

The LASER LINE by United Grinding Technologies produces extremely complex geometries of rotationally symmetric tools and indexable cutting inserts in just one clamping. With five CNC machine axes and three overlaid laser beam guide axes, a total of eight CNC axes are available. Complex geometries including cuts and cavities are mastered with precision.

Another important application of laser machining in diamond tools is etching of chipbreakers. The stringy "birds nest" that forms around tools and workpieces when the formed chip fails to break causes shortened tool life, broken tools, poor finish, and costly machine downtime. The problem is especially troubling when turning aluminum and other nonferrous materials. To combat this, customized laser-etched 3D chip breaker geometries can be etched into PCD and CVD diamond inserts and cutting tools.

Laser cutting has the potential to replace or complement grinding and EDMing in economical production of diamond tools. Today, the common systems available in the market have pulse durations in the nanosecond range or longer. This creates some problems in achieving high quality cutting edges. Diamond is prone to crack when machined with nanosecond-laser pulses. Moreover, quality and precision of laser-shaped surfaces do not meet the requirements for the use as cutting tools without further postprocessing. The use of pulse duration in the picosecond range reduces the tendency of diamond to form cracks, but does not eliminate cracks completely. The use of lasers in femtosecond range eliminates cracks formation. As such, the quality of the machined surface and head-affected zone reduce significantly. The cost of picosecond- and femtosecond-lasers machines is very high (in the range of $1,000,000).

However, two important points should be remembered. First, the material-removal rate with nanosecond-laser pulses is comparable with EDMing and grinding, whereas that with the picosecond- and femtosecond-lasers is considerably lower. Second, the quality of the polished surface cannot be achieved with laser machining.

REFERENCES

Asmussen, J. and Reinhard, D.K. 2002. *Diamond Films Handbook*. New York: Marcel Dekker.

Barnard, A. 2000. *The Diamond Formula: Diamond Synthesis: A Gemological Perspective*. Boston, MA: Butterworth-Heinemann.

Boyd, F.R. and Gurney, J.J. 1986. Diamonds and the African lithosphere. *Science* 232:472–477.

Bruton, E. 1970. *Diamonds*. Philadelphia, PA: Chilton Book.

Bundy, F.P. 1989. Pressure-temperature phase diagram of elemental carbons. *Physica A* 156:169–178.

Burstein, E. 2003. A major milestone in nanoscale material science: The 2002 Benjamin Franklin Medal in Physics presented to Sumio Iijima. *Journal of the Franklin Institute* 340(3–4):221–242.

Chenguang, L. 1992. Composition, morphology and distribution of RE element containing phases in cemented carbide. *International Journal of Refractory Metals and Hard Materials* 11(5):295–302.

Chrenko, R.M., Tuft, R.E., and Strong, H.M. 1977. Transformation of the state of nitrogen in diamond. *Nature* 270:141–144.

Davis, J.R. (ed.) 1995. *Tool Materials (ASM Specialty Handbook)*. Metals Park, OH: ASM.

Davis, R.F. 1993. *Diamond Films and Coating*. Park Ridge, NJ: Noyes Publications.

Dismukes, J.P., Gaines, P.R., Witzke, H., Leta, D.P., Kear, B.H., and Behal, S.K. 1988. Demineralization and microstructure of carbonado. *Material Science and Engineering A* 105/106:555–563.

Evans, T. and Wild, R.K. 1965. Plastic bending of diamond plates. *Philosophical Magazine* 12(117):479–489.

Field, J.E. (ed.) 1992. *The Properties of Natural and Synthetic Diamond*. London, U.K.: Academic Press.

Grodzinski, P. 1953. *Diamond Technology*, 2nd edn. London, U.K.: N.A.G. Press.

Hall, H.T. 1960. Ultra-high-pressure, high-temperature apparatus: The 'belt'. *Review of Scientific Instruments* 31:125–131.

Hall, H.T. 1970. Sintered diamond: A synthetic carbonado. *Science* 169:868–869.

Hall, H.T. 1982. Diamonds, synthetic. In *Encyclopedia of Chemical Processing and Design*, Vol. 15, McKetta, A.W.C.J.J. (ed.). New York: Marcel Dekker. pp. 410–435.

Hannay, J.B. 1879. On the artificial formation of the diamond. *Proceedings of the Royal Society of London* 30(200–205):450–461.

Jeynes, C. 1978. Natural polycrystalline diamond. *Industrial Diamond Review* (January) 39:15–28.

Krebs, R.E. 2006. *The History and Use of Our Earth's Chemical Elements: A Reference Guide*, 2nd edn. London, U.K.: Greenwood Press.

Lammer, A. 1988. Mechanical properties of polycrystalline diamond. *Materials Science and Technology* 4:949–955.

Lassner, E. and Schubert, W.-D. 1999. *Tungsten: Properties, Chemistry, Technology of the Element, Alloys, and Chemical Compounds*. New York: Kluwer Academic.

Liu, A.Y. and Cohen, M.L. 1989. Prediction of new low compressibility solids. *Science* 245:841–842.

Melton, C.E. 1985. *Ancient Diamond Time Capsules*. Hull, GA: Melton-Giardini.

Moissan, H. 1884. Nouvelles expériences sur la reproduction du diamant. *Comptes Rendus de l'Académie des Sciences (Paris)* 118:320–341.

Neikov, O., Naboychenko, S., Gopienko, V.G., and Frishberg, I.V. 2009. *Handbook of Non-Ferrous Metal Powders: Technologies and Applications*. Oxford, U.K.: Elsevier.

Partington, J.R. 1962. *A History of Chemistry*. London, U.K.: Macmillan & Co.

Pierson, H.O. 1999. *Handbook of Chemical Vapor Deposition: Principles, Technology and Applications*, 2nd edn. New York: Noyes Publications.

Prikhna, A.I. 2008. High-pressure apparatuses in production of synthetic diamonds (review). *Journal of Superhard Materials* 30(1):1–15.

Purniawan, A. 2008. Deposition and characterization of polycrystalline diamond coated on silicon nitride and tungsten carbide using microwave plasma assisted chemical vapour deposition technique. Mechanical Engineering, Universiti Teknologi Malaysia, Johor Bahru, Malaysia.

Royère, C. 1999. The electric furnace of Henri Moissan at one hundred years: Connection with the electric furnace, the solar furnace, the plasma furnace? *Annales pharmaceutiques françaises* 57(2):116–130.

Smith, C.P., Bosshart, G., Ponahlo, J., Hammer, V.M.F., Klapper, H., and Schmetzer, K. 2000. GE POL diamonds: Before and after. *Gems & Gemology* 36:192–215.

Smith, N.R. 1974. *User's Guide to Industrial Diamonds*. London, U.K.: Hutchinson Benham.

Spear, K. and Dismukes, J. 1994. *Synthetic Diamond: Emerging CVD Science and Technology*. Pennington, NJ: John Wiley & Sons.

Uhrenius, B., Brandrup-Wognsen, H., Gustavsson, U., Nordgren, A., Lehtinen, B., and Manninen, H. 1991. On the formation of impurity-containing phases in cemented carbides. *International Journal of Refractory Metals and Hard Materials* 10(1):45–55.

Upadheaya, G.S. 1998. *Cemented Tungsten Carbides. Production, Properties, and Testing*. Westwood, NJ: Noyes Publications.

von Platen, B. 1962. A multiple piston, high pressure, high temperature, apparatus. In *Modern Very High Pressure Techniques*, Wentorf, R.H. (ed.). Washington, DC: Butterworths.

Wilks, J. and Wilks, E. 1991. *Properties and Applications of Diamond*. Oxford, U.K.: Butterworth-Heinemann.

Appendix C: Basics of the Tool Geometry

The major objective of this appendix is to familiarize potential readers with the basic notions and definitions used in the analysis of the tool geometry.

C.1 BASIC TERMS AND DEFINITIONS

The geometry and nomenclature of cutting tools, even single-point cutting tools, are surprisingly complicated subjects (Astakhov 1998/1999, 2010). It is difficult, for example, to determine the appropriate planes in which the various angles of a single-point cutting tool should be measured; it is especially difficult to determine the slope of the tool face. The simplest cutting operation is one in which a straight-edged tool moves with a constant velocity in the direction perpendicular to the cutting edge of the tool. This is known as the 2D or orthogonal cutting process illustrated in Figure C.1. The cutting operation can best be understood in terms of orthogonal cutting parameters, so the basic cutting terminology and the theory of metal cutting were developed for orthogonal cutting.

Although representation of basic terms shown in Figure C.1 is simple and rather straightforward, it is not always easy to correlate the presented terms with real machining operations such as with turning shown in Figure C.2. Figure C.3 helps to correlate the terminology used in orthogonal cutting and turning.

C.1.1 Workpiece Surfaces

In orthogonal cutting (Figure C.1), the two basic surfaces of the workpiece are considered:

- Work surface is the surface of the workpiece to be removed by machining.
- Machined surface is the surface produced after the cutting tool pass.

In many practical machining operations, an additional surface is considered. The transient surface is the surface being cut by the major cutting edge (Figure C.3a). Note that the transient surface is always located between the work surface and the machined surface. Besides simple shaping, planning, and broaching (excluded helical), the cutting edge does not form the machined surface in many real cutting operations. As clearly seen in Figure C.3b, the machined surface is formed by the tool corner and minor cutting edge. Unfortunately, not much attention is paid to these two important components of the tool geometry although their parameters directly affect the integrity of the machined surface including the surface roughness and machining residual stresses. Misunderstanding of the previously discussed matter causes a great mismatch in the results of known modeling of the cutting process and reality as the tribological conditions on the tool flank face (over the tool–workpiece tribological interface) including forces, stresses, and temperatures are considered to be important in the formation of the integrity of the machined surface. As clearly seen in Figure C.3, it is not so as these tribological conditions affect the formation of the transient surface.

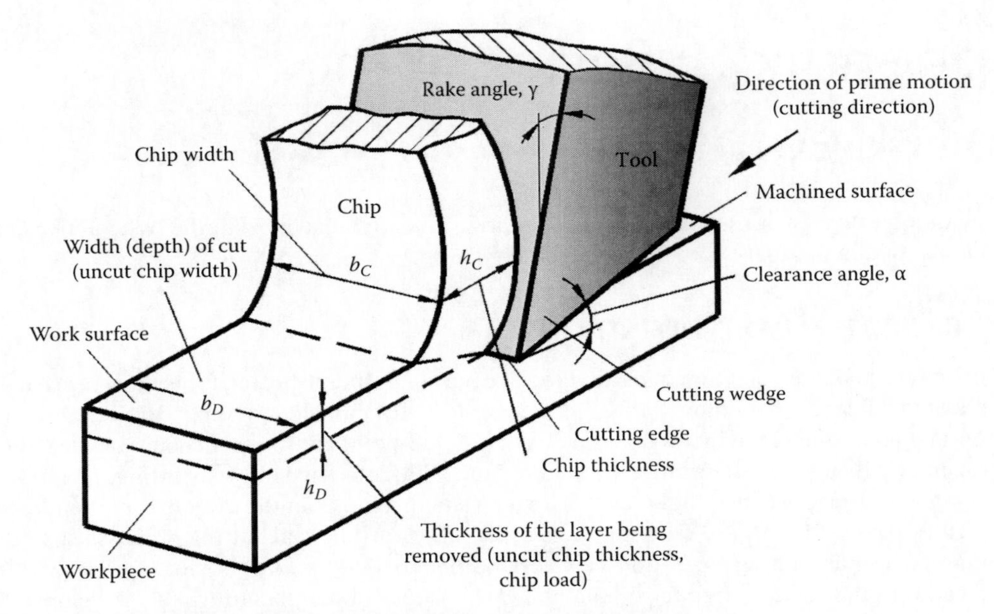

FIGURE C.1 Visualization of the basic terms in orthogonal cutting.

FIGURE C.2 Arrangements of the components in turning.

C.1.2 Tool Surfaces and Elements

The design components of the cutting tool are defined as follows:

- *Rake face* is the surface over which the chip, formed in the cutting process, slides.
- *Flank face* is the surface(s) over which the machined surface passes.
- *Cutting edge* is a theoretical line of intersection of the rake and the flank surfaces.
- *Cutting wedge* is the tool body enclosed between the rake and the flank faces.
- *Shank* is the part of the tool by which it is held.

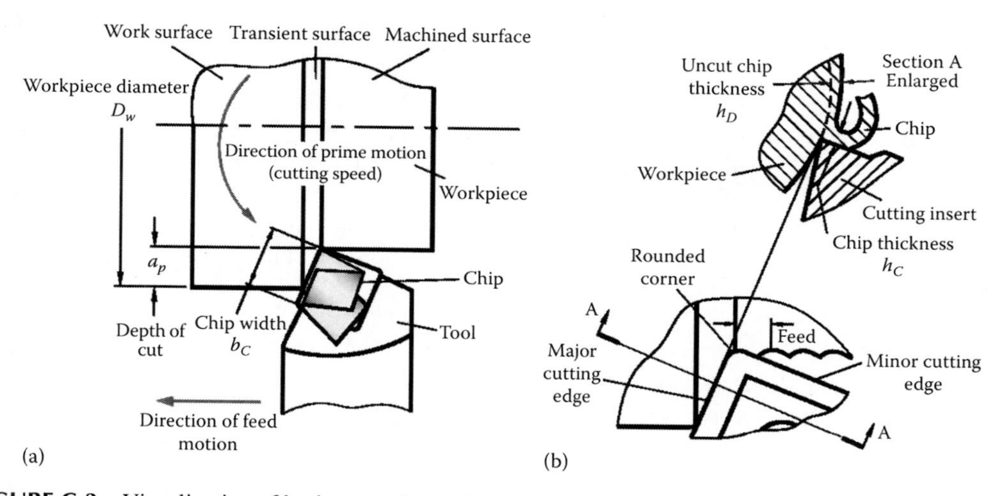

FIGURE C.3 Visualization of basic terms in turning: (a) general view and (b) enlarged cutting portion.

Understanding the previously introduced terms is important for further considerations. As introduced, the cutting edge is a *line*, in general, a 3D line that does not have any 'meat' attached. Therefore, it cannot have strength, temperature, wear, chipping, and other physical characteristics of a 3D body although these characteristics are routinely prescribed to the cutting edge in the professional and even in scientific literature. In reality, the cutting wedge and its corresponding surfaces (the rake and flank) should be considered in the modeling of the cutting process, cutting tool design, and optimization of machining operations.

C.1.3 Types of Cutting

Orthogonal cutting is that type of cutting where the straight cutting edge of the wedge-shaped cutting tool is at right angle to the direction of cutting as shown in Figure C.4a. The additional distinctive features of orthogonal cutting are as follows:

- The cutting edge is wider than the width of cut.
- No side spread of the layer being removed occurs on its transformation into the chip.
- Plane strain condition is the case, that is, a single *slice* (by a plane perpendicular to the cutting edge) of the model shown in Figure C.4a can be considered in the analysis of the chip formation model.

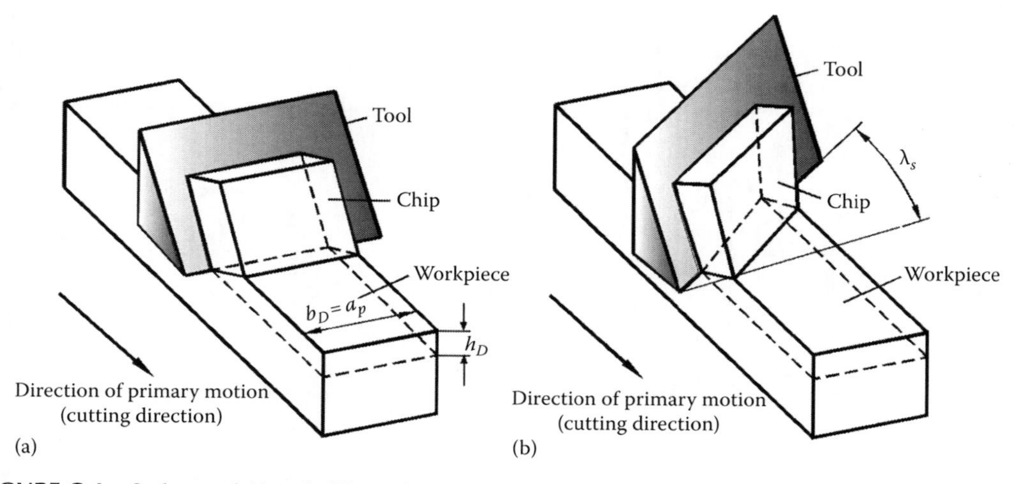

FIGURE C.4 Orthogonal (a) and oblique (b) cutting.

- The cutting edge does not pass the previously machined surface by this cutting edge, so there is no influence of the previous cutting passes on the current pass. This is not the case in many real machining operations, for example, drilling operations, because the temperatures and machining residual stresses built on the previous pass might significantly affect the cutting conditions on the current pass. Moreover, this influence depends on many cutting parameters such as the rotational speed of the workpiece (which defines the time difference between two successive positions of the cutting edge and the intensity of the residual heat) and axial feed (which defines the machining residual stresses left from the previous pass of the cutting tool).

Oblique cutting is that type of cutting where the straight cutting edge of the wedge-shaped cutting tool is not at right angle to the direction of cutting. Figure C.4 illustrates the difference between orthogonal and oblique cutting. In orthogonal cutting (Figure C.4a), the cutting edge is perpendicular to the direction of the primary motion, while in oblique cutting (Figure C.4b), it is not. The angle that the straight cutting edge makes with the direction of the cutting speed is known as the tool cutting-edge inclination angle λ_s. The plastic deformation of the layer being removed in oblique cutting is more complicated than that in orthogonal cutting (Shaw 2005).

Free cutting is that type of orthogonal or oblique cutting when only one cutting edge is engaged in cutting. Although this definition is widely used in the literature on metal cutting (Zorev 1966; Armarego and Brown 1969; Shaw 2005), it does not provide the proper explanation to the idea of free cutting. For example, if a cutting edge is not straight, it does not perform free cutting. In contrast, a number of cutting edges can be simultaneously engaged in cutting in surface broaching, but each edge is engaged in free cutting. In the definition, *free* means that the elementary chip flow vectors from each point of the cutting edge are parallel to each other and do not intersect any other chip flow vectors. An example of free cutting is shown in Figure C.5a. If more than one adjacent cutting edges are involved in cutting (Figure C.5b shows an example of two cutting edges) or when the cutting edge is not straight (Figure C.5c), the chip flows formed at different cutting edges or at different points of the same cutting edges cross each other, creating chip flow interface and thus causing greater chip deformation and cutting force than in free cutting.

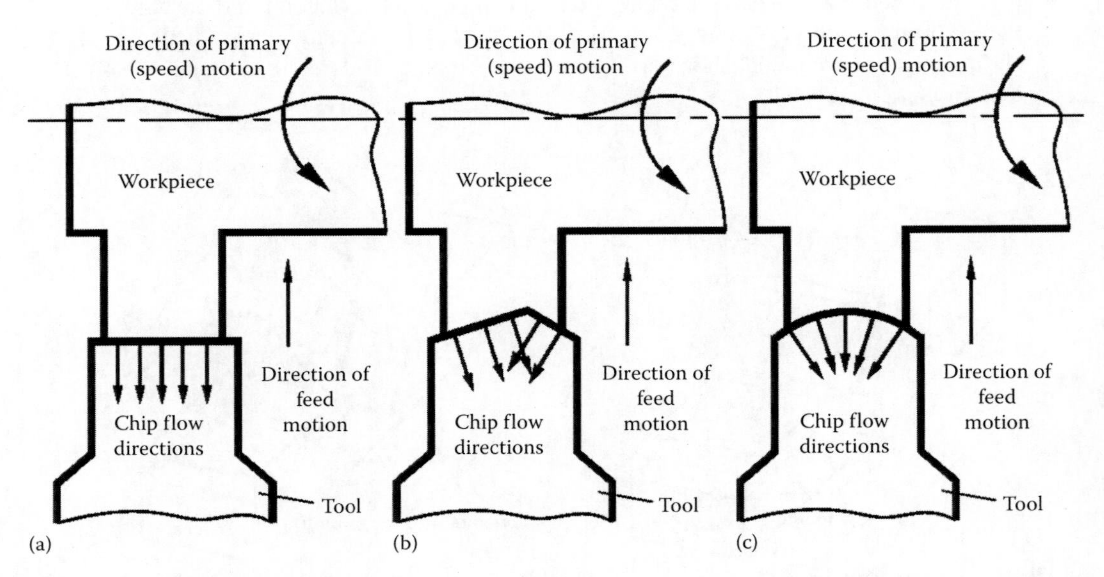

FIGURE C.5 Model showing (a) free and (b) nonfree cutting with two cutting edges and (c) nonfree cutting with a rounded cutting edge.

C.2 SYSTEMS OF CONSIDERATION OF TOOL GEOMETRY

Two basic standards provide definitions of the cutting tool geometry: ISO 3002-1 (1982)/Amd 1:1992 "Basic quantities in cutting and grinding. Part 1: Geometry of the active part of cutting tools - general terms, reference systems, tool and working angles, chip breakers" and ANSI B94.50 (1975) "Basic nomenclature and definitions for single-point cutting tools. 1975 (reaffirmed 1993)." Both geometry standards discuss two systems of consideration of the cutting tool geometry, namely, the tool-in-hand and tool-in-use systems (hereafter, T-hand-S and T-use-S, respectively). The former relates to the so-called static geometry, while the latter is based on the consideration of tool motions with respect to the workpiece. In the author's opinion, however, these two systems are insufficient for the proper consideration of cutting tool geometry. Another system, namely, the tool-in-machine system (Astakhov 1998/1999, 2010) (hereafter, T-mach-S), should also be considered.

The introduction of an additional system of consideration may be thought by some as a kind of overcomplicating of the cutting tool geometry and its practical applications so that it is suitable only for ivory academicians as it has little practical value at the shop floor level. In the author's opinion, the opposite is actually the case. Namely, misunderstanding the tool geometry in the previously mentioned systems leads to improper selection of the tool geometry parameters and prevents meaningful optimization of practical machining operations. This is particularly true for drilling tools as discussed in Chapter 4.

C.2.1 Tool-in-Hand System

C.2.1.1 T-Hand-S Reference Planes

The cutting tool geometry includes a number of angles measured in different planes. Although the definitions of the standard planes of consideration of the tool geometry are the same for all systems of consideration, these planes are not the same in these systems. This is because a set of the standard planes in each particular system is defined in a certain coordinate system. Thus, it is of crucial importance to set the proper coordinate system in each system of consideration as the very first step in any geometry analysis.

Because the angles and other geometric features may vary from point to point along the cutting edge of a tool, it is necessary to locate the reference system at whatever point one desires to be able to define the tool geometry. Therefore, each plane in any of the previously discussed systems of consideration is defined with respect to a selected point on the cutting edge.

Each plane is provided with a symbol consisting of P with a suffix indicating the plane's identity (e.g., P_r, the tool reference plane) when a point of the major cutting edge is considered. When it is necessary to distinguish clearly a plane passing through a selected point on the minor cutting-edge, the appropriate symbol bears a prime (e.g., P'_s, the tool minor cutting edge plane).

When the cutting edge, face, or flank is curved, the tangents or tangent planes to the selected point should be used in the reference system of planes.

Figure C.6 establishes the basis for the T-hand-S for a selected point on the major cutting edge. In this figure, the vector of the cutting speed $\mathbf{v}$ is placed in the assumed direction of the primary motion, known as the direction of the cutting velocity (customarily referred to as the cutting speed), and vector of the feed velocity $\mathbf{v}_f$ is in the assumed direction of the cutting feed. The summation of these two vectors gives the resultant cutting speed vector $\mathbf{v}_e$ as shown in Figure C.6.

Once the assumed directions of the primary and feed motions are established for the selected point of the cutting edge, two basic reference planes can be defined:

- *Tool reference plane* P_r as to be perpendicular to the assumed direction of the primary motion. Note that both tool geometry standards (ISO 3002-1 [1982]/Amd 1 and ANSI B94.50 [1975]) define this plane incorrectly.
- *Assumed working plane* P_f as to be perpendicular to the reference plane P_r and containing the assumed direction of feed motion.

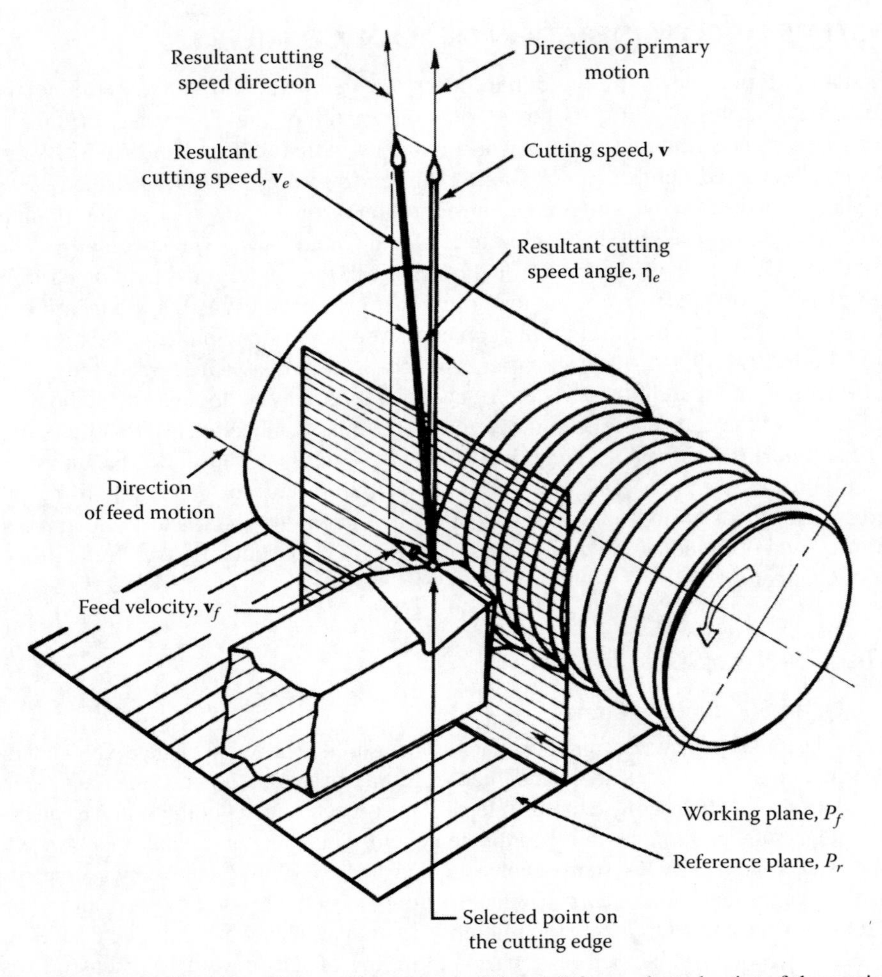

FIGURE C.6 Assumed directions of the tool motions in T-hand-S for a selected point of the cutting edge.

Note that both previously mentioned standards define the tool reference plane as passing through the point of consideration. In the author's opinion, it is not necessarily. It is defined as to be perpendicular to the direction of the primary motion, but the exact location of the plane is a matter of consideration convenience. To the contrary, the working plane should always pass through the considered point, that is, its location is unique as strictly defined by two perpendicular vectors that must lie in this plane.

Figure C.7 shows graphical representation of the following planes:

- *Tool cutting-edge plane* P_s is perpendicular to P_r, and contains the major cutting edge. If the major cutting edge is not straight, then the tool cutting-edge plane should be determined for each point on the curved cutting edge, thus being the plane that is tangent to the cutting edge at the point of consideration and perpendicular to the reference plane.
- *Tool orthogonal plane* P_o is defined as a plane through the selected point on the cutting edge and perpendicular both to the tool reference plane P_r and to the tool cutting-edge plane P_s.
- *The tool normal plane* P_n is defined as the plane normal to the cutting edge at the selected point on the cutting edge.

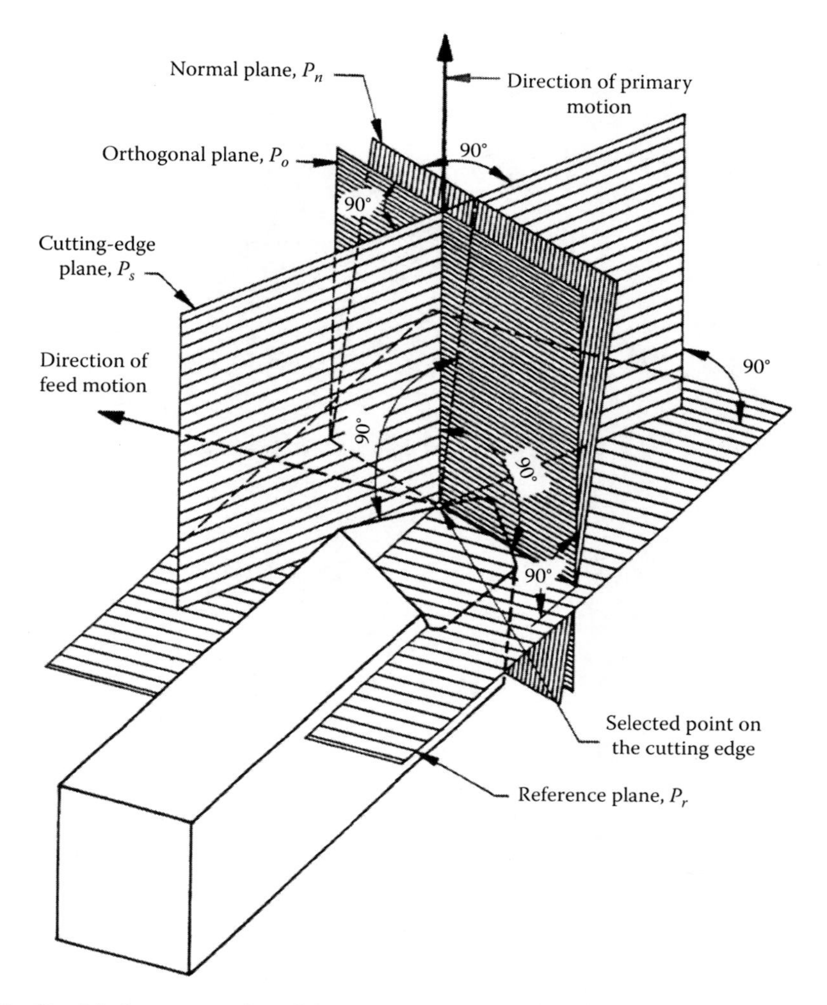

FIGURE C.7 Graphical representation of the tool cutting edge, orthogonal and normal planes.

Figure C.8 shows graphical representation of *the tool back plane* P_p. As can be seen, this plane is defined as a plane through the selected point on the cutting edge and perpendicular both to the tool reference plane P_r and to the tool working plane P_f.

Similarly to points of the major cutting edge, a system of reference planes can be defined for points of the minor cutting edges. Figure C.9 shows graphical representation of the tool reference, cutting edge, and normal planes for the selected point of the minor cutting edge.

C.2.1.2 T-Hand-S Coordinate System and Angles

Figure C.10 sets the tool coordinate system in T-hand-S. The origin of this coordinate system is always placed at a selected point on the cutting edge. The z-axis is always in the assumed direction of the primary motion, while the x-axis is in the direction of the assumed direction of the feed motion. The y-axis is perpendicular to the z- and x-axes to form a right-hand Cartesian coordinate system. It is extremely important not to associate this coordinate system with the actual holders, with the location of this holder in the machine, and with the actual speed and feed directions, assuming that a cutting element is a single-point cutting tool considered later in this section. The corresponding transformations to the geometry in T-use-S (through T-hand-S and T-mach-S) is then

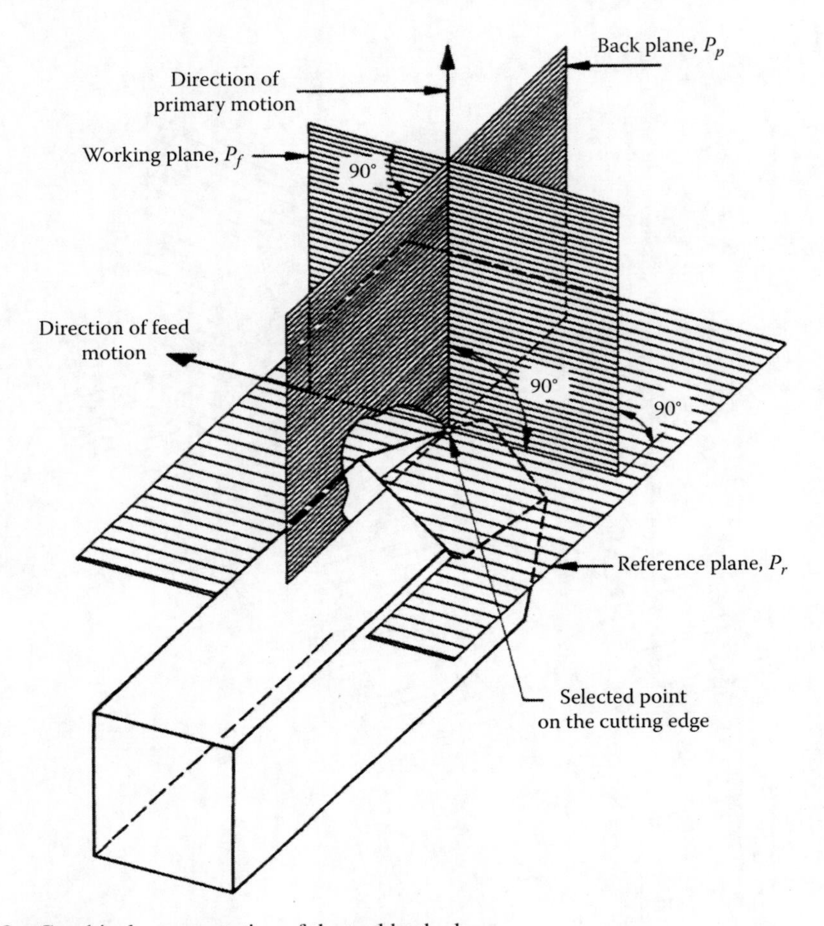

FIGURE C.8 Graphical representation of the tool back plane.

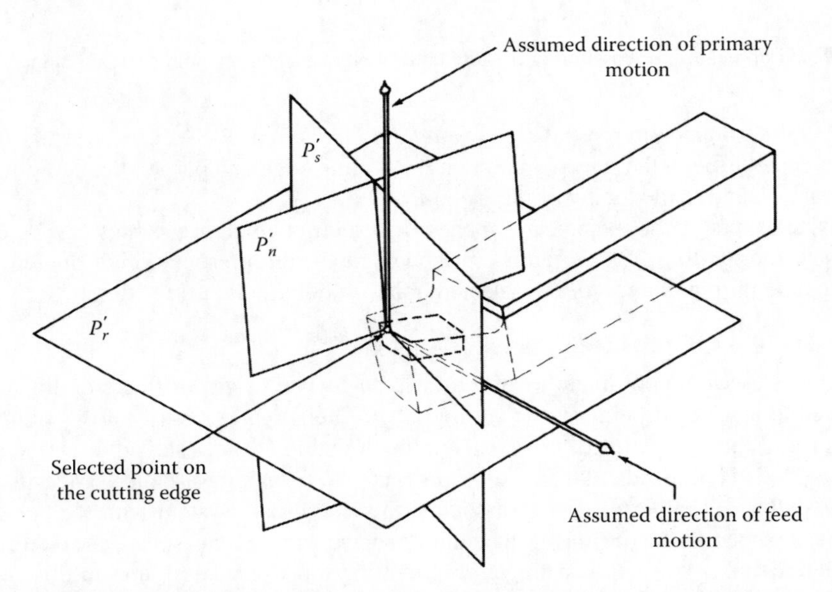

FIGURE C.9 Graphical representation of the tool reference, cutting edge, and normal planes for the selected point of the minor cutting edge.

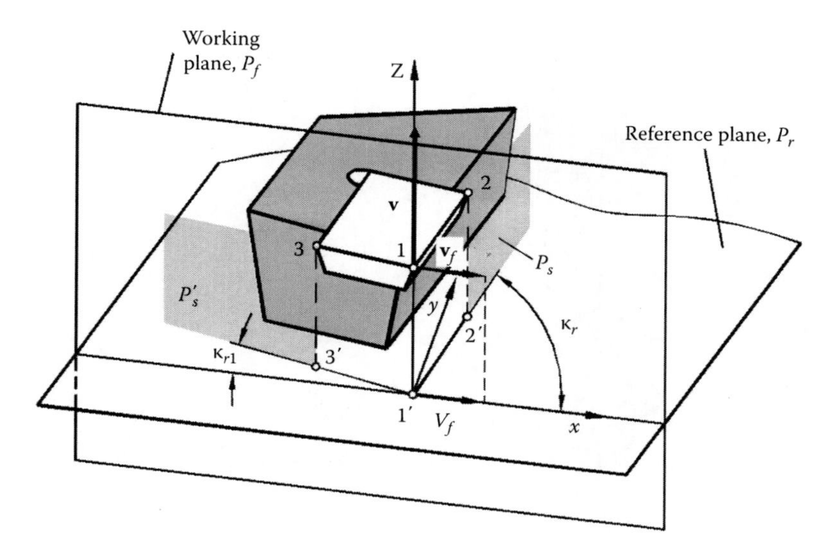

FIGURE C.10 Tool-in-hand coordinate system and graphical representation of the tool cutting edge angles of the major and minor cutting edges.

accomplished, accounting for the actual location of this cutting element in the tool holder, location of this holder in the machine, and speed and feed directions in actual machining.

In Figure C.10, line 1–2 is the major cutting edge, and 1–3 is the minor cutting edge. For the sake of simplifying the current consideration, these cutting edges are shown as straight lines. The set of tool angles can be defined for each of these cutting edges as shown in Figure C.10.

The following angles are defined for a considered point on the major cutting edge:

- *Tool cutting edge angle* κ_r as the acute angle between the projection of the major cutting edge into the reference plane and the x-direction. In another more strict definition, it is the angle measured in the reference plane P_r between the lines of intersection of the tool working plane P_f and the tool cutting-edge plane P_s with the reference plane. Angle κ_r is always positive, and it is measured in a counterclockwise direction from the position of the assumed working plane P_f in the manner shown in Figure C.10.
- *Tool minor cutting edge angle* κ_{r1} as the acute angle between the projection of the minor (end) cutting edge into the reference plane and the x-direction. In another more strict definition, it is the angle measured in the reference plane P_r between the lines of intersection of the tool working plane P_f and the tool minor cutting-edge plane P'_s with the reference plane. Angle κ_{r1} is always positive (including zero), and it is measured in a clockwise direction from the position of the assumed working plane.

The geometry of a cutting element is defined by certain basic tool angles, and thus, precise definitions of these angles are essential. A system of tool angles in T-hand-S is shown in Figure C.11. The rake, wedge, and flank angles are designated by γ, β, and α, respectively, and these are further identified by the subscript of the plane of consideration.

The definitions of basic tool angles in T-hand-S for point a on the cutting edge are as follows:

- *Tool approach angle* ψ_r (the lead angle as termed by ANSI B94.50 [1975]) is the acute angle that the tool cutting-edge plane P_s makes with the tool back plane P_p and is measured in the tool reference plane P_r as shown in Figure C.11.
- *Tool included angle* ε_r is the angle between the tool cutting-edge plane P_s and the tool minor cutting edge plane P'_s measured in the tool reference plane P_r as shown in Figure C.11.

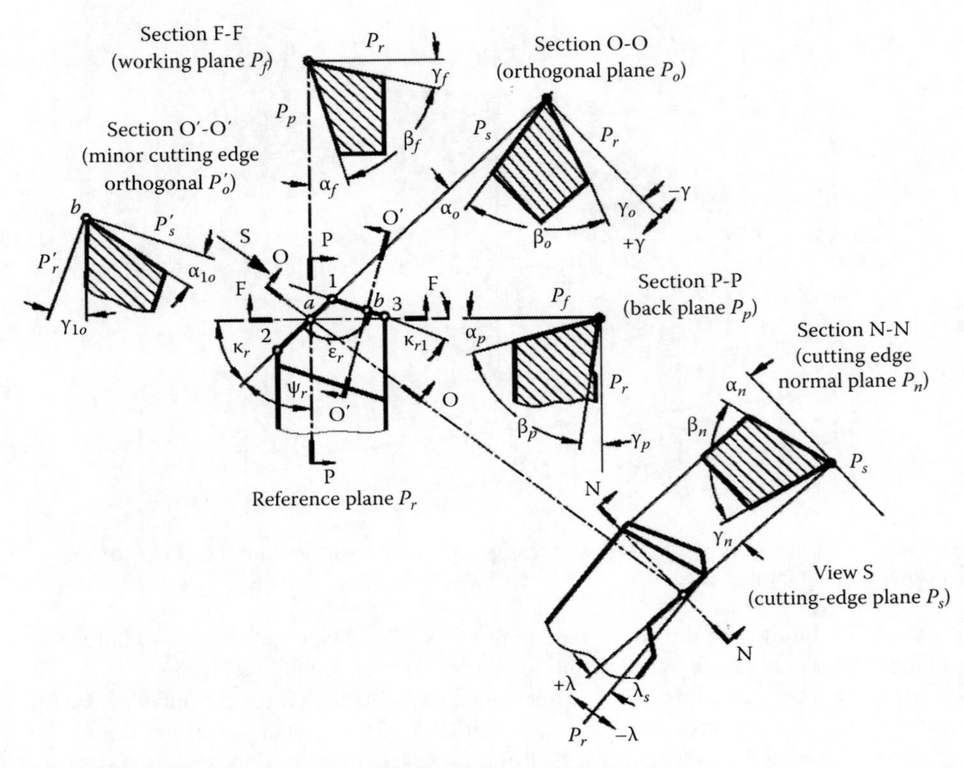

FIGURE C.11 System of tool angles in the defined planes in T-hand-S.

- *The rake angles* γ_f, γ_p, γ_o, γ_n are defined in the corresponding reference planes. The rake angle is the angle between the reference plane drawn through the selected point on the cutting edge (point a in Figure C.11) and the intersection line formed by the considered (orthogonal, back, or working) reference plane and the tool rake face. The rake angle is defined as always being acute when looking across the rake face from the selected point and along the line of intersection of the face and the corresponding reference plane. Angle γ_f is known as the tool side rake, γ_p is known as the tool back rake, γ_o is known as the tool orthogonal rake, and γ_n is known as the normal rake. The sign of the rake angles is well defined as shown in Figure C.11 for the orthogonal rake angle, that is, if the intersection line formed by the considered (orthogonal, back, or working) reference plane and the tool rake face is above the reference plane, then the rake angle is negative; if it is below, the rake angle is positive; and if it coincides with the trace of the reference plane, it is zero or neutral.
- *The clearance angles* α_f, α_p, α_o, α_n are defined in a similar way to the rake angles. The clearance (angle) is the angle between the tool cutting-edge plane P_s and the intersection line formed by the tool flank plane and the considered reference plane as shown in Figure C.11. Angles α_f, α_p, α_o, α_n are clearly defined in the corresponding planes as shown in Figure C.11. Angle α_f is known as the tool side clearance, α_p is known as the tool back clearance, and α_n is known as the normal clearance.
- *The wedge angles* β_f, β_p, β_o, β_n are defined in the corresponding reference planes. The wedge angle is the angle between the two intersection lines formed as the corresponding reference plane intersects with the rake and flank faces. For all cases, the sum of the rake, wedge, and clearance angles is 90°, that is, $\gamma_p + \beta_p + \alpha_p = \gamma_n + \beta_n + \alpha_n = \gamma_o + \beta_o + \alpha_o = \gamma_f + \beta_f + \alpha_f = 90°$.

- *Cutting edge inclination angle* λ_s is defined as the angle between the cutting edge and the tool reference plane P_r measured in the tool cutting-edge plane P_s drawn through a considered point of the cutting edge. This angle is defined as always being acute and positive if the cutting edge, when viewed in a direction away from the selected point at the tool corner being considered, lies on the opposite side of the reference plane from the direction of the primary motion. This angle can be defined at any point of the cutting edge. The sign of the inclination angle is well defined in Figure C.11.

The same angles can be defined for the minor cutting edge in the corresponding reference planes. Section O′–O′ represents the orthogonal plane drawn through a point b on the minor cutting edge. In this section, the clearance angle of the minor cutting edge (i.e., its point b) α_{o1} is defined as the angle between the tool minor cutting-edge plane P_s' and the intersection line formed by the tool flank plane and the minor orthogonal plane P_o', while the rake angle of the minor cutting edge (i.e., its point b) γ_{o1} is defined as the angle between the tool minor reference plane P_r' and the intersection line formed by the tool rake plane and the minor orthogonal plane P_o'.

Simple relationships exist among the considered angles in T-hand-S (Astakhov 2010):

$$\tan\gamma_n = \cos\lambda_s \tan\gamma_o \tag{C.1}$$

$$\tan\alpha_p = \frac{\tan\alpha_o}{\cos\kappa_r} \tag{C.2}$$

$$\tan\alpha_f = \frac{\tan\alpha_o}{\sin\kappa_r} \tag{C.3}$$

$$\cot\alpha_n = \cos\lambda_s \cot\alpha_o \tag{C.4}$$

$$\tan\lambda_s = \sin\kappa_r \tan\gamma_p - \cos\kappa_r \tan\gamma_f \tag{C.5}$$

$$\tan\gamma_f = \tan\gamma_o \sin\kappa_r + \tan\lambda_s \cos\kappa_r \tag{C.6}$$

$$\tan\gamma_p = \tan\gamma_o \cos\kappa_r - \tan\lambda_s \sin\kappa_r \tag{C.7}$$

$$\tan\gamma_o = \cos\kappa_r \tan\gamma_p + \sin\kappa_r \tan\gamma_f \tag{C.8}$$

$$\cot\alpha_o = \cos\kappa_r \cot\alpha_p + \sin\kappa_r \cot\alpha_f \tag{C.9}$$

It must be stated, however, that Equations C.5 through C.9 apply only when the cutting edge angle κ_r is less than or equal to 90°.

Example C.1

Problem

The optimal cutting performance of a single-point tool for turning was found when this tool has the following geometry: normal clearance angle $\alpha_n = 12°$, normal rake angle $\gamma_n = 8°$, cutting edge inclination angle $\lambda_s = 10°$, and tool cutting edge angle $\kappa_r = 60°$. Find the corresponding angles in T-hand-S in the orthogonal, back, and working planes that were used in tool design and manufacturing.

Solution

The clearance angle in the orthogonal plane calculates using Equation C.4 as

$$\alpha_o = \arctan(\cos\lambda_s \tan\alpha_n) = \arctan(\cos 10° \tan 12°) = 11.82°$$

The rake angle in the orthogonal plane calculates using Equation C.1 as

$$\gamma_o = \arctan\left(\frac{\tan\gamma_n}{\cos\lambda_s}\right) = \arctan\left(\frac{\tan 8°}{\cos 10°}\right) = 8.12°$$

The clearance angle in the assumed working plane calculates using Equation C.3 as

$$\alpha_f = \arctan\left(\frac{\tan\alpha_o}{\sin\kappa_r}\right) = \arctan\left(\frac{\tan 11.82°}{\sin 60°}\right) = 13.59°$$

The clearance angle in the back plane calculates using Equation C.2 as

$$\alpha_p = \arctan\left(\frac{\tan\alpha_o}{\cos\kappa_r}\right) = \arctan\left(\frac{\tan 11.82°}{\cos 60°}\right) = 22.71°$$

The rake angle in the assumed working plane calculates using Equation C.6 as

$$\gamma_f = \arctan(\tan\gamma_o \sin\kappa_r + \tan\lambda_s \cos\kappa_r) = \arctan(\tan 8.12° \sin 60° + \tan 10° \cos 60°) = 11.96°$$

The rake angle in the back plane calculates using Equation C.7 as

$$\gamma_p = \arctan(\tan\gamma_o \cos\kappa_r - \tan\lambda_s \sin\kappa_r) = \arctan(\tan 8.12° \cos 60° - \tan 10° \sin 60°) = -4.65°$$

C.2.2 Tool-in-Machine System

The previous considerations of the cutting tool geometry are related to T-hand-S where the tool corner (point 1 in Figure C.10) and the axis of rotation of the workpiece are located on the same horizontal reference plane P_r. This is not always the case in practical machining as shown in this section for simple machining operations. Therefore, another, nonstandardized by previously mentioned tool geometry standards but extremely important for drilling tools, system referred to as the tool-in-machine (T-mach-S) is introduced. The symbol used for a reference plane and/or tool angle bears an additional suffix *m* for *machine* (e.g., P_{rm}, the tool reference plane in T-mach-S) to distinguish it from the corresponding T-hand-S plane (e.g., P_r, the tool reference plane in T-hand-S).

Figure C.12 shows idealized cases for turning and boring. As can be seen, the vector of the cutting speed **v** (the cutting speed direction) is perpendicular to the reference plane P_r. As the tool reference plane in T-hand-S P_r coincides with the tool reference plane P_{rm} in T-mach-S and the tool

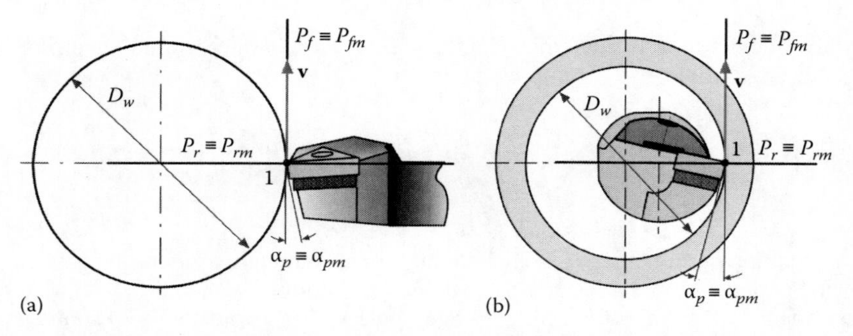

FIGURE C.12 Tool corner point 1 and the axis of rotation locate at the same reference plane: (a) turning and (b) boring.

assumed working plane P_f coincides with the working plane P_{fm} in T-mach-S, the tool geometry parameters in T-mach-S are the same as in T-hand-S. Figure C.12 shows an example for the back clearance angle in turning and boring.

In reality, however, the position of point 1 may not be as shown in Figure C.12, which can change angles assigned in T-hand-S. Corner point 1, after being installed in the machine, is often found to be shifted in the vertical direction as shown in Figure C.13. In many axial tools, the cutting edge(s) is intently located with a certain shift from this reference plane P_r (i.e., in twist drills). This shift causes changes in the cutting angles that should be accounted for. The reason for this change is explained as follows.

As clearly seen in Figure C.13, being always perpendicular to the radius of rotation (radius 0–1), the vector of the cutting speed **v** (the cutting direction) is located as a certain angle τ_{ad} to the reference plane P_r is T-hand-S. As the reference plane in any system of consideration must be perpendicular to the vector of cutting speed **v**, the reference plane in T-mach-S P_{rm} is introduced as shown in Figure C.13. Naturally, this plane is at angle τ_{ad} to the reference plane in T-hand-S P_r. As the working plane must contain both directions of the primary and feed motions, this plane in T-mach-S for the considered case, referred to as working plane in T-mach-S P_{fm}, is also rotated by angle τ_{ad} with respect to P_f as shown in Figure C.13.

The surplus angle τ_{ad} is calculated (Granovsky and Granovsky 1985; Astakhov 1998/1999) using geometry models shown in Figure C.13 as

$$\tau_{ad} = \arctan \frac{h_o}{D_w/2} \tag{C.10}$$

where h_o is the vertical shift of point O in the tool back plane. Its sense is clearly seen in Figure C.12.

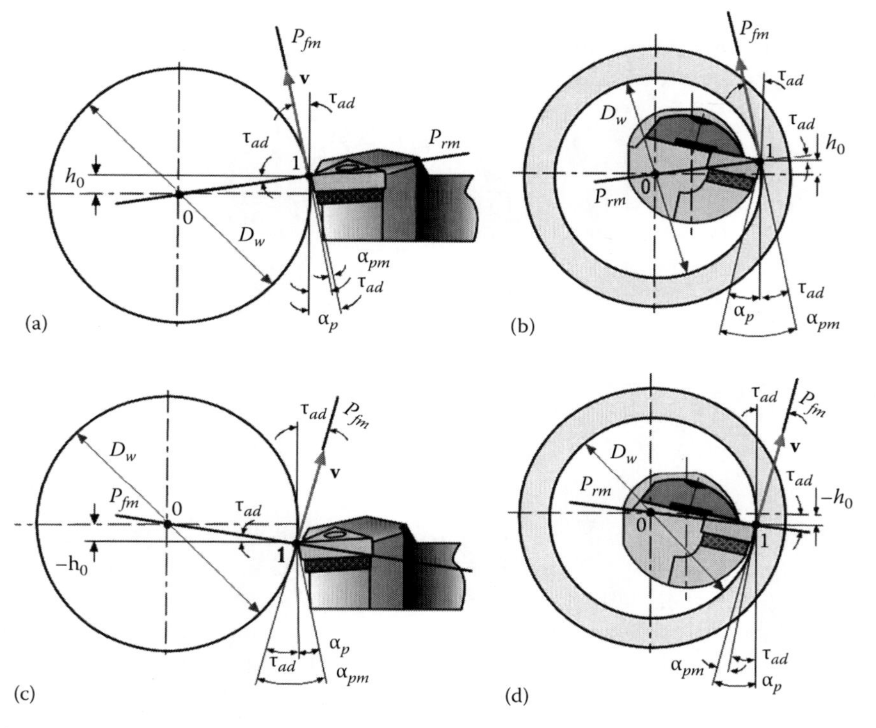

FIGURE C.13 Vertical shift of tool corner 1 with respect to the reference plane in T-hand-S: scenarios: (a) positive shift h_o (above the T-hand-S reference plane) in turning, (b) positive shift h_o (above the T-hand-S reference plane) in boring, (c) negative shift $-h_o$ (below the T-hand-S reference plane) in turning, and (d) negative shift $-h_o$ (below the T-hand-S reference plane) in boring.

Figure C.13 shows the modification of the T-hand-S back clearance angle (α_p shown in Figure C.12) for four basic scenarios: (a) positive shift h_o (above the T-hand-S reference plane) in turning, (b) positive shift h_o (above the T-hand-S reference plane) in boring, (c) negative shift $-h_o$ (below the T-hand-S reference plane) in turning, and (d) negative shift $-h_o$ (below the T-hand-S reference plane) in boring.

It is obvious from the models shown in Figure C.13 that in the case of positive shift h_o (above the T-hand-S reference plane) in turning (Figure C.13a), the T-mach-S back clearance angle α_{pm} of point 1 is greater than the T-hand-S back clearance angle α_p shown in Figure C.12a and is calculated as

$$\alpha_{pm} = \alpha_p - \tau_{ad} \tag{C.11}$$

In the case shown in Figure C.13b, that is, in boring, this angle becomes smaller than α_p:

$$\alpha_{pm} = \alpha_p + \tau_{ad} \tag{C.12}$$

In the case of negative shift $-h_o$ (below the T-hand-S reference plane) in turning (Figure C.13c), the T-mach-S back clearance angle α_{pm} of point 1 is smaller than the T-hand-S back clearance angle α_p shown in Figure C.12 and is calculated as

$$\alpha_{pm} = \alpha_p + \tau_{ad} \tag{C.13}$$

In case of boring (Figure C.13d), however, the opposite is true, that is,

$$\alpha_{pm} = \alpha_p - \tau_{ad} \tag{C.14}$$

The other angles in T-mach-S are also modified. The T-mach-S tool cutting edge angle κ_{rm} when $\lambda_s = 0$ is calculated as (Astakhov 2010)

$$\kappa_{rm} = \arctan(\tan\kappa_r \cos\tau_{ad}) \tag{C.15}$$

In most practical cases, however, particularly in drilling tools, $\lambda_s \neq 0$. In this case (Astakhov 2010),

$$\kappa_{rm} = \arctan\left(\frac{\tan\kappa_r \cos(\tau_{ap} - \tau_{ad})}{\cos\tau_{ap}}\right) \tag{C.16}$$

where τ_{ap} is the modification angle calculated as

$$\tau_{ap} = \arctan\frac{\tan\lambda_s}{\sin\kappa_r} \tag{C.17}$$

Equation C.16 is general as it is valid for any sign of the inclination angle λ_s (including $\lambda_s = 0$) and vertical shift h_o. In fact, when $\lambda_s = 0$ and thus $\tau_{ap} = 0$, Equation C.16 becomes the same as Equation C.15. When $h_o = 0$ (no vertical shift), then $\tau_{ad} = 0$ according to Equation C.10 so that $\kappa_{rm} = \kappa_r$ according to Equation C.15. It also follows from Equation C.16 that when the inclination angle λ_s and vertical shift h_o are of opposite signs, then $\kappa_{rm} < \kappa_r$, while when these signs are the same, then $\kappa_{rm} > \kappa_r$.

The T-mach-S cutting edge inclination angle is calculated as (Astakhov 2010)

$$\lambda_{sm} = \arctan(\sin\kappa_{rm} \tan(\tau_{ap} - \tau_{ad})) \tag{C.18}$$

Equation C.18 is valid for any sign of the inclination angle λ_s (including $\lambda_s = 0$) and vertical shift h_o.

The T-mach-S orthogonal rake angle is calculated as

$$\gamma_{om} = \arctan\left(\tan\kappa_{rm}\tan\lambda_{sm} + \frac{\tan\gamma_{pm}}{\cos\kappa_{rm}}\right) \tag{C.19}$$

Equation C.19 is valid for any sign of the inclination angle λ_s (including $\lambda_s = 0$) and vertical shift h_o. When $h_o = 0$ (no vertical shift), then $\tau_{ad} = 0$, $\kappa_{rm} = \kappa_r$, and $\lambda_{sm} = \lambda_s$ so that $\gamma_{om} = \gamma_o$.

The T-mach-S orthogonal clearance angle is calculated as

$$\alpha_{om} = \arctan(\tan\alpha_{pm}\cos\kappa_{rm}) \tag{C.20}$$

Equation C.20 is valid for any sign of the inclination angle λ_s (including $\lambda_s = 0$) and vertical shift h_o. When $h_o = 0$ (no vertical shift), then $\tau_{ad} = 0$ and thus $\kappa_{rm} = \kappa_r$ and $\lambda_{sm} = \lambda_s$ so that $\alpha_{om} = \alpha_o$ according to Equation C.20.

Figure C.14 shows deviations $\Delta_{T\text{-}mach\text{-}S}$ of the tool cutting edge angle $\Delta\kappa_r$, orthogonal rake angle $\Delta\gamma_o$, orthogonal clearance angle $\Delta\alpha_o$, and cutting edge inclination angle $\Delta\lambda_s$ in T-mach-S from those in T-hand-S as functions of the tool cutting edge angle when a single-point cutting tool having normal rake angle, $\gamma_n = 10°$; normal clearance angle, $\alpha_n = 8°$; and cutting edge inclination angle, $\lambda_s = 10°$, is installed with $h_o = 2$ mm. The diameter of the workpiece $D_w = 30$ mm. As can be seen, the cutting tool inclination angle changes significantly, while deviations of the orthogonal rake and clearance angles diminish in the region of the widely used tool cutting edge angle in turning tools. Combined with high accuracy of tool installation on CNC machines (small h_o's), the difference in angles in T-hand-S and T-mach-S is insignificant, so it is routinely ignored, and thus, T-mach-S is not considered in the professional and scientific literature as well as in tool geometry standards. However, this is not nearly the case for many drilling tools as shown in Chapter 4 where the discussed differences are of prime concern.

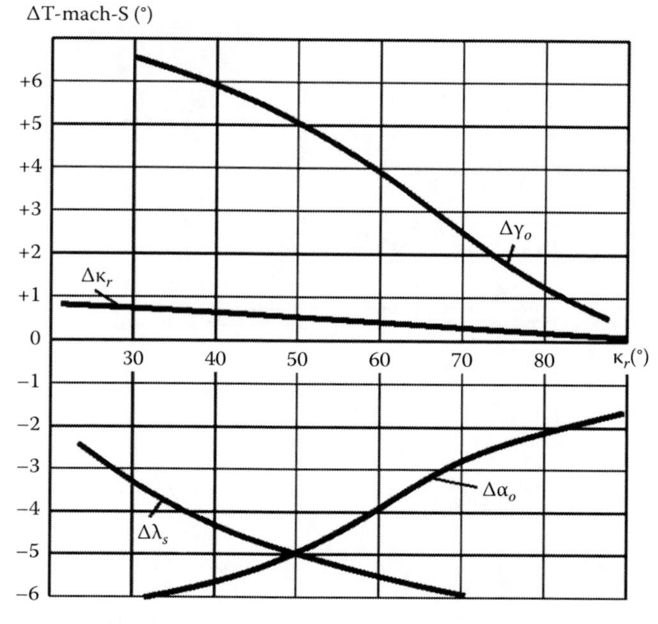

FIGURE C.14 Changes of the T-hand-S angles in T-mach-S as functions of the tool cutting edge angle.

C.2.3 Tool-in-Use System

C.2.3.1 T-Use-S Reference Planes

The T-use-S considers the geometry of the cutting tool accounting for machining kinematics. As the parameters of the tool geometry are affected by the actual resultant motion of the cutting tool relative to the workpiece, a new system, referred to as the tool-in-use (T-use-S), and the corresponding tool angles, referred to as the working angles, are established in this new system.

Figures C.15 and C.16 show graphical representation of the following planes:

- *Working reference plane* P_{re} defined as a plane through the selected point on the cutting edge and perpendicular to the resultant cutting direction.
- *Working plane* P_{fe} defined as a plane through the selected point on the cutting edge and containing the resultant cutting direction and perpendicular to the working reference plane P_{re}.
- *Working back plane* P_{pe} defined as a plane through the selected point on the cutting edge and perpendicular both to the working reference plane P_{re} and to the working plane P_{fe}.
- *Working cutting-edge plane* P_{se} defined as a plane tangent to the cutting edge at the selected point and perpendicular to the working reference plane P_{re}. This plane thus contains the resultant cutting direction.
- *Working orthogonal plane* P_{oe} defined as a plane through the selected point on the cutting edge and perpendicular both to the working reference plane P_{re} and to the working cutting-edge plane P_{se}.

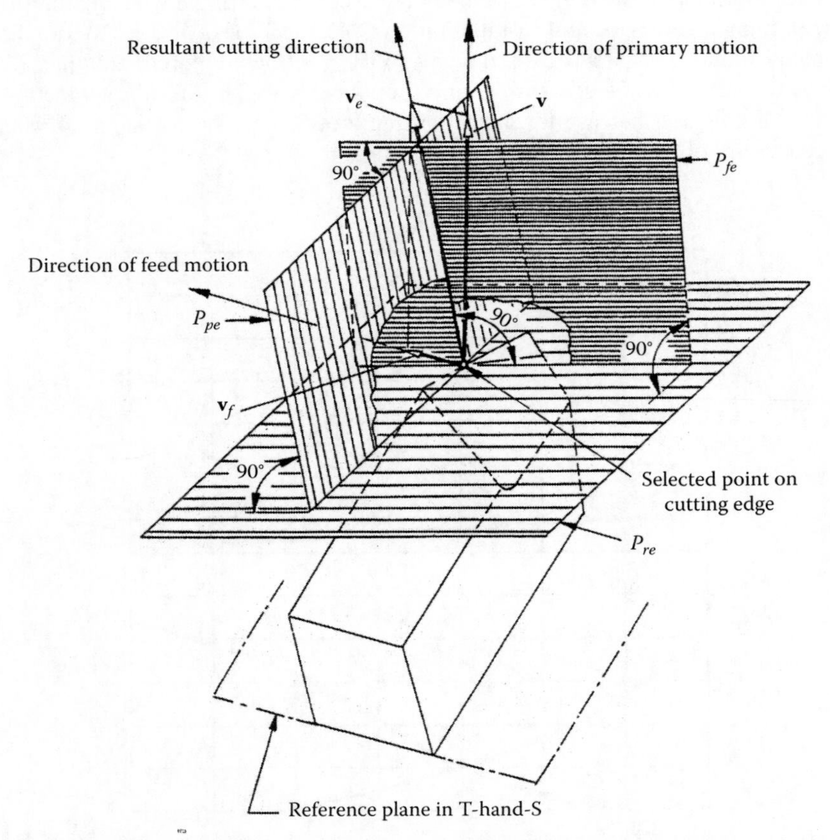

FIGURE C.15 Representation of the reference, working, and back planes in T-use-S.

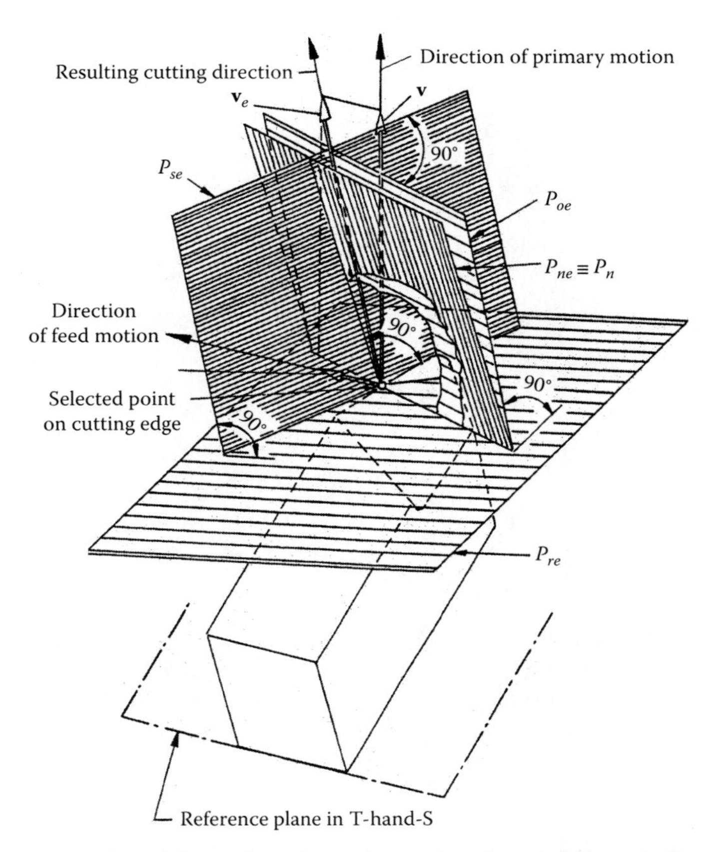

FIGURE C.16 Representation of the cutting edge, orthogonal, and normal planes in T-use-S.

- *Cutting edge normal plane* P_{ne}. According to ISO and ANSI standards (ISO 3002-1 Basic quantities in cutting and grinding. Part 1: Geometry of the active part of cutting tools—General terms, reference systems, tool and working angles, chip breakers 1982; ANSI B94.50 [1975] Basic nomenclature and definitions for single-point cutting tools 1975 [reaffirmed 1993]), this plane is identical to the cutting edge normal plane defined in the T-hand-S, that is, $P_{ne} \equiv P_n$. In the author's opinion, this notion is incorrect as it is not based on the physics of the metal cutting process. This physics implies that the proper rake and clearance angles in the T-use-S are measured in a plane containing the resultant direction of chip flow (Astakhov 2010).

C.2.3.2 T-Use-S Angles

A set of angles is defined in T-use-S. These angles have the prefix w (*working*) in their title. Because working angles may vary from point to point along the cutting edge, the definitions of the angles refer always to the angles at the selected point on the cutting edge. When the cutting edge, face, or flank is curved, the tangents or tangent planes through the selected point are used in the reference systems of planes used to define the angles.

Each angle is specified, where appropriate, with reference to a particular cutting edge on the tool, depending upon the location of the selected point on the cutting edge. The title of the angle may include an indication of whether the selected point is located on the major or minor cutting edge (e.g., at a selected point on the major cutting edge, there is the tool normal rake, and at a selected point on the minor cutting edge, the corresponding angle is termed the tool minor cutting edge normal rake).

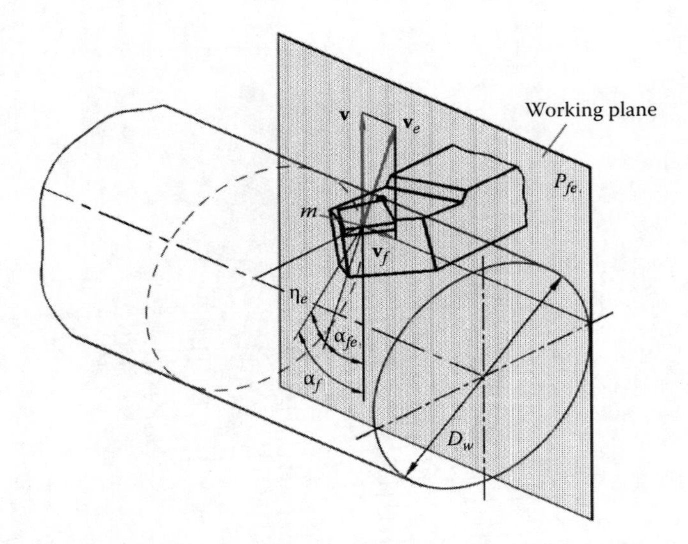

FIGURE C.17 Showing the sense of the clearance angle α_{fe} in T-use-S.

The angles of the cutting tool in T-use-S are defined in these planes in the same manner as in T-hand-S with respect to the T-use-S reference planes. To comprehend the concept of working planes, consider a simple model shown in Figure C.17. As can be seen, the clearance angle in the working plane α_{fe} in T-use-S is smaller than that in T-hand-S by angle

$$\tan\eta_s = \frac{v_f}{v} = \frac{f}{\pi D_w} \tag{C.21}$$

Figure C.18 shows a stricter model. It visualizes the sense of working rake γ_{fe} and clearance α_{fe} angles in T-use-S considered in the working plane P_{fe}. As can be seen, the three planes in T-hand-S pass through selected point m on the cutting edge, namely, the tool reference plane P_r, tool cutting-edge plane P_s, and working plane P_f. As such by definition, tool rake γ_f and clearance α_f angles in T-hand-S are as shown in Figure C.18. In T-use-S, the reference plane P_{re} is perpendicular to the resultant cutting direction, that is, rotated counterclockwise by angle η_e with respect to P_r.

It follows from Figure C.18 that the working rake γ_{fe} and flank α_{fe} angles are calculated as

$$\gamma_{fe} = \gamma_f + \eta_e \tag{C.22}$$

$$\alpha_{fe} = \alpha_f - \eta_e \tag{C.23}$$

As can be seen, angle γ_{fe} becomes larger and α_{fe} becomes smaller in T-use-S than those in T-hand-S.

In many practical cases of turning and boring, angle η_e is small. For example, in turning with the following parameters, $D_w = 40$ mm, $f = 0.3$ mm/rev, angle $\eta_e \approx 0.14°$, which is much smaller than common tolerances on tool angles. Therefore, T-use-S angles are considered only in some special cases of machining with high feeds, that is, in thread cutting (Astakhov 2010).

C.3 IMPORTANCE OF THE INTRODUCED ANGLES

The introduced angles that constitute the cutting tool geometry have functional importance. Each angle is of concern when a certain function of the cutting tool is considered, that is, the cutting force, direction of chip flow, and theoretical surface roughness of the machined

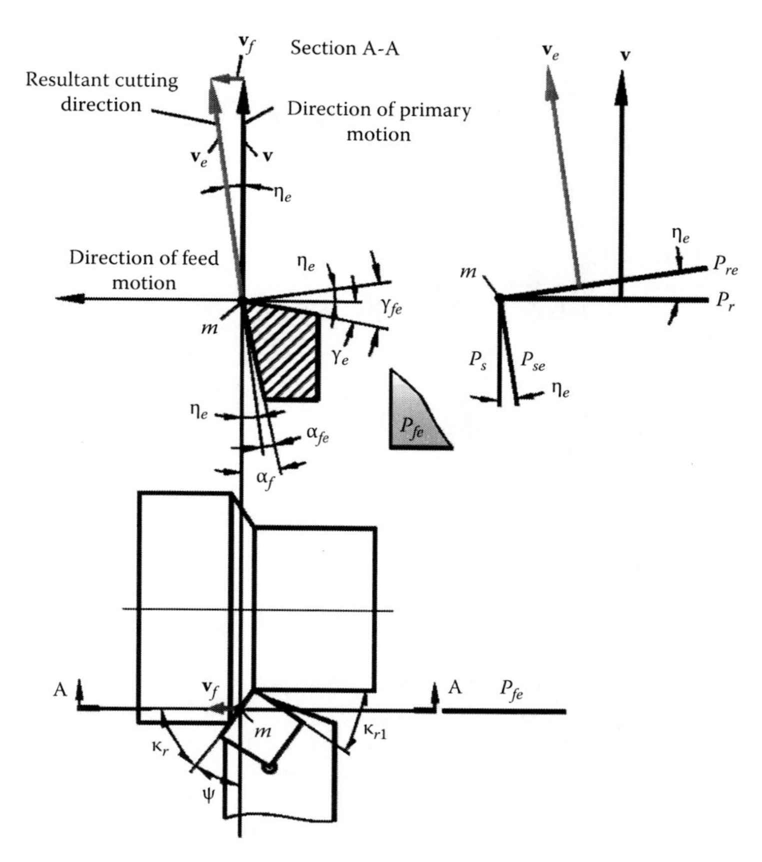

FIGURE C.18 Sense of working rake γ_{fe} and clearance α_{fe} (in the working plane) angles in T-use-S.

surface. However, one angle stands out among other angles—it is *the clearance angle that is the major distinguishing feature of the cutting tool.*

C.3.1 Clearance Angle

C.3.1.1 Introduction

Understanding the proper terminology is a key issue in the design of optimal machining processes and cutting tools including drilling tools. This section provides methodological help on the matter, discussing the basics of splitting, shearing, and metal cutting in terms of their similarities and principal differences in order to distinguish the process of metal cutting among other closely related manufacturing processes.

What unifies these processes is that they all aim to physical separate a part of the workpiece by applying an external force (forces). They differ, however, by the direction of the external force application, by the mechanical properties of the work material to resist such a separation, and, what is the most important, by the power required to accomplish the process.

C.3.1.2 Splitting

The easiest way to explain splitting as a separation process is to use wood splitting as an example. The first thing to understand about wood splitting is that it is different from wood cutting. Wood is a natural polymer—parallel strands of cellulose fibers held together by a lignin binder. These long chains of fibers make the wood exceptionally strong—they resist stress and spread the load over the length of the board. Furthermore, cellulose is tougher than lignin. It is easier to split a board with the grain

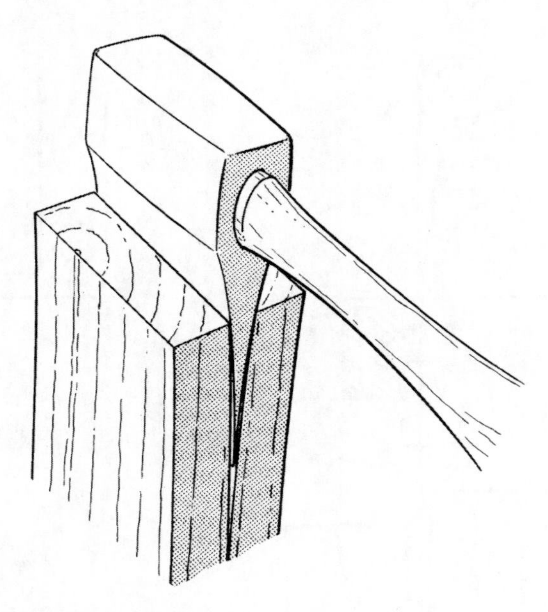

FIGURE C.19 Wood splitting by a hatchet.

(separating the lignin) than to break it across the grain (separating the cellulose fibers). This makes wood an anisotropic material, that is, its properties depend on the direction in which they are measured and its strength depends on the direction of load application. When one splits wood, he or she is separating the cellulose fibers from each other. This is why wood can only be split along the grain.

The tool pictured in Figure C.19 is a maul or its lighter version—hatchet. It differs from an axe in that it is designed exclusively for splitting wood. It has a much wider head, which means that it does a better job of pushing the cellulose fibers apart and breaking the wood into pieces. It is heavier, which means it hits the wood harder than the axe and busts it up. Finally, it has a sharp edge, but not a thin one, which means that it does not get stuck in the wood like an axe will.

For very tough and/or large pieces of wood, splitting wedges are used (Figure C.20a). A wedge is used like this. First, the wedge is *set* in the wood by gently tapping it in with the maul held in one hand or a small sledge, known as a *single jack*, can be used for this task. Basically, it is desired for the wedge to dig in enough that it stands up on its own. It is sometimes helpful to hit the wood with the maul in

(a)

(b)

FIGURE C.20 Splitting wedge (a) and wood splitting (b) by two splitting wedges.

order to make a divot in which one then starts the wedge. Once the wedge is set, one begins to hit it with the back side of the maul (or a dedicated sledge hammer) and drive it into the wood (Figure C.20b).

Figure C.21 shows a force model for wood splitting. In terms of mechanical engineering, a wedge is a simple machine consisting of a solid block of metal shaped as an inclined plane. It converts motion in one direction into a splitting motion that acts at right angles to the wedge faces. In this model, a wedge is set in the wood and is loaded by dividing (axial) force F_a. This force results in two normal (separating) forces N_w acting to push aside two sides of the wood. As such, each normal force creates a bending moment with respect to the tip of crack A ahead of the cutting edge. As a result, a crack runs ahead of the cutting edge so that the wood splits into two parts. The running crack provides some space for further cutting edge penetration. Therefore, the wood splits by tensile stress so that the major strength characteristic involved is the tensile strength of the wood (the work material) along fibers. The separating force N_w generated dividing force F_a is calculated as

$$N_w = \frac{F_a}{2\sin(\beta/2)} \quad \text{(C.24)}$$

where β is the wedge angle as shown in Figure C.21.

The wood, however, *fights back*—as wedge penetrates, the normal reaction forces from the wood results in two friction forces F_{fr}. These two forces increase with the wedge penetration, causing a necessity of increasing the splitting force to continue splitting.

A simple analysis of the model shown in Figure C.21 reveals the following:

- The work material fails by a bending moment that the normal forces created at the tip of the crack. Thus, the tensile strength of the wood along the fibers is the major mechanical characteristic of the work material.
- Normal and friction forces act on the both faces of the splitting wedge.
- The cutting edge does not actually cut—it serves as the stress concentrator promoting the crack ahead of it.
- The wedge face–wood contact area is large and it increases with the wedge penetration. This creates great friction forces.

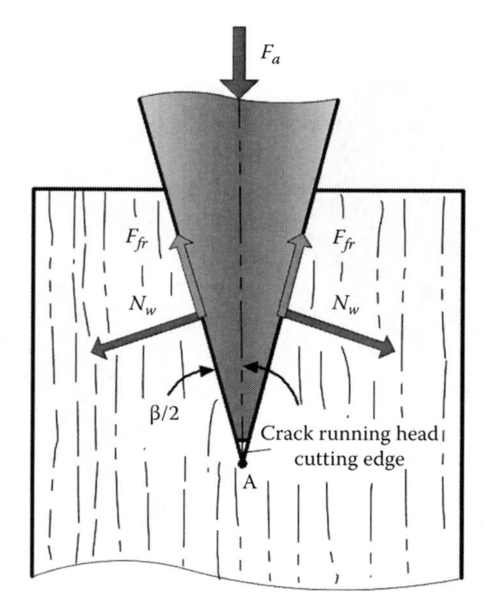

FIGURE C.21 Force model for splitting wood by a splitting wedge.

- Splitting is not a continuous process. Rather, it is a cyclic process consisting of two stages: (1) edge perpetration with no splitting (the normal forces on its side grow) and (2) tensile failure in the region of point A causing crack propagation when the normal forces become large enough to exceed the tensile strength of the work material. At certain crack length, its growth stops as the stress at its tip becomes less than the tensile strength of the work material. To continue the process, further edge penetration is needed.
- Optimization of the process can be thought of in terms of the work material and tool design. Some species split much easier when green; others seem to split easier when dry. For many types of wood, splitting is the easiest when firewood is frozen, that is, when its strain at fracture is at minimum. The wedge angle β can be optimized for a given wood properties/condition to reduce the severity of friction on the sides of the wedge. Such an optimized wedge angle reduces the contact length (area) on the sides while promoting crack propagation as shown in Figure C.22. The side (rake) faces of the tool can be modified. Figure C.23a shows the traditional modification with anti-return slots on the side faces to prevent the wedge slipping out of the log. Figure C.23b shows modern modification with inclined (helical) slots. Besides addressing anti-return problem, such slots reduce the contact length (area), and thus, friction forces reduce.

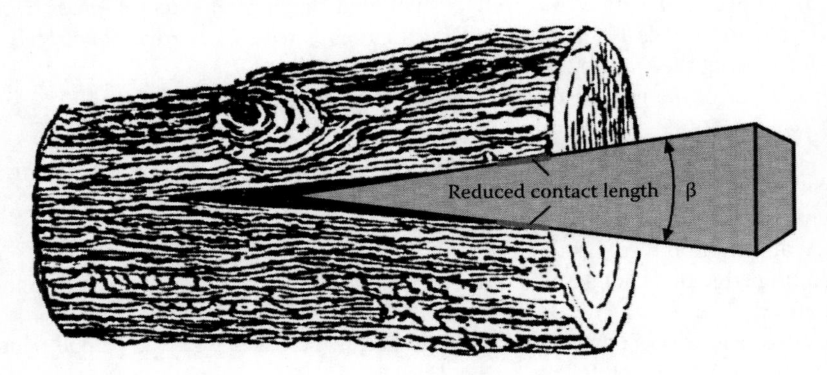

FIGURE C.22 Optimized wedge angle leads to reduced contact length (area).

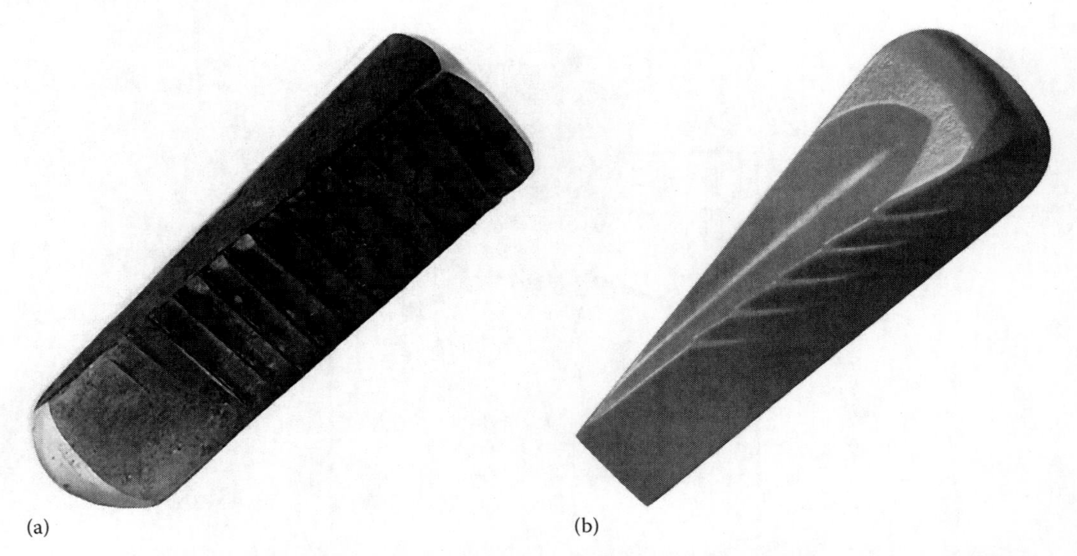

FIGURE C.23 Modification of the side faces of the splitting wedge: (a) tradition modification and (b) modern modification.

FIGURE C.24 Slicing bread with a knife.

Slicing bread with a knife is another use of splitting (Figure C.24). When a knife is pressed or chopped straight through an object, that object is essentially being pinched in half. One might ask why a sharp knife is more effective at this than a dull knife. This is a matter of physics—a transfer of kinetic energy, to be specific. When a knife is sharp, its edge is narrower than the edge of a dull knife. This means the point of contact between the knife and the object to be cut is much smaller. The weight of the knife combined with the force one uses in pressing down is exerted over a much smaller point of contact, magnifying that force dramatically so that the sharp knife cuts through the object more cleanly and with less force than a dull one.

C.3.1.3 Shearing

Shearing is the deformation and then separation of a material substance in which parallel surfaces are made to slide past one another. In shearing, one layer of a material is made to move on the adjacent layer in a linear direction due to action of two parallel forces F_{sh} located at distance a_{cl} known as the clearance distance as shown in Figure C.25. A typical example of shearing is cutting with a pair of scissors (Figure C.26). Scissors are cutting instruments consisting of a pair of metal blades connected in such a way that the blades meet and cut materials placed between them when the handles are brought together. Figure C.27 shows the difference in cutting with a knife and scissors, that is, between separating by splitting and shearing.

Shearing operations use the same principle of material separation as shown in Figure C.28. Many sheet-metal parts are made from a blank of suitable dimension that is first removed from a large sheet or coil using a variety of manufacturing processes called shearing operations as they are all based on the shearing process. In these operations, the sheet is cut by subjecting it to shear

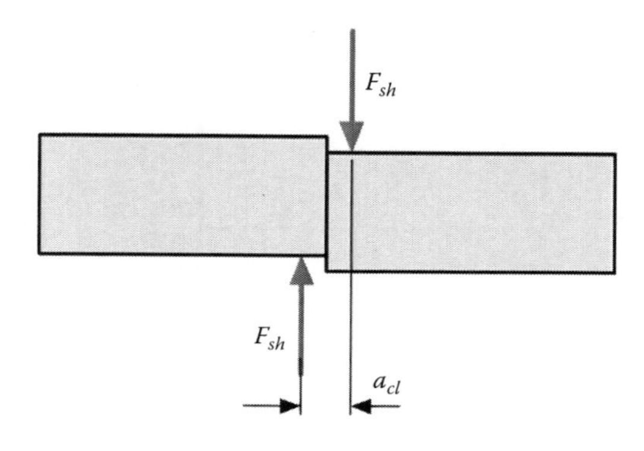

FIGURE C.25 Shearing.

FIGURE C.26 Cutting with a pair of scissors.

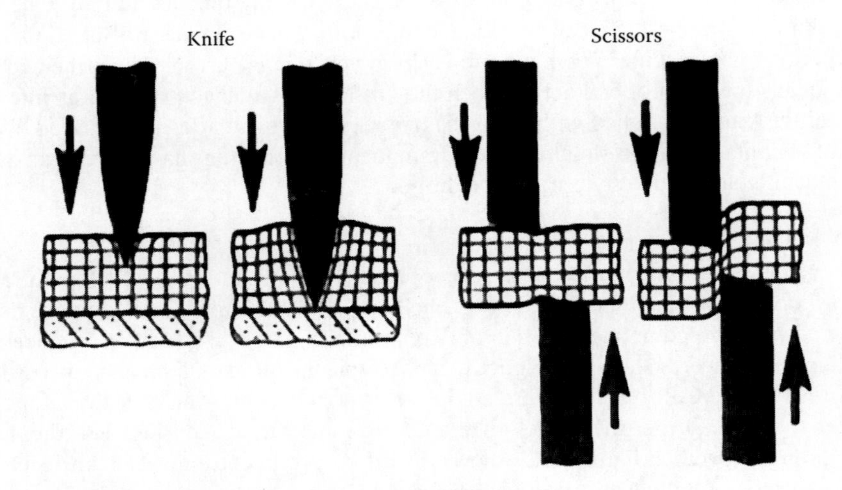

FIGURE C.27 Showing the difference in cutting with a knife and scissors.

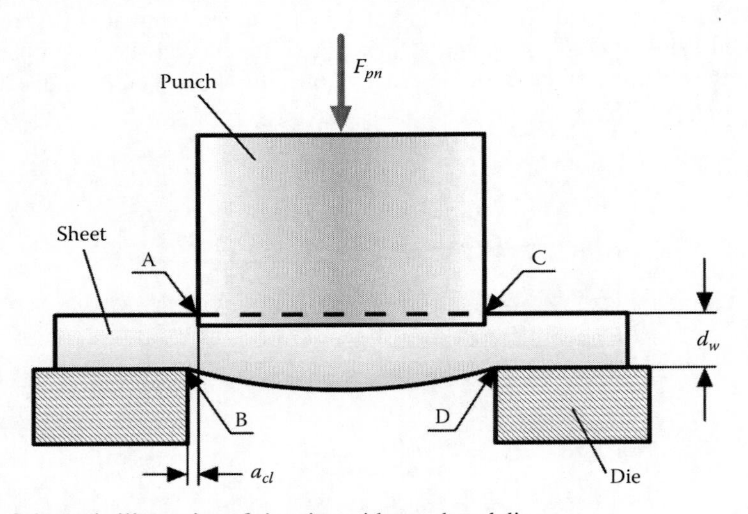

FIGURE C.28 Schematic illustration of shearing with punch and die.

stress typically between a punch and a die as shown in Figure C.28. Shearing usually starts with the formation of the shear planes and then cracks on both the top and bottom edges of the workpiece (A and B, and C and D in Figure C.28). These cracks eventually meet each other and separation occurs. The rough fracture surfaces are due to these cracks. The particularities of the punch wear and quality of blanks depend on the clearance a_{cl} between the punch and die. The smooth and shiny surface on the hole is the result of the burnishing of the flank edge of the punch.

Generally, the punching force calculates as the product of the shear strength of the work material and the shearing area (Vidosic 1964), that is,

$$F_{pn} = A_{sh}s_s = L_{pn}d_w s_s \quad \text{(C.25)}$$

where

A_{sh} is the shearing area
s_s is the ultimate shear strength of the work material
L_{pn} is the length or perimeter of cut
d_w is the thickness of sheet being sheared.

Comparison of splitting and shearing reveals the following similarities:

1. The contact between the work material and the tool takes place on both working faces, that is, on the conditional rake and flank faces, which causes excessive plastic deformation of the work material and great friction forces/losses over the contact areas.
2. A great force should be applied. In splitting, the ultimate tensile stress should be achieved to start the crack ahead of the tool. In shearing, as it follows from Equation C.25, the punching force can be significant because the length or perimeter of the cut can be great. Thus, the punch force builds up rapidly during shearing because the entire thickness is sheared at the same time. As a result, multiton presses of great dimensions are used for punching.

C.3.1.4 Metal Cutting

The process of metal cutting is defined as a forming process, which takes place in the components of the cutting system that are so arranged that the external energy applied to the cutting system causes the purposeful fracture of the layer being removed. This fracture occurs due to the combined stress including the continuously changing bending stress causing a cyclic nature of this process. The most important property in metal cutting studies is the system time. The system time was introduced as a new variable in the analysis of the metal cutting system, and it was conclusively proven that the relevant properties of the cutting system's components are time dependent. The dynamic interactions of these components take place in the cutting process, causing a cyclic nature of this process (Astakhov 2006).

It follows from this definition that the presence of bending moment in the deformation zone and relatively small size of this zone distinguish metal cutting among other closely related manufacturing processes and operations. The bending moment forms the combined state of stress in the deformation zone that significantly reduces the resistance of the work material to cutting. As a result, metal cutting is the most energy-efficient material removal process (energy per removed volume accounting for the achieved accuracy) compared to other closely related operations.

The positive clearance angle is the major distinguishing feature of the cutting tool, that is, it is the only feature that distinguishes the cutting tool from those tools used in closely related manufacturing process as in splitting and shearing (Astakhov 2010). In other words, other angles of the tool geometry can assume virtually any signs and values that only affect optimality of the cutting process, while the clearance angle should always be positive—it is the only condition of existence of the cutting process.

The previous statement is explained using the model shown in Figure C.29. This figure presents the simplest example of orthogonal cutting where the primary motion is straight having velocity **v**. As shown in Figure C.29a, a square bar stock is used as a tool. The side face of this bar stock is used as the rake face having cutting edge *ab*, and thus, the rake angle γ is zero for the sake of simplicity of the current considerations. The square face of this bar stock *ABCD* is used as the flank face, so the clearance angle α_n is also zero. The tool thus formed is set to cut the chip having chip thickness h_D and width $b_d = a_p$.

When the primary motion with velocity **v** is applied to the tool, the chip begins to form as shown in Figure C.29b. However, the normal cutting process will not last. As soon as flank face *abcd* comes into contact with the machined surface, severe rubbing takes place, which makes cutting process impossible due to severe friction force on the flank face, frictional vibrations, and high temperature as a result of heat generation on rubbing. To make the cutting process stable, a positive clearance angle $\alpha_n > 0$ should be applied to the cutting tool in the manner shown in Figure C.29c.

Although it is obvious that the cutting process is unconditionally impossible when $\alpha_n < 0$, one may argue, however, that the discussed negative consequences of the zero clearance angle

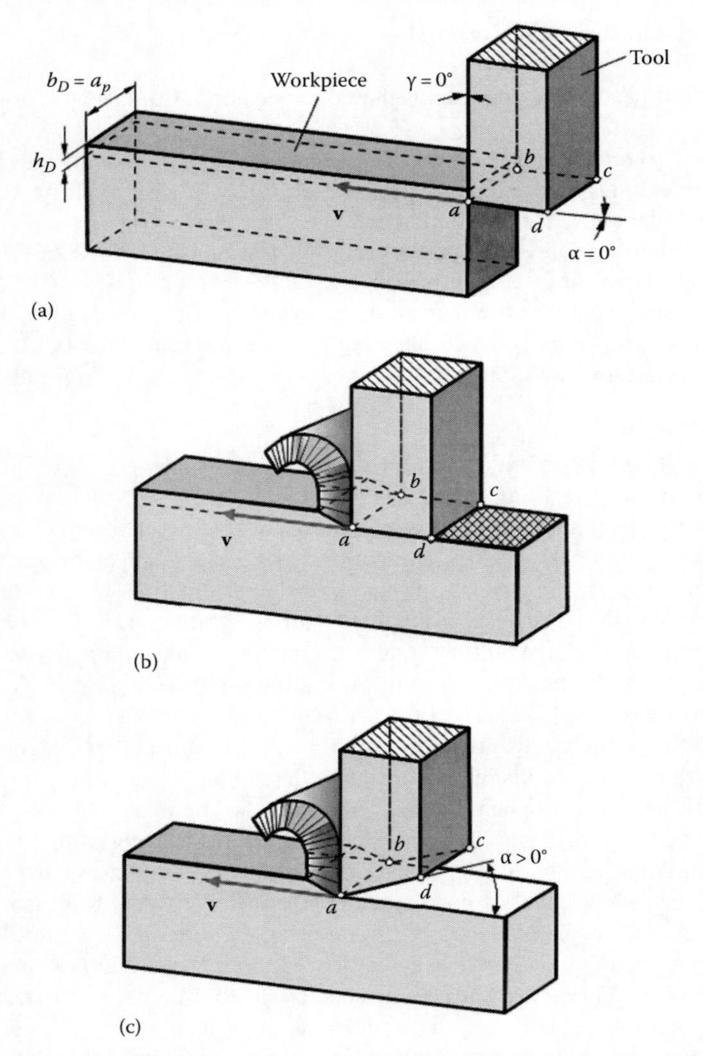

FIGURE C.29 Formation of a cutting tool from a square block: (a) initial arrangement, (b) attempt to cut with zero clearance able, and (c) proper cutting with positive clearance angle.

can be reduced by generous lubricating of the rubbing surfaces, reducing the rubbing area, etc., so that while it may not be optimal, the cutting process is possible with the zero clearance angle under these modified conditions. In reality, however, it is not possible in practical machining due to two primary reasons: (1) kinematics of machining and (2) work material elasticity. These are explained as follows.

A widely used arrangement in practical studies of orthogonal cutting is tuning the end of a tube. The end of a tube is cut in a lathe by a tool having a zero inclination angle. The diameter of the tube is selected to be much greater than the thickness of its wall so that the cutting velocity does not change significantly along the cutting edge, and thus, it can be regarded to be constant. Let us consider what happens if a cutting tool having a zero clearance angle is used in this operation.

Figure C.30 shows a simple model of a tube end turning with a tool having a zero clearance angle. A simple comparison of this model and that shown in Figure C.29 reveals that the great difference between them is the presence of the feed motion in tube end tuning. The feed motion changes the whole picture as it results in an inclined machined surface as the tool moves now not in the direction of the primary motion but in the direction of the resultant motion. This causes the interference between the tool flank face having a zero clearance angle and the machined surface as shown in Figure C.30. No matter how small the feed velocity $\mathbf{v}_f$ is, this interference prevents cutting. This also follows from Figure C.18 where the clearance α_f in T-hand-S should be greater than angle η_e to achieve a positive clearance angle α_{fe} in T-mach-S, that is, to make cutting kinematically possible.

As mentioned earlier, the second reason why practical machining is not possible with zero clearance angle is work material elasticity. When a material is cut, it deforms elastically and then plastically under the action of the cutting force. Once the tool passes, the force is removed. As a result, the work material springs back due to its elastic recovery. The concept and example of elastic recovery are discussed in Chapter 4 (Section 4.5.9.5) in great detail.

As the work material is cut, the cutting tool causes its elastic and then plastic deformation. When the tool passes a certain region on the workpiece leaving the machined surface, elastic deformation causes elastic recovery because such deformation is recoverable (Chapter 4, Section 4.5.9.5). As such, when the load that caused the deformation is removed, the material will spring back.

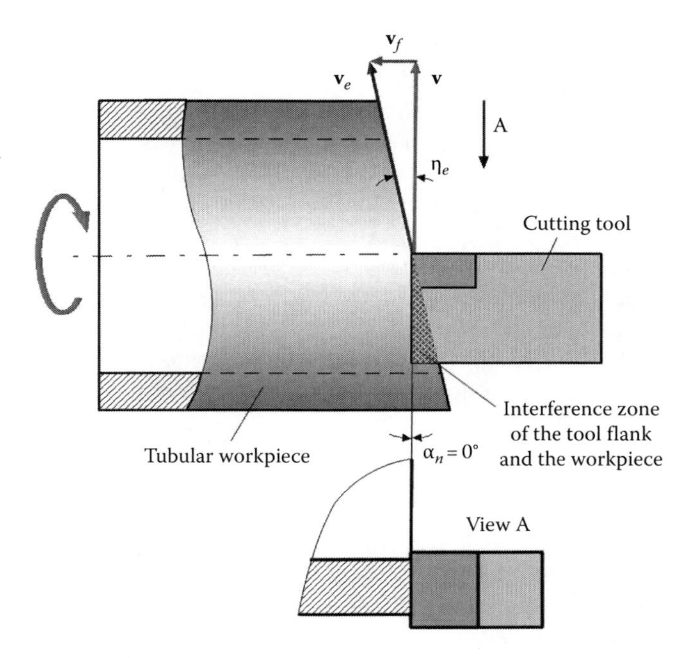

FIGURE C.30 Model of tube end tuning with a tool having a zero clearance angle.

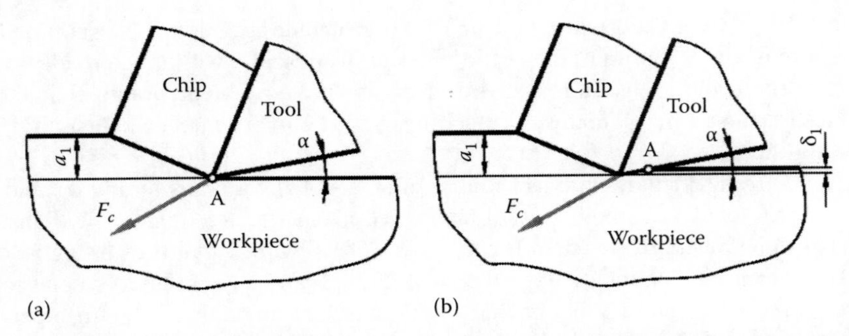

FIGURE C.31 Flank face contact: (a) with no springback and (b) with springback of the work material.

Figure C.31a shows a commonly considered model with no elastic recovery (springback) of the work material. In this model, point A (a point on the cutting edge) is the last point of contact of the tool in its motion relative to the workpiece. In reality, however, a model shown in Figure C.31b is the case where the elastic recovery δ_1 of the work material after removing the cutting force F_c takes place so that the machined surface contacts the tool flank face along certain length that ends at point A. As discussed in Chapter 4 (Section 4.5.9.5), the greater strength of the work material, the higher elastic recovery δ_1; the lower the Young's modulus, E, the higher elastic recovery δ_1. As a result, flank wear is a predominant wear in cutting tools.

As discussed earlier, angle η_e is rather small for many practical machining operations, so the selection of the optimum flank angle is trade-off between reduction of friction over the flank contact area due to elastic recovery of the work material (followed by the wear land on the flank face) and the properties of the cutting wedge. The latter include its strength and thermal properties (e.g., heat conduction). Therefore, there are two opposite trends taking place when the clearance angle increases:

- The contact on the tool flank due to springback of the work material decreases. This leads to a reduction in the tool–workpiece contact area. Because the mean shear stress at the tool flank interface is constant (Astakhov 2006), this leads to the corresponding reduction of heat due to friction. As a result, the flank temperature decreases that leads to increased tool life and improves the quality of machined parts.
- The wedge angle β decreases. As such, the strength of the region adjacent to the cutting edge decreases that causes possible chipping of the cutting edge. Moreover, as the wedge angle decreases, heat dissipation (heat sink) through the tool decreases that causes higher tool wedge temperatures with the maximum on the contact interfaces. These factors lower tool life.

As a direct result of such contradictive effects, the influence of the flank angle on tool life always has a well-defined maximum as shown in Figure C.32. This figure also shows that the optimal (in terms of tool life) flank angle increases when the uncut chip thickness decreases. This suggests that the flank angles of the cutting tool designed for finishing operation should be increased to achieve greater tool life.

C.3.2 Rake Angle

The rake angle comes in three varieties: positive, zero (sometimes referred to as neutral), and negative as shown in Figure C.33a through c, respectively. There is a great body of experimental and numerical modeling results dealing with the influence of the value and sign of the rake angle on the machining process. In the author's opinion, the role and importance of the rake angle in metal cutting is not well understood because these available data are contradictive and often misleading.

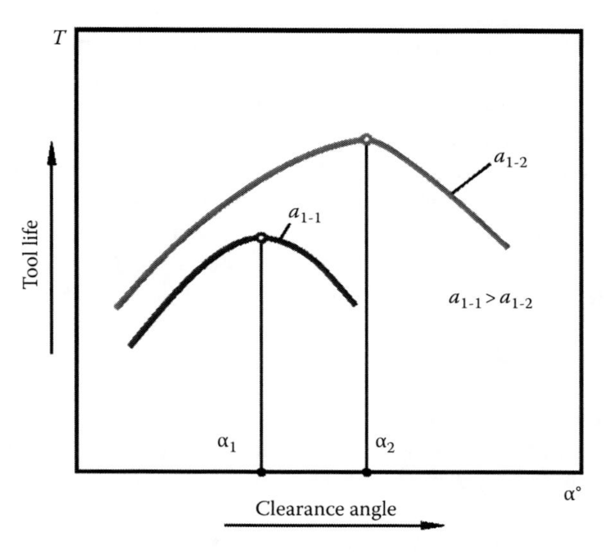

FIGURE C.32 General idea of the influence of the flank angle on tool life.

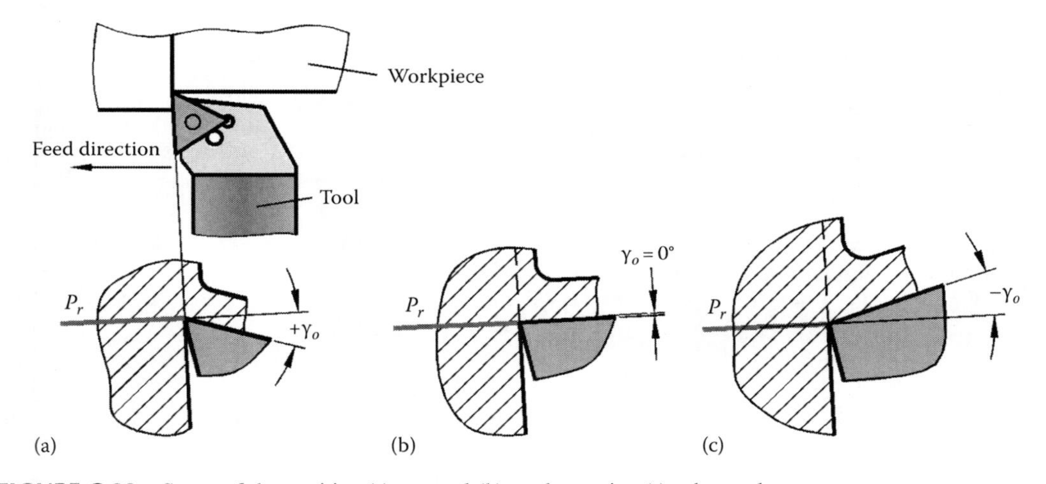

FIGURE C.33 Sense of the positive (a), neutral (b), and negative (c) rake angles.

Moreover, the available studies did not concern with the system consideration of the influence of the rake angle on the various outcomes of the cutting process. Rather, one outcome parameter is normally considered (e.g., the cutting force), while others (e.g., tool life) are ignored. Using these data, a practical tool/process designer cannot make an intelligent selection of the proper rake angle for a given application.

As mentioned earlier, Shaw (1988) argued that the specific cutting energy (and thus the cutting force) decreases about 1% per degree increase in the rake angle, while Dahlman et al. (2004) showed that by controlling the rake angle, it is possible to generate tailor-made machining residual stresses in the product. Gunay et al. (2005) in their experimental study found that a change in the rake angle from 0° to +2.5° resulted in a 2% reduction of the cutting force while a change from −2.5° to 0° resulted in a 3.4% reduction. Tetsuji et al. (1999) in their test on rock cutting found that the cutting force of the bit with a 20° rake angle decreased about 30%–80% (depending upon other machining parameters), compared to that of the bit with a −20° rake angle. Moreover, an increase in cutting force with the cutting depth becomes lower with increase in the rake angle. Gunay et al. (2006) carried out a detailed experimental study of the influence of the rake angle in machining of AISI 1040 steel. They found a very small influence, which diminishes at high cutting speed. Saglam, Yaldiz, and Unsacar carried out an extensive research program on machining of AISI 1040 steel bars hardened to HRC 40 (Saglam et al. 2007) in order to reveal the effect of tool geometry. A system consideration of the major geometry parameter was attempted as the interinfluence of these parameters was considered. It was found that an increase in the rake angle noticeably reduces the cutting force while the cutting temperature increases. It was also found that the influence of the rake angle depends on the tool cutting edge angle. More dramatic influences of the rake angle on the cutting force and temperature were found for high cutting speeds.

A complete analysis of the influence of the rake angle on the cutting process as well as recommendation for the selection of the so-called effective rake angle was presented by the author earlier (Astakhov 2010). It is shown that

- Rake angle has significant effect on the amount of work of plastic deformation in metal cutting
- The effect of the rake angle is more profound at low cutting speeds although it is still significant at moderated and high cutting speeds.

Two important effects of the rake angle should be considered as relevant to drilling tools: (1) influence of the rake angle on the force direction and (2) influence of the rake angle on the tool–chip contact length.

C.3.2.1 Influence of the Rake Angle on the Tool–Chip Contact Length

In metal cutting, the tool–chip contact length known as the length of the tool–chip interface determines major tribological conditions at this interface such as temperatures, stresses, and tool wear (Astakhov 2006). Moreover, all the energy required by the cutting system for chip removal passes through this interface. Therefore, it is of great interest to find out a way to assess this length.

To deal with the problem, the Poletica criterion (*Po*-criterion) is introduced (Poletica 1969) as the ratio of the contact length l_c to the uncut chip thickness h_D:

$$Po = \frac{l_c}{h_D} \quad \text{(C.26)}$$

It was found that for a wide variety of work materials, this criterion can be calculated through the CCR ζ as

$$Po = \zeta^{k_r} \quad \text{(C.27)}$$

where $k_r = 1.5$ when $\zeta < 4$ and $k_r = 1.3$ when $\zeta \geq 4$.

Because CCR reduces with an increase in the rake angle as shown in Figure C.34 (Zorev 1966), the rake angle directly affects the tool–chip contact length. Although this effect may vary for practical combinations of the work and tool materials, machining regimes, and many other particularities of the machining system, the general trend is still the same, that is, the greater the rake angle, the shorter the length of the tool–chip contact. Reductions of CCR and tool–chip contact length with the rake angle have opposite effects on the outcomes of the cutting process.

Reduction of CCR reduces the work of plastic deformation, which is the major contributor to the energy spent in metal cutting (Astakhov 2010). Therefore, less energy should pass through the tool–chip interface that reduces the portion of the normal stress due to plastic deformation and thus normal force on this interface. In turn, this reduces tool chipping due to high normal stresses in the region of the rake face adjacent to the cutting edge.

Reduction of the tool–chip contact length, however, increases the tool–chip normal stresses (Astakhov 2006). The total effect (the reduction of the normal stresses due to reduction of plastic deformation of the work material and increase of these stresses due to reduction of the tool–chip contact length) depends on many particularities of a given machining operation.

Reduction of the tool–chip contact length may also affect the friction conditions at the tool–chip interface. Analyzing numerous experimental results, Poletica concluded (Poletica 1969) that although the mean shear stress τ_c at the tool–chip interface can be correlated with many mechanical properties of the work material, the best fit seems to be achieved with the ultimate tensile strength, σ_{UTS} (Astakhov 2006). The following empirical relation shows good correlation with available experimental data:

$$\tau_c = 0.28\sigma_{UTS} \tag{C.28}$$

Therefore, as the shear stresses at the tool–chip interface remain the same, the reduction of the contact length reduces the friction force at the tool–chip interface, that is, a smaller amount of heat is generated due to the friction at this interface that, in turn, should result in lower contact temperatures. Although this is true, the location of the maximum temperature at this interface shifts toward the cutting edge as the contact length decreases. As cross-sectional area of the cutting wedge becomes smaller, higher temperatures occur.

C.3.2.2 Influence of the Rake Angle on Force Direction

In professional literature for practical tool designers, a notion *positive/negative rake angle* (sometimes referred to as positive/negative tool geometry) is widely debated (Destefani 2002). In the author's opinion, the notion *positive/negative tool geometry* is an atavism that came from *ancient* time when carbides, as tool materials, were rather brittle especially when used on the old, nonrigid, underpowered machines with excessive spindle runout. As such, brittle carbides chipped when positive rake angles were used. The reason for that is as follows. When cutting with a positive rake angle as shown in Figure C.33a, the interaction of the tool rake face with the

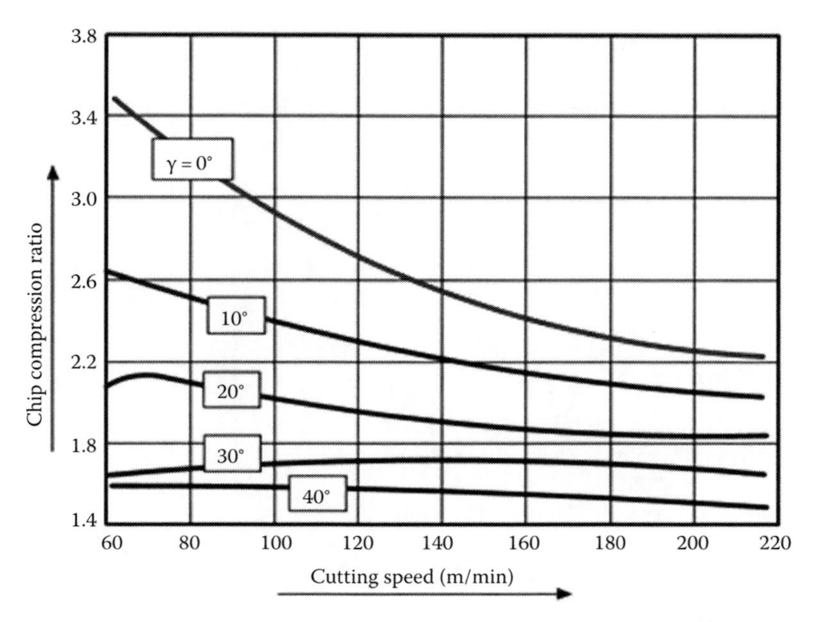

FIGURE C.34 Influence the rake angle on CCR for the range of the cutting speed used for carbide tools. Free cutting with $a_p = 6$ mm, $h_d = 0.15$ mm. Work material: AISI steel 4130.

moving chip results in certain distributions of the normal and tangential stresses over the tool–chip contact length l_c (Astakhov 2004). These distributions can be represented in terms of the corresponding resultant normal N and tangential F forces acting on the rake face and thus on the cutting wedge. As can be seen, in the case of the tool with a positive rake angle (Figure C.34a) the normal force N causes bending of the tip of the cutting wedge. The presence of the bending may cause wedge chipping if the tool material is not strong enough. However, the tool–chip contact area reduces with the rake angle, so the point of application of the normal force shifts closer to the cutting edge that reduces the bending moment. On the contrary, when cutting with a tool having a negative rake angle (Figure C.33b), the normal force N acting on the tool rake face does not cause the mentioned bending. Instead, it results in compression of the tool material. Because tool materials have very high compressive strength, the strength of the cutting wedge is higher.

A special case of machining with a high rake angle should be considered here as it is relevant to drilling tools. It was noticed that when a twist drill is used for enlarging previously drilled holes in relatively soft work materials, such as brass, copper, and Babbitt, the drill jumped ahead of the feed into the hole, which caused vibration.

To understand why this happens, consider a simplified force model for machining with a tool having a high rake angle as shown in Figure C.35. In cutting, the radial component F_p of the resultant force R normally pushes the tool out of the workpiece. However, it may not be the case in machining with a tool having a high rake angle. As follows from the model shown in Figure C.35, the radial force is calculated as

$$F_p = F_f \cos\gamma - R\sin\gamma + F_q \tag{C.29}$$

where

- F_f is the friction force over the tool–chip interface
- F_q is the force on the tool flank land that depends on the flank angle, tool wear, MWF, and other cutting parameters (Astakhov 2006). This force can be accounted fairly well when its specific value of 30–60 N/1 mm of the cutting edge length is considered.

The first component $F_f \cos\gamma$, which pushes the tool away from the workpiece, decreases with the rake angle, while the second component $R \sin\gamma$, which pulls the tool into the workpiece, increases.

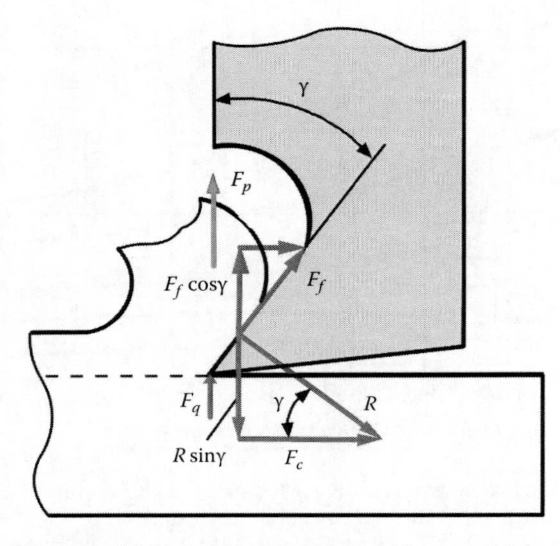

FIGURE C.35 Simplified force model for machining with a tool having a high rake angle.

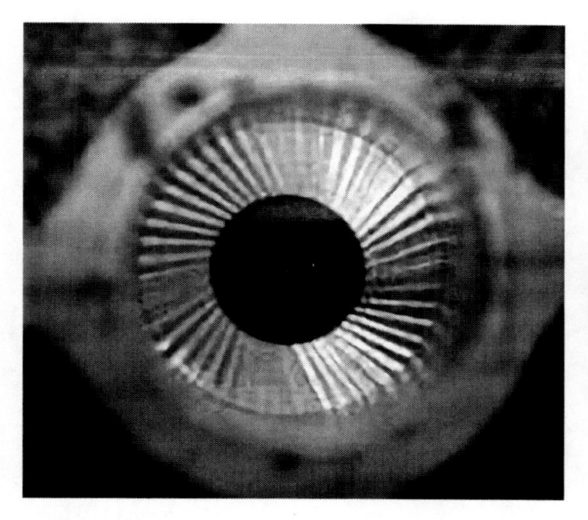

FIGURE C.36 Chatter marks on the bottom of the hole being enlarged with a twist drill.

Therefore, as the rake angle increases and a sharp cutting tool is used (small F_q), the radial force F_p can be directed into workpiece, which is the root cause of the described phenomenon (chatter in drilling). Its typical appearance is shown in Figure C.36.

The era of twist drill cutting edge modification began almost at the same time with the wide use of these tools. Having solved the problem with chip transportation due to a helical chip flute, twist drills brought another challenge to drilling, which is the high rake angle at the drill periphery while that close to the drill center is low. When a twist drill was used for enlarging previously drilled holes, the drill jumped ahead of the feed into the hole, which caused vibration, poor quality of drilled holes, drill breakage, etc. The reason for that is high rake angles in the regions adjacent to drill periphery as discussed in Chapter 4. The common solution of this problem is shown in Figure 4.56.

C.3.3 Tool Cutting Edge Angle

C.3.3.1 Components of the Cutting Force and Uncut Chip Thickness

The tool cutting edge angle is often considered as the most important angle of the tool geometry as it has multifaceted influence on practically all aspects of the metal cutting process and greatly affects the outcomes of a machining operation. This is because it defines the magnitudes of the radial and tangential components of the cutting force and, for given feed and cutting depth, it defines the uncut chip thickness, width of cut, chip flow direction, and tool life. The physical background of this phenomenon can be explained as follows: when κ_r decreases, the chip width increases correspondingly because the active part of the cutting edge increases. The latter results in improved heat removal from the tool and hence tool life increases. For example, if tool life of a high-speed steel face milling tool having $\kappa_r = 60°$ is taken to be 100%, then when $\kappa_r = 30°$, its tool life is 190%, and when $\kappa_r = 10°$, its tool life is 650%. Even more profound effect of κ_r is observed in the machining with single-point cutting tools. For example, in rough turning of carbon steels, the change of κ_r from 45° to 30° leads sometimes to a fivefold increase in tool life.

The reduction of κ_r, however, has its drawbacks. One of them is the corresponding increase of the radial component of the cutting force. As discussed in Chapter 1, the cutting force R is a 3D vector that is normally resolved into three orthogonal components, namely, the power (tangential, toque) component F_c, axial component F_f, and radial component F_p. For simplicity, these are often

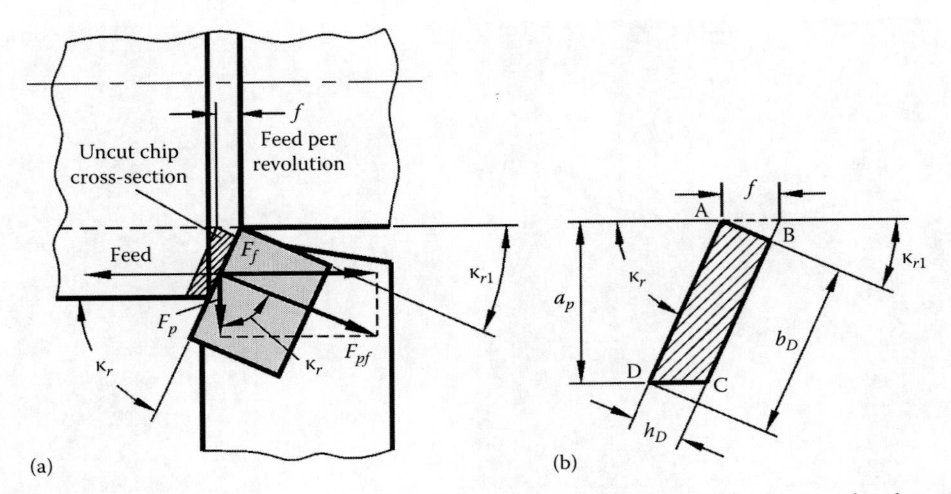

FIGURE C.37 Forces in the reference plane (a) and parameters of the uncut chip cross-sectional area (b).

called tangential, axial, and radial forces, respectively. Figure C.37a depicts the component F_{pf} in the reference plane. As can be seen, the radial F_p and axial F_f forces are actually components of F_{pf}. As follows from Figure C.37a, the radial and axial forces are related as

$$\frac{F_f}{F_p} = \tan \kappa_r \tag{C.30}$$

As follows from Equation C.30, lowering κ_r from 45° to 20° results in more than a twofold increase in the radial force that increases the bending force acting on the workpiece and thus may reduce the accuracy of machining because the rigidity of the workpiece varies along its length. When the workpiece is machined between centers, an increased radial force causes the so-called barreling, and when the workpiece is clamped only in the chuck, tapering may occur. Besides, because lowering κ_r leads to an increased radial force, this force often causes vibrations so that the advantages of small cutting edge angle may become not too profound.

The tool cutting edge angle has direct influence on the uncut (undeformed) chip cross-sectional shape, uncut chip cross-sectional area, and thus on the uncut chip thickness, which is by far the most important parameter of a machining operation because it determines (for a given work material) the cutting force, plastic deformation of the work material in its transformation into the chip, allowable feed, tool–chip contact length, etc. Therefore, the correlations between κ_r and the uncut chip thickness (known in practice of metal cutting as the chip load) should be established.

In orthogonal cutting, the concept of the uncut chip thickness is self-obvious as it is equal to the thickness of the layer that has been removed (Figure C.1). Figure C.37b helps to comprehend the concept of the uncut chip thickness in the simplest case of turning. As seen, the uncut chip cross-sectional area is represented by polygon ABCD. Side AD is formed by the major cutting edge, side AB is formed by the minor cutting edge, and side DC is a part of the workpiece surface to be machined. The uncut chip thickness h_D is the thickness of the layer to be removed per one revolution of the workpiece as measured perpendicular to the cutting edge. As it follows from Figure C.37b,

$$h_D = f \sin \kappa_r \tag{C.31}$$

It follows from Equation C.31 that under a given feed f, the uncut chip thickness can be varied in a wide range by changing angle κ_r. When this angle becomes zero, the uncut chip thickness is also zero (no cutting), and when $\kappa_r = 90°$, the uncut chip thickness reaches its maximum, becoming $h_D = f$ as in orthogonal cutting. Therefore, $h_D \in (0, f)$.

The following equation for the uncut chip area correlates the feed f and the depth of cut a_p, with the uncut chip thickness h_D and nonorthogonal chip width b_D:

$$a_p f = h_D b_D = A_D \tag{C.32}$$

where A_D is the uncut chip cross-sectional area.

C.3.3.2 Minor Cutting Edge

The role of the minor cutting edge and its influence on the cutting force and power consumption are seldom considered in the literature on metal cutting. At best, the influence of the tool minor cutting edge angle κ_{r1} is mentioned in the consideration of the theoretical roughness of the machined surface or the geometric component of roughness. In the author's opinion, the term *minor* probably misled many researchers in the field causing a common perception that, besides the microgeometry of the machined surface, this cutting edge does not affect the cutting process to any noticeable degree.

Zorev provided a detailed analysis of the chip formation by the minor cutting edge (Zorev 1966). He studied the velocity hodograph, associated plastic deformation, and flows in this region. Using the results of this study, one can visualize the chip cross-sectional area cut by the minor cutting edge with the aid of Figure C.38. Figure C.38a shows a hypothetical single-point cutting tool having $\kappa_{r1} = 90°$; that is, practically, this tool does not have a minor cutting edge. Figure C.38b shows the cross-sectional area ABC of *a tooth* of the surface profile left after this surface was machined by this tool. Real cutting tools have the minor cutting edge with $\kappa_{r1} \ll 90°$ so that the surface profile left by the cutting tool is ACD while area ADB is cut by the minor cutting edge as shown in Figure C.38c.

C.3.4 Inclination Angle

Although the sense and sign of the inclination angle λ_s is clearly shown in Figure C.11 (View S) and it is defined earlier as the angle between the cutting edge and the reference plane as viewed in the tool cutting-edge plane, experience shows that there are certain difficulties and confusions in understanding this angle and its influence on the cutting process. Figure C.39 aims to clarify the

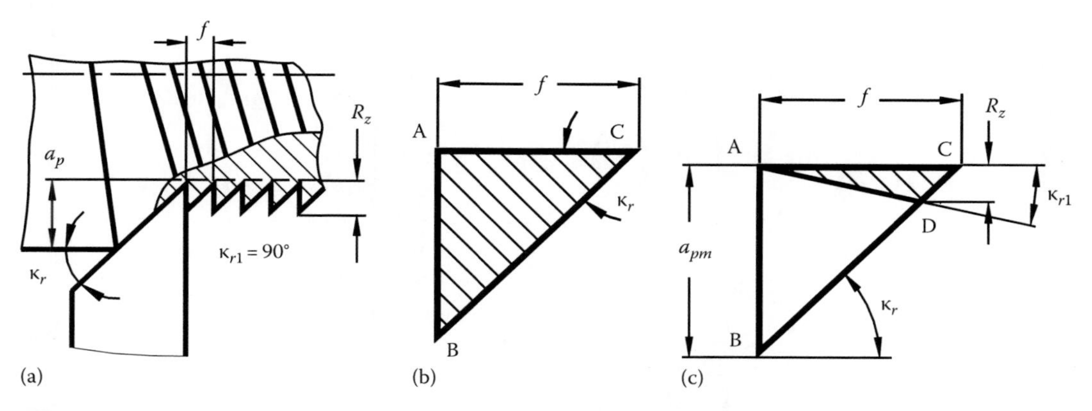

FIGURE C.38 The cross-sectional area of the chip cut by the minor cutting edge: (a) hypothetic tool having a 90° tool cutting edge angle of the minor cutting edge, (b) the cross section of the surface roughness left after cutting when the tool minor cutting edge angle is 90°, and (c) geometrical model to calculate the cross-sectional area of the chip cut by the minor cutting edge when the tool minor cutting edge angle is less than 90°.

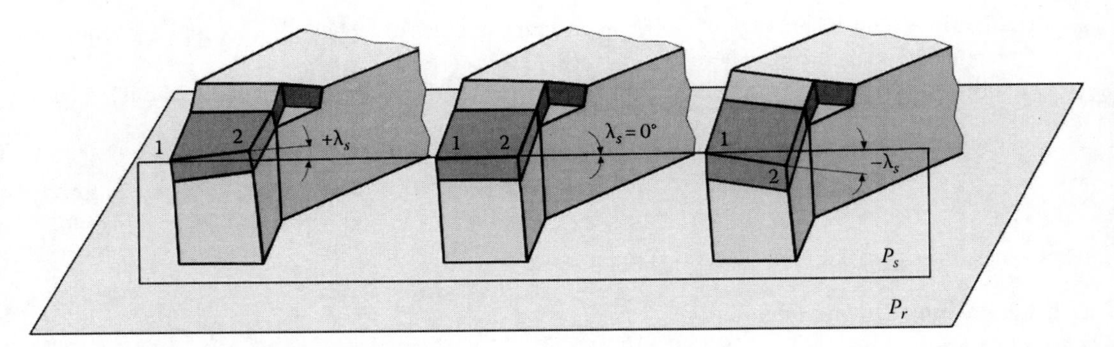

FIGURE C.39 Sign of the inclination angle.

issue. The inclination angle λ_s is measured in the cutting-edge plane P_s, which is perpendicular to the reference plane P_r and contains the cutting edge 1–2. Numbers 1 and 2 designate the ends of the cutting edge. As such, if point 1 (the tool corner) locates below point 2, then the inclination angle λ_s is positive; if points 1 and 2 are at the same level, then $\lambda_s = 0$; and when point 1 locates above point 2, then the inclination angle λ_s is negative.

The sign of the inclination angle defines the chip flow direction as shown in Figure C.40. When λ_s is positive, the chip flows to the right and can potentially damage the machined surface. When λ_s is negative, the chip flows to the left. When $\lambda_s = 0$, the chip flow direction in the reference plane is entirely determined by the tool cutting edge κ_r.

In the literature on metal cutting, the influence of this angle is rarely discussed as this influence is complicated so that the system approach is needed to reveal this influence. DeVries (1992) presented an analysis on the influence of the inclination angle on the chip flow angle (direction) showing that according to previous researches, this angle is close to the inclination angle. DeVries analyzed the influence of the inclination angle on the cutting force using the model of the cutting force derived using the improper Merchant's force model modified for the case of oblique cutting. He did not find any influence of the inclination angle in its range of 0°–20°.

In standard tuning tools with indexable inserts, the inclination angle λ_s cannot be readily distinguished because it is rather small for standard single-point tools. This angle forms in the T-mach-S when the insert is placed in a tool holder and the tool holder is mounted in the machine. As such, the inclination of the base face in the working plane γ_f and in the back plane γ_p is indicated in catalogs. For a common tool holder with $\kappa_r = 60°$, $\gamma_f = -6°$ and $\gamma_p = -6°$. Using Equation C.5, one can calculate the most common inclination angle used in turning as

$$\lambda_s = \arctan(-\sin\kappa_r \tan\gamma_p - \cos\kappa_r \tan\gamma_f) = \arctan(-\sin 60° \tan(-6°) - \cos 60° \tan(-6°)) = 2.2° \quad \text{(C.33)}$$

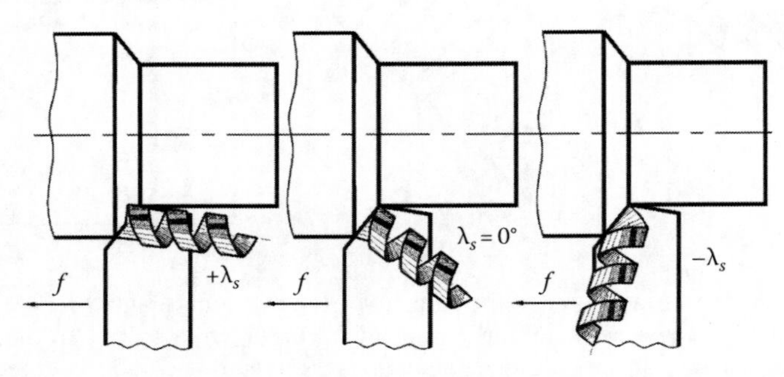

FIGURE C.40 Influence of the sign of the inclination angle on the direction of chip flow.

In the author's opinion, the influence of the inclination angle in other practical machining operations is significant as far as the width of cut is great. In other words, the inclination angle affects the cutting process when this angle is able to change the state of stress in the deformation zone. As explained in great detail by Zorev (1966) and pointed out by Shaw (1984), it happens when the chip flow direction changes significantly due to inclination angle. Unfortunately, instead of being reevaluated for new developments in tool materials and experimental techniques, these results are simply forgotten in modern theory and practice of metal cutting and tool design.

C.3.5 Relevant Angles: Direction of Chip Flow

A simple yet logical and important question about the cutting tool angles is about their relevancy to tool performance. On the one hand, the obvious answer relates to pure geometrical considerations as follows: they should (1) assure tool free penetration into the workpiece, (2) provide the desired direction of the components of the cutting force and theoretical surface roughness (the tool cutting edge angles of the major and minor edges), (3) assure the desired direction of chip flow (the cutting tool inclination angle), and so on. On the other hand, the selected angles should assure a certain desired tool life, machining residual stress, chip shape, and so on, that is, they should be related to the mechanics and physics of cutting. While the mentioned geometrical considerations are rather simple and well studied, the latter is not, so it presents challenges in the tool design, implementation, and failure analysis. Unfortunately, the available literature on metal cutting provides little help on the matter.

Among many important issues in the correlation of the tool geometry parameters with tool performance at the physical level is the plane of measurement of the rake and clearance angles. Although being very important in tool design and manufacturing, the previously introduced reference systems and corresponding rake and clearance angles in these systems are purely geometrical. In the author's opinion, the following should be taken into account in the consideration of the relevant rake and clearance angles:

1. The rake angle should be considered in the plane containing the vector of the direction of true chip flow (the chip flow plane P_{ch}) because along this direction, (a) the chip velocity relative to the rake face is at maximum; (b) although the chip deformation could vary along the chip width particularly when the active part of the cutting edge is great, the average chip deformation, and thus the work of plastic deformation of the layer being removed, is the greatest; and (c) the tool–chip contact length is the greatest.
2. The clearance angle should be considered in the direction where the sliding speed over the tool–workpiece interface and the previously discussed springback (elastic recovery) of the machined surface are the greatest. Unfortunately, these two conditions are seldom fulfilled simultaneously, that is, in one plane of measurement. Moreover, it often happens that one of these two conditions is fulfilled in the chip flow plane P_{ch}. This might present a problem in a 2D modeling of the cutting process including FEM. When the length of the active part of a single-point tool edge and inclination angle of this edge are not great, which is the case in many practical machining operations where this length does not normally exceed a few millimeters and inclination angle is no more than 7°, the difference is not of appreciable significance so that the clearance angle can be safely considered in the chip flow plane. However, this is not nearly the case for some cutting tools, for example, for drills as considered in Chapter 4.

Therefore, the determination of the chip flow direction and the corresponding chip flow plane are important.

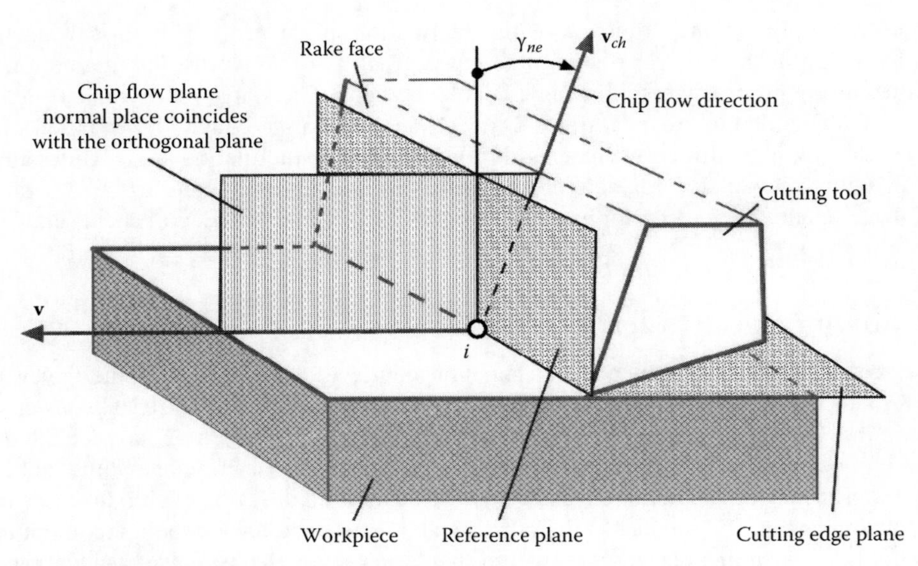

FIGURE C.41 Representation of the chip flow direction in orthogonal cutting.

In orthogonal cutting (Figure C.4a), the cutting edge inclination angle λ_s is zero. Figure C.41 shows the T-use-S tool geometry in orthogonal cutting for a selected point *i* of the cutting edge. As previously defined, the reference plane is perpendicular to the cutting speed vector and passes through point *i*, the orthogonal plane is perpendicular to the reference plane and passes through point *i*, and the cutting-edge plane is perpendicular to the reference plane and contains the cutting edge. As can be seen in Figure C.41, the cutting edge plane coincides with the machined surface.

As the cutting edge is perpendicular to the cutting speed vector, the normal plane coincides with the orthogonal plane. As previously defined, the normal rake angle is measured in the normal plane between the two intersection lines that this plane makes with the reference plane and the rake face. The normal clearance angle is measured in the normal plane between the two intersection lines that this plane makes with the flank face and the cutting-edge plane. The chip flow direction is perpendicular to the cutting edge as shown in Figure C.41 so that orthogonal, normal, and chip flow planes are coincident.

The normal rake and clearance angles should be considered in the cutting mechanics and correlated with the physics of cutting because the chip velocity and the contact length of the tool-chip interface the rake face are greatest in the direction of chip flow and because the machined surface springback and sliding velocity at the tool–workpiece interface are greatest compared to any other directions of consideration.

In oblique cutting (Figure C.4b), the cutting edge inclination angle λ_{se} is not zero, which makes its consideration a bit more complicated compared to orthogonal cutting. Figure C.42 shows T-use-S tool geometry in oblique cutting for a selected point *i* of the cutting edge. As previously defined, the reference plane is perpendicular to the cutting speed vector and passed through point *i*. The angle that the reference plane makes with the cutting edge is the previously defined to be the cutting edge inclination angle λ_{se}. The normal plane is perpendicular to the cutting edge and passes through point *i*, the orthogonal plane passes through point *i* and is perpendicular to the reference plane and to the projection of the cutting edge into the reference plane, and the cutting-edge plane is perpendicular to the reference plane and contains the cutting edge. As can be seen in Figure C.42, the cutting edge plane coincides with the machined surface.

As in orthogonal cutting, the normal rake angle is measured in the normal plane between the two intersection lines that this plane makes with the reference plane and the rake face. The normal clearance angle is measured in the normal plane between the two intersection lines that this plane

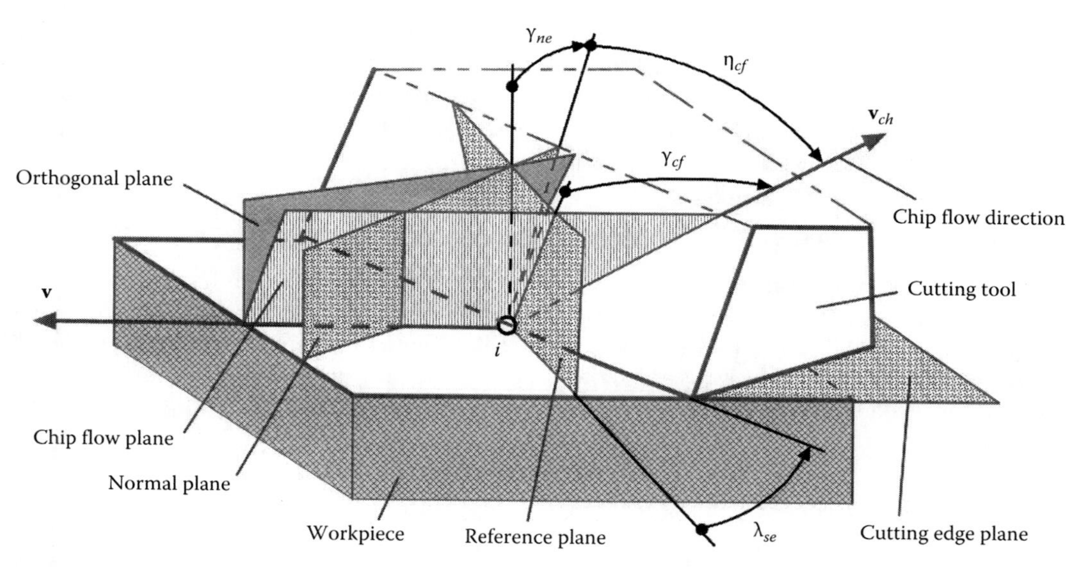

FIGURE C.42 Representation of the chip flow direction in oblique cutting.

makes with the flank face and the cutting-edge plane. An important question would be about the relevance of these normal angles to the cutting physics considerations.

As stated earlier, the rake angle should be considered in the plane containing the vector of the direction of true chip flow. Figure C.42 shows the chip flow direction represented by the vector of chip flow $\mathbf{v}_{ch}$. In the rake face plane, the angle between this direction and the intersection line of the normal plane and the rake face plane is universally known as the chip flow angle η_{cf} (Stabler 1964; Shaw 1984). The chip flow plane is defined as a plane that passes through point i and contains the vector of chip flow $\mathbf{v}_{ch}$. In this plane, the rake angle γ_{cf} is measured between two intersection lines, which the chip flow plane makes with the reference plane and the rake face. As shown in (Astakhov 1998/1999), the rake angle γ_{cf} and the chip flow angle η_{cf} are correlated as

$$\gamma_{cf} = \arcsin(\sin\eta_{cf}\sin\lambda_{se} + \cos\eta_{cf}\cos\lambda_{se}\sin\gamma_{ne}) \tag{C.34}$$

The determination of the actual chip flow angle requires a great number of carefully prepared tests (Astakhov 2006). To simplify many practical calculations, the Stabler rule (Stabler 1964) is normally used. According to this rule, $\eta_{cf} = \lambda_{se}$. If this is substituted into Equation C.34, it becomes

$$\gamma_{cf} = \arcsin[1 - \cos^2\lambda_{se}(1 - \sin\gamma_{ne})] \tag{C.35}$$

This angle should be considered as the relevant rake angle. However, the earlier considerations are valid only for oblique free cutting, which rarely happens in reality. As such, Equation C.35 becomes more complicated (Astakhov 1998/1999).

The determination of the relevant clearance angle is not simple and straightforward as that for the relevant rake angle. It is always a temptation to measure the relevant rake and clearance angles in one reference plane (the chip flow plane) because in this plane, (1) stress, strain, temperature, and so on fields in the workpiece ahead of the cutting edge are continuous so that a 2D modeling including FEM using commercial FEM packages is significantly simplified; (2) the springback of the work material is the greatest as the deformation of the work material is the greatest in this plane; and (3) the clearance angle in this plane is smaller than that in the normal plane so that the length of the tool–workpiece interface due to springback of the work material is the greatest. However, the sliding velocity (the velocity over the tool–chip interface) in this direction is not the greatest. For the case considered in Figure C.42, the greatest sliding speed is in the orthogonal plane, so it appears that the orthogonal clearance angle should be considered. In practice, however, the inclination angle for most of single-point tools is small so is the difference between the normal and orthogonal clearance angles.

C.4 CUTTING EDGE

As discussed earlier, the cutting edge is a line of intersection of the rake and flank surfaces, and thus, it is treated so at the macro level. A logical question about why a sharp edge is made round/chamfered, that is, dull, using a wide-spread edge preparation (EP) technology has to be answered.

The manufacturing practice, however, shows that EP sometimes works well but sometimes leads to catastrophic consequences for the tool. Therefore, it is necessary to answer a series of simple and practical questions:

1. What is exactly EP? Technology and cost.
2. What is exactly a *sharp tool*?
3. Why does EP sometimes work and sometimes not work?
4. What is the optimal EP parameters for given cutting conditions? How does it correlate with the other parameters of the tool geometry, design, and its manufacturing quality?

In the author's opinion, any talk/presentation/text/all of earlier texts/whatsoever about EP is a pure speculation unless the answers to the previously posted questions are provided in clear and concise manner.

C.4.1 SHARPNESS

As discussed earlier, the cutting edge is a line of intersection of the rake and flank surfaces, and thus, it is treated so at the macro level. As shown further, it may not be the case in reality. To understand how far a common model of chip formation with a perfectly sharp cutting edge is from reality, one first needs to understand the optimal cutting conditions with a perfectly sharp cutting edge. A simple model for this case is shown in Figure C.43a. The particularities of this model were discussed by the author earlier (Astakhov 1998/1999, 2006, 2010). In this model, the action of the normal force N on the rake face at a certain distance from the cutting edge creates the bending

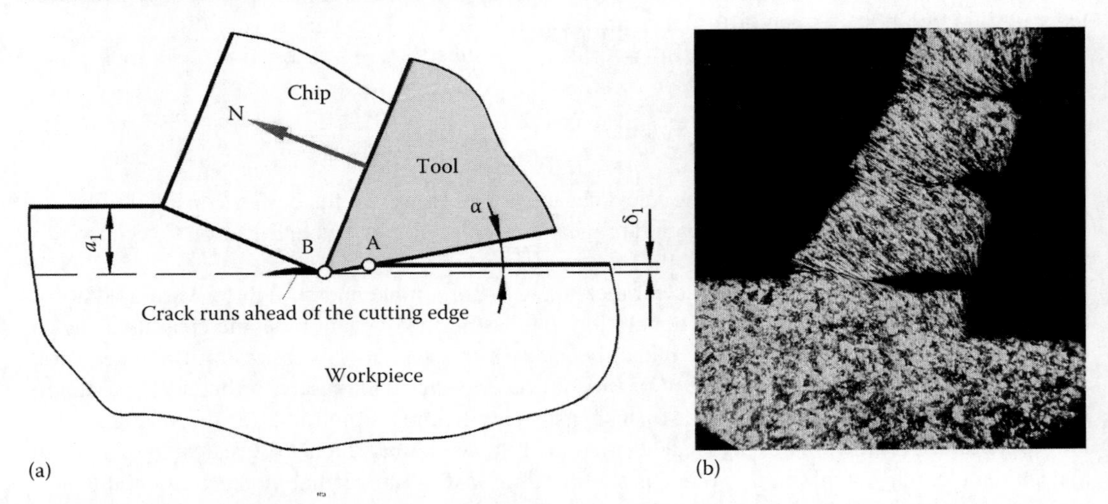

FIGURE C.43 Cutting with a perfectly sharp cutting edge: (a) model and (b) partially formed chip.

moment ahead of this edge that combines with the compressive load applied by the cutting force. These two together create the so-called combined (complex) state of stress in the zone ahead of the cutting edge that, when optimized, minimizes the resistance of the work material to tool penetration (i.e., cutting). In machining of not very ductile materials, a visible crack forms ahead of the cutting edge as shown in Figure C.43b. The sharper the edge, the higher probability of formation of a visible crack as this edge serves as a stress concentrator.

The elastic recovery (springback) of the work material δ_1 is at minimum as the minimum cutting force is required at these optimized cutting conditions. As a result, the flank contact area (represented by segment AB in Figure C.43a) is small provided that the clearance angle is optimal. This maximizes tool life in terms of reduction of the flank wear.

In reality, however, the cutting edge is not a line but rather a transitional surface between the rake and flank surfaces that could be, in general, of any shape. Figure C.44 shows examples of common roughness of the flank and rake surfaces. Note that these pictures are taken from the top-of-the-line drills made for the most advanced automotive plant. Therefore, it can be safely assumed that surface roughness of other drilling tools is even much worse. This explains the earlier statement

FIGURE C.44 Examples of common coarse roughness of the flank and rake surfaces: (a) flank face in the region of the chisel edge, (b) the primary flank along the major cutting edge, (c) rake face, and (d) primary flank face adjacent to the drill corner.

about the existence of the transition surface between the rake and flank surfaces instead of a sharp cutting edge. Even if it does not exist (visibly) for a new tool, it is formed just after machining of a few holes because of microfracturing of asperities due to rough rake and flank faces. As a result, a model shown in Figure C.43a is not adequate to many practical drilling operations. Therefore, the cutting edge should be represented by a surface in any modeling of metal cutting with real-world drilling tools.

It is convenient for modeling to represent this surface by a cylindrical surface having radius R_{ce} as shown in Figure C.45. The first stage in comprehending the influence of the cutting edge roundness is the consideration of forces and contact stresses acting on the tool flank face as related to R_{ce}. Figure C.46 shows a simple model. As can be seen, the cutting tool has the cutting edge radius R_{ce}. Due to this radius, the total uncut chip thickness h_D is *physically* separated at point A into two work material flows in its transformation into the chip: the actual uncut chip thickness h_{Da} and the layer of thickness h_1 to be burnished by the round part adjacent to the tool flank face. In technical terms, physical separation of a material into two or more parts is referred to as fracture, so the fracture stress corresponding to the state of stresses at point A should be achieved to accomplish the discussed separation separation. This leads to an increase cutting force compare to the tool with the sharp cutting edge shown in Figure C.43.

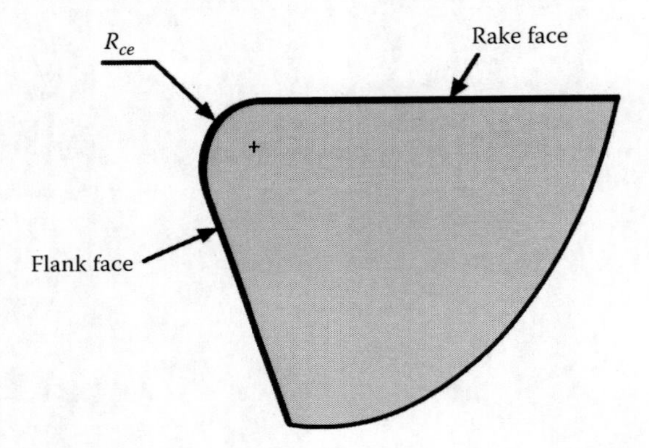

FIGURE C.45 Radius of the cutting edge.

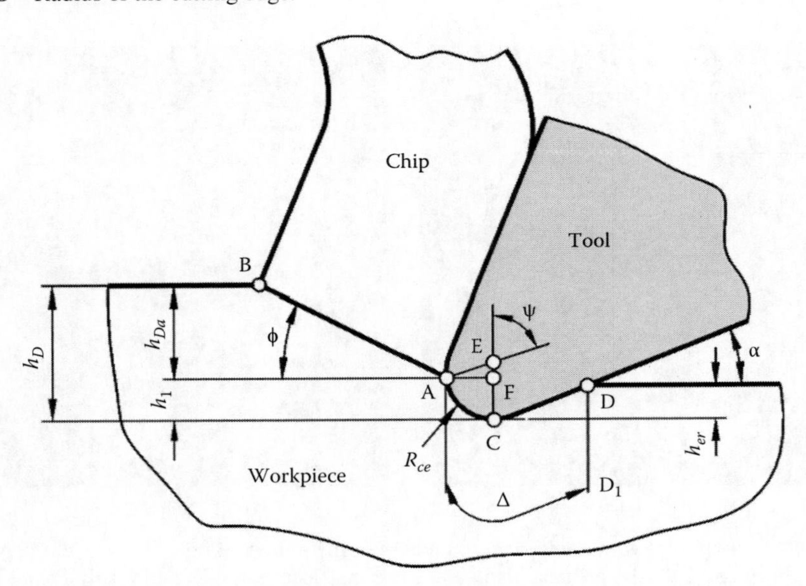

FIGURE C.46 Simple model of chip formation with a rounded cutting edge.

The arc distance between points A and D designated as Δ consists of curved AC and straight CD parts. It is calculated as

$$\Delta = \cup AC + CD = R_{ce}\psi + \frac{h_{er}}{\sin\alpha} \tag{C.36}$$

where
 ψ is the central angle corresponding to arc AC
 h_{er} is elastic recovery (springback) of the machined surface.

Because $\psi = \arccos(1 - h_1/R_{ce})$, then it follows from Equation C.36 that

$$\Delta = R_{ce}\left(\arccos\left(1 - \frac{h_1}{R_{ce}}\right) + \frac{h_{er}}{R_{ce}\sin\alpha}\right) \tag{C.37}$$

It is known (Silin 1979) that the cutting process ceases and the layer to be removed undergoes plastic deformation similar to burnishing when

$$\frac{h_1}{R_{ce}} \le 0.5 - \frac{\tau_{in}}{\sigma_{YT}} \tag{C.38}$$

where
 σ_{YT} is the yield strength of the work material
 τ_{in} is the strength of adhesion bonds at the tool–workpiece interface determined using results of adhesion tests (Shuster 1988).

The strength of adhesion bonds depends on mutual adhesion properties of the tool and work materials as well as on the surface roughness of the tool.

Combining Equations C.37 and C.38, one can obtain

$$\frac{\Delta}{R_{ce}} = \arccos\left(0.5 + \frac{\tau_{in}}{\sigma_{YT}}\right) + \frac{h_{er}\left(0.5 - (\tau_{in}/\sigma_{YT})\right)}{h_1\sin\alpha} \tag{C.39}$$

The plastic deformation of the surface layer can be characterized by the burnishing factor $m_b = h_1/h_{er}$, which according to Poletica (1969) can be approximated by the CCR ζ. As a result, Equation C.39 becomes

$$\frac{\Delta}{R_{ce}} = \arccos\left(0.5 + \frac{\tau_{in}}{\sigma_{YT}}\right) + \frac{\left(0.5 - (\tau_{in}/\sigma_{YT})\right)\sin\varphi}{\cos(\varphi-\gamma)\sin\alpha} \approx 1.25\sqrt{\frac{Br}{\sin\alpha}} \tag{C.40}$$

where Br is a similarity criterion, referred to as the Briks criterion (Silin 1979; Astakhov 1998/1999),

$$Br = \frac{\cos\gamma}{\zeta - \sin\gamma} \tag{C.41}$$

The experimental results showed that $h_1 \approx 0.5R_{ce}$ is a good approximation when cutting ductile materials (Silin 1979). As such, Equation C.40 becomes

$$\Delta = R_{ce}\left(\frac{\pi}{3} + \frac{0.5\sin\varphi}{\cos(\varphi-\gamma)\sin\alpha}\right) \tag{C.42}$$

As follows from Equations C.41 and C.42, the contact length on the tool flank is a function of the tool rake angle and CCR.

Using the earlier considerations and the model shown in Figure C.46, one can obtain expressions for h_1 and h_{er} as

$$h_1 = R_{ce}\left(1 - \frac{1}{\sqrt{1+Br^2}}\right) \tag{C.43}$$

$$h_{er} = \frac{R_{ce}\left(1-\sqrt{1/1+Br^2}\right)Br}{\cos\gamma + Br\sin\gamma} \tag{C.44}$$

According to Poletica (1969), the stress distribution at the flank contact surface is as follows:

$$\tau_{cf}(x) = \tau_y \exp\left[-3\left(\frac{x}{\Delta}\right)^2\right] \tag{C.45}$$

where
τ_y is the yield shear strength of the work material
x is the distance from the cutting edge.

Integrating Equation C.45, one can obtain the mean shear stress at the tool flank interface

$$\tau_{cf} = 0.505\tau_y \tag{C.46}$$

which is in excellent agreement with experimental results obtained by Zorev (1966) and Chen and Pun (1988).

The foregoing analysis allows obtaining the expression to calculate the friction force at the tool–workpiece interface as

$$F_{fF} = 0.625\tau_y R_{ce} b_1 \sqrt{\frac{Br}{\sin\alpha}} \tag{C.47}$$

where b_1 is the true uncut chip width, which calculates depending on the tool geometry (Astakhov 2010). In a simple case of turning, it is equal to the width of the chip (Figure C.37).

It directly follows from Equation C.47 that the friction force at the tool flank is directly proportional to the radius of the cutting edge. Moreover, a simple comparison of the model shown in Figures C.43a and C.45 clearly shows that a sharp cutting edge *is always better than a honed/chamfered one* in terms of metal cutting as its performance results in a lower cutting force and temperature. If one has any doubt about this statement, let him hone the edges of his razor (e.g., Gillette Fusion) and then try to shave.

C.4.2 Edge Preparation

C.4.2.1 Idea

Cutting edge preparation (hereafter, CEP) is a modification of the cutting edge (considered as a theoretical line) into a transitional surface between tool rake and flank surfaces (planes). In other words, CEP is a technology of applying a defined radius to the cutting edge to improve quality of the transitional surface between the rake and the flank faces and thus improve cutting edge quality.

Although as suggested by the current analysis and confirmed by experimental results (Thiele et al. 2000; Tugrul Özel et al. 2005), a sharp cutting edge is better than a honed one in terms of cutting forces and temperatures; multiple experimental studies and author's experience reveal the following:

- Tool life of high-speed steel, carbide, PCD, and PCBN tools (single point, twist and straight-flute drills, reamers, milling tools, etc.) increases when the proper (optimal hone radius) EP is used (Agnew 1973; Mayer and Stauffer 1973; Rech 2006; Biermann and Terwey 2008; Cheung et al. 2008).
- The size and surface finish and the process stability (spiraling, chatter marks) in machining of aluminum alloys are much better when a suitable EP is used. Even small hand honing by a diamond file can improve these parameters noticeably.

These well-known results are explained as follows. As mentioned previously, defects of the cutting edge are present in nearly all cutting tools (Rodriguez 2009). They are mainly the result of finish EDMing and grinding. Considered by many as microscopic, these defects are not small as many practitioners believe. Figure C.44 exemplifies this statement. These defects include microcracks, burrs, burns, and serrated cutting edges that lead to great variations in tool performance particularly in demanding applications. In the author's opinion, increased adhesion of the work and tool materials is the major damaging factor. Adhesion between tool and work materials always results in the formation of the BUE, which grows in each cycle of chip formation (Astakhov 1998/1999). The problem is in its size and strength of adhesion bonds. The poorer the topology of the cutting edge, the greater these two factors. When BUE reaches a certain height (depending on the strength of adhesion bonds), it is periodically removed by the moving chip. Such a removal does the damage as BUE takes away pieces of the tool material. Figure C.47 shows a graphical example of the damaged cutting edge due to this process. This issue is discussed in Chapter 2 where adhesion tool wear is considered in detail and a *fairy tale* about the so-called *protective action* of BUE is completely dismantled.

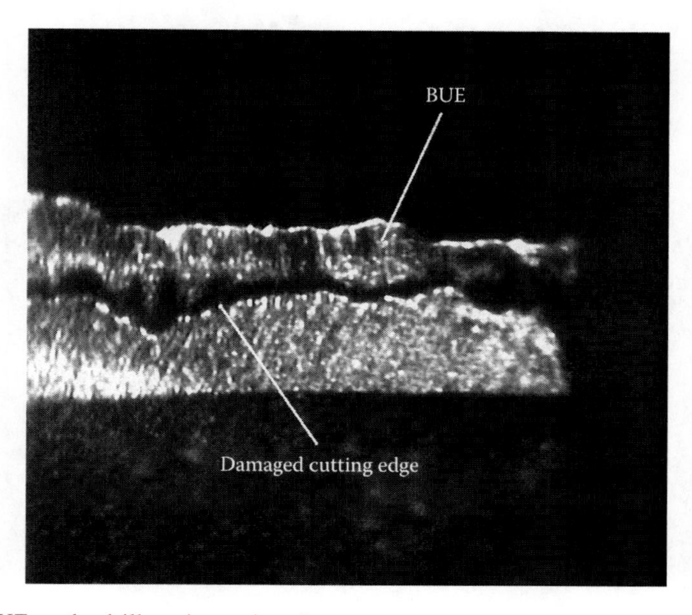

FIGURE C.47 BUE on the drill cutting major edge.

The available information and author's personal experience show that EP works because it

1. Significantly improves the microfinish on the tool–chip and tool–workpiece contact interfaces that reduces adhesion forces over these interfaces (Astakhov 2006).
2. Heals surface microdefects such as cracks and voids in the vicinity of the cutting edge left by the grinding wheel as shown in Figure C.48 and C.49. These defects are critical because they cause micro- and then macrochipping of the cutting edge. EP just *heals* these defects (Komarovsky and Astakhov 2002) as ductile micro-cutting takes place even on superhard tool materials (Jahanmir et al. 1999; Zhong 2003; Kang et al. 2006) (Figures C.48 and C.49).
3. Reduces tool vibration (Astakhov 2011).
4. Improves coatability of the cutting tools. As well known (e.g., Bouzakis et al. [2009]), the application of PVD coating on a sharp edge results in a very high internal stress. As a result, such a coating breaks away and peels off very shortly after starting cutting. For this reason, EP is mandatory for tools to be coated. Figure C.50 shows an example of coating properly applied on a rounded cutting edge.

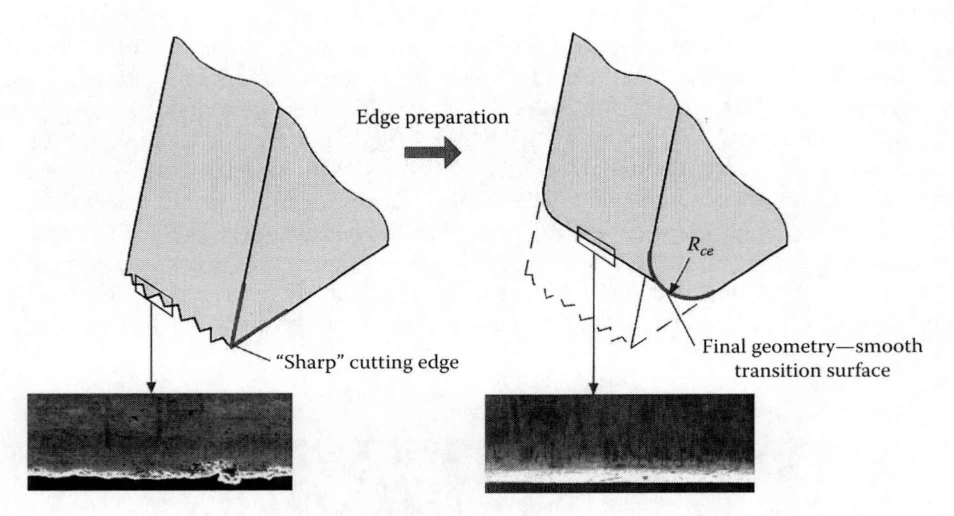

FIGURE C.48 Essence of the CEP technology.

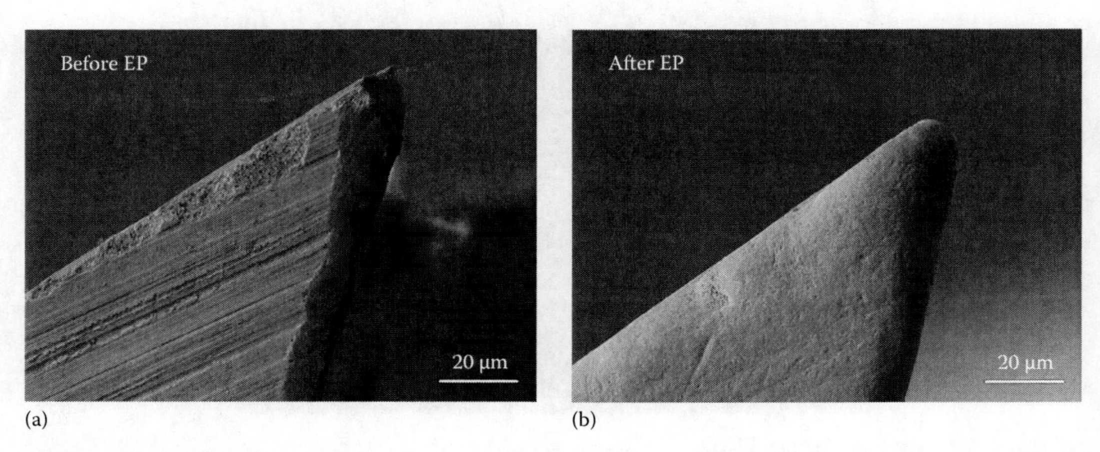

FIGURE C.49 Example of EP results: the cutting edge (a) before EP and (b) after EP by a drag finishing machine by Otec Co.

FIGURE C.50 Tool carbide cutting edge coated by layers of CVD diamond coating.

The next logical question is when EP works. The foregoing analysis suggests that EP is always a trade-off between the factors that cause an increase in tool life as better microfinish of the contact surfaces (less adhesion between the tool and work materials) and better coating and strength of the cutting wedge and those that cause a decrease in tool life as higher friction forces and temperatures on the contact surfaces. As a result, there should be the optimal solution, in terms of tool life, radius of the cutting edge.

The foregoing analysis and the model shown in Figure C.45 suggest that *the radius of the cutting edge R_{ce} should be judged against the uncut chip thickness h_D*. For characterization of cutting tool sharpness, the relative tool sharpness (hereafter, RTS) of the cutting edge was introduced as RTS = h_D/R_{ce} (Outeiro and Astakhov 2005). The minimum value of this ratio that corresponds to negligibly small influence of the cutting edge radius on the cutting process is referred to as the critical relative tool sharpness RTS_{cr}. Zorev (1966) suggested the following empirical rule: the radius of the cutting edge does not affect the cutting process if RTS is equal to or more than 10. In many practical machining operations, however, RTS is less than 10. As a result, the radius of the cutting edge should be considered as a significant factor in the modeling of the cutting process. For example, if one tries to evaluate the influence of the cutting feed and the parameters of the cutting tool geometry, the discussed RTS_{cr} should be always kept in mind.

The radius of the cutting edge and thus RTS affect both the contact stresses (force) on the tool flank and tool geometry as discussed by the author earlier (Astakhov 2010). For example, RTS affects heat partition in the cutting system. Figure C.51 shows an example of energy partition in the cutting system. As seen, the amount of thermal energy transported by the chip (Q_c), conducted into the workpiece (Q_w), and conducted into the tool (Q_t) directly depends on RTS (Outeiro and Astakhov 2005). As RTS decreases, more heat goes into the workpiece causing machining residual stresses and into the tool causing lower tool life. Moreover, when machining with shallow uncut chip thicknesses, a small RTS causes a shift of the region of maximum tool temperatures from the rake face into the flank face that causes excessive flank wear (Astakhov 2010).

The radius of the cutting edge may also affect the rake angle. As shown by the author earlier, this angle is calculated as

$$\gamma_{n1} = \begin{cases} \arcsin\left(\dfrac{h_D}{R_{ce}} - 1\right) & \text{if } h_D < R_{ce} \cdot (1 + \sin\gamma_n) \\ \gamma_n & \text{if } h_D \geq R_{ce} \cdot (1 + \sin\gamma_n) \end{cases} \tag{C.48}$$

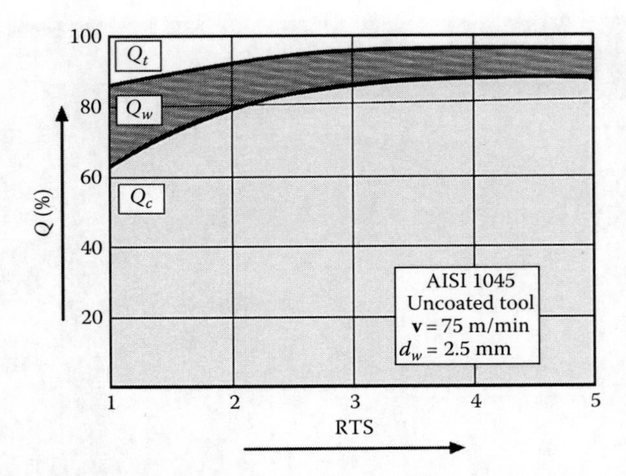

FIGURE C.51 Influence of RTS on energy balance when machining AISI 1045 steel.

Available information on the theory and application of EP allows drawing the following important conclusions:

1. ISO 1832: 2004/2005 and ANSI B212-4 both define various types of EP to be included in the cutting insert designations. The detailed information was presented by the author earlier (Astakhov 2010). Among multiple available forms of EP, more than 80% of honed cutting tools receive a radius hone (Figure C.52a), which is centrally located on the cutting corner of the tool, that is, the extent of EP $L_{ep} = R_{ce}$. Tools with this type of hone are used for general applications. A half-parabolic shape is known as a *waterfall* or *reverse waterfall*, depending on its orientation to the rake and flank surfaces. With a waterfall-shaped hone, the EP is skewed toward the top side of the tool as shown in Figure C.52b where normally for the waterfall EP, its size along the rake face is twice greater ($2L_{ep}$) than that along the flank face (L_{ep}). The main benefit of a waterfall-shaped hone is that the honing process leaves more tool material directly under the cutting edge, which further strengthens the corner (Shaffer 2000).
2. Although the optimal RTS of equal to or greater than 10 is most desirable for cutting tools, practical RTS for drilling tools is selected from the range of 6–10. Note that the optimal RTS is very sensitive to the parameters of the drilling system, so it varies significantly depending upon a particular machining system, cutting tool, tool and work materials, etc. What is the most important, however, is that this optimum lies within a rather narrow range. An example is shown in Figure C.53a (Cheung et al. 2008) for steel drilling. In drilling of aluminum alloys with PCD drills, the window of opportunity is even smaller as shown in Figure C.53b. As can be clearly seen, even small deviation from the optimal cutting edge radius to either side results in steep reduction in tool life. This explains a great

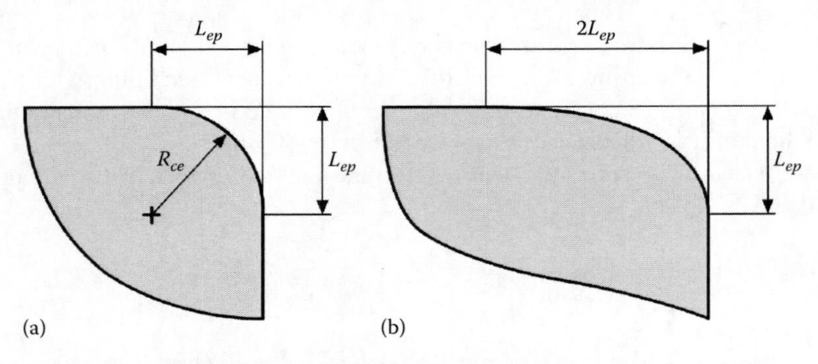

FIGURE C.52 Common shape of EP: (a) radius hone shape and (b) waterfall-shaped hone.

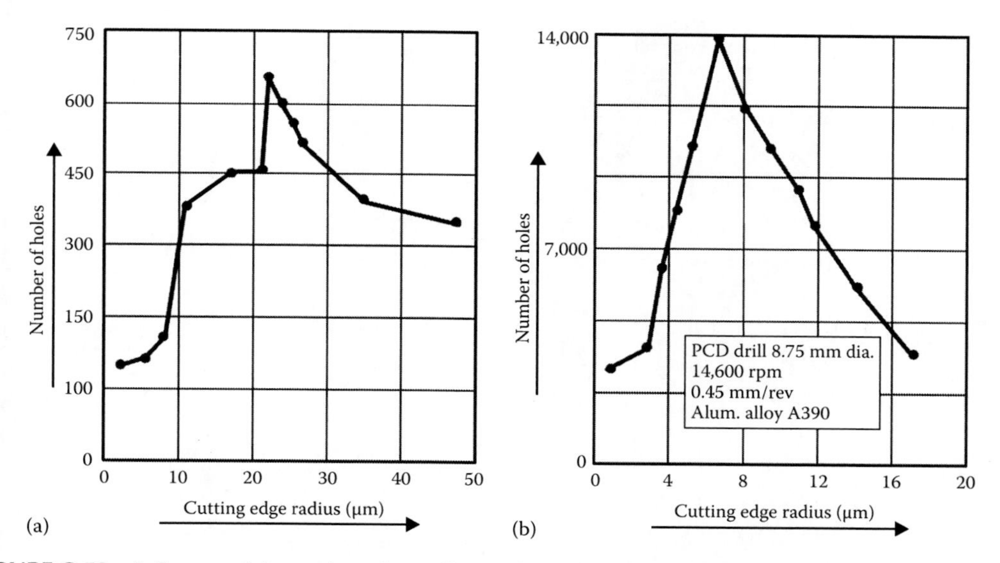

FIGURE C.53 Influence of the cutting edge radius on the tool life in drilling: (a) twist drills in steel drilling and (b) PCD drills in high-silicon aluminum alloy A390 drilling.

scatter in EP application results as the optimal radius and its allowable deviation range (tolerance) are not normally established and thus are not a part of the tool drawing. Therefore, the optimum radius of CEP should be determined for critical applications.

3. EP should be kept at minimum in machining of high-strength work material of low thermoconductivity and great strain hardening, for example, titanium alloys.

C.4.2.2 Technology and Cost

There are three basic methods of EP: mechanical, thermal, and chemical. Each method includes a number of technologies developed. Figure C.54 lists these methods and associated technologies. The basic description and details of these technologies are discussed by Rodriguez (2009). Each technology developer/EP machine manufacturer shows various sales presentations proving that a particular technology/machine is the best. These presentations/brochures/catalogs are generously supported by high-quality SEM images *before and after*, showing significant improvement in the quality of the transition surface and rake and flank contact areas.

A rather great variety of available technologies makes it not simple for a cutting tool engineer/manufacturing/application specialist to pick up one that seems to be the right choice for a given application or for a product line of a particular tool company. To the first approximation, a comparison presented by Platit Co. (shown in Table C.1) can be used. Figure C.55 shows a typical manufacturing cost structure for common drilling tools (Rodriguez 2009).

C.4.2.3 Metrology of EP

The proper metrology has to be an inherent part of any application of EP technology although this is not nearly the case in reality. Such a metrology allows one to

1. Compare the results of EP, that is, compare the quality of the cutting edge before and after EP application
2. Assure that the optimized parameters of EP are actually applied
3. Assure the consistency of EP technology or, alternatively, compare different EP technologies in order to select the most suitable for a given application
4. The rule of thumb is as follows: if a proper metrology of EP is not available, it is better not to apply EP at all.

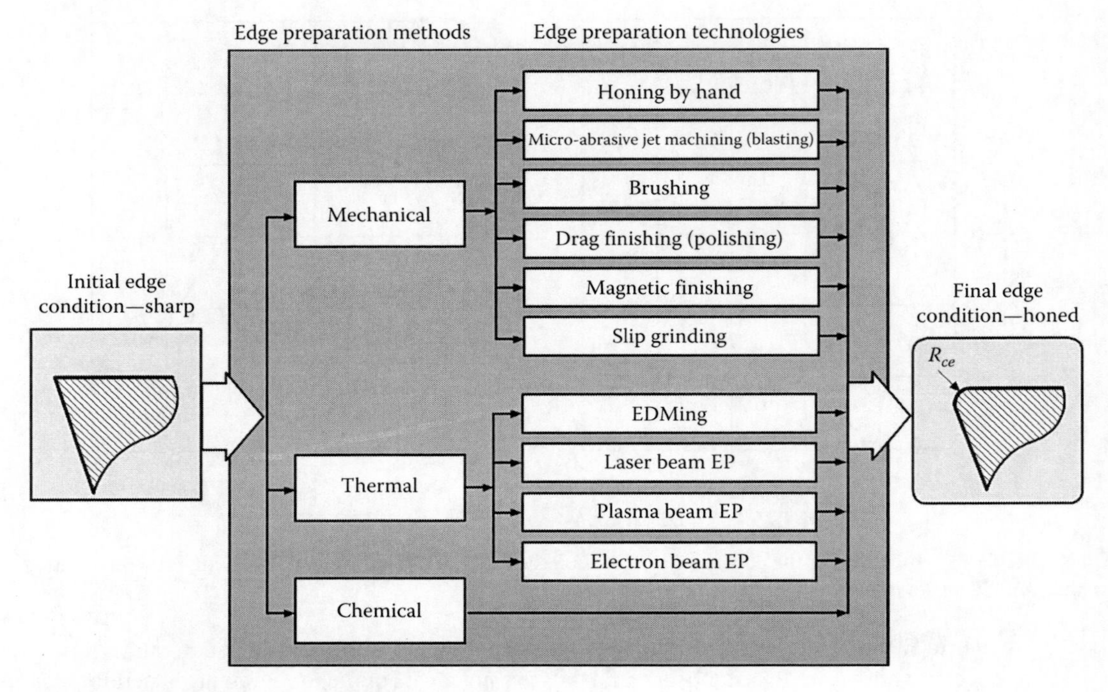

FIGURE C.54 Basic methods and technologies of EP.

TABLE C.1
Qualitative Comparison of Various EP Technologies

Criteria/ Features	Honing by Hand (with Diamond File)	Brushing	Drag Finishing	Micro-Blasting (Dry)	Micro-Blasting (Wet)	Water Jet	Magnetic Finishing
Quality	Best	Good	Good	Medium	Good	Good	Good
Repeatability	Depending on operator	Good	Good	Medium	Good	Good	Good
Flexibility	Very high	High	Medium	Medium	High	Medium	High
Productivity	Low	Medium	Medium	Medium	High	Very high	Medium
Cost	Salary only	High	Medium	Low	Medium	Very high	High
Availability of standard machine		Yes	Yes	Yes	Yes		Yes
Flute polishing		Yes	Yes	Yes	Yes		Limited in depth
Coating droplet removal		Yes	Yes	Yes	Yes		Yes
Special features	Typical for very small batch of tools		Problems with small tool diameter	Residual abrasive material on tool surfaces	High compressed air consumption	Large-scale production, corrosion problems	Demagnetizing needed

Source: Platit Co., Selzach, Switzerland.

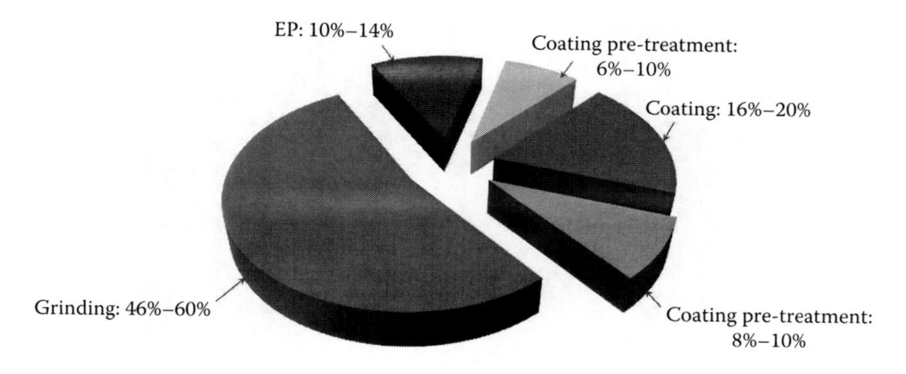

FIGURE C.55 Manufacturing cost structure for common drilling tools.

A comparison of the quality of the cutting edge before and after EP application can be carried out using SEM to obtain images similar to images shown in Figure C.49. Being useful for the research purposes, this way is totally not feasible as an inspection procedure in toolmaking for obvious reasons. The use of universal optical microscopes at high magnification results in unclear images not suitable for the proper assessment. Therefore, specially designed and calibrated microscopes, optical and ultrasonic, should normally be used. According to the author's experience, the portable inspection machine *promSkpGo* by Zoller Co. (Figure C.56) intended for a process-oriented measurement of the cutting edge rounding can be used for fast and accurate assessment of EP.

A specially designed universal drill holding fixture shown in Figure C.57 allows fast and proper orientation of the drill's cutting edge to be investigated with respect to the microscope optical axis. As can be seen, the drill shank as its datum (see Chapter 7) is positioned in a V-block and held tightly by a spring lever. A spherical joint and graduated dials allow final adjustment of the drill position. Using this fixture, any desirable edge of the drill can be properly oriented and inspected.

At the first stage of the quality assessment, the drill cutting edge and the adjacent rake and flank faces are observed in the manner shown in Figure C.58. The following can be seen in this figure:

- EP did not improve the surface roughness of the rake and flank faces in the manner as intended (see Figure C.49).
- The width of EP is not uniform. As such, the maximum material removal by EP took place at the drill corner.

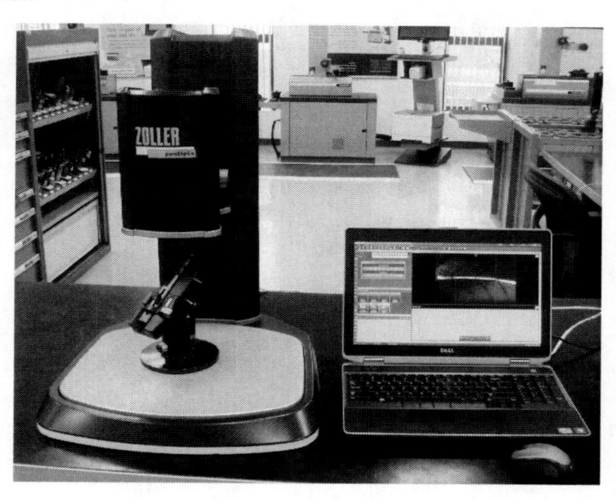

FIGURE C.56 Portable inspection machine *promSkpGo* by Zoller Co. for the assessment of the quality of the cutting edge.

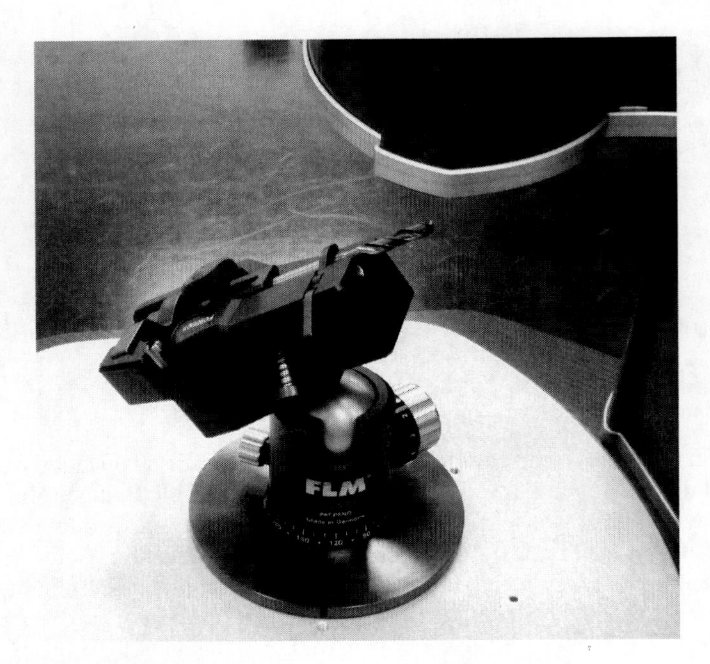

FIGURE C.57 Universal drill holding fixture.

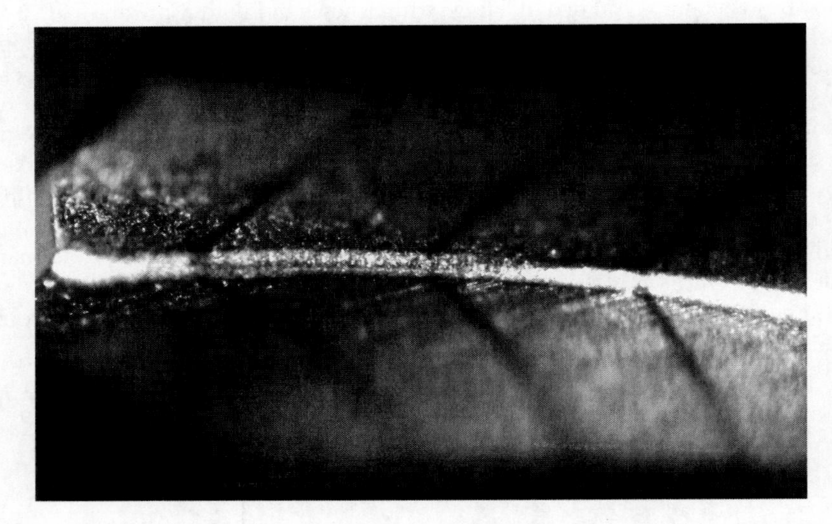

FIGURE C.58 Qualitative assessment of the cutting edge and the adjacent rake and flank faces.

The black shadow lines indicate the boundaries, and the middle of the region of the cutting edge is scanned for quantitative assessment.

The next step is digitizing of the selected region of the cutting edge and assessment of its radius distribution as shown in Figure C.59. As was expected from the earlier analysis of visual appearance of the cutting edge (Figure C.58), the variation of its radius is great. Figure C.60 shows the summary report of the quality of EP. It includes the maximum radius of the cutting edge approximated digitally, the distribution of this radius over the length of the selected region, and the roughness of the cutting edge (termed as *chipping*). Using this report, one may conclude that the quality of EP is very low so that the application of such an EP to the drill will result in lower tool life than that for the same drill with no EP.

Figure C.61 shows an example of improved EP. As can be seen, the waterfall-type (see Figure C.52b) EP is applied. The desirable range of the cutting edge radius (the tolerance on this radius)

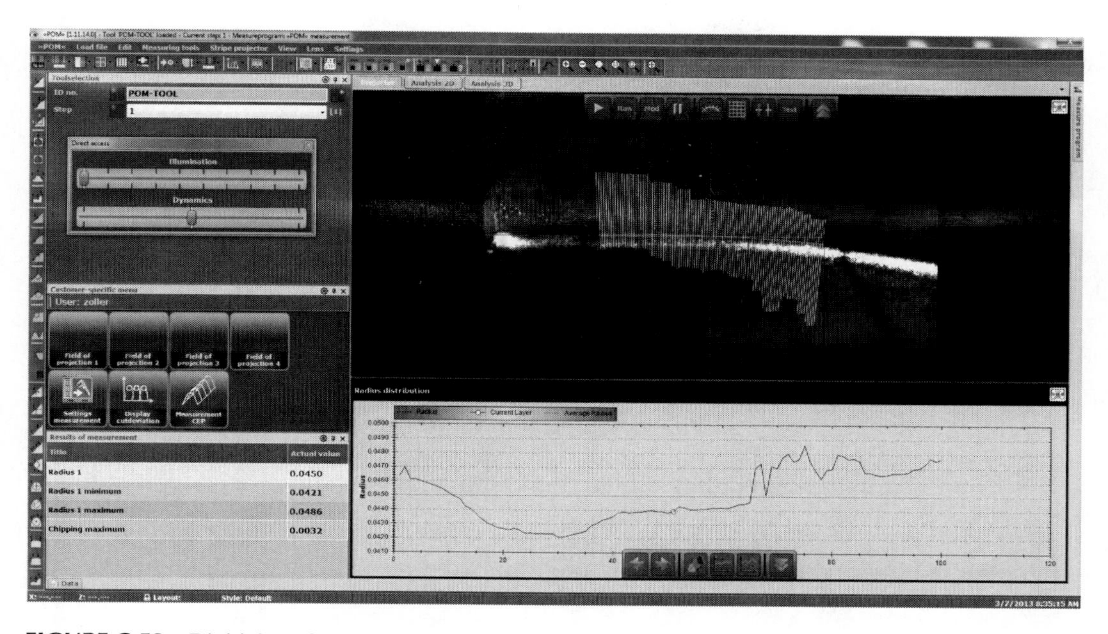

FIGURE C.59 Digitizing of the selected region of the cutting edge and assessment of its radius distribution.

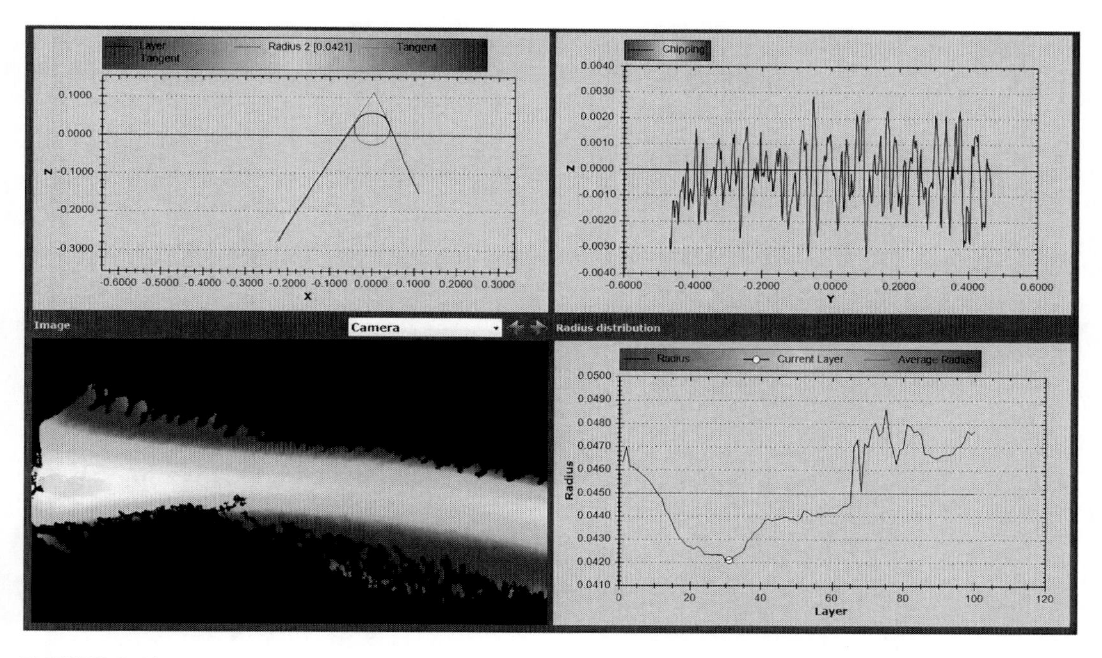

FIGURE C.60 Summary report of the quality of EP.

is set as its lower and upper boundaries. The variation of this radius over the cutting edge is small and just below the mean value preset in the evaluation program. Being much better than that shown in Figure C.60, the roughness of the cutting edge is still too high that clearly indicates that the roughness of the flute and flank grinding should be improved to obtain the maximum advantage of EP in HP drilling.

Figure C.62 shows a more detailed evaluation of the quality of EP. The parameters of EP in any selected section over the inspection region can be evaluated individually. Moreover, the rake,

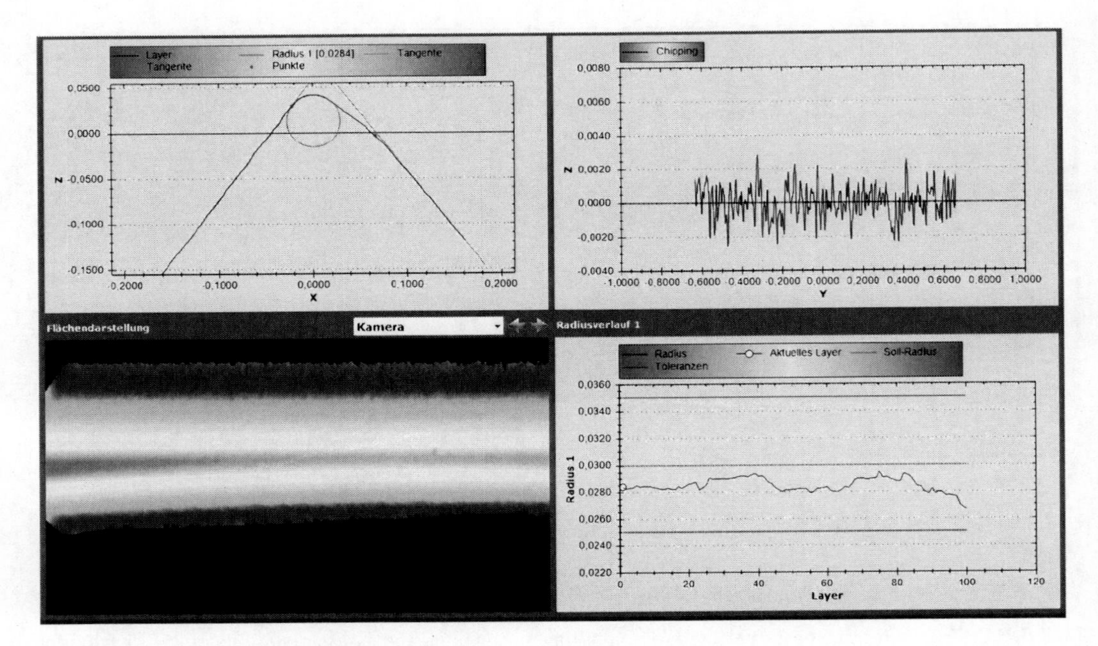

FIGURE C.61 Example of improved EP.

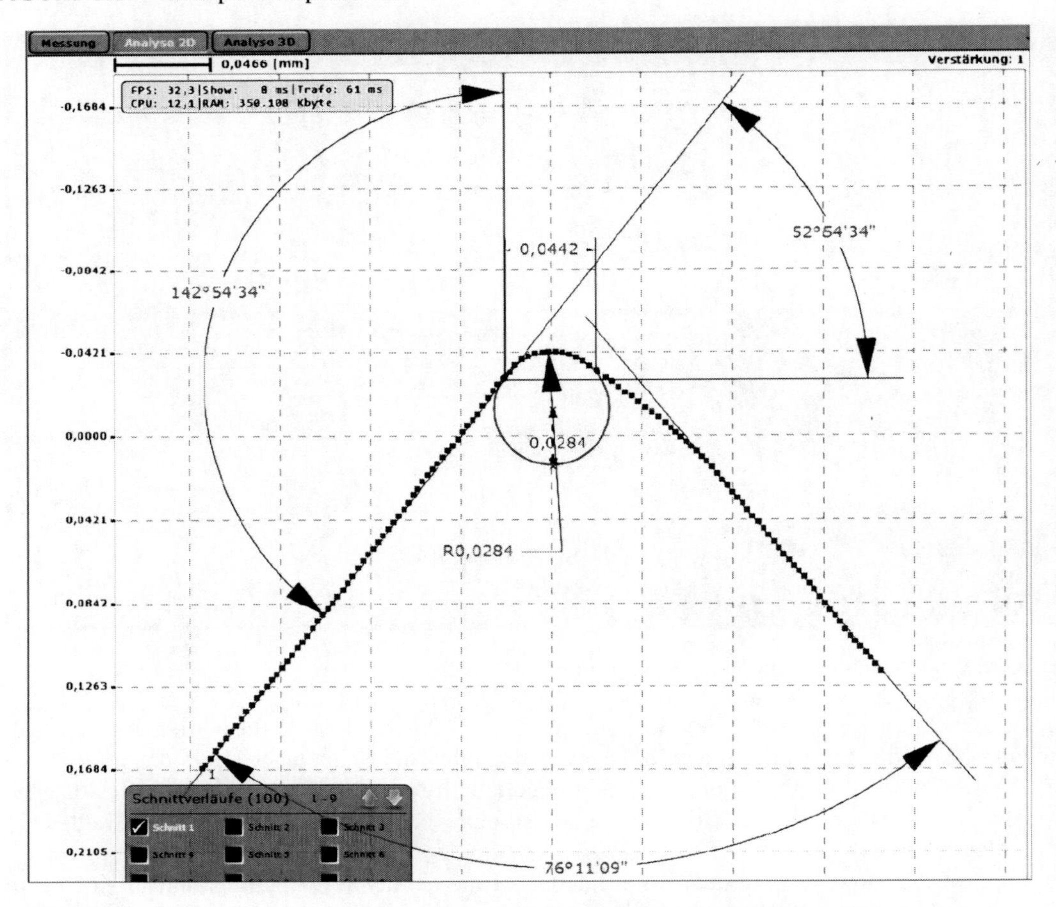

FIGURE C.62 Parameters of EP and tool geometry in the selected section of the region.

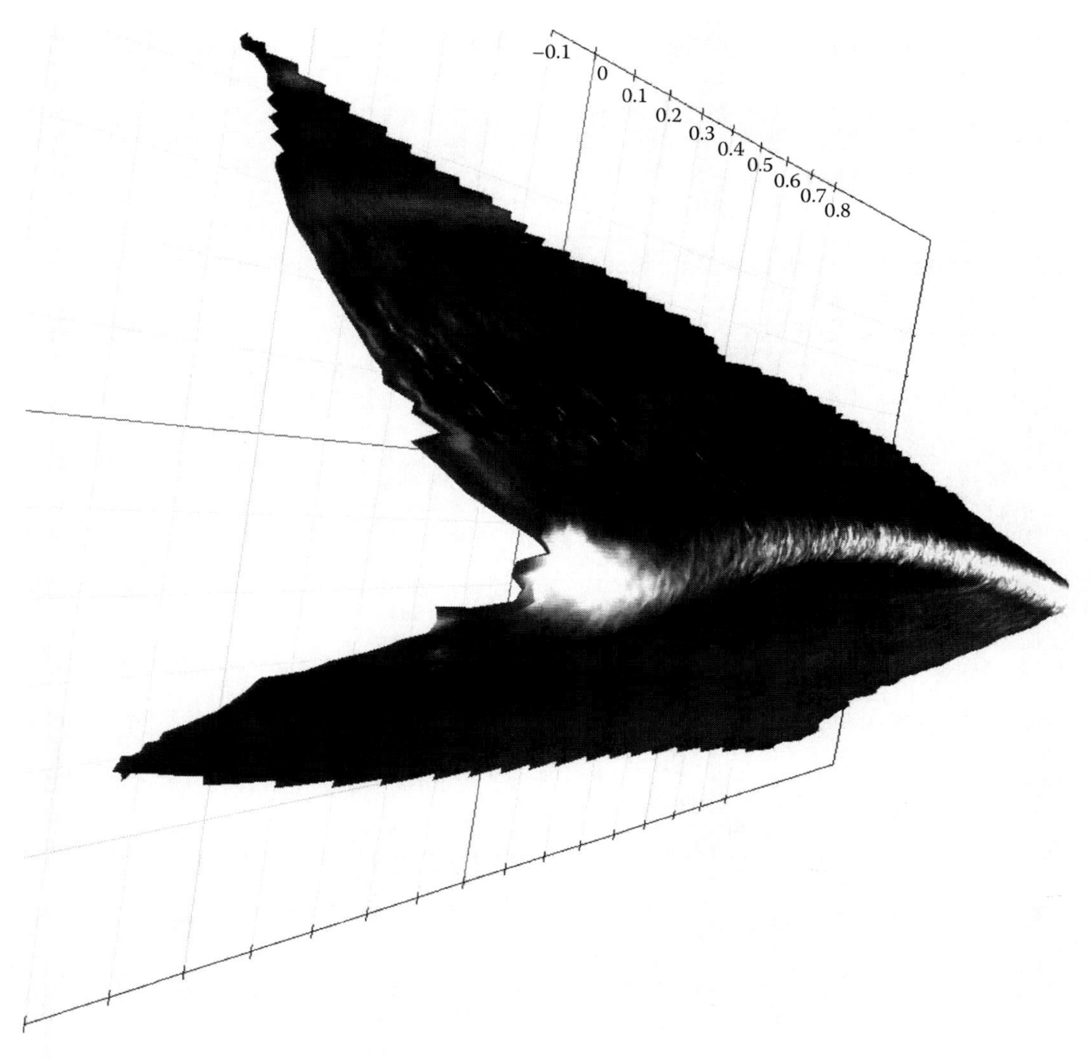

FIGURE C.63 Topography of the whole cutting edge.

clearance, and wedge angles can also be measured as shown in the figure. Figure C.63 shows topography of the whole cutting edge in 3D. This image can be rotated in 3D, so a detailed examination of any region of the cutting edge can be accomplished.

Note that when the drill corner is included into consideration, its rounding is the greatest. It is clearly seen in Figure C.63. In the author's opinion, this is the weakest point of EP for HP drills as sharp drill corners are of prime importance in HP drilling. Honing by hand (with diamond file) is listed as the best EP technology in Table C.1 simply because an experienced EP operator never touches drill corners. No other EP technology available today can provide protection for drill corners.

C.4.2.4 Final Recommendation on EP

Some application recommendations are as follows:

1. As with coating, EP should not be attempted to use to solve tooling or other drilling-system-related problems.
2. Optimization of the drilling tool geometry, selection of its tool material for a given application, proper MWF application, and so on should be accomplished before considering EP.

3. EP machine/technology provider is not responsible for improving tool life and other parameters of a drilling operation. In other words, the parameters of EP should be defined by drilling tool manufacturer/user. EP machine/technology provider is responsible for assuring these parameters are within the tolerance assigned.
4. As EP works only in a rather narrow range of EP parameters (i.e., shape, radius), these parameters should be determined experimentally together with allowable ranges of their variation.
5. The tool manufacturer/user (if EP is used by the user) should have a reliable means to control EP parameters.
6. The optimized EP parameters and their tolerances should be imbedded in the tool drawing properly.

REFERENCES

Agnew, J. 1973. The importance and methods of carbide edge preparation. SME paper MR73-905.

American National Standard ANSI B94.50. 1975. Basic nomenclature and definitions for single-point cutting tools. 1975 (reaffirmed 1993).

Armarego, E.J. and Brown, R.H. 1969. *The Machining of Metals*. Upper Saddle River, NJ: Prentice Hall.

Astakhov, V.P. 1998/1999. *Metal Cutting Mechanics*. Boca Raton, FL: CRC Press.

Astakhov, V.P. 2004. *Tribology of Metal Cutting*. New York: Marcel Dekker.

Astakhov, V.P. 2006. *Tribology of Metal Cutting*. London, U.K.: Elsevier.

Astakhov, V.P. 2010. *Geometry of Single-Point Turning Tools and Drills. Fundamentals and Practical Applications*. London, U.K.: Springer. pp. 1–78.

Astakhov, V.P. 2011. Turning. In *Modern Machining Technology*, Davim, J.P. (ed.). Cambrige, U.K.: Woodhead Publishing.

Biermann, D. and Terwey, I. 2008. Cutting edge preparation to improve drilling tools for HPC processes. *CIRP Journal of Manufacturing Science and Technology* 1:76–80.

Bouzakis, K.-D., Gerardisa, S., Katirtzogloua, G., Makrimallakisa, S., Bouzakisa, A., Cremerd, R., and Fussd, H.-G. 2009. Application in milling of coated tools with rounded cutting edges after the film deposition. *CIRP Annals—Manufacturing Technology* 58(1):61–64.

Chen, N.N.S. and Pun, W.K. 1988. Stresses at the cutting tool wear land. *International Journal of Machine Tools and Manufacture* 28:79–92.

Cheung, F.Y., Zhou, Z.F., Geddam, A., and Li, K.Y. 2008. Cutting edge preparation using magnetic polishing and its influence on the performance of high-speed steel drills. *Journal of Materials Processing Technology* 208:196–204.

Dahlman, P., Gunnberg, F., and Jacobson, M. 2004. The influence of rake angle, cutting feed and cutting depth on residual stresses in hard turning. *Journal of Materials Processing Technology* 47(2):181–184.

Destefani, J. 2002. Cutting Tools 101: Geometries. *Manufacturing Engineering* 129(10):25–42.

DeVries, W.R. 1992. *Analysis of Material Removal Processes*. New York: Springer-Verlag.

Granovsky, G.E. and Granovsky, V.G. 1985. *Metal Cutting (in Russian)*. Moscow, Russia: Vishaya Shkola.

Gunay, M., Korkut, I., Aslan, E., and Seker, U. 2005. Experimental investigation of the effect of cutting tool rake angle on main cutting force. *Journal of Materials Processing Technology* 166:44–49.

Gunay, M., Seker, U., and Sur, G. 2006. Design and construction of a dynamometer to evaluate the influence of cutting tool rake angle on cutting forces. *Materials and Design* 27:1097–1101.

International Standard ISO 3002-1. 1982. Basic quantities in cutting and grinding. Part 1: Geometry of the active part of cutting tools—General terms, reference systems, tool and working angles, chip breakers.

Jahanmir, S., Ramulu, M., and Koshy, P. (eds.) 1999. *Machining of Ceramics and Composites Manufacturing Engineering and Materials Processing*. Boca Raton, FL: CRC Press.

Kang, X., Junichi, T., and Akihiko, T. 2006. Effect of cutting edge truncation on ductile-regime grinding of hard and brittle materials. *International Journal of Manufacturing Technology and Management* 9(1–2):183–200.

Komarovsky, A.A. and Astakhov, V.P. 2002. *Physics of Strength and Fracture Control: Fundamentals of the Adaptation of Engineering Materials and Structures*. Boca Raton, FL: CRC Press.

Mayer, J.E. Jr. and Stauffer, D.G. 1973. Effects of tool hone and chamfer on wear life. ASME techical paper MR73-907.

Outeiro, J.C. and Astakhov, V.P. 2005. The role of the relative tool sharpness in modelling of the cutting process. Paper presented at the *Eighth CIRP International Workshop on Modeling of Machining Operations*, Chemnitz University of Technology, Chemnitz, Germany. pp. 517–523.
Poletica, M.F. 1969. *Contact Loads on Tool Interfaces (in Russian)*. Moscow, Russia: Mashinostroenie.
Rech, J. 2006. Influence of cutting edge preparation on the wear resistance in high speed dry gear hobbing. *Wear* 261:505–512.
Rodriguez, C.J.C. 2009. Cutting edge preparation of precision cutting tools by applying micro-abrasive jet machining and brushing, Kassel University, Kassel, Germany.
Saglam, H., Yaldiz, S., and Unsacar, F. 2007. The effect of tool geometry and cutting speed on main cutting force and tool tip temperature. *Materials and Design* 29:101–111.
Shaffer, W.R. 2000. Getting a better edge. *Cutting Tool Engineering* 52(3):44–48.
Shaw, M.C. 1984. *Metal Cutting Principles*. Oxford, U.K.: Oxford Science Publications.
Shaw, M.C. 1988. Metal removal. In *CRC Handbook of Lubrication: Theory and Practice of Tribology, Volume II: Theory and Design*, Booser, E.R. (ed.). Boca Raton, FL: CRC Press.
Shaw, M.C. 2005. *Metal Cutting Principles*, 2nd edn. Oxford, U.K.: Oxford University Press.
Shuster, L.S.H. 1988. *Adhesion Processes at the Tool–Work Material Interface (in Russian)*. Moscow, Russia: Mashinostroenie.
Silin, S.S. 1979. *Similarity Methods in Metal Cutting (in Russian)*. Moscow, Russia: Mashinostroenie.
Stabler, G.V. 1964. The chip flow law and its consequences. Paper presented at the *Proceedings of the Fifth International MTDR Conference*, University of Birmingham, Birmingham, U.K. pp. 243–251.
Tetsuji, O., Hirokazu, K., and Shigeo, M. 1999. Influence of clearance and rake angles on the cutting force of rock. *Journal of NIRE (National Institute for Resources and Environment)* 2(3):73–80.
Thiele, J.D., Melkote, S.N., Peascoe, R.A., and Watkins, T.R. 2000. Effect of cutting-edge geometry and workpiece hardness on surface residual stresses in finish hard turning of AISI 52100 steel. *Journal of Manufacturing Science and Engineering* 122(4):642–649.
Tugrul Özel, T., Hsu, T.-K., and Zeren, E. 2005. Effects of cutting edge geometry, workpiece hardness, feed rate and cutting speed on surface roughness and forces in finish turning of hardened AISI H13 steel. *The International Journal of Advanced Manufacturing Technology* 25(3–2):262–269.
Vidosic, J.P. 1964. *Metal Machining and Forming Technology*. New York: Roland Press.
Zhong, Z.W. 2003. Ductile or partial ductile mode machining of brittle materials. *The International Journal of Advanced Manufacturing Technology* 21(8):579–585.
Zorev, N.N. (ed.) 1966. *Metal Cutting Mechanics*. Oxford, U.K.: Pergamon Press.

Index

D

G

H

N

S